Modern Diesel Technology: Heavy Equipment Systems

Robert Huzij, Angelo Spano,
Sean Bennett

Australia • Brazil • Japan • Korea • Mexico • Singapore • Spain • United Kingdom • United States

Modern Diesel Technology: Heavy Equipment Systems
Robert Huzij, Angelo Spano, Sean Bennett

Vice President, Career and Professional Editorial: Dave Garza

Director of Learning Solutions: Sandy Clark

Executive Editor: David Boelio

Managing Editor: Larry Main

Senior Product Manager: Sharon Chambliss

Editorial Assistant: Lauren Stone

Vice President, Career and Professional Marketing: Jennifer McAvey

Executive Marketing Manager: Deborah S. Yarnell

Senior Marketing Manager: Jimmy Stephens

Marketing Specialist: Mark Pierro

Production Director: Wendy Troeger

Production Manager: Stacy Masucci

Content Project Manager: Cheri Plasse

Art Director: Benj Gleeksman

Technology Project Manager: Christopher Catalina

Production Technology Analyst: Thomas Stover

© 2009 Delmar, Cengage Learning

ALL RIGHTS RESERVED. No part of this work covered by the copyright herein may be reproduced, transmitted, stored or used in any form or by any means graphic, electronic, or mechanical, including but not limited to photocopying, recording, scanning, digitizing, taping, Web distribution, information networks, or information storage and retrieval systems, except as permitted under Section 107 or 108 of the 1976 United States Copyright Act, without the prior written permission of the publisher.

For product information and technology assistance, contact us at
Professional & Career Group Customer Support, 1-800-648-7450

For permission to use material from this text or product,
submit all requests online at **cengage.com/permissions.**
Further permissions questions can be e-mailed to
permissionrequest@cengage.com.

Library of Congress Control Number: 2008926708

ISBN-13: 978-1-4180-0950-2

ISBN-10: 1-4180-0950-4

Delmar
Executive Woods
5 Maxwell Drive
Clifton Park, NY 12065
USA

Cengage Learning is a leading provider of customized learning solutions with office locations around the globe, including Singapore, the United Kingdom, Australia, Mexico, Brazil, and Japan. Locate your local office at **international.cengage.com/region**

Cengage Learning products are represented in Canada by Nelson Education, Ltd.

For your lifelong learning solutions, visit **www.cengage.com/delmar**

Visit our corporate website at **www.cengage.com**

Notice to the Reader
Publisher does not warrant or guarantee any of the products described herein or perform any independent analysis in connection with any of the product information contained herein. Publisher does not assume, and expressly disclaims, any obligation to obtain and include information other than that provided to it by the manufacturer. The reader is expressly warned to consider and adopt all safety precautions that might be indicated by the activities described herein and to avoid all potential hazards. By following the instructions contained herein, the reader willingly assumes all risks in connection with such instructions. The publisher makes no representations or warranties of any kind, including but not limited to, the warranties of fitness for particular purpose or merchantability, nor are any such representations implied with respect to the material set forth herein, and the publisher takes no responsibility with respect to such material. The publisher shall not be liable for any special, consequential, or exemplary damages resulting, in whole or part, from the readers' use of, or reliance upon, this material.

Printed in the United States of America
5 6 7 11

Contents

Preface

ABOUT THIS TEXT

Modern Diesel Technology (MDT): *Heavy Equipment Systems* has been developed to ensure the availability of an up-to-date, system-specific text book to heavy equipment technician programs across North America. One of the objectives of the authors was to package relevant off-road equipment content within a single text book so that post-secondary institutions and students could rely on a single resource to cover the relevant subject matter in a user-friendly format. As teachers in a post secondary environment, we found it had become necessary to use many different text books and equipment manuals, to cover the components and circuits found on modern heavy equipment and college program syllabi. It has been a number of years since any non-proprietary heavy equipment text books have been written, making those available extremely out of date. One of the challenges we faced with this book was to limit the content to areas that were not now adequately covered in any one book. This edition consists of 19 chapters dealing with subject matter specific to off road equipment. Because there are other text books available that adequately cover subject matter such as diesel engines, and electronic fuel injection, we decided not to include these topics in this book so we could focus on subjects that are not currently available in a single volume.

TECHNOLOGICAL CHANGES

During the development of this book, changes in technology have taken place at an astounding rate. The components and subsystems on modern equipment have progressed to a point were they are now controlled with pilot-operated circuits integrated into the electrical control systems of the equipment. The challenge we face as educators in this field in keeping up with technological changes has become a daunting one. This text book is an attempt at bringing together some of the technical changes that have taken place during the past 20 years into a one-book format.

The contents include the fundamentals in each subject area and progresses to more advanced components and circuits specific to off road equipment. In many cases off road equipment uses components and circuits that are unique to this field. The way in which we elected to break down subject matter in order to make delivering and studying the material easier for both teachers and students is shown below.

- Safety
- Introduction to hydraulics
- Hydraulic system components
- Hydraulic system schematics
- Hydraulic system maintenance and diagnostics
- Hydraulic brakes
- Spring applied hydraulically released brake systems
- Air and Air over hydraulic brake systems
- Introduction to power trains
- Mechanical clutches
- Mechanical transmissions
- Powershift transmissions
- Hydrostatic drive and hydraulic retarder systems
- Driveline systems
- Heavy duty drive axles
- Final drives
- Steering systems
- Tracks and under carriage systems
- Suspension systems

THE OFF ROAD EQUIPMENT INDUSTRY

The off road industry in North America has progressed to a point where in our view it has produced much better equipment reliability and longevity. One example of this progress is the changes that have taken place in off road equipment brake systems. Modern brake systems on equipment today, with proper maintenance, may last for the life of the equipment. That was not the case in the past. A paradigm shift in thinking led to an evolution in equipment design, which maintains North American manufacturers at the forefront of innovation. The role educators play in continuing to instill work ethics and practices to students entering the trade cannot be overstated. Many sectors of the North American off road industry face technical skills shortages. This text book attempts to address the learning outcomes that are required in this field today. Technicians who are now working with heavy equipment technology understand the importance of staying current in their field, and are encouraged to emphasize this to newcomers to the industry. Learning is and always has been a lifelong process that must be passed on to aspiring technicians both in school and on the job.

Robert Huzij
Angelo Spano
Sean Bennett

ABOUT THE AUTHORS

Bob Huzij is currently the program coordinator at Cambrian College in Sudbury, Ontario, Canada, for Motive Power Programs which includes the Heavy Equipment Technician Post Secondary two-year program, as well as the HDET (Heavy Duty Equipment Technician) apprenticeship program. Bob is the Motive Power Curriculum advisory representative on the Ontario provincial Heavy Equipment Advisory Committee. Bob is currently the chair of the Motive Power Provincial Advisory Committee, responsible for the provincial curriculum standards for motive power trades in Ontario.

In the past Cambrian College was the lead college for the development of new motive power provincial curriculum, where Bob held the position of program chair for the Heavy Equipment Technician apprenticeship, Farm Equipment technician, and Powered Lift Truck technician curriculum. He also held the position of program chair for the development of Ontario provincial trade's exemption exams for Heavy Equipment, Farm Equipment technician, and the Powered Lift truck Technician. Prior to working for Cambrian College, he was part of a team responsible for the development and delivery of heavy equipment apprenticeship program in private industry that had approximately 280 heavy equipment technicians and apprentices. Bob ran his own company for a number of years that was responsible for the development of equipment technical training manuals for one of the large heavy equipment manufacturer in Canada.

Angelo Spano, a certified Heavy-Duty Equipment technician, and dual-certified adult educator, has been involved in the Heavy-Duty Equipment field since 1979. A graduate of Centennial College, Scarborough, Ontario, Mr. Spano received his Certificate of Qualification in 1984. After having worked for Canadian National Railways, Condrain Construction (the largest sewer and water main company in Canada), and a CASE Equipment dealership, Mr. Spano joined the faculty of Centennial College in January of 1999, as a sessional instructor. He excelled at sharing his passion, and in August, 1999, became a full-time instructor in the Heavy-Duty and Truck/Coach department. Between 2002 and 2005, along with teaching at Centennial College, Mr. Spano conducted two six-week training programs for the Moa nickel mine in Cuba, sat on the committee for Curriculum Development, and became one of the coordinator's for the Heavy-Duty and Truck/Coach department. Finally, in 2006, Mr. Spano helped develop a co-op program in his department at Centennial College. He is currently co-chair of the Health & Safety Committee, sits on the Power Lift Truck Advisory Committee, and continues to coordinate the Heavy Duty co-op program.

Sean Bennett is currently on faculty at Centennial College in Toronto where he is the Program Coordinator for Truck and Diesel apprenticeship and post-secondary programs. His industry background was with a Freightliner dealership and Mack Trucks Incorporated. Sean has authored numerous text books on diesel engine, truck, heavy equipment, and automotive technology in a body of work that includes the market leading *Heavy Duty Truck Systems* and *Truck Engines, Fuel, and Computerized Management Systems*. In addition he has chaired motive power curriculum development projects for the province of Ontario.

Acknowledgements

The authors and publisher wish to thank the following individuals who provided feedback during the production process:

Robert L. Burkholder, Gregory Poole Equipment Co., Raleigh, NC
Pat Dillard, Nashville Auto-Diesel College, Nashville, TN
Cole Eddy, Nashville Auto-Diesel College, Nashville, TN
Christopher W. Gillette, Case New Holland, New Holland, PA
Randall E. Hall, Canadore College, North Bay, Ontario, Canada
Jim Hoyle, Scott Construction Equipment Co., LaVergne, TN
Scott Jones, Nashville Auto-Diesel College, Nashville, TN
Bobby L. Leatherman, Nashville Auto-Diesel College, Nashville, TN
Todd Moss, Nashville Auto-Diesel College, Nashville, TN
Preston Nelson, Case New Holland, New Holland, PA
Jamie Nunley, Nashville Auto-Diesel College, Nashville, TN
Charles Shropshire, Nashville Auto-Diesel College, Nashville, TN
Phillip Welch, Thompson Machinery, LaVergne, TN

Instructor's CD

An instructor's CD has been prepared to accompany this textbook. The CD includes an Instructor's Guide which includes an answer key to the end of chapter review questions, a PowerPoint presentation, and an image gallery containing images from the book.

CHAPTER

Safety

Learning Objectives

After reading this chapter, you should be able to

- Evaluate potential danger in the workplace.
- Describe the importance of maintaining a healthy personal lifestyle.
- Outline the personal safety clothing and equipment required when working in a service garage.
- Distinguish between different types of fire and identify the fire extinguishers required to suppress small scale fires.
- Outline some procedures required to use shop jacking and hoisting equipment safely.
- Recognize the potential danger when using different types of shop equipment and the importance of using exhaust extraction piping.
- Identify what is required to work safely with chassis electrical systems and shop mains electrical systems.
- Outline the safety procedures required to work with oxyacetylene cutting and welding equipment and how to safely use arc welding stations.

Key Terms

chain hoist
cherry picker
come-alongs
scissor jack
single-phase mains
static charge
static discharge
three-phase mains
Underwriters Laboratories (UL)

INTRODUCTION

The mechanical repair trades are physical by nature, and those employed as technicians probably have higher than average levels of personal fitness. Technicians are required to work safely when handling heavy equipment and to safely handle materials that can be hazardous.

Modern living requires that we assess risk on a minute-by-minute basis and then strategize how to handle a given situation to avoid potential danger. After all, the simple act of crossing a city street requires some planning. There may be a law that prohibits jaywalking so the first decision you make is whether or not you observe that law. Most cities (in North America) accord the pedestrian some rights over vehicles when it comes to crossing a street, but to be eligible for those rights, you may have to observe such things as crosswalk paths and stop/go lights. The result is that you get to choose whether you make the act of crossing a busy city street one of indescribable adventure and danger or a relatively safe procedure that millions successfully perform daily.

The same can be said of working safely in a repair shop environment. There are a multitude of rules and regulations, most of them posted for maximum exposure. But ultimately, it is up to the individual whether he or she chooses to observe those rules. The great thing about the free world we live in, is you get to decide. Nevertheless, if you are planning on a career in the heavy service repair industry, you should attempt to stack your decisions in favor of your ultimate safety.

A major heavy equipment manufacturer monitored accidents in its over a five-year period in one of its assembly plants and came up with the following conclusion: a line-production employee's risk of serious injury (defined as one that required some time off work) during the first year of employment, was equal to that of years two through six combined. In simple terms: if you can survive your first year injury-free, thereafter, your risk diminishes significantly.

Teachers of mechanical technology often complain that it is difficult to teach safe work practices to entry-level technicians. The difficulty arises from the fact that entry-level motive power students may be motivated to learn transportation technology but tend to turn off when asked to learn about the health and safety issues that accompany working life. The sad news is that it sometimes difficult to teach safe work practices to persons who have never been injured and may harbor an illusion that they are immune from injury. On the other hand, an injured person probably has acquired, with the injury, some powerful motivation to avoid a repeat.

Shake Hands with Danger!

"Shake Hands with Danger!" is the title of a film about workplace safety made by Caterpillar many years ago. It uses some painfully realistic shop scenarios to drive its message home. The title of the film is the chorus in a catchy, drawled C&W ballad that repeats itself many times over during the film, usually just before someone is about to sustain a deadly or limb-severing experience. The hard message of the film is that most lapses in safety are caused by stupidity. This is accurate, and the great feature of this cautionary and often humorous presentation is that helps us identify some of the dangerous shortcuts we take on a daily basis without giving a thought to the consequences. Every technician working with heavy on- and off-highway equipment should see this film at least once during his or her career. I know of one dealership that runs this film once a year to all their service personnel and they claim a 50% lower overall accident rate as a result.

A Healthy Lifestyle

Repairing on- and off-highway heavy equipment certainly requires more physical strength than working at a desk all day, but it would be a tough call to say it was in itself a healthy occupation. Lifting a clutch pack or pulling a high load on a torque wrench for sure requires some muscle power but does not parallel lifting weights in a gym where the repetitions, conditions, and movements are carefully coordinated to develop muscle power. Jerking on a torque wrench while attempting to establish final torque on main caps during an in-chassis engine job can tear muscle as easy as develop it. It pays to think about how you use your body and to use your surroundings to maximize leverage and minimize wear and tear. Make a practice of using hoists to move heavier components, even if you know you could manually lift the component: you may believe it is macho to manhandle a 150-pound clutch pack into position, but all it needs is a slight twist of the back while doing so and you can sustain an injury that can last a lifetime. There is nothing especially macho about hobbling around, suffering with the throes of chronic back pain for years.

Part of maintaining a healthy lifestyle means eating properly and making physical activity a component of everyday living. What this means for each individual will differ. Team sports are not just for teenagers, and whether your sport is hockey, baseball, basketball, or football, there are plenty of opportunities to compete at all ages and at a range of levels. If team sports are not your thing, there are many individual pursuits that you can explore. Working out in a gym, hiking, and canoeing are good for your mind as well as your body, and even golf gets you outside and walking. Because of the physical nature of repair technology, it makes sense to routinely practice some form weight conditioning, especially as you get older.

PERSONAL SAFETY EQUIPMENT

Personal safety equipment covers a range of safety apparel we use to protect our bodies. Some of this personal safety equipment, such as safety boots, should be worn continually in the workplace, whereas other equipment, such as hearing protection, may be worn only when required—for instance, when noise levels are high.

Safety Boots

Safety boots or shoes are required footwear in a repair shop. In most jurisdictions, there is legislation mandating the use of safety shoes with steel shanks, steel

Figure 1-1 UL-approved safety boots.

toes, and **Underwriter's Laboratories (UL)** *http://www.ul.com* certification, but beyond any legal requirements, common sense dictates the use of protective footwear in a shop environment. Given the choice, especially in a heavy equipment service and repair facility, safety boots (see **Figure 1-1**) are a better choice than safety shoes because of the additional support and protection to the ankle area.

There is a range of options when it comes to selecting a pair of safety boots, cost being a major factor. If you are going to work on a car under a tree over a weekend, it may be that a low-cost pair of safety boots will suffice, but for the professional mobile equipment technician who wears this footwear daily for the lifetime of the boot, it pays to invest a little more. Generally, the higher the cost of the footwear, the longer it will last, and, more importantly, the greater the comfort level.

Safety Glasses

Many shops today require all their employees to wear safety glasses while on the shop floor. This is really just common sense. Eyes are sensitive to dust, metal shavings, grinding and machining particulate, fluids, and fumes. They are also more complex to repair than feet.

Perhaps the major problem, when it comes to making a habit of using safety glasses, is the poor quality of most shop-supplied eyewear. Shops supply safety glasses because liability issues mandate their availability. Low-cost, mass-produced, and easily scratched plastic safety glasses may meet the employer's obligation to make safety glasses available to employees but are unlikely to encourage the daily use of eye protection.

The solution is to not to depend on your employer to provide this essential safety apparel. Get out of the mindset that safety glasses should be free. Free safety glasses are uncomfortable and, although they may protect eyes from flying particulate, may actually impair vision. Spend a little more and purchase a good quality pair of safety glasses. Even if you do not normally wear eyeglasses, after a couple of days, you will forget you are wearing them. **Figure 1-2** shows three different methods of protecting your eyes in the workplace.

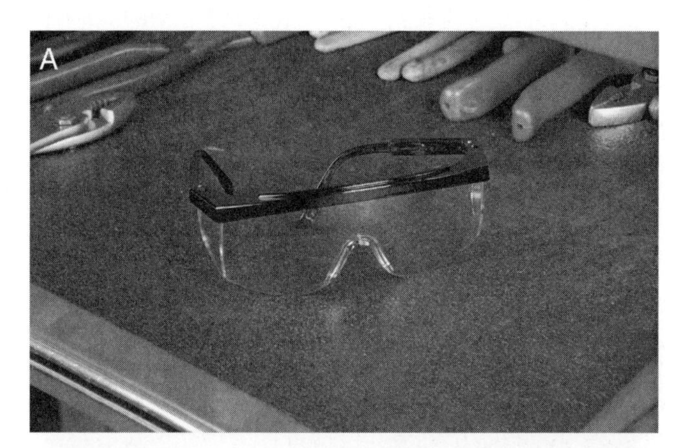

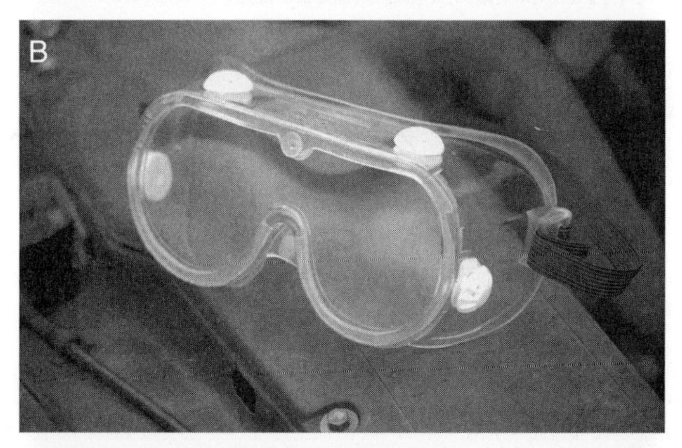

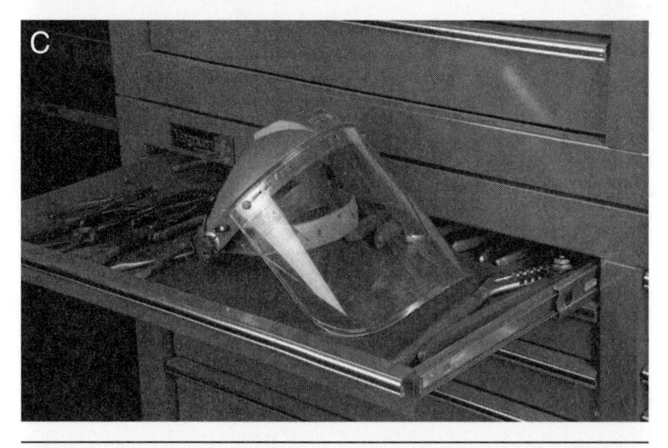

Figure 1-2 (A) safety glasses, **(B)** splash goggles, and **(C)** face shield. (*Courtesy of Goodson Shop Supplies*)

Hearing Protection

Two types of hearing protection are used in shops. Hearing muffs are connected by a spring-loaded band and enclose the complete outer ear. This type of hearing protection is available in a range of qualities, determined by the extent to which they muffle sound. Cheaper versions may be almost useless, but good quality hearing muffs can be the most effective when noise levels are high and exposure is prolonged.

A cheaper and generally effective alternative to hearing muffs is a pair of ear sponges (plugs). Each sponge is a malleable cylindrical or conical sponge that can be shaped for insertion into the outer ear cavity: after insertion, the sponge expands to fit the ear cavity. The disadvantage of hearing sponges is that they can be uncomfortable when used for prolonged periods. **Figure 1-3** shows some typical shop hearing protection devices.

For high noise level activities, such as riveting or operating dynamometers, the use of both hearing muffs and sponges should be considered; however, you should be aware that insulating the brain from sound can be disorientating, resulting in loss of balance.

CAUTION *Damage to hearing is seldom incurred by a single exposure to a high level of noise; it is something that more typically results from years of exposure to excessive and repetitive noise levels. Protect your hearing! And note that hearing can be as easily damaged by listening to music at excessive volume as by exposure to buck riveting.*

Gloves

A wide range of gloves can be used in shop applications to protect the hands from exposure to dangerous or toxic materials and fluids. The following are some examples.

CAUTION *Never wear any type of glove when using a bench-mounted, rotary grinding wheel: there have been cases where a glove has been snagged by the abrasive wheel, dragging the whole hand with it.*

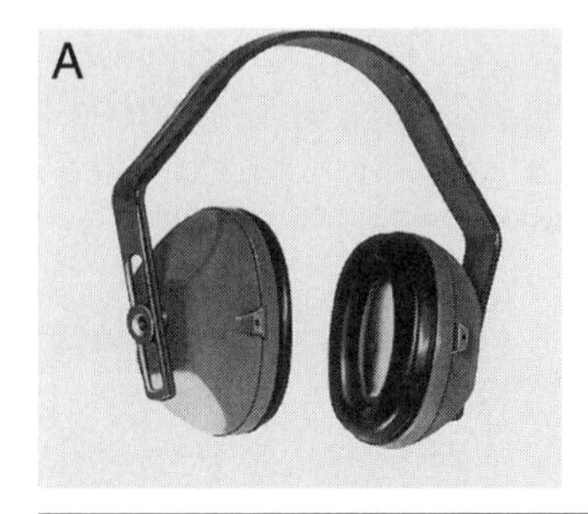

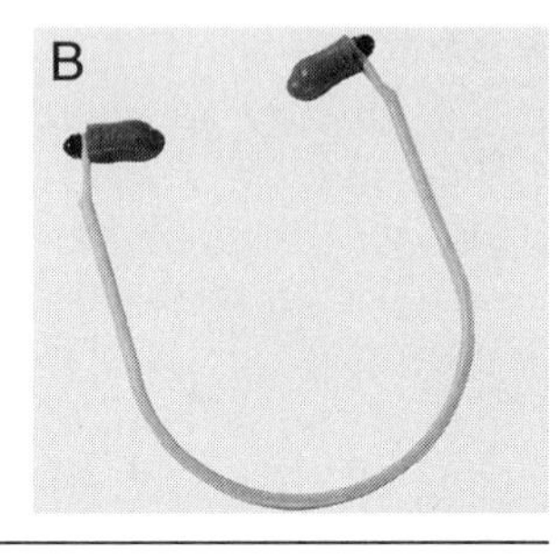

Figure 1-3 Typical **(A)** ear muffs and **(B)** ear plugs. (*Courtesy of Dalloz Safety*)

Vinyl Disposable. Most shops today make vinyl disposable gloves available to service personnel. These protect the hands from direct exposure to fuel, oils, and grease. The disadvantage of vinyl gloves is that they do not breathe, and some find the sweating hands that result uncomfortable. Most shop-use vinyl gloves today are made of thin gossamer that minimizes the loss of tactile awareness.

Cloth and Leather Multipurpose. A typical pair of multipurpose work gloves consists of a rough leather palm and a cotton back. They can be used for a variety of tasks ranging from lifting objects to general protection from cold when working outside. This category of work glove can also provide some insulation for the hands when performing procedures such as buck riveting. You should cease to use this type of glove after its saturation with grease or oil.

Welding Gloves. Welding gloves are manufactured from rough cured leather. They are designed to protect the hands from exposure to the high temperatures created in welding and flame cutting processes. Such gloves should be task-specific. The rough leather they are made from absorbs grease and oil easily, reducing their ability to insulate. Avoid using welding gloves rather than tools to handle heated steel because the gloves will rapidly harden and require replacement.

Some shops recommend the use of welding gloves when performing heavy suspension work because of their ability to protect fingers. If you perform both welding and suspension work in your job, make sure you use separate gloves for each: the gloves you use for suspension work will absorb oil and grease, after which they should not be used for welding.

Dangerous Materials Gloves. Gloves designed to handle acids or alkalines should be used for the specific task only. Gloves in this category are manufactured from nonreactive, synthetic rubber compounds. Care should be taken when washing up after using this type of glove.

CAUTION *Never wear leather gloves to handle refrigerants: leather gloves rapidly absorb liquid refrigerant and can adhere to the skin.*

Back Care

Back injuries are said to affect 50% of repair technicians, at some point in their careers, seriously enough for them to have to take some time off work. A bad back does not have to be an occupational hazard. Most of us begin our careers in our twenties when we have sufficient upper body strength to be able to sustain plenty of abuse. As we age, this upper body strength gradually diminishes and bad lifting practices can take their toll.

The best strategy for taking care of your back is to observe some simple rules for lifting heavy items. These rules can be summarized as follows:

- Keep your back erect while lifting.
- Keep the weight you are lifting close to your body.
- Bend your legs, and lift using the leg muscles.

Figure 1-4 shows the correct method of using your back while lifting.

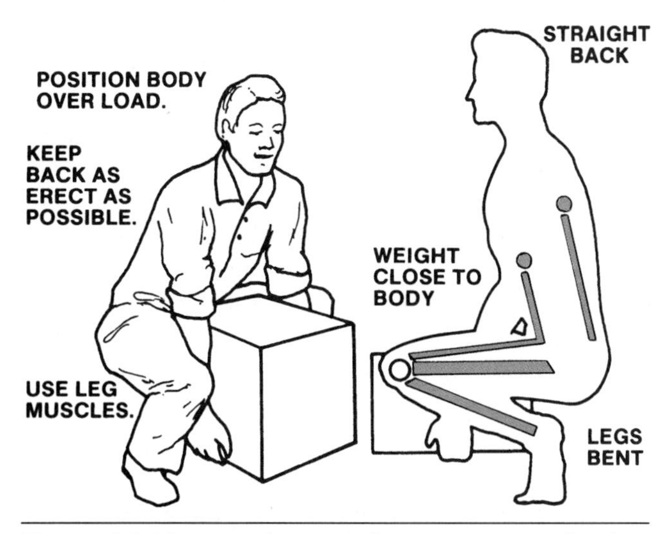

Figure 1-4 Use your leg muscles—never your back—when lifting any heavy load.

Back Braces. A back brace may help you avoid injuring your back. Wearing a back brace makes it more difficult to bend your back, "reminding" you to keep your back straight when lifting. You may have noticed that the sales personnel in one national hardware and home-goods chain are all required to wear back braces. As a mobile equipment technician, you will be required to use your back for lifting, so you should consider the use of a back brace. Body shape plays a role when it comes to back injuries: if you are either taller than average height or overweight, you will be more vulnerable to back injuries.

Coveralls

Many shops today require their service employees to wear a uniform of some kind. These may be work shirts and pants, shop coats, or coveralls (**Figure 1-5**). Uniforms have a way of making service personnel look professional. The uniform of choice in truck, bus, and heavy equipment service facilities should be coveralls, given the nature of the work. The coveralls should preferably be made out of cotton for reasons of comfort and safety. When ordering cotton coveralls for personal use, remember to order at least a size larger than your usual nominal size: unless otherwise treated, cotton shrinks when washed.

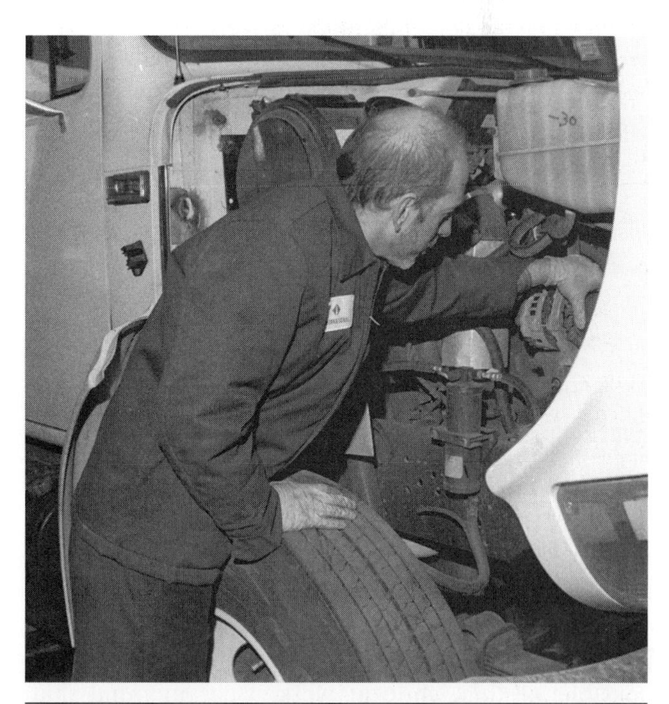

Figure 1-5 Shop coats can be worn to protect clothing.

CAUTION *Avoid wearing any type of loose-fitting clothing when working with machinery. Shop coats, neckties, and untucked shirts can all be classified as loose-fitting clothing.*

When artificial fibers are used as material for coveralls, they should be treated with fire retardant. Cotton smolders when exposed to fire (that is, when it is not saturated with oil, fuel, or grease, when it becomes highly flammable). Cleanliness is essential: oily shop clothing not only looks unprofessional, it can be dangerous! Artificial fibers, when not treated with fire retardant, melt when exposed to high temperatures and may fuse to the skin. You should note that, even when treated with fire retardant, some artificial fibers will burn vigorously when exposed to direct flame for a period of time.

Butane Lighters

There are few more dangerous items routinely observed on the shop floor than the butane cigarette lighter. The explosive potential of the butane lighter is immense, yet it is often stored in a pocket close to where it can do the most amount of damage. A chip of hot welding slag will almost instantly burn through the plastic fuel cell of a butane lighter. Owners of these devices will often compound the danger they represent by lighting torches with them. If you have to have a lighter on your person while working, purchase a Zippo!

CAUTION *Note that, even when treated with fire retardant, some artificial fibers will burn vigorously when exposed to direct flame for a period of time.*

Hair and Jewelry

Long hair and personal jewelry produce some of the same safety concerns as loose-fitting clothing. If it is your style to wear long hair, it should be secured behind the head and you should consider wearing a cap. Because of the recent trend to wear more body jewelry, you should consider removing as much as possible of this while at work. Bear in mind that body jewelry is often made of conductive metals, so you should consider both the possibility of snagging jewelry and creating some unwanted electrical short circuits.

FIRE SAFETY

Service and repair facilities are usually subject to regular inspections by fire departments. This means that obvious fire hazards are identified and neutralized. Although it should be stressed that fire fighting is a job for trained professionals, any person working in a service shop environment should be able to appropriately respond to a fire in its early stages. This requires some knowledge of the four types of fire extinguisher in current use.

Fire Extinguishers

Fire extinguishers are classified by the types of fire they are designed to suppress. Using the wrong type of fire extinguisher on certain types of fire can be extremely dangerous and actually accelerate the spread of the fire you are attempting to control. Every fire extinguisher will clearly state the types of fire it is designed to suppress using the following class letters. Your role as a technician in the suppressing a fire is to assess the risk required and only intervene if there is minimal risk.

Class A. A Class A fire is one involving combustible materials such wood, paper, natural fibers, biodegradable waste, and dry agricultural waste. A Class A fire can usually be extinguished with water. Fire extinguishers designed to suppress Class A fires use foam or multipurpose dry chemical, usually sodium bicarbonate.

Class B. Class B fires are those involving fuels, oil, grease, paint and other volatile liquids, flammable gases, and some petrochemical plastics. Water should not be used on Class B fires. Fire extinguishers designed to suppress Class B fires make use of smothering: they use foam, dry chemical, or carbon dioxide. Trained fire personnel may use extinguishers such as Purple K (potassium bicarbonate) or halogenated agents to control fuel and oil fires.

Class C. Class C fires are those involving electrical equipment. First intervention with this type of fire should be to attempt to shut off the power supply: assess the risk before handling any switching devices. When a Class C fire occurs in a vehicle harness, combustible insulation and conduit can produce highly toxic fumes. Great care is required when making any kind of intervention in vehicle chassis or building electrical fires. Fire extinguishers designed to suppress electrical fires use carbon dioxide, dry chemical powder, and Purple K.

Class D. Class D fires are those involving flammable metals: some metals, when heated to their fire point, begin to vaporize and combust. These metals include magnesium, aluminum, potassium, sodium, and zirconium. Dry powder extinguishers should be used to suppress Class D fires. **Figure 1-6** provides a visual guide

	Class of Fire	Typical Fuel Involved	Type of Extinguisher
Class A Fires (green)	**For Ordinary Combustibles** Put out a Class A fire by lowering its temperature or by coating the burning combustibles.	Wood Paper Cloth Rubber Plastics Rubbish Upholstery	Water*[1] Foam* Multipurpose dry chemical[4]
Class B Fires (red)	**For Flammable Liquids** Put out a Class B fire by smothering it. Use an extinguisher that gives a blanketing, flame-interrupting effect; cover whole flaming liquid surface.	Gasoline Oil Grease Paint Lighter fluid	Foam* Carbon dioxide[5] Halogenated agent[6] Standard dry chemical[2] Purple K dry chemical[3] Multipurpose dry chemical[4]
Class C Fires (blue)	**For Electrical Equipment** Put out a Class C fire by shutting off power as quickly as possible and by always using a nonconducting extinguishing agent to prevent electric shock.	Motors Appliances Wiring Fuse boxes Switchboards	Carbon dioxide[5] Halogenated agent[6] Standard dry chemical[2] Purple K dry chemical[3] Multipurpose dry chemical[4]
Class D Fires (yellow)	**For Combustible Metals** Put out a Class D fire of metal chips, turnings, or shavings by smothering or coating with a specially designed extinguishing agent.	Aluminum Magnesium Potassium Sodium Titanium Zirconium	Dry powder extinguishers and agents only

**Cartridge-operated water, foam, and soda-acid types of extinguishers are no longer manufactured. These extinguishers should be removed from service when they become due for their next hydrostatic pressure test.*

Notes:

(1) Freezes in low temperatures unless treated with antifreeze solution, usually weighs over 20 pounds (9 kg), and is heavier than any other extinguisher mentioned.

(2) Also called ordinary or regular dry chemical (sodium bicarbonate).

(3) Has the greatest initial fire-stopping power of the extinguishers mentioned for class B fires. Be sure to clean residue immediately after using the extinguisher so sprayed surfaces will not be damaged (potassium bicarbonate).

(4) The only extinguishers that fight A, B, and C classes of fires. However, they should not be used on fires in liquefied fat or oil of appreciable depth. Be sure to clean residue immediately after using the extinguisher so sprayed surfaces will not be damaged (ammonium phosphates).

(5) Use with caution in unventilated, confined spaces.

(6) May cause injury to the operator if the extinguishing agent (a gas) or the gases produced when the agent is applied to a fire is inhaled.

Figure 1-6 Guide to fire extinguisher selection.

Figure 1-7 Store combustible materials in approved safety cabinets. (*Courtesy of the Protectoseal Company*)

to selecting fire suppression equipment, and **Figure 1-7** shows a cabinet used to safely store combustibles.

SHOP EQUIPMENT

There is an extensive assortment of shop equipment that technicians should become familiar with: some of this can be dangerous if you are not trained how to use them. Make a practice of asking for help if you are not familiar with how to operate any equipment you are not familiar with.

Lifting Devices

Many different types of hoists and jacks are used in heavy equipment shops. These can range from simple pulley and chain hoists to a variety of hydraulically actuated hoists. Weight-bearing chains on hoists should be routinely inspected (this is usually required by law): abrasive wear, deformed links and nicks are reasons to place the equipment out of service. Hydraulic hoists should be inspected for external leaks before using, and any drop-off while in operation should be reason to take the equipment out of service. Never rely on the hydraulic circuit alone when working under equipment on a hoist: after lifting, support the equipment using a mechanical sprag or stands. Remember that this also applies when working on any kind of hydraulically actuated devices on heavy equipment.

CAUTION *Never rely on a hydraulic circuit alone when working underneath raised equipment. Before going under anything raised by hydraulics, make sure it is mechanically supported by stands or a mechanical lock.*

Jacks. Many types of jack are used in heavy equipment service facilities. Before using a jack to raise a load, make sure that the weight rating of the jack exceeds the supposed weight of the load. Most jacks used in service repair shops are hydraulic, and most use air-over-hydraulic actuation because this is faster and requires less effort. Bottle jacks are usually hand-actuated and designed to lift loads of up to 10 tons: they are so named because they have the appearance of a bottle. Air-over-hydraulic jacks capable of lifting up to 30 tons are also available.

Using hydraulic piston jacks should be straightforward: they are designed for a perpendicular lift only. The jack base should be on level floor and the lift piston should be located on a flat surface on the equipment to be lifted. Never place the lift piston on the arc of a leaf spring or the radius of any suspension device on the mobile equipment. After lifting the equipment, it should be supported mechanically using steel stands. It is acceptable practice to use a hardwood spacer in conjunction with a shop jack: make sure it is exactly level. Whenever using a jack, ensure that the vehicle being jacked can roll either forward or backward: parking brakes should be applied and wheel chocks used on the axles not being raised.

Cherry Pickers. Cherry pickers come in many shapes and sizes. Light-duty cherry pickers can be used to raise a heavy component, such as a cylinder head, from an engine while heavy-duty cherry pickers (see **Figure 1-8**) can lift a large bore diesel engine out of a chassis. Most cherry pickers have extendable arms. As the arm is lengthened, the weight the device can hoist reduces significantly. Take care that the weight you are about to lift can be sustained by the cherry picker without toppling.

CAUTION *As the arm of a cherry picker is lengthened, the weight it can lift reduces significantly. Ensure that the weight you are about to lift can be sustained by the cherry picker without toppling.*

Figure 1-8 Typical heavy-duty cherry picker.

Figure 1-9 Shop exhaust extraction piping.

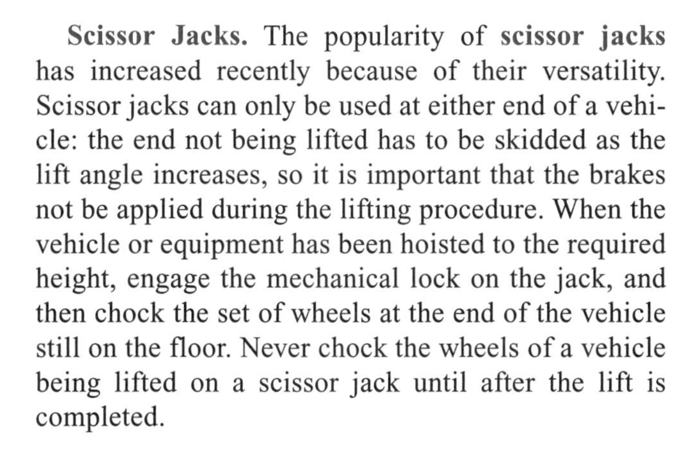

Scissor Jacks. The popularity of **scissor jacks** has increased recently because of their versatility. Scissor jacks can only be used at either end of a vehicle: the end not being lifted has to be skidded as the lift angle increases, so it is important that the brakes not be applied during the lifting procedure. When the vehicle or equipment has been hoisted to the required height, engage the mechanical lock on the jack, and then chock the set of wheels at the end of the vehicle still on the floor. Never chock the wheels of a vehicle being lifted on a scissor jack until after the lift is completed.

Chain Hoists. These are often called chain falls. **Chain hoists** can be suspended from a fixed rail or a beam that slides on rails or be mounted on any of a number of different types of A-frames. Chains hoist in shops in most jurisdictions are required to be inspected periodically. An inspection on a mechanical chain hoist will inspect chain link integrity and the ratchet teeth and lock. Electromechanical units will require an inspection of the mechanical and electrical components. Where a chain hoist beam runs on rails, brake operation becomes critical: some caution is required when braking the beam because aggressive braking can cause a pendulum effect on the object being lifted.

Come-alongs. Come-alongs describe a number of different types of cable and chain lifting devices that are hand-ratchet actuated. They used both to lift objects and to apply linear force to them. When used as a lift device, come-alongs should be simple to use, providing the weight being lifted is within rated specification. However, come-alongs are more often used in service shops to apply linear force to an object or component: great care should be taken, ensuring that the anchor and load are both secure and that the linear force does not exceed the weight rating of the device.

GENERAL SHOP PRECAUTIONS

Because every service facility is different, the potential dangers faced in each shop will differ. In this section, we will outline some general rules and safety strategies to be observed in service and repair shops.

Exhaust Extraction

Diesel engines should be run in a shop environment using an exhaust extraction system: in most cases this will be a flexible pipe or pipes that fit over the exhaust stack(s). Take care when climbing up to fit an exhaust extraction pipe over the vertical exhaust stack(s): use a ladder when you cannot get a secure foothold elsewhere. When parking mobile heavy equipment in and out of service bays, park the unit in the bay and shut the engine off. Avoid running an engine without the extraction pipe(s) fitted to the stack(s). **Figure 1-9** shows the exhaust extraction piping used in a diesel engine testing and repair facility.

WARNING *Diesel engine exhaust fumes are known by the State of California to cause respiratory problems, cancer, birth defects, and other reproductive harm in humans. Avoid operating diesel engines unless in a well-ventilated area. When starting an engine outside a shop, warm the engine before driving it into the shop to reduce the contaminants emitted directly into the shop while parking the unit.*

Workplace Housekeeping

Sloppy housekeeping can make your workplace dangerous. Clean up oil spills quickly. You can do this by applying absorbent grit: this not only absorbs oil but makes it less likely that a person could slip and fall on an oil slick. Try to organize parts in bins and on benches when you are disassembling components: besides adding to the danger potential of your work environment, it indicates a lack of professionalism to see parts and tools strewn all over the shop floor.

Components Under Tension

On mobile heavy equipment, numerous components are under tension, sometimes deadly tensional loads. Never attempt to disassemble a component that you suspect is under high tensional load unless you are exactly sure of how to go about. Consult service literature and ask more experienced co-workers when you are unsure of a procedure. Some examples of components that are under tensional load are

- Air spring brake chambers
- Roll-up van doors
- Elliptical suspension spring packs
- Suspension torsion springs
- Electronic unit pump (EUP) assemblies on some engines
- Roll-up garage doors

Compressed Fluids

Fluids in both liquid and gaseous states can be extremely dangerous when proper safety precautions are not observed. Because of residual pressures in circuits, equipment does not have to be actually running to represent a serious safety hazard. Compressed fluids can be generally divided into pneumatic and hydraulic equipment, but you will see when you read through the next section that many fluid power systems make use of both compressed air and hydraulics.

Pneumatics Safety

Compressed air is used on some heavy equipment chassis both in brake and auxiliary systems. Additionally, compressed air is used to drive both portable and non-portable shop tools and equipment. Most shop powered handtools use pneumatic pressure such as the 1/2-inch drive impact wrench shown in **Figure 1-10.**

WARNING *Always wear safety glasses when coupling and uncoupling or using pneumatic tools.*

Some examples of chassis systems that use compressed air:

- Air brake circuits and actuators
- Pneumatic suspensions
- Pilot control systems on tankers
- Air start systems

Some examples of shop equipment that use compressed air:

- Pneumatic wrenches
- Pneumatic drills
- Shop air-over-hydraulic presses
- Air-over-hydraulic jacks
- Air-over-hydraulic cylinder hoists

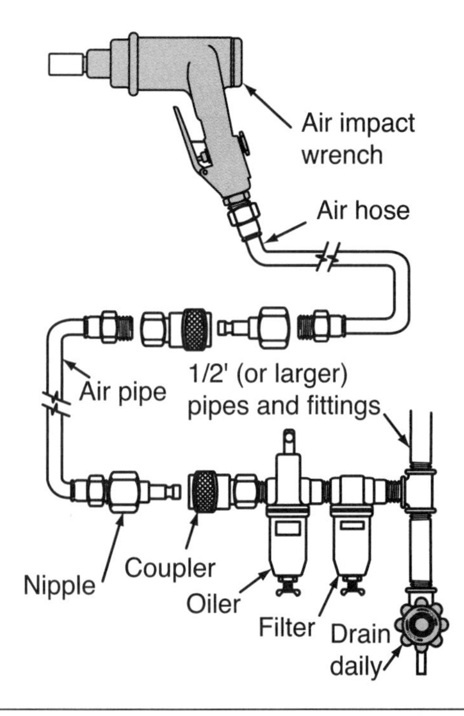

Figure 1-10 Typical setup for a 1/2-inch drive impact gun.

Hydraulic System Safety. Vehicle and shop hydraulic systems use extremely high pressures that can be lethal when mishandled. Never forget that idle circuits can hold residual pressures, and many circuits use accumulators. The rule when working with hydraulic circuits is to be absolutely sure about potential dangers before attempting to disassemble a circuit or component.

WARNING *Always wear safety glasses when working close to shop or vehicle hydraulic circuits and check service literature before attempting a disassembly procedure. Ask someone if you are not sure rather than risk injury.*

Some examples of chassis systems that use hydraulic circuits are

- Wet-line kits for dump and auxiliary circuits
- High-pressure fuel management circuits
- Automatic transmission control circuits
- Air-over-hydraulic brake circuits
- Clutch control circuits

Some examples of shop equipment using hydraulic circuits are

- Jacks and hoists
- Presses
- Bearing and liner pullers
- Suspension bushing presses

CHASSIS AND SHOP ELECTRICAL SAFETY

Mobile heavy equipment vehicles today can use numerous computers, all networked to a central data backbone using multiplexing technology. The computers used to control chassis subsystem components including the engine, transmission, brakes, dash electronics, collision warning systems, satellite communications systems, function on low-voltage electrical signals and use thousands of solid-state components. While some of these electronic subcircuits are protected against transient voltage spikes, others are not. An unwanted high-voltage spike caused by static discharge or careless placement of electric welding grounds can cause thousands of dollars worth of damage.

Some heavy equipment chassis are equipped with electrical isolation switches. These should be opened any time major service or repair work is performed on a vehicle. When any type of electric welding is performed on a chassis, ensure that the ground clamp is placed close to the work. Placing a welding ground clamp on the front bumper when you are welding at the rear of the chassis is not only capable of causing electronic damage but also of taking out bearings and journals in the major powertrain components of the vehicle. Although electricity can be relied on to take the shortest path to complete a circuit, sometimes it will experiment with determining which is the shortest path. Pulsing electricity through crankshaft journals and transmission bearings causes arcing that results in costly damage.

CAUTION *Whenever performing electric arc welding or cutting on a chassis, make sure you place the ground clamp as close to the work area as possible to avoid creating chassis electronic or arcing damage.*

Static Discharge

When you walk across a plush carpet, your shoes "steal" electrons from the floor. This charge of electrons accumulates in your body, and when you go to grab a door handle, this excess of electrons discharges itself into the door handle, creating an arc as it does so. This is known as **static discharge.** Accumulation of a **static charge** is influenced by factors such as relative humidity and the type of footwear you are wearing. Getting a little zap from the static charge that can accumulate in the human body is seldom going to produce any adverse effects to human health but it can damage sensitive solid-state circuits.

Picture a fuel tanker transport running down an interstate. In the same way your body steals electrons from a carpet, so does the tanker steal electrons from the atmospheric air it is forcing itself through; however, the charge differential that can be accumulated by the tanker is much greater and can exceed 50,000 volts. This type of charge differential can be highly dangerous and produce a spark that can easily ignite fuel vapors. This potential danger accounts for the legal requirement to ground out a tanker chassis before undertaking any load or unload operation.

Static Discharge and Computers. Static charge accumulation in the human body can easily damage computer circuits. Because some pieces of equipment today have a dozen, sometimes more, computer-controlled circuits, it is important for technicians to understand the effects of static discharge. The reason that static discharge has not caused more problems than it has in the service repair industry is due primarily to

- Technicians' footwear of choice, usually rubber-soled boots.
- The tendency of shop floors to be concrete-surfaced rather than carpeted.

Neither of the above factors is conducive to static charge accumulation, but technicians should remember that any carpeted flooring is conductive, and due precautions should be taken.

That said, it is good practice, when troubleshooting requires you to access electronic circuits, to use a ground strap before separating sealed connectors, connecting breakout Tees/boxes, or accessing the data bus beyond just connecting an electronic service tool (EST) to it. A ground strap "electrically" connects you to the device you are working on, so that an unwanted static discharge into a shielded circuit is unlikely. Special care should be taken when working with modules that require you physically remove and replace solid-state components such as PROM chips from the motherboard.

Chassis Wiring and Connectors

Every year, millions of dollars worth of damage to mobile equipment is created by service technicians who ignore OEM precautions regarding working with chassis wiring systems. Perhaps the most common abuse is puncturing wiring insulation with test lights and digital multimeter (DMM) leads. When you puncture the insulation on copper wiring, in an instant that wiring becomes exposed to both oxygen (in the air) and moisture (relative humidity). The chemical reaction almost immediately produces copper oxides that then react with moisture to form corrosive cupric acid. The acid begins to eat away the wiring, first creating high resistance, and ultimately consuming the wire. The effect is accelerated when copper-stranded wiring is used because the surface area over which the corrosion can act is so much greater.

CAUTION *Never puncture the insulation on chassis wiring. Read the section that immediately precedes this if you want to know why!*

The sad thing about this type of abuse is that it is so easily avoided. There are so many ways that a repair technician can access wiring circuits using the correct tools, it is just stupidity not to use them. Use breakout Tees, breakout boxes, and test lead spoons.

Breakout Tees. A breakout Tee allows you to access an energized circuit. Most breakout Tees allow you to access a single wire circuit: to "Tee" into the circuit, you separate it at a connector, then connect the breakout Tee. This allows you to test the circuit status while it is electrically active.

Breakout Box. A breakout box is a troubleshooting device required by some OEMs. A breakout box does the same thing as a breakout Tee except that, when it is connected into the circuit, you can check the status of a large number of circuit wires and/or terminals. A troubleshooting tree in interactive diagnostic software will often direct a technician to install a breakout box into a circuit, usually at a connector block, then perform a sequence of DMM driven tests, the results of which may have to be entered into the diagnostic computer.

Test Lead Spoons. Test lead spoons have the potential to save so much unnecessary expenditure, it is surprising that they are not distributed free of charge to anyone working on mobile equipment. Test lead spoons are designed for insertion into weatherproof connectors and terminal blocks without creating damage to either the terminal or the wiring. Get a set from your tool supplier.

Circuit Test Lights. Test lights have few uses on modern equipment and have the potential to inflict damage on electrical wiring when misused. A circuit test light consists of a sharp spike intended to abuse wiring insulation, a lightbulb, a wire, and a ground clamp. If you connect the ground clamp on the tool to a good ground, then any time the spike contacts a voltage differential, the bulb illuminates. Some are specified for working on electronic circuits, and these will illuminate with much smaller voltage potentials. Although this category of circuit test light will not create a current overload in a sensitive circuit, the spike can create all kinds of damage to connector blocks and wiring insulation.

CAUTION *Be aware of the damage you can create using a circuit test light. Many can cause current overloads in sensitive electronic circuit, and even those rated for working on electronics can damage insulation, insulated connectors, and connector blocks.*

Tech Tip: Maybe you should not dispose of your circuit test light in the garbage, after all you paid good money for it. Instead, place it in some remote recess of your toolbox so that the next time you are confronted with an equipment wiring problem, you can better resist the temptation to set about vandalizing its electrical circuits!

Mains Electrical Equipment

Mains electrical circuits, unlike vehicle electrical circuits, operate at pressures that can be lethal, so you have to be careful when working around any electrical equipment. Electrical pressures may be **single-phase main** operating at pressure values between 110 and 120 volts *or* three-phase, operating at pressures between 400 and 600 volts. In some jurisdictions, repairs to mains electrical equipment and circuits are required to undertaken by certified personnel. If you undertake to repair electrical equipment, make sure you know what you are doing! **Figure 1-11** shows a typical three-phase electrical outlet.

Take extra care when using electrically powered equipment when the area you are working in is wet. And remember that a transient spike of AC voltage driven through a chassis data bus can knock out electronic equipment networked to it. Electrical equipment can also be dangerous around vehicles because of its potential to arc and initiate a fire or explosion.

Figure 1-11 High-voltage three-phase electrical outlet.

CAUTION *Do not undertake to repair mains electrical circuit and equipment problems unless you are qualified to do so.*

Some examples of shop equipment using single-phase mains electricity are

- Electric hand tools
- Portable electric lights
- Computer stations
- Drill presses
- Burnishing and broaching tools

CAUTION *Take care when using trouble lights with incandescent bulbs around volatile liquids and flammable gases: these are capable of creating sufficient heat to ignite flammables. Many jurisdictions have banned the use of this type of trouble light, and they should be never used in garages in which gasoline, propane, and natural gas-fueled vehicles are present. Best bet: use a fluorescent-type trouble light in rubber-insulated housing!*

Some examples of shop equipment using high-voltage, three-phase electricity are

- Most welding equipment
- Dynamometers
- Lathes and mills
- Large shop air compressors

OXYACETYLENE EQUIPMENT

Technicians use oxyacetylene for heating and cutting probably on a daily basis. This equipment is used less commonly for braising and welding. Some basic instruction in the techniques of oxyacetylene equipment safety and handling is required. The following information should be understood by anyone working in close proximity with oxyacetylene equipment.

Acetylene Cylinders

Acetylene regulators and hose couplings use a left-hand thread. The regulator gauge working pressure should *never* be set at a value exceeding 15 psi (100 kPa): acetylene becomes extremely unstable at pressures higher than 15 psi. The acetylene cylinder should always be used in the upright position. Using an acetylene cylinder in a horizontal position will result in the acetone draining into the hoses.

It is not possible to determine with any accuracy the quantity of acetylene in a cylinder by observing the pressure gauge because it is in a dissolved condition. The only really accurate way of determining the quantity of gas in the cylinder is to weigh it and subtract this from the weight of the full cylinder, often stamped on the side of the cylinder.

CAUTION *It is a common malpractice to set acetylene pressure at high values. Check a welder's manual for the correct pressure values to set for the equipment and procedure you are using.*

CAUTION *Never operate an acetylene cylinder in anything but an upright condition. Using acetylene when the cylinder is horizontal results in acetone exiting with the acetylene can destabilize the remaining contents of the cylinder.*

Oxygen Cylinders

It should be noted that oxygen cylinders tend to pose more problems than acetylene when exposed to fire. They should always be stored in the same location in a service shop when not in use (this should be identified to the Fire Department during an inspection) and not left randomly on the shop floor.

Oxygen regulators and hose fittings use a right-hand thread. The cylinder pressure gauge will accurately indicate the oxygen quantity in the cylinder, meaning that the volume of oxygen in the cylinder is approximately proportional to the pressure.

Oxygen is stored in the cylinders at a pressure of 2,200 psi (15 MPa) and the hand wheel actuated valve "forward-seats" to close the flow from the cylinder and "back-seats" when the cylinder is opened: it is important to ensure, therefore, that the valve is fully opened when in use. The consequence of not fully opening the valve is leakage past the valve threads. **Figure 1-12** shows a high-pressure oxygen cylinder.

CAUTION *Never use oxygen as a substitute for compressed air when cleaning components in a shop environment. Oxygen can combine with solvents, oils, and grease, resulting in an explosion.*

Figure 1-12 High-pressure oxygen cylinder.

Regulators and Gauges

A regulator is a device used to reduce the pressure at which gas is delivered: it sets the working pressure of the oxygen or fuel. Both oxygen and fuel regulators function similarly in that they increase the working pressure when turned clockwise (CW). They close off the pressure when backed out counterclockwise (CCW).

Pressure regulator assemblies are usually equipped with two gauges. The cylinder pressure gauge indicates the actual pressure in the cylinder. The working pressure gauge indicates the working pressure, and this should be trimmed using the regulator valve to the required value while under flow.

Hoses and Fittings

The hoses used with oxyacetylene equipment are usually color-coded. Green is used to identify the oxygen hose, and red identifies the fuel hose. Each hose connects the cylinder regulator assembly with the torch. Hoses may be single or paired (Siamese). Hoses should be routinely inspected and replaced when defective. A leaking hose should never be repaired by wrapping with tape. In fact, it is generally bad practice to consider repairing welding gas hoses by any method: simply replace when they fail.

Fittings couple the hoses to the regulators and the torch. Each fitting consists of a nut and gland. Oxygen fittings use a right-hand thread, and fuel fittings use a left-hand thread. The fittings are machined out of brass that has a self-lubricating characteristic. Never lubricate the threads on oxyacetylene fittings.

Torches and Tips

Torches should be ignited by first setting the working pressure under flow for both gases, then opening the fuel valve *only* and igniting the torch using a flint spark lighter. Set the acetylene flame to a clean burn (no soot), then open the oxygen valve to set the appropriate flame. When setting a cutting torch, set the cutting oxygen last. When extinguishing the torch, close the fuel valve first, then the oxygen: the cylinders should be shut down finally and the hoses purged. **Figure 1-13** shows an oxyacetylene station with a cutting torch.

Welding, cutting, and heating tips may be used with oxyacetylene equipment. Consult a welder's manual to determine the appropriate working pressures for the tip/process to be used. There is a tendency to set gas working pressure high. Even when using a large heating tip often described as a rosebud, the working pressure of both the acetylene and the oxygen should be set at 7 psi. (50 kPa). **Figure 1-14** shows a typical cutting torch.

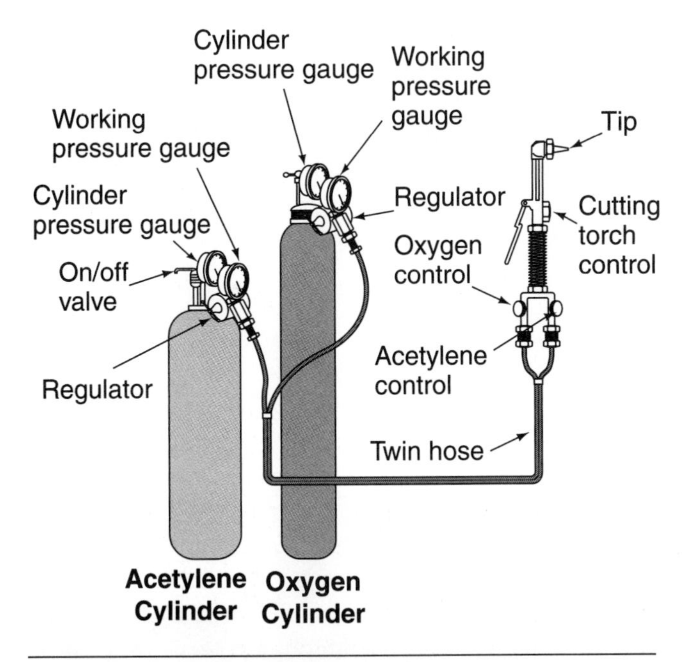

Figure 1-13 Oxyacetylene station set up with a cutting torch.

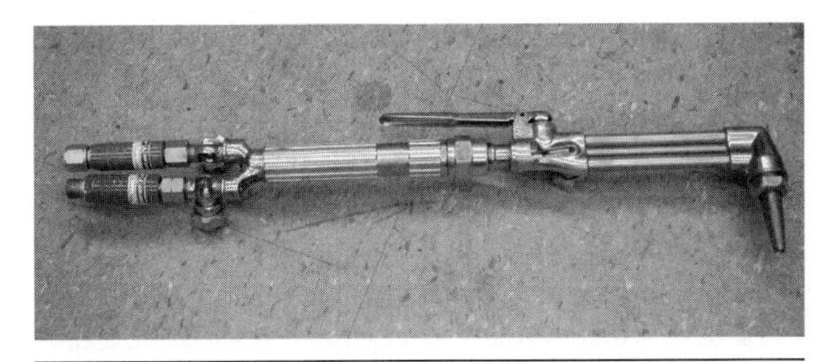

Figure 1-14 Oxyacetylene cutting torch.

Backfire

Backfire is a condition where the fuel ignites within the nozzle of the torch, producing a popping or squealing noise: it often occurs when the torch nozzle overheats. Extinguish the torch and clean the nozzle with tip cleaners. Torches may be cooled by immersing in water briefly with the oxygen valve open.

Flashback

Flashback is a much more severe condition than backfire: it takes place when the flame travels backward into the torch to the gas-mixing chamber and beyond. Causes of flashback are inappropriate pressure settings (especially low-pressure settings) and leaking hoses/fittings. When a backfire or flashback condition is suspected, close the cylinder valves immediately beginning with the fuel valve. Flashback arresters are usually fitted to the torch and will limit the extent of damage when a flashback occurs.

Eye Protection

Safety requires that a #4 to #6 grade filter be used whenever using an oxyacetylene torch. The flame radiates ultraviolet light that can damage eyesight. Even when UV rated sunglasses are not sufficient protection.

Oxyacetylene Precautions

Take the following precautions when working with oxyacetylene:

- Store oxygen and acetylene upright in a well-ventilated, fireproof room.
- Protect cylinders from snow, ice, and direct sunshine.
- Remember that oil and grease may ignite spontaneously in the presence of oxygen.
- Never use oxygen in place of compressed air.
- Avoid bumping and dropping cylinders.

- Keep cylinders away from electrical equipment where there is a danger of arcing.
- Never lubricate the regulator, gauge, cylinder, and hose fittings with oil or grease.
- Blow out cylinder fittings before connecting regulators: make sure the gas jet is directed away from equipment and other people.
- Use soapy water to check for leaks: NEVER use a flame to check for leaks.
- Thaw frozen spindle valves with warm water: NEVER use a flame.

Adjustment of the Oxyacetylene Flame

To adjust an oxyacetylene flame, the torch acetylene valve is first turned on and the gas ignited. At the point of ignition, the flame will be yellow and produce black smoke. Next, the acetylene pressure should be increased by using the torch fuel valve. This will increase the brightness and reduce the smoking. At the point the smoking disappears, the acetylene working pressure can be assumed to be correct for the nozzle jet size used. Now, the torch oxygen valve is turned on. This will cause the flame to become generally less luminous, and an inner blue luminous cone surrounded by a white-colored plume is formed at the tip of the nozzle. The white-colored plume indicates excess acetylene. As more oxygen is supplied, this plume reduces until there is a clearly defined blue cone with no white plume visible. This indicates the *neutral* flame used for most welding and cutting operations.

Oxidizing Flame. If after setting a neutral flame, the oxygen supply is increased, the blue cone will become smaller and sharper in definition and the outer envelope will become streaky. This is known as an *oxidizing flame* indicating excess oxygen. For most welding procedures an oxidizing flame should be avoided, but in some special applications, a slightly oxidizing flame is required.

Carburizing Flame. A *carburizing flame* is indicated by the presence of a white plume surrounding an inner blue cone. A carburizing flame should also be avoided in most welding operations, although a very slight carburizing flame is used in certain special applications.

Electric Arc Welding

Electric arc welding and cutting processes are used extensively in truck and heavy equipment service garages. Arc welding stations either receive or generate high-voltage charges, then condition them to lower-voltage, high-current circuits. Before attempting to use any type of arc welding equipment, make sure you receive some basic instruction and training. The following types of welding station are non-specialized in application and are found in many repair shops:

- Arc welding station: uses a flux-coated, consumable electrode, often known as *stick welding.*
- Metal inert gas (MIG): uses a continuous reel of wire which acts as the electrode, around which inert gas is fed to shield the weld from air and ambient moisture.
- Tungsten inert gas (TIG): uses a non-consumable tungsten electrode surrounded by inert gas, and filler rods are dipped into the welding puddle created.
- Carbon-arc cutting: arc is ignited using carbon electrodes to melt base melt while a jet of compressed air blows through the puddle to make the cut.

Figure 1-15 shows a typical arc welding electrode holder. Because less training and faster weld speeds are possible with MIG welding processes, stick arc welding is less commonly used in today's shops. However, when specialty alloy and tempered materials are required to be welded, arc welding electrodes (sticks) are commonly selected.

ONLINE TASKS

1. Use an Internet search engine to identify some personal safety equipment suppliers. Check the features offered in safety equipment such as
   - Safety glasses
   - Safety boots
   - Hearing protection
   - Coveralls
   - Gloves

Figure 1-15 Arc welding electrode holder.

2. Paste this url into your favorites/bookmarks folder: *http://www.osha.gov/*. This is one of the best maintained, government-operated Web sites: it is up to date with an abundance of useful links.

Shop Tasks

1. Take a tour of the shop garage making a note of every type of fire extinguisher. List the types of fire each will suppress. Note the location of fire hazards and the assigned place to store combustibles such as oxyacetylene equipment, fuels, window-washer concentrate, and grease and waste barrels.
2. With the guidance of an instructor, run through the process required to change an oxyacetylene torch body to the fuel and oxygen hoses. Do the same with the attachments such as the cutting head, welding head, and rosebud (heating) head.
3. Under the guidance of an instructor, set the pressures on a oxyacetylene torch for cutting, ignite the fuel, and set the oxygen to produce a neutral flame.
4. Identify the welding equipment in your shop/garage. Identify the types of electrode used with arc welding equipment and interpret the AWS codes (such as E-7018). Do the same with MIG welding wire, and identify the shielding gas used.

Summary

- To work safely in a service garage environment, it is important to evaluate potential dangers and respond appropriately.
- Working as a service technician is more physical than many occupations, so it is important to maintain healthy personal lifestyle.
- Personal safety clothing and equipment such as safety boots, eye protection, coveralls, hearing protection, and different types of gloves are required when working in a service garage.
- Technicians should learn to distinguish between the four different types of fire and identify the fire extinguishers required to suppress them.
- Jacks and hoists are used extensively in service facilities and should be used properly and inspected routinely.
- The potential danger of exhaust emissions should be recognized so that equipment, when inside, is run with exhaust extraction piping.
- It is important to identify what is required to work safely with chassis electrical systems because of the costly damage caused by simple errors.
- Shop mains electrical systems are used in portable power and stationary equipment and can be lethal if not handled properly.
- Oxyacetylene cutting and welding equipment is used extensively in service facilities and technicians should understand how the equipment is set up.
- Care should be taken when using any of the arc welding processes because of the potential for personal injury (electric shock) and equipment damage (high-voltage spikes).

Review Questions

1. Which of the following best describes the term, “Shake hands with Danger”?

   A. a way of life in any modern repair shop
   B. a safety video presentation made by Caterpillar
   C. a sure way of avoiding accidents
   D. something that only beginners do

2. Experience has shown that a worker is more likely to be injured:

   A. during the first day on the job
   B. during the first year of employment
   C. during the second to fourth year of employment
   D. during the year before retirement

3. When lifting a heavy object, which of the following should be true?
   A. Keep your back erect while lifting.
   B. Keep the weight you are lifting close to your body.
   C. Bend your legs and lift using the leg muscles.
   D. All of the above.
4. What is Purple K?
   A. a new type of stimulant
   B. a dry powder fire suppressant
   C. a toxic gas
   D. a type of paint
5. Which of the following is usually a requirement of a safety shoe or boot?
   A. UL certification
   B. steel sole shank
   C. steel toe
   D. all of the above
6. What type of gloves are recommended by some shops when working with heavy suspension components?
   A. synthetic rubber
   B. vinyl disposable
   C. leather welding gloves
   D. Gloves should never be worn when working with suspension components.
7. Which of the following is false?
   A. Never wear gloves when using a bench grinder.
   B. Wear leather gloves when working with refrigerant.
   C. Disposable vinyl gloves help protect hands from exposure to oil and fuel.
   D. Welding gloves harden when repeatedly exposed to high temperatures.
8. Which type of fire can usually be safely extinguished with water?
   A. Class A
   B. Class B
   C. Class C
   D. Class D
9. When attempting to suppress a Class C fire in a chassis, which of the following is good practice?
   A. Disconnect the batteries.
   B. Use a carbon dioxide fire extinguisher.
   C. Avoid inhaling the fumes produced by burning conduit.
   D. All of the above.
10. When buck riveting steel panels, which of the following personal protection items are required?
   A. safety glasses
   B. hearing protection
   C. work gloves
   D. all of the above
11. What sealing compound should be used on brass oxyacetylene fittings?
   A. nothing
   B. never-seize
   C. multipurpose thread sealant
   D. Teflon® tape
12. When oxyacetylene hoses are color-coded, which color is used to indicate the oxygen hose?
   A. blue
   B. green
   C. red
   D. black

13. To tighten an acetylene regulator fitting, which of the following should be true?

    A. Turn CW.
    B. Turn CCW.
    C. Torque oxygen fitting first.
    D. Ensure the regulator valve is open.

14. Which type of oxyacetylene flame should be set for most heating and welding procedures?

    A. neutral
    B. oxidizing
    C. carburizing
    D. white

15. What is the maximum safe acetylene flow pressure that should be set at the regulator?

    A. 5 psi
    B. 15 psi
    C. 105 psi
    D. 150 psi

16. Technician A says that many jurisdictions have banned the use of trouble lights with incandescent bulbs because they can ignite flammable substances. Technician B says that fluorescent trouble lights should be used in service garages. Who is correct?

    A. Technician A only
    B. Technician B only
    C. Both A and B
    D. Neither A nor B

17. Which of the following compressed gases in a cylinder tends to be the most dangerous in the event of a fire?

    A. oxygen
    B. acetylene
    C. propane
    D. argon

18. When performing arc welding on the rear of a piece of mobile heavy equipment, where should the ground clamp be fixed?

    A. on the front bumper
    B. on the battery ground post
    C. on the battery positive post
    D. as close to the weld area as possible

19. What is the minimum level of eye protection required when using an oxyacetylene cutting torch?

    A. UL-rated safety goggles
    B. any shades with 100% UV protection
    C. lens with grade 4-rated filter
    D. lens with grade 12-rated filter

20. Technician A says that it is okay to work on a defective mains electrical system providing you have learned to identify the three color codes used in the wiring. Technician B says that only qualified electricians should work on three-phase, high-voltage circuits and components. Who is correct?

    A. Technician A only
    B. Technician B only
    C. Both A and B
    D. Neither A nor B

CHAPTER

2 Introduction to Hydraulics

Learning Objectives

After reading this chapter, you should be able to:

- Explain fundamental hydraulic principles.
- Define *fluids* as they apply to pneumatics and hydraulics.
- Explain the terms hydrostatics and hydrodynamics.
- Interpret Pascal's Law and Bernoulli's Principle.
- Apply the laws of hydraulics.
- Define laminar flow.
- Explain how pressure and flow are used to manage outcomes in a hydraulic circuit.
- Describe an open-circuit hydraulic system.
- Describe a closed-circuit hydraulic system.
- Calculate force, pressure, and area.
- Identify some of the hydraulic circuits used in heavy equipment.

Key Terms

Bernoulli's Principle
closed-center
flow
fluid
force
head
head pressure
hydrodynamics
hydrostatics
laminar flow
manometer
open-center
Pascal's Law
pressure
prime mover
ram
reservoir
spool valve
Torricelli's tube
work

INTRODUCTION

Fluid power is the practice of using gases or liquids in confined circuits to transmit power or multiply human or machine effort. A **fluid** is any substance that has fluidity, that is, it is able to flow and yield to alter shape. For our purposes, any gas or liquid can be classified as a fluid. In mobile hydraulics, we use both pneumatics and hydraulics in different ways. Here are some examples:

Pneumatic circuits:

- Air brakes
- Air suspensions
- Wiper motors
- HVAC controls
- Air starters

Hydraulic circuits:

- Hydraulic brakes
- Power steering systems
- Powershift and automatic transmissions
- Fuel systems
- Wet-line kits
- Torque converters
- Lift gates

The term *hydraulics* is used to specifically describe fluid power circuits that use liquids, usually especially formulated oils, in confined circuits to transmit force or motion. The practice of hydraulics predates recorded history and is thought to have played a role in the construction of the Egyptian pyramids. In this chapter we will focus on gaining an understanding of basic hydraulic circuits. A heavy equipment technician is required to have a good knowledge of hydraulics to properly understand steering and brake systems, hydraulic auxiliary equipment, engine operation, fuel systems, and many types of transmissions.

Although hydraulics may have been used by the ancient Egyptians and Romans, it did not really become a science until Blaise Pascal (France 1623–1662). Pascal is so important to the understanding of modern hydraulics that most programs of study begin with what we know today as **Pascal's Law,** which states:

Pressure applied to a confined liquid is transmitted undiminished in all directions and acts with equal force on all equal areas, at right angles to those areas.

FUNDAMENTALS

Although we do not often differentiate between them, hydraulics can be divided into two branches, which we know as **hydrostatics** and **hydrodynamics.** Hydrostatics is the science of transmitting force by pushing on a confined liquid. In a hydrostatic system, transfer of energy takes place because a confined liquid is subject to pressure. A good example of a hydrostatic circuit is that used to actuate a boom cylinder. Hydrodynamics is the science of moving liquids to transmit energy. A water wheel or turbine can both be classified as a hydrodynamic device, but the best example on mobile equipment chassis is the torque converter used to impart drive to a powershift transmission.

- Hydrostatics: low fluid movement with high system pressures.
- Hydrodynamics: high fluid velocity with lower system pressures.

Pressure

Pressure is force applied to a specific area. When a confined fluid is subject to pressure, the force applied to the area of confinement will be uniform throughout (Pascal's Law). When a liquid confined in a vessel is pushed on, the pressure that results acts evenly on all of the walls of the vessel. This characteristic makes it possible to transmit force or "push" through pipes. In hydraulics, we use liquids rather than gases as hydraulic fluids, because liquids are not easily compressed. Because of this incompressibility, we can relay action almost instantaneously, so long as the circuit is full of liquid and contains no air. Pressure is usually expressed in pounds per square inch or kilopascals.

Creating Pressure. Pressure is created when there is resistance to flow in a circuit. If we get a bicycle pump that has a plunger with a section area of 1 sq. in. and is equipped with a pressure gauge and we draw into that pump a column of water, then trap the water into the pump with a finger, we can say the following:

- If no force is applied to the pump plunger, the gauge will read zero.
- If 10 pounds of linear "push" is applied to the pump, the gauge will read 10 psi.
- If 15 pounds of linear "push" is applied to the pump, the gauge will read 15 psi.
- If the finger closing the pump outlet is removed, the water will be displaced as force is applied to the pump.

We can conclude from this that the pressure of a confined liquid is always in proportion to the applied force. And if the restriction is removed (in our example here, the finger), gauge pressure drops to zero.

Atmospheric Pressure. The Earth's atmosphere extends upward for around 50 miles (80 km) and we know that air exerts a force because of its weight. A column of air measuring 1 sq. in. (6.45 cm) extending 50 miles (80 km) into the sky would weigh 14.7 pounds (6.67 kg) at sea level as shown in **Figure 2-1.** We express

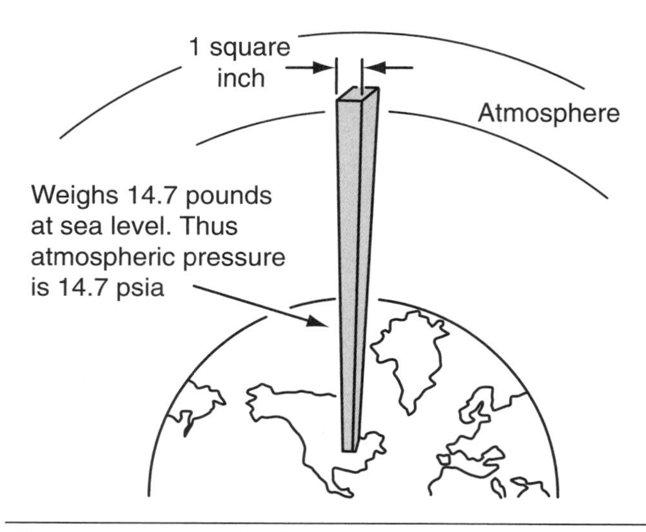

Figure 2-1 Atmospheric pressure.

atmospheric pressure at sea level as 14.7 psia (pounds per square inch absolute) or 101.3 kPa, its metric equivalent. If we stood on a high mountain, the column of air would measure less than 50 miles (80 km) and the result would be a lower weight of air in the column. Similarly, if we were below sea level (in a mine for instance), the weight of air would be greater in the column. In North America, we sometimes use the term *atm* (short for *atmosphere*) to describe a unit of measurement of atmospheric pressure: the Europeans use the unit *bar* (short for *barometric* pressure). The relative equivalents of these values appear a little later in this chapter.

Force. **Force** is push or pull effort. The weight of one object placed upon another exerts force on it proportional to its weight. If the objects were glued to each other and we lifted the upper, a pull force would be exerted by the lower object proportional to its weight. Force does not necessarily result in any work done. If you were to push on the rear of a parked railway locomotive, you could apply a lot of force but that effort would be unlikely to result in any movement of the locomotive. The formula for force (*F*) is calculated by multiplying pressure (*P*) by the area (*A*) it acts on.

$$F = P \times A$$

Table 2-1 shows the mathematical relationship between force, pressure, and sectional area.

Table 2-1: MATHEMATICAL RELATIONSHIP BETWEEN FORCE, PRESSURE, AND SECTIONAL AREA

To Calculate	Use Formula
Pressure	$P = \frac{\text{force}}{\text{area}}$
Force	$F = \text{pressure} \times \text{area}$
Area	$A = \frac{\text{force}}{\text{pressure}}$

Head Pressure. If you have ever drained fuel out of a tank through its drain valve, you will have observed that the fuel runs out of the tank faster when the tank is full and progressively slower as the tank empties. This occurs because the pressure that forces the fuel from the tank is defined by the weight of the fuel in the tank, and as the tank empties, less weight results in lower pressure. We express this as **head pressure.** By definition, **head** is the vertical distance between two levels in a fluid and it can be expressed using linear or pressure values. For instance, a 10 foot (3.05 m) head of water is equivalent to 4.4 psi (30 kPa). The weight of a fuel or hydraulic oil is less than that of water so we would have to first reckon its specific gravity and then calculate the head required to move it.

Vacuum. A perfect vacuum is a volume that has been evacuated of all matter. That means that no air (mixture of nitrogen and oxygen molecules) can be present in a true vacuum. A true vacuum is difficult to achieve, but we tend to use the word to describe any pressure condition lower than atmospheric pressure. Partial vacuums are used in hydraulics so that when a pump creates a low-pressure area on its inlet side, the static head pressure of our atmosphere acts on the **reservoir** (fluid storage tank) oil to push it toward the pump. In this way, pumps are used to move fluid through circuits whether they are on a wet-line kit, steering gear circuit, or fuel subsystem.

Pressure Scales. There are a number of different pressure scales used today but all are based on atmospheric pressure. One unit of atmosphere is the equivalent of atmospheric pressure and it can be expressed in all these ways:

1 atm = 1 bar (European)
= 14.7 psia
= 29.92 inches Hg (inches of mercury)
= 101.3 kPa (metric)

However, the above values are not precisely equivalent to each other:

1 atm = 1.0192 bar
1 bar = 29.53 inches Hg
= 14.503 psia
1" Hg = 13.6 inches H_2O @ 60°F

An Italian named Evangelista Torricelli discovered the concept of atmospheric pressure. He inverted a tube filled with mercury into a bowl of the liquid and then observed that the column of mercury in the tube fell until atmospheric pressure acting on the surface balanced against the vacuum created in the tube, as shown in **Figure 2-2.**

At sea level, vacuum in the column in **Torricelli's tube** would support 29.92 inches of mercury. Mercury is usually abbreviated to its elemental symbol Hg. To this day, we use manometers to measure pressure values both below and above atmospheric. To convert 1 psi to its equivalent in inches of mercury, 1 psi is equal to 2 inches Hg (or more precisely, 2.036 inches Hg).

Manometer. A **manometer** is a single tube arranged in a U-shape mounted to a linear measuring scale as shown in **Figure 2-3.** It may be filled to the zero on the calibration scale with either water (H_2O) or mercury (Hg), depending on the pressures range to be measured.

A manometer can measure either push or pull on the fluid within the column. Actual examples of how we use manometer readings are as follows:

- Vacuum restriction on a diesel engine fuel subsystem primary circuit is specified in inches of mercury ("Hg).

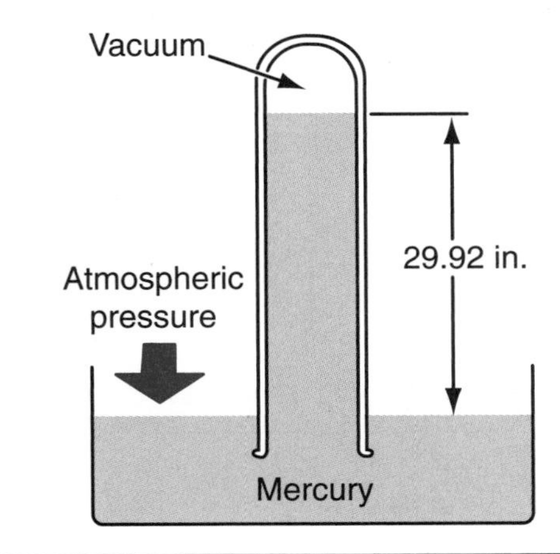

Figure 2-2 Toricelli: Atmospheric pressure expressed in inches of Mercury (Hg).

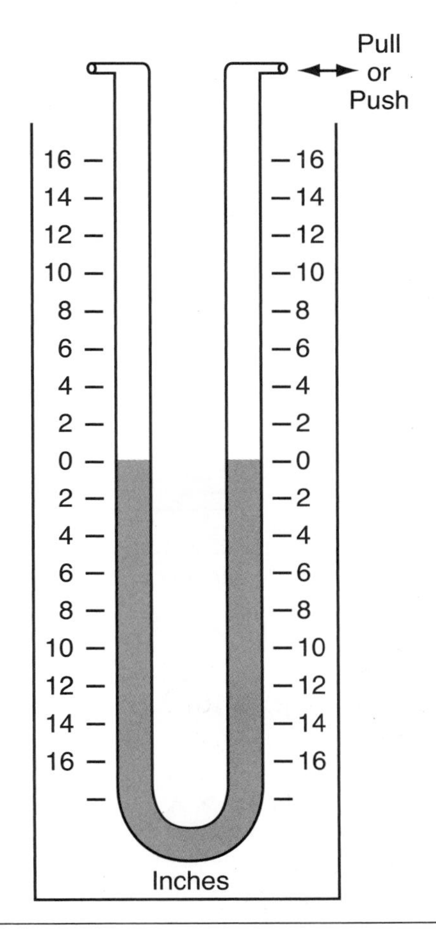

Figure 2-3 Manometer shown at atmospheric pressure: Medium in column can be either H_2O or Hg.

- Inlet restriction of a diesel engine air filter system is specified in inches of water ("H_20).

Note that an inch of mercury is equivalent to 13.6 inches of water, so it is important that you do not mix the two scales. In practice, we tend to use pressure gauges that read values below atmospheric pressure rather than manometers: these pressure gauges display values in inches of mercury on a rotary scale.

Absolute Pressure. Absolute pressure uses a scale in which the zero point is a complete absence of pressure. Gauge pressure has atmospheric pressure as its zero point. A gauge therefore reads zero when exposed to the atmosphere. To avoid confusing absolute pressure with gauge pressure, absolute pressure is expressed as psia. Gauge pressure is usually expressed as psi or psig.

HYDRAULIC LEVERS

Hydraulic levers can be used to demonstrate Pascal's Law. Pressure equals force divided by the sectional area

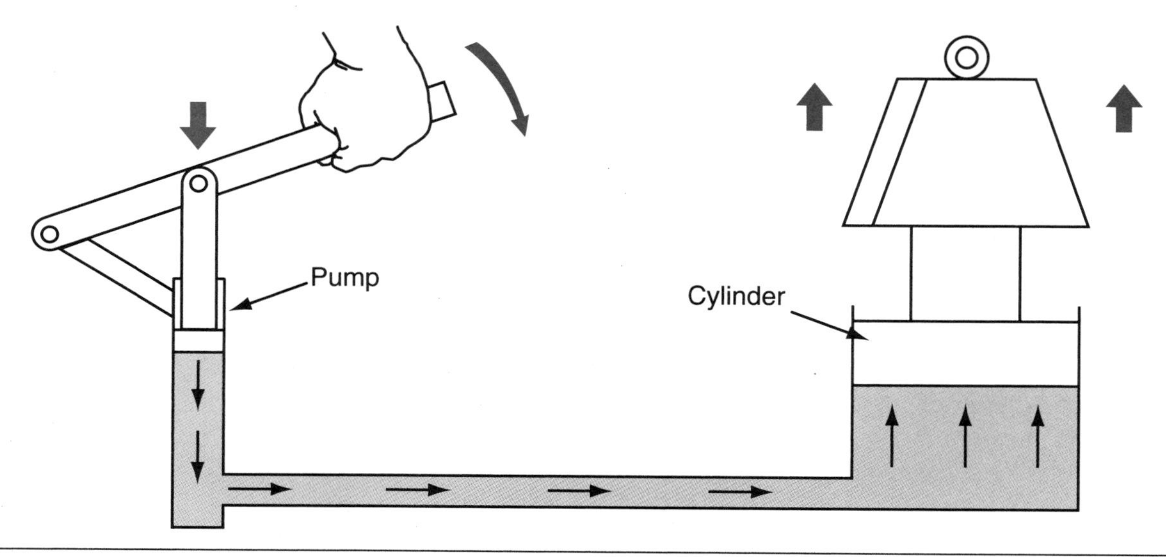

Figure 2-4 Simple hydraulic lever: Pressure multiplies force.

it acts on. Similarly, force equals pressure multiplied by area.

$$F = P \times A$$

So if we want to calculate the force exerted by a cylindrical ram you would use the following formula:

Force exerted = hydraulic pressure (psi)
Force exerted × cylindrical area (diameter^2 × 0.7854)

If we build a simple hydraulic circuit such as that shown in **Figure 2-4,** we have the building blocks of a hydraulic jack.

A variation on the hydraulic jack principle can be seen in the type of hydraulic circuit shown in **Figure 2-5,** the basis of a hydraulic ram. A hydraulic ram is a linear actuator in which hydraulic force is converted into mechanical energy. The arrangement shown in Figure 2-5 consists of a pair of unequally sized cylinders connected by a pipe. In the figure, one of the cylinders has a sectional area of 1 sq. in. and the other, 50 sq. in., so the following would be true:

- Applying a force of 2 pounds on the piston in the smaller cylinder would lift a weight of 100 pounds supported on a piston in the larger cylinder.
- Applying a force of 2 pounds on the piston in the smaller cylinder produces a circuit pressure of 2 psi, because the force is being applied to a sectional area of 1 sq. in.
- The circuit potential is 2 psi and because this acts on a sectional area of 50 sq. in. it can raise 100 pounds.

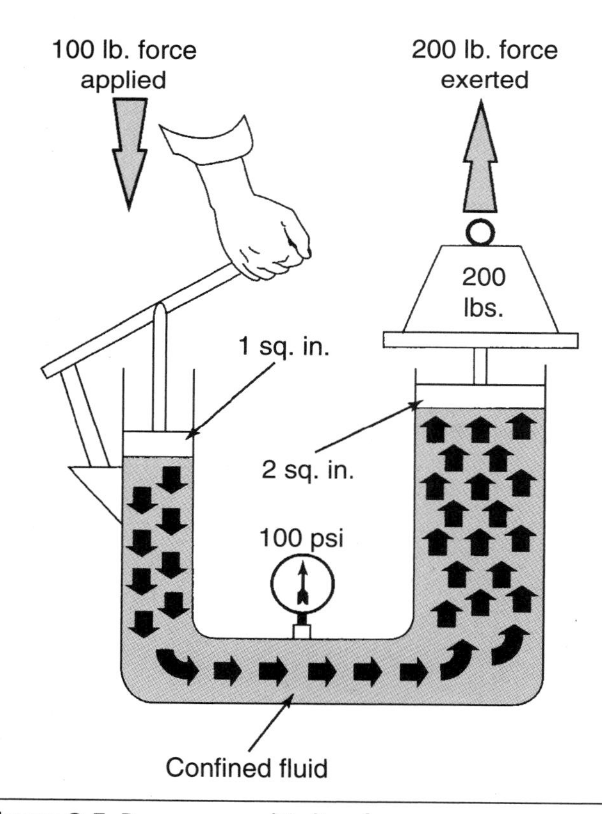

Figure 2-5 Pressure multiplies force.

- Accordingly, if a force of 10 pounds was to be applied to the smaller piston, the resulting circuit pressure would be 10 psi, and the circuit would have the potential to raise a weight of 500 pounds.

This principle is applied every time we use a hydraulic jack. A hydraulic jack is a simple hydraulic lever as is a hoist lift or forklift mast **ram.**

There has to be a trade-off in this type of circuit just as there is with any mechanical lever. When 2 pounds of force is applied to the smaller cylinder to raise a 100 pound weight 1 inch in the larger cylinder, the piston in the smaller piston must move through 50 inches as shown in **Figure 2-6.** This movement is accomplished using multiple strokes and a check valve in typical hydraulic circuits.

Flow

Flow is the term we use to describe the movement of a hydraulic medium through a circuit. Flow occurs when there is a difference in pressure between two points. In a hydraulic circuit, flow is created by a device such as a pump. A pump exerts push effort on a fluid. Flow rate is the volume or mass of fluid passing through a conductor over a given unit of time: an example would be gallons per minute (gpm).

Measuring Flow. Flow can be measured in two ways: velocity and flow rate. The velocity of a fluid in a confined circuit is the speed at which it moves through it. It is measured in feet per second (fps). Flow rate is the volume of fluid that passes a point in a hydraulic circuit in a given time. It is measured in gallons per minute (gpm). Flow rate determines the speed at which a load moves. Fluid velocity and flow rate have to be considered when sizing the hydraulic hoses and lines that connect hydraulic circuit components.

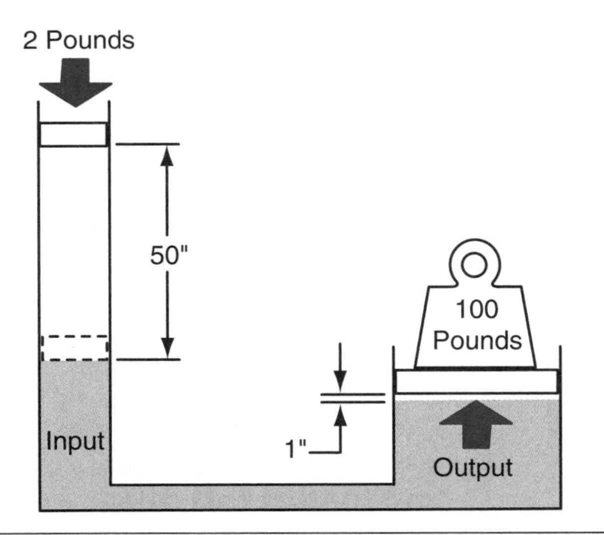

Figure 2-6 Applied force and distance in a hydraulic circuit.

If a small diameter pipe opens into a larger diameter, then we can conclude:

- A constant flow rate will result in lower velocity when the diameter increases.
- A constant flow rate will result in higher velocity when the diameter decreases.
- The velocity of oil in a hydraulic line is inversely proportional to its cross-sectional area.

We should also add that lower fluid velocities are generally desirable to reduce friction and turbulence in the fluid.

The same will work for hydraulic cylinders. Given an equal flow rate, a small cylinder will move faster than a larger cylinder. If the objective is to increase the speed at which a load moves, then

- Decrease the size (sectional area) of the cylinder.
- Increase the flow to the cylinder (gpm).

The opposite would also be true, so if the objective was to slow the speed at which a load moves, then

- Increase the size (sectional area) of the cylinder.
- Decrease the flow to the cylinder (gpm).

Therefore, the speed of a cylinder is proportional to the flow it is subject to, and inversely proportional to the piston area.

Pressure Drop. In a confined hydraulic circuit, whenever there is flow, a pressure drop results. Again, the opposite applies. Whenever there is a difference in pressure, there must be flow. Should the pressure difference be too great to establish equilibrium, there would be continuous flow. In a flowing hydraulic circuit, pressure is always highest upstream and lowest downstream. This is why we use the term *pressure drop.* A pressure drop always occurs downstream from a restriction in a circuit.

Flow Restrictions. Pressure drop will occur whenever there is a restriction to flow, and it will occur downstream from the restriction. In hydraulics, a restriction in a circuit may be unintended, such as a collapsed line, or intended, such as a restrictive orifice. The smaller the line or passage that hydraulic fluid is forced through, the greater the pressure drop. The energy lost because of a pressure drop is converted to heat energy.

Work. **Work** occurs when effort or force produces an observable result. In a hydraulic circuit, this means moving a load. To produce work in a hydraulic circuit, we must have flow. Work is measured in units of force multiplied by distance, for example, pound-feet.

Work = Force × Distance

Energy. There are many forms of energy, and energy is simply the capacity to perform work. In a hydraulic circuit, the objective is to transfer energy. We transfer energy from one form to another and from one point to another. The idea is to accomplish this as efficiently as possible and not waste too much by transforming it into heat.

In a typical hydraulic circuit, mechanical energy is required to drive a hydraulic pump to create flow and kinetic energy potential in the fluid. Fluid under pressure is the potential energy of a hydraulic circuit. This energy can then be reconverted to mechanical energy to move a load. This mechanical energy is kinetic energy, or the energy of motion. The prime source of the energy used in a hydraulic system may be the heat value of the fuel used in an engine or the electrical energy in a battery. The term **prime mover** is used to describe the machine that creates the mechanical energy required to power a hydraulic pump, although it is sometimes used to describe the pump itself.

Bernoulli's Principle. **Bernoulli's Principle** states that if flow in a circuit is constant, then the sum of the pressure and kinetic energy must also be constant. As we said before, when fluid is forced through areas of different diameters, the result is changes in fluid velocity. Fluid flow through a large pipe will be slow, but when the large pipe reduces to a smaller pipe, the fluid velocity will increase. **Figure 2-7** shows a hydraulic circuit consisting of two cylinders connected by a pipe, with force being applied to one of the cylinders. Bernoulli proved that in such a circuit, when force is applied to cylinder on the left, the pressure measured at the connecting pipe will be less than that measured at either

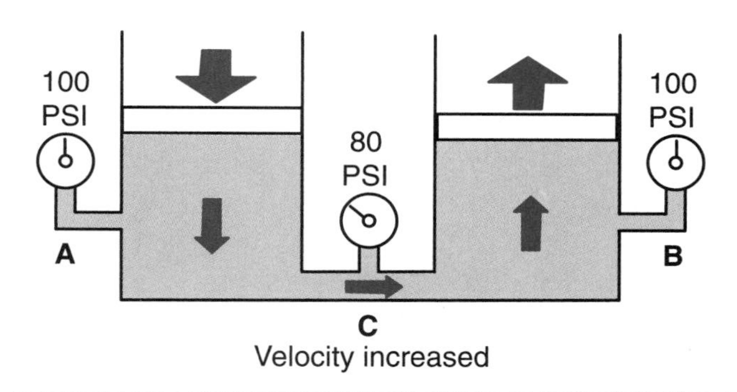

Figure 2-7 Bernoulli's Principle.

cylinder. This happens because the increase in fluid velocity that occurs in the connecting pipe means that some of the pressure energy has been converted to kinetic energy. In the cylinder on the right, assuming that there have been no frictional losses, the pressure should return to the same value as the cylinder on the left because the kinetic energy has been converted back into pressure.

Laminar Flow. Flow of a hydraulic medium through a circuit should be as streamlined as possible. Streamlined flow is known as **laminar flow.** Laminar flow is required to minimize friction. Changes in section, sharp turns, and high flow speeds can cause turbulence and crosscurrents in a hydraulic circuit, resulting in friction losses and pressure drops.

TYPES OF HYDRAULIC SYSTEM

Hydraulic systems can be grouped into two main categories:

- Open-center systems
- Closed-center systems

The primary difference between open-center and closed-center systems has to do with what you do with the hydraulic oil in the circuit after it leaves the pump.

Open-Center Systems. In an **open-center** system, the pump runs constantly, which means that oil flows through the circuit continuously. Furthermore, the pump used is a fixed displacement pump so it produces continuous flow when it is driven. A valve is used to manage the circuit, and when this valve is in its "open" or neutral position, fluid is allowed to return to the reservoir. An example of an open-center hydraulic system is the power-assisted steering used on highway vehicles.

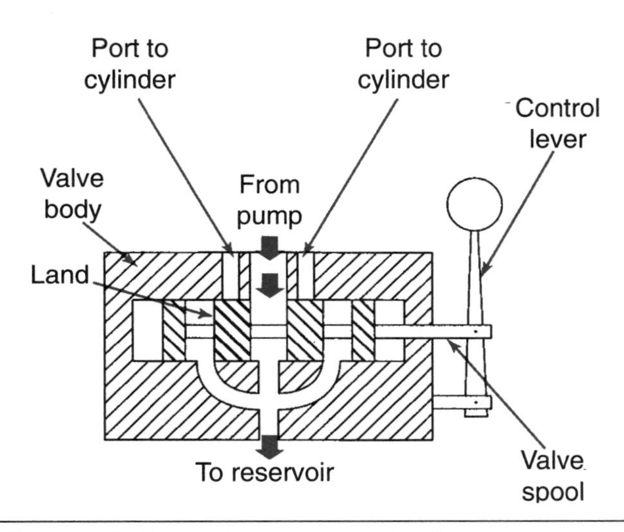

Figure 2-8 Opened-center spool valve.

Figure 2-8 shows an example of an open-center **spool valve.** A spool valve consists of a stationary cylindrical bore within which a spool or piston moves with the objective of directing flow.

Closed-Center Systems. In a **closed-center** system, the pump can be "rested" during operation whenever flow is not required to operate an actuator. This means that the control valve blocks flow from the pump when it is in its "closed" or neutral position. A closed-center system requires the use of either a variable displacement pump or of proportioning control valves. When a variable displacement pump is used, the pump continues to rotate but it ceases to move fluid when rested. Closed-center systems have many uses on agricultural and industrial equipment in applications such as bucket forks and lift rams.

Calculating Force. In hydraulics, force is the product of pressure multiplied by area. For instance if a fluid pressure of 100 psi acts on a piston sectional area of 50 sq. in. it means that 100 pounds of pressure acts on each square inch of the total sectional area of the piston. The linear force that results can be calculated as follows:

Force = 100 psi × 50 sq. in. = 5000 pounds

Summary of Basics. This concludes the introduction to some of the basic principles of hydraulics. Having some knowledge of the language of hydraulics is important to understanding not only the hydraulic circuits used on heavy equipment systems, but also transmissions, steering, brakes, suspensions, engines, and fuel systems.

ONLINE TASKS

Unlike many areas of motive power technology, the Internet can be the source of plenty of useful information on the basics of hydraulics. Here are some tips:

1. Log on to the *http://www.howstuffworks.com* site, **and by using some of the key words in this chapter, see what you come up with.**
2. Use a search engine and punch in Blaise Pascal. See what you can discover about this genius that has nothing to do with hydraulics.
3. Check out this url: *http://hydraulics.eaton.com.* See if you can navigate to any information that explains some of the basic principles discussed in this chapter.

SHOP TASK

Select a piece of mobile equipment in your shop and identify how fluid power principles are used in its various circuits. In this exercise, do not limit yourself to identifying only hydrostatic implement circuits.

Summary

- Fluid power is the practice of using gases or liquids in confined circuits to transmit power or multiply human or machine effort.
- A fluid is any substance that has fluidity: any gas or liquid can be classified as a fluid.
- Fundamental hydraulic principles include Pascal's Law, Bernoulli's Principle, and how force, pressure, and sectional area are used in hydraulic circuits to produce outcomes.
- Pascal's Law states that pressure applied to a confined liquid is transmitted undiminished in all directions and acts with equal force on all equal areas, at right angles to those areas.
- Hydrostatics is the science of transmitting force by pushing on a confined liquid. In a hydrostatic system, transfer of energy takes place because a confined liquid is subject to pressure.

- Hydrodynamics is the science of moving liquids to transmit energy.
- The formula for force (*F*) is calculated by multiplying pressure (*P*) by the area (*A*) it acts on.
- Flow occurs when there is a difference in pressure between two points. In a hydraulic circuit, flow is created by a device such as a pump.
- In a hydraulic circuit, whenever there is flow, a pressure drop results. The opposite also applies: whenever there is a difference in pressure, there must be flow. When the pressure difference is too great to establish equilibrium, continuous flow results.
- Bernoulli's Principle states that if flow in a circuit is constant, then the sum of the pressure and kinetic energy must also be constant.
- An open-center system is one in which the pump runs constantly, resulting in continuous oil flow through the circuit: a valve is used to manage the circuit and when this valve is in its "open" or neutral position, fluid is allowed to return to the reservoir.
- A closed-center system is one in which the pump can be "rested" whenever flow is not required to operate an actuator: this means that the control valve blocks flow from the pump when it is in its "closed" or neutral position.
- A closed-center system requires the use of either a variable displacement pump or of proportioning control valves.
- A typical simple hydraulic circuit consists of a reservoir, pump, valves, actuators, conductors, and connectors.

Review Questions

1. Which of the following best describes the operating principles used to move a forklift
   - A. mast ram
   - B. hydrostatics
   - C. hydrodynamics
   - D. hydrology
   - E. hydromechanics
2. Which of the following is a correct specification of atmospheric pressure at sea level?
   - A. 14.7 inches of Hg
   - B. 29.8 psi
   - C. 101.3 kPa
   - D. 29.8 inches of H_2O
3. Convert a manometer reading of 12 inches Hg into pounds per square inch (psi).
   - A. 6 psi
   - B. 14.7 psi
   - C. 24 psi
   - D. 101.3 psi
4. When two unequally sized, hydraulic cylinders in the same circuit are subjected to equal flow, which of the following should be true?
   - A. The larger cylinder will raise less weight.
   - B. The smaller cylinder will move faster.
   - C. The larger cylinder will require more pressure.
   - D. The smaller cylinder will raise more weight.
5. When there is a fixed restriction to flow in a hydraulic circuit, what should happen to the pressure downstream from the restriction?
   - A. remains unchanged
   - B. reduces
   - C. increases
   - D. surges
6. In an open-center hydraulic circuit, which of the following should be true when describing pump operation?
   - A. runs constantly
   - B. produces higher operating pressures
   - C. manages to have rest cycles
   - D. produces lower operating pressures

7. When the application of force produces movement, what results is known as

   A. horsepower
   B. power
   C. work
   D. friction

8. When a diesel engine is used to drive a hydraulic pump that powers a lift ram, which of the following is correctly known as the *prime mover*?

   A. the diesel engine
   B. the hydraulic pump
   C. the lift ram
   D. the load being moved

9. When a vented reservoir is used in a hydraulic system, what force moves the fluid through the circuit upstream from the hydraulic pump?

   A. vacuum
   B. atmospheric pressure
   C. hydraulic pressure
   D. return pressure

10. Which of the following best describes what we call *laminar flow* in a hydraulic circuit?

    A. restricted flow
    B. streamlined flow
    C. aerated fluid
    D. contaminated fluid

11. Which of the following is a measure of fluid *flow*?

    A. inches of Hg
    B. inches of H_2O
    C. gallons per minute (gpm)
    D. head

12. When a hydraulic force of 1000 psi acts on a sectional area of 50 sq. in., how much linear force is generated?

    A. 50 pounds
    B. 5,000 pounds
    C. 50,000 pounds
    D. 500,000 pounds

13. Who is responsible for the law of hydraulics that states, "pressure applied to a confined liquid is transmitted undiminished in all directions and acts with equal force on all equal areas, at right angles to those areas"?

    A. Newton
    B. Pascal
    C. Bernoulli
    D. Watt

14. Technician A says that substances in a liquid state are classified as fluids. Technician B says that the air we breathe is by definition a fluid. Who is correct?

    A. Technician A only
    B. Technician B only
    C. Both A and B
    D. Neither A nor B

15. Which of the following is a measure of *flow rate*?

    A. psi
    B. gpm
    C. BHP
    D. lb/ft

CHAPTER

3 Hydraulic System Components

Learning Objectives

After reading this chapter, you should be able to

- Identify the key components required in simple hydraulic circuits.
- Identify the types of reservoirs and accumulators used in hydraulic circuits.
- Describe the function of valves used in hydraulic circuits.
- Explain the operating principles of the different types of valves used in hydraulic circuits.
- Define the term *positive displacement.*
- Identify fixed and variable displacement pumps.
- Outline the operating principles of external and internal gear pumps.
- Outline the operating principles of balanced and unbalanced vane pumps.
- Outline the operating principles of axial and radial piston pumps.
- Outline the operating principles of hydraulic motors.
- Describe the construction of hydraulic conductors and couplers.
- Install hydraulic fittings, couplers, and hoses using the correct procedure.
- Outline the properties of hydraulic fluids.
- Interpret both SAE and ISO viscosity grades.

Key Terms

accumulator
cam ring
couplers
directional control valve
dryseal couplers
flow control valve
linear actuators
quick-connect coupler
quick-release coupler
pilot-operated relief valve
positive displacement
ram
reservoir
rotary actuators
spool valve
swashplate
viscosity
viscosity index (VI)

HYDRAULIC COMPONENTS

Most of the components in a hydraulic circuit are interrelated in much the same way as components in an electrical circuit. The components in the circuits described in this chapter will be consistent with those found in mobile hydraulic systems.

RESERVOIRS

A **reservoir** in a hydraulic system has the following roles:

- Stores hydraulic oil.
- Helps keep oil clean and free of air.
- Acts as a heat exchanger to help cool the oil.

A reservoir is typically equipped with:

- **Filler Cap.** This is the means of replenishing oil in the hydraulic circuit. Depending on the system, the filler cap may be equipped with a vent or be held under pressure.
- **Oil Level Gauge** or **Dipstick.** This provides a means of verifying that there is sufficient oil in the reservoir.
- **Outlet and Return Lines.** These provide a means of conducting the oil out of and back into the reservoir. Although the lines can be positioned on top of or at the sides of a reservoir, the end of each line should be located near the bottom within the reservoir. If an oil return line discharges above the fluid level in a reservoir, foaming can result.
- **Baffle(s).** The function of a baffle or baffles in a reservoir is to separate the return oil from that being drawn out by the pump. However, baffle(s) may be unnecessary on some systems, depending on how the lines and filters are arranged in the circuit.
- **Intake Filter.** The intake filter is often just a screen or strainer located in series with the system oil filter.
- **Drain Plug.** Drain plugs provide a means of dumping the hydraulic system oil during system service. Some drain plugs are magnetic to attract and hold small metal particles that pass through the system.

HYDRAULIC JACKS

A hydraulic jack consists of a simple hand-actuated pump plunger, a reservoir, two check valves, and an output piston. Referencing figure 2-4, when the pumping plunger is pulled up, a partial vacuum is created in the plunger chamber. This causes the atmospheric pressure acting on the oil in the reservoir to push fluid into the plunger chamber through the inlet check valve. While the pumping plunger is being pulled upward through its stroke, load induced pressure acting on the output piston holds the outlet check valve closed.

When the plunger chamber has been charged with oil (plunger at the top of its stroke), the pressure in the plunger chamber should be close to the atmospheric pressure in the reservoir. When downward force is applied to the pumping plunger, as soon as the pressure in the plunger chamber exceeds that in the output piston cylinder, the outlet check valve opens, permitting flow and pressure to charge the output cylinder.

Output flow of a simple hydraulic jack is determined by the volume displaced in a single pump cycle. The force developed is defined by the force potential applied to the pumping plunger. This simple hydromechanical device forms the basis of hydraulic floor jacks and single-acting hydraulic cylinders used on equipment.

ACCUMULATORS

A mechanical coiled spring is a simple **accumulator.** When the spring is compressed, it "stores" energy and can also be used to dampen shock and buffer pressure rise. Accumulators used in hydraulic circuits perform the following roles:

- Store potential energy.
- Dampen shocks and pressure surges.
- Maintain a consistent pressure.

In an actual hydraulic circuit, an accumulator is usually used for a specific function in one of the foregoing roles. An accumulator that stores potential energy is often used to back up a hydraulic pump in the event of pump failure or at system start-up. For instance, in large off-highway dump trucks using hydraulically assisted brakes, in the event of oil supply failure, an accumulator is used to feed several charges of oil into the circuit to effect an emergency stop. Accumulators can also be used to dampen shock loads, control pressure surges, and ensure smooth operation of a hydraulic circuit. Although it uses a self-contained circuit, any vehicle shock absorber is a type of accumulator. In terms of operating principles, accumulators generally fall into one of three categories:

- Gas-loaded
- Weight-loaded
- Spring-loaded

Gas-Loaded Accumulators

Although liquids tend *not* to compress, gases do compress; therefore, they can be used to load a volume of oil under pressure. In this type of accumulator, the gas (usually an inert gas such as nitrogen) and hydraulic oil occupy the same chamber but are separated by a piston, diaphragm, or bladder. When circuit pressure rises, incoming oil to the chamber compresses the gas: when circuit pressure drops off the gas in the chamber expands, forcing oil out into the circuit. Most gas-loaded accumulators are pre-charged with the compressed gas that enables their operation. **Figure 3-1** shows a sectional view of a typical gas-loaded accumulator.

Weight-Loaded Accumulators

A weight-loaded accumulator uses a gravity piston and cylinder. Weight is loaded to the piston so that it acts on hydraulic oil in the accumulator chamber. When circuit pressure rises, oil is unloaded into the accumulator chamber thus raising the piston and the weight it supports. When circuit pressure drops off, gravity forces the weight down, thereby charging oil to the circuit.

Spring-Loaded Accumulators

A spring-loaded accumulator functions much like the weight-loaded version except that the weight is replaced by a mechanical spring. This type of accumulator is used to buffer pressure surges in clutch pistons in automatic transmissions.

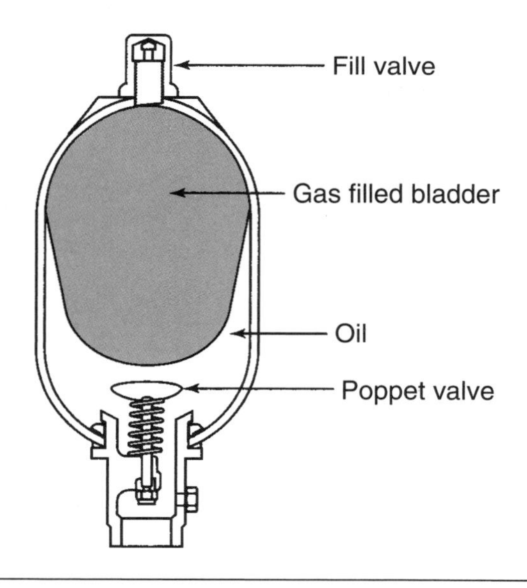

Figure 3-1 Hydraulic accumulator. (*Courtesy of Mack Trucks*)

PUMPS

Pumps create flow. They aspirate fluid at an inlet and displace it to an outlet. Displacement can take place in two ways: non-positive and positive. A non-positive displacement pump picks up fluid and moves it. Examples of this are water wheel or the coolant pump used on a typical diesel engine. Non-positive displacement pumps can be classified as hydrodynamic: if the outlet of this type of pump is blocked, it will mark time.

A **positive displacement** pump not only creates flow but it also backs it up. Fluid picked up by a positive displacement pump is sealed and held while the pump turns. Positive displacement pumps are classified as hydrostatic. They have a positive seal between the inlet and outlet so if the outlet ever became blocked, something would have to give. Most hydraulic circuits use only positive displacement pumps, and we can divide these into two general categories:

- Fixed-displacement pumps
- Variable-displacement pumps

Figure 3-2 shows a comparison between positive and non-positive pumping principles.

When studying pumps, you should remember that hydraulic pumps only create flow: they do not create pressure. Pressure results by creating resistance to flow.

Fixed-Displacement Pumps

A fixed-displacement pump will move the same amount of oil each cycle with the result that the slug volume picked up by the pump at its inlet equals the slug volume discharged to its outlet per cycle. This means that the speed of the pump determines how much hydraulic oil is moved. Fixed-displacement pumps are often used as an assist in a subsection of a hydraulic circuit and to move diesel fuel through a fuel subsystems.

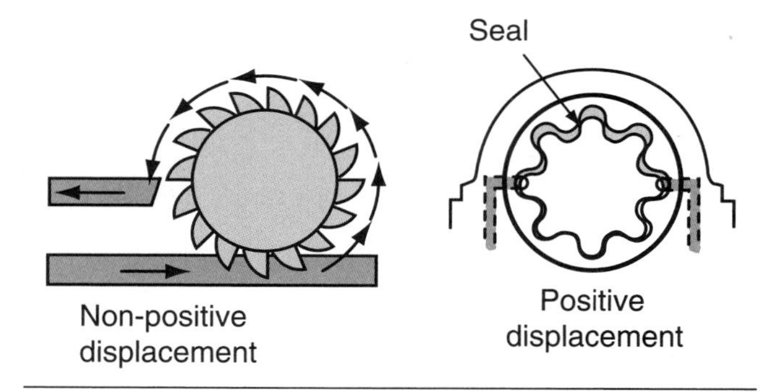

Figure 3-2 Non-positive and positive displacement pumping principles.

Variable-Displacement Pumps

Variable-displacement pumps are positive displacement pumps designed to vary the volume of oil they move each cycle, even when they are run at the same speed. They use an internal control mechanism to vary the output of oil, usually with the objective of maintaining a constant pressure value and reducing flow when demand for oil is minimal.

Gear Pumps

Gear pumps are widely used in mobile hydraulics because of their simplicity. They are also used to move fuel through diesel fuel subsystems and as engine lube oil pumps. Gear pumps are always fixed displacement pumps. Three types of gear pumps are used:

- External gear
- Internal gear
- Rotor gear

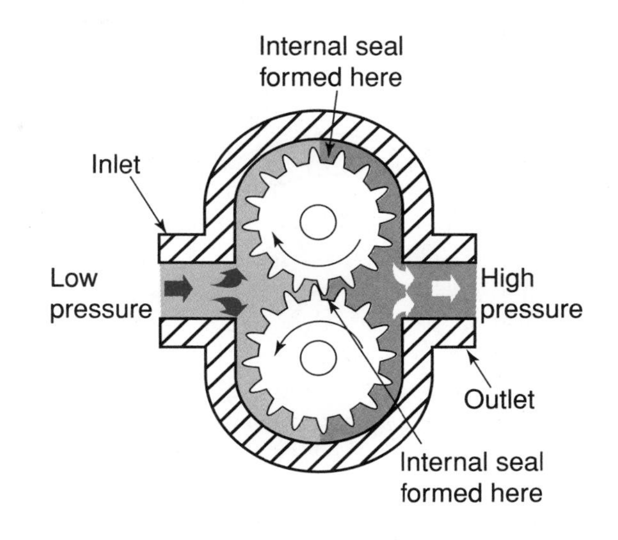

Figure 3-3 External gear pump. (*Courtesy of Mack Trucks*)

External Gear Pumps. Two close-fit, intermeshing gears are driven within a housing. One of the gears is a drive shaft, this drives the second because it is in mesh with it. As the gears rotate, oil from the inlet is trapped between the teeth and the housing. This oil is then moved around the outside of the gear teeth to the outlet of the pump housing. The close mesh of the gear teeth forms a seal that prevents oil from backing up to the inlet. Oil is discharged through the outlet side of the pump housing from which the circuit is charged. This principle is demonstrated in **Figure 3-3. Figure 3-4** shows a cutaway of a typical gear-type pump.

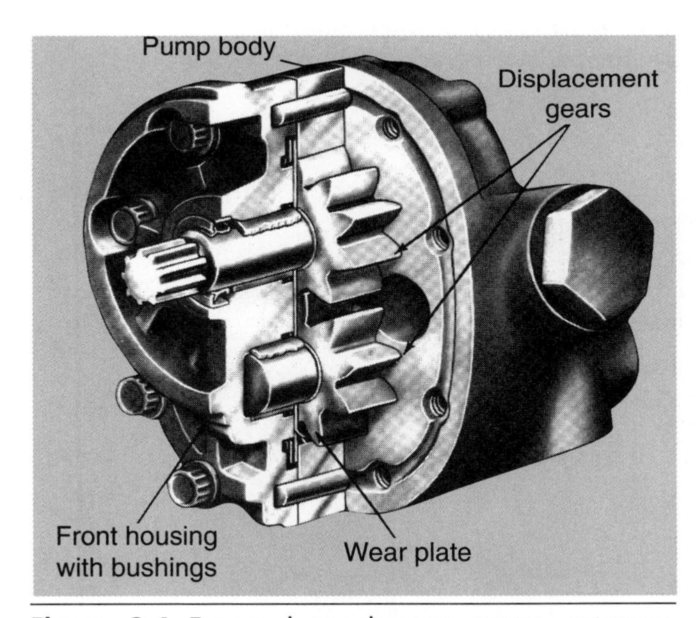

Figure 3-4 External tooth-gear pump cutaway. (Reproduced by permission of Deere & Company, John Deere Publishing, Moline, IL. Hydraulics © 1999. All rights reserved. Visit our website: http://www. deere.com/publications)

Internal Gear Pumps. The internal gear pump also uses a pair of intermeshing gears, but now a spur gear rotates within an annular internal gear, meshing on one side of it. Both gears are divided on the other side by a crescent-shaped separator. The driveshaft turns the spur gear that meshes with the annular internal gear driving. Both gears will rotate in the same direction. As the gear teeth come out of mesh, oil from the inlet is trapped between their teeth and the separator, and this oil is unloaded to the outlet. As with the external gear pump, when the gear teeth mesh, a seal is formed to prevent backup. **Figure 3-5** shows the operating principle of an internal gear pump, and **Figure 3-6** is a cutaway showing the internal components.

Rotor Gear Pumps. A rotor gear pump is a variation of the internal gear pump. An internal rotor with external lobes rotates within an outer rotor ring with internal lobes. No separator is used. The internal rotor is driven within the outer rotor ring. The internal rotor has one less lobe than the outer rotor ring, with the result that only one lobe is fully engaged to the rotor

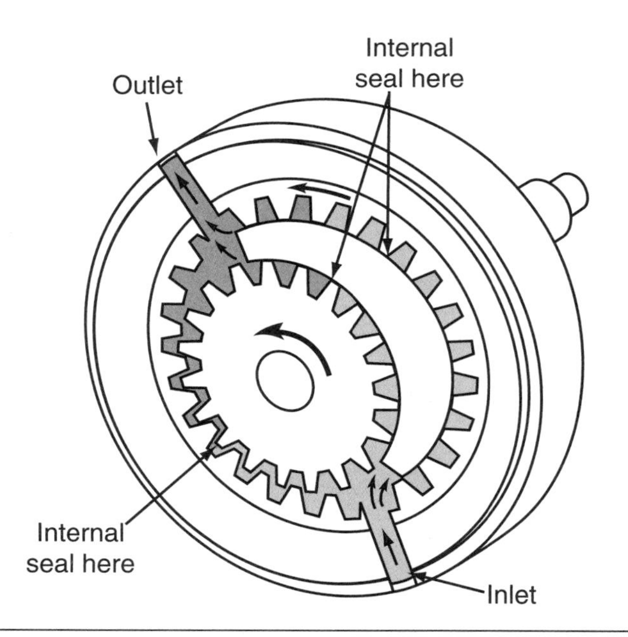

Figure 3-5 Operating principle of an internal-gear pump. (Reproduced by permission of Deere & Company, John Deere Publishing, Moline, IL. Hydraulics © 1999. All rights reserved. Visit our website: http://www.deere.com/publications)

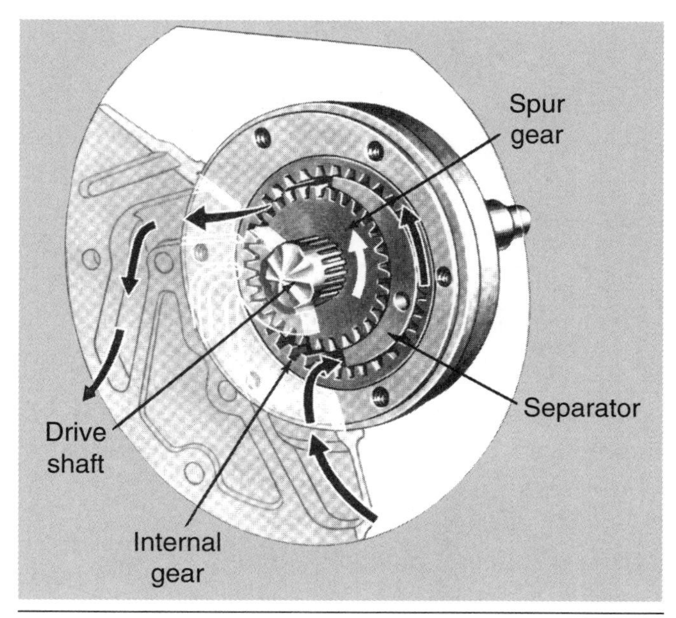

Figure 3-6 Internal-gear pump components. (Reproduced by permission of Deere & Company, John Deere Publishing, Moline, IL. Hydraulics © 1999. All rights reserved. Visit our website: http://www. deere.com/publications)

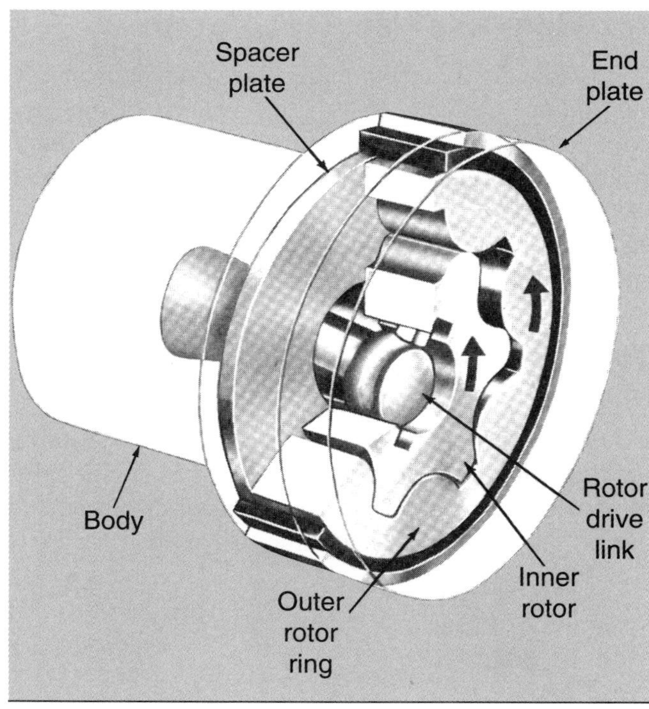

Figure 3-7 Rotor-type pump. (*Courtesy of John Deere*)

ring at any given moment of operation. As the lobes on the internal rotor ride on the lobes on the outer ring, oil becomes entrapped: as the assembly rotates, oil is squeezed out to the discharge port. **Figure 3-7** shows the operating principle of a rotor gear-type pump.

Vane Pumps

Vane pumps are also used extensively in hydraulic circuits. Mobile equipment power-assisted steering systems commonly use vane pumps. Vane pumps function using a slotted rotor that is fitted with sliding vanes: the rotor turns within a stationary liner known as a **cam ring.** The two types of vane pumps are:

- Balanced
- Unbalanced

Balanced Vane Pumps. In a balanced vane pump, the rotor turns within a stationary liner. The rotor is fitted with slots within which sliding vanes are inserted. When the rotor is driven, centrifugal force throws the vanes outward so they contact the liner walls. As the vanes follow the contour of the liner wall, fluid from the pump inlet is trapped between the crescent-shaped "chambers" formed between vanes. These chambers are continually expanding and contracting as the rotor

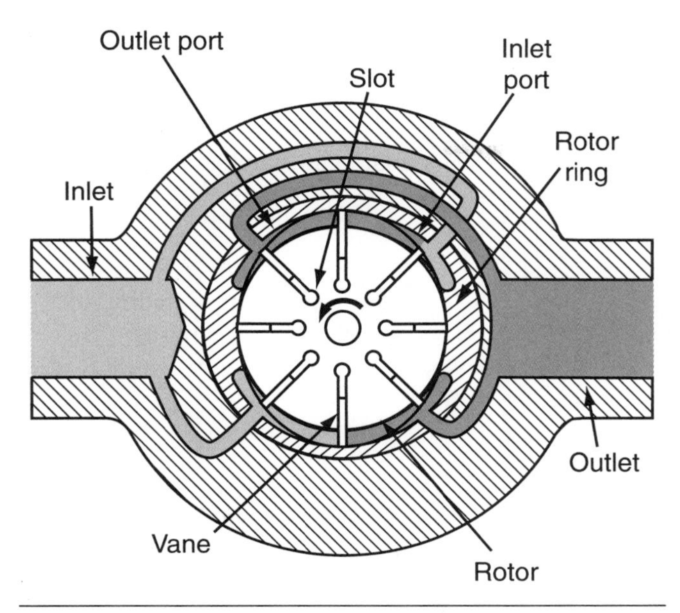

Figure 3-8 Balanced vane pump. (Reproduced by permission of Deere & Company, John Deere Publishing, Moline, IL. Hydraulics © 1999. All rights reserved. Visit our website: http://www.deere.com/publications)

turns: oil from the inlet is trapped in the chamber formed by two vanes. Turning the rotor causes the chamber to contract into the register with the outlet. This action repeats itself twice per revolution because there are a pair of both inlet ports and discharge ports. Balanced vane pumps are always fixed displacement. **Figure 3-8** shows the operating principle of a balanced vane pump.

Unbalanced Vane Pumps. The unbalanced vane pump uses the same principle as the balanced version with the exception that the operating cycle only occurs once per revolution. The location of the axis of the rotor is offset and has only one inlet and one outlet port. In operation, inlet oil becomes entrapped between a pair of vanes when they are expanded and in register with the inlet port. As the rotor turns, the chamber is contracted before the rotation of the pump brings the outlet port into register with the chamber. The disadvantage of the unbalanced vane pump is the radial load placed on one side of the rotor as each chamber formed between the vanes is contracted between fill and discharge: this can cause the bearing to fail. No such force acted on the opposite side of the pump because the inlet oil was under little or no pressure. **Figure 3-9** shows the operating principle of an unbalanced vane pump.

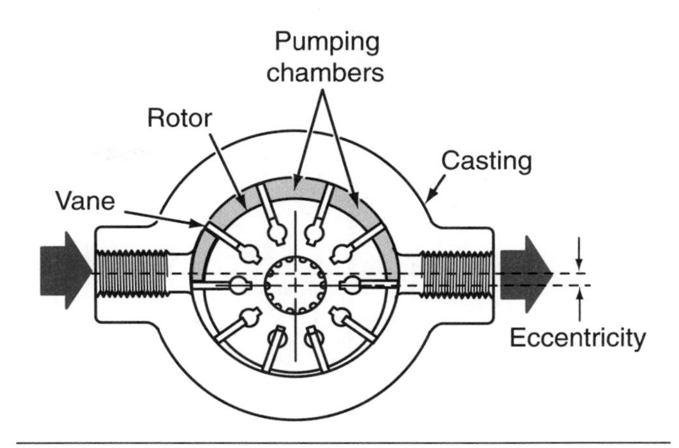

Figure 3-9 Unbalanced vane pump.

Piston Pumps

There is a wide variety of piston pumps, beginning with the simplest pumps and progressing to some of the more complex pumps used in hydraulic circuits. There are three general types of piston pumps:

- Plunger pumps
- Axial piston
- Radial piston

Plunger-type pumps are seldom found on hydraulic circuits but axial and radial piston pumps are used on systems that demand high flow and high-pressure performance.

Plunger Pumps. A bicycle pump is an example of a plunger pump, as are the fuel hand-priming pumps used on many diesel fuel systems. A plunger reciprocates within a stationary barrel. Fluid to be pumped is drawn into the pump chamber formed in the barrel on the outward stroke of the plunger. This fluid is then discharged on the inboard stroke of the plunger.

Axial Piston Pumps. In an axial piston pump, the pumping elements are arranged in parallel with the pump driveshaft axis. A cylinder block houses the pumping elements, which consist of a piston and barrel bores machined into the cylinder head. The base of each piston rides against a tilted plate known as a **swashplate** or wobble plate. The swashplate does not rotate, but in some instances, the tilt angle can be controlled. Fluid is charged to each pump element as the piston is drawn to the bottom of its travel. As the cylinder head rotates (because the piston follows the tilt of the swashplate), it is driven upward through its stroke, forcing fluid out of the chamber through its outlet. **Figure 3-10** shows a

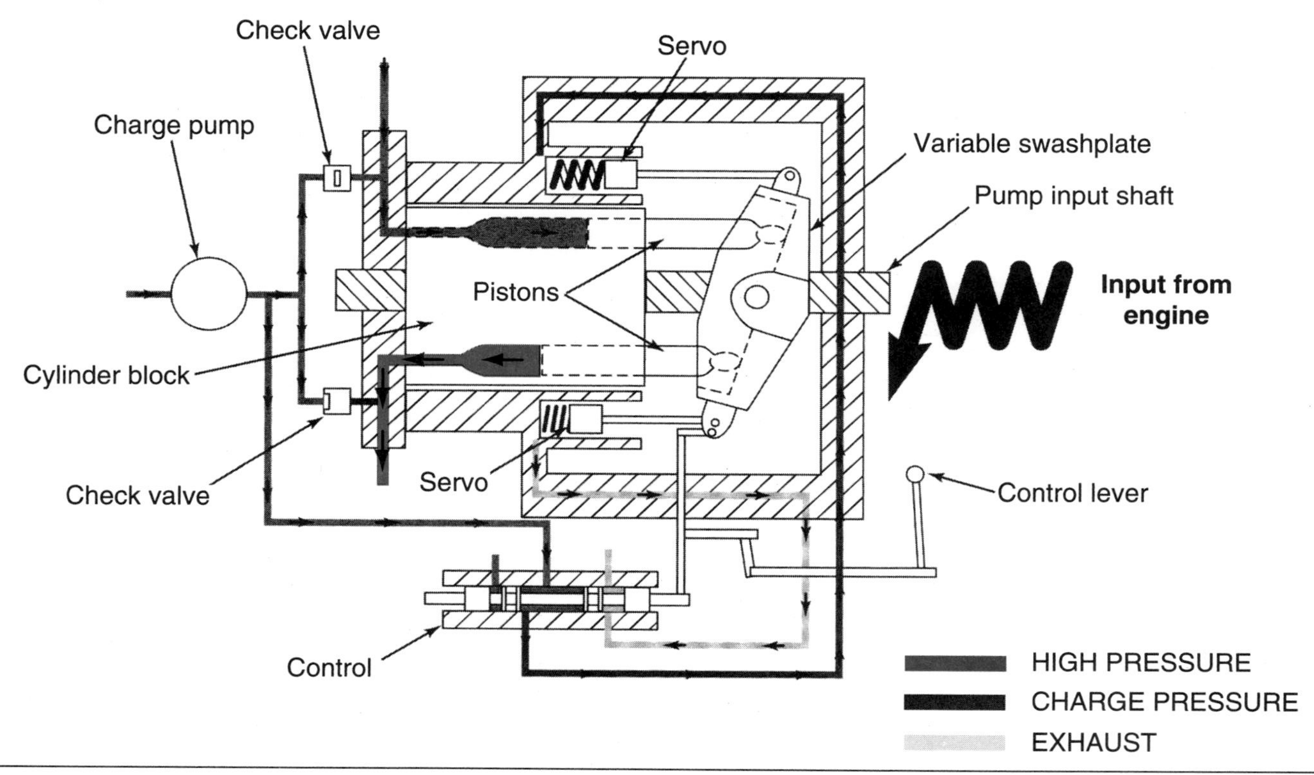

Figure 3-10 Sectional view of a swashplate pump: Note the means of varying the swash angle.

sectional view of a variable-displacement, swashplate pump: in a fixed-displacement version, the angle (pitch) of the swashplate cannot be altered.

If the angle of the swashplate were fixed (it is in some), the pump would be of the fixed-displacement type. When the swashplate angle is movable and controlled by a servo, the distance the pistons are able to move back can be either increased or decreased, and the pump would be classified as variable displacement. The farther the pistons travel, the greater the amount of oil that can be displaced per pump cycle. In **Figure 3-11**, it is shown how varying the swash angle can be used to alter the displacement of the pump.

Radial Piston Pumps. Radial piston pumps are capable of high pressures, high speeds, high volumes, and variable displacement. However, they cannot reverse flow. Radial piston pumps operate in two ways:

- Rotating cam
- Rotating piston

Rotating Cam. In a rotating cam pump, radially opposed pistons are located in a fixed pump body: a central driveshaft is machined with an eccentric cam. As the cam profile is rotated, it forces the pistons outward to effect a pumping stroke. Each piston is loaded by spring force to ride the cam profile. This means that the outward stroke of the piston is effected by cam profile and the inboard stroke by spring force. Oil inlet and outlet passages are located at either end of each pump element. Spring-loaded valves in the inlet and outlet ports permit oil to flow into and out of the piston bores.

During the inlet stroke, the piston is returned by the spring that loads it to ride the cam profile, creating low pressure in the piston bore. This low pressure in combination with oil pressure unseats the inlet valve, allowing the pump chamber to be charged with oil. When the pump chamber is filled with oil, the inlet valve is closed by its spring.

After the pump chamber has been charged, the cam rotates and acts on the piston, driving it outward through its pumping stroke: oil in the pump chamber is

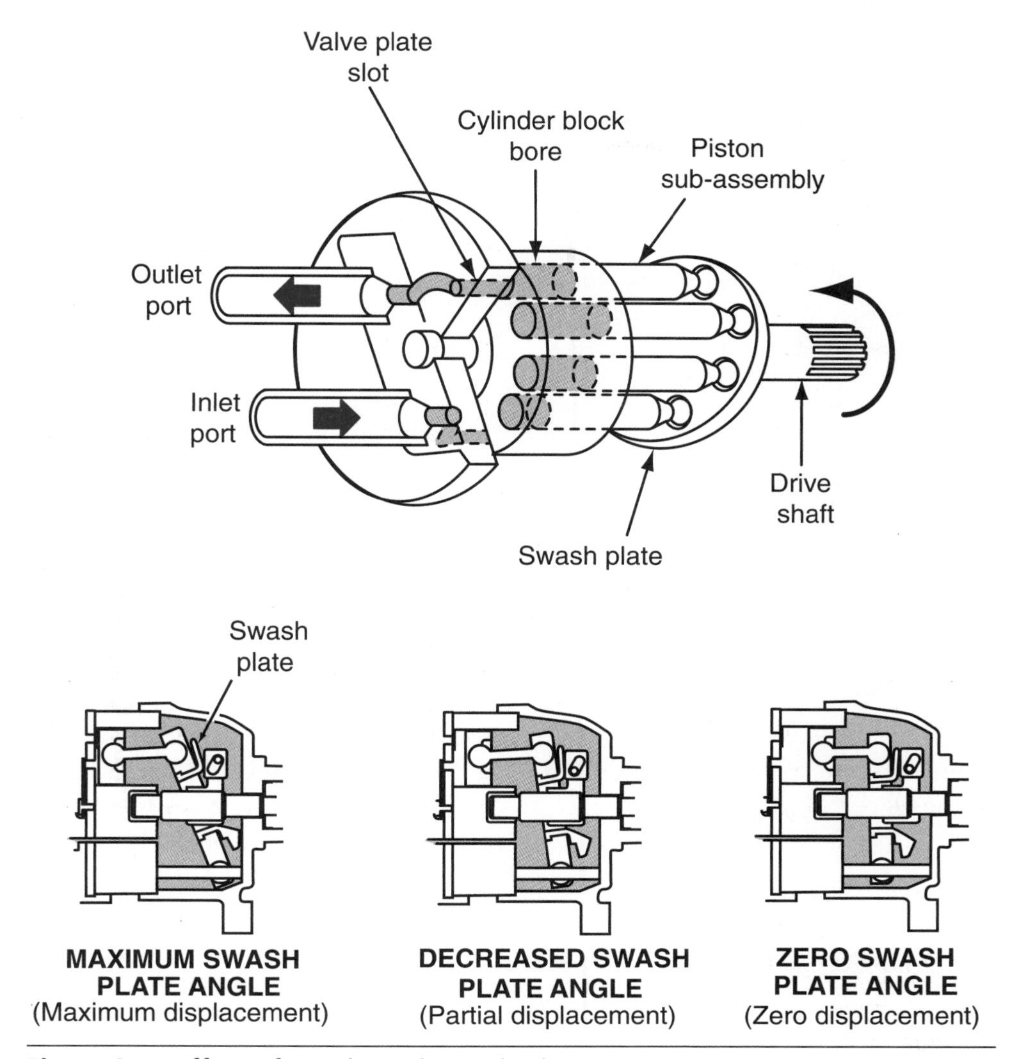

Figure 3-11 Effect of swash angle on displacement.

pressurized and unseats the discharge valve, unloading it into the outlet circuit. The pumping cycles of the pistons work sequentially as the cam rotates: this produces a continuous flow of oil. Four-, six-, and eight-piston versions of the pump are common. Oil output depends on the rotational speed of the pump (assuming it is of the fixed-displacement type).

A common mobile application of cam-actuated radial piston pumps, is the high-pressure pump used on recently introduced common rail, diesel fuel injection. **Figure 3-12** shows the components used in a radial piston pump.

When a variable-displacement rotating cam pump is required, a stroke control mechanism has to be used. This functions to hold the pistons away from the actuating cam profile, either shortening or eliminating the piston effective stroke. A stroke control valve admits oil into the center of the pump so that it acts on and holds off the pistons: the result is that while the camshaft continues to turn, pump chamber output can be held at nearly zero. When the hydraulic circuit requires pump outlet flow once again, hydraulic pressure at the center of the pump is dropped. This allows the piston springs to return the pistons to ride the cam profiles, and pumping action resumes.

Rotating Piston. A rotating piston pump functions similarly to an unbalanced vane pump. Radially reciprocating pistons are arranged within an eccentric cylinder. As the cylinder assembly rotates, the pistons are loaded by centrifugal force against the pumping housing wall. This permits the piston bores to be filled with oil from inlet ports, and as the cylinder is rotated, the pistons are driven inward displacing the oil to outlet ports.

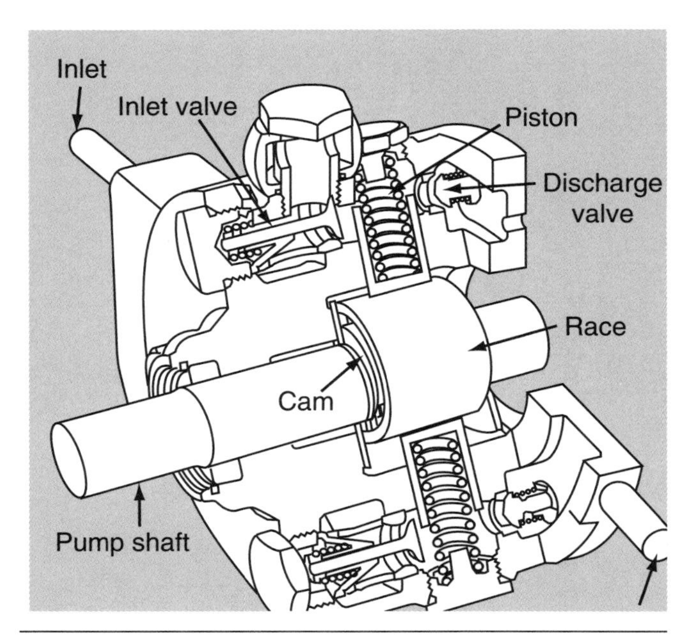

Figure 3-12 Rotating cam-type of radial piston pump. (Reproduced by permission of Deere & Company, John Deere Publishing, Moline, IL. Hydraulics © 1999. All rights reserved. Visit our website: http://www.deere.com/publications)

Variable displacement is achieved by adjusting the axis of the outer housing to that of the cylinder. This alters piston travel and, therefore, the volume of oil displaced per cycle.

Pump Performance

Hydraulic pumps are usually rated by delivery rate or, in other words, the volume it can displace over a given time period. This is usually expressed in gallon per minute (gpm) at a specific rpm, such as 1500 or 1200 rpm. This specification should be accompanied by a rating of the back-pressure a pump can sustain while producing the gpm rating. As pump delivery pressure increases, internal leakage will also increase with the result that usable volume decreases.

In a fixed-displacement pump, flow is related to pump speed. The faster the rotational speed, the more oil pumped. This means that pump ratings should be specified with flow, pressure, and rpm values.

Pump Drives. On both on- and off-highway mobile heavy equipment, hydraulic pumps are usually driven either directly or indirectly by the vehicle engine. Power-take-off (PTO) devices may be driven directly by the engine or from gearing in the transmission. Alternatively, electrical power produced by the chassis electrical system can be used to drive electric motors located anywhere in the equipment or on devices towed by the equipment. For instance, several remotely located electric motors (this is a common alternative use for diesel engine electric cranking motors) may be used to actuate a number of independent hydraulic circuits.

VALVES

Valves are used to manage flow and pressure in hydraulic circuits. There are three basic types of valves used in hydraulic circuits. The general operating principles of these three categories of valves are shown in **Figure 3-13.**

Pressure Control Valves

Hydraulic systems are designed to operate at specific working pressures, so pressure control valves are used to regulate pressure in the system or in sections of a system. Many different types can be used. Examples are pressure-reducing and safety-relief valves. The purpose of these control valves is to separate two circuits working at different pressures. They can also act as safety-relief valves to limit the maximum working pressure of a specific system or circuit.

Counterbalance Valves. Counterbalance valves are used to maintain back-pressure in a circuit or component. They are designed to block a fluid flow path until pilot pressure (direct or remote) overcomes the spring tension setting of the valve and opens the spill circuit. These are often used in a double-acting lift ram to prevent sudden or uncontrolled dropping.

Relief Valves. Two general types of pressure-relief valve are used in hydraulic circuits:

Directly-operated. A directly-operated relief valve consists of two operating ports, a ball poppet, and a tension spring. The tension spring loads the ball poppet against the valve seat, giving the valve a normally closed operating status. When hydraulic pressure acting on the ball poppet is sufficient to overcome the spring pressure, the ball unseats and permits fluid to spill back to the reservoir, bypassing the circuit, and dropping circuit pressure. When system pressure returns to normal, spring pressure once again loads the poppet onto the valve seat. Directly-operated relief valves may be

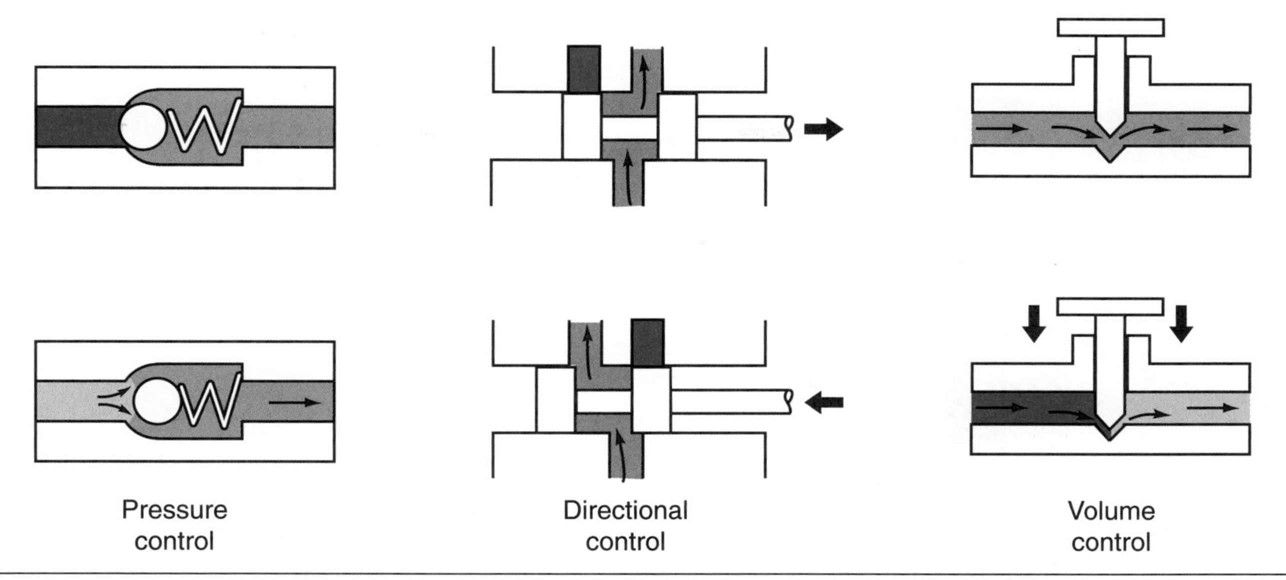

Figure 3-13 Three types of hydraulic valves. (Reproduced by permission of Deere & Company, John Deere Publishing, Moline, IL. Hydraulics © 1999. All rights reserved. Visit our website: http://www.deere.com/publications)

adjustable: a screw located behind the tension spring permits adjustment of the valve opening pressure. **Figure 3-14** shows an illustration of a directly-operated relief valve.

Pilot-operated. A **pilot-operated relief valve** is used to handle large volumes of fluid with little pressure differential. It consists of a main relief valve, through which the main pressure line passes, and a smaller relief valve that acts as a trigger to control the main relief valve. The valve body has a secondary port that directs flow back to the reservoir when the main relief valve opens. A passage in the valve body directs fluid flow to the pilot valve: as inlet pressure increases, so does pressure in the pilot passage. When inlet pressure equals the pilot valve setting, it opens, releasing oil trapped behind the main relief valve and opening it. Excess oil then spills through the discharge port, preventing further pressure rise. At the moment inlet pressure drops below the valve setting pressure, the valves close. **Figure 3-15** shows an illustration of a pilot-operated relief valve.

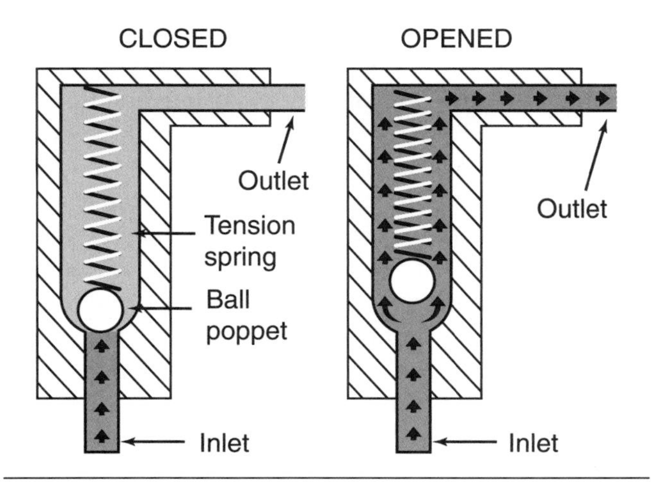

Figure 3-14 Directly-operated pressure-relief valve. (*Courtesy of Mack Trucks*)

Flow Control Valves

Flow control valves are used to regulate the speed of an actuator. These function by defining a flow area and can be classified as adjustable and non-adjustable. They are rated by operating pressure and capacity. They can be used in meter-in or meter-out applications. There are two general types of flow control valve:

- Bleed-off
- Fixed orifice

When meter-in is used to control flow, the valve is placed in series between the control valve and the actuator. This method of flow control is used in hydraulic circuits that continually resist pump flow, such as that used to actuate a dump box ram.

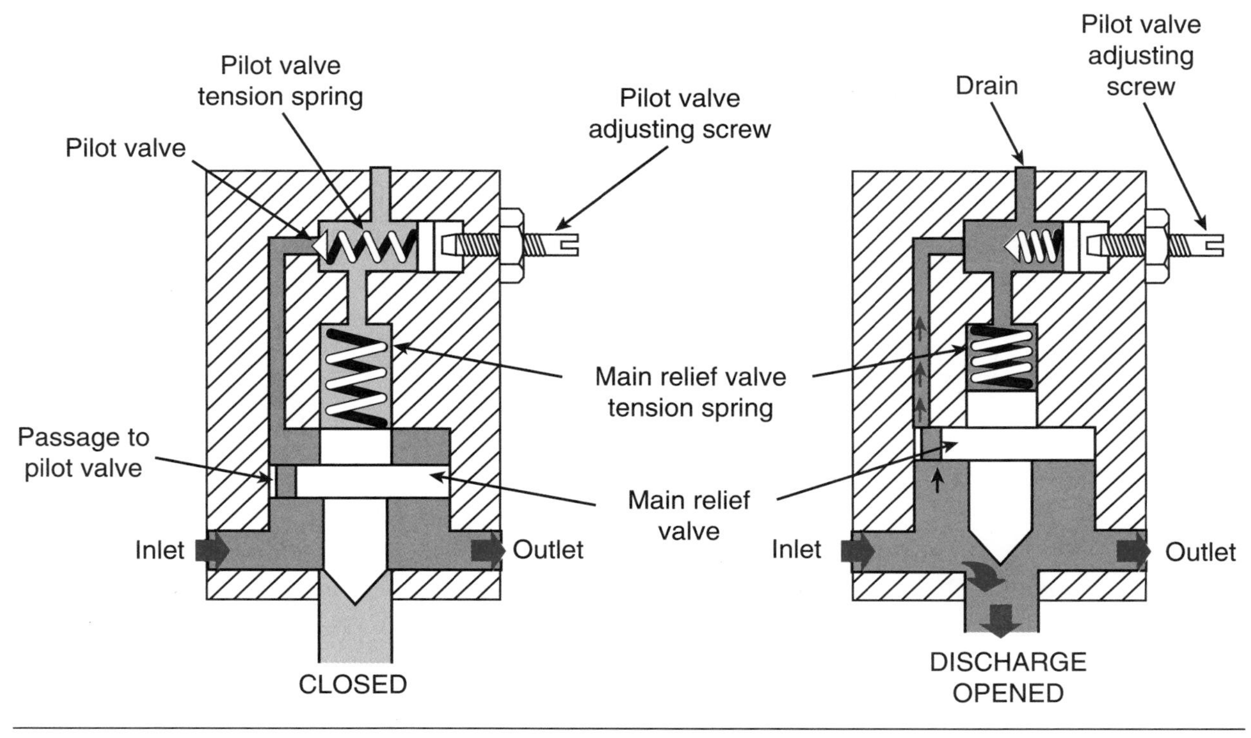

Figure 3-15 Pilot-operated pressure-relief valve. (*Courtesy of Mack Trucks*)

A meter-out flow control valve is required when an actuator has the potential to run away. The meter-out flow control valve is placed between the actuator and the reservoir to control flow away from the actuator.

A bleed-off flow control valve is placed between the pump outlet and the reservoir, and it meters diverted flow as opposed to working flow. The result is that metered flow is directed back to the reservoir at the load pressure rather than the relief valve pressure. Bleed-off control valves are the least precise means of controlling flow.

A simple fixed orifice (fixed restriction) valve in a hydraulic circuit can be used to control flow when it is located in series to control actuator speed by either diverting or slowing flow.

Directional Control Valves

Directional control valves direct flow of oil through a hydraulic circuit. They include

- Check valves
- Rotary valves
- Spool valves
- Pilot valves

The center section of Figure 3-13 shows the operating principles of directional control valves.

Check Valves. A check valve uses a spring-loaded poppet that seats and unseats. They are designed to open to permit flow in one direction but close to prevent flow in the other.

Rotary Valves. A rotary valve uses a rotary spool that turns to open and close oil passages and thereby allows the option of routing of oil to different subcircuits. They are commonly used as pilots for other valves in hydraulic systems with numerous subcircuits.

Spool Valves. Spool valves use a sliding spool within a valve body to open and close hydraulic subcircuits. They are the primary type of directional control valves used in hydraulic circuits and function to start, actuate, and stop most activity in a typical circuit. A spool valve consists of a body and a cylindrical spool machined with annular recesses and lands. Figure 2-8 in Chapter 2 shows a typical spool valve. Though there is no limit to the number of lands that can be used in a spool valve, two-, four-, and six-land spools are most common. Each pair of recesses machined into a spool valve is known as a stack. The spool recesses direct oil to and from sections of a hydraulic circuit, while the lands are used to seal off sections of a circuit. Spool valves may be controlled manually, hydraulically, pneumatically, or electrically. Spool valves are used extensively in hydraulic systems and automatic transmissions.

When spool valves are used for directional control, they can be classified by characteristic or function as follows:

- Number of spool positions
- Number of flow paths in extreme positions
- Flow routing in the center position
- Method of shifting the spool
- Method of retracting the spool

Table 3-1 shows the directional control options possible in spool valves by type.

Pilot Valves. Pilot valves may be controlled mechanically, hydraulically, or electrically. The idea is to locate this type of control valve close to the function it controls, thereby simplifying the hydraulic circuit. Pilot valves may use poppet, rotary, or spool operating principles and are used extensively in hydraulic circuits and automatic transmissions. Figure 3-15 shows a typical hydraulically actuated pilot valve.

ACTUATORS

Hydraulic actuators convert the fluid power from the pump into mechanical work. In mobile hydraulic systems, actuators can be grouped as hydraulic cylinders and hydraulic motors. A hydraulic cylinder is a **linear actuator.** A hydraulic motor is a **rotary actuator.**

Hydraulic Cylinders

There are two types of hydraulic cylinder:

- Piston
- Vane

Although they will be described briefly here, vane type hydraulic cylinders are nearly obsolete.

Piston-type Cylinders. In hydraulics, both single-acting and double-acting cylinders are used. In both types, a piston is moved linearly within the cylinder bore when it is subjected to pressurized oil. In application, a dump box ram requires only a single-acting cylinder while the boom on an excavator requires a double-acting cylinder.

Table 3-1: SPOOL VALVE DIRECTIONAL CONTROL OPTIONS

Type	Classification	Operation
Path of flow	Two-way	2 flow paths in two extreme spool positions
Path of flow	Four-way	4 flow paths in two extreme spool positions
Control	Manually operated	Hand actuated spool shift
Control	Pilot actuated	Hydraulic pressure shifts spool
Control	Solenoid actuated	Solenoid action shifts spool
Control	Solenoid controlled, pilot actuated	Solenoid shifts pilot spool which directs pilot flow to the main spool
Position	Two-position	Spool has 2 extreme positions of dwell
Position	Three-position	Spool has 2 extreme positions plus center position
Spring	Offset spring	Spring returns spool to normal offset position when shifter force is released
Spring	Center spring	Spring returns spool to center position when shifter force is released (always 3-position)
Spool	Open center Closed center Tandem center Partially closed center Semi-open center	The common types of spool valve categorized by flow pattern when the spool is in its center position (3-position valves) or crossover position (2-position valves)

Single-acting Cylinders. In a single-acting cylinder, hydraulic pressure is applied to only one side of the piston. Single-acting cylinders may be either outward- or inward-actuated. When an outward-actuated cylinder has hydraulic pressure applied to it, the piston and rod are forced outward to lift the load. When the oil pressure is relieved, the weight of the load forces the piston and rod back into the cylinder. One side of a single-acting cylinder is dry. The dry side must be vented so that when oil pressure on the pressure side is relieved, air is allowed to enter, thus preventing a vacuum. In a outward-actuated single-acting cylinder, the "dry side" or vent chamber is on the rod side of the piston.

When an inward-actuated cylinder has hydraulic pressure applied to it, the rod is pulled inward into the cylinder. This means that oil pressure acts on the rod side of the piston, and the vent chamber is on the opposite side as shown in **Figure 3-16**.

A seal on the piston prevents oil leakage from the wet side to the dry side of the cylinder, and a wiper seal is used to clean the rod as it extends and retracts. To prevent dirt from entering the dry side of the cylinder, a porous breather is used in the air vent.

A **ram** is a single-acting cylinder in which the rod serves as the piston. The rod is slightly smaller than the cylinder bore. A shoulder on the end of the rod prevents the rod from being expelled from the cylinder. A ram-type cylinder eliminates the need for a dry side because oil fills the entire inner chamber when actuated.

Double-acting Cylinders. Double-acting cylinders can provide force in both directions, which means that oil pressure can be applied to either side of the piston within the cylinder. When oil pressure is applied to one end of the cylinder to extend it, oil from the opposite side returns to the reservoir. This will be reversed when the cylinder is retracted. Good examples of double-acting cylinders are those required to actuate the swing arm and bucket on a backhoe application.

Double-acting cylinders may be balanced or unbalanced. In the balanced double-acting cylinder, the piston rod extends through the piston head on both sides, giving an equal surface area on which hydraulic pressure can act, on both sides of the piston. In an unbalanced double-acting cylinder, a piston rod is located on only one side of the piston. The result is that there is more surface area on which hydraulic pressure can act on the side without the rod. It should be noted that the actual balance or unbalance of these cylinders depends on load. If the loads are not equal on either side, the balances will vary. **Figure 3-17** shows a sectional view of a typical double-acting cylinder.

Vane-type Cylinders. Vane-type cylinders may be found in some much older hydraulic systems. A vane-type cylinder provides rotary motion. It consists of a cylindrical barrel, within which a shaft and vane rotate when pressurized oil enters. As the shaft vane rotates, it closes off an inlet passage in the top plate, leaving only a small orifice to discharge the oil: this slows the rotating vane as it comes to the end of its stroke. Most vane-type cylinders are double-acting. Pressurized oil is directed through one chamber to swing left, the other to swing right. Double-acting vane-type cylinders can be used in applications such

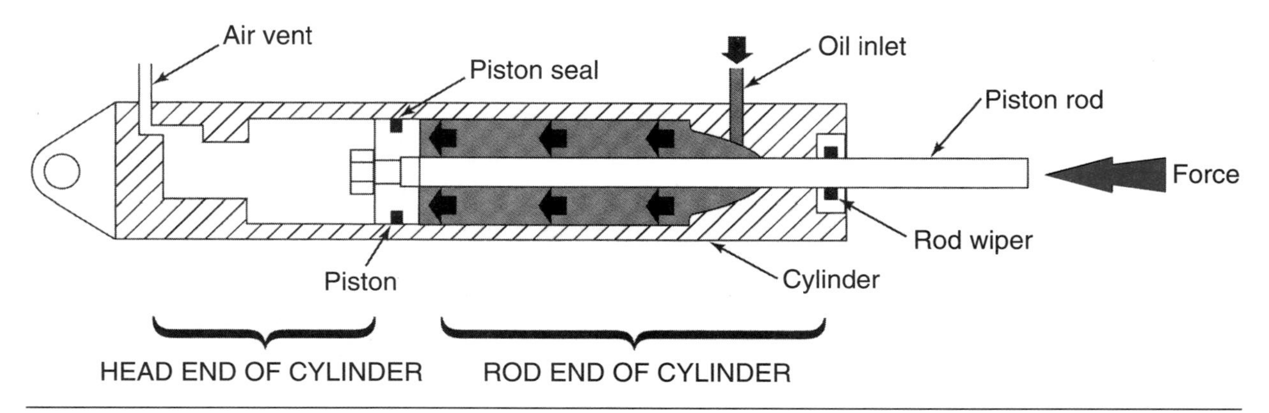

Figure 3-16 Single-acting cylinder. (*Courtesy of Mack Trucks*)

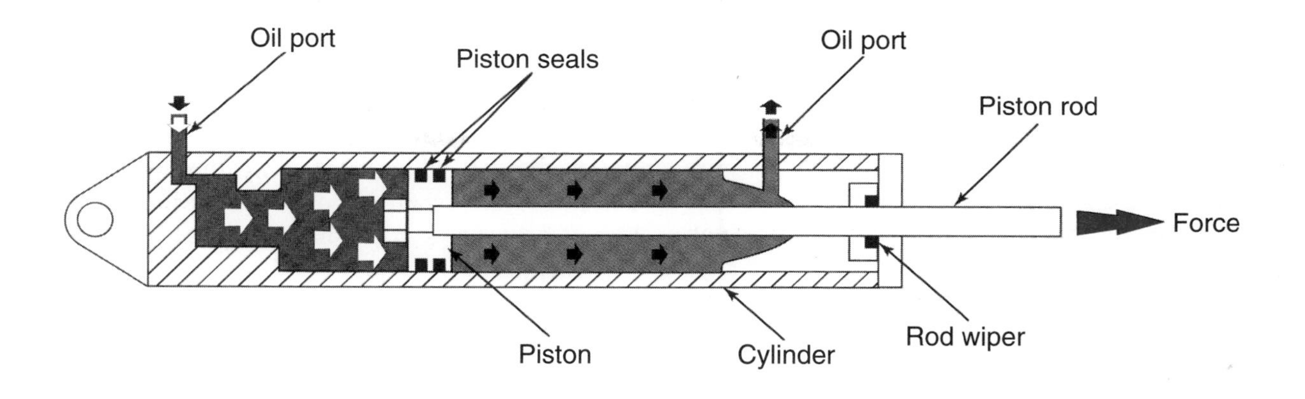

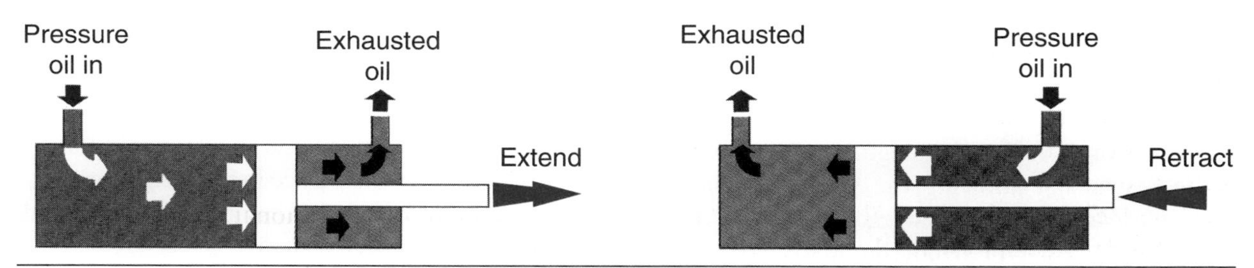

Figure 3-17 Double-acting cylinder. (*Courtesy of Mack Trucks*)

as backhoes because they enable a boom and bucket to be swung rapidly from trench to pile: an alternative to a double-acting vane-cylinder would be to use a pair of cylinders.

HYDRAULIC MOTORS

Hydraulic motors function opposite from hydraulic pumps:

- Pump: draws in oil and displaces it, converting mechanical energy into fluid force.
- Motor: oil under pressure is forced in and spilled out, converting hydraulic energy into mechanical energy.

A pump drives its fluid, while a motor is driven by fluid. **Figure 3-18** shows this principle.

The work output of a motor is in the form of twisting force that we describe as torque. Like pumps, motors can be either fixed displacement or variable displacement. In a fixed-displacement motor, the speed is varied by managing the input flow of oil: this type of motor usually has a fixed torque rating. Variable-displacement motors can have both variable speeds and output torque. Input flow and pressure usually remain constant, so the speed and torque are managed by altering the displacement per cycle. There are three categories of hydraulic motor:

- Gear motors
- Vane motors
- Piston motors

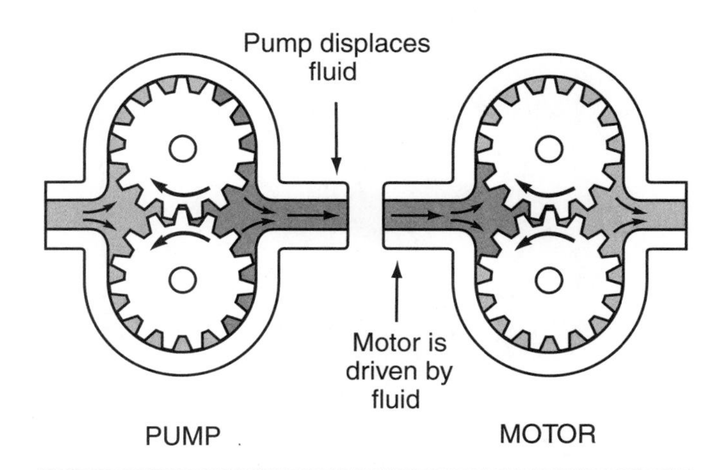

Figure 3-18 Hydraulic pump versus motor principles. (Reproduced by permission of Deere & Company, John Deere Publishing, Moline, IL. Hydraulics © 1999. All rights reserved. Visit our website: http://www. deere.com/publications)

All hydraulic motors rotate: they are driven by incoming hydraulic oil under pressure.

Gear Motors

Gear motors are similar to gear pumps except that they function in reverse. There are two categories of gear motor:

- External gear
- Internal gear

External Gear. An external gear motor is driven by pressurized hydraulic oil forced into the pump inlet. This oil acts on a pair of intermeshing gears, turning them away from the inlet with the oil passing between the external gear teeth and the pump housing. Hydraulic oil is then discharged to the pump outlet at a lower pressure. An output or driveshaft is integral with the axis of one of the two gears, allowing the motor to convert hydraulic potential to mechanical energy. **Figure 3-19** shows an external gear-type hydraulic motor.

Internal Gear. An internal gear motor is similar to an internal gear pump in that an external lobed rotor is driven within an internal lobed rotor ring. The axis of the internal rotor is offset from that of the rotor ring. Pressurized hydraulic oil is delivered to the motor inlet, charging the cavity formed between the inner and outer rotor lobes and forcing both to rotate. This action discharges the oil at lower pressure through the outlet. The motor driveshaft is connected to the inner rotor. **Figure 3-20** shows an internal gear-type hydraulic motor.

Vane Motors

Vane motors use both balanced and unbalanced operating principles and function similarly to vane pumps operating in reverse. However, vane-type hydraulic motors are normally equipped with springs to load the vanes into the liner wall: this is necessary because incoming oil at the inlet is at a high enough pressure (especially at start-up) to overcome any amount of centrifugal force developed by pump rotation and prevent the vanes from contacting the pump liner or rotor ring. Once in motion, the combination of centrifugal force and oil pressure is usually sufficient to load the vanes into the liner wall.

A balanced version of a vane-type motor (see **Figure 3-21**) will locate a pair of diametrically opposed inlet and outlet ports to prevent offset loading of the output shaft connected to the rotor. Balanced vane

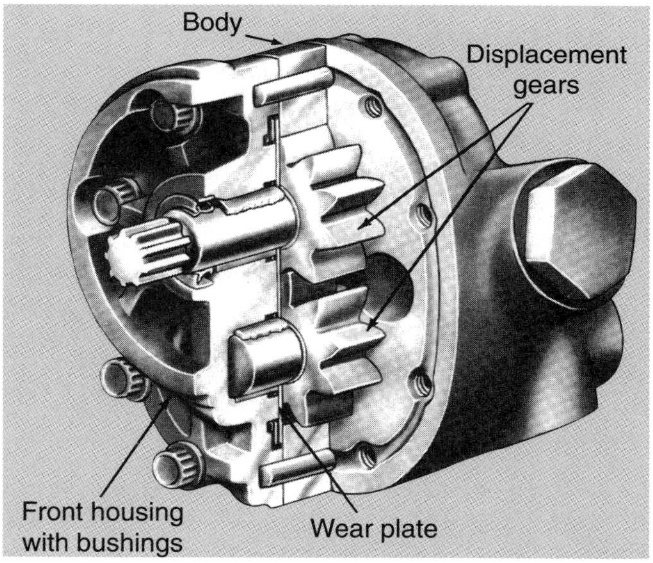

Figure 3-19 External-gear motor. (Reproduced by permission of Deere & Company, John Deere Publishing, Moline, IL. Hydraulics © 1999. All rights reserved. Visit our website: http://www.deere.com/publications)

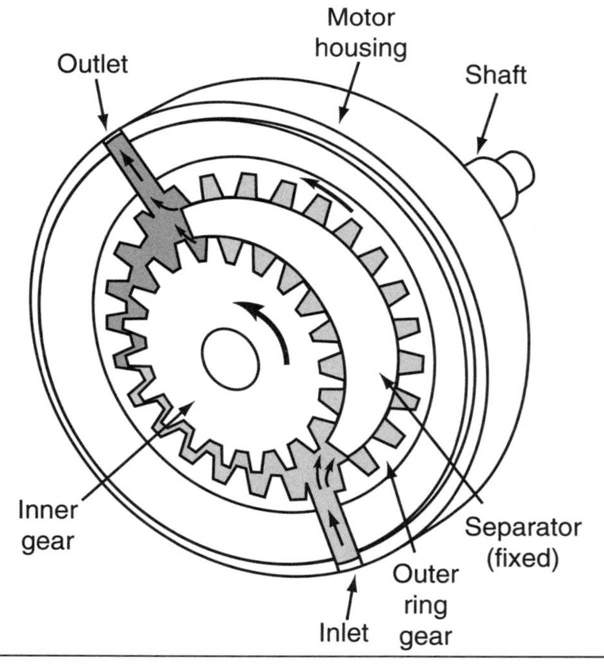

Figure 3-20 Internal-gear motor. (Reproduced by permission of Deere & Company, John Deere Publishing, Moline, IL. Hydraulics © 1999. All rights reserved. Visit our website: http://www.deere.com/publications)

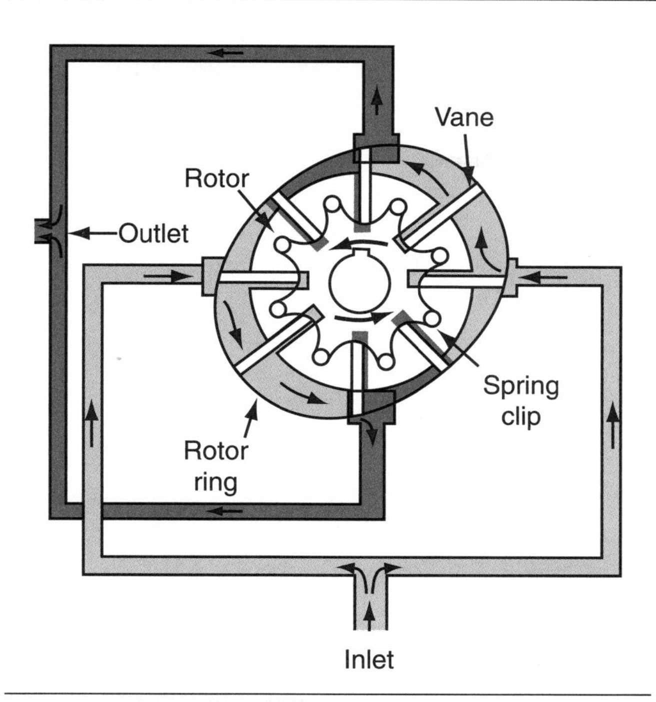

Figure 3-21 Balanced vane motor in operation. (Reproduced by permission of Deere & Company, John Deere Publishing, Moline, IL. Hydraulics © 1999. All rights reserved. Visit our website: http://www. deere.com/publications)

motors are only capable of fixed displacement, but they operate with higher efficiencies than gear motors and rotational direction can be reversed simply by switching the direction of fluid flow.

Piston Motors

Because piston motors are more complex and require more maintenance than gear and vane motors, they tend to be used in applications where higher speeds and pressures are required, such as for hydrostatic drive transmissions. The two types used are:

- Axial piston
- Radial piston

However, in mobile hydraulic applications, it is unusual to find anything but axial piston motors; when radial piston motors are used, it is often in offshore equipment.

Axial Piston. Axial piston motors are often used as hydrostatic drive units on off-highway equipment. The operation of this pump is shown in **Figure 3-22.**

The endcap contains ports A and B, each of which can act as either inlet or outlet ports, depending on the desired direction of rotation. The pistons reciprocate in

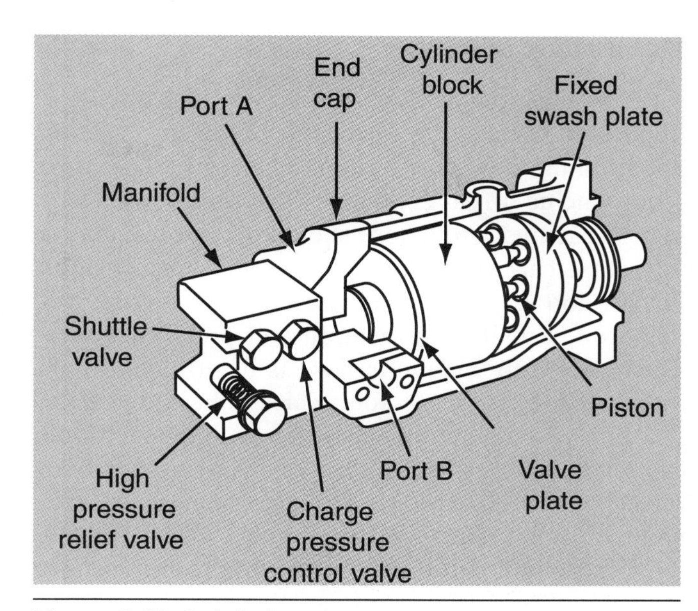

Figure 3-22 Axial piston motor.

bores machined into the cylinder block rotor. As fluid enters port A, the piston bore is charged with high-pressure oil that forces the piston against a fixed-angle swashplate. This allows the piston to slide down the swashplate face, turning the cylinder block that in turn rotates the output shaft. As the cylinder block continues to turn, other piston bores align with inlet port A, allowing each to be similarly actuated in sequence. Oil is discharged at lower pressure through outlet port B as it comes into register at the end of each piston stroke.

Most axial piston motors are fixed displacement and not reversible. In variable-displacement applications of axial piston motors, the swashplate angle can be adjusted by using an arm and lever assembly. The greater the swashplate angle to the shaft, the larger the amount of oil that has to be displaced, resulting in a decrease in motor output speed. In other words, swashplate angle determines motor displacement per cycle.

CONDUCTORS AND CONNECTORS

In any hydraulic circuit, the hydraulic medium has to conducted from the various components plumbed into the circuit. In mobile hydraulic equipment, hoses tend to function best as hydraulic conductors because they:

- Allow for movement and flexing
- Absorb vibrations
- Sustain pressure spikes
- Enable easy routing and connection on the chassis

Hydraulic Hoses

The selection of hoses in a hydraulic circuit is important because the hose play a major role in determining how the system performs. The flow requirements of a conductor determine what the internal diameter of a hose should be. For instance, a hose that can withstand the pressure specification but has too small internal diameter will restrict circuit flow, causing overheating and pressure losses. The size of any hydraulic hose is determined by its inside diameter. This is sometimes indicated as *dash size* in 1/16-inch increments: each dash number indicates 1/16 inch. A #4 dash hose is equivalent to 4/16 inch or 1/4 inch. **Table 3-2** shows some dash sizes and their nominal equivalents.

A large internal diameter hose has to be stronger to sustain the working pressures of a hydraulic circuit. Another consideration for hose selection is that the hose must be compatible with the hydraulic fluid used in the system. There are four general types of hoses used in hydraulic circuits:

- Fabric braid
- Single-wire braid
- Multiple-wire braid (up to 6 wire braid)
- Multiple-spiral wire (up to 6 wire spirals)

These types of hydraulic hoses are shown in **Figure 3-23.**

Table 3-2: DASH SIZE EQUIVALENTS

Dash Size	Nominal diameter
# 4	1/4 inch
# 6	3/8 inch
# 8	1/2 inch
# 10	5/8 inch
# 12	3/4 inch

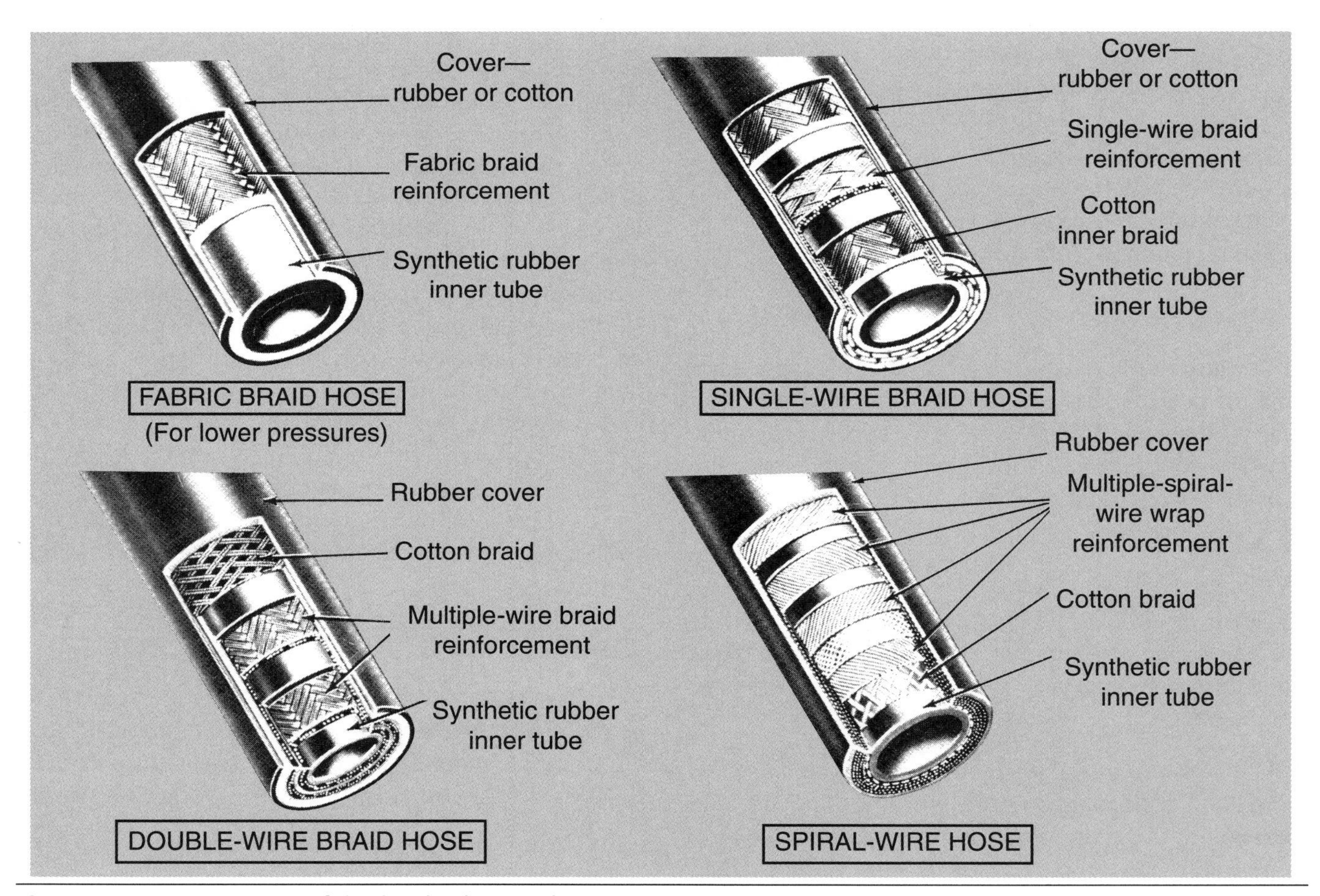

Figure 3-23 Four types of hydraulic hoses. (Reproduced by permission of Deere & Company, John Deere Publishing, Moline, IL. Hydraulics © 1999. All rights reserved. Visit our website: http://www.deere.com/publications)

Fabric Braid. Fabric braid hoses are used for petroleum base hydraulic and fuel oils at low pressures. There are various categories of fabric braid hose, depending on application. They are often used on the low-pressure side of a fuel subsystem or the line that connects a hydraulic reservoir to a pump. They are also commonly used in return circuits. Typically, fabric braid hoses are constructed using a synthetic inner tube with fiber reinforcing, sometimes with a spiral wire to prevent collapse. The outer wrapping can be either synthetic rubber or fiber braid. If fabric braid hose is to be used at lower than atmospheric pressures, the maximum low-pressure specification must be observed. Ratings up to 40 inches Hg inlet restriction are available, but the specified rating drops as the inside diameter increases. Fabric braid line is not recommended for use as pressure lines in mobile hydraulic circuits.

Single-wire Braid Hose. Single-wire braid hose is commonly used as the conductor for the pressure side of diesel fuel subsystems and for some hydraulic circuits. The inner tube is usually manufactured with synthetic rubber reinforced with a single braid of high tensile steel wire. The outer wrap consists of a synthetic rubber or braided cotton, designed to be oil and abrasion resistant. Single-wire braid is often used in truck, farm, and heavy equipment hydraulic circuits and fuel subsystem applications.

Multiple-wire Braid Hose. Double-wire braid hose is commonly used as the conductor for high-pressure hydraulic circuits. The inner tube is usually manufactured with synthetic rubber reinforced with at least two braids of high tensile steel wire. The outer wrap again consists of a synthetic rubber or braided cotton, designed to be oil and abrasion resistant. Double-wire braid is often used in hydraulic circuits using higher pressure. Up to six wire braids may be used, but spiral-wire hose is more common in high-pressure circuits.

Spiral-wire Hose. Spiral-wire braid hose is used high-pressure hydraulic circuits and is the preferred option in systems that have to sustain high pressure surges. The inner tube is usually manufactured with synthetic rubber reinforced with multiple spirals of high tensile steel wire. The outer wrap consists of a synthetic rubber or braided cotton, designed to be oil and abrasion resistant.

Installing Hoses

Some care should be taken when installing hydraulic hoses because improper installation can result in rapid failure. Ensure that the OEM specified working pressures are observed when replacing hoses: **Table 3-3** identifies the maximum working pressures of some common hydraulic hose types. **Figure 3-24** provides some tips on routing hydraulic hoses. The following precautions should be observed:

- **Taut hose.** There should always be some slack when a hose is installed to prevent straining even when the coupled ends do not move in relation to each other. Taut hoses tend to bulge and weaken in service, and hoses under pressure can fractionally shrink.
- **Loops.** Angled fittings should be used to avoid creating loops: this should reduce the total length of hose required.
- **Twisting.** Hoses should not be subjected to twisting either during installation or in operation.

Table 3-3: HOSE WORKING PRESSURES

Hose Size	Working Pressure: Single-wire Braid	Working Pressure: Multiple-wire Braid	Working Pressure: Multiple-spiral Wire
1/4 inch #4	3000 psi	5000 psi	NA
3/8 inch #6	2250 psi	4000 psi	5000 psi
1/2 inch #8	2000 psi	3500 psi	4000 psi
5/8 inch #10	1750 psi	2750 psi	3500 psi
3/4 inch #12	1500 psi	2250 psi	3000 psi
1 inch #16	800 psi	1875 psi	3000 psi
1 1/2 inch #24	500 psi	1250 psi	3000 psi
2 inch #32	350 psi	1125 psi	2500 psi

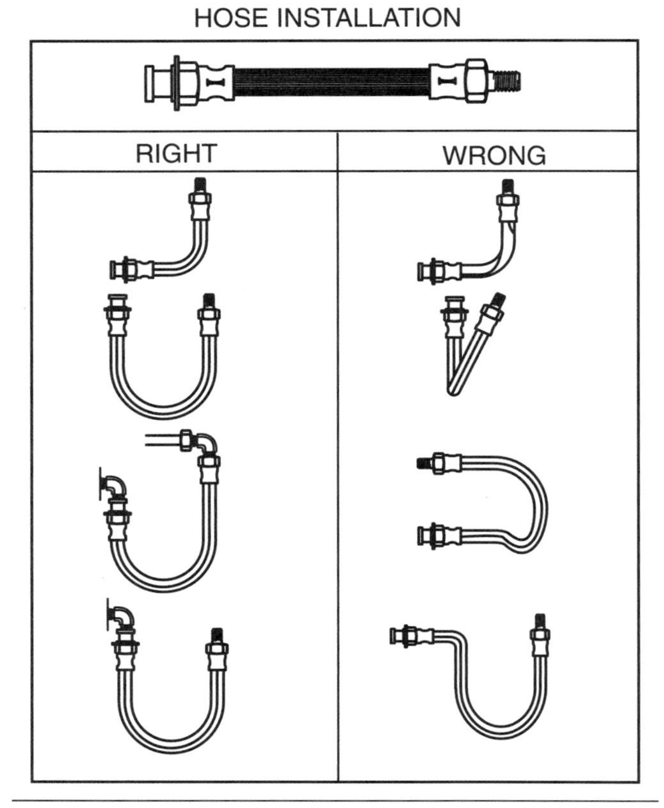

Figure 3-24 Hydraulic hose routing options.

When installing hoses, tighten the fitting on the hose, not the hose on the fitting.

- **Rubbing.** Use hose clamps and brackets to ensure that hoses do not contact moving parts on the chassis.
- **Heat.** Hoses should be prevented from contacting hot surfaces, such as engine components. Shield hoses when necessary.
- **Sharp bends.** Bends choke down flow and can lead to premature failure. The maximum bend radius permitted depends on the type of hose used, but generally, tighter bends are permissible in low-pressure systems.

Couplers (Connectors)

Hydraulic hose **couplers** (also known as connectors and fittings) are made of steel, stainless steel, brass, or fiber composites. Hose couplers or fittings can either be reusable or permanent. Hose fittings are installed at the hose ends and the mating end consists of either a nipple (male fit) or a socket (female fit). Some examples of couplers are shown in **Figure 3-25.** Adapters are separate from the hose assembly and are used to couple hoses to other components such as valves, actuators or pumps.

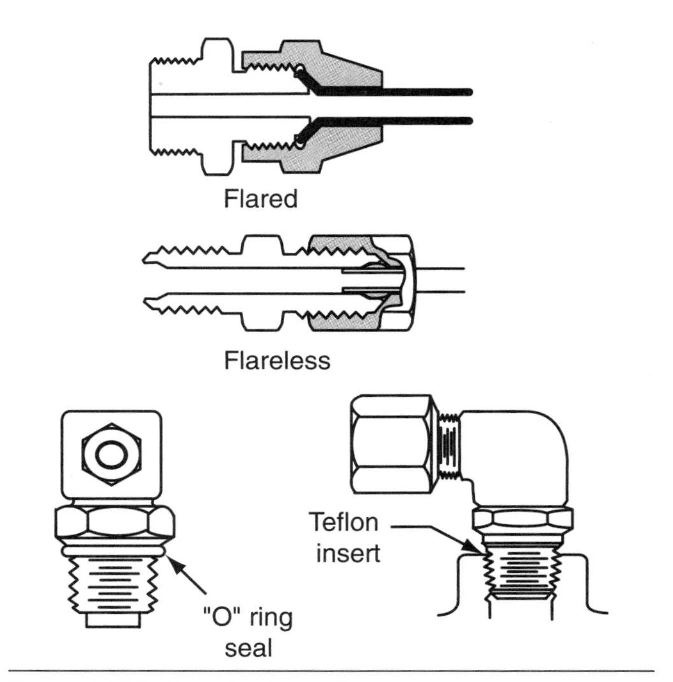

Figure 3-25 Assorted couplers.

Permanent Hose Fittings

When a hydraulic hose is fitted with permanent hose fittings, the fittings are discarded with the hose in the event of a hose failure. These fittings are crimped or swagged onto the hose. When a hose fitted with permanent hose fittings fails, the hose must be replaced. If it cannot be replaced as an assembly, one must be made up using stock hose cut to length and fitted with either crimp-type or reusable fittings. **Figure 3-26** shows some standard permanent hose couplers.

Reusable Hose Fittings

Reusable fittings are common in repair shops: a hose assembly station, some stock hose, and an assortment of reusable fittings can replace many of the hundreds of different types of hoses used on various pieces of equipment. When a hose with reusable fittings wears out, the fittings can be removed and assembled onto new stock hose. Reusable fittings are usually screwed onto hose, although some types of low-pressure hose may use press fits.

Assembling Hose Fittings

Figure 3-27 shows some hose-to-fitting assembly methods. Because of the high pressure used in many hydraulic circuits, the OEM assembly procedure must be precisely observed.

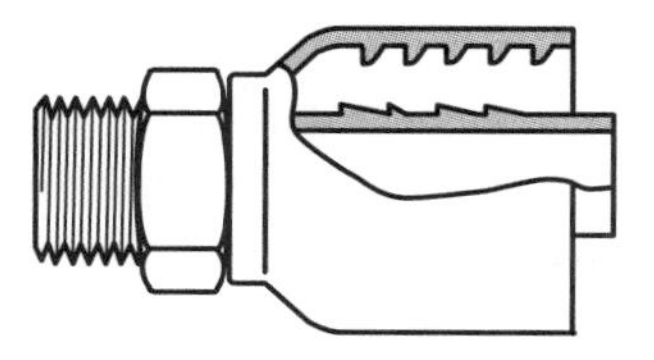

1. SAE straight Thrd. "O" ring

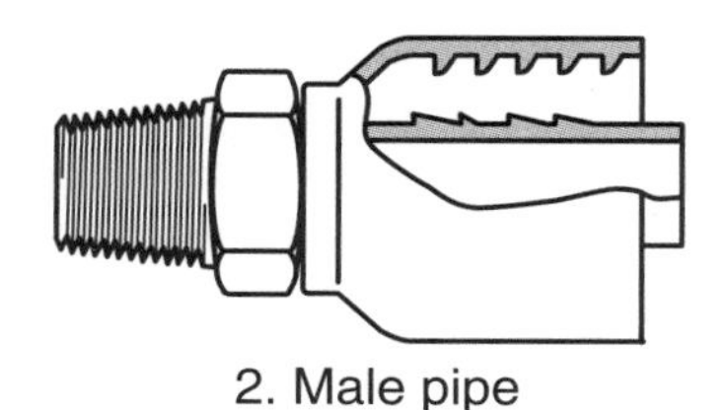

2. Male pipe

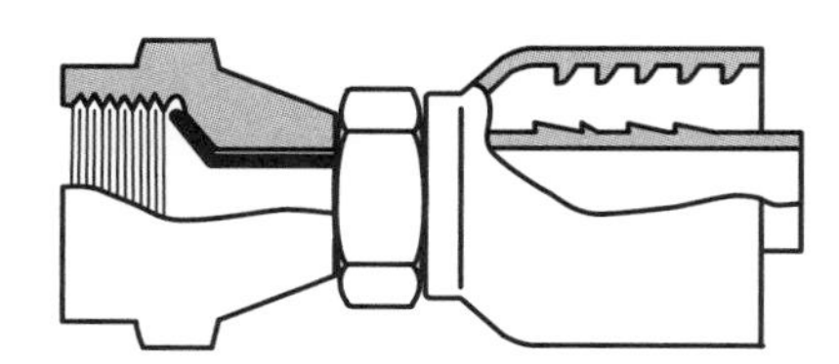

3. 37° JIC

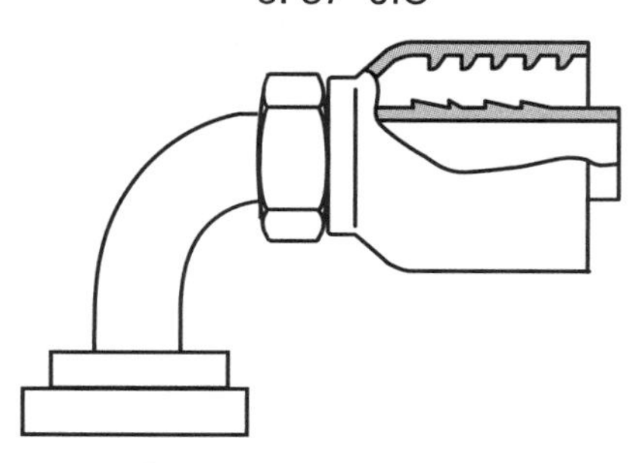

4. Code 61 / 62 SAE flanges

Figure 3-26 Permanent hose couplers.

CAUTION *Separation of a fitting and hose at high pressure can be dangerous! Never reuse a suspect fitting and observe the manufacturer's assembly procedure to the letter.*

Sealing Fittings. Fittings can be sealed to couplers in the following ways:

- Tapered threads
- O-rings
- Nipple and seat (flare)

When making hydraulic connections, ensure that the coupler fittings are compatible with each other.

Adapters. An adapter is separate from the hose assembly. Adapters have the following functions:

- Couple a hose fitting to a component
- Connect hydraulic lines in a circuit
- Act as a reducer in a circuit
- Connect a pair of hoses on either side of a bulkhead

Coupling Guidelines

When making hydraulic connections, the following guidelines should be observed:

- Torque the fitting on the hose, not the hose on the fittings.
- Couple male ends before female ends.
- Ensure the sealing method of each fitting to be coupled is the same.
- Use 45- and 90-degree elbows to improve hose routing.
- Use hydraulic pipe seal compound on the male threads only (on thread-seal unions).
- Use two wrenches when tightening unions to avoid twisting hose.
- Never over-torque hydraulic fittings.

Tech Tip: When tightening the fittings on a pair of hydraulic couplers, always use two wrenches to avoid twisting hoses or damaging adapters.

Pipes and Tubes

Pipes used in hydraulic circuits are generally made from cold drawn, seamless, mild steel. The pipe should never be galvanized because the zinc can flake off and plug up hydraulic circuits. Tubing can also be used, and it has the advantage of being able to sustain some flex, hence its use in vehicle brake systems. Tubing should be manufactured from cold drawn steel if used in moderate- to high-pressure circuits. When used in low-pressure circuits, copper or aluminum tubing may be used.

Pipe Couplers. Pipe joints are made by threading the OD and then turning it into a tap-threaded bore. The standard threads used in vehicle applications are:

- National standard tapered thread (NPT)
- National standard dryseal taper pipe thread (NPTF)

An NPT coupling is sealed by the interference contact between the flanks of the mating threads and,

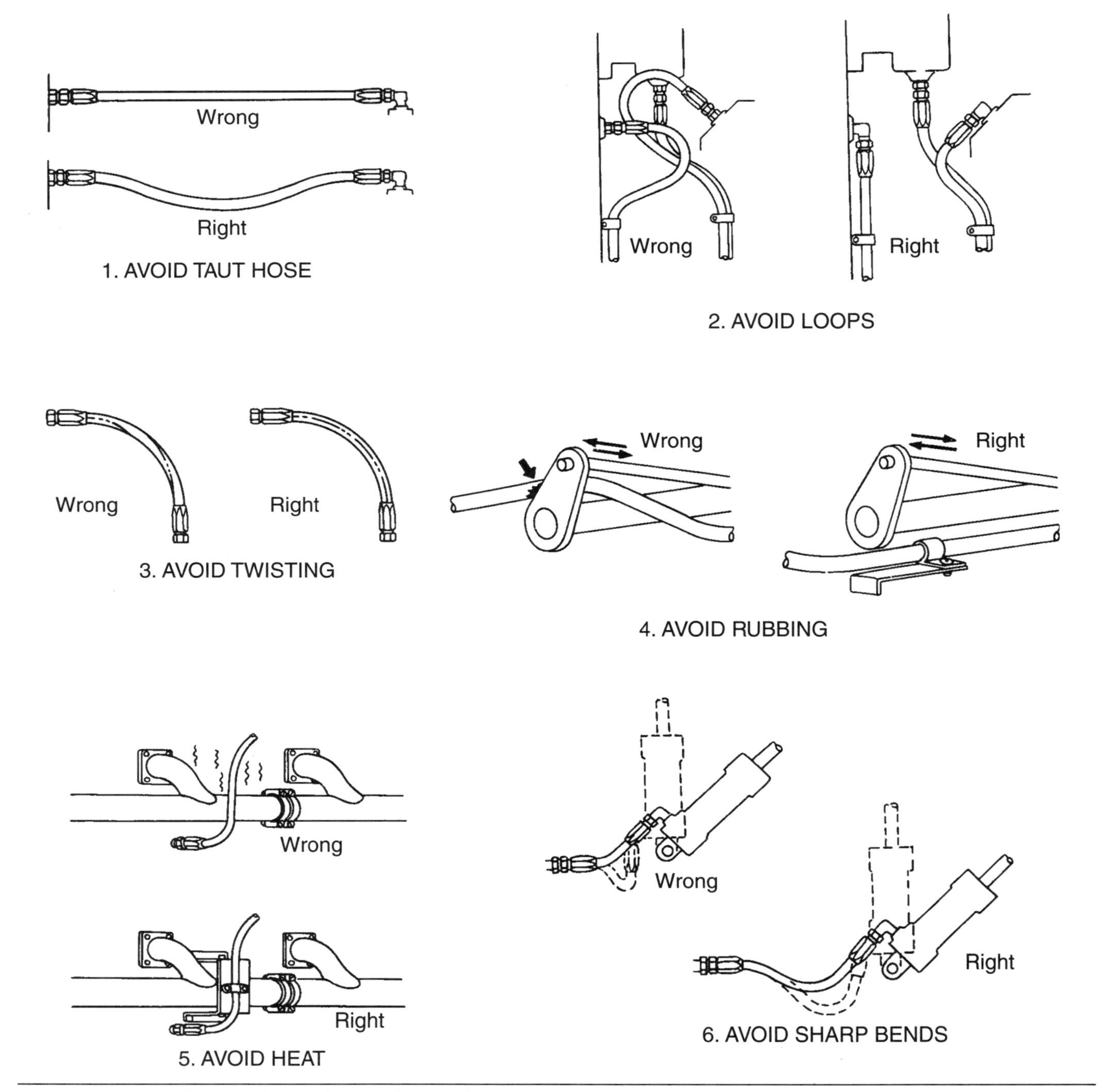

Figure 3-27 Installing hydraulic hoses. (Reproduced by permission of Deere & Company, John Deere Publishing, Moline, IL. Hydraulics © 1999. All rights reserved. Visit our website: http://www.deere.com/publications)

consequently pipe sealing compound is required to ensure a seal. In an NPTF dryseal coupling, the roots and crests of the opposing threads contact before the flanks contact, and a seal is created by crushing the thread crests. **Dryseal couplers** are intended for one-off usage and usually do not require thread seal compound.

Tech Tip: When coupling pipe connectors, remove burrs from both the male and female threads, then apply sealing compound to the male threads only.

Tube Couplers. Tube couplers are designed so that the fitting can be tightened while the tube remains stationary. Two types of tube coupling seal are used: flared and flareless. Flared fittings seal by the metal-to-metal crush created when the flared mating surfaces contact between a pair of fittings. Flareless fittings seal by compression, either with or without a ferrule.

Flared Fittings. Flared fittings are used in thin wall tubing and the flare angle is either 45 degrees (SAE standard), or 37 degrees (joint industry committee or JIC standard). The following types of flare fittings are used:

- **Inverted flare.** A 45-degree flare is rolled to the inside of the fitting body. Commonly used in hydraulic brake circuits.
- **Two-piece flare.** A tapered nut aligns and seals the flared end of the tube.
- **Three-piece flare.** A three-piece flare fitting consists of a body, a sleeve and a nut that fits over the tube. The sleeve free floats, permitting clearance between the nut and tube and aligning the fitting. When tightened, the sleeve is locked without imparting twist to the flared tube.
- **Self-flaring.** These fittings use a wedge-type sleeve: as the sleeve is tightened, the fitting is forced into the tube end, spreading it into a flare. This type of fitting combines the advantages high strength and fast assembly.

CAUTION *Never attempt to cross-couple SAE and JIC fittings: the result will be to damage both.*

Flareless Fittings. Flareless fittings are commonly used in vehicle fluid power circuits, especially pneumatic circuits. They require no special tools for assembly and are reusable. There are three basic types:

- **Ferrule fittings.** These consist of a body, a compression nut, and ferrule. A wedge-shaped ferrule is compressed into the fitting body by the compression nut, creating a seal between the tube and the body.
- **Compression fittings.** These are used with thin-walled tubing and seal by crimping the end of the tube to form a seal.
- **O-ring fittings.** The principle is similar to ferrule type fittings except that a compressible rubber-compound O-ring replaces the ferrule. As the fitting nut is torqued, the O-ring is compressed, forming a seal between the tube and the fitting body. Several different types of O-ring are used, including round section, square section, D-section, and steel-backed.

Tech Tip: When tightening tube fittings, torque only until snug. Over-torquing any kind of line fitting is a direct cause of leaks. Use two wrenches when torquing fittings, one to hold the fitting body and the other to turn the compression nut.

Quick-connect Couplers. When hydraulic lines have to be frequently connected and disconnected, **quick-release couplers** are used. Quick-release couplers are simple connect/disconnect devices with an integral positive seal that allows coupling and uncoupling without admitting air to the circuit. A typical application is a wet-line kit on a dump trailer that has to be connected to multiple tractor units. A quick coupler is a self-sealing device that shuts off flow when disconnected. This eliminates the need to bleed a system each time a disconnect/connect takes place. **Quick-connect couplers** consist of a male and female coupler. There are four types:

- **Double poppet.** A double poppet quick connector has a self-sealing poppet in both the male and female halves. When coupled each poppet is force off its seat, allowing oil flow through the coupling. When disconnected, the poppets in both the male and female halves are closed by spring force, sealing both sides of the disconnected circuit. A connected double poppet coupler is locked into position by a ring of balls that are held into position by a spring loaded, sleeve. When the sleeve is retracted, spring pressure on the ring of balls is relieved, permitting the disconnect.
- **Sleeve and poppet.** Sleeve and poppet quick-connect couplers have a self sealing poppet in one half and a tubular valve and seal in the other half of the coupling. This type of quick coupler provides faster sealing on disconnect, minimizing the entry of air or oil leakage from the coupling.
- **Sliding seal.** Sliding seal couplers use a sliding gate the covers a port on each coupling half at disconnect. It tends to be prone to some leakage during connect and disconnect.
- **Double rotating ball.** Double-rotating ball quick connectors couple by inserting the line plug into the coupler body while turning a locking valve

lever. The action of rotating the locking lever forces back a pair of sealing balls, permitting circuit flow. When disconnected, the locking lever is turned and spring pressure loads the sealing balls into recesses in the locking valve lever, sealing the circuit. Most double rotating ball quick-couplers have an automatic lock release that permits both immediate sealing and hose release in the event of an unintended hydraulic circuit separation. For instance, if such a union is made between a tow unit and a trailer, separation will occur without damage to the hydraulics of either circuit.

HYDRAULIC FLUIDS

Hydraulic fluid is the medium used to transmit force through a hydraulic circuit. Most hydraulic fluids are refined from petroleum-base stocks and are seldom compatible with vehicle brake fluids. Hydraulic fluids used in hydraulic systems are usually specialty hydraulic oils, but engine and transmission oils can also be used. Always check when adding to or replacing hydraulic oil. Synthetic hydraulic oils are commonly used in today's hydraulic circuits because they have wider temperature operating ranges and offer greater longevity.

Hydraulic oils must perform the following:

- Act as hydraulic media to transmit force.
- Lubricate the moving components in a hydraulic circuit.
- Resist breakdown over long periods of time.
- Protect circuit components against rust and corrosion.
- Resist foaming.
- Maintain a relatively constant viscosity over a wide temperature range.
- Resist combining with contaminants such as air, water and particulates.
- Conduct heat.

When minimum operating temperatures are specified for a hydraulic system, fluid operating temperatures, rather than ambient temperatures, are referenced. However, because fluid viscosity increases as temperature drops (the oil thickens), minimum operating temperatures must at least respect the lowest ambient temperatures. If hydraulic oil thickens to the point that the pump inlet becomes restricted, the pump starves for oil, resulting in cavitation and pump damage.

Maximum operating temperatures influence the types of seals used in the circuit components because some seals can harden at high temperatures. Hydraulic fluids thin as temperature increases, resulting in lower lubricity: this results in a loss of lubrication at high load points in the circuit.

Viscosity

Viscosity is a rating of a liquid's resistance to fluid shear, but it is also a measure of its resistance to flow. Petroleum oils tend to thin as temperatures increase and begin to gel as temperatures decrease. In a hydraulic circuit, if the viscosity of hydraulic fluid is too low (fluid is thin), the chances of leakage past seals and poor surface lubrication increase. If viscosity is too high (fluid is thick), more pump force is required to push fluid through the circuit and the operation of actuators may be sluggish.

There are two measures of viscosity: absolute (also known as dynamic) and kinematic (which factors the density of the oil). In hydraulics, kinematic viscosity tends to be used to rate oils.

SAE Viscosity Grades. The viscosity rating of hydraulic fluids is graded using SAE codes similar to engine or transmission lubricants. Lower SAE numbers are used to denote low-viscosity oils and high SAE numbers are used to denote high-viscosity oils. This means that SAE 10W hydraulic oil will flow more readily than SAE 30 graded oil at a given temperature.

ISO Viscosity Grades. International Standards Organization (ISO) is commonly used to grade hydraulic oil viscosity. ISO 32 is the most commonly used viscosity grade, and this is approximately equivalent to SAE 10 grade. Lighter ISO grades such as ISO 15 and 22 tend to be used in extreme cold weather applications, while the heavier grades, 46, 68, and 100 are used in industrial high-pressure applications. The correct viscosity for a specific hydraulic system depends on the following variables:

- Starting viscosity at the minimum predicted ambient temperature.
- Desired viscosity range when the system is subjected to predicted extreme duty and load cycles.

Viscosity Index

Viscosity index (VI) is used to rate the oil's change in thickness as temperature changes. If oil thickens at low temperatures and becomes very thin at high temperatures, it has low VI. When oil can retain a relatively consistent viscosity through a wide temperature range, it can be described as having high VI. In a hydraulic circuit, the hydraulic medium is required to be thick enough to prevent leakage while also offering low flow resistance through the circuit, so high VI is essential.

ONLINE TASKS

1. Use any Internet search engine to research hydraulic oil specifications. By using a specific piece of equipment, your geographic location, and the operating environment, outline the reasons you have selected a particular oil.
2. Check out this url: *http://www.hydraulic-supply.com/*. You may have already come across this when undertaking the online exercise outlined in Chapter 2. Bookmark it.

SHOP TASKS

1. Select any hydraulic hose on a piece of equipment in your shop. Without removing it, identify the hose and coupler specifications to the extent that you can either order (from parts) or make up an equivalent hose using in-house assembly equipment.
2. Select a piece of mobile equipment in your shop. Attempt to identify all the pumps in the system by type. Note the location of each pump and its prime mover. Identify any hydraulic motors in the system.

Summary

- The potential energy of a hydraulic circuit is pressurized hydraulic fluid.
- Hydraulic pumps convert mechanical energy into hydraulic potential.
- The prime mover of a hydraulic circuit in mobile heavy equipment is the vehicle engine either directly by driving the pump, or indirectly, by means of a PTO or chassis electricity.
- Common pumps used in mobile equipment are internal and external gear, unbalanced and balanced vane, and axial and radial piston pumps.
- Valves are used manage flow and direction through a hydraulic circuit.
- Actuators, such as hydraulic cylinders and motors, convert hydraulic potential into mechanical movement.
- Hydraulic oil is used to store and transmit hydraulic energy through a hydraulic system.
- Hydraulic oil viscosity is rated by SAE and ISO standards.
- VI rates an oil's ability to change in thickness as temperature changes. If oil thickens at low temperatures and becomes very thin at high temperatures, it has low VI.

Review Questions

1. When a hydraulic pump discharges the same slug volume it picks up per cycle, it should be described as:
   - A. variable displacement
   - B. constant cycle
   - C. fixed displacement
   - D. non-positive cycle
2. How is hydraulic oil routed through a typical external gear pump?
   - A. through the center between the intermeshing gears
   - B. around the outside between gear teeth and pump body
   - C. through the inlet to the gear output shaft
   - D. around the idler gear shaft to the outlet port

3. A vane-type hydraulic pump has one inlet port and one outlet port it is classified as:

   A. balanced
   B. unbalanced
   C. variable-displacement
   D. twin-cycle

4. Technician A says that a hydraulic cylinder is a linear actuator. Technician B says that a hydraulic motor is a rotary actuator? Who is correct?

   A. Technician A only
   B. Technician B only
   C. Both A and B
   D. Neither A nor B

5. Which of the following is not a function of a hydraulic accumulator?

   A. dampens pressure surges
   B. stores potential energy
   C. maintains consistent circuit pressure
   D. charges hydraulic actuators

6. When applied to hydraulic flow in a circuit, what does the term *laminar flow* mean?

   A. high turbulence
   B. vortex flow
   C. streamlined flow
   D. flow through elbows

7. If a hydraulic pressure of 1000 psi acts on a piston measuring 25 square inches, what amount of linear force will result?

   A. 25 pounds
   B. 250 pounds
   C. 2,500 pounds
   D. 25,000 pounds

8. Which type of gear pump uses a crescent-shaped separator located between and external spur gear and an annular internal gear?

   A. internal gear
   B. external gear
   C. rotor gear
   D. triple gear

9. Which of the following is used in an axial piston-type hydraulic pump?

   A. rotating cam
   B. swashplate
   C. radial piston
   D. compressor type

10. What specification(s) is normally used to rate hydraulic pump performance?

    A. flow in gpm
    B. pressure in psi
    C. revolutions per minute (rpm)
    D. A, B, and C

11. What is usually used to drive mobile equipment hydraulic pumps?

    A. electricity
    B. engine
    C. transmission
    D. A, B, and C

12. Which of the following is the primary function of a flow control valve?

    A. control the speed of an actuator
    B. control circuit pressure
    C. manage circuit resistance
    D. switch circuit outcomes

13. What type of actuator would be required to actuate an excavator boom?

    A. single-acting
    B. double-acting
    C. separate lift and return rams
    D. vane-type motor

14. What force is used to turn a hydraulic motor?
    A. electrical
    B. diesel engine
    C. transmission
    D. hydraulic
15. What type of hydraulic hose should be able to sustain the highest working pressures?
    A. spiral wire
    B. single-wire braid
    C. double-wire braid
    D. fabric braid
16. When comparing a section of 3/8-inch single-wire braid hydraulic hose to a 1/2-inch single-wire braid, which of the following should be true?
    A. The 3/8-inch hose will have a higher specified flow rate.
    B. The 3/8-inch hose will have a higher specified working pressure.
    C. The 1/2-inch hose will have a lower specified flow rate.
    D. The 1/2-inch hose will have a higher specified burst pressure.
17. Which of the following threads used in hydraulic fittings is described as dryseal?
    A. UNF
    B. UNC
    C. NPT
    D. NPTF
18. Which of the following should be true of a hydraulic fluid described as being of high viscosity?
    A. thickens as temperature increases
    B. more pump force is required to push it through the circuit
    C. has better resistance to cold weather gelling
    D. performs better through a wide temperature range
19. What seat angle is required by SAE standard fittings?
    A. 35 degrees
    B. 37 degrees
    C. 45 degrees
    D. 60 degrees
20. Technician A says that hydraulic tubing is generally manufactured from seamed, mild steel tube stock. Technician B says that hydraulic tubing should be galvanized. Who is correct?
    A. Technician A only
    B. Technician B only
    C. Both A and B
    D. Neither A nor B

CHAPTER

Hydraulic Symbols and Schematics

Learning Objectives

After reading this chapter, you should be able to

- Identify typical hydraulic circuit symbols.
- Interpret simple hydraulic schematics.

Key Terms

American National Standards Institute (ANSI)

conductors

International Standards Organization (ISO)

SYMBOLS AND SCHEMATICS

When working with vehicle hydraulic schematics, it is important to be able to interpret standard symbols. Symbols are two-dimensional figures that roughly approximate the components they represent in a hydraulic schematic. The symbols used in hydraulic schematics are standardized by the **American National Standards Institute (ANSI)** and by the **International Standards Organization (ISO).** This helps makes life easier because most manufacturers today use either ISO or ANSI symbols in their schematics. The graphics used in hydraulic schematics consist of shapes and marks.

Shapes and Marks

Symbols for the critical components on a hydraulic schematic will be explained in some detail in this section. However, the basic shapes used in hydraulic schematics are

- Circle/semicircle—indicates a pump or motor symbol.
- Square (or envelope)—represents one position or path through a valve. Two squares together indicate a two-position valve.
- Diamond—indicates a component that conditions fluid in the circuit, such as a filter or heat exchanger.
- Rectangle—indicates a hydraulic cylinder or reservoir.

A symbol is drawn by using one of the four basic shapes above and adding the appropriate marks. Typical marks include

- Solid lines—indicate the flow route for hydraulic fluid.
- Dashed lines—indicate a pilot line that connects a control circuit to a slave circuit.
- Dotted lines—indicate a return or exhaust circuit.
- Center lines—enclose assemblies.
- Arrows—show the direction of fluid flow or rotational direction of pumps and motors.
- Arcs—show points of adjustment in the circuit such as the flow control valve.

Conductor Symbols. Hydraulic pipes, hoses, and tubes that carry fluid in a hydraulic circuit are known as **conductors.** They are represented by lines on a schematic. Solid lines indicate inlet, pressure, or return circuits. Broken lines with long dashes usually indicate pilot or control lines, while lines of short dashes indicate leakage oil. An arced line (that is not semicircular) between two dots indicates a flexible hose. **Figure 4-1** shows how conductors are typically represented on a hydraulic schematic.

A semicircular loop on one line crossing another indicates that the two lines are not hydraulically connected. However, sometimes crossing (but not hydraulically connected) lines simply cross each other. In the event that a pair of intersecting lines are hydraulically connected, a dot is shown at the intersection point. These three arrangements are shown in **Figure 4-2.**

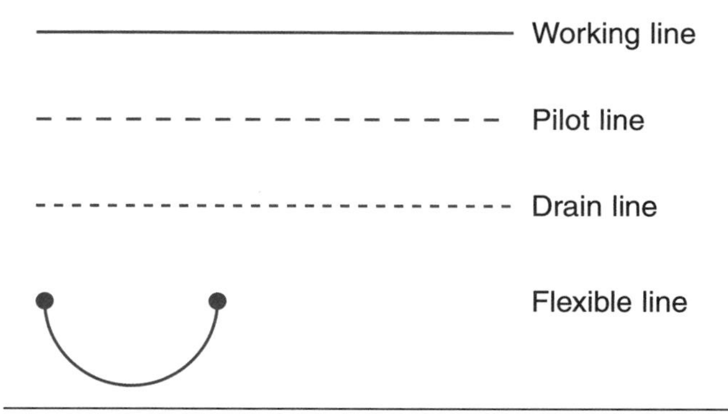

Figure 4-1 Conductor symbols.

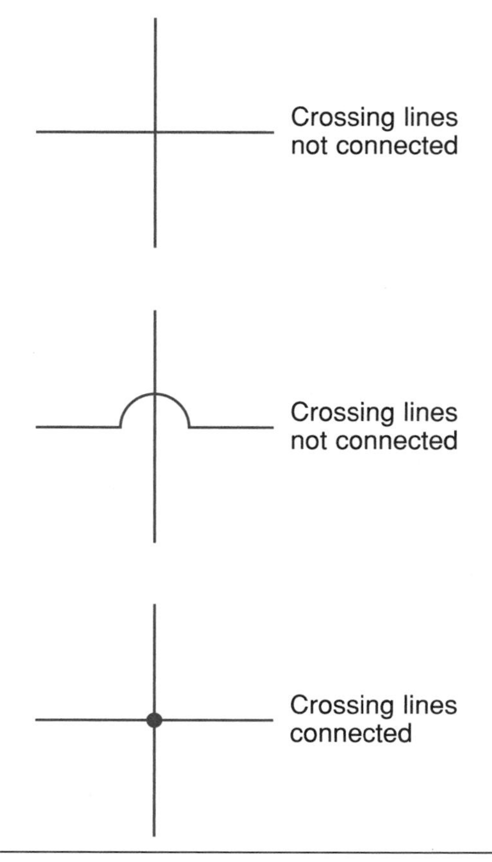

Figure 4-2 Intersecting lines.

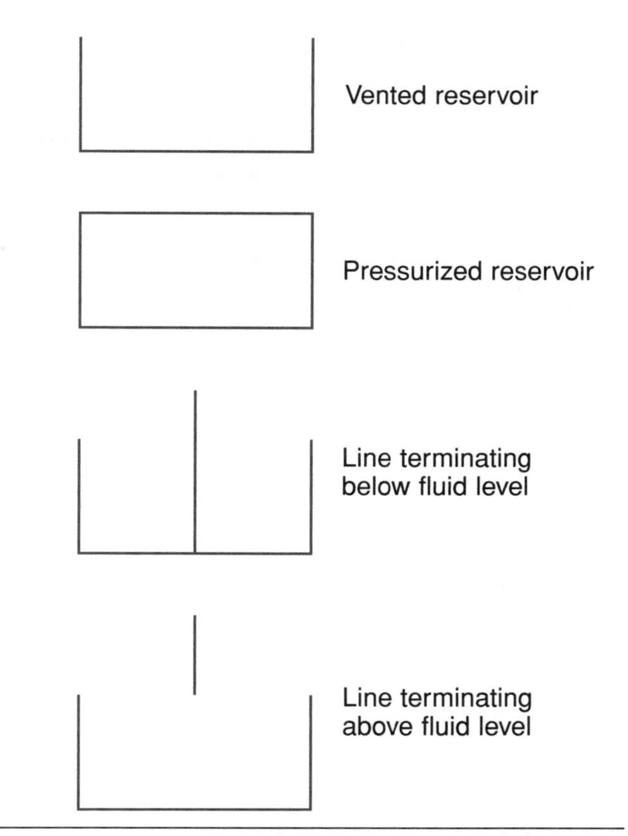

Figure 4-3 Reservoir symbols.

Reservoir Symbols. A horizontal rectangle is used in schematics to represent a reservoir. A pressurized reservoir is indicated when the rectangle is drawn fully enclosed as indicated in **Figure 4-3.** A vented-to-atmosphere reservoir is indicated when the rectangle is open at the top as indicated in Figure 4-3.

The way in which lines are shown entering a rectangle symbol is also of significance. A line terminating either above the rectangle or before its base indicates that return flow is above the fluid, whereas one that is drawn contacting the base of the rectangle indicates a line that terminates below the fluid level. All these symbols are shown in Figure 4-3.

Lines connected to a reservoir are almost always drawn from the top, regardless of where the connection is made on the actual component. Every reservoir has at least two hydraulic lines connected to it and can have many more. Because numerous components in a hydraulic circuit depend on a single reservoir, the reservoir symbol is often repeated rather than creating a scramble of return lines all over the schematic. In most cases, the reservoir symbol is the only one to be repeated on a schematic: you can regard it in the same way as a ground symbol is regarded on an electrical schematic.

Pump and Motor Symbols. Pump and motor symbols are similar in appearance. Both use a circle. A hydraulic pump is indicated when the small triangle at the output line (within the circle) is shown pointing outward (**Figure 4-4**). A hydraulic motor is indicated when the small triangle at the inlet line (within the circle) is shown pointing inward (**Figure 4-5**). When hydraulic pumps and motors are fixed-displacement, the circle that represents them is shown with just the inlet and outlet lines contacting the circle. When either is variable-displacement, an arrow is drawn through the circle as shown in Figure 4-4.

Hydraulic motors appear almost identical to the symbols shown in Figure 4-4, except the direction of the triangles is reversed as shown in Figure 4-5.

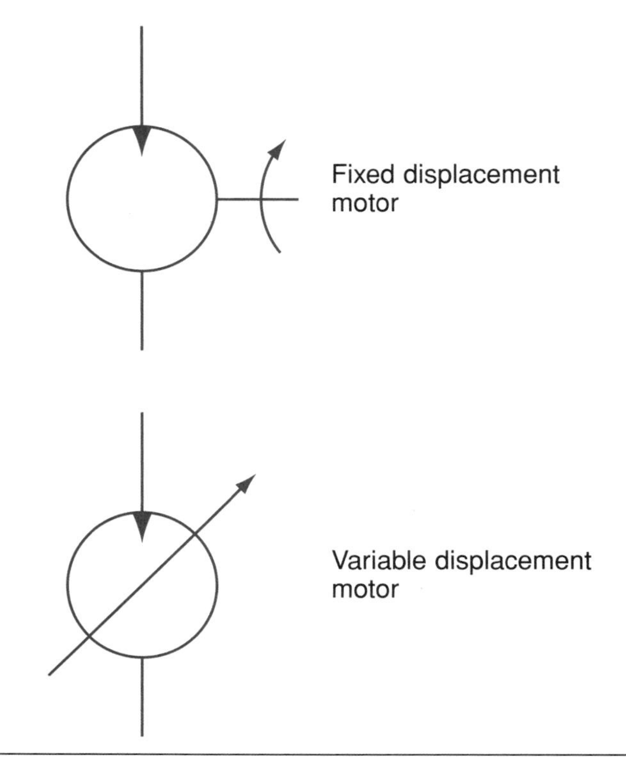

Figure 4-5 Fixed- and variable-displacement hydraulic motor symbols.

Valve Symbols

When valves are represented in hydraulic schematics, they tend to be a little more tricky. We will take a look at some simple valves used in schematics, beginning with a pressure relief valve.

Pressure Relief Valves. Pressure relief valves are required in even the simplest hydraulic circuits: they usually drain to the reservoir when tripped. A dashed

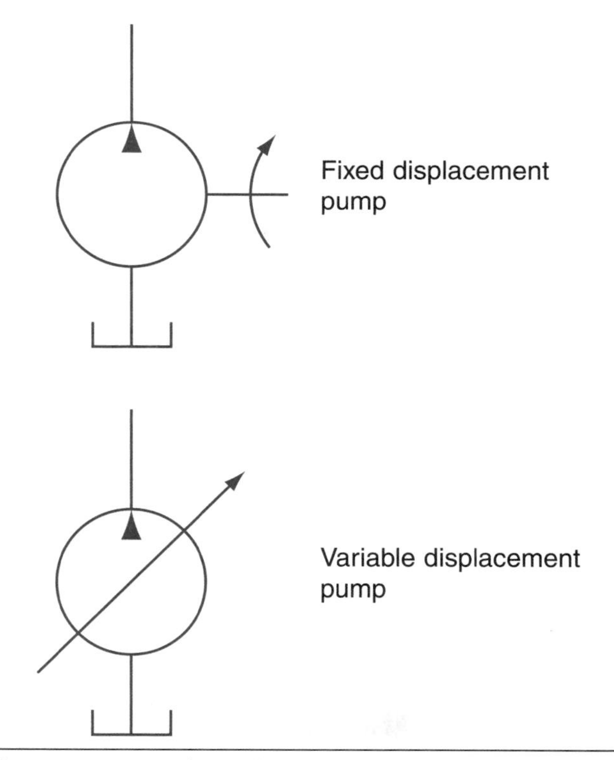

Figure 4-4 Fixed- and variable-displacement pump symbols.

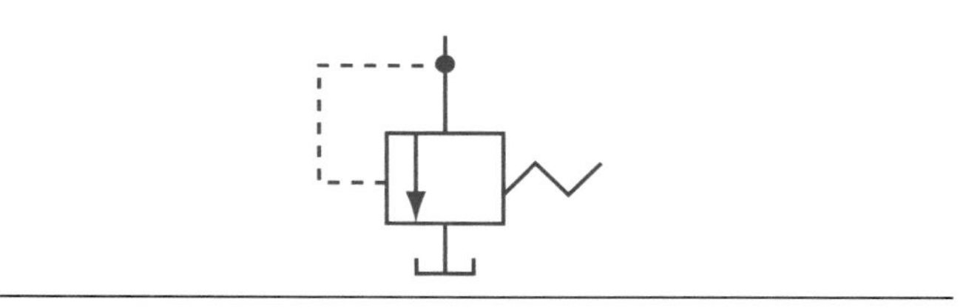

Figure 4-6 Pressure-relief valve symbol.

line as shown in **Figure 4-6** represents a pilot line: pressure in this line opposes spring force on the opposite side of the valve. When pilot pressure exceeds the mechanical force of the spring, the spool shifts, opening the valve and returning fluid to the reservoir. This relieves the pressure on the pump side, reducing pilot pressure and allowing spring force to return the relief valve to its normal operating status. Relief valves are used to define maximum circuit pressure.

Directional Control Valve Symbols. Directional control valves are used to manage fluid flow in a hydraulic circuit. There are many types, but generally these valves are classified by number of valve positions and number of ports. The number of ports in a valve is usually referred to as way(s) so a valve with three ports is known as a three-way valve. When a spool valve is used for directional control it consists of a body, a spool, and a means of actuating the spool. When attempting to interpret hydraulic schematics, remember that the actuator always acts on the spool: moving the spool is going to open and close the passages in the valve body.

To understand how valves are represented in hydraulic schematics, you should note the following:

- Boxes with arrows show the flow of fluid when the valve is shifted.
- A box without an arrow indicates flow (if any) in the neutral position.
- A box without an arrow shows the number of ports (ways) in the valve.

Figure 4-7 shows a two-position, two-way valve. It is shown with a mechanical actuator indicated on the right side of the figure and a spring return to neutral on the left side of the figure. We can also tell that this valve is normally-closed (NC) in operation because both ports are shown to be blocked in neutral: arrows on the port symbols would indicate that they were open. This schematic shows a valve that would typically be used as a safety gate: when the gate is not closed, then no flow is possible in the circuit that connects to it.

Figure 4-8 shows a more complex type of direction control valve that is electrically actuated. This valve has three positions indicated by its three larger, center boxes, and four ports (ways), indicated in the center box. It is a closed-center valve, meaning that when the spool is neutral, all the ports are blocked. The small rectangles with diagonal lines on each end identified as C1 and C2 are electrically actuated coils. The spool is shifted by energizing the coils. Springs on either side of the valve, depicted under the coils, return the valve to the center position when the coils are not energized.

In Figure 4-8, the following alpha symbols are used:

- C1—left side coil
- C2—right side coil
- A—connects to an external device
- B—connects to an external device
- P—pressure feed
- T—tank (reservoir)

In operation, when C1 is energized the valve shifts, directing pressure to the B port, and drains the A port to the reservoir. The reverse occurs when C2 is energized: this feeds pressure to the A port and drains the B port to the reservoir.

In **Figure 4-9,** another three-way, four-position directional control valve is shown. It is also electrically actuated (note the coils on either side) with springs to return the spool to the center position when the coils are not energized. This type of directional control valve can be used to control a hydraulic cylinder: it is integrated into the schematic shown a little later in Figure 4-13.

Cylinder Symbol. Cylinders are commonly used actuators in mobile hydraulics technology. When pressure is charged to one side of a cylinder, output movement occurs, providing the applied pressure acting on the sectional area of the cylinder is sufficient to overcome the load. A larger diameter cylinder applies more force than a smaller diameter cylinder, assuming equal pressure is applied to both. The symbol for a cylinder is easy enough to identify and is represented in **Figure 4-10.**

Filter, Cooler, and Accumulator Symbols. Filters and strainers are represented in hydraulic schematics by

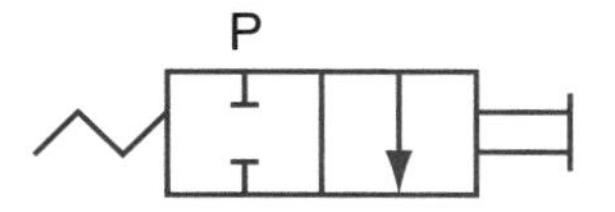

Figure 4-7 Two-position, two-way valve symbol.

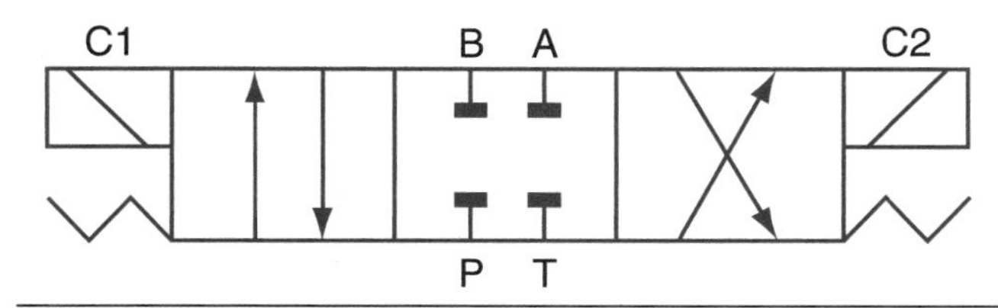

Figure 4-8 Three-position, four-way valve symbol.

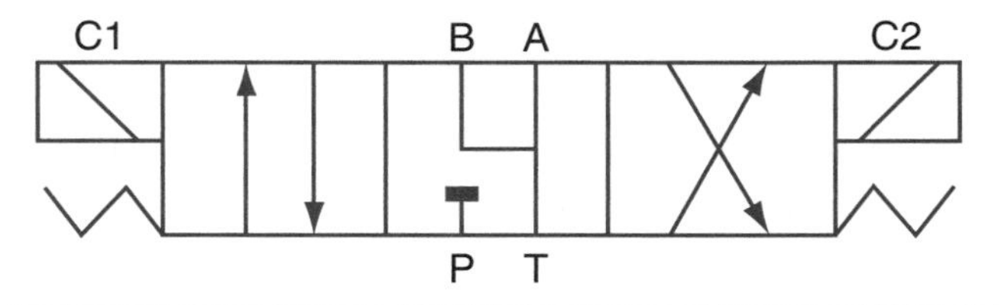

Figure 4-9 Alternate three-position, four-way valve symbol.

Figure 4-10 Symbol for a hydraulic cylinder.

a diamond shape with a vertical broken line through it. The symbol for a heat exchanger or cooler is similar with a solid colored triangle at the top and bottom that are connected by a vertical line. These symbols are shown in **Figure 4-11.**

Accumulators are represented by a rectangle, radiused (rounded) at the corners. The accumulator potential is also indicated in the schematic. Both spring- and gas-loaded accumulators are shown in **Figure 4-12.**

Simple Hydraulic Circuit Schematic

Figure 4-13 incorporates most of the symbols we have introduced so far into a simple hydraulic circuit built to actuate a cylinder. Note how the directional control valve that manages the circuit is integrated into it.

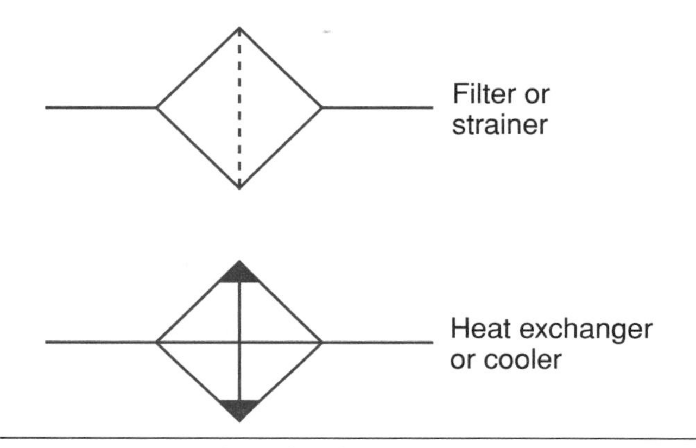

Figure 4-11 Symbols for filters and strainers.

ISO/ANSI symbols

Figure 4-14 shows most of the ISO/ANSI symbols you are likely to come across. You can use this to interpret some typical hydraulic circuits.

Sometimes equipment manufacturers make their schematics a little more user-friendly for technicians less accustomed to working on hydraulic circuits. **Figure 4-15** shows a boom, bucket, and skid control

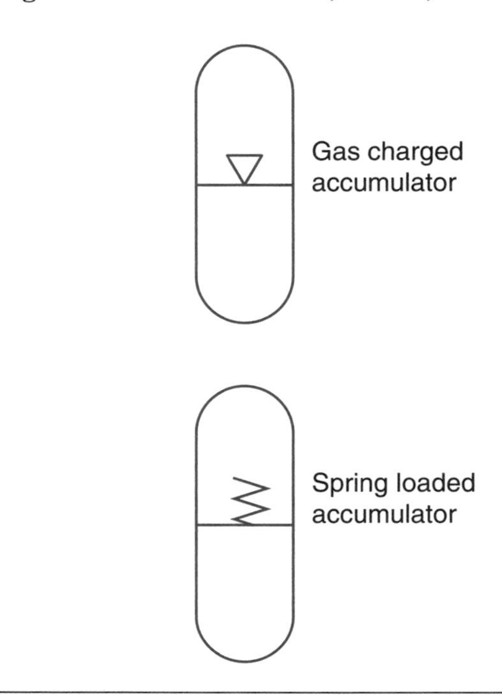

Figure 4-12 Accumulator symbols.

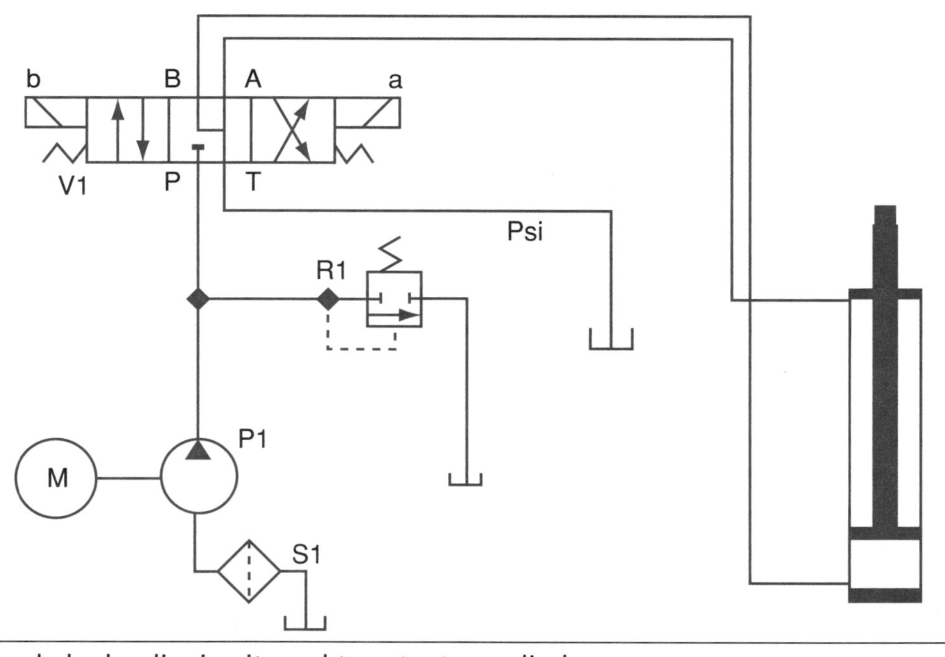

Figure 4-13 Simple hydraulic circuit used to actuate a cylinder.

Lines

Symbol
LINE, TO RESERVOIR ABOVE FLUID LEVEL BELOW FLUID LEVEL
LINE, WORKING (MAIN)
LINE, PILOT (FOR CONTROL)
LINE, LIQUID DRAIN
FLOW, DIRECTION OF HYDRAULIC PNEUMATIC
LINES CROSSING
LINES JOINING
LINE WITH FIXED RESTRICTION
LINE, FLEXIBLE
STATION, TESTING, MEASUREMENT OR POWER TAKE-OFF
VARIABLE COMPONENT (RUN ARROW THROUGH SYMBOL AT 45°)
PRESSURE COMPENSATED UNITS (ARROW PARALLEL TO SHORT SIDE OF SYMBOL)
TEMPERATURE CAUSE OR EFFECT
RESERVOIR VENTED PRESSURIZED
VENTED MANIFOLD

Pumps

Symbol
HYDRAULIC PUMP FIXED DISPLACEMENT VARIABLE DISPLACEMENT

Motors and Cylinders

Symbol
HYDRAULIC MOTOR FIXED DISPLACEMENT VARIABLE DISPLACEMENT
CYLINDER, SINGLE ACTING
CYLINDER, DOUBLE ACTING SINGLE END ROD DOUBLE END ROD ADJUSTABLE CUSHION ADVANCE ONLY DIFFERENTIAL PISTON

Miscellaneous Units

Symbol
ELECTRIC MOTOR
ACCUMULATOR, SPRING LOADED
ACCUMULATOR, GAS CHARGED
HEATER
COOLER
TEMPERATURE CONTROLLER

Figure 4-14 ISO/ANSI schematic symbols. (Reproduced by permission of Deere & Company, John Deere Publishing, Moline, IL. Hydraulics © 1999. All rights reserved. Visit our website: http://www.deere.com/publications)

Miscellaneous Units (cont.)

FILTER, STRAINER	
PRESSURE SWITCH	
PRESSURE INDICATOR	
TEMPERATURE INDICATOR	
COMPONENT ENCLOSURE	
DIRECTION OF SHAFT ROTATION (ASSUME ARROW ON NEAR SIDE OF SHAFT)	

Methods of Operation

SPRING	
MANUAL	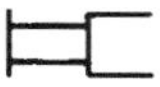
PUSH BUTTON	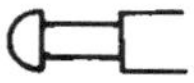
PUSH-PULL LEVER	
PEDAL OR TREADLE	
MECHANICAL	
DETENT	
PRESSURE COMPENSATED	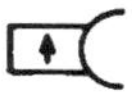
SOLENOID, SINGLE WINDING	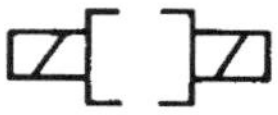
SERVO MOTOR	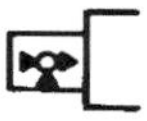

PILOT PRESSURE REMOTE SUPPLY	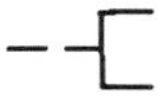
INTERNAL SUPPLY	

Valves

CHECK	
ON-OFF (MANUAL SHUT-OFF)	
PRESSURE RELIEF	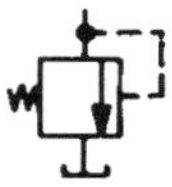
PRESSURE REDUCING	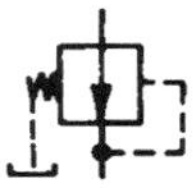
FLOW CONTROL, ADJUSTABLE–NONCOMPENSATED	
FLOW CONTROL, ADJUSTABLE (TEMPERATURE AND PRESSURE COMPENSATED)	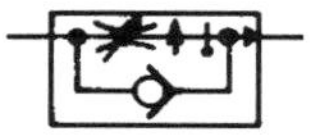
TWO POSITION TWO WAY	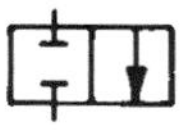
TWO POSITION THREE WAY	
TWO POSITION FOUR WAY	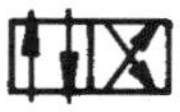
THREE POSITION FOUR WAY	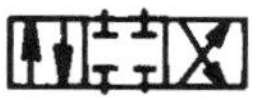
TWO POSITION IN TRANSITION	
VALVES CAPABLE OF INFINITE POSITIONING (HORIZONTAL BARS INDICATE INFINITE POSITIONING ABILITY)	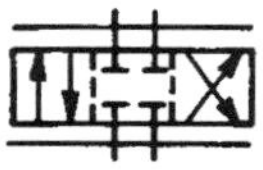

Figure 4-14 (Continued)

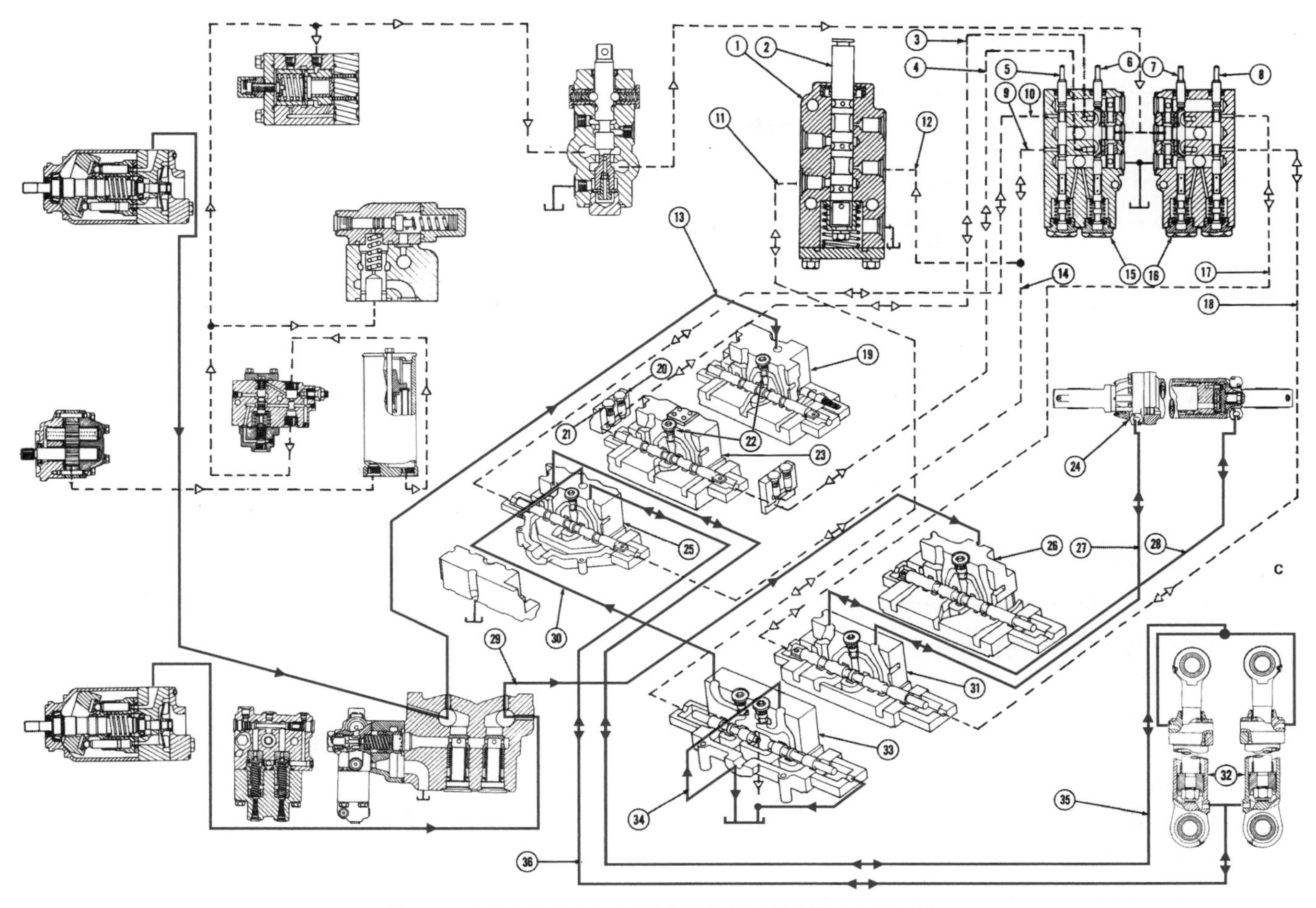

SCHEMATIC FOR BOOM, BUCKET AND STICK CONTROL

For Description of Components Also See PUMP FLOW AND PRESSURE CONTROL

1. Pump selector valve. Manually operated to make a combination of the output of both axial piston pumps.
2. Spool for pump selector valve.
3. Oil line for pilot pressure oil to main control valve for bucket OPEN.
4. Oil line for pilot pressure oil to main control valve for bucket CLOSE.
5. Spool in pilot valve for boom control.
6. Spool in pilot valve for bucket control.
7. Spool in pilot valve for swing control.
8. Spool in pilot valve for stick control.
9. Oil line for pilot pressure oil to main control valve for boom RAISE.
10. Oil line for pilot pressure oil to main control valve for boom LOWER.
11. Oil line for pilot pressure oil to boom crossover valve from pump selector valve.
12. Oil line to pump selector valve from pilot valve for boom control.
13. Oil line for main system oil from the combiner valve (front pump) to the control valve with three stems.
14. Oil line for pilot pressure oil to main control valve for boom RAISE.
15. Pilot valve. Manually operated for control of pilot pressure oil to the main control valves for the boom and bucket.
16. Pilot valve. Manually operated for control fo pilot pressure oil to the main control valves for the stick and swing.
17. Oil line for pilot pressure oil to main control valve for stick IN.
18. Oil line for pilot pressure oil to main control valve for stick OUT.
19. Main control valve for right track motor.
20. Line relief valve (nine).
21. Make-up valve (twelve).
22. Load check valve (eight).
23. Main control valve for bucket cylinder.
24. Cylinder, stick.
25. Main control valve for boom cylinders.
26. Main control valve for left track motor.
27. Oil line for main system oil from control valve for stick to rod end of stick cylinder (stick OUT).
28. Oil line for main system oil from control valve for stick to head end of stick cylinder (stick IN).
29. Oil line for main system oil from the combiner valve (rear pump) to the control valve with five stems.
30. Oil line from the boom crossover valve to the boom valve.
31. Main control valve for the stick cylinder.
32. Cylinders, boom.
33. Control Valve, crossover. When spool (2) in pump selector valve (1) is moved, main system oil from the rear pump goes to control valve (25) through line (30). This puts the oil from the two axial piston pumps (front and rear) together for a faster boom RAISE.

NOTE: The stem in the stick crossover valve does not work.

34. Oil line for the return to tank of oil flow from the swing pump.
35. Oil line for main system oil from control valve for boom to rod end of boom cylinders (boom LOWER).
36. Oil line for main system oil from control valve for boom to head end of boom cylinders (boom RAISE).

Figure 4-15 Boom, bucket, and skid control hydraulic schematic. (*Courtesy of Caterpillar*)

hydraulic schematic used by Caterpillar. As you can see, every component in the circuit is explained using graphics rather than symbols.

HYDRAULIC FORMULAE

The following are some formulae used to spec out hydraulic circuit performance.

Torque (T) in pound-feet
T = HP × 5252 (divided by) rpm

Horsepower (HP)
HP = T × rpm (divided by) 5252

Revolutions per minute (rpm)
rpm = HP × 5252 (divided by) T

Oil flow velocity (V)
V = GPM × 0.3208 (divided by) A
GPM = gallons per minute

A = (pipe id sectional area)

Hydraulic horsepower
HP = psig × GPM (divided by) 1714
psig = pounds per square inch gauge

Cylinder thrust force
T = A × psi

Cylinder piston speed (S)
S = CIM (divided by) A
CIM = oil flow into cylinder
A = piston sectional area

Interesting Facts

- Compressibility of hydraulic oils: 0.5% per 1000 psi applied pressure.
- Compressibility of water: 0.3% per 1000 psi applied pressure.
- Hydraulic power and temperature: 1 watt consumption raises temperature of 1 gallon oil by 1 degree Fahrenheit per hour.
- Hydraulic power: for every 1 HP of drive power, 1 gpm at 1500 psi can be produced.
- Fluid velocity: pressure line at 3000 psi equals equivalent of 25 feet per second.

ONLINE TASKS

1. Use an Internet search engine to locate sites that offer hydraulic schematics and the means to interpret them. **Caution:** Avoid those sites, and there are many, that require payment for a download.
2. Check out this url: *http://www.sae.org/*. Navigate to identify pages that set standards for mobile hydraulic equipment. Explain what a J-standard is.

SHOP TASKS

1. Select a piece of mobile equipment equipped with hydraulic implements. Select one implement, for instance, a bucket control cylinder, and create a schematic that routes the circuit from prime mover, to pump, through control valves, and ending up at the implement. Use the circuit symbols identified in this chapter.
2. Select a piece of mobile equipment in your shop equipped with hydraulic implements and identify each type of control valve in the circuit. In each case, identify the means used to shift the spool in each control valve.

Summary

- ANSI and ISO graphic symbols are used to represent hydraulic components and connectors in hydraulic schematics.
- A circle (sometimes a semicircle) is the symbol used to represent pumps or motors in hydraulic schematics.
- A circle symbol with the triangle pointing outward indicates a hydraulic pump.
- A circle symbol with the triangle pointing inward indicates a motor.
- A variable-displacement pump or motor is indicated by an angled arrow drawn through the circle.
- A square (or envelope) is used to represent one position or path through a valve, while two squares together indicate a two-position valve.
- Diamonds indicate a component that conditions fluid in the circuit such as a filter or heat exchanger.
- Rectangular symbols indicate a hydraulic cylinder or reservoir.

- Hydraulic symbols are drawn on a schematic using one of the four basic shapes and adding the appropriate marks.
- Marks are used on hydraulic schematics to show the flow routing of the circuit being represented.
- Solid lines indicate the flow path for hydraulic fluid.
- Dashed lines indicate a pilot line that connects a control circuit to a slave circuit.
- Dotted lines indicate a return or exhaust circuit.
- Center lines are used to enclose assemblies or subcircuits.
- Arrows show the direction of fluid flow or the rotational direction of pumps and motors.
- Arcs are used to show points of adjustment in the circuit such as flow control valve.

Review Questions

1. Technician A says that a circle symbol on a hydraulic schematic can represent a hydraulic pump. Technician B says that a circle symbol on a hydraulic schematic can represent a hydraulic motor. Who is right?

   A. Technician A only
   B. Technician B only
   C. Both A and B
   D. Neither A nor B

2. What does a circle symbol with the triangle pointing outward indicate?

   A. Gas-loaded accumulator
   B. heat exchanger
   C. hydraulic pump
   D. hydraulic motor

3. Technician A says that dashed lines on a hydraulic schematic indicate a pilot line that connects a control circuit to a slave circuit. Technician B says that dotted lines are used in schematics to indicate a line under constant pressure. Who is right?

   A. Technician A only
   B. Technician B only
   C. Both A and B
   D. Neither A nor B

4. What symbol shape is used to indicate a component that conditions fluid such as a heat exchanger?

   A. circle
   B. rectangle
   C. square
   D. diamond

5. Technician A says that arrows are used to show the direction of fluid flow in hydraulic schematics. Technician B says that arrows are used to show the rotational direction of pumps and motors in hydraulic schematics. Who is right?

   A. Technician A only
   B. Technician B only
   C. Both A and B
   D. Neither A nor B

6. Technician A says that hydraulic oil is not compressible. Technician B says that hydraulic oil is less compressible than water. Who is right?

   A. Technician A only
   B. Technician B only
   C. Both A and B
   D. Neither A nor B

7. Technician A says that an accumulator is represented on a hydraulic schematic by a rectangle, rounded at the corners. Technician B says that a control valve is represented on a hydraulic schematic by an oval shape. Who is correct?

   A. Technician A only
   B. Technician B only
   C. Both A and B
   D. Neither A nor B

8. If equal pressure is applied to a large diameter cylinder and one with a smaller diameter, which of the two will produce the most force?

   A. Both cylinders must produce equal force.

   B. The larger diameter cylinder produces more force.

   C. The smaller diameter cylinder produces more force.

9. A diamond-shaped symbol with solid colored triangles top and bottom connected by a vertical line is used to represent

   A. a filter

   B. a strainer

   C. a heat exchanger

   D. a safety-relief valve

10. A schematic shows a pair of unbroken, intersecting lines with a dot at the intersection point. This usually indicates that

    A. two hydraulic lines cross but do not connect

    B. two hydraulic lines connect

    C. a regulator is represented by the dot

    D. an accumulator is represented by the dot

CHAPTER

Hydraulic System Maintenance and Diagnostics

Learning Objectives

After reading this chapter, you should be able to:

- Work safely around hydraulic equipment.
- Outline the benefits of a preventative maintenance program.
- Describe the features of a sound maintenance program.
- Outline a drain, flush, and refill procedure.
- Identify the effects of unwanted restrictions in hydraulic circuits.
- Connect a hydraulic analyzer into a hydraulic circuit.
- Perform some typical hydraulic analyzer tests.
- Use troubleshooting charts and tables to help identify circuit malfunctions.
- Identify the hydraulic circuits found on a typical excavator.
- Outline some diagnostic tests for the track drive, swing, boom, stick, and bucket hydraulic circuits on an excavator.

Key Terms

aeration
boom circuit
bucket circuit
drift
hydraulic analyzer
pilot
preventative maintenance (PM)
preventative maintenance inspections (PMI)
residual pressure
stick circuit
swing circuit
track circuit

MAINTENANCE

Proper maintenance is a key to having a hydraulic system perform properly and to minimize repair downtime. Good maintenance requires some upfront investment in labor, data tracking, and scheduling, plus some low cost items such as fluid and filters; but the payoff is avoiding costly downtime and major component failure. Properly planned and scheduled maintenance is known as **preventative maintenance (PM).** A key factor of preventative maintenance is the scheduling of a **preventative maintenance inspections (PMI).** Adhering to PMIs in a planned maintenance schedule has been proven to minimize costly unscheduled equipment failures.

The type of preventative maintenance schedule that suits a particular operation varies depending on the equipment and the kind of environment it is operated in.

Any maintenance program should be flexible enough to provide for monitoring the program itself: this type of evaluation allows the program to be adapted as equipment and operating conditions change. Typical maintenance problems result from:

- Low reservoir oil level
- Clogged oil filters
- Contaminated or improper oil in the system
- Dirt in the hydraulic circuit

Preventative maintenance schedules should be observed. Most routine maintenance on hydraulic circuits takes little time and can eliminate expensive component repair and replacement. This section will take you through some typical maintenance procedures.

Safe Practice

Hydraulic circuits are designed to run at high pressures and to support high loads. It is essential that you work safely around chassis hydraulic equipment. Make sure you understand the contents of the safety chapter (Chapter 1) in this book. Here is a quick reminder of some of the basic rules specific to working around hydraulic circuits:

- Never work under any device that is only supported by hydraulics. A raised bucket or dozer blade must be mechanically supported before you work under it. Just as when using a floor jack, you must use some mechanical device for supporting any raised equipment or components.
- Hydraulic circuit components can retain high **residual pressures:** the system does not have to be active for this to be a potential hazard. Try to ensure that pressures are relieved throughout the circuit before opening it up. Crack hydraulic line nuts slowly and be sure to wear both safety glasses and gloves.

Drain, flush and fill

A draining, flushing, and refilling procedure is the only way to remove contaminants and degenerated hydraulic fluid from a hydraulic circuit. The frequency with which this procedure is performed will depend on the severity of the working conditions and, in many cases, simply draining and replacing the hydraulic oil will be sufficient. Sludging oil, solid deposits, and the appearance of varnishing on components are indicators that draining the system should be followed by system flushing before refilling with new hydraulic oil.

Drain System. Gravity drain the system, clean any sediment from the reservoir, and replace the filter elements. Inspect the old oil. If it shows signs of severe degeneration, flushing the system should be considered. Indicators of severe degeneration of hydraulic oil are a burned odor, excessive stickiness, discoloration, and lacquer flakes in oil.

Flush System. Most hydraulic system flushing should be performed with the specified service hydraulic oil. After draining, fill the system reservoir with the specified oil, bleed the hydraulic pump, and then operate the equipment to circulate the oil through the entire system. All the valves should be actuated to ensure that the flushing oil reaches every part of the circuit. Finally, drain and dispose of the flushing oil.

Fill the System. Remembering that the intrusion of dirt into hydraulic systems probably causes more problems than any other single problem, ensure that everything used in the refill procedure is clean. A new filter should be used. Fill the reservoir to the prescribed level with new hydraulic oil. Then run the equipment through its entire working cycle several times to ensure that any air is purged from the circuit. The equipment should run smoothly. Finally, recheck the oil level in the reservoir.

Checking system performance. Before and after repairs, and as part of preventative maintenance, hydraulic systems should be checked. Some OEMs have system-specific checklists and these should be used when available. Typical checks should be performed in the following fields

Check Reservoir. This check should be performed regularly because it can be performed in seconds and can identify small problems before they become major problems. Before opening the system filler cap, clean it and the surrounding area with a clean rag. When dipsticks are used, clean the dipstick with a lint-free cloth before checking the level. Look for

- Oil **aeration.** Foaming oil or bubbles can indicate an air leak in the system, usually on the suction side of the circuit.
- Change in oil level. Reservoir oil level that requires constant topping up usually indicates an external leak in the system.
- Milky oil. This almost always indicates the presence of water in the system. Often system flushing will be required because hydraulic oil and water emulsify under pressure.

Check Oil Lines and Connections. Both low- and high-pressure sides should be checked for leaks and physical damage. Low-pressure side leaks can result in air being pulled into the system, while high-pressure side leaks will produce external leakage. Pinched lines can create restrictions that result in foaming, overheating, and reduction in hydraulic efficiency.

Check Circuit Components. A thorough visual inspection should be made of all of the hydraulic circuit components including the oil cooler, valves, cylinders, pumps, and motors. The circuit should be run through its working cycle to check for leaks and whether the flow is sufficient to run the system. Check the oil cooler fins for plugging up and check the system valves for operation, remembering that wear can cause internal leakage. Motors should not be allowed to run hot. If this happens, check that the oil supply is adequate and that the oil cooler is functioning properly. Also check for leaks around the motor hose fittings, shaft, seals, and body mating surfaces.

Flow and Pressure Testing. Various types of **hydraulic analyzers** are available for testing hydraulic circuits. A hydraulic analyzer typically consists of a pressure gauge(s), flow meter, manual gate (choke) valve, and thermometer. **Figure 5-1** shows one type of hydraulic analyzer. **Figure 5-2** shows how the analyzer is plumbed into a circuit to perform a simple pump output test.

Typically, a hydraulic analyzer checks,

- **Flow.** Flow testing determines whether the pump is producing its specified output.
- **Pressure.** Pressure testing verifies the operation of relief valve and, in closed-center systems, pump performance.

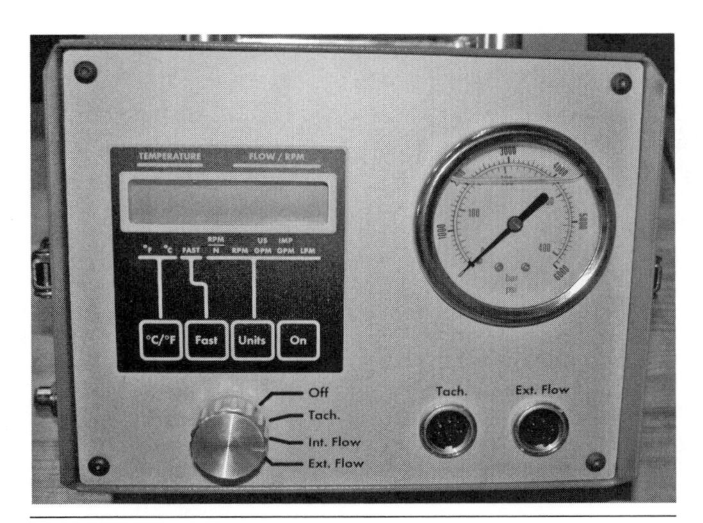

Figure 5-1 Hydraulic system analyzer.

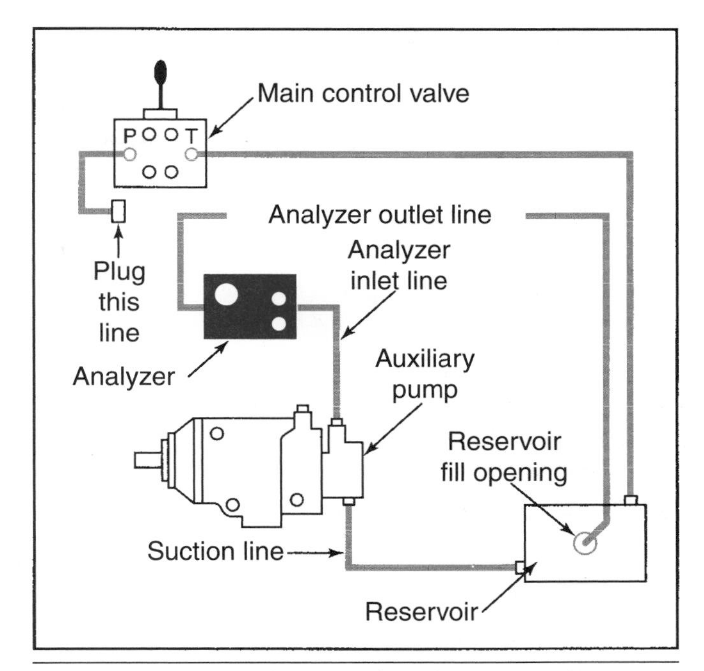

Figure 5-2 Circuit diagram for pump test. (*Courtesy of Mack Trucks*)

- **Temperature.** Hydraulic oil temperature should be measured because system pressure and flow specifications are calculated with the system at operating temperature.
- **Leakage.** Leakage testing identifies leakage in a specific circuit component.

When you are required to test a vehicle hydraulic circuit, observe the OEM test cycle. It is recommended that you record the results of each test step using a procedure such as that shown in **Table 5-1** for a pump performance test and **Table 5-2** for a hydraulic circuit performance test. In our example here, we have used a typical excavator system: most manufacturers use system-specific tables for recording performance test data in their service literature. The swing, boom, stick, and bucket circuits are described in a little more detail later in this chapter.

A typical test cycle consists of the following steps:

1. Install the hydraulic tester into the circuit to be tested, between the pump and the control valve. The pressure line should be connected to the inlet port of the tester.
2. Connect the hydraulic tester outlet (return) to the system reservoir.
3. After installation, check the system oil level.
4. Start the engine and gradually close the hydraulic tester gate valve to load the circuit and bring the temperature up to normal operating value.

Table 5-1: HYDRAULIC PUMP PERFORMANCE TESTS

Record Pump Flow in gallons per minute (gpm)

Pressure	0 psi	500 psi	1000 psi	2000 psi	2500 psi	3000 psi	3500 psi
Flow in gpm							

Table 5-2: HYDRAULIC CIRCUIT TESTS

Record Pump Flow in gallons per minute (gpm)

Circuit Test	0 psi	500	1000	2000	2500	3000	3500
Boom up							
Boom down							
Stick in							
Stick out							
Bucket in							
Bucket out							
Swing CW (right)							
Swing CCW (left)							

5. Open the gate valve and record system flow at zero pressure.
6. Move the system control valve into one of its power positions and gradually close the tester gate valve, recording flow in 250 psi increments from zero up to full system pressure.
7. Repeat the above test in each of the system control valve power positions.

Tech Tip: When performing steps 6 and 7, ensure the system oil temperature is the same. This may require that you allow the oil to circulate and cool between each test.

OEMs identify the test points along with specifications in their service literature. The test points used to diagnose the hydraulic circuits in an excavator are shown in **Figures 5-3** and **5-4.** In this test profile, a tachometer is connected to monitor engine speed. Note the test locations for both the pumps and control valve.

Diagnosing Test Results. You should make sure that you consult the OEM literature when diagnosing hydraulic circuit test results, but here are some typical guidelines used to test a simple hydraulic cylinder type circuit:

- Pump flow at maximum pressure should be at least 75% of the pump flow measured at zero pressure: OEM specifications can require higher percentage values. A pump that fails to meet 75% of this value is probably worn.
- When flow at each pressure value throughout testing is within 75% of pump flow at zero pressure, the hydraulic circuit components are probably functioning properly.
- When pressure drops off before full load, a leaking circuit is indicated. To determine whether the leakage is occurring in the control valve or in the cylinder, disconnect the cylinder return hose and move the control valve to a power position. If oil leaks at the cylinder return, it is probably leaking internally and is defective. If no oil exits the cylinder return line, the control valve may be at fault.
- A relief valve should open within a window of its specified pressure, so check the OEM specifications. A defective relief valve can be identified when the valve opens before system maximum pressure is achieved. If circuit flow suddenly drops off before system pressure is achieved, a defective relief valve is indicated.

VIEW OF PUMPS ON ENGINE

1. Location of plugs for checking system pressure and flow from the rear pump. 2. Location of plugs for checking system pressure and flow from the swing pump. 3. Location of plugs for checking system pressure and flow from the front pump. 4. Location of plug for checking the signal pressure to the front and rear pumps.

Figure 5-3 Test points for system checks. (*Courtesy of Caterpillar*)

Figure 5-4 Control valve pressure test location. (*Courtesy of Caterpillar*)

LEAKAGE

Theoretically, a hydraulic pump should be able to turn a hydraulic motor at the same speed, providing both are of equal displacement. To achieve this would require 100% volumetric efficiency. However, this is not achievable due to circuit leakage factors. Circuit leakage in hydraulic circuits can be classified as:

- Internal leakage
- External leakage

Intended Internal Leakage

Any leakage that is designed into a hydraulic circuit can be classified as internal leakage. Intended internal leakage is used to lubricate plungers and pistons, valve spools, pumps, motor components, and pretty much anything that moves in the circuit. Intended internal leakage can also help "guide" a close-fitting component, such as a plunger, in the bore it reciprocates within. Leak paths are built into certain hydraulic components, such as motors, to provide balanced operation.

Unintended Internal Leakage

Unintended internal leakage can occur when normal wear opens up flow paths in intended internal leakage circuits, or it can result from internal seal failure. Excessive internal leakage slows down actuators and, in the case of a failed seals, can create a leakage path that diverts all of the pump flow around the actuator.

Pressure and Leakage. Whether internal leakage is intended or not, it increases with pressure. In addition, heat is generated at the leakage flow path. Therefore, operating a circuit at higher than specified pressures produces both increased leakage and adds heat to the circuit fluid. While excess internal leakage can result in loss of control or function in a hydraulic circuit, the good news is that the oil is not lost. One way or another, it will be routed back to the reservoir, either by means of the return circuit or by drain passages.

Viscosity and Leakage. The viscosity of hydraulic oil influences internal leakage flow rate. A low viscosity oil flows more readily and therefore increases internal leakage. The viscosity and viscosity index of the hydraulic oil used are key considerations, both in providing the appropriate intended internal leakage and in preventing excessive internal leakage.

Internal Seal Failure. When an internal seal fails, the result can be a complete failure of the hydraulic circuit. Oil continues to flow through the circuit and heat is generated at the location of the seal failure. A sudden rupture of an internal seal can be dangerous in many types of hydraulic circuit.

External Leakage

All hydraulic circuit external leakage is unintended. External leakage presents less of a challenge in troubleshooting because of the mess it makes. Apparently minor external hydraulic leaks should always be promptly repaired because what begins as seal leakage can progress to a rupture. External leakage can usually be sourced to one of the following:

- Improper assembly
- Poor maintenance
- Shock load on an actuator
- Lines subjected to excessive vibration

CAUTION *Pin-holes in hydraulic lines result in high-pressure oil escaping from the system in minute droplets: these oil droplets can penetrate the skin, often with little external evidence, and the result can be blood poisoning. Use a piece of cardboard to search for leaks in hydraulic lines; never use a bare hand.*

TROUBLESHOOTING CHARTS

Most hydraulic circuit failures follow a pattern producing either a sudden or a gradual loss of system performance. Most manufacturers produce system troubleshooting trees or charts to help technicians diagnose malfunctions. The use of electronic management of hydraulic circuits and pressure sensors has facilitated the sourcing of problems, sometimes by guided tree navigation. The following sequence of tables **(Tables 5-3 through 5-14)** identifies some typical hydraulic circuit malfunctions and remedies.

TROUBLESHOOTING PROCEDURE

Any troubleshooting procedure should be specific to the equipment being diagnosed and to a range of other variables such as:

- Hydromechanical controls
- Electromechanical controls
- Electronic (computer) controls
- Equipment OEM

For purposes of study, we will base our troubleshooting profile on an excavator equipped with upper structure swing, boom, bucket, and stick controls, and we will assume that the controls are hydromechanical. The drive and steer systems of the excavator are part of the **track circuit** and are studied later in this textbook. When approaching any troubleshooting procedure, remember that the specified oil flow and pressure are required. Oil flow is a function of prime mover (engine) rpm, and pressure readings indicate the resistance to oil flow within the circuit. In general, you should begin a troubleshooting procedure with a visual check and proceed to performance tests before progressing to instrument testing using hydraulic analyzers. The following section briefly describes the circuits integrated into a typical excavator.

Swing Circuit

The upper structure is mounted on the excavator undercarriage and incorporates the cab, engine housing, and boom assemblies. This upper structure is hydraulically rotated by the **swing circuit** and is identified in **Figure 5-5.**

The swing motor is capable of rotating the upper structure of the excavator through a 360-degree turn. The oil supply for the swing circuit is usually separate from that used to supply the implements and track drive circuits. In troubleshooting profiles it is normal practice to refer to swing direction as CW (right) or CCW (left) rotation. **Figure 5-6** shows a cross-section of the excavator swing drive control.

Boom Circuit

The **boom circuit** is actuated by a pair of synchronized hydraulic cylinders. The cylinders are double-acting and are used to raise or draw back the boom assembly. The excavator boom assembly is shown in **Figure 5-7.**

Boom operation is mastered by a control valve assembly. **Figure 5-8** shows a control valve body that is integrated into a valve assembly consisting of boom control (callout 9), bucket control (callout 8), and right track motor control (callout 4) valves.

Stick Circuit

The **stick circuit** is actuated off the boom and uses a single, double-acting cylinder to vary the bucket pivot

Table 5-3: SYSTEM INOPERATIVE

Cause	Remedy
No oil in system.	Fill system and check for leaks.
Improper oil in system.	Reference OEM specifications. Flush and refill system.
Restricted oil filter.	Change oil and replace filter.
Worn-out pump or circuit components.	Repair or replace defective components. Change oil and flush system.
Leakage in suction line (air in system).	Repair or replace as necessary.
Excessive load.	Check OEM specifications for rated maximum load.
Defective pump drive.	Repair or replace belts and couplings, check alignment, backlash, and belt tension.
Internal leakage.	Check circuit component operation, including relief valve. Repair or replace as necessary.
External leakage.	Visually inspect circuit lines and components. Repair or replace as necessary.

Table 5-4: ERRATIC OPERATION

Cause	Remedy
Aerated system.	Check suction side of circuit for leaks.
Excessively cold oil in system.	Allow sufficient warm-up oil and consider using a winter-rated oil (lower viscosity).
Restricted or damaged circuit components.	Clean and replace. Flush circuit and change oil.
Restricted lines and or filters.	Clean or replace lines and filters as necessary.

Table 5-5: SLUGGISH OPERATION

Cause	Remedy
Oil gelling.	Allow oil some warm-up time. Consider using lower viscosity oil.
Low pump drive speed.	Increase prime mover (drive engine) rpm.
Aerated system.	Check suction side for leaks and repair.
Worn system components.	Check for internal and external leakage. Check for cause of wear. Repair or replace components as necessary.
Restricted filters or lines.	Clean and/or replace filters and lines.
Low oil level.	Check reservoir oil level.
Oil leaks.	Torque fittings and visually check seals and lines.
Improper adjustments.	Check relief valve settings and orifice flow areas per OEM specification.

Table 5-6: SYSTEM OPERATES TOO QUICKLY	
Cause	**Remedy**
Wrong size or improperly adjusted restrictor.	Replace or adjust as necessary.
Prime mover running too fast.	Adjust engine speed to specification.

Table 5-7: FOAMING HYDRAULIC OIL	
Cause	**Remedy**
Improper oil, low oil level,	Check oil specification, correct oil
contaminated oil.	level, flush and refill oil in circuit.
Aerated oil.	Check suction side for air induction.

Table 5-8: OVERHEATED OIL	
Cause	**Remedy**
Oil cycling through relief valve for prolonged period.	Return control valve to neutral position when not in use.
Contaminated, incorrect, or low oil level.	Replace oil and flush circuit, ensuring reservoir oil level is correct.
Prime mover running too fast.	Reduce engine speed.
Restricted lines or filters.	Clean and/or replace lines and filters.
Insufficient heat exchanger ratings or restricted heat exchanger grills.	Clean contamination from heat exchanger air flow passages. Check heat exchanger specifications.
Mechanical malfunction/binding/ lockup of components.	Locate problem and repair or replace.
Excessive internal leakage.	Locate defective component and repair or replace.

Table 5-9: LEAKAGE AT PUMP OR MOTOR	
Cause	**Remedy**
Failed seal(s).	Replace, check shaft alignment.
Loose or failed components.	Tighten and/or replace components.

Table 5-10: NOISY PUMP	
Cause	**Remedy**
Incorrect or contaminated oil, low oil level.	Check OEM oil spec, clean and flush oil, fill to correct level.
High inlet restriction in suction line.	Clean and/or replace suction line.
Worn or damaged pump.	Repair or replace pump.

Table 5-11: HYDRAULIC LOAD DROPS, CONTROL VALVE IN NEUTRAL	
Cause	**Remedy**
Internal or external cylinder leaks.	Repair or replace worn components.
Control valve not centering on release.	Check linkage and for spool binding. Repair as necessary.

Table 5-12: LEAKING CONTROL VALVE	
Cause	**Remedy**
Loose tie-bolts.	Torque to specification.
Worn or damaged seals.	Replace seals.

Table 5-13: BINDING CONTROL VALVE	
Cause	**Remedy**
Tie-bolts too tight.	Back off and torque to specification.
Valve linkage misaligned.	Align linkage.
Internally damaged control valve.	Replace control valve.

Table 5-14: Hydraulic Cylinder Leakage	
Cause	**Remedy**
Damaged or worn seals.	Replace seals.
Damaged rod.	Replace cylinder assembly.

Figure 5-5 Excavator upper structure controlled by swing circuit.

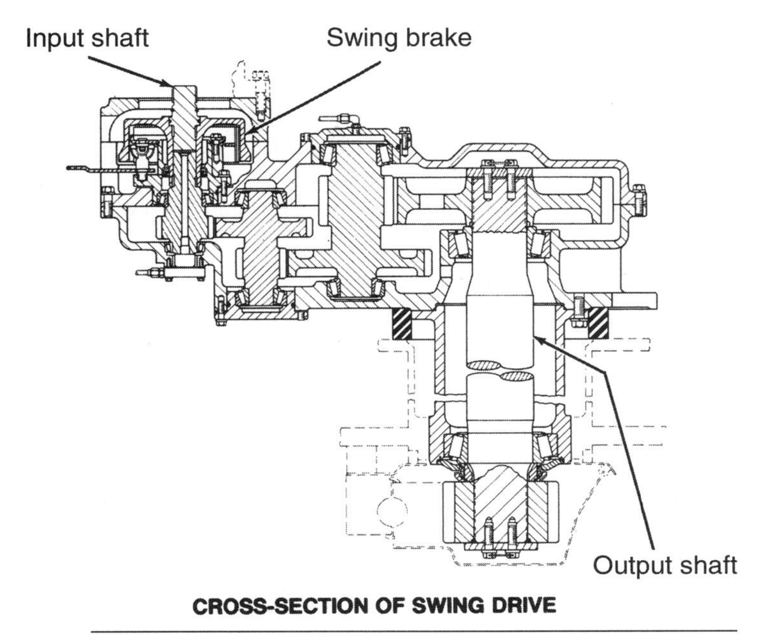

Figure 5-6 Cross-section of swing drive: 1. Input shaft. 2. Swing brake. 3. Output shaft. (*Courtesy of Caterpillar*)

Figure 5-7 Excavator boom, stick and bucket controls. (*Courtesy of Caterpillar*)

angle. You can see this by referencing Figure 5-7 callout 2. The cylinders used as actuators in the bucket, stick, and boom circuits are similar in both design and operation. A sectional view of a typical cylinder is shown in **Figure 5-9.**

Bucket Circuit

The excavator bucket is mounted on a pivot on the stick implement and is moved by a double-acting cylinder actuated by the **bucket circuit** as shown in Figure 5-7, callout 3. The bucket is scooped inward as the bucket cylinder is extended and moved outward to dump as the bucket cylinder is retracted.

Excavator Circuit Features

When troubleshooting a suspected hydraulic circuit in the typical excavator machine we are using as our example, you should keep in mind the following points:

- The swing circuit is usually a closed circuit with a dedicated pump, control valves, relief valve, and motor. The relief valve is located in the pressure supply between the main control valve and the swing motor. The swing circuit will not usually cause the engine rpm to drift below rated when it is actuated by itself.
- The system piston pumps are used to charge the track motors. A typical arrangement uses the front axial piston pump to supply the right track motor and the rear axial piston pump to supply

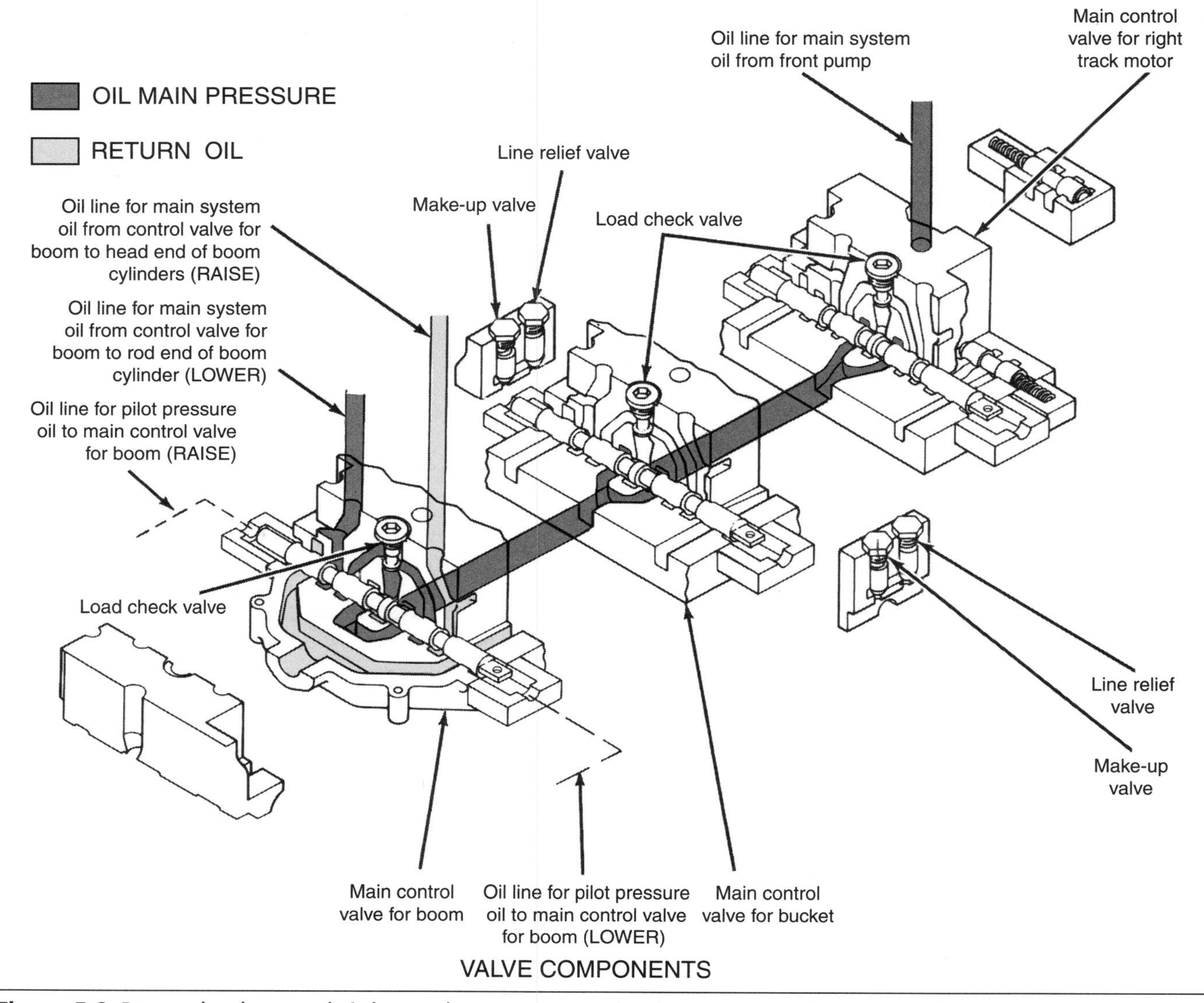

Figure 5-8 Boom, bucket, and right track motor control valve assembly. (*Courtesy of Caterpillar*)

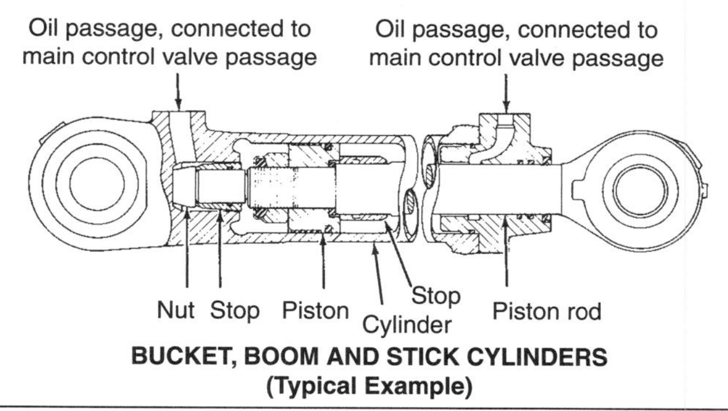

Figure 5-9 Typical boom, stick, and bucket cylinders. (*Courtesy of Caterpillar*)

the left track motor. The common components to both track circuits are the speed and direction control valve, the main relief valve, and the swivel. The speed and direction control valve manages the main control valves (one for each track) by using pilot pressure. Pilot pressure moves the stem in the main control valve that then routes oil pressure from one of the piston pumps to a track motor. Hydraulic oil to and from the track motors must pass through the swivel, the only component in each track circuit that is common to both circuits other than the brake valve. A track control valve is shown in **Figure 5-10.**

Because the brakes are activated by spring force, hydraulic pressure from the brake valve must relieve the spring force before the excavator can be moved.

- A drop in engine rpm or an increase in implement load can cause the piston pumps to decrease output. This is done by a signal to both

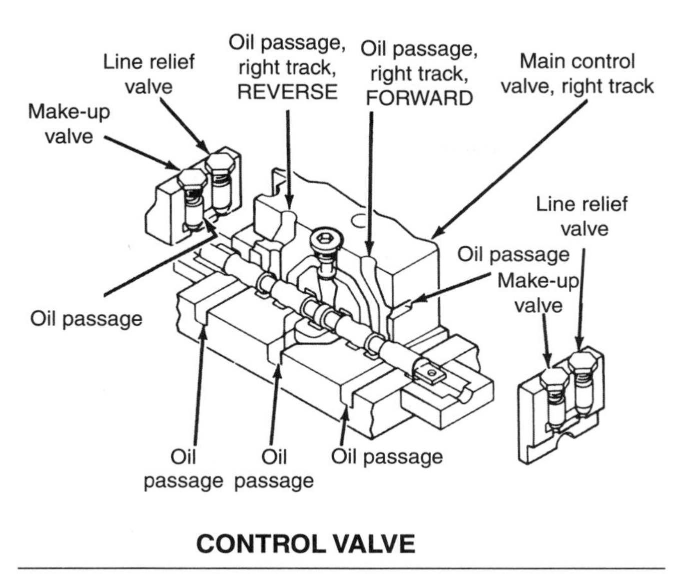

Figure 5-10 Main (right) track control valve. (*Courtesy of Caterpillar*)

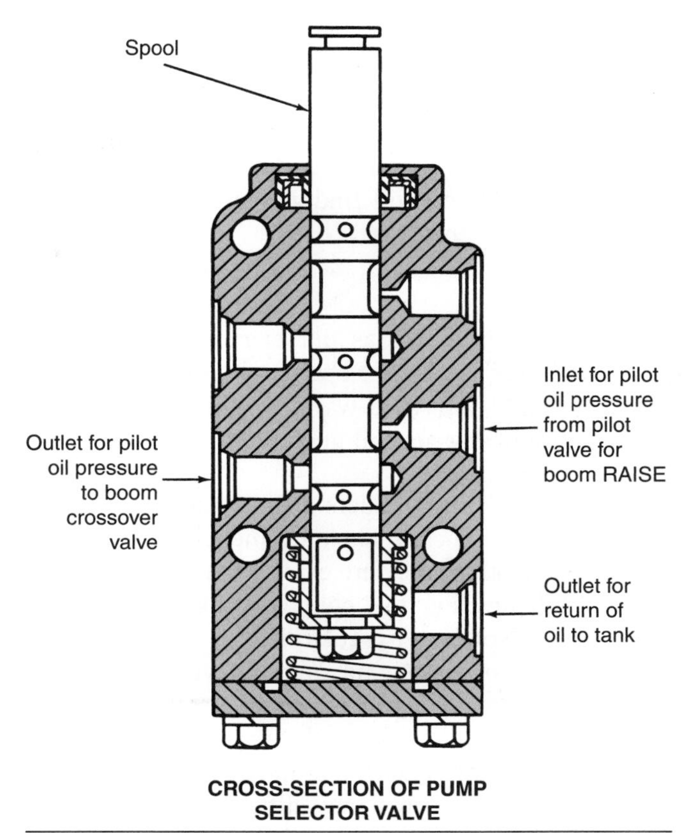

Figure 5-11 Cross-section of a pump selector valve. (*Courtesy of Caterpillar*)

pumps that alters pump swash angle and puts them in stroke-back mode. The 'signal' is sourced from the underspeed valve or the summing valve, which senses a sudden increase in load from the implement. **Figure 5-11** shows a sectional view of a pump selector valve.

- To ensure that the specified signal pressure is sent to the axial piston pumps, the pilot pressure must be correct. Pilot pressure can be adjusted and set by correctly shimming the pilot relief valve. Check this value to specifications and adjust if required.
- When there is a problem that is common to all the hydraulic circuits on the machine, check the engine stall speeds first, using the manufacturer service literature.

Typical Excavator Circuit Problems

Now we will take a look at some typical hydraulic circuit problems, using our typical excavator model. The information in this section references Caterpillar service literature but it has been adapted to share commonality with other OEM hydromechanically controlled systems.

Visual Checks

Begin a visual check of the equipment with the engine turned off and any hydraulic implements neutralized: lower booms and buckets to the ground. Then:

1. Check the reservoir oil level. Slowly loosen the reservoir cap to relieve any residual pressure.
2. Remove and inspect strainers and filter elements for obvious restriction.
3. Closely inspect all lines and hoses checking for kinks, crushing, and leakage.
4. Inspect external control linkages for bends, breakage, and excessive play.

Performance Tests

When performance testing, you are required to start the engine and work the implements. During this phase of troubleshooting, you should be looking for evidence of leakage and obvious malfunctions, such as a valve or actuator that is not working properly. The speed of rod movement or motor output torque can be used to determine the performance of actuators in the system. Use Tables 5-1 and 5-2 located earlier in this chapter to guide you through the tests.

Tech Tip: A key to accurate troubleshooting is developing the ability to navigate through OEM hydraulic schematics. Familiarize yourself with the schematic(s) before attempting to source a problem, and you will find you spend less time locating problems.

Using the excavator as our example, you should raise and lower the boom, open and close the bucket, and swing the boom both ways. Do this several times using the following general guidelines:

1. Observe the cylinders that actuate the boom, stick, and bucket. The movement of each should be smooth as they are extended and retracted.
2. While cycling the cylinders, listen for pump noise. Note any irregularities.
3. Listen for the sound of relief valves tripping. Check the relief valve opening pressures to specification. Relief valves that trip below specification cause system malfunctions: relief valve opening pressures may differ for each implement and the swing circuits.

Typical Problems and Causes

Once again, we will use our typical excavator to look at some possible hydraulic problems and their sources.

Problem: Cylinder extend and retract cycle times exceed those shown in the specifications.

Possible Causes:

1. High idle engine rpm set too low. This results in a reduction of pump output flow. *Note*: The rated engine rpm (full load) is the speed at which maximum horsepower transfer is available—in hydromechanically managed engines, this is often lower than engine high idle speed by 5 to 20%. Check service literature on what is required to adjust the rated speed of a specific engine; this may be programmable (electronically controlled engines) or require governor trim adjustment.
2. Relief valve trip pressure in the pilot system incorrectly (low) set.
3. Damaged piston pumps.
4. Swashplates in the implement pumps seized or set at a minimum angle.

Problem: Cycle time required to extend or retract only one cylinder exceeds that shown in the specifications.

Possible Causes: Note that this type of problem is usually sourced in the control circuit of the affected cylinder.

1. Check the linkage between the control lever and the pilot valve, making sure that it is properly connected and not bent or damaged in any way.
2. Turned spool in the pilot valve. A spool can sometimes turn in its bore, blocking pilot fluid delivery to the main control valve.
3. Spool or housing damage in the main control valve. This type of problem can be caused by the loss of too much oil.
4. Cylinder damage or leaking piston seals.
5. Front piston pump malfunction, if the problem is in the boom or bucket circuits.
6. Rear piston pump, if the problem is in the stick circuit.

Problem: Drop in engine rpm (often close to stall) when the load on the hydraulic system is increased.

This condition can usually happen only when the piston pumps do not stroke back and there is no reduction in their output.

Possible Causes:

1. Low relief valve pressure setting in the pilot system.
2. Severe loss of engine power output.
3. Damaged actuator or servo valve in one or both piston pumps.

Problem: With the engine running at rated rpm, there is insufficient force (pressure) at the implements.

Possible Causes:

1. Low relief valve pressure in the pilot system. This results in modulation at the main control valves, resulting in insufficient pilot pressure to shift the spools in the main control valves to full travel.
2. Main pressure relief valve set too low.
3. Internal pump damage limiting pressure. Check the flow from the case drain on the pump.

Problem: Aerated hydraulic oil.

Possible Causes:

1. Leak in the oil line between the reservoir and the pump.

2. Failure to bleed the hydraulic system after assembly, inspection, or testing (See Photo Sequence 1).
3. Relief valve malfunction causing it to repeatedly open and close.
4. Leakage at the cylinder seals.

Problem: Noisy pump operation.

Note that noise coming from the area in which the pumps are located may not necessarily be caused by a pump. Check the pump drives. Noise from a pump drive can reverberate through the connecting shaft and housings, making it seem as though the noise is in the pump.

Possible Causes:

1. Oil aeration. This could be the cause if all the pumps are noisy.
2. Pump damage. If the noise is sourced from only one pump, it is probably caused by wear or damage to that pump.

Problem: High hydraulic oil temperature.

High oil temperature can be caused by a number of factors. For instance, removing baffles in the control compartment can allow heated air to act on the pumps and inhibit heat exchanger operation. Check the airflow paths on the equipment.

Possible Causes:

1. Oil aeration.
2. Restriction of air flow through the heat exchanger (oil cooler).
3. High ambient air temperatures.
4. Internal restriction in the heat exchanger—restricting the flow of oil through the cooler because of sludging or flow control valve problem.

Problem: Low hydraulic oil temperature.

Low oil temperature can cause problems. The temperature of the oil during operation must enable it to flow freely through the lines and components.

Possible Causes:

1. Cooler flow control valve malfunction permitting excessive oil flow to the cooler.
2. Cooler, reservoir, and other hydraulic components insufficiently protected from low ambient temperatures. Note OEM recommendations for operating equipment in Northern and subarctic conditions.

Problem: Improper modulation of all the implements.

When troubleshooting this condition, remember that the only part of the hydraulic system that is common to all implements is the pilot circuit. A defect in the pilot system can cause surging in the implement circuits and prevent the correct operation of the piston pumps.

Possible Causes:

1. Leakage in the system that prevents the pressure reaching specified values.
2. Low oil temperature.
3. Low opening pressure of the pilot system relief valve.
4. Aerated hydraulic oil.
5. Damaged or worn pilot pump.

Problem: The modulation of one implement is not normal.

When this occurs in just one circuit, check for a root cause that is specific to the malfunctioning circuit.

Possible Causes:

1. Bent or disconnected linkage to a pilot valve spool.
2. Turned or binding spool in a pilot valve.
3. Air in circuit.
4. Binding main control valve spool.

Problem: Machine will not move when either the FORWARD or REVERSE pedals is depressed.

Focus on the components in the track control and drive circuits, including the front and rear piston pumps (one pump for each track motor), main relief valve, line relief valves, swivel, motors (one for each track), make-up valves, main control valves, pilot control valves and overspeed valves. Note that the only components common to both track motors are the main relief valve, the swivel, and the track brake valve (a signal pressure from the track brake valve is required to release the brakes before the machine can be moved).

Possible Causes:

1. Pressure in the pilot system not reaching that specified to release the track brakes.
2. Machine safety lever is located in the SAFE position.
3. Stop has been turned in on the track brake valve, preventing signal pressure from releasing the track brakes.
4. Dirt in the track brake valve.
5. Leaking seal in the brake line section of the swivel.
6. External object preventing the tracks from turning.

Problem: Machine moves but does not steer.

Possible Causes: This problem must be caused by either the steering valve or its linkage. First ensure that the linkage is adjusted properly and not bent, broken, or disconnected. If there is no problem with the linkage, check inside the steering valve to be sure the lever is not disconnected.

Problem: Excessive hydraulic cylinder **drift** (movement off preset position).

Cylinder drift is usually caused by circuit leakage. In stick and bucket circuits you should check the main control valves, line relief valves, makeup valves, and cylinders. In the boom circuit, check these components plus the boom vent valve, boom check and relief valve, and the rate of boom lower valve.

Possible Causes:

1. Leakage in and around the seals on the cylinder pistons.
2. Leakage at the line relief or makeup valves in the main control valve module for the circuit with drift.
3. Spool in the main control valve not correctly centered, usually caused by a broken spring or binding valve spool.
4. Small amount of pilot (signal) pressure at the end of the spool in the main control valve causing an off-center spool and back leakage.
5. In the case of boom cylinder drift, the cause can be any of those above or a defective boom vent valve, boom check valve, and relief valve.

Problem: Boom will not raise or lower, but all other implements operate correctly.

Possible Causes:

1. Stuck spool in the boom vent valve, causing the boom check valve to block and preventing head-end oil to exit the cylinder for boom.
2. Seized spool in the main control valve caused by a defect in the spool or body of the valve.
3. Turned spool in the pilot valve bore.
4. Disconnected, bent, or out-of-adjustment control linkage between the lever and pilot valve.

Problem: High oil level in the final drive housing.

Possible Causes:

1. Excessive oil added to the housing (service problem).
2. Oil leakage from the track motor.
3. Oil leakage from the track brake.

Problem: High oil level in the pump drive housing.

Possible Causes:

1. Excessive oil added to the housing (service problem).
2. Leakage past pump seals.

Problem: Surge results when the boom or stick control valve for the boom or stick is activated.

Possible Causes:

1. Summing valve not working properly.
2. Horsepower output of the engine has been reprogrammed or otherwise altered.
3. Line from the summing valve to the pumps is not open.

Problem: Noisy line relief valves.

Possible Causes:

1. Low relief-valve set pressure.
2. High main relief-valve set pressure.

Bleeding a Hydraulic Pump

After any hydraulic pump replacement, it is imperative that the pump be bled of any air that may be trapped in the inlet hose between the hydraulic reservoir and the pump. This procedure outlines a general procedure used to bleed off the trapped air that may cause any aeration or cavitation within the pump. It is by no means intended to replace specific manufacturer procedures or guidelines. It must also be noted that the equipment using hydraulic pumps is infinitely variable and that this procedure may not be applicable to all applications.

P1-1 This procedure is implemented after a hydraulic pump, inlet hose, or hydraulic reservoir is drained or replaced or when air is suspected to have entered the hydraulic system. The example we are using is on a Caterpillar 248B Skid Steer Loader.

P1-2 This procedure outlines the order to bleed off any trapped air in the inlet hose to the pump after a pump has been reinstalled into the machine. Before any work is started on a machine, proper safety procedures and 'lock-out' systems must be implemented. First, raise the boom high enough to place the safety arm to lock the boom in the raised position. Remove the holding pin and slowly lower the safety arm in place. Note: Do not let the safety arm fall into place as it may damage the cylinder rod.

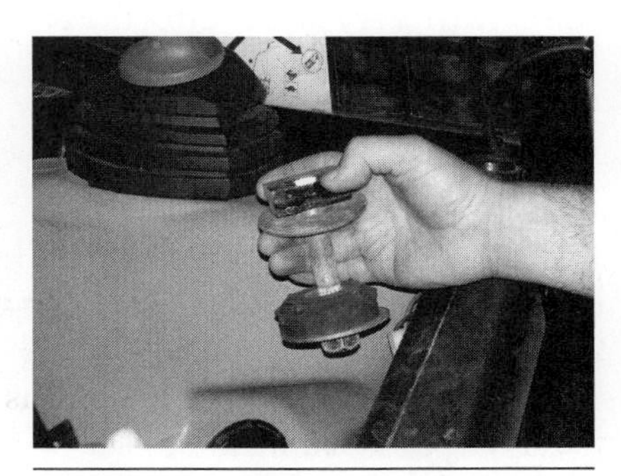

P1-3 Because the pump is located below the operatorís cab, it must be raised to access the pumps and hydraulic components. To raise the cab, first remove the cab bolts located at the front corners of the cab.

P1-4 This machine is equipped with gas-filled cylinders and will raise with very little effort. For machines not equipped with gas-filled cylinders, an over-head crane is necessary to raise the cab.

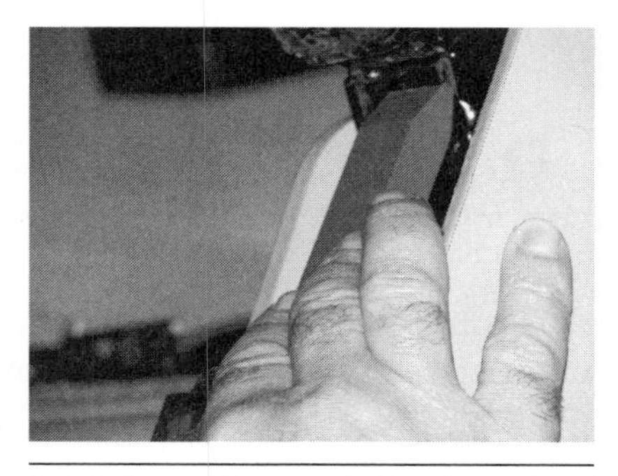

P1-5 Once the cab is in the raised position, make sure the safety latch is in place to prevent any sudden drop of the cab.

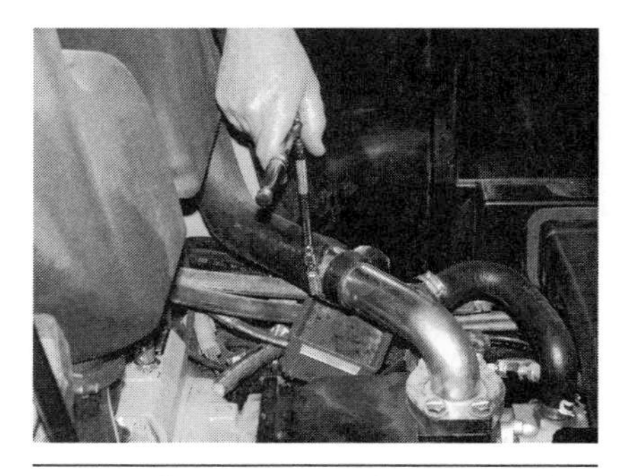

P1-6 After any work is completed and it is necessary to bleed the pump, locate the inlet line to the pump and remove the hose clamp.

P1-7 Insert a dull-tipped screwdriver between the flex-hose and inlet flange fitting of the pump.

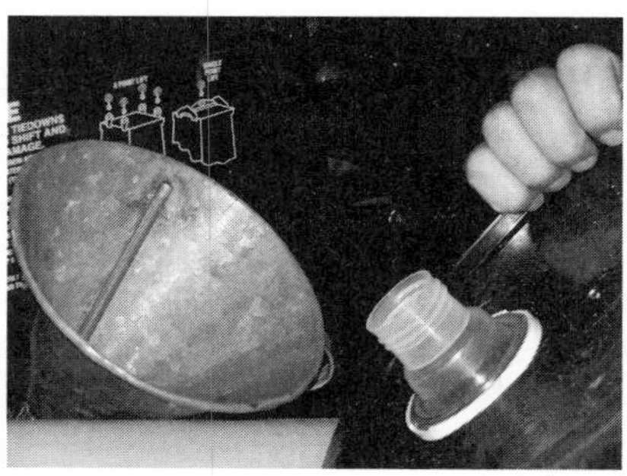

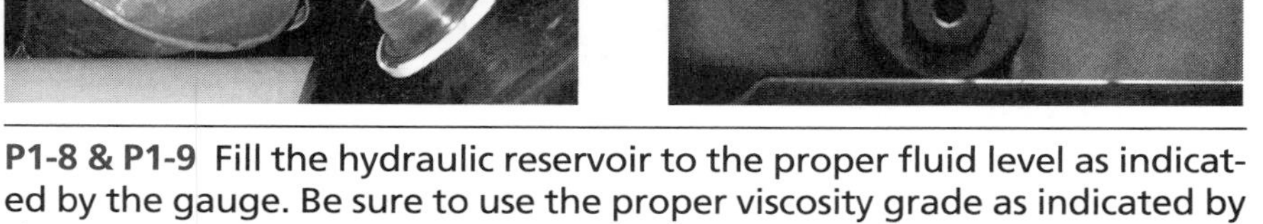

P1-8 & P1-9 Fill the hydraulic reservoir to the proper fluid level as indicated by the gauge. Be sure to use the proper viscosity grade as indicated by the OEM.

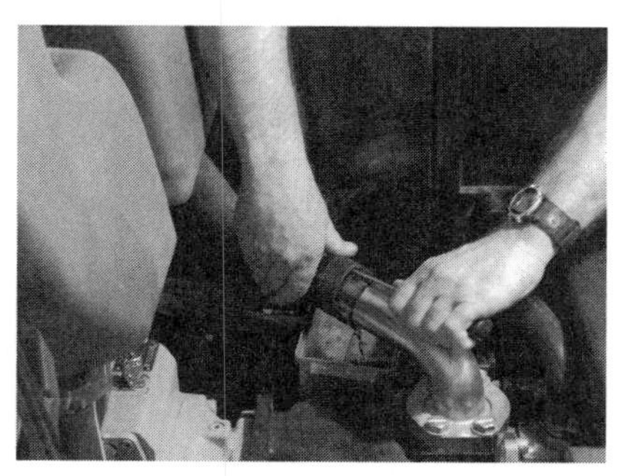

With the reservoir at the full level, oil should start to flow through the pump inlet line, pushing the air out through the opening created by the inserted tool. When the oil flows at a steady rate and there is no evidence of air in the line, remove the tool and reinstall the hose clamp. On certain machines it may be necessary to slightly pressurize the reservoir to allow oil flow to the hydraulic pump. A hydraulic pump may be mounted higher than the reservoir.

The machine is now safe to start at low idle to allow oil to circulate through the system. At this point the technician should be looking for any possible oil leaks and any other abnormal sounds and functions.

It is very important to double-check the hydraulic fluid level after the cab and boom have been lowered to the ground and are in proper checking position.

Alternate Systems

Most hydraulic circuits used on off-highway equipment share the same operating principles no matter where they are manufactured or by which OEM. Once you understand the operation of one piece of equipment, most others can be likened to it. **Figure 5-12** is a complete hydraulic circuit schematic of a simple backhoe loader: you should be able to recognize the various implement circuits identified on the schematic.

ONLINE TASKS

1. Locate each of the following urls and navigate to pages that provide troubleshooting information:
   - *http://www.cat.com/*
   - *http://www.deere.com*
   - *http://www.komatsuamerica.com/*
   - *http://www.jcb.com/*
   - *http://www.liebherr.com/lh/en/*
   - *http://www.groveworldwide.com/na/eng/crane/default.htm*
   - *http://www.casece.com/index.asp?RL=NAE*

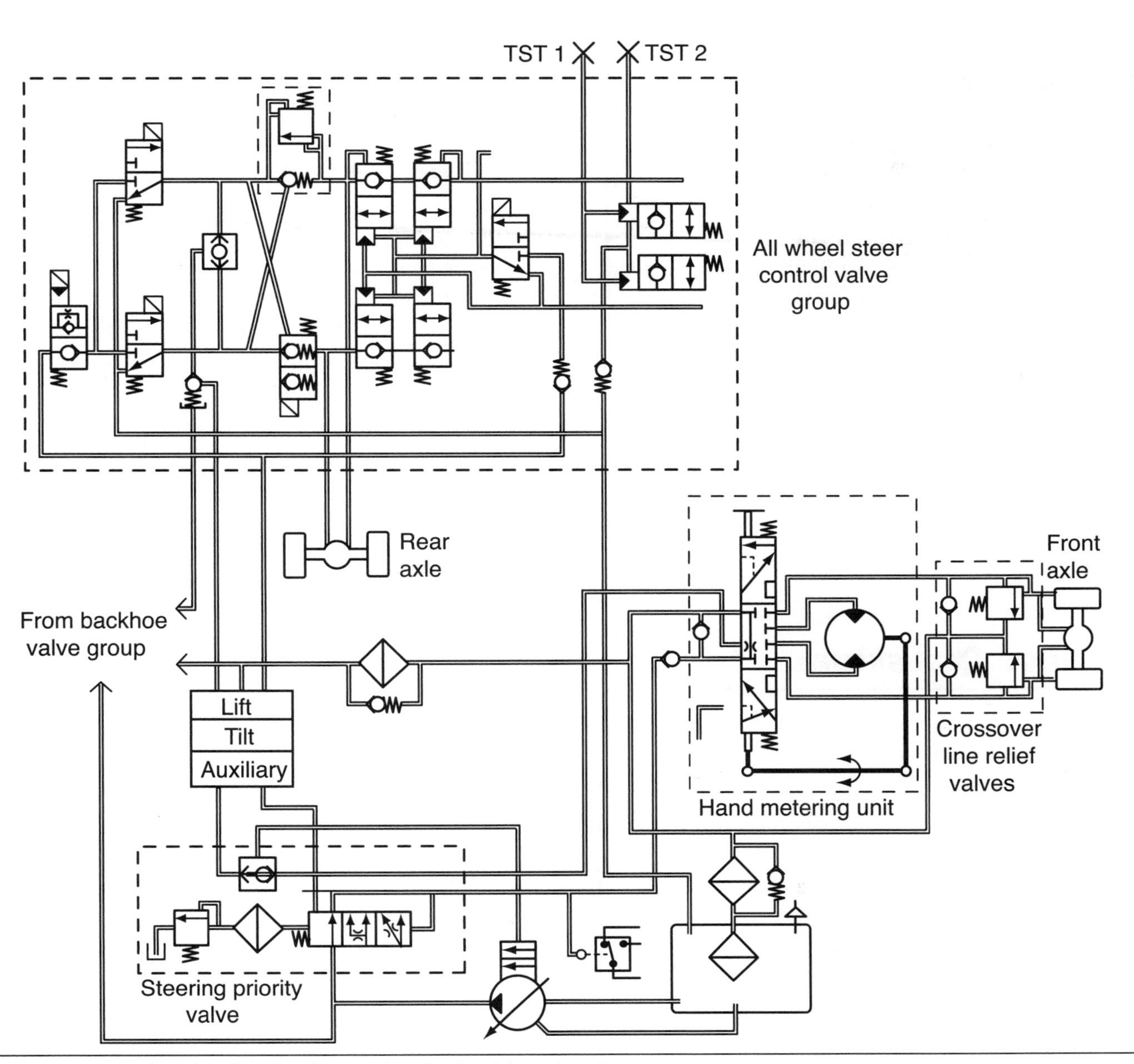

Figure 5-12 Hydraulic circuit schematic of a backhoe loader.

2. To get relevant service and troubleshooting information, you usually have to have access to an OEM data hub, which is limited access and password protected. If your college or workplace has data hub access, select a piece of equipment, identify a hypothetical problem, and download the troubleshooting data. For instance, the Caterpillar site is *http://www.IMAP@cat.com.* However, you will be prompted for passwords at logon, and you will not get any farther unless you have access.

SHOP TASKS

1. Locate the hydraulic schematics that outline the hydraulic circuits on a piece of equipment. Make a copy and identify the location of the pump(s), control valve(s) and actuators. Note all the circuit test locations on your copy.
2. Use OEM service literature to connect a hydraulic analyzer into the hydraulic circuits on a piece of equipment. Complete a troubleshooting test profile such as those used in Tables 5-1 and 5-2 in this chapter.

Summary

- Maintenance procedures on mobile equipment hydraulic systems begin with ensuring the system is clean both inside and outside the circuit.
- Routine replacement of hydraulic fluid, sometimes accompanied by system flushing, is recommended to minimize system malfunctions and downtime.
- Hydraulic circuit testers are used to analyze hydraulic circuit performance.
- A typical general-purpose hydraulic circuit tester consists of pressure gauge(s), flowmeter, flow control valve, and thermometer.
- When hydraulic analyzer test data falls outside specification, use OEM troubleshooting tables and charts to source the problem(s).
- A key to troubleshooting is developing the ability to navigate through OEM hydraulic schematics.
- A typical excavator uses hydraulics in the track drive, swing, and implement circuits: the circuits may be complimentary so, to ensure that you are following the correct procedure, the OEM service literature should be referenced.
- The implement circuit on a typical excavator consists of boom, stick, and bucket circuits: each circuit uses double-acting hydraulic cylinders.

Review Questions

1. A hydraulic pump is flow tested and at zero psi, produces only 60% of the OEM specified flow. Which of the following is the probable cause?

   A. defective pressure relief valve
   B. worn pump
   C. flow should be as specified when a flow restriction is added
   D. flow should be as specified when pressure exceeds 100 psi

2. The boom will not raise or lower on a piece of equipment, but all other implements appear to operate correctly. Which of the following is the most likely cause?

   A. aerated hydraulic oil
   B. out-of-adjustment control linkage between the lever and pilot valve
   C. low pump relief valve setting
   D. oil leakage from track brake

3. A track-driven excavator will not move when either the FORWARD or REVERSE pedals is depressed, but all the implements function normally. Which of the following would be the LEAST likely cause of the problem?

   A. insufficient pilot pressure to release the track brakes
   B. machine safety lever located in the SAFE position
   C. stop fully turned in on the track brake valve
   D. aerated hydraulic oil

4. When operating a backhoe, excessive hydraulic cylinder drift occurs. Technician A says that cylinder drift can be caused by circuit leakage. Technician B says that the main control valves should be checked. Who is correct?

   A. Technician A only
   B. Technician B only
   C. Both A and B
   D. Neither A nor B

5. Technician A says that aerated hydraulic oil can cause a circuit to operate erratically. Technician B says that excessively cold hydraulic oil can cause a system to operate erratically. Who is correct?

   A. Technician A only
   B. Technician B only
   C. Both A and B
   D. Neither A nor B

6. After a PM drain, flush, refill, and filter replacement service, which of the following procedures is recommended?

   A. Do not operate the system for 24 hours after the service.
   B. Static bleed each circuit using a pneumatic bleeder ball.
   C. Run system through several cycles to purge air from system.
   D. Dump and renew the new oil after one hour of operation.

7. A hydraulic bucket load drops when the control valve is moved to neutral. Technician A says that this could be caused by a control valve that binds and does not center. Technician B says this could be caused by a seized mechanical linkage. Who is correct?

   A. Technician A only
   B. Technician B only
   C. Both A and B
   D. Neither A nor B

8. Pressure drops off in a hydraulic implement circuit well before full load is achieved. Which of the following is the more likely cause?

   A. defective relief valve
   B. internal leak in the circuit
   C. seized control valve spool
   D. cold hydraulic fluid

9. The oil level in a final drive housing of a track-driven excavator is excessively high. Technician A says that oil leakage from the track motor could be the cause. Technician B says that oil leakage from the track brake could be the problem. Who is correct?

   A. Technician A only
   B. Technician B only
   C. Both A and B
   D. Neither A nor B

10. The boom on an excavator will not raise or lower, but all the other implements appear to work properly. Technician A says that this could be caused by a seized spool in the main control valve. Technician B says that this could be caused by a turned spool in the pilot valve bore. Who is correct?

    A. Technician A only
    B. Technician B only
    C. Both A and B
    D. Neither A nor B

CHAPTER

Hydraulic Brakes

Learning Objectives

After reading this chapter, you should be able to

- Describe the fundamentals of off-road hydraulic brake systems.
- Identify the major components that make up off-road hydraulic brake systems.
- Describe the principles of operation of off-road hydraulic brake systems.
- Outline the maintenance and repair procedures associated with off-road hydraulic brake systems.
- Identify the basic troubleshooting procedures for off-road hydraulic brake systems.

Key Terms

accumulator
brake application pressure
brake pressure
brake shoes
brake valve
calipers
charge valve
check valve
drum
foot valve
hygroscopic
master cylinder
metering
modulated brake pressure
park brake valve
pressure switch
rotor
servo brake
slave cylinder
system pressure
variable-pressure reducing valve
wheel cylinder

INTRODUCTION

Over the last 30 years, off-road equipment manufacturers have used a wide variety of brake designs. Modern day off-road brake systems have evolved to become what might be considered technological marvels in brake design. Brake systems on most off-road equipment are able to operate in conditions that would overload and wear out a conventional on-road brake system in a very short time. In most cases, there are no paved roads for the equipment to operate on; dirt and gravel roadways are the norm in this environment. Given the grades and roadways, the ability to safely slow down and stop equipment that weighs as much as 200 tons is a daunting task.

In the past both air brakes and air-over-hydraulic brakes were commonly used on off-road equipment, but these have fallen out of favor because of the high maintenance costs associated with these brake designs. Modern equipment design favors the use of hydraulic brake systems that are classified as internal wet disc brakes. These systems use a brake design that has the brakes located inside the axle assembly which keeps the dirt out. The design has evolved so remarkably that these brakes may last for many thousands of hours before they need to be overhauled. It is not uncommon for the brake friction discs to last up to 10,000 hours of operation before they need to be replaced.

FUNDAMENTALS

Brake design has changed dramatically since the invention of the wheel. As equipment speed and weight increased, the brake system design had to keep pace with these changes. Brake systems are used to control the speed of the equipment as well as to bring it to a safe stop when required. No matter what the design of the brake system, the principle of operation remains similar. All brake systems rely on some form of friction to slow or stop the equipment. We know that energy cannot be destroyed, but it can be converted to another form of energy. Kinetic energy of moving equipment is converted to heat energy when the operator depresses the brake pedal. A brake application by the operator forces friction material to rub on a metal surface to slow down or stop the equipment. The friction device is usually located in the drive axle and, depending on the axle design, the brake system can be located either inboard or outboard on the axle assembly.

Early on it was quickly discovered that it made more sense to wear out the friction lining than to wear out the drums or rotors. The early brake systems used on equipment involved the use of complex linkages that took advantage of lever principles to transmit the force of the operator's foot to the wheel ends. As brake systems evolved, the use of hydraulics to transmit the operator's foot pressure to the wheel ends became increasingly popular. Early hydraulic brakes designs used a fluid confined in a hydraulic circuit to transmit the operator foot pressure to a master cylinder, which builds pressure that is proportional to the amount of force the operator applies with his foot. Modern equipment uses some form of hydraulic assist to take the place of the operator's foot pressure (force). The wheel-end clamping force used on these modern systems is determined by how much **modulated brake pressure** is sent to the wheel ends.

The nature of off-road equipment operating conditions requires that the brake systems have a number of safety features built into them. The use of two independent brake circuits on equipment is a must. An independent mechanical park brake is also required to safely park the equipment. In some industries, such as mining, the equipment must be equipped with three independent brake circuits. The third circuit (emergency brake) is a combination of the wheel ends and the park brake circuit.

The complexity of the brake systems on off-road equipment has increased dramatically over the past 10 years. The use of complex hydraulic brake circuits, with variable-displacement pressure-compensated pumps with ABS circuits, is becoming common in off-road equipment. These modern brake circuits monitor the **brake application pressure** and will apply the brakes automatically if the pressure falls below safe operating pressure. Not only will the brakes apply when the **brake pressures** are too low, but many systems will also automatically apply the brakes if there is a loss of electrical power or if transmission clutch or torque pressure falls below safe operating levels. On many types of equipment the brakes cannot be released after an emergency application until the operator activates a brake interlock circuit. This interlock brake circuit ensures that the operator is in the cab when the brakes are released. To gain a full understanding of these brake systems, it is imperative that a technician have a good understanding of basic hydraulics and can read and understand electric/hydraulic schematics. This chapter will cover the basic principles of hydraulic brakes, including shoe-type brakes and external caliper disc brakes, as well as taking an in-depth look at various types of hydraulically applied internal wet disc brake systems.

BASIC HYDRAULIC BRAKE SYSTEM COMPONENTS

To fully understand the complex brake systems that are used on larger off-road equipment, we need to examine the basic brake systems and their components that are still found on some smaller off-road equipment. **Figure 6-1** shows the master cylinder that is typically found on off-road utility vehicles, such as a 4-wheel drive Toyota Land Cruiser, which are used extensively by personnel in the mining industry to get around the job site. Note that the front and rear brakes use separate circuits incorporated into the design of the master cylinder. Each section of a dual master cylinder has its own brake reservoir. This provides a degree of safety; if one of the brake circuits were to fail, the other would still be operational. A warning light located in the operator's compartment alerts the driver to a loss of a brake circuit.

Dual Circuit Master Cylinder

Master cylinder design has not changed much over the last 40 years. The last major change incorporated the use of two independent circuits in the master cylinder,

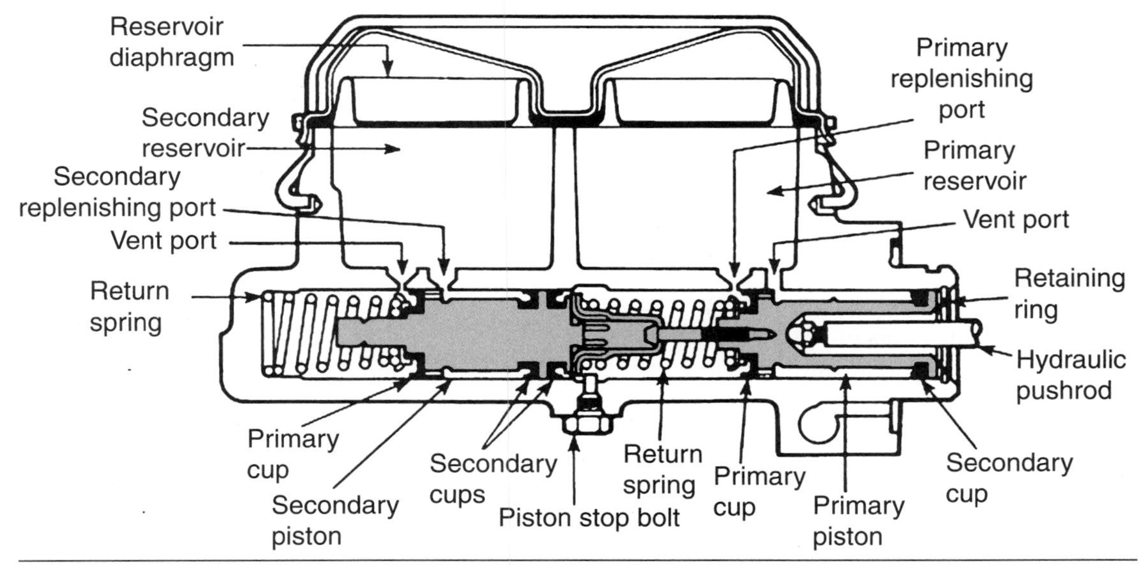

Figure 6-1 Cutaway of a typical master cylinder used on some of the smaller types of off-road equipment.

known as a tandem master cylinder. Since the development of the dual circuit master cylinder, very few changes have taken place in the design. However, the material quality used in the internal components, especially in the seals and cups, has improved over the years. Figure 6-1 shows a cutaway of a typical dual circuit master cylinder. Note that the single bore houses both the primary and secondary pistons for both circuits. They are separated by seals, which prevent the fluid from one circuit from entering the other circuit. If a failure were to occur in either circuit, the other circuit would not be affected.

The master cylinder, as its name suggests, controls the flow of hydraulic brake fluid to the wheel end actuators, which are often referred to as **slave cylinders** or **wheel cylinders** in the manufacturer's service manuals. The actual wheel end forces that are produced by the master cylinder are totally dependent on the force that is applied to the **master cylinder** by the operator. In **Figure 6-1,** each hydraulic chamber is connected to its own reservoir. In the "brakes released" position, the primary and secondary pistons are held in the retracted position by a spring. When the operator activates the brakes, the primary piston is forced to move forward thereby closing the inlet port, which allows brake pressure to build in the primary circuit. The increase in primary brake pressure forces the secondary piston to move forward, closing off the inlet port and causing the pressure in the secondary circuit to increase. When the operator releases the force on the brake pedal, pressure in the brake circuit drops instantly as the pistons retract. The compensating ports located in the piston chambers help vent the brake fluid back into the reservoirs.

Hydrostatic Drive Brake Master Cylinder

Hydrostatic drive equipment often incorporates a conventional master cylinder brake circuit that is capable of dissipating drive pressure from the drive circuit during a brake application. **Figure 6-2** shows a cutaway of this type of master cylinder. The operator controls the brake pedal application through a mechanical brake pedal in the operator's compartment. A boot on the plunger end of the master cylinder seals the unit from the environment. When the brakes are not engaged, oil from the reservoir keeps the valve spool chamber and the brake plunger chamber filled with oil. In the released position, the valve spool blocks the flow of drive signal pressure from returning to the tank. When the operator depresses the brake pedal, the valve spool is forced forward, bleeding off the drive system signal pressure back to the tank. Note that both fluid chambers in the master cylinder are still open to the tank. As the valve spool continues its forward travel it pushes against the brake plunger blocking the forward chamber from the tank side. Continued brake pedal travel builds pressure in this chamber; this pressure applies the brakes.

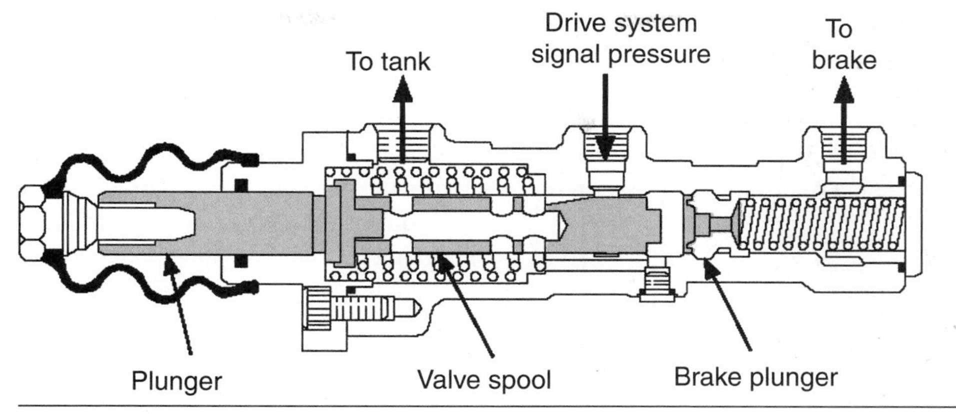

Figure 6-2 Cutaway of a typical master cylinder used on some of the smaller hydrostatic drive equipment.

When the operator releases the brake pedal the valve spool and brake plunger are forced back by return springs into the released position, dissipating brake pressure from the circuit and blocking the port that allows the drive system signal pressure to be bled off to the tank. This design allows the dissipation of drive system signal pressure to coincide with the application of the brakes on this type of drive system. Many equipment manufacturers that use hydrostatic drives incorporate this type of master cylinder design into their equipment.

Dual Master Cylinder

Conventional dual master cylinders are not always suited to all applications of off-road equipment. Independent control of brakes on each side of the equipment requires the use of a specialized type of master cylinder. On many equipment designs, two separate master cylinders are required so that each brake circuit on the equipment can be controlled independently. The master cylinders used on both sides of the brake circuit are identical in design. A separate brake pedal is used on each master cylinder, allowing the operator to apply the brakes to each side of the equipment independently. This design gives the equipment greater steering control when operating in tight quarters.

Both master cylinders have their own supply port, which is connected to a common brake reservoir, or in some cases, dual reservoirs. A return spring in each master cylinder keeps the valve stem in the neutral position when the brakes are not activated.

When the operator activates a brake pedal, the plunger shown in **Figure 6-3** moves to the left, forcing the valve stem to block the flow of oil to the reservoir. This action allows pressure to increase in the brake circuit. At the same time the increase in pressure in the chamber forces the compensator valve to unseat, allowing oil to flow through the bridge pipe to the other master cylinder where it is blocked by the other compensator valve. When both left and right brakes are activated simultaneously, both compensator valves open to allow the pressure to equalize in both brake circuits. This permits equal brake engagement on both sides of the equipment. The compensator valves act as **check valves** when independent brake application is required on either side, and are forced off their seats to equalize the brake pressure when the operator activates both brakes at the same time.

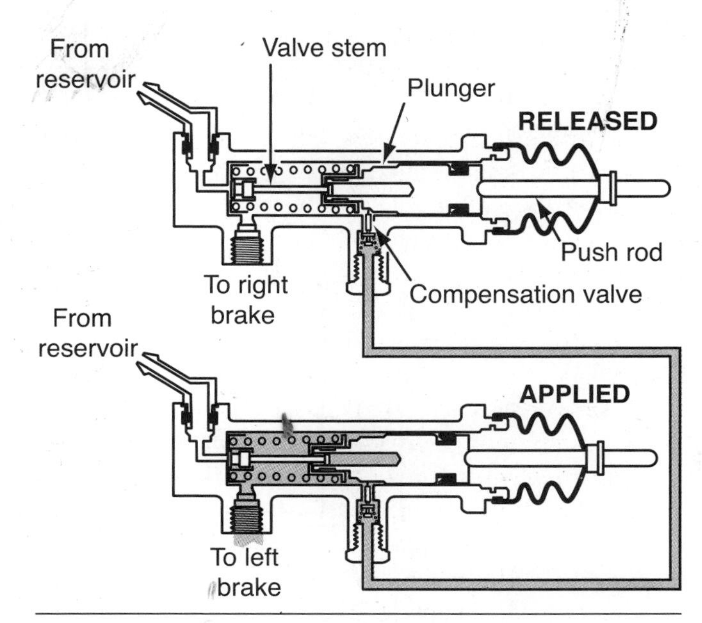

Figure 6-3 Cutaway of a typical dual master cylinder configuration with the brakes applied on the left and right brake circuits independently.

When the brake pedal(s) is released, the return spring forces the plunger and valve stem to return to the neutral position, opening the supply port back to the tank. The pressure in the system forces the plunger back and the pressure in the brake circuit is equalized.

Boost Assist Master Cylinder

On large equipment where brake pressures need to be considerably higher than what can be achieved with normal operator foot pressure, additional hydraulic pressure can be used to multiply brake force without exerting extra pedal effort on the part of the operator. This piston applies additional pressure to the master cylinder plunger to achieve a higher brake pressure than is possible with operator foot pressure. Depending on the equipment manufacturer, the oil pressure used for the power-assist side can come from a separate pump or from a hydraulic circuit that incorporates a reduced pressure circuit. In **Figure 6-4** the master cylinder is shown in the applied position. The master cylinder piston is held in this position by a return spring. In this piston, the boost piston rear seal keeps the inlet oil from entering the master cylinder. The two spring chambers in the master cylinder are open to the reservoir, which keeps the pressure equalized. Since the brake line openings are located in the spring chamber area, they remain filled with oil. Note that when the boost piston is the released position, the center valve is also open to the reservoir.

When the operator applies foot pressure to the brake pedal, the boost piston is forced to move to the left, allowing pressurized oil to enter the chamber behind the boost piston that applies hydraulic pressure to add additional force to the input plunger. The input plunger forces the master cylinder plunger to move to the left, causing the center valve to close the opening to the reservoir. When the center valve closes the opening to the reservoir, the pressure starts to build in the spring chamber in front of the master cylinder plunger. During normal braking action, the flapper valve located on the right side of the boost piston closes. If the operator depresses the brake pedal too quickly, the inlet oil is not able to fill the boost-assist side of the chamber quickly enough.

To ensure smooth pedal effort during a quick brake application, the flapper valve allows the oil to flow from the low pressure side of the boost piston to the inlet side of the boost piston, keeping the inlet side of the piston constantly filled with oil. Once satisfactory pressure is reached on the inlet side of the boost piston, the flapper valve closes. The pressure applied to the boost piston is proportional to the effort exerted by the operator's foot on the brake pedal. Under certain conditions, a small hole in the boost piston allows oil to flow to the reservoir. When the operator applies more force to the brake pedal, the metal-to-metal contact surface between the input piston and the boost piston is increased, slowing down the oil leakage back to the reservoir. During boost piston movement, the oil on the low pressure side of the boost piston is sent to the reservoir. When the normal oil level is exceeded, the excess oil flows back to the tank through an overflow port located in the reservoir.

Continued brake pedal travel allows the pressure in the spring chamber to build and, since this chamber is open to the brake wheel ends, the brakes are applied. In Figure 6-4 note that the center valve seal blocks the

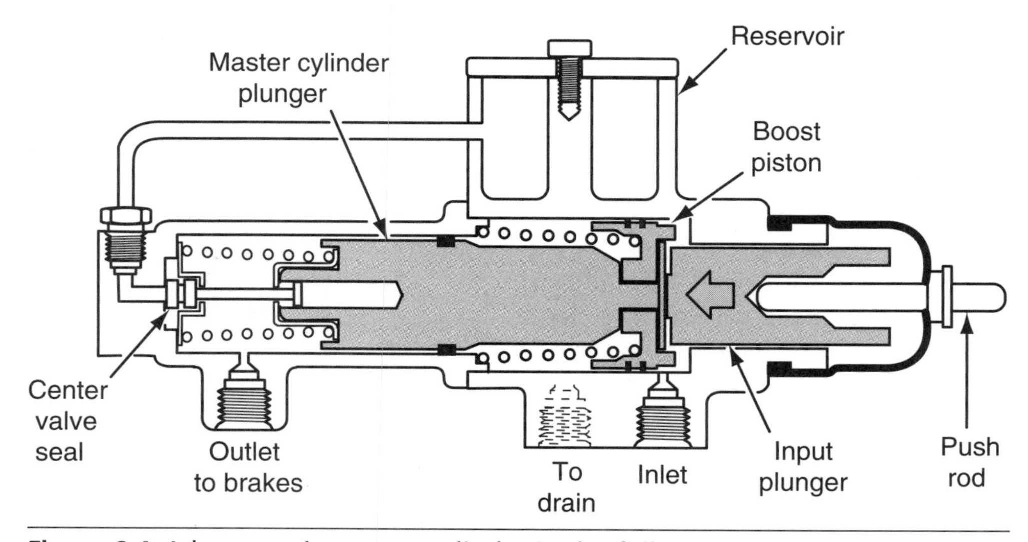

Figure 6-4 A boost-assist master cylinder in the fully engaged position.

flow of oil back to the reservoir when the spring chamber is pressurized. If a brake application becomes necessary when the engine is not running, the operator's foot effort on the pedal still can apply the brakes even though no power assist is available. During initial release of the brake pedal, the boost pressure on the right side of the boost piston moves the input piston back to the right. At this point the pressurized oil from the inlet port flows to the low-pressure side of the boost piston through the hole in the center of the piston. During this step the oil from the inlet port flows to the reservoir until the boost piston returns to its original position which blocks the inlet port.

Hydraulic-Assist Two-Stage Master Cylinder

Equipment manufacturers use a wide variety of master cylinder designs to accommodate equipment component layout. This particular application has the brake pedal connected to a bell crank lever and cable, which controls the movement of the brake booster plunger. When the operator depresses the brake pedal, the booster plunger forces the power piston to move. **Figure 6-5** shows the master cylinder in the applied position; the return spring is keeping the power piston and booster connector against the left side of the master cylinder. The booster connector has a spring inside that keeps the plunger piston in the retracted position. Plunger travel is limited by a snap ring located on the end of the plunger piston. As long as the plunger piston is in the released position, inlet oil can flow into the spring chamber by way of a notch in the plunger piston.

Since the spring chamber is open to the return, oil circulates through the system. Oil can flow to the main chamber of the master cylinder through an orifice located at the left end of the main chamber. During this stage the main chamber and the second stage chambers (as well as the brake lines) are connected hydraulically to the reservoir, keeping the pressure equalized in the system. This design uses two pistons to apply the brakes and has an external reservoir that maintains the fluid necessary to engage the brakes regardless of brake pad wear.

The complex operation of this brake design necessitates that the brake engagement be shown in two steps to clearly explain its operation. The first step, the start of engagement, is explained in detail so that you can fully understand the brake engagement process. A relief valve located in the master cylinder housing limits the amount of pressure available to act on the power piston. Further travel of the main piston blocks the inlet orifice passage to the reservoir; this causes the pressure to build in the main chamber and second stage chamber. At this point, pressure in the main chamber activates the service brakes until the second stage piston seats against the end seal. Once this occurs, oil from the main chamber is blocked from acting on the service brakes, at which point the oil pressure from the second stage chamber of the master cylinder acts to apply the service brakes. Further brake pedal travel allows the pressure to build up in the main chamber until the pressure is high enough to unseat the second stage piston, allowing the oil to flow to the service brakes. An external

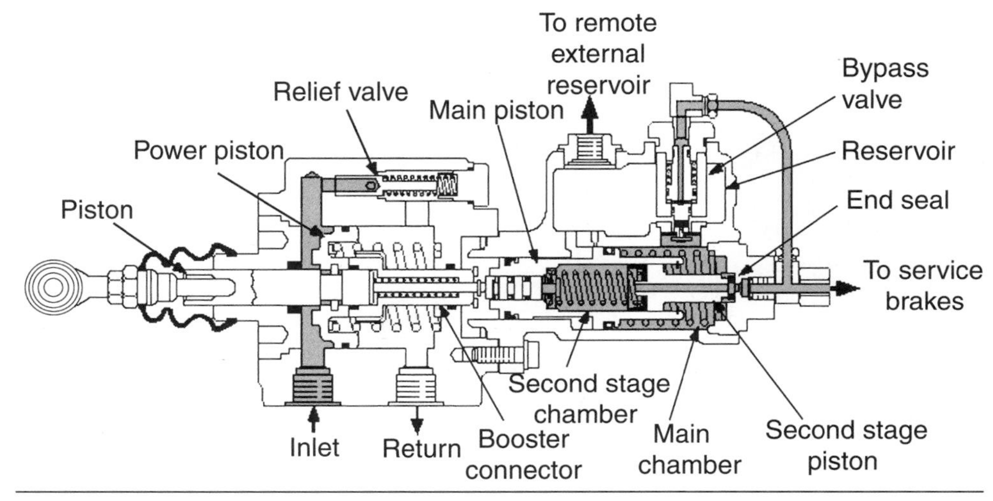

Figure 6-5 A hydraulically assisted master cylinder with the master cylinder in the fully engaged position.

line connected on the brake line side of the master cylinder allows the brake pressure to act on a bypass valve located in the master cylinder housing above the second stage piston chamber. The bypass valve limits the amount of brake pressure that can be developed if the pressure were to rise above a predetermined point, the bypass valve would open, forcing the brake pressure to return to second stage pressure. Brake pressures are generally low at this point initial engagement, to provide smooth control of braking during partial braking.

In Figure 6-5 the master cylinder is in the fully applied position, which forces the fluid to act on the bypass valve. When this pressure reaches a predetermined point the bypass valve opens, forcing the brake engagement to go to second stage engagement. The fluid flows from the main chamber to an external reservoir causing a pressure drop in the main chamber, which forces the second stage piston to seat on the end seal. The pressurized oil in the second stage chamber cannot escape because of the sealing action of the cupped seal on the piston. Because this area (second stage) is approximately 30% of the area of the main chamber, the pressure flowing to the service brakes will be much higher than it was during the first stage of brake application.

When the operator releases the brake pedal, the plunger piston is forced back by spring force. At this point a passage through the power piston opens, allowing the oil behind the power piston to return to the reservoir. The power piston and booster connector are forced back by a return spring, and the pistons in the master cylinder are forced back by a combination of oil pressure and spring pressure. This brake design allows the brakes to be applied by the operator even if the engine is not running. However, the pressure will be limited to the pedal effort since no external oil pressure is available to apply an assist to the power piston. The power piston will bottom out on the booster connector, forcing the main piston to build a limited amount of brake pressure.

Dual Power-Assist Master Cylinder

Modern equipment design has incorporated the use of independent dual brake circuits built into one master cylinder for the front and rear brakes for obvious safety reasons. Failure of a brake circuit that had separate front and rear brake circuits still allows the operator to stop the equipment safely. **Figure 6-6** shows a typical dual power-assist master cylinder in the released position. This design can be broken down into two sections: the power cylinder and the master cylinder. The operation of each section will be described in detail in this chapter.

In the released position, the brake pedal is fully released and return springs keep the internal components in their respective positions. Oil is present in the chamber between the load feedback piston and the plunger. Oil from the inlet also passes around the servo

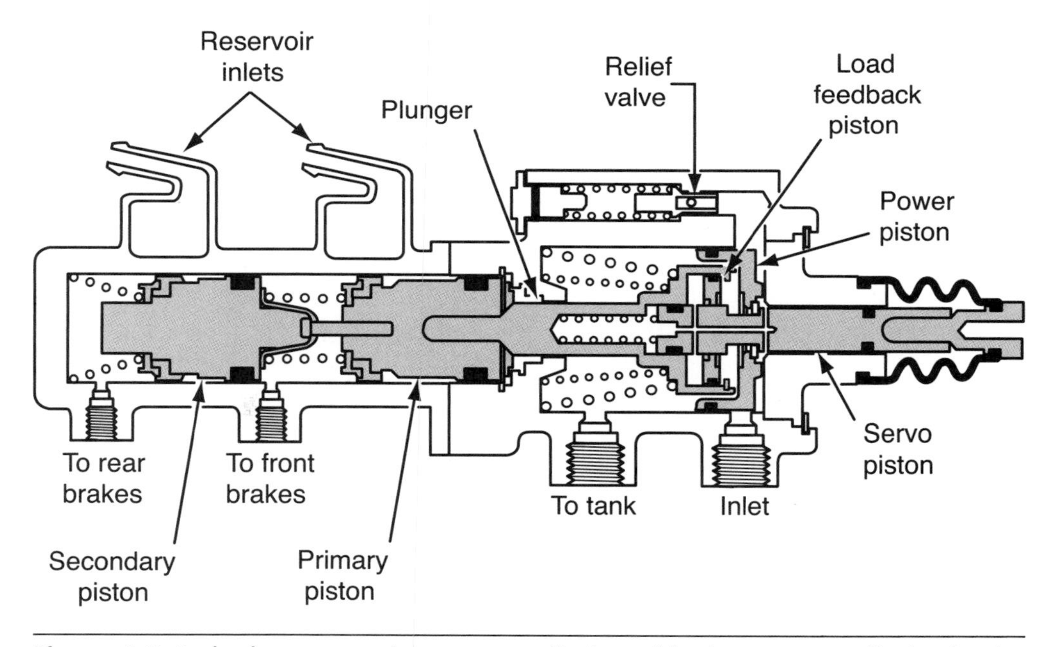

Figure 6-6 A dual power-assist master cylinder with the master cylinder in the fully released position.

piston and flows back to the tank by way of the power cylinder chamber. The master cylinder houses two separate brake compartments, primary and secondary, which are open to the reservoir when the brake assembly is in the released position. This allows the pressure in both circuits to be equalized with the brake line full.

To activate the service brakes the operator depresses the brake pedal; this forces the servo piston to shift to the left, blocking the flow of inlet oil to the tank. This action allows the pressure to increase quickly behind the power piston to increase the assist force that is applied to the plunger, which then acts on the primary piston. A relief valve located in the power cylinder housing limits the pressure that is available to act on the power piston. The power piston also has an area where the pressure acts on a load feedback piston, giving the operator feedback from the brake application. Both primary and secondary pistons in the master cylinder are connected together mechanically.

When the primary and secondary pistons are forced to the left, they block the inlets. This forces the pressure to rise in the two chambers that are applying the brakes. Chamber pressure in the front brake chamber applies pressure to the secondary piston face, forcing the pressure to equalize in both front and rear chambers. When the operator releases the brake pedal the servo piston will return to its original location because the force of the return springs allows oil to flow around the servo piston and back to the tank.

Hydraulic Brake Control Valve

Brake control valves are used to control the flow of fluid to the service brakes on equipment that uses hydraulic brakes to get its pressurized oil supply from a hydraulic circuit. This circuit can be part of a main hydraulic system, or it can be a dedicated brake hydraulic circuit designed just for this purpose. There are a wide variety of **brake valves** in use on off-road equipment to suit many different applications. To gain an understanding of how these brake valves work, a commonly used brake valve design will be explained.

Figure 6-7 shows a cutaway view of a typical hydraulic brake control valve. Note that this valve

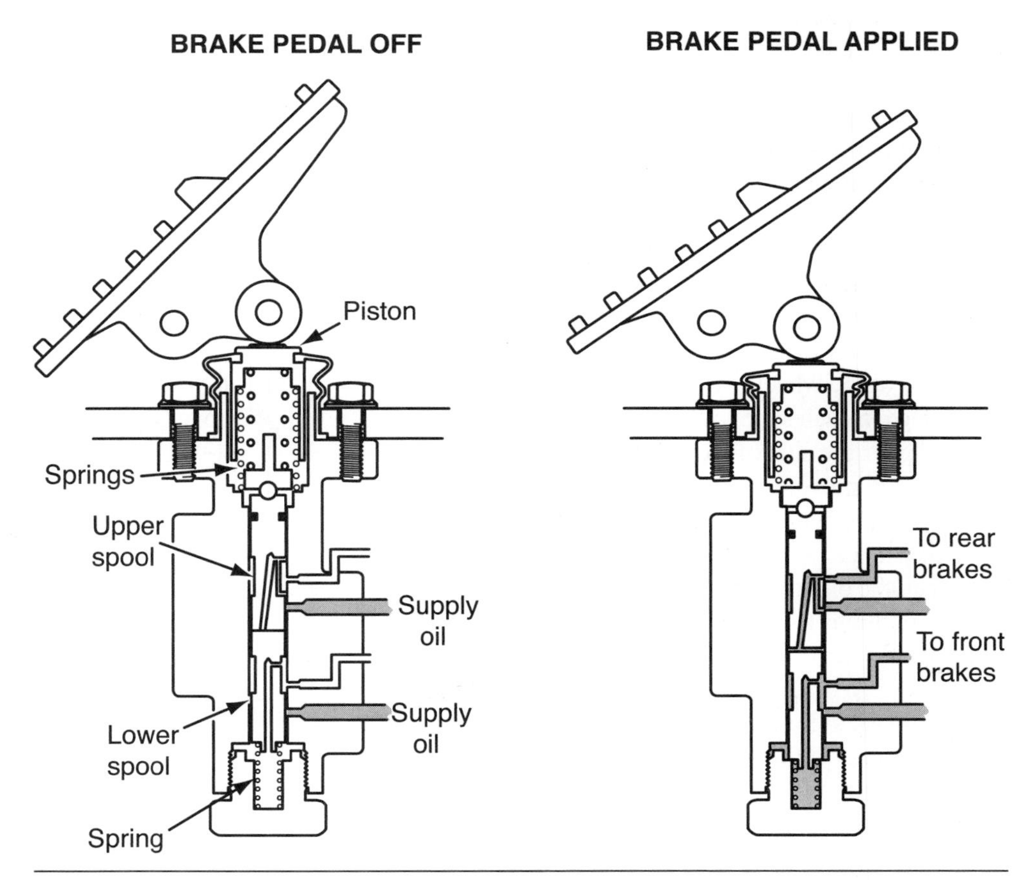

Figure 6-7 A dual brake valve: the valve on the left is in the released position and the one on the right is in the applied position.

incorporates what is commonly referred to as a duel circuit pedal. In other words, it has two independent brake circuits, one on top of the other, in the same valve body. Other designs use two separate valves connected to the same pedal, or two separate pedals (often found on loaders) where have one pedal applies the brakes only and the other brake pedal applies the brakes as well as disengaging the transmission. This design allows the transmission to disengage automatically whenever this pedal is used to apply the brakes during truck loading operations. Another common application for dual pedals is on equipment that requires the brakes to be applied independently on each side of the equipment. The modern backhoe is a good example of equipment that often uses this brake design.

Figure 6-7 shows two supply ports that have pressurized oil present at the inlets. The brake pedal on the left is shown in the released position with no oil flowing through the valve. A spring at bottom of the valve keeps the internal spools centered in the brake valve body. As long as the two internal valves remain in this position, the supply oil is blocked from flowing to the wheel end brakes. The upper spring returns the pedal to the retracted position when the operator releases the pedal.

When the operator steps on the brake pedal, the upper piston forces the top spring to compress, moving the two spools down. When this takes place, the oil from the inlet passages flows to the brakes. Small passages located between the inlet side and the outlet side of the spools act as restrictions to limit the amount of oils that can flow. The more the pedal is depressed, the more oil will flow. This allows the brake pedal to function as a **variable-pressure reducing valve** which limits the pressure flowing to the wheel end brakes. The more the pedal is depressed, the higher the pressure will be at the wheel end brakes.

A small internal passage in each spool allows the oil to flow from the modulated pressure side of the circuit to the underside of the spools. This provides a resistance to the pedal movement and gives the operator physical feedback by providing a slight resistance. When the force on the upper springs is lower than the oil pressure underneath the spools, the pressure forces them to move up. By metering the flow of oil, the brake valve works like a variable-pressure reducing valve.

When the operator releases the brake pedal, the oil pressure below the spools and the lower spring returns the spools to the neutral position by sending the flow of oil from the wheel end brakes back to the reservoir, while at the same time blocking the flow of oil from the inlet side to the brakes.

Dual Pedal Brake Valve

Figure 6-8 shows a typical brake valve that is used on a variety of backhoe loaders or small skid steer loaders that require independent side to side braking. With the pedal in the released position, the return springs keep the modulator spool and both brake selector spools in the released position. The oil flows from the supply side through the modulator valve and back to the reservoir through a small hole drilled through the modulator valve. Both brake spools in the released position allow the brake side to flow back to the reservoir. Two springs are used in the modulator valve; the lower spring is used to return the modulator valve to the released position when the operator releases the brakes.

When the operator begins to apply the brakes to the left side of the equipment as shown in Figure 6-8, the selector spool is forced down, blocking the return to the tank passage. At the same time it opens a passage from the outlet side of the modulator valve to the service brakes on one side. When this occurs the modulator spool moves down, allowing some of the supply oil to flow through a cross-drilled hole into the lower spring cavity of the outlet port. The position of the modulator valve causes a restriction to the lower spring cavity.

When the springs in the modulator valve are further compressed, the valve is forced to shift further; this

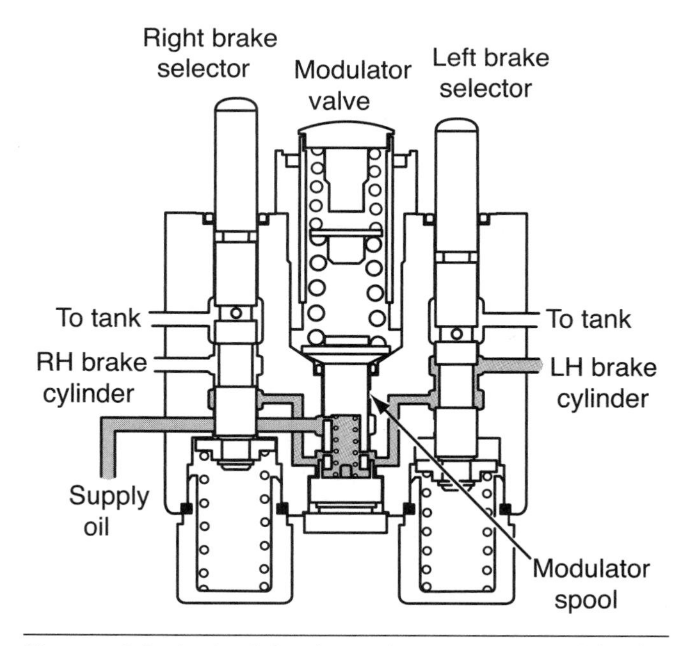

Figure 6-8 A dual brake valve cutaway with the left brake beginning engagement.

increases the oil pressure at the outlet ports. The amount of restriction is determined by the force applied to the brake pedal, which will modulate the pressure acting like a variable-pressure reducing valve. The modulator valve is capable of supplying oil to both brake valves simultaneously, but only the left side is being utilized in this example.

If the operator were to apply both brakes together, the pressure would equalize in both brake circuits just as it would if one or the other brake circuits were to be engaged separately due to the action of the modulator valve. For example, when brake line pressure increases in a brake circuit, the pressure of the oil in the modulator valve rises above the upper spring pressure, forcing it up. This causes more restriction in the oil flow through the cross-drilled hole until the opposing forces are equalized, thus metering the flow of oil to the service brakes to maintain the pressure.

Hydraulic Slack Adjusters

Equipment manufacturers use different means of compensating for brake wear. This chapter explains the operation of one type of hydraulic slack adjuster that CAT uses on some of its equipment. **Figure 6-9** shows the hydraulic slack adjuster cutaway in two different views, one in the released position and the other in the brakes applied position. This type of slack adjuster is designed to compensate for brake wear. The slack adjusters are located in series between the brake valve and the wheel end brakes. Each wheel end has its own hydraulic slack adjuster.

When the operator applies the service brakes, pressurized oil flows into the slack adjuster through the inlet port from the master cylinder or brake valve, depending on the equipment brake circuit design. This slack adjuster will work on either hydraulic brake circuit. Continued wear of brake material will require that more oil be displaced with each brake application as the wear increases. In **Figure 6-10** the oil pressure from the brake valve forces the large piston to the end of the adjuster. The oil now fills the oil passage that leads to the small piston. The small piston will move away from the seat if the pressure in the oil passages is higher that the pressure in the brakes, forcing oil to flow to the brakes.

When the service brakes are released, the oil flow is vented to the inlet port side of the slack adjuster. The lower brake pressure returns the smaller piston to the seat. This forces the service brakes to exhaust the oil from the wheel end brakes through the slack adjuster. This action forces the large piston back into the released position.

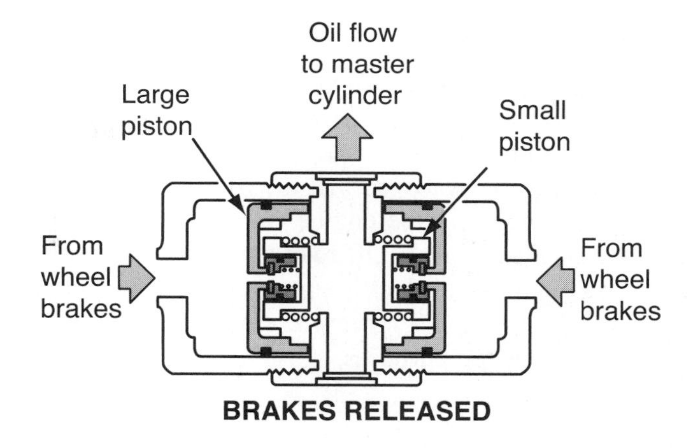

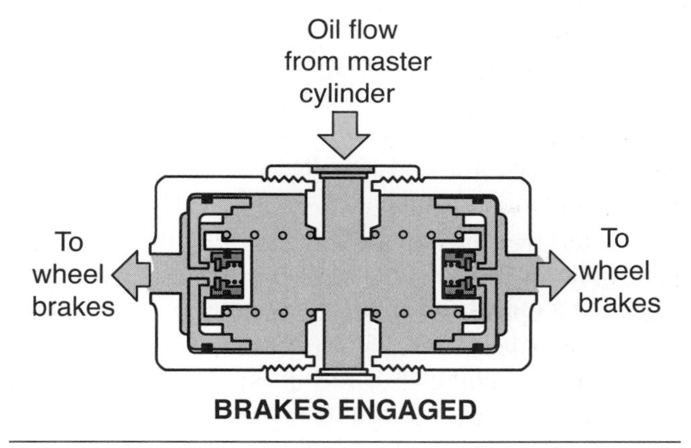

Figure 6-9 Cutaway of hydraulic slack adjusters: one is in the released position and the other is in the applied position.

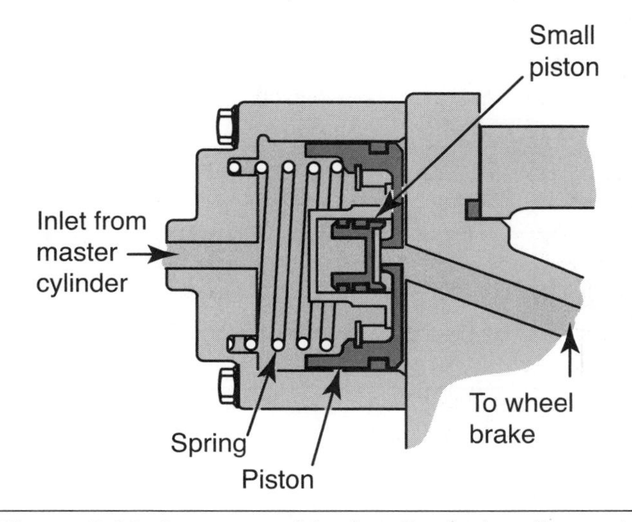

Figure 6-10 Cutaway of hydraulic slack adjuster in the applied position.

Diverter Valve

Many applications on off-road equipment require the use of a diverter valve in the brake circuit for various reasons. Equipment with spring-applied park brakes requires a diverter valve so that the park brakes can be released in the event the engine can no longer be run safely. See **Figure 6-11.** Off-road haulage truck manufacturers often use an electric supplemental drive pump, shown in **Figure 6-12,** to provide emergency hydraulic pressure for towing purposes.

Drum Hydraulic Brakes

Although hydraulic drum brakes are not as popular on off-road equipment today as in the past, they still account for a number of brake systems currently in use. The basic system consists of a drum assembly generally made from cast iron and is bolted or held in place on a rotating assembly that includes the wheel, hub assembly, and drive axle. A backing plate is required to hold the shoes, wheel cylinder, adjusters, and mounting linkage in place. In some of the smaller equipment, the use of a mechanical park brake that is integrated into the rear brakes of the equipment is also possible. The shoes are lined with a friction material that is forced to rub on the inside of the drum when the brakes are actuated.

On off-road equipment, the shoes are generally pinned on the bottom and have a wheel cylinder at the top, which forces the shoes out against the inside of the drum during brake actuation. When the brakes are applied, the friction material converts the kinetic energy of the moving equipment to heat energy. Hydraulic pressure is applied to the wheel cylinders from an actuating device, such as a master cylinder or brake hydraulic system, to force the brake lining against the inside of the drum. **Figure 6-13** shows what takes place when the brake shoe is forced against the drum. The frictional drag that is created by the rotation of the drum will wedge the shoe tighter into the drum: this is known as self-energizing.

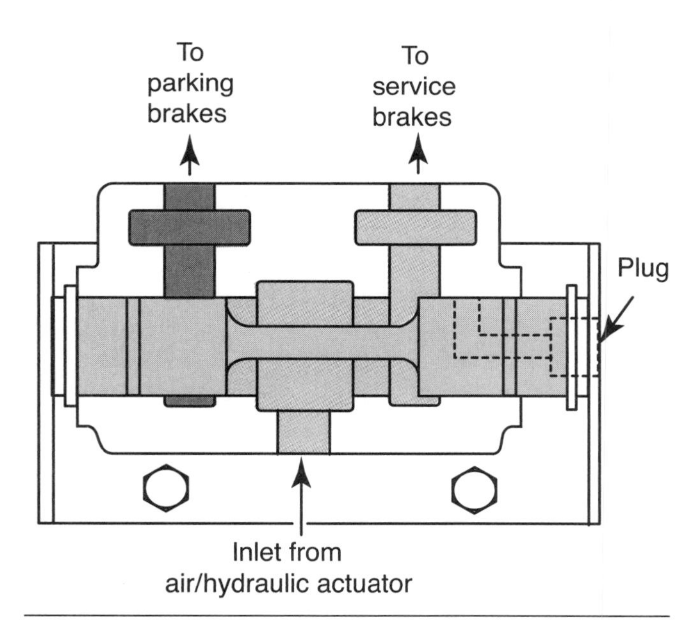

Figure 6-11 Cutaway of a diverter valve used on off-road equipment with spring-applied park brakes.

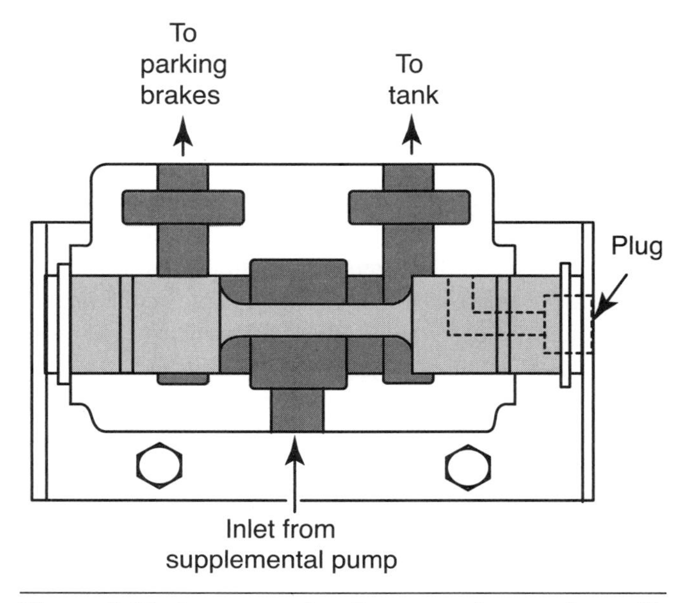

Figure 6-12 Cutaway of a diverter valve used on off-road equipment with a connection for a supplemental electric hydraulic pump.

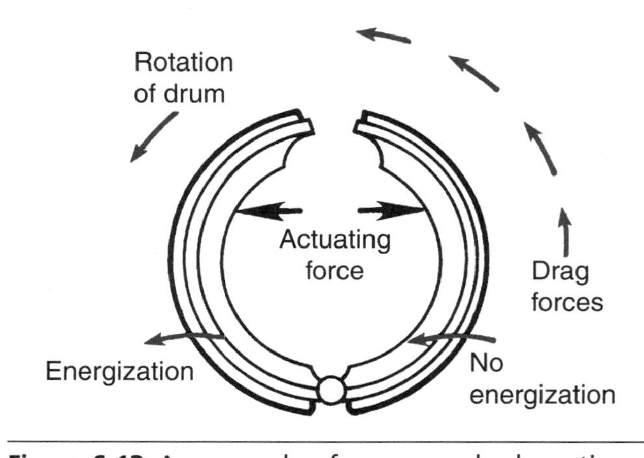

Figure 6-13 An example of non-servo brake action.

In our example of a non-servo drum brake, each shoe is anchored to the backing plate at the bottom to an anchor pin. The wheel cylinder applies the actuating force to the toe of the shoe when the equipment is moving in forward to energize the leading shoe. In this design, this energizing concept only applies to one shoe at a time and is dependent on the direction of travel. When traveling in reverse, the trailing shoe becomes energized. As you can see, this design does not utilize this principle very well because only one shoe can use the energizing concept in each direction of travel.

Some brake designs have an additional feature that uses the servo principle as well as the self-energizing principle. These designs are often found on smaller types of equipment. The principle of operation is shown in **Figure 6-14** where servo action takes place when hydraulic force is applied to the toe ends of the **brake shoes** during forward motion. The primary shoe will react to the rotational forces first because it has the weaker return spring. The shoe lifts off the anchor and is forced against the inside of the drum. The shoe can pivot on the heel of the shoe, causing it to try to rotate with the drum. This action forces the secondary shoe to actuate due to the force the primary shoe exerts on the connecting link forcing it into the drum. This design causes both shoes to work together. The rotational forces, along with the actuating forces, cause the complete shoe assembly to become energized as one unit.

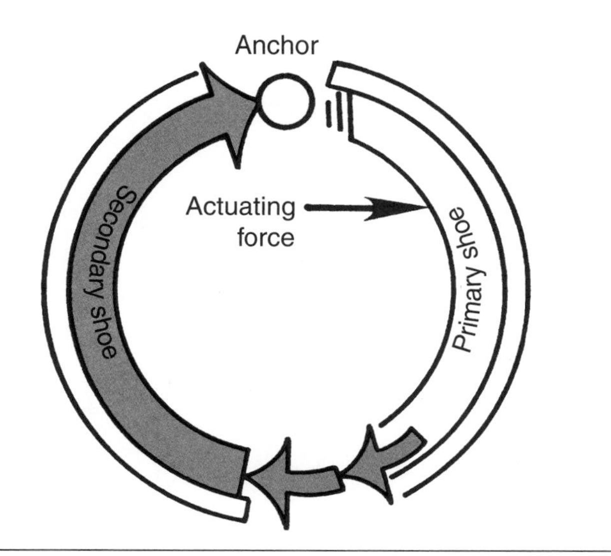

Figure 6-14 The self-energizing concept of a servo action.

Wheel Cylinders

Hydraulic drum brakes, regardless of whether they are a non-servo or **servo brake** design system, must have a wheel cylinder to actuate them. The wheel cylinders use hydraulic pressure from a brake valve to force the shoes out against the drum. There are two styles of wheel cylinders in use on these systems (**Figure 6-15**). The single-acting wheel cylinder has only one piston and actuates one shoe of an anchor pin pivoted design. Double-acting wheel cylinders actuate two shoes at one time, one from each end of the wheel cylinder. A specific amount of brake fluid is sent to the wheel cylinder at one time, and as the brake lining wears out, the wheel cylinder travel becomes inadequate to apply sufficient force to stop the equipment. Regardless of

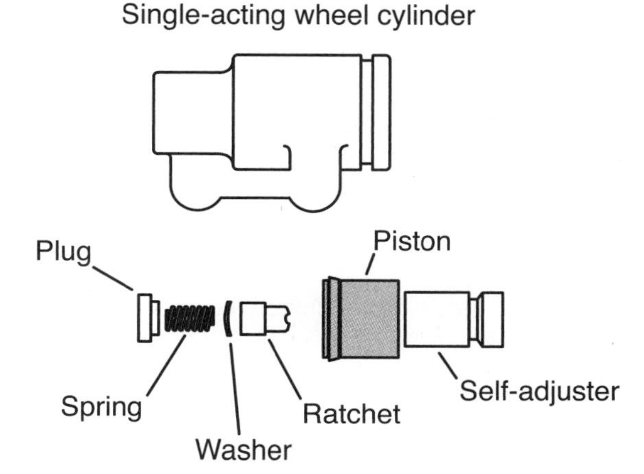

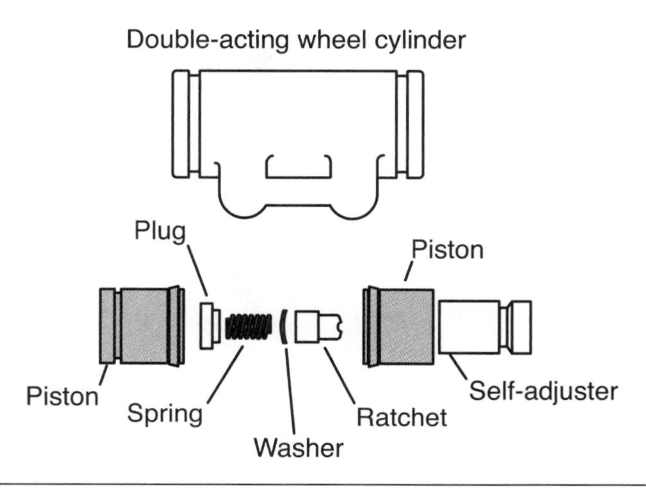

Figure 6-15 Illustration shows the two most common types of wheel actuators in use: single-acting and double-acting wheel cylinders.

design, wheel cylinders rely on cups to seal the pressurized brake fluid inside the housing behind the piston. When the operator releases the brake pedal, the brake shoe retractor springs pull the shoes back toward the anchor pins and into the released position. Note that some of the older off-road drum brake systems used a residual check valve in the master cylinder to maintain a residual pressure of approximately 12 psi in the wheel cylinder. This was required to take up any slack and prevent wheel cup sag, which could lead to leaks or air getting into the wheel cylinder. The residual pressure wasn't enough to apply the brakes; it takes far more pressure to overcome the return spring tension. All wheel cylinders have a rubber boot on the piston end that protects the piston area from dirt and water.

Self-Adjusting Mechanism

The self-adjustment mechanism is necessary to keep the brake shoes close to the drum and to take up the extra clearance caused by drum and shoe wear. The system can only operate in the reverse direction. It will not be able to make any self-adjustments in the forward direction. The adjusting lever, cable, and over travel spring are key components, as shown in **Figure 6-16.** When the brakes are activated in the reverse direction and the brake shoes are adjusted correctly, the cable tension and spring force are equal. These equalized forces keep the adjusting lever from moving. During normal operation, the brake lining and drum will wear and the upper cable travel will increase, causing the tension to decrease. This causes the cable tension to become lower than the spring force. When this occurs, the spring force pulls the adjusting lever back and down, turning the adjusting screw slots. This causes the brake shoes to move out until the tension on the adjusting lever spring and the cable is in balance.

Hydraulic External Caliper Brakes

External hydraulic disc brake systems come in two basic designs: the fixed caliper and the sliding caliper. They are easily identified by the observing whether or not the calipers have pistons on both sides of the rotor. If there are pistons on both sides of the rotor (as shown in **Figure 6-17**), they are the fixed-caliper design. Both brake designs have their own shortcomings and wear characteristics. Regardless of the caliper design, the hydraulic portions of the brake systems are generally identical in operation.

The sliding-caliper design, is easily recognized as shown in **Figure 6-18,** it has hydraulic pistons on one side only. In operation, the sliding caliper pulls the opposite pad and mounting hardware toward the rotor each time the piston on one side is activated. This design is much superior to the drum brake for a number of good

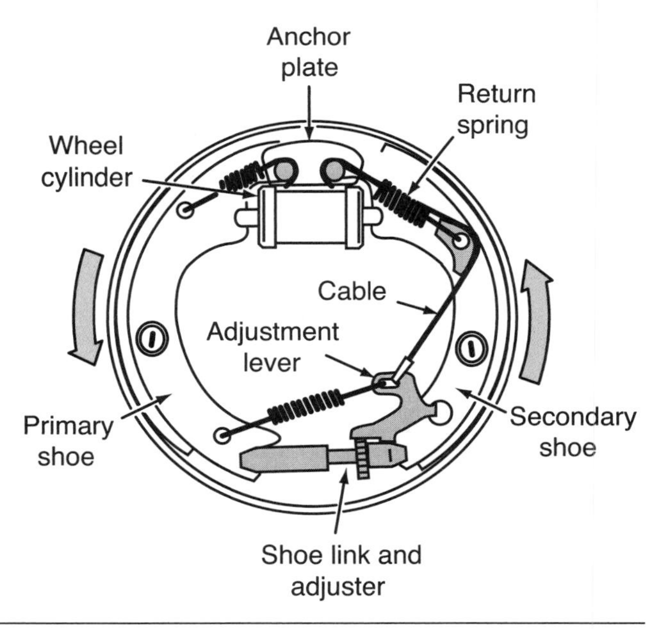

Figure 6-16 The self-adjusting mechanism in the forward direction. No self-adjusting takes place in forward.

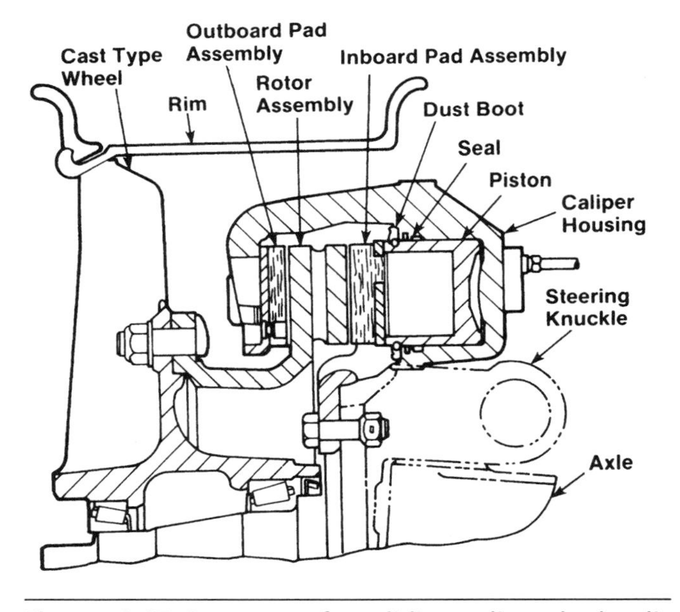

Figure 6-17 Cutaway of a sliding-caliper hydraulic disc brake system. (*Courtesy of International Truck and Engine Corporation*)

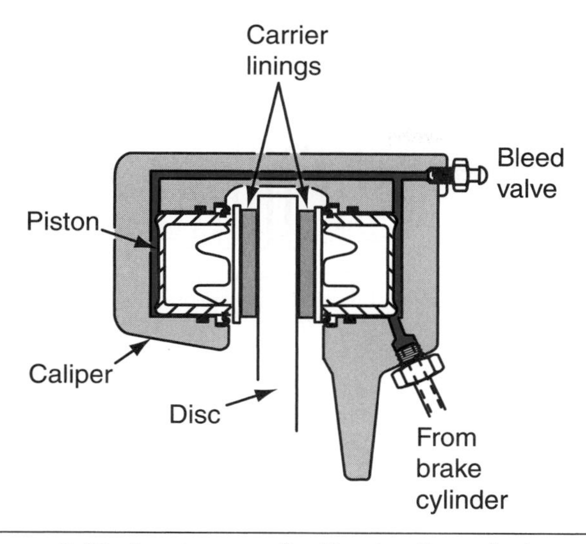

Figure 6-18 A cutaway of a fixed caliper design.

reasons. The most obvious is the difference in how the caliper clamp the **drum** or rotor. The caliper squeezes the rotor, and as it heats up, the **rotor** expands into the caliper. When a drum brake heats up, the drum expands away from the shoes, which can lesson the brake forces. The clamping forces that the caliper generates on to the rotor are not affected by servo action as on a drum brake. As the drums heat up, the drum brakes quickly lose their braking capability; when disc brakes heat up, the brakes don't lose their braking capabilities as quickly because of better cooling. The braking surfaces on the rotors are exposed to the air and can shed heat more quickly than can be achieved by drum brakes. The drum brake friction surfaces are inside the drum and cannot dissipate the heat build up as well as an external disc brake system can. When examining the stopping potential of equipment brake systems, you must consider the clamping forces involved and the swept area of the brake system.

Disc brakes have many obvious advantages over drum brakes when used on off-road equipment. They are more resistant to brake fade, provide better cooling capability, have an acceptable tolerance to dirt and water, require less maintenance, do not need to be adjusted, and have a greater swept area that drum brakes of the same size. Look at the formula shown in **Table 6-1.**

Table 6-1: SWEPT AREA CALCULATION

R_p	Radius to the outside of the pad
R_r	Radius to the outside of the rotor
W_p	Pad Width
Note	If R_p nearly equals R_r, use the following formula to calculate swept area:
Formula	Disc Brake Swept Area = $6.28W_p (2R_r - W_p)$

Note: 6.28 represents $\pi \times 2$ (3.14 × 2)

$$\textbf{Disc Brake Swept Area} = 6.28 \times 4 (2 \times 15 - 4) = 25.12 \times 26 = 653.12 \text{ sq. in. per rotor}$$

When engineers design equipment, they design the components to have a total swept area that factors in the weight, speed, and traction of the equipment. To see how this impacts the brakes on some equipment, we can calculate the swept area of a similar size drum brake system to see what the difference is in swept area between the two designs. The example below shows a similar diameter dimension and pad width to show what the difference looks like.

In our example, we will use the following values to complete the formula:

R_p	15 inches
R_r	15 inches
W_p	4 inches
Note	If R_p nearly equals R_r, use the following formula to calculate swept area:
Formula	Disc Brake Swept Area = $6.28W_p (2R_r - W_p)$

Drum Brake

$$\textbf{Swept Area} = 3.14 \times \text{D (Drum Diameter)} \times \text{L (Pad Width)} = 3.14 \times 30 \text{ inches} \times 4 \text{ inches} = 378.8 \text{ sq. in. per drum}$$

It is not hard to see which brake system has the higher swept area.

Now that we know a little bit about what makes one brake system inherently superior to another, we will examine a few other factors that affect the equipment's stopping ability. A number of factors affect brake performance. Brake performance is limited to the force applied by the piston, the deflection of the drum or rotor, and associated brake hardware. The wear incurred by the brake components, the operating temperatures of the brake components, and tire traction should not become major factors in equipment brake performance. Ideally, only tire traction should be a limiting factor on how well equipment can stop. If any other factor mentioned earlier in the paragraph limits the stopping power of the equipment, the brakes are probably not adequate.

Internal Wet Disk Brakes

Brake systems on off-road equipment were always considered an area that required a lot of maintenance until the introduction of internal wet disk brakes. No matter how well external disc brakes worked, they had one major hurdle that was difficult to overcome. The equipment operated in less than ideal conditions, which often leads to a short brake component life in many applications. The problem stemmed from the fact that the rotors and calipers were exposed to the road grime that the equipment operated in, leading to short brake component life and often higher maintenance costs.

The introduction of the internal wet disc brake on off-road equipment changed that; brakes that were sealed internally in the axle housing, away from the road grime, proved to be the answer to high maintenance costs associated with external disc brakes.

The internal wet disc brake system is highly suited to applications where road grime contamination is extremely likely to occur. Braking is achieved through the use of a number of friction discs that are alternately spaced between number of steel discs that are held stationary by outside teeth or notches splined to the wheel spindle. The friction discs are splined internally to the rotating wheel. The sealed system operates in an oil bath, which cools the brake components as well as providing a means of lubrication. Some manufacturers mount the internal wet disc brake components right next to the differential, and others mount them at the wheel end of the axle. This depends entirely on equipment design and application. Lubrication and cooling on some designs are often integrated with the lubrication and cooling for the axle assembly and require the use of specialized lubricants. In heavy-duty applications, a separate cooling circuit that has its own pump and cooler can be used to dissipate the heat generated by this type of brake system. This brake system requires very little in the way of maintenance. No adjustments are generally required because the design compensates for wear automatically.

When the brakes are in the released position, as shown in **Figure 6-19** a number of return springs (**Figure 6-20,**) keep the discs and plates separated thus allowing the wheels to turn with very little resistance. Depending on equipment application, the brake cavity is either filled with oil or has an independent brake cooling circuit that circulates fluid through the brake component cavity. The steel discs are externally splined to the axle, and the friction discs are splined to the rotating wheel. When the equipment is in motion, the friction discs rotate with the

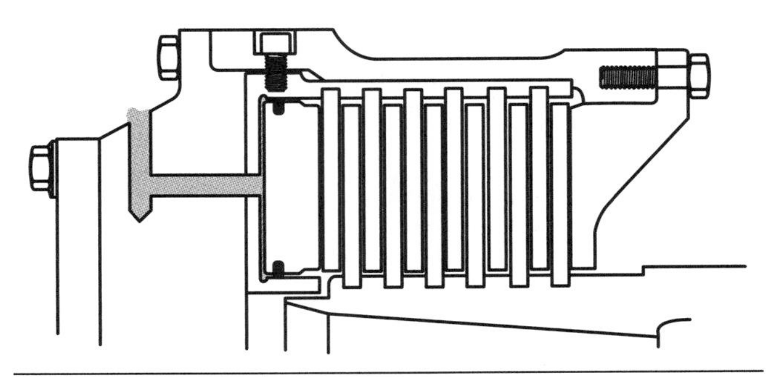

Figure 6-19 The internal wet disc brake in the released position.

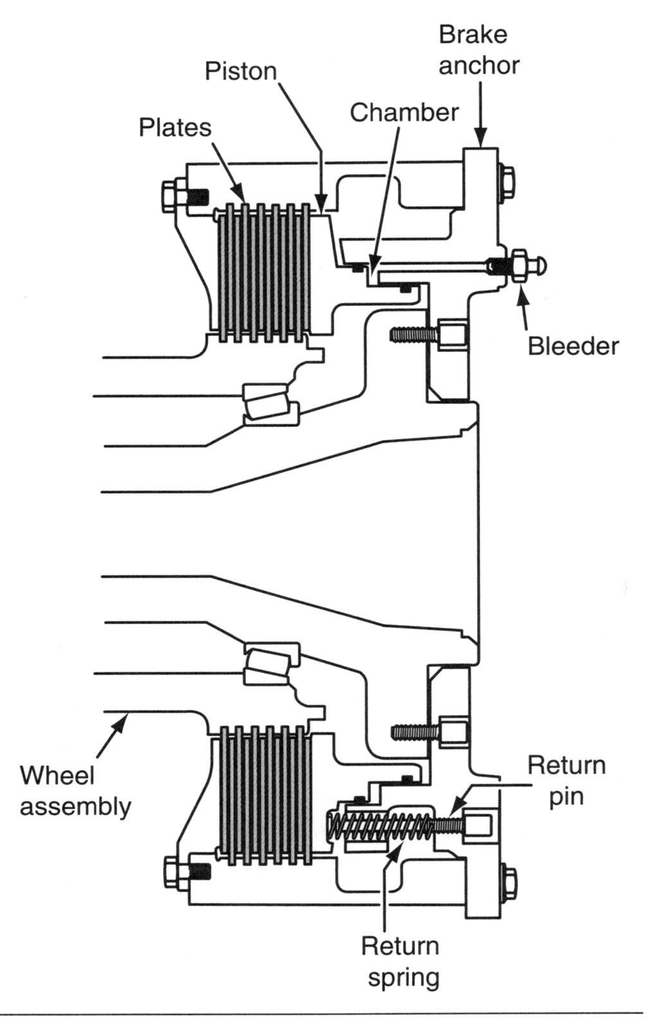

Figure 6-20 Cutaway view of an internal hydraulic wet disc brake assembly.

wheel; but they are not in contact with the stationary steel discs, which are splined to the spindle. A minimal amount of friction is generated as the constant circulation of oil in the housing keeps things well-lubricated.

When the operator applies the brakes, oil is sent to the wheel end through a foot brake valve and enters the wheel end through an inlet port located externally on the inside of the wheel assembly. High pressure oil enters the piston cavity and forces the piston to move to the right, squeezing the rotating friction discs and stationary steel discs together. This squeezing effect of the friction discs, which are splined to the rotating wheels, slows down the equipment. If enough hydraulic force is exerted on the piston, the wheel will stop rotating, bringing the equipment to a stop. The discs are immersed in a constant oil bath, which absorbs the heat energy created by the discs. Heavy-duty application of this design will have a separate cooling circuit that is constantly circulating oil through a cooler and return filter system to keep the oil relatively clean and cool.

Inboard Hydraulic Wet Disk Brake

The internal wet disc brake design has many variations that are based on equipment application. Some off-road equipment applications do not have room in the wheel end to accommodate a wet disc brake, necessitating a design variation of this brake system. Many wheel loaders and farm tractors fall into this category; they have the wet disc brakes located in the final drive right next to the differential. With this design the clutch and plate piston are held stationary by pins fastened to the outer housing of the axle assembly. A friction disc is located between the two stationary members and is splined to the rotating drive axle (sun gear shaft), as shown in **Figure 6-21.** The amount of friction and the number of steel plates used varies and depends on the size of the equipment and its application. When the brakes are in the released position, the piston is held away from the disc by return springs.

When the operator applies the brakes, (**Figure 6-22**) hydraulic oil is sent to the back side of the clutch piston, forcing it to squeeze the disc(s) together between it and the plate. Since the sandwiched friction disc is connected to the wheels, the equipment slows or comes to a complete stop. The brake and differential share the same oil for lubrication and cooling and require the use of a specific type of lubricant. If lubricant must be added when servicing this type of equipment, be sure to use the correct type of lubricant; conventional differential lubricant

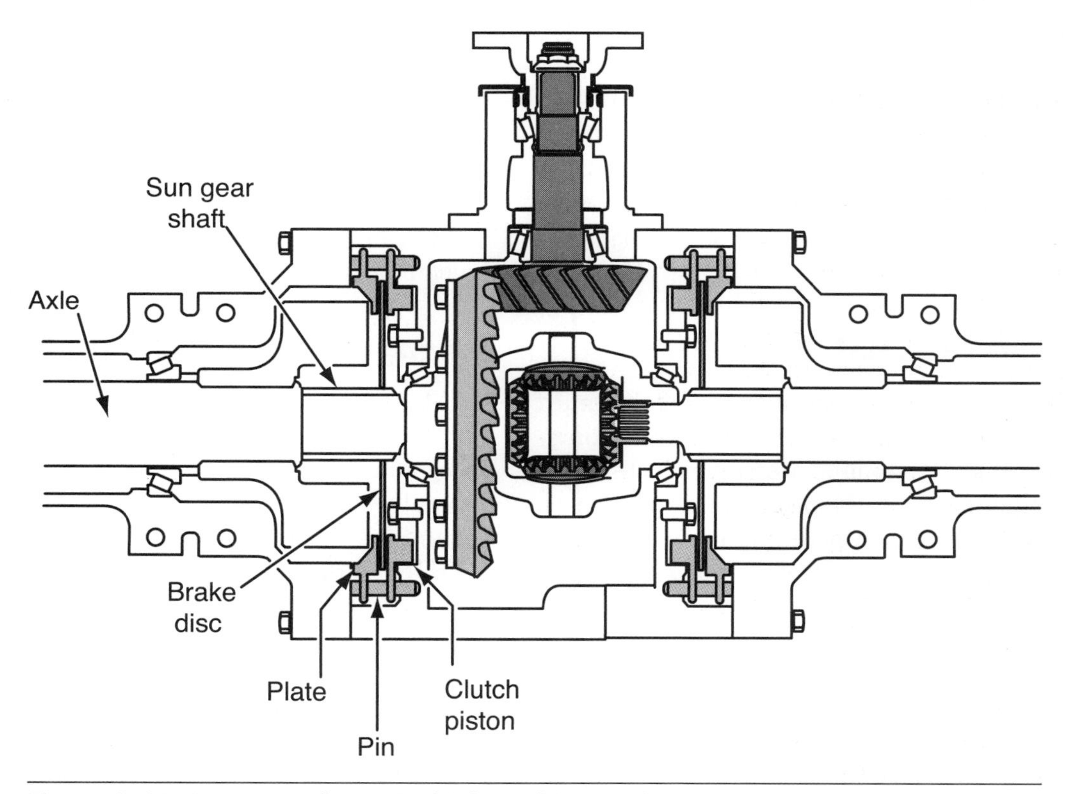

Figure 6–21 Cutaway of a typical inboard internal wet disc brake.

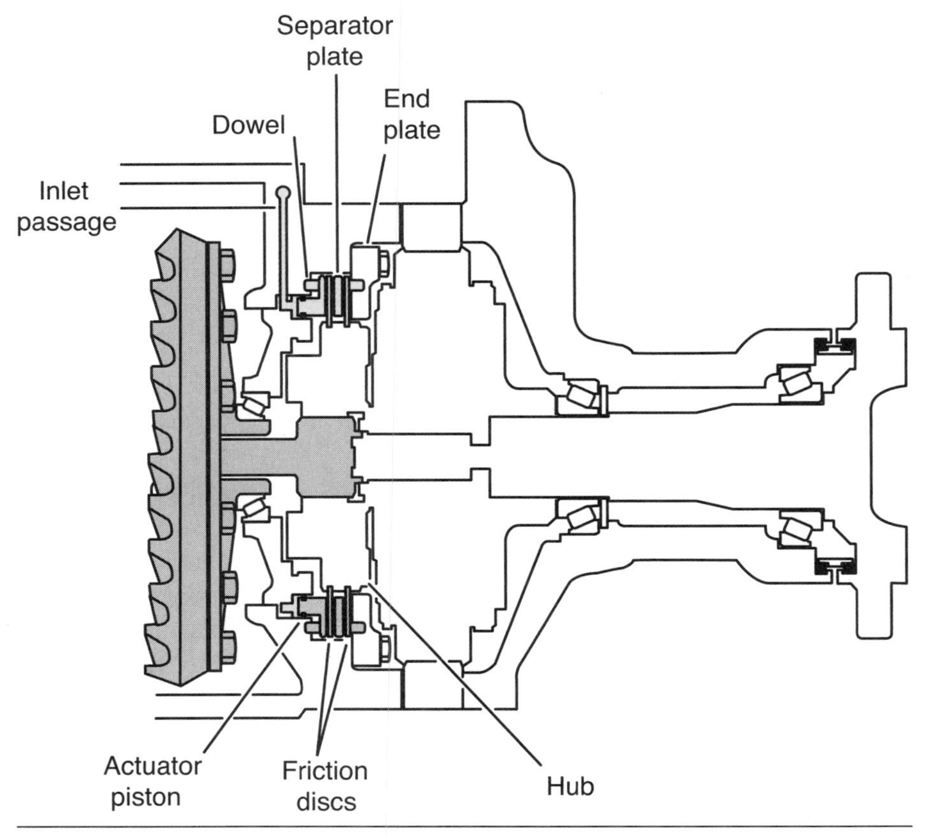

Figure 6-22 Cutaway of a typical inboard internal wet disc brake system in the applied position.

may not be compatible with the brake friction material and could result in early brake failure and severe differential component damage. This brake design does not require any brake adjustment during its life other than periodic oil changes based on the equipment manufacturer's recommendations.

Mechanical Internal Park Brake

Equipment that is placed in the park position must not be parked using hydraulic pressure only for obvious safety reasons. If hydraulic pressure is lost for any reason, the equipment would move and cause property damage or injury to anyone in the immediate area. Manufacturers have come up with many variations of a mechanical park brake on off-road equipment. **Figure 6-23** shows a typical mechanical multi-disc park brake that uses an external mechanical lever to apply the park brake.

This type of mechanical brake has a caliper that locks the bevel pinion to the housing to prevent any movement when parked. It is actuated by a mechanical level on the outside of the housing. The caliper assembly is mounted to the inside of the housing, and the friction discs are mounted to the pinion shaft with internal spines. When the operator sets the park brake, the pads in the caliper squeeze the friction discs to lock them together and prevent the pinion shaft from rotating.

Spring-Applied Park Brake

Modern equipment today generally uses a variation of a hydraulic multi-disc wet disc brake that incorporates a series of extremely heavy-duty springs to apply a mechanical force to the disc. This design requires a constant source of hydraulic pressure to keep the park brake in the released position. By the very nature of the design, it adds a measure of safety to the park brake in that, if the equipment were shut down for any reason or lost hydraulic brake pressure, the park brakes would automatically apply. The park brake example shown in **Figure 6-24** can be found on large articulating haulage trucks and is located on the output transfer drive.

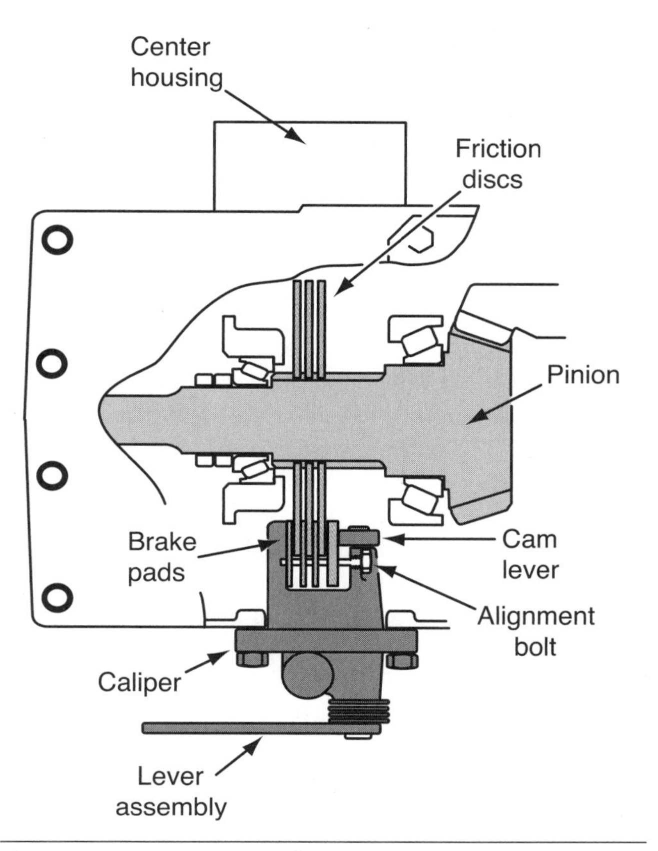

Figure 6-23 Cutaway of a typical internal mechanical park brake system.

The park brake consists of two sets of discs: steel and friction. The steel disks are externally splined to the axle housing and do not rotate; the friction discs are splined to the rotating shaft. A hydraulic piston is located in the assembly with a pressure plate, which uses a number of heavy-duty springs, on one side of it and the hydraulic piston on the other side. See **Figure 6-25.** The large heavy springs force the pressure plate against the piston, which locks the friction discs to the stationary steel disks and prevents any movement of the axle shaft whenever the equipments park brake is engaged. Because this design requires hydraulic pressure to release the brakes, the brakes are automatically applied whenever the equipment is not running. When the operator wants to release the park brake, hydraulic oil must be sent to the hydraulic side of the piston, which will build up enough force if the pressure is sufficient to overcome the force of the springs releasing the park brake. This additional safety feature ensures that the park brakes do not release if the hydraulic pressure is to low. The down side is that the park brakes can drag if not enough pressure is available to fully release them.

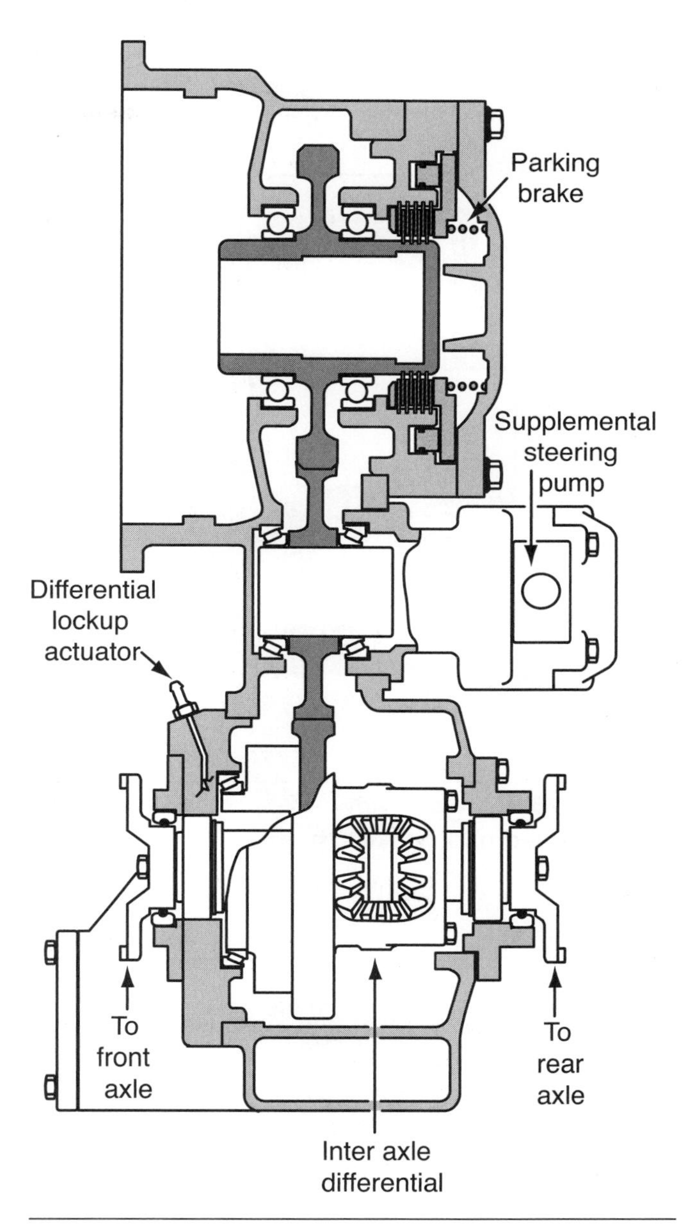

Figure 6-24 Cutaway of a typical internal spring-applied park brake system located on the output of a transfer case drive assembly.

Combination Service/Park Brake

Some manufacturers combine the park brake and service brake in the wheel end brake assembly, shown in **Figure 6-26.** This is accomplished by using a separate circuit and piston to apply the park brake, which uses the same discs and plates that the service brakes use. This lowers cost and complexity, and at the same time increases reliably and safety. This design has a spring-applied wheel end brake in each wheel end of

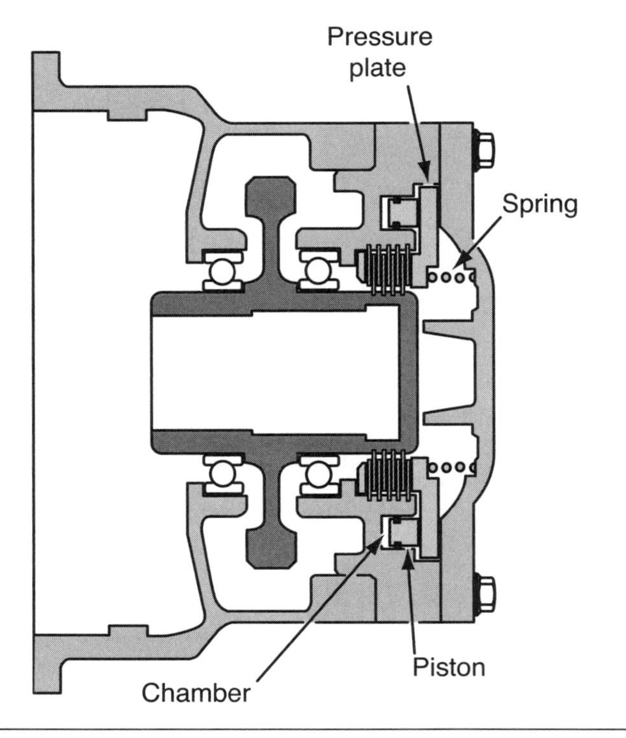

Figure 6-25 Close-up view of a typical internal spring-applied park brake system.

the equipment, making it almost failsafe. It is very unlikely that all four wheel ends would fail at the same time.

The park brake circuit uses a series of heavy springs located around the park brake piston, which is positioned behind the service brake piston. When the operator releases the park brakes, he or she activates the park brake release circuit, which sends hydraulic oil pressure to the hydraulic cavity between the park brake piston and the service brake piston. This forces the service brake piston (the only one to directly contact the discs) to release the force it is exerting on the discs. The oil pressure exerts a force to overcome the multiple spring pressure that is pushing on the park brake piston, which in turn is exerting a mechanical force on the service brake piston. When the operator applies the park brake, hydraulic oil is released from the cavity between the two pistons, allowing the heavy park brake springs behind the park brake piston to apply pressure to the service brake piston and clamping the stationary steel discs and friction discs together to apply the brakes. A cutaway view of this system is shown in **Figure 6-27.**

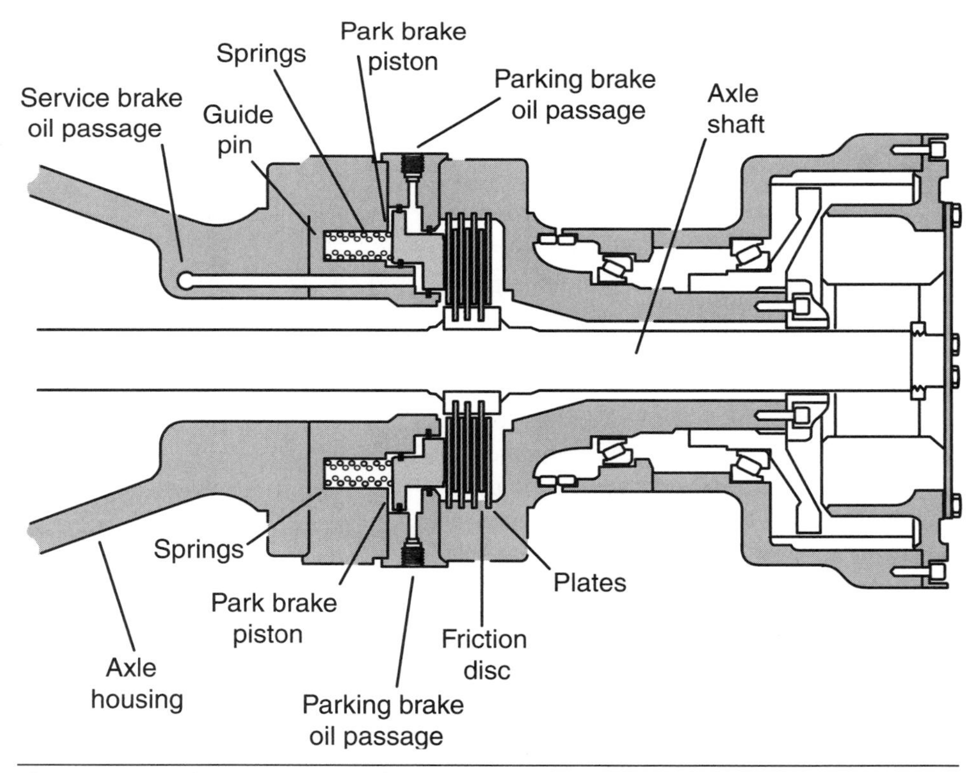

Figure 6-26 Close-up view of a typical internal spring-applied park brake system.

Figure 6-27 Close-up view of typical combination spring-applied park brake/service brake system. (*Courtesy of Caterpillar*)

Hydraulic Brake Fluids

Brake fluids used in a brake system must have certain characteristics to be suitable for a heavy equipment application. The fluid must not boil at temperatures that would normally be encountered in a typical brake circuit. It must not change viscosity when cold and should not freeze at subzero temperatures. The fluid should not compress significantly, and it must flow through small brake system passages easily. It should have the ability to protect the brake components from corrosion and be compatible with material commonly found in brake systems. The fluid should not alter its properties significantly over time, and it should be compatible with other types of similar brake fluids. A good fluid should not form any gum or deposits on brake components during its life.

In North America, brake fluid standards are quantified by two groups: the Society of Automotive Engineers (SAE) and the Department of Transportation (DOT). Conventional brake fluid has three categories: DOT 3, DOT 4, and DOT 5. The first two brake fluids, DOT 3 and DOT 4, are formulated from a polyglycol base, which has some unique characteristics. One characteristic, unique to this type of brake fluid, is its ability to absorb water (hygroscopic). This is a desirable characteristic that ensures that if any water gets into the brake system it will quickly be absorbed by the brake fluid. This prevents corrosion from forming and also prevents the water from freezing in the brake lines during cold weather operation. The down side to a brake fluid absorbing water is that it lowers the boiling point over time. Any hygroscopic brake fluids that have been sitting on a shelf must be completely sealed in their respective containers, or they will absorb any moisture that is present in the air and quickly become unusable. One of the most important qualities a brake fluid can have is its resistance to boiling. Let's use DOT 3 to demonstrate how water in the brake fluid lowers the boiling point. DOT 3 brake fluids have a dry boiling point of approximately 400 degrees F. (205 degrees C). Once the brake fluid absorbs some water, the boiling point will quickly fall to around 284 degrees F (140 degrees C). It's easy to see that the wet brake fluid would pose a problem when the brake fluid gets hot. Once boiling occurs in brake fluid in a wheel caliper, the gas bubbles formed are compressible and cause a pressure drop, significantly reducing the stopping ability.

DOT 5 brake fluid is formulated from a silicone base and does not absorb water, in other words it is non-hygroscopic. When using this type of fluid in a brake circuit, it must be kept free from water; any water entering the brake system will cause corrosion. The other negative feature is that it is slightly compressible. This makes it unsuitable for brake systems that utilize ABS circuits. Whenever any service work is to be performed on a brake system, identify the type of brake fluid in use so that any fluid that may need to be added to the brake system is compatible with the existing fluid. Refer to **Table 6-2.**

Table 6-2: DOT BRAKE FLUID BOILING POINT PARAMETERS

	DOT 3	DOT 4	DOT 5
Dry Boiling Point Degrees F	401°F	446°F	500°F
Degree C	205°C	230°C	260°C
Wet Boiling Point Degree F	284°F	311°F	356°F
Degree C	140°C	155°C	180°C

HYDRAULIC BRAKE CIRCUITS

Modern off-road equipment design engineers have taken advantage of basic hydraulic principles when designing brake circuits to safely control the speed and stopping ability of equipment. Conventional hydraulic brake circuits use fluid that is confined in a brake hydraulic circuit to transmit hydraulic force to the wheel end brakes. Distance from the master cylinder to the wheel ends is often considerable. This makes the use of hydraulic principles an excellent choice to transmit force over long distances. With minor exceptions, this concept has not changed much over the years. Actuation of the master cylinder by the operator builds up hydraulic pressure in the master cylinder which is transmitted to a slave cylinder. The slave cylinder is often boost assisted on heavy equipment brake systems to help increase the force at the wheel ends that is required to slow down or stop the equipment safely. The force applied to the wheel end brakes is directly proportional to the pressure applied by the master cylinder to the slave cylinder.

Basic Brake Circuit

To fully understand the operation of complex hydraulic brake circuits you must first understand the basic brake circuits that are currently used on off-road equipment. The basic circuit uses a master cylinder to supply control pressure that is used to operate the slave cylinder. With the brake pedal in the released position, the return spring keeps the plunger in the master cylinder in the neutral position. The check valve at the end of the master cylinder is open to the reservoir. Another check valve located at the end of the slave cylinder is also open to the reservoir as well as to the wheel end brakes. This design ensures that the pressure is equalized in the brake circuit and all lines are kept full of fluid. A return spring forces the power piston over to the left side of the brake valve. The primary piston blocks the inlet port for the boost pressure from entering the rear of the power piston when in the released position.

To apply the service brakes (**Figure 6-28**), the operator applies pedal pressure to the master cylinder, which immediately closes the port to the reservoir and starts to build pressure in the master cylinder. The resulting pressure acts on the inlet port of the boost cylinder and forces the primary piston, power piston, and slave piston to move to the left, closing of the check vale at the end of the slave cylinder chamber that is open to the reservoir. This allows the inlet in the boost cylinder to open, sending hydraulic pressure behind the power piston. The boost pressure behind the piston assists in developing the pressure necessary to apply the brakes.

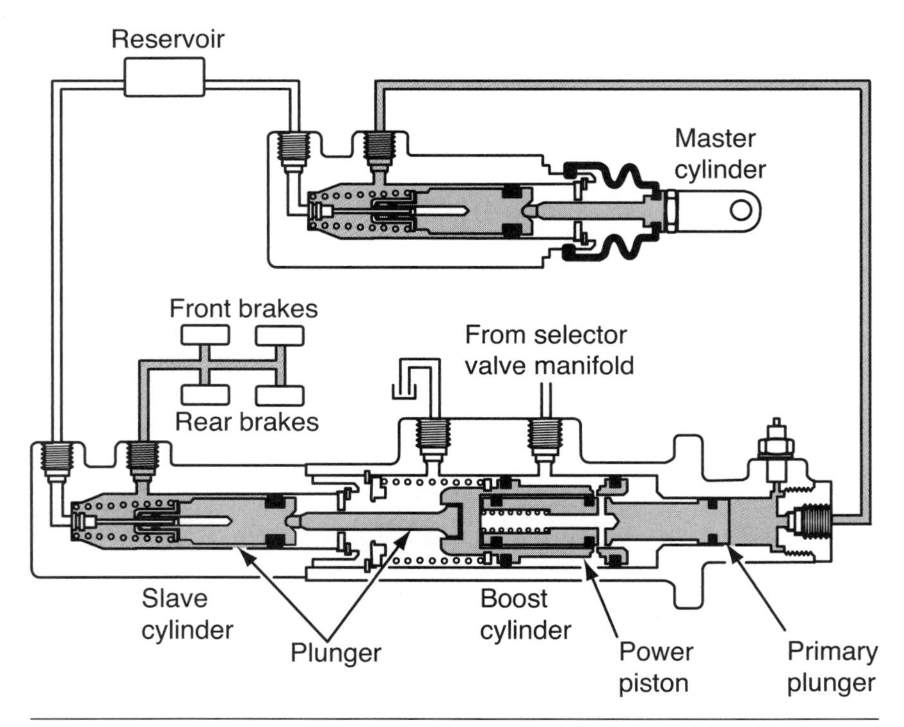

Figure 6-28 Cutaway view of a typical dual hydraulic service brake circuit with the brakes applied.

Basic Hydraulic Brake Circuit

This section explains the operation of a hydraulically-applied internal wet disk brake system used on off-road equipment. This system uses the hydraulically-controlled internal wet disk brakes that are ideally suited for use on off-road equipment operating in an extremely abrasive environment such as a quarry or mining operation. The components that are used in this type of brake circuit consist of a few basic components commonly found in a hydraulic circuit and a few specialized components that are unique to this type of brake circuit. An overview of how this brake system works is explained and a detailed explanation of each component follows. A detailed explanation of a few common hydraulic brake circuits is also provided along with a brief discussion of basic diagnostic and repair procedures.

Figure 6-29 shows a pictorial view of a basic hydraulic brake circuit. When the operator activates the tandem foot brake valve, oil flows to the brake wheel ends to apply the multi-disc brakes. A hydraulic pump provides a constant flow of oil to the brake circuit, sending the oil flow to the **charge valve** that regulates the hydraulic brake pressure in the circuit. An **accumulator** charge valve is used to maintain adequate brake pressure in the brake circuit by controlling the flow of pressurized oil in the brake circuit. Once maximum pressure is achieved in the brake circuit (kick-out pressure), the charge valve redirects the pump flow back to the tank.

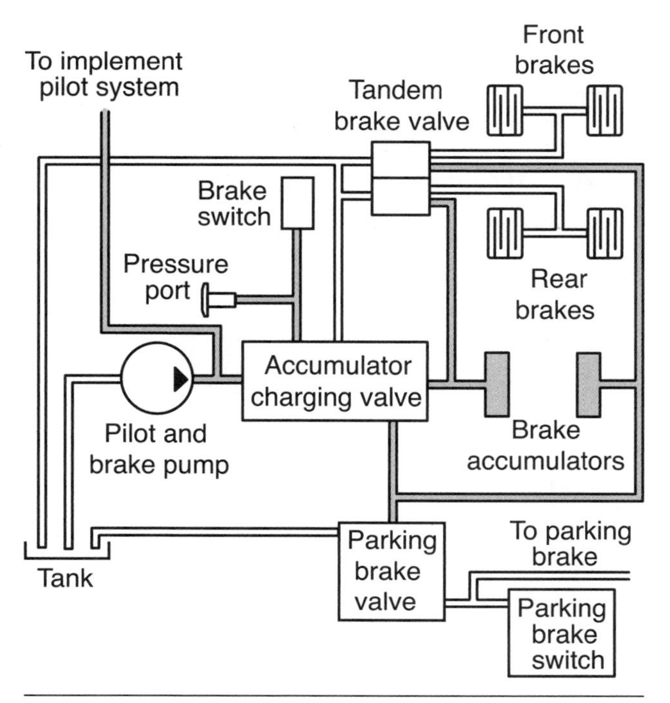

Figure 6-29 Pictorial view of a typical basic hydraulic brake circuit.

When the oil pressure falls below a predetermined amount (kick-in pressure), the charge valve redirects the flow of oil back into the brake circuit, charging the circuit back up to the kick-out pressure.

Two accumulators are used in this basic brake circuit: one for the front service brakes and one for the rear service brakes. The pre-charged nitrogen accumulators store hydraulic energy, which is used by the brake circuit to apply the brakes. In the event of an engine failure, the accumulators store enough energy to allow the brakes to be applied. A separate **park brake valve** is used to activate the spring-applied park brakes. Note that two independent circuits are used for the service brakes: one for the front brake and one for the rear. A third circuit is used to activate the park brakes. This design gives the equipment three independent brake circuits, making it extremely reliable and safe.

The park brake consists of a spring-applied hydraulically-released brake that can be mounted on any component that is mechanically connected to the wheels. Many equipment manufacturers mount it on the output shaft of the transmission or transfer case or even on the differential input shaft. The park brake in this circuit (**Figure 6-30**) has an electric hydraulic **pressure switch** that monitors the brake pressure; if it were to fall below a predetermined pressure the pressure switch would send electrical power to the park brake solenoid to activate the park brake. The park brake solenoid valve cuts off the oil flow to the spring-applied park brake. When the park brake is in the applied position, the park brake solenoid is not energized. The park brake does not receive any oil in this position; the oil is allowed to drain back to the tank through the park brake valve.

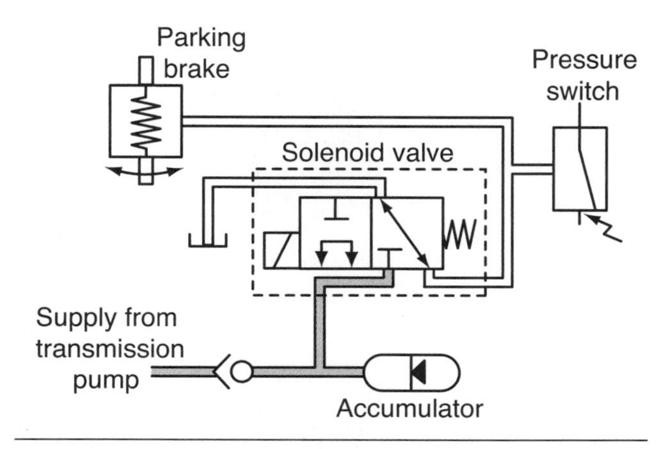

Figure 6-30 Schematic of the park brake circuit.

In many applications, off-road equipment brakes are used in what we call a severe-duty application. In other words, the brakes are used to a point where they require a cooling circuit to keep the brake oil temperature within safe limits. In these circumstances, an extra cooling circuit can be designed into the equipments brake system to cool the oil. In **Figure 6-31** oil is circulated through the wheel end brake clutch pack during operation, continually removing any contamination formed as a result of disc wear. This continuous flow of oil also cools the brake components; before it enters the wheel ends it is filtered and cooled. In this instance, the brake cooling circuit has a separate pump dedicated for this purpose. Some equipment manufacturers use excess oil from the main hydraulic circuit to cool the brakes. Pressures in the brake cooling circuits are generally low (5 to 25 psi) in comparison to the brake activation pressures.

Brake Components

Brake Pump. A hydraulic pump is necessary to supply oil for an internal wet disc brake circuit. A gear pump with an output of approximately 5 to 12 gpm (gallons per minute) is generally used for this purpose because of its low cost. The pump is driven directly by the engine, or in some cases it is mounted to the torque converter and driven by gearing in the torque converter (**Figure 6-32**). The brake pump requires very little in the way of maintenance other than making sure it gets a clean, unrestricted supply of hydraulic oil.

Accumulator. The piston-type accumulator used on brake systems is pre-charged with dry nitrogen. The amount of pre-charge can vary considerably from application to application. The amount of pre-charge

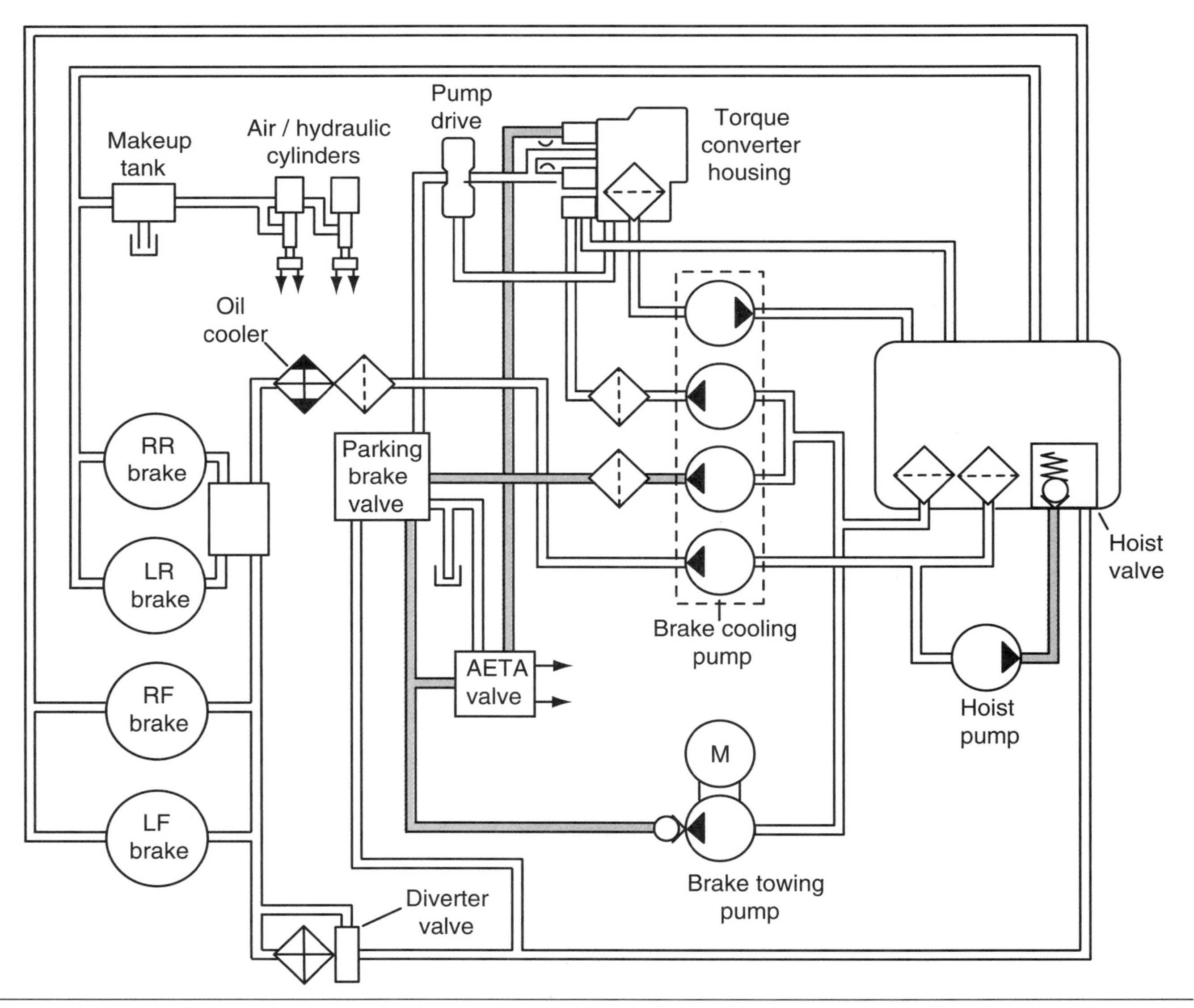

Figure 6-31 Schematic of a typical brake cooling circuit.

and the charge procedure is covered in the maintenance and troubleshooting section in this chapter. The accumulator contains a free-floating piston that has high-pressure O-rings, which separate the dry nitrogen on one side and hydraulic oil on the other side. Its purpose is to store pressurized oil that can be used to apply the brakes when required. See **Figure 6-33** for the location of two brake accumulators.

Since hydraulic oil does not compress measurably, the pressurized hydraulic oil loses its pressure quickly when released into the brake system. Nitrogen gas on one side of the accumulator piston helps maintain the hydraulic pressure in the accumulator circuit so that a number of brake applications can be used before the pressure gets low enough to start the accumulator charge process. Without any nitrogen pressure in the

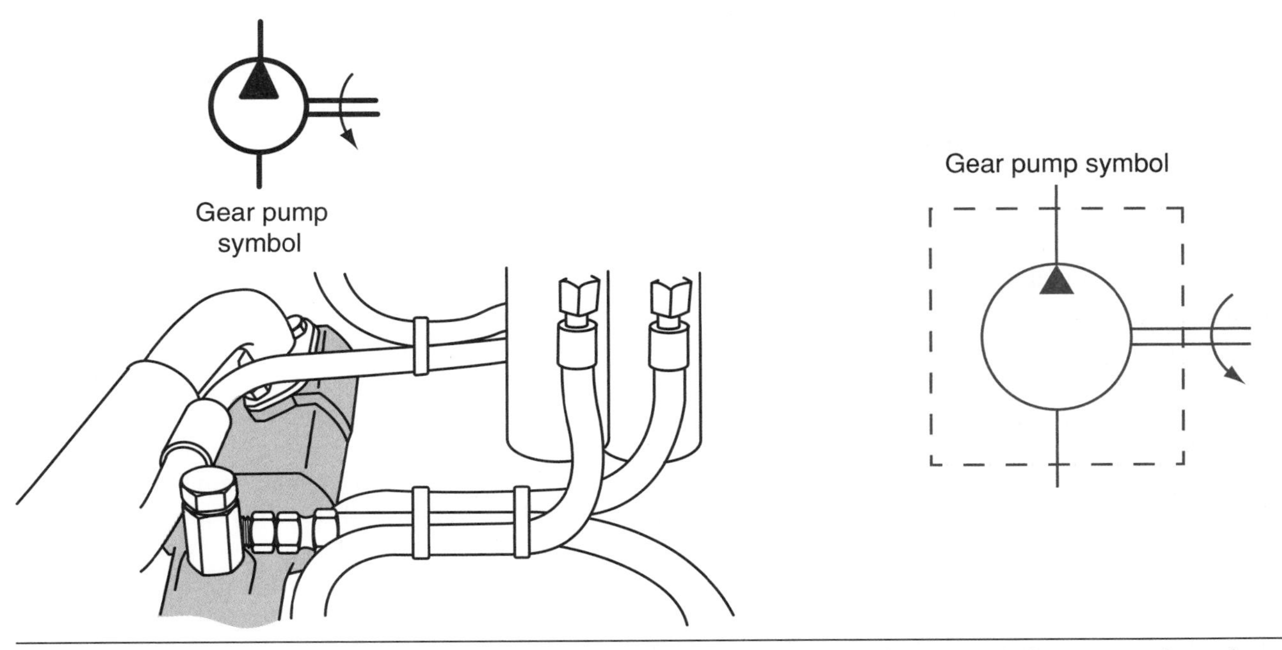

Figure 6-32 Schematic of a typical brake pump and its location. In this case, it is mounted to the torque converter.

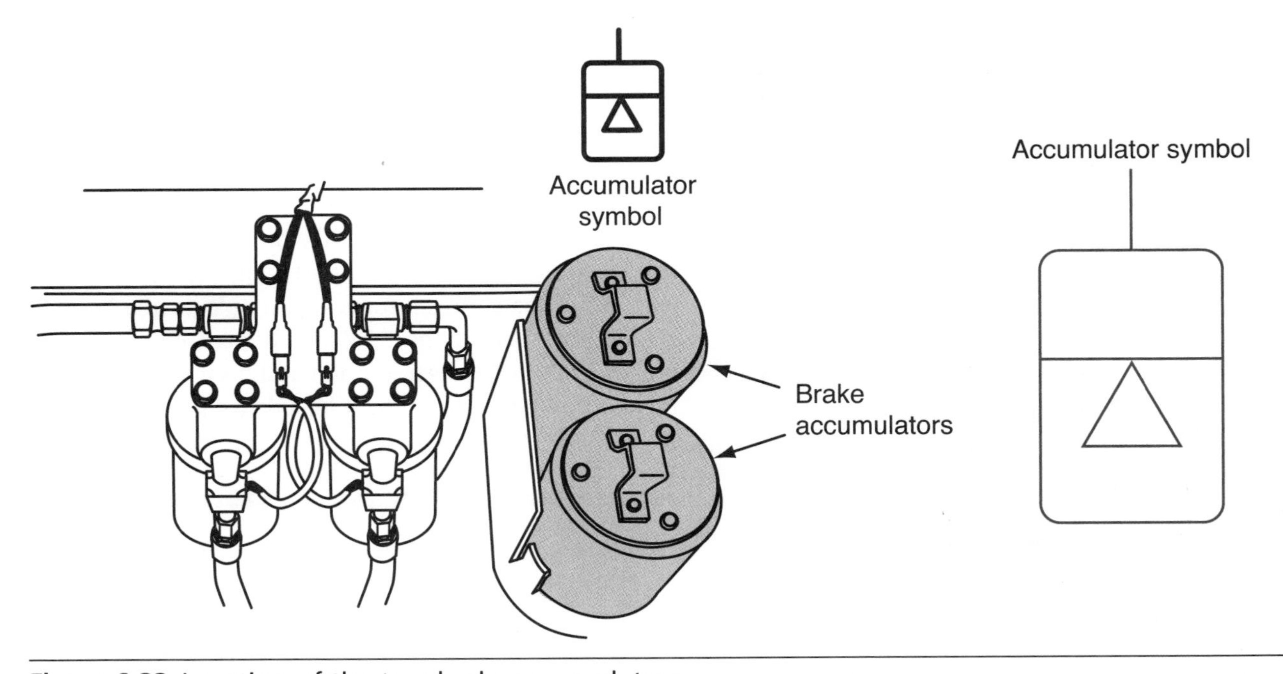

Figure 6-33 Location of the two brake accumulators.

accumulator, the brake pressure will fall almost immediately. In the charged mode, the pressure on both sides of the accumulator equalizes. When the operator applies the brakes, some of the hydraulic pressure is used up. The higher pressure on the nitrogen side of the piston forces the piston down, increasing the pressure on the hydraulic side until both are equal. The accumulators should be checked for nitrogen pre-charge on a regular interval. The piston seals can develop a leak, causing oil to enter the gas side of the accumulators or gas to enter the oil side. If the pre-charge is low, it is an indication that the nitrogen gas is leaking into the oil side.

Brake Charge Valve. The brake charge valve (**Figure 6-34**) supplies oil flow when required to the brake circuit accumulators. The purpose of the charge valves is to maintain a predetermined amount of pressure in the brake accumulator circuit when the engine is running. Refer to **Figure 6-35.** The flow control valve (8) regulates the flow of oil into the brake circuit. When the oil pressure requirements are met, the flow control valve senses the pressure and sends the flow of oil back to the tank. An inverse shuttle valve (12) in the charge valve is connected to the accumulators by two separate lines. The shuttle valve can sense the pressure in either accumulator and charge one or the other whenever the pressure falls below a predetermined amount.

The charge valve contains a number of specialized internal components, an inverse shuttle valve (12), a pressure relief valve (6), cut-in and cut-out pressure valves (11), a flow control valve (8), a flow control orifice (9), and a check valve (10). The pressure relief valve limits the maximum pressure inthe brake circuit. A flow control valve regulates the oil flow to the brake

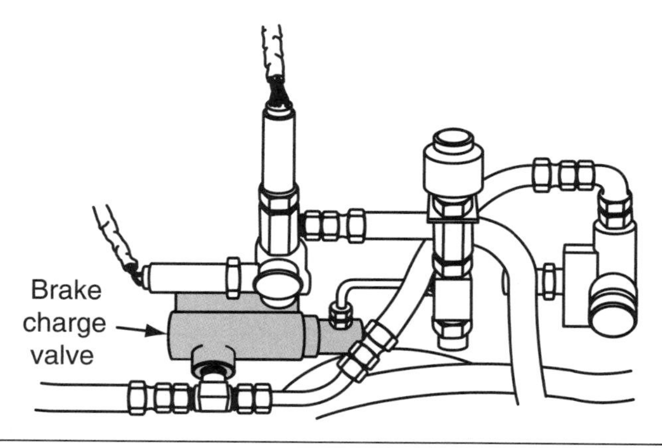

Figure 6-34 Location of a brake charge valve.

system, allowing any excess oil to flow back to the tank. A check valve stops theoil flow from accumulator side of the circuit from returning to the tank.

The cut-in and cut-out pressure circuit has a spool valve that is capable of sensing low charge pressure (cut-in pressure). When this occurs, the spool shifts, allowing the oil to charge the accumulator circuit. A flow control valve also shifts at the same time to stop the flow of oil to the inverse shuttle valve. This allows the inverse shuttle valves to charge the two accumulators separately. A port (7) connects to two brake accumulator pressure switches: one pressure switch activates a low brake pressure light in the operator's compartment, and the other pressure switch sets the park brake if the brake pressure falls below a predetermined amount.

Brake Control Valve. When the operator steps on the brake pedal, as shown in **Figure 6-36,** the upper piston forces the top spring to compress, moving the two spools down. When this takes place, the oil from the inlet passages flows to the brakes. Small passages located between the inlet side and the outlet side of the spools act as restrictions, limiting the amount of oils that can flow to the brakes. The more the pedal is depressed, the more oil will flow. This allows the brake pedal to function as a variable-pressure reducing valve by limiting the pressure flowing to the wheel end brakes. The more the pedal is depressed, the higher the pressure will be at the wheel end brakes. A small internal passage in each spool allows the oil to flow from the modulated pressure side of the circuit to the underside of the spools. This provides a resistance to the pedal movement, which gives the operator feedback by providing a slight resistance. When the force on the upper springs is lower than the oil pressure underneath the spools, it forces them to move up. By metering the flow of oil, the brake valve' act like a variable-pressure reducing valve.

When the operator releases the brake pedal (shown on the left of Figure 6-36), the oil pressure below the spools and the lower spring returns the spools to the neutral position by sending the flow of oil from the wheel end brakes back to the reservoir and at the same time blocking the flow of oil from the inlet side to the brakes.

Accumulator Pressure Switch

The accumulator pressure switch is a hydraulic electric valve that monitors the accumulator pressure and triggers a park brake application when the pressure in the accumulators gets below a predetermined value. A

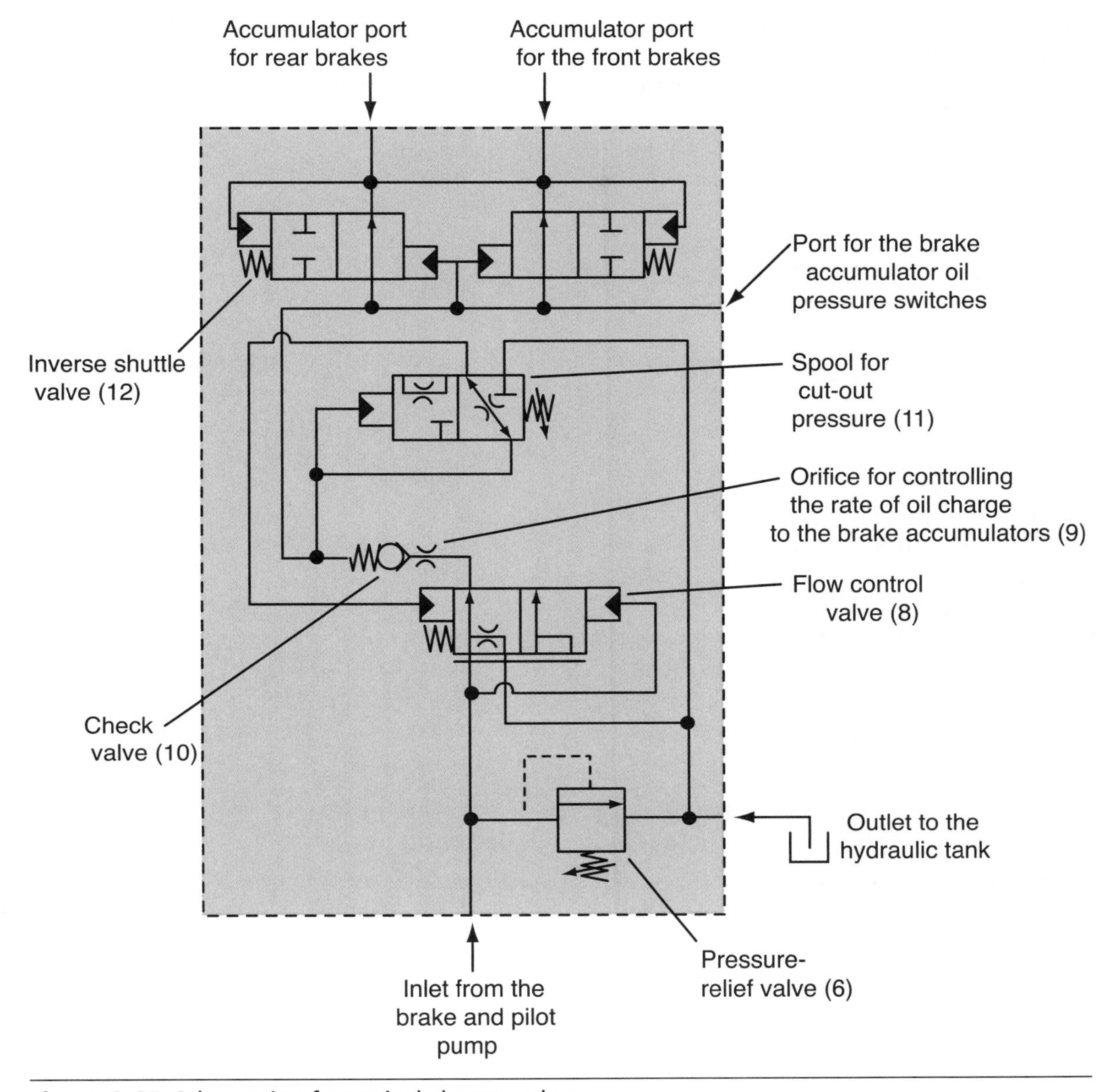

Figure 6–35 Schematic of a typical charge valve.

second accumulator pressure switch turns on a low accumulator pressure light in the operator's compartment when the pressure gets below a predetermined pressure. The brake accumulator pressure switch will prevent the operator from releasing the brakes before a safe predetermined accumulator pressure is reached. Refer to **Figure 6-37.**

After the operator starts the engine, the charge valve will charge the brake accumulators until the pressure reaches the low limit of the pressure switch. Once the pressure reaches the pressure switch's low limit value, the electrical contacts open in the pressure switch and turn off the low brake pressure indicator light in the operator's compartment. The accumulator pressure switch also triggers a signal to the power train ECM that allows the park brakes to be released. In the event of a low brake pressure problem during operation, the accumulator pressure switch will turn on a brake impending indicator light in the operator's compartment and apply the park brake once the pressure falls below the switch's low limit value. The operator will not be able to release the park brake until the safe accumulator pressure is above the low limit value. When this occurs, the operator will have to cycle the park brake switch on and off before the brakes can be released. An electrical interlock circuit prevents the accidental release of the park brake should it be applied inadvertently by a loss of brake pressure.

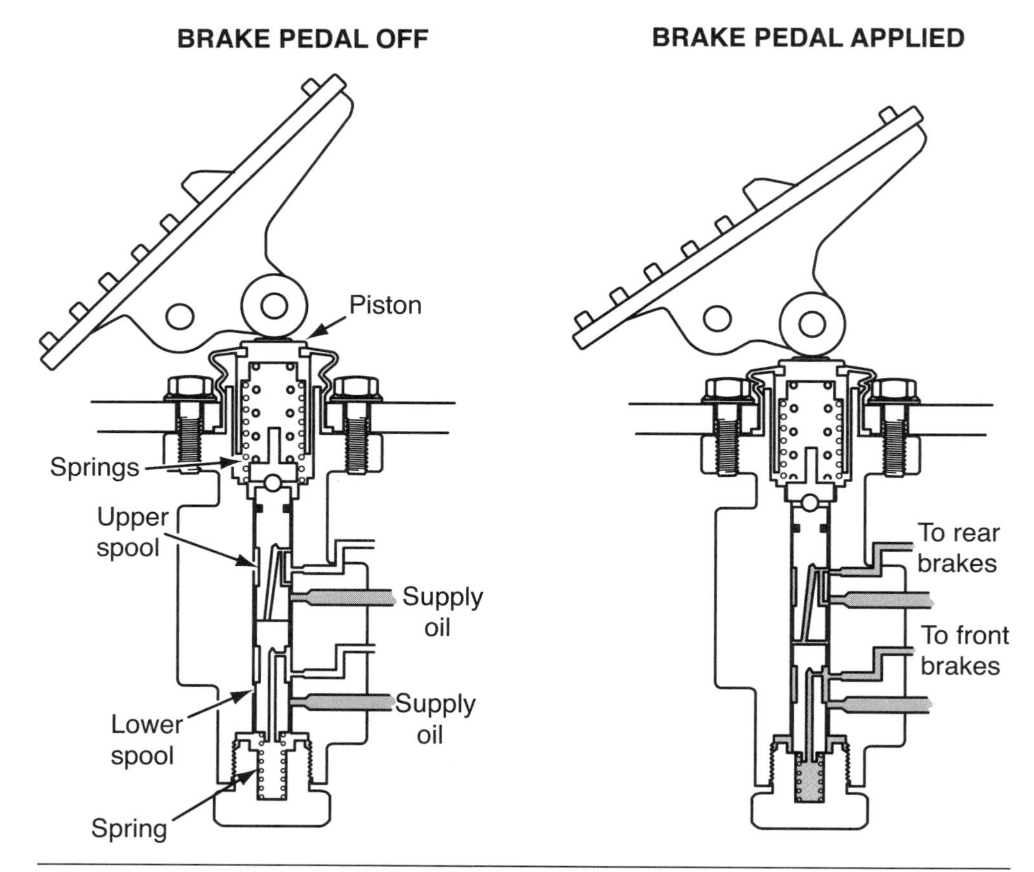

Figure 6-36 Cutaway of two brake valves. The one on the left is in the released position. The one on the right is in the applied position.

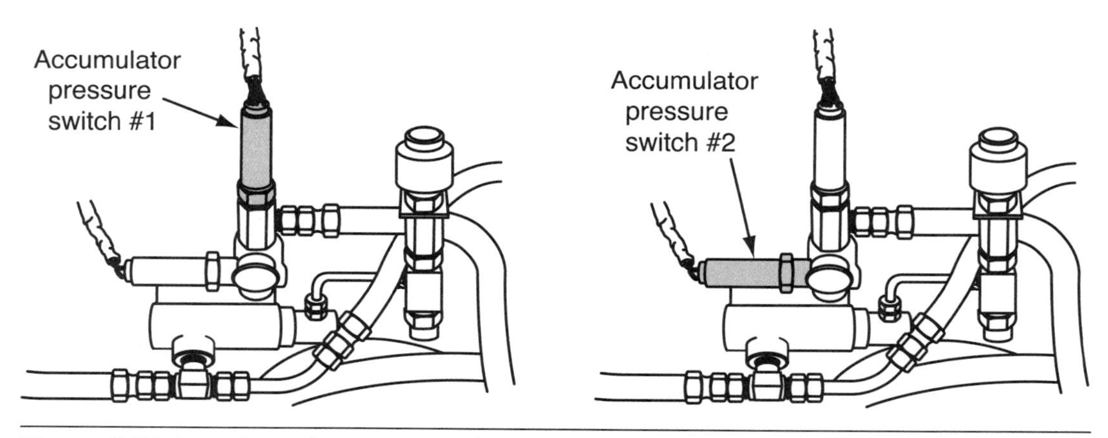

Figure 6-37 Location of two accumulator pressure switches. The one on the left is for the indicator light in the operator's compartment. The one on the right applies the park brake if the pressure gets below a predetermined amount.

Park Brake Control Valve

The park brake valve control is an electric hydraulic valve that is activated by an electric control button in the operator's compartment. The valve shown in **Figure 6-38** controls the flow of oil to the spring-applied park brake. The park brake circuit cannot release the park brake until brake pressure is above a predetermined minimum pressure. In the event that the brake pressure were to fall below the predetermined safe operating pressure, the park brake would be set automatically by the accumulator brake pressure switch.

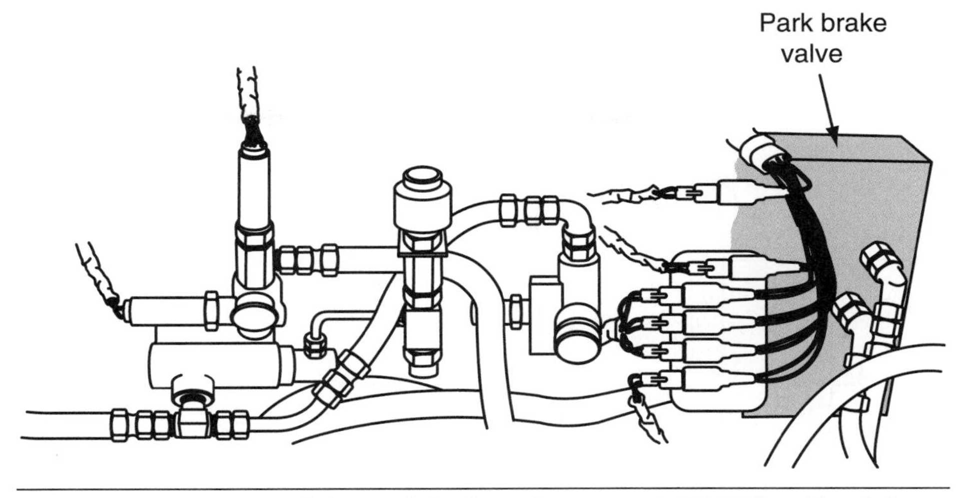

Figure 6–38 Location of the park brake valve on an R1700G load haul dump loader.

When the park brake switch in the operator's compartment is set to the "off" position, the solenoids on the park brake control valve are energized and the valve allows the oil to flow to the park brakes, disengaging them. When the operator activates the park brake, the solenoids on the park brake valve are de-energized. This also takes place when the operator turns the key switch to the off position or if brake pressure falls below a predetermined value. This allows the oil to drain from the park brakes back to the tank, activating the spring-applied park brakes.

Figure 6-39 shows the schematic view of the park brake valve in the on position. Solenoid (2) is de-energized, and the flow of oil from the pilot relief valve at port (5) is blocked by the valve spool (10). In this position the park brakes are applied by spring pressure. Oil from the park brakes is vented back to the tank through ports (9) and (7) through valve spool (10). When the operator releases the park brakes, solenoid (2) is energized and opens the port at valve spool (10), allowing oil to flow from the pilot port relief valve at port (5) to port (9) releasing the spring-applied brakes.

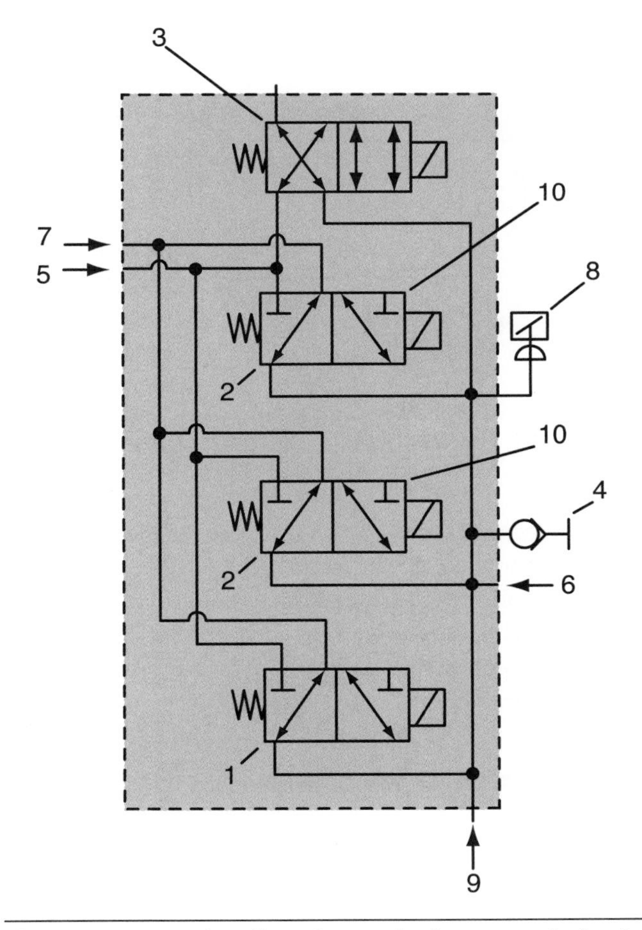

Figure 6-39 Hydraulic schematic for a park brake valve from an R1700G load haul dump loader.

Service Brake Operation

The service brakes on this equipment are a hydraulically-applied internal wet disc brake system. The park brake is a spring-applied oil pressure released system. Both systems use the same wheel end brake components to apply the brakes. Since the park brakes are spring engaged, the service brakes cannot be applied when the park brake is engaged because of the design of the system (**Figure 6-40**). After the engine is started the brake/pilot pump (54) sends oil flow to check valve (42) and on to the brake accumulator charge valve (59). The charge valve will charge the accumulators (17, 18) until

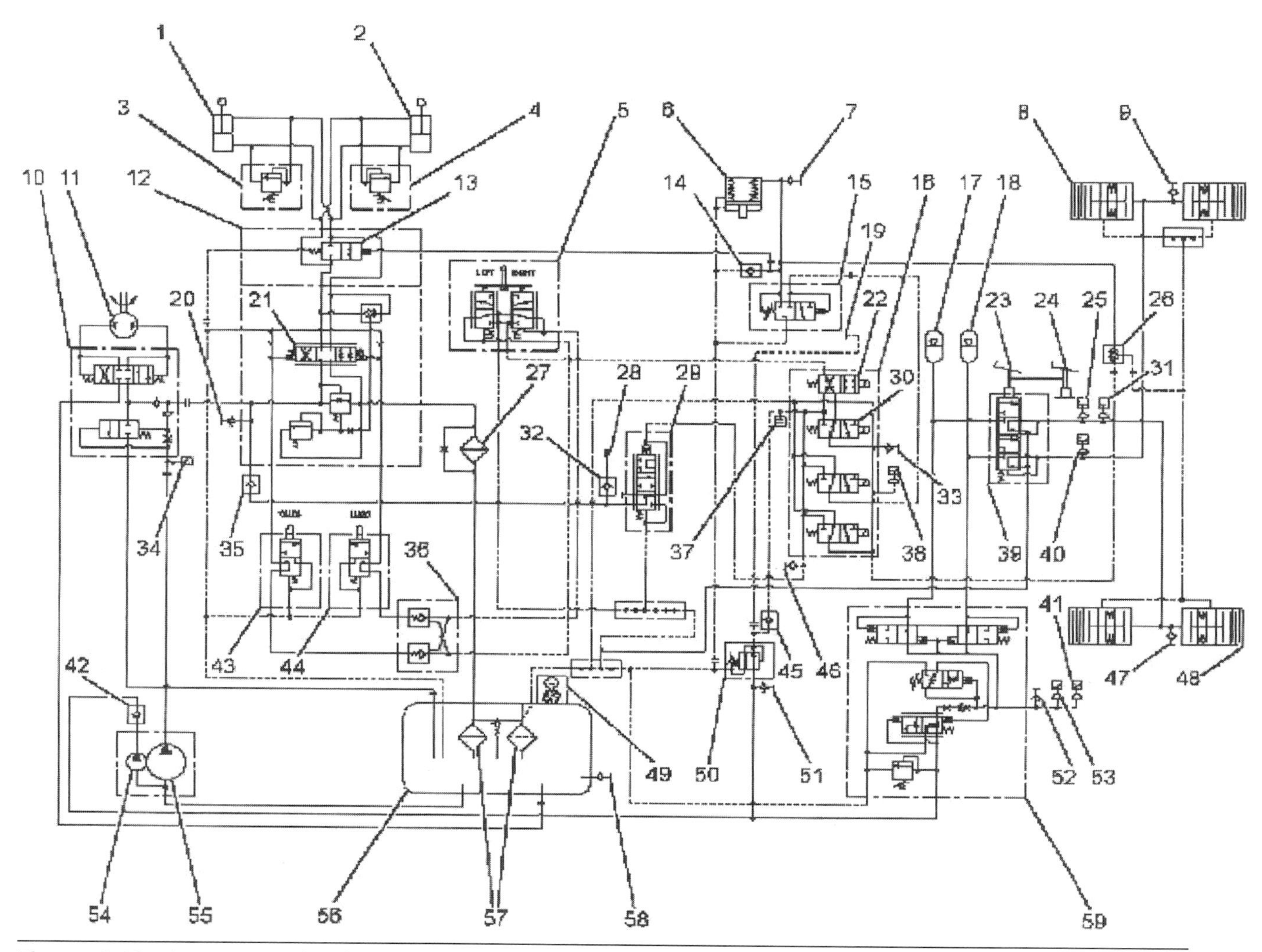

Figure 6-40 The R1700G loader brake schematic. *(Courtesy of Caterpillar)*

the predetermined cut-out pressure is reached. The oil then flows to the service brake valve (39). When the operator applies the service brake pedal, modulated oil pressure is directed to the wheel end brakes. The farther the pedal is depressed, the more pressure is sent to the wheel end brakes.

Park Brake Operation

When the operator applies the park brake (Figure 6-40), the park brake solenoids are de-energized, which causes the park brake control valve (16) to redirect the flow of oil from the wheel end brake to the hydraulic tank. The park brake springs force the pistons to apply pressure to the friction discs thus stopping the equipment.

When the operator releases the park brake, the park brake solenoids are energized to redirect the oil flow from the tank to the wheel end park pistons, which force the piston to retract against spring pressure and releasing the park brakes. As long as there is oil pressure behind the park brake pistons opposing the park brake spring pressure, the park brakes will remain in the released position. If for any reason brake pressure or engine power is lost, the park brakes will apply immediately.

TESTING/ADJUSTING AND TROUBLESHOOTING

This section provides examples of service and diagnostic procedures that are normally carried out by qualified technicians. When performing this type of work, always consult the manufacturer's shop manual for the specific procedures to follow. Before any service or testing can be performed on any equipment, a technician must carefully read over the section of the shop manual that covers the service or testing procedures to be performed. Failure to follow the manufacturer's recommended procedures can lead to serious injury or equipment damage. Unless you have the proper training to carry out any service work or diagnostic testing, the work is best left to qualified technicians.

Performing service work on hydraulic systems requires that technician follow established safety procedures when working with pressurized hydraulic circuits. Disposal of hydraulic fluids must always be carried out in accordance with local laws and established approved, safe practices.

Tech Tip: High pressure oil can remain in a hydraulic brake circuit for an indefinite period of time after the engine is stopped. Failure to properly release any stored pressure can result in serious injury or death if manufacturers' recommended procedures are not followed. Never use your hands, even if you are wearing gloves, to detect any hydraulic leaks in a hydraulic circuit; use a cardboard for this purpose. High-pressure fluid escaping from a pin-sized hole can cause severe injury or death. High-pressure oil can penetrate the ski, causing severe injury. Always seek medical attention immediately if this type of injury is suspected.

Before any work can begin, the equipment must be parked on level ground and have approved wheel chokes installed on the both sides of the tires. Be sure that any implements attached to the equipment are lowered to the ground. The steering frame lock must be installed along with any appropriate lockouts or tags that are required. If any steering lock mechanisms or transmission locks are present on the equipment, ensure that they are engaged. To relieve any hydraulic pressure that may be in the main hydraulic system, turn on the key switch and activate the steering and dump hoist controls to relieve any pressure in the system. Depress the brake pedal several times to relieve the brake **system pressure.** Depress the hydraulic tank relief valve on the hydraulic tank, if it is a pressurized hydraulic tank, to release the pressure. Be sure you have a suitable container to hold any fluid that may be released during the work procedures that you are about to follow. Visually inspect the complete brake system for any signs of oil leakage or damaged components. Before any testing takes place, be sure the oil is at operating temperature. As with all diagnostic procedures, the first place to start is at the beginning; in the case of a hydraulic system, that beginning means the tank. Check the oil level; look for any signs of air in the oil (aerated oil). Place a small amount of oil in a clear container and visually examine it for air bubbles. *Note*: This must be performed immediately after the equipment is shut down; otherwise, the bubbles will disperse on their own in a few minutes. Inspect the filter element for excessive dirt; use a magnet to determine if the particles are metallic. Check the condition of all hydraulic lines and fittings for any obvious damage. In diagnostics, there are always several causes to any known problem that you may encounter. Before performing specific tests, read over the troubleshooting section in the shop manual to review any common problems or symptoms before beginning diagnostic procedures. Ensure that the accumulator nitrogen charge pressure is correct. Be sure that the oil you are using is the correct viscosity for the operating conditions and temperatures that the equipment is operating under. The following troubleshooting chart outlines the most common brake system problems encountered on internal wet disc brake systems. Use **Table 6-3** as an example of a diagnostics procedure. Always refer to the manufacturer's service manual for the specific equipment that you are diagnosing.

Brake System Pressure Release

This procedure is an example of the steps that should be taken to release the brake system pressure; always consult the equipments service manual for the specific procedures that are to be used on the equipment that you are servicing. Not all procedures are identical and vary from equipment to equipment. They may even be different on the same type of equipment that was produced at a different location or time. The equipment must be parked on level ground and have approved wheel chokes installed on both sides of the tires. Be sure that any implements attached to the equipment are lowered to the ground. The steering frame lock must be installed along with any appropriate lockouts or tags that are required. If any steering lock mechanisms or transmission locks are present on the equipment, ensure that they are engaged.

Procedure

- Set the park brake and stop the engine before beginning.
- Make at least 80 applications to the service brake to dissipate the accumulator hydraulic pressure. Observe the brake pressure gauge on the dashboard of the equipment. Note that there is a gauge for each brake circuit, front and rear. The gauges should read 0 after all the pressure is relieved with the brake pedal depressed.

Table 6-3: COMMON BRAKE SYSTEM PROBLEMS

Problem	Possible Cause
Service brakes are slow to apply.	• Nitrogen gas charge is low • Hydraulic fitting or lines are leaking • Brake discs are worn beyond limits • Aeration of the brake fluid • Damaged hydraulic brake lines
Accumulator pressure will not reach cut-out pressure at engine idle.	• Low brake pump output, excessive brake pump wear
No front service brakes.	• Brakes lines to the front brakes are pinched or twisted • Internal damage to the front brake components • Faulty service brake control valve
No rear service brakes.	• Brakes lines to the rear brakes are pinched or twisted • Internal damage to the front brake components • Faulty service brake control valve
Park brake does not engage.	• Faulty park brake control switch • Sticky park brake solenoid control valve
Service brakes do not hold. No warning system activation.	• Internal brake discs worn beyond permissible tolerances
Service brakes do not apply evenly.	• Uneven brake disc wear • Brake control valve is sticking during use • Low brake pressure to either brake circuit
Excessive force required to depress the brake control valve.	• Sticking brake control valve due to oil contamination
Service brakes not releasing completely.	• Wheel end brake component damage • Sticking brake control valve due to oil contamination • Obstruction under the brake pedal; dirt or rocks preventing the pedal from releasing completely
Brake pedal kick-back during brake application.	• Air in the brake system

Brake Pump Flow Test

The brake pump test is designed to determine the condition of the pump and its ability to pump oil. Pump flow tests are conducted at two different pump speeds to access the pump's condition. The difference between pump flow of two different operating pressures is called flow loss. The specifications used in this example (**Table 6-4** and **6-5**) are for reference purposes and do not represent actual flows for any given pump. Always refer to the manufacturer's shop manual for the equipment in question.

To perform this test on the equipment, a flow meter must be installed to measure the pump output. An example of actual flow loss is shown in **Table 6-6.**

Brake Accumulator Charge and Test Procedure

This procedure is an example of the steps that should be taken to check and charge a brake accumulator. Always refer to the equipments service manual for the specific procedures that are to be used on the equipment that you

Table 6-4: DETERMINING PUMP FLOW LOSS

Method of determining flow loss
Pump Flow at 690 kPa (100 psi) Pump Flow at 6900 kPa (1000 psi) Flow Loss
Example of Determining Flow Loss
217.6 L/min (57.5 US gpm) 196.8 L/min 52 US gpm) Loss = 20.8 L/min (5.5 US gpm)

are servicing. Not all procedures are identical and vary from equipment to equipment. They may even be different on the same type of equipment that was produced at a different location or time. The equipment must be parked on level ground and have approved wheel chokes installed on both sides of the tires. Be sure that any implements attached to the equipment are lowered to the ground. The steering frame lock must be installed along with any appropriate lockouts or tags that are required. If any steering lock mechanisms, such as transmission locks, are used on the equipment, ensure that they are engaged.

Tech Tip: The only gas that can be used to charge accumulators is dry nitrogen gas. No other gases are permitted for this procedure; some gases, such as oxygen, are highly explosive and should never be used for this purpose. Be sure that the nitrogen bottles have the right connections to match the hoses and gauges that are going to be used for this procedure. Before beginning this procedure, make sure that all the hydraulic pressure is dissipated from the brake system.

To ensure that the nitrogen charge pressure is correct, the hydraulic pressure must be totally dissipated with the accumulator piston bottomed out in the accumulator. To ensure that the correct pressure is present in the accumulator, a temperature correction chart should be consulted before beginning this procedure. In our example (**Table 6-7**) we will use 800 psi as our desired charge pressure at 21 degrees C (70 degrees F).

To check the nitrogen charge in the accumulator, you must connect a 0 to 2000 psi (0 to 13800 kPa) pressure gauge to the connection on the top of the accumulator. (Refer to the service manual for the equipment being serviced for the correct procedures.) If the reading on the gauge is high, release some of the excess nitrogen. If the pressure is low, follow the general steps in the service manual for the equipment being serviced to charge the accumulator.

An example of steps used to charge an accumulator is shown below.

Procedure

- Connect the appropriate hose to the nitrogen cylinder to valve (3) as shown in **Figure 6-41.**
- Close valve (3) and open the nitrogen cylinder. Adjust screw (1) 0 on the regulator (9) until the pressure on gauge (7) is at the correct charge pressure. *Note*: Be sure to compensate for ambient temperature variations.

Table 6-5: DETERMINING PUMP FLOW LOSS

Method of Determining Flow Loss				
Flow Loss (L/min or US gpm) Pump Flow at 690 kPa (100 psi)	×	100	=	Percent of Flow Loss
Example of Determining Flow Loss				
(20.8 L (5.5 gpm)) (217.6 L/min (57.5 US gpm))	×	100	=	9.5%

Table 6-6: ACTUAL PUMP FLOW LOSS

Example of Actual Pump Flow Loss				
Pump Flow at 6900 kPa (1000 psi) Pump Flow at 690 kPa (100 psi)	×	100	=	Percent of Flow Loss

Table 6-7: PRESSURE/TEMPERATURE RELATIONSHP

Accumulator Pre-charge Pressure/ Temperature Relationship for 5500 kPa (800 psi)	
−7 degrees C (20 Degrees F)	5000 kPa (725 psi)
−1 degrees C (30 Degrees F)	5105 kPa (740 psi)
4 degrees C (40 Degrees F)	5235 kPa (760 psi)
10 degrees C (50 Degrees F)	5340 kPa (775 psi)
16 degrees C (60 Degrees F)	5420 kPa (785 psi)
21 degrees C (70 Degrees F)	5500 kPa (800 psi)
27 degrees C (80 Degrees F)	5605 kPa (815 psi)
32 degrees C (90 Degrees F)	5735 kPa (830 psi)
38 degrees C (100 Degrees F)	5840 kPa (845 psi)
43 degrees C (110 Degrees F)	5920 kPa (860 psi)
49 degrees C (120 Degrees F)	6000 kPa (875 psi)

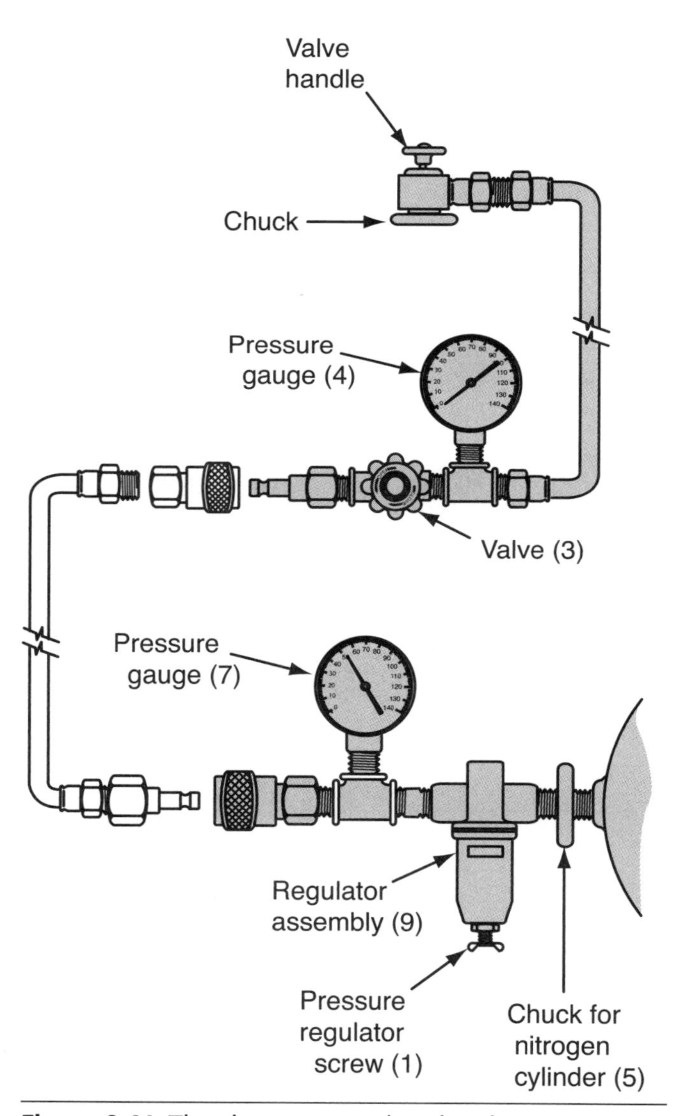

Figure 6-41 The charge procedure hardware required to charge a brake.

- Open the valve to the accumulator (3) and proceed to charge the accumulator. Once the pressure is stabilized in the accumulator, close valve (3). Pressure on gauges (4) and (7) should be the same.
- Repeat the above procedure if the pressure does not match.
- Close valve (3) and close the nitrogen bottle valve.
- Remove the chuck and test equipment and reinstall the protective cap onto the accumulator.

When installing a new or reconditioned accumulator perform the following steps to purge the accumulator of any air trapped inside. Then perform the steps to charge an accumulator listed above. All of the air must be purged out from the nitrogen side of the accumulator before any nitrogen is introduced into the accumulator. Pour approximately 2 quarts (1.9 L) of SAE 10W oil into the nitrogen side of the accumulator. This is necessary to lubricate the top piston seal and ensure that all of the air is removed from the nitrogen side of the accumulator.

Procedure

- Install the accumulator on the equipment.
- Refer to Figure 6-41 to identify the equipment that is needed to connect the nitrogen cylinder with the appropriate gauges and adapters to properly charge the accumulator.
- Install the chuck (5) to the accumulator nitrogen charge valve with the other end of the charge hose pointing into a suitable container capable of holding the oil from the nitrogen side of the accumulator.

- Start and run the engine until it reaches normal operating temperature. As the charge valve charges the bottom section of the accumulator, the air and oil in the top of the accumulator will be forced out into the container. When the oil stops flowing out the hose close valve (3) completely.
- Stop the engine and release all the brake system pressure as outlined in the service manual under "Releasing Brake System Pressure."

Now follow the steps to charge the accumulator to complete the procedure.

Accumulator Charge Valve Test Procedure

This following procedure is an example of the steps that must be used to check accumulator charge valve operation. Always consult the equipment service manual for the specific procedures that are to be used on equipment that you are servicing. Not all procedures are identical and may vary from equipment to equipment; in some cases they may even be significantly different for the same model of equipment that was manufactured at a later or earlier time. The equipment must be parked on level ground and have approved wheel chokes installed on both sides of the tires. Be sure that any implements attached to the equipment are lowered to the ground. The steering frame lock must be installed along with any appropriate lockouts or tags that are required. If any steering lock mechanisms, such as transmission locks, are present on the equipment, ensure that they are engaged.

Procedure

- Ensure that the engine is not running before proceeding.
- Depress the service brake pedal until all the pressure is removed from the brake accumulators. The number of times that you must depress the brake pedal will vary from equipment to equipment. In our example, we will use 75 applications.
- Connect a pressure gauge (0–25000 kPa or 0–3600 psi) to gauge connection pressure tap (2) shown in **Figure 6-42.**
- Observe the pressure reading before starting the engine. The pressure should be 0 psi.
- Start the engine and build up the charge pressure until it reaches cut-out pressure. Refer to the manufacturer's service manual for the correct specifications. In our example, we will use 14480 kPa (2100 psi). Stop the engine at this point.

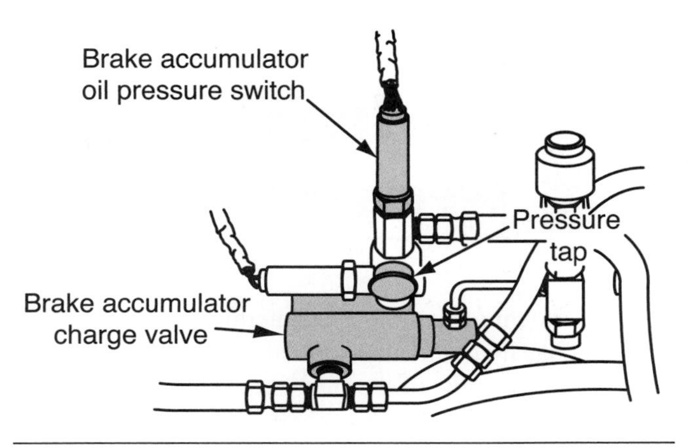

Figure 6-42 Location where a gauge must be connected to check accumulator charge valve operation on a CAT R1700G load haul dump loader. (1) Brake accumulator oil pressure switch for the alert indicator, (2) Pressure tap (brake accumulator oil), (3) Brake accumulator charging valve.

- Turn the key to the "on" position without starting the engine.
- Apply the service brakes repeatedly and count the number of times you apply the brakes while checking the pressure on the service gauge.
- Once the gauge gets down 8300 kPa (1200 psi) the accumulator oil pressure gauge on the dashboard should come on. If this occurs as described, then the accumulator low oil pressure alert is working properly.
- Apply the service brakes and continue counting the brake applications while observing the gauge pressure. At one point the pressure will drop off rapidly; this is an indication of how much nitrogen pre-charge pressure is actually in the accumulator.
- Add up the number of brake applications it takes to get to the point where the gauge dropped off rapidly. Check the manufacturer's shop manual to determine the number of applications that is correct for the equipment you are servicing. In our example, we will use 75 brake applications as the minimum amount.
- If the cut-out pressure is not correct refer to the manufacturer's shop manual for the adjustment procedures. Adjustment procedures will vary from equipment to equipment. Not all charge valves have a cut-out pressure adjustment that can be readily performed in the field. It may be necessary to replace the charge valve in some cases (See Photo Sequence 2).

2 Hydraulic Brakes Charge Valve Pressure Check

The following example is a general procedure used to check brake charge valve pressure and operation on equipment that has hydraulically-applied internal wet-disc brakes. It is not intended to replace the shop manual and is only shown as an example of the steps required to perform this type of procedure. Always refer to the manufacturer's technical documentation for the actual step-by-step instructions for this type of work. Before beginning this procedure, be sure to ensure that the oil flow for the brake circuit is within specifications. Restriction to flow in any hydraulic circuit determines the correct pressure values.

CAUTION: *Never use your hands to check for leaks in any high-pressure hydraulic circuit. Use a small piece of cardboard. A pin-size leak will penetrate the skin. If oil is accidentally injected into the skin, immediate medical attention is required. Oil in a hydraulic circuit can become extremely hot during operation. This can lead to severe burns if accidentally splashed onto yourself or other personnel in the work area.*

P2-1 Depending on the employer, some form of tagging or lockout procedure must be used to protect the technicians while work is being performed on the equipment.

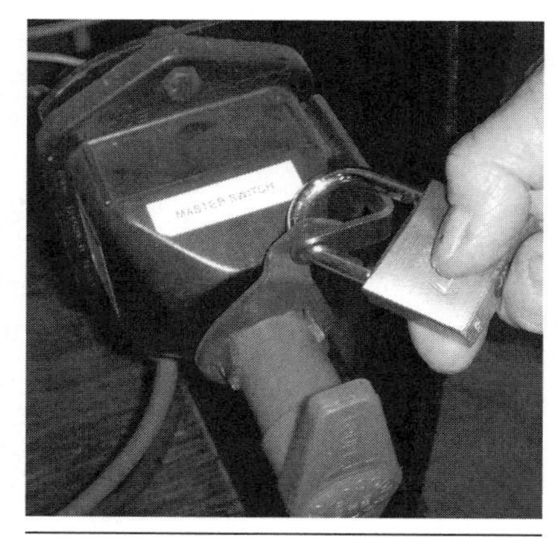

P2-2 Some employers require you to actually lock out the main power disconnect on the equipment. This lock can only be installed and removed by the technician who performing the work on the equipment. It must be removed on completion of the work.

P2-3 Proceed by lowering the brake system pressure by activating the brake valve as many times as necessary to get the brake pressure to go to zero. The foot brake valve is located in the operator's compartment.

P2-4 Connect a suitable gauge to the fitting. A 0 to 3000 psi (20700 kPa) hydraulic pressure gauge must be used. At this point, the gauge should read zero after it is connected and tightened.

P2-5 Have someone start the engine and observe the reading on the pressure gauge as it increases. A short pause in the oil pressure rise indicates the accumulator nitrogen precharge pressure. The pressure will continue to rise to a point where it will pause again. This pause point is the cut-in pressure. Continue observing the pressure rise until it reaches its maximum level. This is known as the cut-out pressure. Now refer to the manufacturer's specifications for the equipment in question for the correct values.

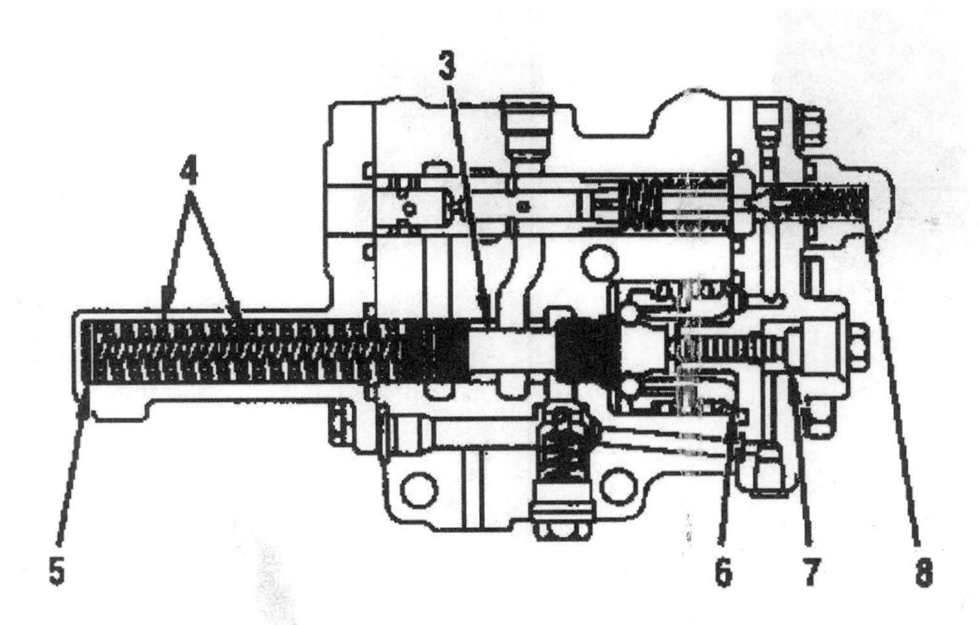

P2-6 If the cut-out pressure is not within specifications, the following adjustments can be made. Dissipate all the oil pressure in the brake circuit as outlined previously. The brake charge valve must be removed from the equipment to make the necessary adjustments.

- The cover for spring 4 in the diagram must be removed to gain access to the shim pack.
- Add or remove shims according to requirements. One shim added or removed will change the pressure by approximately 33 psi (230 kPa).
- Reinstall the brake charge valve and recheck the pressures as outlined previously.
- After the cut-out pressure is within specification, run the equipment and apply the brake pedal as many times as it takes to get the pressure to the point where it will start to charge back up. Observe the pressure at this point. This is the cut-in pressure.
- If the pressure difference between the cut-in and cut-out pressure is not within specifications, shims must be added or removed in location (6).
- Remove the charge valve and add or remove the correct amount of shims to achieve the correct spread in pressures.
- Reinstall the charge valve and retest the pressures for correct values. Reinstall any additional hardware that was removed to access the brake charge valve.

ONLINE TASKS

1. Use a suitable Internet search engine of your choice to research the use of hydraulic brake oil specifications. Identify a piece of equipment in your particular location, paying attention to the operating temperature changes that exist in your area. Outline the reasoning behind selecting this particular type of hydraulic oil.
2. Check out this url: *http://www.hydraulic-supply.com/html/productline/mfgprod/vickers-hydraulics.htm* to discover important facts and classifications of hydraulic oil used on off-road hydraulic brake circuits.

SHOP TASKS

1. Select a piece of equipment in your shop that has a hydraulically-applied brake system and identify the components that make up the brake circuit. Identify the type of brake hoses and lines that are used and the manufacturers' specifications for these components.
2. Select a piece of mobile equipment in your shop. Identify the type and model of the equipment and list all the components and the corresponding part numbers of a typical hydraulically-applied brake circuit. Use the correct parts book for that particular piece of equipment.

Summary

- All brake systems rely on friction to slow or stop the equipment. We know that energy cannot be destroyed but must be converted to another form of energy. With brake systems we convert the kinetic energy of moving equipment to heat energy when the brake pedal is depressed to slow down or stop the equipment.
- The use of two independent brake circuits on equipment is a must on heavy equipment. In addition, the use of an independent mechanical park brake is also required to safely park the equipment. In some industries such as mining, the equipment is required by law to be equipped with three independent brake circuits.
- On many types of off-road equipment applications, the use of two separate master cylinders is required so that each brake circuit on the equipment can be controlled independently.
- On large equipment where brake pressures need to be considerably higher than what can be achieved under what would be considered normal operator foot pressure, additional hydraulic pressure can be used to multiply brake force without exerting extra pedal effort on the part of the operator.
- Modern equipment design has incorporated the use of independent dual brake circuits built into one master cylinder for the front and rear brakes for obvious safety reasons.
- On most off-road equipment, the shoes are generally pinned on the bottom and have a wheel cylinder at the top that forces the shoes out against the inside of the drum during brake actuation.
- Hydraulic drum brakes, regardless of whether they are a non-servo or servo brake design system, must have a wheel cylinder to actuate them. The wheel cylinders use hydraulic pressure from a

brake valve to force the shoes out against the drum.

- Some of the older off-road drum brake systems used a residual check valve in the master cylinder to maintain a residual pressure of approximately 12 psi in the wheel cylinder. This was required to take up any slack and prevent wheel cup sag, which could lead to leaks or air getting into the wheel cylinder.
- The self-adjustment mechanism is necessary to keep the brake shoes close to the drum and take up the extra clearance caused by drum and shoe wear. The system can only operate in the reverse direction. It will not be able to make any self-adjustments in the forward direction.
- External hydraulic disc brake systems come in two basic designs; the fixed-caliper and the sliding-caliper design.
- Disc brakes have many obvious advantages over drum brakes on off-road equipment. They are more resistant to brake fade, provide better cooling capabilities, have an acceptable tolerance to dirt and water, require less maintenance and do not need to be adjusted, and have a greater swept area that drum brakes.
- When engineers design off-road equipment, they take into account the total swept area of the brake system and the weight, speed, and traction of the equipment when determining brake pressures along with swept area of the brake design.
- The internal wet disc brake system is highly suitable for applications where road grime and contamination are extremely likely to occur. Braking is achieved through the use of a number of friction discs that are alternately spaced between number of steel discs, which are held stationary by outside teeth or notches that are splined to the wheel spindle.
- Lubrication and cooling on some inboard wet disc brake systems are often integrated with the lubrication and cooling for the axle assembly and require the use of specialized lubricants. In heavy-duty applications, a separate cooling circuit that has its own oil pump and cooler can be used to dissipate the heat generated by this type of brake system.
- Inboard hydraulic wet disc brakes allow the brake and differential to share the same oil for lubrication and cooling, and require the use of a special type of lubricant that is compatible with the brake lining.
- When equipment is left in the park position, it must not be parked on hydraulic pressure alone but must use a mechanical park brake.
- Many mechanical park brake consists of two sets of discs. Steel disks are externally splined to the housing and do not rotate. Friction discs are splined to a rotating shaft. A pressure plate has a number of heavy-duty springs on one side of it and the hydraulic piston on the other side. Either one can lock the friction and steel discs together.
- In North America, brake fluid standards are quantified by two groups: the Society of Automotive Engineers (SAE) and the Department of Transportation (DOT). Conventional brake fluid has three categories: DOT 3, DOT 4, and DOT 5.
- Any hygroscopic brake fluids that have been sitting on the shelf must be completely sealed in their respective containers to prevent any moisture that is present in the air from being absorbed by the fluid.
- DOT 3 brake fluids have a dry boiling point of approx 400 degrees F (205 degree C). Once the brake fluid absorbs some moisture, the boiling point will quickly fall to around 284 degrees F (140 degrees C).
- DOT 5 brake fluid is formulated from a silicone base and does not absorb water, in other words it is non-hygroscopic. It must be kept free from water because any water entering the brake system will cause corrosion whereever it has settled.
- Two accumulators are used in a typical off-road brake circuit; one for the front service brakes and one for the rear service brakes. Nitrogen pre-charged accumulators store hydraulic energy, which is used by the brake circuit to apply the brakes. In the event of an engine failure, the accumulators store enough energy to allow the brakes to be applied.
- The park brake on off-road equipment consists of a spring-applied hydraulically-released brake that can be mounted on any component that is mechanically connected to the wheels. Many equipment manufacturers mount them on the output shaft of the transmission or transfer case or even on the differential input shaft.
- A hydraulic pump is used to supply oil for the brake circuit on equipment with internal hydraulic wet disc brake systems. A gear pump with an output of approximately 5 to 12 gpm is often used for this purpose because of its low cost.

- The accumulator contains a free-floating piston that has high-pressure O-rings, which separate the dry nitrogen on one side from the hydraulic oil on the other side. Its purpose is to store pressurized oil that can be used to apply the brakes when required.
- The purpose of the accumulator charge valve is to maintain a predetermined amount of hydraulic pressure in the brake accumulator circuit when the engine is running.
- The brake pedal functions as a variable-pressure reducing valve, limiting the pressure flowing to the wheel end brakes. The more the pedal is depressed, the higher the pressure will be at the wheel end brakes. A small internal passage in each spool allows the oil to flow from the modulated pressure side of the circuit to the underside of the spools. This provides a resistance to the pedal movement, which gives the operator feedback by providing a slight resistance.
- The accumulator pressure switch is a hydraulic electric valve that monitors the accumulator pressure and triggers a park brake application when the pressure in the accumulators gets below a predetermined value.
- The operator will not be able to release the park brake until the safe accumulator pressure is above the low limit value. The park brake switch must be cycled on and off before the brakes can be released. An electrical interlock prevents the accidental release of the park brake should it be applied inadvertently by a loss of brake pressure.
- When the park brake switch in the operator's compartment is set to the "off "position, the solenoids on the park brake control valve are energized and the valve allows the oil to flow to the spring-applied park brakes, disengaging them. When the operator activates the park brake, the solenoids on the park brake valve are de-energized.
- The service brakes on most off-road equipment are generally hydraulically-applied internal wet disc brakes. The park brake is a spring-applied oil pressure released system. On some systems both brake circuits use the same wheel end brake components to apply the brakes.
- On spring-applied park brake systems, as long as there is oil pressure behind the park brake pistons opposing the park brake spring pressure, the park brakes will remain in the released position. If for any reason brake pressure or engine power is lost, the park brakes will apply immediately.
- High-pressure oil can remain in a hydraulic brake circuit for an indefinite period of time after the engine is stopped. Failure to properly release any stored pressure can result in serious injury or death. Manufacturers' recommended procedures must be followed. Never use your hands, even if you are wearing gloves, to detect any hydraulic leaks in a hydraulic circuit: use a cardboard for this purpose.
- The brake pump test is designed to determine the condition of the pump and its ability to pump oil. Pump flow tests are conducted at two different pump speeds to access the pump's condition. The difference between pump flow of two different operating pressures is called flow loss.
- The only gas that can be used to charge accumulators is dry nitrogen gas. No other gases are permitted for this procedure; some gases such as oxygen, are highly explosive and should never be used for this purpose.
- To ensure that the nitrogen charge pressure is correct, the hydraulic pressure must be totally dissipated with the accumulator piston bottomed out in the accumulator. To ensure that the correct pressure is present in the accumulator, a temperature correction chart should be consulted before beginning this procedure.
- All of the air must be purged out from the nitrogen side of the accumulator before any nitrogen is introduced into the accumulator. Pour approximately two quarts (1.9 L) of SAE 10W oil into the nitrogen side of the accumulator. This is necessary to lubricate the top piston seal and ensure that all of the air is removed from the nitrogen side of the accumulator.
- Charge valve adjustment procedures will vary from equipment to equipment. Not all charge valves have a cut-out pressure adjustment that can be readily performed in the field. It may be necessary to replace the charge valve in some cases.

Review Questions

1. What does the term *hygroscopic* mean?

   A. designed to absorb moisture
   B. designed to resist moisture absorption
   C. designed to act as a hydraulic medium
   D. designed to resist expansion when subjected to temperature rise

2. Which of the following describes a typical hydraulic brake system?

   A. A hydraulic brake pedal cannot multiply force.
   B. Application force is multiplied hydraulically by the wheel cylinders.
   C. Actual braking force at the wheels is not dependent on mechanical force.
   D. Application force is multiplied hydraulically by the operator foot.

3. Which of the following describes the function of a tandem master cylinder during normal operation is true?

   A. Primary piston is actuated hydraulically.
   B. Secondary piston is actuated mechanically.
   C. There is a single bore.
   D. Primary piston is actuated mechanically.

4. Where does the hydraulic pressure that operates hydraulic boosters come from?

   A. power steering pump
   B. reserve electric motor and pump
   C. dedicated hydraulic pump
   D. all of the above

5. Which brake shoe in a non-servo drum brake assembly is energized during braking when a vehicle is traveling in a reverse direction?

   A. leading shoe
   B. both shoes
   C. trailing shoe
   D. neither shoe

6. What is the most common type of parking brake used on off-road equipment with hydraulic brake systems?

   A. canister brake
   B. driveline-mounted, hydraulic applied drum brake
   C. cable and lever actuated secondary shoe
   D. driveline-mounted, mechanically applied brake

7. When the operator depresses the brake pedal of the typical loader with hydraulic brakes, which of the following occurs?

   A. Hydraulic pressure actuates the mechanical booster assemblies.
   B. Hydraulic fluid flows through the quick release valves.
   C. Modulated hydraulic pressure actuates hydraulic wheel end brake assembly.
   D. Modulated hydraulic pressure actuates the diverter valve.

8. When replacing brake pads on a loader, what should be done before forcing a caliper piston back intoits bore?

   A. Open the bleed nut and drain the displaced fluid.
   B. Open the master cylinder lid to allow the fluid to return to the reservoir.
   C. Bleed the master cylinder first.
   D. Remove the brake line to drain the caliper.

9. What type of discs are used on a wet disc multi-disc brake assembly on heavy equipment?
   A. steel and friction discs
   B. friction discs only
   C. steel discs and brass discs
   D. brass discs and friction discs
10. What is the purpose of the brake pump in a hydraulic brake system?
   A. supply oil to the accumulator and the pilot circuit
   B. supply oil to hydraulically-appliedspring release brakes
   C. supply oil to the spring-applied hydraulic release brakes
   D. supply oil to the park brake system calipers
11. What happens when the internal seal on the spindle of a hydraulically-applied wet disc wheel end leaks?
   A. Hydraulic oil will enter the differential.
   B. Hydraulic oil will enter the wheel end.
   C. The differential oil will enter the wheel end brakes.
   D. Hydraulic oil will leak externally.
12. What advantage does a multi-disc brake system have over conventional brake systems?
   A. more swept area
   B. less moving components
   C. braking surface area is minimized
   D. less cooling required
13. On off-road brake systems, what does the term *charge valve cut-out pressure* mean?
   A. The charge valve stops supplying of the accumulator.
   B. The charge valve begins to supply the pilot circuit.
   C. The charge valve begins to supply the accumulator.
   D. The charge valve stops charging the reservoir.
14. What type of device is the accumulator pressure switch in a wet disc brake system?
   A. hydraulic/electric
   B. air brake
   C. electric/air
   D. air/hydraulic
15. What is the purpose of the accumulator(s) in a hydraulic applied wet disc brake system?
   A. to store hydraulic energy for emergency brake application
   B. to apply the park brakes if engine power were lost
   C. to release the driveline park brake when brake pressure is low
   D. to apply the service brakes

CHAPTER

Spring-Applied Hydraulically-Released Brakes

Learning Objectives

After reading this chapter you should be able to

- Describe the fundamentals of off-road spring applied hydraulically-released brake systems.
- Identify the major components that make up spring-applied hydraulically-released brake systems.
- Describe the principles of operation of spring-applied hydraulically-released brake systems.
- Outline the maintenance and repair procedures associated with spring-applied hydraulically-released brake systems.
- Identify the basic troubleshooting procedures for spring-applied hydraulically-released brake systems.

Key Terms

accumulator
application pressure
brake pressure
brake pump
charge valve
check valve
foot valve
friction discs
park brake valve
pilot pressure
pressure switch
reverse modulation
sequence valve
steel discs
system pressure
wheel end pressure

INTRODUCTION

Off-road equipment brake systems have developed into a reliable braking systems that are safe and efficient compared to what they were in the past. Modern brake systems incorporate complex hydraulic circuits to get the job done. This chapter covers one of the most popular brake systems in use today: the spring-applied hydraulically-released wet disc brake system.

Brake systems on off-road equipment must be able to operate in extremely abrasive conditions that would wear out conventional external brake systems very quickly (**Figure 7-1**). The roadways that equipment operates on are extremely abusive to a brake system. Dirt and gravel roadways are common in this environment. To safely slow down and stop equipment that can weigh as much as 200 tons is a requirement in this environment. The grades and roadways the equipment must operate on place a severe-duty status on the brake system. The constant exposure to abrasive dirt quickly wears out the fiction lining.

In the past external hydraulic brakes (both drum and disc) were common on off-road equipment; they are seldom used on new equipment because of the high costs associated with maintaining these systems. Many manufacturers favor the use of internal spring-applied

Figure 7-1 Typical excavator that utilizes a spring-applied hydraulically-released brake system. (*Courtesy of Caterpillar*)

hydraulically-released wet brake systems. These systems incorporate a brake design with the brakes located inside the axle assembly; this keeps abrasive road dirt out. The design has evolved over the years to be so reliable that the brakes usually last for thousands of hours of operation before needing to be overhauled. Another major advantage is that these brakes do not require any mechanical adjustments. With proper maintenance, it is not uncommon to see the **friction discs** last 10,000 hours before needing replacement.

FUNDAMENTALS

Brake design concepts have changed considerably over the years. Equipment speed and weight increased and brake system design was forced to keep pace. Brake systems on off-road equipment are designed for extremely high braking capacity; the brake system must control the speed of the equipment as well as bring it to a safe stop when required. Brake systems have always relied on friction to slow or stop the equipment. Kinetic energy of the equipment is converted to heat energy when the equipment is slowed down or brought to a stop. When the operator applies the brakes, friction material is forced to exert pressure on a corresponding metal surface to slow down the equipment. These friction devices are located in the drive axles and, depending on axle design, the brake systems can be located either inboard or outboard on the axle assembly.

It quickly became apparent that it made more economical sense to wear out the friction lining rather than to wear out the rotors. As brake systems evolved, the use of mechanical spring pressure and hydraulics became increasingly popular. Early hydraulic brake design used a fluid confined in a hydraulic circuit to transmit the pressure required for braking. The operator's foot created the mechanical force on the master cylinder. Modern equipment uses some form of hydraulic assist to make the task of applying the brakes effortless. The wheel end clamping forces used by a spring-applied hydraulically-released brake is determined by the size and number of springs used in the wheel ends to apply mechanical force to the friction lining.

Because of the extremely abrasive conditions in which off-road equipment operate, the brake systems must have a high level of reliability designed into the system. The concept of having four independent spring-applied brakes on equipment is safer than using conventional hydraulically-applied wheel end brakes. With hydraulically-applied brakes, if a hydraulic **brake pressure** loss occurs, it is possible to lose some braking potential. This is not the case with spring-applied hydraulically-released service brakes; a brake pressure loss would immediately and automatically apply the service brakes with no potential for any brake loss.

The use of an independent mechanical brake is also a requirement to safely park the equipment. In some industries, such as mining, the use of three independent brake circuits is not an option; it is a requirement. A third additional circuit (called emergency brakes) is a combination of the wheel ends and the park brake circuit.

The complexity of the brake systems on off-road equipment has increased dramatically over the past 10 years. The use of complex hydraulic brake circuits, which can incorporate variable-displacement pressure compensated pumps, is becoming common. These modern brake circuits monitor **application pressures** and apply the brakes automatically if the pressure falls below safe operating limits. Not only will the brakes apply when the brake pressures become low, but they will also apply if there is a loss of electrical power or if transmission clutch or torque pressure falls below safe operating levels.

Many manufacturers implement a brake interlock circuit; the brakes cannot be released after an emergency application unless the operator manually resets a brake interlock circuit. This interlock brake circuit ensures that the operator is *in* the operator's compartment before the brakes can be released. To understand these brake systems, it is important that you have a good understanding of basic hydraulics and can read and understand electric/hydraulic schematics. Chapter 6, Hydraulic Brakes, covers the basic principles of hydraulic brakes including shoe-type brakes and external caliper disc brakes.

BASIC BRAKE SYSTEM COMPONENTS

A spring-applied hydraulically-released brake system (**Figure 7-2**) shares similar components that are used in a hydraulically-applied brake circuit. The use of a multi-disc wheel end brake design is like the ones used in hydraulically-applied brakes. The main difference is the large heavy springs that are used in the spring to actually apply the brakes. Hydraulic pressure is used to keep the brakes in the released position in this design.

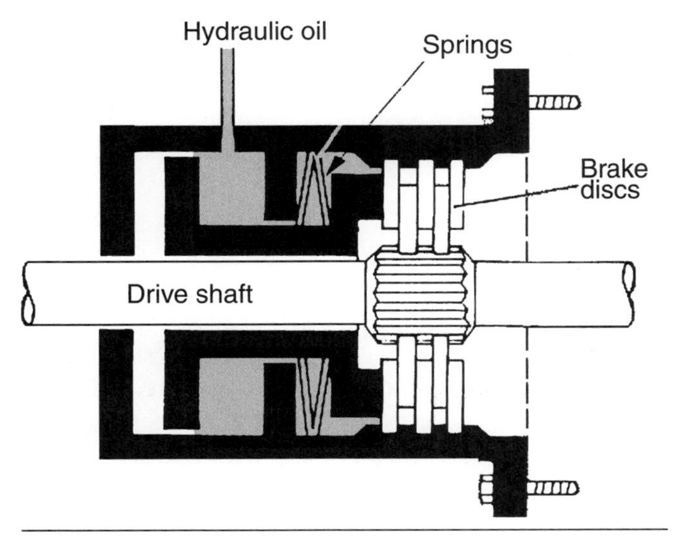

Figure 7-2 Cutaway of a spring-applied hydraulically-released brake system used on an excavator final drive.

Brake Pump

The brake pump (**Figure 7-3**) that is used in this type of brake circuit is generally a heavy-duty gear-type pump that is capable of supplying a continuous oil flow of up to 19 gpm (gallons per minute) or 72 L/m (Liters per minute) at 2400 rpm. In many applications, this pump also supplies the oil flow for the brake cooling system, as well as the equipment's pilot circuit. Gear pumps are used because of their low cost and ability to operate in less than ideal operating conditions. The pumps are generally mounted to the torque converter and are located in a high heat area. In some applications manufacturers are using variable-flow piston pumps, which can be de-stroked and thus eliminating the need for a **charge valve** in the brake circuit. The downside to using a variable-displacement pump for the brake circuit is the higher cost when compared to a gear pump. The upside is the cost saving of not having to use a charge valve in the brake circuit to regulate brake pressure.

Hydraulic Accumulators

Although there are different accumulator designs, those used in brake systems are all similar in operation. There are three basic designs as shown in **Figure 7-4** that can be found in brake systems. The piston-type pneumatic accumulator is probably the most popular design in use, with the bladder-type generally coming in second. A third type, the diaphragm-type is also found on some types of equipment. These accumulators all perform the same function: they store hydraulic energy that can be used by the brake circuit when required. The size of the accumulators depends on the design of the brake system; the larger the equipment, the larger the accumulator capacity.

Piston-type accumulators, as shown in **Figure 7-5** consist of a cylindrical tube that have a free-floating piston inside to keep the hydraulic oil and nitrogen separated. The diaphragm-type of accumulator uses a synthetic rubber diaphragm that separates the hydraulic oil from the nitrogen gas. The bladder-type of accumulator (**Figure 7-6**) has a metal shell that is fitted with synthetic flexible bladder. The bladder is pre-charged with nitrogen, and the metal shell holds the hydraulic oil. A poppet valve, located at the bottom hydraulic port, keeps bladder contained in the accumulator when the hydraulic pressure is discharged. The use of accumulators on a brake circuit allows the design engineers

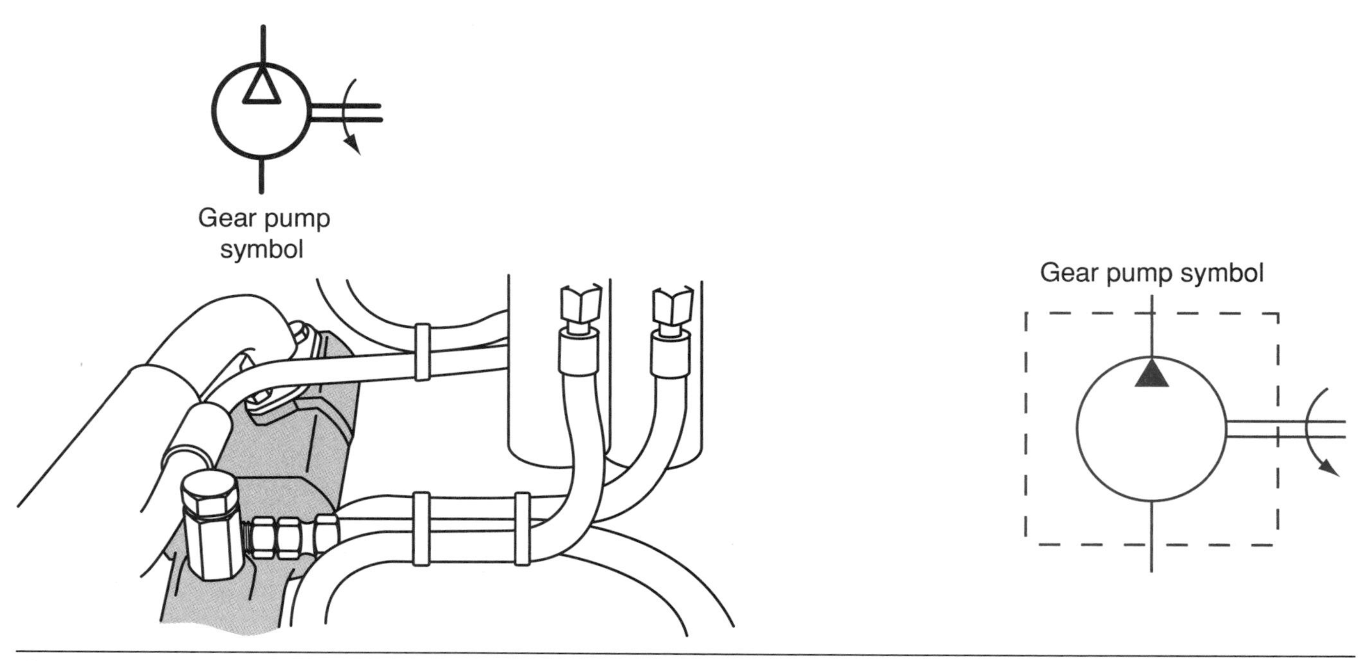

Figure 7-3 Schematic of a typical brake pump. It is mounted to the torque converter.

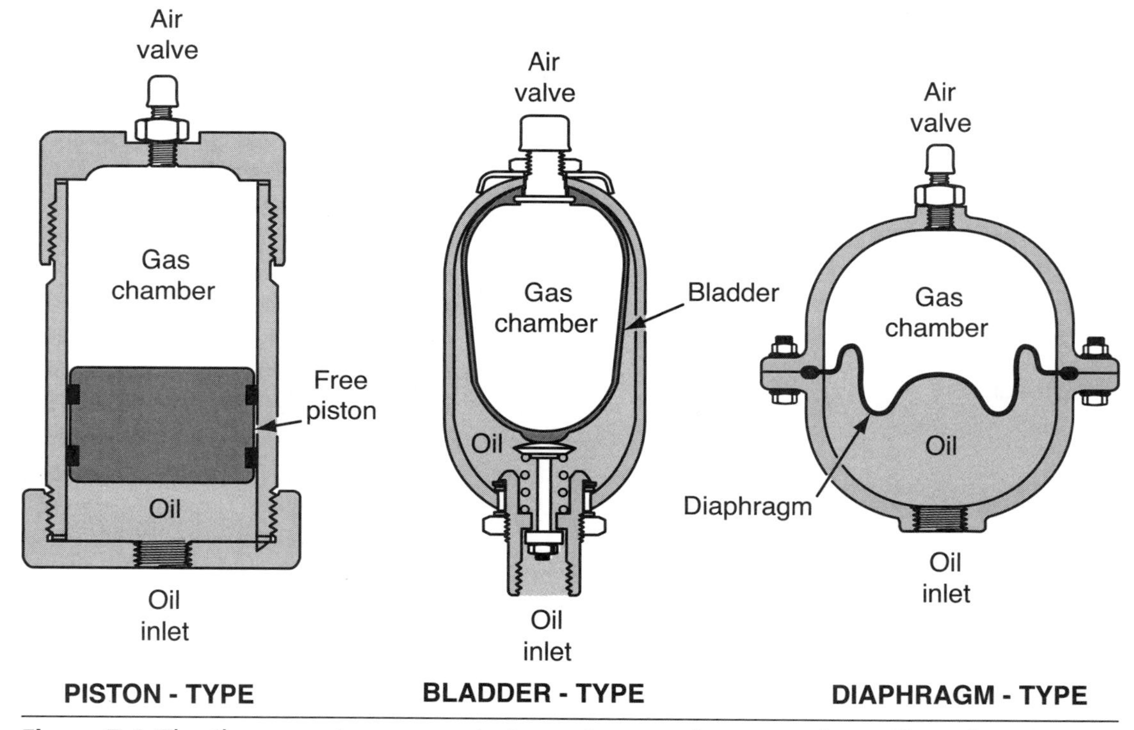

Figure 7-4 The three most common designs of accumulators used on off-road equipment brake circuits.

to use a much smaller hydraulic pump to keep the system charged, preventing excessive pump operation that requires additional horsepower to turn.

Only approved connectors are to be used to charge accumulators. Always refer to the manufacturer' s recommended tools and procedures before attempting to work on these components. Only dry nitrogen gas is allowed for use as a charging agent in an accumulator; under no circumstances should air or oxygen be used because of its explosive nature.

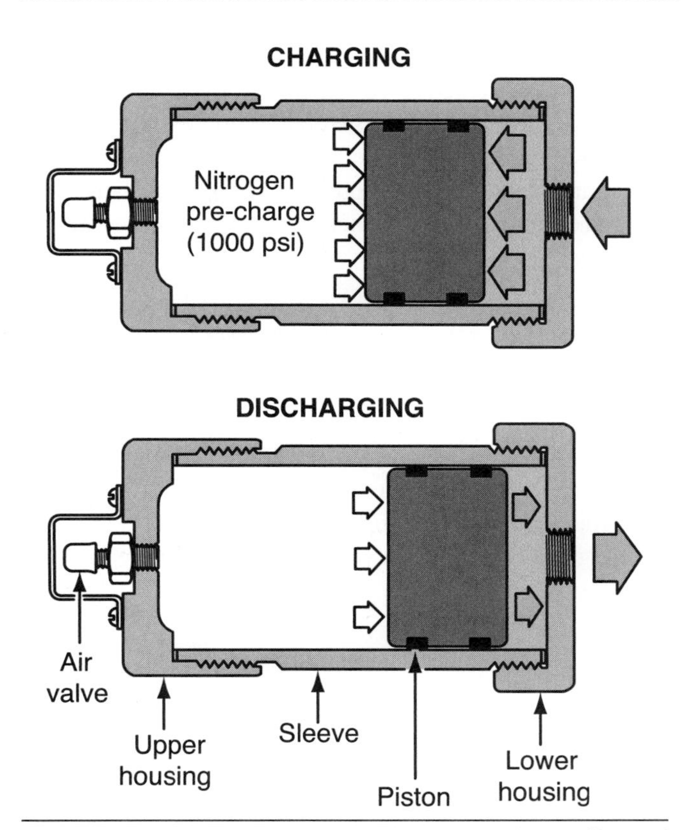

Figure 7-5 Piston-type accumulators used on off-road equipment.

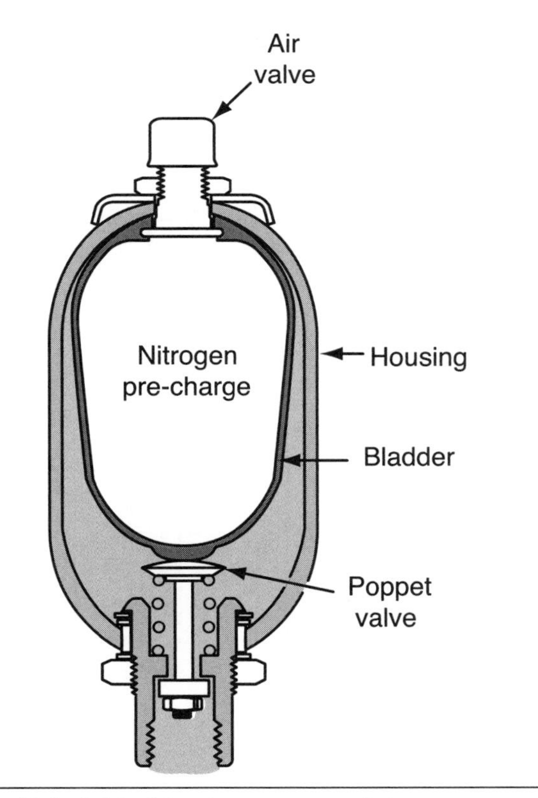

Figure 7-6 A bladder-type accumulator.

Tech Tip: The accumulator is a high-pressure vessel and must be treated accordingly. Under no circumstance should any welding or brazing be done on the accumulator casing. If physical damage to the casing is discovered it must not be used. If damage occurs during operation, the pressure must be bled off before the accumulator is replaced.

- Always read the manufacturer's documentation that comes with the new replacement accumulators. Some manufacturers ship the accumulators with a low pre-charge (10 bar/150 psi) to prevent damage during shipping.
- Under no circumstances should tire valves be used to replace the gas valve in an accumulator. Tire valves are not designed to withstand the high pressures used in an accumulator.

Piston-Type Accumulators

Piston-type accumulators have a free-floating piston that keeps the hydraulic oil and high-pressure nitrogen separated. The piston has a high-pressure seal that is replaceable. Bladder-type accumulators (Figure 7-6) have a flexible bladder that is not serviceable in the field. In most cases, when these accumulators develop a leak they are bled down and replaced. Accumulators (shown in Figure 7-4) must be checked for pre-charge on a regular basis usually during routine maintenance and servicing. Low accumulator pre-charge will cause excessive cycling of the charge valve, leading to high oil temperatures. Always refer to the manufacturer's service procedures before attempting to service a brake system that uses accumulators. The high pressures used in the accumulator can cause severe injury or death if not handled properly. Many manufacturers recommend that the accumulator pre-charge pressure be checked after approximately three months in service, then after six months of service, and then annually.

Figure 7-7, shows the location of the brake charge valve. The charge valve sends pressurized oil to the oil side of the accumulator. As the pressure rises in the oil side of the accumulator, it forces the piston to compress the 1000 psi (83 bar) pre-charge. As the piston retracts, the pressure in the pre-charge rises with the oil pressure and equalizes to the cut-out pressure. When the brakes are operated the pressure will drop with each brake application. The higher pressure above the piston on the gas side will force the piston down to try to maintain

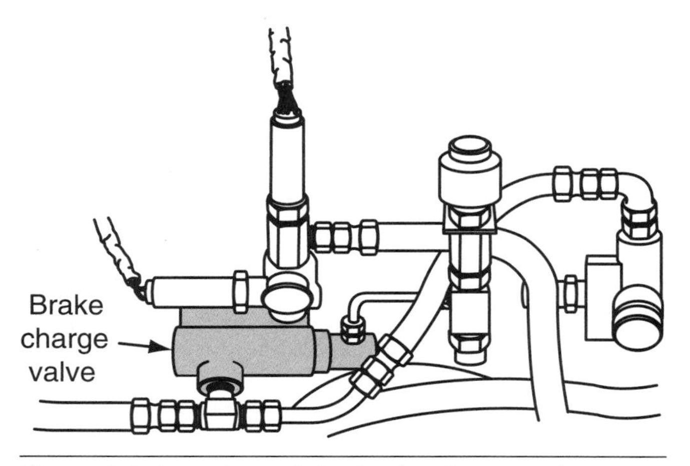

Figure 7-7 Location of the brake charge valve.

equal pressure on both sides. This continues until the cut-in pressure is reached at 1600 psi (110 bar). At this point the charge valve will start the charge process over again. The stored energy in the accumulator allows the operator to maintain an adequate oil pressure reserve to ensure that full brake release can be achieved at any time during operation.

Accumulator Charge Valve

The accumulator charge valve, shown in the schematic in **Figure 7-8,** is used to regulate oil flow from the brake pump and to charge the accumulators, which store the energy for use in the brake circuit during equipment operation. The charge valve controls the rate of charge as well as the cycle rate for the brake circuit.

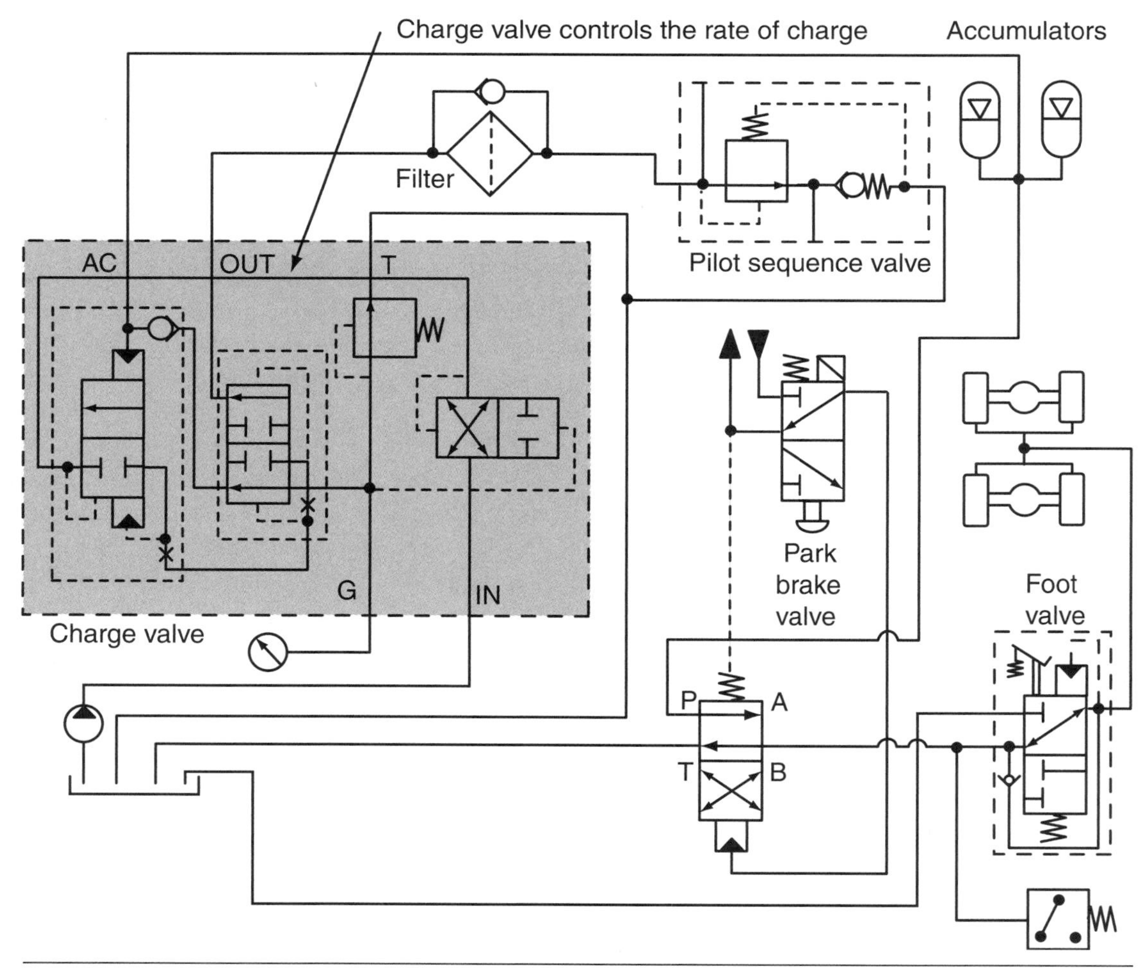

Figure 7-8 Typical location of a brake charge valve in the schematic.

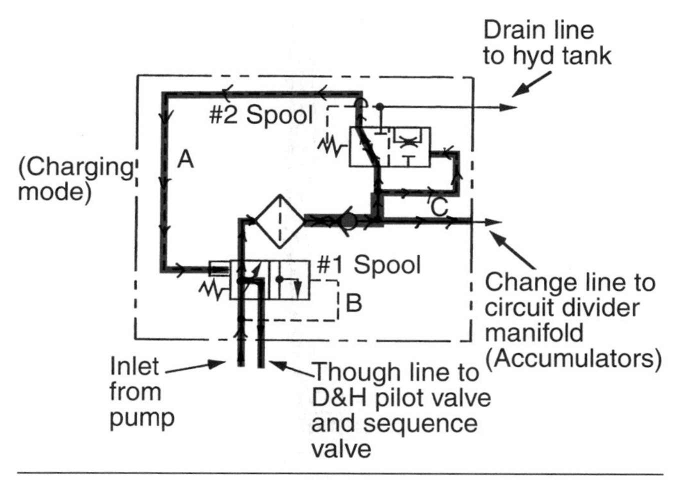

Figure 7-9 Schematic of a charge valve in the charge position.

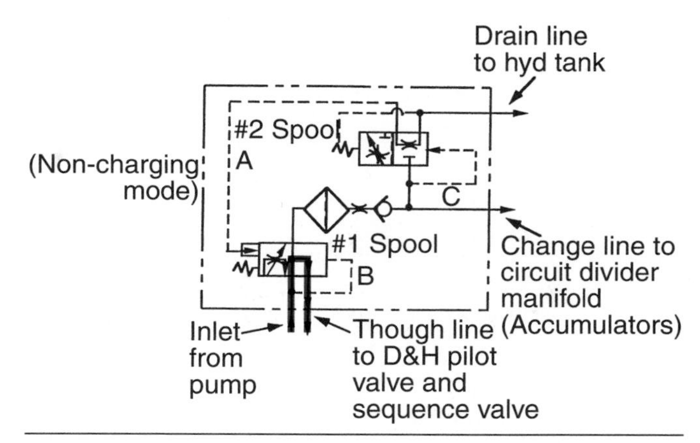

Figure 7-10 Schematic of a charge valve in the fully charged position.

In the fully charge position the oil from the brake pump flows to the dump/hoist and steering pilot circuit. During operation, as shown in **Figure 7-9,** when the pressure falls below a predetermined value (cut-in pressure), the charge valve diverts the oil flow to the accumulators until they are fully charged (cut-out pressure). The charge pressure values are not the same on every type of equipment, but they are generally similar. When the accumulators are fully charged, the oil is diverted to the **pilot pressure** valve and then to the brake cooling circuit.

In **Figure 7-10** the charge valve has completed charging the accumulators to 2000 psi (138 bar) cut-out pressure. The spring force on spool #2 has been overridden by the oil pressure in line C and shifts the spool so that the oil flow is blocked through line A. Oil pressure from the pump inlet through line B shifts spool #1 over to send all the oil through a non-restricted port to the dump/hoist/steering pilot circuit.

After a series of brake applications, the accumulator pressure is depleted to where the pressure falls below the cut-in pressure 1600 psi (110 bar) in our example. This shifts spool #1 over, which restricts the flow of oil to the dump/hoist pilot circuit. The oil from the brake pump now flows through an internal filter and unseats the internal **check valve** sending the oil flow to charge the **accumulators.** Spool #2 shifts to permit the oil to flow through line A, which helps keep spool #1 in the accumulator charge position. The charging continues until the pressure in the accumulators reaches cut-out pressure, 2000 psi in our example. Once the charge cut-out pressure is achieved, the pressure in line B shifts spool #1 over, sending the pump flow to the dump/hoist/steering pilot circuit.

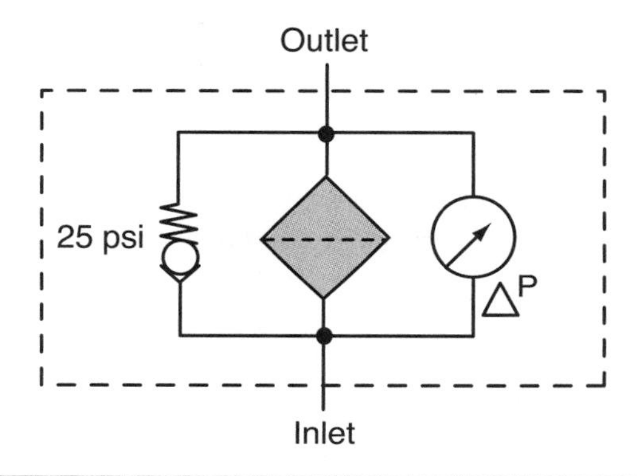

Figure 7-11 Schematic of an in-line hydraulic filter used in the brake circuit.

Hydraulic Return Filter

The hydraulic filter in the brake circuit provides partial flow filtration for the hydraulic oil. The filter, as shown in **Figure 7-11** has a 25 psi (1.7 bar) bypass relief valve built into the filter. The filter will go into bypass whenever there is a 25 psi difference in pressure across the filter. The filter is also equipped with a visual restrictor gauge to make it easy for the operator to check for dirty filter elements.

Pilot Pressure Sequence Valve

In our example in **Figure 7-12** the pilot relief valve is a non-adjustable full relief valve that is preset to maintain 200 psi (14 bar) for the dump/hoist/steering pilot circuit.

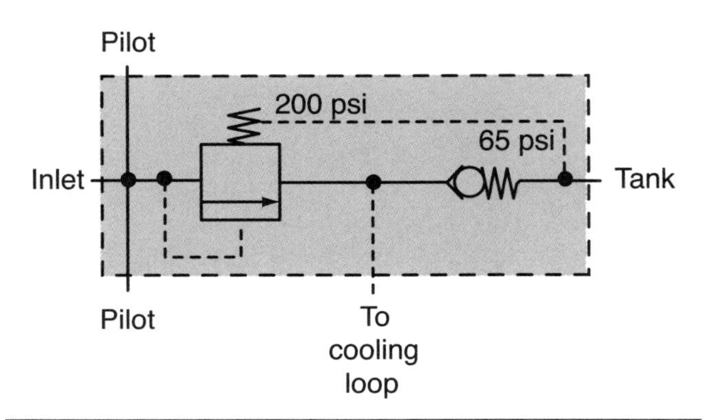

Figure 7-12 Schematic of a pilot pressure sequence valve.

The oil flow for this circuit comes from the accumulator charge valve and in through the inlet port. The inlet and pilot ports are connected together internally. The pilot sequence valve is a closed-center valve that regulates the pilot pressure to 200 psi. It is used to operate the dump/hoist/steering valves indirectly through a pilot circuit. There are no provisions made for adjustment to the pilot pressure in this valve. The excess oil left over from the pilot circuit flows to the brake cooling circuit. The brake cooling circuit only receives oil after the pilot circuit flow has been satisfied. An internal check valve ensures that the pressure in the brake cooling circuit never gets above 65 psi (4.5 bar). This is necessary to protect the oil cooler from excessive pressure. Any excess oil will open the internal check valve and send excess oil back to the hydraulic tank. This check valve has no provisions for adjustment or replacement. If the valve fails to deliver the correct pressure, the complete sequence valve must be replaced.

Accumulator Pressure Switch

The accumulator **pressure switch** in our example is a piston-type pressure switch. It has two sets of contacts that open and close simultaneously when the pressure changes in the brake circuit. In our example, the pressure switch is set to 1400 psi (196 bar). When the pressure falls below 1400 psi (196 bar), the normally open (NC) contacts open and the normally closed contacts (NC) close. The normally closed contacts are connected to an indicator light on the dash in the operator's compartment and alerts him or her of an impending emergency brake application. In **Figure 7-13** the normally open contacts control the power to the park brake solenoid. If power is interrupted to the park brake solenoid, the park brake remains in the applied position. When the accumulator pressure falls below 1400 psi (196 bar), the normally open contact interrupts the flow of power to the park brake solenoid causing the brakes to be applied.

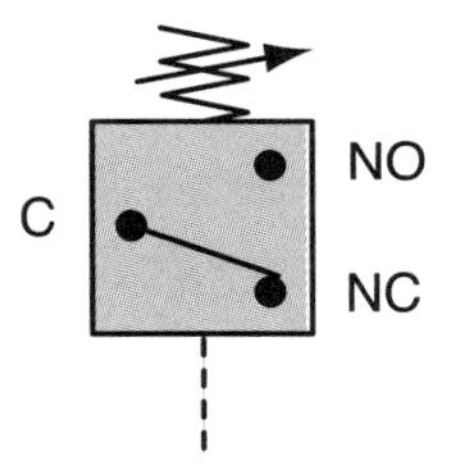

Figure 7-13 Schematic of an accumulator pressure switch.

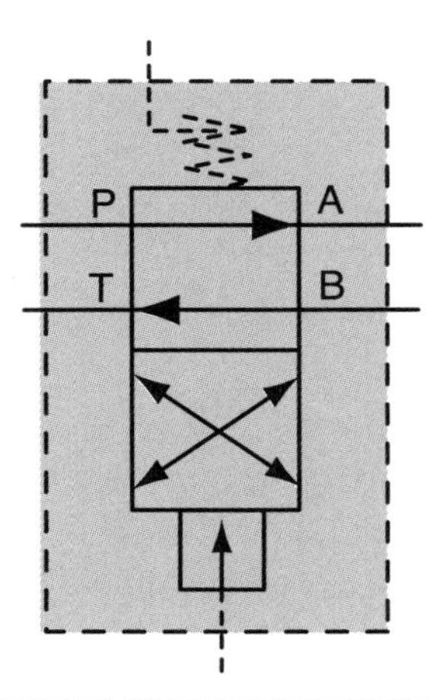

Figure 7-14 Schematic of a park brake selector valve for a spring-applied hydraulically-released brake system.

Park Brake Selector Valve

The park brake selector valve is a four-way two-position hydraulic valve that is pilot-operated by the park brake control valve. It is also used to apply the park brakes in the event of clutch pressure loss or if the park brake is applied by the operator. In **Figure 7-14** oil pressure is present at port P during the brakes applied mode, but it cannot flow through port A because it is deadheaded with a plug. When the operator activates the park brake control valve and releases the brakes, oil flows to the pilot-operated port on the selector valve. This pilot port is connected to the clutch pressure in our example, and if the engine is running and we have sufficient clutch pressure at the pilot port, the selector valve will release the brakes. This feature prevents the brakes from being released when there is no clutch pressure. The spring on the other side of the selector valve will not allow the selector valve to release the brakes.

If clutch pressure is present at the pilot port on the selector valve, the park brake control valve will remain in the released position, allowing oil to flow from port P to port B and through the foot-operated brake valve to the wheel ends to release the spring-applied brakes. If clutch pressure were to be lost while the equipment was operating, the spring in the end of the selector valve would shift the spool down and allow the oil to flow back to the tank from the wheel ends thus applying the spring brakes.

Foot-Operated Brake Control Valve

The brake control valve, as shown in **Figure 7-15** used on a spring-applied hydraulically-released brake system is a closed-center (closed to tank), open-to-pedal **reverse modulation** hydraulic brake valve. Unlike conventional brake systems that use hydraulic pressure to apply the brakes, spring-applied brakes require oil pressure to keep the brakes released, so the brake pedal in this application allows oil to flow to the wheel ends whenever the pedal is in the released position.

The foot-operated brake valve limits the brake **wheel end pressure** to 1500 psi (103 bar) in our example, while the brake **system pressure** will vary from 1600 (110 bar) to 2000 psi (138 bar). Once 1500 psi (103 bar) is achieved at the wheel ends, the pressure backs up through a pilot line to the top of the spool, forcing the spool by overcoming the spring on the bottom of the spool and shutting off oil flow to the wheel ends. If the pressure falls below 1500 psi (103 bar) the spring at the bottom of the **foot valve** pushes the spool up, allowing oil pressure to increase. By varying the spring tension with the foot pedal, reverse modulation is achieved because oil pressure will always try to balance the varying foot pedal spring tension. When the operator depresses the foot pedal, the spool is forced downward stopping the input flow and gradually allowing the wheel end oil to return to the tank. The more the operator depresses the pedal, the more pressure will be lost at the wheel ends. This applies the brakes further. Completely depressing the foot pedal to the floor releases all the wheel end oil pressure back to the tank and applies the spring brakes. The spring-applied brake pedal works in reverse when compared to the hydraulically-applied foot brake valve. The foot-operated brake pedal in Figure 7-15 has two separate brake circuits: one for each of the front and rear brakes. The previous explanation was given for one side of the brake pedal. The system works exactly the same way for the other side and is controlled by the same foot pedal.

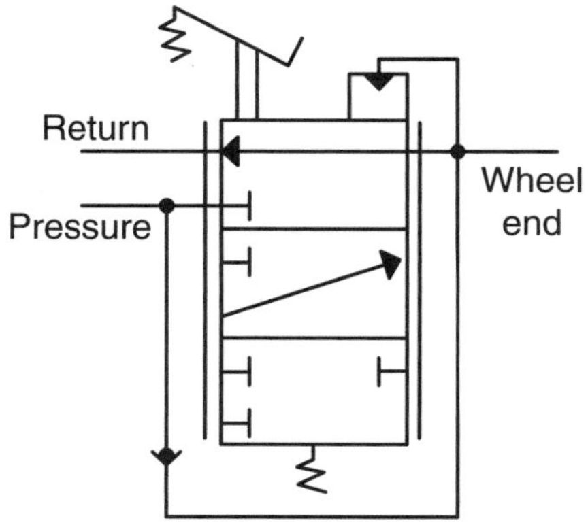

Figure 7-15 Schematic of a spring-applied brake foot control valve.

Park Brake Solenoid

An electrically controlled park brake solenoid is used to apply the park brakes in this example. The operator applies the park brakes by de-energizing the solenoid through a park brake button on the dash of the equipment. In **Figure 7-16,** when the solenoid is de-energized the spool shifts, allowing oil to flow from port P to port A and from port B to port T. When the operator energizes the solenoid, the brakes are released, allowing the oil to flow from port B to port T and from port A to port T.

Driveline Park Brake

Off-road equipment often incorporates a separate mechanical park brake, shown in **Figure 7-17,** that is primarily designed for parking the equipment, although on some equipment it is part of the emergency brake circuit. These units are often found mounted to a driveline on the equipment. Park brakes on off-road equipment are always mechanically applied and hydraulically released. This makes the use of the park brake extremely safe since no hydraulic pressure is required to park the equipment.

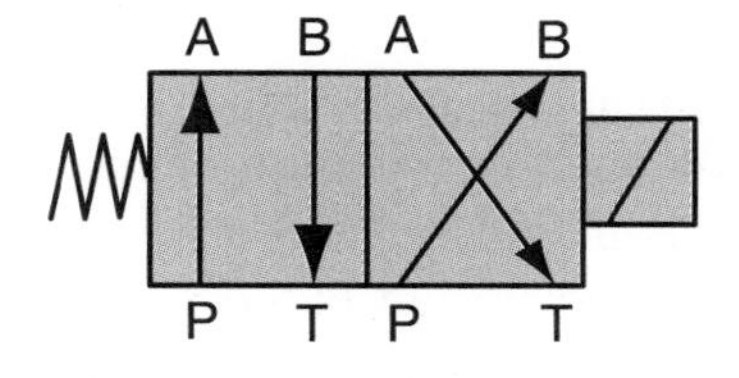

Figure 7-16 Schematic of a park brake solenoid.

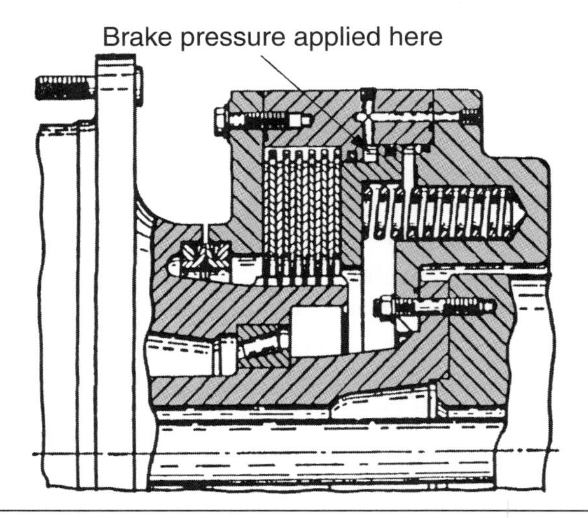

Figure 7-17 Cutaway of a spring-applied hydraulically-released brake system used on an excavator final drive.

The example that will be explained in this section is no exception to this rule; it is mounted to the front driveline of the equipment. Two sets of discs are used in this arrangement. The friction discs are splined to the drive shaft and rotate with it. The stationary discs are splined to a housing that contains the rest of the park brake components. Large, heavy internal springs are used to squeeze the two sets of discs together; these springs can develop clamping pressures that can exceed 35,000 lbs. per square inch. When the park brake is engaged, the stationary **steel discs** prevent the friction discs from rotating. With the equipment running, the park brakes are released by hydraulic oil that is directed by the park brake solenoid to the back of an internal piston in the brake assembly located in the housing. The piston compresses the brake application springs, allowing the friction discs to rotate with the driveline. Due to the design of driveline brake assemblies, it does not require full system brake pressure to release the park brakes; a hydraulic pressure-regulating valve is generally used to adjust the release pressure to suit the application.

Hydraulic Oil Cooler

Hydraulic oil coolers are necessary on off-road equipment applications so that the oil temperature be maintained at a safe temperature. In many severe-duty applications, the oil temperature can exceed safe operating temperatures. There are two designs of oil coolers that are generally used for this purpose: the oil-to-air and the liquid-to-liquid designs. Regardless of the type used, cleanliness is a must; and the heat exchangers cooling fins must be kept clean. On newer equipment the oil temperature is monitored and the operator is alerted that the oil temperature is too high; on older equipment it is often up to the operator to spot problems with the oil temperature. The most effective way to clean the oil cooler is to use a high-pressure wash or steam. A good cleansing agent should be sprayed onto the cooler fins and allowed to soak before pressure washing. After a thorough cleaning the coolers can be dried with compressed air. Drying out the coolers is necessary because dirt will stick to the wet surface of the fins much quicker than when the fins have been properly dried. The pressure that flows through the brake cooling circuit, as shown in **Figure 7-18,** is regulated by an additional check valve that limits the pressure in the cooling circuit.

Spring-Applied Hydraulically-Released Wet Disc Brake

The spring-applied hydraulically-released brake system is used by many manufacturers of off-road equipment. Although the brake system uses hydraulic components and pressures that are similar to hydraulically-applied wet

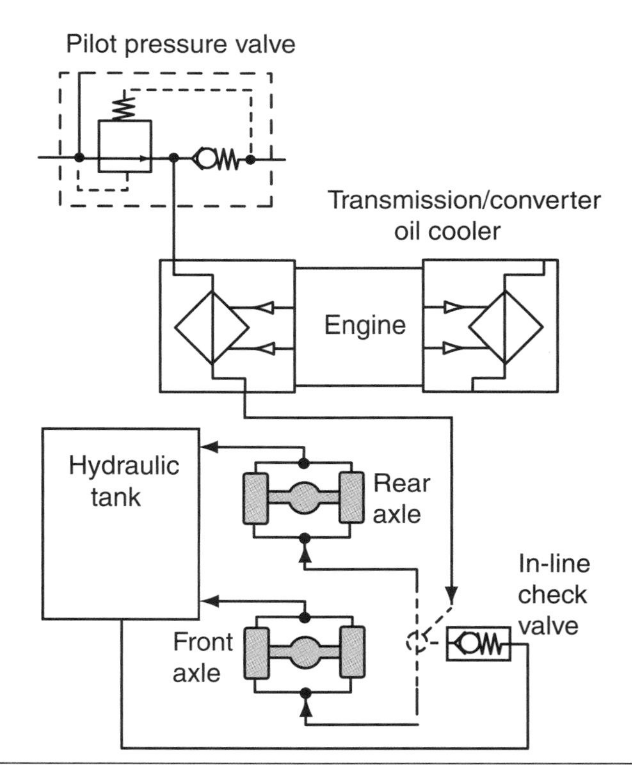

Figure 7-18 Schematic of a typical brake cooling circuit used on a spring-applied hydraulically-released brake.

disc brakes, the system differs in one significant way: the brake pressures are reverse modulated. In other words, the pressures are reduced in the brake circuit when the brakes are applied. Conventional hydraulically-applied service brakes operate by increasing the pressure in the wheel ends to apply the brakes. The brake components used in the spring-applied service brake system look similar to the components that are used in conventional hydraulically-applied service brake systems, but differ in the way they actually function.

Conventional brake valves let pressures into the brake circuit when the brakes are applied; spring-applied service brakes let pressure out of the brake circuit when the brakes are applied. When checking brake pressures in the spring-applied hydraulically-released service brake circuit, you must remember that the pressure is gradually reduced when the brakes are applied; and when fully applied, there is no brake pressure in the wheel ends—the friction discs are clamped to the stationary steel discs by mechanical spring pressure. The spring pressures in this type of brake system can exceed 40,000 lbs. per wheel end. Before attempting to service a spring-applied brake system, read over the disassembly procedure to ensure that the work will be performed correctly. Failure to follow these manufacturer's recommended procedures can result in serious injury or death.

In **Figure 7-19** the brakes are shown in the applied mode. When sufficient brake pressure is sent into the cavity between the piston and the housing, the pressure will push back against the brake application springs until the friction discs are free to rotate with the axle. The brake system is designed to monitor the brake pressure so that in the event of low brake pressure, the operator will see an indicator light up on the dashboard warning him of an impending emergency brake application. The brakes are designed to go on automatically in the event that the pressure goes below a safe, predetermined value.

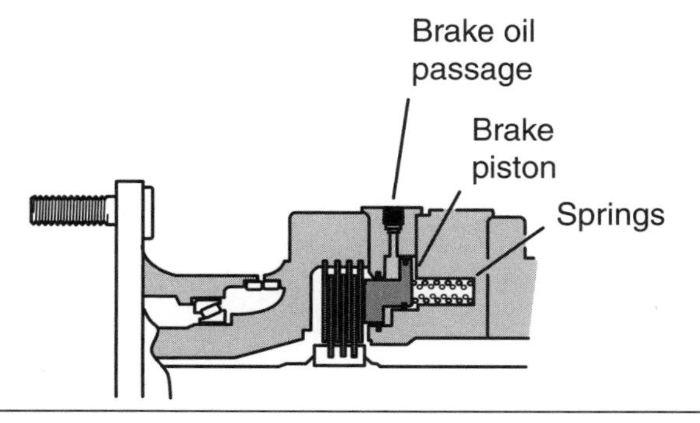

Figure 7-19 Cutaway of a typical spring-applied wet-disk brake.

This type of brake system is inherently much safer that the conventional hydraulically-applied wet disc brake system because each wheel end brake on a spring-applied system can act independently of any other. If a problem occurs in any of the four wheel ends that results in a brake pressure loss, the springs in that particular wheel end can apply the brakes. There are anywhere from 15 to 40 brake application springs in each wheel end brake assembly, so it would be highly unlikely to have a brake application failure on equipment with this type of system. Even if a few brake application springs were to break in a wheel end there would be no adverse effect on the stopping ability of the equipment.

SPRING-APPLIED BRAKE SYSTEM OPERATION

A spring-applied brake system, as shown in **Figure 7-20** operates in reverse when compared to conventional brakes. Each wheel end brake is capable of exerting as much as 80,000 lbs. of force from the spring pressure in the wheel ends. The pressure required to fully release the spring brakes completely is generally in the neighborhood of 1500 psi (103 bar). Figure 7-20 shows the general layout of the brake circuit by using a combination of schematic and pictorial views.

With the equipment running, the pump forces oil to flow to the accumulator charge valve where it is directed to the two accumulators until they are fully charged. Once the accumulators are fully charged, the charge valve redirects the oil to a filter before entering the brake pilot sequence valve. The pressure-reducing sequence valve allows 200 psi (14 bar) to flow to the dump/hoist/steering pilot circuit. The remainder of the oil is sent to the brake cooling circuit. When the operator activates the park brake control valve, oil flows to the park brake selector valve, allowing oil from the accumulators to flow to the brake control valve. The oil flows through the brake valve when the pedal is in the released position and then on to the wheel ends where it forces the hydraulic brake piston to retract the brake application springs thereby releasing the service brakes.

To activate the service brakes, the operator steps on the brake pedal. This gradually reduces the brake pressure in the wheel ends and applies the brakes. The return oil from the foot valve is redirected back to the tank. The more force the operator applies to the brake pedal, the greater the pressure drop in the service brake circuit. Unlike a conventional brake pedal, the spring-applied brake pedal operates in reverse. It increases the pressure

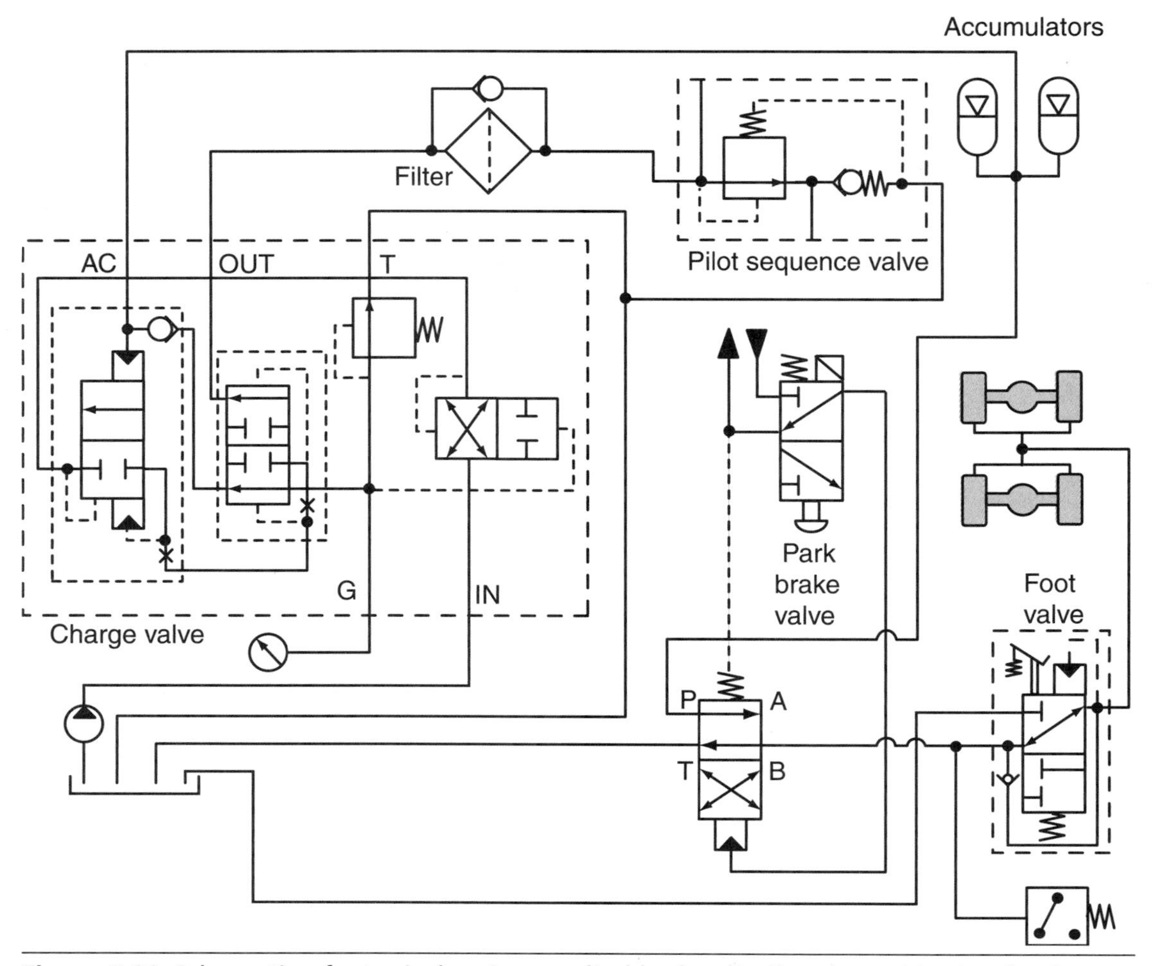

Figure 7-20 Schematic of a typical spring-applied hydraulically-released brake circuit.

to the wheel end brakes when it releases the brakes and decreases the pressure when it applies the brakes.

To park the equipment, the operator must activate the park brake control valve (**Figure 7-21**) that is located in the operator's compartment. This action stops the oil flow to the park brake selector valve. The spool in the park brake selector valve shifts and allows the wheel end oil pressure to return to the tank through the foot valve. This blocks the oil flow from the accumulators and prevents it from flowing to the wheel end brakes. The brake application springs in the wheel ends can now apply the wheel end brakes.

To release the park brake, the operator activates the **park brake valve** which redirects oil to the park brake selector valve. This shifts the spool in the selector valve and allows oil from the accumulators to flow through the foot valve to the wheel end, releasing the brakes. In the event that brake pressure was lost for any reason, the worst thing that would happen is that the brakes would apply. To prevent the operator from operating the equipment with low brake pressure, which could cause the brakes to drag, a brake pressure switch monitors accumulator brake pressure. If brake pressures fall below a predetermined amount, a light illuminates on the dashboard in the operator's compartment.

Brake Cooling Circuit

In **Figure 7-22** the oil flow comes in from the charge valve "out" port and goes through a filter before entering the pilot pressure sequence valve. The pilot pressure sequence valve is considered a full flow relief valve that will maintain a predetermined amount of pressure for the pilot circuit. In our example, this pressure is 200 psi (14 bar). This pressure is generally not field-adjustable. Once pilot pressure is obtained, excess oil flows past a check valve and limits the pressure in the cooling circuit in our example to 65 psi (4.5 bar). The lower pressure is necessary to protect the cooler from excessive pressure. The cooler lowers the temperature of the cooling

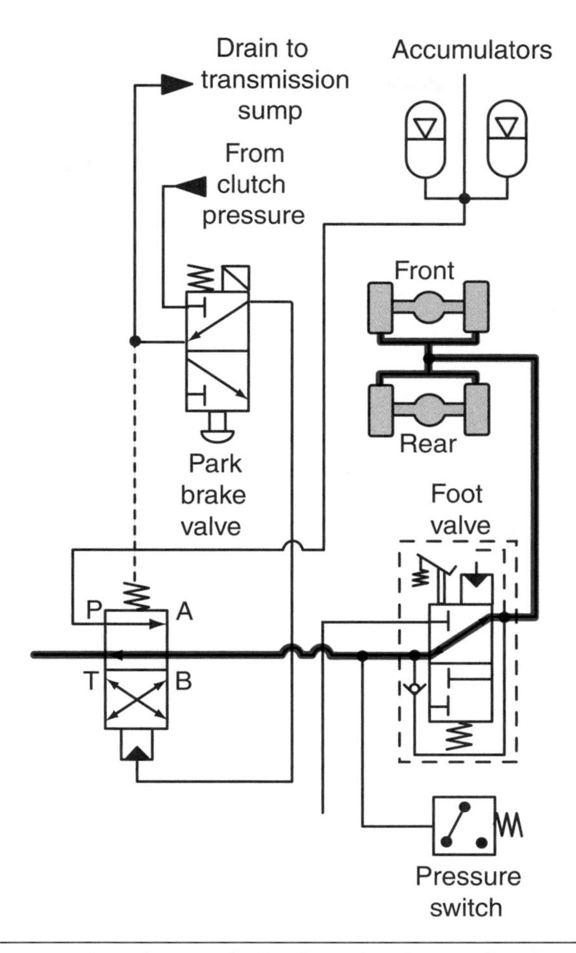

Figure 7-21 The park brake circuit in the brakes applied position.

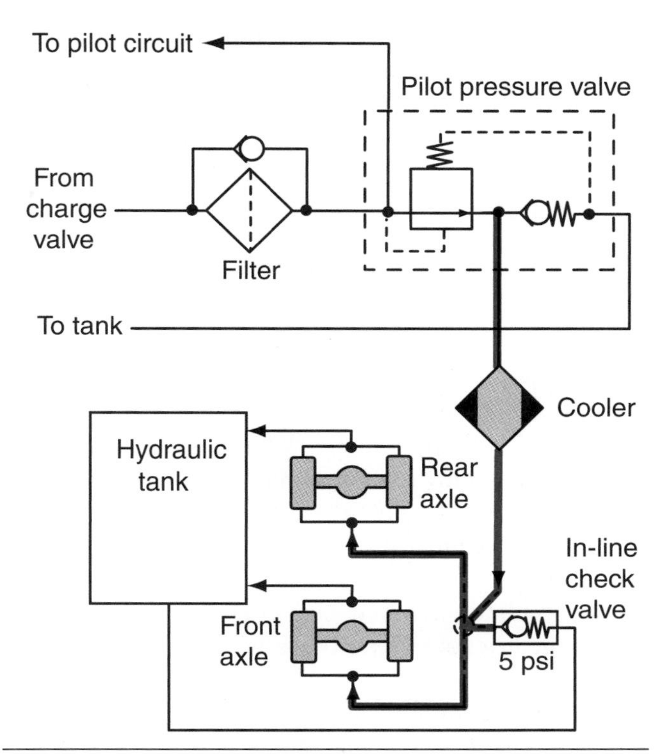

Figure 7-22 A brake cooling circuit from a spring-applied hydraulically-released brake circuit.

oil before it enters the wheel ends. The oil then flows to the wheel end cooling circuit where another check valve, rated at 5 psi (.3 bar) in our example, prevents pressures higher than the check valve rating on the downstream side of the wheel end. The cooling oil flows around the inside of the brake discs compartment absorbing heat from the discs before returning to the tank. Overpressurization of the wheel end cooling circuit may result in a blown wheel end seal. Some manufactures route the oil flow through the park brake circuit to dissipate extra heat before returning it back to the tank.

MAINTENANCE

The cleanliness of the hydraulic system cannot be overemphasized. To get a long trouble-free life out of the brake components, it is very important to maintain the hydraulic system according to the manufacturer's recommendations. Excessive heat is the enemy in all cases when it comes to hydraulic circuits. Regular filter change intervals should be adhered to and followed up with periodic oil analysis. Following the manufacturer's recommended maintenance procedures will reduce the number and severity of common problems that develop over time because of poor maintenance practices. By following the prescribed maintenance practices, small problems can be detected and corrected before a major break down occurs.

One of the most common problems encountered in equipment hydraulics is maintaining the correct level of oil in the tank. Plugged or dirty hydraulic filters, both on the suction side and the pressure side, must be cleaned and replaced as prescribed. Oil aeration is another big problem; loose fitting hydraulic suction lines or O-rings that do not seal properly on suction lines can create unnecessary suction leaks that introduce air into the oil and can lead to excessive cavitation of the oil. The misuse of the correct type and grade of oil is another area of neglect. Operators will sometimes dump whatever is handy into the hydraulic tank, thinking oil is oil. As simple as these problems sound, they are in fact the biggest cause of hydraulic system problems.

Problems usually start when the system is opened up for servicing or repairs. It is extremely difficult to maintain a clean environment in the field. Cleanliness should always be on a technician's mind as the single most important issue. Keeping the dirt and contamination out of the hydraulic system is essential; even the small particles that are hard to see with the naked eye can score valves and plug orifices. Using clean containers to transport oil to the job site is of the utmost importance. When performing work on a job site you should always clean up the work area before opening a hydraulic system. This will minimize the chance of contamination. When changing or adding oil, make sure the area around the tank opening is clean and the container that is being used does not have any contamination on the top that could fall into the tank.

Accumulator maintenance often requires that the pre-charge be checked periodically. Most accumulators are serviceable in the field. Always read over the manufacturer's recommended service procedures before attempting to service an accumulator. In general when checking the pre-charge, if you observe that the pressure is higher on the gas side than it should be, then it is likely that the seals on the piston are not sealing effectively and oil will be present in the gas side. If the pressure on the gas side is lower than normal, check for an external leak at the gas charge fitting or cylinder end seals.

Tech Tip: High-pressure oil can remain in a hydraulic brake circuit for a long period of time after the equipment is parked. Failure to properly release any stored pressure can result in serious injury or death if the manufacturer's recommended procedures are not followed. Never use your hands, even if you are wearing gloves, to detect any hydraulic leaks in a hydraulic circuit. Use a cardboard for this purpose. High-pressure oil can penetrate the skin, causing severe injury. Always seek medical attention immediately if this type of injury is suspected. When checking nitrogen pre-charge in an accumulator, always follow the manufacturer's recommended procedures for the system that you are servicing. There are significant differences in the way the systems behave when checking accumulator pre-charge, not following recommended procedures may result in serious injury or even death.

Tech Tip: The only gas that should be used to charge accumulators is dry nitrogen gas. No other compressed gases are permitted for this procedure; other types of gases, even compressed air, can lead to explosions and should never be used. Be sure that the nitrogen bottles have the correct connections to match the hoses and gauges that are going to be used for this procedure. If there is any visible physical damage to the accumulator, it must be replaced immediately.

CAUTION *Before attempting to replace the accumulator, make sure to discharge all pressure in the accumulator following the manufacturer's recommended procedures.*

TESTING AND ADJUSTING

Testing and Recharging

This procedure is only an example of the steps that should be taken to check and charge a bladder-type brake accumulator. Always refer to the equipment service manual for the specific procedures that are to be used on the equipment that you are servicing. Not all procedures are identical and vary from accumulator to accumulator. They may even be different on the same type of equipment that was produced in a different location or time. The equipment must be parked on level ground and have approved wheel chokes installed on both sides of the tires. Be sure that any implements attached to the equipment are lowered to the ground. The steering frame lock must be installed along with any appropriate lockouts or tags that are required. If any steering lock mechanisms, such as transmission locks, are used on the equipment, ensure that they are engaged.

To check the nitrogen pre-charge in the accumulator(s), all the oil pressure must be dissipated in the brake circuit and the piston must bottom out at the lower end cap. Once you are confident that the fluid side of the accumulator is free from pressure, you are ready to begin.

Procedure

- Before attempting to remove the protective guard from the top of the accumulator, that protects the gas valve (**Figure 7-23**), clean the surrounding area with compressed air.

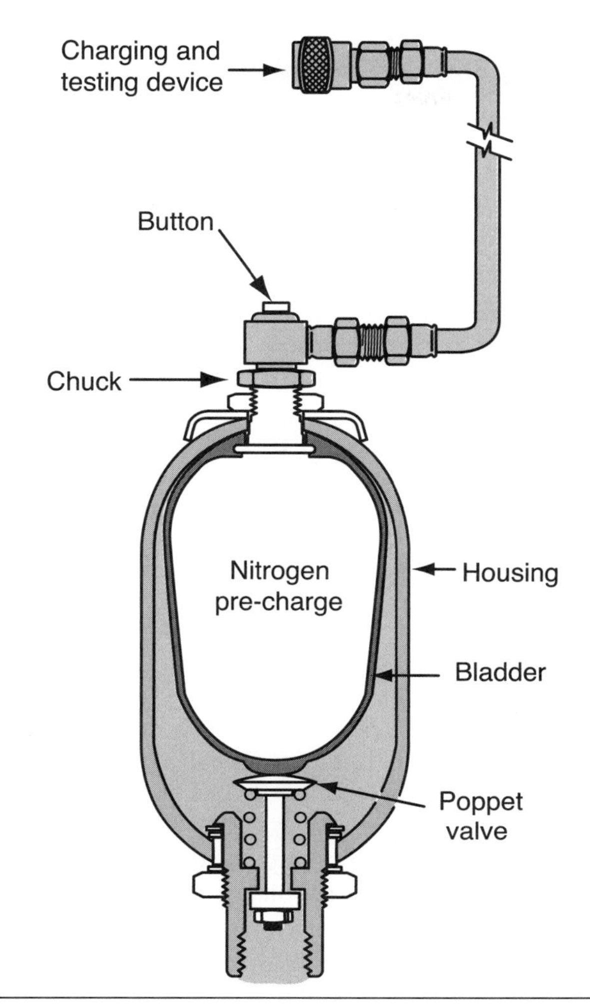

Figure 7-23 A typical bladder-type accumulator along with the connectors and hardware used to charge it.

- Make sure the appropriate connector and test gear are selected before attempting to connect the gas valve on the accumulator. *Not all connectors have the same threads.* Once the connection is tightened and checked for leaks, open the appropriate test valve and record the pressure on the gauge. This reading will be the actual pre-charge in the accumulator.
- As in all cases when performing this type of service work, refer to the manufacturer's specifications for the correct pre-charge. *Always refer to the manufacturers temperature correction chart, it may be necessary to compensate for temperature differences.* Checking and pre-charging the accumulator in cold or hot ambient temperatures without compensating for temperature difference will result in either overcharging or undercharging the accumulators. Once the pre-charge charge levels are correct, remove the test gear and replace the protective cap.
- Connect the nitrogen bottle to the adapter hose with the appropriate gauge. Now close the bleeder valve.
- Next, open the shut-off valve on the nitrogen bottle and slowly pre-charge the accumulator. Monitor the charge pressure carefully while performing this step. *When charging bladder or diaphragm accumulators, never allow the pre-charge to exceed specifications as this could lead to bladder or diaphragm damage.*
- If the pre-charge is too high, you can release some of it by carefully opening the bleeder valve. Pre-charge pressure can change significantly because of temperature differences. Always allow the gas temperature to stabilize before rechecking the pre-charge and removing the adapter and nitrogen bottle. See **Table 7-1** for examples of the pressure/temperature relationship.
- Reinstall the protective cap and check for leaks by applying a soap and water mixture. Any leaks will form small bubbles around the leak area. Leaks must be corrected before the equipment is returned to service.

Brake Performance Test

Performance brake tests must be performed anytime the brakes are serviced. Brake tests must be performed in a suitable area that will not compromise the safety of anyone in the immediate area. Ensure that there is adequate room, both front and rear, to allow for any unintended equipment movement. The equipment must be tested on level ground. This procedure, as with any drivability testing, should only be attempted by qualified personal. The equipment must be brought up to operating temperature with the brake system fully charged.

Procedure

- Release all brakes before beginning any tests. To test the service brakes, depress the brake pedal fully and observe that the brake gauges are showing the correct pressures. Refer to the equipment specifications in the shop manual for the correct values.
- The equipment must be placed in the appropriate gear for this test. Refer to the equipments technical documentation for the correct gear selection and

Table 7-1: PRESSURE/TEMPERATURE RELATIONSHIP

Accumulator Pre-charge Pressure/Temperature Relationship for 5500 kPa (800 psi)	
–7 degrees C (20 degrees F)	5000 kPa (725 psi)
–1 degrees C (30 degrees F)	5105 kPa (740 psi)
4 degrees C (40 degrees F)	5235 kPa (760 psi)
10 degrees C (50 degrees F)	5340 kPa (775 psi)
16 degrees C (60 degrees F)	5420 kPa (785 psi)
21 degrees C (70 degrees F)	5500 kPa (800 psi)
27 degrees C (80 degrees F)	5605 kPa (815 psi)
32 degrees C (90 degrees F)	5735 kPa (830 psi)
38 degrees C (100 degrees F)	5840 kPa (845 psi)
43 degrees C (110 degrees F)	5920 kPa (860 psi)
49 degrees C (120 degrees F)	6000 kPa (875 psi)

rpm range. Not all brake tests on equipment are checked the same way.

- With the service brakes applied, depress the accelerator until the test rpm is reached. Observe and record the rpm in gear. The equipment should not move.

CAUTION *Do not keep the equipment in this operating condition for more than 20 seconds at a time.*

- On equipment that incorporates an emergency park brake circuit, locate the test switch on the dashboard in the operator's compartment that will be used for this procedure. This park brake test switch test drains the hydraulic brake fluid from the spring-applied park brake circuit while the emergency brakes are in the released position. This provision allows the operator to isolate the park brake circuit so it can be checked separately.
- Place the equipment in the correct gear for this test and slowly increase the test rpm until the desired rpm is reached. Refer to the manufacturer's technical documentation for the specifications. Do not allow the converter to stall for more than 20 seconds at a time without a cool down period.
- The equipment should not move during any of the brake tests. If the equipment does not pass and of the appropriate tests it must be repaired before being placed back into service.

Accumulator Charge Valve Test Procedure

This procedure is to be used as an example of procedure that must be followed to check accumulator charge valve operation. Always refer to the equipment shop manual for the specific procedures that are to be used on the equipment that you are testing. Not all procedures are the same and may vary from equipment to equipment; in some cases they may even be different for the same model of equipment that was manufactured at a later or earlier date.

As with any brake tests, the equipment must be parked on level ground and have approved wheel chokes installed on both sides of the tires. Any implements attached to the equipment must be lowered to the ground. The steering frame lock must be installed along with any lockouts or tags that are required. If any steering lock mechanisms, such as transmission locks, are present on the equipment, ensure that you engage them.

Procedure

- Ensure that the engine is shut down for this next step. Continually depress the service brake pedal until all hydraulic pressure is depleted from the brake accumulators. The number of times that you depress the brake pedal will vary from equipment to equipment.
- Connect a pressure gauge (0 to 25000 kPa or 0 to 3600 psi) to an appropriate location in the charge valve circuit. The connection point for the gauge will vary; always refer to the equipment service manual for the correct location.

- Observe and record the pressure reading before starting the engine. The pressure should be 0 psi. If there is still pressure showing on the gauge, repeat the bleed down procedure again.
- Start the engine and allow the charge pressure to build up until it reaches cut-out pressure. Once maximum charge pressure is showing on the test gauge, shut down the engine. Observe and record the values shown on the test gauge for comparison later. Compare them to the manufacturer's specifications in the shop manual.
- The next step requires that the key to be in the off position. Apply the service brakes repeatedly and note the number of times you apply the brakes. Each time the brake pedal is depressed, the pressure on the test gauge will drop slightly.
- As soon as the gauge reaches the low accumulator oil pressure level, the accumulator oil pressure indicator light on the dashboard should come on. If this occurs, then the accumulator low oil pressure warning system is working correctly. Refer to the equipment service manual for the correct low accumulator pressure value.
- Apply the service brakes continually while counting the brake applications and observing the test gauge pressure. At one point the pressure will drop off rapidly; this is an indication of how much nitrogen pre-charge pressure is actually in the accumulator.
- Calculate the number of brake applications it took to get to the point where the gauge dropped off rapidly. Check the manufacturer's service manual to determine the correct pre-charge value.
- Start the engine and run the equipment until the brake cut-out pressure is reached. If the cut-out pressure reading is not correct, refer to the manufacturer's shop manual for the adjustment procedures. Adjustment procedures will vary from equipment to equipment. Not all charge valves have a cut-out pressure adjustment that can be readily performed in the field. It may be necessary to replace the charge valve in some cases.

Brake Pump Flow Test

To test the flow capabilities of brake pump, a standard hydraulic pump test is used to verify the pump's condition and its ability to pump oil under pressure. This pump flow test is conducted at two different pump speeds to access the pump's condition. The difference is calculated between the pump flows at two different operating pressures and is referred to as flow loss. Hydraulic pumps lose efficiency when they wear internally, and a large pumping loss will be present if the pump has excessive wear. The specifications used in this example (**Tables 7-2** and **7-3**) are for reference purposes only and do not represent actual flows for any specific pump. Refer to the manufacturer's technical documentation for the equipment being tested.

To perform this particular test on the equipment, a suitable flow meter must be installed to the pump outlet to measure the pump output. An example of actual flow loss is shown in **Table 7-4.**

Table 7–2: DETERMINING PUMP FLOW LOSS

Example of Determining Pump Flow Loss
Pump Flow at 690 KPa (100 psi)
Pump Flow at 6900 KPa (1000 psi)
Flow Loss
Example of Determining Flow Loss
217.6 L/min (57.5 US gpm)
196.8 L/min 52 US gpm)
Loss = 20.8 L/min (5.5 US gpm)

Table 7-3: DETERMINING PUMP FLOW LOSS

Example of determining pump flow loss		
Loss (L/min or US gpm)	× 100	= Percent of Flow Loss
Pump Flow at 690 kPa (100 psi)		
Example of determining pump flow loss		
(20.8L (5.5 gpm))	× 100	= 9.5 %
(217.6 L/min (57.5 US gpm))		

Table 7-4: ACTUAL PUMP FLOW LOSS

Example of Calculations Used to Determine Actual Pump Flow Loss		
Pump Flow at 6900 kPa (1000 psi)	× 100	= Percent of Flow Loss
Pump Flow at 690 kPa (100 psi)		

Checking Internal Friction Disc Wear

This test is generally performed after the equipment fails a brake test. One of the wear conditions that will lead to the equipment inability to pass a brake test is excessive wear to the friction disc lining. On some spring-applied brake designs, it is almost impossible to accurately determine the friction disc wear without dismantling the wheel ends. On brake systems that have no access ports to the wheel end disc chambers that provide a place to check wear easily, you must eliminate all other possibilities that could lead to a failed brake test first, and then be prepared to dismantle the wheel ends.

Fortunately, most models of equipment have an access port as shown in **Figure 7-24** in the wheel end that allows the friction discs to be checked for wear. The procedures explained in this chapter are shown as examples only. To perform these checks on equipment, always refer to the equipment shop manual test procedure for the equipment in question. The procedures vary somewhat from equipment to equipment.

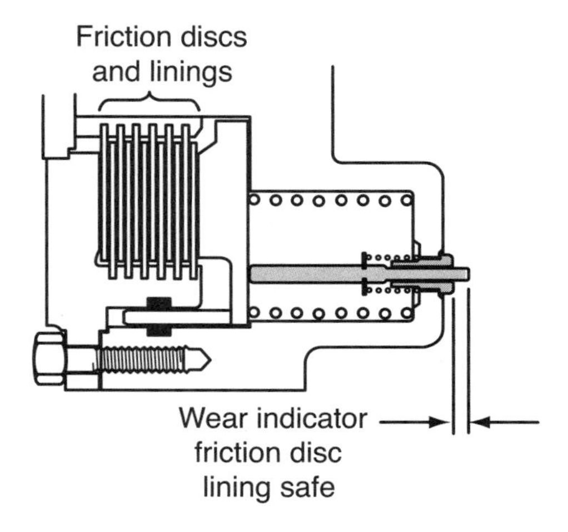

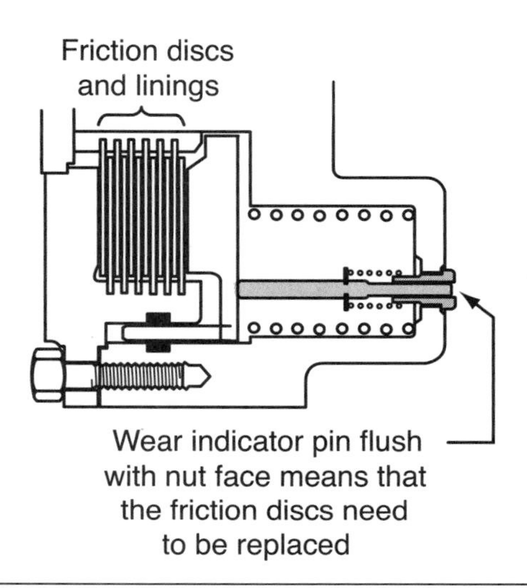

Figure 7-24 Location of the inspection port used to perform a disc brake wear check on a spring-applied hydraulically-released brake system.

CAUTION *Before performing any checks or adjustments to the brake system, the equipment must be parked on level ground and have approved wheel chokes installed on both sides of the tires. Any implements attached to the equipment, such as a bucket, must be lowered to the ground. The steering frame lock must be installed along with any lockouts or tags that are required. If any steering lock mechanisms, such as a transmission lock, are present on the equipment, ensure that they are engaged.*

The following checks can only be performed correctly while the equipment is shut down with all power turned off and with the brakes in the applied position. Since the brakes are spring-applied, they will automatically be in the applied position when the engine is powered down. When performing these tests be sure to provide a container to capture any fluids that might be lost during any adjustments to the brake system. Follow approved fluid disposal procedures according to local regulations and employer policies.

Visual Inspection. Check over the brake system for oil leaks that can cause brake system performance problems and will inevitably lead to brake component wear. Inspect the axle housing breather plugs; these can lead to a build up of pressure in the axle housing, which often leads to oil seal leakage. During normal operation, the friction discs will wear and cause the oil to become contaminated with friction lining material. On some brake systems that do not use an auxiliary cooling circuit, the replacement of the oil at regular suggested service intervals is highly recommended. With this design, the oil that lubricates and cools the brake lining is contained in the wheel end and does not get circulated through a filter and cooled as in some other designs. If excessive brake lining wear is discovered, the oil and friction discs should be replaced as soon as possible. On some equipment the oil in the complete axle assembly must be replaced. In many cases, whenever a friction disc change is performed on a wheel end brake, both wheel end friction discs must be replaced at the same time.

Procedure to Check Friction Disc Wear

- Locate the wear indicator plug on the wheel end and clean the area completely before removing the inspection plug.
- Refer to the equipment technical documentation to be sure that the correct pin is selected to perform the next step. Failure to use the correct pin can lead to inaccurate measurements and false assumption.
- Place the wear indicator pin into the access hole until it bottoms out on the reaction plate. If the pin fits flush or is below the housing face, the friction discs need to be replaced. If the pin projects above the housing face, as shown in **Figure 7-25,** then the friction discs are still good.
- This wear check procedure must be performed on all four wheels to determine the condition of the friction discs. When replacing the inspection plug, make sure to replace the seal or O-ring as this compartment is normally filled with oil.

On most brake designs it is recommended that the air be bled from the spring-applied brake wheel ends any time the brakes are disassembled. Air in the piston cavity will prevent the brakes from releasing completely, which can lead to excessive friction disc wear. Always refer to the manufacturer's recommended procedures for the equipment in question. The example given in this section requires that a minimum of two individuals perform the air purge. As always, before any service work is performed, the equipment must be parked on level ground and have approved wheel chokes installed on both sides of the tires. Any implements attached to the equipment must be lowered to the ground. The equipment steering frame lock must be installed along with any lockouts or tags that are required. If any steering lock mechanisms, such or transmission locks, are present on the equipment, ensure that you engage them. Failure to follow the manufacturer's recommended procedures for performing repairs or tests on the equipment can lead to serious injury or death. To perform any work on brake systems, the individuals performing the repairs or adjustments must be qualified technicians who have read over the manufacturer's recommended procedures before attempting to make any adjustments or checks to the brake system. Under no circumstance should anyone perform any repairs to a brake system without the proper training.

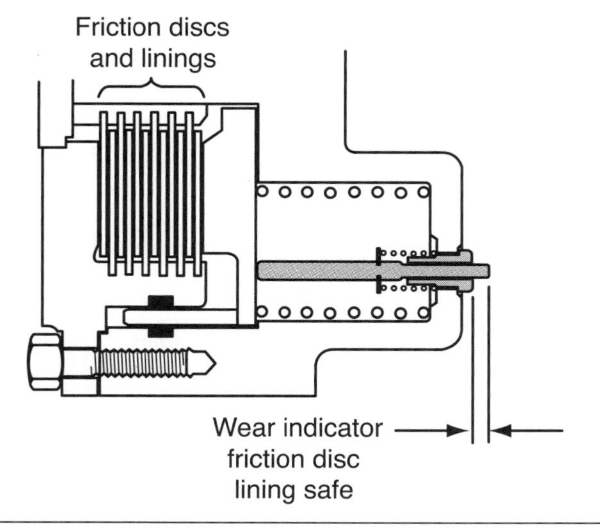

Figure 7-25 Location of and procedure used to determine friction disc wear on a spring-applied hydraulically-released brake system.

Procedure to Bleed Air from the Brakes

- Ensure that the hydraulic reservoir is full before beginning an air purge.
- Locate the air purge screw on the wheel end in question. Have someone remain in the operator's compartment during the procedure. Start the engine and release the park brake. The engine must remain running during the air purge procedure so that the brake accumulator pressure is maintained at or near maximum pressure.
- To perform this type of procedure, a clear tube that fits onto the purge valve snugly is mandatory. Obvious reasons are that the oil can be transferred to a container without spillage and air bubbles are easily seen in the clear tubing during the air purge procedure.
- With the engine running, park brakes released, and someone in the operator's compartment at all times during this procedure, open the bleed screw slowly and look for air bubbles in the fluid. Be sure to use a suitable container to capture the oil. Keep bleeding the wheel end until the oil flow is free from air bubbles.
- Close the purge valve and apply the park brakes. Now release the park brake and continue to bleed the air as before until the fluid is free from air. Perform this procedure to each wheel end, one at a time, until all the air is bled from the brake system. Failure to remove all the air from the wheel ends could prevent the brakes from releasing fully, causing premature wear to the friction discs.

Pressure Testing the Brake Cooling Circuit

The brake cooling circuit on spring-applied hydraulically-released brakes can develop internal leaks that can result in hydraulic oil flowing into the axle housing. A leak in the brake cooling circuit will result in oil over-filling the axle housing and eventually running out of the axle housing vent. If not found during routine axle servicing, the high oil level will generally cause a seal failure in the axle housing face seals due to excessive oil pressure build up. To locate the source of the oil leak, each individual brake cooling circuit in each wheel end must be pressure tested individually. A low air pressure regulator is required along with the appropriate fitting to mate to the cooling inlet and outlet ports on each wheel end as shown in **Figure 7-26.** It is advisable to pressure test each wheel end individually after any service has been performed to the internal brake components. A leak found at this point is easily repaired compared to finding it after the equipment has been placed back in service.

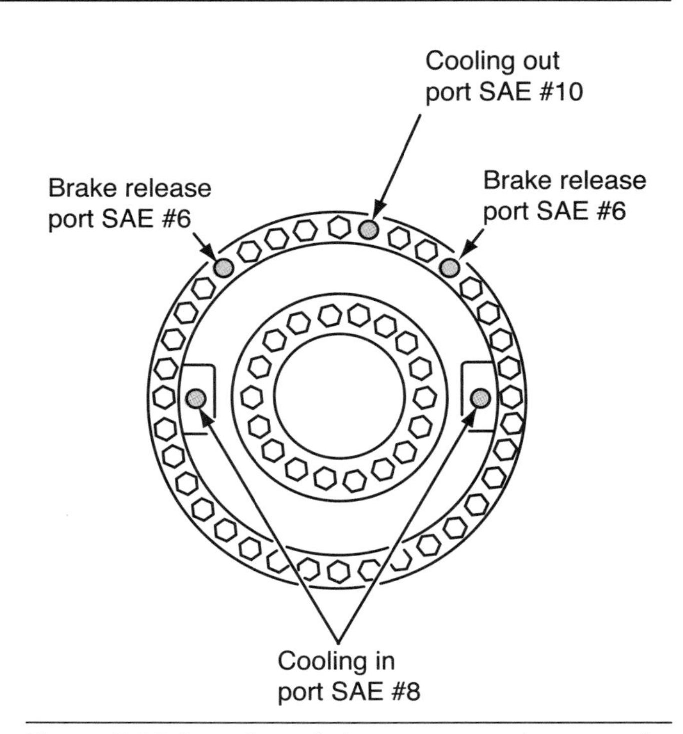

Figure 7-26 Location of the test ports that must be used to pressure test the brake circuit.

Tech Tip: Before performing any service or repairs to a brake system, the equipment must be parked on level ground and have approved wheel chokes installed on both sides of the tires. Any implements attached to the equipment such as a bucket, must be lowered to the ground. The steering frame lock must be installed along with any lockouts or tags that are required. If any steering lock mechanisms, such as transmission locks, are present on the equipment, ensure that you engage them. Failure to follow the manufacturer's recommended procedures can result in personal injury or death.

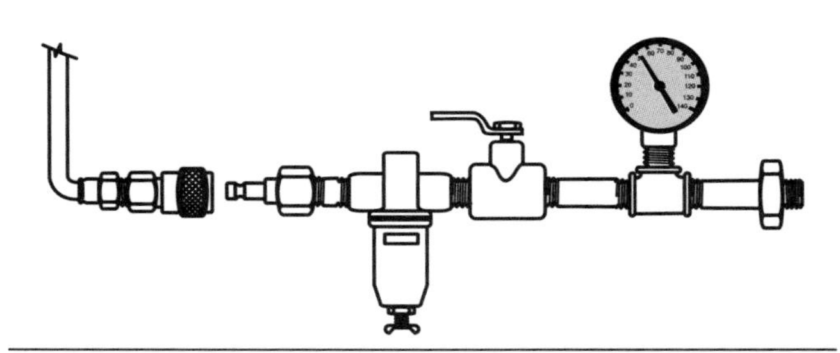

Figure 7–27 Equipment required to perform a pressure test on the wheel end brake cooling circuit.

Procedure to Test for Internal Leaks

- The test equipment for this procedure as shown in **Figure 7-27,** consists of a 30 psi pressure gauge, air shut-off valve, air regulator, pipe tee, pipe nipple, reducing bushing, and an air line connection.
- This test must be performed with the engine shut down and the park brakes applied. Install a blind plug to seal the cooling outlet port of the wheel end (Figure 7-27). Connect the pressure test equipment, as shown in **Figure 7-28** to the inlet port of the cooling circuit.
- Set the regulator to 12 psi and slowly open the air valve until 12 psi of air is showing on the gauge.

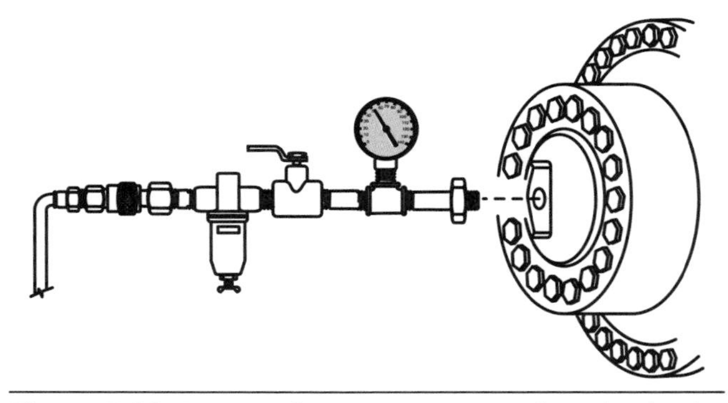

Figure 7-28 Test equipment connected to the brake cooling circuit.

- Close the air valve between the air regulator and the gauge and observe the pressure on the gauge.
- The air pressure should hold for approximately 15 seconds. If it fails to hold, the wheel assembly must be disassembled and repaired. Test both wheel ends on each axle assembly to determine which one is leaking. It is quite possible that both wheel ends are leaking internally.

Repairing The Brake Assembly

Table 7-5 indicates the parts and quantities needed to overhaul a wheel end brake.

The following is shown as an example of the general steps that are used to disassemble a spring-applied wheel end brake assembly once it has been removed from the equipment. The example used is not intended to replace the equipment service manual. Always refer to the specific manufacturer's technical documentation when performing service work to brake systems. This actual service work must also include the service manual for the axle assembly that is being serviced. As with any service manual procedures, following the steps in the exact sequence given is very important. The large springs used in this brake system pose a hazard to anyone who does not follow the correct disassembly procedures as outlined in the equipment manufacturer's service documentation. Application spring tension in this type of brake system can reach 250,000 lbs. of force. Failure to follow the manufacturer's recommended disassembly procedures can lead to personal injury or death.

Having all the necessary tools on hand before attempting to perform this type of repair work is essential. The illustrations used in this example are generic and may not be exactly the same as the equipment that you are repairing. Proper disassembly of the front cover and backing plate on these types of brake systems is essential to prevent component damage or personal injury. Proper cleaning of the brake assembly is required before any disassembly work can proceed.

Procedure

- Note that the front cover on the brake assembly, as shown in **Figure 7-29** is under spring pressure for approximately the first inch of bolt travel. The spring pressure can be neutralized by applying approximately 1500 psi to the brake inlet piston port with a portable hydraulic power pack. Although it is not necessary to do this, it makes the removal of the brake front cover somewhat easier. If no portable power pack is available, the front cover can still be removed by loosening the front cover bolts in a criss-cross pattern, half a turn at a time, until the spring pressure is dissipated.
- The next step requires the removal of the quad ring seal as shown in **Figure 7-30.** Be sure to carefully inspect the quad ring and corresponding groove in the housing for any signs of damage.

Table 7-5: PARTS LIST

Description	Qty.	Description	Qty.
No. 1 Multi-disc brake assembly	1	No. 16 Outer piston seal assembly	1
No. 2 Cooling outlet plug	1	No. 17 Outer cover capscrews	12
No. 3 Cooling outlet plug O-ring	1	No. 18 Outer cover washers	12
No. 4 Bleeder screw	2	No. 19 Brake outer cover	1
No. 5 Plug	2	No. 20 Friction disc	3–6
No. 6 O-ring	2	No. 21 Reaction plate	3–6
No. 7 Actuating fluid plug	1	No. 22 Inner piston seal	1
No. 8 Actuating plug O-ring	1	No. 23 Brake piston	1
No. 9 Plug	2	No. 24 Wear indicator pin	1
No. 10 Brake housing	1	No. 25 Indicator pin O-ring	2
No. 11 Inlet O-ring	1	No. 26 Indicator guide O-ring	1
No. 12 Outlet O-ring	1	No. 27 Indicator guide	1
No. 13 Brake apply spring	15	No. 28 Warning sticker	2
No. 14 Outer cover O-ring seal	1`	No. 29 Sticker cover	2
No. 15 Piston pressure ring	1	No. 30 Sticker cover screw	2

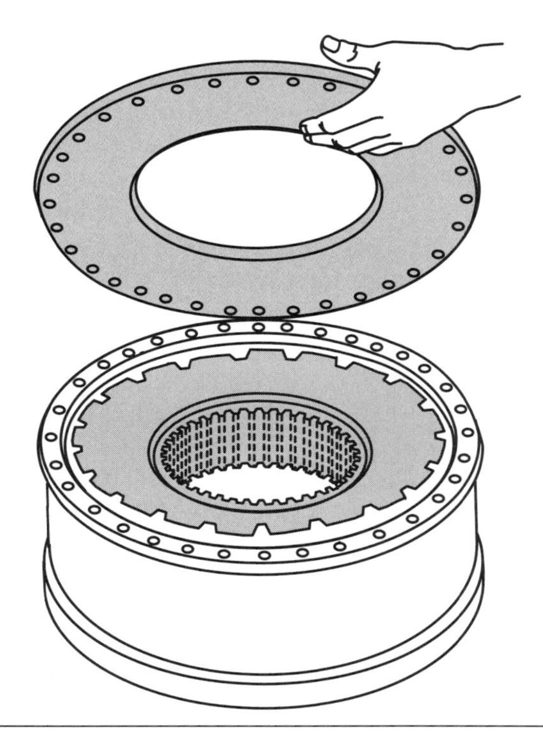

Figure 7-29 Front cover removed from the spring applied-brake assembly.

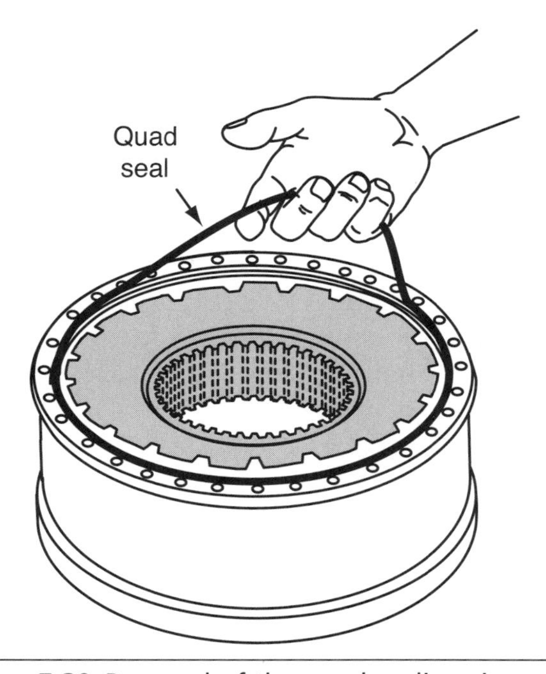

Figure 7-30 Removal of the quad sealing ring.

- Carefully remove the steel and friction discs from the housing and make a mental note of the number of discs and the sequence in which they are installed in as shown in **Figure 7-31.**

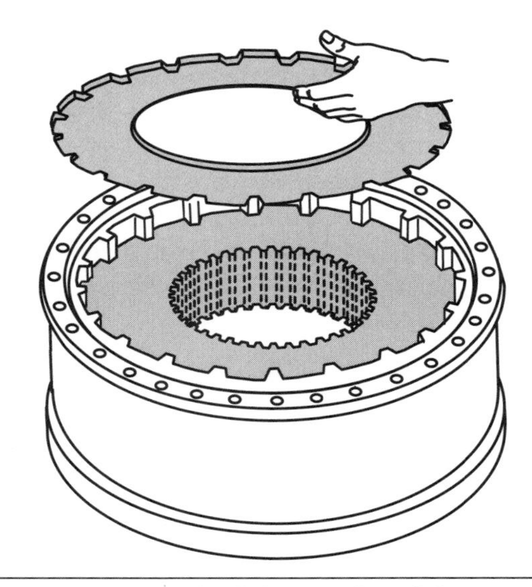

Figure 7-31 Removal of the steel and friction discs.

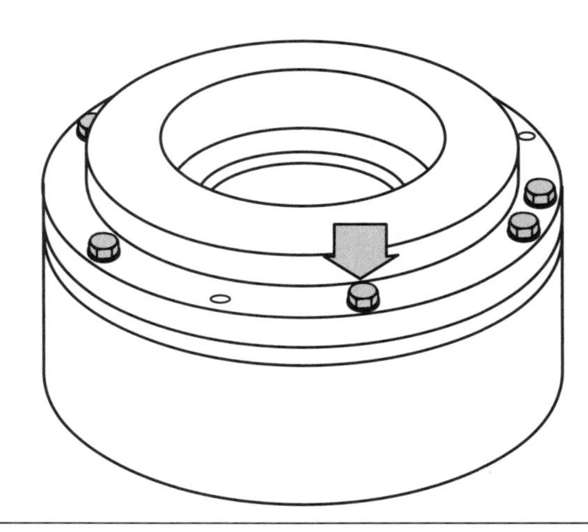

Figure 7-32 Backing plate bolts to release the application spring tension.

- The next step requires that the brake assembly be turned over to gain access to the backing plate bolts. Caution must be exercised with the following steps. The backing plate bolts are under a great deal of spring force and should never be removed with an impact. The backing plate bolts must be removed in the correct order to prevent any stress to the blots.
- The next step requires the removal of every second bolt from the backing plate cover, as shown in **Figure 7-32,** slowly releasing the backing plate by turning each remaining bolt a quarter of a turn at a time in a criss-cross pattern until the

backing plate is fully released. Remove all but three bolts from the assembly. The three bolts are left in place to keep the assembly from falling apart when it is flipped over.

- After flipping the brake assembly over, remove the remaining three bolts from the housing and separate the housing from the backing plate. Remove the springs from the brake housing as shown in **Figure 7-33.**
- Next, remove the O-rings from the backing plate and the port O-rings from the brake housing as shown in **Figure 7-34.**
- Using a soft blow hammer, gently tap the piston to remove it from the housing as shown in **Figure 7-35.** Be careful that the brake piston does not fall during the final removal stages from the housing; it could easily be damaged. Carefully remove the outer piston seal and expander from the piston and examine the piston and sealing rings for any damage or wear. Next, remove the inner seal and expander ring from the brake housing and inspect the parts for wear or damage.

Clean all the brake parts that will be reused and inspect them for any signs of wear or damage that may affect braking performance. Be sure that all parts are free from nicks or burrs and that no cracks are visible on any of the components. Only the correct OEM replacement parts should be used to reassemble the brake assembly. The seals are delicate and should be handled with extreme care during the installation process. Follow the manufacturer's recommended installation procedure during assembly.

Reassemble the components in the reverse order of disassembly, paying particular attention to the way the O-rings and expander rings are installed on the piston and inner housing. **Figure 7-36** shows the correct orientation of the piston seals once correctly installed.

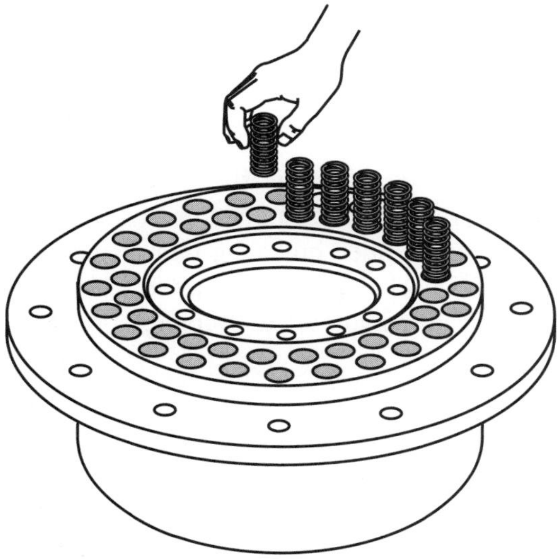

Figure 7-33 Brake spring being removed from the brake housing.

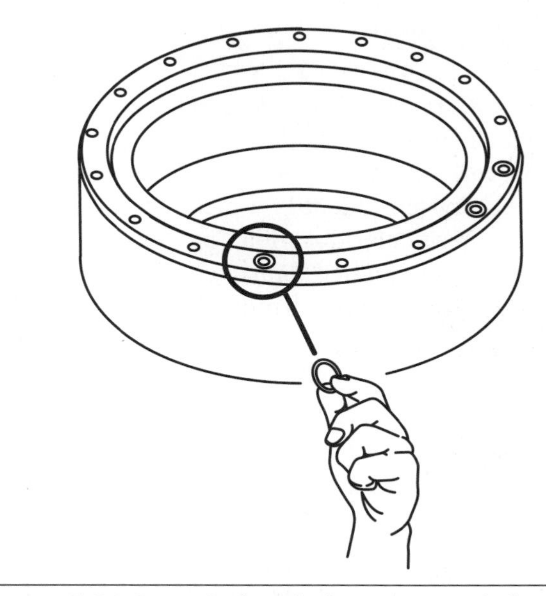

Figure 7-34 Port O-ring being removed from the brake housing.

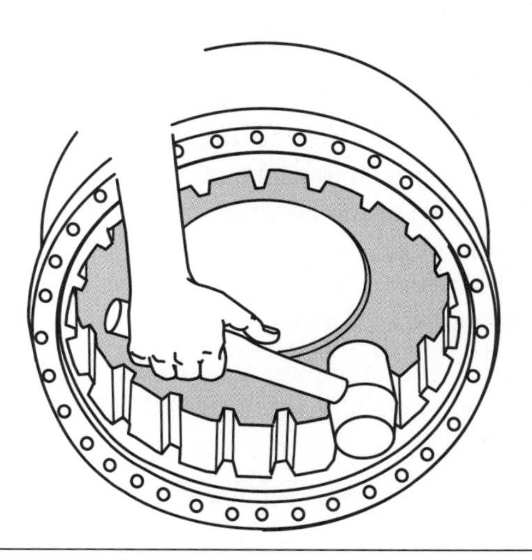

Figure 7-35 Brake piston removal procedure.

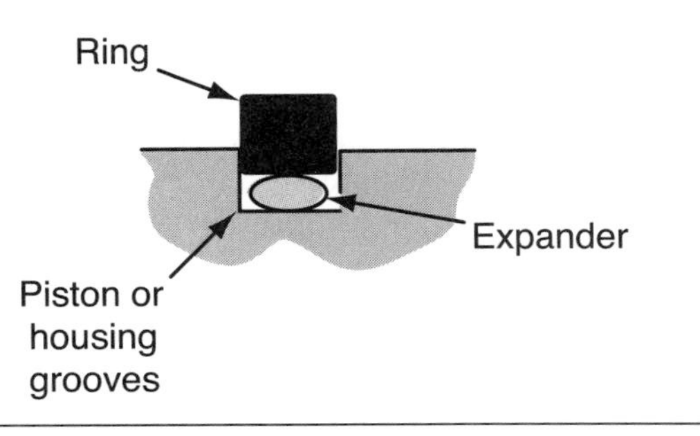

Figure 7-36 Correct orientation of the sealing rings on the brake piston.

ONLINE TASKS

1. Use a suitable Internet search engine of your choice to research the heat dissipation in an off-road spring-applied hydraulically-released brake system. Identify a piece of equipment in your particular location, paying attention to the spring-applied brake cooling circuit and how heat energy from the spring-applied hydraulically-released brake system is dissipated. Outline the reasoning behind the type of cooling circuit used on this particular piece of equipment.
2. Check out this url: *http://www.newtechbrake.com/newt/pages/about.html.* Look for information on new brake technology.

SHOP TASKS

1. Select a piece of equipment in your shop that has a spring-applied hydraulically-released brake system and identify the components that make up the brake circuit. Identify the type of cooling circuit used; note how oil is directed to this circuit and what cooling pressures are used.
2. Select a piece of mobile equipment in your shop. Identify the type and model of the equipment and list all the components and the corresponding part numbers of a typical spring-applied hydraulically-released brake circuit. Identify and use the correct parts book for the particular piece of equipment being worked on.

Summary

- Brake systems on off-road equipment must be able to operate in extremely abrasive conditions that would quickly wear out conventional external brake systems. The roadways that equipment operates on are considered extremely abrasive to a brake system.
- Many off-road equipment manufacturers favor the use of internal spring-applied hydraulically-released wet brake systems. These systems incorporate a brake design that has the brakes located inside the axle assembly, which keeps abrasive road dirt out.
- Brake systems convert the kinetic energy of the moving equipment to heat energy when the equipment is slowed down or brought to a stop by the operator applying the brake. Friction material exerts pressure on a corresponding metal surface to slow down the equipment.
- The wheel end clamping forces used by spring-applied hydraulically-released brakes are dependent on the size and number of springs used in the wheel ends to apply mechanical force to the friction lining.
- Many manufacturers implement a brake interlock circuit in which the brakes cannot be released after an emergency application unless the operator manually resets the brake interlock circuit. This interlock brake circuit ensures that the operator is in the operator' compartment before the brakes can be released.
- Modern spring-applied hydraulically-released brake circuits apply the brakes when brake pressures become to low, and also apply the brakes if there is a loss of electrical power or if transmission clutch or torque pressure falls below safe operating levels.
- A spring-applied hydraulically-released brake system shares similar components that are used in a hydraulically-applied brake circuit. The use of a multi-disc wheel end brake design is very similar to the ones used in hydraulically-applied brakes.
- Brake pumps that are used on a spring-applied hydraulically-released brake circuit are generally heavy-duty gear-type pumps that are capable of supplying a continuous oil flow of up to 19 gpm (gallons per minute) or 72 L/m (Liters per minute) at 2400 rpm.
- The diaphragm-type of accumulator uses a synthetic rubber diaphragm that separates the hydraulic oil from the nitrogen gas.

- The bladder-type of accumulator has a metal shell that is fitted with synthetic flexible bladder. The bladder is pre-charged with nitrogen and the metal shell holds the hydraulic oil. A poppet valve is located at the bottom hydraulic port that keeps the bladder contained in the accumulator when the hydraulic pressure is discharged.
- The accumulator is a high-pressure vessel and must be treated accordingly. Under no circumstance should any welding or brazing of any kind be done on the accumulator casing. If physical damage is discovered to the casing, it must not be used.
- Under no circumstances should a tire valve be used to replace the gas valve in an accumulator. Tire valves are not designed to withstand the high pressures used in the accumulator.
- Piston-type accumulators have a free-floating piston that keeps the hydraulic oil and high-pressure nitrogen separated. The piston has a high-pressure seal that is replaceable.
- The stored energy in the brake accumulator allows the equipment to maintain adequate oil pressure reserve, ensuring a full brake release can be achieved at all times during operation.
- The accumulator charge valve is used to regulate oil flow from the brake pump to charge the accumulators that store the energy for use in the brake circuit during equipment operation. The charge valve controls the rate of charge as well as the cycle rate for the brake circuit.
- The hydraulic filter in the brake circuit provides partial flow filtration of the hydraulic oil. The filter has a 25 psi (1.7 bar) bypass relief valve built into the filter. The filter will go into bypass whenever there is a 25 psi difference in pressure across the filter.
- The pilot sequence valve used in a spring-applied hydraulically-released brake circuit is a closed-center valve that regulates the pilot pressure to 200 psi and is used to operate the dump/hoist/steering valves indirectly through a pilot circuit.
- The accumulator pressure switch is a piston-type pressure switch that has two sets of contacts that open and close simultaneously when the pressure falls below a predetermined amount in a brake circuit.
- The park brake selector valve is a four-way two-position hydraulic valve that is pilot-operated by the park brake control valve. It is also used to apply the park brakes in the event of clutch pressure loss or when the operator wants to apply the park brake manually.
- Unlike conventional hydraulic brake systems which use hydraulic pressure to apply the brakes, spring-applied brakes require oil pressure to keep the brakes released. The brake pedal in this application allows oil to flow to the wheel ends whenever the pedal is in the released position.
- Park brakes on off-road equipment are always mechanically applied and hydraulically released. This makes the use of the park brake extremely safe because no hydraulic pressure is required to park the equipment.
- Hydraulic oil coolers are necessary on many off-road brake applications so that the oil temperature can be maintained at a safe temperature. In many severe-duty applications, the oil temperature could exceed safe operating temperatures if no additional cooling is used.
- When checking brake pressures in the spring-applied hydraulically-released service brake circuit, you must remember that the pressure is gradually reduced when the brakes are applied, and when fully applied there is no brake pressure in the wheel ends. The friction discs are clamped to the stationary steel discs by mechanical spring pressure.
- A spring-applied brake system operates in reverse when compared to conventional brakes. Each wheel end brake is capable of exerting as much as 80,000 lb of force from the spring pressure in the wheel ends.
- To activate the service brakes in a spring-applied hydraulically-released brake circuit the operator depresses the pedal to gradually reduce the brake pressure in the wheel ends applying the brakes. The return oil from the foot valve is redirected back to the tank. The more force the operator applies to the brake pedal, the greater the pressure drop in the service brake circuit.
- To prevent the equipment from operating with low brake pressure, which could cause the brakes to drag, a brake pressure switch monitors the accumulator brake pressure.
- The pilot pressure sequence valve used in a brake circuit is considered a full flow relief valve that will maintain a predetermined amount of pressure for the pilot circuit.

- One of the most common problems encountered in equipment hydraulics is maintaining the correct level of oil in the tank. Plugged or dirty hydraulic filters both on the suction side and pressure side, must be cleaned and replaced per the manufacturer's recommendations.
- Keeping the dirt and contamination out of the hydraulic system is essential; even the small particles that are hard to see with the naked eye can score valves and plug orifices.
- Never use your hands, even if you are wearing gloves, to locate hydraulic leaks in a hydraulic circuit. Use a cardboard for this purpose. High-pressure oil can penetrate the skin causing severe injury. Always seek medical attention immediately if this type of injury is suspected.
- To check the nitrogen pre-charge in a accumulator(s), all the oil pressure must be dissipated in the brake circuit and the piston must bottom out at the lower end cap.
- The only gas that should be used to charge brake accumulators is dry nitrogen gas. No other compressed gases are permitted for this procedure; other types of gases, even compressed air, can lead to explosions and should never be used.
- If there is any visible physical damage to the accumulator, it must be replaced immediately. *Note:* Before attempting to replace the accumulator make sure to discharge all the pressure in the accumulator, following the manufacturer's recommended procedures.
- Checking and pre-charging the accumulator in cold or hot ambient temperatures without compensating for temperature difference will result in either overcharging or undercharging the accumulators.
- Performance brake tests must be performed anytime the brakes are serviced. Brake tests must be performed in a suitable area that will not compromise the safety of anyone in the immediate area. Ensure that there is adequate room, both front and rear, to allow for any unintended equipment movement. The equipment must be tested on level ground.
- To perform a park brake test, depress the park brake test switch that is located on the dashboard in the operators compartment. This test drains the hydraulic brake fluid from the spring-applied park brake circuit while the emergency brakes are in the released position. This test allows isolation of the park brake circuit so it can be checked separately.
- A brake pump flow test is conducted at two different pump speeds to access the pumps condition. The difference is calculated between the pump flows at two different operating pressures and is referred to as flow loss.
- Most wet disc brake systems have an access port in the wheel end that allows the friction discs to be checked for wear.
- On internal wet disc brake designs, it is recommended that the air be bled from the spring-applied brake wheel ends anytime the brakes are disassembled. Air in the piston cavity will prevent the brakes from releasing completely, which can lead to excessive friction disc wear.
- The brake cooling circuit on spring-applied hydraulically-released brakes can develop internal leaks that could result in hydraulic oil flowing into the axle housing. A leak in the brake cooling circuit will result in oil over-filling the axle housing and eventually running out of the axle housing vent.
- To locate the source of the oil leak in a spring-applied brake cooling circuit, each individual brake cooling circuit in each wheel end must be pressure tested individually. A low air pressure regulator is required along with the appropriate fitting to mate to the cooling inlet and outlet ports on each wheel end.
- To test the cooling system chamber for leaks, the air pressure should hold for approximately 15 seconds. If it fails to hold, the wheel assembly must be disassembled and repaired. Test both wheel ends on each axle assembly to determine which one is leaking. It is quite possible that both wheel ends are leaking internally.
- The large springs that are used in the spring-applied brake system pose a hazard to anyone who does not follow the correct disassembly procedures as outlined in the equipment manufacturer's service documentation. Application spring tension in this type of brake system can reach 250,000 lbs. of force. Failure to follow the manufacturer's recommended disassembly procedures can lead to personal injury or death.
- Only the original OEM replacement parts should be used to overhaul any brake assembly. The seals are delicate and should be handled with extreme care during the installation process.

Review Questions

1. What does the term cut-in pressure refer to on a spring-applied hydraulically-released brake system?

   A. the pressure at which the accumulator charge valve starts to charge
   B. the pressure at which the accumulator charge valve stops to charge
   C. the pressure at which sequence valve starts to supply oil to the cooling circuit
   D. the pressure at which the pilot circuit starts to supply oil to the sequence valve

2. What is the first function that the pilot sequence valve performs on a spring-applied hydraulically-released brake system?

   A. provide oil to the accumulator circuit
   B. control the cut-in pressure
   C. control the cut-out pressure
   D. provide oil flow for the steer/hydraulic control circuit

3. What purpose does the 65 psi check valve serve in the pilot sequence valve?

   A. provide a higher pressure oil flow for the brake cooling circuit
   B. provide a lower pressure oil flow for the brake cooling circuit
   C. provides a lower pressure for the pilot circuit control circuit
   D. maintains a safe pressure in the wheel ends

4. What could be responsible for the brake charge valve cycling too often in a spring-applied hydraulically-released brake system?

   A. high accumulator pre-charge pressure
   B. low accumulator pre-charge pressure
   C. high wheel end brake pressure
   D. high pilot circuit pressure

5. What is the purpose of the accumulator pressure switch?

   A. monitor the pilot pressure
   B. monitor the cooling pressure
   C. monitor brake pressure
   D. monitor cut-out pressure

6. What component controls the wheel end pressure in the spring-applied hydraulically-released brake circuit?

   A. brake valve
   B. spool 2 in the charge valve
   C. pilot sequence valve
   D. pressure regulating valve in the charge valve

7. If a charge valve was not used in a spring-applied hydraulically-released brake circuit, what type of oil pump must be used?

   A. gear pump
   B. variable-displacement piston pump
   C. fixed-displacement piston pump
   D. gerotor pump

8. What feature does a low-pressure return filter have when used in a spring-applied hydraulically-released brake system?

   A. bypass valve
   B. pressure regulator
   C. full flow with no bypass
   D. two check valves

9. What component receives oil first from the brake pump in a spring-applied hydraulically-released brake system?

   A. accumulator charge valve
   B. accumulators
   C. pilot sequence valve
   D. brake valve

10. What regulates the pressure that flows through the wheel ends of the brake circuit on a spring-applied hydraulically-released brake system?

    A. accumulator pressure switch
    B. pilot sequence valve
    C. check valve
    D. brake valve

11. How is brake spring pressure regulated in the spring-applied brake circuit?

    A. by the brake valve
    B. by the charge valve
    C. by the accumulators
    D. by the pilot sequence valve

12. What purpose does the accumulator serve in a spring-applied hydraulically-released brake system?

    A. prevents the brake pump from cycling too often
    B. provides emergency brake capabilities
    C. ensures that the brakes will always apply
    D. allows the brakes to be applied even if the engine is not running

13. What must be done first when work has to be performed on a spring-applied hydraulically-released brake system?

    A. release the accumulator hydraulic pressure
    B. release the emergency brakes
    C. drain the hydraulic tank
    D. release the accumulator pre-charge

14. Which pressure is manually adjustable on a spring-applied brake system charge valve?

    A. cut-in pressure
    B. cut-out pressure
    C. both A and B
    D. service brake pressure

15. What is the purpose of the park brake pressure switch in a spring-applied brake system?

    A. turn on a light in the operator's compartment
    B. turn off a light in the operator's compartment
    C. activate the park brake
    D. activate the charge circuit

16. What is the location of the safety relief valve in a spring-applied brake system?

    A. inside the pilot sequence valve
    B. inside the charge valve
    C. inside the brake valve
    D. before the charge valve

CHAPTER

Air and Air-Over-Hydraulic Brake Systems

Learning Objectives

After reading this chapter, you should be able to

- Identify types of off-highway equipment using air brakes.
- Identify the components of an air brake system.
- Explain the operation of a single- and dual-circuit air brake system.
- Identify the major components of an air compressor.
- Describe the operation of desiccant- and aftercooler-type air dryers.
- Outline the operating principles of the valves and controls used in an air brake system.
- Explain the operation of an air brake chamber.
- Outline the functions of the hold-off and service circuits in spring brake systems.
- Describe the operation of S-cam and wedge-actuated drum brakes.
- Describe the operating principles of manual and automatic slack adjusters.
- List the components and describe the operating principles of an air disc brake system.
- Describe the major components and operation of parking and emergency braking systems.
- Identify the major components of an air-over-hydraulic braking system.
- Describe the operation of a typical air-over-hydraulic ABS system.
- Outline some typical maintenance and service procedures performed on air-over-hydraulic brake systems.
- Interpret air-over-hydraulic system schematics.

Key Terms

aftercooler
air dryer
air/hydraulic master cylinder
air-over-hydraulic intensifier
brake chambers
brake shoes
check valve
coefficient of friction
compressor
couplers
dash control valves
disc brakes
double check valve
drum brakes
dryer reservoir module (DRM)

dual-circuit application valve
foot valve
master cylinder
modulating valve
pop-off valve
proportioning valves
ratio valve
relay valves
safety pressure-relief valve
supply tank
system pressure
treadle valve
two-way check valve
wet tank
wheel cylinder

INTRODUCTION

Air and air-over-hydraulic brakes are not commonly used in contemporary off-highway heavy equipment but there remain a few applications. Nevertheless, technicians working on off-highway heavy equipment should have a good understanding of both air and air-over-hydraulic brakes because, unlike highway trucks, off-highway equipment tends to have a much longer operational life. Some examples of equipment using air brakes are:

- Scrapers
- Wheel loaders
- Articulated compactors
- Articulated rock/mining/quarry trucks
- Off-highway construction dump trucks
- Off-highway logging trucks and trailer/multi-trailer trains

The first part of this chapter will focus on air-only brake systems and the final part on air-over-hydraulic brakes. Because an air-over-hydraulic brake system has a high degree of component commonality with air-only and hydraulic-only brakes, it makes sense to study them last.

AIR BRAKE SYSTEMS

Air brake operation is similar in some ways to hydraulic brake operation, except that compressed air is used in place of hydraulic fluid to actuate the brakes. The potential energy of an air brake system is in the compressed air itself. The retarding effort delivered to the wheels has nothing to do with mechanical force applied to the brake pedal. The source of the potential energy of an air brake system is the vehicle engine, which drives an air **compressor.** The compressor discharges compressed air to the vehicle air tanks.

The brake pedal, actuated by the operator, meters (precisely measures) some of this compressed air out to the **brake chambers.** Brake chambers are used to convert air pressure into mechanical force to actuate foundation brakes. Foundation brakes consist of shoes and drums or rotors and calipers that convert the mechanical force applied to them into friction. In other words, in an air brake system, braking effort at the wheel is achieved in the same way as in a hydraulic brake system. The force that actuates the foundation brakes is converted to friction, which retards movement. The friction is converted to heat energy, which then must be dissipated (transferred) to the atmosphere.

While air brakes are used almost exclusively on highway heavy-duty trucks and trailers, they are not so often found on today's off-highway equipment unless it is on off-highway tractors and trailers, or multi-trailer trains of the type used in logging. Air brake systems have some advantages over hydraulic brakes when equipment has to be coupled and uncoupled by operators such as when a tractor unit is required to couple to a trailer or towed unit. A manufacturer that builds on-highway trucks engineered with air brake systems often opts to use air brakes in their off-highway product. However, because air brakes use pressurized air, which is compressible, they have some disadvantages when compared with hydraulic brakes, which use a nearly incompressible liquid medium. When air brakes are used in off-highway heavy equipment, it is usually on older equipment.

Advantages of air brakes over hydraulic brakes are:

- Air is limitless in supply so minor leaks do not result in brake failures.
- The air brake circuit can be expanded easily so that trailers can be coupled and uncoupled from the tractor circuit by a person with no mechanical knowledge.
- In addition to providing the energy required to brake the vehicle, compressed air is also used as the control medium. That is, it is used to determine when and with how much force the brakes should apply in any situation.
- Air brakes are effective even when leakage has reduced their capacity, meaning that in the event of air leakage, the system can be designed with sufficient fail-safe devices to bring the vehicle safely to a standstill.

The disadvantages of air brake systems when compared with hydraulic brakes are:

- Slower control response times.
- Slower actuation times.
- Lower pulse frequencies when used in anti-lock braking system (ABS) systems.

Overview of Air Brake Basics

On-highway air brake systems are governed by legislation introduced in 1975 called Federal Motor Vehicle Safety Standard No. 121 (known as FMVSS 121 in the United States, CMVSS 121 in Canada). Over the years, it has been modified in small ways to keep up to date with technology, but from the beginning it required all highway vehicles using air brakes to use a dual-circuit application circuit. However, FMVSS 121 does not apply to vehicles that run exclusively off-highway. This means that there are air-brake equipped heavy equipment vehicles that use a single-circuit brake system though, in reality, they are few in number. When a dual-circuit brake system is used, the system is divided into two subcircuits known as primary and secondary circuits. Throughout this chapter, we will reference FMVSS 121 because this benchmarks a degree of safety that may or may not be adhered to in off-highway vehicles.

Off-highway equipment using air brakes is not mandated to use computer-controlled, ABS. When ABS is used, it is important to note that it does not significantly alter the basic brake system that will be introduced in this chapter. **Figure 8-1** shows the layout of a current air brake system on a Caterpillar 775D quarry truck. If you are familiar with air brake circuits on highway equipment, you will note that the off-highway system shown here is similar. It is a dual-circuit system, but it is simpler in general design: for instance, there is no **supply tank** in the system shown here. This is an OEM preference. In some cases, a supply tank is used on off-highway equipment. As we guide you through air brake system components, remember that you will see a broader range of components in off-highway equipment because the statutory standards that apply to on-highway equipment do not apply. In describing air brakes in this chapter, we will assume that the equipment is heavily optioned, but this is not always the case.

Figure 8-2 shows the layout of the air brake system used on a Caterpillar 621G on an articulating wheel tractor and scraper unit. Note that the layout here is more similar to that used on a highway tractor and semi-trailer combination.

AIR BRAKE SUBSYSTEMS

While there is no obligation for off-highway equipment to have separate service and parking brake systems, combined with a dual-service application circuit, it makes sense to do this for safety reasons. Just as in the case of dual-circuit hydraulic brakes, dual-circuit pneumatic service brakes provide redundancy in the event of the complete loss of a circuit. In addition, having a means of mechanically applying parking brakes provides a safety factor in the event of a complete loss of system air pressure.

The term *service brakes* is used to describe the running brakes on a vehicle: the running brakes are applied by a **foot valve** or hand valve. The term *parking/emergency brakes* is used to describe the mechanically-applied parking brakes on a vehicle: these are controlled by the operator using dash valves.

Assuming that a vehicle is designed to couple to a trailer or other type of articulated towed unit, a complete air brake system is made up of the following subsystems:

- The air supply circuit. The air supply circuit is responsible for charging the air tanks with clean, dry, filtered air. It consists of an air compressor, a governor, an **air dryer/aftercooler,** a supply tank, a low-pressure switch, and a system safety (pop-off) valve.
- Primary circuit. In an articulated combination, the primary circuit is usually responsible for actuating the tractor rear wheel brakes and actuating the trailer service brakes when applied from the foot valve. The foot valve is also known as a **treadle valve** or **dual-circuit application valve.** The primary circuit components consists of a primary tank, the foot valve, a pressure gauge, and quick release or rear axle **relay valves.**
- Secondary circuit. In an articulated combination, the secondary circuit is usually responsible for actuating the front axle service brakes and the trailer brakes when actuated by the trailer hand control valve on the tractor. It consists of a secondary tank, a foot valve, a pressure gauge, and quick release or relay valves.
- **Dash control valves** and the parking/emergency circuit. The control circuit consists of two or three dash-mounted valves, which manage the parking and emergency circuits. The parking/emergency circuit consists of spring brake assemblies, and the interconnecting plumbing.
- Trailer circuit. The trailer service and parking/emergency brakes are controlled by the tractor unit. When tractor and trailer air systems are required to be routinely connected and disconnected, they are coupled by a pair of hoses and **couplers.** The trailer brake circuit consists of couplers, air hoses, air tanks, pressure protection valves, and relay valves.

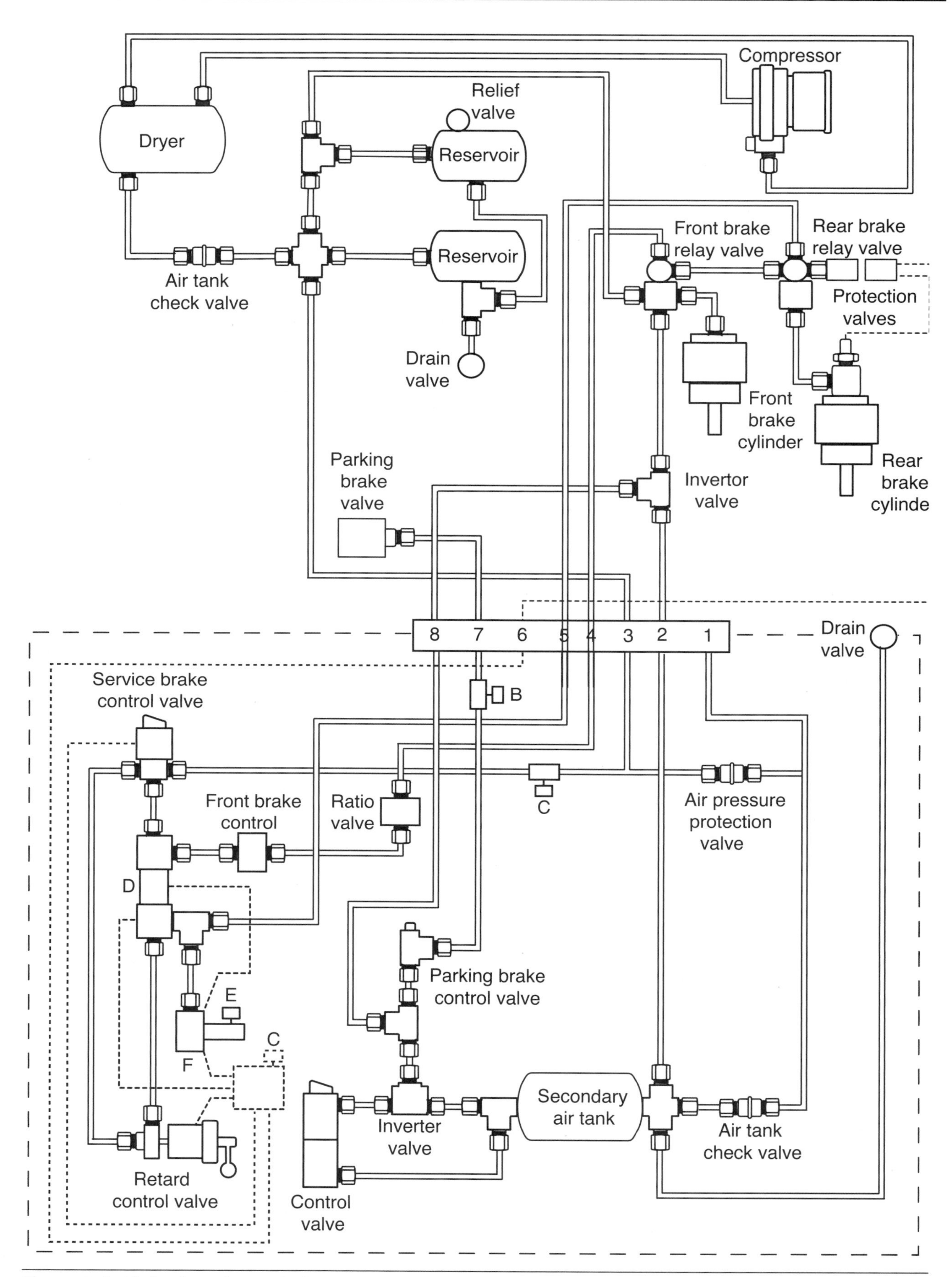

Figure 8-1 Air brake schematic from a Caterpillar 775D quarry truck.

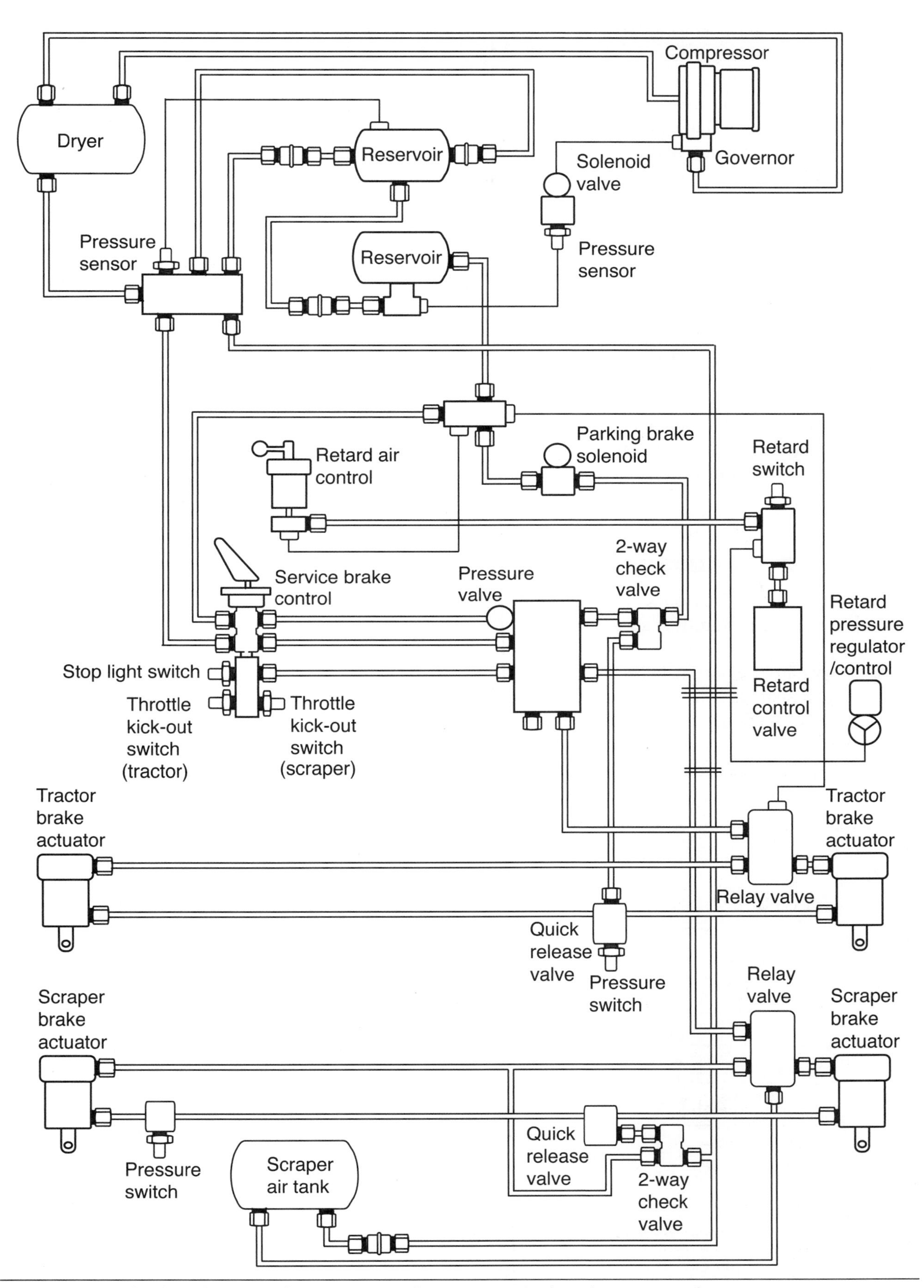

Figure 8-2 Air brake schematic from a Caterpillar 621G articulated tractor and scraper.

- Foundation brakes. The function of foundation brakes is to convert the air pressure supplied by the application circuits (primary and secondary) into the mechanical force required to stop a vehicle. The parking and emergency brakes may also use the foundation brakes to effect mechanical application of the brakes. Foundation brakes consist of brake actuators (chambers or rotochambers), slack adjusters, S-cams or wedges, shoes, pads, linings, drums, and discs. Additionally vehicles equipped with ABS will have wheel speed sensors mounted on the wheel assemblies.

Note the location of the dual-circuit air brake system components in the figures used in this book. Our approach to the study of air brake systems begins by introducing and explaining the operation of each of the subcircuits, starting with the air supply circuit and working through to the foundation brakes. Next, a more detailed description of the critical air brake components and valves is provided.

THE AIR SUPPLY CIRCUIT

The air supply circuit is responsible for charging the air tanks with clean, moisture-free air to be used by the brake system and to supply the remaining air requirements of the vehicle. The potential energy of an air brake system is compressed air. The air is compressed by the supply circuit and then distributed and stored in reservoirs (tanks) at pressures typically around 125 psi (860 kPa). Valves controlled by the operator can then deliver compressed air to the brake system components. Most equipment today is designed to have their parking/emergency brakes fully applied by mechanical force in the event that there is no system air pressure. Therefore, air is required first to release the parking/ emergency brake circuit and then to effect service braking when needed. **Figure 8-3** shows a typical air supply circuit.

Air Compressors

The air compressor is driven by the engine. Compressors may be either single- or multiple-cylinder. An air compressor drive usually comes directly from an engine accessory drive gear. In some cases, a compressor may be controlled by a clutch, and in light-duty and some older equipment, it may be belt-driven. In this textbook, we will make the assumption that all compressors are gear-driven by the engine.

The components of an air compressor are much like those of an engine. The compressor crankshaft is supported by main bearings, which are lubricated with engine oil. Connecting rods are attached to the crank throws and support pistons with rings to seal them in the cylinder bore. When the crankshaft is rotated by the engine accessory drive, the pistons reciprocate in bores machined in the compressor housing. The connecting rod bearings and cylinder walls also are lubricated by engine oil. The engine oil is pumped directly into the compressor crankshaft. The spill and throw-off oil from the bearings drains to a sump. The oil level in the sump is determined by the location of the oil return line. Because the oil pumped into the compressor is allowed to

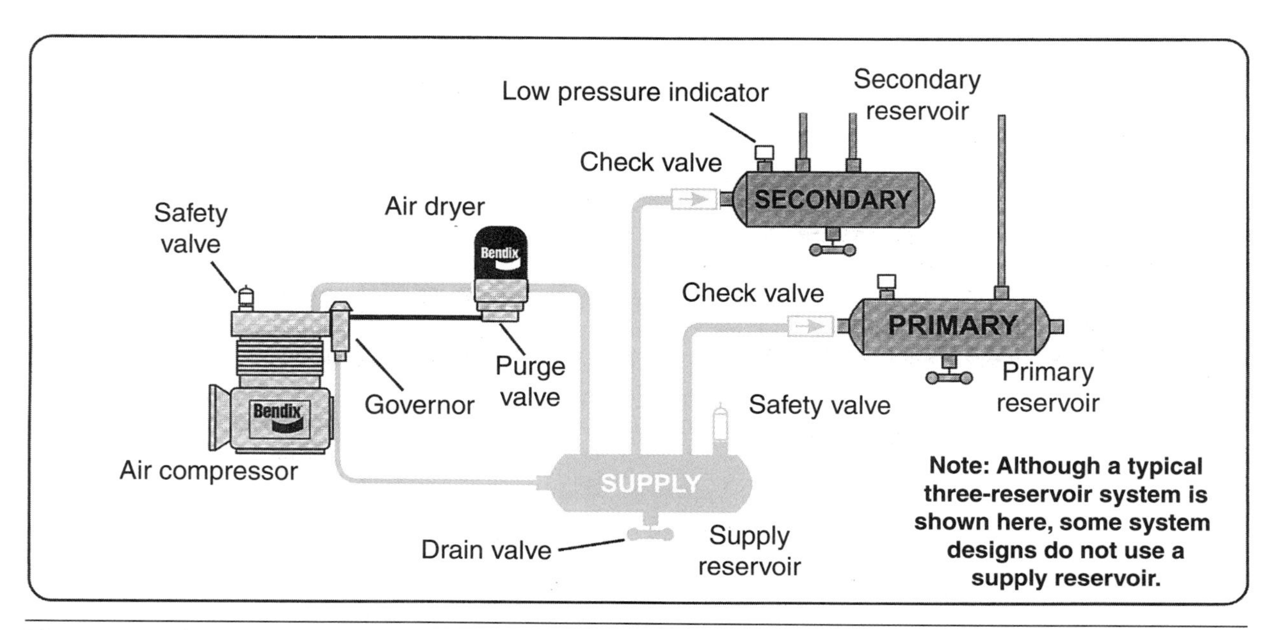

Figure 8-3 Air supply circuit components. (*Courtesy Bendix Commercial Vehicle Systems*)

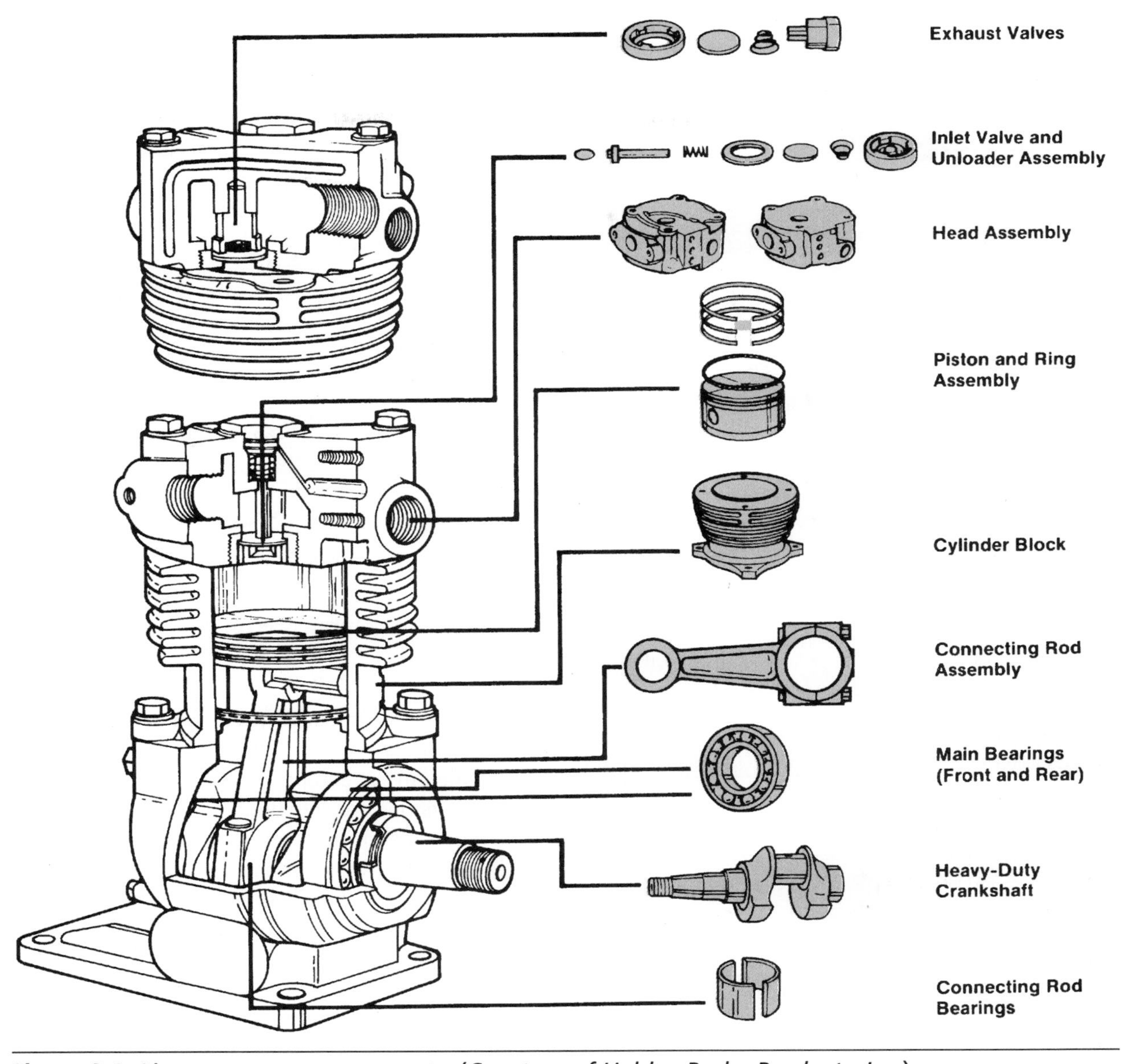

Figure 8-4 Air compressor components. (*Courtesy of Haldex Brake Products, Inc.*)

drain back from its sump to the engine crankcase, the lubricating oil in the compressor is constantly changing.

The cylinder head in the compressor contains valves that are designed to open in one direction only. An inlet valve is designed to open when low pressure is created in the cylinder. A discharge valve is designed to open only when cylinder pressure is greater than the pressure in the discharge line.

Compressor Operation. When the engine is run, it rotates the compressor crankshaft, causing the pistons to reciprocate in the cylinder bores. The internal components of the compressor are shown in an exploded view in **Figure 8-4.** On the downward stroke of the piston, low pressure is created in the cylinder. This low pressure unseats the inlet valve, which allows filtered air to be pulled into the cylinder. When the piston reaches the bottom of its stroke, it stops, then reverses. As the piston is driven upward, it pressurizes the air charge, closing the inlet valve. When the pressure in the cylinder exceeds the pressure on the outlet side of the discharge valve, the discharge valve opens. When the discharge valve is open, the compressed air in the cylinder is unloaded into the discharge line toward the system supply tank. This operating cycle is shown in **Figures 8-5** and **8-6.**

Compressor Cycles. Because most compressors are direct-driven by the vehicle engine, the compressor

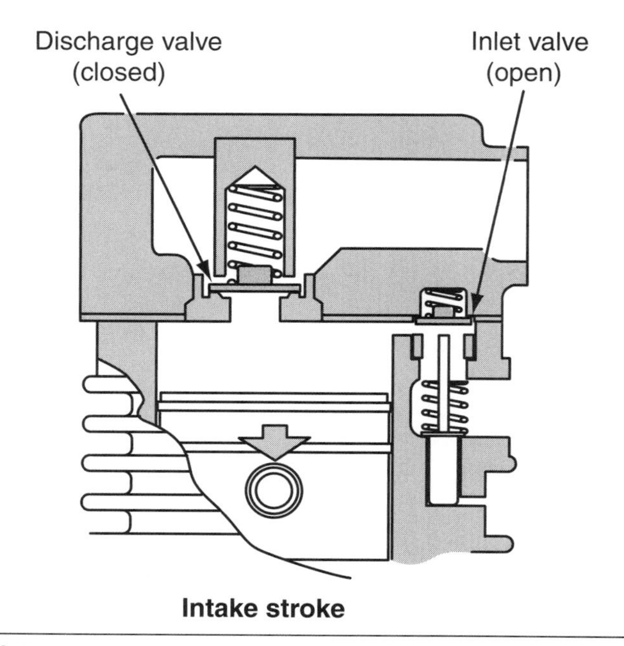

Figure 8-5 Air compressor operation: intake.

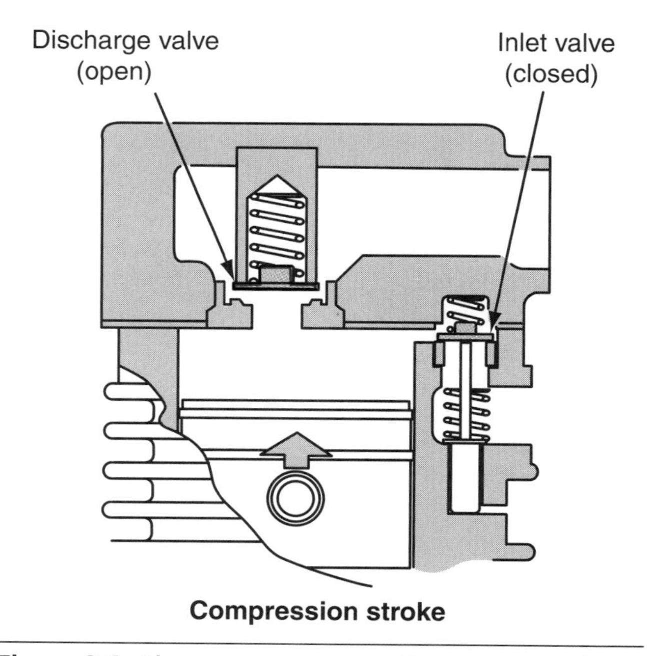

Figure 8-6 Air compressor operation: compression.

is continuously rotated. However, while being driven it may be in either a loaded cycle or unloaded cycle. These cycles are controlled by an air governor. When the compressor is in loaded cycle, it compresses the air and delivers it to the discharge circuit. The governor senses pressure in the supply tank (if equipped) or primary tank (when there is no supply tank) by means of a small gauge signal line. When the governor signals the compressor into unloaded cycle, an air signal is delivered from the governor unloader port to unloader pistons located in the compressor cylinder head. When the unloader pistons are actuated, they hold the inlet valves open. Each piston then pulls in air on its downstroke and immediately blows it back out of the inlet valve on the upstroke. In other words, no compression takes place. The unloaded cycle means that the compressor is being rotated but is not performing the work of compressing air.

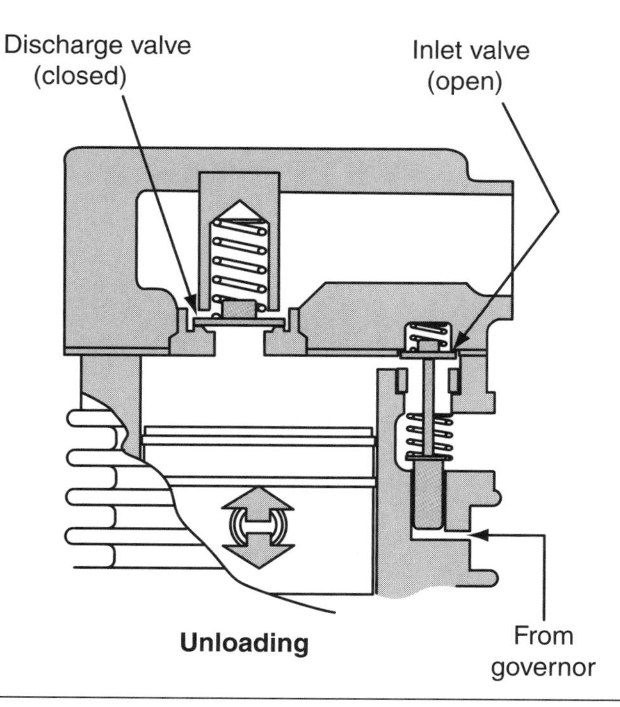

Figure 8-7 Air compressor unloading.

Governor cut-in occurs when the **system pressure** has dropped below a specified value and requires recharging. The compressor goes into loaded cycle at governor cut-in. The compressor goes into unloaded cycle at governor cut-out. This is the specified maximum system pressure. **Figure 8-7** shows the operation of the unloader mechanism from a typical air compressor.

Governors

The functions of the air governor are to monitor system air pressure and manage the loaded and unloaded cycles of the compressor. The governor assembly can be mounted directly to the air compressor (not generally recommended because of higher temperatures close to the compressor) or remotely, often on the firewall of the engine compartment. **Figure 8-8** shows a typical mounting location of an air governor.

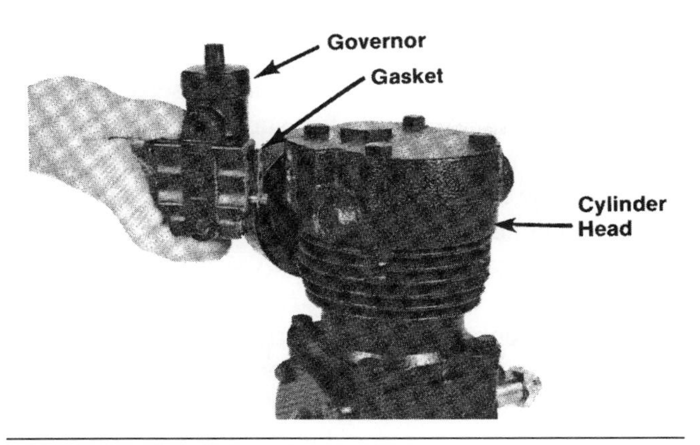

Figure 8-8 Example of a firewall mounted governor. *(Courtesy of Haldex Brake Products, Inc.)*

Air pressure in any air brake system must be carefully managed at what is called system pressure. System pressure falls within the range of pressures between governed pressure (maximum system pressure) and compressor cut-in pressure (minimum system pressure). Governed pressure is usually set between 120 and 130 psi on most vehicles, but you should note the following, which may or may not be adhered to by manufacturers of heavy off-highway equipment:

- FMVSS 121 specifies that setting governor cut-out pressure at any value between 115 and 135 psi is acceptable.
- The Technical and Maintenance Council (TMC) of the American Trucking Association recommends that governor pressure be set at 125 psi.

The governor measures system pressure via a signal line that runs from the supply tank and acts on the lower sectional area of the governor piston. The governor spring acts in opposition to this piston. Governor spring tension is adjusted to define the governor cut-out pressure value. When the system pressure achieves the specified maximum, an air signal is sent from the governor unloader port to the unloader pistons in the compressor cylinder head. When the unloader pistons are actuated, the inlet valves are held open and the compressor is no longer capable of compressing air.

As system air is used by the vehicle air circuits, system pressure drops. When the system pressure drops to between 20 and 25 psi below governed pressure, the governor spring forces the governor piston to its seated position. This triggers an exhaust valve to dump the air signal to the compressor unloader pistons, allowing the inlet valves to seat once again. Governor cut-in occurs at the moment the unloader signal to the compressor is exhausted by the governor and the compressor returns to its effective cycle.

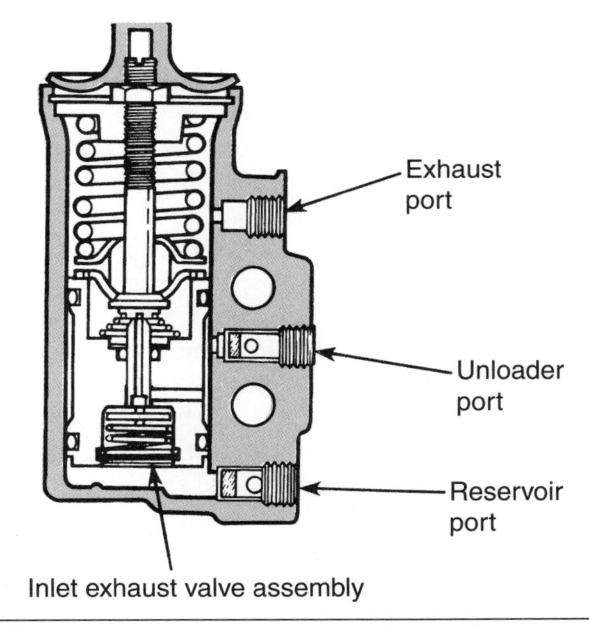

Figure 8-9 Cutaway view of a governor showing the connection ports. *(Courtesy of Bendix Commercial Vehicle Systems LLC. All rights reserved)*

System Pressure. The working air pressure on an air brake equipped vehicle is referred to as system pressure. It should be noted that system pressure is not a specific value on any system; rather, it is any value between the governor cut-in and governor cut-out values. For instance, on an air brake system with governor cut-out set at 125 psi, system pressure would indicate any air pressure value between 125 psi and 100 psi. This definition of system pressure is important when it comes to testing system performance with diagnostic pressure gauges. **Figure 8-9** shows a cutaway view of a typical governor, identifying the connection ports.

Tech Tip: Because FMVSS 121 requires governor cut-out be set within the range of 115 psi to 135 psi, most off-highway heavy equipment OEMs also adhere to this specification. Governor cut-in is usually set to occur at no more than 25 psi less than governor cut-out pressure. Should the difference between governor cut-out and cut-in be less than 20 psi, the result would be too-frequent cycling of the compressor loaded and unloaded cycles.

Air Dryer

An air dryer is a common option on an air brake system, and most equipment today uses one, especially when operating in humid or cold conditions. **Figure 8-10** shows the location of an air dryer in the supply circuit. The air compressed by the compressor is either atmospheric or ported boost air from the turbo, both of which contain some percentage of vaporized moisture. This moisture can harm air system valves if it is not removed. The air dryer is located between the compressor discharge and the supply tank. Compressed air is heated to temperatures up to 300°F (150°C) or higher when using a boost air source. Air dryers use two methods of separating moisture from the compressed air.

The air dryer functions primarily as a heat exchanger that is, an **aftercooler,** to reduce the air temperature, dropping it to a value cool enough to condense any vaporized moisture it receives. Actually, first-stage cooling of the hot compressed air occurs in the discharge line that connects the compressor with the dryer. In some older applications, copper lines were used for this purpose. Current steel-braided discharge lines still play a role in cooling compressed air. They do not achieve this as efficiently as copper line, but flexible steel-braided line does outlast copper.

Air dryers often use external fins to increase cooling efficiency. When compressed air enters the dryer, it first passes through an oil filter that removes trace quantities of oil that have been discharged by the compressor. Next, the air is passed through a desiccant pack designed to absorb water. Some air dryers use only one of these principles to remove moisture from air, whereas others use both. Water and oil separated and condensed in the dryer drain to a sump. Dry air exits the dryer and is routed to the supply tank. **Figure 8-11** shows a cutaway view of a typical air dryer.

The water and oil that drain to the sump must be discharged or purged from the dryer. A signal from the unloader port of the governor (the same signal that "unloads" the compressor at governor cut-out) is used to trigger a purge valve in the air dryer sump. Any liquid in the sump is blown out during the dryer purge cycle.

Some dryers are equipped with an electric heater to prevent winter freeze-up. A thermostat controls the heater. The dryer shown in Figure 8-13 is equipped with a heater. It should be noted that air dryers are designed to separate the normal trace quantities of oil discharged by a compressor. A compressor that is pumping oil will rapidly destroy the dryer desiccant packs. Desiccant packs require replacement, largely dependent on how much contaminant they have been exposed to. **Figure 8-12** shows some of the many different types of air dryers used on modern equipment.

Dryer Reservoir Module. A **dryer reservoir module (DRM)** is an integrated reservoir, dryer, governor, and four-valve pressure protection pack that has become more common. The components that make up this assembly all function as they do in a nonintegrated assembly. The third photograph from the bottom in Figure 8-12 shows one type of DRM unit.

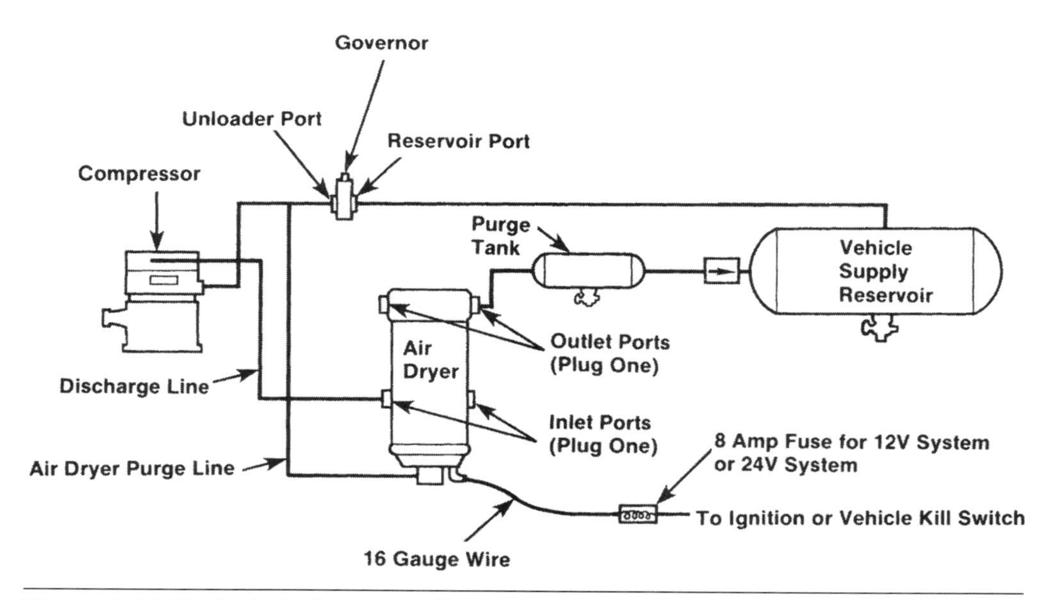

Figure 8-10 Typical air dryer installation. (*Courtesy of Bendix Commercial Vehicle Systems LLC. All rights reserved*)

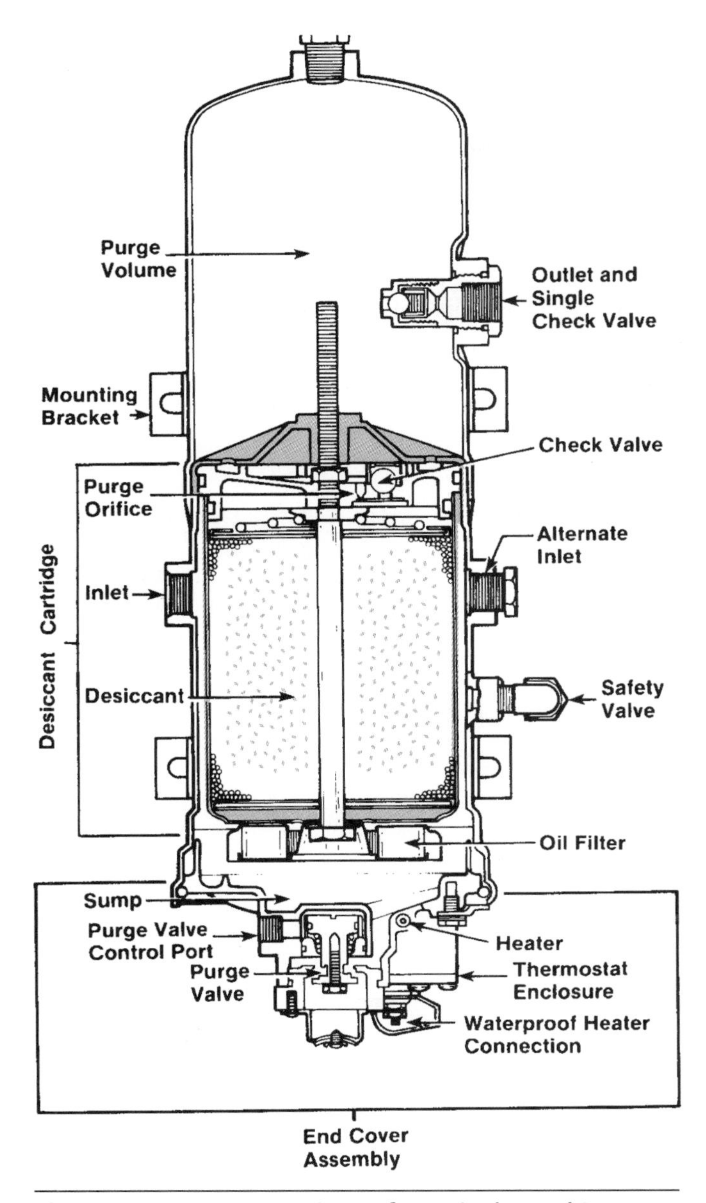

Figure 8-11 Cutaway view of an air dryer. (*Courtesy of Bendix Commercial Vehicle Systems LLC. All rights reserved*)

Tech Tip: Air dryers must be fitted with a safety valve. Because the system safety valve is located in the supply tank, if the hose from the dryer to the supply tank gets plugged (with ice or contaminants) or kinked, the governor will not cut out the compressor effective cycle, and the resulting high pressure could explode the air dryer.

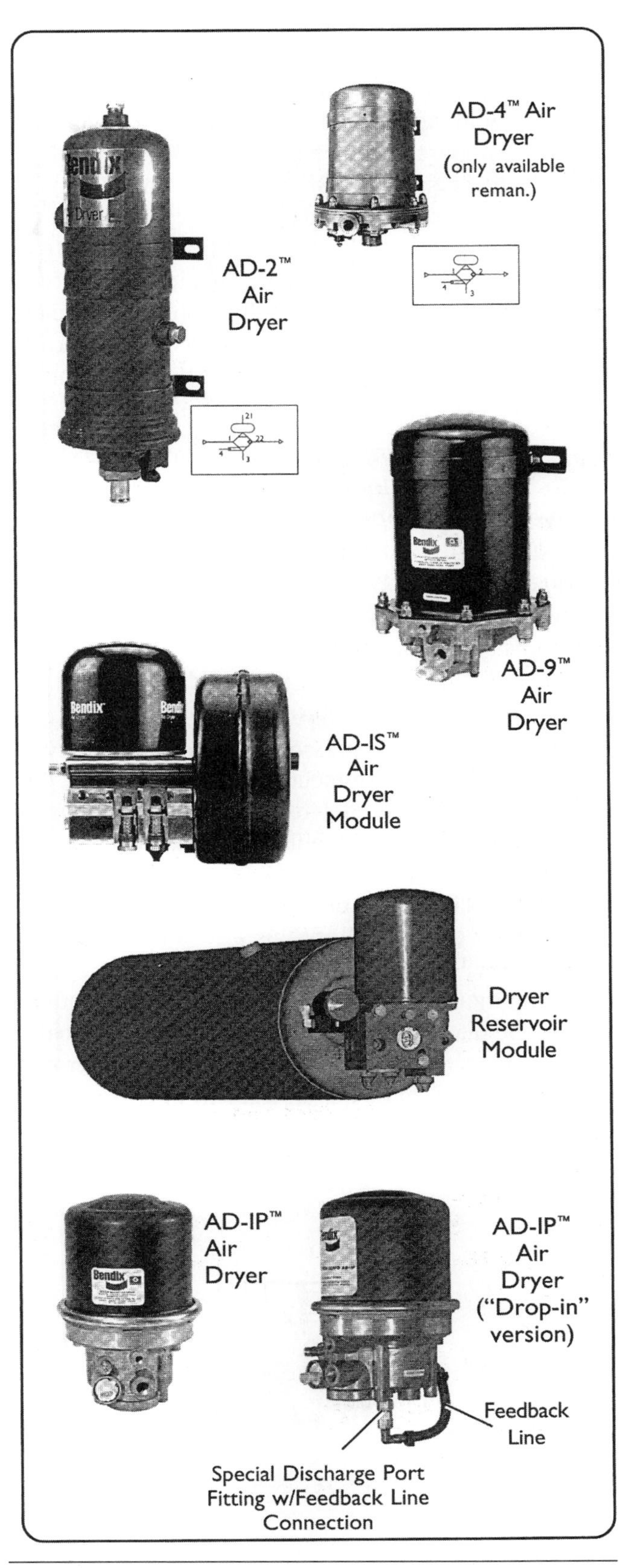

Figure 8-12 Assortment of air dryers used on mobile equipment air systems. (*Courtesy of Bendix Commercial Vehicle Systems LLC. All rights reserved*)

Alcohol Evaporators

Not all of the moisture in the air can be removed. Alcohol evaporators are used in some systems to help prevent freeze-ups. Alcohol vapor is introduced into the system by the alcohol evaporator. This forms a solution with water and lowers its freezing point. Alcohol evaporators should be located downstream from the air dryer. It is important to use only alcohols intended for use in air brake systems. Brake valves contain moving parts and most are lubricated on assembly to last for the life of the component. Pure alcohol (methyl hydrate or methyl alcohol) can dry out lubricants in brake valves, causing premature failure. Alcohols sold specifically for use in brake systems contain brake valve lubricant. **Figure 8-13** shows a typical alcohol evaporator unit.

WARNING *Alcohol evaporators MUST be located downstream from the air dryer in the supply circuit. Alcohol will turn air dryer desiccant into mush if pumped through the system.*

Supply Tank

The supply tank (or *reservoir*) is the first tank to receive air from the compressor. It is not mandatory to equip off-highway equipment with a supply tank, but when used, it is located downstream from the compressor. The supply tank is sometimes referred to as the **wet tank** because, in a system with no air dryer, this is where the air cools and condensing moisture collects. If a compressor discharged oil, this also would collect in the wet tank.

In common with all other air tanks on a mobile chassis, the internal walls of the supply tank are coated with an anticorrosive liner. The anticorrosive liner protects the steel walls of the air tank from corrosion by moisture. The liner itself, however, is susceptible to the carbon coking that can be caused by a compressor pumping oil.

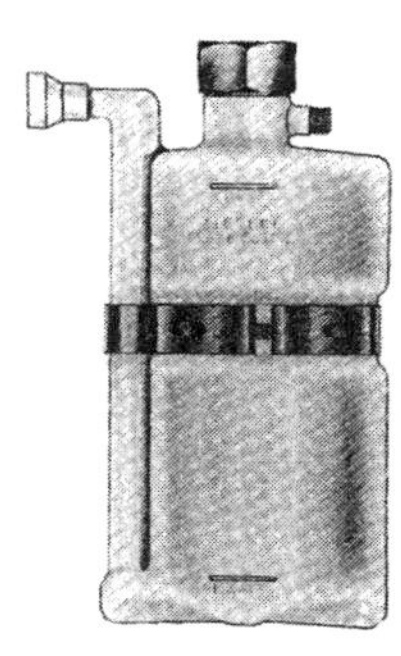

Figure 8-13 Alcohol evaporator. (*Courtesy of Bendix Commercial Vehicle Systems LLC. All rights reserved*)

Reservoirs must be equipped with draining devices, usually known as dump valves. A draining device can be as simple as a hand-actuated drain cock or an electrically-heated automatic drain valve. Many reservoirs are pressure-protected by single **check valves.** A single check valve is a one-way device that will allow a tank inlet to be charged with air but will not discharge through the inlet. **Figure 8-14** shows some typical air tanks, dump valves, single check valves, and low-pressure indicators.

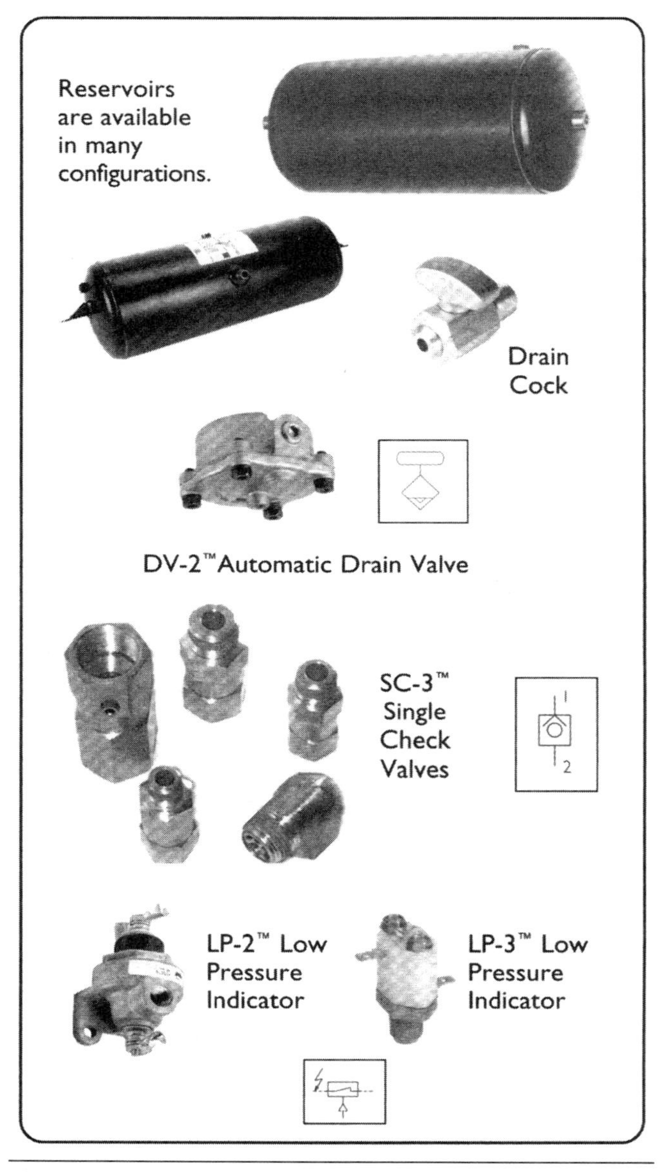

Figure 8-14 Typical air reservoirs, dump valves, check valves, and low pressure indicators. (*Courtesy of Bendix Commercial Vehicle Systems LLC. All rights reserved*)

The maximum working pressure of the tank is usually 150 psi. To ensure that this pressure is not exceeded, a **safety pressure-relief valve** is fitted. This is often known as a **pop-off valve.** Should something such as a debris-restricted signal hose cause the governor to malfunction and not signal compressor cut-out, the safety valve would trip at 150 psi to prevent damage to the chassis air circuits. **Figure 8-15** shows a typical safety pressure-relief valve.

Moisture Ejectors. Automatic moisture ejectors (see Figure 8-14) respond to changes in tank pressure and expel any moisture that has collected in the tank. They are sometimes known as spitter valves or automatic dump valves. All moisture ejectors must be able to manually open. As mentioned earlier, some moisture ejectors are equipped with thermostatically actuated heaters that prevent winter freeze-up.

Other Supply Circuit Valves. Other types of valves can be used in the supply circuit. Note that the same valves may be used elsewhere in the circuit. **Figure 8-16** shows some of these auxiliary valves. Their functions are briefly described here:

- **Double check valve:** Used when a component or circuit must receive or be controlled by the higher of two possible sources of air pressure.
- **Pressure protection valve:** A normally closed, pressure-sensitive control valve. Some are externally adjustable. Can be used to delay the charging of a circuit or reservoir, or to isolate a circuit or reservoir in the event of a pressure drop-off.
- **Pressure reducing valve:** Used in any part of a circuit that requires a constant set pressure lower than system pressure.

Tech Tip: Air tanks are pressure vessels. Air tanks are required to be hydrostatically tested after assembly. They should never be repair-welded. Even if the weld is sound, the heat causes the interior anticorrosive liner to melt and peel away from the tank wall. Failed air tanks always should be replaced.

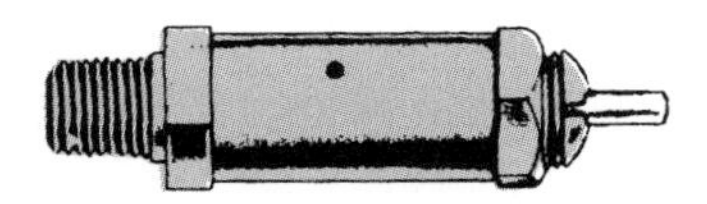

Figure 8-15 Safety pressure-relief valve, also known as a pop-off valve. (*Courtesy of Bendix Commercial Vehicle Systems LLC. All rights reserved*)

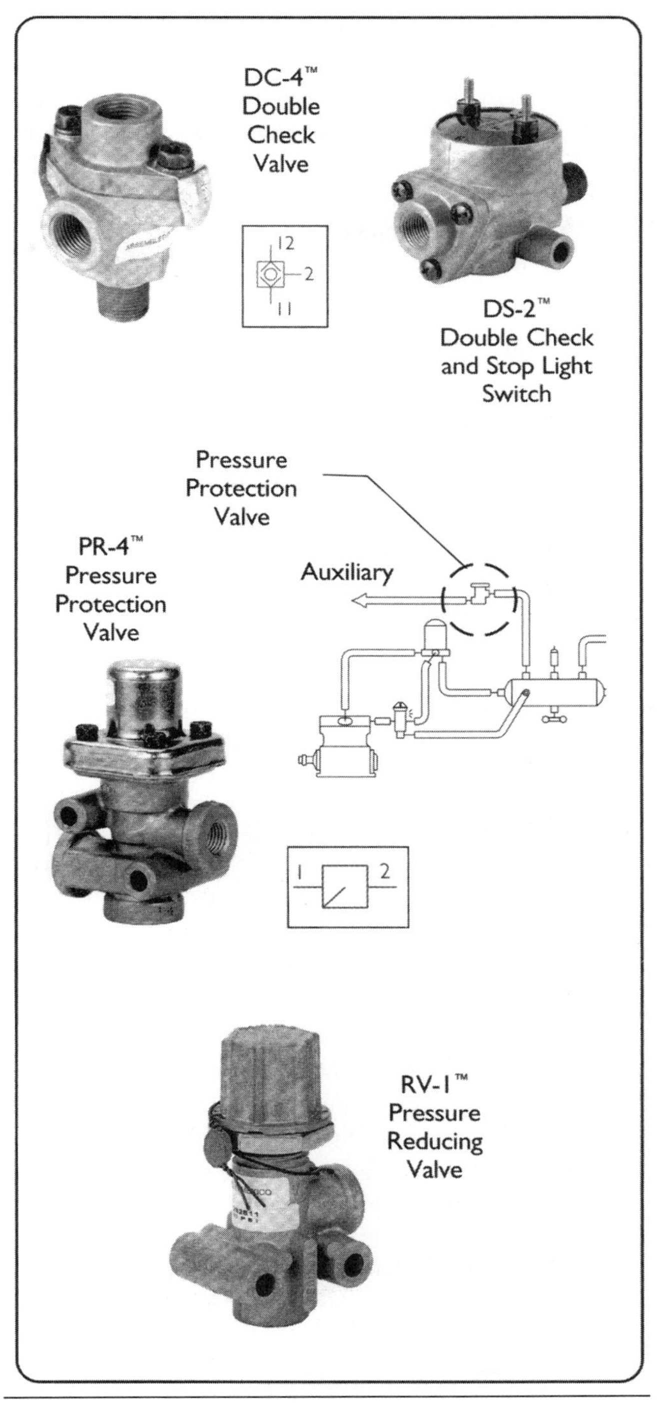

Figure 8-16 Auxiliary air system valves. (*Courtesy of Bendix Commercial Vehicle Systems LLC. All rights reserved*)

Low Pressure Indicator

In an air brake system, if the system air pressure drops too low, the brakes will cease to function properly. A low-pressure indicator is a pressure-sensing electrical switch that closes at any time the pressure drops below its preset value. The low-pressure indicator should be specified to close at 50% governor cut-out pressure. In a typical system, this will be at approximately 60 psi. When the switch closes, a light and a buzzer are actuated.

PRIMARY CIRCUIT

On a typical tractor/trailer combination, the primary service circuit is responsible for actuating the brakes over the tractor drive-axle tandem and foot-actuated applications of the trailer service brakes. This is why the primary circuit sometimes is referred to as the rear brake circuit. The secondary circuit is responsible for actuating the tractor front axle brakes and hand valve-actuated applications of the trailer service brakes. This explains why it is sometimes known as the front brake system.

Primary Tank

The primary tank receives air directly from the supply tank when equipped: in off-highway systems not equipped with a supply tank, air from the supply circuit is routed directly to the primary tank. If the primary and secondary tanks are supplied by the supply tank in series, the air must pass through the primary tank first. The supply air enters the primary tank by means of a one-way check valve.

This means that compressed air can only travel one way from the supply tank. For instance, if the supply tank pressure were to drop below that of the primary tank, air could not drain back. The one-way check valve helps isolate the primary circuit from the remainder of the chassis air system. Should a severe leak take place elsewhere in the system, the primary circuit air supply would be pressure-protected. Like all air tanks, the primary tank must be equipped with a drain cock or an automatic moisture ejector.

The primary tank has two main functions. The first is to make a supply of air at system pressure available to the **relay valves** for direct application of the service brakes. The second is to make a supply of air at system pressure available to the foot valve. The foot valve and primary circuit relay valves are plumbed directly into the primary tank.

Primary circuit pressure should be monitored by a dash gauge. A signal line supplies this gauge either directly from the tank or indirectly through the foot valve manifold. The gauge used in a specific chassis may be a single unit with separate, color-coded needles to indicate primary and secondary circuit pressure, or it may be two separate gauges, each identified as primary and secondary. The primary circuit is generally green; on a single-unit, two-needle gauge, the green needle would indicate the primary circuit pressure.

The primary tank also supplies air to the dash control valve assembly, the push-pull valves, or the modular dash control valve assembly. This supply is ensured by a **two-way check valve,** so that in the event of an upstream primary circuit failure, air would be routed to the dash control valves from the secondary tank by shuttling a two-way check valve. In older systems that used a two-way check valve supplied from the primary and secondary tanks, if a line ruptured downstream from the two-way check valve, the air in both circuits would be lost. In the current modular dash control valve assembly, only one circuit is capable of discharging.

Role of the Foot Valve in the Primary Circuit

Foot valves are shown in **Figure 8-17, Figure 8-18,** and **Figure 8-19.** The primary portion of the foot valve is actuated mechanically. This means that it is directly connected to the treadle. A supply of air from the primary reservoir is made available to the primary supply port at system pressure. Travel of the primary piston will determine how much air is metered out to actuate whatever components are in the primary circuit. Also, the air metered out by the foot valve is used as a signal to actuate relay valves on both the tractor and the trailer. When the foot valve is actuated, the primary circuit air does the following:

- It actuates or signals the brake valves plumbed into the primary circuit.
- It actuates the relay piston in the foot valve, that is, it actuates the secondary circuit.
- It reacts against the applied foot pressure to provide "brake feel."

Brake Valves in the Primary Circuit

In most cases, relay valves will be used to actuate the brakes on the primary circuit. These usually will be the rear axle brakes. The relay valve signal is delivered from the foot valve using a small-gauge signal line. The small-gauge line is used to reduce lag. Signal lag tends to be greatest when the volume of compressed air is greatest. The relay valve is also connected directly to a closer supply of compressed air, either by a large-gauge air hose or sometimes directly coupled to the tank itself.

The relay valve is actuated by the signal pressure and is designed to deliver a much larger volume from the local supply of air to the service brake chambers it supplies. The signal pressure delivered to the relay valve and

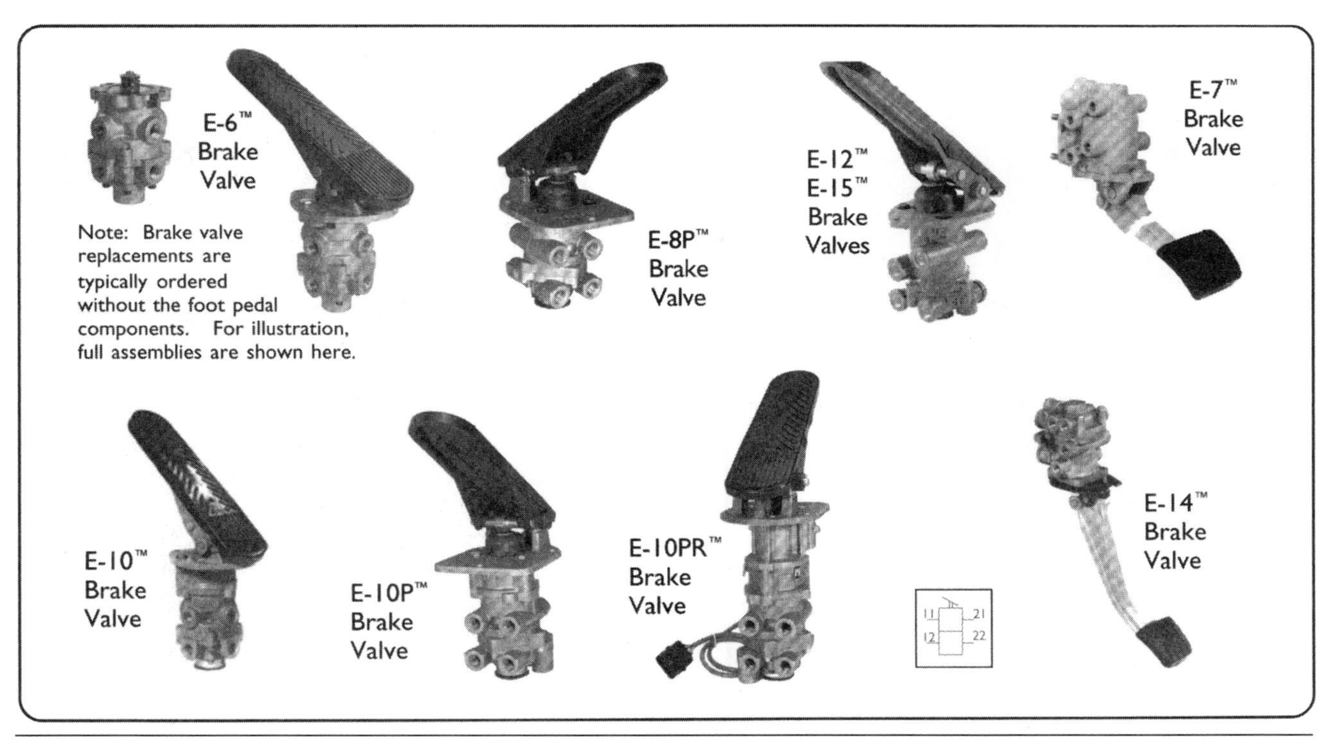

Figure 8-17 Types of dual-circuit brake foot pedal control valve: They are also called application or treadle valves. (*Courtesy of Bendix Commercial Vehicle Systems LLC. All rights reserved*)

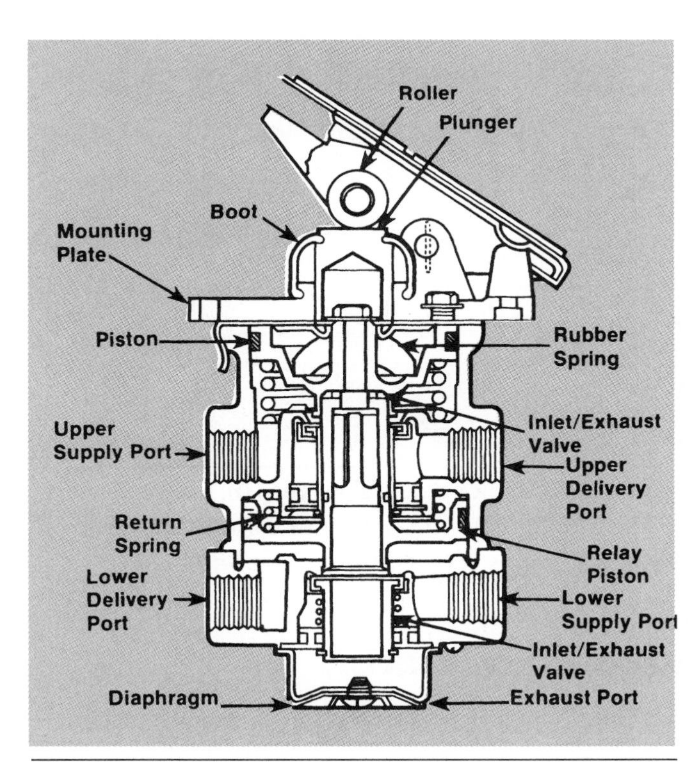

Figure 8-18 Cutaway view of a typical dual circuit brake valve. (*Courtesy of Bendix Commercial Vehicle Systems LLC. All rights reserved*)

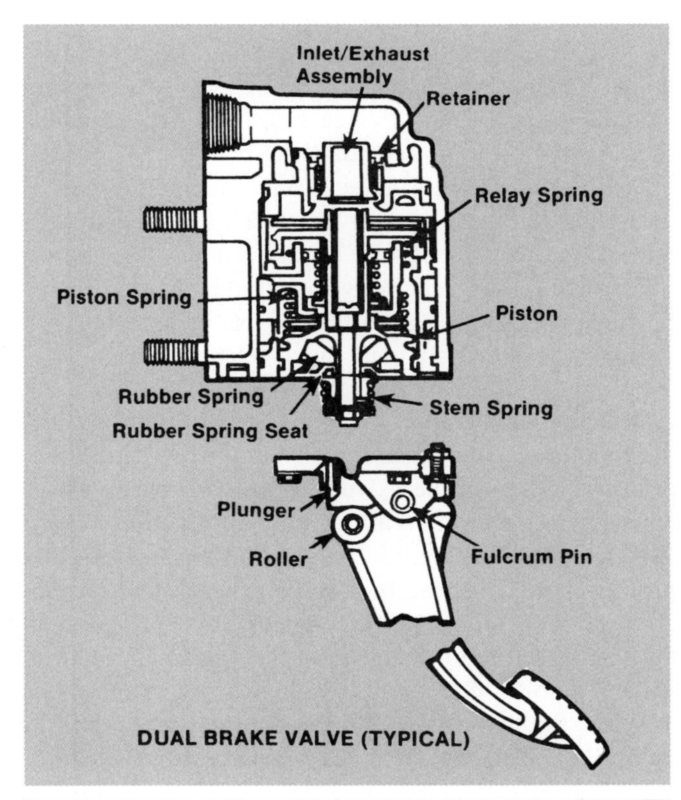

Figure 8-19 Suspended pedal brake valve. (*Courtesy of Bendix Commercial Vehicle Systems LLC. All rights reserved*)

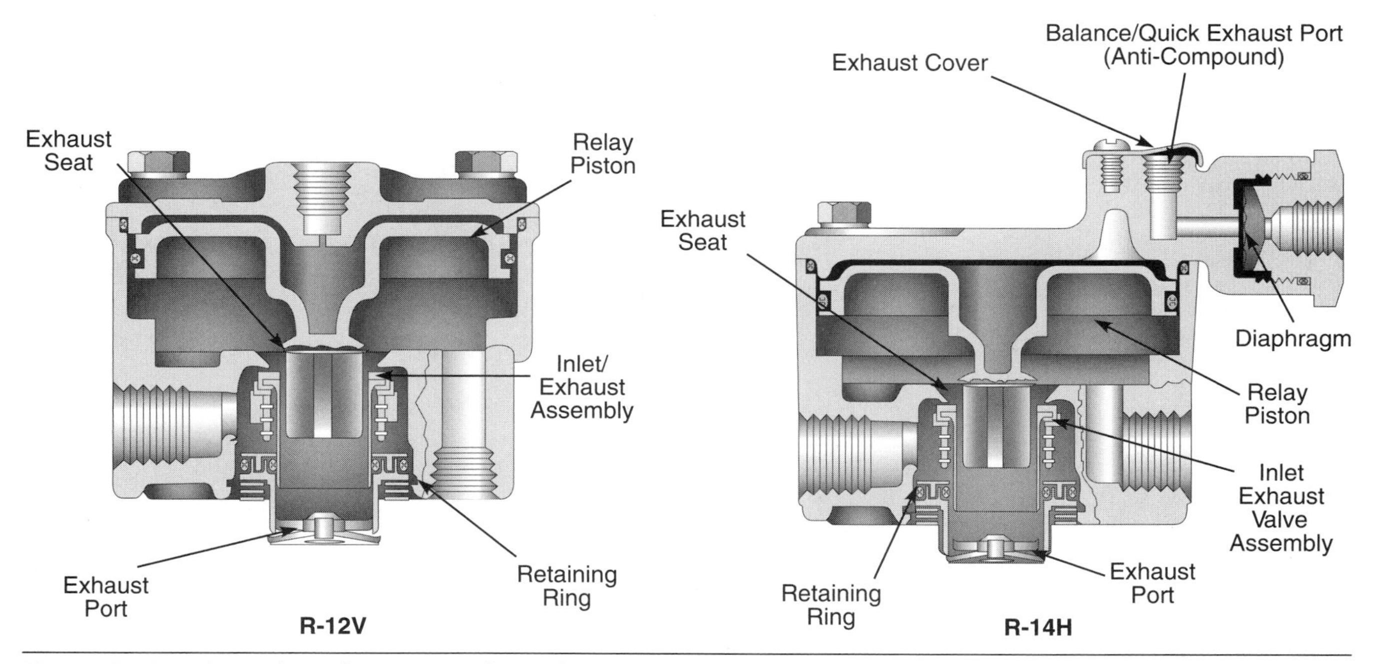

Figure 8-20 Relay valves. (*Courtesy of Bendix Commercial Vehicle Systems LLC. All rights reserved*)

the application pressure it delivers to the brake chambers are designed to be identical in most brake systems. The low volume of signal air (in relation to the application air) speeds up application and release times. **Figure 8-20** shows a couple of common types of relay valve.

Some systems, mostly much older systems, use quick-release valves in place of the relay valves to route air to the tractor rear service brakes. When a quick-release valve is used to actuate the rear tractor service brakes, the air that acts on the service diaphragms has to be routed through the foot valve. This results in extended lag time. To optimize actuation and release times for service brakes, relay valves tend to be used in place of quick-release valves on tractor rear service brakes. **Figure 8-21** shows a typical quick-release valve.

Trailing Unit Primary Circuit Brake Signal. When the foot valve is depressed, a signal is routed from the primary portion of the valve to a two-way check valve downstream from the **trailing unit application valve.** Whenever the trailing unit application signal from the primary portion of the foot valve exceeds that on the opposite side of the two-way check valve, it shuttles the valve to route the signal to the tractor protection valve. The bias of the two-way check valve is designed to select the highest source pressure and route that to its outlet. Whenever the signal pressure from the foot valve is greater than the signal pressure from the trailer application valve, the trailing unit service application signal is provided by the primary circuit.

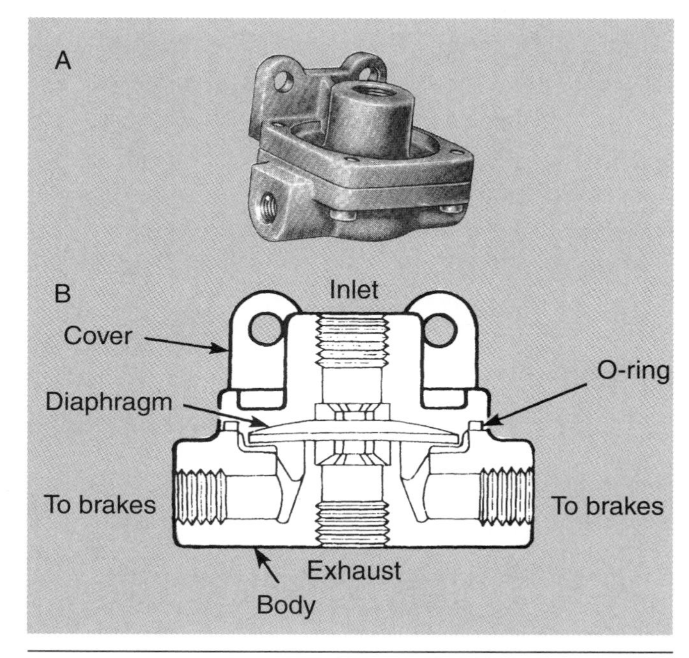

Figure 8-21 Quick-release valve: (A) External view, (B) Sectional view. (*Courtesy of Bendix Commercial Vehicle Systems LLC. All rights reserved*)

SECONDARY CIRCUIT

The secondary service circuit is usually responsible for actuating the brakes over the tractor steering axle and trailing unit application valve-sourced applications of the trailing unit service brakes. This is why the secondary circuit is sometimes referred to as the front brake circuit.

Secondary Tank

The secondary tank can receive air directly from the supply tank, or air may be routed to it indirectly through the primary tank. Air in the secondary tank is pressure-protected by a one-way check valve. This one-way check valve isolates the secondary circuit from the rest of the chassis air system, meaning that loss of compressed air elsewhere in the circuit will not affect the secondary circuit air supply. Like all air tanks, the secondary tank must be equipped with a drain cock or automatic moisture ejector.

The secondary tank has two main functions. The first is to make a supply of air at system pressure available to the relay valves or quick-release valves for direct application of the service brakes. The second is to make a supply of air at system pressure available to the secondary or relay portion of the foot valve. The foot valve and secondary circuit relay valves are plumbed to the secondary tank by means of air hoses.

Secondary circuit pressure is required to be monitored by a dash gauge. A signal line supplies this gauge either directly from the tank or indirectly through the foot valve manifold. If a single-unit gauge with separate needles to indicate primary and secondary circuit pressures is used, the secondary circuit needle indicator is red. When a two-gauge dash system is used, the secondary circuit gauge is identified as such. The secondary tank is connected to a two-way check valve supplying the dash control valve assembly. However, under normal operation, the two-way check valve prioritizes the primary circuit feed to the dash valve assembly. This means that secondary circuit air will only be used to supply the dash valve assembly in the event of a primary circuit failure.

Role of the Foot Valve in the Secondary Circuit

Again, make sure you understand the description of the dual-application foot valve earlier in this chapter. When it is operating normally, the secondary or relay portion of the foot valve is actuated pneumatically, that is, by air pressure. Whatever air pressure value is metered by the primary portion of the foot valve to actuate the primary circuit is also metered to actuate the relay portion of the valve. The relay portion of the foot valve is designed to act exactly as a relay valve. This means that it will replicate the signal pressure value from the primary portion of the foot valve, sending out exactly that value using secondary tank source air to actuate the brakes on the secondary circuit.

In the event of total failure of the primary circuit air supply, the foot valve primary piston will travel downward until it mechanically contacts the relay piston. Because the relay piston meters air according to the distance it travels, it functions as usual, metering secondary source air to actuate the secondary circuit brakes. In the event of a total failure of the secondary circuit air, no air will be metered out to actuate the secondary circuit when the relay piston is displaced.

Remember, whenever a failure occurs in either the primary or secondary circuits, the operator is alerted to the condition audibly and visually and is required to bring the unit to a standstill and wait for mechanical assistance.

Brake Valves in the Secondary Circuit

Relay valves or quick-release valves may be used to actuate the vehicle brakes on the secondary circuit. In most cases, the secondary circuit is responsible for the front service brakes. In many older systems, because the front brakes were located close to the foot valve, they were actuated directly by the relay portion of the foot valve. In other words, the air directed out of the relay portion of the foot valve was the air that was actually used to charge the front service brake chambers. This air was routed to the chambers by means of a quick-release valve. This provided a short application lag that actually helped ensure balanced application timing because the air had to travel farther to reach the rear brakes. Because a quick-release valve was used, release timing lag was exactly what it would be if a relay valve was used.

Manually-switched front axle limiting valves have not been used for a number of years, but they may sometimes be seen on older and off-highway vehicles. These operate simply by limiting the application pressure to the front axle brakes by 50% whenever the *slippery road* setting is selected. When air directed to the front axle service brakes passes through an automatic **ratio valve,** it is proportioned to limit application pressure under conditions of less severe braking but to permit full front wheel braking under severe braking conditions.

Trailing Unit Secondary Circuit Brake Signal. The trailer application signal from either a dash- or steering column-mounted trailer application valve (also known as a trolley valve) is sourced from the secondary circuit. A supply of secondary circuit air at system pressure is available at the trailer application valve. When the lever on the valve is actuated, air is metered to the trailer service application signal line. Downstream from this application valve is a two-way check valve. The two source supplies to the two-way check valve are the trailer application valve and the foot valve. The bias of the two-way check valve is designed to select the highest source pressure and route that to its outlet. Whenever the signal pressure from the trailer application valve is greater than the signal pressure from the foot valve, the trailer service application signal is provided by the secondary circuit.

Service Brake Chambers. Service brake chambers are single brake chamber devices and should be differentiated from single hold-off chamber, dual-chamber spring, and rotochamber brake assemblies. When the single chamber on a service brake chambers is charged with air, it actuates the brakes on the unit: the light duty spring in the chamber functions to retract the brakes only. **Figures 8-22** and **8-23** show the operation of service brake chambers.

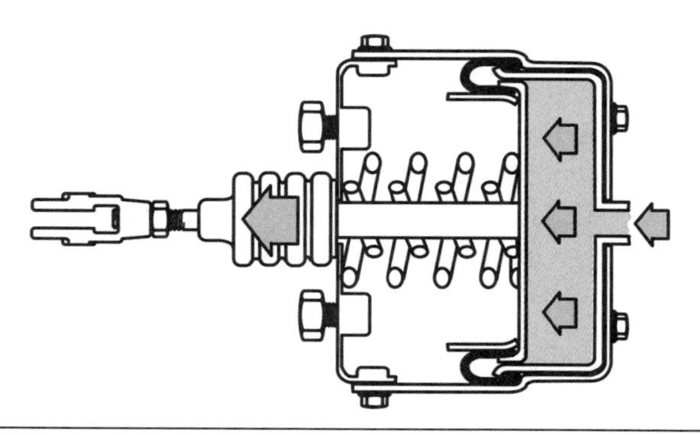

Figure 8-22 Service chamber subject to service brake pressure.

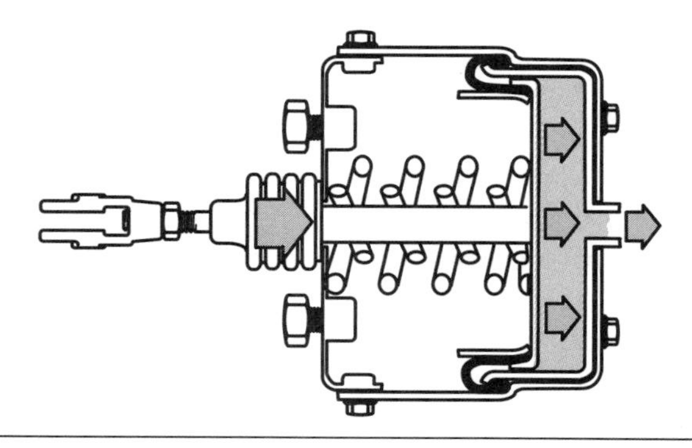

Figure 8-23 Service chamber discharging service brake pressure.

DASH CONTROL AND THE PARKING/EMERGENCY CIRCUIT

The parking and emergency circuits in an air brake system consist of dash-mounted hand control valves, spring brake chambers, and self-applying, fail-safe features. Because all highway vehicles equipped with air brakes must have a means of mechanically applying the parking brakes, so do most off-highway vehicles. Also, because air brakes are vulnerable to loss of braking capability when air leakage occurs, a fail-safe mechanically-actuated emergency brake system is required. On many off-highway, air brake equipped vehicles, there is a single dash control button.

Single Hold-Off Chamber Brakes

Single hold-off chamber brakes are designed to function as simple, single chamber park and service braking units. These units are equipped with a powerful spring that applies mechanical force to the slack adjuster to apply the brakes. To release the unit, hold-off air is applied to the brake chamber. Service braking is effected by modulating the hold-off pressure to the unit. **Figure 8-24** shows a single hold-off chamber with air applied to the hold-off diaphragm, releasing the braking force.

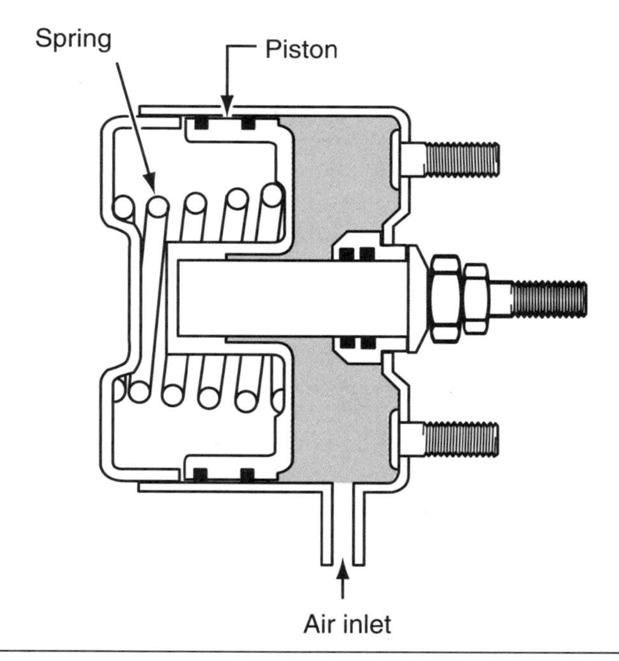

Figure 8-24 Single hold-off brake chamber in its released position.

Spring and Rotochamber Brakes

Spring brake and rotochamber assemblies are dual-chamber brake actuators designed to perform both service and parking/emergency brake function; they are used by most air brake equipped vehicles today. Service applications are performed by air pressure, whereas the parking/emergency function is performed mechanically; that is, by powerful springs. The two chambers are known as service and hold-off (emergency) chambers. Each chamber receives its own air supply that is directed to act on a diaphragm. When the service chamber receives air pressure, it performs service braking. The hold-off diaphragm must have air supplied to it in order to release—that is, cage—the main spring. Without air in the hold-off chambers, a vehicle with a properly functioning brake system will not move. **Figure 8-25** shows selection of spring brake and rotochamber assemblies used on equipment.

When the chassis system air pressure is zero, there is no air acting on the hold-off diaphragms on the equipment, so the main springs are fully applied to the spring chamber pushrod. This means the system is in park mode. When air pressure builds in the system and the parking brakes are released, air is directed to the hold-off chambers and the force of the main springs is "held-off" the brake chamber pushrod. At least 60 psi must be present in all the spring brake hold-off chambers before the vehicle parking brakes can be considered to be released.

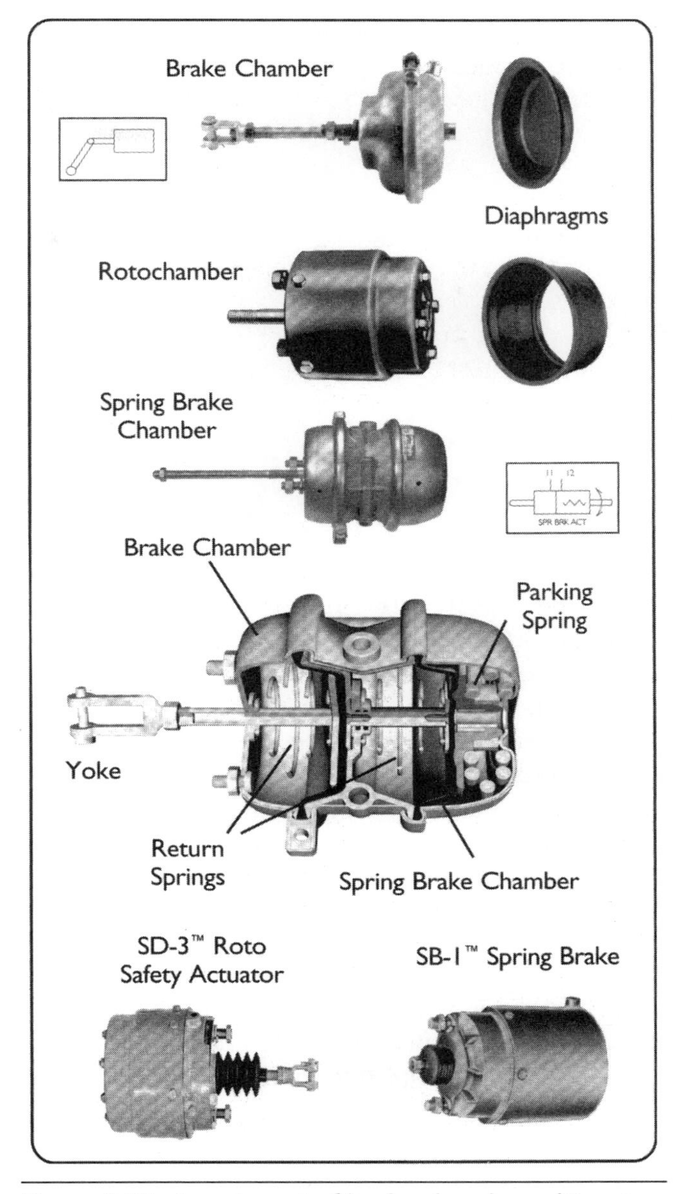

Figure 8-25 Assortment of brake chambers. (*Courtesy of Bendix Commercial Vehicle Systems LLC. All rights reserved*)

Dash Control Valves. Dash control valves are push-pull air valves that enable the operator to release and apply the vehicle parking brakes and supply the trailer with air pressure. Specifically, these valves manage pressure in the hold-off circuit of the equipment. Most of these valves are pressure-sensitive, so they will automatically move from the applied to exhaust position if the supply pressure drops below a certain minimum. In many off-highway applications, hold-off circuit air is controlled by a single dash valve such as that shown in **Figure 8-26.**

The dash valve assembly usually receives its air supply from the primary tank, but it may also be backed up with air from the secondary circuit by means of a two-way check valve. The two-way check valve receives air from two sources, in this case the primary and secondary reservoirs; and it is designed to shuttle to output whichever source arrives first or is at the higher pressure. Under normal conditions, primary circuit air is output by the two-way check valve, but in the event of a primary circuit failure, secondary circuit air is routed to the dash control valve assembly. The more recently introduced modular control valve assembly is equipped with a two-way check valve combined with pressure protection for each air pressure source. This means that if an air line failed downstream from the modular control valve, only one of the source air circuits will bleed down.

Three types of dash valve configuration can be used on off-highway equipment: one-, two-, and three-valve systems. The type depends on how the equipment is configured. Remember that their function is to manage

Figure 8-26 Dash parking brake control button. (*Courtesy of Caterpillar*)

air in the hold-off circuit. The function, color, and shape of each of the three valves may or may not conform to that required by on-highway FMVSS 121 legislation. In the description here, we will reference the FMVSS 121 characteristics of the dash control valves.

The shape of each valve is important because it enables it to be identified by touch. In on-highway applications the shank size of each valve is different, so it is not possible to install an incorrect knob on a valve. A single-valve system may be used on equipment that is not required to be coupled to trailing units and will be known as a system park valve. This is usually a push-pull valve, but toggling-type valves are also occasionally used. When air brakes are used on a tractor unit designed to haul a trailer, the system must be equipped with a minimum of two valves: the system park valve and the trailer supply valve. A three-valve system may also be used. A three-valve system incorporates the system park and trailer supply valves and adds a tractor park valve. This third dash control valve isolates the tractor park function from any trailing unit brakes. All three valves are described in the following paragraphs and can be observed in the air system schematics used in this chapter.

System Park. This is the only valve on the dash of a single-dash valve system. If it conforms to FMVSS 121, the dash control knob for system park should be diamond-shaped and yellow. It masters the lead vehicle parking brakes, that is, it manages hold-off air. If used in a multi-dash valve system, it also masters the trailer supply valve and tractor park valve. In older systems, neither of these valves could be actuated until the system park valve had been actuated. When the system park valve is pushed inward, it directs air to the spring brake hold-off chambers in the tractor brake system. The tractor hold-off chambers are charged with air that is routed from the system park valve and directed through quick-release valves before being delivered to the hold-off chambers.

When the system park valve is pulled out (exhausted), both the tractor and trailing unit parking brake systems are put into parking mode—that is, the spring brakes are applied by exhausting all hold-off air. In applications using the current modular-control valve assembly, the trailing unit (trailer supply valve pushed in) may be supplied with air with the tractor brake spring brakes applied, but, in these, trailer supply air is immediately cut when the system park valve is exhausted.

When the system park valve is pulled outward, the air supply to the trailing unit supply valve and the tractor hold-off circuit is exhausted. The trailer(s) immediately goes into park mode. The air in the hold-off circuit up to the quick-release valves is exhausted at the system park dash valve and the air in the hold-off chambers is dumped at the quick-release valve exhaust. Because the system park valve masters the trailer supply valve and, when used, the tractor park valve, the whole tractor/trailer combination is immediately put into park mode.

Trailing Unit(s) Supply. The trailing unit supply valve is usually octagonal in shape and red, consistent with that required in on-highway applications. Unless a modular control valve is used, the trailer supply valve can only be actuated after the system park valve has been actuated. When this happens, air is directed from the valve to the trailing unit supply port of the tractor protection valve, and, from there, directed by means of a hose and couplers to the trailing unit air circuit(s). In more recent applications using a modular-control valve assembly, the trailer may be charged with air before the system park valve is depressed; however, when both valves are depressed, pulling the system park valve out also will cut the trailer air supply. The immediate result is that the trailer supply valve pops out and the trailing unit(s) parking brakes are applied.

Whenever the trailing unit(s) supply valve is actuated (pushed in) and the system pressure is above 60 psi, air is directed into the trailing unit(s) spring brake hold-off chambers, releasing the parking brakes. When the system park valve is pulled out, the trailer supply air is cut, applying the trailer parking brakes. **Figure 8-27** shows some examples of dash circuit control valves.

Tractor Park. The tractor park valve, when used, is usually round and blue. It is only used in dash three-valve control systems and is mastered by the system park valve. In a dash three-valve system, when the system park valve is depressed, air is made available to the trailer supply and tractor park valves. Both these valves must be depressed before the rig is taken out of park mode. However, to apply the parking brakes in the whole rig, only the system park valve has to be pulled. You may come across these valves in some older air brake systems, especially off-highway construction dump trucks.

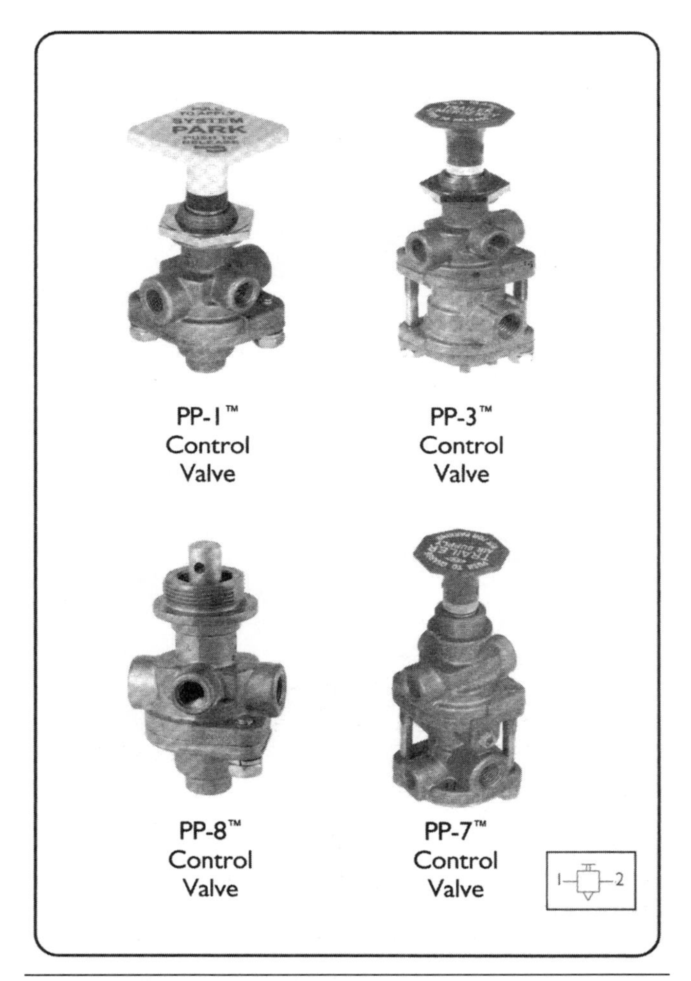

Figure 8-27 Dash control valves: Note the shape of the FMVSS 121 compliant knobs, which may or may not be used in off-highway applications. *(Courtesy of Bendix Commercial Vehicle Systems LLC. All rights reserved)*

Tech Tip: Make sure you understand the function of each dash control valve in the system. Block the wheels and get into a piece of equipment with a fully charged air system. Use the controls and observe the effect.

FOUNDATION BRAKES

So far, we have dealt with the components that manage the vehicle braking. The foundation brakes are the components that perform the braking at each wheel on the end of the axles. They consist of a means of actuating the brake, a mounting assembly, friction material, and a drum or rotor onto which braking force can be applied. Foundation brakes can be divided into three types: the S-cam type, disc, and wedge brake assemblies. **Figure 8-28** shows all three types.

S-cam Foundation Brakes

In an S-cam brake, the air system is connected to the foundation brake by a slack adjuster. The actuator is a brake chamber. Extending from the brake chamber is a pushrod. A clevis is threaded onto the pushrod, and a clevis pin both connects the pushrod to slack adjuster and allows it to pivot. The slack adjuster is splined to the S-camshaft, which is mounted in fixed brackets with bushing-type bearings.

When force from the brake chamber pushrod is applied to the slack adjuster, the S-camshaft is rotated. The S-camshaft extends into the foundation assembly in the wheel and its S-shaped cams are located between two arc-shaped **brake shoes.** The foundation assembly is mounted to the axle spider, which is bolted to a flange at the axle end. The brake shoes may either be mounted on the spider, with a closed anchor pin that acts as a pivot, or open-anchor mounted, meaning that they pivot on rollers onto which they are clamped by springs. Opposite the anchor end of the shoe is the actuating roller that rides on the S-cam. The shoes are lined with friction facings on their outer surface. Attached to the wheel assembly and rotating around the fixed brake shoes is a drum. The drum is integral with the wheel hub and rotates with it. When brake torque is applied to the S-camshaft, the S-shaped cams are driven into actuating rollers on the brake shoes, forcing the shoes into the drum. This action is shown in **Figure 8-29.**

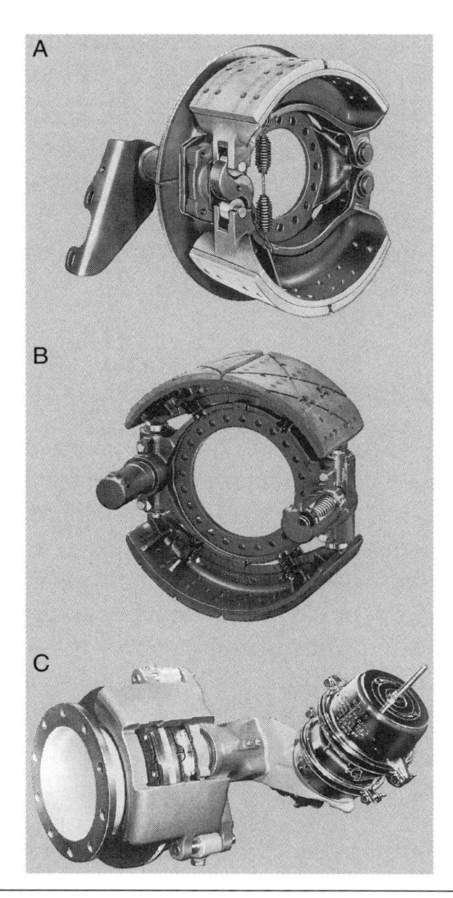

Figure 8-28 Foundation brake types: (A) S-cam brake, (B) Wedge brakes, (C) Disc brakes. (*Courtesy of ArvinMeritor*)

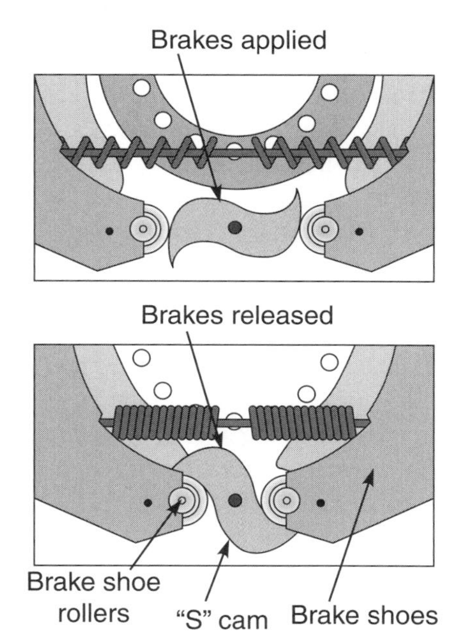

Figure 8-29 S-cam brake operation. (*Courtesy of ArvinMeritor*)

When the shoes are forced into the drum, the friction that retards the movement of the drum is created. In mechanical terms, the energy of motion is converted into friction, which must be dissipated by the drum as heat. Some cam-actuated brakes use flat cam geometry. This means that the shape of the actuating cam is flat instead of being S-shaped. Both types are shown in **Figure 8-30.**

S-Camshafts. The function of the S-camshaft is to convert the brake torque applied to it by the slack adjuster into the linear force that spreads the brake shoes and drives them into the brake drum. The slack adjuster is splined to the S-camshaft and retained by washers and snap rings. Two different spline configurations, known as coarse spline and fine spline, are used. They are also oriented left or right. They must never be interchanged.

S-camshafts are supported on the axle housing and on the brake spider by nylon bushings that permit the shaft to rotate to transmit brake torque to the shoes. The bushings should be lubricated with chassis grease on a routine basis.

CAUTION *S-cams are oriented with left and right cam profiles and must never be interchanged.*

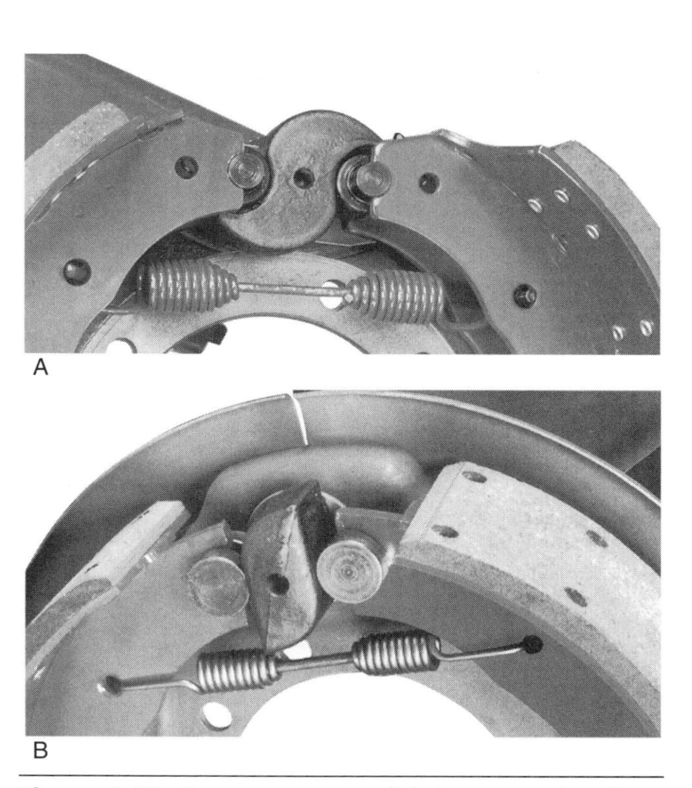

Figure 8-30 Cam geometry: (A) S-cam and rollers, (B) Flat cam and roller. (*Courtesy of ArvinMeritor*)

Spiders. The brake spider is bolted or welded to the axle end and provides a mounting for the foundation brake components. It must be tough enough to sustain both the mechanical forces and the heat to which it is subjected. The brake shoes are mounted to the spider by means of either open or closed anchors. The anchor is the pivot end of the shoe.

Brake Shoes. Brake shoes are arc-shaped to fit to the inside diameter of the drum that encloses them. They are mounted to the axle spider assembly. Each shoe pivots on its own anchor pin retained in a bore in the brake spider. Today, most shoes are of an open-anchor design. This means that the anchor end of the shoe does not enclose the anchor pin that forms the pivot. In most cases, open-anchor shoes are retained at the anchor end by a pair of springs that clamp the shoes to each other, forcing them against their individual anchors. On the actuation end of the brake shoes, a roller is retained by a clip spring or pin, as shown in **Figure 8-31.**

A single retraction spring loads the brake shoe rollers onto the S-cam profile. The function of the retraction spring is to hold the shoes away from the inside surface of the brake drum when the brakes are not being applied. When the brakes are applied, the S-cam rotates, acting on the brake shoe rollers and forcing the brake shoes into the drum. The major advantages of open-anchor brake shoes are lower maintenance and faster brake jobs.

Closed-anchor shoes operate identically to open-anchor shoes. However, on the anchor end of the shoe, a pair of eyes encloses the spider-mounted anchor pin on either side. The disadvantage of the closed-anchor brake shoe design is that more maintenance time required to perform brake jobs, mainly because the anchor pins tend to seize in their spider bores.

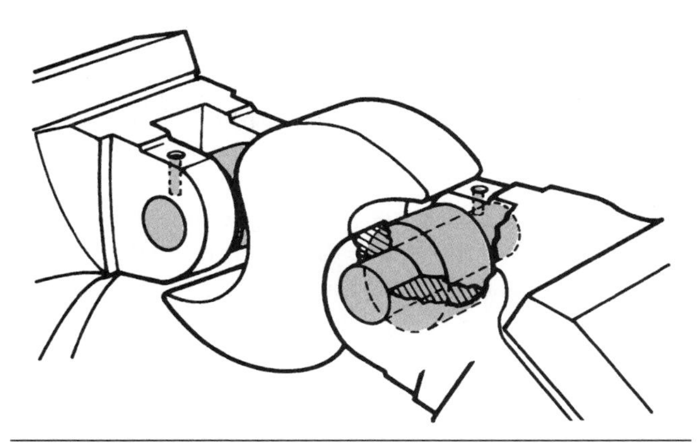

Figure 8-31 Brake shoe rollers. *(Courtesy of ArvinMeritor)*

Friction faces are attached to the shoes. These are graded by their **coefficient of friction** rating, which is given a letter code. Coefficient of friction describes the aggressiveness of any friction surface. For instance, sheet ice has lower friction resistance than an asphalt surface, so it can be said to have a lower coefficient of friction. However, the letter codes used on brake shoe friction linings are too general to be used as a reliable guide for replacement.

Ensure that the OEM recommended friction linings are used when relining shoes. It also should be noted that coefficient of friction alters with temperature, generally decreasing as temperature increases. The friction facing is either riveted or bolted to the shoe. The general practice today is to replace the whole shoe with one with new friction lining attached at each brake job. This reduces the overall time required for the brake job. **Figure 8-32** shows how a typical S-cam foundation brake is assembled.

Edge Codes. Letters imprinted on the edge of heavy-duty friction facings are used to describe the coefficient of friction (See **Table 8-1**). The first letter indicates the cold coefficient of friction, the second the hot coefficient of friction. The farther down in alphabetical sequence, the higher the coefficient of friction—that is, the more aggressive the lining. So, a G rating indicates a higher coefficient of friction than an F rating.

Remember, because of the wide range of friction coefficients covered by each letter rating, most industry opinion is that this is an inadequate rating of actual brake performance. Two letters are always used to rate any lining. For instance, a lining code rated FF would have both cold and hot coefficient of friction ratings within the range of 351–450.

Table 8-1: LETTER GRADED EDGE CODES

Letter	Coefficient of Friction Rating
D	0.150–0.250
E	0.251–0.350
F	0.351–0.450
G	0.451–0.550
H	Over 0.551

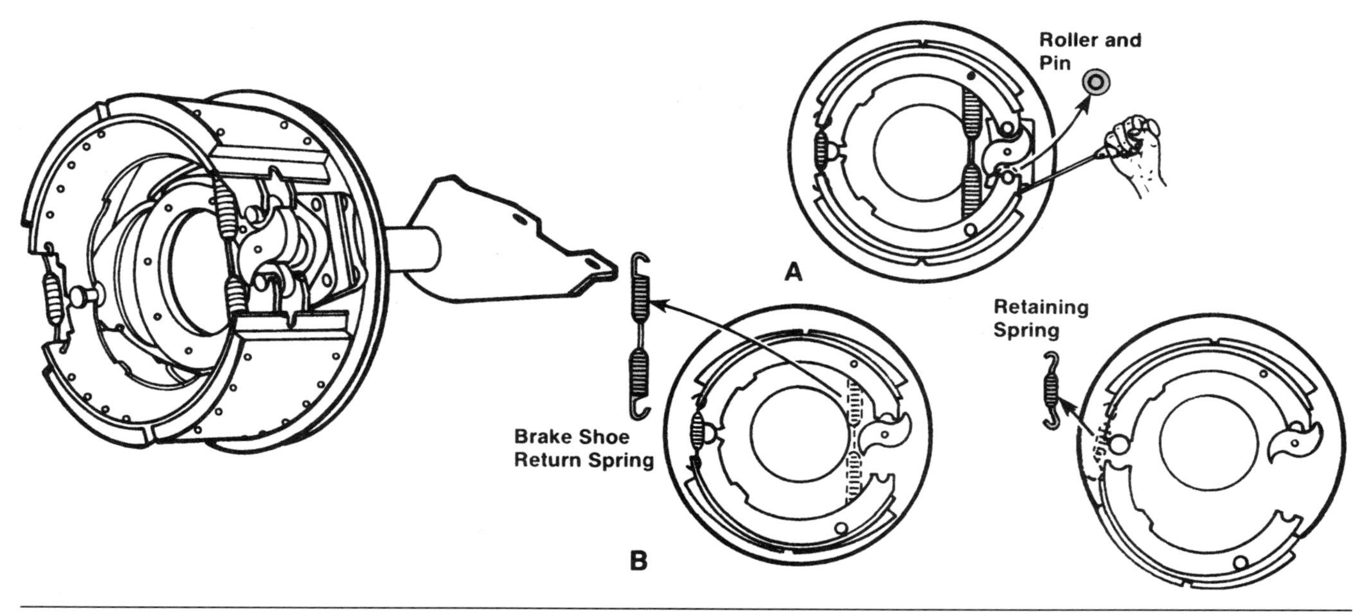

Figure 8-32 Assembling S-cam foundation brakes. (*Courtesy of ArvinMeritor*)

Brake Drums. Brake drums are manufactured from cast alloy steels or fabricated steel. The two have different coefficients of friction, so they should never be mixed. Brake drums are manufactured in several different sizes. The friction facings used in brake linings today, which do not use asbestos, tend to be harder. This means that the friction facings on the shoes last longer but can also be much tougher on brake drums. It has become routine in some applications for the brake drums (along with the shoes) to be changed at each brake job. Brake drums tend to harden in service. After they show heat checks (small cracks) and heat discoloration (blue shiny areas), they should be replaced rather than machined to oversize.

Brake drums are prone to out-of-round distortion when they are stored for any length of time; so it is good practice to machine drums when new to ensure they are concentric. It is not regarded as good practice to machine brake drums after they have been in service because of the tempering effect caused by extreme heat. When drums are replaced, the correct size and weight class should be observed to ensure the brakes are balanced. Drums of different weights will run at different temperatures. When drums are machined, ensure that the maximum machine and service dimensions are observed. For instance, on a 16.5-inch drum, the maximum recommended service dimension is 0.120 inch and the maximum machining dimension is 0.090 inch.

S-Cam Foundation Brake Hardware. The S-cam foundation brake hardware includes all the springs, anchor pins, bushings, and rollers on the spider assembly. It is good practice to replace everything at each brake overhaul. The springs used in the foundation brake assemblies must sustain high temperatures, which causes them to lose their tension over time; it is critically important that these be changed. Cam rollers should be lubricated on the roller-pocket race and never on the cam contact face. Lubricating the cam roller-pocket race helps avoid flat spots on the cam contact face. The cam contact faces should never be lubricated because the roller is designed to rotate with the S-cam when brake torque is applied to it.

Wedge Brake Systems

In this brake design, the slack adjuster and S-camshaft are replaced by a wedge-and-roller mechanism that spreads the shoe against the drum. They may be single- or double-actuated. In a single-actuated wedge brake system, the shoes are mounted on anchors. A single wedge acts to spread the shoes, forcing them into the drum. Dual-actuated wedge brakes tend to be used in heavy-duty applications. The dual-actuated wedge brake assembly has an actuating wedge located at each end of the shoe. The actuating principle of a wedge brake assembly is shown in **Figure 8-33.**

The brake chambers in a wedge foundation brake system are attached directly to the axle spider. The wedge-and-roller actuation mechanism is enclosed within the actuator and chamber tube. When air is delivered to the

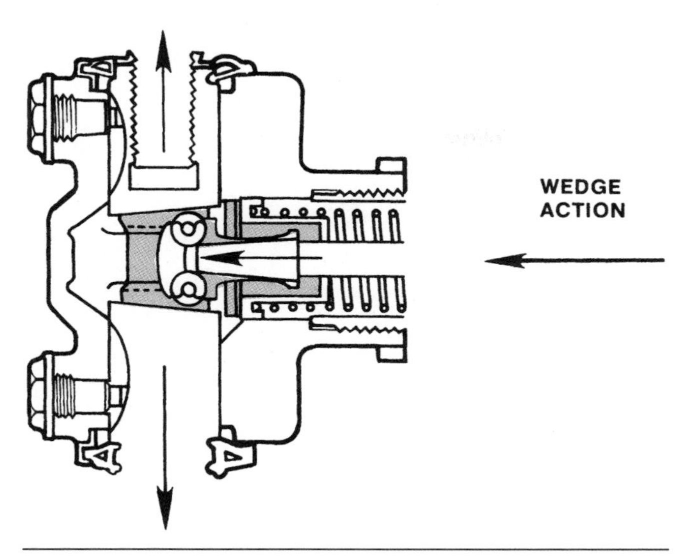

Figure 8-33 Wedge-actuated brake system. (*Courtesy of ArvinMeritor*)

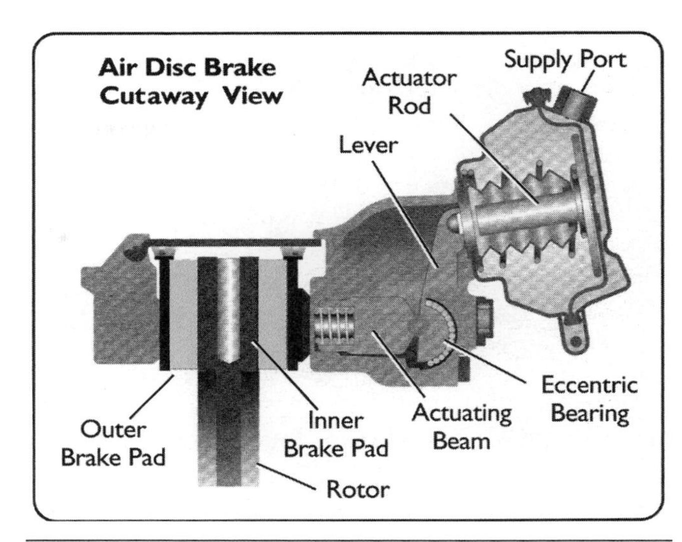

Figure 8-34 Air disc brake cutaway. (*Courtesy of Bendix Commercial Vehicle Systems LLC. All rights reserved*)

service chamber, the actuating wedge is driven between rollers that in turn act on plungers that force the shoes apart. When the service application air is exhausted from the service chambers, retraction springs pull the brake shoes away from the drum. The actuating wedges usually are machined at a 14-degree pitch, but they also may be machined at 12 or 16 degrees. They must be carefully checked on replacement.

A self-adjusting mechanism is standard on all wedge brakes, and this is contained in the wedge brake actuator cylinder. It consists of a serrated pawl and spring that engages to helical grooves in the actuator plungers. A ratcheting action prevents excessive lining-to-drum clearance occurring as the friction linings wear. Despite the fact that dual-actuated wedge brakes have higher braking efficiencies than S-cam brakes, they tend to be rarely used today. They are seen by many as having a much higher overall maintenance cost. Of note is the tendency of the automatic adjusting mechanism and plungers to seize when units are not frequently used. This often necessitates the disassembly of the foundation brake assembly to repair the problem.

Air Disc Brakes

Air **disc brakes** have been proven to have higher efficiencies than **drum brakes;** they also have higher initial, operating, and maintenance costs. **Figure 8-34** shows a cutaway view and operating principle of air disc brakes. Although the operating principles remain the same, the actuating mechanisms used differ somewhat between manufacturers.

The air disc brake assembly consists of a caliper assembly, a rotor, and an actuator mechanism. The actuator mechanism is a brake chamber, usually acting on an eccentric lever to amplify its force and convert the linear force produced by the chamber into brake torque to a pair of threaded tubes that act on the pads. The objective of an air disc brake system is to apply clamping force to a caliper assembly within which the pads are mounted. Running between the friction pads of the caliper is a rotor that is bolted to, and therefore rotates with, the wheel assembly. When the power beam is actuated, the threaded tubes force the brake pads into the rotor. Because the caliper is floating, when the inboard-located pad is forced into the rotor, the outboard friction pads are simultaneously forced into the outboard side of the rotor with an equal amount of force. There are also some older air-actuated disc brakes that use brake chambers that act on a slack adjuster to actuate the caliper assembly.

Air disc brakes provide the advantage of better brake control and higher braking efficiency than S-cam actuated brakes. They also are immune to brake fade and boast a slight overall weight advantage over S-cam brake assemblies. Air disc brakes are an excellent option in severe service applications where braking frequency and loads are at their highest. **Figure 8-35** shows a typical air disc foundation brake assembly.

Figure 8-35 Air disc brake foundation assembly. (*Courtesy of ArvinMeritor*)

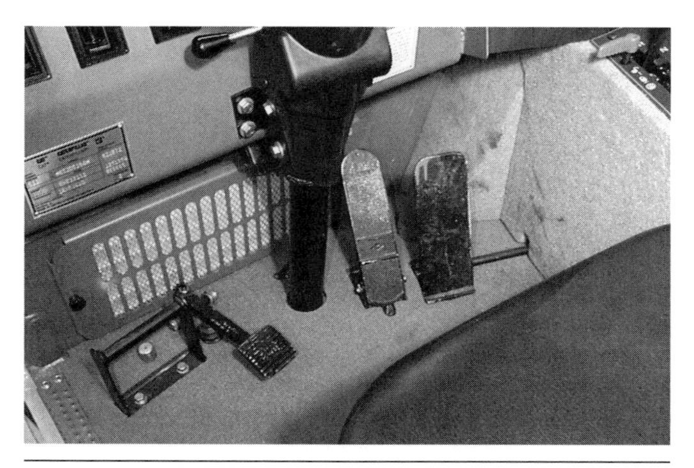

Figure 8-36 Location of a foot valve in a typical cab. (*Courtesy of Caterpillar*)

SERVICE BRAKE MANAGEMENT

This section will describe how an air brake system is managed by taking a closer look at some components. The foot valve is key to understanding air brake management. For instance, a foot valve plays a role in all service braking, so it has a role in both primary and secondary circuit application pressure management. The description of the foot valve provided in this section outlines all of its functions, so some information will be repeated from earlier sections.

Foot Valves

The term foot or treadle valve is used to describe the **dual-circuit application valve** required on all current highway truck brake systems and almost all off-highway air brake systems. Understanding how the foot valve operates is essential because it masters both the primary and secondary service brake circuits on an air brake system.

The foot valve is responsible for service brake applications. Service brake applications cover all vehicle braking except for parking/emergency braking. During the day-to-day operation of a vehicle, it is necessary to apply the brakes with exactly the right amount of force required to safely stop a vehicle. Too little force can result in the vehicle not stopping in time, whereas too much force can lock the wheels. The foot valve is the operator's means of managing the service brakes in an air brake system. It plays the same role as the brake pedal in your car. **Figure 8-36** shows a typical location of a foot valve in a cab.

In a hydraulic brake system, the application pressure delivered from the brake pedal is always associated with the amount of mechanical force applied by the foot, even in a power-assist system. Because of this relationship with the amount of pressure applied and the braking force, the operator is provided with what is known as "brake feel." The potential energy of an air brake system is compressed air. This means that the mechanical force applied to the brake pedal has no direct relationship to the resulting brake force. Instead, a different means of providing the operator with brake feel is required. In other words, the foot valve in an air brake system must be made to at least *feel* like a brake pedal in a hydraulic brake system. That is, the pedal should produce increased resistance as braking force increases. The main function of the foot valve is to provide the operator with a means of precisely managing the correct amount of brake force for each stopping requirement.

The foot valve is divided into two sections (see Figure 8-18 and Figure 8-19). The section closest to the treadle (foot pedal) is known as the primary section and is responsible for primary circuit braking. The section further on from the treadle is known as the relay section and is responsible for secondary circuit braking.

The primary section of the foot valve receives a supply of air from the primary tank at system air pressure. The relay section of the foot valve receives a supply of air directly from the secondary tank at system air pressure. When the operator depresses the brake pedal, the supply air from each section of the foot valve is metered out to apply the brakes on each circuit, primary and secondary. The air from one circuit never comes into direct contact with the other when in the foot valve.

Foot Valve Operation

The primary section of the foot valve is actuated mechanically; that is, by foot pressure. This causes the plunger under the treadle to move the primary piston. Any primary-piston travel off its seat results in the closing of the exhaust valve and then, depending on primary-piston travel, air being metered from the primary supply port out to the primary circuit service brake components. This metered air performs two other essential functions. First, it is directed downward to act on the relay (secondary) piston. The relay portion of the foot valve is responsible for actuating the secondary circuit brakes. It is designed to operate exactly like a relay valve. System pressure from the secondary tank is constantly available at its supply port. The air pressure value metered out to actuate the primary circuit brakes is also directed downward to actuate the relay piston. Like most relay valves, the foot valve relay piston is designed so the actuating signal pressure moves the piston to meter out pressure exactly equal to the signal pressure value to the secondary circuit.

Brake feel is an essential feature of an air brake system. Brake feel is provided by directing whatever air pressure value has been metered out to the primary circuit to act against the applied foot pressure. This means that the farther the treadle travels, the more foot resistance is felt, because the air pressure below it is greater. The idea is to provide air brakes with the feel of hydraulic brakes. In hydraulic brakes, whether power-assisted or not, actual braking force is always proportional to the mechanical force applied to the brake pedal. In air brakes, because the potential energy of the brake system is compressed air, the treadle valve simply meters out the required amount of compressed air to make a stop.

Primary Circuit Application: Normal Operation. Air from the primary reservoir is available to the primary section of the foot valve at system pressure. When the treadle on the foot valve is depressed, air from the primary circuit is metered out to:

1. Actuate the primary circuit brakes.
2. Act as a signal to actuate the relay portion of the foot valve.
3. Act against the applied foot pressure to provide brake feel.

If the treadle on the foot valve is depressed so that 20 psi of air is metered out to actuate the primary circuit brakes, 20 psi also will act on the relay piston to actuate the secondary circuit brakes and to act against the applied foot pressure to provide the operator with brake feel. In normal operation, the primary circuit is actuated mechanically.

Secondary Circuit Application: Normal Operation. When the primary circuit is actuated, primary circuit air is directed to enter the relay piston cavity. This air pressure moves the relay piston close the secondary exhaust valve and then to allow air to flow out of the secondary supply port to be metered out to actuate the secondary circuit brakes. This means that, in normal operation, the secondary circuit is actuated by air pressure.

Releasing Brakes. When the operator releases the brakes, the foot valve spring causes the primary piston to retract, opening the exhaust valve. Primary signal air (the air acting on the relay piston) and the air acting to provide brake feel are simultaneously exhausted through the primary exhaust valve. When primary circuit air acting on the relay piston is exhausted, the relay piston spring retracts the relay piston, exhausting the secondary circuit signal.

Circuit Failures

The main reason for having a dual-circuit service application system is to provide a fail-safe in the event of one circuit failure. This type of system redundancy is mandated by on-highway applications but for obvious reasons is desirable in off-highway applications. When one circuit fails, the companion circuit is designed to bring the vehicle to one safe stop. Total failures of one of the two circuits are rare, but not unknown. Physical damage to either of the circuit air tanks is an example of such a failure.

In the event of a failure of either circuit, the operator should bring the equipment to a standstill and await mechanical assistance. When such a failure occurs, the operator is alerted to the condition by gauge readings, warning lights, and a buzzer. As a fail-safe for both the primary and secondary circuits, the spring brakes or rotochambers bring the vehicle to a standstill.

Primary Circuit Failure. In the event of a primary circuit failure, the foot valve treadle will depress the primary piston, which will not meter any air because none will be available at the primary supply port. This means that no air will be metered out to actuate the primary circuit brakes, to actuate the relay piston, or to provide brake feel. The operator's foot on the treadle will now travel farther with little resistance until the primary piston physically contacts the relay piston. The relay piston will now be actuated mechanically and, according to the distance it travels, meter out air to actuate the secondary circuit brakes. The secondary circuit will operate normally, except that instead of being actuated by air from the primary circuit, it will be actuated mechanically.

Secondary Circuit Failure. In the event of a secondary circuit failure, the primary circuit will operate normally with no loss of brake feel. The air metered into the relay piston cavity will move the relay piston, but no secondary circuit air will be available at the supply port, so none will be metered out to the secondary circuit service brakes.

Front Wheel Limiting Valves

A front-wheel, limiting valve (also known as a "wet-dry" switch) is used in some, usually older, air brake systems. This manually-actuated valve was switched from the cab and reduced the application pressure to the front axle service brakes to 50% of application pressure when toggled to the *wet road* setting. More recent air brake systems tend not to use a manually-switched limiting valve, replacing it with a ratio-modulated valve known as an automatic **ratio valve. Figure 8-37** shows both external and sectional views of ratio valves.

An automatic ratio valve limits front axle braking under light- and medium-duty service applications. Typically, the valve functions as follows. The ratio valve will not crack until the application pressure exceeds 10 psi. With application pressures between 10 and 40 psi, the front axle service brakes receive 50% of the application pressure. Between 40 and 60 psi, the ratio valve graduates the ratio of air delivered to the front axle service brakes to 100%. With any application pressure greater than 60 psi, the front brakes receive 100% of the application pressure. This means that, under severe braking, the front axle brakes receive 100% application pressure.

Automatic ratio valves are considered safer than front-wheel limiting valves because of the importance of the front axle brakes under any severe (panic) braking condition. Most research and analysis of panic brake performance has shown that the operator can maintain better control when the front axle service brakes are provided with 100% application pressure. Theoretically, steer axle ratio valves are not required on ABS. However,

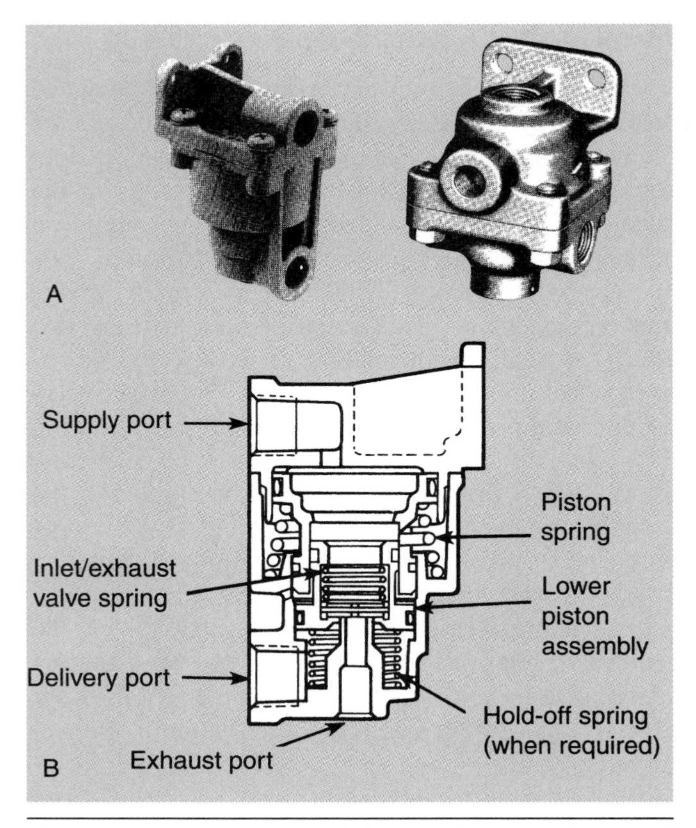

Figure 8-37 Ratio valves: (A) Typical ratio valves, and (B) Sectional view of a ratio valve. (*Courtesy of Bendix Commercial Vehicle Systems LLC. All rights reserved*)

because ABS brake systems are required to meet performance standards in the event of an ABS failure, they still appear on some chassis.

Spring Brake Control Valve

A spring brake control valve is a combination valve that is used to manage both service and parking/emergency brake functions to spring brake chambers. This valve also may incorporate the functions of an inversion valve, regulate spring brake chamber hold-off pressure, and have anti-compounding features. **Figure 8-38** shows a typical multifunction **modulating valve** that can perform up to four functions:

1. Limits hold-off pressure to the spring brakes
2. Provides quick release of air from the hold-off chamber providing faster actuation of spring brakes
3. Prevents compounding by having an anti-compounding feature
4. Provides inversion valve braking in the event of full system pneumatic failure

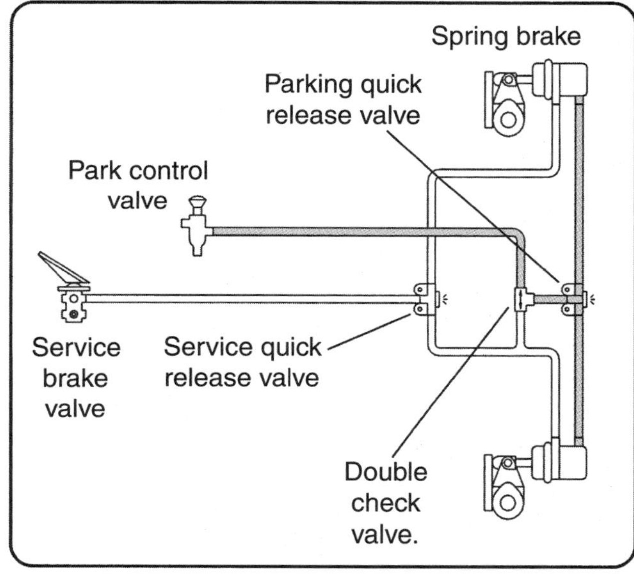

Figure 8-38 Spring brake control valves and hold-off circuit. (*Courtesy of Bendix Commercial Vehicle Systems LLC. All rights reserved*)

Check Valves

A check valve routes air flow in one direction and, in most cases, prevents any backflow. A couple of different types of check valves are used in air brake systems to perform tasks such as pressure protection and priority routing.

One-Way Check Valves. These devices could not be simpler. They permit air to travel in one direction only. They are often located at the inlet to a tank or subcircuit, in which case they will act as pressure-protection devices. This means that air can enter the tank or subcircuit but not exit through the check valve.

Figure 8-39 Double check valve. (*Courtesy of Haldex Brake Products, Inc.*)

Two-Way Check Valves. Two-way check valves can be used in several areas of an air brake circuit. They are disc or shuttle valves. The two-way check valve receives air from two sources and has a single outlet. The valve core is designed to shuttle so the higher-pressure source air is always delivered to the outlet. In other words, the two-way check valve will transmit the higher-pressure source air to the outlet. It is recommended that these valves be mounted horizontally. **Figure 8-39** shows a typical double- or two-way check valve.

Stop Lamp Switches

The stop lamp switch is designed to close an electrical circuit when line pressure is sensed. Its function is to illuminate the brake lights on the vehicle. It is rated by the amount of air pressure required to close the electrical circuit. It can be located at various points in the service application circuit.

Brake Chambers

Brake chambers convert the compressed air back into the mechanical force required to stop a vehicle. A simple service brake chamber consists of a two-piece housing within which are a diaphragm, a pushrod, and a retraction spring. The two sections of the housing are clamped together so that the diaphragm forms an airtight seal. The pushrod is linked to the foundation brakes by means of a slack adjuster.

When compressed air acts on the flexible rubber diaphragm, force is applied to the pushrod plate. The amount of linear force is proportional to the amount of air regulated by the operator using the service brake application valve. **Figure 8-40** shows the principle behind the actuation of foundation brakes by a single-chamber actuator. The spring in this assembly is relatively light-duty and functions simply to return (retract) the pressure plate after a brake application.

The amount of linear force delivered by a brake chamber can be calculated by knowing the sectional area of the diaphragm and the amount of air pressure applied to it (see **Figure 8-41**). A common size of diaphragm is 30 square inches. If a 100-psi service application was applied to a 30-square-inch brake chamber, the linear force developed can be calculated as follows:

Linear Force = 30 sq. in. × 100 psi = 3,000 lbs

Spring Brake Operation. All spring brakes have a means of being mechanically caging. Caging a spring brake compresses the main spring. A spring brake should be mechanically caged before it is either removed or replaced. Most spring brakes are sold with a cage bolt assembly. Caging of a spring brake consists of inserting the cage bolt lugs into slots in the internal cage plate, rotating the bolt, and then compressing the spring using the cage nut and washer. A caged spring brake should be regarded as armed and dangerous.

A summary of spring brake operation follows:

- No air pressure in either the hold-off or service circuits: brake mechanically applied by the full force of the main spring in the spring brake assembly. This condition is used to mechanically "park" the brake or to effect emergency braking in the event of total loss of air.

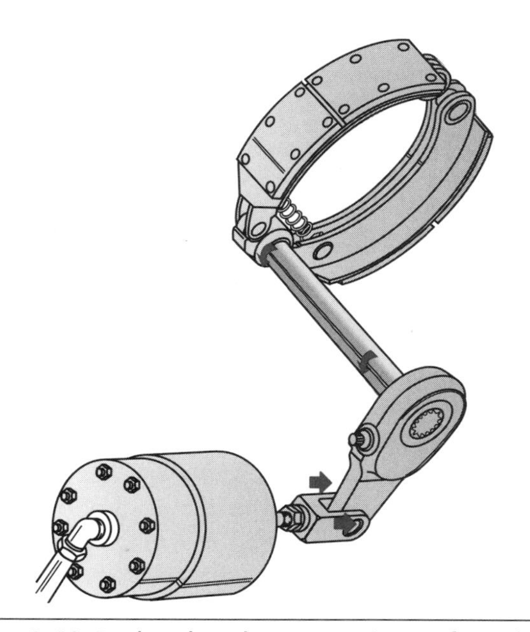

Figure 8-40 Brake chamber actuating a foundation brake assembly.

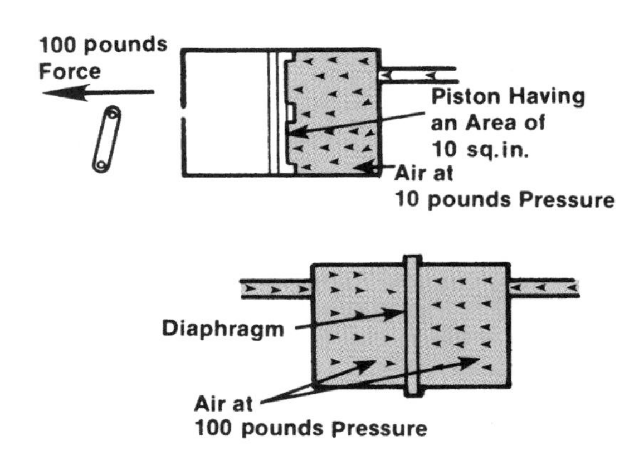

Figure 8-41 Force delivered when compressed air is used as potential energy. (*Courtesy of Bendix Commercial Vehicle Systems LLC. All rights reserved*)

- Air pressure in hold-off circuit, none in service chamber: parking brake is released and the vehicle is capable of being moved. To release the parking brakes and move the vehicle, air pressure greater than 60 psi must be present in spring brake hold-off chambers.
- Air pressure in hold-off circuit, service application pressure in service chamber: service brakes applied at the service brake application pressure value. This condition exists at any time the vehicle is mobile and the service brakes are applied.
- No air in hold-off circuit, service application pressure in the service chamber: braking is compounded by having spring and air pressure applied to the foundation brake assembly simultaneously. Compounding of brake application pressures can damage the foundation brake components. Most systems have anti-compounding valves to prevent this condition.
- Spring brakes are mechanically caged with a cage bolt, nut, and washer assembly, often bolted to the side of the spring brake assembly.

CAUTION *Spring brake chambers should be regarded as being potentially lethal. Always observe the OEM precautions when working on or around spring brake chambers.*

Rotochambers

Rotochambers use the same principles as typical spring brake chambers, except that in place of the pancake diaphragm, a rolling-type diaphragm is used.

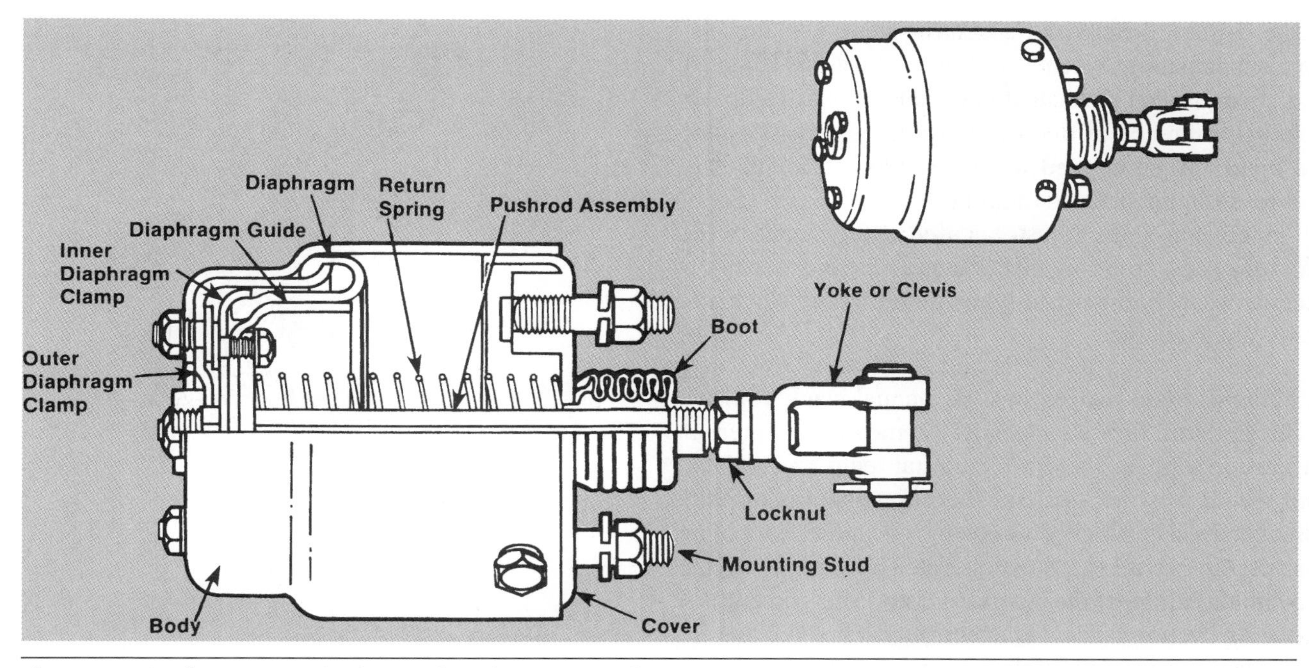

Figure 8-42 Shows a typical rotochamber assembly.

This produces a constant output force throughout the pushrod stroke and provides a much longer stroke. Rotochambers are available in a number of sizes and provide a wide range of output forces. They are more likely to be found in off-highway equipment where their larger physical size and heavier weight are not factors. **Figure 8-42** shows a typical rotochamber assembly.

Slack Adjusters

The linear force produced by the brake chamber must be converted into brake torque (twisting force). Rotation of the S-cam forces the shoes into the drum to effect braking. Slack adjusters are used to connect the brake chamber with the S-cam shaft. **Figure 8-43** shows the roles of the slack adjuster in the actuation of the brakes. First, the slack adjuster is the mechanical link between the brake chamber and the foundation brake assembly. It is attached to the brake chamber pushrod by means of a clevis and pin and to the S-camshaft by means of internal splines on the slack adjuster and external splines on the S-camshaft. The slack adjuster is also a lever. As a lever, it amplifies the force applied to it by the brake chamber. The amount of leverage potential depends on the distance between the clevis pin that connects it to the chamber pushrod and the fulcrum formed at the centerline of the S-cam. This is known as the leverage factor. For instance, if the leverage factor of

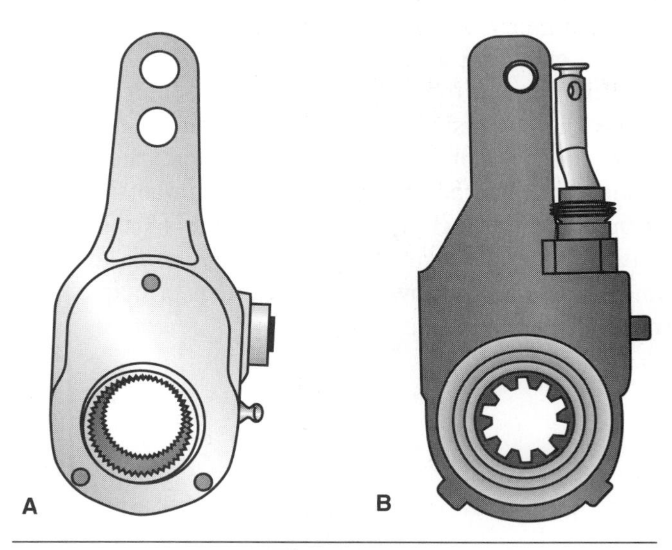

Figure 8-43 (A) Manual and (B) Automatic slack adjusters.

a slack adjuster is 6 inches or 0.5 foot, and a linear force of 3,000 pounds is applied to it by the brake chamber, then the brake torque can be calculated as follows:

Brake Torque = Force × Distance
= 3,000 lb. × 0.5 foot = 1,500 lb.-ft. torque

Figure 8-43 shows a manual and an automatic slack adjuster. The leverage factor of a slack adjuster is critical to the correct operation of a brake system. For this reason, it is important whenever a slack adjuster is changed that the clevis pin be located in the correct slack adjuster hole. A difference of one-half inch can drastically alter the brake torque applied to the foundation brake. This will result in unbalanced braking.

Finally, the slack adjuster is used as the means of removing excess free play that occurs as the brake friction facings wear. Two general types of automatic slack adjuster are available.

Manual Slack Adjusters. A manual slack adjuster is located on the S-camshaft by means of internal splines on a gear. There are external teeth on the gear mesh with a worm gear on the adjusting wormshaft. The wormshaft is locked to position by a locking collar that engages to a hex adjusting nut. This locking collar prevents rotation of the wormshaft and sets the adjustment. As friction face wear occurs, slack results; that is, the chamber-actuated pushrod will have to travel farther to apply the foundation brake. This slack is removed by adjusting the brakes.

The brake is adjusted by depressing the locking collar and engaging a closed wrench (usually 9/16 inch) over the hex on the wormshaft. The wrench can now rotate the wormshaft to remove the slack. After the brake has been properly adjusted to obtain the correct amount of pushrod free play, the adjustment is set by reengaging the locking collar. **Figure 8-44** shows the critical components of a typical manual slack adjuster, and **Figure 8-45** shows the adjustment principle of a manual slack adjuster.

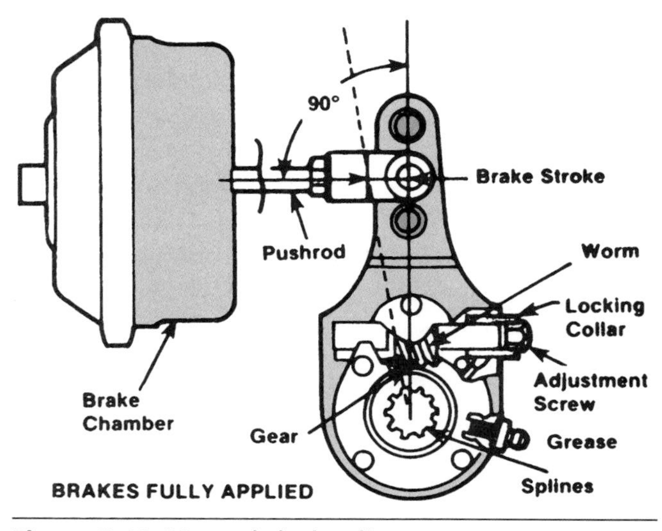

Figure 8-44 Manual slack adjuster.

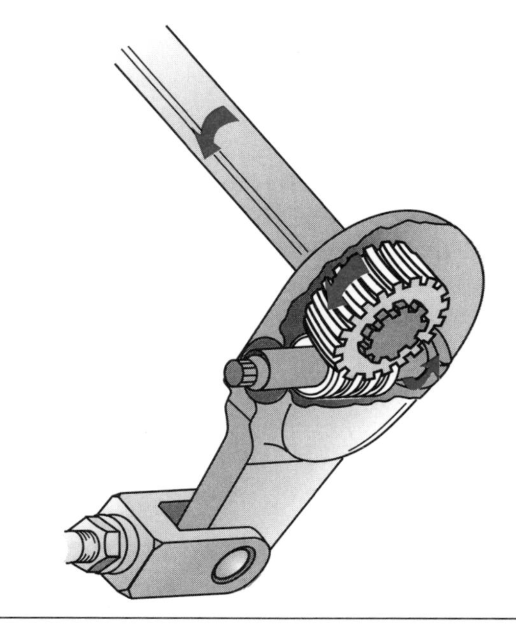

Figure 8-45 Adjustment principle of a manual slack adjuster. (*Courtesy of Caterpillar*)

Stroke-Sensing Automatic Slack Adjusters. The stroke-sensing automatic slack adjuster responds to increases in pushrod stroke to actuate an internal cone clutch and rotate an adjusting wormshaft. This wormshaft meshes with the external teeth of the gear engaged to the splines of the S-camshaft. This type of slack adjuster operates on the assumption that if the applied stroke dimension is correctly maintained, the drum-to-lining clearance will be correct. **Figure 8-46** and **Figure 8-47** show cutaway and external views of an automatic slack adjuster.

Clearance-Sensing Automatic Slack Adjusters. Clearance-sensing automatic slack adjusters respond only to changes in the return stroke dimension, so they function almost opposite to stroke-sensing models. They "adjust" only when the return stroke dimension increases; if the applied stroke dimension becomes excessive, no adjustment will occur. Clearance-sensing automatic slack adjusters work on the assumption that if the shoe-to-drum clearance is normal, the stroke should be within the required adjustment limits.

Tech Tip: Slack adjusters without a grease fitting should not be assumed to be functional. Slack adjusters should be lubricated with a low-pressure grease gun.

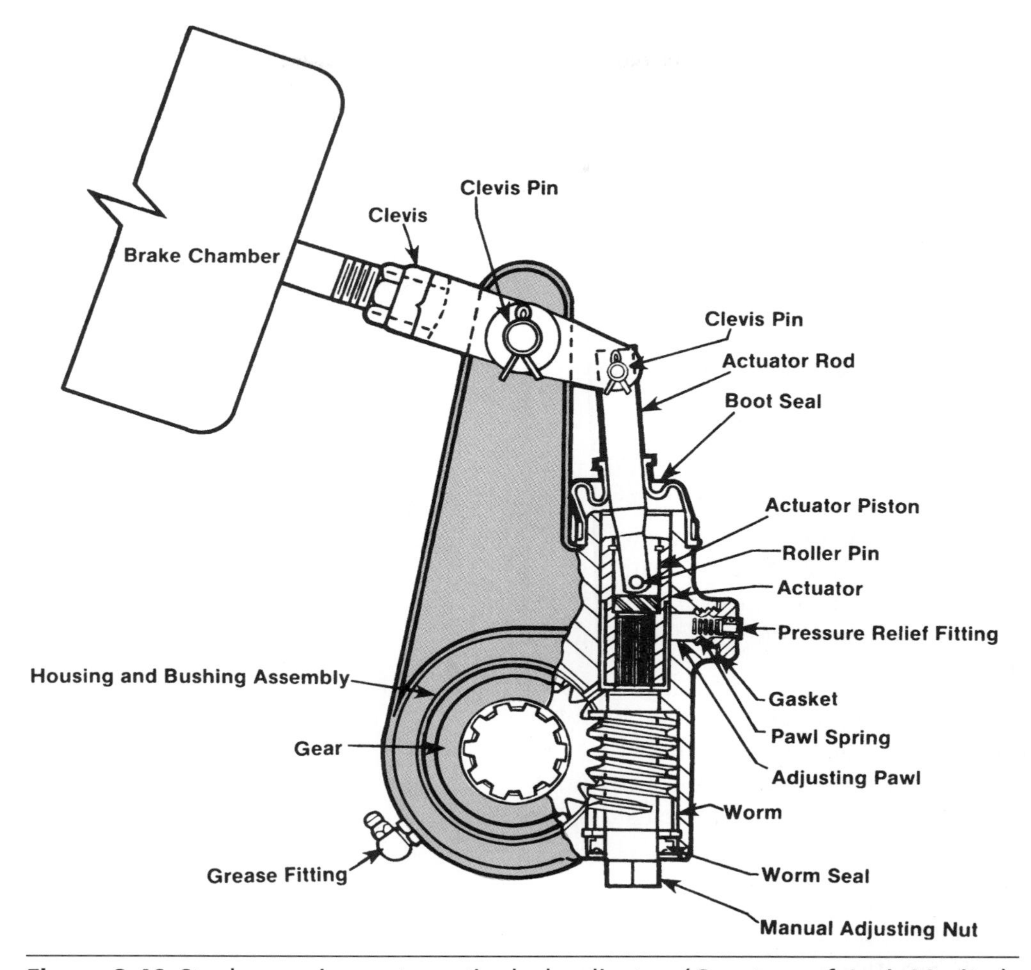

Figure 8-46 Stroke sensing automatic slack adjuster. (*Courtesy of ArvinMeritor*)

Figure 8-47 Automatic slack adjuster. (*Courtesy of ArvinMeritor*)

Gauges

Air pressure gauges in air brake circuits are used to display air pressure in the primary and secondary circuits and to give the operator an indication of what the system and application pressures are. Either individual gauges or a two-in-one gauge can be used to display the system pressure in the primary and secondary circuits.

When a two-in-one gauge is used, a green needle indicates primary circuit pressure and a red needle indicates secondary circuit pressure. If a loss of pressure occurs in either of the circuits, the operator is alerted to the condition after the pressure drops below 60 psi. The alert is usually visible (gauges/warning lights) and audible (buzzer). Some systems also have an application pressure gauge. An application pressure gauge tells the operator the exact pressure value of a service application.

Mechanics of S-cam, Disc, and Wedge Brakes

The performance of S-cam, disc, and wedge brakes varies significantly, and this can produce problems in a tractor and trailing or multi-trailer combination in which the foundation brakes are mixed. For instance, a tractor equipped with S-cam brakes coupled to a trailing unit equipped with wedge brakes will often cause trailer overbrake/tractor underbrake because, in moderate to severe braking, double-actuated wedge brakes have superior brake mechanics. A tractor unit equipped with air disc brakes hauling a trailer with S-cam brakes can cause the opposite problem because of the higher mechanical efficiency of disc brakes over an S-cam actuated drum.

AIR-OVER-HYDRAULIC BRAKES

Air-Over-Hydraulic Components

Air-over-hydraulic brake systems combine both air and hydraulic principles to braking. The objective is to combine the advantages of both pneumatic and hydraulic brake systems to optimize braking. Although there are a number of variations of air-over-hydraulic brake systems used on equipment, the following statements tend to be true of all systems:

- Pneumatic Circuit: This is used as the control circuit. It masters a hydraulic slave circuit.
- Hydraulic Circuit: This is the actuation circuit. Hydraulic force is therefore used to actuate the foundation brakes.

The schematic shown in **Figure 8-48** is a simplified Caterpillar air-over-hydraulic brake system identifying the air, hydraulic, and electrical components.

Figure 8-49 is a working schematic of the air-over-hydraulic brake system used on a Caterpillar 950E wheel loader. The components of the system are identified in the text that follows the figure.

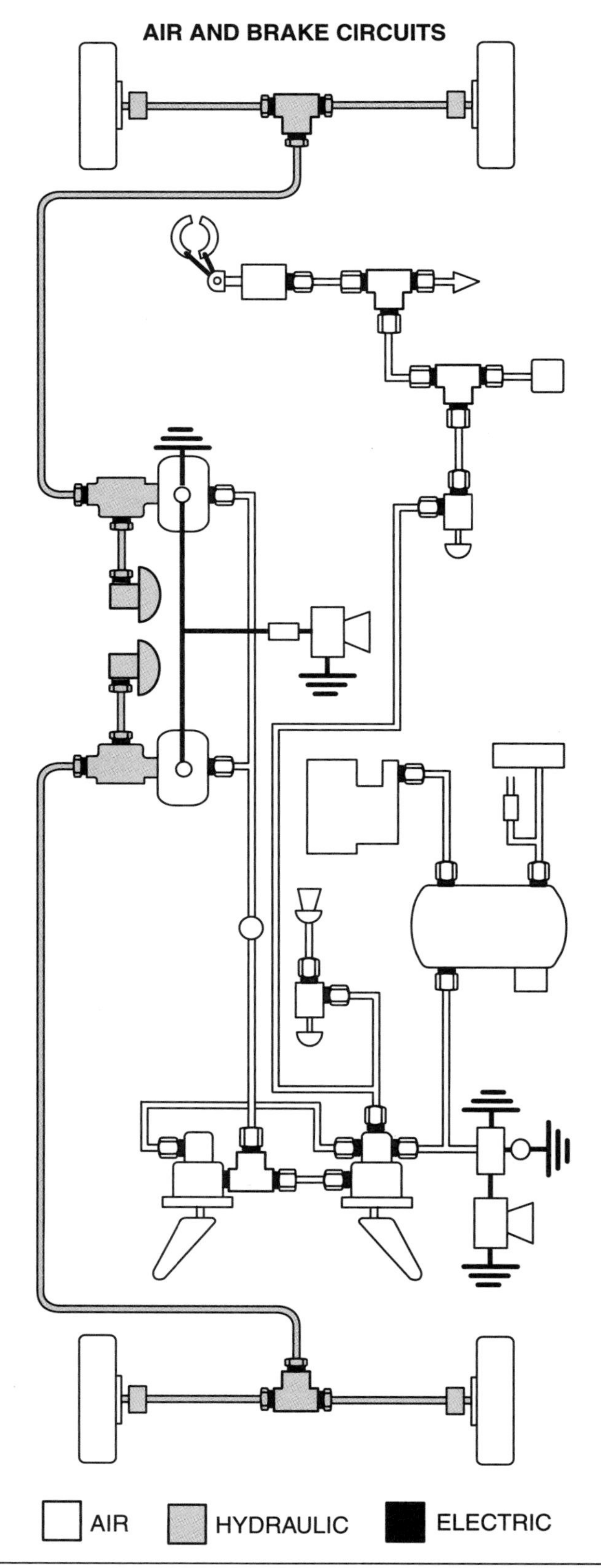

Figure 8-48 Simplified air-over-hydraulic brake system showing all the key components.

Wheel Loader Brake Circuit

Referencing Figure 8-49, you can see that the wheel brakes on this wheel loader are caliper-actuated disc brakes. Each of the four wheels on the vehicle is equipped with a disc brake assembly, and in addition, there is a secondary/parking brake located on the front driveshaft. On this unit, the secondary/parking brake uses a drum and shoe principle: the brake chamber that actuates it is spring-actuated (default) and air-released.

The wheel disc brakes are controlled by a pair of pedals in the operator's station. When the right pedal is depressed by the operator's boot, only the wheel brakes are actuated. When the left pedal is depressed, the wheel loader transmission is first disengaged, then the wheel brakes are actuated. On releasing the left

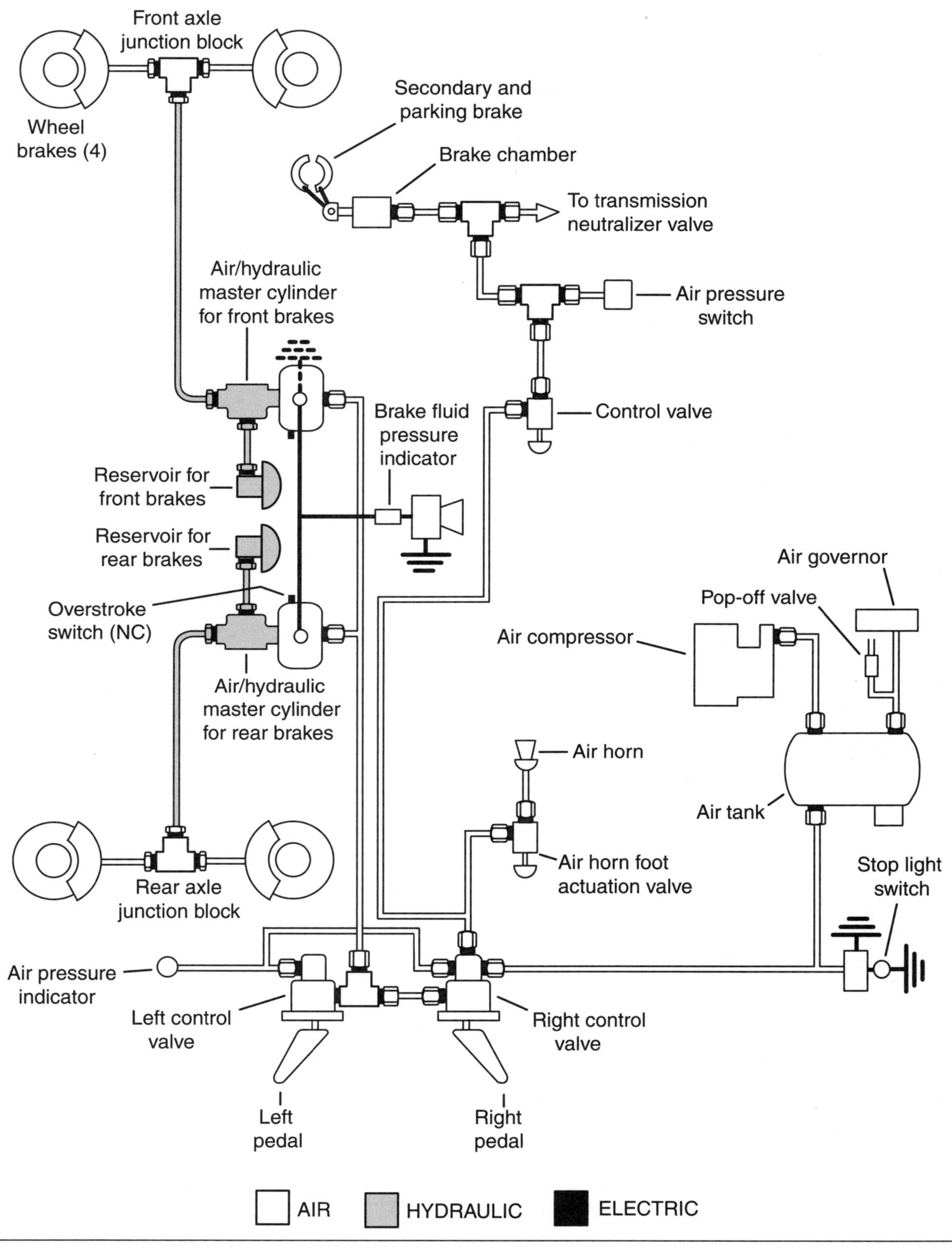

Figure 8-49 Air-over-hydraulic schematic of a Caterpillar 950D wheel loader.

pedal, the transmission is engaged before the brakes are released on the unit to prevent rolling when on an incline.

The secondary/parking brake on the unit is controlled manually by a valve located underneath the steering wheel. Because air pressure acts to release the secondary/parking brake, it will begin to apply (spring pressure) automatically if system air pressure gets too low. The brake system on this wheel loader can be subdivided into four separate circuits:

1. Air supply for the air control system.
2. Air circuit for the secondary/parking brake circuit.
3. Air control circuit for the wheel brakes.
4. Hydraulic slave circuit for the wheel brakes.

Operating Principles

There are several variations of air-over-hydraulic brakes used on mobile equipment. In general, the control circuit is managed by air pressure and the application circuit by hydraulics. Depressing the brake pedal moves the application valve plunger. This allows modulated system pressure air to flow through quick-release valves and to be directed to a combination **air/hydraulic master cylinder.** Some OEMs refer to the air/hydraulic **master cylinder** as an **air-over-hydraulic intensifier.** An air/hydraulic master cylinder divides the air and hydraulic sections of the circuit. In addition, separate air/hydraulic master cylinders may be used for the front and rear brakes. Air pressure applied to the booster moves an actuating piston, which creates hydraulic pressure for the foundation brake actuators.

The actual amount of hydraulic pressure applied to the foundation brakes depends on the modulated (by the foot valve) air pressure applied to air/hydraulic master cylinder. The resulting hydraulic actuation pressure is routed to the **wheel cylinders,** which effect vehicle braking in the same manner as in a hydraulic brake system.

Hydraulic pressure forces the pistons of each wheel cylinder outward. Outward movement of the pistons applies brake pressure on the brake drum or disc **calipers.** A brake-lining wear switch may be mounted in each air/hydraulic master cylinder to indicate excessive travel of the air piston. Depending on the type of equipment, a front-wheel limiting valve can be used on the front axle brakes. This limiting valve proportions fluid pressure during light braking, both to balance the braking and reduce brake lining wear. A load **proportioning valve (LPV)** is integrated into some systems and is used to modulate hydraulic (application) pressure to front and rear brakes based on vehicle loads. This can control suspension dive on rough terrain on an unloaded piece of equipment.

Air Control Circuit

The general setup of the air control circuit is similar to that required on an air brake equipped vehicle. Typically it consists of the following components:

- Engine driven compressor
- Air dryer circuit
- Governor
- Air storage reservoirs
- Foot (treadle) valve
- Quick-release valve(s)
- Air/hydraulic master cylinder (air-over-hydraulic intensifier)

Most modern air-over-hydraulic systems are dual-circuit systems and may incorporate an anti-lock brake system (ABS). Dual-circuit systems provide the brake circuit with a safety redundancy factor, meaning that a sudden leak in one portion of a system should not result in a total loss of braking. Other fail-safe features include the usual buzzers and lights that warn of low air pressure in the system and overstroke in the air/hydraulic master cylinder. When the air/hydraulic master cylinder overstroke alert is illuminated on the operator interface, the usual cause is hydraulic circuit leakage or low hydraulic oil level. **Figure 8-50** shows the air control circuit components of a typical air-over-hydraulic brake system; it does not show the location of the air/hydraulic (master) cylinders on the schematic.

Hydraulic Foundation Brake Circuit

Air-over-hydraulic systems may use either a drum- or disc-based foundation brake. The arrangement may be either on all four wheels or mixed disc and drum systems. There are other variations, however. Some systems can use a mixture of hydraulic disc and full air S-cam drums. When a mixed air-only plus an air-over-hydraulic system is used, a delay valve is required to keep the hydraulic brakes from coming on before the slower-acting air brakes. The hydraulic foundation brake circuit typically consists of the following components:

- Air/hydraulic master cylinder
- Metering and proportioning valves
- Wheel cylinders and calipers
- Drum or disc foundation assemblies

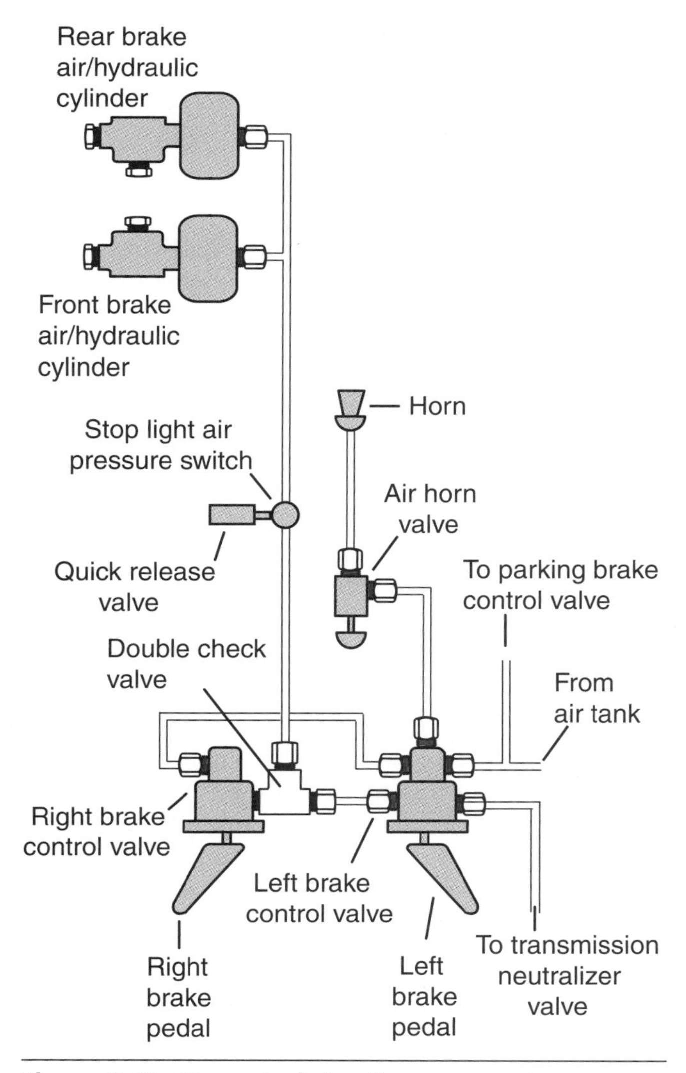

Figure 8-50 Air control circuit components.

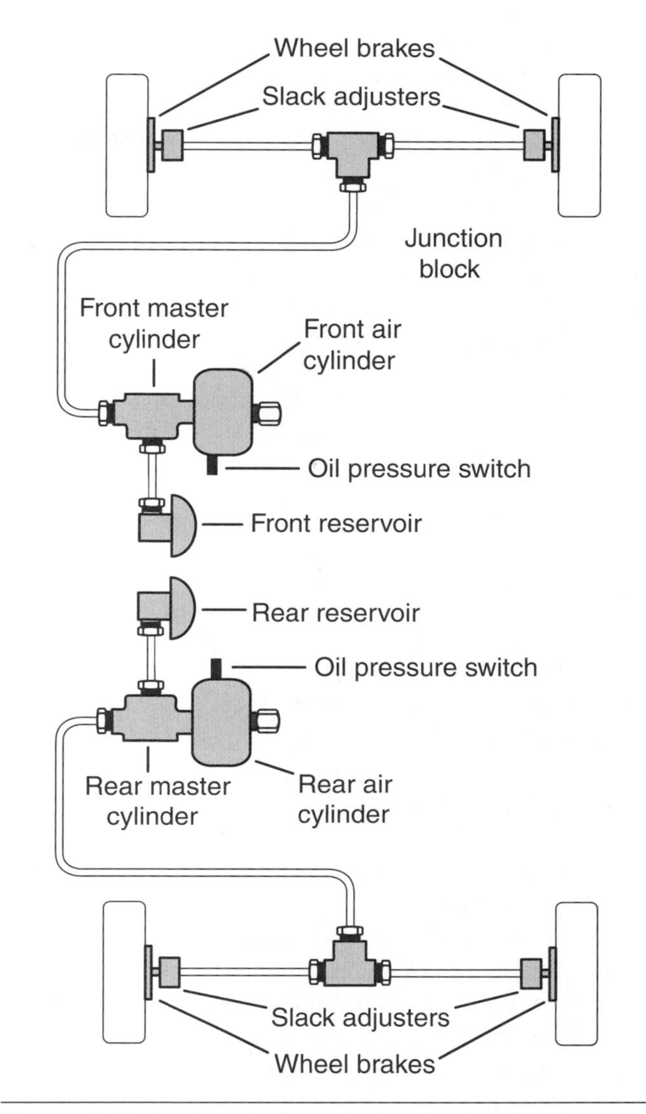

Figure 8-51 Hydraulic foundation brake circuit.

Figure 8-51 shows a typical hydraulic foundation brake circuit. Some OEMs refer to this as a hydraulic slave circuit. Note the position of the front and rear air/hydraulic master cylinders in the circuit: these units divide the air control (pneumatic) and hydraulic actuation circuits.

Secondary/Parking Brake

A common type of parking brake on air-over-hydraulic systems is the spring brake canister brake. Its parking brake action is obtained by forcing a wedge assembly between the plungers in the wheel cylinder on each rear wheel. Engagement of the wedge assembly by the spring brake expands the plunger in the wheel cylinder, forcing the brake shoes into the brake drum. Cable-actuated parking brakes are also found in some air-over-hydraulic systems. This type of unit is mounted at the rear of the transmission so that it acts on the rear wheels through the driveshaft. The parking brake is incorporated in each rear wheel and is cable-controlled from the cab area. For a transmission-located parking brake to function, both rear wheels must be firmly on the ground.

Also commonly used on air-over-hydraulic brake systems is a spring-actuated secondary/parking brake that uses system air pressure on the unit to hold off the spring pressure in the air chamber. This type of secondary/parking brake is controlled manually by an operator dash valve. Because air pressure acts to release the secondary/parking brake, it will begin to apply (spring pressure) automatically if system air pressure gets too low, providing an operational safety factor. **Figure 8-52** shows as

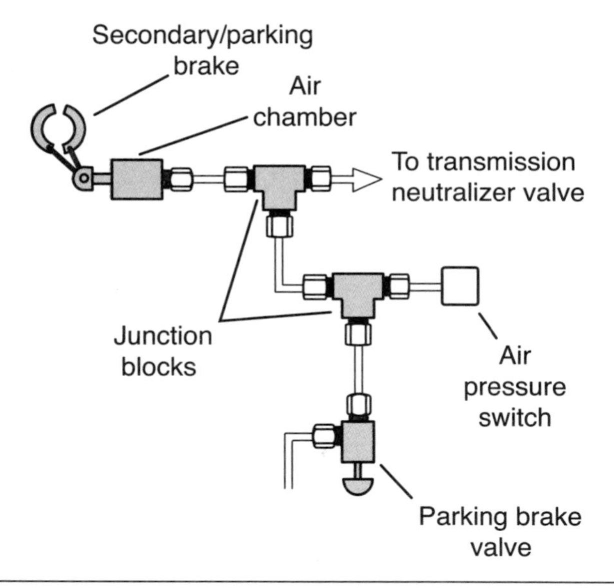

Figure 8-52 Schematic for a spring-apply/air-release secondary/parking brake.

simplified secondary/parking brake circuit that uses a spring-apply/air-release principle.

Air-Over-Hydraulic Abs

ABS can be incorporated into an air-over-hydraulic brake system, and when it is, it functions within the hydraulic slave circuit of the system. Because of this, it operates similarly to ABS in a hydraulic-only air brake circuit. Because the ABS is integrated into the hydraulic section of the circuit, the cycling frequency is high speed, similar to hydraulic-only systems. ABS management may be data bus driven, enabling directional stability in poor traction conditions.

Bleeding An Air-Over-Hydraulic System

To bleed a typical air-over-hydraulic system used on a two axle vehicle, follow the following procedure:

Procedure

1. Adjust the air pressure in the air reservoir to that specified by the OEM. If the air pressure drops below this specified pressure during the bleeding operation, start the engine.
2. When the air pressure is at the specified system pressure, stop the engine.
3. Bleed the air from each air bleeder screw in the following order:
   - Front brakes and air/hydraulic master hydraulic cylinder plug.
   - Right front wheel air bleeder screw.
   - Left front wheel air bleeder screw.
   - Rear brakes and air/hydraulic master hydraulic cylinder plug.
   - Right rear wheel air bleeder screw.
   - Left rear wheel air bleeder screw.
4. Fill the brake reservoir with fluid. Refill frequently during bleeding so no air enters the brake lines.
5. Attach a vinyl tube to the bleeder screw. Insert the other end of the tube into a bottle partially filled with brake fluid.
6. Depress the brake pedal and at the same time turn the bleeder screw to the left.
7. Tighten the bleeder screw when the pedal reaches the full stroke. Release the pedal.
8. Repeat until no bubbles are observed.

ONLINE TASKS

1. Access the Bendix website at *http://www.bendix.com/bendix/products* and navigate through it. This site contains much useful general and service information on air brake systems. Although much of this information applies to on-highway equipment you should nevertheless find it beneficial.
2. Because Caterpillar still manufactures equipment with air brake systems, you can obtain useful information on their online service information system (SIS). If your college or workplace has SIS data hub access, select a piece of equipment (begin with a Cat 775D quarry truck), and download the brake information and schematics. The Caterpillar SIS site is at *http://www.IMAP@cat.com*. However, you will be prompted for passwords at logon, and you will not get any farther unless you have access.

SHOP TASKS

1. Locate a piece of equipment equipped with air brakes and manual slack adjusters. Adjust brake-stroke travel to the manufacturer's specifications.
2. Locate a piece of equipment equipped with air brakes and automatic slack adjusters. Adjust brake-stroke travel to the manufacturer's specifications.
3. Locate a piece of equipment equipped with air-over-hydraulic brakes and, either using the procedure outlined earlier in this chapter or an OEM procedure, bleed the brake circuit using the correct sequence.

Summary

- An air dual-circuit brake system is composed of supply circuit, primary circuit, secondary circuit, parking/emergency control circuit, trailer circuit, and foundation brake assemblies. Off-highway equipment is not required to meet the redundancy standards of highway vehicles, so their air brake systems may be simplified.
- Air compressors are single-stage, reciprocating-piston air pumps that are either gear- or belt-driven.
- Air dryers are used to help eliminate moisture and contaminants from the air system.
- Dual-circuit application or foot valves, ratio valves, quick-release valves, relay valves, dash control valves, double-check valves, and check valves are some of the critical valves of an air brake system.
- The potential energy of compressed air is changed into mechanical force in an air brake system by slack adjusters and brake chambers.
- The most common type of foundation brake assembly used on current air brake systems is the S-cam type.
- Slack adjusters multiply the force applied to them by the brake chamber into brake torque.
- Brake torque applied to the S-camshafts results in the shoes being forced against the drum.
- Air disc brakes operate by using an air-actuated caliper to squeeze brake pads against both sides of a rotor.
- Wedge brakes use a drum and a pair of shoes, and air-actuated wedges are used to force the shoes against the drum.
- Service braking applies to the running brakes on equipment. Service braking is managed by the operator using a treadle or foot valve.
- The parking/emergency system manages the hold-off circuit. The hold-off circuit is uses system air pressure to "hold-off" the mechanical park system of a piece of equipment, usually spring brake or rotochamber units.
- Spring brakes and rotochambers are twin chamber units used for air-actuated service braking and mechanically-actuated parking/emergency braking.
- In a typical air-over-hydraulic brake system, the control circuit can be compared with the control circuit in typical air-only brake systems, while the actuation circuit can be compared with that in a typical hydraulic brake circuit.
- The air control circuit in air-over-hydraulic brake systems is charged with compressed air by a compressor with the system-pressure operating window managed by a governor. The service brakes are operator-actuated by a foot (treadle) valve. The modulated air that exits the foot valve is delivered to an air-actuated, hydraulic master cylinder.
- When the hydraulic master cylinder receives a pneumatic signal from the foot valve, it forces brake fluid out to actuate the foundation brakes.
- Air-over-hydraulic brakes may use disc- or drum-type foundation brakes.
- A typical air-over-hydraulic ABS would add sensors, an electronic control module, and a modulator to the system—all of which function similarly to those in hydraulic-only brake.

Review Questions

1. Which of the following air brake system components functions as a lever?

   A. the brake chamber
   B. the rotor
   C the caliper
   D. the slack adjuster

2. Which of the following components sets the system cut-in and cut-out pressures?

   A. the service reservoir
   B. the governor
   C. the air dryer
   D. the supply tank

3. What is the function of a system safety pop-off valve?

   A. It releases excess pressure from the compressor.

   B. It pressure-protects air in the supply tank.

   C. It alerts the operator when air pressure has dropped below system pressure.

   D. It relieves pressure from the supply tank in the event of governor cut-out failure.

4. The function of the foot valve or dual-circuit application valve is to:

   A. increase the flow of air to the service reservoirs

   B. reduce the flow of air from the service reservoirs

   C. meter air to the service brake circuit

   D. meter air to the brake hold-off chambers

5. The potential energy of an air brake system is:

   A. compressed air

   B. foot pressure

   C. spring pressure

   D. vacuum over atmospheric

6. Which of the following components is a requirement on every air tank?

   A. an air dryer

   B. a safety valve

   C. a drain cock

   D. a pressure protection valve

7. The function of an air brake chamber is to:

   A. convert the energy of compressed air into mechanical force

   B. multiply the mechanical force applied to the slack adjuster

   C. reduce the air pressure for a service application

   D. increase the air pressure for parking brakes

8. An air brake-equipped vehicle is moving at 20 mph and the operator makes a service brake application. Which of the following should be true in all of the spring brake chambers on the vehicle?

   A. Only the hold-off chambers are charged with air.

   B. Only the service chambers are charged with air.

   C. Both the service and hold-off chambers are charged with air.

   D. Neither the service nor the hold-off chambers are charged with air.

9. Technician A says that automatic slack adjusters can improve brake balance and reduce downtime and maintenance costs. Technician B says that automatic slack adjusters should not be adjusted after installation. Who is correct?

   A. Technician A

   B. Technician B

   C. Both A and B

   D. Neither A nor B

10. Which of the following determines the amount of leverage gained by a slack adjuster?

    A. the slack adjuster length

    B. the pushrod stroke length

    C. the S-cam geometry

    D. the number of internal splines

11. S-cam type brake shoes, retraction, and fastening hardware are mounted to:

    A. a spider

    B. a backing plate

    C. the drum

    D. the axle spindle

12. What actuates the relay piston in a dual-circuit foot valve on an air brake system when the valve is operating normally?

    A. foot pressure
    B. air pressure
    C. spring pressure
    D. mechanical pressure

13. Technician A says that air disc brakes usually produce higher mechanical braking efficiencies than equivalent S-cam brakes. Technician B says that the reason wedge brakes produce higher mechanical braking efficiencies than equivalent S-cam brakes is that they are usually double-actuated. Who is right?

    A. Technician A only
    B. Technician B only
    C. Both A and B
    D. Neither A nor B

14. In the event of a total failure of the primary circuit in a vehicle equipped with dual-circuit air brakes, how is the relay/secondary circuit actuated in the foot valve?

    A. pneumatically
    B. hydraulically
    C. mechanically
    D. cannot be actuated

15. Which of the following is an advantage of a rotochamber when compared with a spring brake chamber?

    A. lightweight
    B. less costly
    C. smaller in size
    D. longer stroke

16. In a typical air-over-hydraulic brake system, depressing the treadle causes which of the following?

    A. Hydraulic fluid flows through quick release valves.
    B. Air pressure forces the pistons of each wheel cylinder outward.
    C. An LPV redistributes air pressure to front and rear brakes.
    D. Air pressure is sent to the air/hydraulic master cylinders assemblies.

17. Which of the following terms is used by some OEMs to describe an air/hydraulic master cylinder?

    A. air-over-hydraulic intensifier
    B. slave valve
    C. hydraulic calipers
    D. treadle valve

18. If the air/hydraulic master cylinder overstroke alert illuminates, which of the following is the more likely cause?

    A. low air pressure
    B. governor improperly set
    C. low hydraulic fluid
    D. excessively high hydraulic pressure

CHAPTER

9 Introduction to Powertrains

Learning Objectives

After reading this chapter, you should be able to

- Identify and define the various types of gears used in heavy-duty equipment applications.
- Calculate gear ratios, gear speed, gear pitch, and torque multiplication.
- Explain the relationship between torque and speed through various gear combinations.

Key Terms

amboid gear
backlash
direct drive
drive gear
driven gear
gears
gear pitch
gear ratio
helical cut
herringbone gear
hypoid gear
laws of leverage
overdrive
pitch diameter
plain bevel gear
planetary gear
rack-and-pinion gear
spur
straight cut spur
synchronized
torque
underdrive
worm gear

INTRODUCTION

The four main purposes of a powertrain are to

1. Connect and disconnect power from the engine.
2. Modify speed and torque.
3. Provide a means for reverse.
4. Equalize power distribution to the drive wheels.

An engine provides the torque required to propel a vehicle. A clutch or torque converter provides the means to connect or disconnect the torque produced by the engine to the transmission. Gear ratios within a transmission allow for different speed and torque ratios needed to propel the vehicle under varying load conditions. The transmission provides a means of reversing engine input rotation for reversing the direction of the vehicle. In many heavy-duty equipment applications there are an equal number of forward and reverse speeds. Differentials equalize power distribution to the final drives. See **Figure 9-1.** In four- or six-wheel drive applications, the power distribution is to both left- and right-hand final drives as well as front to rear final drives as shown in **Figure 9-2.**

GEAR FUNDAMENTALS

The purpose of **gears** or gearing, is to relay a rotating (twisting) force known as **torque.** Almost all gears need to be supported by a shaft to allow an axis for them to rotate, either by rotating independently from each other or together as a unit (see **Figure 9-3**).

Figure 9-1 Front-end loader showing the components of a powtertrain. (*Courtesy of Caterpillar*)

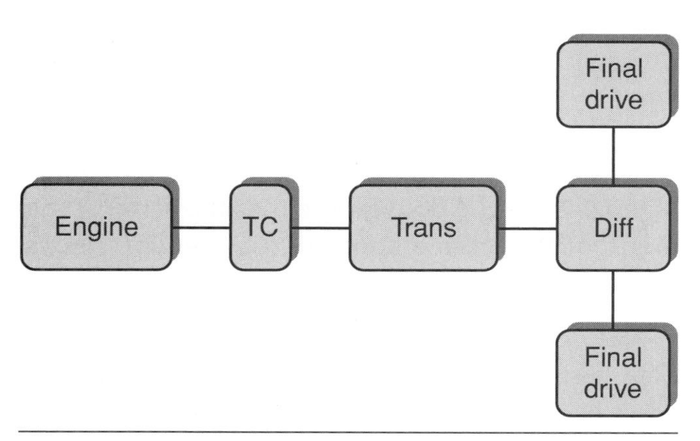

Figure 9-2 The engine produces the torque needed to propel a vehicle and the various components that transfer the torque to the final drives. (*Courtesy of Caterpillar*)

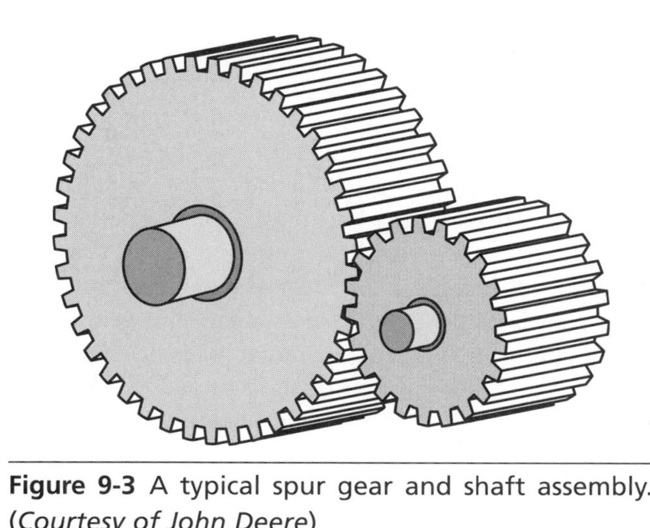

Figure 9-3 A typical spur gear and shaft assembly. (*Courtesy of John Deere*)

Gears working together can act as a lever to multiply torque or to directly transfer torque to another shaft through mating gears. They can also provide a speed increase above the original input speed. This speed increase cannot be done without a penalty. Because gears act as a lever the same **Laws of Leverage** principles apply (**Figure 9-4**).

A longer lever provides a mechanical advantage and a greater amount of torque is achieved. You can liken this to a small-diameter gear driving a large-diameter gear. The action is inversely proportional: if a large diameter gear is driving a small diameter gear, torque reduction occurs with a speed increase. To define these principals further, we must understand how torque is achieved. As stated earlier, torque is a twisting force. It is calculated by multiplying the force applied to a lever by the length of the lever. As shown in **Figures 9-5** and **9-6,** a pipe wrench is attempting to rotate a pipe. By applying a force of 100 pounds to the lever, which in this case is the pipe wrench, at a distance of 12 inches we

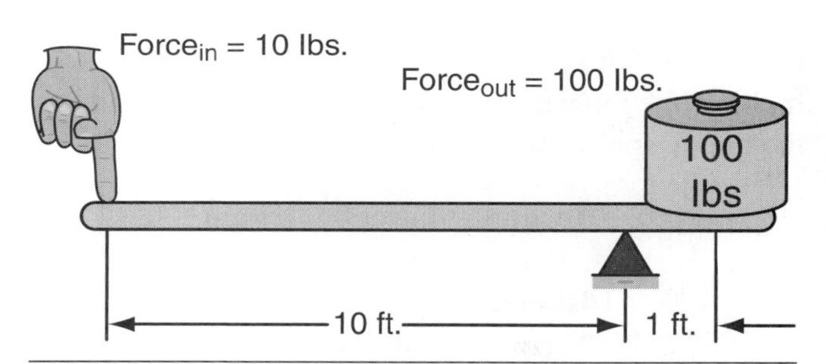

Figure 9-4 Applying a small force at the end of a long lever will produce a larger force on the small lever. (*Courtesy of Vickers Incorporated*)

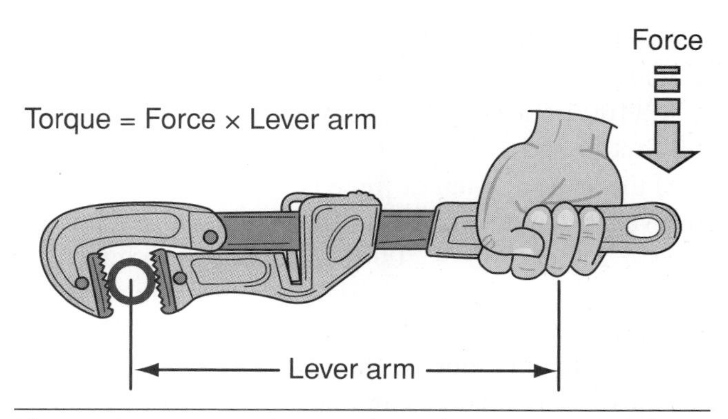

Figure 9-5 Torque is equal to the amount of force multiplied by the length of the lever. (*Courtesy of Vickers Incorporated*)

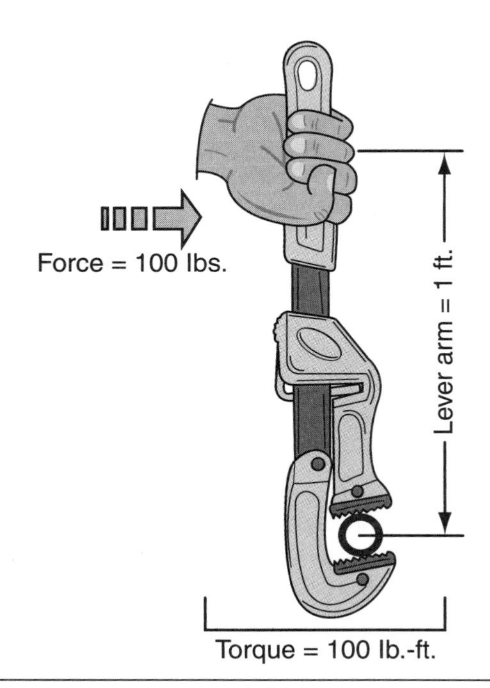

Figure 9-6 100 pounds of force applied to a 1-foot lever equals 100 foot-pounds of torque. (*Courtesy of Vickers Incorporated*)

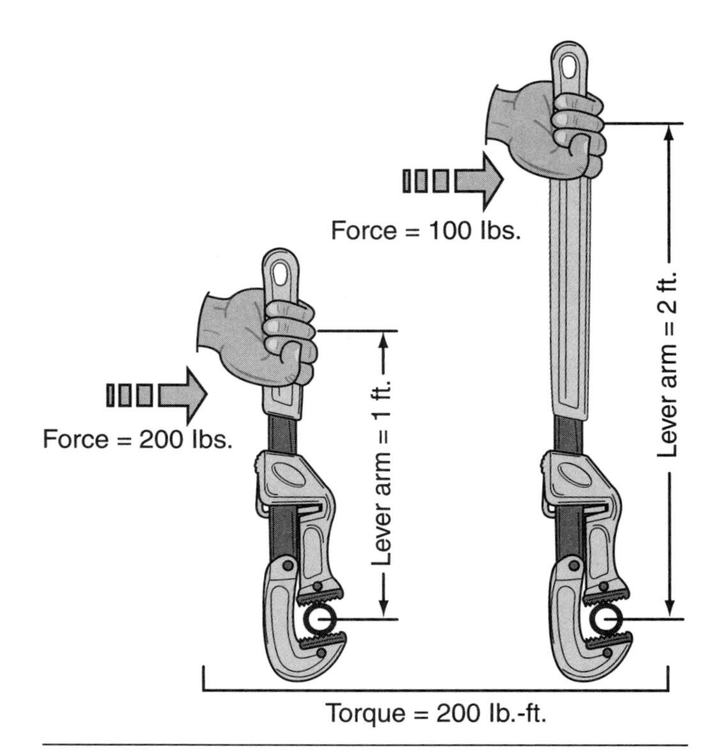

Figure 9-7 Torque doubles if either the force or the length of the lever is doubled. (*Courtesy of Vickers Incorporated*)

multiply the force by the distance of the lever. Therefore the torque applied would be calculated as follows:

Torque = Force × Distance
Force = 100 pounds
Distance = 1 foot
Torque = 100 pounds × 1 foot
Total Torque = 100 pounds foot or foot-pounds

There are two ways of increasing torque: by increasing the force applied, or by increasing the length of the lever. If we increase the force applied to 200 pounds with a 1 foot lever, a total torque of 200 foot-pounds is achieved. If the lever is extended to 2 feet, less force (100 pounds) is required to achieve a torque of 200 foot-pounds (**Figure 9-7**).

If the same amount of force is applied to the extended lever, a torque increase will result: 200 pounds of force × 2 foot lever = 400 foot-pounds of torque.

Gears come in a large variety of types and styles to suit the needs of the application. They are usually classified by the type and/or the design of the teeth. The two most common types are the **straight cut spur** and **helical cut.** Straight cut spur teeth are arranged parallel to the supporting shaft. **Spur** gears in mesh will always have at least one pair of teeth in mesh. This tooth design has lower torque capability than helical cut teeth and tends to be noisier than other gear tooth designs. Straight cut spur teeth under load create radial loads on their respective supporting shafts. Radial loading, which is a force perpendicular (90 degrees) to the shaft, occurs when gears push away from each other (Figures 9-3 and 9-8).

Helical cut gear teeth are cut at an angle or helix from the parallel of the supporting shaft. This tooth design has greater torque capability and is quieter because there is always more than one pair of teeth in mesh and the helical design offers greater surface area for tooth contact. The other contributing factor to greater torque capability occurs because the paired teeth on the parallel plain are in full mesh; the paired teeth approaching have started to mesh, the teeth that are coming out of mesh are still in partial mesh. Overall, this produces increased tooth contact and less clattering of gear teeth. Similar to the straight cut spur gear teeth, helical cut gears also produce radial loads on their supporting shafts, but because of their helical design they also produce a side thrust to their respective supporting shafts. This is because helical gears tend to not only push away from each other, but also to push away at an angle parallel to their shafts as shown in **Figure 9-8.**

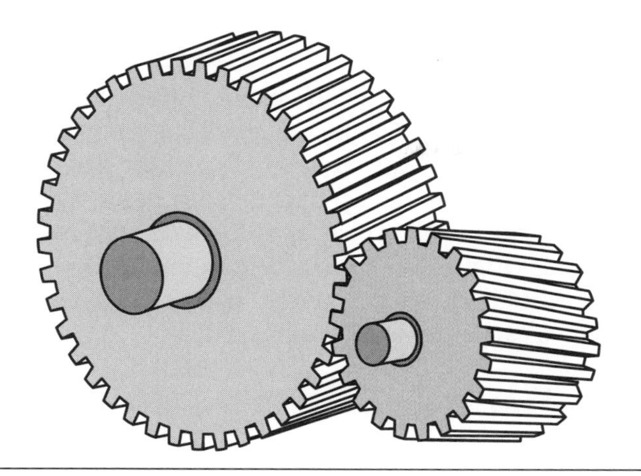

Figure 9-8 Helical spur gear set. (*Courtesy of John Deere*)

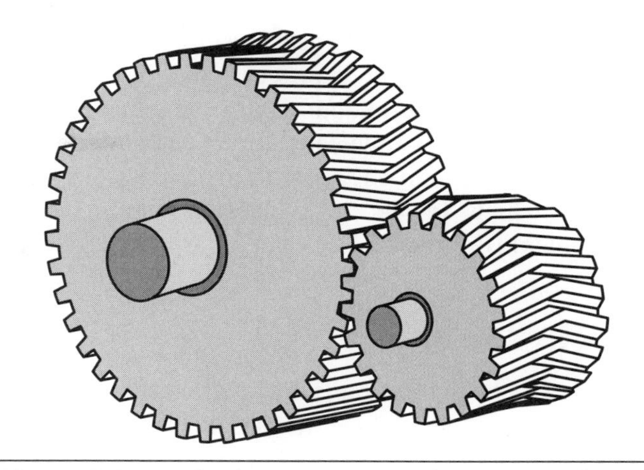

Figure 9-9 Herringbone gear set. (*Courtesy of John Deere*)

Herringbone Gears

Herringbone gears have double helical-cut gear teeth extending away from each other in opposite directions from the tooth centerline. This design provides the high torque capability of traditional helical cut teeth without the side thrusting. Side thrust is eliminated because, as each double helical comes into mesh, one side thrust cancels out the other one. Although side thrusting is eliminated, the gears still produce a radial load on the shaft (**Figure 9-9**).

Plain Bevel Gears

Plain bevel gears consist of a smaller **pinion gear** matched to a larger **crown gear.** The pinion gear tooth design is a straight cut spur tapered to the edge of the shaft; similarly crown gear teeth are tapered or angled to a smaller thickness at the outer edge as shown in **Figure 9-10.** The pinion gear axis is placed directly on the centerline and 90 degrees to the crown-gear axis. This allows the rotating force of the pinion gear to alter its drive by 90 degrees.

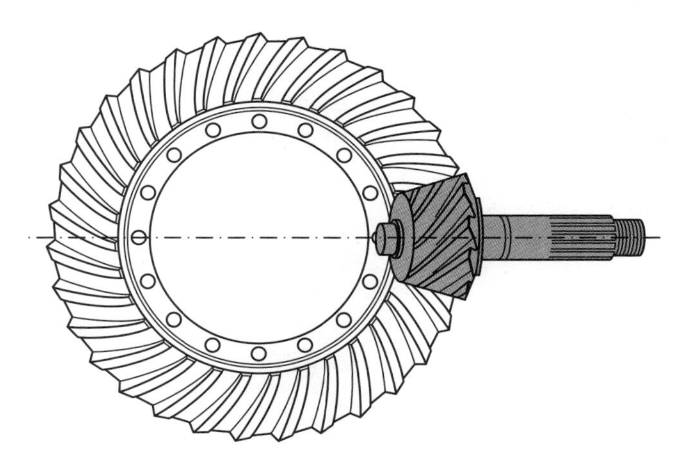

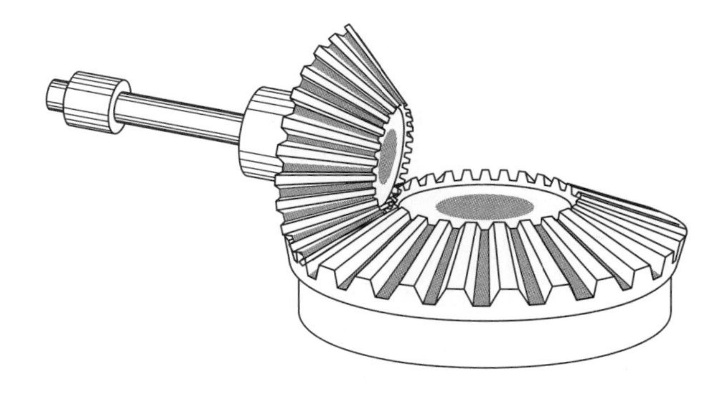

Figure 9-10 Plain bevel gear set. (*Courtesy of Arvin Meritor*)

Spiral Bevel Gears

Spiral bevel gears are similar to the plain bevel gear arrangement with a smaller pinion gear matched to a larger crown gear. It also provides the means to alter the direction of drive input on the pinion gear by 90 degrees; the pinion gear axis is still placed on the centerline axis of the crown gear. The difference between the two configurations is in the gear design. Plain bevel gear teeth are similar to a straight spur gear. However, the spiral bevel gears are of a helical-convex design, meaning that instead of traveling from its outer circumference to its center circumference in a straight line, the teeth have a convex shape and are placed angularly to the center of the gear. The advantage of this design over the plain bevel gear is similar to the relationship of the straight spur gear and helical gear designs. A greater gear tooth contact is achieved and, therefore, a greater torque capability. The spiral design also allows more than one tooth to be in mesh simultaneously, which results in a quieter running gear arrangement (**Figure 9-11**).

Hypoid Gears

Hypoid gears resemble spiral gears in gear design. However, hypoid gears have the axis of the pinion gear, (the smaller of the two gears) below the centerline of the crown gear axis. Again, the purpose of this design is to achieve a greater tooth contact area, which results in greater torque capabilities. Hypoid gears also allow more than one gear tooth to be in mesh simultaneously for a smoother, quieter running gear set. Hypoid gear arrangements typically have the convex side of the crown gear as the forward thrust side of tooth contact (**Figure 9-12**).

Amboid Gears

Amboid gears resemble hypoid gears in every aspect except that the pinion gear axis is mounted above the centerline of the crown gear axis. This also allows greater gear contact and, in many cases, has the concave side of the crown gear tooth as the leading thrust side (**Figure 9-13**). This configuration allows for high-driveline clearances and is widely used for the rear-drive carrier on tandem-drive axles.

Planetary Gears

Planetary gears are based on the concept of our solar system. Think of the sun as located in the center of our solar system with all the planets rotating around it. As shown in, **Figure 9-14** a planetary gear arrangement consists of three different gears: a sun gear, a ring gear, and a set of planetary pinion gears. Like our solar system, the sun gear is located in the center of the gear arrangement. The planetary gears

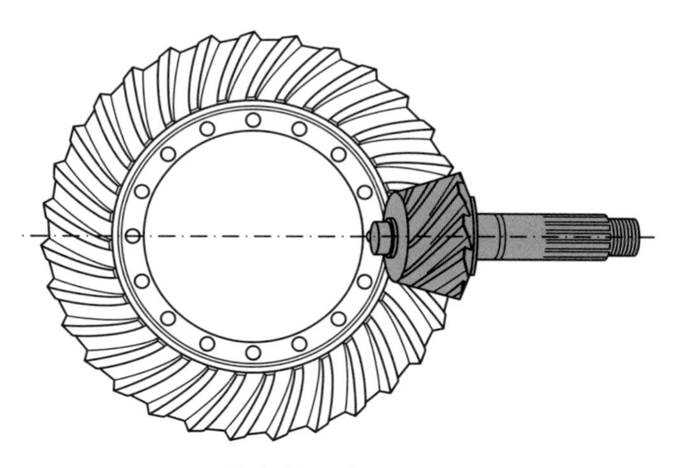

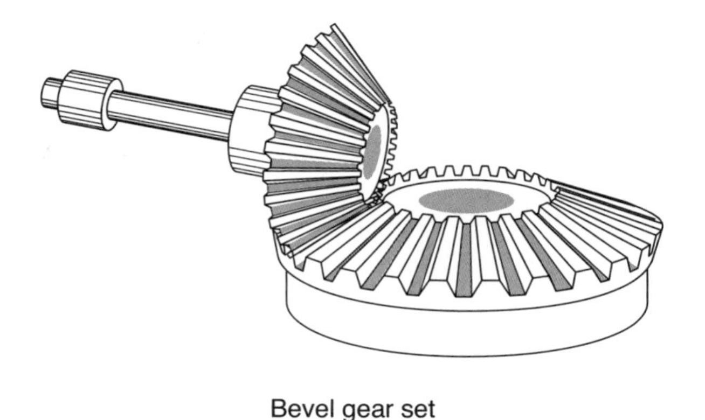

Figure 9-11 Spiral bevel gear set. (*Courtesy of Arvin Meritor*)

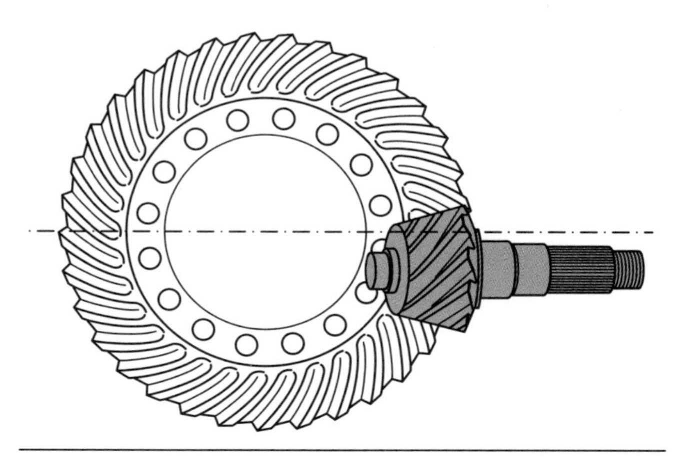

Figure 9-12 Hypoid gearing arrangement. (*Courtesy of Arvin Meritor*)

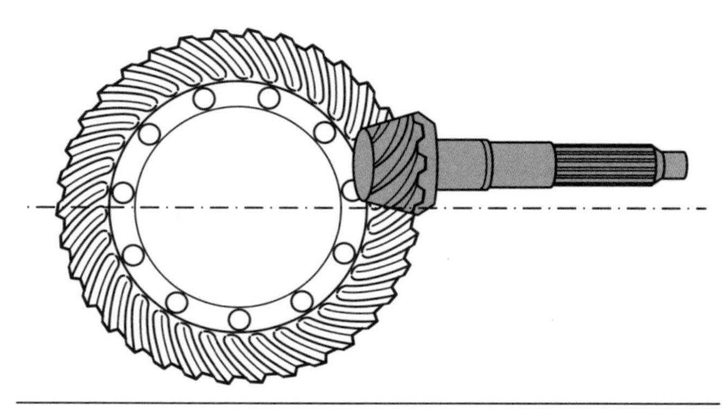

Figure 9-13 Amboid gearing arrangement. (*Courtesy of Arvin Meritor*)

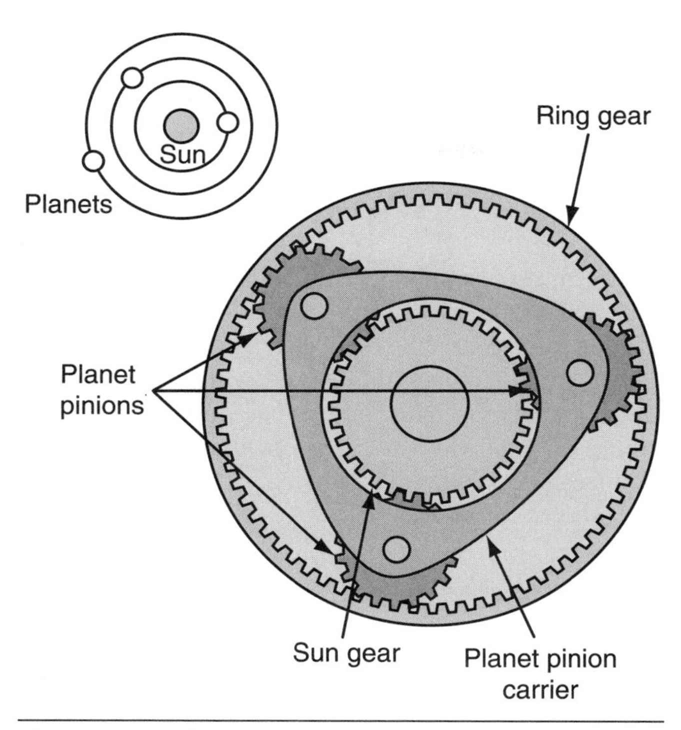

Figure 9-14 Planetary gear sets are based on the design of our solar system. (*Courtesy of John Deere*)

(like the planets) are in mesh with and rotate around the sun gear; they are supported and spaced evenly around the circumference of the sun gear by a planetary carrier. The carrier not only supports the planetary gears, but also gives them an axis to rotate on. The ring gear (an internal gear) is meshed with all the planetary gears and supports or holds together the entire gear set. Planetary gears are used for a wide variety of applications, from transmissions to differentials to final drives. This is because one planetary gear set can have as many as seven different gear-ratio configurations and the integrated gear design is capable of high torque and speeds. This will be covered in greater detail in Chapter 12, "Powershift Transmissions."

Worm Gears

Worm gears are used where high-torque/low-speed applications are required. They consist of a screw gear or worm gear meshed with an external gear (**Figure 9-15**). They also provide the means to alter the drive input of the worm gear by 90 degrees, similar to the crown and pinion gear designs. This gear application can be seen in many steering gears and gearboxes used with cable winching on cranes.

Rack-and-pinion Gears

Rack-and-pinion gears are similar to worm gears in their design except that, instead of a worm gear in mesh with an external gear, it uses a rack with a straight gear-tooth design meshed with an external gear as shown in **Figure 9-16**. Rack-and-pinion gears are used to convert rotary motion into linear motion or vice versa, depending on the input or drive component. For example if the external gear on a rotating shaft is the input or **drive gear**, it will cause the rack (output or

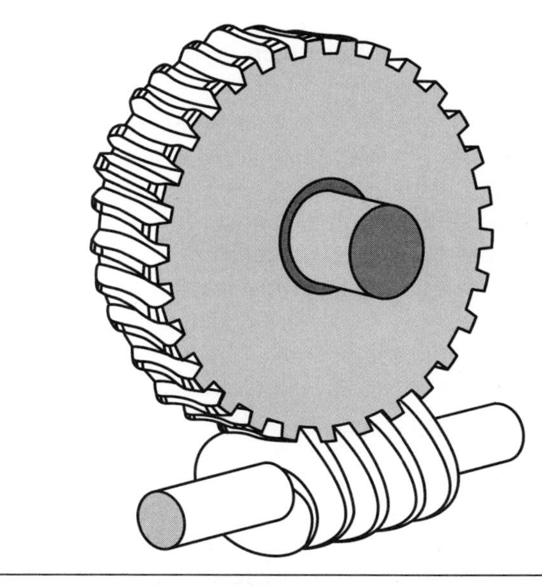

Figure 9-15 Worm gear arrangement. (*Courtesy of John Deere*)

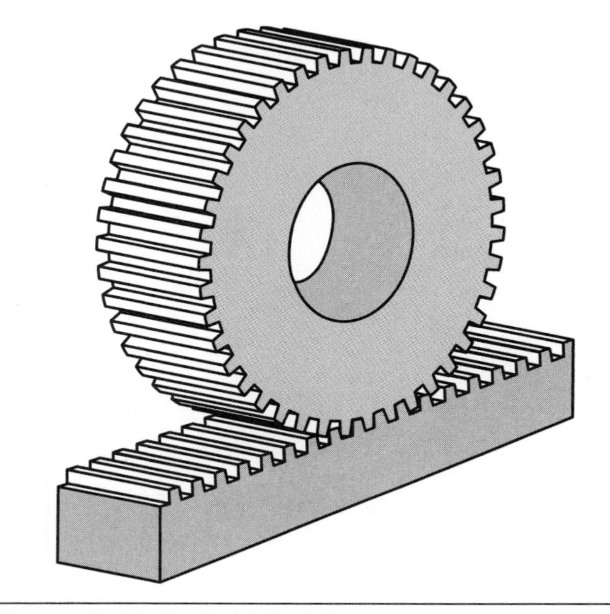

Figure 9-16 Rack-and-pinion gear arrangement. (*Courtesy of John Deere*)

driven gear) to move in a linear direction. If the rack is the input gear, moving in a linear direction, it will cause the pinion gear to rotate on its axis or to rotate the shaft to which it is fixed.

GEAR GEOMETRY

Gear geometry is an important factor in gear design; it provides long service life with nominal wear. **Figure 9-17** shows basic gear tooth nomenclature. Outside diameter refers to the gears' outermost diameters, measured directly across the axis from tooth edge to tooth edge. The root diameter is the measurement across the axis within the root or base of the gear teeth. **Pitch diameter** is also the measurement across the axis of the gear, but it is measured approximately between the outside diameter and the root diameter (known as the pitch line). The area within the outside diameter and pitch diameter is referred to as the addendum and the area within the root diameter and the pitch diameter is referred to as the dedendum. The tooth contact area is the tooth flank. The fillet radii is the machined radius between gear teeth at the root of the gear tooth.

Gear Pitch

Gear pitch is a factor of gear design, all gears in mesh are engineered to gear pitch. Gear Pitch refers to the number of teeth per given unit of pitch diameter. Here is a simpler example of gear pitch calculation: **Figure 9-18** shows a gear with 36 teeth and a pitch diameter (measured from the pitch line, not the outside diameter) of 6 inches. If you divide the number of teeth by the pitch diameter, the answer equals the gear pitch. In this example, 36 (number of teeth) divided by 6 (pitch diameter) equals 6; therefore, the gear pitch is equal to 6.

The important thing to remember is that gears in mesh, regardless of their differences in diameters, must have the same gear pitch. If a larger gear (72 teeth, pitch diameter of 12 inches) is meshed with a gear having a pitch diameter of 6, the larger gear must also have the same gear pitch. In this case, you divide 72 teeth by 12 inches, producing a gear pitch of 6. Congratulations! The gears mesh perfectly.

Gear Backlash

Gear **backlash** is the clearance, or free play, between two gears in mesh. Gears are designed to roll at their pitch diameters (**Figures 9-19** and **9-20**). Gear teeth are strongest at this point and can transfer the greatest amount of force. If there is too much clearance or backlash between the gears, the pitch diameters are extended from each other and cause the addendum sections of the gear teeth to transfer the force. Since the tooth design is not meant to transfer the total force at this point, excessive tooth loading can cause premature failure. If there is too little backlash, this can cause the teeth to contact the root diameters, which will also result in premature gear failure.

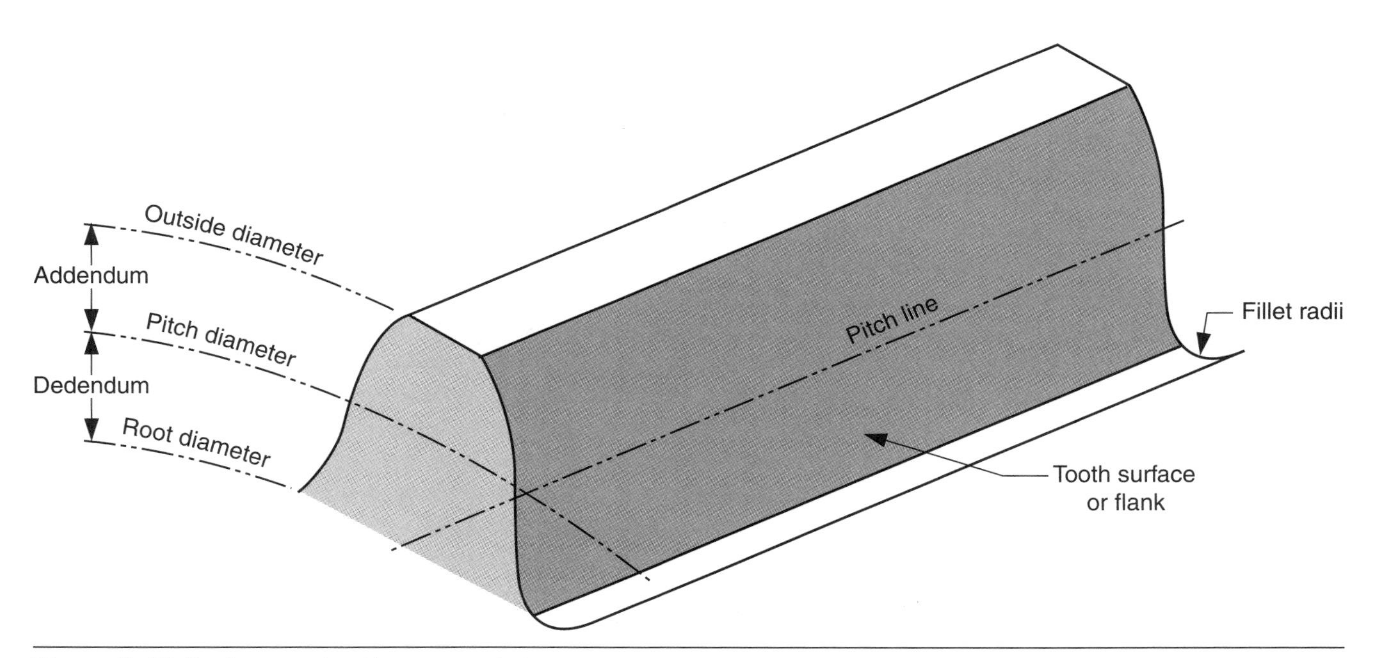

Figure 9-17 Basic gear tooth nomenclature.

Gear Tooth Contact

The amount of gear tooth contact is dictated by the gear tooth design. **Figure 9-21** shows the three stages of gear tooth contact: 1. Gears coming into mesh, 2. Gears in full mesh, and 3. Gears coming out of mesh. Gears coming into mesh contact the addendum portion of the driven gear and the dedendum portion of the driving gear as shown on reference point 1. The tooth loading (LT) or load transfer is relatively low because the gear teeth in full mesh (reference point 2) support the major force, and the teeth coming out of mesh (reference point 3) also support a portion of the tooth loading. The teeth in full mesh are designed to transfer their full load at their respective pitch diameters in a rolling action. There is no sliding motion between the gear teeth when compared to the teeth coming into and out of mesh.

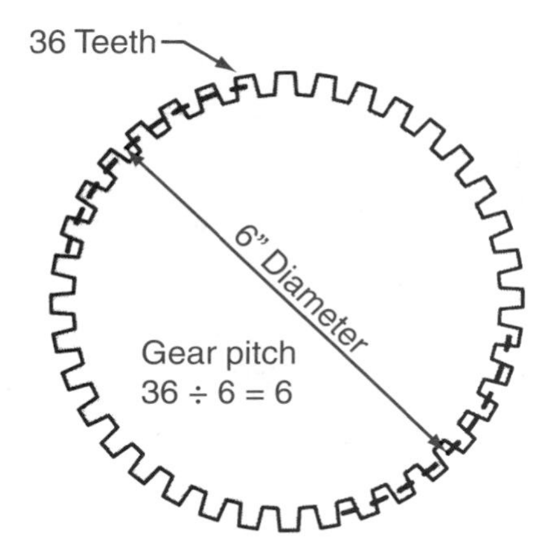

Figure 9-18 Calculating gear pitch.

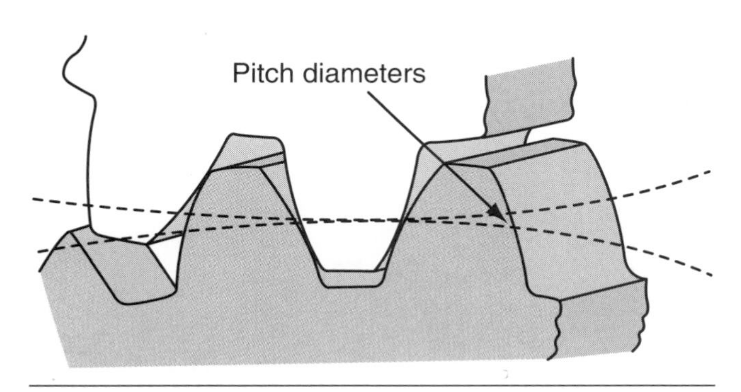

Figure 9-19 Gears rolling at their pitch diameters. (*Courtesy of John Deere*)

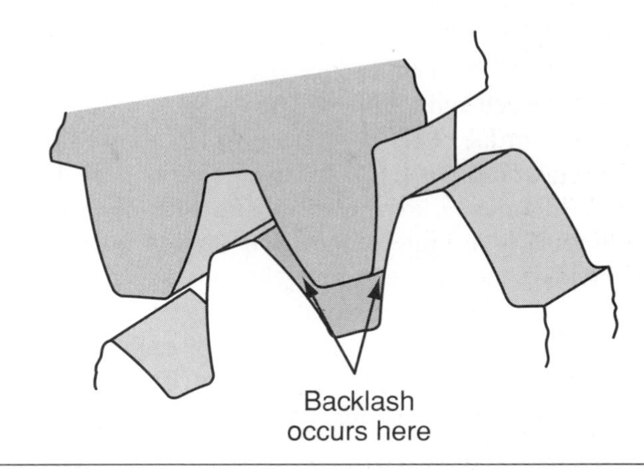

Figure 9-20 Clearance between two gears in mesh is referred to as backlash. (*Courtesy of John Deere*)

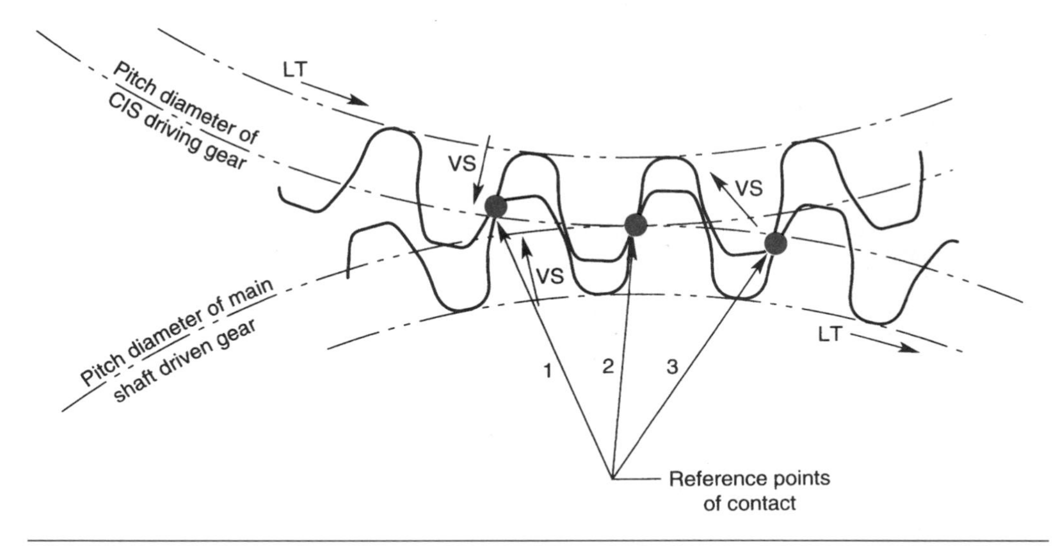

Figure 9-21 Three stages of gear tooth contact: 1. Gears coming into mesh, 2. gears in full mesh, and 3. gears coming out of mesh.

GEAR SPEED AND TORQUE

As mentioned earlier, gears are used to transfer torque from one gear to another, and that speed and torque are affected inversely and proportionately. This section will show how this effect takes place.

Gear Ratios

Gear ratio describes the speed relationship between two gears in mesh. The speed ratio is the number of rotations of the input gear, compared to the number of rotations of the output gear. This relationship of speed is dictated by the number of teeth on each gear and the speed at which they are rotating. To calculate gear ratios, you must always determine the input member (drive gear) from the output member (driven gear). **Figure 9-22** shows two gears of equal size and equal number of teeth. Because there is no difference between the two gears, the output speed and torque will be the same as the input speed and torque. This is referred to as "direct drive" having a gear ratio of 1:1. If the two gears in mesh are of different diameters, as shown in **Figure 9-23,** a small gear (input) is driving a large gear (driven or output). Since it takes the smaller input gear more speed (revolutions) to complete one revolution of the larger output gear, a speed reduction occurs. This type of gear ratio is referred to as an underdrive. Inversely, if a large gear (input) is driving a small gear (driven or output), a speed increase occurs because the large gear takes less than one revolution to rotate the driven (small) gear one revolution. This type of gear ratio is referred to as an overdrive as shown in **Figure 9-24.**

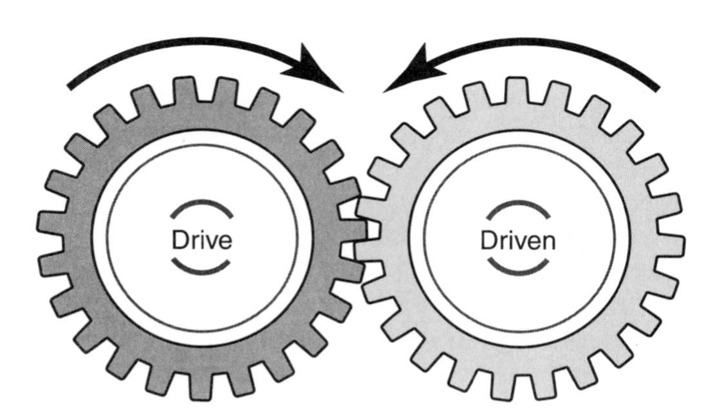

Figure 9-22 When gears have and equal number of teeth in mesh, the input speed will be equal to the output speed.

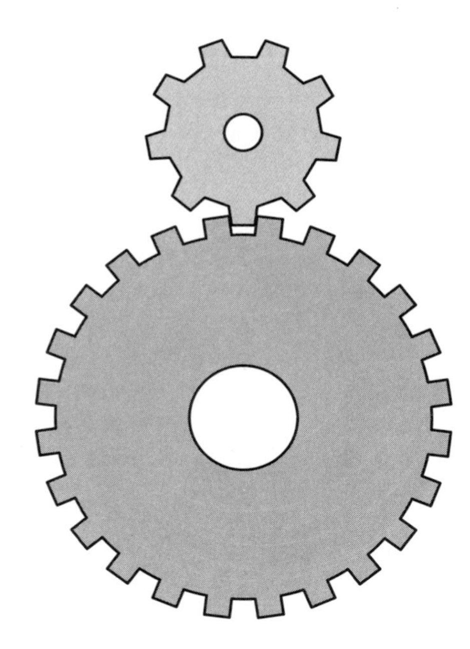

Figure 9-23 A small gear driving a large gear will rotate slower, with greater torque. (*Courtesy of John Deere*)

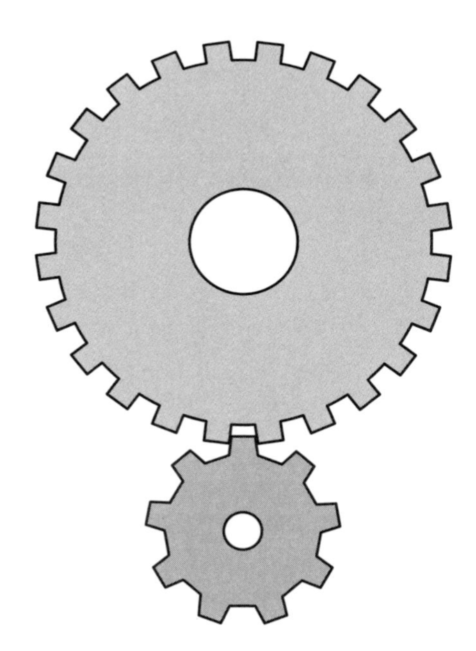

Figure 9-24 A large gear driving a small gear will rotate faster, with less torque. (*Courtesy of John Deere*)

Gear ratios can be calculated either by the relationship of the number of teeth on the input gear and the number of teeth on the output gear, or by the differential of speeds between the input gear and output gear. To calculate a gear ratio based on the relationship of the number of teeth between the input and output gears, you simply divide the number of teeth on the drive gear by the number of teeth on the driven gear. This is an example of an underdrive gear ratio when the drive gear has 25 teeth and the driven gear has 125 teeth.

$$\frac{\textbf{Number of teeth on driven gear}}{\textbf{Number of teeth on drive gear}} = \frac{\textbf{125 teeth}}{\textbf{25 teeth}} = \textbf{5}$$

Therefore on this combination of gears, for every 5 revolutions of the drive gear the driven gear will rotate one revolution. Expressed as an underdrive gear ratio this equals **5:1.**

Note: Whenever the number of teeth on the drive gear is lower than the number of teeth on the driven gear an under-drive will result.

An example of an overdrive gear ratio is as follows: the drive gear has 80 teeth and the driven gear has 50 teeth. Using the same formula, it is calculated as follows:

$$\frac{\textbf{Number of teeth on driven gear}}{\textbf{Number of teeth on drive gear}} = \frac{\textbf{50 teeth}}{\textbf{80 teeth}}$$

$$= \textbf{0.625}$$

Therefore, on this combination of gears, for every 0.625 revolutions of the drive gear, the driven gear will rotate one revolution. Expressed as an overdrive gear ratio, this equals **0.625:1.**

Note: Whenever the number of teeth on the drive gear is greater than the number of teeth on the driven gear, an overdrive will result.

A direct drive combination has the same number of teeth on the drive gear as on the driven gear. There is no speed or torque differential.

To calculate gear ratios through speed differentials, the formula is opposite of that using the number of teeth on each of the drive or driven gears. Divide the speed in rpm (rotations per minute) of the drive gear or shaft by the rpm of the driven gear or shaft. In the following example, if the drive (input) gear is rotating at a speed of 200 rpm and the driven (output) gear is rotating at a speed 40 rpm. The formula will read as follows:

$$\frac{\textbf{Speed/rpm of drive gear}}{\textbf{Speed/rpm of driven gear}} = \frac{\textbf{200 rpm}}{\textbf{40 rpm}} = \textbf{5.0}$$

This results in an under-drive gear ratio of **5.0:1.**

Note: Whenever the speed of the drive (input) gear is higher than the speed of the driven (output) gear, an underdrive results.

The next example shows an **overdrive** gear ratio using the speed differentials of the following gears and shafts. Drive (input) gear speed is 100 rpm; driven (output) gear speed is 120 rpm.

$$\frac{\textbf{Speed/rpm of drive gear}}{\textbf{Speed/rpm of driven gear}} = \frac{\textbf{100 rpm}}{\textbf{120 rpm}} = \textbf{0.83}$$

This results in an overdrive gear ratio of **0.83:1.**

Note: Whenever the speed of the drive (input) gear is lower than the speed of the driven (output) gear, an overdrive results.

Because gear speeds are proportionate to their respective gear ratios, calculations of either speed (rpm) and or ratio can be calculated by combining the two formulas by using at least 3 of the 4 known data and a "cross multiplication "calculation.

For example calculate the output speed of a gear combination with the drive (input) gear having 25 teeth and the driven (output) gear having 125 teeth with an input speed of 200 rpm. The formula is as follows:

$$\frac{\textbf{Number of teeth on driven gear}}{\textbf{Number of teeth on drive gear}} = \frac{\textbf{Speed/RPM of drive gear}}{\textbf{Speed/PRM of driven gear}}$$

Placing the known data in its appropriate slot

$$\frac{\textbf{125}}{\textbf{25}} = \frac{\textbf{200}}{\boldsymbol{X}}$$

Therefore,
Driven (output) gear X

$$= \frac{\textbf{25} \times \textbf{200}}{\textbf{125}} = \frac{\textbf{5000}}{\textbf{125}}$$

Driven (output) gear $X = 40$ RPM

This result indicates that this gear combination, with the input gear rotating at a speed of 200 rpm through the gear ratio of 5:1, has the output gear rotating at a reduced (underdrive) speed of 40 rpm.

Using the same example of gear teeth and speeds, change the known data to: Drive (input) gear has 25 teeth rotating at a speed of 200 rpm and the output gear is rotating at a reduced speed of 40 rpm. The formula is as follows:

$$\frac{\textbf{Number of teeth on driven gear}}{\textbf{Number of teeth on drive gear}} = \frac{\textbf{Speed/RPM of drive gear}}{\textbf{Speed/PRM of driven gear}}$$

Place the known data in the appropriate slot

$$\frac{X}{25} = \frac{200}{40}$$

Therefore,
Number of teeth on driven (output) gear X

$$= \frac{25 \times 200}{40} = \frac{5000}{40}$$

Number of teeth on Driven (output) gear X

$$= 125 \text{ teeth}$$

Speed and Torque Relationship

We now understand how the speed of two gears in mesh is affected by the diameter and number of teeth on a drive gear versus the diameter and number of teeth on a driven gear. Gears act as levers to transfer torque from the engine to the drive wheels through different gear configurations to obtain maximum efficiency from the engine speed and torque limitations. Speed and torque are inversely proportional, meaning that by gaining speed through a given gear ratio as in an overdrive, we reduce torque, and inversely, by reducing speed (as in an underdrive) we increase torque. Torque is always transferred through the center of the shaft delivering it. If a gear is attached to the shaft delivering the rotational force, it must deliver this torque at an extension equal to half the pitch diameter of the gear. Therefore, the torque delivered to the end of the gear is equal to torque divided by half the pitch diameter. Torque must be divided by the pitch diameter because we are trying to transfer this torque at an extension away from the center of the driving shaft. This is much like trying to lift a pail of water at an extended arm's length compared to lifting it straight up beside you.

Tech Tip: The drive gear of any gear combination will always reduce the original input torque value. The increase in torque is dictated by the diameter of its mating driven gear.

Because torque is transferred to the gear pitch diameter and the driven gear receives this torque at its equivalent pitch diameter, the driven gear will transfer torque to the center of its shaft equal to the torque times half the pitch diameter. The torque must be multiplied because the driven gear is now acting as a lever and fulcrum assembly as shown in **Figure 9-25.** When the fulcrum is farthest from the end of the lever (see Figure 9-4), the force is multiplied equal to the length of that lever. A torque increase occurs at the center of its shaft. If the fulcrum is placed closer to the end of the lever, that distance reduces the multiplication factor proportionately to the distance.

Figure 9-26 shows how torque is transferred through two gears of equal size and equal number of teeth. The input torque of the drive gear is equal to 10 foot-pounds at a distance of 1 foot. Therefore, 10 foot-pounds is divided by the lever distance of 1 foot, which is equal to 10 pounds of force. This 10 pounds of force is relayed to the drive gear's outer pitch diameter. Because this force is acting at an extension of 1 foot to the center of the driven gear, we multiply the 10 pounds of force by the 1-foot lever, which is equal to 10 foot-pounds of torque.

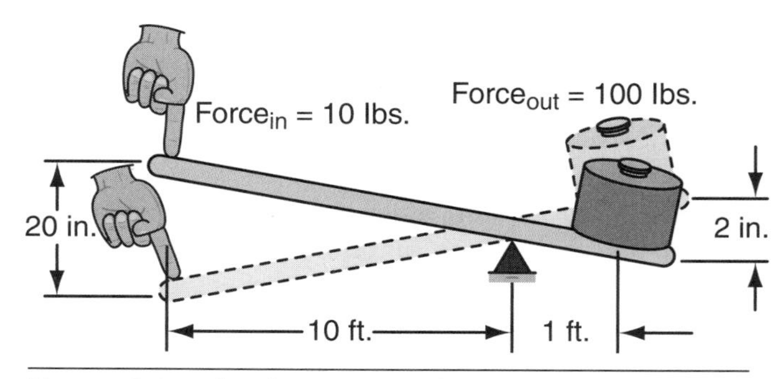

Figure 9-25 The force gained by the length of the lever is proportional to the distance loss. (*Courtesy of Vickers Incorporated*)

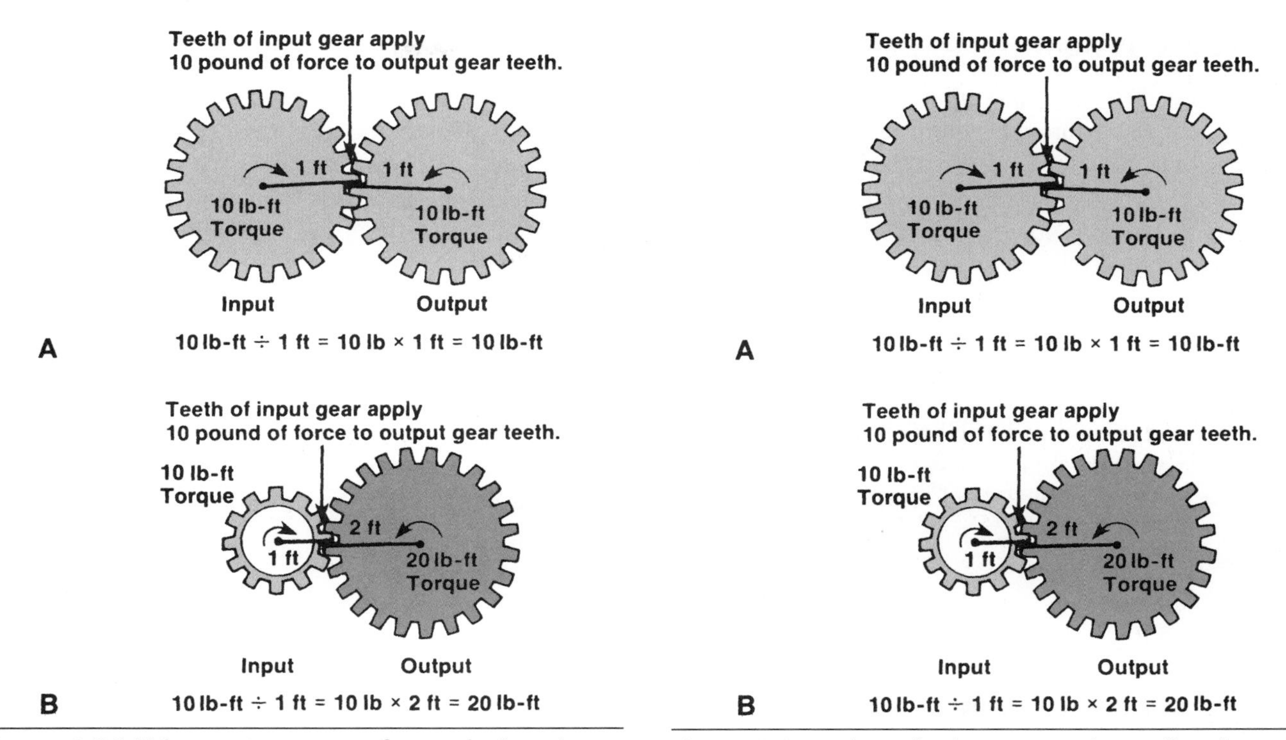

Figure 9-26 When gears are of equal size, input torque equals output torque.

Tech Tip: Idler gears do not change speed or torque ratios between drive and driven gear sets.

Figure 9-27 shows how torque is increased by using a drive gear that is half the diameter of the driven gear. The same input torque of 10 ft-lbs divided by the 1-foot distance of the drive gear equals 10 lbs of force. This 10 lbs of force is relayed to the drive gear's outer pitch diameter. Because this force is acting at an extension of 2 feet to the center of the driven gear, we multiply the 10 lbs of force by the 2-foot lever, which is equal to 20 ft-lbs of torque. Torque increase is doubled with a speed reduction and decreased to half the input speed due to the driven gear being twice the diameter of the drive gear.

Here is a third example. The drive gear is twice the diameter of the driven gear such as an over-drive gear ratio. Given: The drive gear has a pitch radius of 2 feet with a torque of 10 ft-lbs applied to it. The torque transfer to the driven gear is equal to 10 ft-lbs divided by 2 feet, which is equal to 5 lbs of force. This 5 lbs of force is applied to the driven gear at a distance of 1 foot. Therefore 5 lbs of force is multiplied by 1 foot, which equals 5 ft-lbs of torque. The torque is reduced by half and the speed has doubled due to the drive gear being twice the diameter of the driven gear.

Figure 9-27 When the input gear is smaller than the output gear, a torque increase occurs.

Tech Tip: Speed and torque are transferred equally because the gears in mesh are of equal size.

Idler Gears

External gears in mesh rotate in opposite directions. Idler gears are used to transfer rotation in a specific direction without changing the gear ratio. **Figure 9-28** shows an idler gear used to transfer gear rotation of a driven gear in the same direction of the drive gear. Idler gears can also be used to transfer rotational force in place of chain drives or belt drives.

Figure 9-28 Idler gears are used to transfer a specific gear direction without changing speed or torque.

ONLINE TASKS

1. Access the Caterpillar online service information system (SIS) through your college or workplace data hub. Find the final drive gear ratio of a 988G-model front-end wheel loader. The Caterpillar SIS site is at *https://login.cat.com* however, you will be prompted for a password at logon and you will not get any farther unless you have access.
2. Access the Web site *http://auto.howstuffworks.com*. Although this Web site is primarily based on automotive technology, it has a very wide knowledge base of basic principles which apply to most mechanical trades. Use the search function to find "How Helical Gears Work."

SHOP TASKS

1. Locate a transmission partially disassembled and calculate the gear ratio of any two gears in mesh. Note which gear is the driving gear and which gear is driven. Identify the type of gearing used within the transmission, e.g.: helical, spur, herringbone, etc.
2. Locate a differential partially disassembled and identify the type of gearing used, e.g.: hypoid or amboid. With the same gear set, calculate the final drive ratio.
3. Locate a gear within a disassembled transmission and calculate the gear pitch as outlined in this chapter.

Summary

- Gears relay a rotating force known as torque.
- Gears can inversely provide a speed increase or torque increase.
- A variety of gear designs are used in heavy equipment dependent on the application of the equipment.
- Gear geometry nomenclature is as follows: outside diameter, root diameter, pitch diameter, pitch line, addendum, dedendum, tooth flank, and fillet radius.
- Gear pitch refers to the number of teeth per unit of pitch diameter.
- Gear backlash refers to the clearance between two gears in mesh.
- The three stages of gear contact are: coming into mesh, full mesh, and coming out of mesh.

Review Questions

1. A hypoid gearing arrangement has its pinion gear mounted:
   A. below the centerline of the ring gear.
   B. above the centerline of the ring gear.
   C. in line with the center of the ring gear
   D. none of the above
2. The term *gear backlash* refers to:
   A. the amount of load placed on two gears in mesh
   B. the amount of clearance between two gears in mesh
   C. the amount of axial clearance between two gears in mesh
   D. the amount of end play on the transmission output shaft
3. A gear with a "gear pitch" of 10 will only mesh with a:
   A. 10-inch gear
   B. 5-inch gear
   C. 20 gear pitch diameter
   D. gear having the same gear pitch

4. Spur gears cut at an angle or spiral:
   A. have a greater load capability
   B. run more quietly than helical gears
   C. both A and B
   D. none of the above
5. What is the gear ratio of two external gears in mesh if the driven gear has 24 teeth and the drive gear has 18 teeth?
   A. 0.75:1
   B. 1.33:1
   C. 4.32:1
   D. 43.2:1
6. In Question 5, the gear ratio is:
   A. an overdrive
   B. an underdrive
   C. a direct drive
   D. an under-overdrive
7. Gear pitch refers to:
   A. the number of teeth per inch around the gear circumference
   B. the number of teeth per inch around the gear circumference divided by the number of teeth
   C. the gear width
   D. the number of teeth around the circumference of the gear divided by the gear diameter
8. Helical gears:
   A. are cut at an angle or spiral
   B. have a greater load capability
   C. run more quietly than spur gears
   D. all of the above
9. The three main components in a planetary gear set are:
   A. the sun gear, ring gear, and planetary gears
   B. the sun gear, pinion gear, and planetary gears
   C. the sun gear, ring gear, and pinion gear
   D. the pinion gear, ring gear, and planetary gears
10. How many gear ratios can be achieved through a planetary gear set?
    A. five
    B. six
    C. seven
    D. eight

CHAPTER

10 Mechanical Clutches

Learning Objectives

After reading this chapter, you should be able to

- Identify and define the various types of clutches and flywheels used in heavy-duty equipment applications.
- List the components of various types of clutch assemblies.
- Identify and define the various types of clutch linkages used in heavy-duty equipment applications.
- Give a failure analysis of clutch and flywheel premature wear.

Key Terms

Band-type clutch
centrifugal clutch
ceramic facing
coefficient of friction
coil spring
cone-type clutch
dampened disc
diaphragm or Belleville-type spring
disk-and-plate-type clutch
driven member
driving member
dry-type clutch
dynamic friction
electro-kinetic friction
electro-magnetic clutch
expanding shoe-type clutch
friction
friction disc
flywheel run-out
hydraulically-released clutch
mechanical linkage clutch
organic facing
over-center clutch
pneumatically-controlled clutch
pressure plate
pull-type clutch
push-type clutch
release bearing
rigid disc
slave cylinder
static friction
tube-type clutch
wet-type clutch

INTRODUCTION

There are many types of clutches used in heavy equipment applications today, and every machine uses at least one type of clutching mechanism. These will be described in this chapter.

It is often necessary to allow an engine to run without the vehicle moving. The main purpose of a clutch is to allow a means of connecting and disconnecting powerflow or torque from a driveline. It also provides a means of getting a vehicle moving from a "stand still" position (**Figure 10-1**). The engine produces the

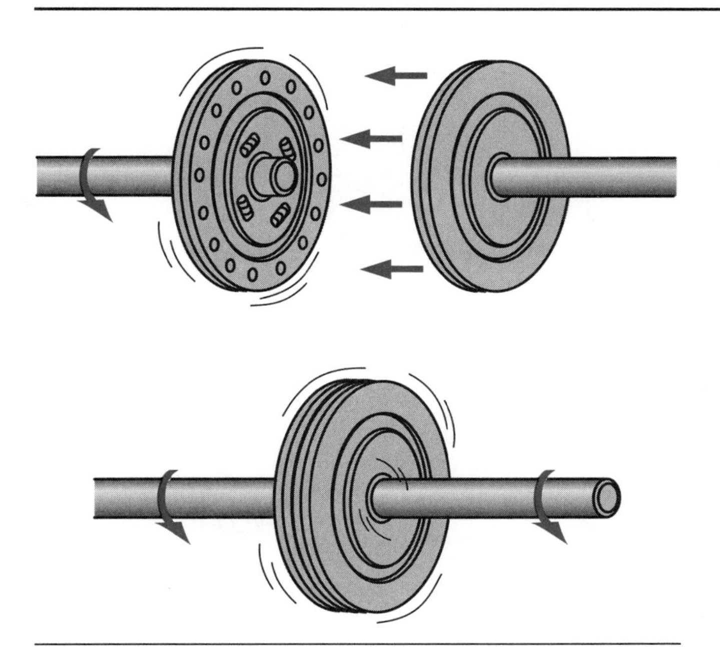

Figure 10-1 The main purpose of the clutch is to connect and disconnect rotational torque.

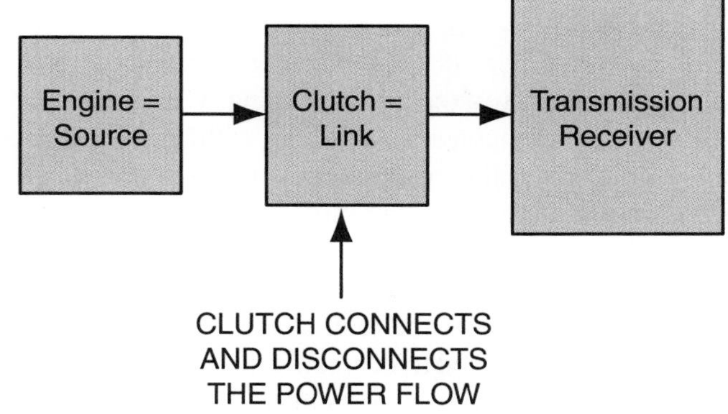

Figure 10-2 A clutch is the link between an engine source and a transmission receiver. (*Courtesy of John Deere*)

rotational force or torque needed to accomplish work. A clutch assembly can act as the link between the engine and transmission to supply the rotational force to the transmission when required as shown in **Figure 10-2.** An example of this type of clutching application is the very popular **disk-and-plate-type-clutch.**

This type of clutch uses one or more discs splined to an output shaft that is sandwiched between the flywheel and a pressure plate. Because the engine is running and the vehicle is in a "stand still" position, the coupling phase is achieved through the friction created between the flywheel clutch disc and clutch pressure plate when squeezed together. It must also provide a positive drive with no slippage between the two members once the unit is completely coupled. Other types of clutches are used to provide a drive in a single direction only, and freewheel in the other. The application of power transfer, whether from the engine to the transmission or within the transmission itself and on to the final drives, will dictate the type and style of clutch to be used for a specific machine. Cranes use a variety of clutch assemblies not normally found in commercial or earth-moving vehicles, but they work under the same principals. These types of clutches will also be described.

To understand how clutches work, we must first understand the principals of friction. **Friction** is the resistance to relative motion between two surfaces in contact with each other. To apply this meaning to a clutch assembly, we must consider the components of a clutch and the surfaces where the friction is created. First, the flywheel (driving member), which is in motion, provides one contact surface. Next, the clutch disc (driven member) is attached to the driveline and located between the flywheel and the pressure plate, which is initially stationary. Finally, the pressure plate, which is bolted to the flywheel, acts as the second driving member and contact surface. The pressure plate is also the component that is actuated to apply or release itself against the **friction disc.**

When the clutch disc, or friction disc as it is sometimes called, is sandwiched between the flywheel and the pressure plate, a calculated amount of friction is necessary to start the clutch disc moving. This is referred to as **static friction.** Static friction is the force that resists a stationary object to the degree of initial motion. In respect to a clutch application, the static coefficient of friction applies to the force restricting the movement of an object (the clutch disc) that is stationary on a relatively smooth, hard surface (engine flywheel). The load on the vehicle, the vehicle weight, and any gear ratio within the drivetrain determine the amount of static friction against the clutch disc. Once the point of static friction is overcome, the stationary clutch disc starts to move. It takes considerably less friction to bring the clutch disc to the same speed as the flywheel and pressure plate because it is not starting from a stationary point. This is known as **dynamic or kinetic friction.**

Tech Tip: The Kinetic coefficient of friction applies to the forces restricting the movement on an object that is in motion on a relatively smooth hard surface.

Friction generates a great amount of heat that must be absorbed and dissipated every time a clutch function occurs. For this reason, manufacturers of clutch components must engineer the components to resist the heat produced and allow a means for cooling. Without proper cooling, these components would fail within a very short time frame. Careful consideration is taken when selecting proper materials for the clutch components which must withstand the heat generated and the torsional forces that are produced during normal functions of the clutch. For this reason, the total **coefficient of friction** ratio must be calculated to overcome the static and kinetic forces as well as the heat generated from the friction.

To calculate the coefficient of friction between two materials in contact with each other we can apply this formula: The coefficient of friction (u) is equal to, the amount of force (f) to move the material or object divided by the weight (w) of the material or object. An example of this is expressed as follows. Consider a 200-pound block of wood resting on a cast-iron surface. If it takes 100 pounds to move the wooden block on the cast-iron surface, we divide the force it takes to drag or push the block along the cast-iron surface by the weight of the wooden block. Therefore (f) 100 lbs divided by (w) 200 lbs equals a coefficient (u) ratio of **0.5:1.** Now consider a different material and comparison. Let's look at a 200-pound block of bronze on the same cast-iron surface. If it requires only 40 pounds of force to move the bronze block, we divide the force required to move the block (40 lbs) by the weight of the block (200 lbs), which equals to a coefficient ratio of **0.2:1** as shown in **Figure 10-3.** Evaluating these two materials on the same surface, we can clearly see that wood has a higher friction value or coefficient ratio than bronze on a cast-iron surface. The higher the ratio value, the higher the friction between those two surfaces. To consider these materials in a clutch application, wood would grab faster and harder than bronze because of the material's coefficient value.

This is not always the ideal application for a clutch. There must be some allowable slippage for a smooth power transfer from the engine to the drivetrain and to allow for heat absorption and dissipation. For this reason, clutch manufactures use many materials or combine materials to achieve these results.

CLUTCH AND FLYWHEEL DESIGN

Clutch and flywheel designs vary from manufacturer to manufacturer. This chapter describes the basic principles of all types of clutches and flywheels and discusses some of the more popular clutches used in the industry.

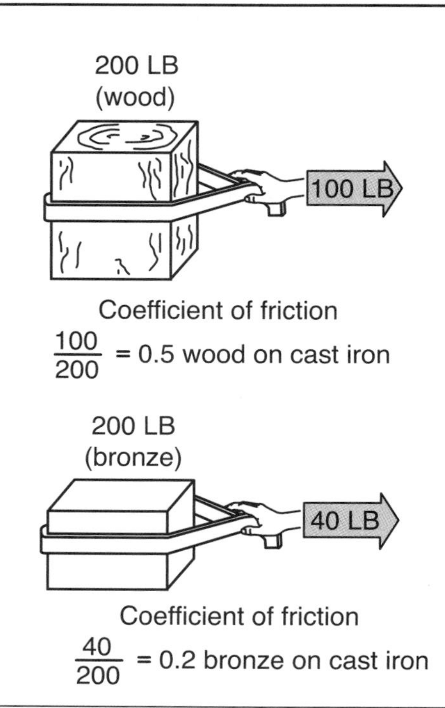

Figure 10-3 The coefficient of friction is the ratio of the force required to move an object divided by the weight of the object.

Disk-and-Plate-Type Clutch Design

There are many design features to disk-and-plate-type clutches. Some of these clutches utilize a single friction disc and as many as five friction discs, depending on the application and style of the clutch assembly. Some utilize a diaphragm spring or coil-type springs for the **pressure plate** clamping force and some a **Push-type** or **Pull-type** method to release or disengage the clutch assembly. Disk-and-plate-type clutches can also be of the **wet-type** or **dry-type** design.

Friction Disc Design

Clutch discs or friction discs, are designed with a friction material on both sides of the disc; one surface will contact the flywheel and the other will contact the pressure plate. The friction material is either riveted or bonded to the disc. The center hub of the disc is splined to accommodate the input shaft of the transmission. When the clutch assembly is engaged (the disc clamped between the pressure plate and flywheel), engine torque is transferred from the flywheel to the friction disc to the splined input shaft of the transmission and is equal to the engine speed.

Because of the high-torque loading these discs are susceptible to, two design types are used: **ridged discs** and **dampened discs.** Ridged discs transfer engine torque directly to the disc hub from the clutch disc that is riveted to it. This design relies solely on slippage between the two facings to start the disc rotating before complete engagement. This design, shown in **Figure 10-4** can cause driveline torsional shock loading with rapid clutch engagement. Ridged discs do not absorb any driveline or torsional vibrations created through various working angles of U-joints or improper gear selection within the transmission. Ridged discs are more susceptible to cracking and they can aid to U-joint failures on commercial on-highway vehicles.

Dampened discs are designed to partially absorb or dampen the torque transfer from the flywheel to the clutch disc and on to the transmission input shaft. Dampened discs are designed with coil-type springs placed between a two-piece hub (see **Figure 10-5**). This design allows some deflection to occur between the initial stationary disc and the rotating flywheel when the two components first come into contact. This action will compress the springs, absorbing the initial torque load of the engine. During regular driving conditions, this clutch design can also absorb vibrations that occur from U-joints and drivelines or from incorrect gear selection. The number of springs and the spring force will dictate the amount of dampening effect of each disc design.

Another identifying feature of clutch discs is their **friction facing.** The type of facing determines the friction characteristics of the disc. There are two basic types of friction facings used: **organic** and **ceramic.** See **Figure 10-6.** Organic facings use a combination of materials such as glass, mineral, wool, cork, and carbon. Asbestos is not normally used, although it may be combined with other materials in very low percentages to give the material greater heat absorption characteristics. Organic material offers the best drivability factor because of its ability to slip slightly during engagement. Under extreme loads or operating conditions such as stop-and-go, it is more likely to experience fade because of its lower coefficient factor as the temperature increases with the increased demand or load.

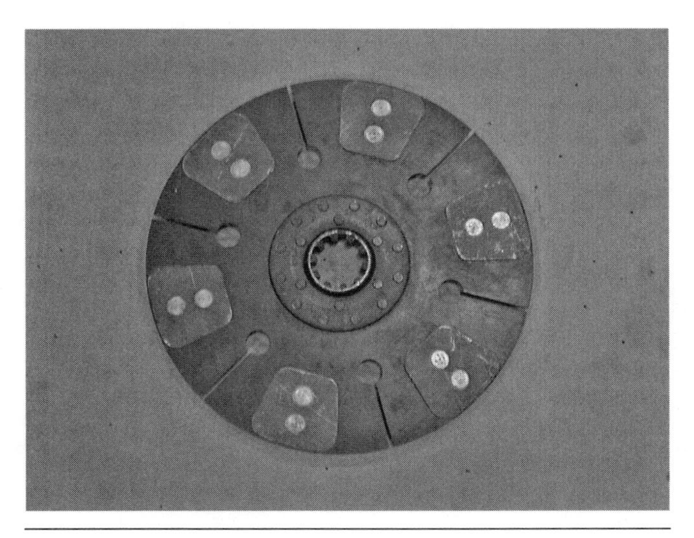

Figure 10-4 Rigid-type clutch disc with ceramic facing.

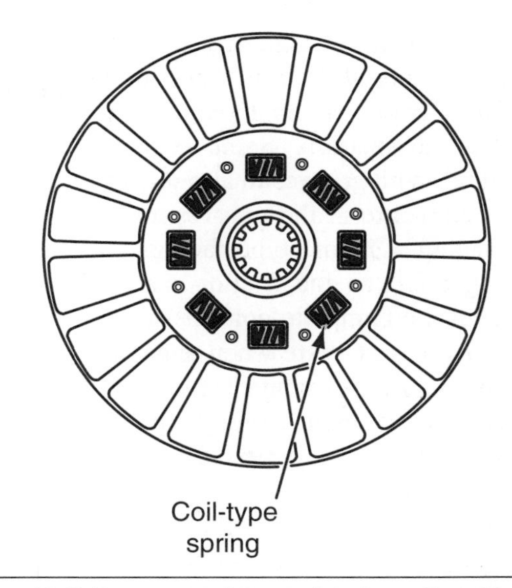

Figure 10-5 Dampened-type friction disc. Note the coil springs within the center hub.

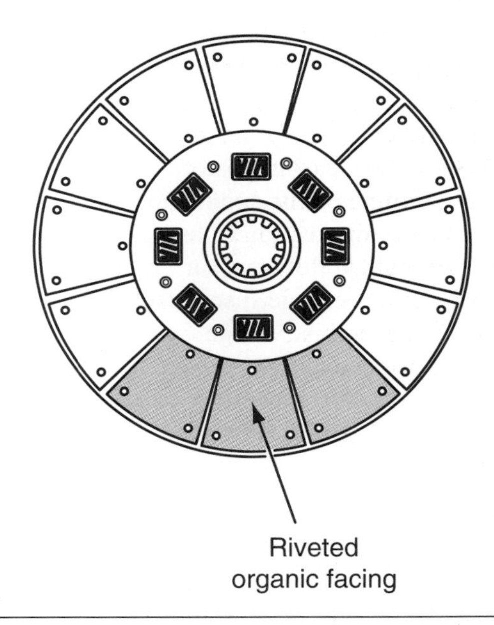

Figure 10-6 Organic facing on dampened-type disc.

Organic facings are either riveted or bonded to the steel backing plate of the disc. Riveting the disc to the plate offers a more secure mounting; bonding offers greater heat dissipation. Some organic facings are manufactured with grooves to allow any particles from wear to accumulate in the grooves and be thrown off the disc facing. Organic type facings are "full face." Full facing refers to the friction material coming in contact with the complete contact surface area of the flywheel or pressure plate when engaged. The greater amount of surface area contact, the higher the friction coefficient factor. Organic-type facings have, a lower coefficient than ceramic-type facings, which is another reason for increasing the surface area of the disc (**Figure 10-7**).

Ceramic facings are made of ceramic and copper or iron materials. These materials combined together give this type of facing a greater coefficient of friction, as well as better heat absorption and dissipation characteristics. These characteristics offer less slippage and wear during engagement of the clutch. This makes it an ideal disc for off-road applications or when the clutch is subjected to heavy loads or frequent stop-and-go situations. The "pads," as they are referred to, are riveted individually to the disc and do not cover the full face of the disc. Depending on the application and torque loads of the clutch, the disc can have as many as six pads or as few as two pads of different surface areas. Much like the organic discs, ceramic discs also come in either a rigid or a dampened disc design.

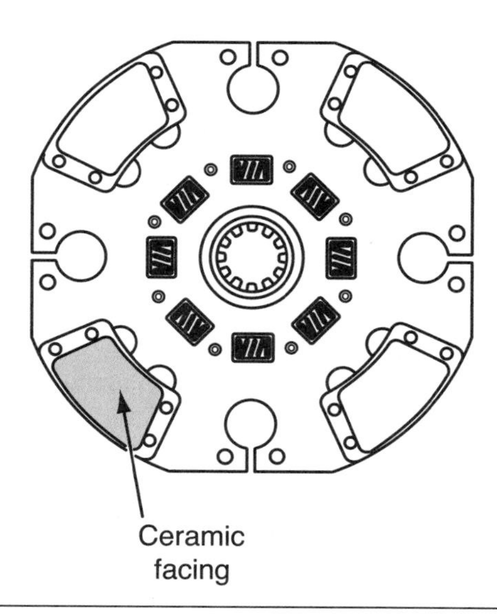

Figure 10-7 Ceramic-type facings on a dampened disc.

Pressure Plate Design

The pressure plate assembly (**Figure 10-8**) has the responsibility of supplying the clamping force necessary to keep the friction disc clamped tightly against the flywheel without slipping. It is directly bolted to the flywheel and assists in transferring engine torque to the disc. It must also have the capability of releasing the clamping force and moving away from the disc to allow the disc to free wheel. The pressure plate is allowed to move within the clutch cover and still provide a positive drive because the pressure plate is either equipped with lugs that will travel along a channel within the clutch cover, or mounted with flexible steel straps that will flex as the pressure plate moves toward or away from the disk. Guide pins may be used

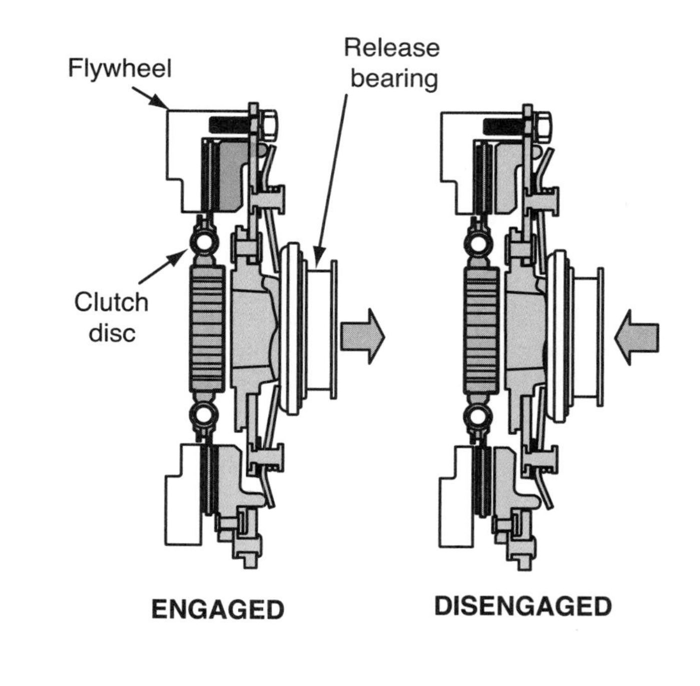

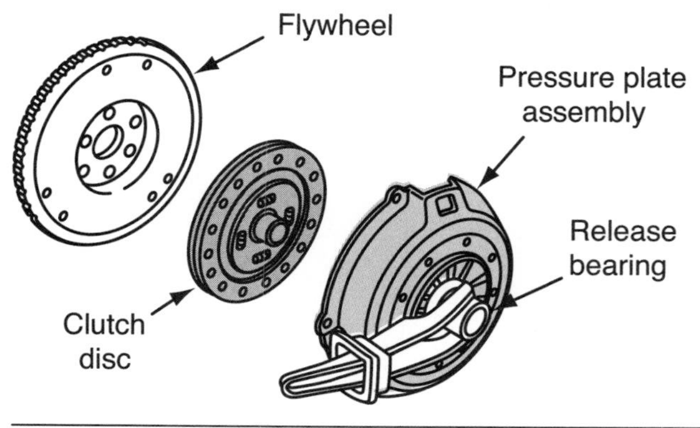

Figure 10-8 Push-type pressure plate assembly.

to allow the pressure plate to slide along during any clutch action.

The clamping force is achieved by using springs. Two types of spring design are used: coil-type springs and diaphragm-type springs. Coil-type springs are progressive springs. As they are compressed, an increasingly greater amount of force is required to keep compressing the spring to its maximum point. The number of springs used on a pressure plate is dictated by the amount of clamping force required for a particular clutch application (**Figure 10-9**). There can be as few as six springs and as many twelve or more. Because **coil springs** are progressive-type springs, disengagement of the clutch by the operator pushing the clutch pedal also becomes progressively harder. The force required to disengage the clutch at the pedal must be carefully calculated to prevent operator fatigue.

Diaphragm springs, or **Belleville-type springs** as they are sometimes referred to, are one-piece beveled springs with slots cut into the center circumference. These slots allow the spring to compress more uniformly as it travels inward during disengagement of the clutch. This design has a constant spring force throughout its compressed length, making it a desirable feature for pressure plates. Because the diaphragm spring is uniform, release levers are not required within the pressure plate assembly. The **release bearing** contacts the spring at its center point where the slots are cut into the spring. Pivot pins that secure the spring to the pressure-plate housing act as the pivot point to pull the pressure plate away from the disc (**Figure 10-10**). Release levers are used to compress the springs on pressure plates by utilizing coil-type springs to move the pressure plate away from the disc.

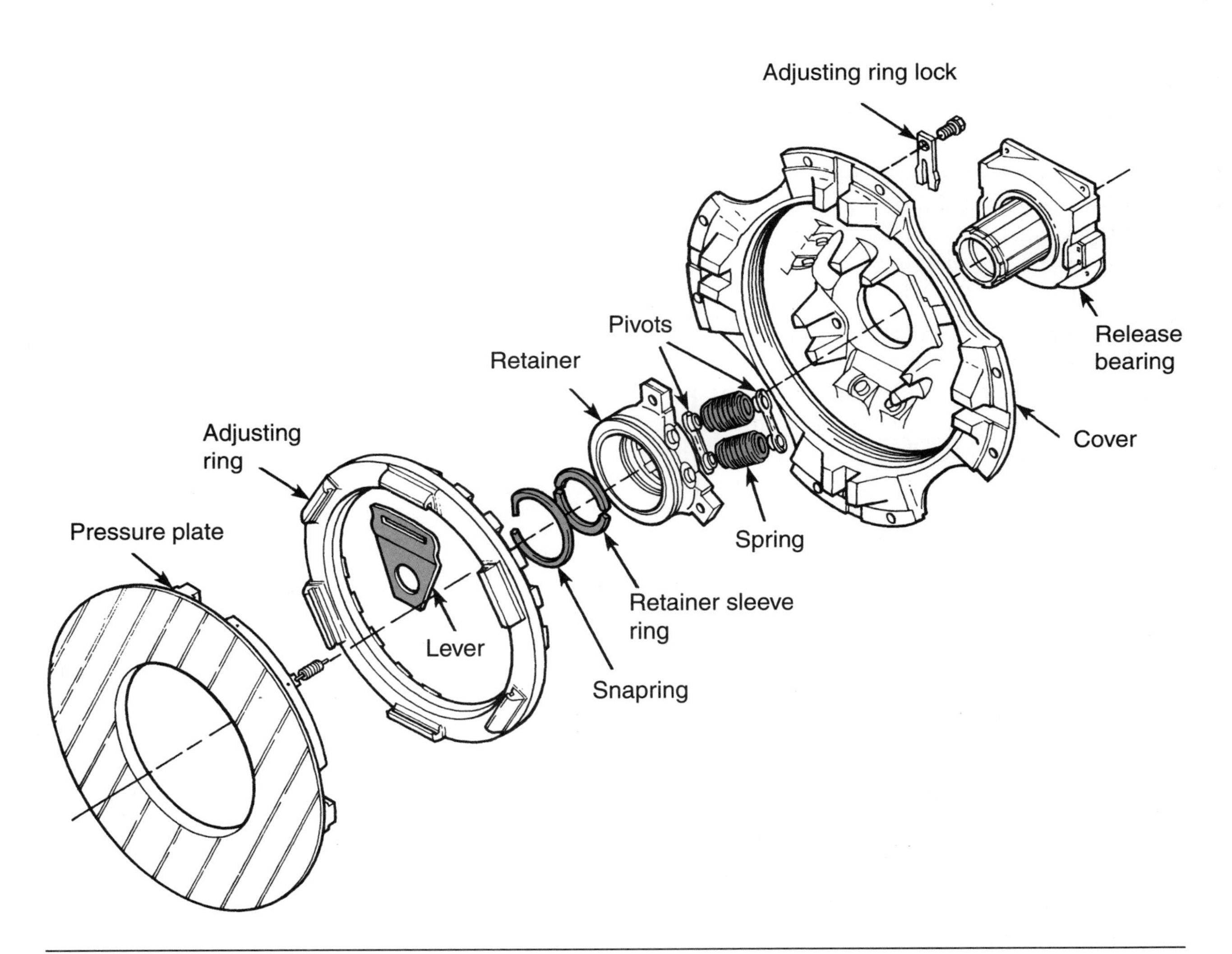

Figure 10-9 Pressure plate assembly utilizing coil-type springs. (*Courtesy of Eaton Corp.—Eaton Clutch Div.*)

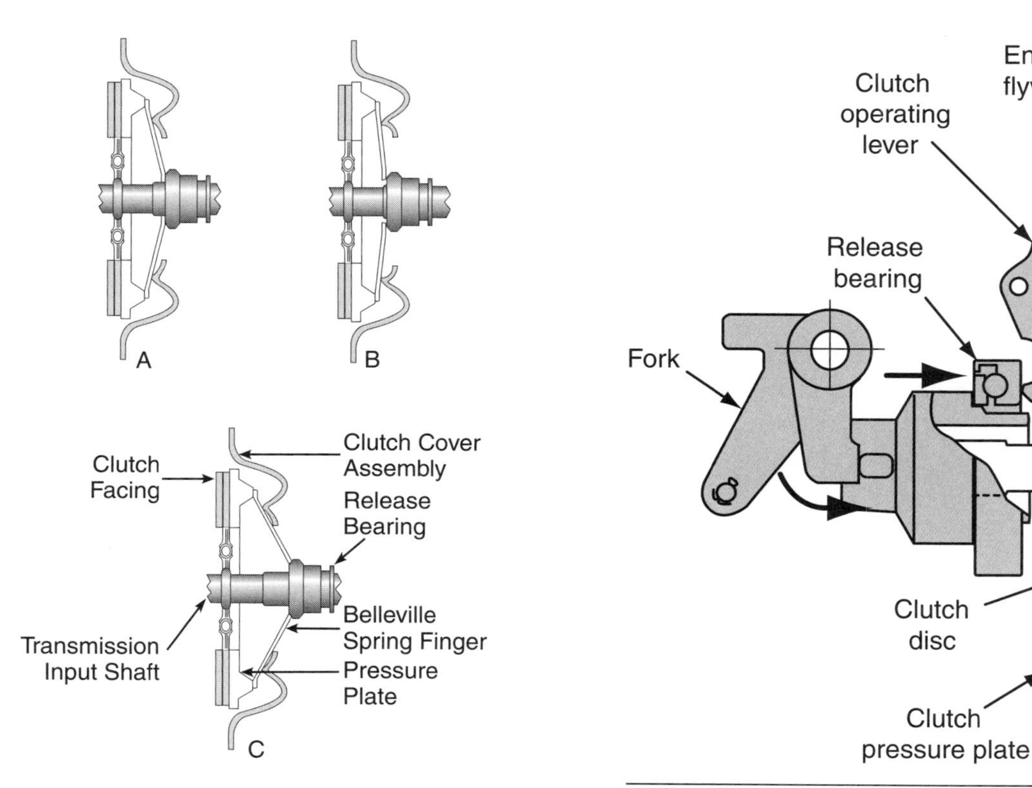

Figure 10-10 Diaphragm clutch operation: (A) New, engaged position; (B) Released position; (C) Worn, engaged position.

Figure 10-11 Release bearing travels toward the flywheel against the clutch operating levers. This causes the pressure plate to move away from the flywheel and disengage the clutch.

On **push-type clutches,** the release levers act as the pivotal point for the release bearing to push on the ends of the levers, or "fingers" as they are sometimes called and to pull the pressure plate away from the disc (**Figure 10-11**). Push-type clutches so-named because of the direction the release bearing moves in relation to clutch disengagement and flywheel location. When the release bearing moves toward the flywheel to disengage the clutch, it is described as a push-type clutch. If the release bearing moves away from the flywheel to disengage the clutch, this describes a "pull type clutch." Pull-type clutches will be covered later in this chapter.

The pressure plate is normally made of cast iron or cast steel alloys depending on the manufacturer and the application necessary. Cast-iron pressure plates offer good heat dissipation qualities, whereas cast steel alloy pressure plates offer longer life and durability. In some applications where engine rpm can be very high, the release fingers are equipped with concentric weights. These weights assist clutch springs to apply additional pressure against the pressure plate at higher rpm. During normal driving conditions, the clutch springs keep enough force on the pressure plate to prevent the clutch disc from slipping between it and the flywheel. At higher rpm it is possible to create enough centrifugal force within the pressure plate to reduce the spring force against the pressure plate. By implementing small weights into the release fingers of the clutch housing, the same centrifugal force created at higher rpm will cause the weights to move outward, apply additional force to the pressure plate, and prevent slippage of the clutch assembly.

Clutch Release Bearings

The clutch is operated by a pivoting lever or fork and must actuate a rotating clutch assembly. This is accomplished by the release bearing. The release bearing is designed to travel freely along the transmission input shaft, which also splines the clutch discs. The hub of the release bearing has a collar designed to accommodate a fork mechanism to control the release and application of the clutch assembly. The fork, or yoke assembly as it is sometimes called, pivots on a shaft either in the clutch or in the transmission housing and holds the bearing in place (**Figure 10-12**). The outer bearing shell incorporates a ball-type bearing, which

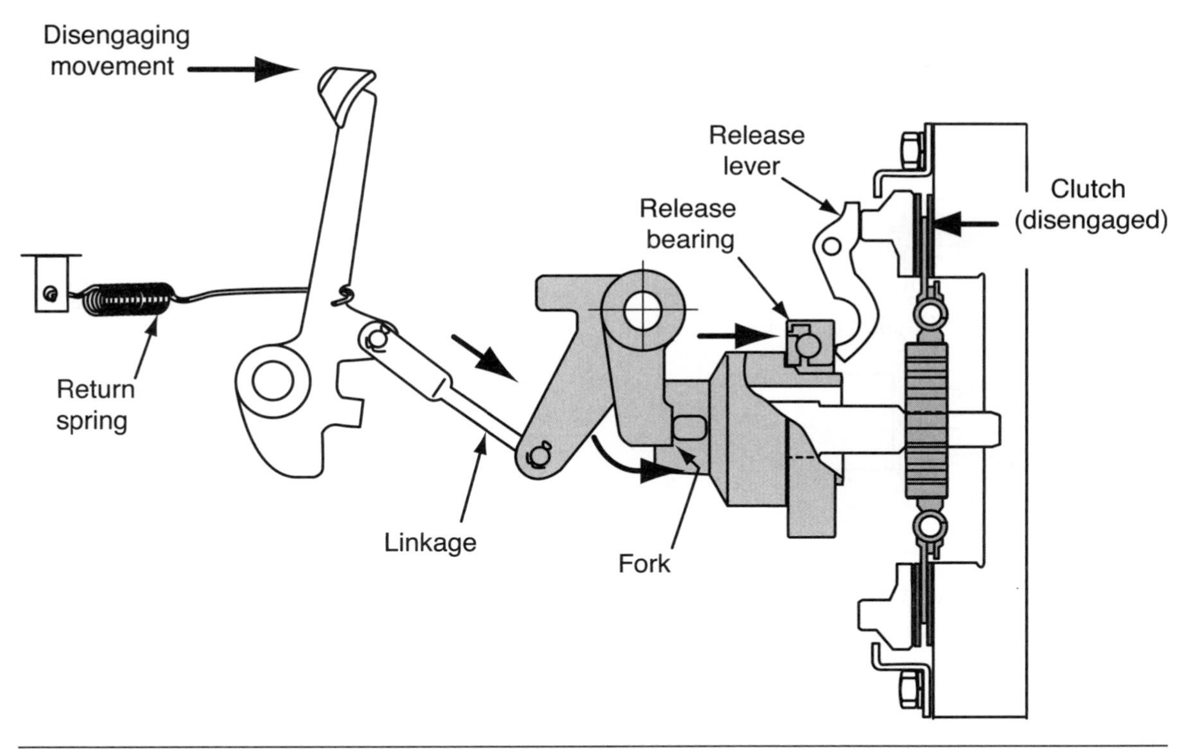

Figure 10-12 When the clutch pedal is depressed, the clutch fork to pivots on the clutch shaft and pushes the clutch release bearing into contact with the clutch release levers.

can rotate once the clutch fork slides the bearing into contact with the release fingers or levers of the clutch assembly. The clutch fork fitted to the hub of the release bearing prevents the hub from rotating. Since the bearing must come in contact with the release fingers each time the clutch pedal is depressed, the rotating clutch assembly will rotate the bearing at engine speed. Improper adjustment of the release bearing can result in premature bearing and/or clutch failure. Push-type clutches need a clearance adjustment between the release bearing and clutch levers specified by the manufacturer of the unit. Some release bearings are sealed and are not serviceable. However, some manufacturers attach a grease line from the clutch housing to the bearing for periodic greasing. If this access is not provided, the transmission inspection cover must be removed to locate the grease fitting directly on the release bearing. These types of bearings have a much greater service life, if properly maintained, than non-serviceable or sealed bearings.

Multiple-Disc Design

As mentioned earlier, some clutch assemblies utilize a single-disc or multiple-disc design. To achieve greater coefficient factors because of higher or greater torque loads on larger equipment and high horsepowered engines friction discs must be able to withstand these forces. Using a ceramic disc design is one method of achieving this demand, but if the demand is even greater than a ceramic disc can provide, other means must be considered. In order to meet these higher loads, friction discs must be manufactured with a greater amount of surface area.

One possible method of increasing the surface area is to increase the diameter of the discs and pressure plates, or we can implement multiple discs and plates within one clutch assembly as shown in **Figure 10-13.** This design allows the manufacturer to keep the diameter of the assembly to a minimum and still achieve a clutch assembly with greater torque capabilities. Utilizing multiple discs within one clutch assembly requires plates between the discs for contact areas. These plates are known as intermediate plates or separation plates. The plates must act as part of the pressure plate assembly and must rotate with the clutch housing to provide the transfer of torque from the engine. Lugs located around the outside diameter allow the plates to fit into channels within the clutch housing for easy servicing. The additional clutch disc will spline to the

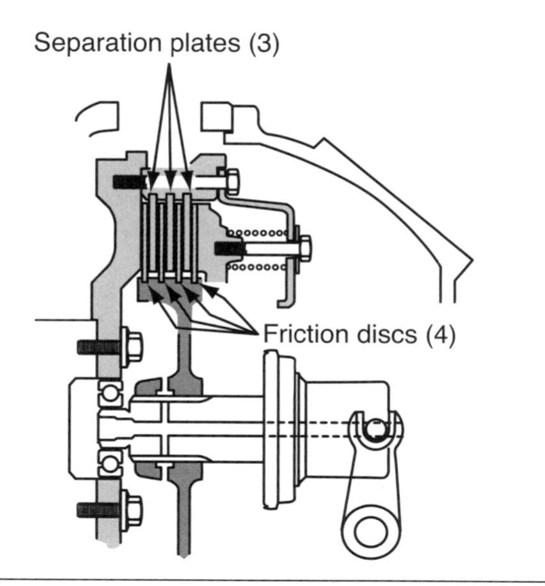

Figure 10-13 Push-type clutch utilizing multiple discs and plates. (18) 3 separation plates and (17) 4 friction discs.

transmission input shaft with the other disc separated by the intermediate plate. Commercial vehicles have been effectively using a two-disc design for many years.

These clutch assemblies, like the single-disc design, are also designed with either a diaphragm spring or coil-type springs for the pressure plate clamping force. Multiple-disc clutch designs can be either **push-** or **pull-type.** A **pull-type clutch assembly,** as discussed earlier, pulls the release bearing away from the flywheel to release the pressure to disengage the clutch assembly. The pressure plate assembly is significantly different from that of the push-type clutch assembly. The pressure plate floats within the cover by means of lugs or guide pins (similar to the push-type clutch) with small return springs attached to the clutch cover to help pull the plate away from the discs when the clutch is being disengaged. An adjusting ring is threaded into the clutch cover to provide a means of adjusting the assembly and also to affix the levers that come in contact with the pressure plate. These levers provide a mechanical advantage to multiply the force applied to the pressure plate by the clutch springs.

The retainer and release bearing assembly controls the levers. The levers are fitted into a collar on the retainer, the retainer is held in place by a snap ring on the release-bearing sleeve and the sleeve is fitted to the release bearing. (See **Figure 10-14.**) This assembly is allowed to float or move along the transmission input shaft as in the push-type clutch.

Clutch assemblies utilizing coil-type springs place the springs on an angle between the clutch cover and the retainer. Coil-type springs are referred to as a progressive springs, meaning that it requires more force to compress the spring a greater distance. This angle-spring design allows the springs, when compressed, to move to a position almost perpendicular to the input shaft, allowing an even force to move the springs away from the pressure plate as shown in **Figure 10-15.** This is a significant factor in driver fatigue considering the number of clutch applications that must be made during the course of the day.

Clutch assemblies utilizing diaphragm springs always provide a more uniform pressure. They also utilize the retainer and levers to multiply the spring force against the pressure plate. The only significant difference between the two types is the springs used. Because the release bearing, sleeve, and retainer are assembled as one unit the internal race of the bearing is constantly rotating with the clutch assembly and the outer bearing cover is stationary with the fork assembly. This means that this bearing is in constant rotation with engine speed whether the clutch is disengaged or not. Because of the high duty cycle of these bearings, they must be lubricated at proper maintenance intervals or premature bearing failure will occur (**Figure 10-16**).

The adjusting ring used to keep a specified gap between the release bearing and release fork is rotated within the clutch housing to change the relative position between the two components as the clutch discs wear. An adjusting-ring lock will secure the position of the ring. Some clutch manufacturers use self-adjusters that will automatically adjust the clutch as it wears. They can sense pedal travel and adjust the clutch accordingly.

Clutch Brakes

Many pull-type clutches implement clutch brakes. The purpose of transmission clutch brakes is to stop the rotation of the transmission input shaft, thus stopping the transmission countershafts and gears. Because most commercial heavy-duty trucks use non-synchronized transmissions, the input shaft and countershafts must be stopped from rotating to prevent gear clashing when placing the transmission of a stationary vehicle into first or reverse gears.

Clutch brakes can stop the rotation of the input shaft because they are splined to the shaft by two tangs that run the length of the input shaft and rotate with the

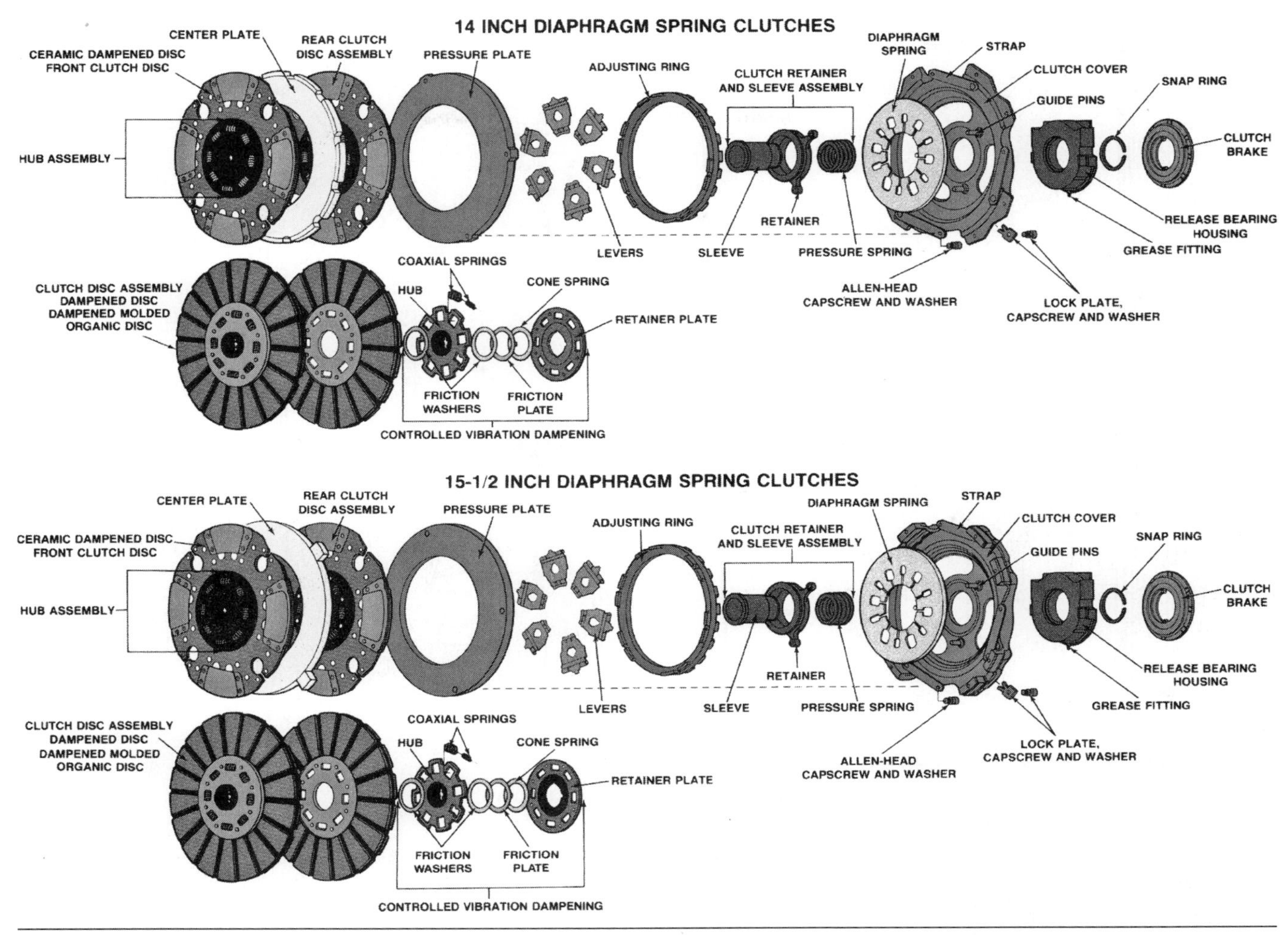

Figure 10-14 Dual-disc, pull-type clutch assembly in a diaphragm spring design. (*Courtesy of Arvin Meritor*)

shaft. This shaft is located between the clutch-release bearing and transmission housing. When the driver depresses the clutch pedal, the release bearing first disengages the clutch. Because of inertial forces of the clutch disc and the transmission shafts and gears, they will still rotate for a few seconds after the clutch has been disengaged. Within the last inch of pedal travel, the release bearing squeezes the rotating clutch brake against the stationary transmission input shaft support bearing housing, thus stopping it and the input shaft. With the transmission shafts and gears stationary, the driver is able to select a gear without any clashing or grinding of gears.

Vehicles equipped with clutch brakes should not have the clutch pedal fully depressed when shifting gears while in motion. This can severely damage the clutch brake.

There are three basic types of clutch-brake designs:

1. **Conventional clutch brake.** Similar to a clutch disc, the clutch brake is a round metal disc with organic fiber discs bonded to either side of the metal disc (**Figure 10-17**). The tangs on the metal disc will drive the clutch brake with the rotation of the transmission input shaft. The bonded fiber discs on one side will come into contact with the release bearing cover and those on the opposite side will make contact with the machined transmission input shaft bearing cover. A conventional two-piece clutch brake design is available by aftermarket suppliers. This two-piece design allows replacement of the clutch brake without removal of the transmission. (See **Figure 10-18.**)

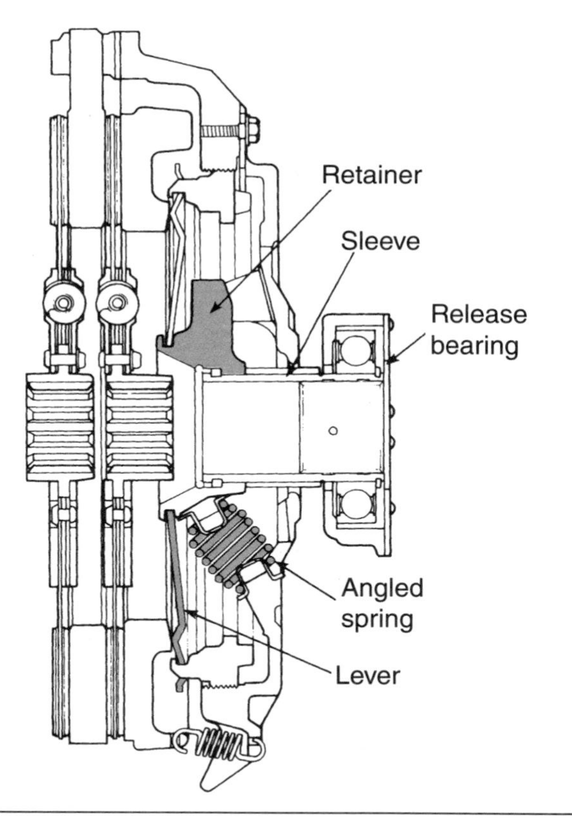

Figure 10-15 The coil springs in this clutch are angled between the cover and a retainer. Levers multiply the spring force against the pressure plate. (*Courtesy of Eaton Corp.—Eaton Clutch Div.*)

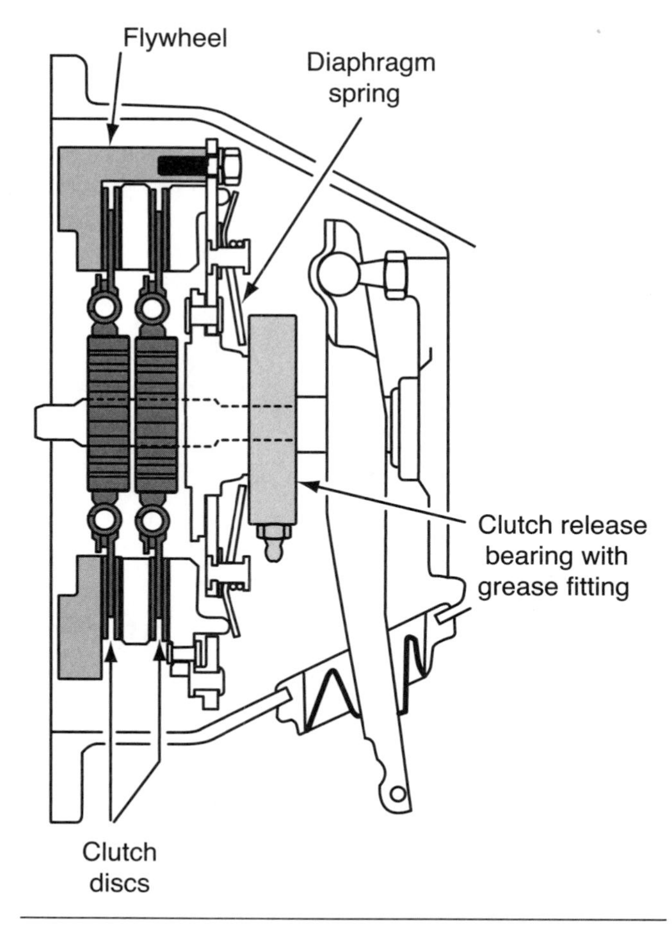

Figure 10-16 Clutch release bearing shown with grease fitting. High duty cycle of this bearing requires scheduled lubrication.

2. **Limited torque clutch brake.** This type of clutch brake can be engaged when the vehicle is moving (**Figure 10-19**). It allows the transmission input shaft to slow down for better synchronization when up-shifting to higher gear ratios. It will still provide non-clashing of gears when selecting first or reverse.
3. **Torque limiting clutch brake.** This type of clutch brake is designed for selecting first and reverse gears only when the vehicle is stationary. Its design feature is to limit the amount of torsional load placed between the rotating transmission input shaft and the stationary components (**Figure 10-20**). The torque limiting clutch brake has a hub that locates the tangs inside a cover with Belleville washers that are designed to slip when a torsional load of approximately 20 to 25 foot-pounds of torque is reached. This design reduces overloading of the clutch brake, resulting in greater service life.

Wet-Type Clutches

Most wet-type clutches work under the same principal as any traditional dry-type clutch. As mentioned earlier, clutches can generate such a tremendous amount of heat that it must be able to dissipate, which is the reason wet-type clutches were manufactured. By spraying or submersing oil onto the clutch surfaces, the generated heat is absorbed by the oil and carried away to a sump or oil cooler to be cooled and recirculated back to the clutch assembly. These types of clutches tend to be used in high duty cycle applications where there is frequent clutching and loads are high. One application of this type of clutch assembly is used by Volvo motor graders.

As **Figure 10-21** shows, the clutch shaft (1) has two cross drillings and one rifle drilling through the center of the shaft. Transmission oil is pumped through port A to the cross- and rifle-drilled clutch shaft. The oil then escapes through the second cross drilling and sprays a

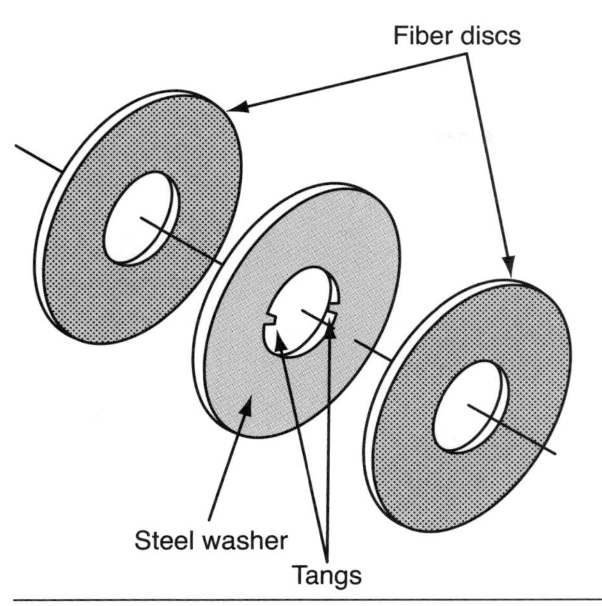

Figure 10-17 A conventional clutch brake. The fiber discs are bonded to either side of the steel washer. The tangs are splined to the master splines on the transmission input shaft.

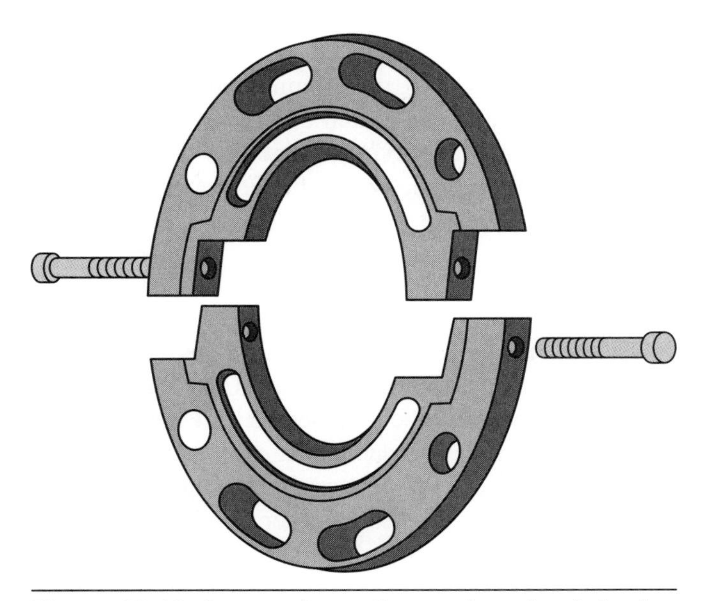

Figure 10-18 A two-piece aftermarket clutch brake, designed to be installed without removing the transmission.

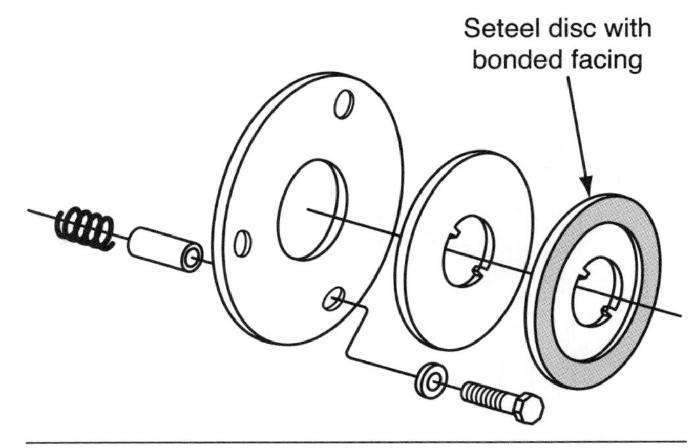

Figure 10-19 Limited torque clutch brake.

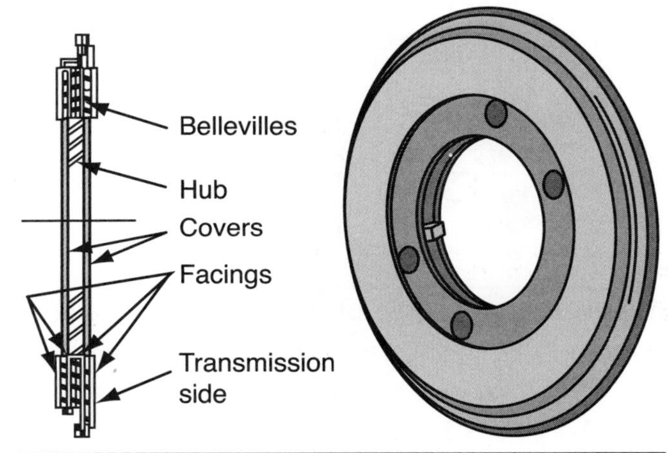

Figure 10-20 Torque limiting clutch brake.

continuous flow of oil to the clutch discs and plates (17 and 18). (Note that a multiple clutch disc design is used in this application.) This keeps the clutch assembly at a constant operating temperature. The oil is also allowed to travel out to the end of the clutch shaft to lubricate pilot bearing (14). As the oil comes in contact with the clutch components it absorbs any heat generated that is hotter than the oil. Because the clutch assembly is turning when the engine is running, centrifugal force also acts on the oil (including the resale bearing) causing it to be thrown around and to lubricate other components within the housing. This constant flow of oil extends the service life of all the clutch components.

The oil then drops to the bottom of the clutch housing and drains through passage B back to the transmission sump for cooling. In wet-type clutches, the oil reduces the coefficient of friction factor between the clutch discs and plates. Therefore, careful consideration must be taken when choosing the friction material on the discs and the type of oil used within the system. Some oils contain high amounts of Molybdenum disulfides that can cause the clutch to slip and wear prematurely. Original equipment manufacturer (OEM) specified oils should be used.

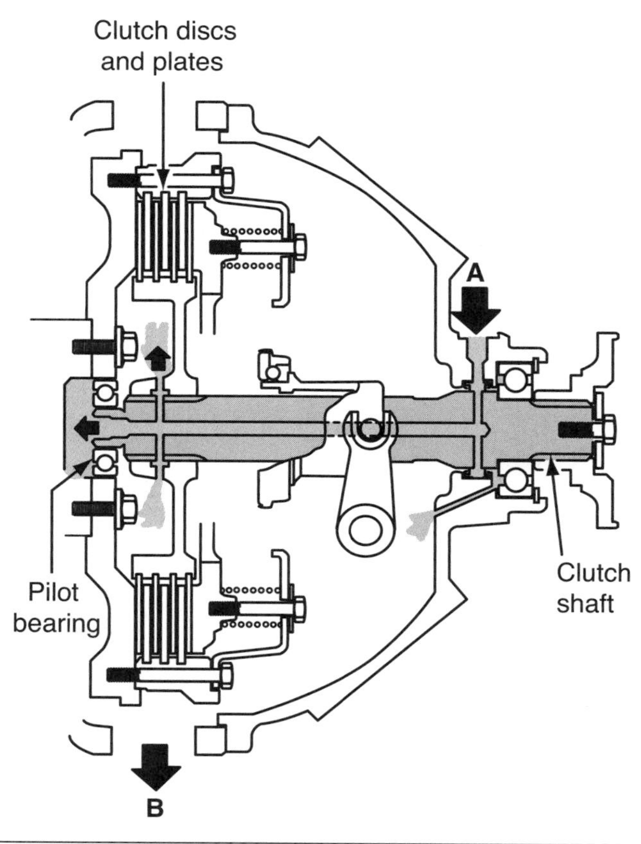

Figure 10-21 Wet-type clutch used in Volvo graders.

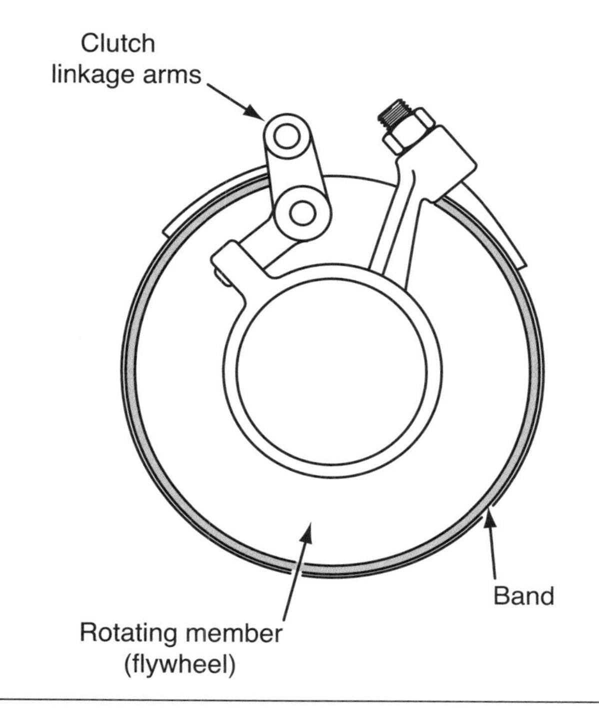

Figure 10-22 Band-type clutch.

Band-Type Clutch

A **band-type clutch** consists of a flywheel with the outside surface acting as the contact area for the friction material. Unlike an engine flywheel that utilizes the face surface, the band surrounds the outside flywheel surface with the friction material that is either bonded or riveted to the band as shown in **Figure 10-22.** Either the flywheel or the band can act as the driving member. When the band is squeezed closer to the flywheel, the friction created between the two units will cause them to rotate together thus transmitting the rotational force. These types of clutches can accommodate high torque loads because of the large surface areas of the flywheel and band. Although not as popular today, modified versions of these clutches are utilized in some conventional cranes.

This design is also utilized as a band-type brake. In this case, the band is a stationary member and, when it is squeezed against the flywheel, causes the flywheel to slow and eventually stop with a greater amount of force applied to the flywheel.

Overrunning Clutches

As the name implies, overrunning clutches allow the **driven member** of the clutch assembly to overrun or rotate faster than the **driving member** if allowed to do so. There are three basic types of overrunning clutches.

1. Roller
2. Cam or Sprag
3. Spring

The roller-type consists of an inner driving member or shaft, an outer driven hub, and a set of rollers placed between the two members as shown in **Figure 10-23.** The number of rollers is dependent upon the application of the clutch. The rollers act as the medium to transfer the torque from one shaft to the other. The inner shaft has a set of ramps that force the rollers to ride up when rotated and make contact with the outer driven hub. Because the area is reduced between the drive and driven members, the rollers makes contact with both surfaces, locking them together and causing them to rotate as one. If the drive member is forced to slow down or speed up, the rollers will move down the ramps and lose contact with the hub or driven member. This causes a "freewheel" or overrunning condition. In some applications, springs are utilized to assist the rollers up the ramps. Careful attention must be taken when servicing so as not to place the

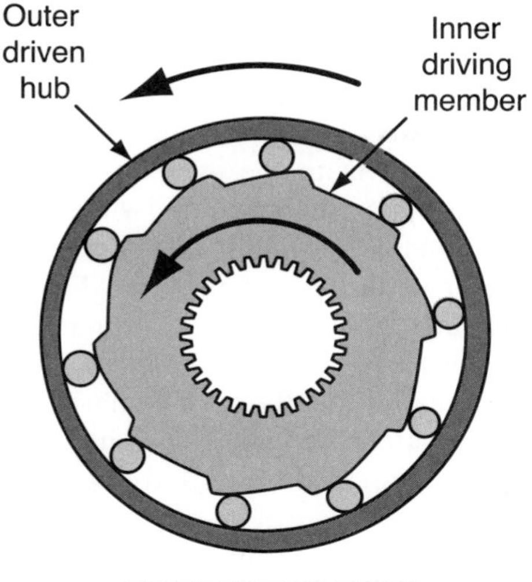

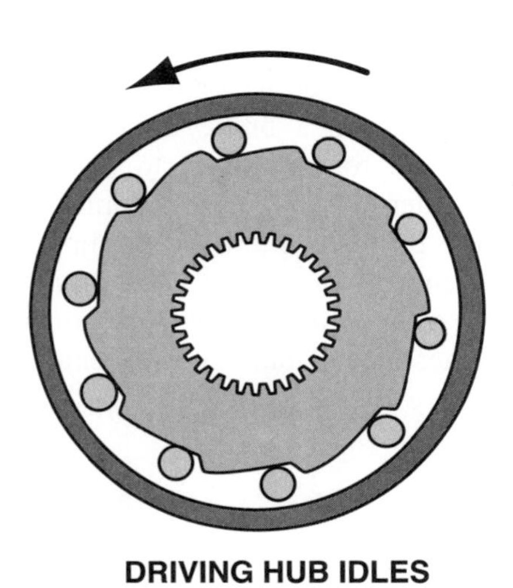

Figure 10-23 Overrunning clutch.

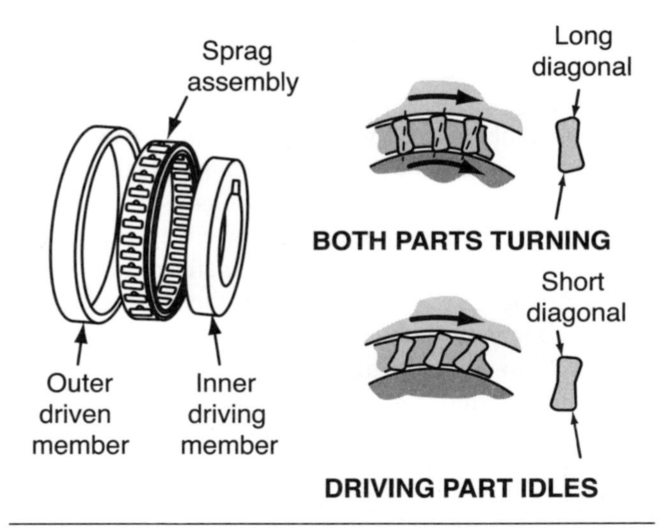

Figure 10-24 Sprag-type overrunning clutch.

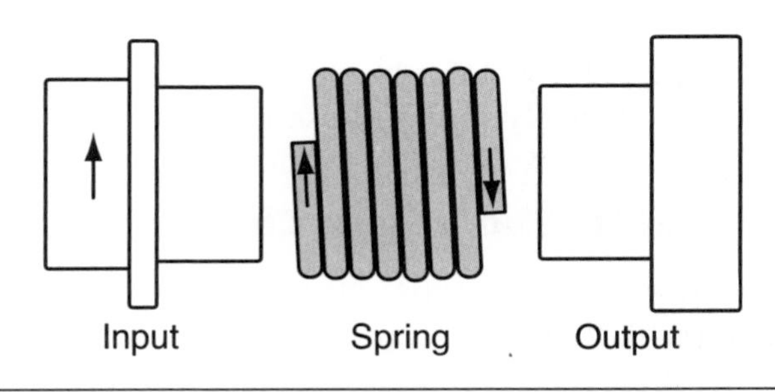

Figure 10-25 This clutch consists of a spring wrapped with a slight interference fit around two adjacent hubs, an input drive hub, and an output driven hub.

springs on the wrong side of the roller. This would cause the clutch unit to freewheel in both directions.

The cam or sprag-type clutch works much like the roller type. They both incorporate drive and driven members, but instead of rollers, sprags or cams are placed between the two members (**Figure 10-24**). When the drive member starts to rotate, it causes the sprags to rotate slightly and wedge against the driven member, causing it to rotate with the drive member and transfer rotational torque. If the driven member were forced to travel faster than the drive member, the driven member would force the sprags to rotate back to their original position, causing an overrunning or freewheel condition.

Spring-type overrunning clutches are used in lower torque applications. This type of clutch uses a coil spring between the driving member and driven member. (See **Figure 10-25.**) When a coil spring is twisted, its diameter is reduced proportional to the amount of twisting. Because the spring is located between the drive and driven member, the twisted spring will clamp itself to the driven member, forcing it to rotate together as a coupled unit. When the driven member is allowed to rotate faster than the drive member, the spring loosens its hold and disengages from the drive member, causing it to overrun the drive member.

Cone-Type Clutches

The **cone-type clutch** is made up of a cone and cup assembly much like the tapered roller bearing cup and cone assembly. Unlike the tapered roller bearing assembly that is trying to reduce friction between two components, the cone-type clutch is trying to achieve friction between the cup and cone to cause them to couple together as one unit. Like any other type of clutching mechanism, it must consist of a drive member and a driven member. Dependent on the application, either the cup or cone can be the drive member for the driven member. A throw-out bearing, or mechanism, is used to move the components in and out of engagement, and supporting shafts are used for the transfer of torque. As shown in **Figure 10-26,** the cup is the drive member and the cone is the controlling driven member that will be moved in and out of the coupling phase by the throw-out bearing. Independent shafts support each member. The cup has a smooth machined tapered surface and the cone incorporates the clutch lining material which is usually made up of ceramic compounds. As the throw-out bearing moves the cone into contact with the cup, the cup immediately starts to transfer torque to the cone, eventually causing them to rotate as one unit. This type of clutching mechanism tends to clutch hard and fast and is ideally used in the synchronization of manually shifted transmissions.

Expanding Shoe-Type Clutches

Expanding shoe-type clutches consist of a drum-and-shoe assembly much like the drum-and-shoe assembly of a brake system. The difference between the two assemblies is that one is trying to achieve a braking effect through the friction created between the stationary shoes and the rotating drum, and the other is trying to transfer rotational torque from one component to the other through the friction created between the drum-and-shoe assemblies. The drum-and-shoe assemblies can act as either the drive member or the driven member. **Figure 10-27** shows the mechanical linkage expanding shoe clutch. Modern day clutch assemblies of this type are typically hydraulically or pneumatically actuated. Conventional operating cranes implement this type of clutching system for free-fall hoists or winches.

As the release bearing is moved toward the drum, it causes the linkage to force the shoes outward into contact with the drum. The shoes are typically lined with an organic fiber pad that has a high coefficient of friction ratio as well as good heat absorption capabilities. The drum, similar to a brake drum has a smooth, machined surface for the shoes to come in contact with. On the outer circumference, cooling fins are cast as part of the drum to allow for cooling. When the shoes apply enough pressure to the drum, they will rotate together as one unit. These types of mechanical linkage clutches incorporate an over-center linkage as shown in **Figure 10-28.**

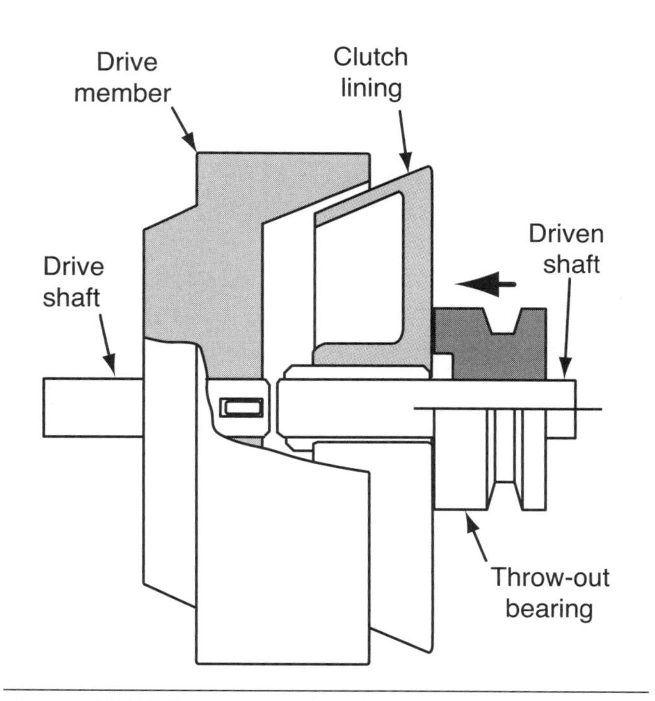

Figure 10-26 Cone-type clutch.

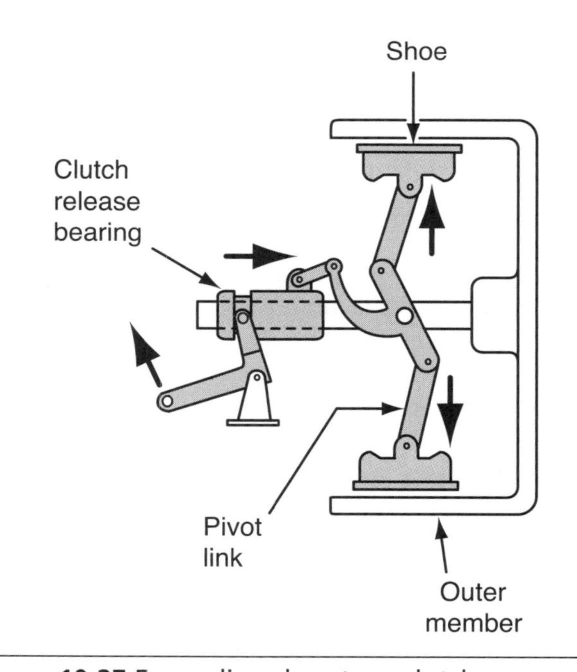

Figure 10-27 Expanding shoe-type clutch.

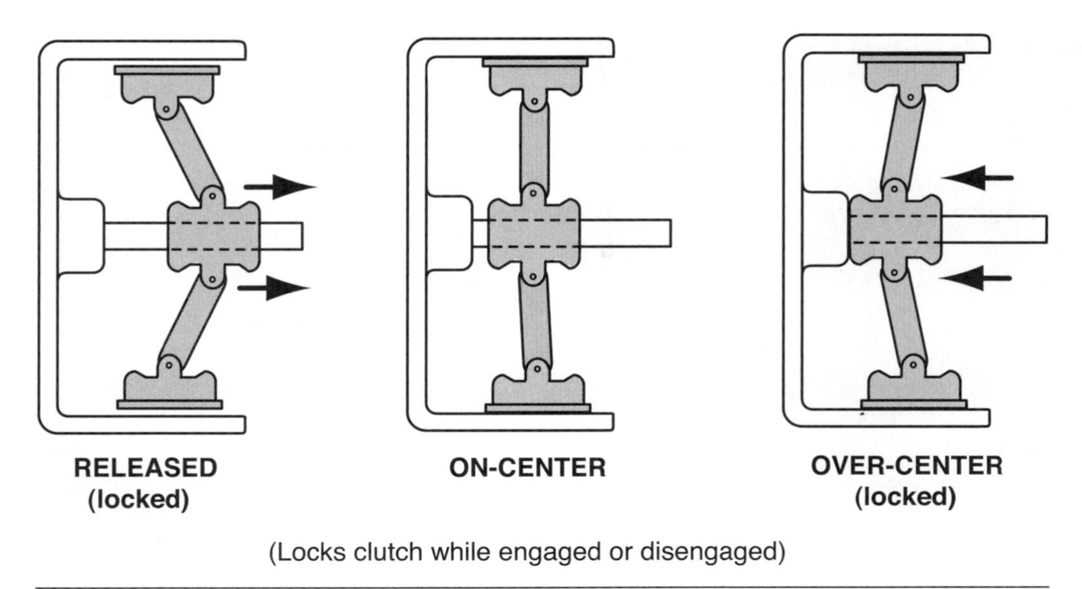

Figure 10-28 Expanding shoe-type clutch with over-center linkage.

When the linkage is perpendicular to the shaft, it is applying the greatest amount of force to the drum. If this position is surpassed, there is less force being applied to the drum but still enough to achieve a 1:1 ratio. The advantage to this over-center position is that the linkage is theoretically locked. To disengage the clutch, the lever must surpass the perpendicular position which takes a greatest amount of force.

A common mechanical over-center clutch still produced by Rockford P.T.O. is the disk-type over-center clutch. See the photographs shown in **Figures 10-29, 10-30, 10-31,** and **10-32.** This clutch operates very similar to a traditional disc-and-plate clutch except that the lever will lock in an over-center position to keep the clutch engaged. A splined friction disc is sandwiched between the plates and acts as the driven member. The friction disc teeth are placed on the outer circumference of the disc and spline to a pot-type flywheel with internal teeth or splines. The flywheel face surface is not used to sandwich the disc between it and the pressure plate. A traditional fork-and-release bearing is used to engage and disengage the clutch. An adjusting ring moves one pressure plate closer or farther away from the friction disc to obtain a designated amount of foot-pounds (ft-lbs) of torque to engage the clutch in the over-center position. This measurement should be tested and adjusted (per OEM recommendation schedules) to adjust for wear and possible clutch slippage.

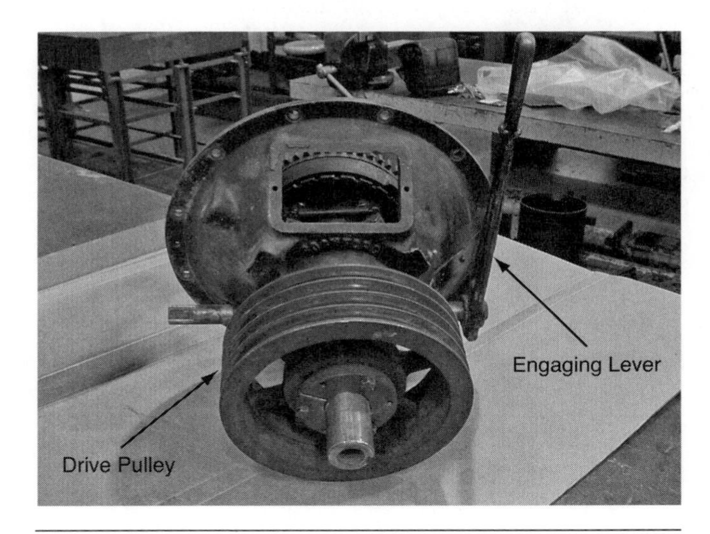

Figure 10-29 Rockford over-center disc-type clutch.

Centrifugal-Type Clutches

Centrifugal clutches are of three basic types.

1. Vane
2. Disc
3. Shoe

The vane-type centrifugal clutch, shown in **Figure 10-33,** consists of a driven outer hub assembly and a set of vanes set into a slotted rotor, which is designated as the driving member. The vanes are able to move freely within the slots of the rotor. As the

Figure 10-30 Rockford over-center disc-type clutch.

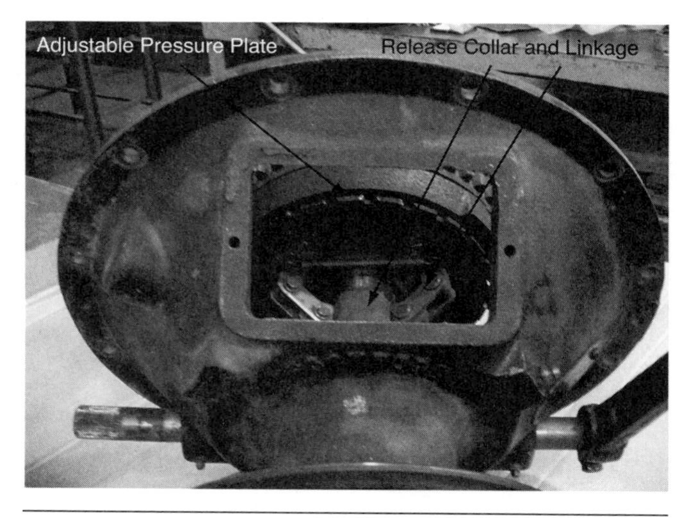

Figure 10-31 Rockford over-center disc-type clutch in released position.

rotor starts to rotate, the centrifugal force on the vanes moves them outward toward the outer hub. When there is enough centrifugal force, dictated by the rotational speed of the rotor and the weight of the vanes the outer driven member will start to rotate. As the resistance diminishes proportional to the speed increase of the driven member, the drive member and driven member may achieve a 1:1 ratio. There is never a positive lock-up with this type of clutching system, but it is very useful when a degree of slippage is necessary.

The disc-type centrifugal clutch consists of a driving inner hub, a clutch basket (driven hub), a set of friction discs and plates similar to those in a standard

Figure 10-32 Rockford over-center disc-type clutch in "locked over-center" position.

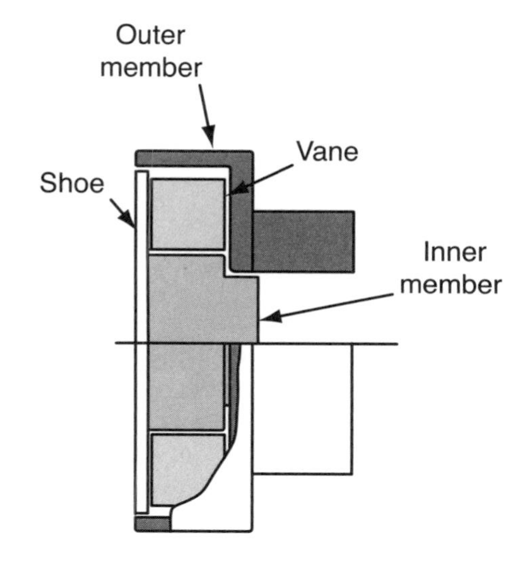

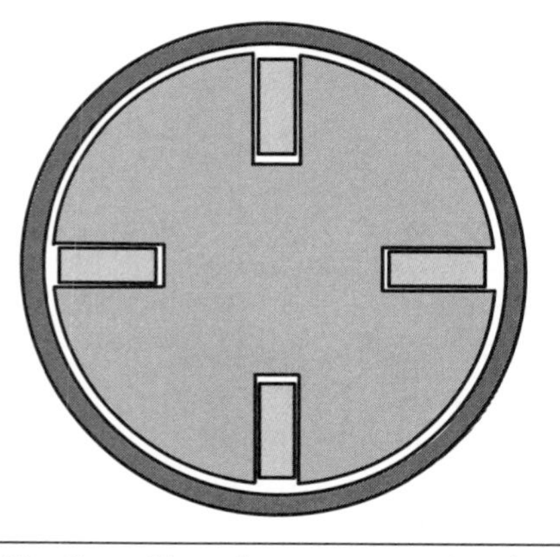

Figure 10-33 Centrifugal vane-type clutch.

disc-and-plate clutch, a set of centrifugal weights and release springs located around the face of the clutch assembly, and a contact pressure plate. The inner driving hub is splined to accommodate the friction discs, the pressure plate, and the centrifugal weights and springs. When the driving hub increases in speed, the centrifugal force acting against the weights causes them to be thrown outward. The weights are attached to a lever mechanism that will force them to pivot outward as the centrifugal force increases. As the weights pivot, the lower mechanism pushes the pressure plate into contact with the friction discs and plates. The clutch plates have tangs that fit into the clutch basket. As the clearance between the friction discs and plates is reduced, the discs drive the plates and basket to accommodate a driven component. **Figure 10-34** shows a sprocket as the driven component. The return springs and specific centrifugal weights dictate the rpm and speed of clutch application. These springs and weights can be adjusted to fit the application. Centrifugal disc-type clutches are used in light-duty applications. They are not able to reach a positive lock-up condition. See **Figures 10-34,** and **10-35.**

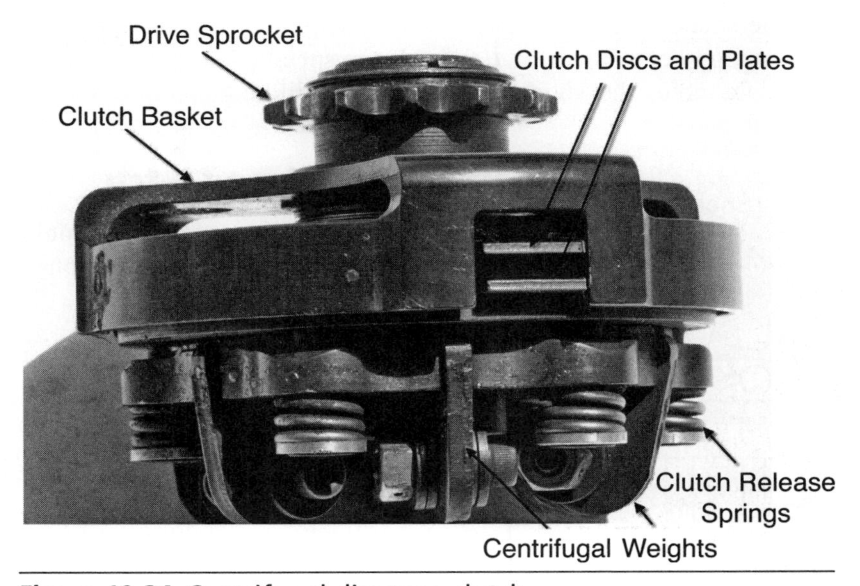

Figure 10-34 Centrifugal disc-type clutch.

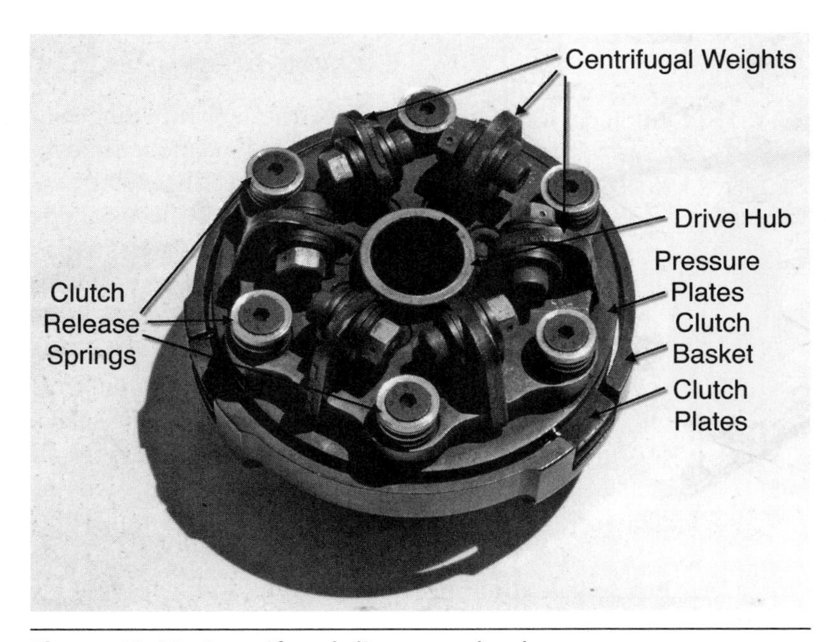

Figure 10-35 Centrifugal disc-type clutch.

The shoe-type centrifugal clutch is very similar to the expanding shoe-type clutch except for the method of engagement. As we discussed earlier, the expanding shoe-type clutch can be actuated by means of a mechanical lever, a hydraulic cylinder, or even an air cylinder.

The centrifugal shoe-type clutch relies on centrifugal force acting on the shoes to expand and contact the driven drum component. As the **Figure 10-36** shows, the inner driving hub rotates at engine speed. There are four sets of shoes driven by the drive pins that also guide the shoes to travel outward toward the drum. A restraining spring keeps the shoes retracted when the driving hub is stationary. When the driving hub rotates, the shoes need to develop enough centrifugal force to overcome the retaining spring force; the shoes then travel outward from the drive pins, contact the drum, and begin to transfer the rotational force. As with the other types of centrifugal clutches, a 1:1 ratio is rarely achieved depending on the load.

All centrifugal clutches have what is referred to as an engagement rpm. This is the speed at which the drive and driven members first come in contact. There is very little power transfer at this stage of engagement. Heat generation and component wear is at its greatest point. As the rpm increases, the power transfer capability increases to the rate of the square of the specific rpm. An example is as follows: if the drive portion of the clutch is rotating at 2,000 rpm the driven component will have a kilo-watt power equal to 4. If the speed is increased to 4,000 rpm 16 kilo-watts of power will be generated.

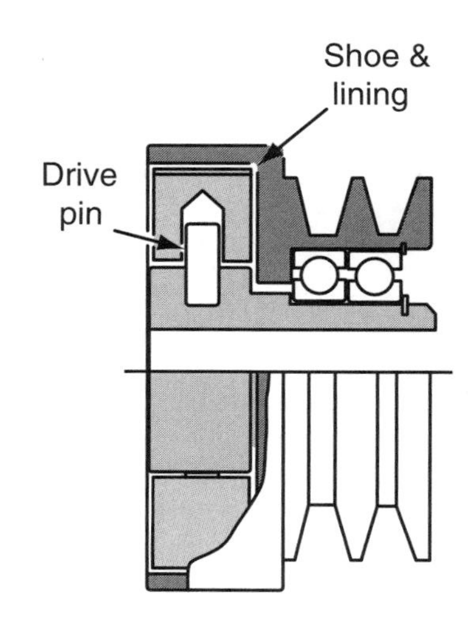

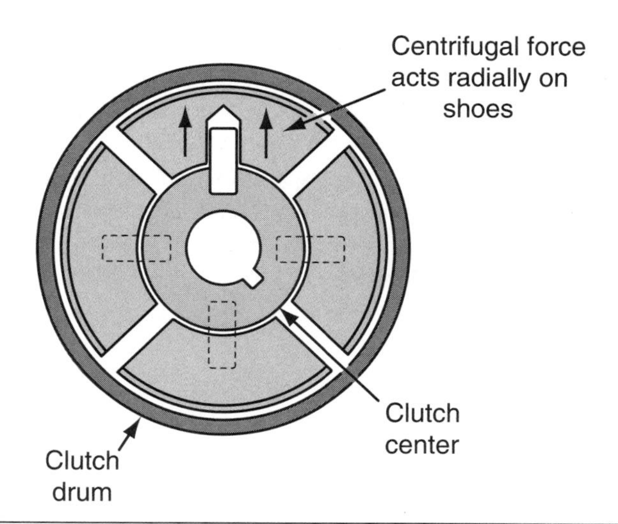

Figure 10-36 Centrifugal shoe-type clutch.

Tube-Type Clutch

A tube-type clutch is a multi-disk-and-plate clutch used in heavy-duty applications. As in most multi-disk-and-plate clutches, the disk will be splined to the drive hub and the plates to the driven hub. An air tube located within the clutch housing and pressure plate can be filled with air which forces it to expand and sandwich the disks and plates, locking the drive and driven members together as shown in **Figure 10-37.** Because air is being used to apply the clutch, it automatically has a cushioning effect during application. Regulating the air pressure and flow can also determine the speed and rate of clutch application.

Electro-Magnetic Clutch

Electro-magnetic clutches use a field coil to produce a strong magnetic field to pull an armature disc to connect a drive. The clutch assembly consists of a rotor assembly (usually a V-belt pulley) with bearings to separate the drive and driven sections. The assembly has a stationary field coil with electrical leads and an armature disc with plate springs to retract that disc, which also connects it to the inner hub as shown in **Figure 10-38.** When the circuit is completed within the field coil, a strong magnetic field is produced through the rotor assembly and armature disc. Since the rotor assembly is made of a heavier casting, the magnetic field will travel through the rotor without disturbing it. The magnetic field also travels through the armature disc, pulling it into contact with the rotor assembly. With the rotor and armature now rotating

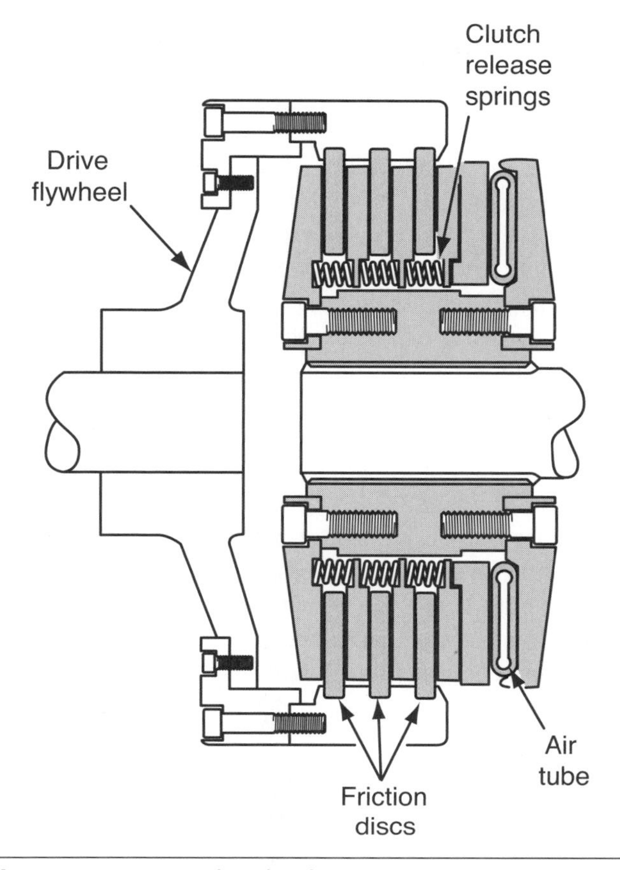

Figure 10-37 Air tube clutch.

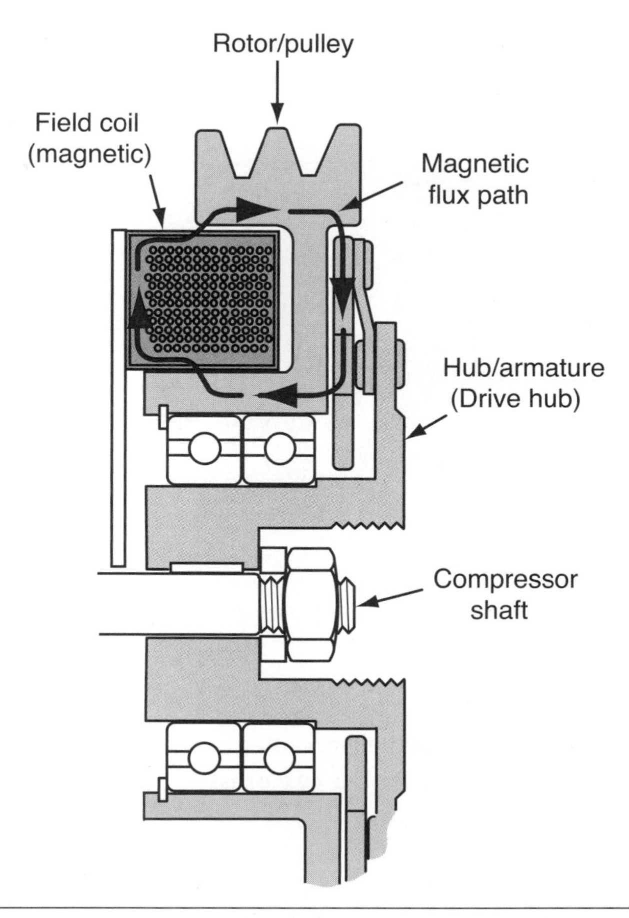

Figure 10-38 When the field coil is energized, it produces a magnetic field through the rotor assembly and armature plate. This causes the two components to rotate together.

together, the connection to the inner hub by the return springs attached to the armature transfers the powerflow to the inner hub. See **Figure 10-39.**

For an excellent example of an electric clutch think of an air conditioning compressor. The rotor pulley assembly is the drive hub, receiving its power from the crankshaft by means of V-belts. When the field coil is activated, the inner driven hub is connected to the compressor input shaft, causing it to rotate. Many manufacturers incorporate a capacitor in to the circuit to absorb any induced voltage when the field circuit collapses when switching off the clutch assembly.

CLUTCH LINKAGE

Many of the previously described clutches must have a means of engaging and disengaging them. Although centrifugal clutches engage from the centrifugal force produced when they rotate at a particular rpm and electric clutches require an electromagnetic field, most disc-and-plate clutches require manual control from the operator or driver of the vehicle. Therefore, most clutches have some form of linkage configuration setup in the operator cab. There are three basic forms of linkage used to control the engagement and disengagement of clutches.

- Mechanical
- Hydraulic
- Pneumatic

Mechanical Linkage Clutches

Mechanical linkage clutch (**Figure 10-40**) consists of a clutch pedal and metal rod linkages that connect through a series of mechanical levers to control the

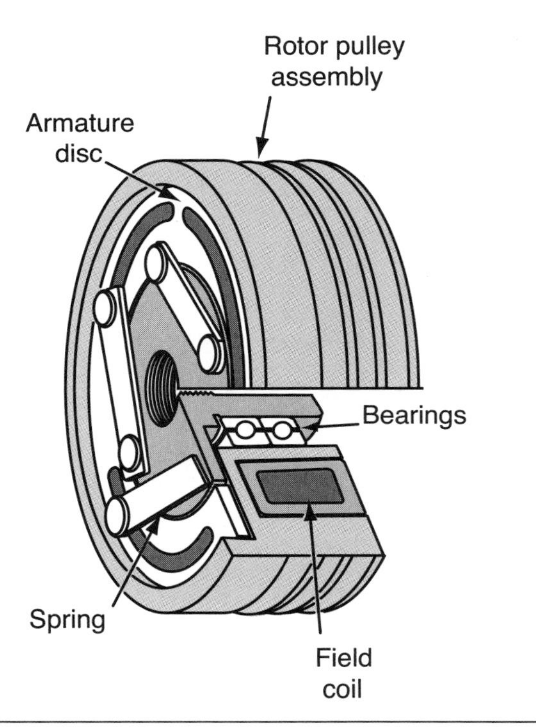

Figure 10-39 Typical electro-magnetic clutch used to drive an air-conditioning compressor.

movement of the clutch fork and release bearing. The length of the levers provides the mechanical advantage and stroke length necessary for the operator to overcome the spring force within the clutch assembly and disengage the clutch. The amount of force required to release or disengage the clutch is determined by the length of the clutch pedal, the length of mechanical levers, and of course, the size and number of springs within the clutch assembly. A return spring is used at the clutch pedal to ensure that the linkage has completely returned after the operator removes his foot from the pedal (clutch engaged). This spring also keeps the release bearing and the clutch release levers, or fingers, from staying in contact with each other once the clutch is re-engaged.

Lighter duty vehicles use mechanical cable instead of rods and levers. Cable does not provide any mechanical advantage unless used with levers. However, cable is much easier to manipulate and easier to locate in unfamiliar equipment (**Figure 10-41**). You should be aware that over an extended period of time cable can become frayed and break apart. If not maintained properly, it can seize and bind within its insulated cover. The

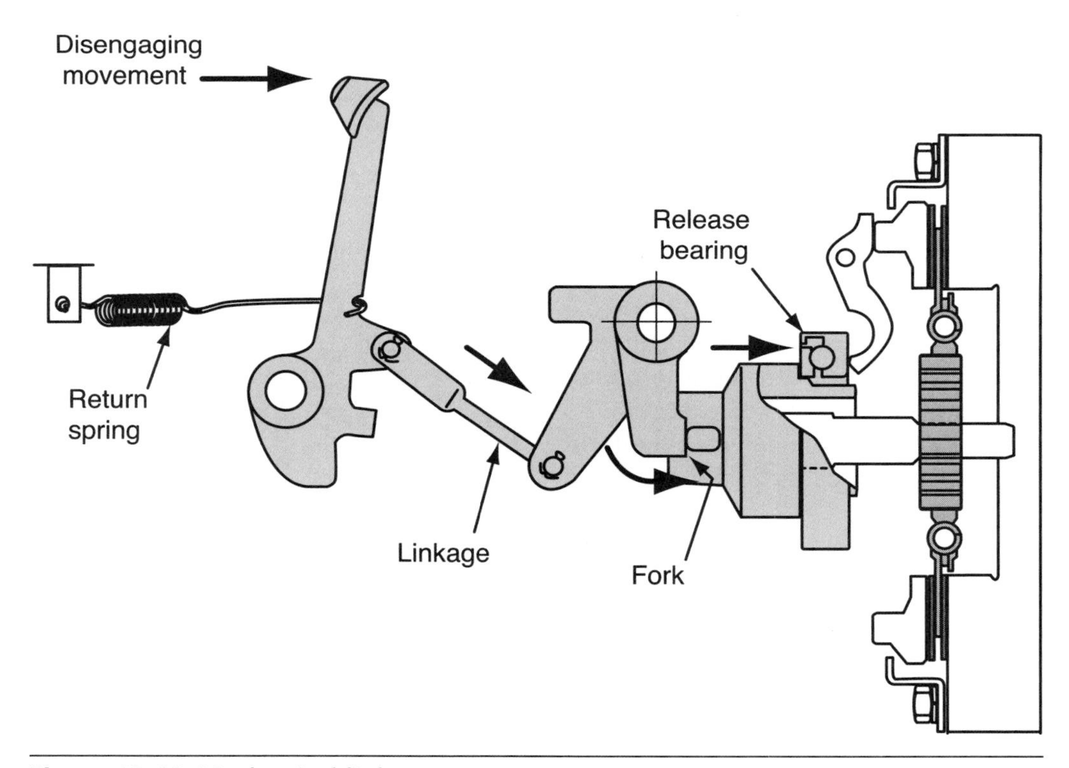

Figure 10-40 Mechanical linkage.

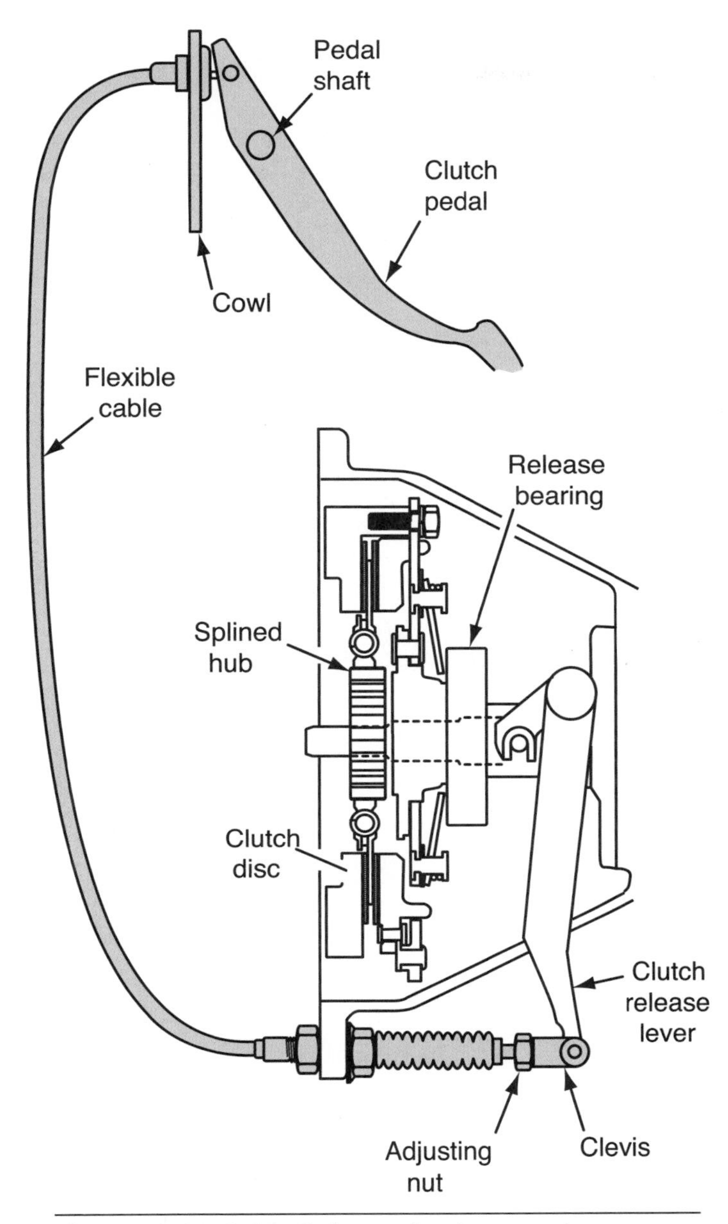

Figure 10-41 Cable linkage clutch control.

manufacturers maintenance procedures should always be followed for proper care and service.

Hydraulically-Released Clutches

Hydraulically-released clutches are very popular in nearly all clutch applications. This is because hydraulic fluid is as versatile as a cable but can also increase or multiply the force applied to the clutch fork without the use of mechanical levers. Hydraulically-released clutches as shown in **Figure 10-42** implement a master cylinder, a **slave cylinder**, and a reservoir to store the hydraulic fluid. On larger, heavy-duty applications, a hydraulic or pneumatic booster may be used to assist clutch pedal movement from the operator.

The master cylinder is equipped with a sealed actuating piston controlled by the clutch pedal. On many master cylinders, it also acts as the reservoir for the system. The reservoir can also be mounted remotely for easy access for servicing; it will then be connected to the master cylinder by means of a hydraulic line. When the clutch pedal is depressed, the sealed actuating piston within the master cylinder starts to displace the oil trapped within it. The oil has no other place to travel except through the line to the slave cylinder. The slave cylinder also contains a sealed piston that is connected to the clutch lever. The resistance at the slave cylinder (from trying to release the clutch) causes pressure to build within the slave cylinder, the line, and the master cylinder. Continuous pedal movement will then cause more oil to move and the pressure to build until it is sufficient to overcome the force and release or disengage the clutch. When the clutch pedal is released, the pressure diminishes within the system. The oil returns to the reservoir and the clutch reapplies. A return spring is sometimes used to ensure that the slave cylinder piston is completely retracted and there is no contact between the release bearing and clutch fingers.

Hydraulic boosters assist the actuation of the master cylinder with a piston that is hydraulically applied by the regulated movement of the clutch pedal. A hydraulic pump supplies the oil flow to the booster control valve (located within the booster) and returns oil either to the reservoir or to a secondary hydraulic circuit that will return the oil. This is shown in **Figure 10-43.** The control valve (integral to the booster) is controlled by the clutch pedal, which simultaneously controls the clutch master cylinder. When the operator starts to depress the clutch pedal, the hydraulic control valve diverts some oil to the booster cylinder proportional to the rate of the clutch pedal movement. When the operator releases the clutch pedal, the control valve returns to its original position allowing the oil within the booster cylinder to return to the reservoir. This release of oil causes the master cylinder piston and the slave cylinder piston to retract and the clutch to reapply (**Figure 10-44**). Because the clutch pedal is always connected to the master cylinder, only with some difficulty can the clutch be disengaged without the engine running.

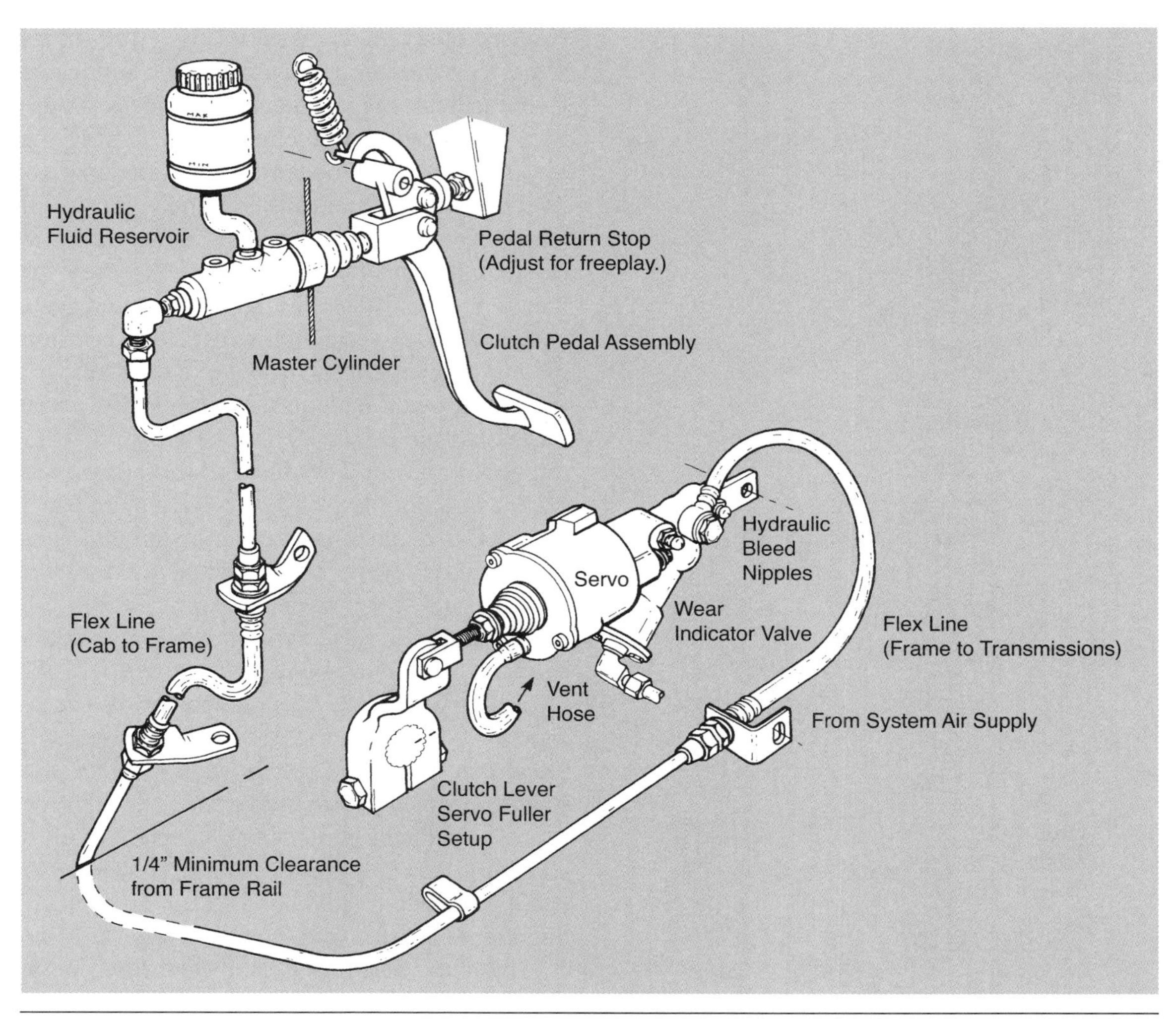

Figure 10-42 Components used in one type of hydraulic clutch circuit.

Pneumatically-Controlled Clutches

Pneumatically-controlled clutches shown in **Figure 10-45,** work under principles similar to hydraulically-controlled clutches. They incorporate an air supply from the air reservoir, a clutch treadle valve, a slave cylinder, and connecting airlines. When the operator depresses the clutch pedal, compressed air from the reservoir is directed through the treadle valve to the slave cylinder. The slave cylinder has a sealed piston connected to the clutch lever; as air is supplied, the pressure builds and the piston starts to move within the cylinder and release the clutch. The treadle valve acts as an air-regulating device controlled by the relative position of the clutch pedal. The farther the pedal is depressed, the more air pressure that must be supplied to overcome the clutch springs and completely release the clutch. The opposite applies when releasing the clutch pedal or re-engaging the clutch. The air must be controlled in a finite manner so engagement and disengagement of the clutch is smooth and even. A booster is not required with this system because the air pressure is not produced at the treadle valve or foot pedal as in the hydraulically-controlled clutch. The treadle valve relays stored pressure from the air reservoir to the slave cylinder.

FLYWHEEL DESIGN

An engine flywheel is a round, heavy, balanced cast-steel or cast-iron component bolted to the end of the engine crankshaft. The purpose of the engine flywheel is not only to store kinetic energy to keep an engine rotating between power-strokes but also to provide a cou-

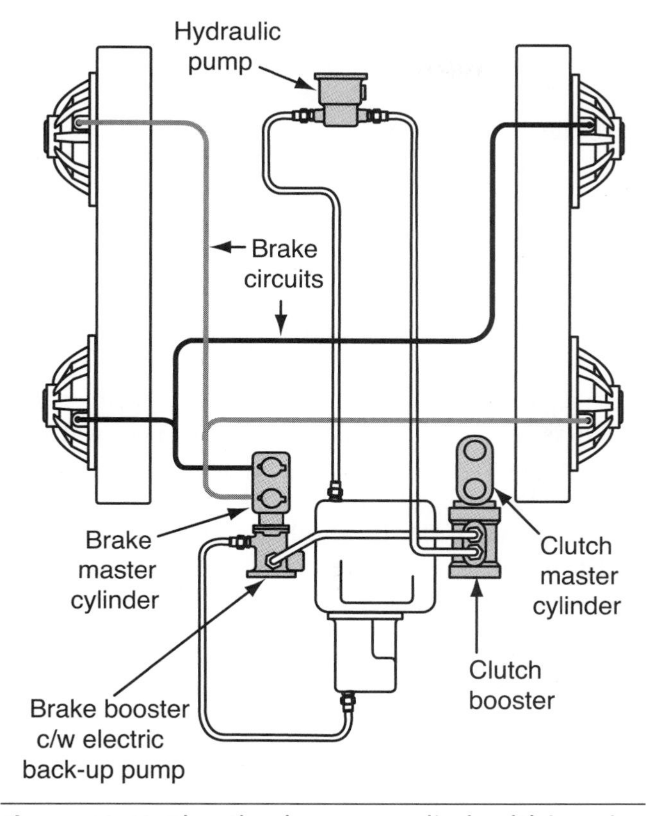

Figure 10-43 The Clutch master cylinder (3) is assisted by a clutch booster (4). Oil is supplied by the hydraulic pump (5). Note that hydraulic boosters are used for both clutch and brake master cylinders.

pling surface for a clutch assembly, torque converter, or any other means of transferring engine torque to a transmission or drive unit. It also acts as the driven gear for the starter motor to crank or "turn over" the engine. The flywheel is fitted with a ring gear that is press-fit around the circumference of the flywheel. This ring gear is a relatively inexpensive replaceable component of the flywheel.

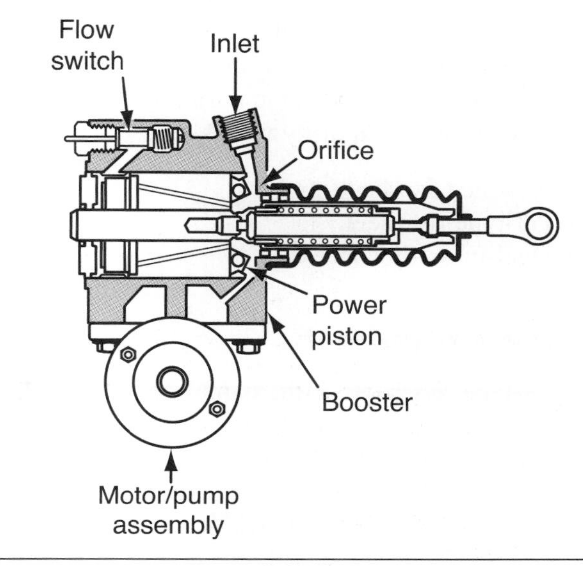

Figure 10-44 Clutch booster.

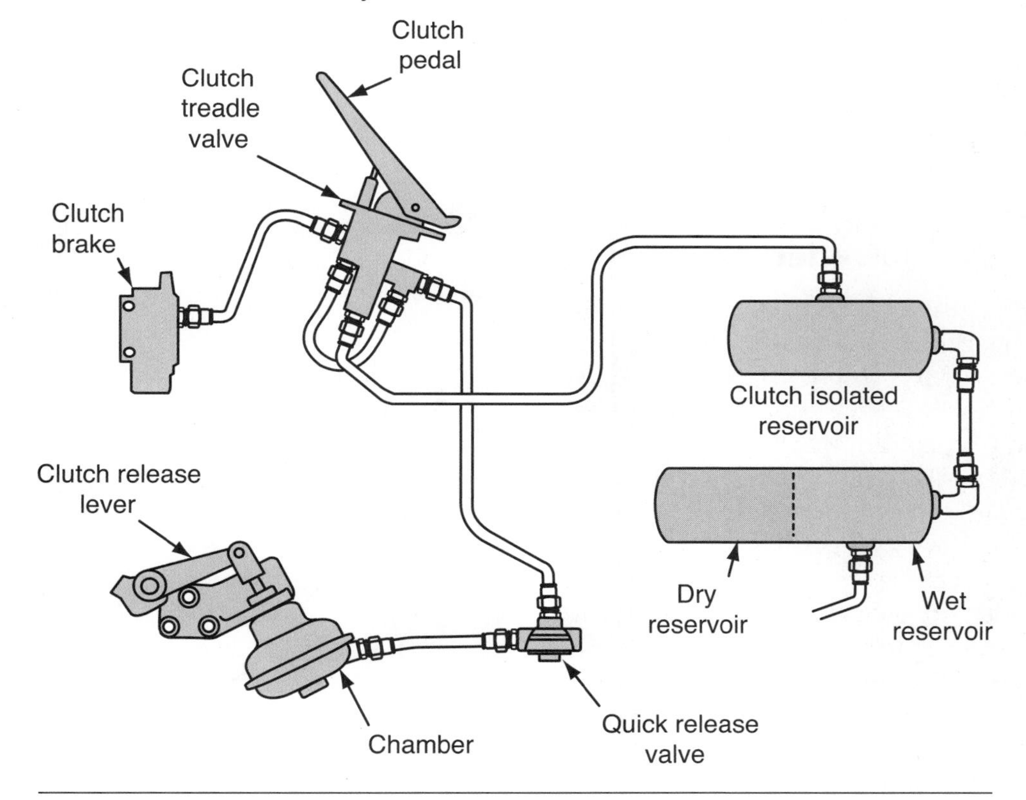

Figure 10-45 Air clutch control.

Damage to the ring gear normally occurs during improper engagement of the starter motor gear. It is considered a good practice to remove and inspect an engine flywheel when a clutch replacement is scheduled on a vehicle. Flywheels should be inspected for thermal cracks or heat cracks, ring gear condition, and warpage. In many cases if a misalignment is suspected between the engine and transmission, an "out-of-round" measurement can be taken. This procedure will be discussed later in this chapter.

Machinery that utilizes a disk-and-plate clutch assembly to drive a transmission normally incorporates one of three basic types of flywheels.

- Flat-face flywheel
- Pot-type flywheel
- Dual-mass flywheel

Flat-Face Flywheel

A flat-face flywheel, as the name indicates, is flat-faced, meaning that the side of the flywheel opposing the engine crankshaft is a smooth flat-machined surface to which a clutch assembly can be bolted. This flywheel, shown in **Figure 10-46,** will have threaded holes around the outer perimeter to bolt the clutch pressure plate. Some flywheels allow for three or four different orientation options that the pressure plate can be bolted to. Because both the flywheel and pressure plate are balanced units, it is not imperative to locate the clutch assembly in the original boltholes.

When servicing this type of flywheel, if thermal cracks are evident on the face of the flywheel, they can be removed by machining the entire face of the flywheel. It is critical not to machine the flywheel to a thickness less than the manufacturer's specifications. If the flywheel is at its maximum point of machining and thermal cracks are still visible, the flywheel should be replaced. If the flywheel is not replaced, thermal cracks can progress deeper into the face of the flywheel and cause it to break apart. If the flywheel is machined to a thickness less than the manufacturer's specifications, the flywheel may warp during high temperatures created from clutch operation.

The center of the flywheel is machined to accommodate a pilot bearing. The purpose of the pilot bearing is to support the input shaft of the transmission when it is coupled to the flywheel and engine bell housing. It is considered good practice to replace the pilot bearing when replacing a clutch assembly.

Pot-Type Flywheel

The pot-type flywheel, in fact resembles a cooking pot. Unlike the flat-face flywheel that is completely flat, the pot-type flywheel is a deeply recessed or counterbored flywheel as shown in **Figure 10-47.** In the trucking industry, this flywheel is used with 14-inch dual-disc

Figure 10-46 Flat-faced flywheel.

Figure 10-47 Pot-type flywheel.

clutch assemblies. The pressure-plate assembly used with this flywheel is significantly different from that used on a flat-face flywheel. Because of the dual-disc clutch assembly, a separation plate is placed between the two friction discs. This separation plate has lugs that allow it to float within the pressure plate used on flat face-flywheels, **Figure 10-48.** The pressure-plate used with pot-type flywheels eliminates the high sides of the pressure-plate housing with a lighter flatter housing.

The separation plate that was part of the clutch housing must now be placed within the pot-type flywheel. The flywheel incorporates replaceable lugs that the separation plate will fit into. This flywheel design has a tendency to overheat the clutch and disc assemblies and clutch particles may accumulate within the flywheel because the clutch discs and separation plate are completely enclosed within the flywheel when the pressure plate is assembled. See **Figure 10-49.**

The replaceable lugs are press-fit into the flywheel and held in place with two set screws. When servicing the flywheel, these lugs must be inspected for wear. When replacing these lugs, removal of both set screws is

Figure 10-48 Pot-type flywheel.

Figure 10-49 Pot-type flywheel, discs, and separation plate located within the flywheel.

necessary: the first set screw locates the lug in place, and the second set screw prevents the first set screw from loosening. The pot-type flywheel must also be inspected for thermal cracks and ring gear wear. Because the flywheel face is recessed within the flywheel depth, it is difficult to machine. If machining of the flywheel face is necessary, the top bolting face must also be machined to the same amount as the face. If this is not done the original flywheel depth will be greater, reducing clutch spring force between the flywheel face, the clutch discs, and the pressure plate, which could result in clutch slippage and premature failure of the friction discs.

Pot-type flywheels used in heavy-duty equipment are slightly different than those used in the trucking industry. Although clutches are rarely used in heavy equipment to propel a vehicle, they are still equipped with a flywheel. Many machines use a torque converter, which is coupled or bolted to the flywheel, to drive the transmission. The pot-type flywheel used in heavy equipment utilizes an internal spline or teeth that a torque converter housing with matching splines can be coupled to without bolting them together. When the transmission is bolted to the engine bell housing, the torque converter and flywheel splines must be aligned for the coupling to occur. See **Figure 10-50.**

Typically, a wet housing is used: the engine and transmission are bolted together as a sealed unit with oil flow circulating within the flywheel housing. This prevents the flywheel and torque converter from oxidizing and also prevents wear between the coupling splines. No machining is necessary because the splines transfer the engine torque to the torque converter. The speed differentials occur within the torque converter, not the flywheel and friction disc. Inspection of the ring gear and internal drive splines is still necessary when servicing the engine or torque converter.

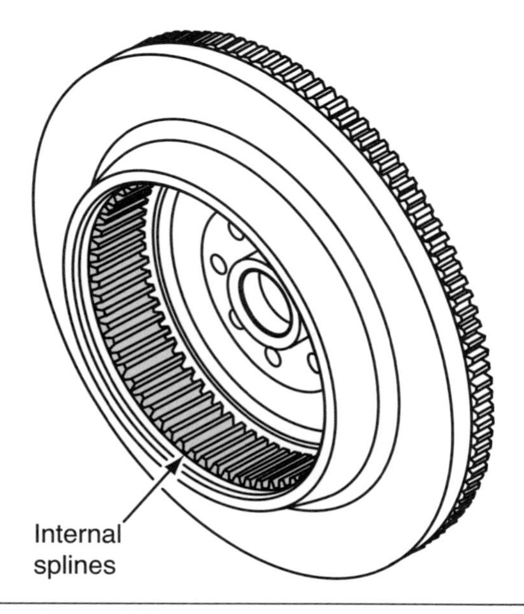

Figure 10-50 Off-road equipment, pot type flywheel with internal spline.

Dual-mass Flywheel

Many engine manufacturers are developing their engines to produce its maximum torque output at lower engine rpm. Because the torque rise occurs at much lower engine speeds, greater torsional vibrations occur. The dual-mass flywheel was developed to control the torsional vibrations created by the power pulses of these diesel engines and to provide a smoother delivery of torque and speed to the transmission. Unlike the traditional single-piece flywheel the dual-mass flywheel is split into two separate masses (**Figure 10-51**). The first mass, which is bolted to the engine crankshaft, transfers the engines mass moment of inertia, while the second mass increases the mass moment of inertia of the transmission. The two masses are linked by a spring dampening system to absorb the power pulses created during high torque loads of the engine. See **Figure 10-52.**

Because of this dampening system, many manufacture's eliminated the need for a dampened-type friction disc when used with a manual transmission. Manufacturers claim that transmission shifting becomes easier because of the lower mass to be synchronized, results in less wear of the transmission synchronizers. Any torsional vibrations that occur when driving are normally absorbed by the dampened-type friction disc when used with a traditional flywheel. The dampening springs in the dual-mass flywheel absorb these torsional vibrations and control any driveline surging.

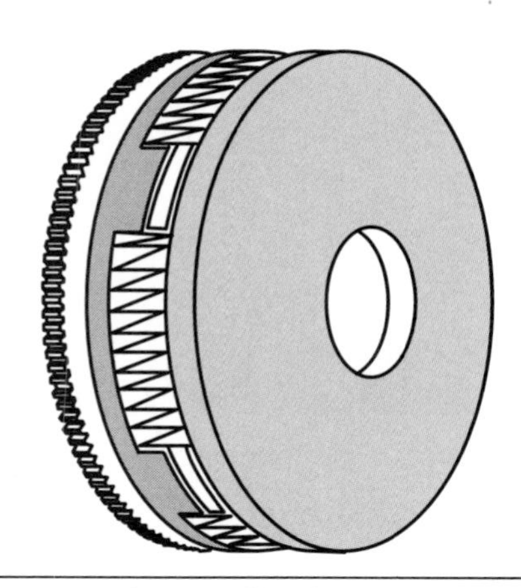

Figure 10-51 Dual-mass flywheel.

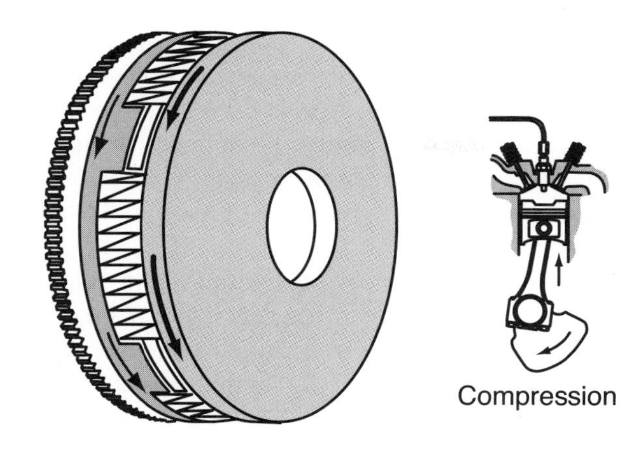

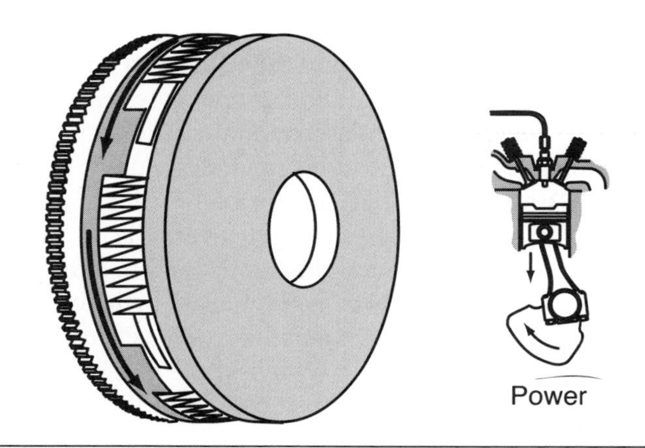

Figure 10-52 A dual-mass flywheel absorbs engine power pulses.

CLUTCH AND FLYWHEEL MAINTENANCE AND REPAIR

The following maintenance and repair procedures are for disc-and-plate-type clutches.

Push-Type Clutch Adjustment

Push-type clutch adjustments vary with different clutch manufacturers. On push-type clutches, as the clutch discs wear, the clearance between the release bearing and the release fork diminishes; therefore, constant adjustment is necessary. If this adjustment is not made, the release bearing will stay in contact with the clutch release fingers and the release bearing will be forced to rotate with the clutch assembly at *all* engine speeds during *all* driving conditions. Premature failure of the release bearing will occur.

Clutch adjustment should be made

- When a clutch replacement has been made.
- At regularly scheduled maintenance intervals.
- When clutch pedal free-play has diminished.

Clutch pedal free-play should not be less than 1.5 inches on heavy-duty applications. On light-duty applications, this adjustment may be less. OEM specifications should always be referred to before making any clutch adjustments.

Pull-type clutch adjustment is discussed in the section on "Clutch Diagnosis and Repair" later in this chapter.

Clutch Adjustment on Mechanical Linkage

When clutch adjustment with mechanical linkage is necessary, you must first ensure that there are no bends or binding within the linkage or adjustment rods. An adjustment rod located at the clutch release fork lever is the typical method of adjustment with most push-type clutches. Before any adjustment is made, all rods and levers should be inspected for wear within the pins and yokes, and they should be properly lubricated (**Figure 10-53**). Any bent or binding linkage and rods should be repaired prior to any adjustments.

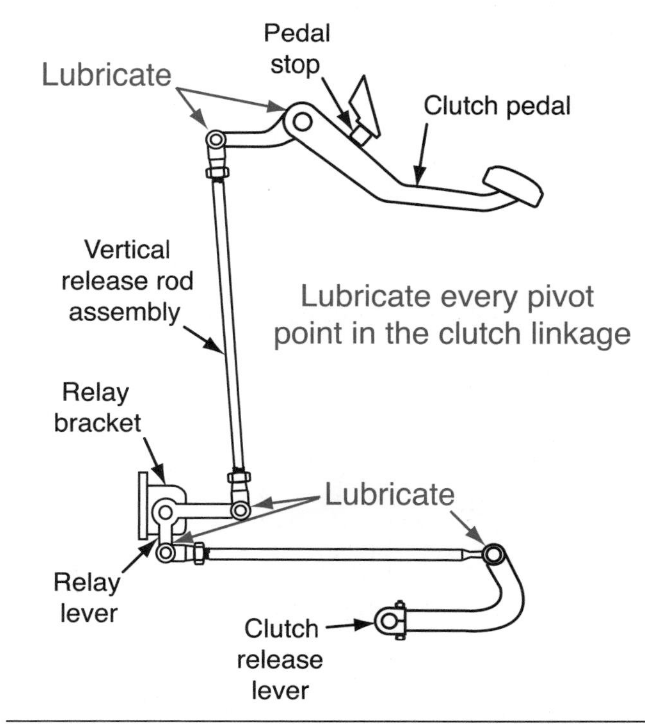

Figure 10-53 Clutch mechanical linkage, lubrication points, and inspection for bent or binding linkage.

Loosen the locking nut on the adjustment rod before adjusting the linkage length to achieve proper clutch pedal free-play. Remove the yoke pin from the clutch release lever and rotate the yoke clockwise or counterclockwise (to shorten or lengthen the adjustment rod) to achieve 0.5 inches of clutch pedal free-play. Rotating the yoke on the adjustment rod should be done in 180- or 360-degree increments. The yoke and pin should be reinstalled on the clutch release lever, and clutch pedal free-play should be measured until proper adjustment is achieved. When the correct free-play is reached, all pins should be secured with new cotter pins and the locking nuts tightened. Free play should be rechecked to guarantee that proper adjustment has been reached.

Hydraulically-Released Clutches

Hydraulically-released clutches (**Figure 10-54**) normally do not require adjustments unless the slave cylinder is equipped with an adjusting rod. Before any adjustment is made, hydraulic fluid in the reservoir should be checked. If the level is low, inspect for any hydraulic leaks in the slave cylinder, hydraulic lines, and master cylinder. These repairs must be made *before* any clutch adjustment is made. Many manufacturers of hydraulically-released clutches equip the master cylinder with a weep hole, which must be inspected for oil leakage. If hydraulic oil is found, the master cylinder is leaking back through the main seal of the master cylinder. Many times this must be inspected from inside of the cab. If a hydraulic booster is used with the master cylinder, leaks may not always be noticed because the oil will be trapped between the master cylinder and the power booster. Master cylinders with weep holes eliminate this problem.

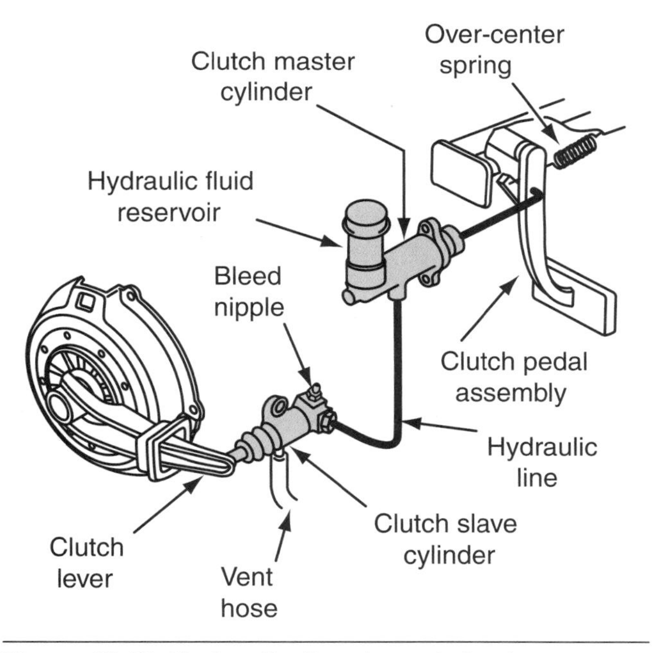

Figure 10-54 Hydraulically-released clutch.

Regular maintenance of hydraulically-released clutches is imperative to proper clutch operation. OEM specifications typically recommend replacing the brake fluid, which is used for hydraulically-released clutches every two to three years. Refer to OEM specifications for proper maintenance intervals. A DOT 3 or DOT 4 brake fluid is used. Newer fluids are constantly being produced, so OEM specifications should always be followed.

Replacing Fluid of Hydraulically-Released Clutches. When replacing hydraulic fluid, proper care and maintenance should be taken for containment of the oil; any oil spillage should be absorbed and cleaned up immediately. Brake fluid is very corrosive to paint and can cause skin irritation. When draining hydraulic oil, open all bleed screws and remove hydraulic lines to allow oil to completely drain from both master cylinder and slave cylinder. To refill hydraulic systems attach all lines and close the bleed screws. Fill the master cylinder to its maximum capacity. Now, open the bleed screw at the slave cylinder and allow the air to drain by gravity. Remember to continuously check oil level in the master cylinder. Do not allow the oil level to drop below the plunger height, as this will introduce air into the system.

The following procedure is used for bleeding the master cylinder and slave cylinder using a pressure ball method, as shown in **Figure 10-55.**

1. Make sure the bleeder ball reservoir is full.
2. Connect bleeder to the master cylinder. The bleeder ball should be pressurized between 7 and 10 psi. This pressure will vary among pressurized ball manufacturers.
3. Wipe any excess dirt from the slave cylinder bleeder screw. Connect a clear flexible line from the slave cylinder bleeder screw. This line should drain into a separate reservoir that is partially filled with clean brake fluid. Before the bleeder ball valve is opened, the slave cylinder bleed screw must be opened.
4. The pressure ball valve should be opened slowly to allow brake fluid from the pressure ball to

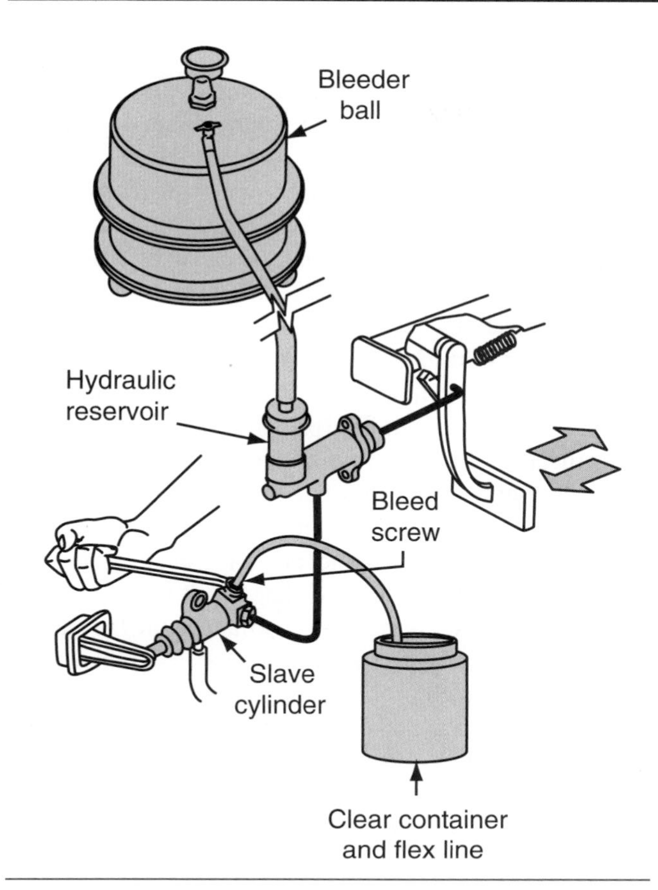

Figure 10-55 Hydraulically-released clutch. Using a bleeder ball method to bleed air from the system.

enter the master cylinder and push any air trapped between the master cylinder and the slave cylinder out through the slave cylinder bleed screw and connecting hose into the separate reservoir. Bubbles will be seen in the separate reservoir if there has been air in the system.

5. When the clear oil flows throughout the separate reservoir and no air bubbles are seen, the slave cylinder bleeder screw should be closed and step 4 should be repeated to make sure there is no residual air in the system. This step can be repeated two or three times as needed to remove all air from the system.

Manual Bleeding Procedure

1. Remove any dirt and debris around the bleed screw. Loosen the slave cylinder bleed screw. Attach a rubber flexible line to slave cylinder bleed screw.
2. Fill master cylinder with the proper fluid to its maximum level. Carefully watch the bleed screw and the line attached to it drain into a separate reservoir. When oil is exiting at a steady rate with no air bubbles, close the bleed screw.
3. Refill the master cylinder to the maximum level and close the master cylinder. Depress the clutch pedal three or four times and hold in the clutch released position. A second person may be required to open the slave cylinder bleed screw. Allow any trapped air to bleed out into the separate reservoir. When the bleed screw has been closed, release the clutch pedal slowly to allow fresh oil to enter the master cylinder. This procedure should be repeated until all air is removed from the system.

Tech Tip: When using the manual bleeding procedure, remember that the master cylinder must be kept to its maximum oil level.

Clutch Adjustment on Cable-Type Linkage

The clutch cable may require periodic adjustment to maintain the correct pedal free-play. Most clutch cables utilize an adjusting nut at the release arm end of the cable as shown in **Figure 10-56.** This adjustment extends or reduces the cable length to allow the clutch lever to move forward or rearward until the free-play required between the release bearing and clutch fingers is achieved. For clutch cables that have two adjusting nuts (the cable being allowed to float within its

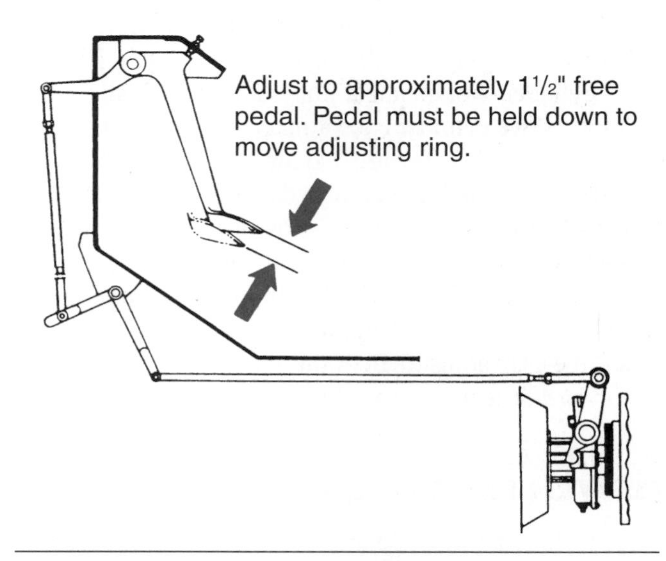

Figure 10-56 Clutch cable-type linkage.

bracket) both nuts must be loosened. When both adjusting nuts are turned in the same direction, the cable housing moves forward or rearward allowing an adjustment for free-play. Depending on the type and style of linkage, to increase the clutch pedal free-play, shorten the length between the clutch release lever and clutch cable bracket. To decrease clutch pedal free-play, extend the length between the clutch release lever and clutch cable bracket. When proper adjustment has been reached, the adjusting nuts should be retightened without altering the extension length. Free-play should always be rechecked after any adjustments have been made.

Air Released Clutches

Air released clutches incorporate a clutch pedal treadle valve, an air slave cylinder, and appropriate airlines and fittings. The treadle valve works similar to that of an air brake treadle valve. The farther the pedal is depressed, the greater the amount of air delivery. Air from the reservoir is piped directly to the clutch treadle-valve air inlet. The treadle-valve delivery port is piped to the clutch slave cylinder. Reservoir air can only flow to the delivery port when the pedal is depressed. The delivery port is open to the exhaust port in the pedal-released position only.

When the pedal is depressed, an internal valve first closes the exhaust port and opens a passage between the inlet port and delivery port, allowing air to flow to the slave cylinder and disengaging the clutch. The treadle valve has a holding stage at which the air pressure delivered equals that of the pressure applied to the pedal. In this position the exhaust port stays closed and the air inlet port is closed thus preventing any further air delivery and trapping the air between the slave cylinder and the treadle valve. This metering of air is very crucial to clutch control when engaging and disengaging the clutch. The air slave cylinder works under the same principal as the hydraulic slave cylinder except that air pressure is substituted for hydraulic pressure. No air leaks should be present when there is no clutch pedal movement or when depressing the clutch pedal. A slow metering of exhaust air should be present when re-engaging the clutch (clutch pedal releasing). Depending on the manufacturer, some slave cylinders incorporate an adjusting rod for clutch pedal free-play adjustment or the slave cylinder is adjusted forward or rearward to achieve proper free-play.

CLUTCH DIAGNOSIS AND REPAIR

A technician must be able to quickly and accurately diagnose and repair clutch problems. Every clutch failure happens for a reason. A good technician must always determine the cause of failure and then use whatever means necessary to correct failure and avoid their reoccurrence. Most clutches provide many hours of dependable service. However, the service life of clutch components varies greatly with operating conditions, operator abuse, or misuse of the clutch.

Operator abuse is commonly the cause of premature clutch failure; "riding the clutch" is a term used when an operator causes the clutch to slip or ride to achieve a slower wheel-speed-to-engine-speed ratio needed to perform an operation not intended for the machine. This causes the clutch assembly (primarily the friction disc) to overheat. Holding the vehicle on a hill by riding the clutch is another situation that can result in premature clutch failure. Resting the foot on the clutch pedal can cause the release bearing to stay in contact with the release fingers on the pressure plate (push-type-clutches) and cause the release bearing to rotate with the clutch assembly at all engine speeds. A good technician will always verify clutch problems by gathering information, such as talking to the customer, test driving the unit, listening for unusual noises, checking clutch pedal free-play, and checking the service manual troubleshooting chart. Using this information plus a good working knowledge of clutch operating principles can direct the technician to a preliminary failure analysis.

Pull-Type Clutch Adjustment

On pull-type clutches, as the clutch discs wear the clearance between the release bearing and the release fork diminishes; therefore, constant adjustment is necessary. If the adjustment is not made, the release bearing and the release fork will stay in contact, which causes the pressure plate to diminish the amount of spring force being applied to the clutch disc and flywheel. If clutch adjustment is not properly maintained, premature failure of the friction disc and possible failure of the pressure plates will occur. Clutch adjustment should be made

- When clutch replacement has been made.
- At regularly scheduled maintenance intervals.
- When clutch pedal free-play has diminished.

Clutch pedal free-play should not be less than 1 to 2 inches according to OEM specifications. See **Figure 10-57.** Pull-type clutches require an internal adjustment of the pressure plate as well as a linkage adjustment.

Adjusting Spicer and Rockwell Assemblies. The internal adjustment for both Rockwell and Spicer pull-type clutches is very similar because both require rotation of the internal adjustment ring located within the clutch pressure plate (**Figure 10-58**). Lipe and Valeo

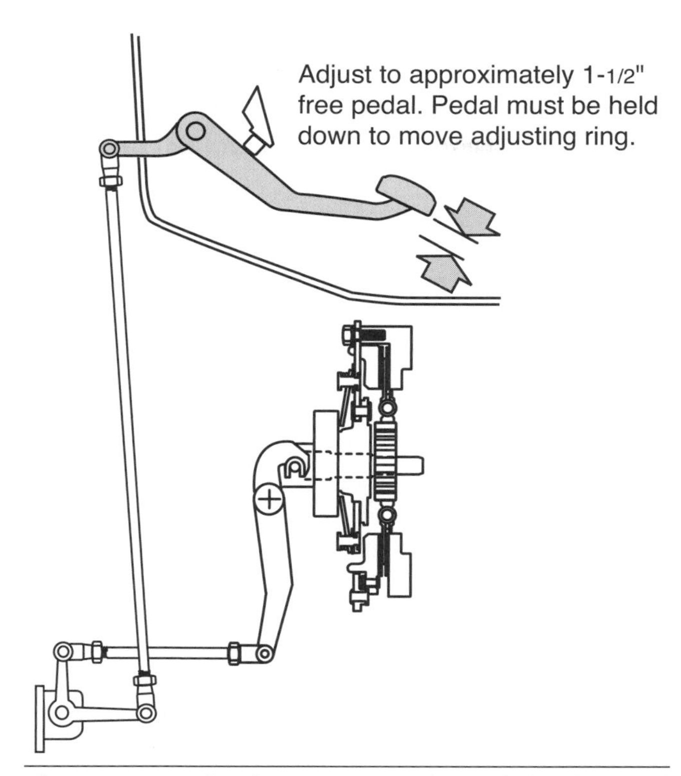

Figure 10-57 The first 1 to 2 inches of clutch pedal travel will take up the clearance between the release bearing and fork.

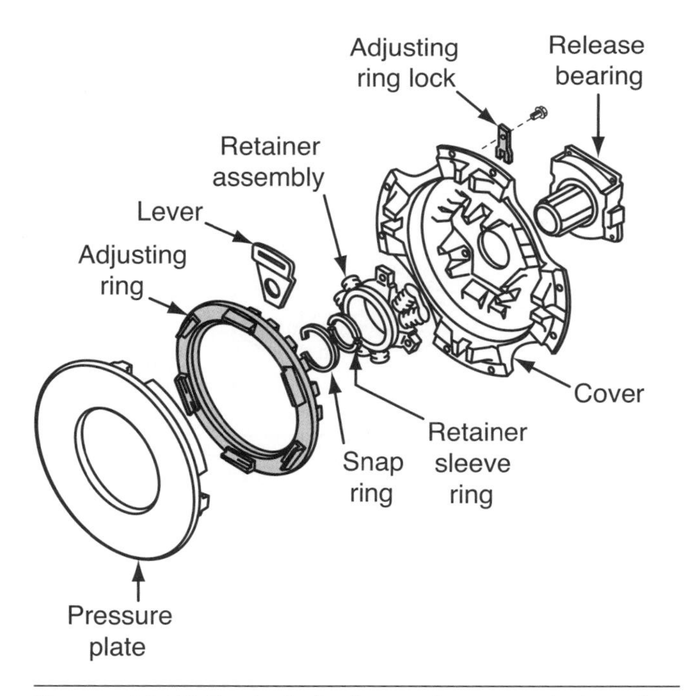

Figure 10-58 Exploded view of an angled spring clutch assembly with adjusting ring.

models require an adjustment to the release-bearing collar as shown in **Figure 10-59.** The internal adjustment on pull-type clutches is an adjustment to set the

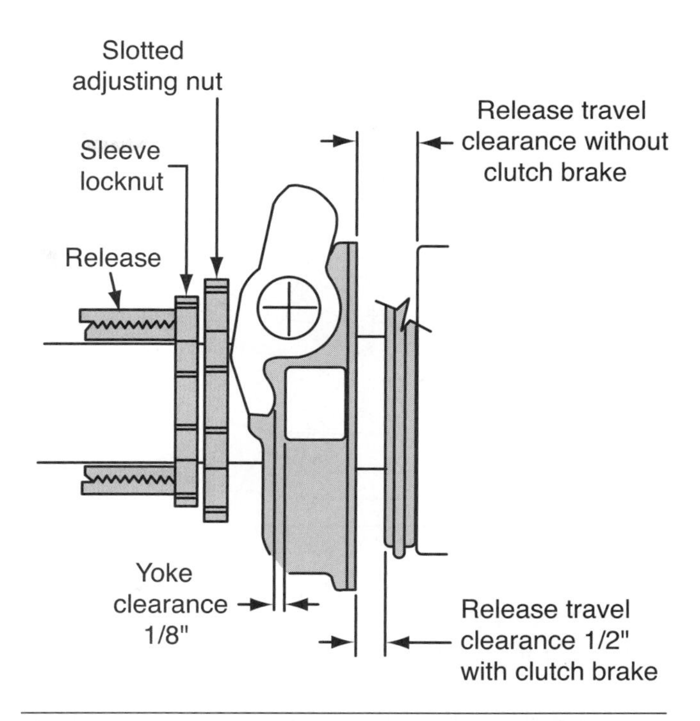

Figure 10-59 Lipe and Valeo adjustment made with the adjusting nut and sleeve collar.

clearance between the clutch release bearing and the clutch brake. The external clutch linkage adjustment sets the clearance between the release-bearing fingers and the release-bearing wear pads.

1. Adjustments to both the Spicer and Rockwell pull-type assemblies must be made with the clutch pedal depressed (clutch released). Remove the inspection cover from the bottom side of the transmission bell housing.
2. Rotate the engine manually to locate the clutch adjustment slot at the bottom side of the transmission bell housing and clutch cover (**Figure 10-60**).
3. With the clutch pedal released (clutch engaged), check the release-bearing clearance between the release fork and the release bearing (**Figure 10-61**).

Note: For clutches using a clutch brake with non-synchronized transmissions, the clearance check should be done between the release bearing and the clutch brake located between the release bearing and the transmission housing. This clearance should be between 1/2 inch and 9/16 of an inch or 12.7 to 14.28 mm. With pull-type clutches that do not use a clutch brake, the clearance should be checked between the release bearing and the clutch cover.

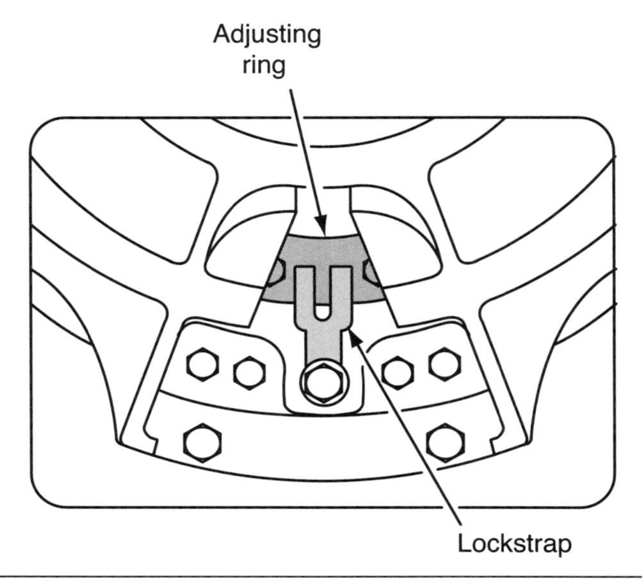

Figure 10-60 Locate the clutch adjustment slot and lock strap through the clutch inspection cover.

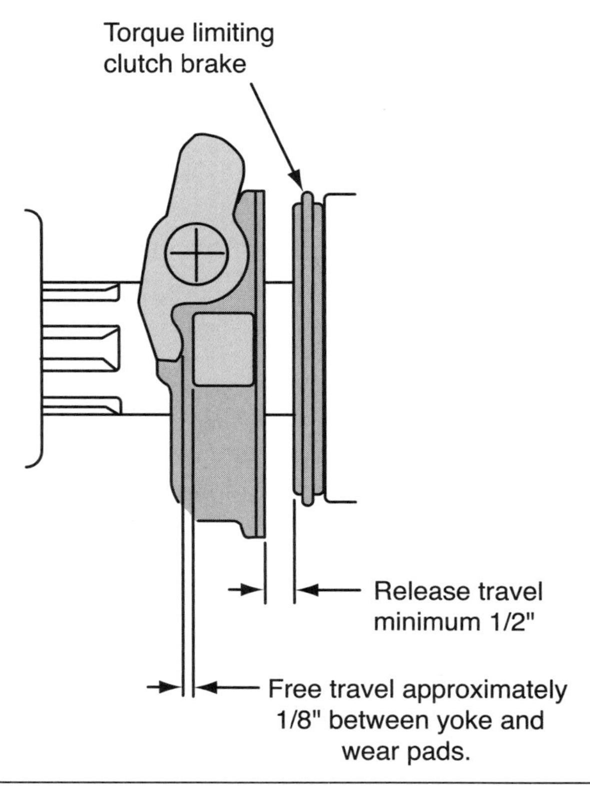

Figure 10-61 Checking release bearing clearance on a pull-type clutch.

4. Locate the lock-plate bolt through the transmission bell housing. This must first be removed before you can adjust the adjustment ring on the clutch pressure plate. An adjusting ring tool can be installed in place of the locking nut bolt to adjust the adjusting ring either clockwise or counterclockwise as shown in **Figure 10-62.** Rotating the internal adjusting ring clockwise moves the release bearing away from the flywheel and toward the transmission. Rotating the adjustment ring counterclockwise moves the release bearing toward the flywheel. Remember that this adjustment must be made with the clutch pedal down (clutch released) otherwise, trying to adjust the adjustment ring will only rotate the engine. Full clutch pressure is applied against the adjustment ring when the clutch is engaged (clutch pedal up). When a proper clearance has been reached, relocate the locking plate and lock-plate bolt, and torque to proper specification. Some clutch models provide an easy-to-reach, quick adjustment method. This permits an adjustment of free travel without the use of special tools or the need to remove any bolts. Eaton/Fuller quick adjust uses a 5/8-inch, 6-point socket to rotate the bolt, which in turn will rotate the adjustment ring. The quick-adjustment bolt has an integrated self-locking mechanism within the 6-point bolt; this locking mechanism engages at each quarter turn of the bolt. See **Figure 10-63.** When and adjustment is

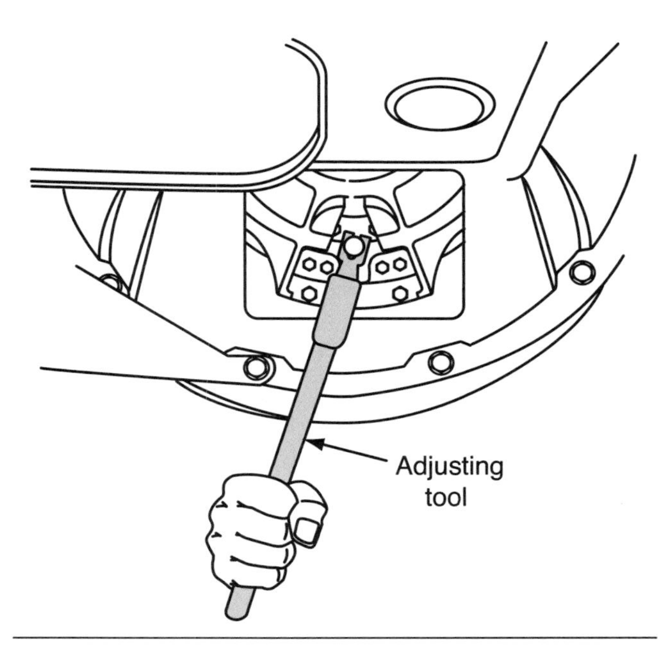

Figure 10-62 Special adjusting ring tools are available for adjusting the clutch.

Figure 10-63 Bolt must be depressed to adjust clutch. Always ensure that the bolt has locked after adjustment has been made.

necessary, the socket wrench must depress this spring-loaded lock before and attempt is made to turn the adjusting screw. Depressing the lock will allow rotation of the adjustment bolt for the quick adjustment. Always ensure that the locking mechanism has returned to the lock position after the clutch adjustment has been made.

5. When the adjustment ring is rotated, each notch represents approximately 0.020 inch or 0.5 mm per notch. When the correct clearance is reached between the release bearing and clutch cover or the release bearing and transmission cover, return the locked plate into position, making sure that it is located between the notches of the adjustment ring, and torque the locking bolt to the proper specified torque. Forgetting to install the locking strap will result in the clutch assembly loosening and clutch failure will occur.

   Adjustment of the ring must be done in increments and then rechecked with the clutch pedal in the up position (clutch-engaged position). Checking clutch adjustment with the clutch pedal depressed will result in an incorrect setting of the release-bearing clearance.

Tech Tip: When using the adjusting ring to check clearance between release bearing and clutch housing or release bearing and transmission housing, the clutch pedal must be released (pedal in the up position) to check the clearances.

6. Check the clearance between the release fork and the release bearing. This clearance should be approximately 1/8 inch or 3 mm. (Refer to Figure 10-62.) If adjustment is necessary, adjust the external clutch linkage to obtain this clearance. Clearance between the release fork and release bearing should amount to clutch pedal free-play of approximately 1½ to 2 inches of travel (**Figure 10-64**). If the clutch assembly is equipped with a clutch brake, clutch brake squeeze should be checked at this point. To check clutch brake squeeze adjustment, depress the clutch pedal 1 inch before the end of its travel (**Figure 10-65**). The clutch brake and release bearing should just start to come into contact at this point. The last 1 inch of clutch pedal travel is the recommended clutch brake squeeze. Lipe and Valeo pull-type clutch adjustments are similar to adjustments for the Spicer and Rockwell heavy-duty pull-type clutches. The inspection and pre-checks can be applied to this type of assembly. The two main adjustments include clutch release-bearing

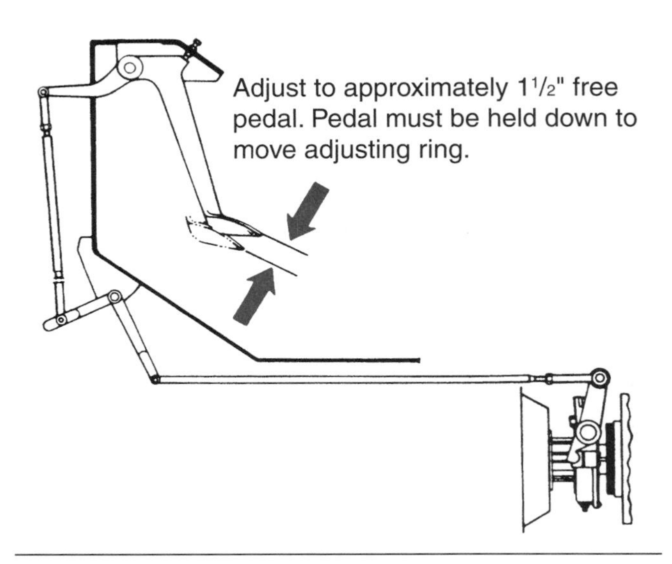

Figure 10-64 Clutch pedal free-play will make up the clearance between the release fork and release bearing.

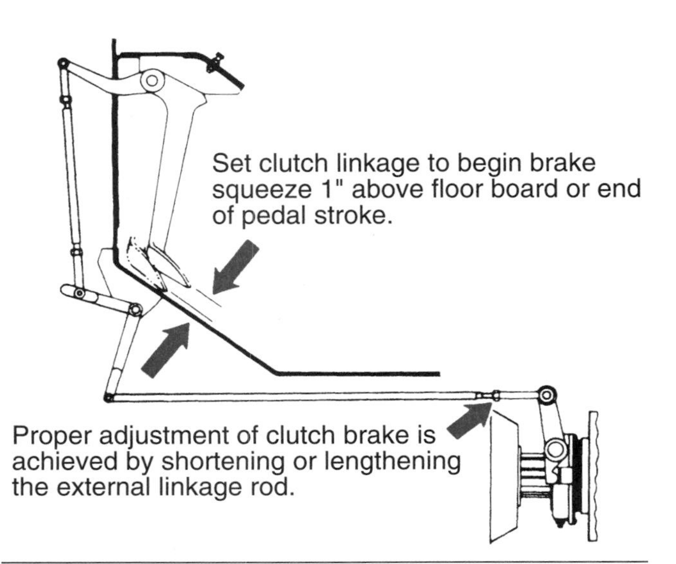

Figure 10-65 Clutch brake squeeze should be set to the last 1 inch of pedal travel.

travel and clutch pedal linkage free-play. Lipe and Valeo use a sleeve collar and locking nut for clutch adjustments. The bearing sleeve collar uses a special C-hook wrench to unlock the locking collar and adjust the sleeve collar to move the pressure plate toward the flywheel or away from the flywheel to obtain proper clearance. Note that this adjustment must be made with the pedal depressed (clutch released). This is always the first adjustment made on pull-type clutches. Clutch fork and release-bearing clearance should be made next, and all other clutch pedal linkages should be made after these adjustments.

7. Clutch brake squeeze adjustment is made by first checking the pedal height for proper travel. Using a 0.010-inch (0.25-mm) feeler gauge between the release bearing and clutch brake, allow it to be trapped between the two components while depressing the clutch pedal to the floor. Slowly release the clutch pedal and note the distance at which the feeler gauge is released from the release bearing and clutch brake. No adjustment is necessary if the pedal release length is between 3/4 and 1 inch. If this measurement is incorrect, lengthen or shorten external linkage, but without altering release fork, to release bearing adjustment to greater than the specified distance.

PUSH- AND PULL-TYPE CLUTCH TROUBLESHOOTING

Typical Problems and Causes

Problem: Clutch slips on engagement or clutch slips while driving.

Possible Causes:

1. Worn or damaged clutch components such as pressure plate and friction disk.
2. Oil or grease on clutch linings.
3. Clutch-linkage and release-bearing adjustment incorrect.

Remedies: Adjust clearance between release bearing and release fork. Any work or damaged clutch component should be replaced. If there is oil or grease on linings the discs must be replaced and the source of oil or grease must be repaired to prevent repeat occurrences.

Problem: Noisy Clutch.

Possible Causes:

1. Worn or damaged linkage.
2. Worn or damaged release bearing.
3. Worn or damaged clutch pressure-plate springs.

4. Friction disc dampening springs broken or loose.
5. Damaged pilot bearing.

Remedies: If a release bearing is noisy, lubricate the release bearing. If the condition persists, the release bearing must be replaced. Inspect the clutch housing for loose springs. If any are found, the clutch assembly should be replaced. Inspect friction disc dampening springs and linings. The disc must be replaced if it is damaged or worn. Inspect pilot bearing for smooth rotation and excessive play. A damaged pilot bearing must be replaced. It is considered good practice to replace the pilot bearing whenever the clutch assembly is removed.

Problem: Clutch Pedal Hard to Operate.

Possible cause:

1. Damaged hub on release-bearing assembly.
2. Worn or damaged clutch linkage.
3. Improperly lubricated linkage, i.e. cables.

Remedies: If there is a damaged hub on a release bearing, replace the release-bearing assembly. If there are any damaged or worn linkages, lubricate and replace the damaged linkage and recheck condition.

Problem: Vibrating Clutch or Pulsating Clutch Pedal.

Possible cause:

1. Worn or damaged splines on transmission input shaft.
2. Pressure plate or clutch assembly out of balance.
3. Engine flywheel out of balance.
4. Worn or damaged splines in hub of clutch discs.
5. Flywheel run-out incorrect.

Remedies: Inspect splines on transmission input shaft, if damaged or worn the transmission input shaft must be replaced. Note that splines on the friction-disc assembly must also be checked for wear. It is very likely that the splines on the friction-disc hub will be worn if the splines on the transmission input shaft are worn. If the pressure plate and assembly are out of balance, remove and check the balance of the clutch cover and pressure-plate assembly. If the condition persists, then the clutch assembly should be replaced. If the flywheel is out of round, follow the procedures for checking the flywheel run-out and make corrective measures (whether machining or alignment) to have the flywheel trued to the engine and transmission housing.

Problem: Clutch Grabs Quickly.

Possible Causes:

1. Broken discs.
2. Broken friction disc facing.
3. Warped pressure plate.
4. Oil or grease on disc facing.
5. Excessive flywheel run-out.

Remedies: Inspect the disc facings for any broken components; these must be replaced if they are flawed. With a straight edge and feeler gauge, inspect for a warped pressure plate it should be replaced if it exceeds specifications. Inspect for oil and grease on disc facings. The disc should be replaced and the source or oil and grease must be repaired to prevent the same symptoms from reoccurring. Excessive flywheel run-out should be tested and proper run-out adjustments made.

Problem: Clutch Noisy in Released Position Only.

Possible Cause:

1. Worn or damaged release bearing.

Remedies: Inspect release bearing for excessive wear or play. Replace the release bearing. Lubricate the release bearing, and if the condition still exists, the release bearing must be replaced.

Problem: Transmission Grinds When Shifting into First or Reverse With Clutch Pedal Depressed.

Possible cause:

1. Clutch linkage improperly adjusted.
2. Clutch brake worn.
3. Clutch not releasing completely.

Remedies: Adjust clutch linkage and inspect clutch brake. If worn it should be replaced. If the clutch is not releasing completely check the clutch adjustment and adjust to proper specifications.

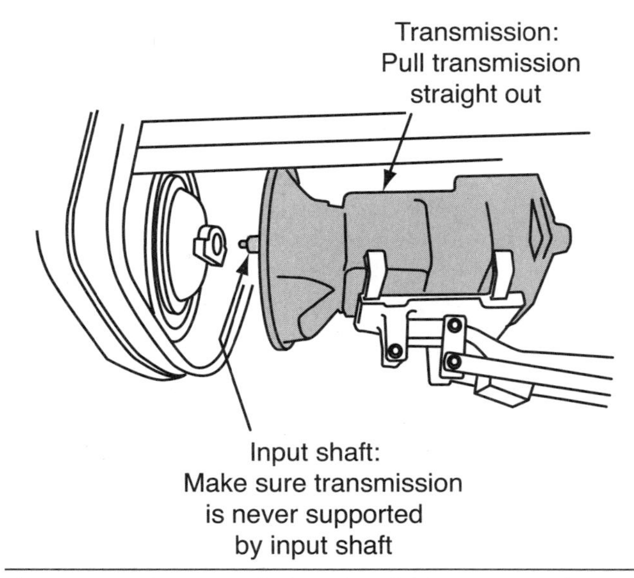

Figure 10-66 Transmission should be removed parallel to the engine without any weight being supported by the input shaft.

TRANSMISSION AND CLUTCH REMOVAL

The following is a general procedure for removing the transmission and clutch assembly. Always consult your manufacturer's service manual for proper removal instructions and specifications. Because of the diversity of clutches used in heavy equipment, it is difficult to describe any one manufacturer's procedure for the removal of the transmission and clutch from a specific machine. The following lists provide general steps to be taken.

Removing the Transmission

1. Remove the transmission shift lever from the transmission.
2. Remove the driveshaft from the transmission output shaft yolk.
3. Disconnect all electrical connections from the transmission.
4. Disconnect all air lines connected to the transmission.
5. Remove any return springs located on the clutch lever of the transmission and disconnect clutch linkage.
6. On hydraulically-released clutches, disconnect the pushrod and spring from the release fork.
7. Remove the hydraulic slave cylinder from the bracket on the transmission. To prevent the slave cylinder piston from exiting the slave cylinder, use a wire or a support mechanism to support the cylinder in place.
8. Use a proper transmission jack to maintain the transmission alignment and to support the transmission assembly. Do not allow the rear of the transmission to drop lower than the front and do not allow the transmission to hang unsupported on the splined hubs of the friction discs; this could cause the discs or the transmission input shaft to bend and/or distort the clutch discs. This will, in turn, cause poor clutch operation and/or future clutch problems.
9. Remove all bolts that attach the transmission to the brackets on the frame of the machine. Make note of the alignment of the transmission, and be sure that it does not hang by the input shaft in the pilot-bearing bore of the flywheel. The clutch assembly and pilot bearing will be damaged if supported by the transmission input shaft alone.
10. Remove all bolts and washers that attach the transmission bell-housing to the engine. Take note when using the transmission jack, that the transmission is still level and no weight is being placed on the transmission input shaft. Pull the transmission bell-housing straight out from the engine. Lower the transmission jack and remove the transmission from the vehicle.
11. With the transmission removed, if the vehicle is equipped with a clutch brake, remove the clutch-brake assembly and inspect it for wear.

Removing the Clutch from the Flywheel

1. Install a clutch alignment tool through the release bearing and clutch disc and into the flywheel pilot bearing. The alignment tool will support the clutch assembly during removal. See **Figure 10-67.**
2. Using the release-bearing release tool, pull the bearing back approximately 1/2 inch and install two 3/8-inch spacers between the clutch cover and the release bearing; release the tool so that the release bearing is supported by the 3/8-inch spacers. These spacers will relieve the spring load of the clutch cover and allow for reinstalla-

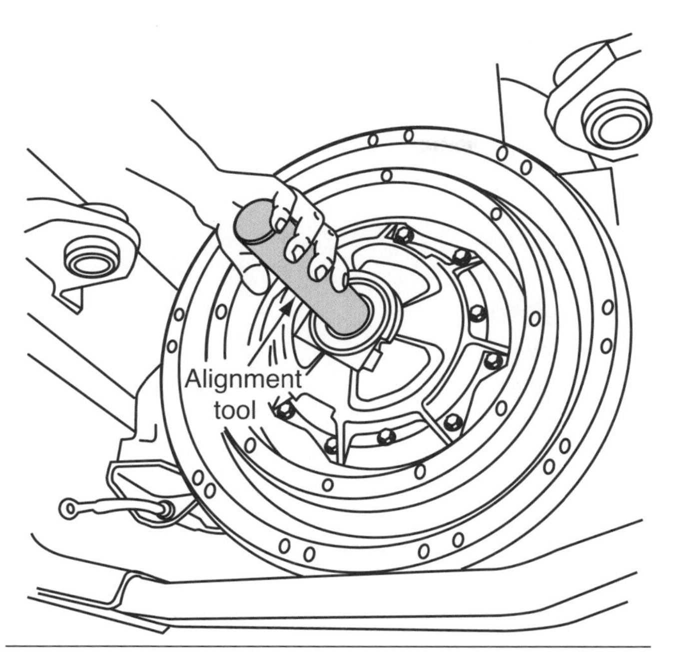

Figure 10-67 Install clutch alignment tool to keep clutch assembly supported during removal.

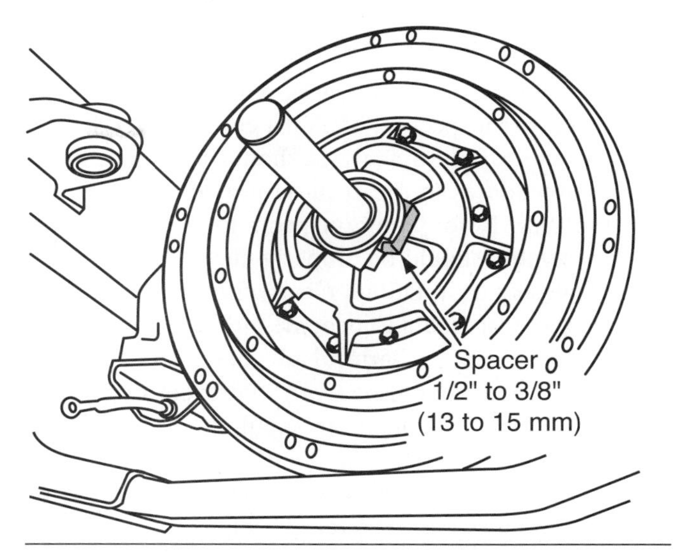

Figure 10-68 Install spacer to relive spring pressure on flywheel.

tion and easy removal of the clutch assembly. See **Figure 10-68**.

3. Loosen all the clutch bolts around the flywheel in a crisscross pattern. Do not remove them. Only loosen them until the clutch spring force has been diminished.
4. Remove two top clutch assembly bolts and install two 3/8-inch diameter bolts approximately 2 ½–3 inches long to be used as guide studs in the holes.
5. Clutch assemblies can be very heavy and manual removal is not advised. Failure to properly support the clutch assembly can result in immediate danger. Specialty tools such as a clutch caddy can be used to facilitate proper removal and installation of heavy-duty clutch assemblies. It is recommended that these tools be properly used for the installation and removal of clutch assemblies.
6. Remove the remaining bolts around the clutch assembly. Remove the clutch pressure plate and cover assembly together with the clutch discs.
7. With the clutch assembly removed, inspect the flywheel.
8. Remove the pilot bearing (**Figure 10-69**). Note that every time the clutch assembly is serviced or the engine is removed, the pilot bearing in the flywheel should be replaced. An internal puller or slide hammer can be used to remove the pilot bearing. The pilot bearing is an inexpensive component. It should be replaced on a regular basis.

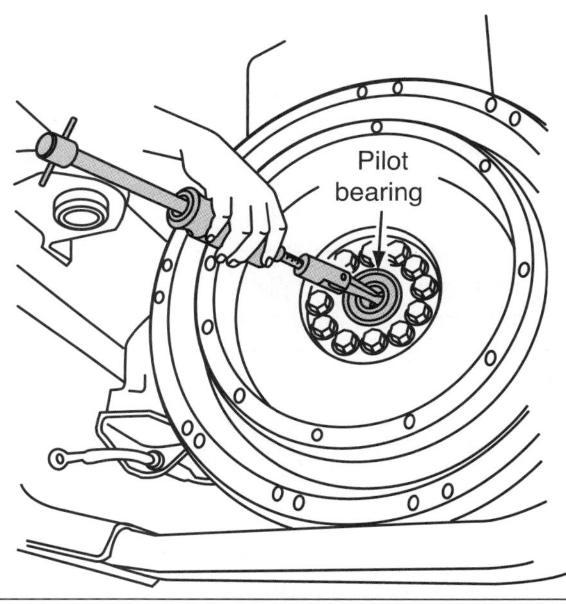

Figure 10-69 Pilot bearing removal using a slide hammer.

Tech Tip: Clutch assemblies can be very heavy. Manual removal is not advised.

Inspecting the Clutch

After removing the clutch assembly from the flywheel, a proper inspection of all the components should be made and a failure analysis should be determined. Correct procedures to repair these problems should be implemented. Before any inspection is started, the clutch assembly should be properly cleaned of dirt and clutch fibers or dust. Do not clean with pressurized air or cause the dust fibers to become airborne. Inhalation of dust fibers can be harmful to your health. Proper safety equipment should be worn to prevent inhalation of dust particles. A damp cloth should be used to wipe down the clutch assembly.

Inspect the pressure plate machined surface first to determine whether it is to be reused (**Figure 10-70**).Then inspect the pressure plate for heat discoloration, thermal cracks, scoring or signs of chattering. Check the pressure plate for signs of warpage by using a precision straightedge and feeler gauge. The pressure plate requires resurfacing if warpage is greater than 0.004 inch (0.10 mm). Thermal cracks greater than 0.060 inch (1.52 mm) deep will also require resurfacing providing the pressure plate thickness is not jeopardized. Check the OEM manual for proper specification limits. Minor heat discolorations (blue to dark brown in color) can be removed with a fine emery cloth and cutting oils. If clutch assembly is used with an intermediate plate, inspect the intermediate plate as you would the pressure plate. Warpage should not exceed 0.002 inch (0.05 mm). Resurfacing the intermediate plate is recommended providing the minimum allowable thickness is not exceeded. See **Figures 10-71** and **10-72.** Inspect drive slots or lugs (depending on the clutch style) on the intermediate plate for damage and excessive wear. If there are any cracks starting from the drive slots or lugs, the intermediate plate should be replaced. Worn drive slots or lugs can cause clutch rattle.

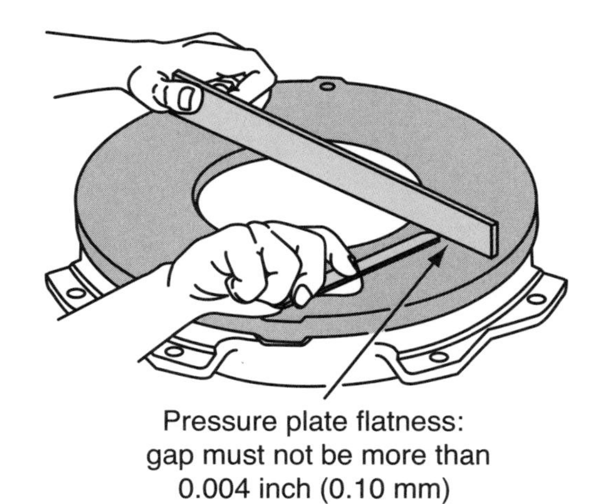

Figure 10-70 Measuring pressure plate flatness with a straight edge and feeler gauges.

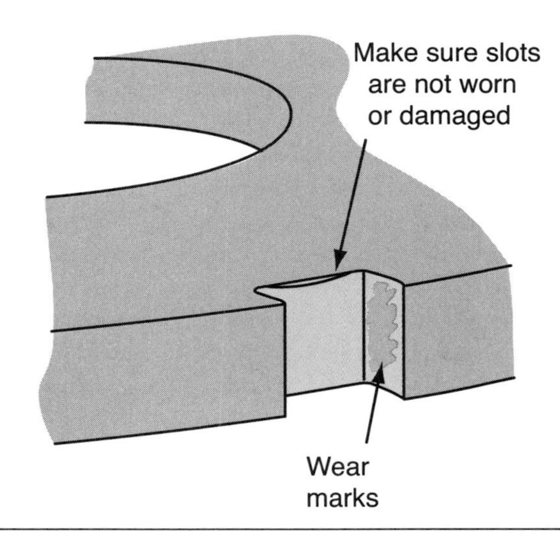

Figure 10-71 Inspecting intermediate plate on 14-inch clutches used with pot type flywheels.

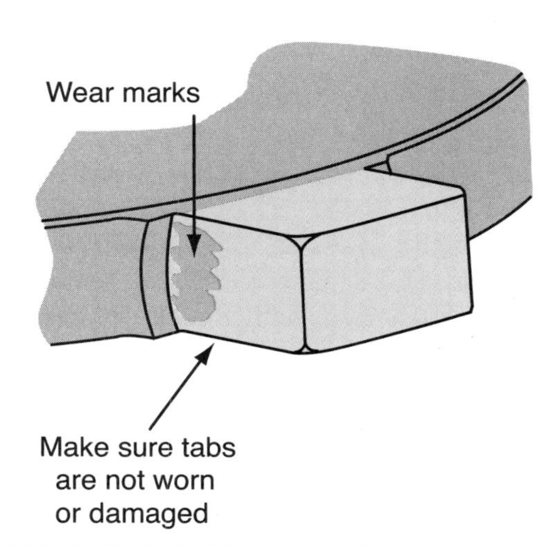

Figure 10-72 Inspecting intermediate plate on 15 $^1/_2$-inch clutches used with flat faced-flywheels.

Inspect the release bearing for smooth operation and that no grinding or binding is felt when rotated by hand. The bearing should have little to no play when checking thrust or radial play. The release bearing can be easily replaced by removing the snap-ring retainer and using a three-leg puller to pull the bearing off the sleeve assembly as shown in **Figures 10-73** and **10-74.** To install a new bearing, place the clutch assembly on a hydraulic press supporting the clutch cover assembly, and slowly press in the new bearing until it reaches the sleeve collar. Reinstall the retaining snap ring.

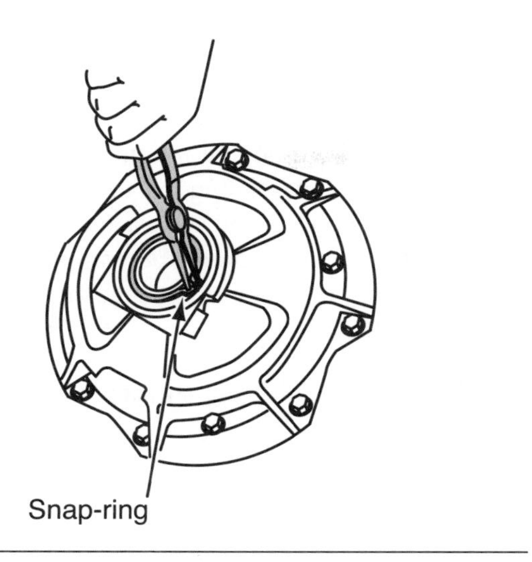

Figure 10-73 Release bearing, snap ring removal.

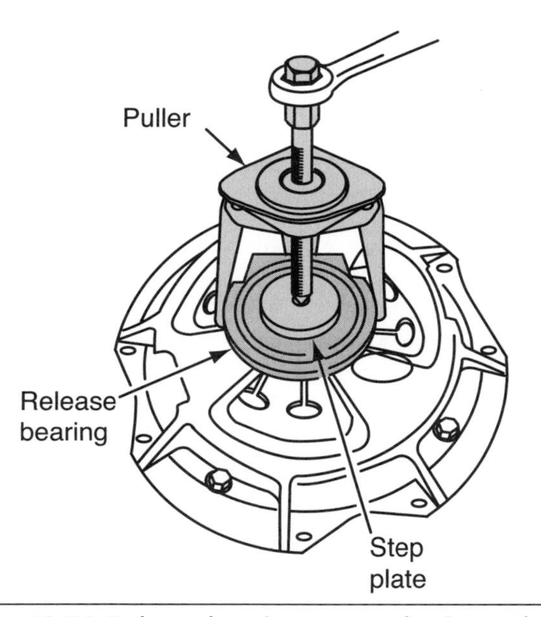

Figure 10-74 Release bearing removal using a three-leg puller.

Inspect the pressure-plate assembly for any broken or distorted springs, inspect levers for cracks and distortion, and inspect the clutch cover boltholes for elongation, if operated with clutch housing loose. It is not a common practice to rebuild the clutch assembly during a RE and RE (REmove and REplace), but it is a good practice to inspect the clutch assembly to determine the probable cause of failure and to remedy the cause so as not to have a repeated failure occur.

Inspect friction discs for glazing, excessive wear should be measured with a micrometer (**Figure 10-75**) and OEM specifications should be referred to. Check for worn hub splines, oil or grease saturation, signs of overheating, loose rivets on ceramic pads, and warped or beveled discs. If a dampened disc is used, inspect the springs for cracks, breaks, and looseness. Friction discs should be replaced if any of these items are noticed on the discs. In the case of oil or grease saturation, inspect the engine rear crankshaft seal, or transmission input shaft seal, for oil leakage. Inspect release bearing for over-greasing. Do not attempt to clean the friction discs of any oil or grease contamination by using heat to burn off the petroleum residue. This will only damage the friction discs and reduce their effective coefficient of friction. The engine flywheel must be removed to properly inspect the rear engine crankshaft seal.

Inspecting the Transmission and Engine Bell-Housing

It is imperative that the driveline, from the engine flywheel to the transmission, be uniform. Any misalignment between the two components will result in vibrations and eventually component failures. This is the reason for inspecting the matting surfaces of the engine and transmission housings. Many engine manufacturers use aluminum housings to keep the weight of components down. When an aluminum flywheel housing is used with a cast-iron transmission housing, the potential for the soft metal to wear is much greater than the harder cast-iron surface if the matting surfaces were to come loose. In addition, when two dissimilar metals are used, a galvanic reaction can take place between aluminum and cast-iron. These are the

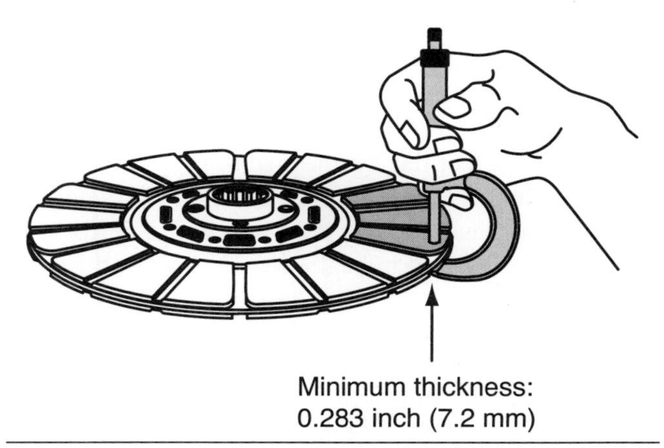

Figure 10-75 Measuring clutch disc for wear. Always refer to OEM specifications.

main reasons for inspecting the housings when the transmission is removed from the engine. The primary points of inspection on the flywheel and transmission housing are

- Bolt holes—damaged or elongated threads.
- Housing surface—worn or damaged from vibration or loose mounting.
- Cracked housing.

If boltholes or threads are damaged, thread, or sleeve, repair kits are available. It is recommended that this procedure be done at a local machine shop. Any misalignment of repair threads or sleeves can lead to mismatched alignment of engine and transmission. If thread repairs cannot be made, the housing should be replaced.

Worn or damaged housing surfaces can be machined or planed provided housing specifications are not surpassed. The greatest amount of wear generally occurs between 3 o'clock and 8 o'clock positions. If the OEM specifications cannot be retained, the housing should be replaced.

Many attempts are made to weld cracked housings. Some of these are successful; many fail and crack again. Most OEM specifications do not recommend that weld repairs be made to flywheel or transmission housings. Replacement housings can be very expensive, and many repair shops attempt to make this repair to keep the cost at a minimum. A technician should keep in mind that a second failure may result in removal of transmission and housings a second time and could result in even greater cost to the customer.

Checking Flywheel Run-Out

Flywheel run-out should be inspected whenever a clutch replacement is made. It is possible that excessive flywheel run-out can be the cause of premature clutch or transmission failure. With many engine rebuild shops, it is also possible that a different flywheel housing was installed on the original engine. A different housing may affect the flywheel run-out of the engine. Flywheel run-out should be checked after every engine rebuild. Four basic run-out tests should be performed using an accurate dial indicator.

Flywheel Face Run-Out (*Axial Eccentricity*) Test. When mounting a dial indicator to the flywheel housing, place the dial indicator needle as close to the flywheel outer edge of the clutch disc contact surface as possible. Push the crankshaft against the thrust bearing to prevent any inaccurate readings. Zero the dial indicator. Rotate the engine flywheel slowly by hand or by a manual cranking tool. Record your readings on the dial indicator at four 90-degree points, making note of the highest and lowest reading (+ or –). The difference between your highest and lowest reading is your total run-out. This reading should be compared to the total allowable run-out from your specific OEM Manual (**Figure 10-76**).

Flywheel Housing Face Run-Out (*Axial Eccentricity*) Test. Using a magnetic-base dial indicator mounted on the outer edge of the flywheel, position the needle on the inner edge of the flywheel housing. The needle will have to pass by the threaded boltholes, so be sure to set the indicator to clear the holes when the engine is rotated. Push the crankshaft against the thrust bearing to prevent any inaccurate readings. Zero the dial indicator and record your readings at four positions, 90 degrees apart. The difference between your highest and lowest reading is your total run-out. This reading should be compared to the total allowable run-out from your specific OEM manual. A typical maximum allowable run-out is 0.38 mm/ 0.015 in (**Figure 10-77**).

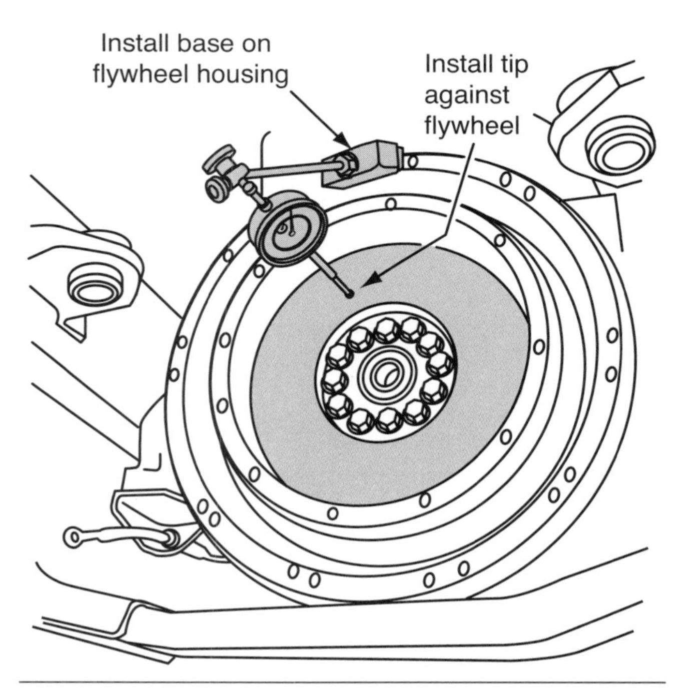

Figure 10-76 Measuring flywheel run-out with dial indicator. Always refer to OEM specifications for run-out limits.

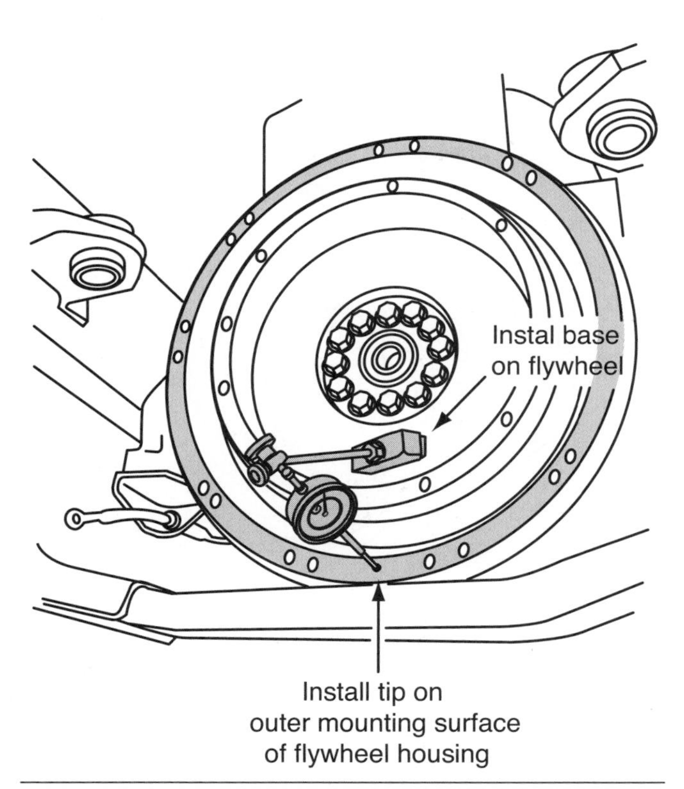

Figure 10-77 Measuring flywheel housing run-out with dial indicator. Always refer to OEM specifications for run-out limits.

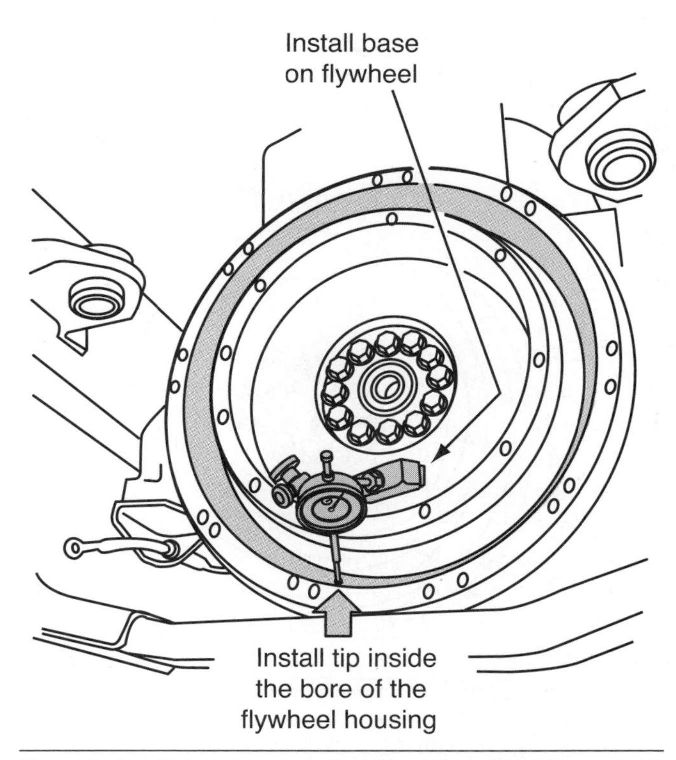

Figure 10-78 Measuring flywheel housing bore run-out with dial indicator. Always refer to OEM specifications for run-out limits.

Flywheel Housing Bore Run-Out (*Radial Eccentricity*) Test. Using a magnetic-base dial indicator mounted on the outer edge of the flywheel, position the needle on the inside edge (bore) of the flywheel housing. Push the crankshaft against the thrust bearing to prevent any inaccurate readings. Zero the dial indicator and record your readings at four positions, 90 degrees apart. Compare the two readings that are 180 degrees apart to determine horizontal and vertical run-out. The difference between your highest and lowest reading between the vertical measurements is your total vertical run-out and the same would apply for your horizontal measurements. This reading should be compared to the total allowable run-out from your specific OEM Manual. A typical maximum allowable run-out is 0.15 mm/ 0.006 inch **(Figure 10-78)**.

Tech Tip: Write the dial indicator measurements with their positive (+) and negative (–) notation. This notation is necessary for making the correct calculations.

Flywheel Pilot Bearing Bore Run-Out Test. When mounting a dial indicator to the flywheel housing, place the dial indicator needle close to the inside surface of the pilot bearing bore. Push the crankshaft against the thrust bearing to prevent any inaccurate readings. Zero the dial indicator and rotate the flywheel one complete revolution, recording your readings at four positions, 90 degrees apart. The maximum allowable run-out is 0.13 mm/ 0.005 inch. Always refer to OEM specifications (**Figure 10-79**).

Checking Crankshaft Endplay

Mount a dial indicator to the flywheel housing. Place the dial indicator needle perpendicular to the flywheel face, as close to the center hub as possible. Push or pull the flywheel against the end of the thrust bearing and zero the dial indicator. Move the flywheel in the opposite direction to the dial indicator setting and record your reading. Compare your reading to the specific engine manufacturer's specifications, and service the crankshaft and thrust bearing as needed.

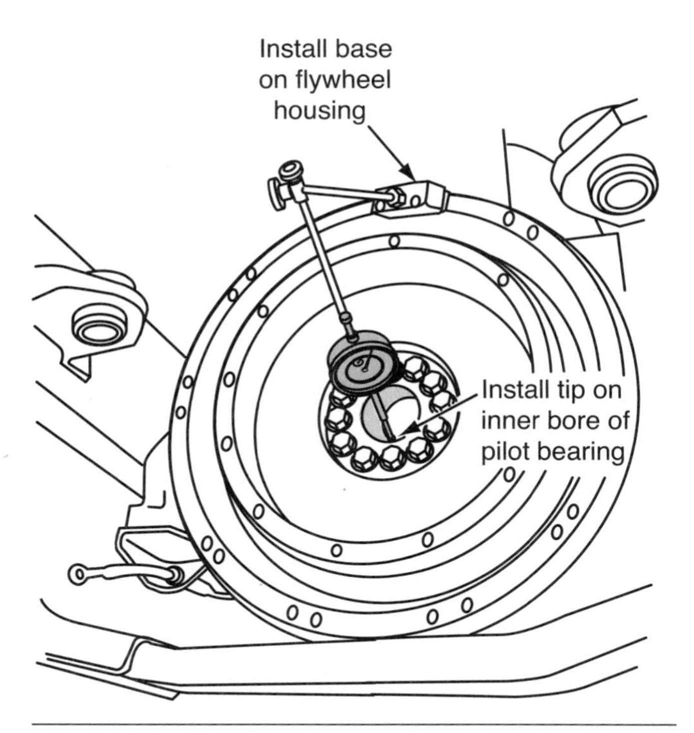

Figure 10-79 Measuring flywheel pilot-bearing bore run-out with dial indicator. Always refer to OEM specifications for run-out limits.

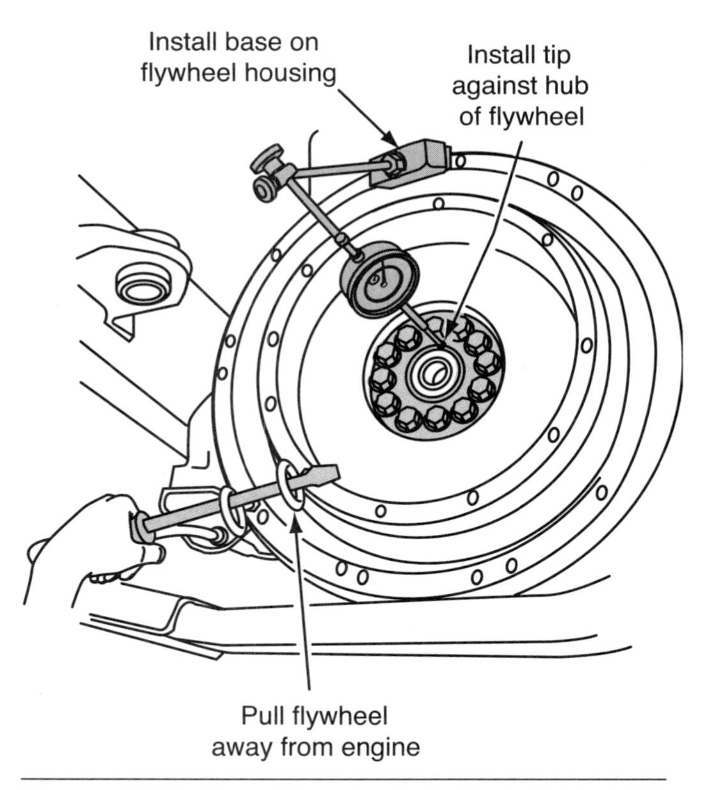

Figure 10-80 Measuring crankshaft endplay with dial indicator. Always refer to OEM specifications for run-out limits.

Endplay specifications can be as low as 0.003 inch or as high as 0.015 inch. For this reason, it is imperative that the original manufacturer's specifications be used (**Figure 10-80**).

ONLINE TASKS

1. Access the industrial powertrain Web site at the following address: *http://industrial-powertrain.com*. This site contains many applications and variations of clutches used in the heavy-duty equipment field. You may find it useful for general and service information on disconnecting clutches.
2. Access the Web site *http://auto.howstuffworks.com*. Although this Web site is primarily based on automotive technology, it provides a very wide knowledge base of basic principles that apply to most mechanical trades. Use the search function to find "How a Clutch Works."

SHOP TASKS

1. Locate a clutch assembly within your shop and identify what type of clutch assembly it is, e.g., pull-type, push-type, etc. Also identify whether it utilizes a diaphragm spring or coil springs to apply clutch pressure.
2. Locate a piece of equipment equipped with a clutch. Identify the type of control that is used to disengage and engage the clutch, e.g., hydraulic or mechanical.
3. Within your shop, locate an engine with the flywheel still mounted in the bell housing. Perform a flywheel run-out test as outlined in this chapter. Refer to the OEM procedure as needed.

Summary

- Pull-type clutches are clutches that disengage when the release bearing moves away from the flywheel.
- Push-type clutches are clutches that disengage when the release bearing moves toward the flywheel.
- Two types of clutch discs used are dampened and ridged.
- Two types of clutch disc material used are organic and ceramic.
- Friction is the resistance to relative motion.
- The coefficient of friction is equal to the amount of force required to move the material or object divided by the weight of the material or object.
- Three types of clutch linkages used are, mechanical, hydraulic, and pneumatic.
- Three basic types of flywheels are flat face, flywheel, pot-type, and dual-mass.
- Two types of clutch pressure plate springs are the diaphragm spring and the coil (coaxial) spring.

Review Questions

1. In which direction does a pull-type clutch disengage?
   A. toward the flywheel
   B. away from the flywheel
   C. pull-type refers to the way it engage
   D. all of the above
2. Two types of clutch discs are:
   A. rigid and ceramic
   B. dampened and organic
   C. ceramic and dampened
   D. all of the above
3. Two types of clutch assembly springs used are:
   A. leaf and coil springs
   B. coil and dampened springs
   C. diaphragm and bevel springs
   D. diaphragm and coil springs
4. The coefficient of friction is
   A. the outward acting force on a rotating body
   B. the ratio of the force required to move an object divided by the weight of the object
   C. the amount of friction produced when two roller bearings are in contact with each other
   D. the force to which a material, mechanism or a component is subjected.
5. Two types of overrunning clutches are
   A. Cone-type and roller-type
   B. Roller-type and sprag-type
   C. Sprag-type and band-type
   D. Band-type and cone-type
6. Which of the following clutch components is splined to the input shaft of the transmission?
   A. flywheel
   B. release bearing
   C. pressure plate
   D. clutch disc
7. Which type of clutch linings grab quicker with less wear?
   A. rigid disc
   B. dampened disc
   C. ceramic disc
   D. organic disc

8. Three types of clutch linkage are:
   A. mechanical, air, and electric
   B. mechanical, air, and magnetic
   C. electric, magnetic, and air
   D. air, mechanical, and hydraulic
9. On a push-type clutch, what happens as the clutch disc wears?
   A. pedal linkage becomes stiff
   B. hydraulic reservoir becomes low
   C. clutch fingers move closer to the release bearing
   D. clutch fingers move away from the release bearing
10. A typical air-conditioning pump clutch uses a:
    A. light-duty push-type clutch
    B. direct-action electric clutch
    C. cone-type clutch
    D. band-type clutch

CHAPTER

Mechanical Transmissions

Learning Objectives

After reading this chapter, you should be able to

- Identify the major components in a typical transmission, including input and output shafts, mainshaft and countershafts, gears, and shift mechanisms.
- Describe the shift mechanisms used in heavy-duty equipment transmissions.
- Trace powerflows from input to output in different gear ratios.
- Define the roles of transfer cases and PTOs in heavy-duty equipment operation.
- Define the principles of single and multiple countershaft transmissions.
- Define the principles of forward/reverse shuttles.
- Explain the importance of using the correct lubricant and maintaining the correct oil level in a transmission.
- List the preventative maintenance inspections that should be made periodically on a standard transmission.
- Describe the procedure for troubleshooting standard transmissions.
- Identify the causes of some typical transmission performance problems, such as unusual noises, leaks, vibrations, jumping out of gear, and hard shifting.

Key Terms

axial load
collar shift
constant-mesh
direct drive
drive gear
driven gear
drivetrain
multiple countershaft
non-constant-mesh
radial load
rotation
shift bar housing
shift fork or yoke
single-countershaft
slave valve
sliding gear
synchronized
synchronizer
transfer case

INTRODUCTION

Whether a vehicle is operated off-highway or on-highway, engine torque is transmitted through the clutch to the input shaft of the transmission. Gears and shafts in the transmission are used to alter the input torque and speed to suit the driving requirements of the vehicle. The transmission output shaft turns the vehicle **drivetrain** downstream from the transmission. The drive-train typically consists of a drive shaft, drive axle carrier, drive axle shafts, and drive wheel assemblies or a track final drive. The different gear ratios used give the vehicle the ability to handle driving conditions that range from a full load stationary start to cruising speeds. A transmission must also provide a reverse capability. Engines always rotate in one direction, so the transmission must be capable of reversing input **rotation** at the output; if this were not possible, a vehicle would only be capable of moving in one direction.

The role of the transmission is to manage the drivetrain. When a vehicle is accelerated or placed under load, the engine applies torque and a speed ratio to the drivetrain components. However, this powerflow may be reversed during deceleration when the drive axle wheels are driving the engine providing what we call engine braking because the engine has become power absorbing rather than power producing. During deceleration, the role of the transmission components is also reversed: output becomes input, and the transmission gear ratio determines how aggressively the engine brakes the drivetrain. In certain applications, the transmission must be capable of providing the same number of forward speeds as reverse speeds or have the necessary means to power special drives such as a PTO (power take-off).

Transmissions may consist of a single housing (main box) or multiple housings. A single-housing transmission may use single or **multiple countershafts.** As the requirement for gear ratios increases, the single housing (main box) can be "compounded" by adding an auxiliary section or sections to it. Compounding is a means of increasing the number of gear ratios in a transmission by the addition of an auxiliary box or multiple countershafts. A compound with two gear ratios can convert a five-speed (ratio) main box into a transmission with up to ten output speeds.

TYPES OF MANUAL SHIFT TRANSMISSIONS

Manual shift transmissions can be classified by

- Type of Gearing Configuration
  - Constant-mesh
  - Non-constant-mesh
- Number of Countershafts
  - Single countershaft
  - Multiple countershaft
- Mode of Shifting
  - Sliding gear
  - Collar shift
  - Synchronized

Transmission Classified by Gearing Configuration

Manual transmissions can be classified as either constant-mesh (**Figure 11-1**) or **non-constant mesh** (**Figure 11-2**). As the names imply, some transmission internal gears are constantly in mesh with each other and never come out of mesh. They are fixed in a position on parallel shafts that do not allow them to move out of mesh. These **constant-mesh** transmissions can use either helical cut gears or straight spur cut gears. Non-constant mesh transmissions have gearing where at least one set of gears on a parallel shaft is allowed to move or slide on its supporting shaft. Non-constant mesh transmissions use only spur cut gears.

Transmission Classified by Number of Countershafts

Most transmissions consist of at least three shafts: the input shaft, which delivers input torque from the engine via the clutch or torque converter; the output shaft, which usually selects the gear ratio and delivers the engine torque to the driveline; and the countershaft, which relays the input shaft torque and speed to the output shaft.

Single-Countershaft Transmissions. Most **single-countershaft** transmissions (**Figure 11-3**) are typically used in lighter-duty or low-horsepower applications. As mentioned earlier, the countershaft must relay input torque from the engine by means of an input shaft that is in constant mesh with it. In Chapter 9 we discussed

Figure 11-1 Constant mesh transmission. All gears are fixed to their respective shafts; these gears do not slide in and out of mesh.

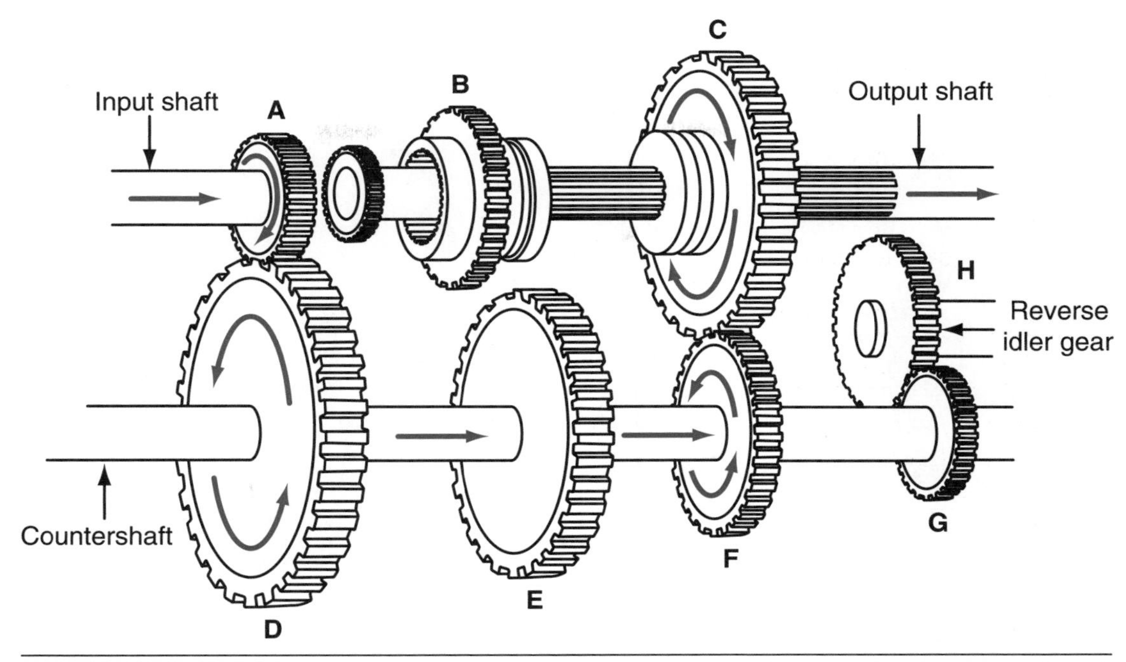

Figure 11-2 A non-constant mesh-transmission. First gear is shown in mesh. Neutral position places first gear C between countershaft gear F and reverse idler gear H. The only gears in constant mesh are input shaft gear A with countershaft gear D and countershaft gear G with reverse idler gear H.

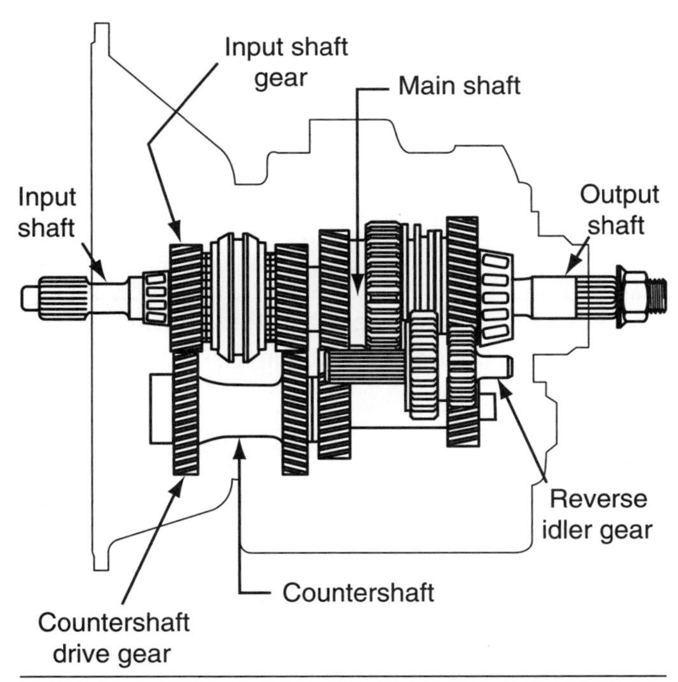

Figure 11-3 A typical single countershaft transmission: 1. Transmission input shaft, 2. Input shaft gear, 3. Countershaft drive gear, 4. Countershaft, 5. Mainshaft, 6. Output shaft.

that external gears in mesh will push away from each other under load, as shown in **Figure 11-4.**

This means that the torque input from the input shaft places a radial thrust onto the countershaft. The countershaft, which is supported by bearings, must be capable of handling this radial thrust. The countershaft also provides different gear ratios for the specific transmission configuration. This means that the torque input can be multiplied even further by the size and number of gear teeth used, increasing the radial thrust placed on the countershaft.

The countershaft gears, together with the output shaft gears will makeup the number of forward and reverse gears of the transmission. For example, five speeds forward and one speed reverse total six gears. Typically, the countershaft has one gear, which relays the input shaft gear-speed and torque, and anywhere from two to six other gears, which determine the different ratios of the transmission to the output shaft. These gears are fixed to the countershaft by either being press-fit or cast as one solid component to the countershaft.

Multiple Countershaft Transmissions. Multiple countershaft transmissions can be described as either

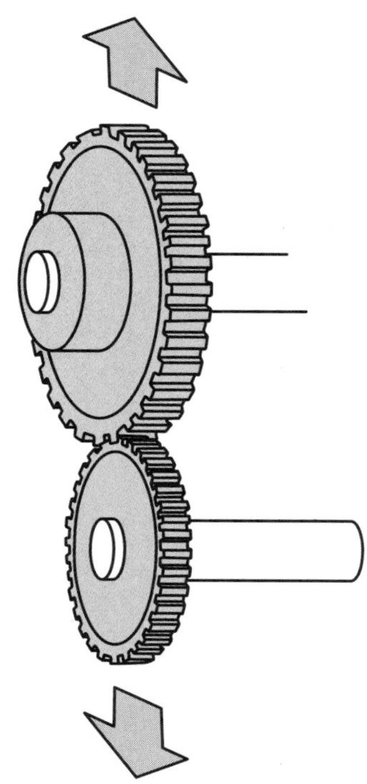

Figure 11-4 Greater resistance to the drive creates greater thrust loads on the mating gears.

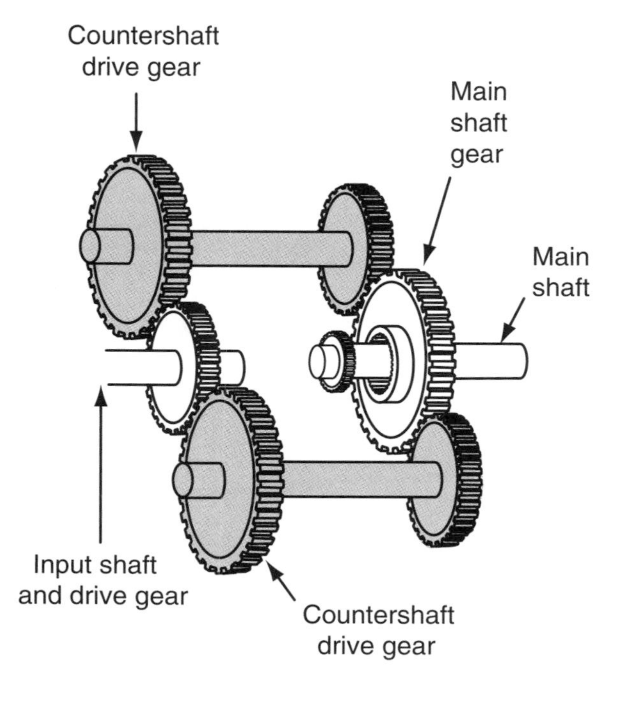

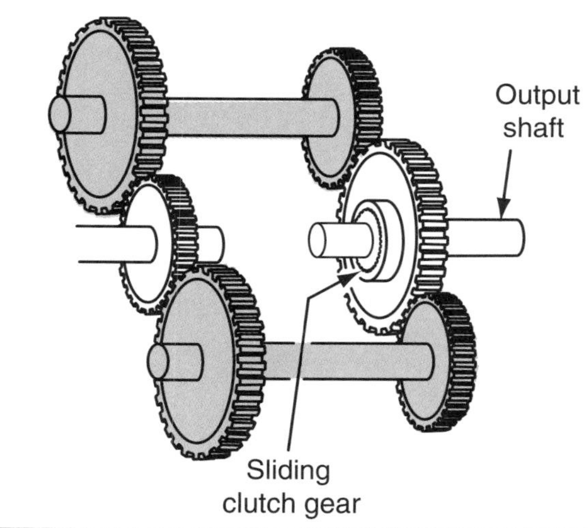

Figure 11-5 Twin countershaft transmission. Input torque is split between two countershafts.

splitting the input shaft torque to the countershafts (as in heavy-duty truck transmissions), or providing for an increase of gear ratios through multiple countershafts as used in many heavy-duty equipment applications. See **Figure 11-5.**

Multiple countershafts, as they are used in heavy-duty truck applications, are subjected to high **radial loads** from the engine and input shaft. Consider a typical truck diesel engine producing approximately 450 hp (horse power) and approximately 1500 ft-lbs of torque being delivered to the transmission input shaft. Providing the input shaft and countershaft gears are the same diameter, the radial load placed on the input shaft, bearing, and gears—together with the countershaft and bearing—is equal to 1500 pounds of force. This force can place a tremendous load on the gearing as well as on the bearings supporting the shafts. If an identical second countershaft were placed at 180 degrees to the input shaft, this would split the torque loading in half and eliminate any radial load on the input shaft. The 1500 pounds of radial force would be reduced to 750 pounds on each countershaft.

Some truck manufacturers utilize three countershafts to split the torque by three. Mack Trucks continue to manufacture transmissions with triple countershafts. **Figure 11-6A** and **Figure 11-6B** show a Mack truck triple countershaft transmission. Note that the countershafts are equally spaced around the mainshafts. This design evenly distributes the torque load around the three countershafts, thereby minimizing deflection and containing gear tooth loading to a minimum.

Multiple countershafts as they are used in heavy-duty equipment applications, must be considered separately

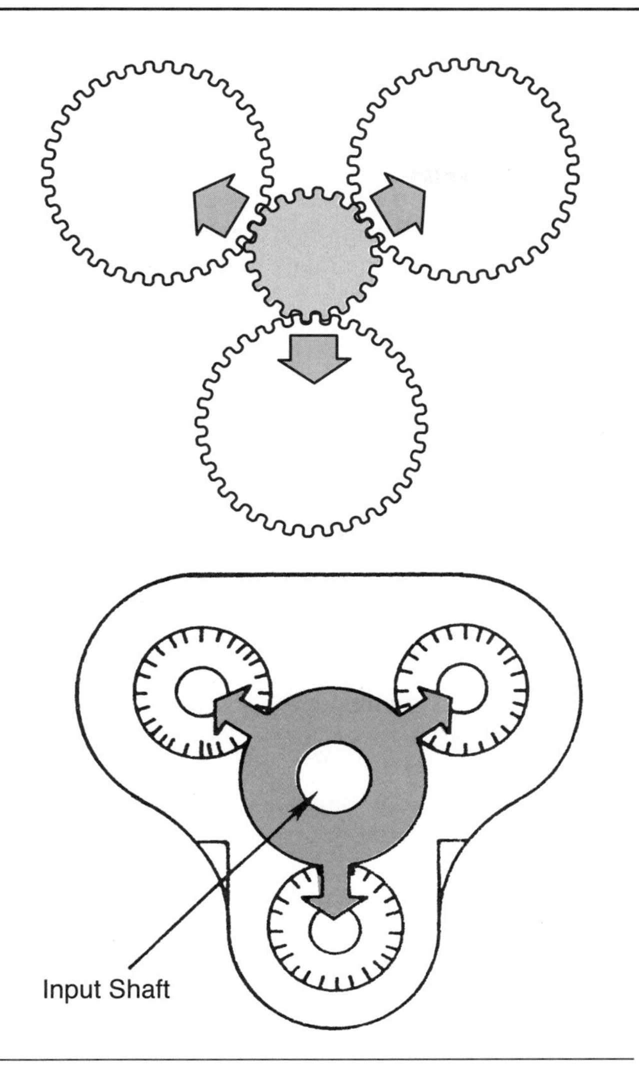

Figure 11-6 (A) A triple countershaft transmission splits the input drive torque by three. (B) A Mack-truck triple-countershaft transmission.

Figure 11-7 This multiple-countershaft transmission has eight forward speeds and four reverse speeds. This transmission does not split the torque between the countershafts. It utilizes them to multiply the number of gears available.

according to the application. Even though manual transmissions are becoming more obsolete in heavy equipment, they are still in use because of the long life expectancy of the equipment. Manual transmissions are typically found in smaller applications of less than 100 hp. The torque loading on smaller horsepower engines eliminates the need for multiple countershafts to split the input torque to the transmission. Large earthmoving equipment (100 hp to 1500 hp) is typically equipped with powershift transmissions or hydrostatic transmissions.

Multiple countershafts can be utilized to multiply gear ratios within a transmission. The countershafts are arranged so that the drive is transferred from one countershaft to the next in a series formation, as shown in **Figure 11-7.** Each countershaft will be messed with only one other shaft, and each shaft will have a different gear configuration to suit the application and number of gears available. As the powerflow travels through each countershaft and gear configuration, the gear ratios can be increased or decreased depending on the size of each gear and the gears selected to provide the powerflow.

Transmission Classified by Mode of Shifting

Standard transmissions are shifted manually, means that the operator makes gear selections on the basis of how he sees the operating requirements of the vehicle in relation to engine load and conditions. There are three general methods of shifting standard transmissions. We can classify these as **sliding gear, collar shift,** and **synchronized shift.** We will first consider sliding gear shift and collar shift. Synchronized shift will be extensively discussed later in this chapter.

Sliding Gear Shift. A transmission that uses a sliding gearshift mechanism has a mainshaft and a countershaft supported parallel to each other. To change from one speed/torque range to another, gears on the mainshaft are moved longitudinally until they engage the desired gear on the countershaft. To effect this movement, spur-cut sliding gears are used: these have a shift collar integral with the gear. When the driver selects a shift, a **shift fork** or **yoke** engages with a collar on the gear and slides it on the mainshaft until it is brought into mesh with a mating gear on the countershaft. Note that these gears are on a continuous spline with the mainshaft as shown in **Figure 11-8.**

Some older, heavy-duty transmissions used this type of sliding gearshift arrangement for all speed ranges. However, this type of shifting is unsynchronized, and creates difficulty for operators who are trying to precisely match the speed of meshing gears. As a result, grinding and gear clash are a problem, especially with inexperienced drivers. Sliding gearshift mechanisms resulted in rapid wear, gear teeth chipping, and fracture. Sliding gearshift transmissions are considered non-constant-mesh transmissions.

In modern standard truck transmissions, the only gear ratios that generally use spur-type sliding gears are first and reverse. The other gears in the transmission use some form of sliding clutch collar or **synchronizer** to engage each gear range. In addition, all non-sliding gears are in constant mesh with their mating countershaft gears. **Figure 11-9** illustrates a five-speed single countershaft truck transmission that uses a combination of sliding gear and collar shift mechanisms to change ratios.

Powerflow. To properly understand how a transmission functions, it is important to understand powerflows. Powerflow is a term that describes the mechanical torque path through a transmission. For example, the powerflow through a transmission in a specific gear ratio is the sequential routing of the torque flow path from the input shaft and gears to the countershaft and gears and on to the specific gear that is engaged on the mainshaft or output shaft. From time to time we will reference powerflows through some of the transmissions we study, beginning here.

First Gear. Using the callout numbers in **Figure 11-9,** let's walk through how shifts are made in this simple transmission. You should be able to identify callout 18 as the first and reverse sliding gear and callout 17 as its integral shift collar. When the shift fork or yoke (this is not shown, but it engages with the shift collar 17) is moved by the gearshift lever in the cab, it slides the collar and gear

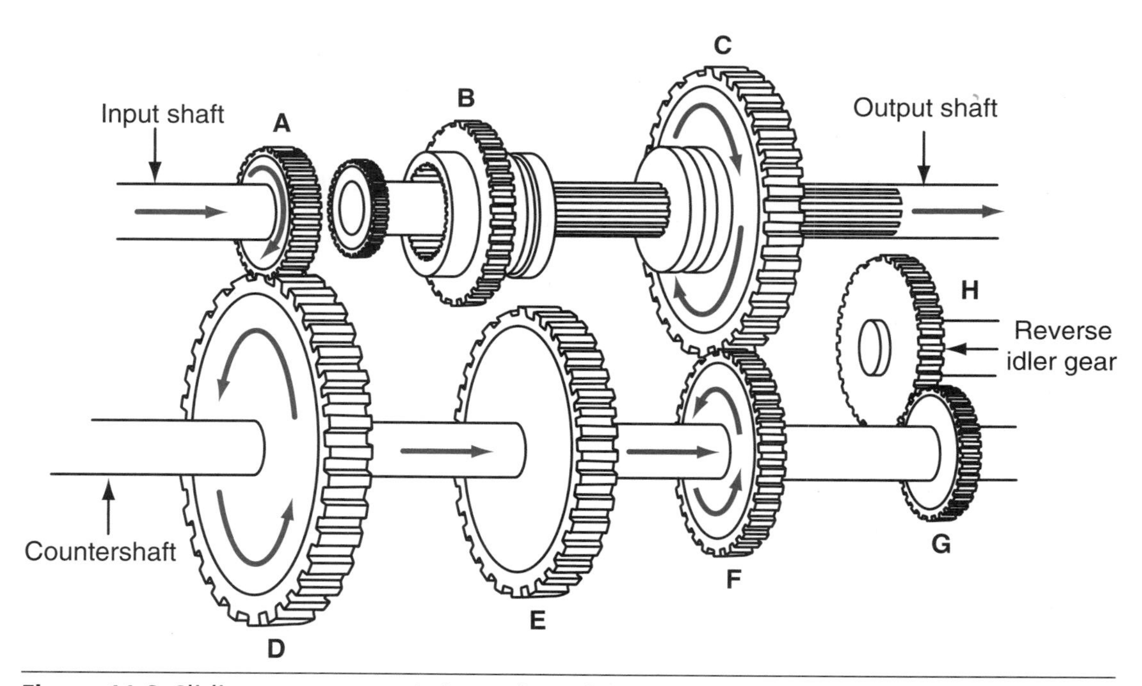

Figure 11-8 Sliding gear transmissions allow the mainshaft gears to slide into mesh with the countershaft.

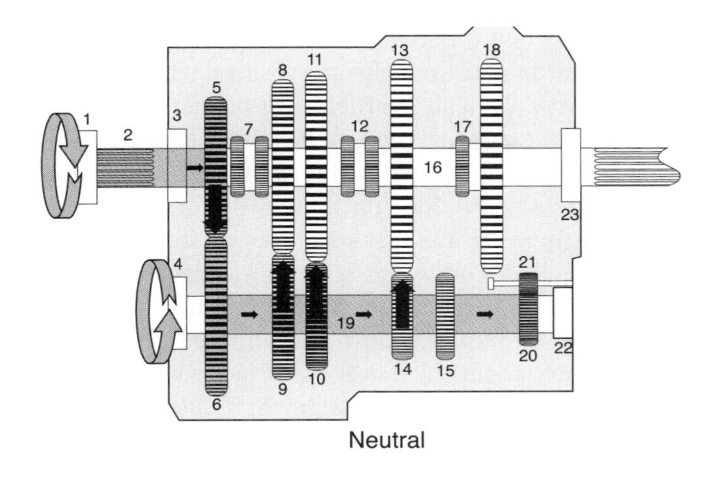

Figure 11-9 Sliding gear (18) slides along mainshaft (16) to engage either first gear (15) or reverse idler gear (21). Gears 5, 8, 11, and 13 are always in mesh with their mated gear.

either to the front or to the rear of the transmission housing. Sliding it forward (to the left) engages the first and reverse sliding gear on the mainshaft with the first gear (callout 15) on the countershaft. This results in directing powerflow through the first gear as shown in **Figure 11-10.** The torque path flows from the engine flywheel, through the clutch plate splines to the transmission input shaft (2), through the input shaft gear (5) and to the countershaft **driven gear** (6). Powerflow is then transmitted through the countershaft to the first gear (15), and up to the first and reverse sliding gear (18) on the mainshaft (16). Because the first and reverse sliding gear is splined to the mainshaft, powerflow is directed through the mainshaft and out to the vehicle driveline.

Reverse Gear. When reverse is selected, the shift lever is moved from neutral to reverse. The shift fork forces the first and reverse sliding gear backward (to the right) until it engages with the reverse idler gear (21). The reverse idler gear is required to allow the first and reverse sliding gear to rotate in the same direction as the reverse gear (20) on the countershaft. Powerflow (**Figure 11-11**) now runs from the input shaft (2) to the input shaft gear (5) and countershaft driven gear (6), then down the countershaft to gears (20) and (21). From the first/reverse sliding gear (18), torque is transferred to the mainshaft (16) and out to the driveline.

Collar Shift. We can now take a closer look at collar shift operation in the basic transmission first shown in Figure 11-9. This transmission uses collar shifting to control the engagement of second, third, fourth, and fifth gears. In a collar shifting arrangement, all gears on the countershaft are fixed to the countershaft. On the other hand, the mainshaft gears are free to freewheel (float) and do so around either a bearing or bushing. These bearings or bushings are positioned on a nonsplined section of the mainshaft. The mainshaft gears are in constant mesh with their mating countershaft gears. This means they turn at all times when the input shaft is driven.

Study the neutral powerflow path of the five-speed transmission as shown in Figure 11-9. The input shaft

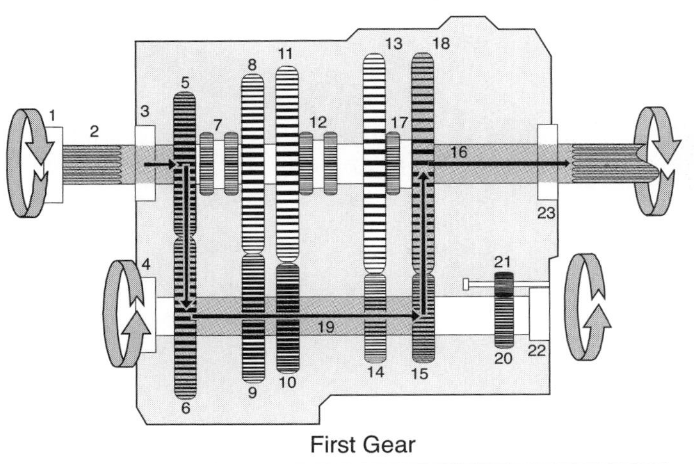

Figure 11-10 First gear powerflow in a typical five-speed transmission.

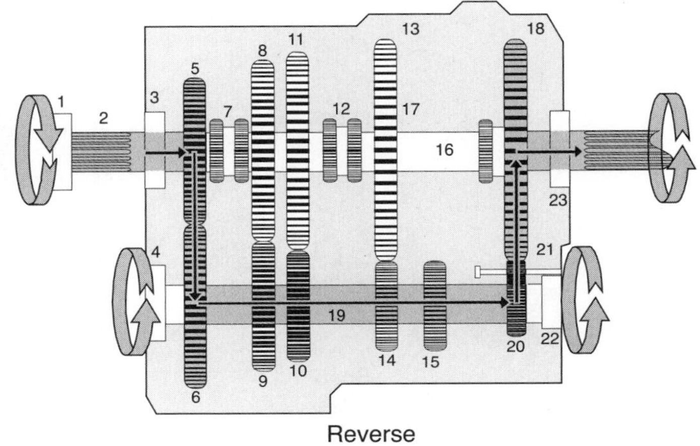

Figure 11-11 Reverse gear powerflow in a typical five-speed single countershaft transmission.

(2) rotates at engine speed at any time the clutch is engaged. The input shaft gear (5) is integral with the input (clutch) shaft so it has to rotate with it. The input shaft gear meshes with the countershaft driven gear (6), so it, the countershaft, and all the gears fixed to the countershaft also have to rotate.

The countershaft gears transfer torque to their mating gears on the mainshaft. But mainshaft gears (8), (11), and (13) all freewheel on the mainshaft on bearings or bushings. Because they are freewheeling, they cannot transmit torque to the mainshaft, so it does not turn. Therefore, there is no output to the driveline.

To enable torque transfer to the mainshaft, one of the freewheeling mainshaft gears must be locked to it. This can be accomplished in a number of ways, and one method uses a shift collar and shift gear. The shift gear is internally splined to the mainshaft at all times. The shift collar is splined to the shift gear. In **Figure 11-12,** you can see that the main gears have a short, toothed hub. The teeth on the main gear hub align with the teeth on the shift gear. The internal teeth of the shift collar mesh with the external teeth of the shift gear and hub.

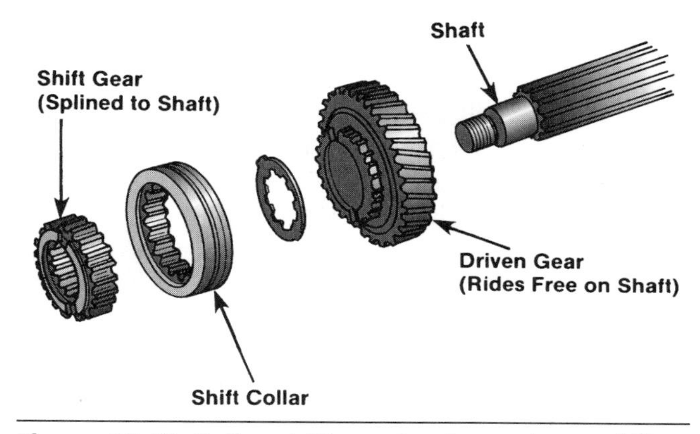

Figure 11-12 Collar shift mechanism using a shifter gear, shift collar, splined hub, and hub-driven gear.

When a given speed range is not engaged, the shift collar simply rides on the shift gear. When the driver shifts to engage that speed range, the shift fork moves the shift collar and slides into mesh with the teeth of the main gear hub. At this moment, the shift collar rides on both the shift gear and main gear hub, locking them together. Power can flow from the main gear to the shift gear, to the mainshaft, and out to the driveshaft.

A second, more common method of locking main gears to the mainshaft does not use a shift gear. Instead, the shift collar is splined directly to the mainshaft. This shift collar, (also called a clutch collar or a sliding clutch), is designed with external teeth. These external teeth mesh with the internal teeth in the main gear hub or body when that speed range is engaged. Most shift collars or sliding clutches are positioned between two gears so they can control two speed ranges, depending on the direction in which they are moved by the shift fork (**Figure 11-13**).

Second Gear Powerflow. To shift from first to second gear, the operator pushes the clutch pedal down to disengage the clutch. The gear select lever is then moved to neutral, disengaging the first and reverse gear (18). Moving the shift lever from neutral to the second gear position causes the second and third shift collar (or sliding clutch) to move (12) back toward the second gear (13) on the mainshaft. The shift collar engages second gear and locks it to the mainshaft. Power flows from (2) to (5) and (6), along the countershaft to (14).

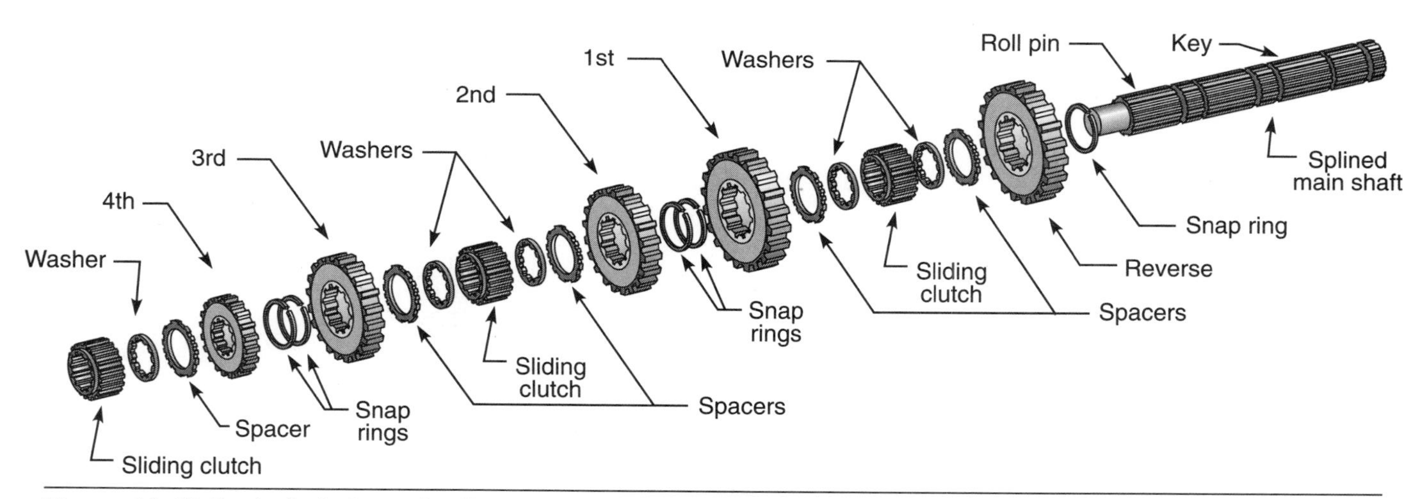

Figure 11-13 Exploded view of a five-speed mainshaft, showing the relative position of the sliding clutches and gears. One sliding clutch controls first and reverse shifts, another controls second and third shifts, and a third controls fourth and fifth gear shifts. Note that the input drive gear is not shown.

Gear (14) drives gear (13), which in turn drives mainshaft output to the driveline (**Figure 11-14**).

Third Gear Powerflow. Once again the driver returns to neutral before shifting to third gear. Moving from neutral to third gear moves the second and third shift collar (or sliding clutch) (12) forward toward the third gear (11), locking it to the mainshaft. Power flows from (2) to (5) and (6), along the countershaft to (10), up to (11), through the shift collar (12) to the mainshaft and out to the driveline (**Figure 11-15**).

Fourth Gear Powerflow. After shifting from third to neutral, the neutral to fourth gearshift causes the shifter fork to move the fourth and fifth shift collar (or sliding clutch 7) into mesh with the fourth gear (8). Power now flows from (2) to (5) and (6), along the countershaft to (9), through (8) and (7) to the mainshaft, and out to the driveline (**Figure 11-16**).

Fifth Gear Powerflow. Shifting from fourth to neutral, then from neutral to the fifth gear causes the shifter fork to move the fourth and fifth shift collar (or sliding clutch 7) into mesh with the input shaft gear (5). This locks the input shaft (2) directly to the mainshaft (16). Input and output speeds are the same. So a 1:1 gear ratio or **direct drive** is established. The powerflow (**Figure 11-17**) is from (2) to (5) through (7) to the mainshaft and out. In fifth gear the clutch shaft gear is still meshed with the countershaft driven gear (6). This means that the countershaft and its gears are all turning. The mainshaft gears (8), (11), and (13) are also

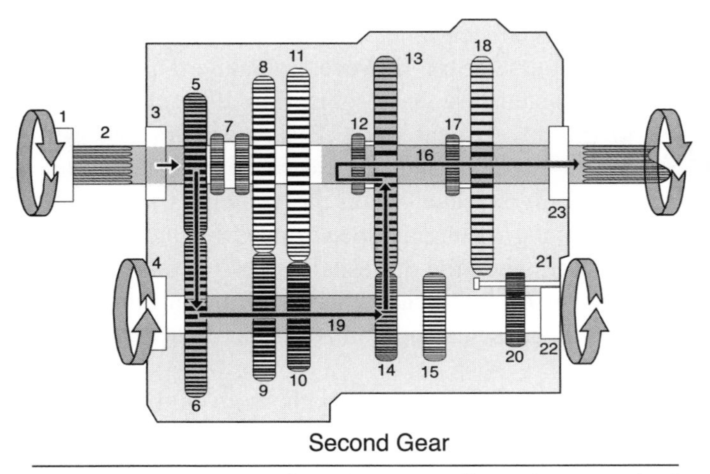

Figure 11-14 Second gear powerflow in a typical five-speed single-countershaft transmission.

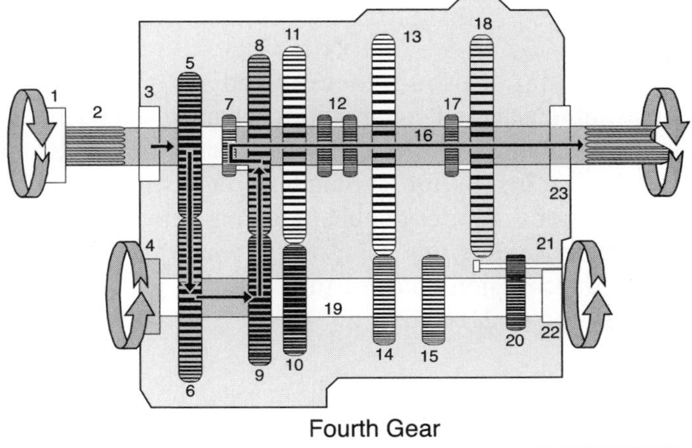

Figure 11-16 Fourth gear powerflow in a typical five-speed single-countershaft transmission.

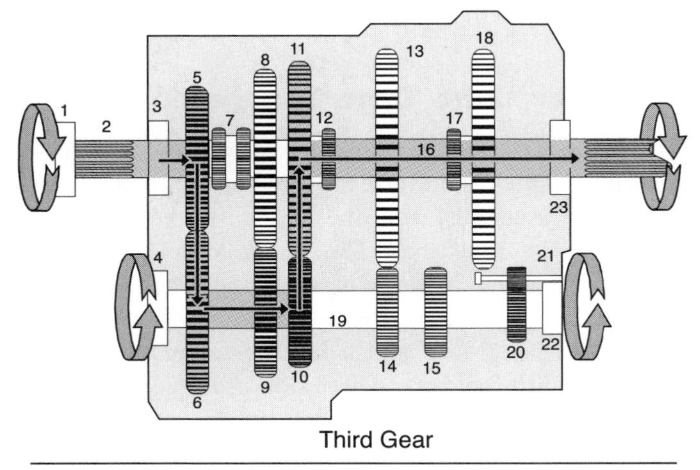

Figure 11-15 Third gear powerflow in a typical five-speed single-countershaft transmission.

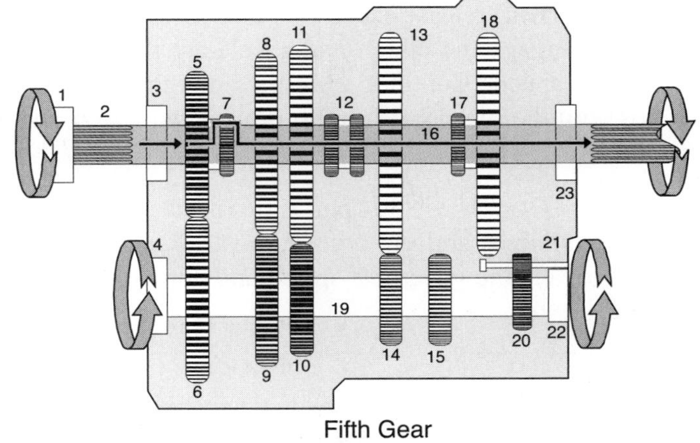

Figure 11-17 Fifth-gear powerflow in a typical five-speed single-countershaft transmission.

freewheeling on the mainshaft, but have no effect on the powerflow.

Take some time to study the powerflow schematics shown in Figures 11-14 through 11-17. Knowing the powerflow paths through the transmission you are working on can make diagnosing and a failure analysis much easier.

Tech Tip: As explained in Chapter 10, a clutch brake is used to stop gear rotation to complete a shift into first or reverse when the vehicle is stationary. The clutch brake is actuated by depressing the clutch pedal completely to the floor. For normal upshifts and downshifts, only partial disengagement of the clutch is needed to break engine torque, in other words, it is not necessary to drive the clutch pedal to the floor to shift.

Operating a Non-Synchronized Transmission. The sliding gear and collar shift mechanisms described in the preceding sections are used on non-synchronized standard transmissions. Shifting non-synchronized transmissions requires double-clutching between gears. We briefly outline this using the example of a seven-speed transmission coupled to an engine with a governed speed of 2100 rpm.

Upshifting. To move away from a standing start with the engine at idle, depress the clutch and move the shift lever to the first gear position. Release the clutch and accelerate the engine to governed speed. When the tach reads governed speed, depress the clutch, release the accelerator pedal, and shift to neutral. Next, engage the clutch and allow engine speed to drop approximately 750 rpm. Then, release the clutch and shift to second gear. Re-engage the clutch and accelerate back up to governed speed. Continue shifting through the gear ratios until the seventh gear ratio is achieved.

Tech Tip: The 750-rpm drop used in the description of shifting procedure varies according to engine-governed speed and torque-rise profile.

Downshifting. To progressively downshift from seventh gear, allow engine speed to drop approximately 750 rpm before depressing the clutch and moving the shift lever to neutral. Now re-engage the clutch and accelerate engine rpm to governed speed. Depress the clutch and move the shift lever into the sixth gear position. Now re-engage the clutch. Continue downshifting in this manner.

SYNCHRONIZED SHIFTING

A synchronizer performs two important functions. Its main function is to bring components that are rotating at different speeds to a common or synchronized speed. In a transmission, a synchronizer ensures that the mainshaft and the mainshaft gear about to be locked to it are both rotating at the same speed. The second function of the synchronizer is to actually lock these components together. If the driver and the synchronizer are both doing their jobs, the result is a clash-free shift. Most standard transmissions having five to seven forward speeds are completely synchronized except for first and reverse gears. However, in many transmissions today, all speeds are synchronized to allow downshifting to the lowest gear ratio. The reason for not synchronizing the first and reverse gears is that they are typically only selected when the vehicle is stationary. Synchronizing of engine speed and driveline speed is not required when the driveshaft is not turning; in fact, a synchronizer could cause hard shifting in these gear positions because a synchronizer needs gear rotation to do its job.

Many transmissions use both main and auxiliary gearing to provide a large number of forward ratios. In these transmissions, synchronizers are normally used only in the two- or three-speed auxiliary gearing. Block, or cone, and pin synchronizers are the most common in heavy-duty applications. Both block and pin-type synchronizers are often referred to as sliding clutches.

Block or Cone Type Synchronizers

Figure 11-18 illustrates a block-type (or cone) synchronizer. The synchronizer sleeve surrounds the synchronizer assembly and meshes with the external splines of the clutch hub. The clutch hub is splined to the transmission output (main) shaft and is held in position by two snap rings. The synchronizer sleeve has a small internal groove and a large external groove in which the shift fork rests. Three slots are equally spaced around the outside of the clutch hub. Inserts fit into these slots and are able to slide freely back and forth. These inserts are designed with a ridge in their outer

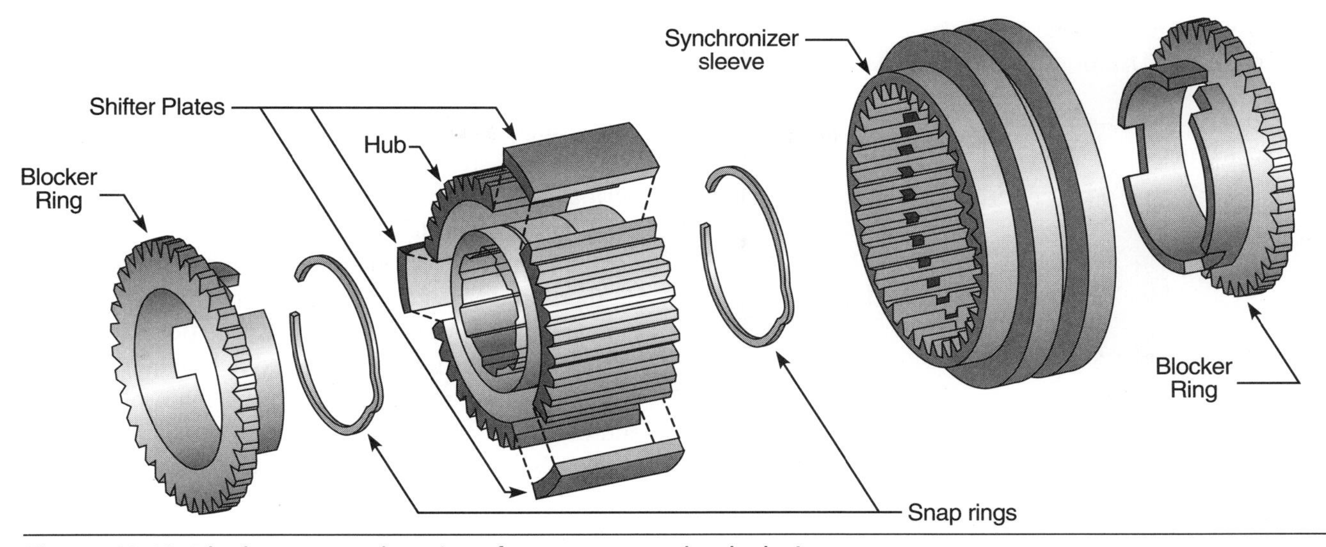

Figure 11-18 Block-type synchronizer features cone-clutch design.

surface. Insert springs hold the ridge in contact with the synchronizer sleeve internal groove.

The synchronizer sleeve is precisely machined to slide on the clutch hub smoothly. The sleeve and hub sometimes have alignment marks to assure proper indexing of their splines when assembling to maintain smooth operation.

Brass or bronze synchronizing blocker rings are positioned at the front and rear of each synchronizer assembly. Each blocker ring has three notches equally spaced to correspond with the three inset notches of the hub. Around the outside of each blocker ring is a set of beveled dogteeth, which are used for alignment during the shift sequence. The inside of the blocker ring is shaped like a cone, as in the cone-type clutch. This coned surface is lined with many sharp grooves.

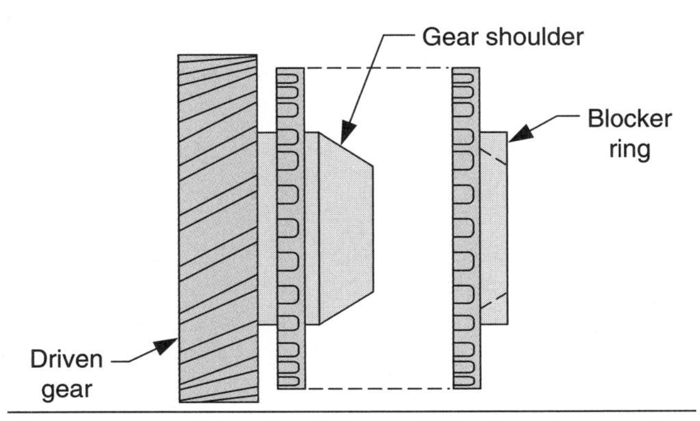

Figure 11-19 Gear shoulder and blocker ring mating surfaces as a cone-type clutch.

The cone of the blocker ring makes up one-half of a cone clutch assembly. The second or mating half of the cone clutch is part of the gear to be synchronized. As you can see in **Figure 11-19,** the shoulder of the main gear is cone-shaped to match the blocker ring. The shoulder also contains a ring of beveled dog teeth designed to align with the dogteeth on the blocker ring.

Operation. In a typical five-speed transmission, two synchronizer assemblies are used. One controls second and third gear engagements; the other synchronizes controls fourth and fifth gear shifting.

When the transmission is in first gear, the second/third and fourth/fifth synchronizers are in their neutral position (**Figure 11-20A**). All synchronizer components are now rotating with the mainshaft at mainshaft speed. Gears on the mainshaft are meshed with their countershaft partners and are freewheeling around the mainshaft at various speeds.

To shift the transmission into second gear, the clutch is disengaged and the gearshift lever is moved to the second gear position. This forces the shift fork on the synchronizer sleeve backward toward the second gear on the mainshaft. As the sleeve moves backward, the three inserts are also forced backward because the insert ridges lock the inserts to the internal groove of the sleeve.

The movement of the inserts forces the back blocker ring, coned friction surface against the coned surface of the second gear shoulder. When the blocker ring and gear shoulder come into contact, the grooves on the blocker ring cone-clutch cut through the lubricant film on the second-speed gear shoulder and a metal-to-metal contact is made (**Figure 11-20B**). The contact generates substantial friction and heat. This is one reason bronze or brass blocker rings are used. A non-ferrous metal such as bronze or brass minimizes wear on the hardened steel gear shoulder. If similar metals were used, wear would be accelerated. This frictional coupling is not strong enough to transmit loads for long periods. But as the components reach the same speed, the synchronizer sleeve can now slide over the external dogteeth on the blocker ring and then over the dogteeth on the second-speed gear shoulder. This completes the engagement (**Figure 11-20C**). Power flow is now from the second-speed gear, to the synchronizer sleeve, to the synchronizer clutch hub, to the main output shaft, and out to the drive shaft.

To disengage the second-speed gear from the mainshaft and shift into third-speed gear, the clutch must be disengaged as the shift fork is moved to the front of the transmission. This pulls the synchronizer sleeve forward, disengaging it from second gear. As the transmission is shifted into third gear, the inserts again lock into the internal groove of the sleeve. As the sleeve moves forward, the forward blocker ring is forced by the inserts against the coned friction surface on the third-speed gear shoulder. Once again, the grooves on the blocker ring cut through the lubricant on the gear shoulder to generate a frictional coupling that synchronizes the gear and shaft speeds. The shift fork can then continue to move the sleeve forward until it slides over the blocker ring and gear shoulder dogteeth, locking them together. Power flow is now from the third-speed gear, to the synchronizer sleeve, to the clutch hub, and out through the mainshaft.

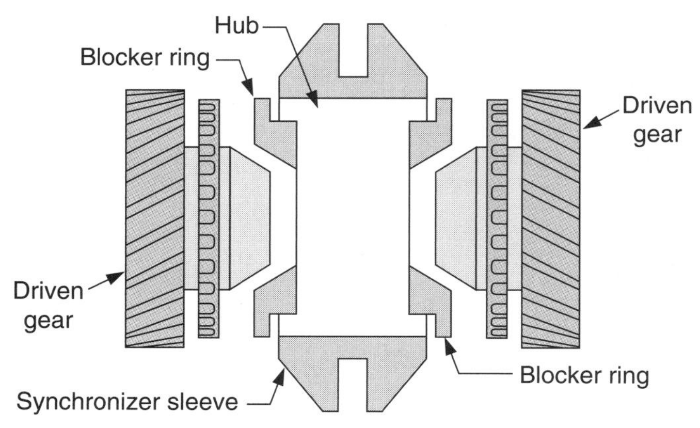

A Synchronizer in neutral position before shift

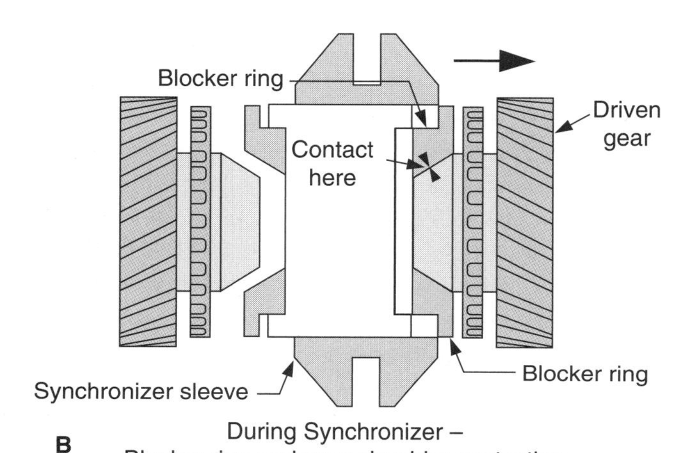

B During Synchronizer – Blocker ring and gear shoulder contaction

C Shift completed – Collar locks driven gear to hub and shaft

Figure 11-20 (A) Block- or cone-type synchronizer in the neutral position. (B) Initial contact of the blocker ring and gear shoulder. (C) Final engagement as the synchronizer sleeve locks the driven gear to the synchronizer hub and shaft.

Pin Synchronizers

The clutch drive gear has both internal and external splines. It is splined to the main (output) shaft and held to it with two snap rings. The sliding clutch gear is loosely splined to the external splines of the clutch drive gear. The sliding clutch gear is made with a flange that has six holes with slightly tapered ends. The shift fork fits into an external groove cut into this flange.

Aluminum alloy blocker rings have coned friction surfaces designed with several radial grooves. Three pins are fastened to each blocker ring, and each pin is machined with two distinctly different diameters. The outer stop synchronizer rings are machined with an internal cone friction surface. The bores of the outer stops have internal teeth that spline to the external teeth on the shoulders of the gears to be synchronized. Finally, the smaller diameter ends of the blocker ring pins rest in the holes of the sliding clutch gear.

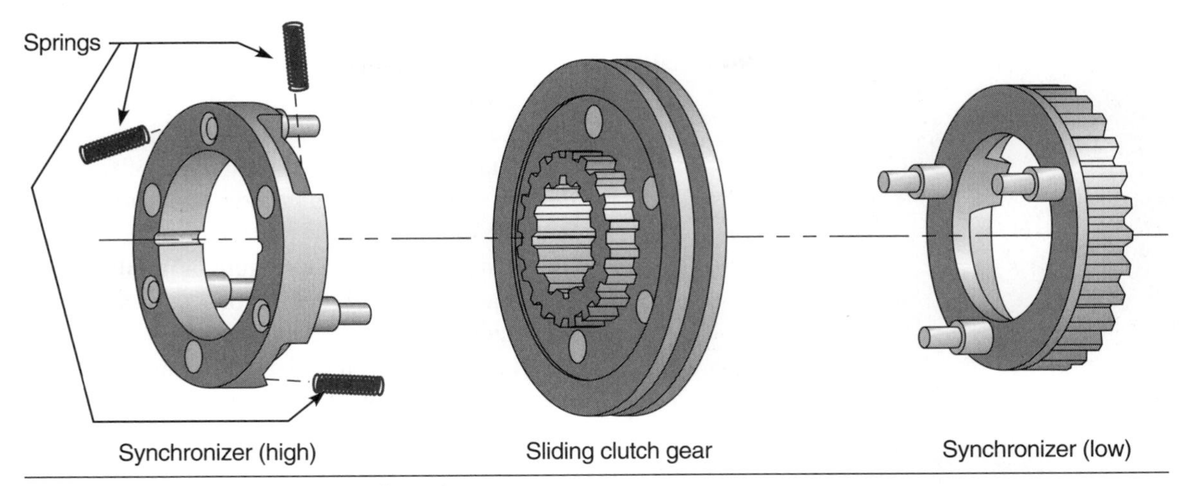

Figure 11-21 A pin-type synchronizer.

Operation. When the pin synchronizer (**Figure 11-21**) is in a neutral (disengaged) position, the clutch drive gear, sliding clutch gear, and the blocker ring all rotate with the main (output) shaft. The gears either side of the synchronizer, as well as the outer rings splined to the gear shoulders, are driven at varying speeds by the countershaft gears. To shift into a pin-synchronized gear, the clutch is disengaged, and the shift lever is moved to the desired speed ratio. This moves the shift fork and the sliding clutch gear toward the gear to be synchronized. The three holes in the sliding clutch gear butt against the side of the larger diameter ends of the three pins on the blocker ring. This forces the blocker ring into the adjoining outer stop ring. The two coned friction surfaces of the blocker and outer rings come into contact. The friction coupling equalizes the turning speed of these components. And because the driven gear is splined to the outer stop ring, it is now turning at mainshaft speed.

As with the block or cone synchronizer, this frictional coupling cannot be held for long. When all components are turning at the same speed, the holes in the sliding clutch gear are able to slide over the larger diameter portion of the pins. The internal teeth of the sliding clutch gear slide over the external teeth of the gear, completing the clutch engagement. To disengage the gear from the shaft, the clutch is disengaged and the shift lever is moved to neutral. This moves the shift fork and sliding clutch gear away from the gear, disengaging the sliding clutch gear teeth from the main gear teeth.

Disc-and-Plate Synchronizers

The disc-and-plate synchronizer uses a series of friction discs alternated with steel plates, not unlike the clutch pack of alternating discs and plates used in many automatic transmissions. The primary difference is that, where automatic transmissions generally use hydraulic pressure to compress the clutch and disc assembly, transmissions using disc-and-plate synchronizers depend upon the physical pressure of the shifter fork during shifting to compress the plates. Various views of the disk-and-plate synchronizer are shown in **Figures 11-22, 11-23,** and **11-24.**

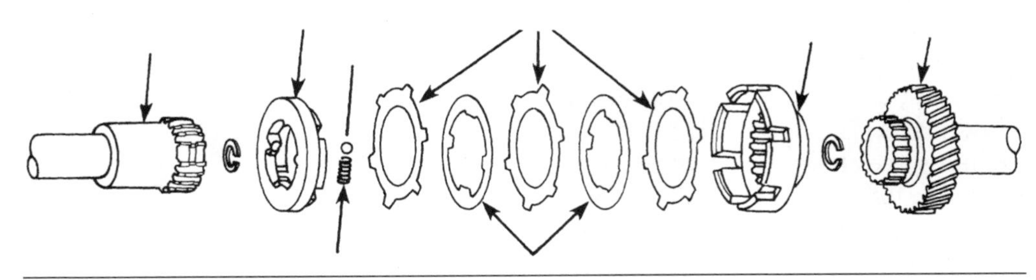

Figure 11-22 Disc-and-plate synchronizer using series of discs and plates. (*Courtesy of Mack Trucks*)

Figure 11-23 Disc-and-plate synchronizer plates with external tangs.

Figure 11-24 Disc-and-plate synchronizer discs with internal teeth.

The discs have external tangs or splines around their outer perimeter. These external splines allow the discs (but not the plates between each disc) to engage with the slots machined inside the synchronizer drum that houses the discs and plates. This means that when the synchronizer drum turns, the clutch discs (with external tangs) turn with it, while the plates (which do not have external splines) do not. The plates do, however, have internal tangs or splines that mate with the blocker ring and input shaft. Because the plates have no external splines and the discs have no internal splines, the only movement that can be transmitted between them comes from clamping friction between the two. In this condition, the synchronizer is in a neutral position, as shown in **Figure 11-25.** If minimal pressure is applied to the disc-and-plate assembly, the amount of friction between the two will be small, as shown in **Figure 11-26** and the amount of movement transferred will be correspondingly small. As more pressure is applied to the disc-and-plate assembly, friction increases and more movement is transferred.

The purpose of the disc-and-plate synchronizer is to match the speed of input gears with output gears during shifting. This occurs when the driver disengages the clutch and moves the gearshift lever, causing the shifter fork to move the synchronizer drum forward. This forward motion causes the discs and plates to rotate the blocker ring around the input shaft until the drive lugs on the blocker ring engage with the outer edge of the detent ball depression cut into the synchronizer gear.

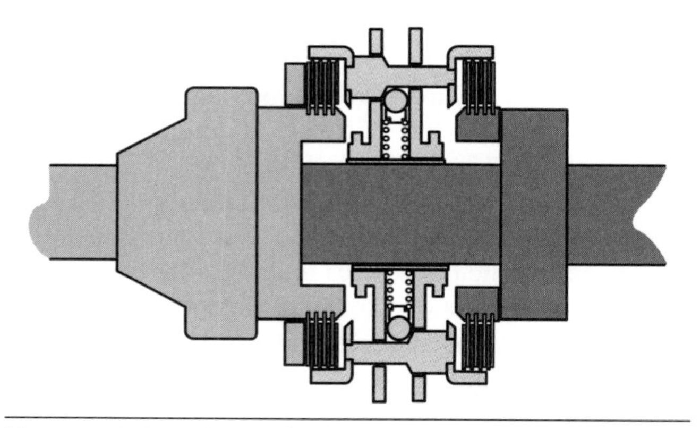

Figure 11-25 Disc-and-plate synchronizer in neutral position.

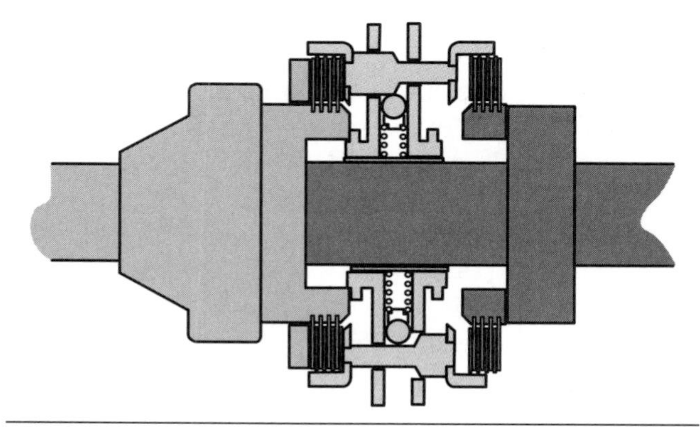

Figure 11-26 Disc-and-plate synchronizer. Initial contact of discs and plates synchronizes mainshaft speed and gear speed and allows meshing of hub gears with internal teeth on gear.

Once the blocker ring has locked into the synchronizer gear, any additional forward motion of the shifter forks will compress the discs and plates together, causing them to rotate at the same speed. The same applies to the rear gear for it to be synchronized to the forward gear as shown in **Figure 11-27.** When the speeds of the input gear and blocker gear are synchronized to the speed of the output gear and both are turning at the same speed, the following should be true:

- Thrust force is released.
- The blocker disengages from its locked position.
- The drum can be moved forward to engage both gears.

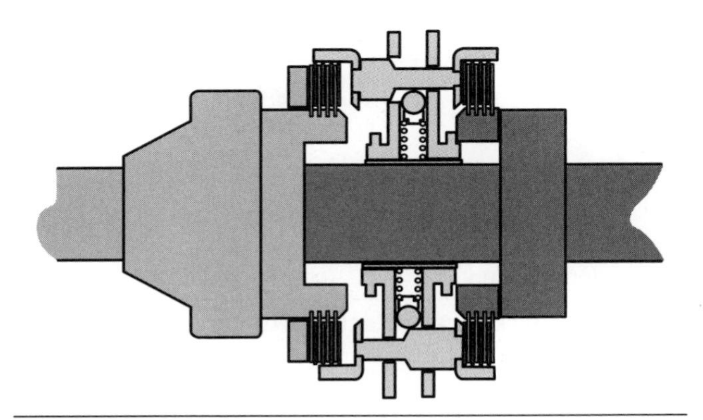

Figure 11-27 Disc-and-plate synchronizer. Initial contact of discs and plates synchronizes mainshaft speed and gear speed and allows meshing of hub gears with internal teeth on gear for opposite gear.

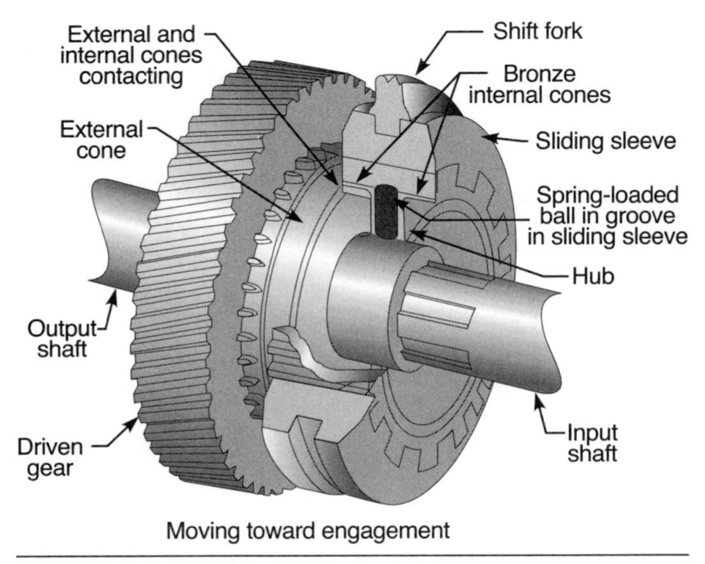

Figure 11-28 A plain synchronizer moving toward engagement.

Plain Synchronizers

A plain synchronizer uses fewer working parts and is similar in design to a block synchronizer (**Figure 11-28**). As is typical with most synchronizers, the hub is internally splined to the mainshaft. Mounted on the hub, and also connected to it, is a sliding sleeve that is controlled by the shift fork movement. In operation, the friction generated between the hub and the gear tends to synchronize their speeds. Pressure on the sliding sleeve prevents the sleeve from moving off the hub and also prevents the teeth on the sleeve from engaging the gear teeth until sufficient pressure has been put on both friction areas to synchronize the components. When in sync, any further movement of the sleeve will allow the sleeve teeth to engage the gear teeth. **Figure 11-29** shows the synchronizer fully engaged.

Driving Synchronized Transmissions

A synchronizer simplifies shifting and helps the driver make clash-free shifts. In operation when a shift is required, the driver should disengage the clutch and move the shift lever toward the desired gear position. When the synchronizer ring makes initial contact with the desired gear, the blockers automatically prevent the shift collar from completing the shift until the gear and mainshaft speeds are matched. At that time, the blocker neutralizes automatically and a clash-free shift is the result. Steady pressure on the shift lever helps the synchronizer do its job quickly. When synchronized, the shift lever moves into gear smoothly and easily.

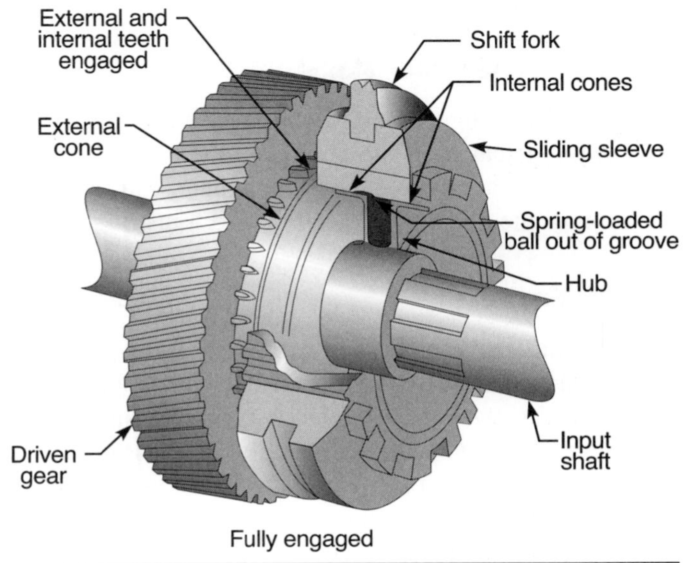

Figure 11-29 A plain synchronizer fully engaged.

If the driver jabs at the synchronizer or teases it by applying and releasing pressure, the synchronizer cannot do its job. It can take up to one or two seconds to match asynchronous speeds, and gentle but constant shift lever pressure produces faster synchronization action. It is possible to override a blocker if the lever is forced into gear. Forcing a transmission into gear defeats the purpose of the synchronizer and can cause gear clash, and ultimately damage the transmission.

When the vehicle is stationary, fully depress the clutch, move the shift lever into first gear, and accelerate to an engine rpm that allows sufficient vehicle momentum to select the next higher gear. Drivers should be aware of the engine torque-rise rpm values because this defines the upper and lower rpm shift limits. Using a progressive shift technique, by not running engine rpm up to governed speed on each shift, saves fuel.

When downshifting, a similar procedure is used. Consider shifting from top (fifth) gear to fourth gear. As the shift point is approached (normally about 100 rpm over the shift point), disengage the clutch and move the shift lever with a steady, even pressure toward fourth gear. The synchronizer will pick up fourth gear, get it up to driveline speed, and allow a clash-free shift from fifth to fourth. After the shift, re-engage the clutch and at the same time accelerate the engine to keep the vehicle moving at the desired speed. If further downshifts are required, continue in a similar manner.

When downshifting into first gear, it is not synchronized and will require a double-clutch operation to complete a clash-free shift. It is possible to double-clutch on all shifts if desired. This aids the synchronizer in its function by getting the engine speed and driveline matched.

SHIFT HOUSING ASSEMBLY

A shift housing assembly consists of a tower assembly and shift bar housing. In some applications some additional linkage is required, this depends on the orientation of the cab with the engine and transmission.

Shift Tower

The shift tower assembly can be direct-actuated as shown in **Figure 11-30.** The shift tower assembly is installed directly on top of the shift bar housing. The gear lever actuating stub floats in clevis recesses in the center of the shift yokes when the transmission is in neutral. The shift yokes are bolted to the shift rails. When a shift is made, the gear lever finger moves the shift rail either forward or backward. Because the shift lever pivots in the tower, the gear lever finger movement is always opposite to the direction the driver moves the lever.

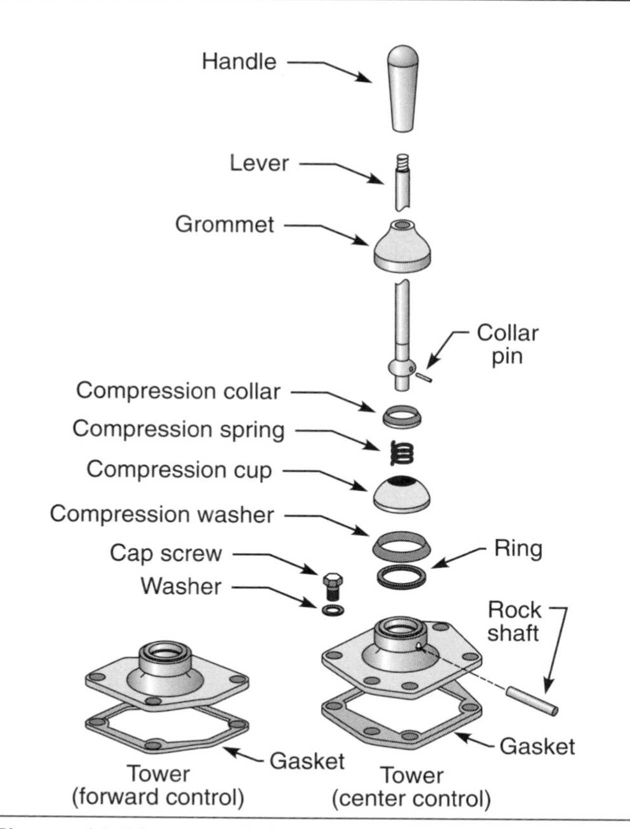

Figure 11-30 Typical direct shifter lever and tower assembly.

Shift Bar Housing

The shift bar housing, also known as the shift rail assembly, houses the components required to convert gear stick movement into range gearshifts within the transmission. There are some design variations between manufacturers, but what you see in **Figure 11-31** is typical. This figure is an exploded view of a shift bar housing used in a seven-speed transmission. **Figure 11-32** illustrates the shift bar housing assembly in a commonly used main five-speed gearbox.

Operation

The gearshift lever in a direct shift tower arrangement fits into slots formed by the shift bracket or yoke blocks. If you look at Figures 11-31 and 11-32, the shift brackets or blocks can either be clamped to a shift rail or integral with shift fork or yoke. The shift bars or rails are designed to slide in the shift bar housing when actuated by the gear lever. The shift forks or yokes are clamped or bolted to the shift bars. When the gearshift lever is moved, it slides the shift

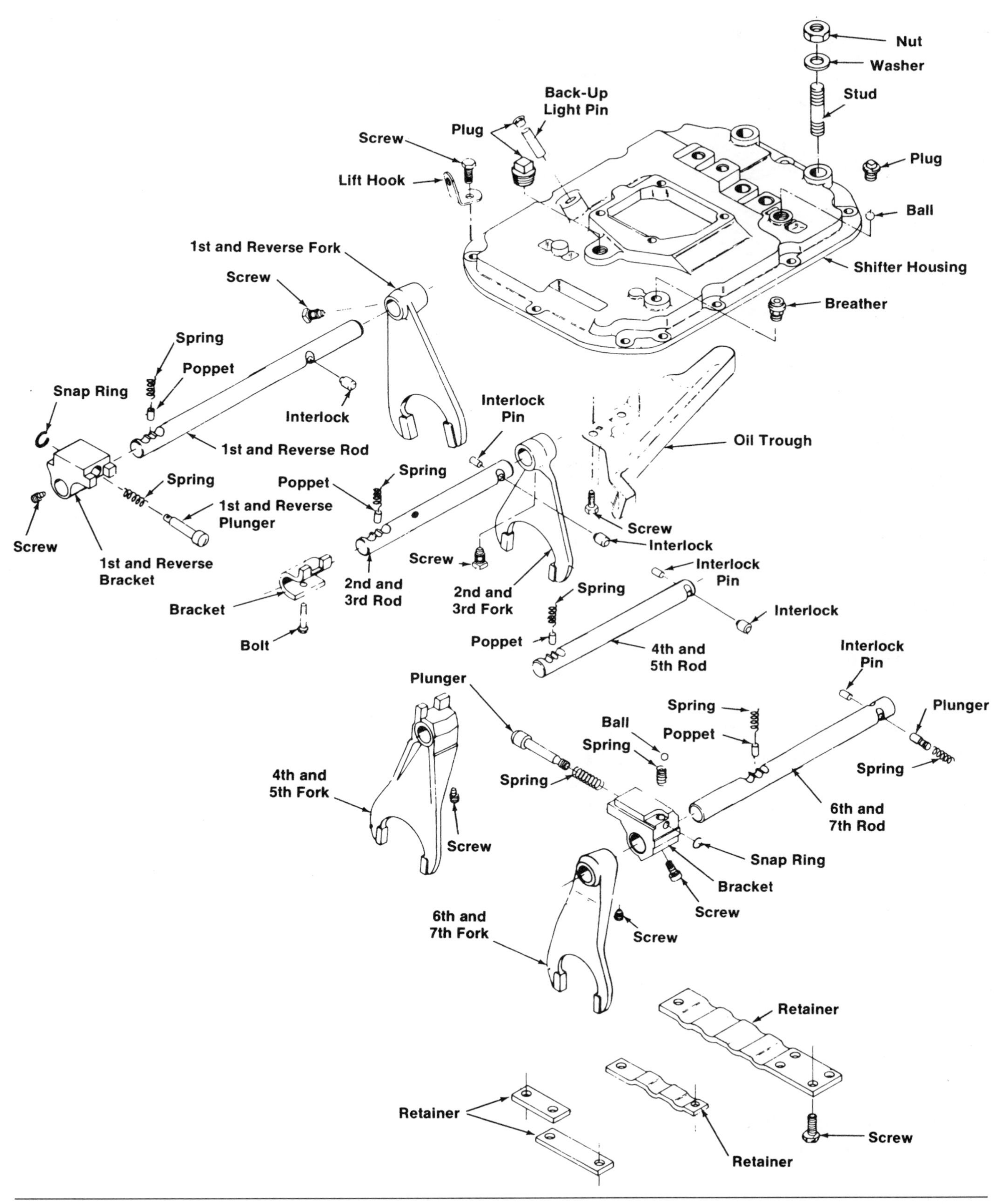

Figure 11-31 Shift housing control components for a seven-speed transmission. (*Courtesy of Eaton Corporation*)

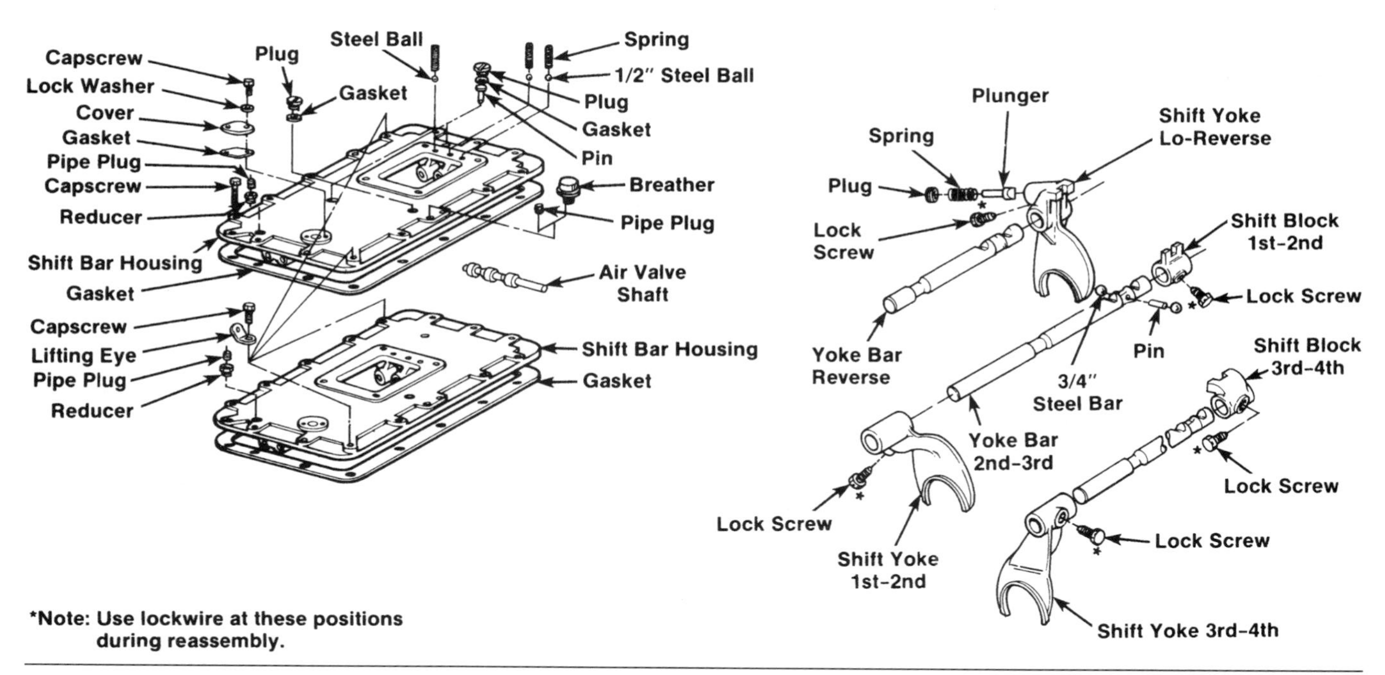

Figure 11-32 Shift housing control components for a five-speed transmission. (*Courtesy of Roadranger Marketing. One great drive train from two great companies—Eaton and Dana Corporation*)

bars and thereby moves the yokes; the shift yoke fits over a shift collar or synchronizer so an actual shift can be made.

After a shift has been made, the shift bar must be held in position so that the selected gear ratio is maintained. To hold the selected gear and prevent unwanted rail movement, the shift bars are machined with recesses; these recesses align with a detent mechanism. The detent mechanism consists of a spring-loaded detent steel ball or poppet: the spring loads the steel ball into the recess in the shift bar. The detent ball holds the shift bar in position and prevents unwanted movement of the other bars. When the driver makes a shift, he must overcome the spring pressure of the detent ball to move the shift bar.

Tech Tip: When troubleshooting a transmission complaint of slipping out of gear, one of the first things you should check is the detent assemblies. Broken springs and seized detent balls can result in unwanted shift rail movement.

FORWARD/REVERSE SHUTTLED TRANSMISSIONS

As discussed earlier, some vehicles require the same number of forward speeds as reverse speeds to accommodate faster cycle times between loads within a loader or bulldozer application. We will look at a wheel loader/backhoe application where this is commonly used. Most manufacturers accomplished this with a forward/reverse shuttle incorporated with a manual transmission (**Figure 11-33**). It should also be noted that this application incorporates a torque converter in place of a clutch. This is to allow faster directional changes, and the torque converter acts as a cushioning device when direction changes are made while still moving in the opposite direction or when an operator forces the machine into a dirt pile with the loader bucket.

Note: It is recommended that the vehicle come to a complete stop before a direction change is made.

Forward/Reverse Shuttle

A forward/reverse shuttle consists of two clutch packs made up of multiple discs and plates splined to an input shaft from the torque converter. The clutch packs are hydraulically engaged and spring released by a piston incorporated within the clutch pack assembly. Either one of the clutch discs or clutch plates can be splined to the input shaft to relay input torque from the engine via the torque converter. Two floating gears (forward and reverse gear) on the input shaft are splined to the opposite disc or plate within each clutch pack

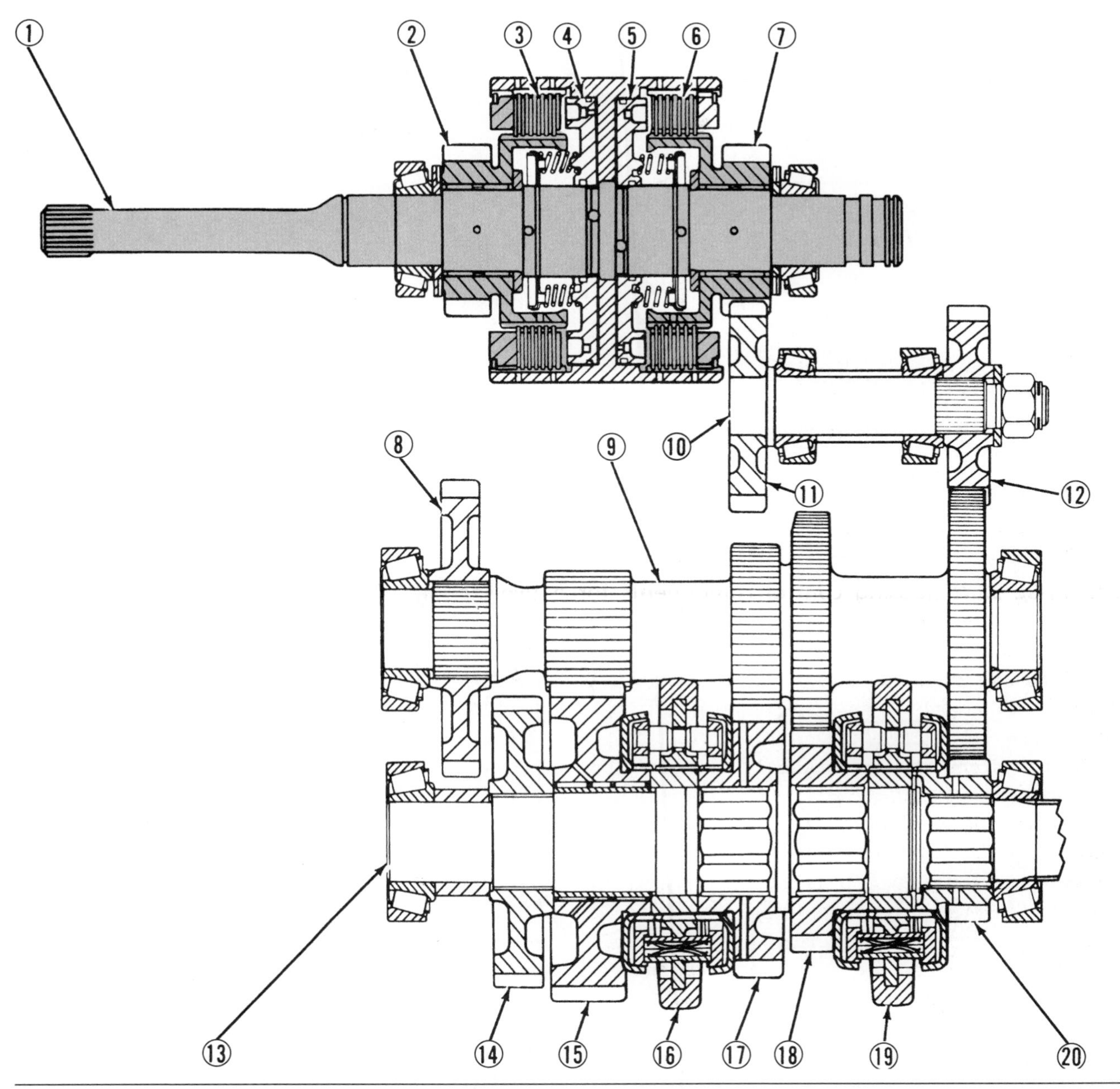

Figure 11-33 Wheel loader/backhoe transmission with four-speed manual transmission and forward/reverse shuttle. (*Courtesy of Caterpillar*)

assembly. For our purposes let's say that the discs are splined to the input shaft and the plates are splined to the forward/reverse floating gears.

Because the clutch packs are spring released, the discs and plates are not in contact with each other. When the engine is started, the torque converter provides rotational torque to the input shaft and the clutch discs that are splined to it. The forward/reverse gears and splined clutch plates are stationary and do not rotate. The forward gear is in constant mesh with the countershaft gears, which are in constant mesh with output shaft gears. The reverse gear is in constant mesh with the reverse idler shaft, which is in constant mesh with the countershaft gears, which in turn, are in constant mesh with the output shaft gears. See **Figure 11-34.**

Forward Direction. In a typical loader/backhoe application the transmission is a manually operated, four-speed, constant-mesh, synchronized, single-countershaft transmission. When the forward/reverse directional control lever (**Figure 11-35**) is selected to the forward drive position, a solenoid controlled valve directs oil pressure to the forward clutch pack piston. It squeezes together the plates and discs that are splined

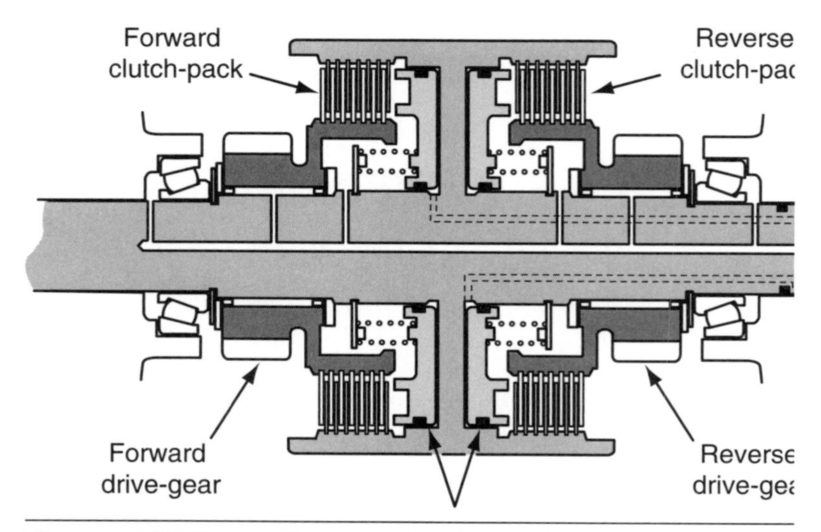

Figure 11-34 Forward/reverse shuttle consisting of two clutch packs to drive either the forward drive-gear or reverse drive-gear.

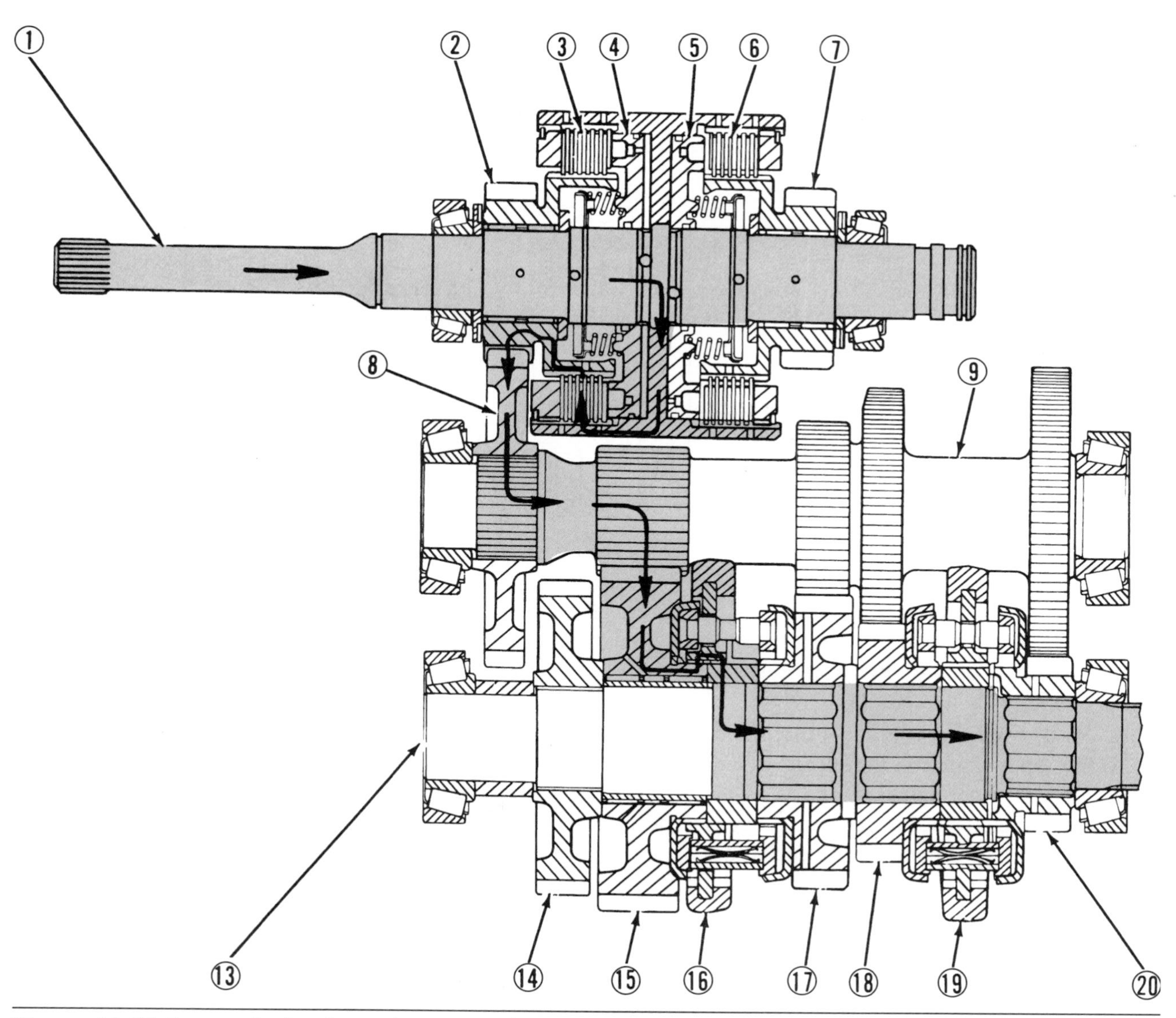

Figure 11-35 Forward clutch pack engaged will drive countershaft and main output shaft gears. (*Courtesy of Caterpillar*)

to the input shaft and forward gear. This causes the forward gear to rotate at the same speed as the input shaft. Because the forward gear is in constant mesh with the transmission countershaft and output shaft gears, they will all rotate within the transmission. The four-speed manual transmission section will drive the vehicle at a selected speed.

Note: No output will be achieved unless a gear/speed selection is made within the manual transmission section. The manual transmission gear speed should be selected before the forward/reverse shuttle. Pre-selecting forward or reverse will cause clashing of the manual transmission synchronizers and gears when attempting to engage the manual section of the transmission. This is equivalent to engaging first gear in a manual transmission without the clutch pedal depressed.

Speed changes can be made on the fly by depressing and holding the clutch disconnect switch located on the shifter handle/knob. The clutch disconnect switch, when depressed, temporarily de-energizes the forward control solenoid valve, causing the forward clutch pack to release. This will place the forward/reverse shuttle in a non-driving neutral condition. Because the transmission is synchronized, the operator can shift to the desired gear or speed without any clashing of gears.

Reverse Direction. When the forward/reverse directional control lever is selected to the reverse drive position, the solenoid-controlled valve directs oil pressure to the reverse clutch pack piston. It squeezes together the plates and discs that are splined to the input shaft and reverse gear. This causes the reverse gear to rotate at the same speed as the input shaft. Because the reverse gear is in constant mesh with the reverse idler shaft and gear, transmission countershaft, and output shaft gears, they will all rotate in the opposite direction within the transmission. Because the input drive is reversed through a separate reverse idler shaft, all the transmission speeds will drive in the reverse rotation (**Figure 11-36**). Gear or speed changes can be achieved in reverse direction in the same manner as the forward speeds. The clutch release switch will neutralize the reverse control solenoid valve while depressed.

Note: During normal working condition the manual transmission section is kept at the desired speed/gear to suit the working condition. The operator shuttles between forward and reverse to achieve directional changes of the machine.

Four-Wheel Drive Option

Many loader/backhoe machines are equipped with four-wheel drive. This gives the machines more traction for loading or for working in very muddy or slippery conditions. These machines are equipped with a front drive axle as well as the regular rear drive axle. Engine torque must now be simultaneously transferred to the front drive axle and rear drive axle at the same transmission speed or gear ratio. This is accomplished by using a transfer gear to equally transfer the engine torque through the transmission to the front and rear drive shafts. See **Figure 11-37.**

Four-wheel drive is not always necessary, so the option to engage four-wheel drive is made available by implementing a clutch pack within the **transfer case** to split the drive between front and rear drive axles.

Transfer Gear and Clutch Pack. The transfer gear is in constant mesh with the transmission output shaft and will rotate in either forward or reverse speed, dictated by the forward/reverse shuttle and manual transmission selected gear. The rear drive axle is the primary drive for the vehicle in this application, so the transfer gear is directly splined to the rear drive yoke and drive shaft.

Note that the transfer gear provides the output from the transmission to the front drive axle and the original transmission output shaft relays the powerflow to the transfer gear. The transfer gear and shaft, together with the main output shaft to the rear drive axle, provide output torque to both drive axles. Refer to Figure 11-37.

A multiple-disc clutch pack made up of friction discs and plates similar to the forward/reverse shuttle is implemented to operate the transfer gear and front output shaft. The friction discs are splined to the transfer gear, and the transfer gear is in constant mesh with the output shaft gear of the rear drive axle. The plates are splined to a hub and the output shaft to the front drive axle. A solenoid-controlled valve, activated by a switch in the cab, directs oil to pressurize or drain the clutch pack when two- or all-wheel drive is selected.

In two-wheel drive mode, powerflow is transmitted to the transfer gear and the output shaft to the rear

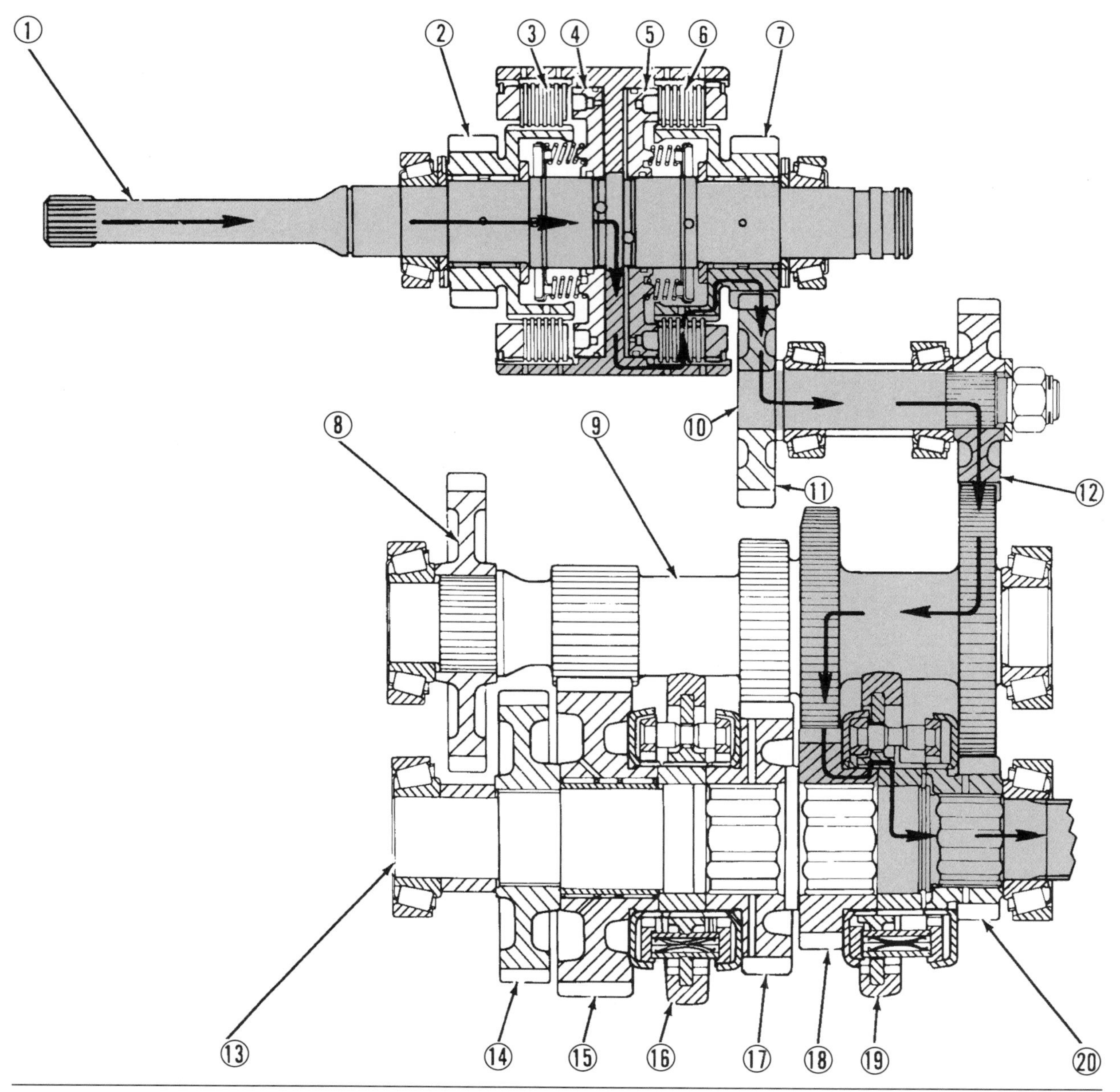

Figure 11-36 Reverse clutch pack engaged will drive reverse idler shaft and gear, countershaft, and main output shaft gears in opposite direction. (*Courtesy of Caterpillar*)

drive axle. The friction discs within the clutch pack will rotate with the output shaft but will not transfer any torque to the front axle, because the clutch pack is disengaged. The clutch-pack plates will rotate at the same speed as the friction discs and transfer gear and shaft, provided that the front and rear drive shafts are rotating at the same speed (all wheels on the ground).

The only time there is a speed differential between the clutch-pack discs and plates is when the vehicles is being propelled forward or reverse with the front wheels off the ground. With the front wheels off the ground, they are not forced to rotate between the friction of the ground and the tire. The clutch plates splined to the forward driveshaft will then remain stationary, and the friction discs splined to the transfer gear and rear output shaft will rotate within the clutch pack. Because there is no synchronization of gears necessary to engage the front-wheel drive axle, all-wheel drive mode can be engaged or disengaged

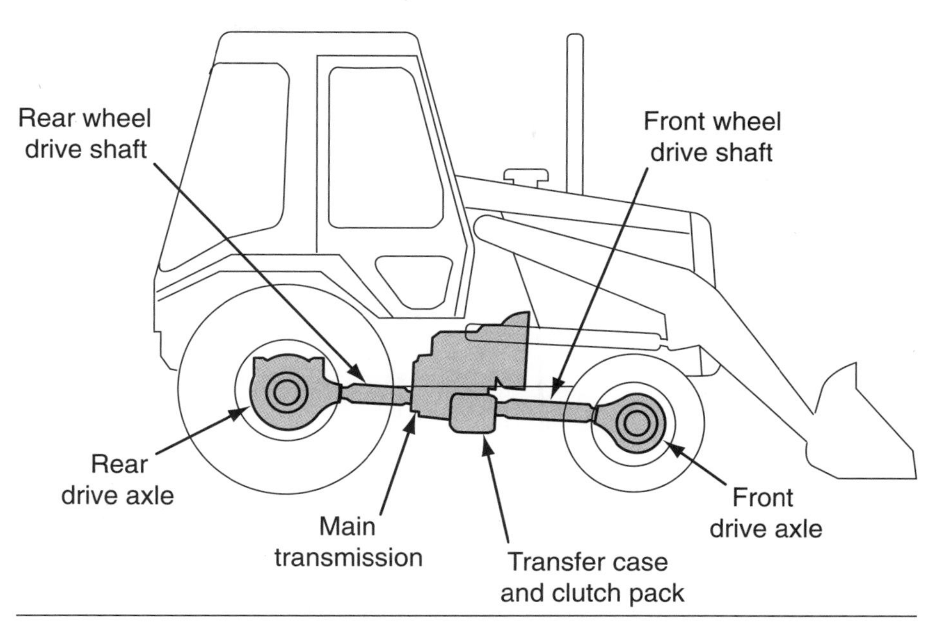

Figure 11-37 Wheel loader with four wheel drive option: Rear drive axle, Rear drive shaft, Main transmission, Transfer gear and clutch pack, Front wheel drive shaft, Front drive axle.

while the vehicle is in motion. See **Figures 11-38** and **11-39.**

There is no speed or gear limitation to switching between two-wheel drive and all-wheel drive. The clutch pack allows a smooth transition between the two modes.

Hydraulic Schematic

Figure 11-40 shows the forward/reverse shuttle hydraulic system with the four-wheel drive option. Oil from the transmission sump is pumped through the transmission circuit using a crescent-type gear pump (20). A screen-type filter is used at the inlet of the pump to contain any large metal fillings that may be drawn into the pump (19). These pumps are typically driven by either the torque converter or the transmission input shaft. In some applications, the pump may be remotely mounted and driven by another driveline source. An in-line filter is used on the outlet side of the pump to contain any smaller particles that may have passed through the suction screen and pump (1). Clean filtered oil is directed to the directional solenoid-controlled selector valve (6) and the AWD (all-wheel drive) solenoid valve (4). In the neutral position, oil passage is blocked at the directional selector valve and oil is forced to open the relief valve (5) and flow toward the torque converter. The torque converter supply oil is also drawn from a restricted orifice located within the relief valve; this continuously allows oil availability to the torque converter. The torque converter inlet relief valve (12) will limit the oil pressure within the torque converter to prevent over pressurizing and damaging the converter. Hot oil from the torque converter (13) is sent to the oil cooler (15) and then directed through the transmission lube circuit to cool and lubricate the bearings. This oil will find its way back to the sump (18) where the transmission pump will start the process over again.

When the operator selects either forward or reverse, the directional-selector solenoid valve shifts to either the forward or reverse position. Transmission pump oil then travels through the selector valve to the appropriate forward or reverse clutch pack (7) and (10) and engages the drive to the transmission countershafts. The relief valve will limit maximum clutch pressure and relay the oil to the torque converter once the clutch pack has been satisfied. When the operator chooses to travel in the opposite direction by shifting the F-N-R (forward-neutral-reverse) selector valve to the reverse position, the direction-selector solenoid valve will energize and position the valve so that pump oil will be directed to the reverse clutch pack and simultaneously drain the forward clutch pack to the transmission sump. When the F-N-R selector is placed in the neutral position, both the forward and reverse clutch packs are allowed to drain to the sump, disconnecting the drive to the transmission countershafts.

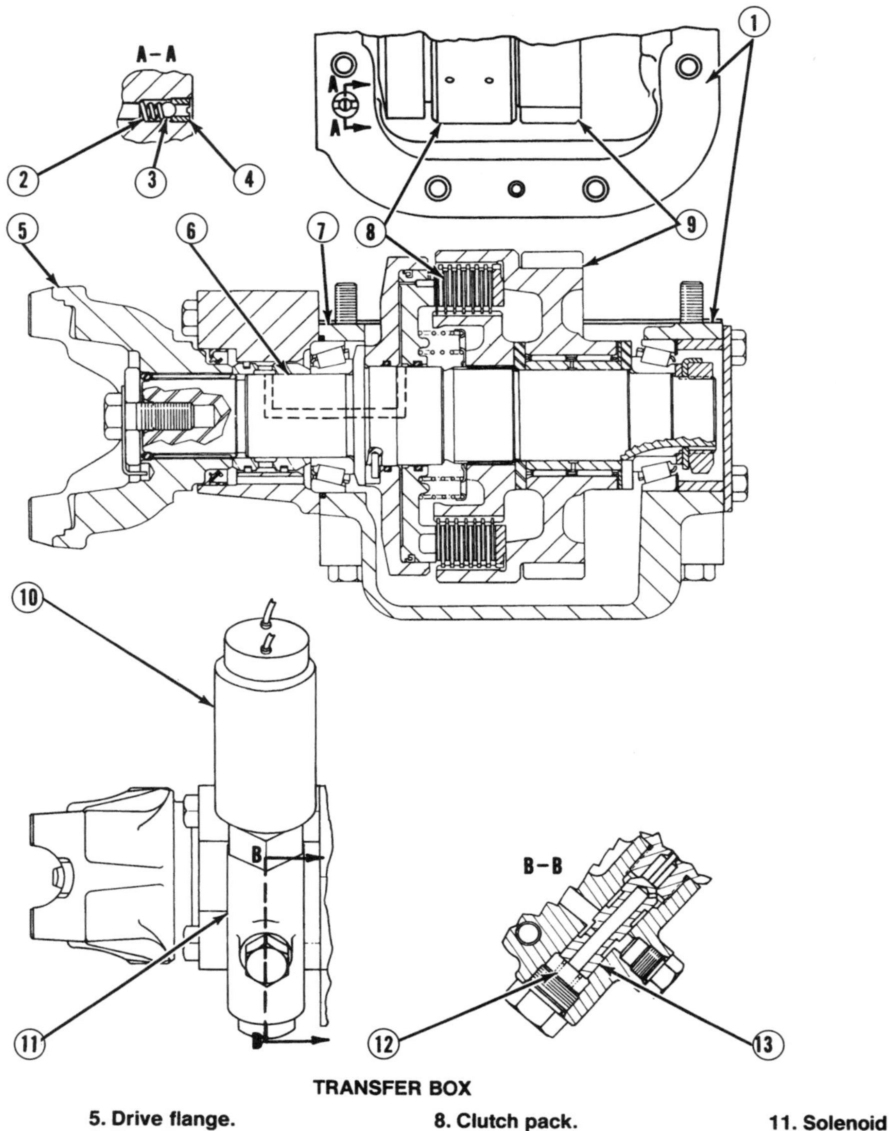

1. Transfer box (drive group).
2. Ball check spring.
3. Ball check.
4. Ball check seat.
5. Drive flange.
6. Transfer drive shaft.
7. Transfer drive group gaskets: two of different thickness.
8. Clutch pack.
9. Transfer drive gear.
10. Solenoid.
11. Solenoid valve.
12. Solenoid valve spring.
13. Solenoid valve spool.

Figure 11-38 Wheel loader transmission with four-wheel drive option. (*Courtesy of Caterpillar*)

When the operator chooses to engage the AWD function, the selector switch in the cab energizes the AWD solenoid valve (4), shifting it to direct pump oil to the AWD clutch pack and engage the front drive axle. The relief valve will also limit maximum clutch pressure to the AWD clutch pack. De-energizing the solenoid valve causes it to shift back to its original position, allowing the clutch pack to drain to the transmission sump and disengaging the drive to the front drive axle.

MECHANICAL TRANSMISSION MAINTENANCE AND REPAIR

Many transmissions used in heavy-duty applications have very diverse maintenance procedures. It is imperative to use and follow OEM-specific maintenance manuals when servicing your equipment. The following maintenance procedures are guidelines to some of the procedures used in the industry today; they are by no means meant to supercede OEM specific

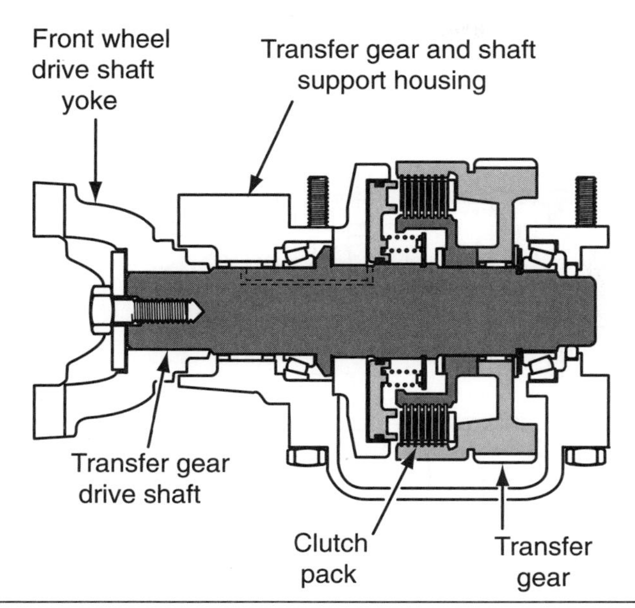

Figure 11-39 Wheel loader with four-wheel drive option: Front wheel driveshaft yoke, Transfer gear driveshaft, Transfer gear and shaft support housing, Clutch pack, Transfer gear.

procedures. Properly operated and maintained, a manual transmission will give hundreds of hours of trouble-free operation. But like all mechanical components, a transmission is subject to wear, fatigue, abuse, and manufacturing defects. Even when operated properly under ideal conditions, adjustments will eventually be necessary and parts will have to be replaced.

Manual Transmission Lubrication

Proper lubrication is one of the keys to a good preventative maintenance program. Maintaining the correct transmission oil level and ensuring the oil is of the OEM specified quality are both critical to ensuring a transmission achieves its expected service life. Following OEM oil change intervals and taking oil analysis samples are also key to a good maintenance program. Because the application of heavy-duty equipment is such that it doesn't accumulate mileage, servicing is based on the number of working hours the machine accumulates or the time period between service intervals. Hour meters log time based from the moment the machine is started until the engine is shut down. To keep an accurate hour meter reading, an engine oil pressure signal or even an alternator pulse activates some hour meters. Direct power activation from the key switch is avoided to eliminate registering hours when the engine is not in operation with the key switch on.

Here are some examples of service charts designations for record keeping:

- Every 10 Service Hours or Daily
- Every 50 Service Hours or Weekly
- Every 250 Service Hours or Monthly
- Every 500 Service Hours or 3 Months
- Every 1000 Service Hours or 6 Months
- Every 2000 Service Hours or 1 Year

Manual transmissions used with standard push- or pull-type clutches will in many cases have a non-pressure lubricated system, meaning that the internal components rely on either a splash lube or the oil's ability to adhere to the components and be carried along as the gears rotate in mesh with each other—in many cases the system relies on both. Transmissions are filled to a designated level and do not use a separate reservoir. Splash lubrication refers to the oil splashing or moving about because the gears and components rotating within the transmission and oil level are throwing the oil outward due to centrifugal force. This will lubricate components that are not within the oil level. Many oils such as "gear-lube" have high adhesion qualities; this allows the oil to stick to components and be carried along from gear to gear, and it also stays with the gears and components when they are not rotating, such as when the unit is parked or not in service.

Standard transmission lubrication requirements will be properly met if the criteria are met:

1. The correct grade and type of oil is used.
2. The oil is changed on schedule.
3. The correct oil level is maintained and it is inspected regularly.

Recommended Lubricants

Only the lubricants recommended by the transmission manufacturer should be used in the transmission. Many transmission manufacturers today prefer to use synthetic lubricants formulated for use in transmissions; they also suggest specific grades and types of transmission oil. (In the past, heavy-duty engine oils and straight mineral oils tended to be more commonly used in transmissions.) Most transmission synthetic lubricants exceed the stated viscosity index grade ratings so they can be expected to perform effectively

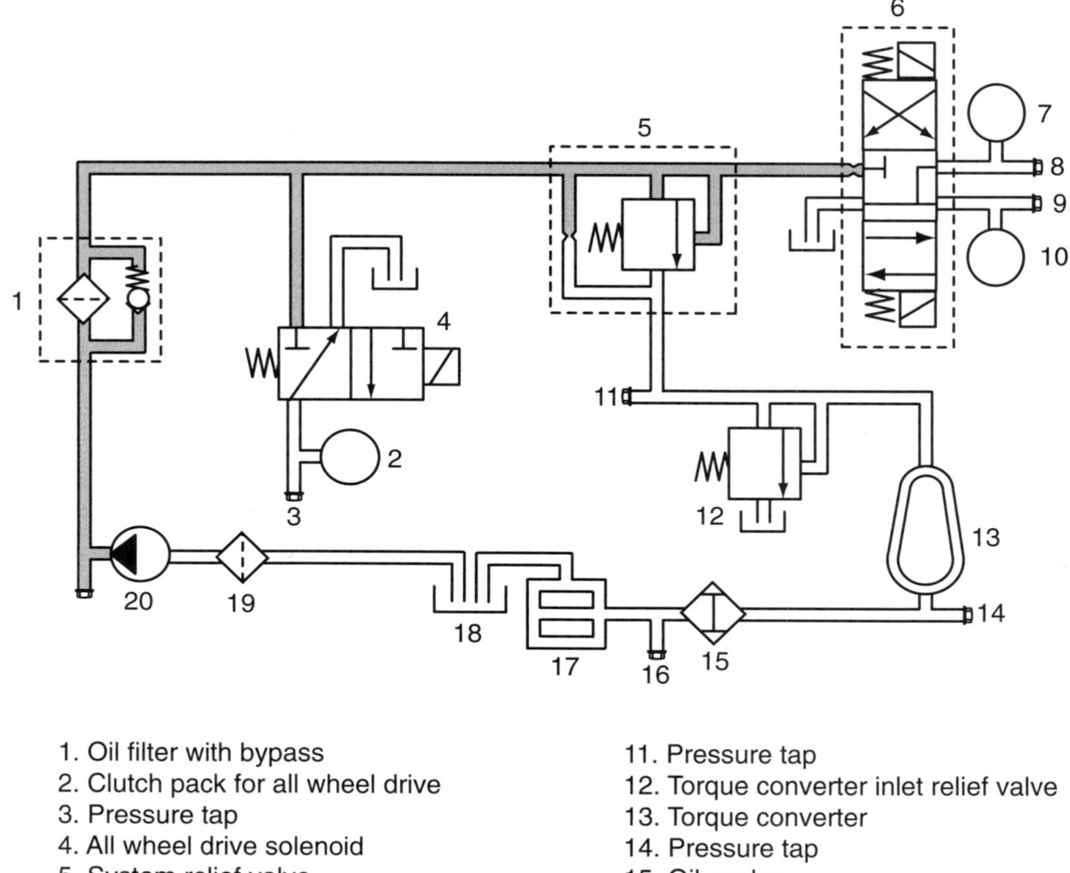

1. Oil filter with bypass
2. Clutch pack for all wheel drive
3. Pressure tap
4. All wheel drive solenoid
5. System relief valve
6. Transmission solenoid valve
7. Reverse clutch pack
8. Pressure tap
9. Pressure tap
10. Forward clutch pack
11. Pressure tap
12. Torque converter inlet relief valve
13. Torque converter
14. Pressure tap
15. Oil cooler
16. Pressure tap
17. Transmission lubrication system
18. Oil reservoir
19. Suction screen
20. Oil pump

Figure 11-40 The transmission hydraulic system in neutral.

through the temperature conditions found from north to south and throughout all seasons.

Although, many transmission manufacturers recommend the use of synthetic lubes in their transmissions, mineral oil gear lubes are still available and are used by some operators.

Mineral-based Lubes. Most manufacturers recommend an early initial oil change after the transmission is placed in service when using mineral-based gear lubes. Although transmission oil is never required to be changed as frequently as engine oil, the first oil change is important because it will flush any cutting debris, created by virgin gears meshing, out of the transmission. A typical transmission oil change is recommended every 1000 service hours or six months of operation.

Synthetic-based Lubes. Modern production machining accuracy has all but eliminated the tendency of a gear set to produce break-in cuttings when new, so the initial drain interval, once considered so important to maximize transmission service life, can be eliminated. Synthetic lubricants are formulated to perform effectively in all temperature conditions found on this continent.

Synthetic lubricants should not be mixed with mineral-based lubes in transmissions. Although mixing of dissimilar lubricants may not produce an immediate failure, the service life of the lube can be greatly reduced. Mixing dissimilar lubricants can sometimes thicken the oil and produce foaming.

CAUTION *Do not add transmission lubricant without first checking what lubricant is already in the transmission. Mineral-based gear oils, mineral-based engine oils, and synthetic gear lubes are all approved for use in transmissions and none of them are particularly compatible. Mixing transmission oils causes accelerated lube breakdown, resulting in lubrication failures.*

Checking Oil Level. The transmission oil level should be routinely checked on a daily basis. This responsibility is normally left with the operator of the equipment. When adding oil to transmissions, care should be taken to avoid mixing brands, weights, and types of oil. When topping up the transmission oil level, it should be exactly even with the filler plug opening as shown **Figure 11-41.** Overfill can cause oil aeration and underfill results in oil starvation to critical components. Some transmissions incorporate a sight-glass for easy viewing; others use a dipstick or a removable plug to check the oil level.

Tech Tip: Transmission oil should be exactly level with the filler plug opening. One finger-digit length equals approximately one gallon of transmission oil; however, testing oil level by dipping your finger into the oil filler hole is not a good practice.

Draining Oil. Drain the transmission when the oil is warm. Remember that it can be hot enough to severely burn your hand. To drain the oil, remove the drain plug at the bottom of the housing. Allow it to drain for at least 10 minutes after removing the plug. Check the drain plug for cuttings and thoroughly clean it before reinstalling. Replace any seal or gasket on the drain plug to prevent any future leaks.

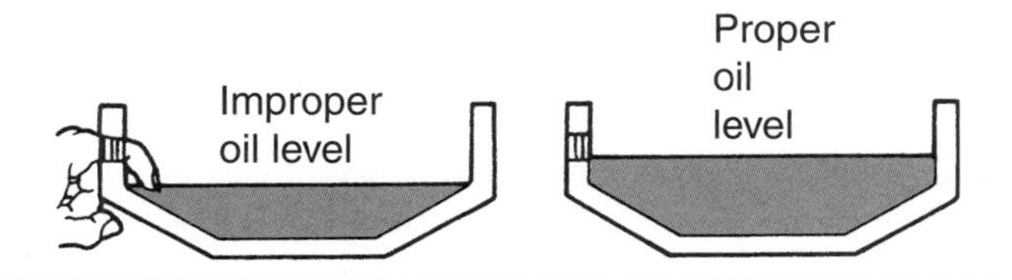

Figure 11-41 Oil level must be level with the bottom of the plug hole.

Refilling Oil. Wipe away any dirt around the filler plug. Then remove the plug from the side of the transmission and refill with the appropriate grade of new oil. Fill until the oil just begins to spill from the filler plug opening. If the transmission housing has two filler plugs, fill both until the oil is level with each filler plug hole.

CAUTION *Do not overfill the transmission. Overfilling usually results in oil breakdown due to the aeration caused by the churning action of the gears. Premature breakdown of the oil results in varnish and sludge deposits that plug up oil ports and build up on splines and bearings.*

Tech Tip: When draining transmission oil, check for metal particles in the oil. Metal particles may indicate excessive wear and warn of an imminent failure. It is not unusual for a newly broken-in transmission to have minute metal particles adhering to a magnetic drain plug.

PREVENTATIVE MAINTENANCE INSPECTIONS

A good preventative maintenance (PM) program can help avoid failures, minimize machine downtime, and reduce the cost of repairs. Often, transmission failure can be traced directly or indirectly to poor maintenance. Taking oil samples as close as possible to the standard change intervals is necessary to receive the full value from oil analysis. You must establish a consistent *trend* of data in order to establish a *pertinent history* of data. Perform consistent oil samplings that are evenly spaced. An increase in the number of oil samples provides a better definition of the trends in data between oil change intervals. More oil samples will allow you to closely monitor wear patterns of the components. This action will ensure that the full life of the transmission is achieved.

Checking for Oil Leaks

Oil leakage in standard transmissions is not common, but when leaks occur, the rear transmission seal (**Figure 11-42**) is a common culprit. A leaking rear main seal can lead to complete transmission failure; it must be repaired. However, it can be difficult to source oil leaks on transmissions, so be sure to identify the leak path before disassembling any components. You can use a checklist to help you identify the source of oil leaks.

The most important thing before beginning a check is to pressure wash the transmission, removing all evidence of oil leaks. Take care not to directly aim the pressure hose into the rear main seal; use a wiper to clean up this area. Now, run the vehicle to operating temperature and re-inspect for leaks.

Transmission Oil Leak Checklist

1. Inspect the speedometer connections. Both mechanical and inductive speedometers connect to the tailshaft assembly. Check the speedometer sleeve torque and ensure that the O-ring or gasket has not failed. Hydraulic thread sealant applied to the threads can sometimes cure this type of leak.
2. Inspect the rear-bearing cover assembly bolts for tightness. These fasteners should be torqued to specification. Thread sealant is required to seal the fastener threads on the rear-bearing cover assembly; leakage will result if this sealant fails.
3. Check the rear-bearing cover gasket and nylon collar. Verify the collar and gasket are installed at the chamfered hole that is aligned near the mechanical speedometer opening. The fastener must be torqued to specification with thread sealant applied to the fastener threads. Use the same procedure to check an electronic tailshaft speed sensor.

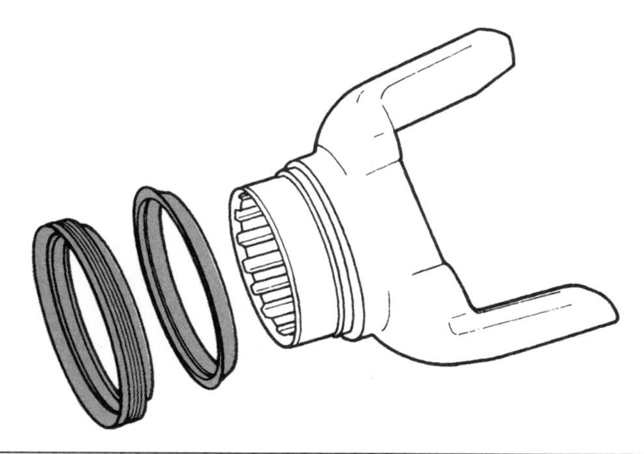

Figure 11-42 Transmission yoke seal. (*Courtesy of Roadranger Marketing. One great drive train from two great companies—Eaton and Dana Corporation*)

4. Check the output yoke-retaining nut for tightness. The retaining nut should be torqued to specification (typically 450 to 600 ft-lbs). You will need a yoke holder to hold the yoke stationary: a torque wrench capable of handling 600 ft-lbs or a lower rated torque wrench used with a 4 to 1 torque multiplier is required to accurately torque the nut.

CAUTION *Over-torquing a transmission output yoke-retaining nut can cause output shaft bearing failure. Do not use a 1-inch air gun to tighten yoke-retaining nuts. It is acceptable to remove the nut using this tool.*

5. Check the PTO manifolds and gaskets, making sure the fasteners are torqued to specification.

CAUTION *Many fasteners used to couple flanges and manifolds to the transmission are bored completely through the housing: these fasteners must have thread sealant applied to the threads. If oil leakage is observed at the PTO covers, replace the PTO cover gaskets and thoroughly clean the fasteners and their mating threads before re-installing.*

6. Check the cover plates and manifolds bolted to the transmission housing, including the front-bearing cover plate, front housing, shift bar housing, rear-bearing cover, and the clutch housing for cracks or breaks. Any or all of these components can be a source of oil leakage.
7. Check the front-bearing cover plate. If the oil return grooves have been damaged by contact with the input shaft, this can cause oil leakage. If the grooves are damaged, you must replace the front-bearing cover. If the front-bearing cover has an oil seal installed, inspect the oil seal. Damaged or worn seals must be replaced to prevent oil leakage, and in the case of the front-bearing cover seal, the transmission will have to be pulled.
8. An oil leak in the front-bearing cover, the auxiliary housing, or in the air breather in the shift bar housing can be caused by a damaged O-ring in the range air shift cylinder. The damaged O-ring bleeds air into the transmission, pressurizing it and resulting in oil leakage.

9. If the transmission is equipped with an oil cooler and oil filter, inspect the connectors, hoses, and filter element to make sure they are tight and leak free.
10. Inspect the oil-drain plug and oil-fill plug for leakage. The oil-drain plug and oil-fill plug should be torqued to specification.

Tech Tip: Always check for a plugged transmission breather when identifying the cause of a transmission oil leak. When transmission oil is raised from cold to operating temperatures, it expands to occupy a greater volume and if the breather is plugged, this can cause seal failure.

Rear Seal Replacement

If your troubleshooting has determined that the rear main oil seal is leaking, the end yoke has to be removed and the seal and slinger replaced. Check the yoke for signs of wear and damage (**Figure 11-43**). Pay special attention to the yoke seal surface. Any wear beyond normal polishing

Tech Tip: Do not attempt to repair a visibly worn or damaged yoke by polishing using crocus cloth. Minor scratches can either track oil under the seal or draw in contaminants.

will require that the yoke to be replaced or re-sleeved. There should be no rust, burrs, nicks, or grooves on the seal surface.

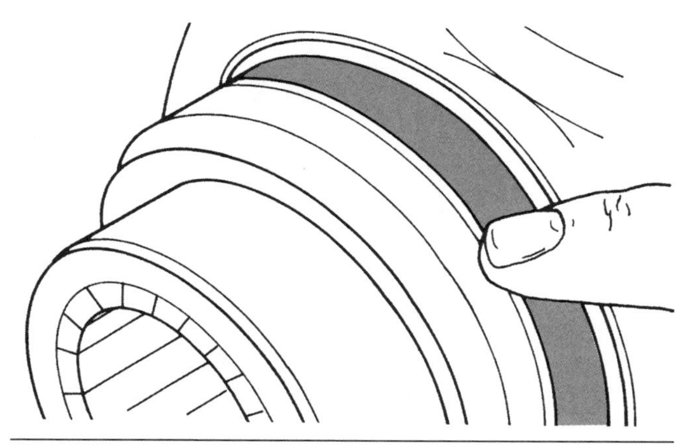

Figure 11-43 Checking the yoke seal for wear and damage. (*Courtesy of Roadranger Marketing*)

Yoke Sleeve Kits. Re-sleeving of yokes is a common practice and cost-effective when weighed against the cost of a yoke replacement. The sleeve fits over the yoke sleeve race with some interference and is made with a hardened steel surface. It is often called a wear ring. Use the wear manufacturer's installation instructions.

Tech Tip: Wear rings that increase the original yoke diameter can cause the new seal to wear more rapidly due to increased seal lip pressure.

Replacement Procedure

1. Block the vehicle wheels and place the transmission in neutral.
2. Disconnect the drive shaft by removing the U-joint from the output yoke.
3. Remove the output yoke nut and use a puller to remove the output yoke.
4. Remove the oil seal from the rear bearing cover.
5. Wipe the bore of the bearing cover with a clean cloth to remove any contaminants.
6. Install the new seal and slinger with the proper driving tool (**Figure 11-44**). Make sure that the seal (or slinger) is not cocked at installation. Even if straightened later, the thin steel cage can be permanently distorted during the installation operation and leakage might occur.
7. Reinstall the output yoke over the transmission output shaft, prevent the yoke from rotating, and torque the nut to specification.

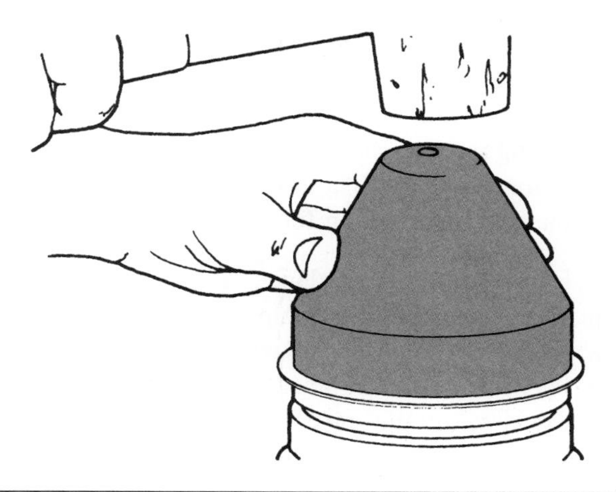

Figure 11-44 Install the seal into the bore with the proper seal driver. (*Courtesy of Roadranger Marketing. One great drive train from two great companies—Eaton and Dana Corporation*)

TROUBLESHOOTING

To isolate the root cause of any transmission problem, you have to follow a troubleshooting procedure. Begin by discussing the problem and its symptoms with the operator. Check any information you can get from the operator with the vehicle history.

Visually inspect the transmission. Look for obvious problems such as broken transmission mounts, fittings, or brackets, and signs of leakage. Road test the vehicle whenever possible. Technicians usually get second- or third-hand reports of trouble experienced with the unit, and these reports do not always accurately describe the actual conditions. Sometimes symptoms may point toward a transmission problem but actually be caused elsewhere in the drivetrain, such as an axle, driveshaft, universal joint, engine, or clutch. This is especially true of complaints of noise and vibration. Road testing the unit yourself reduces the importance of second-hand reports on performance from operators. If the technician is capable of operating the equipment, road testing can be more effective; but riding with an operator can also be informative.

Troubleshooting should always begin with information gathering. Next, you will have to analyze the information and attempt to identify a cause. Nothing is more helpful than experience in troubleshooting and product knowledge, but these skills take time to acquire.

Symptoms

Next, we will take a look at some typical symptoms produced by transmission problems and give you some clues as to what might have caused them.

Noisy Operation. This complaint is common and can be caused by so many different things that you should always begin with a road test. Noise has a way of reverberating through the drivetrain. Here are some non-transmission problems. The noises resulting from these conditions can be mistaken for transmission noise:

- Engine fan out of balance or blades bent
- Defective vibration damper
- Crankshaft out of balance
- Flywheel out of balance
- Flywheel mounting-bolts loose
- Engine rough at idle, producing rattle in gear train
- Clutch or torque converter assembly out of balance
- Engine mounts loose or broken
- PTO engaged
- Universal joints worn out
- Driveshaft out of balance
- Universal joints out of phase or at excessive working angle
- Hanger bearings in driveline dry or out of alignment
- Wheels out of balance
- Tire treads humming or vibrating at certain speeds
- Flywheel housing not concentric

It is your responsibility as a technician to determine the cause, so it is very important that you do not jump to conclusions based on a driver complaint. First, determine the source of the noise. Next, if you are sure it is coming from the transmission, attempt to define the character of the noise. Be aware of the exact conditions that produce the noise: Ask questions. Does the noise only occur in one gear ratio? Does it occur at a specific road speed?

Growling. Low frequency noises such as growling, humming, or grinding are usually caused by worn, chipped, rough, or cracked gears (**Figure 11-45**). As gears continue to wear, a growling/grinding noise becomes more noticeable, usually in gear positions that throw the most load on the worn or damaged gears.

Thumping or Knocking. A thumping or knocking noise can begin as growling. As bearings wear and retainers start to break up, a thumping or knocking noise can be produced. This noise is heard at low shaft speeds in any gear position. Irregularities on gear teeth such as cracks, bumps, and swells can also cause thumping or knocking. Bumps or swells can be identified as highly

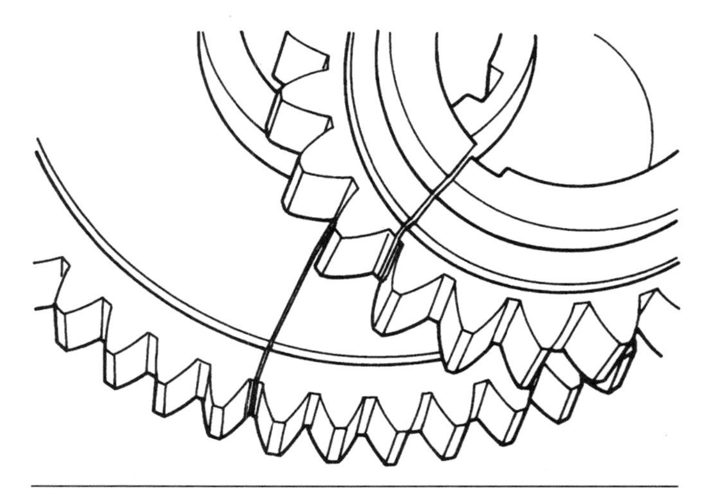

Figure 11-45 Cracked gears can cause such problems as whining, knocking, grinding, or thudding noises. (*Courtesy of Roadranger Marketing. One great drive train from two great companies—Eaton and Dana Corporation*)

polished spots on the face of the gear tooth (**Figure 11-46**) and these are often caused by rough handling during assembly. In most housings, the noise produced will be more prominent when the gear is loaded, making it easier to identify the problem gear. At high speeds, a knocking or thumping noise can turn into a howl.

Rattles. Metallic rattles within the transmission result from a variety of conditions. Engine torsional vibrations are transmitted to the transmission through the clutch. In heavy-duty equipment using clutch discs without vibration dampers (coaxial springs in the hub), rattles may result that are most noticeable in neutral. In general, engine speeds should be 600 rpm or above to eliminate rattles and vibrations noticeable during idle. A cylinder misfire could cause an irregular or lower idle speed, producing a rattle in the transmission. Another cause of rattles is excessive backlash in the PTO unit mounting.

Vibration. Drivetrain vibrations can be very difficult to source. Never forget that they can be produced by any component in the drivetrain so your troubleshooting must involve everything from the engine through to the drive wheel assemblies. Most driveline vibrations are only produced at a specific speed or in a specific gear: these clues can help you determine whether the cause is in the transmission or elsewhere.

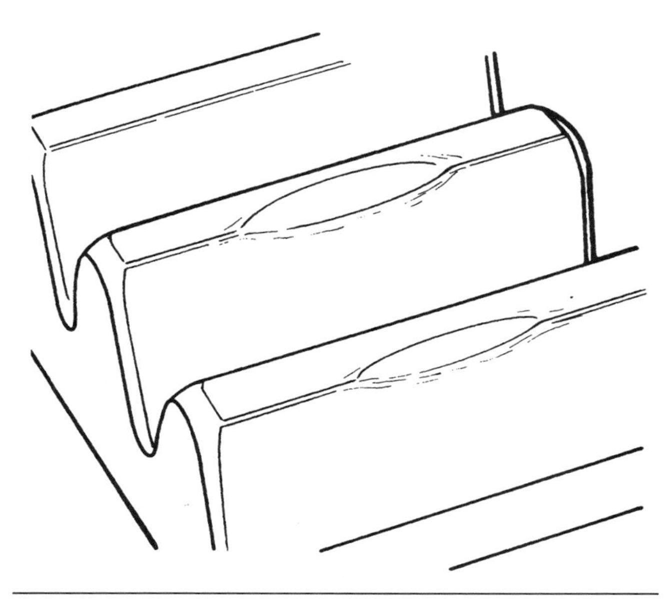

Figure 11-46 Bumps or swells on gear teeth are caused by improper handling of gear before and during assembly. (*Courtesy of Roadranger Marketing. One great drive train from two great companies—Eaton and Dana Corporation*)

If you have isolated the cause of the vibration to the transmission, first note the speeds at which it is most pronounced. Use what you know about the transmission to make your diagnosis. Remember that the input shaft rotates at engine speed when the clutch is engaged, the output shaft rotates at driveshaft speed. Make note of the gear at which the vibration occurs and note whether it occurs only in high range or only in low. Make sure that you have some evidence that the cause is in the transmission before removing the transmission from the vehicle.

Whining. Gear whine is usually caused by insufficient backlash between mating gears, such as improper shimming of a PTO. A high-pitched whine or squeal can also be caused by mismatched gear sets. Such gear sets are identified by an uneven wear pattern on the face of the gear teeth. Pinched bearings with insufficient axial or radial clearance can also produce a whine.

Noise in Neutral. Possible causes of noise while the transmission is in neutral include

- Misalignment of transmission
- Worn or seized flywheel pilot bearing
- Worn or scored countershaft bearings
- Worn or rough reverse idler gear
- Sprung or worn countershaft
- Excessive backlash in gears
- Worn mainshaft pilot bearing
- Scuffed gear tooth contact surface
- Insufficient lubrication
- Use of incorrect grade of lubricant

Noise in Gear. Possible causes of noise while the transmission is in gear include

- Worn or damaged mainshaft rear bearing
- Rough, chipped, or tapered sliding gear teeth
- Noisy speedometer gears
- Excessive endplay of mainshaft gears
- Most of the conditions that can cause noise in neutral

Oil Leaks. Because the transmission in most machines is located in the ground-effect airflow, oil leaks can be difficult to source. Always clean the transmission before attempting to locate the source. Possible causes of oil leakage are

- Oil level too high
- Defective front or rear seals

- Non-shielded bearing used as front or rear bearing cap (where applicable)
- Transmission breather plugged (pressurizing transmission housing resulting in seal failure)
- Fasteners loose or missing from shift tower, shift bar housing, bearing caps, PTO, or covers
- Oil drain-back paths in bearing caps or housing plugged with varnish, dirt, silicone, or covered with gasket material
- Failed gaskets, gaskets shifted or squeezed out of position
- Cracks or porosity in castings
- Drain or fill plugs loose
- Oil leakage from engine
- Speedometer adapter or connections

Gear Slip-Out. When a sliding clutch is moved to engage with a mainshaft gear, the mating teeth must be parallel. Tapered or worn clutching teeth will attempt to *walk* apart as the gears rotate, causing the sliding clutch and gear to slip out of engagement. Slip-out will generally occur when pulling with full power or decelerating with chassis momentum turning the drivetrain.

Any condition that prevents full engagement of a sliding clutch and gear can lead to a slip-out condition. Begin by inspecting the shift tower assembly, ensuring it moves freely and completely into position. If there are problems here, you may have to pull the shift rail housing. Visually check the floorboard opening, operator seat position, and in housings using a remote shift linkage or other causes of binding. Some things to look for are

- Remote shift linkage bent or out of adjustment
- Sloppy ball-and-socket joints
- Worn or loose engine mounts, loose on frame
- Shift rail control-clevis loose on rail or control slot worn
- Taper wear or damage on gear clutch teeth
- Transmission and engine out of alignment caused out-of-specification concentricity at the flywheel housing or bell housing
- Weak or broken detent spring or wear on the detent notch of the yoke bar (**Figure 11-47**)

Vibrations set up by an improperly aligned driveline can add to a slip-out problem. Jump-out occurs when a fully engaged gear and sliding clutch are forced out of engagement. It generally occurs when a force sufficient to overcome the detent spring pressure is applied to the yoke bar, moving the sliding clutch to a neutral position.

Conditions that result in jump-out include

- Heavy and long gearshift levers that swing (over bumpy terrain) enough to overcome detent spring tension (**Figure 11-48**)
- Shift-rail rod detent springs broken, or badly worn detent notches
- Shift-rail bent
- Shift fork pads not square (twisted) with shift-rail

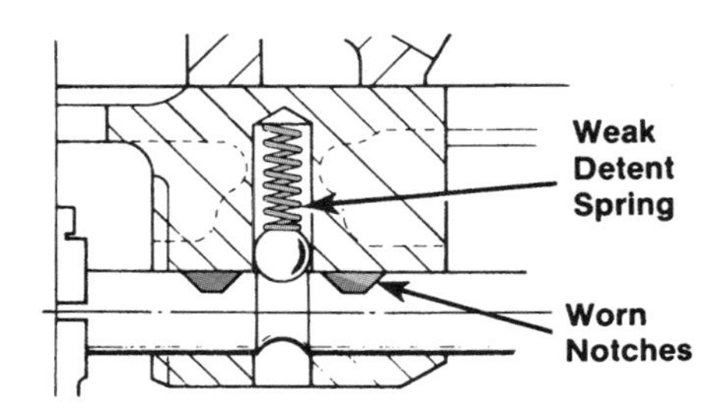

Figure 11-47 Gear slip-out or jump-out can be caused by weak or broken detent springs or worn yoke bar notches. (*Courtesy of Roadranger Marketing. One great drive train from two great companies—Eaton and Dana Corporation*)

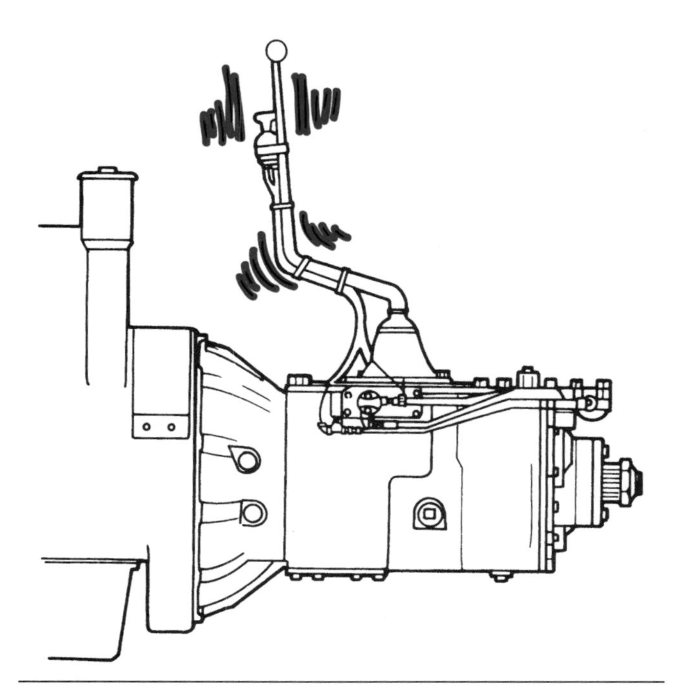

Figure 11-48 When passing over rough surfaces or terrain, the whipping action of a long heavy shifter lever can overcome the detent spring tension, causing the transmission to jump out of gear. (*Courtesy of Roadranger Marketing. One great drive train from two great companies—Eaton and Dana Corporation*)

- Excessive endplay in drive gear, mainshaft, or countershaft; caused by worn bearings or retainers
- Thrust washers worn excessively or missing

Hard Shifting. Hard shifting occurs when the effort required to move a gearshift lever from one gear position to another is excessive. It can be caused by the gearshift linkage or by internal transmission problems. First, you will have to determine whether the problem is caused by the linkage or shift tower assembly, or whether it is a transmission problem. Remove the shift tower from the transmission. Next, you can slide the shift blocks into each gear position by using a pry bar or screwdriver. If the yoke bars slide easily, the problem is probably caused by the linkage assembly. Here are some common causes of hard shifting:

- No lubricant in remote linkage control unit or binding linkage U-joints
- Broken down or overheated transmission lubricant causing varnishing and sludge deposits on splines of shaft and gears
- Badly worn or bent shift rods
- Sliding clutch gears binding on shaft splines
- Clutch teeth burred, chipped, or otherwise damaged
- Clutch pilot bearing seized, rough, or dragging
- Clutch brake engaging too soon when clutch pedal is depressed

Towing Precautions

Vehicles must be towed properly to prevent damage to the transmission and other drivetrain components. When towing a vehicle with a disabled transmission or other problem, standard transmissions require rotation of the main box countershaft and mainshaft gears to provide adequate lubrication. These gears do not rotate when the vehicle is towed with the drive wheels on the ground and the drivetrain connected. Under these conditions if the mainshaft is driven by the rear wheels without lubrication, it can be rapidly destroyed. If the vehicle is to be towed over a long distance, it is recommended that vehicle be placed on a lowboy float and be transported to the repair shop instead of towing the vehicle.

If towing the vehicle is the only option, the procedure to observe should be as follows: Pull the axle shafts or disconnect the driveline. In some applications with planetary hubs, it may only be necessary to remove the sun-gear from the final-drive hubs. This will disconnect the drive between the final-drive hub and the differential.

CAUTION *When the driveline or axle shafts are reinstalled, the axle nuts must be tightened to the correct torque, the axle shafts properly installed (RH and LH), and driveshafts properly phased. Refer to the specific service manual for the correct torque values.*

TRANSMISSION REMOVAL AND DISASSEMBLY

The following is a general procedure for removing and disassembling the transmission assembly. Always consult your manufacturer's service manual for proper removal instructions and specifications. Because of the diversity of transmissions used in heavy equipment machinery, it is difficult to describe any one manufacturer's procedure for the removal of the transmission from a specific machine.

Transmission Removal

1. Remove the chassis battery ground cable.
2. Drain the lubricant from the transmission.
3. Remove the transmission shift lever from the transmission.
4. Scribe a mark on the transmission output yoke, the driveshaft flange, and the output shaft stub on the transmission. Aligning the scribe marks on reinstallation will ensure that these components are correctly installed.
5. Remove the driveshaft from the transmission output shaft yolk.
6. Disconnect all electrical connections from the transmission.
7. Disconnect all air lines connected to the transmission (if equipped).
8. Remove any return springs located on the clutch lever of the transmission and disconnect the clutch linkage.
9. On hydraulically-released clutches, disconnect the pushrod and spring from the release fork.
10. Remove the hydraulic slave cylinder from the bracket on the transmission. To prevent the slave cylinder piston from exiting the slave cylinder, use wire or a support mechanism to support the cylinder in place.
11. Use a proper transmission jack to maintain the transmission alignment and support the transmission assembly. Do not allow the rear of the transmission to drop lower than the front and do not allow the transmission to hang unsupported on the

splined hubs of the friction discs. This could cause the discs or the transmission input shaft to bend and/or distort the clutch discs. This will cause poor clutch operation and/or future clutch problems.

12. Remove the fasteners that attach the transmission to any frame support brackets. Take note of the engine/transmission mounts. If the engine mounts are located on the engine flywheel housing, the transmission mounting bolts are ready to be removed. However, some OEMs place their engine mounts on the transmission bell housing. If this is true, you will have to support the full weight of the engine securely before uncoupling the bell housing.
13. Make note of the alignment of the transmission and that it does not hang by the input shaft in the pilot-bearing bore of the flywheel. The clutch assembly and pilot bearing will be damaged if supported by the transmission input shaft alone.
14. Remove all bolts and washers that attach the transmission bell housing to the engine. Take note that the transmission is still level and no weight is being placed on the transmission input shaft. Pull the transmission bell housing straight out from the engine. Lower the transmission jack and remove the transmission from the vehicle. See **Figure 11-49.**

CAUTION *Some equipment configurations with manual transmissions utilize a torque converter. If this is the case, the torque converter may have to be unbolted from the engine flywheel and be removed with the transmission. An inspection cover on the bottom of the transmission housing will give you access to these bolts. Manually rotating the engine so that the bolts are in the inspection cover area will allow access so that suitable tools may be used to remove them.*

15. Never place any part of your body under any component supported on a hydraulic jack or chain hoist.
16. Use a cherry picker or crane hoist to raise the transmission from the transmission jack and mount it on a transmission overhaul stand.

Transmission Disassembly

It is important that you use the OEM transmission service literature to guide you through the disassembly, inspection, and reassembly procedures. We will use a general approach to the disassembly of a transmission.

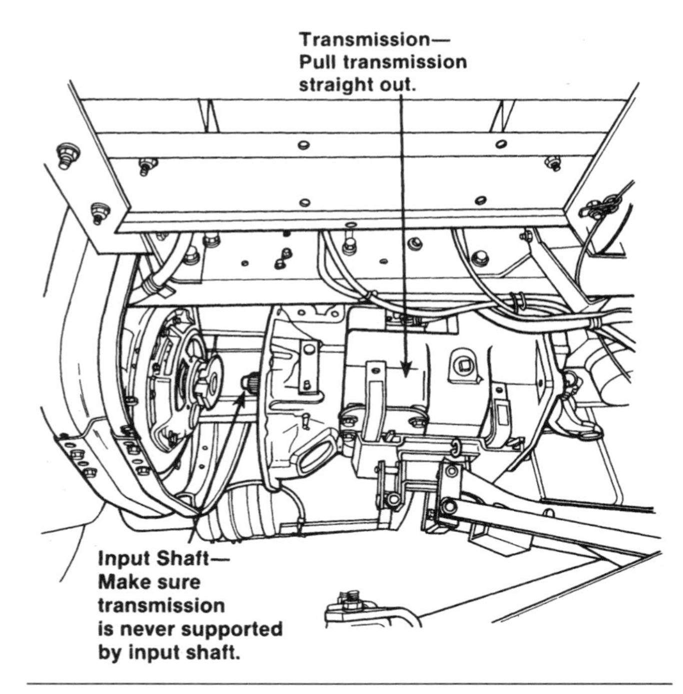

Figure 11-49 When removing the transmission, make sure that it is fully supported and pulled straight out. (*Courtesy of Arvin-Meritor*)

Disassembly Precautions. Whenever disassembling a transmission, follow these general precautions:

- Wash, dry, inspect, and lubricate all reusable bearings as they are removed: you can wrap these in grease paper until ready for reinstallation. Use bearing pullers rather than driving bearings out with a punch.
- Mark housings with Letter or Number Stamps for easier reassembly and locations.
- Work systematically. Arrange removed components neatly on a clean bench in sequential order of removal. This is just good workshop practice and will simplify reassembly. Placing removed parts on clean paper reduces contamination.
- Remove snap rings with snap ring pliers. This allows you to reuse snap rings on reassembly. Using a screwdriver distorts and damages snap rings.
- Work clean. Do not risk dirt contamination when reassembling a transmission. Dirt produces more field failures than any other single cause.
- Be gentle. Never use unnecessary force on shafts, housings, and bearings. Soft face hammers, pry

bars, and bearing pullers make the use of excessive force unnecessary.

Inspection

Failure analysis is the process by which you determine what caused the failure, and it could be the most important step in the overhaul procedure. By the time the transmission is in pieces on the workbench, you should know exactly what caused it to fail. In addition, reusing a questionable component is a high-risk practice that could generate a *comeback* failure, with your employer assuming the cost of the overhaul. When assessing the reusability of transmission components, evaluate the age and application of the equipment. Following are some tips for inspecting transmission components during and after disassembly.

Bearings

The service life of a transmission is often determined by the life of the bearings. Transmission bearings fail due to vibration and dirt. Some of the more prominent reasons for bearing failures are

- Abrasion wear due to dirt
- Fatigue of hard-surfacing on raceways or balls
- Wrong type, gradem or broken-down lubricant
- Lack of lubricant caused by low lubricant levels or bad operating practices
- Vibration breakup of retainer, brinelling of races, fretting corrosion, and seized bearings
- Bearings adjusted too tight or too loose
- Improper fit of shafts or bore
- Acid etch of bearings due to water in lube
- Overloading of vehicle; operator using improper gear selections
- Engine torsionals causing low-frequency vibration that pounds out bearing-hardened surfaces

Tech Tip: Engine torsionals refer to the frequency of torsional pulses delivered to the drivetrain as the force of each cylinder power stroke is unloaded into the crankshaft. Because of today's practice of managing engines at slower speeds and higher torque to produce better fuel economy, lower frequency torsionals are produced and this has caused some transmission failures.

Bearing Inspection. Clean and inspect a bearing as follows:

1. Wash all bearings in clean solvent. Allow the bearings to dry. Check balls, rollers, and raceways for pitting, discoloration, and spalling. Reject any bearings that are questionable.
2. Lightly lubricate bearing and check axial and radial clearances with its race.
3. Reject bearings with excessive clearances; this is an indicator of wear and is usually accompanied by some visual clues.
4. Check the bearing race and shaft fits. The bearing inner bore should be tight on the shaft it is fitted to. Outer races are sometimes interference-fit into the housing bore so check this using the OEM procedure: if an interference-fit bearing race turns in its bore, the housing has to be replaced or repaired.

Analysis of Bearing Failures. We will now take a more detailed look at why transmission bearings fail.

Dirt. More than 90% of transmission bearing failures are caused by abrasive dirt. Dirt usually enters the bearings during reassembly. This is the reason a reconditioned transmission unit has less than half the service life expectancy of a new one. Dirt in a transmission because of poor assembly practices is avoidable, so make a practice of working clean. Dirt can also enter through the seals and/or breather, or even from dirty containers used for addition or change of lubricant.

Soft material such as grain powder, smoke particulate, and light dust forms an abrasive paste within the bearings themselves. The rolling action of the bearing entraps the abrasives, which now work like lapping paste, eroding the hard surfacing. As bearings wear they become noisy. The abrasive action tends to increase rapidly as ground steel particles from the bearing hard-surfacing adds to the abrasives.

Hard, coarser materials can enter the transmission during assembly from hammers, drifts, or power chisels. They can also be produced from within the transmission from raking teeth that cause minute cuttings. These cuttings mixed with the lube produce small indentations in the critical contact areas of bearings and gear teeth.

Fatigue. Bearing fatigue is characterized by flaking or spalling of the bearing race. It is a metallurgical failure of the bearing that can be caused by old age or a manufacturing problem. Spalling is granular weakening of the hardened surface steel that causes it to flake away from the race. Because of their rough surfaces, spalled bearings will run noisy and produce vibration. Normal fatigue failure occurs when bearing service life is at an end: the bearing hard-surfacing appears to be evenly worn through into the base metal below it. All bearings will eventually succumb to fatigue failure.

Lubrication Failures. Bearing failures due to poor lubrication are characterized by discoloration, spalling, and sometimes bearing break-up. Failures can result from low oil level, contaminated lube, improper grade oil, and mixing of incompatible oil types: the use of additives falls into the last category. Lubrication failures are avoidable so ensure that PM practices are adhered to.

Brinelling. A brinelling condition (**Figure 11-50**) can be identified as tiny indentations high on the shoulder or in the valley of the bearing race. It can be caused by improper bearing installation or removal. Driving or pressing unevenly on a bearing can fractionally twist it: any force absorbed by the bearing that is not distributed evenly by the race is the primary cause. This type of damage can be avoided by using correct drivers or pullers.

Fretting. Sometimes, the bearing outer race can pick up the machining pattern of the bearing bore as a result of vibration. This action is called fretting (**Figure 11-51**). A fretted bearing can be mistakenly diagnosed as one that has spun in the bore. Fretting in itself does not require replacement of the bearing.

Contamination. Scoring, scratching, and pitting of bearing contact surfaces can be identified as a contamination failure. It is caused by very fine particles suspended in the lubricant or the use of older extreme pressure (EP) oils. This type of lube contamination will result in abrasive action and polishing of the bearing contact surfaces. It can significantly shorten the life of the bearing.

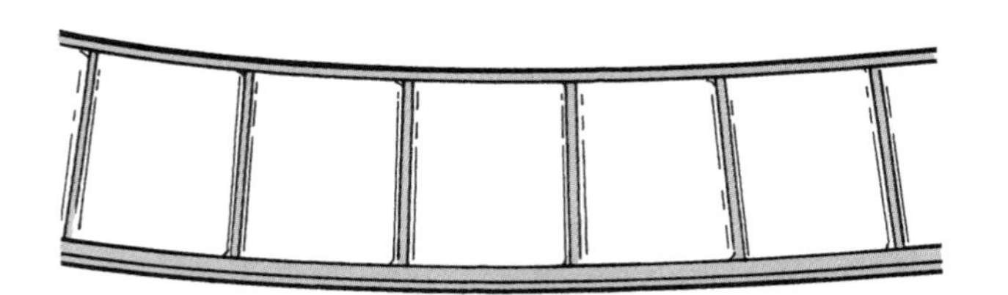

Figure 11-50 Bearing race shown with brinelled effects. (*Courtesy of Roadranger Marketing. One great drive train from two great companies—Eaton and Dana Corporation*)

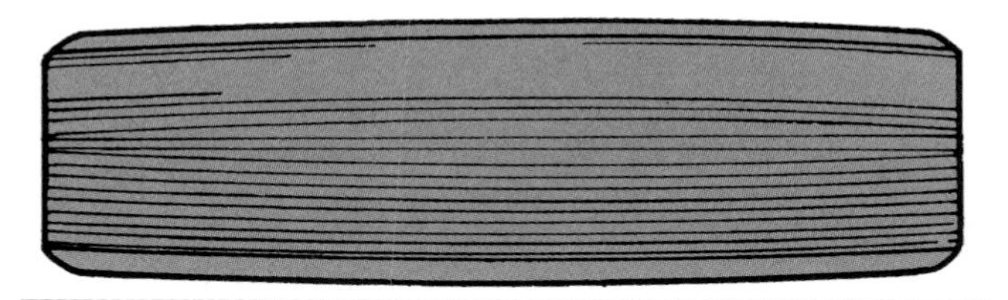

Figure 11-51 A fretted outer race. (*Courtesy of Roadranger Marketing. One great drive train from two great companies—Eaton and Dana Corporation*)

Tech Tip: New bearings should be stored in their shipping wrappers until ready for use. Used bearings should be cleaned in solvent, lubricated, and wrapped in grease paper until ready for installation.

Gears

Visually inspect gear teeth for frosting and pitting. Frosting of a gear tooth is normal and not an indicator of transmission failure. Frosted gears often heal in operation. When the frosting appearance progresses to pitting that can be felt by stroking a fingernail over the surface, the gear will have to be replaced.

Gears with worn clutching teeth should be replaced. Check for taper, reduced length, and chipping of the clutching teeth. Check the axial clearance of each gear. Where excessive clearance is identified, check the gear snap ring, washer, spacer, and gear hub for wear. Check the OEM recommended axial clearance between the mainshaft gears.

Synchronizers. Check the synchronizers for burrs, uneven and excessive wear at contact surfaces, and for metal particles. Examine the blocker pins for wear or looseness. Replace any suspect parts. Remember to check the synchronizers for the auxiliary-drive and low-range gears.

Splines. Check the shaft splines for wear and damage. If the sliding clutch gears, companion flange, or clutch hub have worn into the land walls of shaft splines, you will have to replace the shaft. Check splines and the components that slide on them for binding.

Limit Washers. Visually inspect the contact surfaces of all limit washers. Limit washers define thrust tolerance; if they are scored or reduced in thickness, they should be replaced.

Gray Iron Components. Check all gray iron parts for cracks and porosity. Replace or repair parts found to be damaged. Most OEMs recommended that any housing components found to be defective be replaced. Welding or braising porosity or cracks is generally not recommended.

Clutch Yokes. Check the clutch release components. Replace yokes worn at cam surfaces. Replace the bearing carrier if worn at the contact pads.

Shift Bar Housing. Visually inspect the shift yokes and blocks at the pads and the lever slot. Check the rail action by sliding fore and aft, noting the detent performance. Replace any defective or worn components. Check the yokes for alignment and twisting; replace sprung yokes.

Check lock screws in the yokes and blocks. Tighten and rewire those found loose.

If the housing has been disassembled, check the neutral and engage notches of the shift rails for wear from the detent balls. Each rail should slide smoothly and engage into each detent position with a snapping noise.

Shift Tower. Check the spring tension on the shift lever and replace the tension spring and washer if the lever flops around too freely. If the housing was disassembled, check the spade pin and its corresponding slot, replacing both if worn.

GENERAL REASSEMBLY GUIDELINES

The procedure for reassembling most transmissions reverses the disassembly procedure with a few exceptions. OEM service manuals should be used during reassembly of the transmission. Again, we will use a general approach to the reassembly of the transmission.

Before beginning the reassembly, the transmission housings should be clean. The transmission overhaul kit should contain new gaskets and O-rings. In some applications, manufacturers recommend replacement of all snap rings and C-clips. Make sure all the required replacement parts have been ordered and received.

- Gaskets can be held in place during assembly using a high-tack spray adhesive. As a rule you should avoid using silicone unless instructed by the assembly literature.
- O-ring assembly lubricant is provided in the overhaul kit, and you should use this. This lubricant has a silicone base but does not set.
- Thread sealant should be used on most of the fasteners used in assembling a transmission. Besides clamping components together, transmission bolts more often than not seal the threaded holes they are installed in.
- Check torque instructions. As you reassemble the transmission, refer to the manufacturer's specifications for proper installation and torque of all fasteners, sets crews, lock nuts, and plugs.
- Interference-fit, bearing installation should be done by heating the bearing in a bearing oven, and it should not be heated greater than 375ºF (refer to OEM specifications). After installation of the bearing on the shaft, make sure the bearing has seated completely against the collar by using a punch and hammer to gently tap the bearing race down. A solid feel on the punch will indicate that the bearing is seated.
- Gears, spacers, thrust washers, and snap rings must be properly installed in their right order on the mainshaft.
- Snap rings should be double-checked to guarantee that they are set in their grooves properly.
- Check that there is substantial gearlash between matting gears and that they rotate freely.
- When installing the shift bar housing, make sure the shift forks align with their appropriate gear or synchronizer. Always check that each gear can be selected and that the transmission shafts rotate through each selected gear range.
- Prefill the transmission with proper oil viscosity. The level should be double-checked after final installation.

ONLINE TASKS

1. Access a Service Information System (S.I.S.) specific to a machine with a manual transmission within your shop. Locate the maintenance schedule for the transmission and the type of oil used. Some S.I.S. systems require subscriptions and/or passwords may be required. These must be acquired by your shop or training center.
2. Access the Web site *http://www.roadranger.com.* Find their "Service/Sales Literature" link. Locate the "Hard Shifting Diagnostics Flowchart." This

will give you a systematic approach to troubleshooting Eaton/Roadranger transmission problems.

SHOP TASKS

1. Locate a transmission assembly within your shop and identify the transmission by type of gearing used, mode of shifting, number of forward gears, and number of countershafts used. For example: helical gears/synchronized/single countershaft transmission components.
2. Locate a piece of equipment equipped with a manual shift transmission. Perform an oil level check as outlined in this chapter or to the OEM procedure.
3. Locate a transmission assembly within your shop and calculate the highest gear ratio of that transmission. Is this transmission an overdrive transmission or a direct-drive transmission?

Summary

- Scheduled lubrication services and oil sampling are a key to a good transmission maintenance program.
- Standard transmissions depend on splash lubrication; some of the rotating components contact, pick up, and circulate oil from the oil in the sump. Maintaining the correct oil level is critical for splash lubrication to be effective.
- Only the lubricants recommended by the transmission manufacturer should be used in a transmission. This can be either a gear or engine oil.
- Good preventative maintenance (PM) lowers the incidence of breakdowns, minimizes downtime, and reduces the cost of repairs.
- Leakage in transmission rear seals is a relatively common problem in manual transmissions, but it is one that is easily repaired.
- When diagnosing transmission complaints, it is important that you identify that the transmission is the actual cause of the problem before removing it for repair.
- When disassembling a transmission, each component should be carefully inspected for abnormal wear and damage. Components that are not reusable should be ordered after disassembly, not discovered damaged during reassembly.
- Ensure that a cause of failure is determined before reassembly. If the failure is not confirmed, it is likely to recur within a short period.
- Bearing failures occur because of dirt contamination and poor lubrication. More than 90% of bearing failures are caused by dirt. Cleanliness is critical when repairing and servicing standard transmissions.
- When removing a manual transmission it should be pulled straight out from the engine and not be allowed to be supported from the input shaft.

Review Questions

1. Which of the following indicates that transmission oil in a standard transmission is at the correct level?
   A. visible through the filler hole
   B. reachable with the first finger digit through the filler hole
   C. level with the filler hole
   D. at the high level on the dipstick
2. In a sliding gear transmission, the gears slide along
   A. the spline of the main output shaft
   B. the spline of the countershaft
   C. the spline of the main input shaft
   D. none of the above; the gears are in constant mesh

3. The shift forks in a typical manual synchronized-type transmission are used to move
   - A. mainshaft gears
   - B. countershaft gears
   - C. synchronizers and collars
   - D. balance shaft.
4. The greatest torque transfer in forward speeds occurs
   - A. in an overdrive gear
   - B. in a direct-drive gear
   - C. in the lowest gear
   - D. in the midrange gear
5. A pin-type synchronizer assembly meshes
   - A. on the external side of the drive gears
   - B. on the internal side of the drive gears
   - C. with countershaft only
   - D. none of the above; synchronizers are in constant mesh
6. Which one of the following conditions would be the most likely cause of a transmission jumping out of gear when the truck is moving forward?
   - A. wrong type of lubricating oil used in the transmission
   - B. incorrect transmission lubricating oil level
   - C. worn idler gears
   - D. bent shift fork
7. What component retains a shift rail in position following a shift?
   - A. detent ball
   - B. shift lever
   - C. range cylinder
   - D. dog clutch
8. The component used to ensure that the mainshaft and the mainshaft gear about to be locked to it are rotating at the same speed is known as a
   - A. synchronizer
   - B. sliding clutch
   - C. transfer case
   - D. PTO unit
9. Which of the following statements is/are correct about a gear train?
   - A. transmit speed and torque unchanged
   - B. decrease speed and increase torque
   - C. increase speed and decrease torque
   - D. all of the above
10. In a manual shift transmission, a gear ratio of 1:1 is normally achieved by
    - A. connecting transmission input shaft to countershaft
    - B. connecting transmission input shaft to reverse idler gear
    - C. connecting countershaft to transmission output shaft
    - D. connecting transmission input shaft to transmission output shaft
11. During a routine transmission service, you find small amounts of brass or brass-like metal shavings in the oil. You would suspect which of the following to be the cause
    - A. worn or damaged transmission bearing
    - B. worn or damaged transmission countershaft
    - C. worn or damaged transmission mainshaft
    - D. worn or damaged transmission synchronizer

12. A plain-type synchronizer assembly collar meshes
    - A. on the external side of the drive gears
    - B. on the internal side of the drive gears
    - C. with the countershaft only
    - D. none of the above; synchronizers are in constant mesh
13. Which one of the following conditions would most likely cause either 3rd or 4th gear to jump out of mesh when the vehicle is moving forward?
    - A. wrong type of lubricating oil used in the transmission
    - B. incorrect transmission lubricating oil level
    - C. worn idler gears
    - D. bent shift fork
14. The main purpose of a transmission synchronizer is to
    - A. synchronize the transmission with the engine speed
    - B. synchronize all the transmission speeds
    - C. synchronize or match mainshaft and countershaft rotational speeds
    - D. synchronize or match mainshaft and driveshaft rotational speed
15. In a synchronized constant-mesh transmission, the synchronizer hubs are
    - A. floating on the mainshaft
    - B. splined to the mainshaft
    - C. floating on the countershaft
    - D. splined to the countershaft

CHAPTER

12 Powershift Transmissions

Learning Objectives

After reading this chapter, you should be able to

- Identify the major components in a typical powershift transmission, including clutch packs, planetary gears, shift manifold, and input/output shafts.
- Trace powerflows through planetary gear sets and calculate their gear ratios.
- Explain the importance of using the correct lubricant and maintaining the correct oil level in a transmission.
- Outline the preventative maintenance inspections (PMI) required by powershift transmissions.
- Identify the differences between planetary shift transmissions and multiple-countershaft transmissions.
- Describe procedures for troubleshooting powershift transmissions and torque converters.
- Identify the causes of typical transmission performance problems, such as low power, transmission slipping, and hard shifting.
- Describe hydraulic circuits and oil flows through powershift transmissions.
- Identify the major components of a torque converter and fluid couplings including turbine, impeller/pump, stator, and overrunning clutch.
- Describe the differences between a torque converter and a fluid coupling.
- Describe power and oil flow through a torque converter.

Key Terms

fluid coupling
impeller/pump
multiple countershaft transmission
planetary carrier
planetary gear set
planetary pinion gears
planetary shift transmission
powershift
ring gear
rotating clutch pack
shift valves
stationary clutch pack
stator
sun gear
torque converter lockup clutch modulating valve
torque multiplication
turbine
vortex oil flow

INTRODUCTION

This chapter introduces powershift transmission concepts that include both planetary and multiple-countershaft types as they are used in heavy-duty equipment. Powershift transmissions shift under full power from the engine. In a manual shift transmission engine torque must be disconnected before a gear selection is completed. A powershift transmission—by means of clutch packs and planetary gearing or countershaft gears in constant mesh—shifts without interrupting drive torque or synchronizing gears. Most manufacturers use a fluid coupling with powershift transmissions. A fluid coupling, such as a torque converter, can absorb variations in a torque load during shifting, but in some instances, a clutch is also used to input engine torque to a powershift transmission. When a driveline clutch is used, it is limited to starting the vehicle from a stationary position. Depending on the application of the vehicle, many manufacturers have eliminated the use of both the torque converter and/or the clutch assembly. Instead, slow and smooth oil modulation to the clutch packs allows engine torque to be transmitted directly to the first gear clutch pack and planetary gear set. Powershift transmissions do not need to break torque from the engine during shifts while the vehicle is in motion.

This chapter will also cover the various types and styles of torque converters and fluid couplings used with many powershift transmissions. The laws of simple planetary gearing and compound planetary sets will also be discussed in detail.

TORQUE CONVERTERS/ FLUID COUPLINGS

Powershift transmissions are usually driven by a fluid coupling, often a torque converter. Torque converters and other types of fluid couplings utilize a hydrodynamic principal to transfer engine torque to the transmission, without a direct mechanical link between the two components. The term *hydrodynamic* derives from the Greek words "Hydro" (meaning water, or in this case oil) and "Dynamic" (meaning motion). Engine torque is transferred to the transmission by moving fluid at high velocity onto a driven component that absorbs energy and drives the transmission. This drive median between the engine and transmission is a hydraulic fluid; typically a specialty transmission fluid is used. This is in contrast to a clutch assembly, in which the drive median between the engine and transmission is a mechanical connection represented by friction disc(s). **Figure 12-1** demonstrates the hydrodynamic principal by using a garden hose to direct water at a paddle wheel. The water striking the paddle wheel at a high velocity causes it to absorb the energy and rotate. Because the paddle wheel is directly connected to the drive wheel, the drive wheel is forced to rotate. Speed and torque transfer to the drive wheel is directly proportional to the volume and velocity of the water striking the paddle wheel.

A torque converter is a type of **fluid coupling** that allows the engine to rotate independently of the transmission. (Torque converter/fluid coupling differences

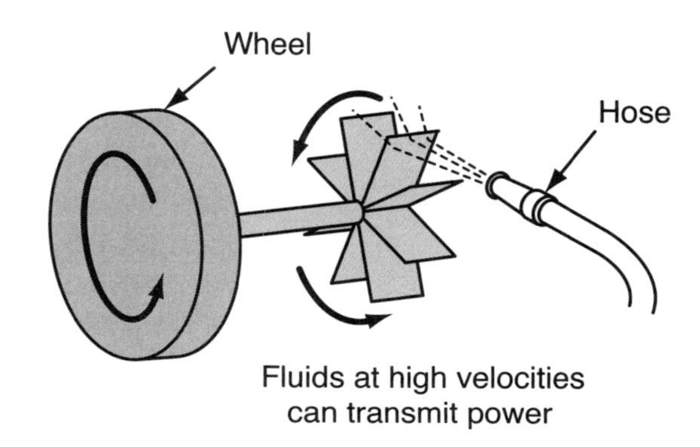

HYDROSTATIC DRIVE
High pressure +
low velocity

TORQUE CONVERTER
Low pressure +
High velocity

One part can drive another
by force of air or oil

Figure 12-1 Principles of fluid coupling.

will be covered later in this chapter.) If the engine is turning slowly, such as in idling, torque transferred through the torque converter is low, requiring only light brake effort to hold the vehicle stationary. Additional brake pressure would be required to prevent the vehicle from moving if the engine was accelerated. As an engine accelerates and pumps more fluid through the torque converter, greater hydrodynamic force is created, resulting in higher torque transfer.

Fluid Couplings

A fluid coupling consists of a pump or impeller, the component that moves the fluid, and a **turbine,** the component that absorbs the energy of the fluid and transfer it to the transmission.

Figure 12-1 shows how air can be used to transfer energy from one fan to another. As the fan speed increases so does the velocity of the air. The velocity of the air coming in contact with the apposing fan blade will absorb this increased energy, forcing the blades to rotate and increase in speed. The same basic principal applies to a fluid coupling that utilizes oil. **Figure 12-2A** shows a pump and turbine rotation together in a fluid coupling.

Figure 12-2B shows a fluid lying level in a bowl. This bowl represents the pump or impeller of a fluid coupling. When the bowl is forced to rotate, centrifugal force acts against the fluid and is forced outwards, as shown in **Figure 12-2C.** When a turbine is placed close to the pump, the fluid is transmitted against the turbine and causes it to rotate with the pump/impeller as shown in **Figure 12-2D.** The fluid then returns to the pump/impeller and is redirected to the turbine. To make the fluid coupling more efficient, vanes are placed on both the pump/impeller and the turbine. These vanes increase the efficiency of the fluid coupling by directing the fluid flow against the vanes which provide a positive surface area to absorb the force. The vanes on the pump/impeller are shaped so that they direct the fluid more efficiently to the turbine. **Figure 12-3** show a typical fluid coupling with pump/impeller and turbine.

The engine drives the pump/impeller of a fluid coupling. Once the engine has started, fluid is directed to the turbine within the fluid coupling. The velocity at which the turbine rotates is dictated directly by engine speed and the resistance of the turbine. When the engine is idling and there is no resistance on the turbine, the two components will rotate at approximately the same speed. If a light load is placed on the turbine, there will be a greater speed differential between the two components. The greatest speed differential is achieved by

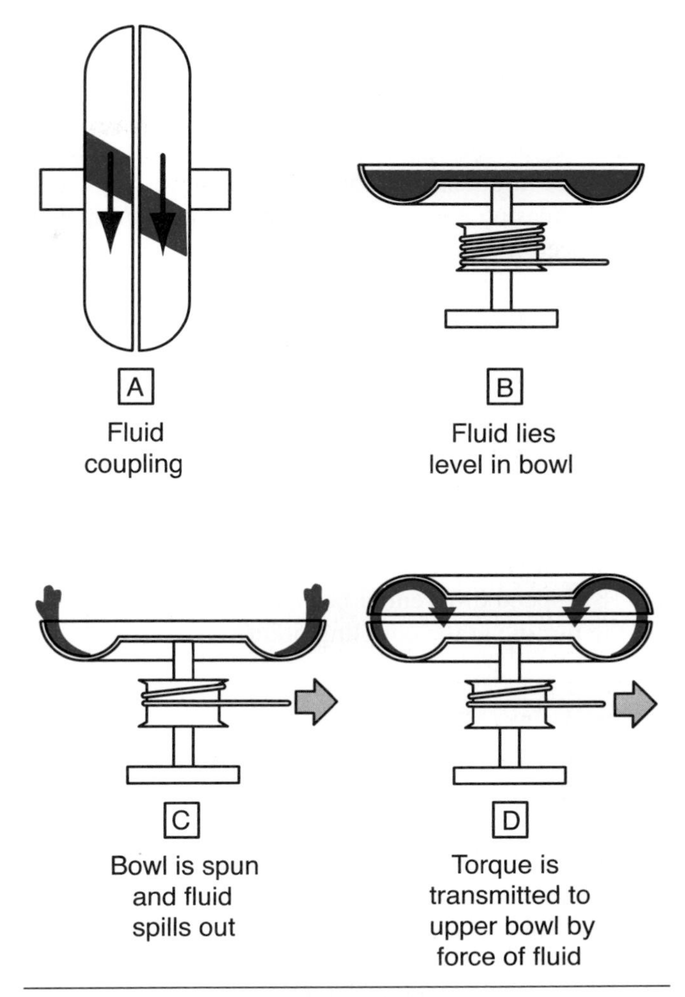

Figure 12-2 Operation of a fluid coupling.

Figure 12-3 A sectioned fluid coupling drive. (*Courtesy of Caterpillar*)

stopping the turbine. The pump/impeller displaces fluid to the stationary turbine. Very simply put, this is what happens:

- The turbine is stationary, so fluid oil becomes very turbulent and looses efficiency.
- Increasing engine speed increases pump/impeller speed which increases fluid velocity to the turbine.
- The turbine can now overcome the resistance and will start to rotate.
- The load is overcome by the increasing speed, so the turbine increases in speed to a point of almost equal speed.
- The fluid coupling has become more efficient because the fluid is less turbulent.

This is the coupling point. However, note that fluid couplings are not 100% efficient. There is approximately a 10% speed differential between the pump/impeller and the turbine at the coupling point.

Torque Converters

Torque converters work under the same principals as a fluid coupling, but with some modifications to make them more efficient. They are equipped with a pump/impeller and turbine much like a fluid coupling, but they are also equipped with a **stator.** Split guide rings, shown in **Figure 12-4,** are designed into the turbine and impeller so that the fluid has an unrestricted fluid path from the **impeller/pump** to the turbine.

A stator is a component used as a fluid lever to assist oil flow within the torque converter. The principals to how a stator works are as follows.

Figure 12-5A, shows fluid directed to a flat surface. The energy of fluid in motion is absorbed by the flat surface and is quickly dissipated. There is only a small

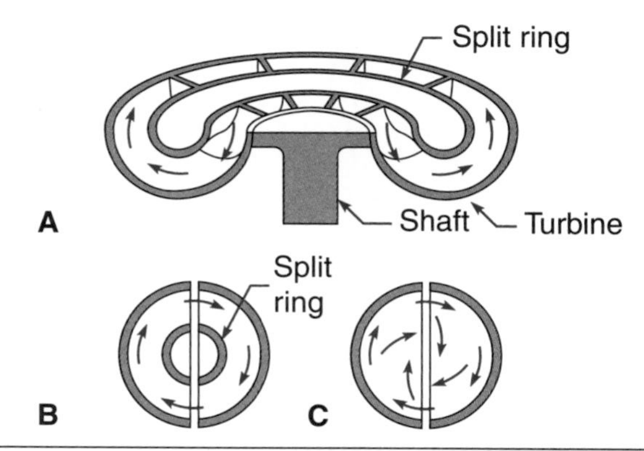

Figure 12-4 Typical fluid coupling with pump/impeller and turbine.

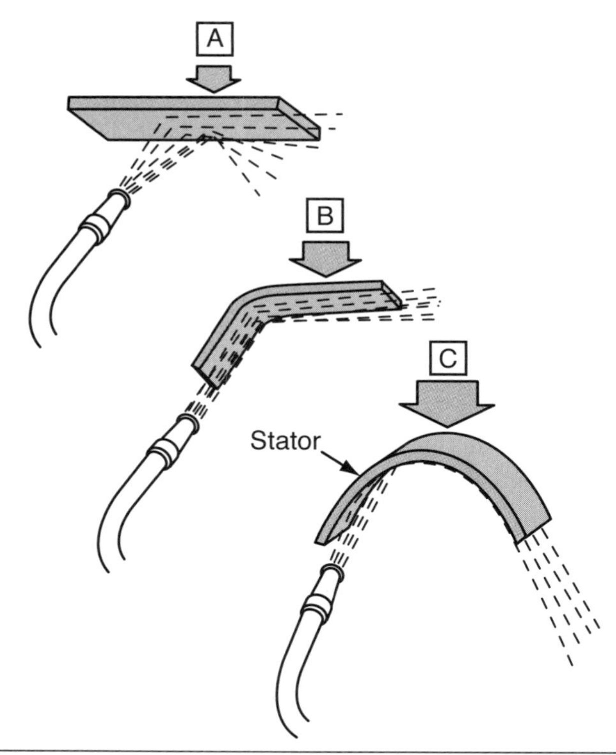

Figure 12-5 (A) Energy of fluid in motion is absorbed by the flat surface and quickly dissipated. (B) Surfaces with a slight curvature allow fluid to flow more smoothly. (C) Fluid energy is completely absorbed by the surface area, increasing the force acting upon it.

amount of force felt on the surface. **Figure 12-5B** shows the surface with a slight curvature; the fluid flows more smoothly. There is an increase of energy absorbed over the previous surface, but with less turbulence. The fluid continues to have velocity and acts on the surface area. **Figure 12-5C** shows a surface area with a greater curvature. The fluid energy is completely absorbed by the total surface area, increasing the force acting upon it. The fluid still has velocity that can be utilized elsewhere.

A stator is located in the center of the torque converter between the pump and impeller, as shown in **Figure 12-6.** When the pump/impeller are rotating and the turbine is stationary, the fluid within the fluid coupling can be turbulent. A stator redirects the fluid, after it has come in contact with the stationary turbine, back to the pump/impeller at an increased velocity. By redirecting the fluid in the same direction as the pump/impeller rotation and with an increased velocity, the stator increases the velocity of the fluid directed to the turbine, thereby increasing the force acting on the turbine. This provides an increase in torque output from the turbine/torque converter. This fluid regeneration is known as **torque multiplication.** The fluid

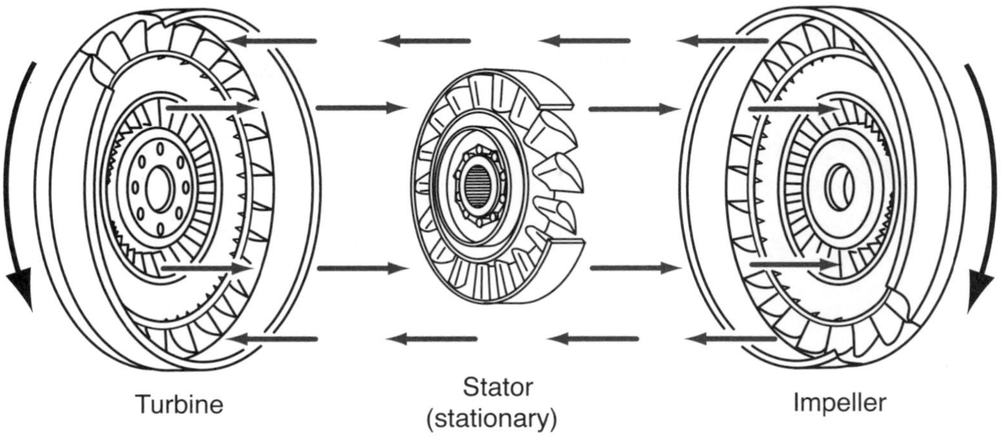

OIL FLOW IN TORQUE CONVERTER

Turbine

Stator
(stationary)

Impeller

Figure 12-6 A stator located between two components can redirect fluid back to the impeller.

regeneration causes the oil to flow in a pattern known as "vortex flow." This vortex flow pattern is a continuous spiral from the pump to the turbine, through the stator, and back to the pump/impeller. The split guide rings assist with vortex flow.

Maximum vortex flow occurs during peak torque converter stall. This is a condition where the turbine has stopped rotating due to an increased load on the vehicle and the pump/impeller is driven at its maximum rpm. The greatest amount of torque multiplication occurs at this point, making the torque converter more efficient than a fluid coupling. This efficiency, however generates a tremendous amount of heat; it is not recommended that a torque converter remain in a stall condition for any extended period. When the torque converter is in a stall condition with the engine idling, the transmission engaged, and the brakes applied, vortex flow is at a minimum and will not overheat the torque converter. If the operator releases the brakes and starts to accelerate the engine, the velocity of the fluid increases, thus increasing vortex flow. This causes the turbine to rotate and drive the transmission and any attached wheels/tracks (depending on application). As the turbine speed increases, the vortex spiral becomes wider and torque multiplication decreases. Soon the turbine speed will be close to pump/impeller speed, the vortex flow spiral will have completely diminished, and the fluid will have changed to "rotary flow." The transformation between vortex flow and rotary flow occurs because, as turbine speed increases, centrifugal force acting on the fluid becomes greater. The fluid trying to return to the pump/impeller becomes harder to achieve. Within the turbine, this force is equal to the fluid flow from the pump/impeller. Vortex flow has completely diminished, the fluid travels in a rotary direction together with pump and turbine. See **Figure 12-7.**

The torque output of a torque converter is infinitely variable between turbine stall and the coupling point at rotary flow. The loads dictate the torque output of the torque converter.

A torque converter is very efficient at maximum vortex flow because it can multiply input torque from the engine. However, efficiency decreases as rotary flow is

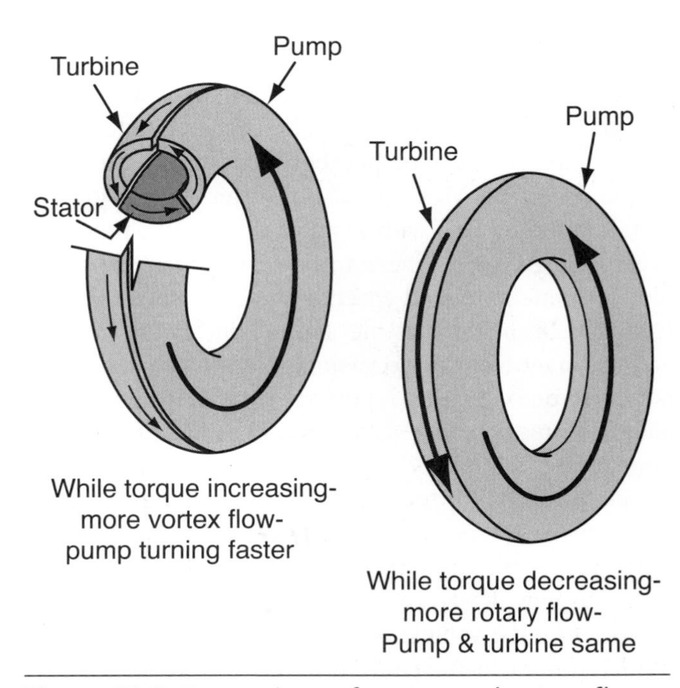

Figure 12-7 Comparison of vortex and rotary flows.

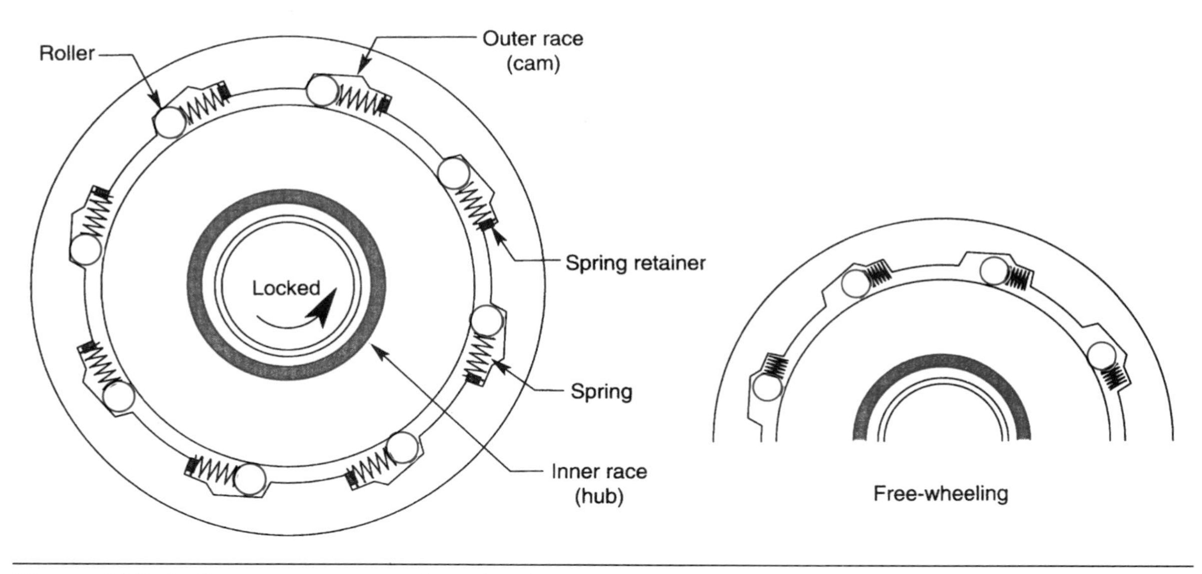

Figure 12-8 Roller-type overrunning clutch.

achieved. The stator, which helps increase vortex flow, can create resistance at rotary flow when the fluid is trying to rotate as a single mass. For this reason many torque converter stators are equipped with an overrunning clutch.

An overrunning clutch on the stator allows the stator to be stationary during vortex flow and to freewheel when rotary flow is achieved. See **Figure 12-8.** Another efficiency factor of the torque converter is the speed differential between the pump and turbine. There can be as much as a 10% loss between input (pump) and output (turbine) speed. Therefore, many torque converters are equipped with a lockup clutch.

A lockup clutch, as shown in **Figure 12-9,** locks the turbine and the torque converter housing together to achieve a 1:1 ratio output between the engine and torque converter. Most torque converters with a lockup clutch are equipped with a stator that incorporates an overrunning clutch. These torque converters are completely immune to stall speed and lockup speed issues. It should be noted that the application of the vehicle and the manufacturer's decision determines whether or not the torque converter is equipped with a lockup clutch and/or a freewheeling stator. Most bulldozer applications use a stationary stator with no lockup feature because most bulldozers work under stall conditions and rarely reach rotary flow for any extended time. Rock trucks or scrapers that may have to carry a load for long distances may utilize a lockup feature in the torque converter. Once the vehicle is in top speed, rotary flow is at its maximum; locking the torque converter makes the vehicle more efficient to run.

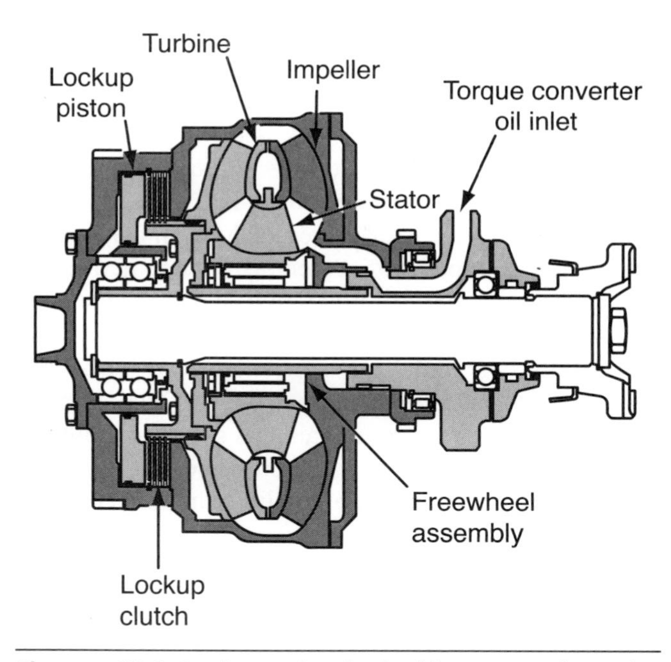

Figure 12-9 Lockup clutch locking together the turbine and torque converter housing.

Torque Divider Torque Converter. A torque divider provides both a hydraulic and a mechanical connection by using a planetary gear set from the engine to the transmission. The torque converter provides the hydraulic connection, while the planetary gear set provides the mechanical connection. During operation, the planetary gear set and the torque converter work together to provide torque multiplication as load on the machine increases.

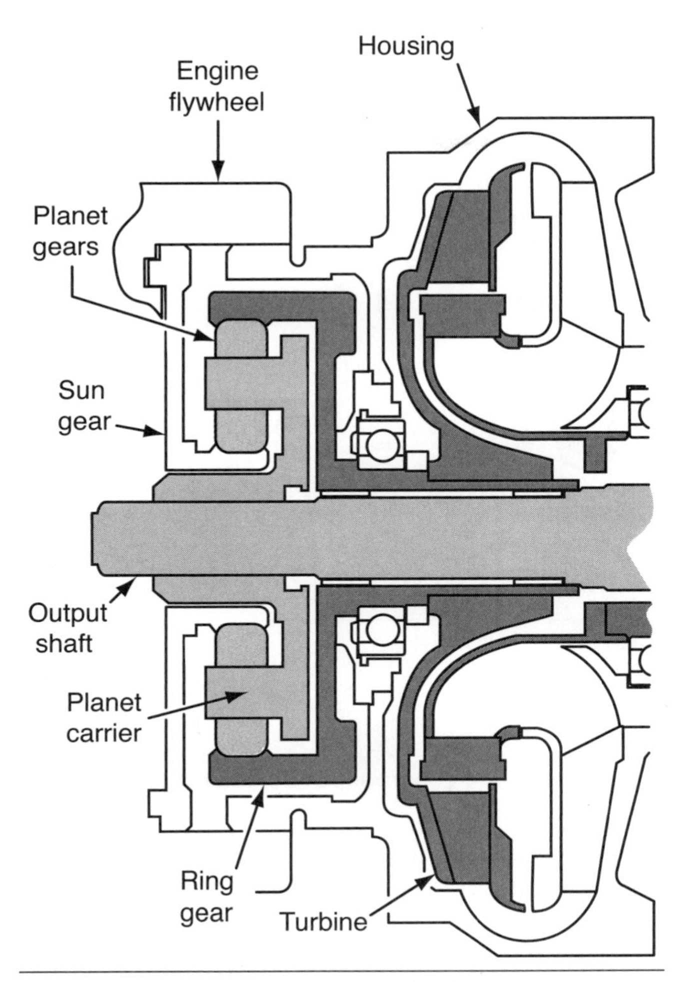

Figure 12-10 Torque divider with planetary gear set.

Figure 12-10 shows a typical torque divider as is used in the Caterpillar D8R. The pump/impeller, torque converter housing, and **sun gear** are on a direct mechanical connection to the engine flywheel. The turbine and the ring gear are connected and rotate together. The output shaft is splined to the **planetary carrier.** The stator is stationary; it is not equipped with an overrunning clutch. Because the sun gear and the pump/impeller are connected to the flywheel, they will always rotate at engine speed. As the impeller rotates, it directs fluid against the turbine blades, causing the turbine to rotate. The turbine and **ring gear** are connected so the turbine rotation causes the ring gear to rotate. During no-load conditions, the components of the planetary gear set rotate as a unit, and the planetary gears do not rotate on their shafts. The torque converter is in a rotary flow condition. Therefore, output speed ratio between the engine and torque converter is 1:1.

When the operator places a load on the machine, the output shaft slows down. A decrease in output shaft speed causes the rpm of the planetary carrier to decrease. Decreasing the planetary carrier rotation causes the planetary gears to rotate, which decreases the rpm of the ring gear and the turbine. The torque converter goes into a vortex flow condition and the torque splits with the torque converter (multiplying the torque hydraulically) and the planetary gear set (multiplying the torque mechanically).

An extremely heavy load can stall the machine. If the machine stalls, the output shaft and the planetary carrier will not rotate. This torque converter stall condition causes the ring gear and turbine to rotate slowly opposite to the direction of engine rotation. Maximum torque multiplication is achieved just as the ring gear and turbine begin to turn in the opposite direction. Avoid holding this machine in a stall condition for any extended period.

During all load conditions, the torque converter provides 70% of the output and the planetary gear set provides the remaining 30%. The size of the planetary gears establishes the torque split between hydraulic torque and mechanical torque.

Impeller Clutch Torque Converter. An impeller clutch-type torque converter is used to change the output of the torque converter by allowing the impeller to slip between the impeller and the torque converter housing (**Figure 12-11**). Because the torque converter can multiply input torque from the engine, undesirable conditions can occur at the final drives of a wheel loader application. In first gear, wheel loaders produce enough torque to cause wheel slippage or wheel spin. If a machine is working in rock conditions, wheel spin can cause tires to be cut and torn apart. The impeller clutch allows the operator to minimize wheel spin by implementing a control feature when the left brake pedal is depressed. Reducing the output torque of the torque converter also provides more engine power for the hydraulic system.

Operation of the Impeller Clutch. An impeller clutch solenoid valve controls the fluid flow to the impeller clutch pack. The impeller clutch solenoid is activated by the powertrain ECM (Electronic Control Module). Refer to Figure 12-11. When the impeller clutch solenoid valve is not energized by the powertrain ECM, oil flows to passage (12) from carrier (14). Oil in passage (12) forces clutch piston (10) against clutch plates (8) and clutch discs (9). Clutch piston (10) and clutch plates (8) are connected to clutch housing (11) with splines. Clutch discs (9) are connected to adapter (18) with splines. Adapter (18) is fastened to the impeller (7) with

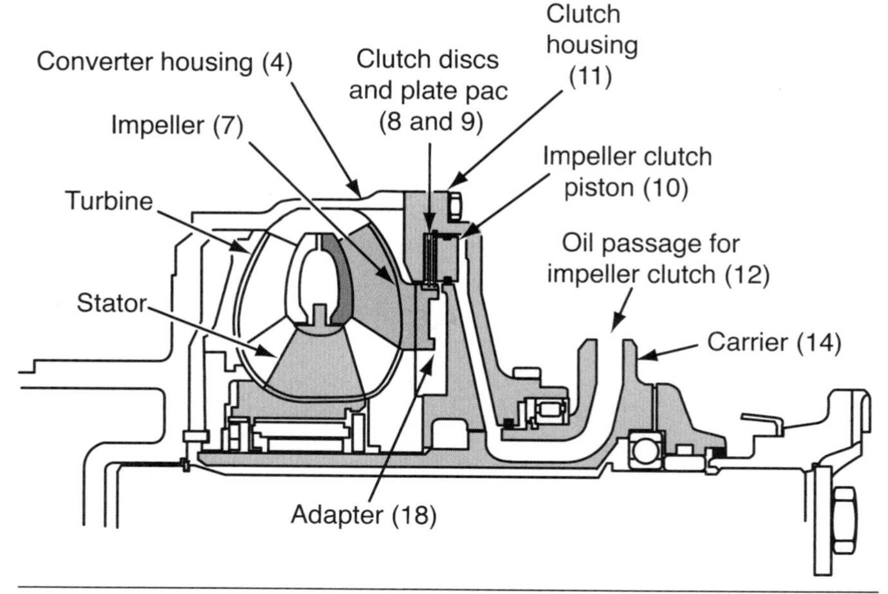

Figure 12-11 Impeller clutch torque converter.

bolts. The friction between clutch discs (9) and clutch plates (8) causes the impeller to rotate at the same speed as converter housing (4). This is the maximum torque output from the torque converter drive. As the amount of current to the solenoid increases, oil pressure to clutch piston (10) is decreased by draining some of the oil to the sump. The friction between the clutch plates and the clutch discs decreases. This causes the impeller to slip from the torque converter housing. When the impeller slips, less oil is forced to the turbine. With less force on the turbine, there is less torque output from the converter turbine shaft.

When the amount of current to the solenoid is at the maximum, there is minimum oil pressure against clutch piston (10). The clutch plates and the clutch discs now have only a small amount of friction and causes the impeller to slip within the torque converter housing. Therefore, the impeller forces only a small amount of oil toward the turbine. There is a minimum amount of torque at the converter output shaft.

The impeller clutch modulates oil pressure at each directional shift, which allows the impeller clutch to absorb energy during all directional shifts. This reduces the amount of energy that is absorbed by the directional clutch packs when an operator shifts transmission direction while the machine is still moving in the opposite direction or under acceleration. This results in an easier shift and less wear of the transmission clutch packs.

The following conditions affect the operation of the impeller clutch:

- The position of the left service brake pedal
- Directional shifts
- Engine speed
- Selected gear
- Direction of rotation of the torque converter output shaft.
- Torque converter output speed
- Position of the rim-pull selector switch in reduced position
- Position of the reduced rim-pull on/off switch

Left Service Brake Pedal. The left service brake pedal controls the amount of brake pressure used to apply the service brakes. It also controls the amount of pressure that actuates the impeller clutch. The impeller clutch is positioned between the engine and the torque converter. By using the left brake pedal, the operator can divert engine power to the implement hydraulic circuit without putting the transmission in the neutral position. As the operator depresses the pedal, the impeller clutch pressure drops quickly to a working pressure. The pressure is then modulated to a reduced pressure, causing the impeller to slip within the torque converter. This sequence occurs for ten degrees of pedal travel. Depressing the left pedal past the 10-degree point applies the service brakes.

Shift Modulation of the Impeller Clutch. When shifting between forward and reverse, the powertrain ECM will override the setting of the left brake pedal. The ECM reduces the pressure of the impeller clutch to create smoother transitions. When this shift in direction is detected, the powertrain ECM reduces the circuit pressure of the impeller clutch to an impeller clutch-hold pressure. If lockup for the transmission directional clutch occurs, the ECM will increase the circuit pressure of the impeller clutch to maximum pressure, providing maximum torque output from the torque converter. If the left brake pedal is not depressed the impeller clutch pressure is not regulated and pressure automatically increases to maximum. The transmission speed sensors and the torque converter output speed sensor are used to determine when the transmission directional clutches have locked up and increase the impeller clutch pressure to maximum.

Function of Reduced Rim-pull Switches. The reduced or maximum rim-pull switch (20) (shown in **Figure 12-12**) and the reduced rim-pull selector switch provide four levels of reduced rim-pull. Reduced rim-pull is enabled only when the transmission is in first speed forward. The reduced or maximum rim-pull switch is a two-position toggle switch found on the vehicle console. The reduced rim-pull selector switch is located in the right-hand console above the operator. In maximum position, the rim-pull switch commands maximum rim-pull. Maximum rim-pull is provided regardless of the position of rim-pull selection switch. The reduced rim-pull selector switch works with the reduced or maximum rim-pull switch in order to change the torque output of the torque converter. The powertrain ECM increases the amount of current to the impeller clutch solenoid valve to reduce the impeller clutch pressure, which results in reduced rim-pull. The position of the reduced rim-pull selector switch regulates the amount of current that is sent to the impeller clutch solenoid valve. The reduced rim-pull indicator lamp is on when reduced or maximum rim-pull switch is in the reduced position and the transmission is in first speed forward. When the transmission is shifted to second speed or above, the reduced rim-pull indicator lamp is off and reduced rim-pull is disabled.

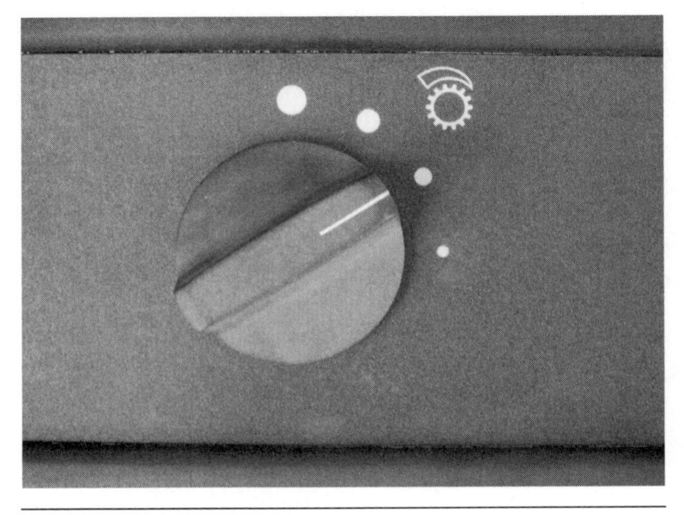

Figure 12-12 Reduced or maximum rim-pull switch.

Modulating Valve (Torque Converter Impeller Clutch). The impeller clutch solenoid valve is a three-way control valve that modulates pressure proportionally. When the powertrain ECM increases the amount of current to the solenoid, the impeller clutch pressure is reduced. When the amount of current from the powertrain ECM is at zero, the impeller clutch pressure is at the maximum, locking the impeller to the torque converter housing.

PLANETARY GEARING

The term *planetary gearing* was introduced because the gears operate similar to our solar system. A simple planetary gear set consists of four main components. A sun gear is located in the center of the gear set (much like the sun is the center of our solar system) and planetary gears, a planetary carrier and the ring gear rotate around it.

A planetary carrier provides the support and centralization of the planetary gears as they rotate around the sun gear. Thus, the planetary gears are allowed to rotate around the sun gear as well as rotate on their own axis within the carrier. The carrier becomes the input, the output or stationary member of the gear set depending on the transmission configuration.

The ring gear provides the final component; it supports the complete planetary gear set and is meshed with all the **planetary pinion gears** on their outer perimeter. Planetary gears are always in constant mesh with each other. Because the gearing is enclosed and self-supporting, planetary gears can handle higher torque loads than regular external gears in mesh. Gear forces are divided equally among the sun, planetary pinions, and ring gears.

Planetary gear sets are compact and can provide many different gear ratio combinations and directional changes. Compounding planetary gear sets increases these combinations even further. When two or more planetary gear sets are linked together by any of the planetary components to complete a desired amount of gear ratios they are called compound planetaries. Heavy equipment transmissions have a multitude of varying transmission configurations so it is critical that a technician understand planetary gearing and how it works.

Principles of Planetary Gearing

There are some basic laws of planetary gearing. For any planetary gear set to work there must be at least one input drive member and one stationary or held member to provide an output. Any of the three main components (sun gear, planetary carrier or ring gear) can be the input, the output, or the held member of the gear set. If this combination cannot be provided, a neutral or non-driving condition occurs. When any combination of held member and input member is created, the third component automatically becomes the output of that planetary gear set. Because of the gearing configuration of planetary gears and depending on which component or member is held or driven, a torque or speed increase can be produced. There are seven different gear ratio combinations that are possible through a single planetary gear set. External gears in-mesh rotate in opposite directions and internal gears in-mesh rotate in the same direction (**Figure 12-13**). Planetary gear sets implement a combination of both principles. The sun gear and planetary pinion gears are both external gears and, because they are in constant mesh with each other, they rotate in opposite directions. The ring gear is an internal gear and is in constant mesh with the planetary gears; they will rotate in the same direction.

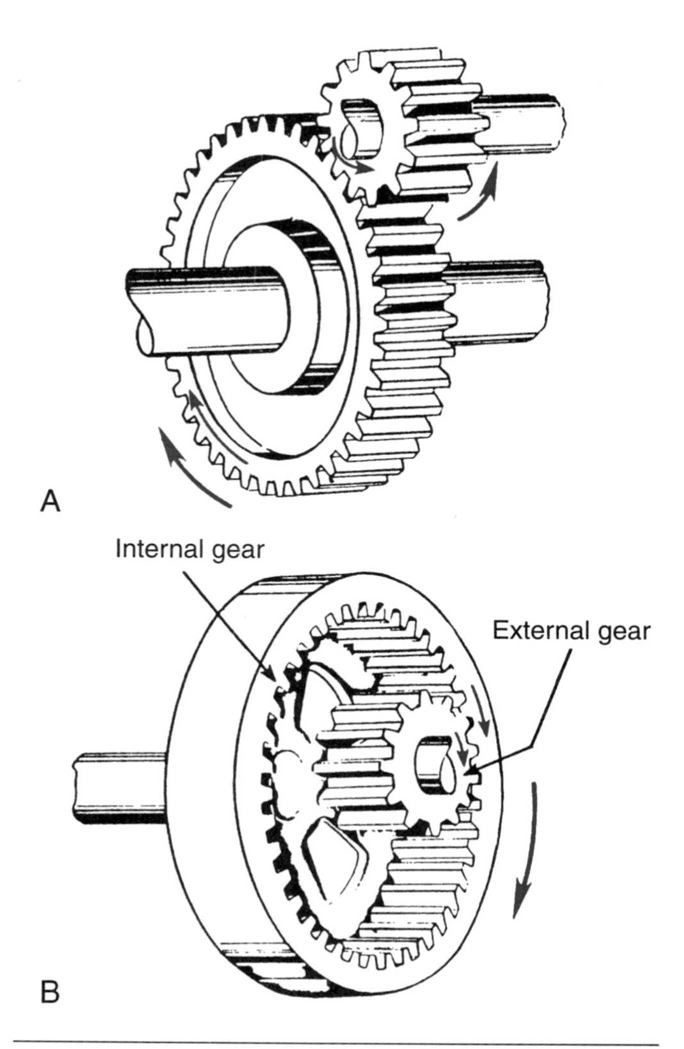

Figure 12-13 Comparison of external and internal gears in mesh.

To calculate gear ratios with planetary gears the same formula applies as with external gears in-mesh, except for calculating the number of gear teeth on the planetary gears. Because the planetary gears are supported by the planetary carrier, there can be as few as three or as many as five planetary pinion gears. This is dictated by the torque loading or application of the transmission. The greater the number of planetary pinion gears within the gear set the greater the amount of torque the gear set can withstand. The formula for calculating gear ratios is Drive into Driven (number of gear teeth on drive member, divided into the number of gear teeth on the driven member) Driven Gear ÷ Drive Gear. Because there are three different members in a planetary gear set that can either be the drive or the driven member, it is important to use the correct number of teeth and component during our calculations. Because we cannot use the number of gear teeth on the individual planetary pinion gears to calculate the gear ratio, we must add the gear teeth of the ring gear plus the gear teeth of the sun gear for the sum of the planetary carrier. For this reason, the planetary carrier will always provide the greatest gear reduction in any planetary gear set when it is the output member.

Figure 12-14 shows a table of eight different combinations that can be achieved through a single planetary gear set. Look at each individual combination and calculate the gear ratio for that combination. To calculate the gear ratios we must first determine the number of gear teeth on each component. See **Figure 12-15.**

Sun Gear	18 Teeth
Ring Gear	43 Teeth
Planetary Carrier	61 Teeth (Sun Gear 18 Teeth + Ring Gear 43 Teeth)

LAWS OF SIMPLE PLANETARY GEAR OPERATION					
Sun Gear	**Carrier**	**Ring Gear**	**Speed**	**Torque**	**Direction**
1. Input	Output	Held	Maximum reduction	Increase	Same as input
2. Held	Output	Input	Minimum reduction	Increase	Same as input
3. Output	Input	Held	Maximum increase	Reduction	Same as input
4. Held	Input	Output	Minimum increase	Reduction	Same as input
5. Input	Held	Output	Reduction	Increase	Reverse of input
6. Output	Held	Input	Increase	Reduction	Reverse of input
7. When any two members are held together, speed and direction are the same as input. Direct 1:1 drive occurs.					
8. When no member is held or locked together, output cannot occur. The result is a neutral condition.					

Figure 12-14 Laws of planetary gearing.

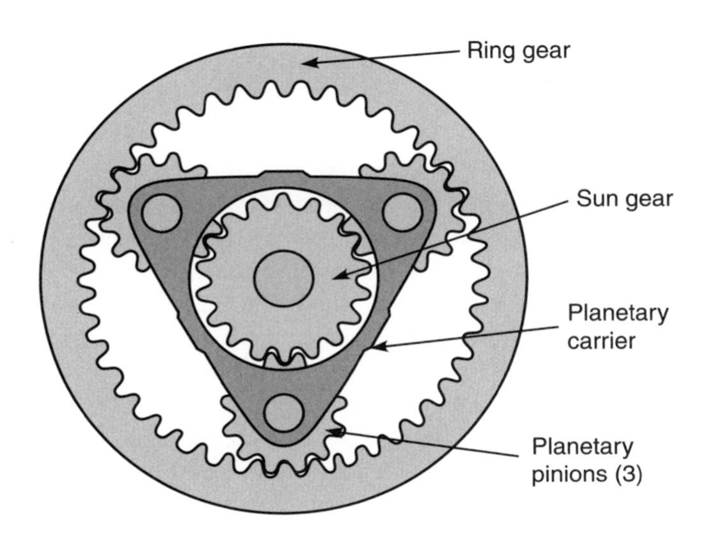

Figure 12-15 Calculating gear ratios in a planetary gear configuration.

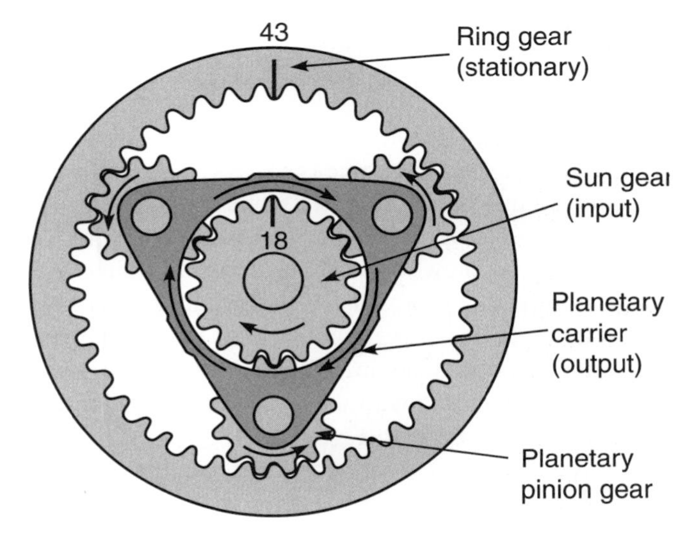

Figure 12-16 Gear ratio: maximum speed reduction/greatest torque increase.

The following eight sections correspond to the table in Figure 12-15.

1. Maximum Speed Reduction. For maximum speed reduction of a simple planetary gear set, the sun gear must be the input member and the ring gear must be the stationary or held member. This automatically makes the carrier the output member. According to the chart this will produce the maximum torque increase and cause the carrier to rotate in the same direction as the input or sun gear. Refer to **Figure 12-16.**

To calculate the gear ratio, divide the number of gear teeth of the input (sun gear 18 teeth) into the number of gear teeth of the output (planetary carrier 61 teeth).

61 teeth ÷ 18 teeth = Gear ratio of 3.3888∞: 1

2. Minimum Speed Reduction. The planetary carrier is still the output, but we reverse the roles of the sun gear and ring gear. The sun gear is now the stationary member and the ring gear is the input. Because the carrier is still the output, the direction of rotation is the same as the input with a minimized gear reduction. This is the second lowest gear ratio with a torque increase achievable through a simple planetary gear set. Refer to **Figure 12-17.**

To calculate the gear ratio, divide the number of gear teeth of the input (ring gear 43 teeth) into the number of gear teeth of the output (planetary carrier 61 teeth).

61 teeth ÷ 43 teeth = Gear ratio of 1.4186: 1

3. Maximum Speed Increase. To achieve a maximum speed increase, the gear with the highest number of teeth

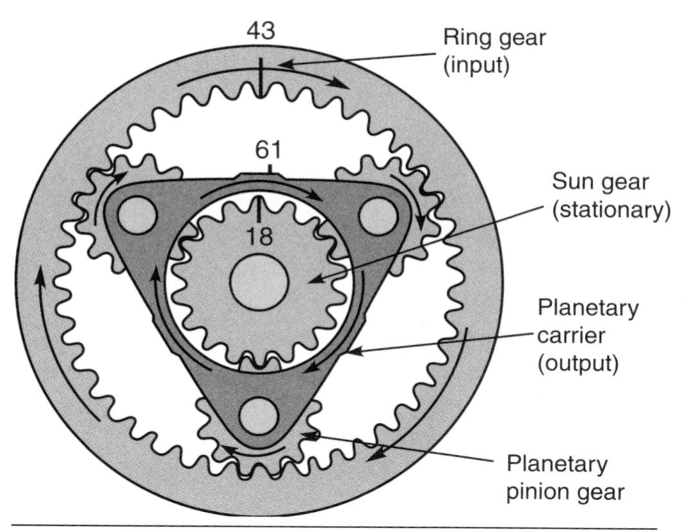

Figure 12-17 Gear ratio: minimum speed reduction/torque increase.

must be the input. The same principle applies to planetary gearing. The planetary carrier is always the component with greatest number of teeth on a planetary gear set, so it must be the input drive member to achieve an overdrive ratio. Depending on which is the held member, either the sun gear or the ring gear will achieve the higher overdrive ratio. Figure 12-15 shows combination 3, with the sun gear as the output and the ring gear held. This achieves the highest speed increase with the direction of rotation the same as the input. Refer to **Figure 12-18.**

To calculate the gear ratio, divide the number of gear teeth of the input (planetary carrier 61 teeth) into the number of gear teeth of the output (sun gear 18 teeth).

18 teeth ÷ 61 teeth = Overdrive gear ratio of 0.2951: 1

4. Minimum Speed Increase. The planetary carrier remains the input and we reverse the roles of the sun and ring gears. Because the planetary carrier is still the input member and is the largest gear, an overdrive gear ratio is the outcome. With the sun gear as the held gear, the ring gear automatically becomes the output member. This will result in a slightly lower overdrive condition with the direction of rotation being the same as the input. Refer to **Figure 12-19.**

To calculate the gear ratio. divide the number of gear teeth of the input (planetary carrier 61 teeth) into the number of gear teeth of the output (ring gear 43 teeth).

43 teeth ÷ 61 teeth = Over-drive gear ratio of 0.7049: 1

5. Speed Reduction Reverse Direction. To achieve a reverse direction from the input of a simple planetary gear set the planetary carrier must be the held member. To achieve a speed reduction in a reverse direction, the sun gear must be the input, which automatically makes the ring gear the output. Because the planetary carrier is held, the pinion gears, which are allowed to rotate on their axes, will act as idler gears to the sun gear. As the sun gear rotates in a clockwise direction the pinion gears rotating on their axes will rotate in the opposite direction of the sun gear (counterclockwise) because they are external gears. The ring gear, an internal gear,

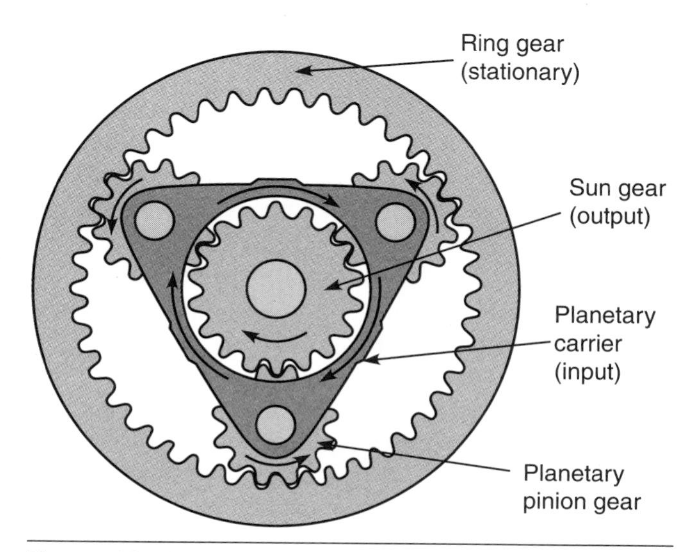

Figure 12-18 Gear ratio: maximum overdrive, speed increase/torque decrease.

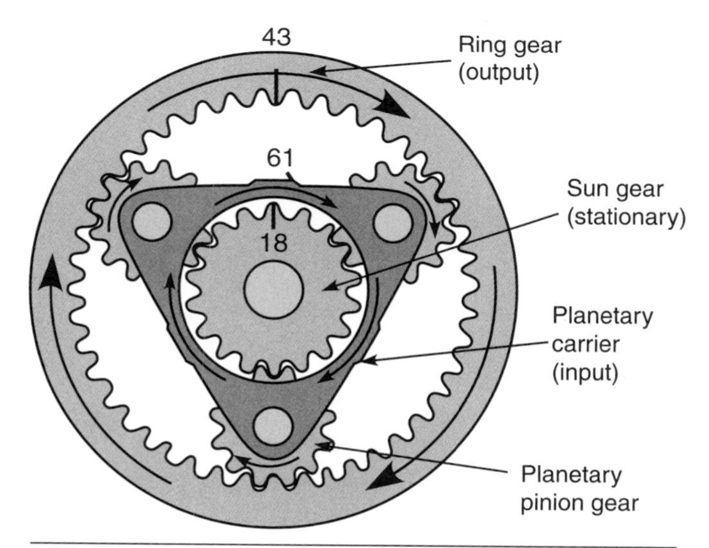

Figure 12-19 Gear ratio: minimum overdrive, speed increase/torque decrease.

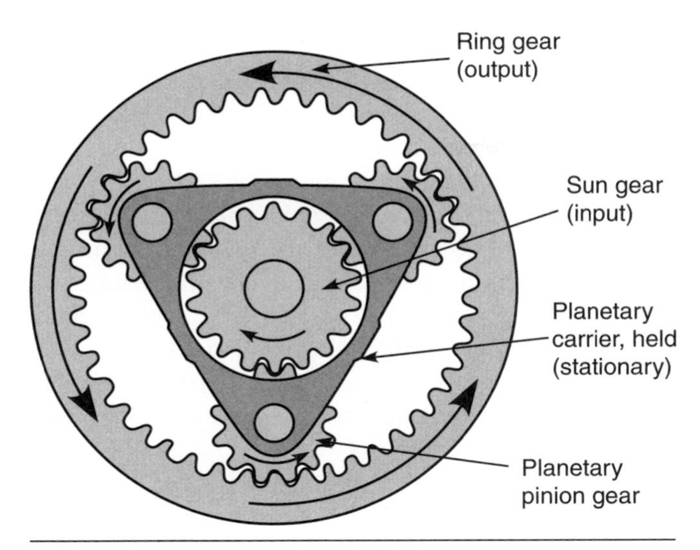

Figure 12-20 Gear ratio: reverse, speed decrease/ torque increase.

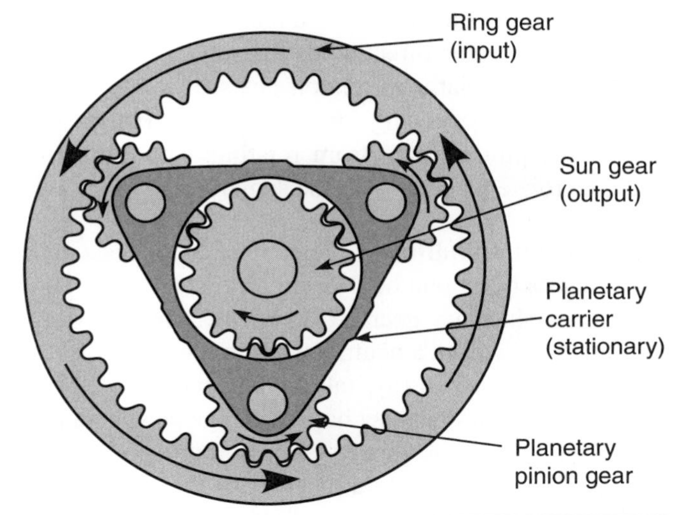

Figure 12-21 Gear ratio: reverse, speed increase/ torque decrease.

will rotate in the same direction (counterclockwise) as the planetary pinion gears and opposite to the sun gear. Reverse direction causes a speed reduction because the sun gear (input) has fewer teeth than the ring gear (output). Refer to **Figure 12-20.**

To calculate the gear ratio, divide the number of gear teeth of the input (sun gear 18 teeth) into the number of gear teeth of the output (ring gear 43 teeth).

43 teeth ÷ 18 teeth = Gear ratio of 2.3888∞ : 1

6. Speed increase Reverse Direction. Remember, to achieve reverse direction, the planetary carrier must be the held member. By reversing the roles of the sun gear and ring gear we can achieve a reverse direction in an overdrive condition. The ring gear is now the input member; which automatically makes the smaller sun gear the output member. With the ring gear as the input member rotating in a counterclockwise direction, the planetary pinion gears again act as idler gears because the planetary carrier is held and the pinion gears are forced to rotate on their axes. They will rotate in the same direction as the input ring gear but will force the output sun gear to rotate in the opposite direction (clockwise). Because the larger ring gear is driving the smaller sun gear, an overdrive condition occurs in reverse direction. Refer to **Figure 12-21.**

To calculate the gear ratio, divide the number of gear teeth of the input (ring gear 43 teeth) into the number of gear teeth of the output (sun gear 18 teeth).

18 teeth ÷ 43 teeth = Reverse overdrive gear ratio of 0.4186 : 1

7. Direct Drive. To achieve a direct-drive combination on a planetary gear set we must lock any two members together so that they can rotate at the same speed in the same direction. By locking two members together the third member is forced to follow along in the same direction and at the same speed. Refer to **Figure 12-22.**

In the example shown, if the sun gear and ring gear are rotated in a clockwise direction at the same speed, the planetary pinions will not rotate on their axes and, therefore, force the carrier to rotate in unison with the ring gear and sun gear at a ratio of 1:1. The sun gear, rotating in a clockwise direction, will try to force the pinion gear

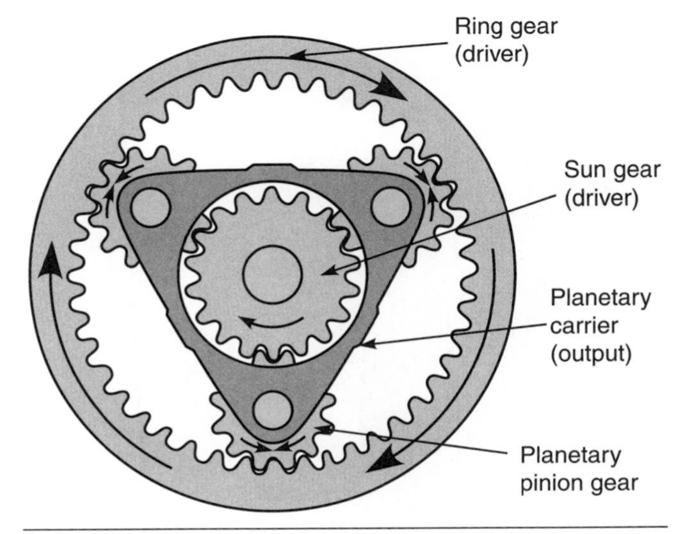

Figure 12-22 Gear ratio: direct drive, speed and direction are the same as the input.

to rotate in a counterclockwise direction. The ring gear, rotating in the same direction as the sun gear, will also try to force the planetary pinion gears to rotate clockwise. These two opposing forces cancel each other and prevent the pinion gears from rotating. The carrier is forced to rotate together with the sun and ring gears.

8. Neutral Condition. There must be at least one input member and one held member to achieve an output from a planetary gear set. If any two members are allowed to freewheel, a neutral condition results. A force (torque in this case) always takes the path of least resistance: if a planetary gear set has the sun gear as the input drive member and the carrier as the output member, the ring gear has to be the held member. If the ring gear is allowed to freewheel, the path of least resistance would be the ring gear, and it would be forced to rotate instead of the carrier. Because the carrier is the output member, the resistance on it would be greater than the resistance on the ring gear. Therefore, the path of least resistance would be the ring gear forcing it to rotate. This combination can occur even when the 1:1 ratio is provided by two inputs or by locking two members together and forcing them to rotate at the same speed. If one input is disconnected or the two drive members are unlocked, the result is the same: the torque will take the path of least resistance and a neutral condition will result. If there are two freewheeling members, neutral will always result. Refer to **Figure 12-23.**

Summary of Planetary Gearing. Here are some simple notes to remember. When you know what the planetary carrier is doing, you will be able to decipher

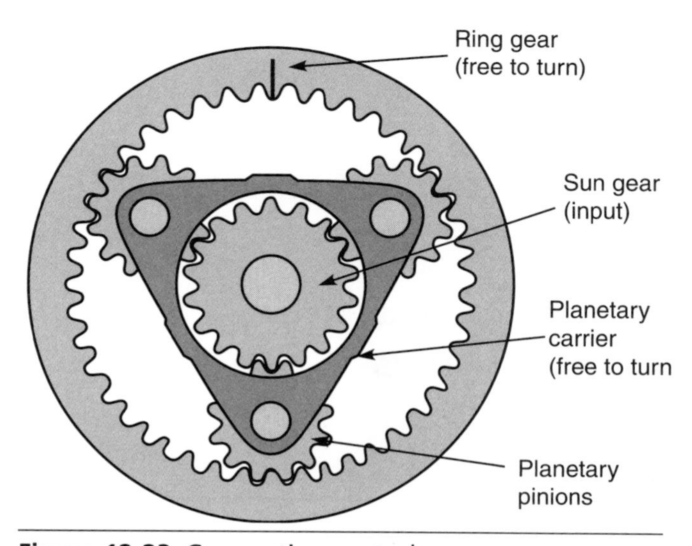

Figure 12-23 Gear ratio: neutral.

speed, torque, or direction of a planetary gear set.

- To calculate the number of gear teeth of the planetary carrier, add the number of teeth on the sun gear to the number of teeth on the ring gear.
- When calculating gear ratios, the carrier is always considered the largest gear in a planetary gear set.
- To calculate gear ratios, divide the number of gear teeth on the drive gear into the number of gear teeth of the driven gear. The held member plays no part in the ratio calculations.
- When the carrier is the output, a speed decrease/torque increase occurs in the same direction as the input.
- When the carrier is the input, a speed increase/torque decrease occurs in the same direction as the input.
- When the carrier is held, one of two reverse directions can occur to the output: One with a speed increase/torque decrease, or one with a speed decrease/torque increase.
- When two inputs move at the same speed or two members are locked to rotate at the same speed, a 1:1 gear ratio occurs.

CLUTCH PACKS

With planetary gear sets, there must be a means of holding certain components in a planetary gear set to produce an output. **Planetary shift transmissions** use a series of clutch packs to hold or provide an input to a particular planetary component to achieve a desired gear ratio through the planetary gear sets. In heavy-duty equipment, the clutch packs used are a multi-disc clutch comprised of friction discs and plates. The friction discs are made of ceramic/bronze compounds, Elastomer compounds, or graphite compounds to withstand high heat. These clutch discs are allocated within the transmission, and the oil provides cooling for these wet discs and plates. Some smaller equipment applications may still utilize organic/paper fibers on there friction discs. Four basic types of friction discs are shown in **Figure 12-24.**

The number of friction discs and plates used within a single clutch pack is strictly dictated by the application of the clutch pack and the torque loading applied to it. For example, first gear clutch pack may utilize 4 discs and 3 plates and second gear clutch pack may only utilize 3 discs and 2 plates of the same diameter. This is because first gear is susceptible to higher torque loads and requires a higher coefficient of friction than second gear.

As like any multi-disc clutch, the friction discs may have their teeth on either the outer edge of the disc or

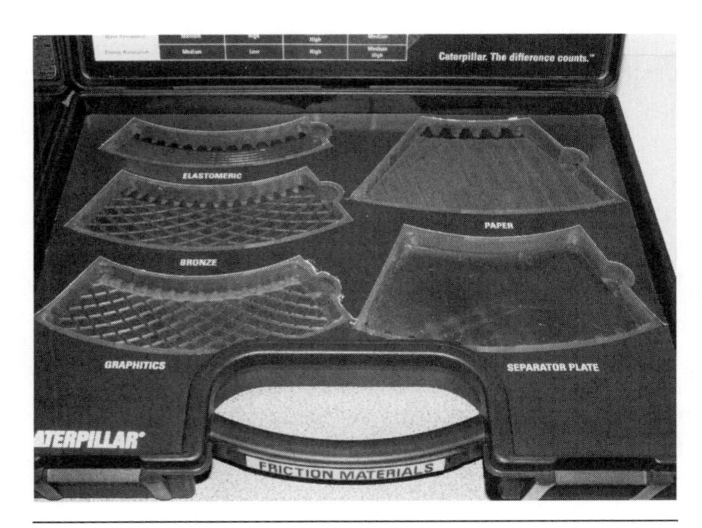

Figure 12-24 Four types of clutch disc material.

on the center edge of the disc. This is dependent on the component (sun gear, ring gear, planetary carrier) that it is either trying to remain stationary or providing a drive. If the friction discs are toothed on the outer perimeter, the friction plates will be toothed on the inner perimeter and vice versa. In many applications, friction plates utilize "dog teeth" or tangs instead of spur-type teeth on their plates. Within each clutch pack, an actuation piston provides the engagement of the clutch pack by means of hydraulic oil pressure. Return springs are typically used to return the actuation piston and release the clutch pack. These components are shown in **Figure 12-25.**

Clutch packs are typically one of two types: rotating clutch packs and **stationary clutch packs.** A **rotating clutch pack,** shown in **Figure 12-26,** provides an input to a planetary component; this means that whenever the clutch pack is engaged, it will transfer engine torque to a specified planetary component and will rotate that component at a specific engine speed.

A stationary clutch pack when engaged, holds a specific planetary component from rotating. See **Figure 12-27.** Typically, stationary clutch packs are toothed to the transmission housing and lock the component to it. For example, if we want maximum torque/minimum speed ratio from a planetary gear set we know that the ring gear must be held and the sun gear must be the input. Therefore a stationary clutch pack is used to lock the ring gear to the transmission housing, holding the ring gear stationary. A rotating clutch pack could then be used to transfer engine torque to the sun gear; the clutch pack will rotate with the sun gear when engaged. With the two clutch packs engaged,

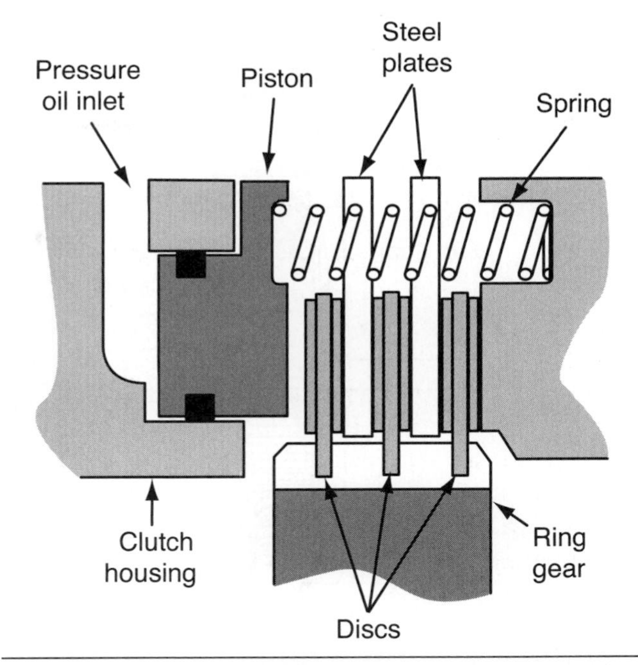

Figure 12-25 Components of a multiple-disc clutch pack.

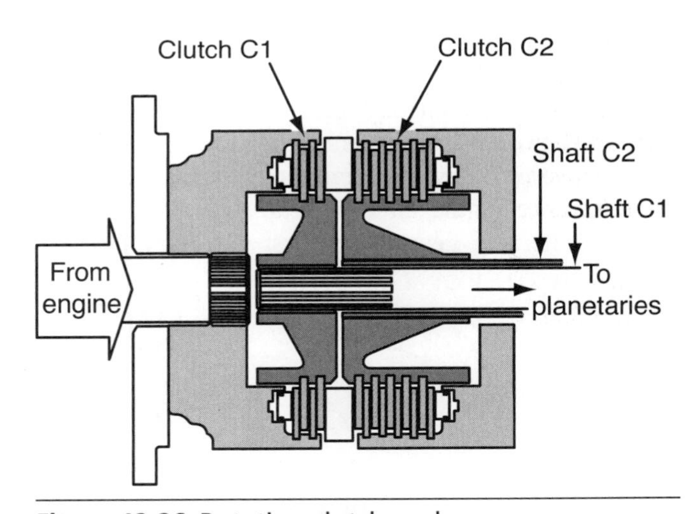

Figure 12-26 Rotating clutch packs.

the planetary carrier automatically produces the output at maximum torque/ minimum speed.

COMPOUND PLANETARY TRANSMISSIONS

Many heavy-duty applications require equipment to have a multitude of forward and reverse speeds. A single planetary gear set may not be able to accomplish this; therefore, compound planetary gear sets are used. Compounding planetary gear sets allows manufacturers

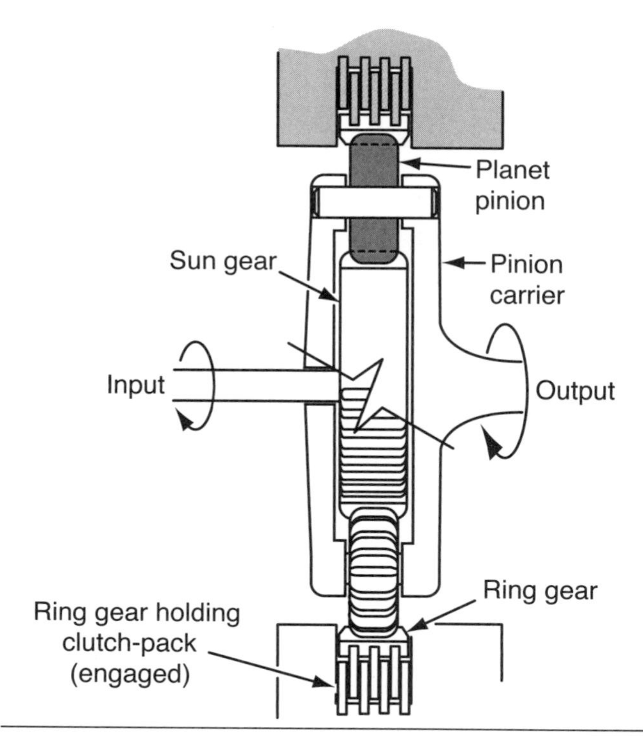

Figure 12-27 Stationary clutch pack.

to design transmissions to meet their specific speed and torque ratios for any application. This is achieved by locating multiple planetary gear sets one behind the other(s). Each gear set transfers engine torque through to the transmission output shaft and each gear set is connected by any one of the three planetary components. For example, the ring gear of the first planetary may be connected to the carrier of the second planetary, and the second planetary sun gear may be connected to the sun gear of the third planetary. An illustration of multiple gear sets is shown in **Figure 12-28.**

The combinations of planetary gear ratios are limited only by the number of planetary gear sets used. In some scraper applications, an eight speed planetary transmission is used. In the heavy-duty equipment industry the configurations of transmissions seem almost infinite. However, we will only describe two popular transmissions used today. It is the responsibility of the technician to become familiar with the specific transmissions. This chapter will provide some basics to understanding powershift transmissions and their principles; it should not be used as a service manual.

Planetary Powerflows

The transmission chosen to describe planetary powerflows is from a Caterpillar® 988G wheel loader. Wheel loaders are very popular in the heavy-duty equipment field, so we will evaluate powerflows and the hydraulic control system using this machine and transmission as an example. The 988G loader has a bucket capacity of approximately 8.2 to 9.2 cubic yards, depending on the application. The operating weight is approximately 110,000 pounds. It is powered by a 3456 engine rated at 475 hp. It has a 4-speed forward, 3-speed reverse, planetary shift transmission that can be operated in automatic or powershift mode. See **Figure 12-29.**

Figure 12-28 Compound planetary transmission using four interconnected planetary gear sets.

Six hydraulically activated clutches (1 rotating clutch pack, 5 stationary clutch packs) are used to achieve this combination with five planetary gear sets. When energized, six electronically controlled solenoids control the oil flow to the clutch packs. A speed clutch and a directional clutch must be engaged to provide powerflow through the transmission. The speed clutch engages before the directional clutch. Number 1 clutch pack and planetary gear set provide reverse direction. Number 2 clutch and planetary gear set provide forward direction. The number 6 clutch pack provides first speed, number 5 clutch pack provides second speed, number 4 clutch pack provides third speed and number 3 clutch pack provides fourth speed. For speed control purposes in reverse, number 3 clutch pack cannot be engaged for fourth speed reverse. The torque converter has an impeller clutch with the option of a lockup clutch as well.

TRANSMISSION

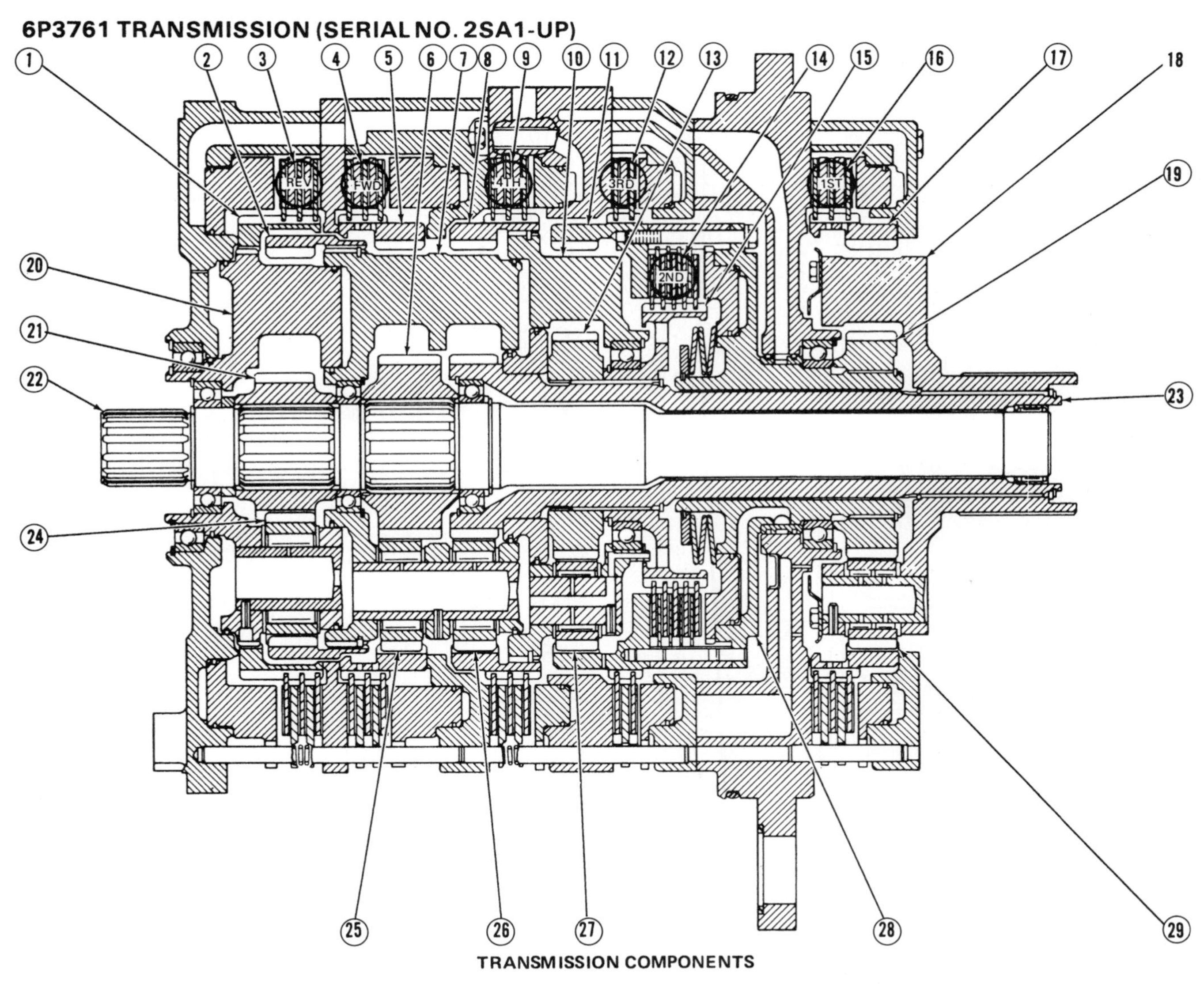

1. Ring gear for No.1 clutch.
2. Coupling gear.
3. No.1 clutch.
4. No.2 clutch.
5. Ring gear for No.2 clutch.
6. No.2 sun gear.
7. No.2 and No.3 carrier.
8. Ring gear for No.3 clutch.
9. No.3 clutch.
10. No.4 carrier.
11. Ring gear for No.4 clutch.
12. No.4 clutch.
13. No.4 sun gear.
14. No.5 clutch.
15. Rotating hub.
16. No.6 clutch.
17. Ring gear for No.6 clutch.
18. No.6 carrier.
19. No.6 sun gear.
20. No.1 carrier.
21. No.1 sun gear.
22. Input shaft.
23. Output shaft.
24. No.1 planetary gears.
25. No.2 planetary gears.
26. No.3 planetary gears.
27. No.4 planetary gears.
28. Housing assembly.
29. No.6 planetary gears.

Figure 12-29 Four-speed forward, three-speed reverse, planetary shift transmission.(*Courtesy of Caterpillar*)

First Speed Forward Power Flow. Refer to **Figure 12-30.** When the operator preselects first speed forward, number 6 and number 2 clutch are engaged. Number 2 clutch is the forward speed clutch, this clutch will stay engaged for all the forward speeds; it holds ring gear (6) stationary. Number 6 clutch is the first speed clutch, which holds ring gear (17) stationary. Input shaft (23) rotates forward drive planetary sun gear (4) at torque converter speed. Because forward drive planetary ring gear is held and the sun gear is the input drive, planetary gears (25) rotate around the ring gear. Because number 2 and 3 planetary carrier (7) occupy the same housing, it is forced to rotate in the same direction as the input shaft and sun gear. The carrier becomes the output at low speed/high torque. Because number 3 sun gear is attached to the output shaft (22), it acts as the stationary member to planetary gear set 3. This will cause the number 3 ring gear (9) to rotate. The number 3 planetary gears will rotate the ring gear, which is splined to number 4 planetary carrier, causing it to rotate. The number 4 sun gear (11) is also splined to the output shaft (22) which acts as a stationary member for the number 4 planetary. This will cause number 4 ring gear (13) to rotate as the output. Number 4 ring gear is fastened to the rotating clutch housing that has sun gear from the number 5 planetary gear set (19) splined to it. Because stationary clutch

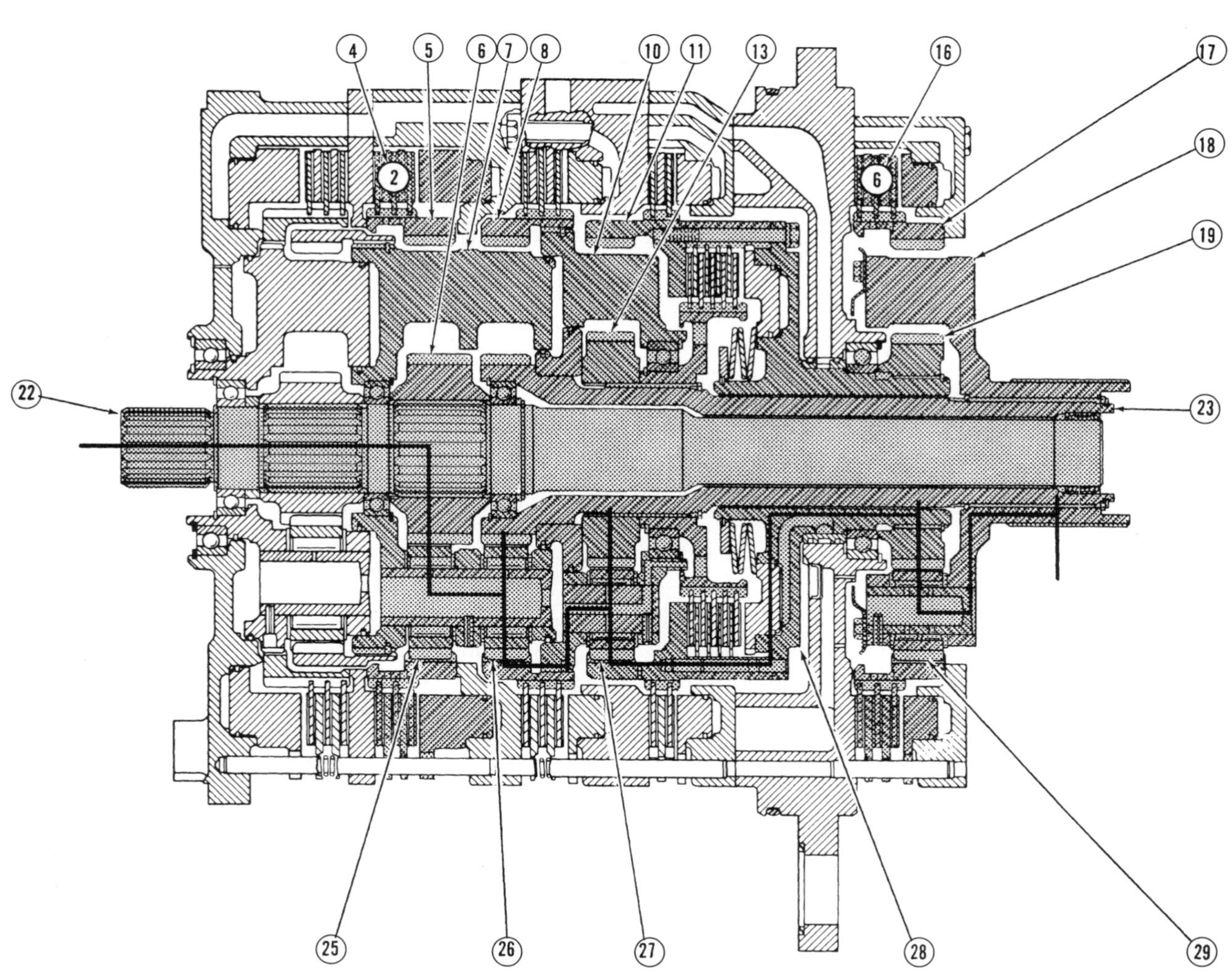

POWER FLOW IN FIRST SPEED FORWARD (No.2 and No.6 Clutches Engaged)

4. No.2 clutch.
5. Ring gear for No.2 clutch.
6. No.2 sun gear.
7. No.2 and No.3 carrier.
8. Ring gear for No.3 clutch.
10. No.4 carrier.
11. Ring gear for No.4 clutch.
13. No.4 sun gear.
16. No.6 clutch.
17. Ring gear for No.6 clutch.
18. No.6 carrier.
19. No.6 sun gear.
22. Input shaft.
23. Output shaft.
25. No.2 planetary gears.
26. No.3 planetary gears.
27. No.4 planetary gears.
28. Housing assembly.
29. No.6 planetary gears.

Figure 12-30 First speed forward with number 2 and number 6 clutch packs engaged. (*Courtesy of Caterpillar*)

pack number 6 (16) is holding number 5 ring gear (17), the planetary carrier becomes the output through output shaft (22). Output shaft (22) and number 5 carrier (18) are splined together and rotate at equal speeds.

As a result, a multiple gear reduction (low speed/high torque) output is divided between number 3 and 4 planetary gear set, and number 5 planetary gear set.

Second Speed Forward Powerflow. Refer to **Figure 12-31.** When the operator selects second speed forward, number 5 and number 2 clutches are engaged. Number 2 clutch is the forward speed clutch, which holds ring gear (6) stationary. Input shaft (23) rotates forward drive planetary sun gear number 4 at torque converter speed. Because forward drive planetary ring gear is held and the sun gear is the input drive, planetary gears (25) rotate around the ring gear. Because number 2 and 3 planetary carrier (7) occupies the same housing, it is forced to rotate in the same direction as the input shaft and sun gear. The carrier becomes the output at low speed/high torque. Because number 3 sun gear is attached to the output shaft (22), it acts as the stationary member to planetary gear set 3. This will cause the number 3 ring gear (9) to rotate. The number

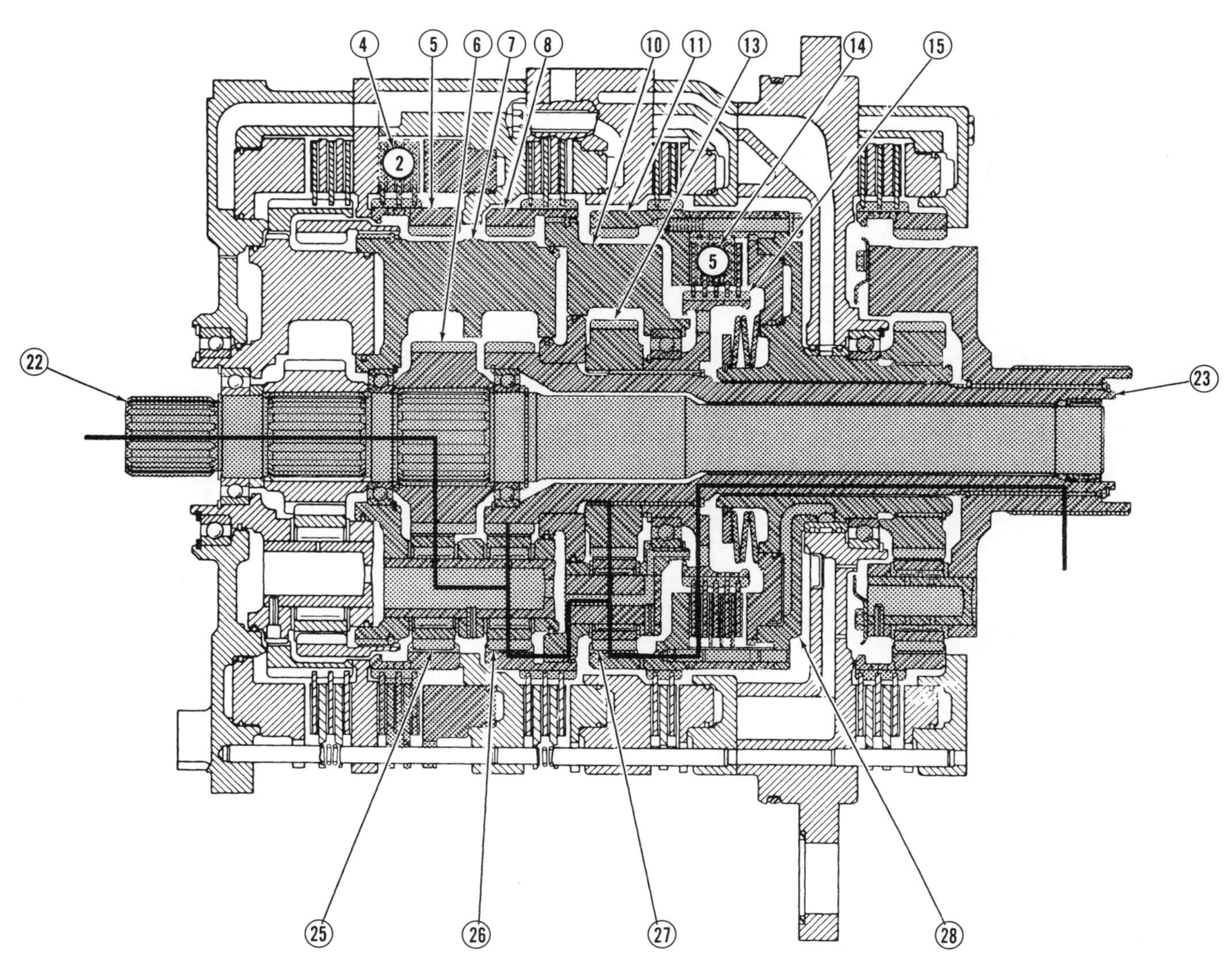

Figure 12-31 Second speed forward with number 2 and number 5 clutch packs engaged. (*Courtesy of Caterpillar*)

3 planetary gears will rotate the ring gear, which is splined to number 4 planetary carrier. Because rotating clutch pack number 5 is engaged, it will prevent number 4 ring gear and sun gear from rotating at different speeds. This causes the input from number 4 carrier to rotate the planetary gear set as one unit or at a ratio of 1:1. The number 4 sun gear (11) is splined to the output shaft (22).

The result is a single reduction (low speed/high torque) through planetary gear set number 3 splitting the torque path with number 4 planetary gear set.

Third Speed Forward Powerflow. Refer to **Figure 12-32.** When the operator selects third speed forward, number 4 and number 2 clutches are engaged. Number 2 clutch is the forward speed clutch, which holds ring gear number 6 stationary. Input shaft (23) rotates forward drive planetary sun gear number 4 at torque converter speed. Because forward drive planetary ring gear is held and the sun gear is the input drive, planetary gears (25) rotate around the ring gear. Because number 2 and 3 planetary carrier (7) occupies the same housing, it is forced to rotate in the same direction as the input shaft and sun gear. The carrier becomes the output at low speed/high torque. Because number 3 sun gear is attached to the output shaft (22), it acts as the stationary member to planetary gear set 3.

THIRD SPEED FORWARD

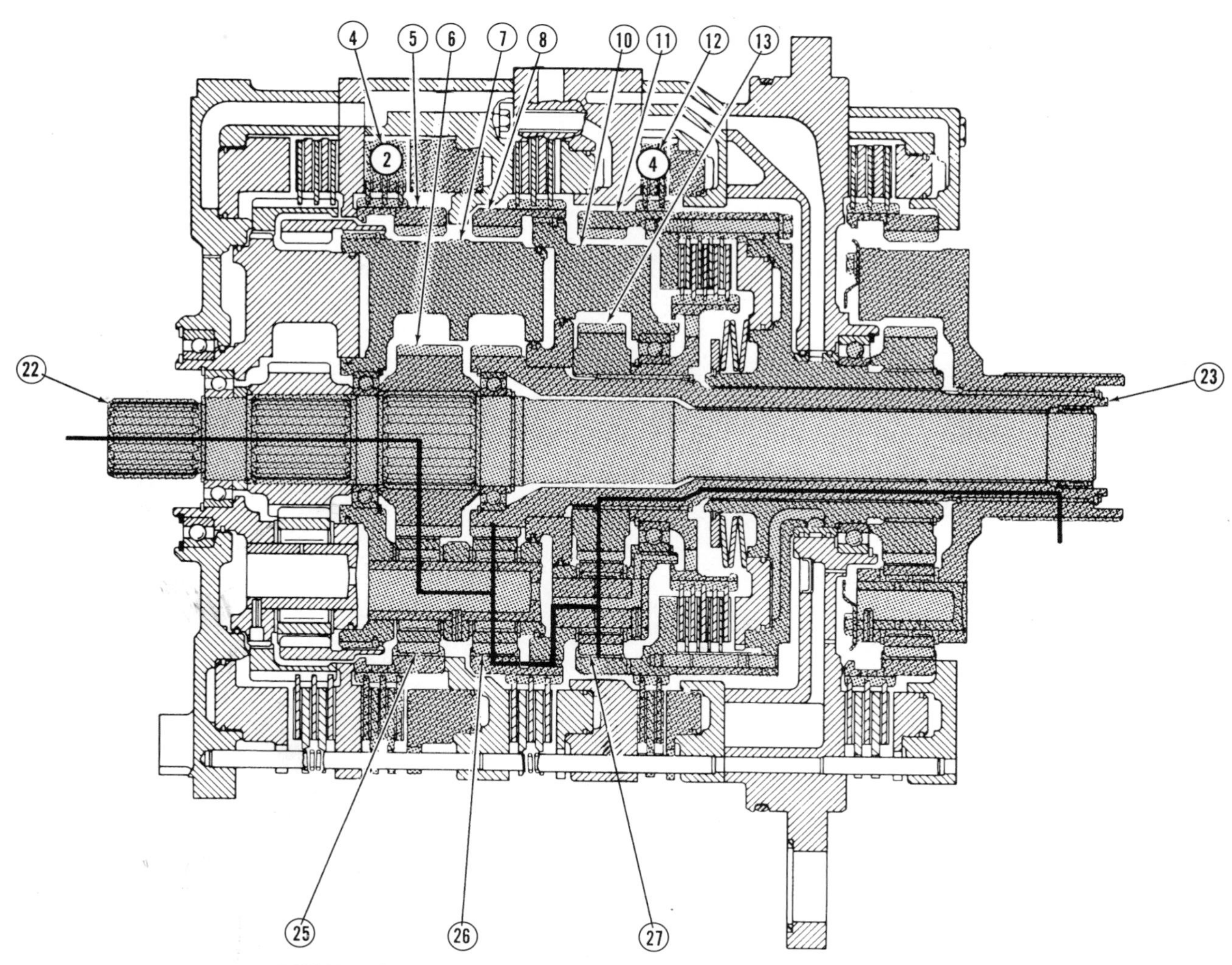

POWER FLOW IN THIRD SPEED FORWARD (No.2 and No.4 Clutches Engaged)

4. No.2 clutch.
5. Ring gear for No.2 clutch.
6. No.2 sun gear.
7. No.2 and No.3 carrier.
8. Ring gear for No.3 clutch.
10. No.4 carrier.
11. Ring gear for No.4 clutch.
12. No.4 clutch.
13. No.4 sun gear.
22. Input shaft.
23. Output shaft.
25. No.2 planetary gears.
26. No.3 planetary gears.
27. No.4 planetary gears.

Figure 12-32 Third speed forward with number 2 and number 4 clutch packs engaged. (*Courtesy of Caterpillar*)

This will cause the number 3 ring gear (9) to rotate. The number 3 planetary gears will rotate the ring gear, which is splined to number 4 planetary carrier. Number 4 planetary carrier becomes the input to its planetary gear set. Because the number 4 clutch pack is engaged, it is holding stationary ring gear (13). Planetary number 4 sun gear becomes the output drive through output shaft (22). Note that number 4 planetary is driving in an overdrive condition compared to its input. The output shaft is still in an underdrive condition due to number 3 planetary providing the input to number 4 planetary at a maximum torque minimum speed condition multiplied by the reduced speed of the rotating sun gear. Because number 3 sun gear is attached to the output shaft and acts as a stationary member for the number 3 planetary gear set, it reduces the output speed of the ring gear even further as the output shaft begins to rotate. This increases torque output from the transmission.

The result is a single reduction (low speed/high torque) through planetary gear-set number 3, splitting the torque path with number 4 planetary gear set in an overdrive condition.

Fourth Speed Forward Powerflow. Refer to **Figure 12-33.** When the operator selects fourth speed forward, number 3 and number 2 clutches are engaged.

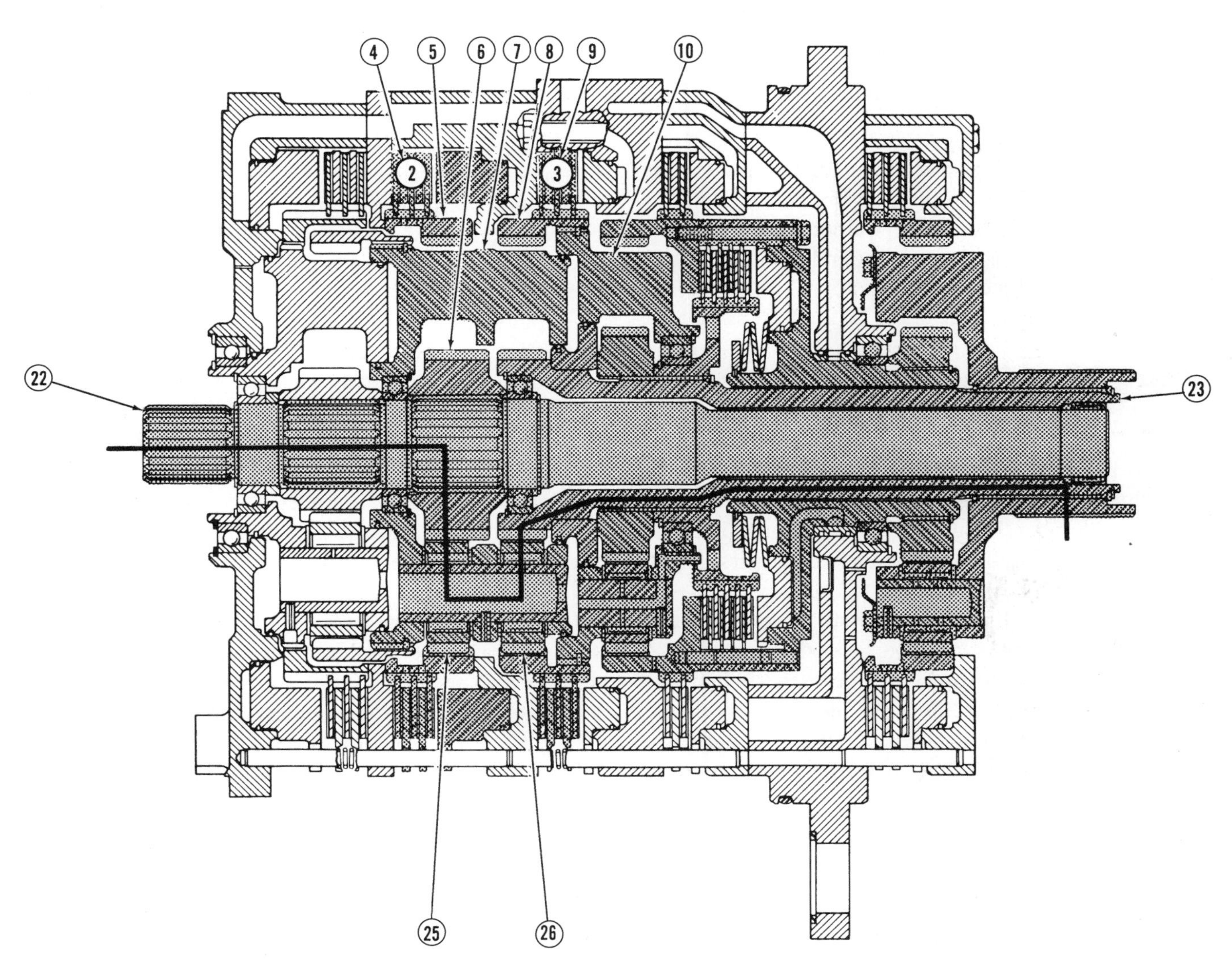

Figure 12-33 Fourth speed forward with number 2 and number 3 clutch packs engaged. (*Courtesy of Caterpillar*)

Number 2 clutch is the forward speed clutch, which holds ring gear number 6 stationary. Input shaft (23) rotates forward drive planetary sun gear number 4 at torque converter speed. Because forward drive planetary ring gear is held and the sun gearis the input drive, planetary gears (25) rotate around the ring gear. Because number 2 and 3 planetary carrier (7) occupies the same housing, it is forced to rotate in the same direction as the input shaft and sun gear. The carrier becomes the output at low speed/high torque. Because number 3 ring gear is held stationary by the number 3 clutch pack, it acts as the stationary member to planetary gear set 3. This will cause the number 3 sun gear to rotate as the output member due to the input of the number 2–3 carrier. It is a direct output with no reduction as the speed of the output shaft increases. This is the highest speed ratio at maximum torque/minimum speed through two compound planetary gear-sets.

Note: Within this specific machine application, no overdrive condition occurs between the transmission input shaft and output shaft. There can be as many as four compound gear reductions for first speed.

First Speed Reverse Powerflow. Refer to **Figure 12-34.** When the operator selects first speed reverse, number 6 and number 1 clutches are engaged. Number 1 clutch is the reverse speed clutch, which holds clutch coupling for number 1 planetary carrier stationary. Number 1 clutch will stay engaged for all the reverse speeds. Number 6 clutch is the first speed clutch, which holds ring gear (17) stationary.

Input shaft (23) rotates reverse drive planetary sun gear (21) at torque converter speed. According to the basic principals of planetary gearing, "whenever the planetary carrier is the held member, a reverse direction from the input occurs." This automatically makes the ring gear (2) the output member (maximum torque/minimum speed) for the number 1 planetary in a reverse direction to the input.

Because ring gear number 1 is interconnected to planetary carrier 2 and 3, they will provide input to the number 2/3 planetary in a reverse direction. Because number 3 sun gear is attached to the output shaft (22), it acts as the stationary member to planetary gear set 3. This will cause the number 3 ring gear (9) to rotate. The number 3 planetary gears will rotate the ring gear, which is splined to number 4 planetary carrier, causing it to rotate. This becomes the input to number 4 planetary. The number 4 sun gear (11) is also splined the output shaft (22), which also acts as a stationary member for the number 4 planetary. This will cause number 4 ring gear (13) to rotate as the output. Number 4 ring gear is fastened to the rotating clutch housing that has sun gear from the number 5 planetary gear set (19) splined to it. Because stationary clutch pack number 6 (16) is holding number 5 ring gear (17), the planetary carrier becomes the output through output shaft (22). Output shaft (22) and number 5 carrier (18) are splined together and rotate at equal speeds.

As a result, a multiple reduction (low speed/high torque) output is divided between number 3 and 4 planetary gear set and number 5 planetary gear set.

Second Speed Reverse Powerflow. Refer to **Figure 12-35.** When the operator selects second speed reverse, number 5 and number 1 clutches are engaged. Number 1 clutch is the reverse speed clutch, which holds clutch coupling for number 1 planetary carrier stationary. Number 5 clutch is the second speed clutch, which holds number 4 ring gear and sun gear together.

Input shaft (23) rotates reverse drive planetary sun gear (21) at torque converter speed. Because the carrier is held, this automatically makes the ring gear (2) the output member (maximum torque/minimum speed) for the number 1 planetary in a reverse direction to the input.

Because ring gear number 1 is interconnected to planetary carrier 2 and 3, they will provide input to the number 2–3 planetary in a reverse direction. Because number 3 sun gear is attached to the output shaft (22), it acts as the stationary member to planetary gear set 3. This will cause the number 3 ring gear (9) to rotate. The number 3 planetary gears will rotate the ring gear, which is splined to number 4 planetary carrier, causing it to rotate. The carrier becomes the output at low speed/high torque. Because number 3 sun gear is attached to the output shaft (22), it acts as the stationary member to planetary gear set 3. This will cause the number 3 ring-gear (9) to rotate. The number 3 planetary gears will rotate the ring gear, which is splined to number 4 planetary carrier. Because rotating clutch pack number 5 is engaged, it will prevent number 4 ring gear and sun gear from rotating at different speeds. This causes the input from number 4 carrier to rotate the planetary gear set as one unit or at a ratio of 1:1. The number 4 sun gear (11) is splined to the output shaft (22).

FIRST SPEED REVERSE

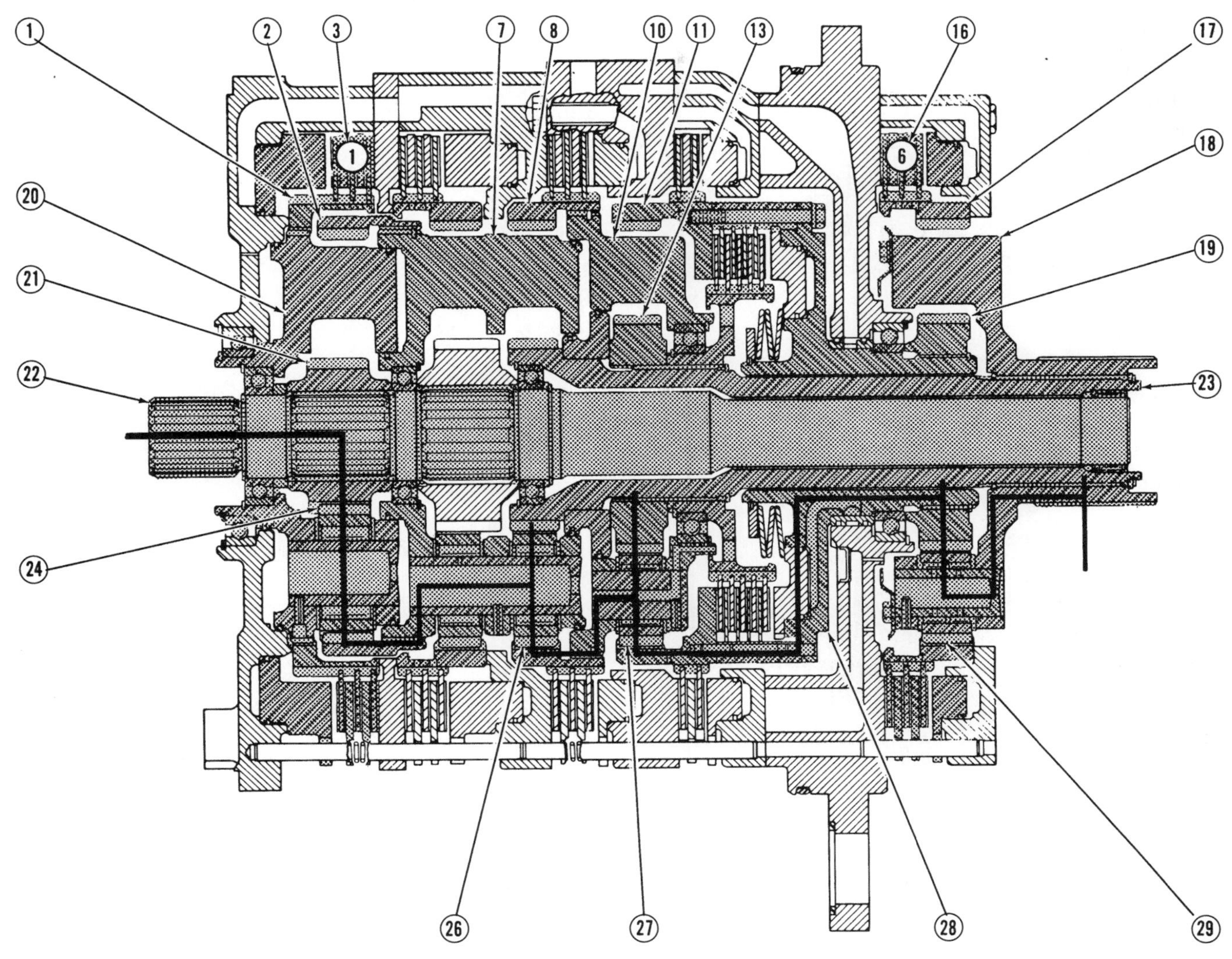

POWER FLOW IN FIRST SPEED REVERSE (No.1 and No.6 Clutches Engaged)

1. Ring gear for No.1 clutch.
2. Coupling gear.
3. No.1 clutch.
7. No.2 and No.3 carrier.
8. Ring gear for No.3 clutch.
10. No.4 carrier.
11. Ring gear for No.4 clutch.
13. No.4 sun gear.
16. No.6 clutch.
17. Ring gear for No.6 clutch.
18. No.6 carrier.
19. No.6 sun gear.
20. No.1 carrier.
21. No.1 sun gear.
22. Input shaft.
23. Output shaft.
24. No.1 planetary gears.
26. No.3 planetary gears.
27. No.4 planetary gears.
28. Housing assembly.
29. No.6 planetary gears.

Figure 12-34 First speed reverse with number 1 and number 6 clutch packs engaged. (*Courtesy of Caterpillar*)

The result is a single reduction (low speed/high torque) through planetary gear set number 3, splitting the torque path with number 4 planetary gear set.

Third Speed Reverse Powerflow. Refer to **Figure 12-36.** When the operator selects third speed reverse, number 4 and number 1 clutches are engaged. Number 1 clutch is the reverse speed clutch, which holds clutch coupling for number 1 planetary carrier stationary. Number 4 clutch is the third speed clutch, which holds number 4 ring gear and sun gear together.

Input shaft (23) rotates reverse drive planetary sun gear (21) at torque converter speed. Because the carrier is held this automatically makes the ring gear (2) the output member (maximum torque/minimum speed) for the number 1 planetary in a reverse direction to the input.

SECOND SPEED REVERSE

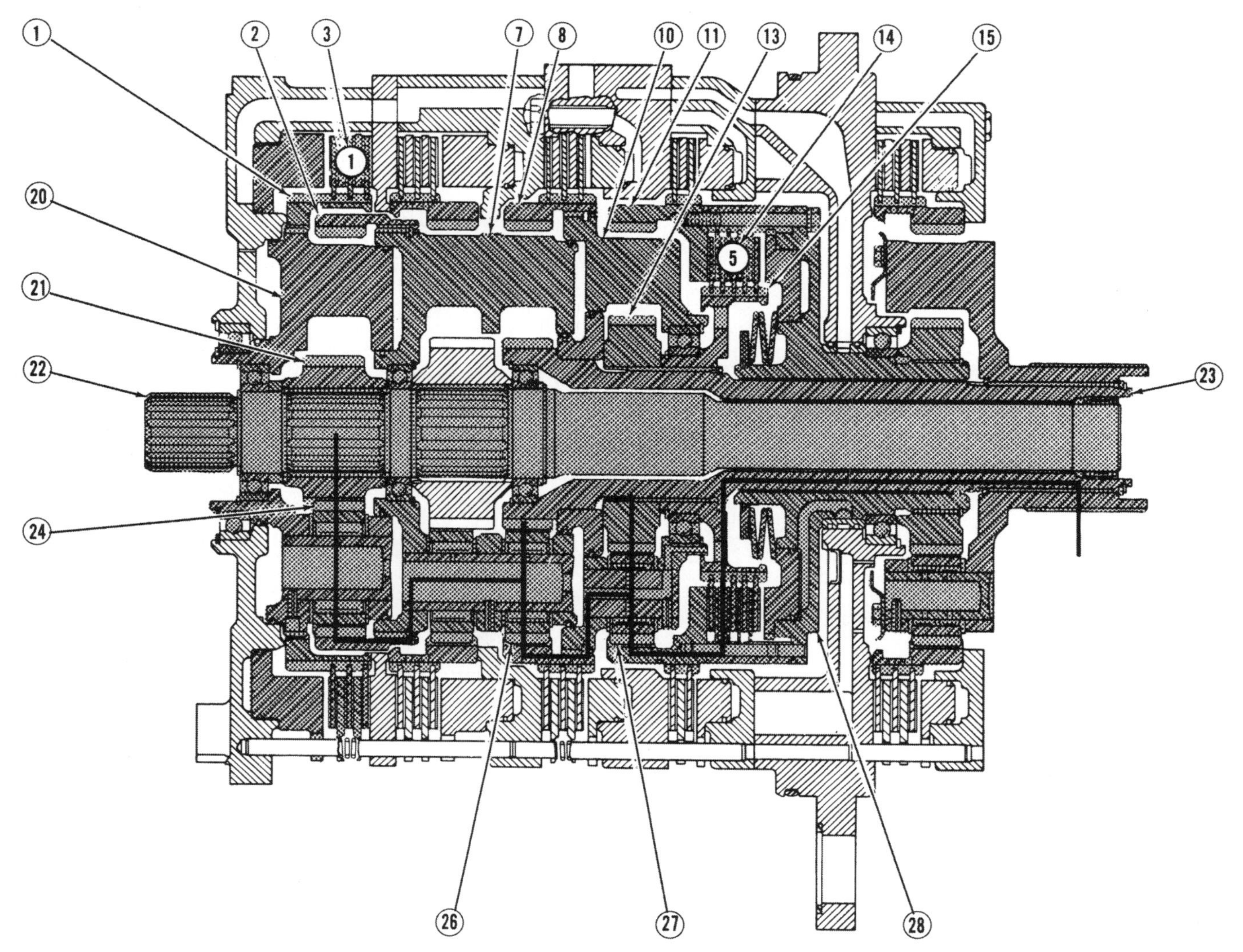

POWER FLOW IN SECOND SPEED REVERSE (No.1 and No.5 Clutches Engaged)

1. Ring gear for No.1 clutch.
2. Coupling gear.
3. No.1 clutch.
7. No.2 and No.3 carrier.
8. Ring gear for No.3 clutch.
10. No.4 carrier.
11. Ring gear for No.4 clutch.
13. No.4 sun gear.
14. No.5 clutch.
15. Rotating hub.
20. No.1 carrier.
21. No.1 sun gear.
22. Input shaft.
23. Output shaft.
24. No.1 planetary gears.
26. No.3 planetary gears.
27. No.4 planetary gears.
28. Housing assembly.

Figure 12-35 Second speed reverse with number 1 and number 5 clutch packs engaged. *(Courtesy of Caterpillar)*

Because ring gear number 1 is interconnected to planetary carrier 2 and 3, they will provide input to the number 2–3 planetary in a reverse direction. Because number 3 sun gear is attached to the output shaft (22), it acts as the stationary member to planetary gear set 3. This will cause the number 3 ring gear (9) to rotate. The number 3 planetary gears will rotate the ring gear, which is splined to number 4 planetary carrier. Number 4 planetary carrier becomes the input to its planetary gear set. Because the number 4 clutch pack is engaged, it is holding stationary ring gear (13). Planetary number 4 sun gear becomes the output drive through output shaft (22). Note that number 4 planetary is driving in an overdrive condition compared to its input. The output shaft is still in an underdrive condition due to number 3 planetary providing the input to number 4 planetary at a maximum torque minimum speed condition multiplied by the reduced

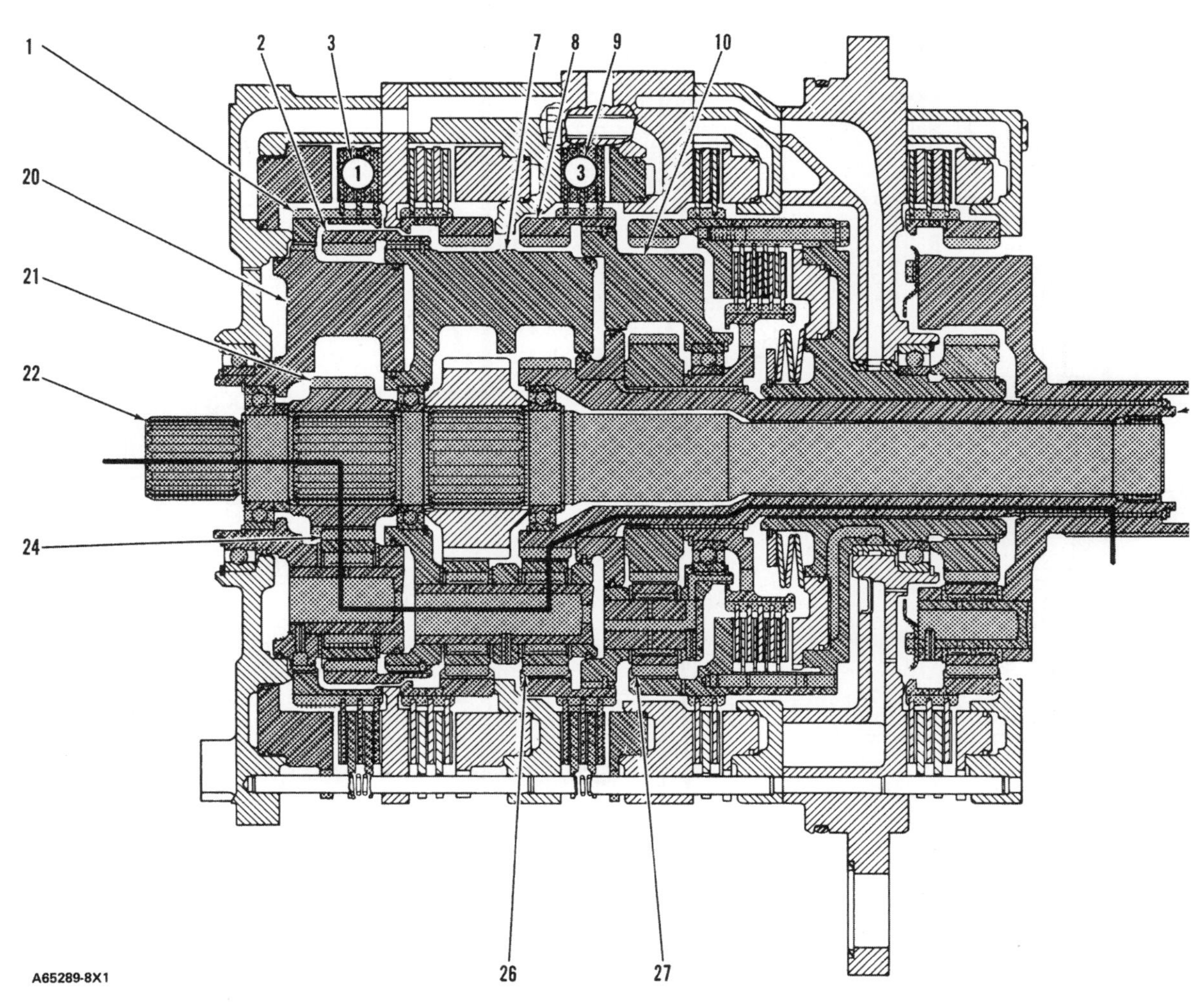

Figure 12-36 Third speed reverse with number 1 and number 4 clutch packs engaged. (*Courtesy of Caterpillar*)

speed of the rotating sun gear. Because number 3 sun gear is attached to the output shaft and acts as a stationary member for the number 3 planetary gear set, it reduces the output speed of the ring gear even further as the output shaft begins to rotate. This increases torque output from the transmission.

The result is a single reduction (low speed/high torque) through planetary gear set number 3, splitting the torque path with number 4 planetary gear set in an overdrive condition.

Transmission Hydraulic System

The transmission hydraulic system is responsible for providing the median for the torque converter, the pressure used to activate all the clutch packs, effective cooling and lubrication within the transmission, and to direct and control oil pressure and flow within the system.

Refer to **Figure 12-37.** The system consists of a positive-displacement external gear pump (8), a transmission filter (1), a transmission hydraulic control relief

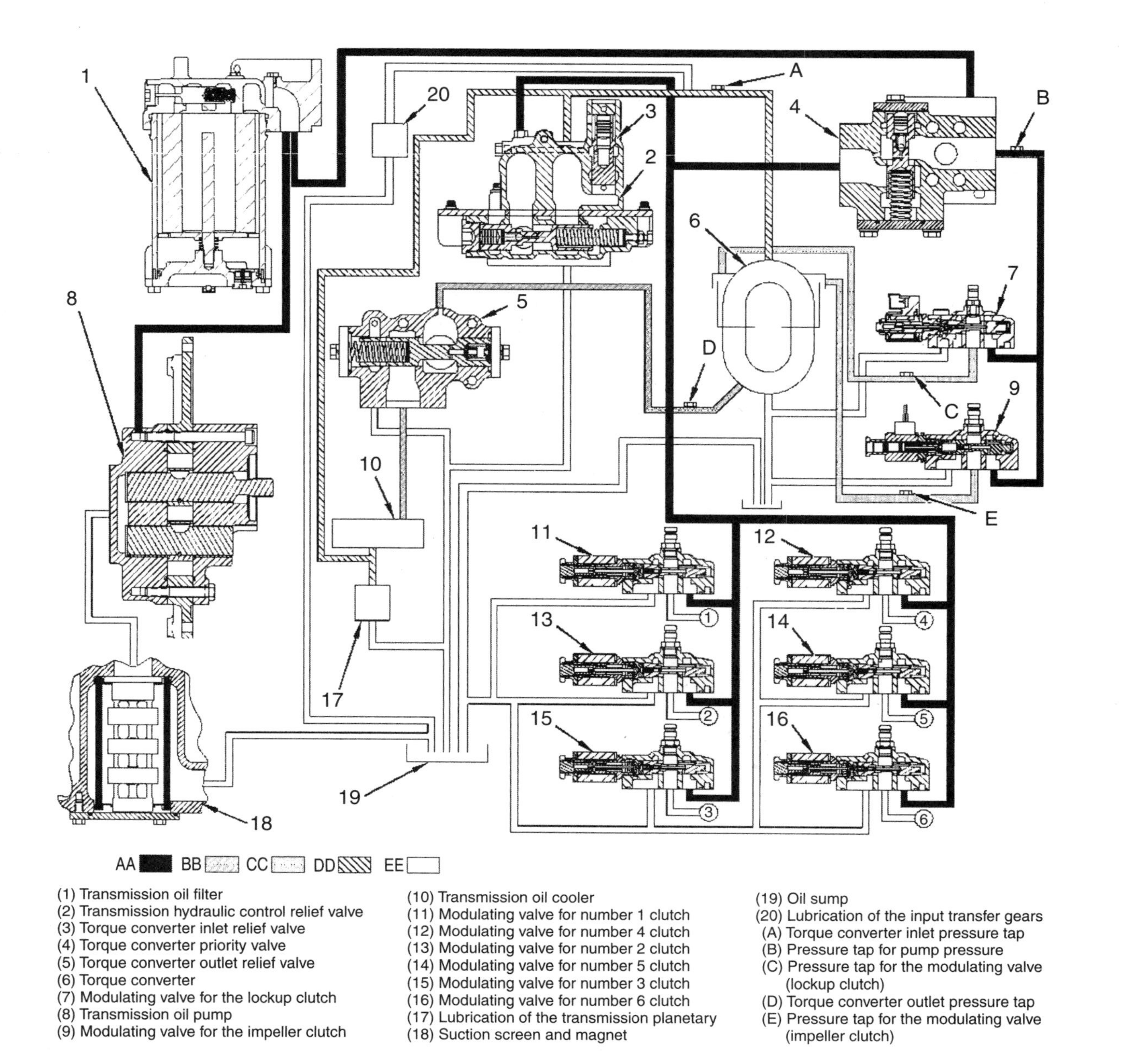

Figure 12-37 Transmission hydraulic control-valve body. (*Courtesy of Caterpillar*)

valve (2), a torque converter priority valve (4), a torque converter outlet relief valve (5), a lockup clutch modulating valve (7), an impeller clutch modulating valve (9), modulating valves for clutch packs (11 through 16), a transmission oil cooler (10), and a pump suction screen and magnet assembly (18). The transmission hydraulic control valve group is mounted on the topside of the transmission planetary gearsets. The transmission housings provide passageways for oil to be directed to the appropriate clutch packs and lubrication paths throughout the transmission.

Hydraulic Oil Flow in Neutral. Refer to **Figure 12-38**. When the engine is started, the transmission pump is driven by the torque converter housing at engine speed. The pump (8) picks up the oil from the transmission sump through the inlet screen and magnet assembly (18). Oil flows first through

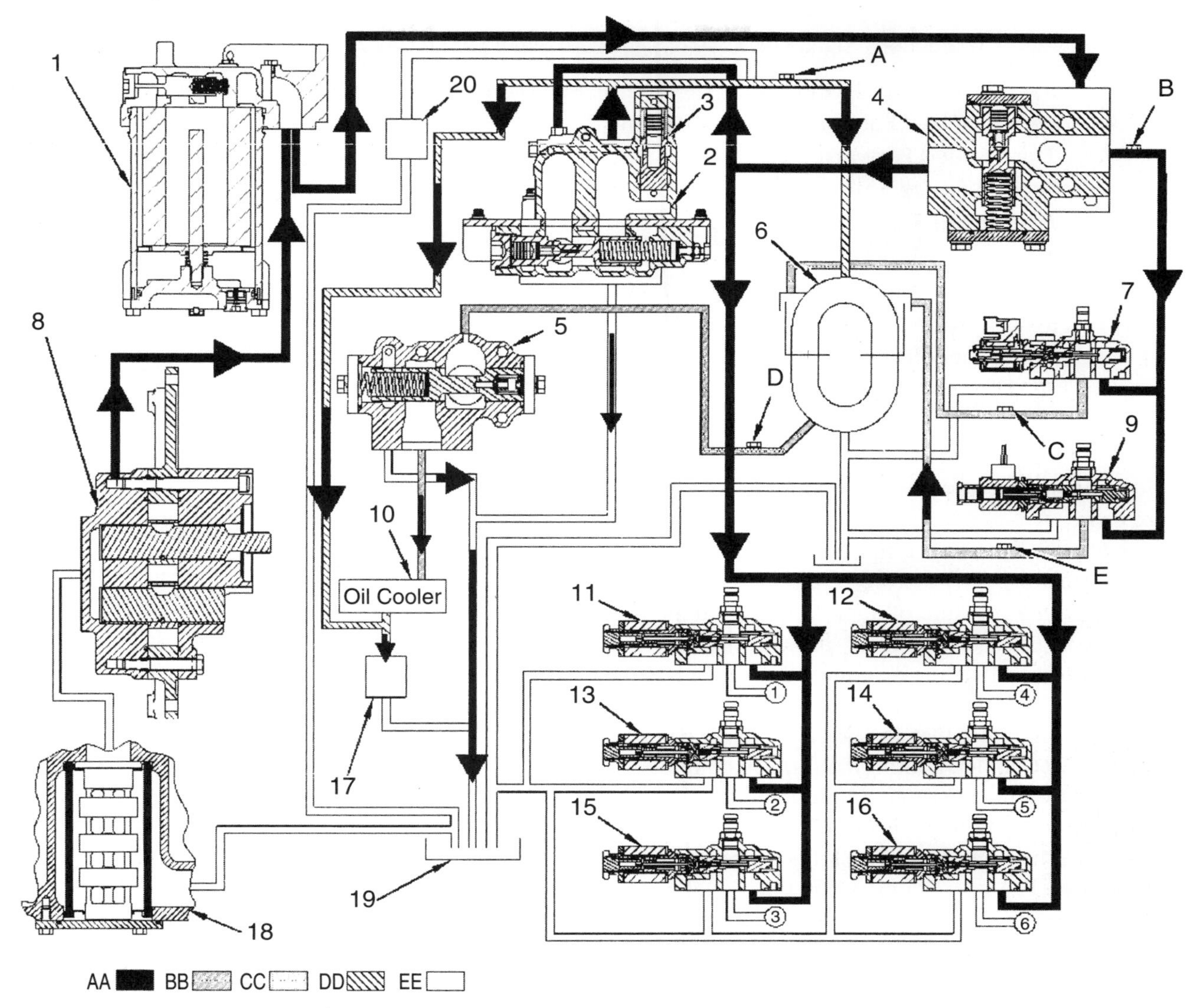

(1) Transmission oil filter
(2) Transmission hydraulic control relief valve
(3) Torque converter inlet relief valve
(4) Torque converter priority valve
(5) Torque converter outlet relief valve
(6) Torque converter
(7) Modulating valve for the lockup clutch
(8) Transmission oil pump
(9) Modulating valve for the impeller clutch
(10) Transmission oil cooler
(11) Modulating valve for number 1 clutch
(12) Modulating valve for number 4 clutch
(13) Modulating valve for number 2 clutch
(14) Modulating valve for number 5 clutch
(15) Modulating valve for number 3 clutch
(16) Modulating valve for number 6 clutch
(17) Lubrication of the transmission planetary
(18) Suction screen and magnet
(19) Oil sump
(20) Lubrication of the input transfer gears
(A) Torque converter inlet pressure tap
(B) Pressure tap for pump pressure
(C) Pressure tap for the modulating valve (lockup clutch)
(D) Torque converter outlet pressure tap
(E) Pressure tap for the modulating valve (impeller clutch)

Figure 12-38 Engine running/transmission in neutral position. (*Courtesy of Caterpillar*)

the oil filter (1) to remove any contaminants and then to the torque converter priority control valve (4). From the priority control valve, the oil is directed to the transmission hydraulic control relief valve (2) and to the clutch solenoid modulating **shift valves** (11 to 16). Clutch modulating valve number 3 is energized in the neutral position and will fill the clutch pack on start-up. This helps lubricate the clutch packs by allowing the ring gears of the directional clutches to rotate within the transmission. Oil is also directed to the lockup clutch modulating valve (7) and the impeller clutch modulating valve (9). The transmission has an individual clutch solenoid modulating valve for each clutch pack. There must be at least two energized solenoids for the transmission to drive: one directional solenoid and one speed solenoid. Refer to **Table 12-1** for the combination of solenoid valves and engaged clutch packs for each forward and reverse speed.

Table 12-1: SOLENOIDS AND CLUTCHES TO SPEED AND DIRECTION

Speed Range and Direction	Energized Solenoids	Engaged Clutches
Fourth Speed Forward	3 and 2	3 and 2
Third Speed Forward	4 and 2	4 and 2
Second Speed Forward	5 and 2	5 and 2
First Speed Forward	6 and 2	6 and 2
Neutral	3	3
First Speed Reverse	6 and 1	6 and 1
Second Speed Reverse	5 and 1	5 and 1
Third Speed Reverse	4 and 1	4 and 1

Courtesy of Caterpillar Tractor Company.

Oil exiting the torque converter may be prone to low pressure drops. A torque converter outlet relief valve is used on the outlet side of the torque converter; this valve is set approximately 50 psi lower than the inlet relief valve to maintain a pressure of approximately 60 psi in the torque converter and prevent possible cavitation within it. Oil from the torque converter can become extremely hot. This hot oil from the outlet relief valve is directed to the oil cooler that can be an air-to-oil cooler or an air-to-water cooler. Exiting cool oil is combined with oil from the transmission hydraulic control relief valve and is directed through the oil lubrication passages within the transmission planetaries, clutch packs, bearings, and shafts. The restrictions inside the oil galleries of the transmission keep the lubrication pressure of the warm oil at approximately 20 psi. This oil is allowed to drain into the transmission sump and is then recirculated through the pump and transmission.

Torque Converter Priority Valve. The torque converter priority valve (4) splits the oil flow between the transmission hydraulic control relief valve and the lockup clutch (7) and impeller clutch modulating valves (9). View Figure 12-38. The priority valve splits the flow with the priority of flow given to the transmission hydraulic control relief valve when a "low flow" condition occurs within the hydraulic system. Because torque converter oil is supplied from the transmission hydraulic control valve, the priority valve will supply oil to it first—this protects the torque converter from a low oil supply and possible damage. During this condition, the lockup clutch and impeller clutch modulating valves will cease to work, but this will not damage the torque converter.

Transmission Hydraulic Control Relief Valve. Refer to **Figure 12-39.** The transmission hydraulic control

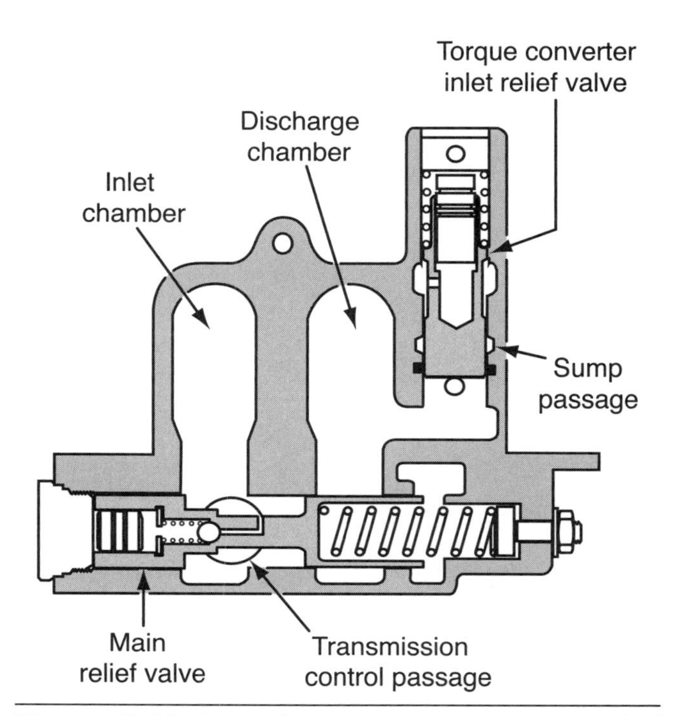

Figure 12-39 Transmission/torque converter relief valve.

relief valve controls the maximum pressure of oil that is supplied to the clutch solenoid modulating valves. Within this valve is the torque converter inlet relief valve (3); it is set lower than the main relief valve pressure of 2620 ± 35 kPa (380 psi) and will maintain an inlet pressure of 1060 ± 100 kPa (154 ± 15 psi) within the torque converter. Although this valve is referred to as a relief valve, it acts as a pressure regulating valve due to the fluctuation of oil flow through the system. The valve tries to maintain a constant pressure regardless of flow rate and resistance to flow. Once the torque converter pressure is reached, excess oil is directed through the oil lubrication passages within the transmission planetaries, clutch packs, bearings, and shafts. Thus separate constant pressures for both the clutch solenoid modulating valves and the torque converter are maintained.

Clutch Solenoid Modulating Valves. Refer to **Figure 12-40.** There are six clutch solenoid modulating valves. Valves 3 and 5 supply oil to the directional clutches: reverse and forward, consecutively. Valves 1, 2, 4, and 6 supply oil to the speed clutches: first, second, fourth, and third, consecutively. The clutch solenoid modulating valves are the outputs of the powertrain electronic control module (ECM). The clutch solenoid modulating valves are proportional solenoid valves. This means that the valves will supply oil to the clutch packs directly proportional to the current supply from the ECM. If there is no current supply, the clutch pressure will be zero; if the current supply is at the maximum, the clutch pressure will be at the maximum.

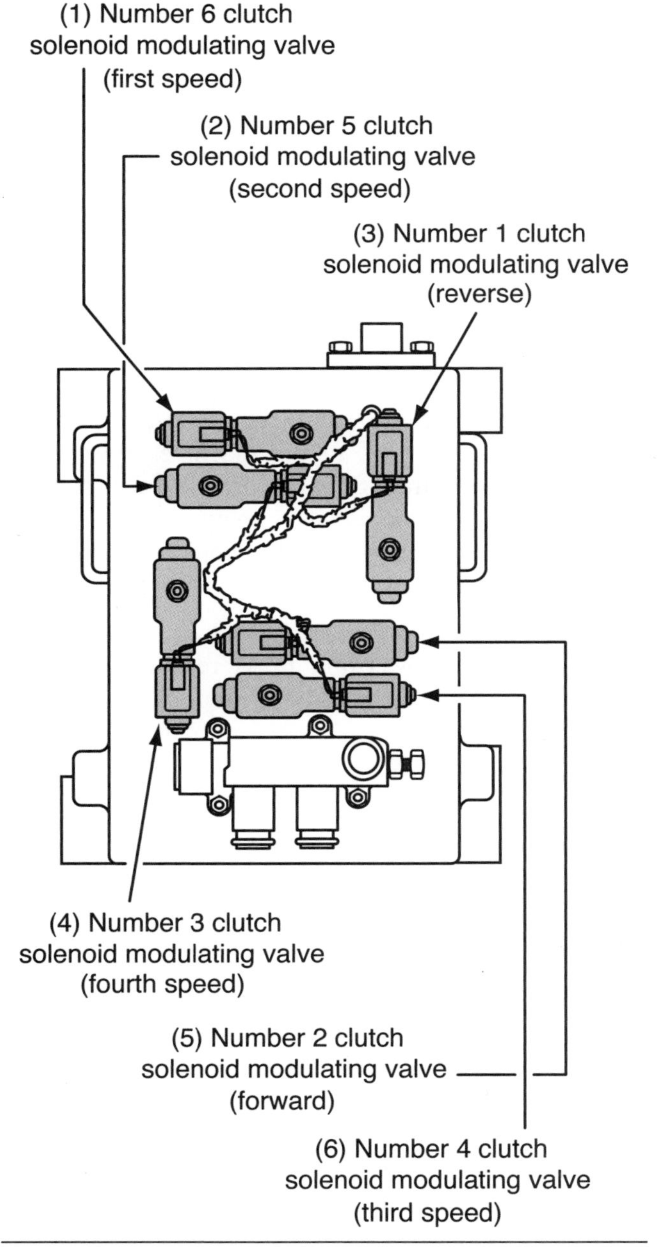

Figure 12-40 Transmission clutch solenoid modulating valves.

The clutch solenoid modulating valves are used by the powertrain ECM to directly modulate the oil pressure that is sent to each individual clutch pack. The ECM sends a pulse width modulated (PWM) signal to vary the current flow to the solenoid valve that controls the plunger travel. The farther the plunger moves within its bore, the greater the amount of oil passage directed to the clutch pack. The powertrain ECM activates the appropriate solenoid valves based on the input from the operator. When the modulating valves are deactivated by the ECM, oil pressure within the clutch pack is drained through the valve back to the transmission sump.

Torque Converter Lockup Clutch Modulating Valve. The **torque converter lockup clutch modulating valve** is a three-way proportional pressure control valve. It is also PWM solenoid, controlled by the powertrain ECM. It functions the same as the modulated shift valves in respect to proportion of pressure to current supply. Refer back to Figure 12-38, component number 7. The lower the ECM current, the lower the pressure signal to the lockup clutch. When the ECM sends a current signal to the solenoid valve, oil is directed to the lockup clutch in the torque converter to lock the turbine to the torque converter housing thus providing an output ratio of 1:1 or direct drive. The torque converter is more efficient at lockup during rotary flow. The ECM can activate the solenoid lockup valve at any speed except first speed. The following conditions must be met before the ECM will activate the lockup solenoid:

- The torque converter output speed is greater than 1400 rpm.
- The torque converter output speed is less than 2120 rpm.
- The left brake pedal must not be depressed.
- The lockup clutch solenoid valve has been deactivated for at least four seconds after a fault for a slipping clutch has been logged.

The lockup clutch will engage when these four conditions are met again in a newly selected gear.

The lockup clutch is disengaged when the torque converter output speed drops below 1250 rpm. This prevents engine lugging and damage to the friction discs and plates of the lockup clutch. Torque multiplication in a vortex flow will resume within the torque converter.

In the event of a machine accelerating down a long slope, the lockup clutch will disengage when the torque converter output speed exceeds 2400 rpm.

Torque Converter Impeller Clutch Modulating Valve. The torque converter impeller clutch modulating valve is a 3-way proportional pressure control valve. It is also a PWM solenoid controlled by the powertrain ECM. It functions the same as the lockup modulated valve in respect to proportion of pressure to current supply except in reverse proportion. The lower the ECM current, the higher the pressure signal to the impeller clutch. Refer back to Figure 12-38, component number 9. When the ECM sends a current signal to the solenoid valve, oil pressure is reduced at the impeller clutch piston. This causes the friction discs and plates to slip between the torque converter housing and the impeller. When the impeller slips, less oil is forced to the turbine. With less oil force on the turbine, there is less torque output from the torque converter, reducing rim-pull and wheel slip.

The following conditions determine what ECM signal is received by the impeller clutch modulating valve:

- The position of the left service brake pedal
- Directional shifts
- Engine speed
- Selected gear
- Direction of rotation of the torque converter output shaft and torque converter output speed
- The position of the rim-pull selector switch in reduced position
- The position of the reduced rim-pull on/off switch

Torque Converter Outlet Relief Valve. The purpose of the torque converter outlet relief valve is to maintain a minimum pressure of approximately 60 psi in the torque converter. As the impeller blades slice through the oil, they create a low-pressure condition within the oil and cause it to vaporize; tiny air bubbles form and implode. When the oil is repressurized, the oil can cavitate components within the torque converter. By pressurizing the torque converter, low-pressure conditions from the impeller are eliminated.

POWERTRAIN ELECTRONIC CONTROL SYSTEM

The powertrain electronic control system is a communication system between the powertrain ECM and the monitoring system. A data link is used to send information back and forth between the two systems. This bidirectional link allows the ECM to input and output information through its circuits and related wiring harness.

Electronic Control Module (ECM)

The powertrain ECM shifts the transmission. The transmission speed and direction control switch send the operator input to the powertrain ECM. The operator input indicates the desired speed selection and direction for the transmission. The powertrain ECM makes decisions based on this input information and the information within the ECM memory. After the powertrain ECM receives the input information and processes it with the information from memory, the powertrain ECM sends a corresponding response via the outputs (modulating valve solenoids).

Inputs. Input devices signal the powertrain ECM of the operating conditions of the machine. The switch inputs of the powertrain ECM are provided with the following signals from the switches: an open, a ground, and a + battery. Sensor inputs provide a constantly changing signal to the powertrain ECM.

Outputs. The powertrain ECM responds to decisions by sending electrical signals through the outputs. The outputs can create an action or provide information to the powertrain ECM.

Input/Output. The bidirectional data link is used to communicate with the other electronic control modules on the machine. The following is a brief example of this shared information:

- The power train ECM receives the harness code input from the monitoring system.
- The powertrain ECM determines the sales model of the machine from the harness code input.
- The powertrain ECM sends the following information to the monitoring system: engine speed, machine ground speed, parking brake switch status, transmission speed selection, transmis-

sion direction selection, and service code of the transmission.

- The monitoring system displays this information for the operator and/or technician.
- The monitoring system also alerts the operator of the machine status by illuminating the corresponding alert indicators or alarms.
- The power train ECM receives service commands from the monitoring system. Service commands can change the operating mode of the powertrain electronic control system. The service command can also direct the powertrain ECM to read or to clear service codes that are stored in the ECM memory.

The powertrain electronic control system has six basic functions:

- Neutral start function
- Manual shift function
- Automatic shift function
- Parking brake function
- Back-up alarm function
- Diagnostic operation

The 988G-wheel loader, shown in **Figure 12-41,** is equipped with a joystick steering control lever. The transmission controller is built into the joystick lever. Forward/reverse direction as well as first, second, third, and fourth speeds can all be selected by pressing the up-shift and down-shift switches or the directional control switches.

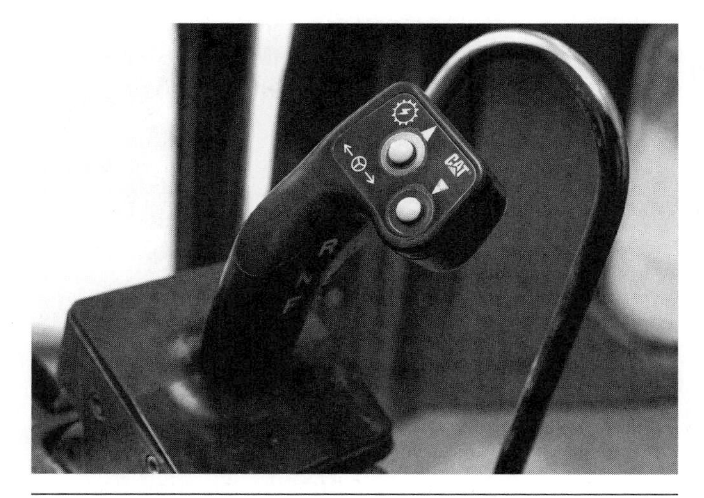

Figure 12-41 Joystick steering and transmission control.

Neutral Start Function

The start relay, an output of the powertrain (ECM), supplies electrical power to the starter solenoid switch that controls the starting motor. When the operator turns the key switch to the start position, the powertrain ECM decides if all the starting conditions are met. These conditions are that the transmission directional control switch is in neutral and the engine is off. To protect the starter motor drive from damage, the ECM will not send a signal to the start relay if the engine is running. When the conditions are met, the ECM will activate the start relay, which will supply battery voltage to the starter solenoid, allowing the starter motor to turn.

Manual Shift Function

The manual shift function allows the operator to use the shift control inputs to select the transmission speed range. In this operational mode, it also allows the operator to change the direction of travel regardless of engine or machine speed. The torque converter impeller clutch allows smooth transitions when directional changes are made.

When an operator decides to up-shift or down-shift, the transmission ECM selects the next higher or lower speed range, depending on which switch is pressed. A new gear is selected on the switch transition from the depressed to released position. Holding one switch in the depressed position will not disable the other switch. The down-shift switch is disabled when the transmission is in first gear. The up-shift switch is disabled when the transmission is in the highest gear. Refer to **Table 12-2** for the shifting sequence.

Automatic Shift Function

The automatic shift function allows the transmission ECM to automatically shift the transmission. The operator must enable this feature by selecting AUTO mode from the Auto/Manual switch (4). There are three auto modes available: 2, 3, and 4. See **Figure 12-42.**

The powertrain ECM evaluates the input that is sent from the engine speed sensor, transmission speed sensors, and the torque converter output speed sensor in order to regulate transmission shifts. The ECM does not allow you to up-shift into a gear that is higher than the gear that is selected with Auto/Manual switch (4). If Auto/Manual is set to the number 3 position, this informs the ECM that a gear speed higher than 3 is not desired and will not allow the transmission to up-shift to fourth

Table 12-2: MANUAL SHIFTING SEQUENCE (988G WHEEL LOADER)

Speed Range and Direction	If REVERSE is selected, then the transmission shifts to the following speed.	If NEUTRAL is selected, then the transmission shifts to the following speed.	If FORWARD is selected, then the transmission shifts to the following speed.	If UPSHIFT is selected, then the transmission shifts to the following speed.	If DOWNSHIFT is selected, then the transmission shifts to the following speed.
4F	3R	Neutral [1](4N)	No Change	No Change	3F
3F	3R	Neutral [1](3N)	No Change	4F	2F
2F	2R	Neutral [1](2N)	No Change	3F	1F
1F	1R	Neutral [1](1N)	No Change	2F	No Change
Neutral [1](1N, 2N, 3N, 4N)	Last[2]	No Change	Last[2]	Next[3]	Prior[3]
1R	No Change	Neutral [1](1N)	1F	2R	No Change
2R	No Change	Neutral [1](2N)	2F	3R	1R
3R	No Change	Neutral [1](3N)	3F	No Change	2R

[1] The speed and the direction that is shown in bold print is the gear readout.

[2] A directional shift out of neutral will cause a speed shift to the last speed that was selected in either forward or reverse.

[3] Speed shifts are allowed when neutral is selected. The gear readout will change the speed that is shown when neutral is selected and a speed shift is made. However, none of the solenoids are activated until forward or reverse is selected.

Courtesy of Caterpillar Tractor Company.

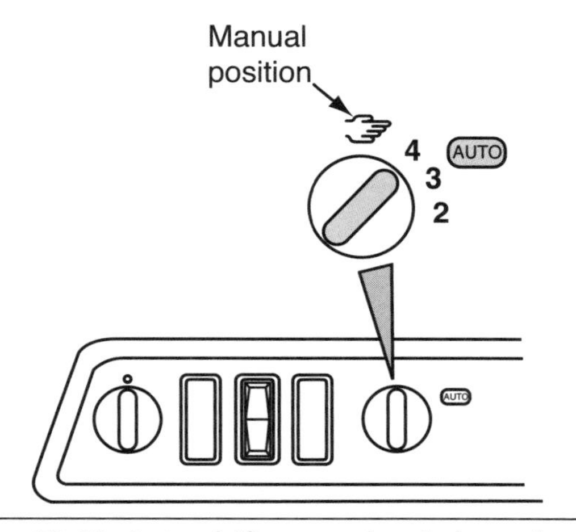

Figure 12-42 Auto shift control switch setting.

speed. There is no automatic downshift from second speed to first speed. The down-shift from second gear to first gear occurs only when the down-shift switch is pressed.

Momentarily pressing the downshift switch instructs the ECM to down-shift the transmission by one gear. Holding down the down-shift switch instructs the ECM to continue to down-shift the transmission as speed decreases. The transmission will remain in the down-shifted gear for three seconds after the down-shift switch is released. Then, automatic shifting is resumed.

Tech Tip: In the automatic mode of auto-shift, the transmission will automatically shift from first gear to second gear. This happens when the operator shifts the direction from forward to reverse. When the operator then shifts from reverse to forward, the transmission will remain in second gear. This feature only functions when the machine is in first gear. A transmission quick-shift can only occur when the operator shifts the direction from forward to reverse. This feature does not affect any directional shift in any of the other gears.

Table 12-3 shows the approximate rpms at which the ECM will initiate a speed shift.

When the key start switch is turned from the "off" position to the "on" position, the powertrain ECM is activated. When the powertrain ECM is initially activated, all the modulating valves (transmission clutch) are de-energized regardless of the position of transmission direction control switch (3). The powertrain ECM then determines if transmission direction control switch (3) is in the "neutral" position. The powertrain ECM disables the transmission if the neutral position is *not* selected; the transmission direction control switch must be returned to the neutral position before a direction can be selected. If the transmission direction control switch is in the neutral position, then the powertrain ECM activates the start relay, allowing the engine to start. The powertrain ECM allows the selected speed to be changed if the transmission direction control switch is in the neutral position when engine is started.

Tech Tip: Transmission direction control switch (3) must be in the "neutral" position when the key start switch is turned to the "start" position. If the transmission direction control switch is not in the neutral position, the start relay will not be activated when the key start switch is turned.

Parking Brake Function

The parking brake function prevents the operation of the machine with the parking brake engaged. Although most equipment has enough torque to drive through the parking brake, operating a machine with the parking brake engaged will cause accelerated wear and lead to possible damage of the parking brake components.

If the operator places the directional switch in forward or reverse with the parking brake engaged, the ECM will place the transmission into neutral by de-energizing all solenoids in the transmission hydraulic control valve. The ECM will activate the warning system. A parking brake indicator lamp will flash and an action alarm will sound to let the operator know he or she is trying to operate the machine with the parking brake engaged. When the parking brake is disengaged, it sends a signal to the ECM indicating the parking brake is released, and the ECM will energize the appropriate solenoids based on the directional and speed switch positions.

Back-Up Alarm Function

The back-up alarm function is an output of the ECM used to alert surrounding workers and operators that the machine is reversing direction. When the operator places the directional control switch into reverse, the ECM activates the back-up alarm relay, which will activate the back-up alarm. Back-up alarms send an

Table 12-3: AUTOMATIC UP-SHIFT AND AUTOMATIC DOWN-SHIFT

Speed Range and Direction Shift	Transmission Speed (rpm) Standard (non-LUC)	Transmission Speed (rpm) Lockup Clutch Enabled	Transmission Speed (rpm) Lockup Clutch Disabled
1F to 2F	344	354	354
2F to 3F	613	717	631
3F to 4F	1086	1269	1116
4F to 3F	999	1180	1027
3F to 2F	564	667	580
2F to 1F	316	326	325
1R to 2R	393	459	404
2R to 3R	701	819	721
3R to 4R	1241	1450	1276
4R to 3R	1141	1349	1174
3R to 2R	645	762	663
2R to 1R	362	427	372

Courtesy of Caterpillar Tractor Company.

audible signal of between 85 and 115 decibels (dB), which should be higher than the ambient noise level. Automated (self-adjusting) and switchable models are available. The switchable model has as many as three settings that are preset by a toggle switch at 97, 107, and 115 dB. The self-adjusting model sounds at 5 decibels above the ambient noise level, to a maximum of 115 dB.

Diagnostic Operation Function

The powertrain ECM detects faults that occur in the input and output circuits within its system. A fault is detected when a signal at the contact of the ECM is outside a specific range. When this occurs, the ECM will record and store the fault information for future reference by the technician to retrieve and repair. If the fault is intermittent, it is still logged as an "intermittent fault" and can be retrieved for future reference. The ECM can provide diagnostic information to assist with troubleshooting the stored faults within the system.

SAE standard J1939, Recommended Practice for a Serial Control and Communications Vehicle Network, incorporates J1587 protocols. The purpose of these recommended practices is to network electronic systems connected to a data bus backbone. This allows data sharing by the controllers connected to the data bus.

Electronic service tools are a requirement for communicating with machine ECMs. Most OEMs are using personal computers (PC) with Windows-based operating systems. Manufacturer-specific programs are loaded onto the PC; a data link connector (6- or 9-pin Deutsch connector) connects the machine data bus and the PC. Together they provide the interface and communication with the ECM. A communication adapter is required between the PC and the machine data bus. Once communication is established with the data bus, the technician can select the ECM he or she wants to communicate with. Each ECM on the data bus is assigned an address known as a MID.

Data Link Connector

The data link connector (19), shown in **Figure 12-43** is located in the right-hand corner of the console near the floor. The connector is used to connect to the data communication equipment that is used for diagnostic purposes.

Figure 12-43 Nine-pin data link connector.

Service codes are used to specify fault paths. Service codes are displayed on the monitoring screen. These codes consist of three identifiers:

- Module Identifier (MID)
- Component Identifier (CID)
- Failure Mode Identifier (FMI)

Module Identifier (MID) The MID is a three-digit code that is shown on the display area of a monitoring screen. The MID is shown for approximately one second before the service code is shown in the same area. The ECM that diagnosed the fault is identified by the MID. A module identifier is assigned to all of the electronic controllers connected to the data bus. The following are examples of some MIDs.

- Monitoring System: MID-030
- Power Train ECM: MID-081
- Implement/Hydraulic ECM: MID-082
- Payload Control System: MID-074
- Engine ECM: MID-036

Component Identifier (CID) Components that are defective are logged by the CID. Examples of these components are the start relay and the reverse solenoid.

Failure Mode Identifier (FMI) The FMI tells the type of failure that has occurred. The FMI is a two-digit code that is shown in the display area. The CID and the FMI are shown together after the MID has been displayed. A decimal point "." precedes the FMI. **Table 12-4** shows an example of FMI codes adapted from the SAE Standards, Section J1587 diagnostics.

Table 12-4: FAILURE MODE IDENTIFIER AND DESCRIPTION

FMI	Failure Description
00	Data valid but above normal operating range
01	Data valid but below normal operating range
02	Data erratic, intermittent, or incorrect
03	Voltage above normal or shorted high
04	Voltage below normal or shorted low
05	Current below normal or open circuit
06	Current above normal or grounded circuit
07	Mechanical system not responding properly
08	Abnormal frequency, pulse, or period
09	Abnormal update
10	Abnormal rate of change
11	Failure mode not identifiable
12	Bad device or component
13	Out of calibration
14	N/A
15	N/A
16	Parameter not available
17	Module not responding
18	Sensor supply fault
19	Condition not met
20	N/A

Courtesy of Caterpillar Tractor Company.

MULTIPLE COUNTERSHAFT TRANSMISSIONS

As mentioned earlier many heavy-duty applications require equipment to have a multitude of forward and reverse speeds. Compound planetary gearing is one method already discussed. Another method of achieving multiple speeds is the use of a series of countershafts to compound the number of ratios required for a specific machine application. Multiple countershafts are set up in a staggered-parallel formation; different gear diameters on each shaft are in constant mesh with each other. In most instances, the external-type gears that float on the countershafts only provide a drive when a clutch pack is engaged. The external gears that are fixed to the countershaft transfer engine torque to the next countershaft. Providing all the appropriate clutch packs are engaged, torque transfer continues through each countershaft until it reaches the transmission output shaft. Rotating clutch packs connected between the countershafts and each independent gear allow for the selection and engagement of each gear ratio by the operator. The number of gear ratios attainable is only limited by the number of countershafts used. In some grader applications, an 8-speed forward, 4-speed reverse, **multiple countershaft transmission** is used. In the heavy-duty equipment industry the configurations of transmissions are many and varied. For this reason, only two popular transmissions are discussed.

MULTIPLE COUNTERSHAFT TRANSMISSION POWERFLOWS

Compound planetary transmissions were discussed earlier. The second transmission chosen is a multiple countershaft powerflow transmission from a Caterpillar® 938G wheel loader. The 938G loader has a bucket capacity of approximately 2.8 to 4.4 cubic. yards, depending on the application. The operating weight is approximately 30,000 pounds. It is powered by a 3126 engine rated at 180 hp. It has a 4-speed forward, 3-speed reverse, multiple countershaft transmission that can be operated in automatic or manual powershift mode. See **Figure 12-44.**

Six hydraulically activated clutches are used to transfer power from the input shaft assembly (7) to the output shaft assembly (12). The torque converter used is a standard type with no lockup or impeller clutch feature. Six electronically controlled solenoids, when energized by the powertrain ECM, control the oil flow to the clutch packs. Unlike a planetary shift transmission that uses stationary and rotating clutch packs, a countershaft transmission uses only rotating clutch packs to provide a drive when engaged. There are six clutches: three directional clutches and three speed clutches.

The directional clutches are:

- Clutch 1 (8) (Forward low)
- Clutch 2 (9) (Forward high)
- Clutch 3 (16) (Reverse)

The speed clutches are:

- Clutch 4 (11) (Second speed)
- Clutch 5 (14) (Third speed)
- Clutch 6 (13) (First speed)

In order for the power to be transferred through the transmission to the output shaft assembly (12),

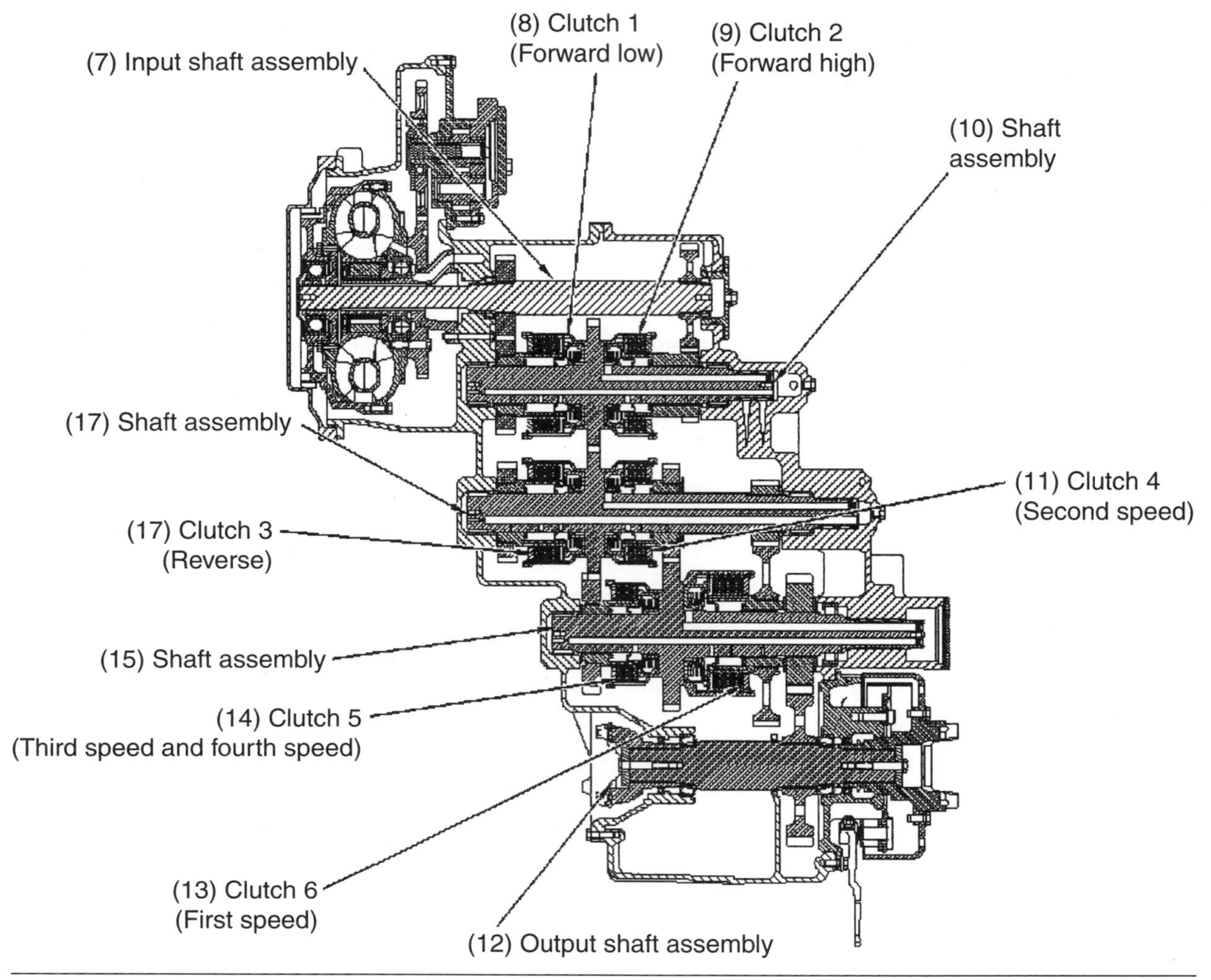

Figure 12-44 Four-speed forward, three-speed reverse, multiple countershaft transmission. (*Courtesy of Caterpillar*)

one direction clutch must be engaged and one speed clutch must be engaged. A modulating valve controls each clutch. The powertrain ECM controls the proper engagement of the clutches by sending a variable PWM signal to the solenoids on the modulating valves.

Figure 12-45 shows a cross-sectional view of the transmission countershaft arrangement with the clutch packs located on each countershaft. The transmission has a total of five shafts: Input shaft assembly (3), Output shaft assembly (33) and three gear shafts. The input shaft assembly (3) is driven by the torque converter turbine shaft. Shaft assembly (9) contains Clutch 1 (4) and Clutch 2 (6), shaft assembly (19) contains Clutch 3 (23) and Clutch 4 (17), and shaft assembly (27) contains Clutch 5 (30) and Clutch 6 (28).

Neutral Position. The powertrain ECM determines the proper clutches that need to be engaged. The proper clutch engagement is based on the inputs to the powertrain ECM. Some of the inputs that are considered are machine speed, engine speed, parking brake, and operator inputs. When the operator turns the start key, the powertrain ECM determines if the transmission is in the "neutral" position and the parking brake is engaged. If these two parameters are met, the ECM will allow the start relay to engage the starter motor to start the engine. Upon start-up, Clutch 5 (third speed) is engaged to allow some lubrication within the transmission and clutch packs prior to machine operation. If the powertrain ECM detects machine movement (parking and service brake disengaged), the powertrain ECM will disengage Clutch 5 and no clutch packs will be engaged.

Refer to **Figure 12-46.** Whenever the transmission direction and speed control lever is moved to the neutral position, or the transmission neutralizer is activated, the solenoid on the modulating valve for Clutch 5 (51) is

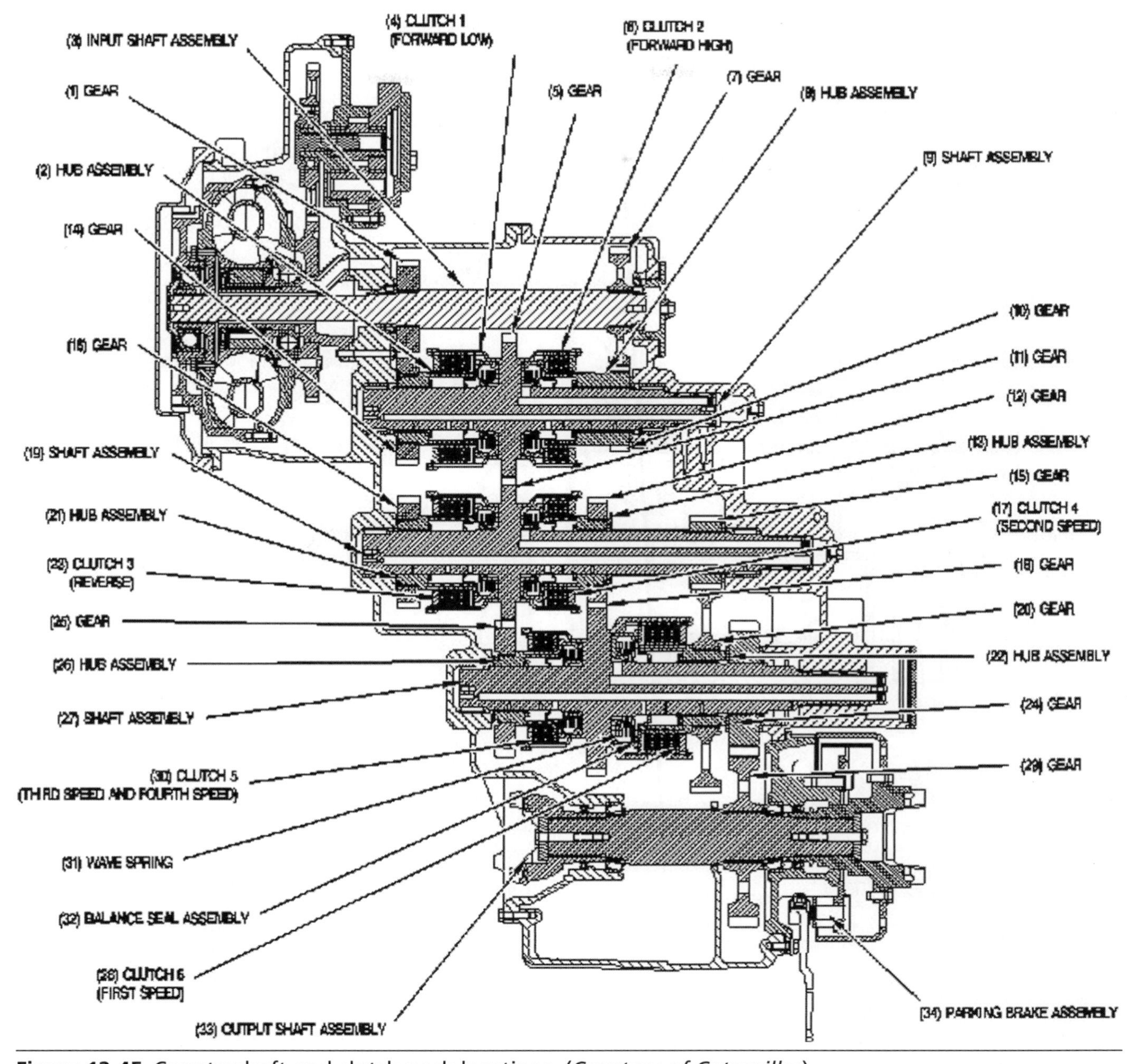

Figure 12-45 Countershaft and clutch pack locations. (*Courtesy of Caterpillar*)

energized. When this solenoid is energized, the modulating valve directs transmission pump oil (51) to Clutch 5 (30) in order to engage the clutch. Simultaneously, the directional and speed clutches (other than third speed Clutch 5) will disengage.

First Speed Forward Powerflow. The powertrain ECM modulates the electrical current that is sent to the selected clutch solenoids based on operator inputs to the ECM. The clutch solenoids are located on the modulating relief valves. Modulating the current to the clutch solenoid controls the clutch pressure by shifting the spool in the modulating valve.

Refer to Figure 12-46. When the transmission direction and speed control lever is moved to the "first speed forward" position, both the solenoid on the modulating valve for Clutch 6 (52) and the solenoid on the modulating valve for Clutch 1 (54) are energized. When these solenoids are energized, the modulating valves direct transmission pump oil (41) to Clutch 1 (4) and Clutch 6

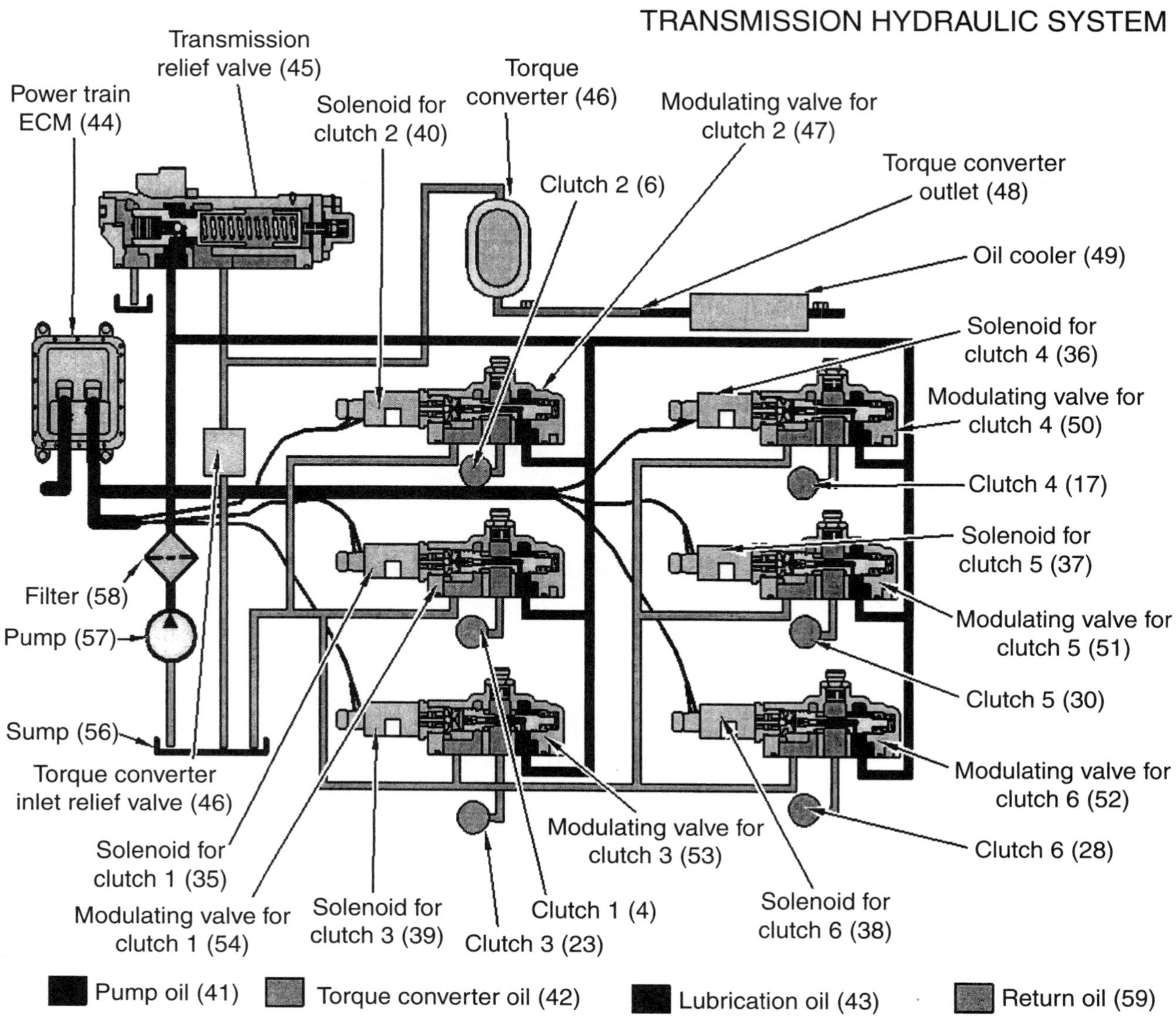

Figure 12-46 Transmission solenoid valve locations. (*Courtesy of Caterpillar*)

in order to engage the clutch packs. The oil in the clutches that are not engaged is sent back to sump (56) and the oil becomes return oil.

With the transmission in the first speed forward position, Clutch 1 (4) and Clutch 6 (28) are engaged. Torque from the engine is transferred through the torque converter to the input shaft assembly (3). Because gear (1) is splined to the input shaft assembly (3), gear (1) turns gear (14) on countershaft (9). Gear (14) is splined to hub assembly (2), which causes the hub assembly to rotate. Because Clutch 1 (4) is engaged, torque is transferred from hub assembly (2) through the engaged Clutch 1 (forward low) to countershaft (9). Gear (5) is part of countershaft assembly (9) and turns gear (11) on countershaft assembly (19). Gear (15) is splined to countershaft assembly (19), which turns gear (20). Gear 15 is considerably smaller than gear (20) which will provide a low speed/high torque condition. Torque is transferred from countershaft assembly (19) through gear (15) and gear (20) to hub assembly (22). First speed Clutch 6 (28) is engaged, and torque is transferred from hub assembly (22) through the engaged Clutch 6 to countershaft assembly (27). Gear (24) is splined to countershaft assembly (27), rotating gear (29), which is splined to output shaft assembly (33). Because gear (24) is smaller than gear (29) a second speed decrease/torque increase occurs through the drive. This final torque increase is transferred from countershaft assembly (27) through gear (24) and gear (29) to output shaft assembly (33). The output shaft provides the means to connect to the front drive axle and rear drive axle.

Second Speed Forward Powerflow. Refer to Figure 12-45. When the transmission speed and direction control lever is placed in the second speed forward position, Clutch 4 (17) and Clutch 1 (4) are engaged. Torque from the engine is transferred through the torque converter to the input shaft assembly (3). Because gear (1) is splined to the input shaft assembly (3), gear (1) turns gear (14) on countershaft (9). Gear (14) is splined to hub assembly (2), which causes the hub assembly to rotate. Because Clutch 1 (4) is engaged, torque is transferred from hub assembly (2) through the engaged Clutch 1 (forward low) to countershaft (9). Gear (5) is part of countershaft assembly (9) and turns gear (11) on countershaft assembly (19). Gear (11) is part of countershaft assembly (19), because Clutch 4 on countershaft assembly (19) (second speed) is engaged; powerflow is transferred through the clutch pack to gear (12). Gear (12) will now drive gear (16) that is in-mesh with it on countershaft (27). Gear (12) is smaller than gear (16) which will still provide a low speed/high torque condition, but not as great as for first speed because of the pitch diameter of the individual gears. Torque is transferred along countershaft assembly (27) through gear (24) and to gear (29), which is splined to output shaft assembly (33). Because gear (24) is smaller than gear (29), a second speed decrease/torque increase occurs through the drive. This final torque increase results in a slightly greater speed increase when compared to first speed.

Third Speed Forward Powerflow. Refer to Figure 12-45. When the transmission speed and direction control lever is placed in the third speed forward position, Clutch 5 (30) and Clutch 1 (4) are engaged. Torque from the engine is transferred through the torque converter to input shaft assembly (3). Because gear (1) is splined to input shaft assembly (3), gear (1) turns gear (14) on countershaft (9). Because Clutch 1 (4) is engaged, torque is transferred from hub assembly (2) through the engaged clutch 1 (forward low) to countershaft (9). Gear (5) is part of countershaft assembly (9) and turns gear (11) on countershaft assembly (19). Because gear (11) is meshed with both gear (5) and gear (25), engine torque is transferred directly from gear (5) to gear (25). Gear (11) acts as an idler between the two meshed gears and provides no additional gear reduction. The gear ratio occurs between the pitch diameters of gear (5) and gear (25). Gear (25) has a larger diameter than gear (25); this results in an overdrive condition between these two gears. With Clutch 5 (30) now engaged, engine torque is transferred along countershaft assembly (27) through gear (24) and to gear (29), which is splined to output shaft assembly (33). Because gear (24) is smaller than gear (29) a speed decrease/torque increase occurs through this drive combination. The final output speed results in a speed increase over that of second gear.

Fourth Speed Forward Powerflow. Refer to Figure 12-45. When the transmission speed and direction control lever is placed in the fourth speed forward position, Clutch 5 (30) and Clutch 2 (6) are engaged. Torque from the engine is transferred through the torque converter to input shaft assembly (3). Because gear (7) is also splined to input shaft assembly (3), gear (7) turns gear (10) on countershaft (9). Gear (7) will drive gear (10) in an overdrive condition because of the larger diameter of gear (7). Because clutch pack (2) is engaged, engine torque is transferred to gear (5) and countershaft (9). Gear (5) is in constant mesh with gear (1), and gear (1) is in constant mesh with gear (25). Gear (1) on countershaft (19) again acts as an idler gear between gears (5) and (25) and relays engine torque directly to gear (5) and countershaft (27). Clutch pack (5) remained engaged between the 3–4 up-shift and now acts as the fourth speed clutch. The speed increase occurs through Clutch 2 (6) forward high, allowing gears (7) and (10) to provide an overdrive gear ratio.

With Clutch 5 (30) engaged, gear (25) will relay engine torque to countershaft (27) and gear (24) and to gear (29) that is splined to output shaft assembly (33). Because gear (24) is smaller than gear (29) a speed decrease/torque increase occurs through this drive combination. The final output speed will result in a speed increase over that of third speed.

First Speed Reverse Powerflow. Refer to Figure 12-45. When the transmission speed and direction control lever is placed in the first speed reverse position, Clutch 6 (28) and Clutch 3 (23) are engaged. Torque from the engine is transferred through the torque converter to input shaft assembly (3). Because gear (1) is also splined to input shaft assembly (3), gear (1) turns gear (16) on countershaft (19). For illustration purposes, countershaft (19) is shown below countershaft (9). In actual fact, it is located parallel to countershaft (9) and gear (16) is splined with gear (1) on input countershaft (3) as described earlier. Because reverse Clutch 3 is engaged on countershaft (19), engine torque is transferred through gears (1) and (6) and clutch pack (3) to gear (15). Gear (15) is in constant mesh with gear (20) on countershaft (27). Because first speed Clutch 6 (28) is engaged, gear (20) will now transfer engine torque to countershaft (27) at a reduced speed/high torque ratio. This is due to the gear ratio provided by diving gear (15) to driven gear (20). Through the first speed Clutch 6 (28) on countershaft (27), engine torque is now transferred to gear (24), which is splined to the countershaft. Gear (24) will provide a further gear reduction with output countershaft (33) and gear (29). Torque is split between the front and rear drive axles to provide reverse direction.

Second Speed Reverse Powerflow. Refer to Figure 12-45. When the transmission speed and direction control lever is placed in the second speed reverse position, Clutch 4 (17) and Clutch 3 (23) are engaged. Torque from the engine is transferred through the torque converter to input shaft assembly (3). Because gear (1) is also splined to input shaft assembly (3), gear (1) turns gear (16) on countershaft (19). Because reverse Clutch (3) and second speed Clutch 4 (17) are engaged on countershaft (19), engine torque is delivered directly to gear (12) on the countershaft. Gear (12) is in constant mesh with gear (16) and will provide a slightly higher gear ratio than first speed reverse. Gears (16) and (24) are directly splined to countershaft (27). This means that engine torque will be transferred directly through countershaft (27), to output countershaft (33). Gears (24) and (29) will provide the last gear ratio reduction to increase engine torque to the output shaft. Torque is then split between the front and rear drive axles to provide second speed reverse.

Third Speed Reverse Power Flow. Refer to Figure 12-45. When the transmission speed and direction control lever is placed in the third speed reverse position, Clutch 5 (30) and Clutch 3 (23) are engaged. Torque from the engine is transferred through the torque converter to input shaft assembly (3). Because gear (1) is also splined to input shaft assembly (3), gear (1) turns gear (16) on countershaft (19). Because reverse Clutch 3 (23) is engaged on countershaft (19), engine torque is transferred through gears (1) and (6) and clutch pack (3) to gear (11). Gear (11) is in constant mesh with gear (25) on countershaft (27). This will provide the highest speed ratio in reverse direction for this particular machine application. Permanently splined gear (24) on countershaft (27) provides a constant gear reduction ratio with gear (29) on output countershaft (33). Torque is then split between the front and rear drive axles to provide third speed reverse.

Table 12-5 shows all forward and reverse speed combinations with the controlling clutch pack. The powertrain ECM controls the modulating valve solenoids that direct pressurized oil to the clutch packs in order to engage the appropriate gear combinations.

Transmission Hydraulic System

The transmission hydraulic system is responsible for providing the median for the torque converter, the pressure used to activate all the clutch packs, effective cooling and lubrication within the transmission, and to direct and control oil pressure and flow within the system. **Figure 12-47** shows all the components that make up the hydraulic system for this transmission.

The tank or reservoir (15) is the sump section of the transmission. A separate reservoir is not used. Trans-

Table 12-5: CONTROLLING CLUTCH PACKS FOR ALL FORWARD/REVERSE COMBINATIONS

	First Speed Forward	Second Speed Forward	Third Speed Forward	Fourth Speed Forward	Neutral	First Speed Reverse	Second Speed Reverse	Third Speed Reverse
Clutch 1	X	X	X					
Clutch 2				X				
Clutch 3						X	X	X
Clutch 4		X					X	
Clutch 5			X	X	X			
Clutch 6	X					X		

Courtesy of Caterpillar Tractor Company.

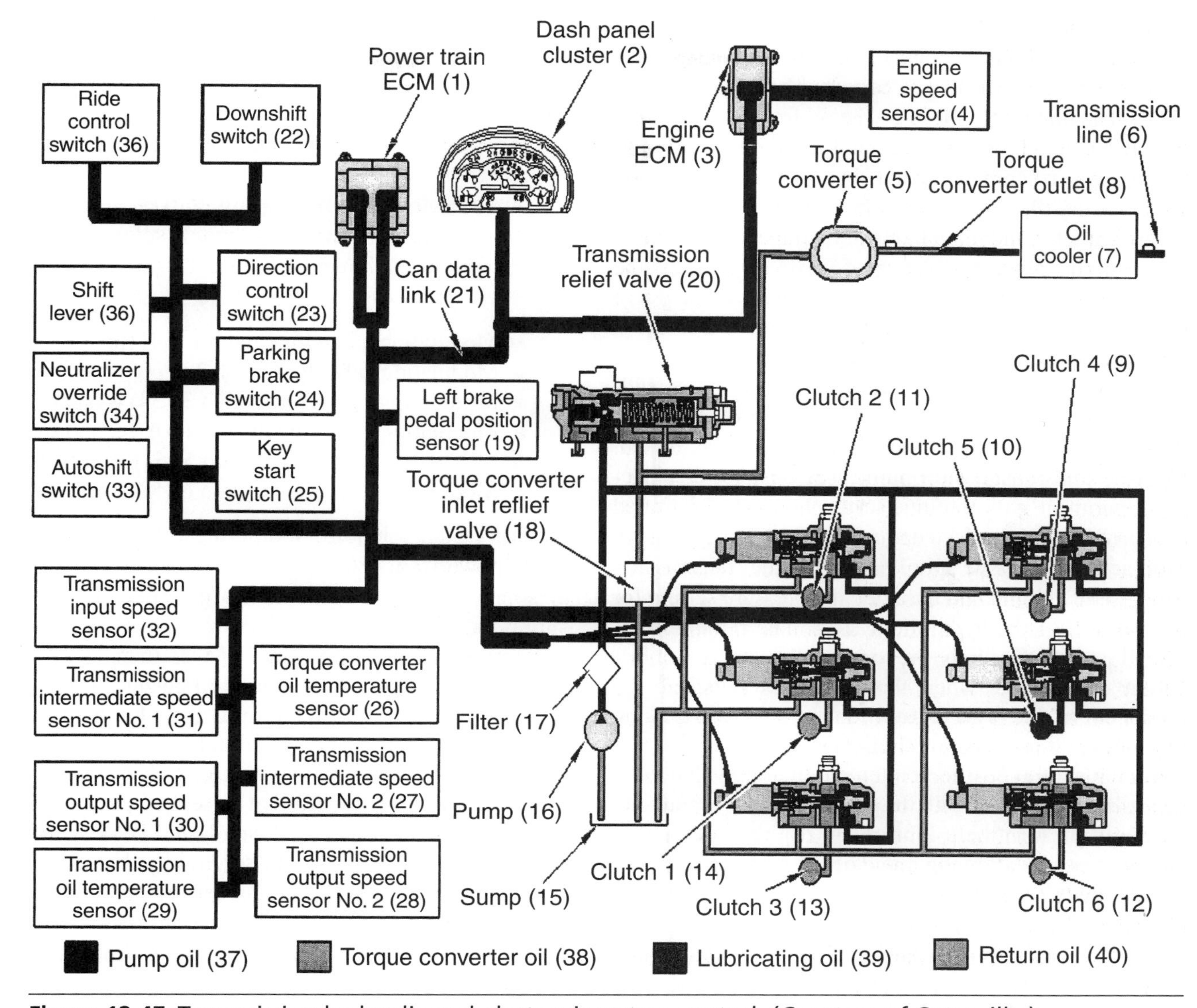

Figure 12-47 Transmission hydraulic and electronic system control. (*Courtesy of Caterpillar*)

mission pump (16) is driven by the torque converter housing and is located to the top-left of the transmission housing. Transmission filter (17) is located on the left side of the machine for easy access when servicing the machine. The transmission relief valve (20) is located with all the modulating shift valves on the top-right section of the transmission. The transmission relief valve is set for 2246 ± 70 kPa (326 ± 10 psi). The torque converter (5) provides the input drive to the transmission and the oil cooler (7) is located to the rear of the machine directly in front of the engine radiator. It is an air-to-oil cooler and will keep the transmission and torque converter between normal operating temperatures of 76°C to 112°C (168°F to 234°F). The torque converter inlet relief valve (18) limits the oil pressure to the torque converter to 965 kPa (140 psi) maximum. This hydraulic schematic represents the clutch packs with clutch ports (9), (10), (11), (12), (13), and (14). Pump oil lines (37) depict high-pressure lines to the shift modulating valves. This oil is used to engage the clutch packs. Torque converter oil lines (38) show oil flow from the transmission relief valve to the torque converter and then to the oil cooler. Oil cooler outlet oil becomes lubrication oil (39) for the transmission bearings, shafts, and clutch packs. Return oil (40) represents oil exiting the clutch packs when the clutches are disengaged. Return oil drains back to the transmission sump where the transmission pump will recirculate it through the system.

Hydraulic Oil Flow in Neutral. Refer to Figure 12-47. When the engine is started, the transmission pump is driven by the torque converter housing at engine speed. The pump (16) picks up the oil from the transmission sump; oil flows first through the oil filter (17) to remove any contaminants and then to the transmission relief valve (20). The transmission relief valve is connected in parallel to the modulating shift valves and in series to the torque converter inlet relief valve. This means the oil must travel through the transmission relief valve before reaching the torque converter relief valve. Oil is then directed to the torque converter and oil cooler.

The transmission modulating valves can be considered open-center valves, which means that oil flow will travel through them when they are not activated. Because the transmission pump is a positive displacement pump, it will continue to displace oil at its rated capacity. It must displace enough oil so that the pressure drop across the valve spool can allow the transmission relief valve to build and maintain the pressure setting for this section of the hydraulic system. The remainder of the oil is used to supply the torque converter and lubrication circuit and, when this circuit is satisfied, the remainder of that oil is drained to the transmission sump through the transmission relief valve.

In the neutral position clutch modulating valve (5) is energized. This will fill the clutch pack on start-up. This will, in turn, help lubrication of the clutch packs and bearings by allowing some of the clutch-pack hubs to rotate while oil pressurizes the system and saturates the components with oil.

The transmission has an individual clutch solenoid modulating valve for each clutch pack that is controlled by the powertrain ECM, based on data input from the operator and machine sensors and switches. The ECM must energize at least two solenoids for the transmission to drive: one directional solenoid and one speed solenoid. Refer to **Table 12-6** which that provides the combination of solenoid valves and engaged clutch packs for each forward and reverse speed.

Clutch Solenoid Modulating Valves. This transmission is equipped with six clutch solenoid modulating valves, as shown in Figure 12-46. They are as follows:

- Modulating valve for Clutch 1 (54) (forward low)
- Modulating valve for Clutch 2 (47) (forward high)
- Modulating valve for Clutch 3 (53) (reverse)
- Modulating valve for Clutch 4 (50) (second speed)
- Modulating valve for Clutch 5 (51) (third speed)
- Modulating valve for Clutch 6 (52) (first speed)

The clutch solenoid modulating valves are outputs of the powertrain (ECM). These valves are proportional solenoid valves. This means that the valves will supply oil to the clutch packs directly proportional to the current supply from the ECM. If there is no current supply, the clutch pressure will be zero, if the current supply is at the maximum, the clutch pressure will be at the maximum. The clutch solenoid modulating valves are used by the powertrain ECM to directly modulate the oil pressure that is sent to each individual clutch pack. The ECM sends a PWM signal to vary the current flow to the solenoid valve that controls the plunger travel. The farther the plunger moves within its bore, the greater the amount of oil passage directed to the clutch pack. The powertrain ECM activates the appropriate solenoid valves based on the input from the operator. When the modulating valves are deactivated by the ECM, oil pressure within the

Table 12-6: ENERGIZED SOLENOIDS AND CLUTCHES TO SPEED AND DIRECTION

Speed Range and Direction	Energized Solenoids	Engaged Clutches
Fourth Speed Forward	5 and 2	5 and 2
Third Speed Forward	5 and 1	5 and 1
Second Speed Forward	4 and 1	4 and 1
First Speed Forward	6 and 1	6 and 1
Neutral	5	5
First Speed Reverse	6 and 3	6 and 3
Second Speed Reverse	4 and 3	4 and 3
Third Speed Reverse	5 and 3	5 and 3

Courtesy of Caterpillar Tractor Company.

clutch pack is drained through the valve back to the transmission sump.

CAUTION *The solenoid coils are not designed to be operated at a 24 direct current voltage (DCV). The powertrain ECM sends a 24-volt PWM signal at a specific duty cycle, which provides an average voltage of about 12 volts to the solenoid coils. Any testing of the solenoid coils should not be made with a 24-DCV supply. The life of the solenoid coils will be drastically reduced. If the solenoid coils must be energized by a source other than the powertrain ECM, 12 DCV should be used to energize the solenoid coils. This will prevent any damage to the solenoids.*

Figure 12-48 shows the components that make up the solenoid modulating valves in the de-energized position. Pump oil (16) flows into the valve body from pump passage (6). Oil then flows into passage (7), through orifice (2), and into chamber (13). Since there is no signal that is sent to solenoid (12), pin (11) cannot hold ball (1) against orifice (10). The oil flows through orifice (10) past ball (1) to the tank passage (9) at a pressure drop close to zero. Oil in tank passage (9) becomes return oil (17). Spring (5) holds shift valve spool (4) to the left. When valve spool (4) is shifted to the left by the spring force, the oil in clutch passage (8) is allowed to flow through tank passage (9). When the oil in clutch passage (8) is vented to the tank, the clutch cannot be engaged. The oil in pump passage (6) is blocked from entering clutch passage (8). Pump oil in passage (7) drains through orifice (10) and to tank passage (9) together with oil from clutch pack passage (8). Even though this oil may enter the clutch passage, the clutch will not engage due to the low pressure passage to the tank. The clutch return springs will force the piston back to drain the oil.

Figure 12-49 shows the components that make up the solenoid modulating valves in the initial energized position. When the powertrain ECM requires a clutch to be engaged, a signal is sent to the solenoid (12). The signal is proportional to the desired clutch pressure. When the signal is sent to the solenoid, pin (11) moves to the right and forces ball (1) toward orifice (2). When the check ball moves toward the orifice, the flow of return oil (18) into tank passage (9) is restricted. This causes oil pressure in chamber (13) to increase and move valve spool (4) to the right against the force of spring (5). This oil becomes the pilot control pressure for the valve spool. When valve spool (4) moves to the

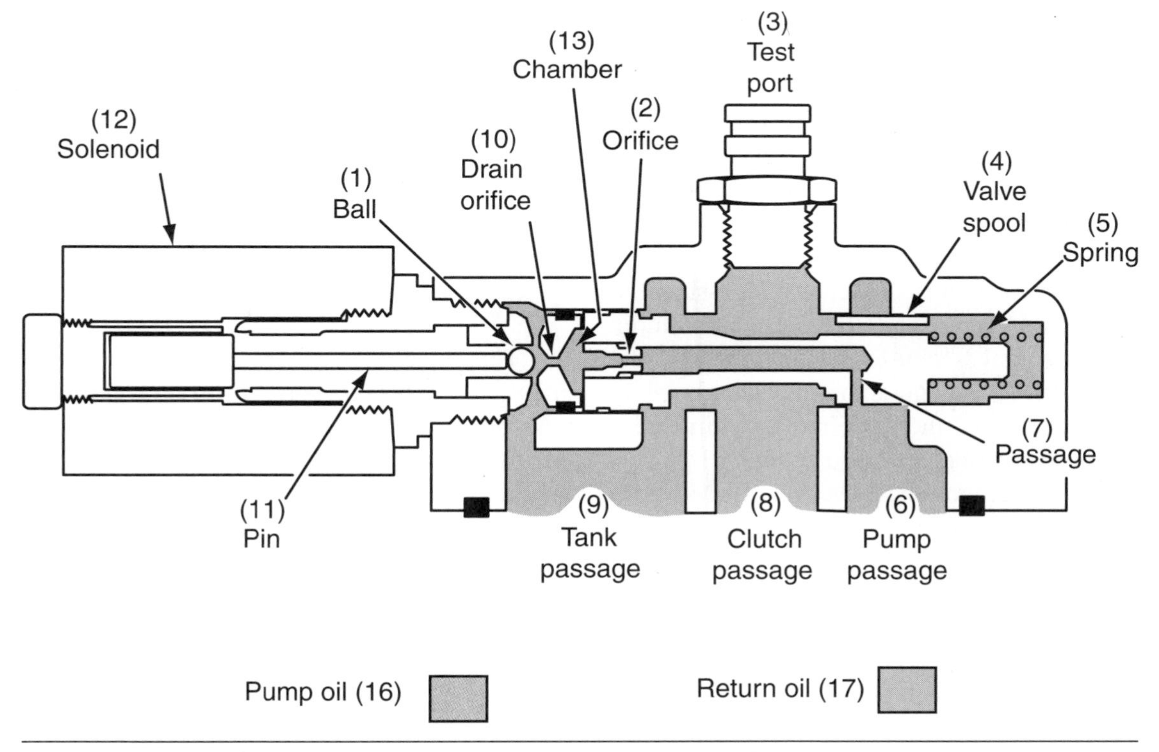

Figure 12-48 Modulating valve with solenoid de-energized.

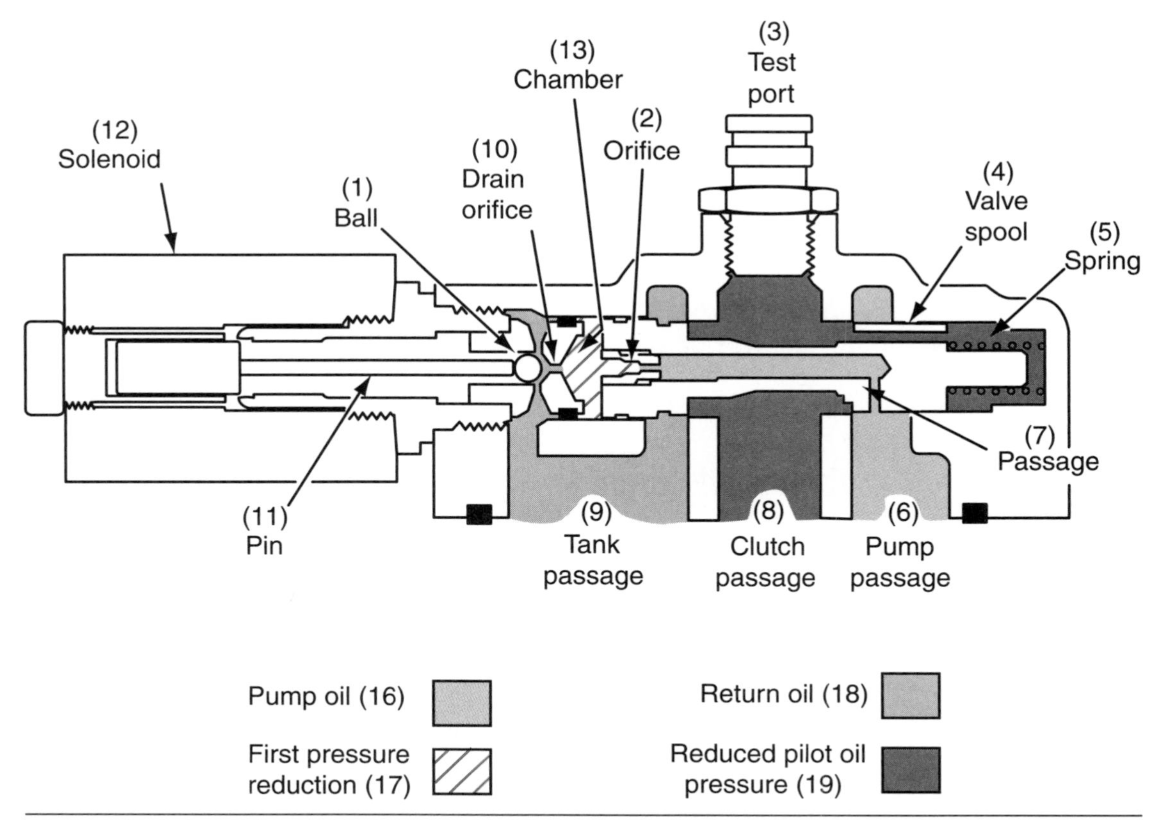

Figure 12-49 Solenoid modulating valve in the initial energized position.

right, pump oil (16) in pump passage (6) enters clutch passage (8). The valve spool blocks the clutch passage from draining to tank passage (9). This causes the clutch pressure to increase due to pump oil flow directed into the clutch passage (8). The oil in the clutch passage also flows through passage (14) into chamber (15). This helps meter and control the oil flow through the valve.

Initially, the powertrain ECM sends a high signal to the solenoid in order to quickly fill the clutch with oil. Then, a reduced signal is sent to the solenoid in order to allow the clutch to engage smoothly. This helps to prevent shock loading of the driveline and to achieve smooth transitions between speed selections.

Once the clutch engages, the powertrain ECM begins to increase the voltage signal to the solenoid to completely engage the clutch pack. (Refer to **Figure 12-50**.) As the signal to the solenoid increases, pin (11) moves to the right and pushes check ball (1) toward orifice (2), completely restricting the flow of oil back to tank passage (9). The oil pressure increase in chamber (13) will push the valve spool to the right. This causes the clutch pressure to increase again. With the clutch pack fully engaged, oil pressure will rise to the transmission relief valve setting. Oil directed through passage (14) together with spring force (5) will meter oil flow through the spool valve to maintain maximum clutch pressure and allow for any leakage within the clutch circuit.

POWERTRAIN ELECTRONIC CONTROL SYSTEM

The powertrain electronic control system is a communication system between the powertrain ECM and the monitoring system. A data link is used to send information back and forth between the two systems. This bidirectional link allows the ECM to input and output information through its circuits and related wiring harness. The data link is also used to communicate with a personal computer through a data link connector.

Whenever there is more than one ECM on a vehicle, networking the ECMs is crucial to finite control of the systems. An engine communicating with a transmission will function more efficiently when one knows what the other is doing. To accomplish this, the ECMs must use a

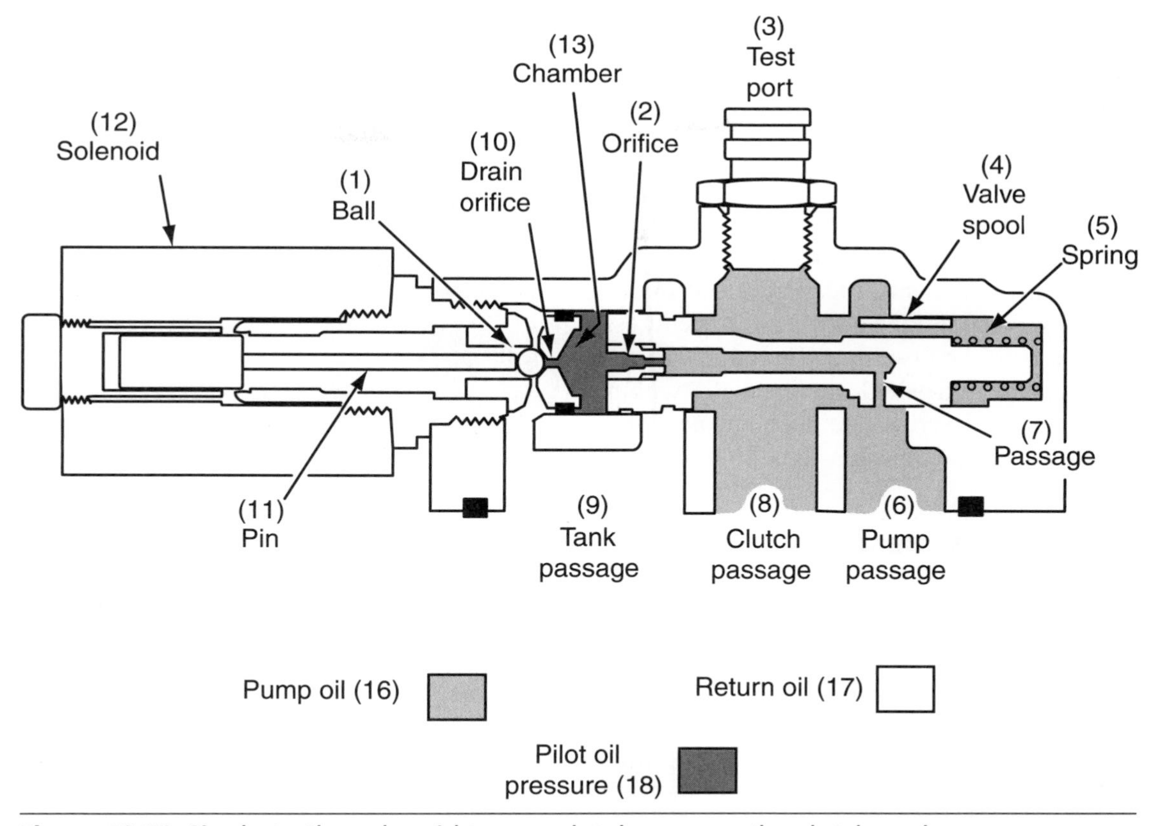

Figure 12-50 Single to the solenoid to completely engage the clutch pack.

common communication or operating protocol. These protocols were discussed earlier in this chapter and are referred to as SAE J1939. SAE J1939 is based on the Controller Area Network (CAN) data link layer. Most diesel-powered off-road vehicles use the J1939–71 application profile for the powertrain. In many machine applications when there are multiple ECMs, such as engine, transmission, hydraulics, payload control, and braking system ECMs, it is important that they all communicate with each other through a CAN network. A CAN network is based on the "broadcast communication mechanism," which is based on a message-oriented transmission protocol. It defines message contents rather than stations and station addresses. Every message has a message identifier, that is unique within the whole network since it defines the content and the priority of the message. This is important when several stations are competing for bus access (bus arbitration). Because of the content-oriented addressing scheme, a high degree of system and configuration flexibility is achieved. It is easy to add stations to an existing CAN network without making any hardware or software modifications to the present stations as long as the new stations are purely receivers. This is referred to as multiplexing. Each CAN message is referenced by a unique number, the PGN (parameter group number). The latest release of the J1939–71 document incorporates several approved additions, and brings the total number of defined messages up to almost 150. This allows for faster communication through the network of ECMs.

The powertrain electronic control system has six basic functions.

- Neutral start function
- Manual shift function
- Automatic shift function
- Parking brake function
- Back-up alarm function
- Diagnostic operation

These functions are identical to the functions described earlier under "Planetary Shift Transmissions," as used by the Cat 988G wheel loader. Please refer to this section for more details. Although the transmissions within these two machines may be different, the ECMs will have similar operating parameters set up within their own electronic control systems. Diagnostic operations will be covered further later in this chapter.

POWERSHIFT TRANSMISSION MAINTENANCE AND REPAIR

Many transmissions used in heavy-duty applications have very diverse maintenance procedures. It is imperative to use and follow OEM. specific maintenance manuals when servicing your equipment. The following maintenance procedures are a guideline to some of the procedures used in the industry today, they are by no means meant to supersede OEM specific procedures. Because we have described two specific transmissions (988G planetary shift and 938G countershaft) within this chapter we will describe the maintenance and repair procedures for these transmissions.

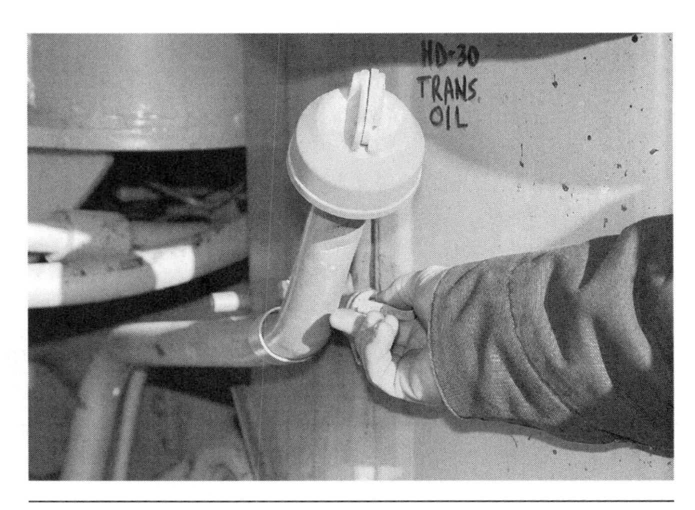

Figure 12-51 Transmission fill tube and dipstick level check.

Powershift Transmission Lubrication

Proper lubrication is one of the keys to a good preventative maintenance program. Maintaining the correct transmission oil level and ensuring that the oil is of OEM specified quality are both critical to ensuring that a transmission achieves its expected service life. Following OEM oil change intervals and oil analysis samples are also key to a good maintenance program. Because the application of heavy-duty equipment is such that it does not accumulate mileage, servicing is based on the number of working hours the machine accumulates or the time period between service intervals. Hour meters log time based from the moment the machine is started until the engine is shut down. To keep an accurate hour meter reading, an engine oil pressure signal or even an alternator pulse activates some hour meters. Direct power from the key switch is avoided to eliminate registering hours when the engine is not in operation with the key switch on.

Here are some examples of service charts:

Maintenance Interval Schedule. All safety information, warnings, and instructions must be read and understood before you perform any operation or any maintenance procedure. Before each consecutive interval is performed, all of the maintenance requirements from the previous interval must also be performed.

The transmission oil level must be checked daily prior to operating the equipment. A dipstick **Figure 12-51** or a sight glass is used in many types of powershift transmissions that make inspection on a daily basis quick and easy. There is no crawling under a machine and no plug to remove to check oil levels as there is on manual shift transmissions. The proper oil viscosity to be used may vary between machine applications and manufacturers.

Selecting the Proper Viscosity Grade. Ambient air temperature is the temperature of the air in the immediate vicinity of the machine. This may differ due to the machine application and from the generic ambient temperature for a geographic region. In order to select the proper oil viscosity, review the regional ambient temperature and the potential ambient temperature for a given machine application. Use the higher temperature of the two in order to select the oil viscosity and use the highest oil viscosity that is allowed for the ambient temperature for that machine. In arctic applications, the preferred method is the use of heaters for machine compartments and a high viscosity oil.

The proper oil viscosity grade is determined by the minimum (lowest) outside ambient temperature. This is the temperature when the machine is started and while the machine is operated. In order to determine the proper oil viscosity grade, refer to the "Min" column in **Table 12-7.** This information reflects the coldest ambient temperature condition for starting and operating a cold machine. Refer to the "Max" column in Table 12-7 to select the oil viscosity grade for operating the machine at the highest temperature that is anticipated.

Machines that are operated continuously, in 24/7 operations (24 hours a day/7 days a week), should use oils that have higher oil viscosity in the final drives and in the differentials. Use of these oils will maintain the highest possible oil film thickness. Since the machines are operated continuously, there is no opportunity for the oil to cool to a dangerous level.

Table 12-7: LUBRICANT VISCOSITIES FOR AMBIENT TEMPERATURES						
Compartment or System	Oil Type and Classification	Oil Viscosities	°C		°F	
			Min	Max	Min	Max
Engine Crankcase	Cat DEO Multigrade	SAE 0W20	−40	+10	−40	+50
	Cat DEO SYN	SAE 0W30	−40	+30	−40	+86
	Cat Arctic DEO SYN	SAE 0W40	−40	+40	−40	+104
	Cat ECF-1	SAE 5W40	−30	+30	−22	+86
	API CG-4 Multigrade	SAE 5W40	−30	+50	−22	+122
		SAE 10W30	−18	+40	0	+104
		SAE 10W40	−18	+50	0	+122
		SAE 15W40	−9.5	+50	−15	+122
Powershift Transmissions	Cat TDTO	SAE 0W20	−40	+10	−40	+50
	CAT TDTO-TMS	SAE 0W30	−40	+20	−40	+68
	Commercial TO-4	SAE 5W30	−30	+20	−22	+68
		SAE 10W	−20	+10	−4	+50
		SAE 30	0	+35	+32	+95
		SAE 50	+10	+50	+50	+122
		TDTO-TMS	−10	+43	+14	+110
Hydraulic Systems	Cat HYDO	SAW 0W20	−40	+40	−40	+104
	Cat DEO	SAE 0W30	−40	+40	−40	+104
	Cat MTO	SAE 5W30	−30	+40	−22	+104
	Cat TDTO	SAE 5W40	−30	+40	−22	+104
	Cat TDTO-TMS	SAE 10W	−20	+40	−4	+104
	Cat DEO SYN	SAE 30	−10	+50	+50	+122
	Cat ECF-1	SAE 10W30	−20	+40	−4	+104
	Cat BIO HYDO (HEES)	SAE 15W40	−15	+50	+5	+122
	API CG-4	Cat MTO	−25	+40	−13	+104
	API CF	Cat BIO HYDO (HEES)	−40	+43	−40	+110
	Commercial TO-4					
	Commercial BF-1	TDTO-TMS	−20	+50	−4	+122
Drive Axles	Cat TDTO	SAE 0W20	−40	0	−40	+32
	Cat TDTO-TMS	SAE 0W30	−40	+10	−40	+50
	Commercial TO-4	SAE 5W30	−35	+10	−31	+50
		SAE 10W	−25	+15	−13	+59
		SAE 30	−20	+43	−4	+110
		SAE 50	+10	+50	+50	+122
		TDTO-TMS	−30	+43	+22	+110

Courtesy of Caterpillar Tractor Company.

WARNING *Do NOT use only the "Oil Viscosities" column when determining the recommended oil for a machine compartment. The "Oil Type and Classification" column MUST also be referred to.*

Tech Tip: SAE 0W oil and SAE 5W oils are not recommended for heavily loaded machines or machines that are operated continuously. Some machine models and/or machine compartments do not allow the use of all available viscosity grades. Therefore, proper selection of oil must be made using OEM recommendations. Some machine compartments allow the use of more than one oil type or viscosity grade; it is recommended that you do not mix these oil types. Also, different brands of oils may use different oil additive packages; it is recommended that different oil brands not be mixed due to chemical incompatibilities.

Transmission Oil Service

WARNING *Care must be taken to ensure that fluids are contained during routine inspection, maintenance, testing, adjusting, and repair of equipment. Be prepared to collect the fluid with suitable containers before opening any compartment or disassembling any component containing fluids that might spill. Dispose of all fluids according to local regulations and mandates.*

When a transmission oil service is due on a machine, it is important to follow these steps:

1. Operate the machine for a few minutes in order to warm the transmission oil.
2. Park the machine on level ground. Lower all implements to the ground and apply slight downward pressure. Engage the parking brake and stop the engine.
3. Remove the transmission oil drain plug and allow the transmission oil to drain into a suitable container. (In some applications, it may be necessary to remove a skidpan or access plate) Inspect the drain plug for any metal or foreign material that might indicate a possible transmission failure. Clean and reinstall the drain plug. See **Figure 12-52.**
4. Replace the transmission oil filter.
5. Use a strap-type wrench to remove the filter element. Dispose of the used filter element properly.

Note: In cases where a canister-type filter element is used, remove bolts from the filter housing cover and lift-off cover. Some filter housing covers are spring-loaded and should be removed carefully. In some canister-type filters, the canister must be rotated or unscrewed to be removed. See **Figure 12-53.** Remove and dispose of filter element. If the filter housing is equipped with a drain valve, this valve should be opened to drain the residual oil contained in the housing. If the housing is not equipped with a drain valve, it is recommended that the oil be extracted using a pump.

Figure 12-52 Transmission drain plug located at the base of the transmission housing. (*Courtesy of Caterpillar*)

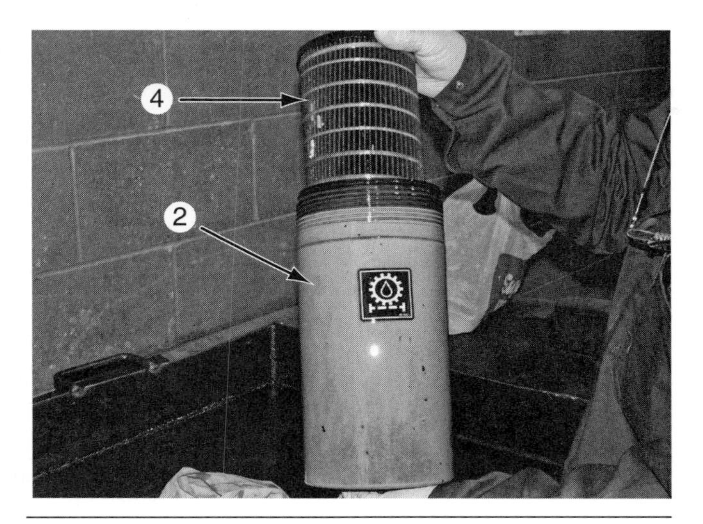

Figure 12-53 Canister filter: (4) disposable filter element, (2) spin-on canister. (*Courtesy of Caterpillar*)

6. Clean the filter-mounting base and make sure that the entire seal is completely removed.
7. Apply a light coat of clean transmission oil to the seal of the new filter element.
8. Install the new transmission oil filter hand tight until the seal of the transmission oil filter contacts the base. Note the position of the index marks on the filter in relation to a fixed point on the filter base.

Tech Tip: The rotation index marks on the transmission oil filter are spaced 90 degrees or 1/4 turn away from each other. When you tighten the transmission oil filter, use the rotation index marks as a guide.

9. Tighten the filter according to the instructions that are printed on the filter. Use the index marks as a guide. As a rule, tighten the filter one-quarter turn after the filter seal has contacted the filter base. You may need to use a strap wrench, or another suitable tool, in order to turn the filter to the amount that is required for final installation. Make sure that the installation tool does not damage the filter.
10. Remove the magnetic strainer cover from the transmission housing. Remove the magnets, the tube, and the screen from the housing and inspect the strainer and magnets for any metal or foreign material that might indicate a possible transmission failure.
11. Wash the tube and the screen in a clean, nonflammable solvent. Use a lint-free cloth, a stiff bristle brush, or pressurized air to clean the magnets.
12. Install the magnets and the tube into the screen. Reinstall the magnetic screen and strainer back into the transmission.
13. Clean the magnetic strainer cover and inspect the seal for damage or flatness. Replace the seal as necessary. Install the magnetic strainer cover and tighten the bolts to proper specifications.
14. Remove the filler cap and fill the transmission with transmission oil.
15. Install the filler cap.

Note: Refer to the section of the operation and maintenance manual on "Lubricant Viscosities and Refill Capacities" for the correct type and amount of oil.

Start and run the engine at low idle with the service brake applied. Slowly operate the transmission control lever from forward to reverse and through each indicated speed in order to circulate the oil within the transmission. Move the transmission control lever to the neutral position, apply the parking brake, and inspect the transmission and transmission filter for leaks with the engine running. Check the transmission oil level and add oil as necessary to obtain proper oil level. See **Figure 12-54.**

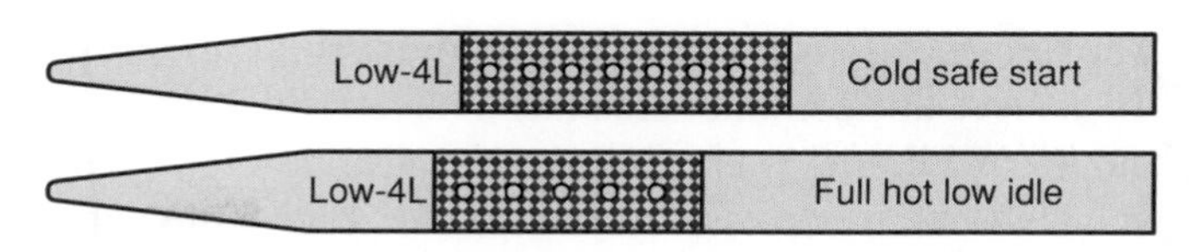

Figure 12-54 Two-sided dipstick with low-idle check on one side and engine safe start on reverse side. (*Courtesy of Caterpillar*)

Note: Certain machines require that the transmission oil level be checked with the engine running, others require that the level be check with the engine stopped. Refer to the operation and maintenance manual for proper procedures to checking oil levels.

In many instances a transmission oil sample is required at every service interval. If an oil sample must be taken, it is recommended that the sample be taken as the oil is draining from the drain plug and not from the filter element. Obtaining an oil sample from the filter will give incorrect readings because the filter contains most of the contaminants. If an oil sample pump is used, an oil sample should be obtained with the engine running and the oil warm.

POWERSHIFT TRANSMISSIONS DIAGNOSTICS

This following section describes some systematic methods of troubleshooting powershift transmissions by using problem and potential cause lists. We will also address troubleshooting and the use of electronic diagnostic equipment. One of the most common complaints a technician receives is the "No Power" complaint. This complaint is one of the hardest problems to solve because it does not lead the technician to a specific component or area of the machine. Therefore, it is critical that the technician involve the operator when trying to isolate this complaint.

Stall Testing

One method of isolating a low power complaint is to perform a "Stall Test" see Photo Sequence 3 on pages 344–345. The torque converter stall test is designed to test the engine, the torque converter, the drive train, and the brake system as a unit. A stall test creates a situation that places the turbine in a stalled or stopped condition with the impeller at maximum engine throttle position.

The torque converter will be at maximum **vortex oil flow.** To achieve this condition, the machine must be kept from moving while the transmission is in a drive mode. The vehicle service brakes must be used to prevent the vehicle from moving and in many cases, it is recommended that wheel chocks be used. Make sure that the service brakes are fully operational and that the transmission oil is at the normal operating temperature of 76 to 112°C (168 to 234°F) when tests are performed.

It is critical to use specific stall test data from the OEM when performing this test. A specific gear selection must be used for a specified time duration. Torque converter overheating and damage can result by not following proper test procedures. This test is best performed outside the shop in an isolated area due to the extreme noise and possible danger involved. Personnel should stay clear of the machine during this test. The transmission auto/manual switch must be in the "manual" position in order to prevent any unexpected shifts during the stall test.

WARNING *To ensure that the transmission oil does not overheat, do not keep the torque converter in a full stall condition for more than 30 seconds. After the torque converter stall test, reduce the engine speed to low idle. In order to cool the oil, move the transmission control to "neutral" position and run the engine between 1200 rpm and 1500 rpm.*

The following test outlines the procedure as it is performed on the Cat 938G® loader, equipped with a hydraulic driven fan, four-speed countershaft transmission with a non-lockup torque converter.

1. Turn the engine start switch to the "off" position.
2. Set the hydraulic fan to the maximum rpm. In order to run the hydraulic fan system at the maximum rpm, the connection to the electro-hydraulic control valve for the hydraulic fan system must be disconnected.

Note: The electro-hydraulic control valve is located on the right side of the engine and below the hydraulic fan pump. Turn the engine start switch to the off position.

3. Disconnect the electrical connection from the connection on the electro-hydraulic control valve.
4. Engage the parking brake and shift the transmission to neutral.
5. Start the engine. Warm the transmission oil to normal operating temperature of approximately 76 to 112°C (168 to 234°F).
6. Fully depress the right-side service brake pedal. Hold the right service brake pedal in the depressed position for the duration of the test. Ensure that the parking brake is applied.
7. Shift the transmission direction and speed control lever to the "fourth speed forward" position. Shift the transmission direction control switch and shift the upshift gear speed switch to the "fourth speed forward" position. Fully depress the accelerator pedal. Allow the engine rpm to stabilize. Then, observe the tachometer and record the result.
8. The correct stall speed for the 938G Series II wheel loader is 2085 ± 65 rpm.

Stall speeds that are lower than the specified rpm indicate incorrect engine performance. Stall speeds that are higher than the specified rpm indicate incorrect transmission performance or incorrect performance of the torque converter. If you must repeat the test, keep the transmission in neutral and wait for two minutes between tests to allow the transmission oil to return to its normal operating temperature.

Problems and Probable Causes. The following six problems and possible causes are provided as examples, but many problems can create diagnostic codes or event codes in the EMS. These codes can help find the root cause of the problem. A technician should be comfortable in his or her ability to retrieve the logged and active codes. Six problems and possible causes are provided as examples

Problem 1: Low stall speed.

Possible Causes:

1. Engine performance is not correct.
2. The oil is cold.

Problem 2: High stall speed in both directions.

Possible Causes:

1. The oil level is low.
2. There is air in the oil.
3. Clutches are slipping.
4. There is torque converter failure.

Problem 3: High stall speed in one gear or in one direction.

Possible Causes:

1. There is a leak in the clutch circuit.
2. There is a clutch failure in the gear that has a high stall speed or in the direction that has a high stall speed.

Problem 4: There is high torque converter pressure.

Possible Causes:

1. There is a restriction inside the torque converter.
2. The oil passage is restricted or the oil cooler is restricted.

Problem 5: There is low torque converter pressure.

Possible Causes:

1. The converter inlet relief valve is stuck.
2. There is low pump flow or pressure.

Problem 6: The torque converter gets too hot.

Possible Causes:

1. There is too much load, which causes the torque converter to slip.
2. The transmission gear is incorrect for the load on the machine. Shift to a lower gear.
3. The oil level is incorrect in the transmission.
4. The water level is low in the engine radiator.
5. There are restrictions in the oil cooler or lines.
6. There is not enough oil to the converter because the converter inlet relief valve is bypassing too much flow.
7. There is not enough oil to the converter because of low pump flow or pressure.

Transmission Problems

Many of the following problems can create diagnostic codes or event codes in the EMS. These codes can help find the root cause of the problem. Technicians should retrieve the logged and active codes.

Four sample problems and possible causes are included here as examples of situations you will encounter. You will be able to diagnose most problems by using resource materials provided by your instructor, shop manuals that are specific to the transmission you are working on, and the manufacturer's technical support documentation.

Problems and Possible Causes

Problem 1: The transmission does not operate in any gear.

Possible Causes:

1. The fuse for the transmission shifter is open or the fuse for the transmission shifter is blown.
2. There is a problem in the electrical circuit. Check ET for active diagnostic codes.
3. There is low system voltage.
4. The transmission shift control unit is faulty.
5. There is low oil pressure.
6. The torque converter has failed.
7. There is mechanical failure in the transmission.
8. The clutch discs and plates are worn too much.
9. The parking brake is engaged.
10. The machine security system is active.
11. The transmission direction and speed control lever and the transmission direction control switch are both in forward or reverse.

Problem 2: The transmission slips in all gears.

Possible Causes:

1. There is low oil pressure.
2. The torque converter has failed.
3. There is mechanical failure in the transmission.
4. The clutch discs and plates are worn too much.

Problem 3: The transmission gets hot.

Possible Causes:

1. There is a low coolant level in the engine radiator.
2. There is too much slippage by the torque converter that is caused by too much load.
3. The transmission gear is incorrect for the load on the machine. Shift to a lower gear.
4. The temperature gauge is faulty.
5. The oil level is incorrect.
6. The oil cooler or lines are restricted.
7. The clutch slips too much. This could be caused by low oil pressure or by damaged clutches.
8. Low oil flow is caused by pump wear or leakage in the hydraulic system.
9. The oil is mixed with air.
10. The torque converter inlet relief valve is stuck open and causes low oil flow through the torque converter.
11. The clutch or clutches are not fully releasing. This can be caused by warped plates or discs. This can

also be caused by a broken return spring or a weak return spring.

Problem 4: There is noise in the transmission that is not normal.

Possible Causes:

1. There are damaged gears in the transmission.
2. There are worn teeth on the clutch plates or on the clutch discs.
3. There are slipping clutch plates and there is noise from the disc.
4. There are other parts that are worn or damaged.

Electronic Diagnostics

Diagnostic software programs are used to access the data bus. The service technician can use diagnostic software to perform troubleshooting and maintenance procedures on the machine. Some of the options that are commonly available are listed below:

- View diagnostic codes
- View active and logged event codes
- View the status of parameters
- Clear active and logged diagnostic codes
- Perform calibration of machine systems
- Program the ECM (Flash)
- Print reports

Using Caterpillar ET to Determine Diagnostic Codes. Connect the Caterpillar Electronic Technician to the machine. Turn the key switch to the "run" position. Start the Cat ET. The Cat ET will initiate communications with the Electronic Control Modules (ECM) on the machine. After communication has been established, the Cat ET will list the Electronic Control Modules. Choose the desired ECM, in this case will be the powertrain ECM. After the diagnostic codes have been determined by the Cat ET, see the test procedure for the corresponding diagnostic code.

Troubleshooting procedures may cause new diagnostic codes to be logged. Therefore, before any procedures are performed, make a list of all the active diagnostic codes in order to determine the system problems. Clear the diagnostic codes that were caused by the procedure when each procedure is complete.

Note: Before performing a procedure, always check that all circuit breakers are not tripped. Repair the cause of any tripped circuit breaker.

A screen, as shown in **Figure 12-55,** is provided in Cat ET for active diagnostic codes. Click the "Diagnostics" button and from the pull-down window choose "active codes." The screen will display the diagnostic codes that are currently active. Active diagnostic information will include a component identifier (CID), a failure mode identifier (FMI), and a text description of the problem.

Logged Diagnostic Codes

A screen is also provided for these codes from the pull-down window. The screen displays the diagnostic codes that are logged or intermittent. These are faults that occur from time to time or are not currently active. The logged diagnostic data includes a CID, a FMI, and a text description of the problem. Also, the logged diagnostic data includes the number of occurrences of the problem and two time stamps. The first time stamp displays the first occurrence of the problem, and the second time stamp displays the most recent occurrence of the problem.

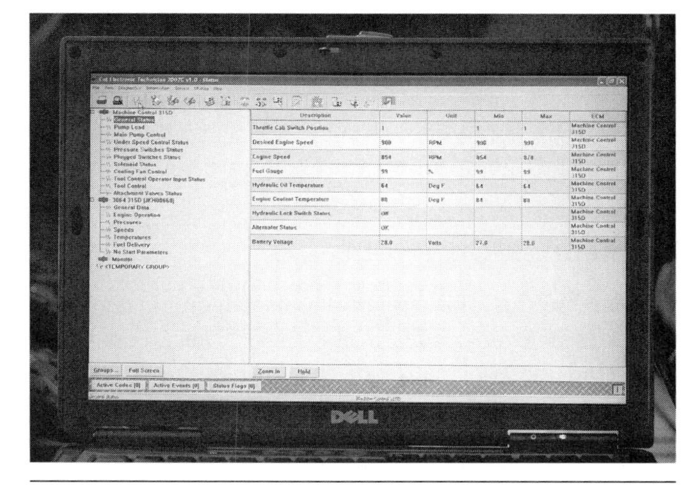

Figure 12-55 Active diagnostic codes shown on computer menu screen.

Diagnostics are logged in non-volatile memory. On power up, the ECM will clear any diagnostic codes that have not been detected or active within the last 150 hours of machine operation.

ONLINE TASKS

1. Access a Service Information System (SIS) specific to a machine with either a planetary or multiple countershaft powershift transmission in your shop. Locate the maintenance schedule for the transmission and the type of oil used. The SIS may require a subscription and or password. These must be acquired by your shop or training center.
2. Access an SIS specific to a machine with either a planetary or multiple countershaft powershift transmission. Locate the "Probable Cause" flowchart. This will give you a systematic approach to troubleshooting specific powershift transmission problems.

SHOP TASKS

1. Locate a powershift transmission assembly in your shop and identify the transmission by type, planetary or countershaft, number of forward and reverse speeds, and number of countershafts, if used.
2. Locate a piece of equipment equipped with a powershift transmission. Perform an oil level check as outlined in this chapter. Refer to the OEM procedure in the appropriate manual.
3. Locate a machine equipped with a powershift transmission in your shop. With the supervision of a trained technician, perform a "Stall Test" as outlined in this chapter or by following OEM procedure. Record your findings and compare them to the OEM specifications.

PHOTO SEQUENCE 3 Transmission/Torque Stall Test Procedure

A torque stall test may be performed anytime the transmission/torque converter circuit is suspected of not meeting performance standards. The test must be performed in a suitable location so as to not compromise the safety of anyone in the immediate area. Ensure that there is adequate room, both front and rear, to allow for any unintended equipment movement. The equipment must be tested on level ground. This procedure should only be performed by qualified personel. The equipment must be brought up to operating temperature with the brake system fully charged, and all fluids in the transmission/torque circuit must be up to the recommended levels and temperatures. A low oil level may produce faulty data and could cause internal damage to the transmission/torque converter.

The following is a general procedure used to check the stall speed on off-road equipment that uses a powershift transmission. It is not intended to replace the manufacturer's recommended procedures and is only an example of the steps required to perform this type of procedure. Always refer to the OEM's technical documentation for the actual step-by-step instructions, as they may vary on different types of equipment.

CAUTION *Never use your hands to check for leaks in a high-pressure hydraulic circuit. Use a small piece of cardboard; a pin-size leak may penetrate the skin. If oil is accidentally injected into the skin, immediate medical attention is required. Oil in a hydraulic circuit can become extremely hot during operation. This can lead to severe burns if accidentally splashed onto yourself or other personnel around the work area.*

P3-1 Depending on the employer some form of tagging or lockout procedure must be used to protect the technicians while work is being performed on the equipment. There are two different procedures used to protect the technician.

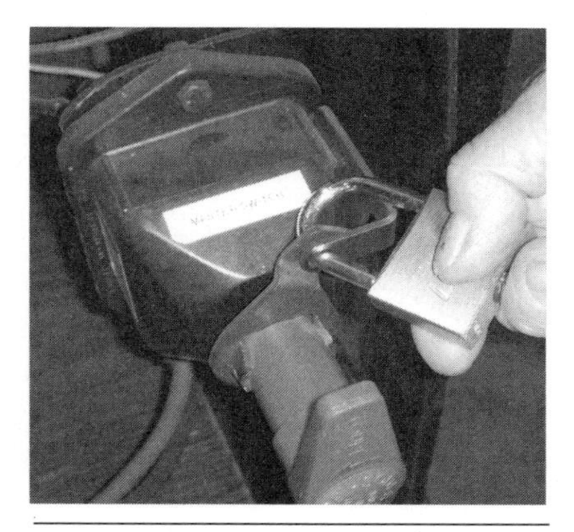

P3-2 Some employers require you to actually lock out the main power disconnect on the equipment. This lock can only be installed and removed by the technician who is performing the work, and it must be removed on completion of the work.

P3-3 Place the appropriate wheel chocks under the tires.

P3-4 Install the lock bar if the equipment has articulating steering.

P3-5 Before performing a torque stall test, the brakes must be able to pass the manufacturer's recommended brake test. Apply the park brake. Ensure that the brake system is fully charged before continuing. Next, depress the brake pedal fully and check the brake pressure gauges for the correct pressure. Refer to the equipment specifications in the shop manual for the correct pressures.

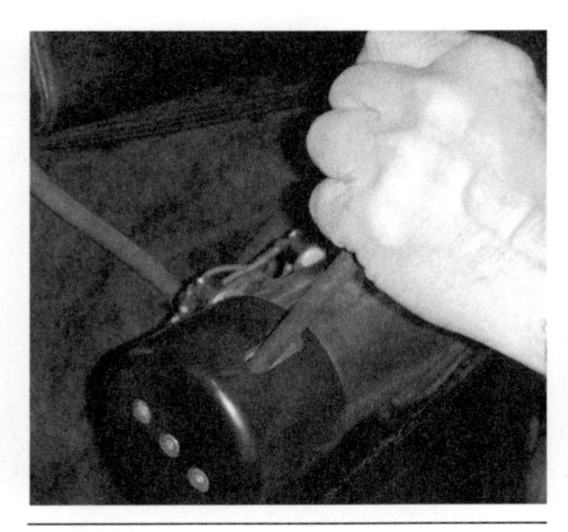

P3-6 The equipment must be placed in the appropriate gear to perform a brake test. Refer to the equipment shop manual for the correct gear selection and rpm range. Not all brake tests on equipment are checked the same way.

Once the brakes have passed the brake test, place the equipment in the correct gear. Refer to the manufacturer's specifications on transmission/torque oil temperature limits before beginning this test. Repeat the stall test, slowly increasing the engine rpm until maximum rpm is reached and record the values. Do not allow the converter to stall for more than 30 seconds at a time without a cooldown period.

Summary

- Two types of oil flow occur in a torque converter: rotary oil flow and vortex oil flow.
- A torque converter is a type of fluid coupling used to transfer engine torque to the transmission.
- A lockup torque converter can provide a direct drive to the transmission.
- A simple planetary gear set consists of a sun gear, a planetary carrier, and a ring gear.
- Depending on which component of a planetary gear set is driven, is held, or is input, torque or speed increases will occur.
- Compound planetary transmissions use combinations of planetary gear sets to produce the required gear ratios for machine applications.
- At each service interval, an oil sample should be taken to monitor transmission condition.
- A multiple countershaft transmission utilizes countershafts in constant mesh to achieve multiple gear ratios.
- The Electronic Control Module (ECM) is a computer that manages the operation of the transmission.
- SAE J1939-compatible hardware and software allow all ECMs to communicate with each other.

Review Questions

1. The three main components in a planetary gear set are
   - A. sun gear, ring gear, and planetary gears
   - B. sun gear, pinion gear, and planetary gears
   - C. sun gear, ring gear, and pinion gear
   - D. pinion gear, ring gear, and planetary gears
2. In planetary transmissions the ring gear is applied or released by
   - A. a brake band
   - B. a cone-type clutch
   - C. a multi-disc clutch pack
   - D. None of the above. The ring gear is never held
3. How many gear ratios can be achieved through a planetary gear set?
   - A. five
   - B. six
   - C. seven
   - D. eight
4. A stator equipped with an overrunning clutch is able to
   - A. freewheel once rotary flow is achieved in the torque converter
   - B. freewheel when vortex flow is achieved in the torque converter
   - C. lock up once rotary flow is achieved
   - D. None of the above. Stators are not equipped with overrunning clutches.
5. A lockup torque converter locks which two components?
   - A. the turbine and stator
   - B. the impeller and stator
   - C. the impeller housing and turbine
   - D. the impeller housing and output shaft
6. Vortex oil flow is
   - A. the flow of oil around the circumference of the torque converter
   - B. the flow of oil across the torque converter
   - C. the flow of oil due to centrifugal force
   - D. the flow of oil through the impeller
7. The purpose of the stator is to
   - A. help pump the oil to the turbine
   - B. redirect oil to the turbine
   - C. redirect oil to the impeller
   - D. help assist rotary flow
8. Fluid couplings are most efficient at
   - A. high speed, no stall
   - B. high-speed, full stall
   - C. low speed, full stall
   - D. all of the above
9. A torque converter stall speed test will provide an indication of
   - A. torque converter condition
   - B. engine performance condition
   - C. transmission performance condition
   - D. all of the above

10. If a torque converter stall speed test is slightly lower than normal, the probable cause could be
    A. a transmission clutch pack may be slipping
    B. engine performance may be lower than normal
    C. overrunning clutch in stator is seized
    D. overrunning clutch in stator is free wheeling
11. When a torque converter stall speed test is much higher than specifications, the probable cause could be
    A. a transmission clutch pack may be slipping
    B. engine performance may be lower than normal
    C. engine performance may be greater than normal
    D. overrunning clutch in stator is seized
12. Torque converter stall speed tests should be performed with
    A. transmission oil cold and slightly lower than normal to allow for thermal expansion
    B. transmission oil at its hottest to simulate working conditions
    C. transmission oil at operating temperature and level full
    D. the converter on a test bench
13. The purpose of the solenoid shift valves is to
    A. relay oil pressure to designated clutches at a preset pressure
    B. develop high enough pressure to engage designated servos or clutches
    C. sense pressure regulator valve signals to determine shift points
    D. shift the selector valve to the desired position
14. Which statement best describes the electro-hydraulic valve body's operation?
    A. The valve body supplies the ECU with all the vital information needed for properly controlling shifts within the transmission.
    B. The valve body pressurizes fluid and provides cooling for hot oil.
    C. The valve allows oil flow through the valve body to the torque converter prior to the clutch pack.
    D. The valve body provides oil flow for lubrication and transmission operation through a series of solenoids controlled by the ECU.
15. Multiple countershaft transmissions differ from planetary transmissions because
    A. countershaft transmissions do not use clutch packs
    B. they are not a powershift transmission
    C. they use countershafts for multiple speed ranges
    D. they divide the input torque to the transmission
16. Multiple countershaft transmissions are limited to
    A. the number of speeds they provide
    B. the amount of torque they can withstand
    C. use with a mechanical clutch
    D. none of the above; they are as variable as planetary transmissions.

17. In a torque converter that uses a freewheeling stator, if the overrunning clutch fails and freewheels in both directions, what problem will this cause?

    A. overheating
    B. low torque output
    C. foaming oil
    D. transmission lockup and the torque converter stall

18. Which component in a torque converter always turns at engine speed?

    A. impeller
    B. turbine
    C. planetaries
    D. stator

Hydrostatic Drive Systems and Hydraulic Retarder Systems

Learning Objectives

After reading this chapter, you should be able to

- Describe the fundamentals of hydrostatic drive systems.
- Describe the components and construction features of hydrostatic drive systems.
- Describe the principles of operation of hydrostatic drive systems.
- Describe the recommended maintenance procedures of hydrostatic drive systems.
- Describe the diagnostic procedures for hydrostatic drives.
- Describe the fundamentals of basic hydraulic retarder systems.
- Describe the components and construction features of basic hydraulic retarder systems.
- Describe the principles of operation of basic hydraulic retarder systems.

Key Terms

case drain
charge check valves
charge pump
charge relief valve
cooling circuit
displacement control valve
fixed displacement
fixed-displacement motor
fixed-displacement pump
friction clutch
high pressure
high-pressure relief valve
hydrodynamic
input retarder
low pressure
metering
output retarder
output shaft
rotor
servo control
shuttle valve
swash plate
turbine shaft
variable displacement
variable-displacement motor
variable-displacement pump

INTRODUCTION

Hydrostatic drive systems, as shown in **Figure 13-1,** are fluid drive systems that are capable of using hydraulic fluid under pressure to convert diesel engine power to fluid power, and then convert it back to mechanical power at the wheels or tracks. This is accomplished by using a positive displacement pump and motor to transfer power through rotary motion to the drive wheels. Although there are some similarities with conventional fluid drives, such as a torque

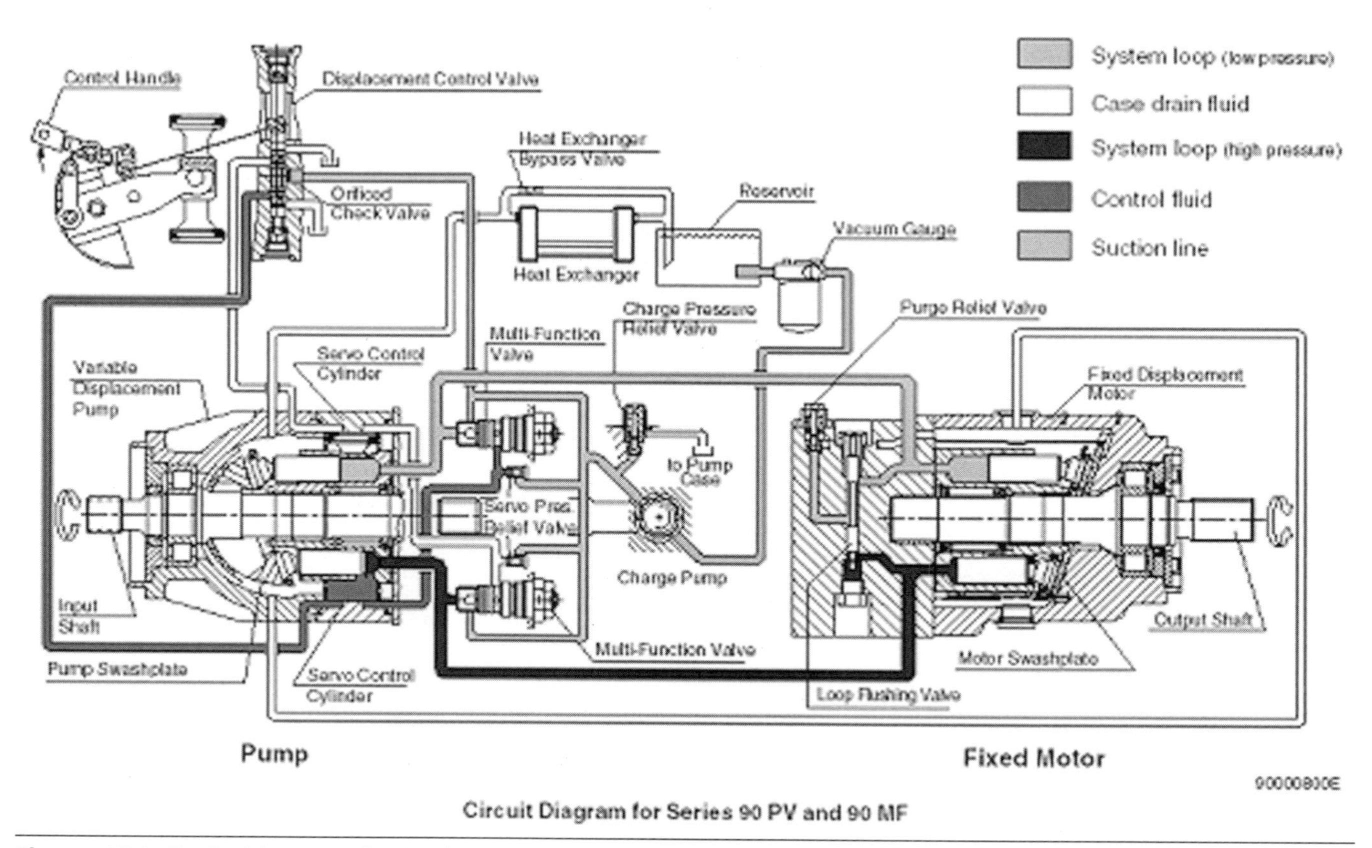

Figure 13-1 Typical layout of a hydrostatic drive system.

converter, the major difference is in how the power is transmitted. The difference between a hydrostatic drive system and a **hydrodynamic** drive system is the velocity of the oil flow and the pressures that are used to transmit power. In a hydrostatic drive system, oil is subjected to **high pressures** at a relatively slow speed. In a hydrodynamic drive system, oil flows at high speed and relatively **low pressure** to provide power to the wheels. In a hydrostatic drive system used on heavy equipment, the hydraulic pump, which is turned by a diesel engine, supplies the movement of oil, while the motor converts the hydraulic energy back to mechanical energy to provide drive. One of the main advantages of using this type of drive is its ability to provide infinite speed and torque range in both directions of travel.

The same rules apply to hydrostatic drive systems as to traditional mechanical drive systems. Torque is reduced as the speed of the motor or drive goes up, and speed is reduced as torque increases. Another advantage of this drive system is its ability to deliver full engine torque at the driven end (drive motor) when the system is at or near maximum pressure. There is considerable flexibility in component placement on these systems. Pumps and motors may be connected together or can be in a separate housing, depending on the design application. A hydrostatic drive can replace a conventional powertrain on a piece of heavy equipment, which is typically made up of a clutch, transmission, and final drive. When equipment uses this drive system design, conventional powertrain components are replaced with a simple hydraulic pump and motor that are hydraulically connected together in a loop, as shown in **Figure 13-2.**

This gives equipment design engineers a lot of flexibility when it comes to placing the hydrostatic drive components in equipment. There are two basic design configurations used in hydrostatic drive systems on heavy equipment: the integral system and the split system.

The *integral* system utilizes a pump and motor that are mounted as a single assembly on the back of a diesel engine. This configuration requires no high pressure hoses to connect the pump and motor together. With this design the output of the motor will usually drive a conventional final drive, replacing the transmission only, as shown in **Figure 13-3.**

The other equally popular design used on heavy equipment, such as a bulldozer, is known as the *split* sys-

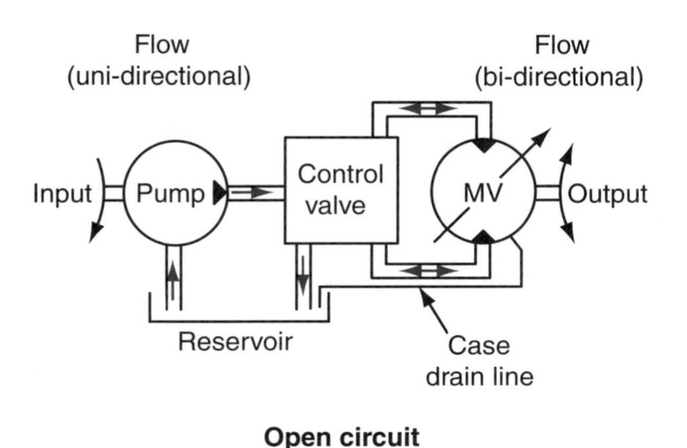

Figure 13-2 Schematic of a open-loop hydrostatic drive system. If the speed increases at the pump, the output power at the motor will also increase.

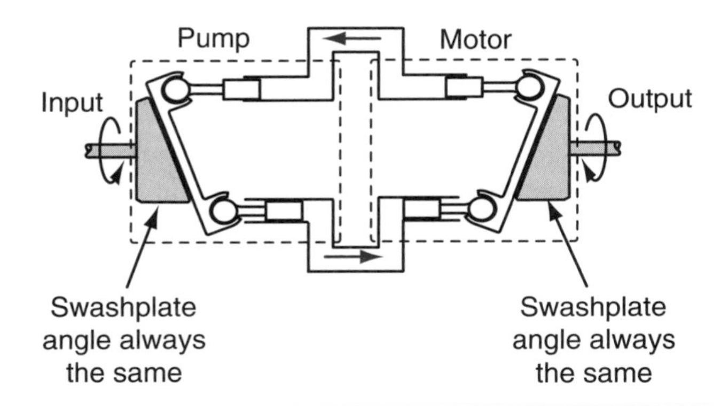

Figure 13-3 A fixed-displacement pump with a fixed-displacement motor. The motor has a fixed swashplate, resulting in constant torque at any given speed. The pump has a movable swashplate that can vary the speed of the pump, resulting in variable speed and constant torque.

tem. This design may have the pump mounted on the back of the diesel engine, and uses a separate motor on each side of the bulldozer to drive the tracks independently. Some equipment manufacturers use two pumps, one for each motor. This design requires the use of high pressure hoses to connect the pump and motors together in a loop.

Hydrostatic drive systems have some unique advantages over conventional drives when used to drive mobile equipment. The drive system is capable of generating a dynamic braking effect, which provides a means of slowing or stopping the vehicle, or even the ability to coast when required. These drive systems are capable of providing available torque on startup that can be as high as 90% of the maximum torque capability of the equipment. This gives equipment design engineers flexibility when it comes to engine placement. The overall weight of the equipment is significantly lower due to the size and weight difference between conventional drives and hydrostatic drives. As for powertrain efficiency, the inertia of the rotating parts is low when compared to convention mechanical drives. These drive systems are capable of having an infinite speed range in forward as well as reverse when they are used with variable-displacement pumps and motors.

FUNDAMENTALS

To gain an understanding of how a hydrostatic drive system actually works, you must understand the basic principles of hydraulics. We must remember a couple of important facts when trying to understand how this system operates. We know that a hydraulic fluid does not compress and will always take the shape of the component it happens to be flowing through. In this chapter we will concentrate on the application of these principles on off-road equipment, especially in their use as a replacement for conventional final drive systems. There are two separate applications on mobile equipment that take advantage of these principles: non-traction drives, which are used to operate things like mast rotators on excavators or cranes; and traction drives, which are used to drive the tracks or wheels. When used as final drives, traction drives can be described as having outputs that are rotary rather that reciprocating in nature.

There are two other factors that we can use to further categorize hydrostatic drive application on mobile equipment. Depending on the application, the equipment design engineer can configure the system to be used as either an open-loop or a closed-loop system. Both basic open-loop and closed-loop systems, as shown in **Figure 13-4,** use pumps, motors, lines, control valves, and reservoirs in their circuits. However, there is a fundamental operational difference in the way the oil is handled after it leaves the motor. In a closed-loop system, as shown in **Figure 13-5,** the oil flows from the pump to the motor and is returned directly to the inlet of the pump. In an open-loop system, the hydraulic oil flows from the pump to the motor and then back to the hydraulic tank.

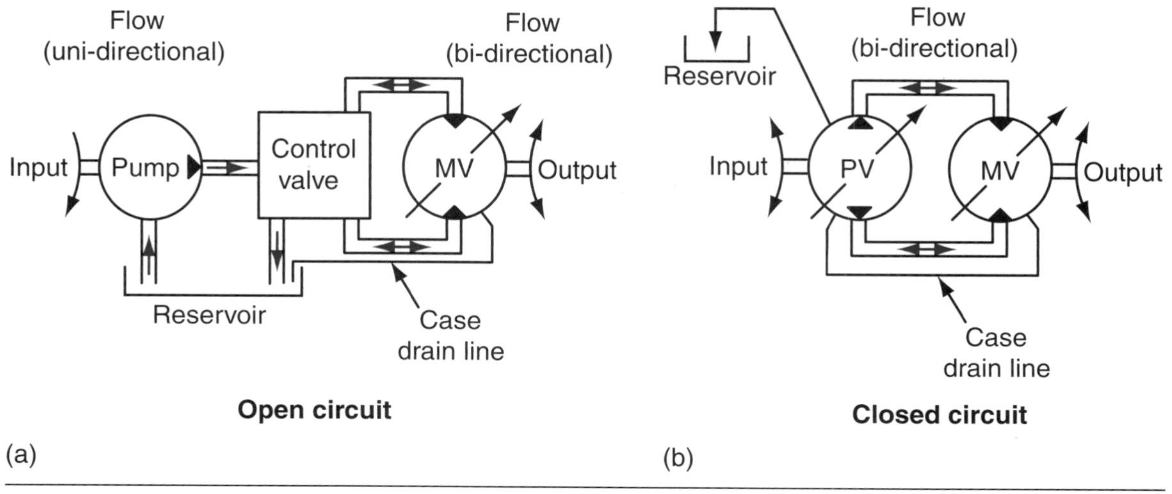

Figure 13-4 Open- and closed-loop hydrostatic drive circuits.

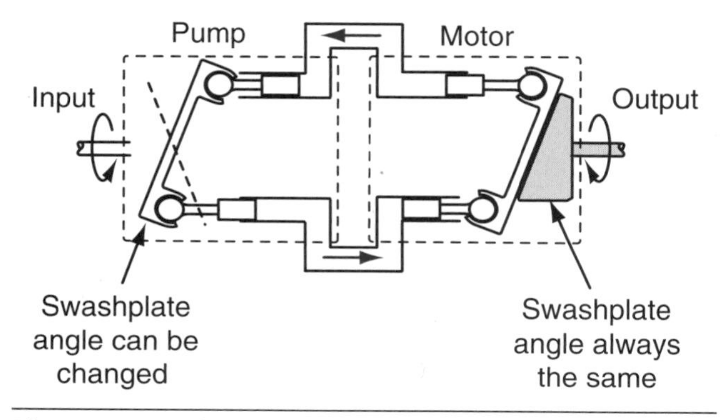

Figure 13-5 This illustration shows a variable displacement pump with a fixed displacement motor. The motor has a fixed swash plate, which results in constant torque at any given speed. The pump has a movable swash plate that can vary the speed of the pump, resulting in variable speed and constant torque.

Depending on equipment design, conventional hydrostatic drive systems typically consist of the following drive configurations:

- A fixed- or a variable-displacement pump, with one or more fixed- or variable-displacement motors.
- You could combine a **fixed-displacement pump** with a **fixed-displacement motor.**
- Or you can combine a fixed-displacement pump to drive a variable-displacement motor, as shown in **Figure 13-6.**
- A **variable-displacement pump** to drive a fixed displacement motor.
- Some manufacturers use a variable-displacement pump with a **variable-displacement motor.**

These components can be combined in many different configurations to suit the application they are to be installed in.

Torque range in a hydrostatic drive system can be explained by comparing what takes place in a conventional equipment powertrain to what takes place in a hydrostatic drive loader. As an example, we can compare the two different drive torque capabilities. Let's look at a hypothetical four-speed standard transmission with a 4:1 ratio in first gear, 3:1 ratio in second

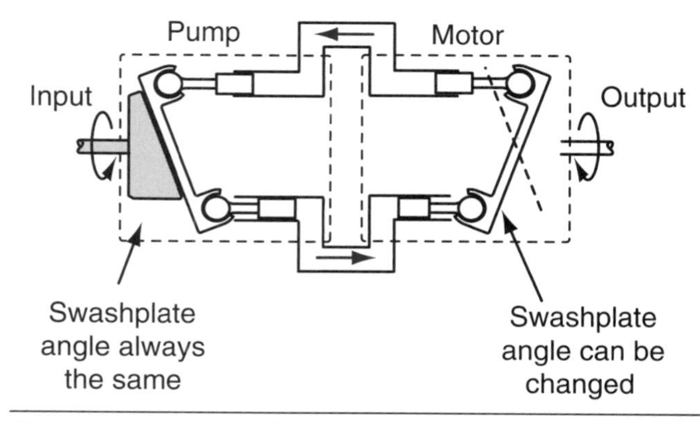

Figure 13-6 How a fixed-displacement pump may be used with a variable-displacement motor. The design is much more complex; the motor stroke must be controlled.

gear, 2:1 ratio in third gear, and a 1:1 ratio in fourth gear. Now let's look at a typical hydrostatic drive loader with a variable-displacement pump that has a flow rate of 4 to 10 cubic inches per minute and a variable-displacement motor with a maximum 10 to 40 cubic inches per minute displacement. We can calculate the ratio as follows:

$$\frac{\textbf{Motor at maximum displacement}}{\textbf{Pump at minimum displacement}} = \frac{\textbf{40}}{\textbf{4}}$$

$$= \textbf{10:1 torque multiplication}$$

$$\frac{\textbf{Motor at minimum displacement}}{\textbf{Pump at maximum displacement}} = \frac{\textbf{10}}{\textbf{10}}$$

$$= \textbf{1:1 torque multiplication}$$

For maximum torque we divided the motor maximum output by the pump minimum output; we got a 10 to 1 ratio. For minimum torque we divided the minimum motor flow rate of 10 cubic inches per minute by the maximum pump displacement of 10 cubic inches per minute for a 1 to 1 ratio. When you compare the two values, we have an overall torque ratio of 10 to 1.

Heavy off-road equipment generally requires a variable range of torque for each type of work to be performed. A wide torque range can be obtained by using a variable-displacement pump and motor. The torque range is always based on the ratio between the pump and motor displacement. If a hydrostatic drive system that consisted of a fixed-displacement motor and pump was used the torque could only be varied by changes in the system pressure.

Design Features of Hydrostatic Drive Systems

We know that pump or motor displacement can be expressed by the volume of oil that is displaced during one revolution of the pump or motor. We have seen how pumps and motors with fixed or variable displacements can be combined to give the output speed and torque that are required on equipment. Although hydrostatic drive systems are considered efficient, they do have speed and torque losses that are inherent to their design. Internal leakage in the pump and motor circuit are required for lubrication and cooling.

These losses become greater whenever the equipment operates at low speeds and high pressure. As the operating speed of the equipment increases, the internal leakage stay relatively the same, increasing the efficiency of the system. Some torque in a hydrostatic drive circuit is lost to friction as well as to high speed and pressure in a circuit. Let's look at how these combinations can be put to use in heavy equipment hydrostatic drive circuits.

Basic Hydrostatic Drive System Components

Charge Pump. The external gear pump is the most common type of pump used to charge a hydrostatic drive circuit. This pump design is ideally suited to this application because of its simplicity and relatively low cost when compared to other pump designs. Variable displacement or high pressure is not required in the charge circuit of a hydrostatic drive, making the fixed-displacement pump a candidate well-suited for this application. The purpose of a **charge pump** is to supply the oil flow for the main pump as well as to provide oil flow for the lubrication and cooling of a typical hydrostatic drive circuit.

Main Pump. The main pump generally used in a hydrostatic drive circuit is a positive displacement pump of either a variable- or a fixed-displacement design. Whether the circuit is a closed- or open-loop circuit dictates whether we would use a fixed- or variable-displacement pump. When selecting the appropriate pump, the application and environment have to be considered.

Factors like pressure and flow rating need to be considered when a design engineer chooses a pump. If the hydrostatic drive circuit is used with a diesel engine as the power source to drive the pump, the options are greater. For example, if you are going to drive the pump with a constant speed source, then you should consider using a fixed-displacement pump with an open-center loop design. This combination allows the oil to flow back to the tank when in the neutral mode, preventing unnecessary heat buildup in the circuit.

The use of a variable-displacement pump (the most efficient pump available) in a hydrostatic drive circuit is determined by various factors; one of them is the operating speed. If the speed of the equipment will vary greatly and require the use of hydraulic power at different operating speeds, then consider using a variable-displacement pump. This pump can decrease or increase the flow of oil in a circuit, with no change in engine speed, just by varying the displacement of the pump.

Drive Motor. The drive motor or motors selected for use in a hydrostatic drive circuit can be of either a variable- or fixed-displacement design. Whether the circuit it a closed- or open-loop circuit determines whether we would use a fixed or variable displacement motor. Again, selection requires consideration of application and environment. The design engineer must consider factors like torque and speed when choosing a motor. If the equipment is powered with a diesel engine as the power source, the options are greater. A motor in a hydrostatic drive circuit is propelled by the flow of fluid sent to it by the main pump. The displacement of the motor will tell us how much flow is required for a specific drive speed. To determine how much load the motor can work with, we need to consider the torque and pressure rating of the motor.

Filter. Contamination of the hydraulic oil must be prevented at all costs, as this will lead to premature component failure. The oil supplied to the charge pump of a hydrostatic drive circuit must be pulled through a full flow suction filter with no bypass, and the filter must be capable of filtering particles as small as 10 microns in size. The cleanliness of the oil used in this drive circuit is critical. It must be free from all contamination because the pumps, valves, and motors used in this circuit are adversely affected by dirty oil. With this filter design, the flow is reduced as the filter becomes clogged, resulting in a corresponding slowdown of circuit response. This will not generally cause any damage to the system in the short term. However, when this occurs, it is an indication that the filter needs replacing. The flow rating of the filter must always be considered when replacing any filter. Do not use a filter that has a lower flow rating than pump specifications demand or poor system performance and/or filter collapse is likely. This will result in serious system damage.

Heat Exchanger. The use of a heat exchanger is a requirement if the operating temperature of the hydraulic oil will exceed 180 degrees at the motor **case drain.** The heat exchanger should be located in the return circuit of the drive circuit. The size of the heat exchanger is determined by the duty cycle of the equipment. To determine what size the heat exchanger should be, calculate the capacity so that it can dissipate approximately 20 to 25% of the input power produced by the circuit. The total system restriction should never result in a case pressure higher than 40 psi at normal operating temperature. A heat exchanger must have a bypass built into the return circuit to prevent back pressure from exceeding the maximum back pressure limits under any operating conditions that may be encountered, such as when the drive oil is cold.

Reservoir. The reservoir size needs to be calculated carefully; the minimum volume of the tank should be half of the charge pump volume measured in gallons per minute (gpm). This volume will ensure that the minimum fluid dwell time is at least 30 seconds at the maximum return flow in the circuit. The suction line outlet on the tank should be located away from the base of the tank to prevent any large particles from entering the pump inlet. A fine mesh screen (100 mesh) should be located on the outlet port of the tank; this will trap any debris from entering the pump. The oil return line to the tank should be positioned so that the fluid has an opportunity to be properly deaerated. To facilitate a suction filter replacement, a shutoff valve should be located at the tank outlet so that the oil flow can be shut off when replacing the filter. A reservoir drain plug or drain valve should be placed in the bottom of the tank to facilitate complete oil draining if the system becomes contaminated.

Hydraulic Lines and Fluids. We must adhere to the guidelines that are provided by the specific equipment manufacturer for the use of a fluid in a hydrostatic drive circuit. Examples of service intervals will be shown for average conditions. If the equipment is operated in conditions that are above the normal ambient levels, then the service life of the oil and filter must be monitored to ensure that oil or filter capabilities, are not exceeded. The selection of hydraulic lines should also be in keeping with equipment manufacturer's recommendations for length, diameter, and pressure capabilities. Consider the bend radii of hydraulic lines that may be exposed to continual flexing during operation, keeping in mind that hoses are also rated for the number of bends that they can sustain during operation.

The type of hydraulic oil used in these systems can be detrimental to the life of the seals. Modern seal materials may include synthetics. It is imperative that a seal made from synthetic material be properly pre-lubricated since the synthetic material does not absorb oil. (Most types of conventional seal material will absorb some oil.) Some seal manufactures recommend that you soak the seal in oil before installation. It should be noted that under no circumstances should seals be installed without pre-lubrication.

REVERSE FLOW CONTROL IN HYDROSTATIC DRIVE SYSTEMS

In order for a hydrostatic drive system to be practical in a heavy equipment application, we need to be able to

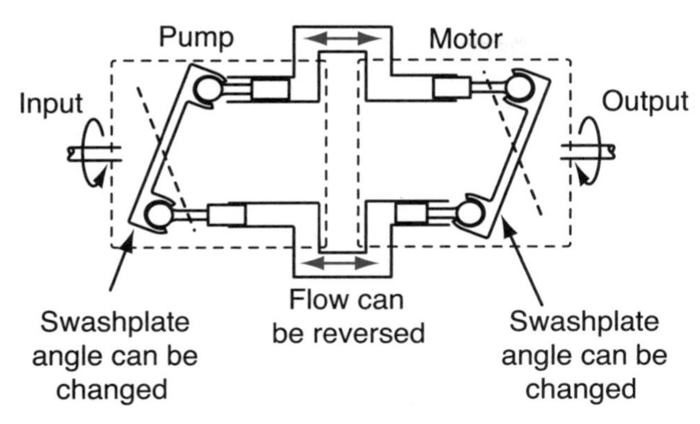

Figure 13-7 A variable-displacement pump may be used with a variable-displacement motor. The system is much more complex as both the pump and motor stroke must be controlled. It has the capability to operate in any of the other three modes that are shown.

reverse the direction of travel without any additional gear train components being added, as shown in **Figure 13-7.** When a heavy equipment design engineer considers the type of pump and motor to use in a particular hydrostatic circuit, he or she will ultimately choose a pump that is capable of reversing or stopping the flow in a circuit without changing the direction of rotation. Reversing or stopping the flow of the pump can be accomplished by moving the swashplate of a variable-displacement pump. Placing the swashplate in a vertical position, where the pistons can no longer reciprocate, stops the oil from flowing to the motor; this is similar to placing a standard transmission in neutral. To reserve the direction of travel, simply reverse the swashplate angle. This action changes the direction of flow; the inlet of the pump now becomes the outlet, and the outlet becomes the inlet, reversing the direction of rotation of the motor.

DISPLACEMENT CONTROL OF HYDROSTATIC DRIVE SYSTEMS

When using a variable-displacement pump or motor, it is generally best to use a displacement control system to control the stroke length of the pistons, as shown in **Figure 13-8.** This system could be as simple as a manual control, or it could be more complex and use a compensator control circuit. On mobile equip-

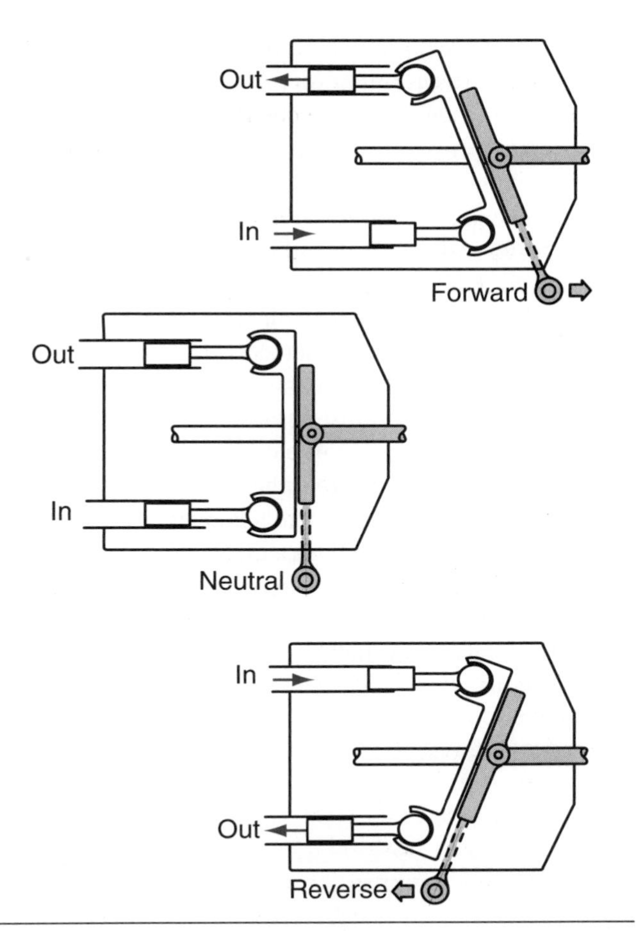

Figure 13-8 The servo piston connection to the reversible swashplate. Any change in the piston position changes the swashplate angle, which varies the amount of oil that is sent to the motor.

ment a **servo control** can be used to change the swashplate angle; this is accomplished by adjusting the stroke length. The displacement controller that is used to vary the swashplate angle can be built into the motor housing or may be remote mounted. The control circuit uses hydraulic pressure to adjust the angle of the swashplate. When the swashplate angle is increased, the stroke is increased, producing an increase in oil flow from the pump without an increase in the engine speed.

One variation of displacement control used by Sundstrand on their series 90 pump is a manual displacement control that uses movement of a control lever to control a spring-centered four-way servo valve, which sends control pressure to either side of a dual-acting servo piston. The servo piston changes the swashplate angle, thus increasing the piston stroke and the output displacement of the pump. This change in flow tends to give a smooth increase in output to a variable-displacement pump.

Newer hydrostatic drive equipment uses a variation of hydraulic displacement control. This control system uses a pressure-control pilot (PCP) valve to control the input signal pressure. The PCP valve uses a 12-volt signal to operate a spring-centered four-way servo valve that sends hydraulic oil to either side of the dual-acting servo valve. The cradle swashplate control is designed so that the angular position of the swash plate is proportional to the electric direct current (DC) input.

Some equipment manufacturers utilize another type of control system, which works something like a conventional automatic transmission. The motor control system uses a 12- or 24-volt electrically operated spool valve that sends pressure to either side of the **displacement control valve,** acting like a conventional forward/neutral/reverse automatic transmission control. When one of the solenoids for forward or reverse is energized, it forces the pump to go to full displacement. These electric control valves are factory preset and, therefore, are not adjustable.

The term *stepless* is often used to describe the operation of a hydrostatic drive. Another common variation in flow control can be used to provide a *stepped* drive. Some manufacturers use two charge pumps mounted in tandem to increase oil flow capabilities on some equipment. One pump is used to provide initial torque to get the piece of equipment moving; as the speed of the equipment starts to increase the pressure in the system will drop off. This starts a second pump to increase the flow to the circuit, simulating a change in gear ratio similar to a conventional transmission.

Circuit Design Features

To protect a hydrostatic drive circuit from dangerous overpressure conditions, an overpressure relief valve, as shown in **Figure 13-9,** is used in each side of the circuit, since either side of the hydrostatic circuit may be subjected to high pressure depending on which direction the equipment is moving. The pressure setting on these safety valves can be very high, generally about 5000 psi. A charge pump, as shown in **Figure 13-10,** is used to provide oil flow for the **cooling circuit,** and to maintain a positive pressure on the low pressure side of the main

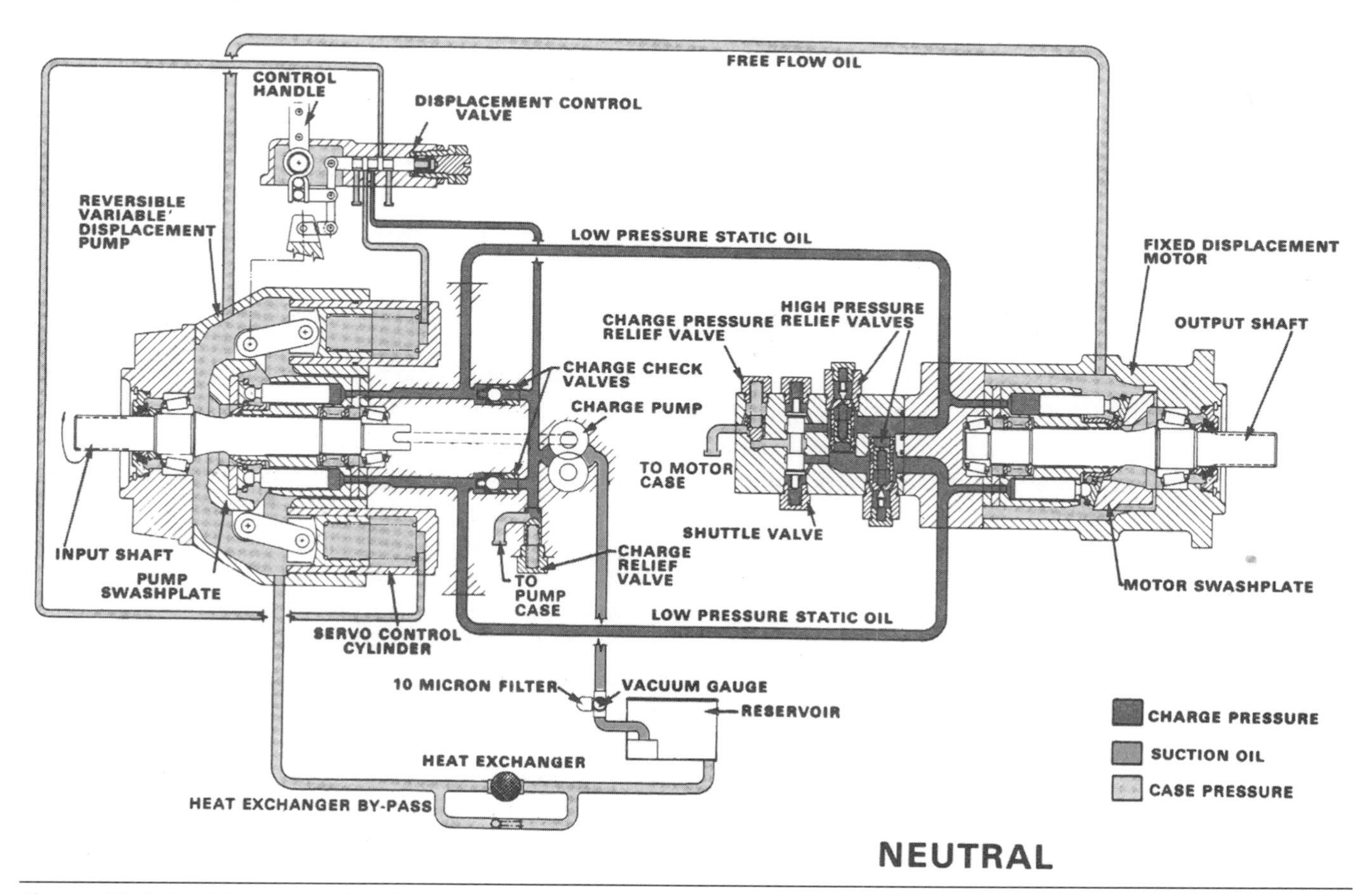

Figure 13-9 In this motor circuit, two high-pressure relief valves as well as the charge pressure relief valve are shown. (*Courtesy Deere & Company*)

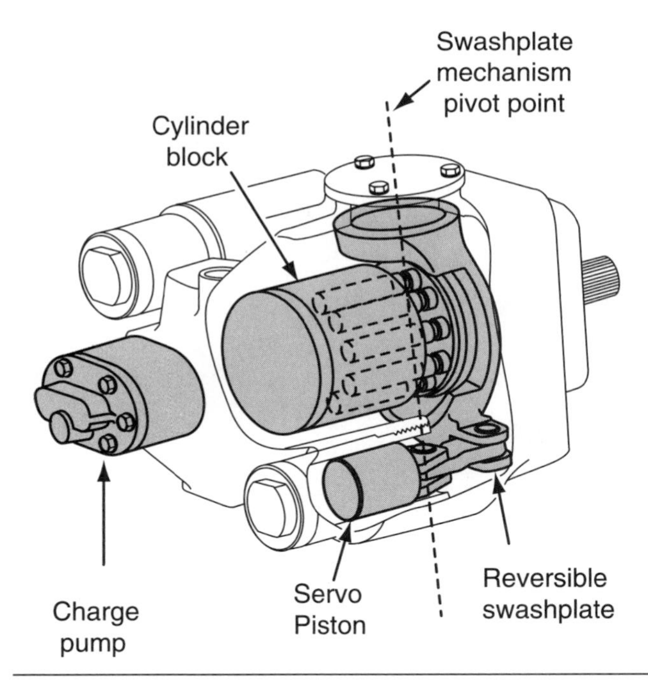

Figure 13-10 A variable-displacement pump that has the charge pump mounted to the back of the housing. The charge pump is driven at the same speed as the main pump.

pump and motor circuit. The control circuit also relies on the charge pump for its supply of oil, as well as for providing makeup oil for internal leakage in the circuit. The charge pump used in this circuit is a fixed-displacement pump that is generally mounted to the main pump and is often driven at the same speed as the main pump.

Cooling Circuit. As with all types hydraulic circuits, heat is generated any time oil is circulated under pressure. This necessitates that all hydrostatic drive circuits provide a way to cool the hydraulic oil. The charge pump pulls filtered hydraulic oil from the reservoir and sends it to the main pump, as shown in **Figure 13-11.**

Note: Some component manufacturers build in an internal filter element for the main pump. The oil must pass through this filter before entering the pump. Hydrostatic drive circuits cannot tolerate any dirt getting into the circuit. Dirt would circulate within the components and quickly cause excessive wear to the pump and motor.

After the pressure in the charge pump circuit reaches charge pump relief pressure, a check valve opens to allow the excess oil to enter the main pump housing to cool and lubricate the internal pump parts. The case drain on the main pump allows the coolant to return to the main reservoir after passing through a cooler to remove any excess heat. The oil cooling circuit used in all hydrostatic drive circuits has a bypass valve located in the cooler that works on pressure differential. If the pressure on the inlet side of the cooler becomes greater than the outlet side of the cooling circuit, the bypass valve allows the oil to bypass the cooler, protecting it from over pressure. When the equipment is operated in the pump neutral condition, the oil from the charge pump is sent to the cooling circuit by the neutral **charge relief valve** and only flows through the pump case, not to the motor case.

Closed-Loop/Open-Loop Hydrostatic Circuits. Although these two circuit designs each use pumps and motors, they differ in the way the oil is handled at the end of each cycle. The closed-loop circuit, as shown in **Figure 13-12A,** sends the oil back to the pump inlet after it leaves the motor. In the open-loop design, the oil leaves the pump and goes back to the hydraulic reservoir before beginning the cycle again. The closed-loop variable-displacement pump circuit in Figure 13-12 shows how the oil is handled after it leaves the pump. The oil is sent through the motor and then returns to the pump. In the open-loop circuit, as shown in **Figure 13-12B,** the oil leaves the pump, travels through a control circuit before traveling through the pump, and then returns to the tank. The charge pump picks up its oil directly from the hydraulic tank, unlike the closed-loop circuit where the return oil for the motor is returned to pump. The variable-displacement closed-loop circuit design is considered the more efficient of the two circuits because the charge pump replenishes the high-pressure circuit with makeup oil and provides oil flow for the cooling and filtration circuit. To get reverse direction with this circuit design, the variable-displacement pump must be an over-center design; this allows the swashplate to move off center in both direction to provide reverse capability.

The open-loop hydrostatic drive circuit is often used where two or more different loads are driven simultaneously by either a fixed- or a variable-displacement pump and with the motor returning the hydraulic oil to the tank. Because this circuit design is often used with two or more loads, it must use a control valve to provide direction and **metering** control. Often this circuit will use a fixed-displacement pump, as it is more cost-effective to incorporate in small equipment design where cost is of importance.

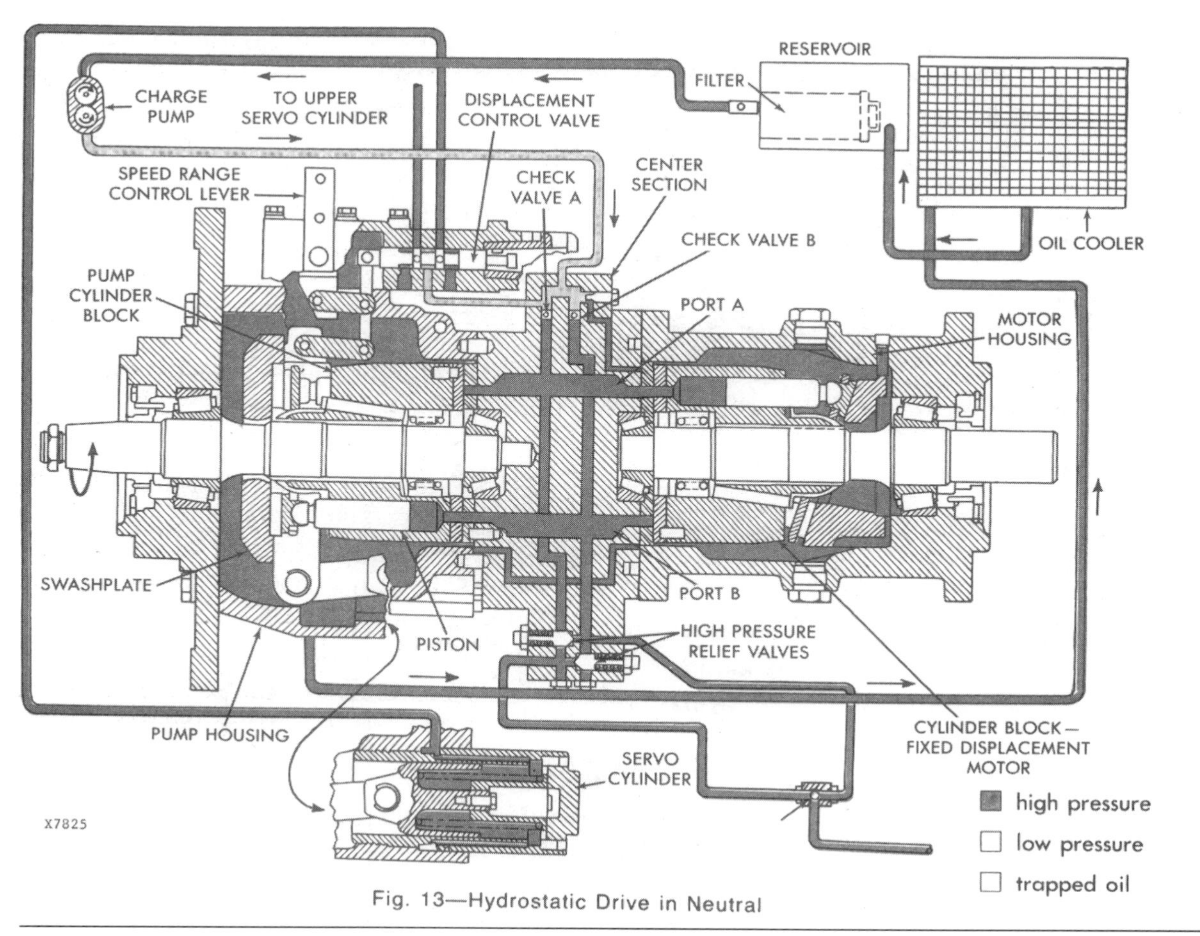

Figure 13-11 Excess charge oil entering the pump case and returning back to the reservoir. (*Courtesy Deere & Company*)

As with any circuit design there are always disadvantages to go along with the advantages. Whenever more that one actuator is used in the circuit at the same time, load interference may become an issue in the circuit, as well as the loss of dynamic braking effect. When operating equipment that uses an open-loop circuit design, the operator must vary the engine power to provide a change in speed.

Operation of a Variable-Displacement Pump/Fixed-Displacement Motor Drive in Neutral

The variable-displacement pump/fixed-displacement motor combination offers infinite speed and direction control. The equipment operator has complete control of speed and direction by using one controller for both speed and direction. A diesel engine provides the power source for the pump. In **Figure 13-13** the engine is running and the control lever is in the neutral position. The displacement control valve is also in neutral position, allowing the servo control ports to send oil flow to the sump. The charge pressure exerts equal pressure on the pump pistons, forcing the swashplate into neutral.

The charge pump pulls oil from the hydraulic reservoir through a 10-micron filter. Hydraulic oil is directed to the low pressure side of the main pump and circulates through the pump and motor housing and back through the heat exchanger before going back to the hydraulic tank. Trapped oil is held in the pump and motor block, as well as in the connecting lines, by the

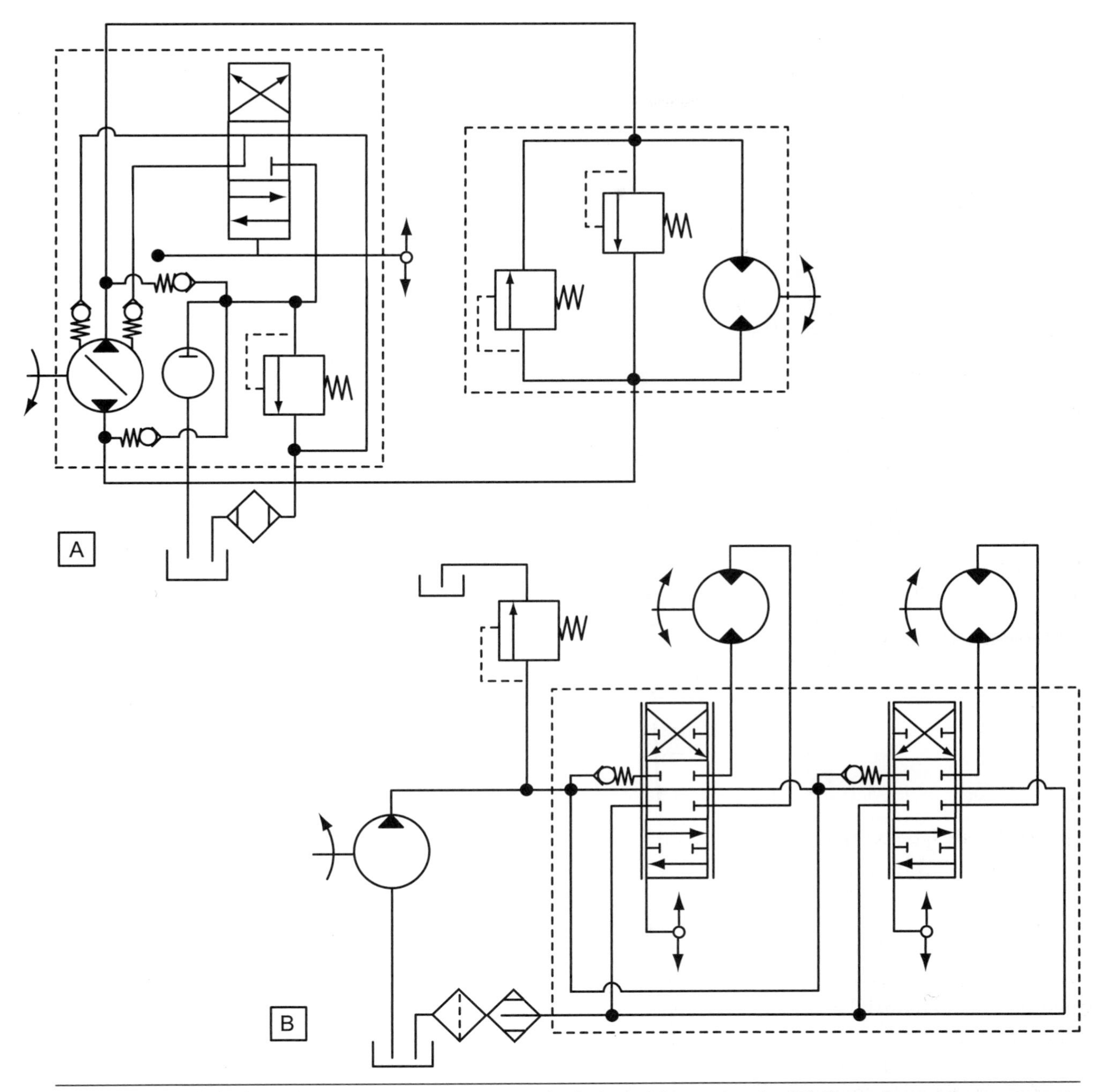

Figure 13-12 (A) The schematic represents a typical closed-loop circuit. (B) The schematic represents a typical open-loop circuit.

two **charge check valves.** Since the control lever is in the neutral position in the pump, the swashplate has de-stroked the pistons in the variable-displacement pump. With the engine running and the pump turning, no oil flows to the motor. However, an oil flow is maintained at low pressure for the control circuit and make-up oil. This compensates for any internal leakage in the pump and motor.

During operation in neutral, the **shuttle valve** in the motor remains centered and allows the charge relief valve to direct the excess oil to flow to the heat exchanger. When operating in this state, the cooling oil flows through the pump and not the motor. The heat exchanger has a bypass, which protects the heat exchanger from high back pressure when operating with cold oil or if the heat exchanger should become plugged internally.

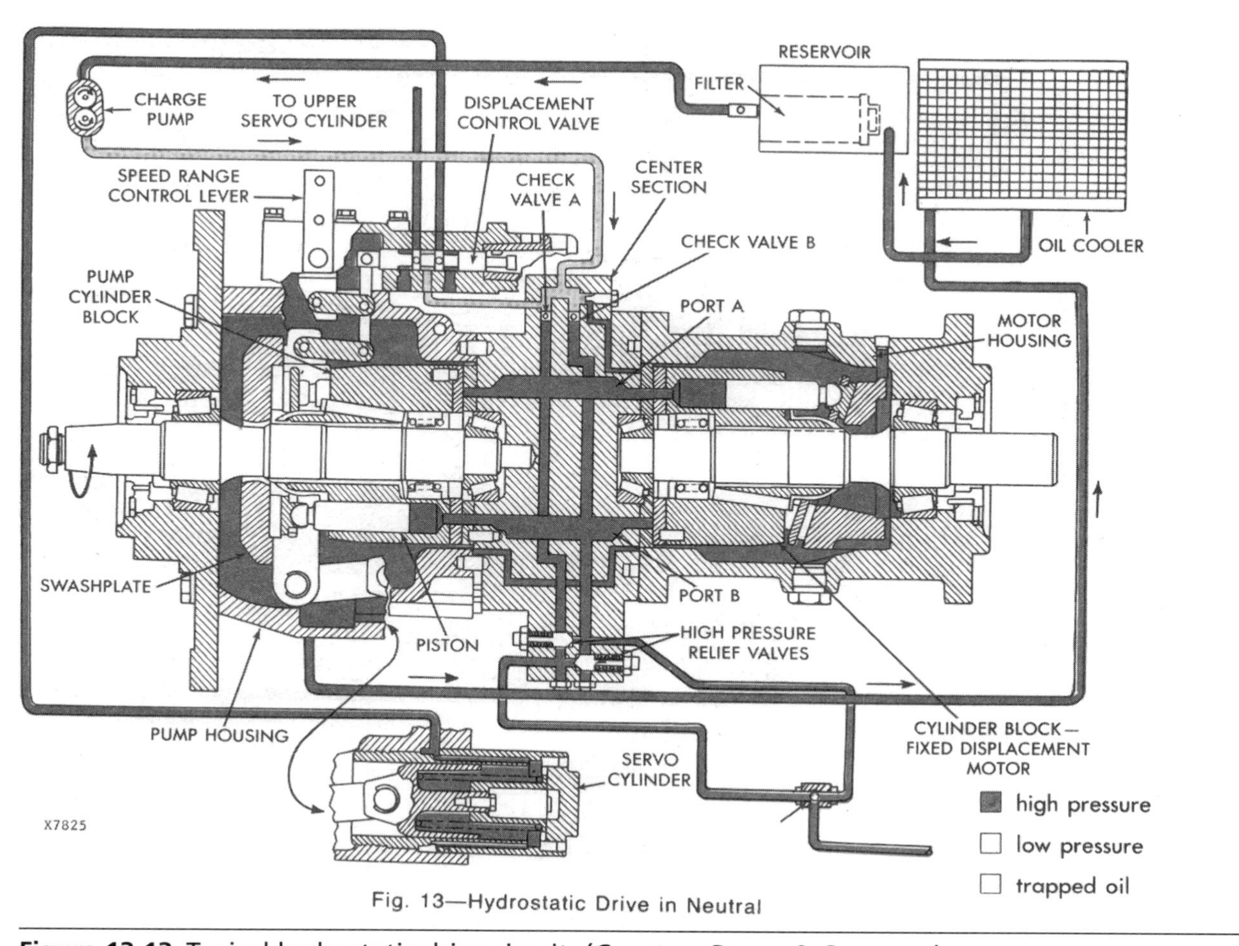

Figure 13-13 Typical hydrostatic drive circuit. *(Courtesy Deere & Company)*

The **high-pressure relief valves,** located in the manifold valve assembly, protect the main circuit from high surge pressures in both main hydraulic lines during periods of quick acceleration. These high-pressure relief valves also protect the main high-pressure circuit from over pressurization due to emergency braking or sudden changes to the load while the equipment is operational.

The shuttle valve is also located in the manifold valve assembly. Its purpose is to provide a circuit connection between either of the two main pressure lines and the charge pressure relief valve when at low pressure.

Displacement Control Valve Operation. In **Figure 13-14,** if we place the control lever in the forward drive position, the displacement control valve is moved to the forward position, allowing the control port to flow oil to the servo piston in the motor. This forces the swashplate against the reverse servo spring to a given position. After the swashplate has moved to the required angle, feedback oil is sent from the forward servo piston in the pump back to the displacement control valve, thus shifting the spool back and leaving just enough oil flow to the forward servo cylinder to maintain the selected swashplate angle.

Operation of a Variable-Displacement Pump/ Fixed-Displacement Motor Drive in Forward

Figure 13-15, a variable-displacement pump/fixed-displacement motor, shows the charge pump pulling oil from the hydraulic reservoir through a 10-micron filter. When the operator shifts the equipment into the forward direction, control pressure oil is directed to the forward servo piston by the displacement control valve.

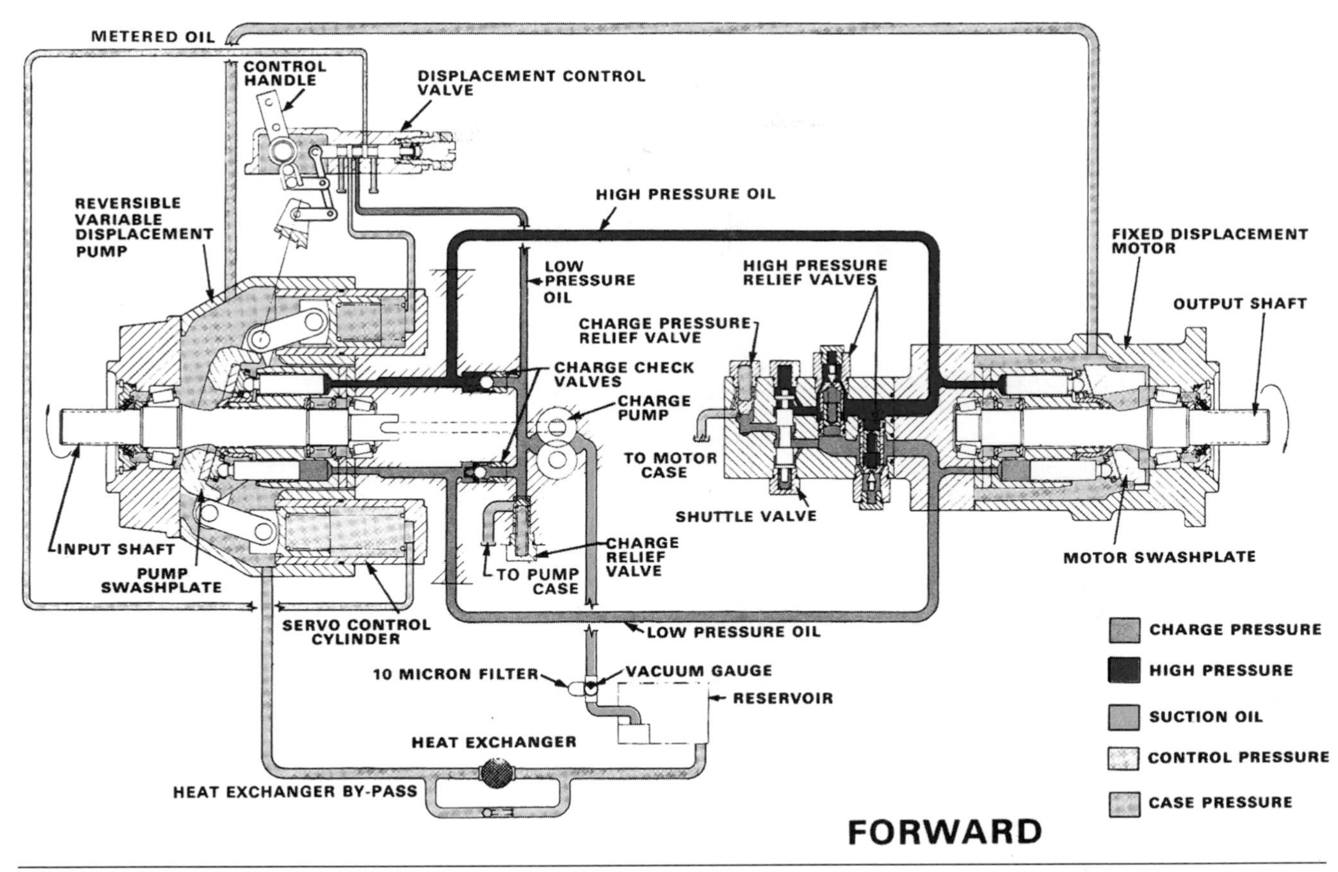

Figure 13-14 Displacement control circuit in forward. *(Courtesy Sauer-Sundstrand Company)*

This moves the swashplate off center, forcing the pump pistons to create oil flow in the circuit. The distance that the pistons reciprocate in the pump depends on the swashplate angle, which will in turn, determine the volume of oil displaced by the pump. **High pressure oil** from the pump is sent to the fixed-displacement motor inlet, forcing the pistons in the motor to rotate in the opposite direction and giving forward direction. Note that the direction of rotation of the pump is always clockwise as viewed from the flywheel end; the direction of the motor is dependent on the direction of travel selected by the operator.

During operation in forward, the motor sends the oil from the outlet side of the motor back to the inlet of the pump. Oil flows in a continuous loop from the pump to the motor; this is where the term *closed-loop* comes from. Excess oil that bypasses the pistons in the motor for lubrication purposes can still flow back to the pump case and out through the heat exchanger to the hydraulic tank. The charge pump will make up any oil that is lost to internal leakage in the circuit. If the system pressure is exceeded while operating in forward, the high pressure relief valve in the motor opens, allowing the oil to bypass the cylinder block assembly on the motor and return to the pump.

Operation of a Variable-Displacement Pump/ Fixed-Displacement Motor Drive in Reverse

Figure 13-16 shows the charge pump pulling oil from the hydraulic reservoir through a 10-micron filter. When the operator shifts the equipment into reverse mode, control pressure oil is sent to the reverse servo piston by the displacement control valve, forcing the swashplate in the pump to shift off center in the opposite direction and forcing the pump pistons to create oil flow in the circuit in the opposite direction. The distance that the pistons reciprocate in the pump depends on the swashplate angle, which determines the volume of oil displaced by the pump. The high pressure oil is

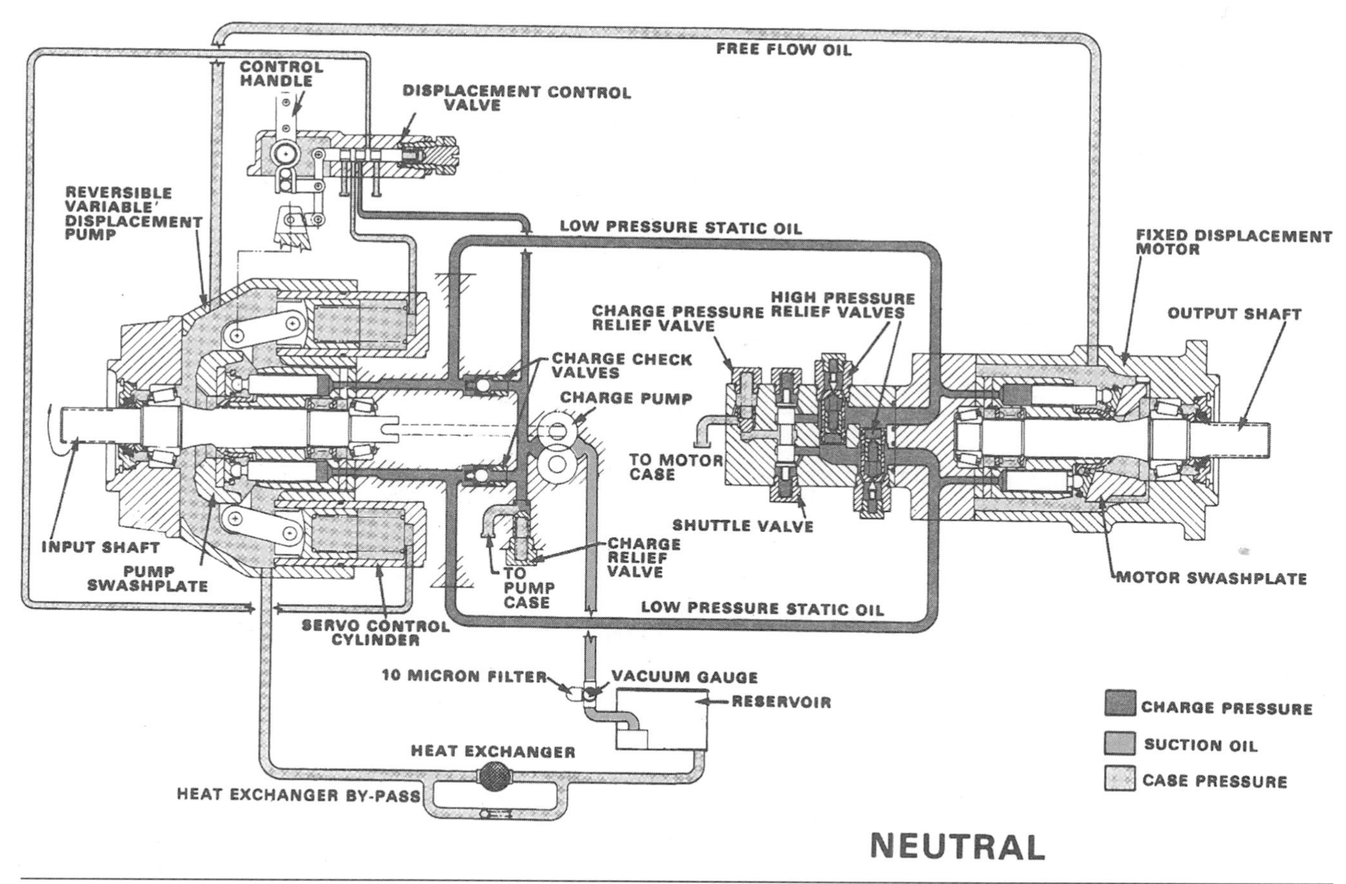

Figure 13-15 Location of 10-micron filter. *(Courtesy Sauer-Sundstrand Company)*

sent to the fixed-displacement reverse motor inlet, forcing the motor to rotate in the reverse direction.

During operation in reverse, oil flow is reversed and the motor sends oil flow back to the opposite side of the pump. Oil flows in a continuous loop from the pump to the motor. Excess oil that bypasses the pistons in the motor for lubrication purposes can flow back to the pump case and out through the heat exchanger to the hydraulic tank. The charge pump will make up any oil that is lost to internal leakage in the circuit. If the regulated system pressure is exceeded while operating in forward, the high pressure relief valve in the motor opens, allowing the oil to bypass the cylinder block assembly on the motor and return to the pump.

Operation of a Variable-Displacement Pump/Variable-Displacement Motor Drive

This combination of pump and motor gives us the greatest flexibility in torque and horsepower combinations. The variable pump/motor combination give us the flexibility to customize the power and torque requirements for different applications. In **Figure 13-17** the charge pump pulls oil from the hydraulic reservoir through a 10-micron filter. When the operator shifts the equipment into the forward direction, control pressure oil is directed to the forward servo piston by the pump displacement control valve and to the motor control valve. Thus moves the swashplate off center and forces the pump pistons to create oil flow in the circuit. The motor pistons can also be stroked to change the distance that the pistons reciprocate in the pump and motor. The result will depend on the swashplate angle in each, which determines the volume of oil displaced by the variable-displacement pump and motor. The high-pressure oil from the pump is sent to the variable-displacement motor inlet, causing the pistons in the motor to rotate in the opposite direction and giving forward direction.

Having the flexibility to vary the swashplate angle in the motor provides the ability to increase or decrease the speed and torque at the **output shaft** of the motor. Note that the direction of rotation of the pump is always clockwise as viewed from the flywheel end; the direc-

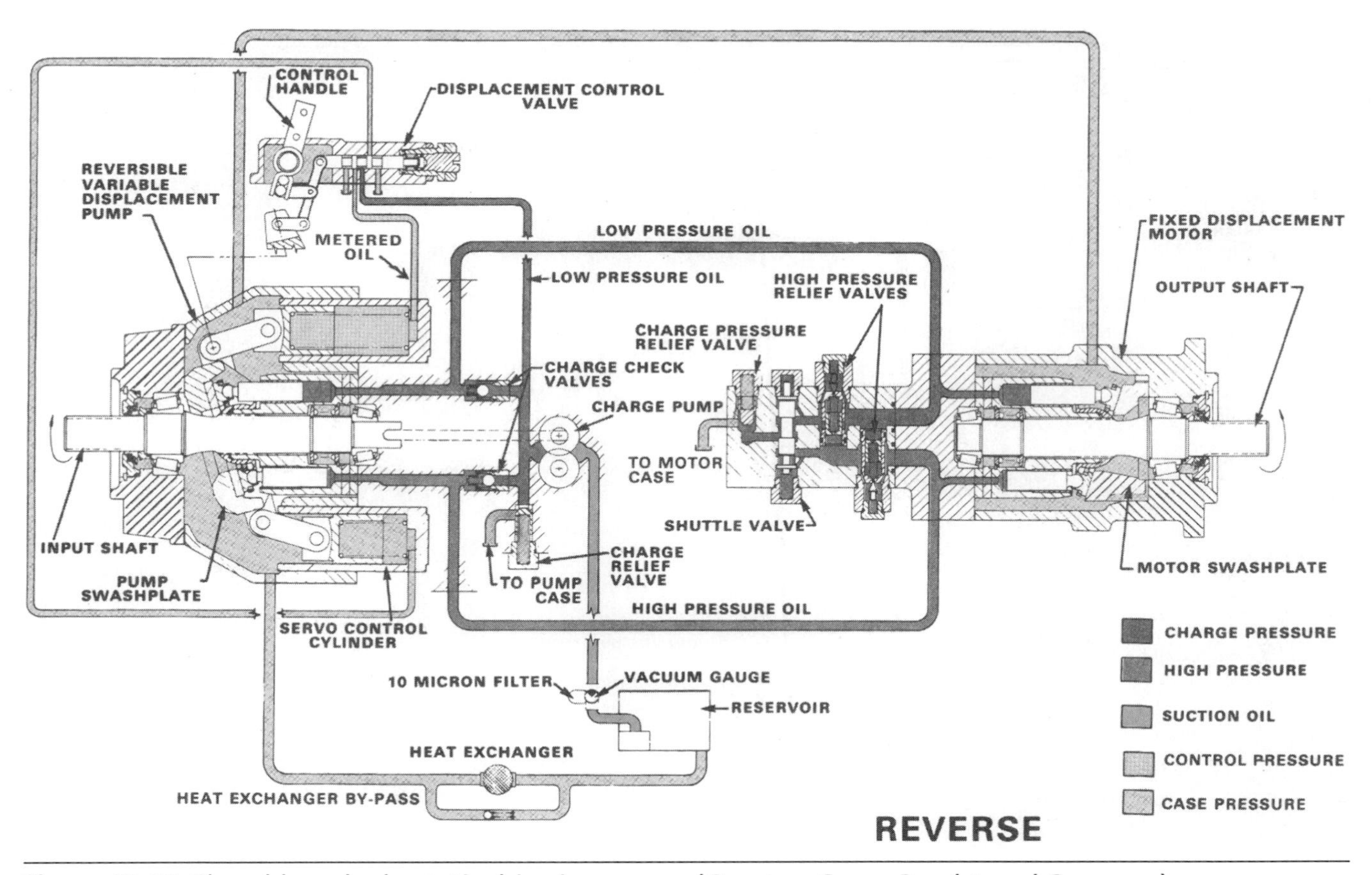

Figure 13-16 Closed-loop hydrostatic drive in reverse. (*Courtesy Sauer-Sundstrand Company*)

tion of the motor rotation is dependent on the direction of travel selected by the operator.

During operation in forward, the motor sends the oil from the outlet side of the motor to the inlet of the pump. Oil flows in a continuous loop from the pump to the motor and back. Excess oil that bypasses the pistons in the motor for lubrication purposes will still flow back to the pump case and out through the heat exchanger back to the hydraulic tank. The charge pump will make up any oil that is lost to internal leakage in the circuit. If the system pressure is exceeded while operating in forward, the high-pressure relief valve in the motor opens, allowing the oil to bypass the cylinder block assembly on the motor and return to the pump.

Operation of a Variable-Displacement Pump/ Fixed-Displacement Motor Drive with Pressure Override

Figure 13-18 shows a pressure override control with a variable-displacement motor driving a fixed-displacement pump. The circuit drive pressure will normally remain below override pressure during normal operation. During this phase of operation, the pressure override valve is in the open position and does not affect the operation of the circuit. During operating conditions where the load exceeds maximum torque available, the pressure override valve takes over the normal displacement control operation. When this occurs, the pressure override control takes over the displacement control of the pump.

In **Figure 13-19,** the pressure override control valve takes control whenever the load requires a system pressure that is higher than override pressure. This valve is a three-way valve that directs charge pressure to the displacement control valve whenever the displacement control valve is in the normally open position. In operation, this valve allows the displacement control pressure to be overridden at a given pressure. System pressure is monitored in both sides of the loop by the override control valve.

In Figure 13-19, the higher pressure is ported into at chamber A, shifting the ball check over, which forces pin B to push the control spool against an internal spring D. As the system pressure increases spool C pushes the spring until the control land E on spool C blocks port F.

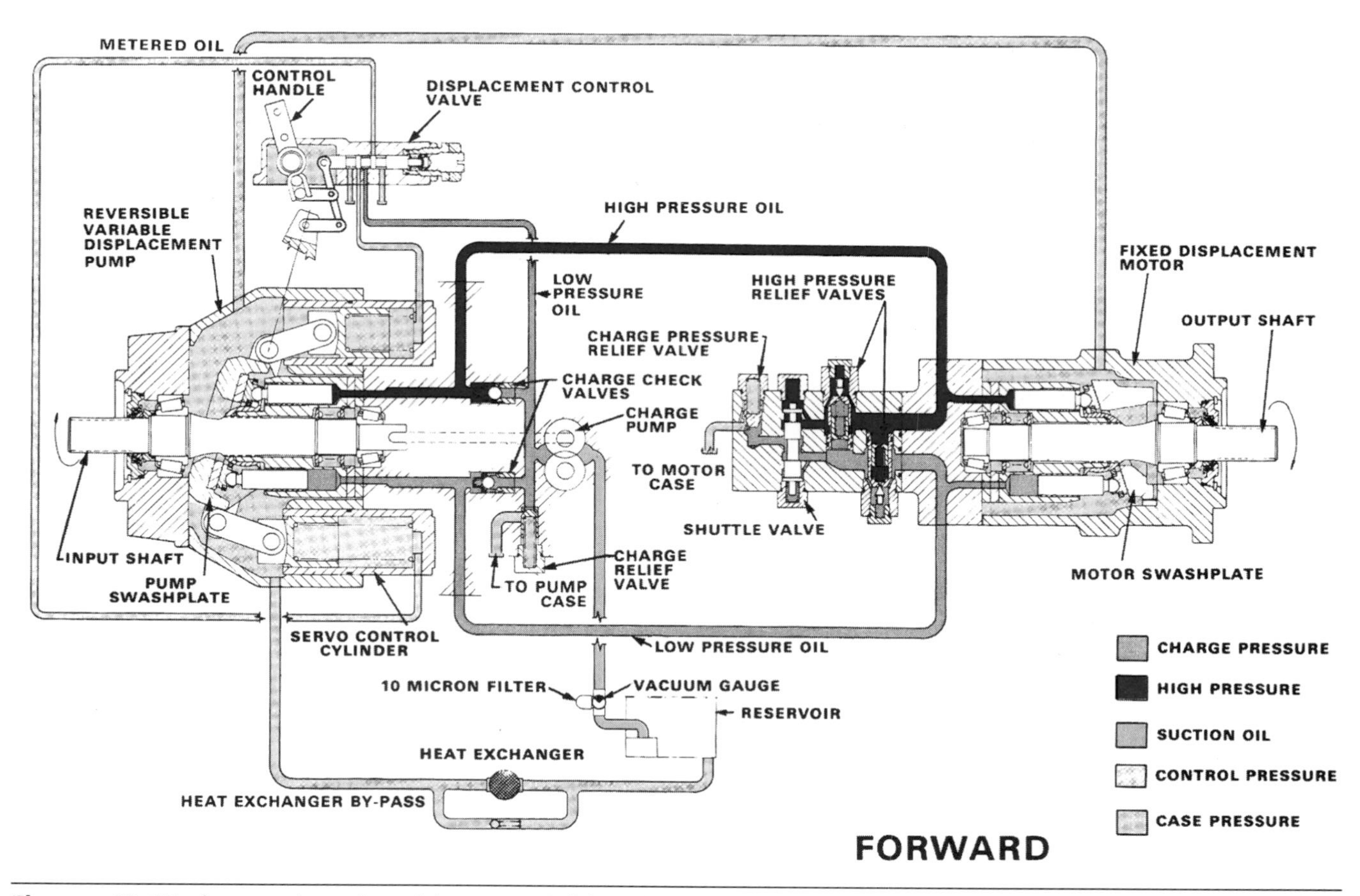

Figure 13-17 shows the closed-loop variable displacement pump and variable-displacement motor combination.

Whenever land E on control spool C blocks port F, the override pressure is in control of the displacement of the pump. The pressure in the override circuit will modulate to maintain servo pressure to ensure that the displacement will always equal a system pressure that is equivalent to override pressure. Adjustments can be made to this valve by changing the spring tension D with the adjustment screw H.

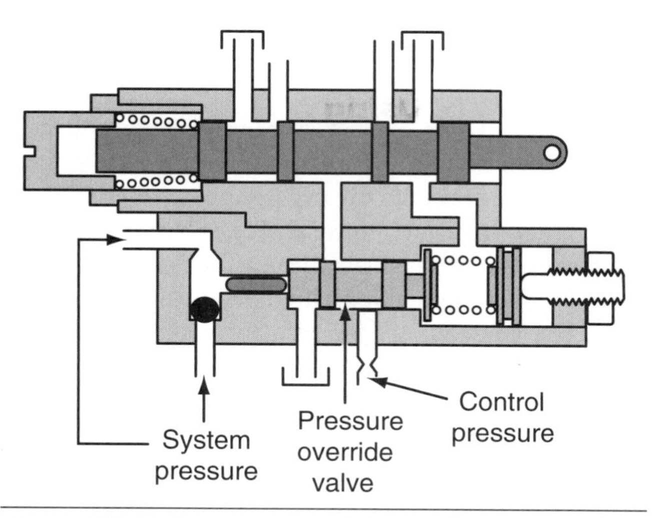

Figure 13-18 Pressure override used on a variable-displacement pump with a fixed-displacement motor.

OPEN-LOOP HYDROSTATIC DRIVE SYSTEMS

An open-loop hydrostatic drive circuit, as shown in **Figure 13-20** (some manufactures refer to it as a hydraulic drive), may be used in heavy equipment where two or more different drive functions or loads are required. This type of drive system may be equipped with a fixed-displacement pump and may use either a fixed- or variable-displacement motor. The motor returns the hydraulic oil to the tank before it is picked up again by the pump. This design, often chosen for multiple loads use, needs a control valve to provide direction and metering control.

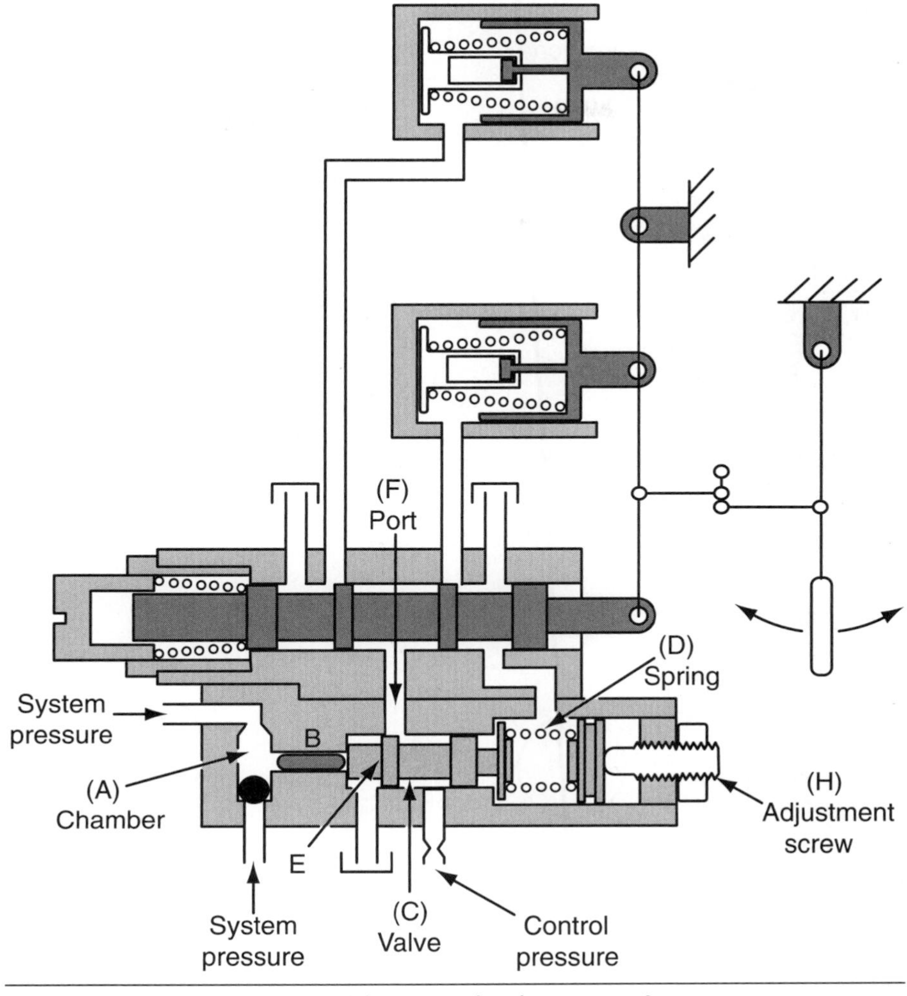

Figure 13-19 Pressure override control valve operation.

On mobile equipment equipped with a diesel engine, we have the additional advantage of having the flexibility of a variable-pump drive speed, which can be used as a form of flow control. Another advantage of using this open-loop circuit design is that it is more cost-effective for smaller equipment where cost is a factor.

As with any circuit design, there are always disadvantages to go along with the advantages. Whenever more that one actuator is activated at the same time in an open-loop circuit, load interference may become an issue in the operation of the circuits. This can also lead to a loss of dynamic braking. When operating equipment that uses the open circuit design, the operator must vary the engine power to provide a change in equipment speed. Depending on the application, the circuit can be used as a source of constant flow to a circuit to provide predictable control; or it can be equipped with a flow control valve to vary the speed of the motor.

Open Circuit Using a Bleed-Off Control

The open circuit design, as shown in **Figure 13-21,** may be divided into three different categories: bleed-off, meter-in, and meter-out control. Bleed-off control uses a variable flow control valve between the pump and motor to allow oil flow back to the tank to control the motor speed. This variation of control will not experience a loss of torque because no pressure drop occurs in this part of the circuit. The motor speed is determined by how much oil is allowed to be bled from the circuit back to tank. This design has one major advantage: The amount of torque required to drive this system is reduced due to the ability of the circuit automatically adjust the pressure to the load.

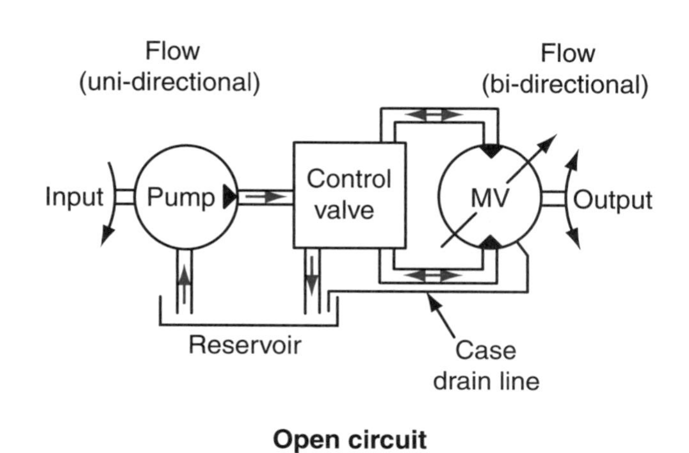

Figure 13-20 Basic open circuit hydrostatic drive.

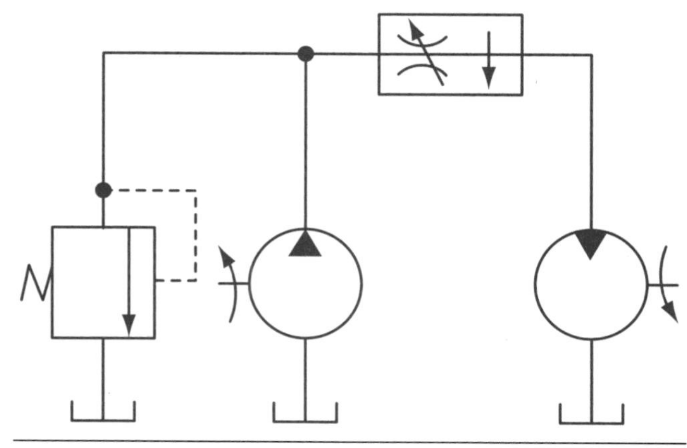

Figure 13-22 Schematic of a meter-in drive circuit.

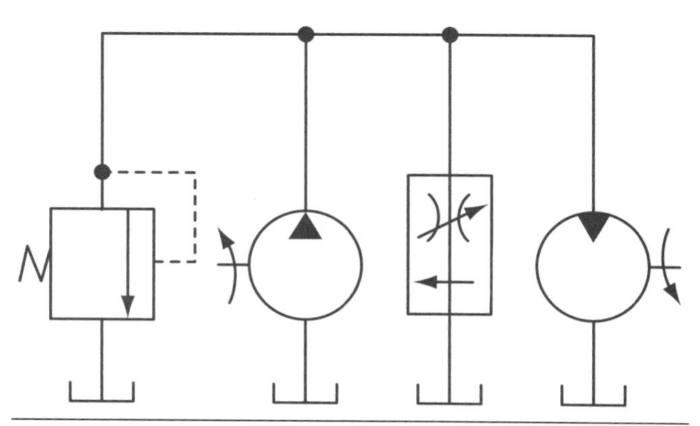

Figure 13-21 Bleed-off drive control circuit.

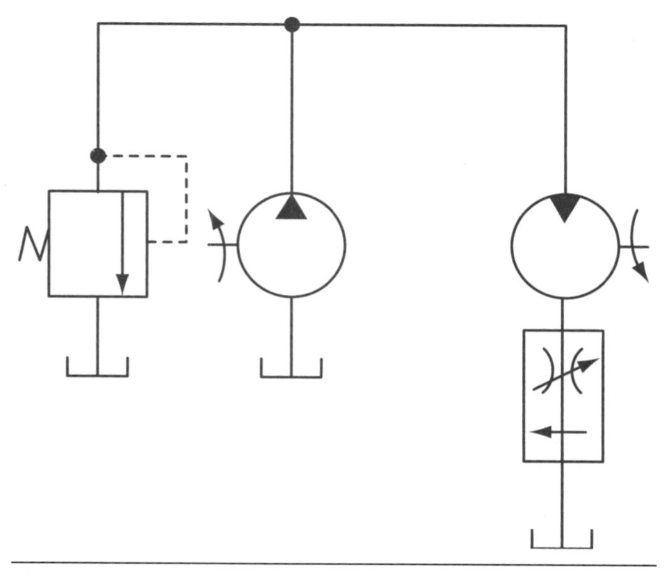

Figure 13-23 Schematic of a meter-out drive control circuit.

Open Circuit Using a Meter-In Control

When a meter in control is used, a variable-flow control valve, as shown in **Figure 13-22,** must be placed between the pump and motor to provide excellent control of the motor output speed. Due to the placement of the variable-flow control valve in this part of the circuit, we experience a pressure drop in this circuit. We also experience a corresponding loss of drive power that is dependent on the pressure drop in the circuit. To compensate for this disadvantage, the circuit can be designed to use a flow control valve that has a built-in relief valve.

Open Circuit Using a Meter-Out Control

The last variation in this type of drive control is the meter-out design, as shown in **Figure 13-23.** It uses a flow control valve in the return line of the circuit to achieve variable motor drive speed. As with the meter-in design, the placement of the variable-flow control valve creates a pressure drop in this circuit thus creating a corresponding loss of drive power that is dependent on the pressure drop in the circuit.

Open Circuit with One Pump and Two Drive Motors

In this design, the pump and motors are not connected directly together; they have a directional control valve in between them. With the engine running and the control valves in the neutral position, the fluid flows from the pump through the control valves and back to the tank. When the operator activates the drive motors, he or she places the shift levers into one of the available positions, allowing oil to flow through the control valves to drive the motors simultaneously in the same direction. The direction of fluid flow determines the direction of motor rotation. If the operator requires an increase in equipment speed, he or she applies engine throttle, which increases pump speed and flow, thereby increasing the speed of the drive motors.

The system pressure is determined by the load on the equipment. In the drive mode, the oil flow from the motors flows through the control valves back to the hydraulic tank. A relief valve is in the circuit to prevent overpressure in the circuit during operation. If this design is used on independent wheel or track drives, the operator can drive one motor in forward and one in reverse, allowing the unit to rotate without any forward or reverse movement. He or she can also choose to rotate one drive at a time, if required, for precise equipment control.

HYDROSTATIC DRIVE SYSTEM MAINTENANCE

Tech Tip: Before you perform any maintenance to hydrostatic drive equipment, take note of any potential hazards that may expose you to the possibility of injury or equipment damage. If work needs to be performed under a piece of equipment, for example under a boom or a part that is supported by hydraulics, it must be blocked mechanically. No one should ever attempt to work under anything that is not supported by a mechanical support. The equipment must be secured with the appropriate wheel chocks to prevent movement if the equipment were inadvertently placed in a condition where freewheeling could occur. Always be aware of the possibility of getting caught up in a pinch point when working with moving or rotating parts.

Before loosening or removing any hoses or components on a hydrostatic drive circuit, you need to make sure that the system pressures have been relieved and that a minimum amount of oil is lost. Some system components may have heavy springs that have stored energy in them. These springs need to be properly removed with the correct tools to prevent component damage or personal injury. A complete review of the manufacturer's service manual for each particular task must be done before work on a job can proceed. Good preventative maintenance practices are critical to maintaining a low-cost service life on hydrostatic drive equipment.

Before we go in to the details about the various maintenance activities that are essential to this type of drive system, look at the basic principles of operation of the various components to gain an understanding of how poor maintenance procedures cause these systems fail over time. The movement of oil from one component to another must be performed with as little internal leakage as possible. Excessive internal leakage will lower the performance capabilities of the equipment and lead to excessive downtime and high repair costs. When the components operate as designed, there is just enough internal leakage to provide adequate lubrication inside the pump, motor, and other components that make up this type of drive system. The oil that is lost to internal leakage flows back through a filter and cooler before being reintroduced back to the main circuit. The components in the hydrostatic drive system can be damaged quickly from too much heat created by excessive internal leakage. Built-in internal leakage in this drive circuit never allows the volumetric efficiency of a hydrostatic drive system to reach 100%.

Consider the type of oil that we are going to use; the wrong grade of oil or oil that is contaminated with dirt quickly increases the internal leakage of the components, and causes a dramatic loss of efficiency in the equipment's operating capabilities. One of the effects of excessive internal leakage is higher-than-normal hydraulic oil operating temperature, which may lead to an accelerated rate of oil breakdown. The natural tendency of any hydraulic oil that is exposed to excessive temperatures is to lose its viscosity. To prevent metal-to-metal contact, a hydraulic oil film must be maintained with a minimum viscosity of 47 SUS at 210°F. Choose the hydraulic fluid carefully; not all hydraulic fluids meet the requirements of equipment manufacturer's specifications. Using the equipment over time with incompatible oil may cause internal wear to take place in the components, which will increase internal leakage above normal levels. Oil that has low viscosity will leak more than an oil with a higher viscosity, so the importance of using oil with the right viscosity index is vital to long system component life.

Keeping dirt out of the hydrostatic circuits and components is easier said that done. When hydraulic oil and filters are scheduled to be replaced, ensure that the external areas around the parts that are going to be disassembled are properly cleaned before removal. Whenever possible, use a high-pressure wash on equipment to thoroughly clean the area to be worked on and to ensure that no water can enter the system before opening up the system. If the hydrostatic drive circuit has to be opened for any reason, it must be properly primed to prevent serious damage to internal components. This damage can start in "low oil" or "no oil" conditions. When equipment is serviced, be sure to replace any O-rings, gaskets, or seals that have to be removed as a result of the work performed.

System Maintenance

Oil replacement in a hydrostatic drive system is dependent on a number of operating parameters. In general, the following guidelines should be considered:

- If the equipment uses a sealed hydraulic tank the recommended, oil change interval can be as long as 2000 hours of operation. This recommended interval may shorten to as little as 500 hours if the system has a vent to atmosphere (open reservoir), such as an air filter on the filler cap. When considering the oil change intervals for equipment that you are going to service, always consult the equipment manufacturer's specifications.
- The maintenance intervals that are developed by equipment owners are often a result of operating experiences and may differ somewhat from the manufacturer's recommendations. An example of this is in the mining industry, where operating conditions would be considered extremely severe compared to the same equipment operating in a logging operation. The service intervals in an underground mining environment necessitate intervals that are often only 200 hours apart.
- Ten-micron filters on hydrostatic drive inlet systems are recommended on most systems.
- The level of oil in the reservoir should be checked daily by the equipment operator.
- Fluid added to the hydraulic tank of a hydrostatic drive circuit should be filtered before it is poured into the tank, and the hydraulic lines and fittings need to be checked daily for leakage and abrasion.
- The heat exchanger should be cleaned externally at a regular interval to maintain safe oil operating temperatures.

Start-Up Procedures after Service. The procedures outlined here are intended for use as general examples of steps to follow; they should not be used for actual start-up procedures on any particular drive system. Always refer to the manufacturer's technical documentation for the equipment that you are working on for the proper procedures and specifications.

Tech Tip: Before we start any hydrostatic drive equipment after servicing or repairing, we need to become familiar with a few general safety precautions. In many cases, when equipment needs to be serviced or repaired, it is not always necessary to remove the parts for repair. Access to the area being serviced should be high-pressure washed before work begins and after it is completed to minimize the chance of contamination and to prevent potential slipping conditions. Washing also minimizes fire hazards from oil on or near hot surfaces.

- After performing certain types of repairs or service work, it may be necessary to have the vehicle raised off the ground on approved safety stands before start-up to prevent possible unexpected equipment movement, which may pose a safety hazard. To determine whether this procedure must be followed, always consult the manufacturer's service manual for the type of equipment and repairs that you are performing.
- When working around areas of the equipment that may pose a fire hazard, only use cleaning agents that are not flammable.
- Use approved wheel chocks on both sides of the drive wheels to prevent accidental movement of the equipment. After performing some types of repairs or service work, starting the equipment for the first time may cause an unexpected loss of drive, which can cause a loss of dynamic braking and lead to an unexpected freewheeling condition on some types of equipment.
- When starting up the equipment for the first time after a repair or service procedure, extreme

caution must be used when dealing with high-pressure hydraulic circuits. Never use your hands to feel for a leak in a hose or component. High-pressure fluid can puncture the skin and get into your blood stream, which may lead to blood poisoning. Always seek medical attention if you suspect this has occurred to you or to another technician.

- If you have serviced a charge pump, main pump or hydraulic motor, the component must be filled with clean, filtered hydraulic oil before it is operated. Check the manufacturer's literature for information as to where and how much fluid is required.
- Ensure that the reservoir is filled to the proper level with the correct type of hydraulic oil.

Note: Depending on the design of the system, if gravity does not fill the charge pump line, then the line must be filled by hand.

- After performing service work to a main pump or motor, it is advisable to disconnect the linkage to the pump control on initial start up to allow the pump to find neutral. Always refer to the manufacturer's recommendations before performing this procedure. Be sure equipment is blocked properly to avoid unexpected movement.
- Before an initial start-up on a hydrostatic drive unit that is powered by an internal combustion engine, one should disable the engine from starting. Use the engine cranking system to prime the hydraulic circuit system. Crank the engine over until a charge pressure gauge mounted on the charge pump circuit registers a pressure build up of at least 20 to 30 psi. This ensures that oil will get to the main components immediately and prevent initial start-up damage from occurring. Refer to the equipment manufacturer's recommendations for this procedure.
- After completing the priming steps, start the engine and let it run for a few minutes at low idle rpm. This will ensure that the system gets filled correctly. Observe the charge pressure to ensure it is stabilized to a normal pressure. Refer to the manufacturer's specifications to obtain the correct values. If charge pump pressure values are not within specifications, shut down the engine to troubleshoot the problem.
- If the charge pump pressure reading is normal, the engine should be shut down and the pump control linkage reconnected. At this time you should also check the oil level in the main hydraulic tank and top up, if necessary.
- You are now ready to run the system through some basic tests. Start the engine and increase the rpm to approximately 1500 to 1800 rpm. Observe the charge pump pressure at this point to ensure it remains within specifications. Refer to manufacturer's specifications for the correct values for the system you are setting up.
- After ensuring that the charge pump pressure is normal, shift the unit back and forth for a few minutes in forward and reverse direction, and observe the charge pump pressure. If the charge pump pressure falls below normal operating pressure while performing this test, shut down the unit and refer to the manufacturer's troubleshooting guide to resolve the problem.
- Although there are some differences in the pressures in hydrostatic drive system designs, the basic pressure test points are obtained in approximately the same locations. Following is a list of test points that should be recorded. Refer to the manufacturer's specifications for test points and values for the equipment you are testing. The values given are examples typical values for the Sundstrand hydrostatic drive and are not necessarily the values for the equipment you are working on.

Tech Tip: Always refer to the manufacturer's technical documentation for the correct specifications for the equipment you are performing work to.

Figure 13-24 shows typical pressure connection points used to test system pressures. These values are only examples of typical pressure values. They should not be used for actual troubleshooting specifications on any drive system.

1. Charge pump pressure at idle and 1800 rpm
2. Pressure test point A: 210 to 240 psi above pump case pressure in neutral at 1000 rpm
3. Pressure test point B: 300 to 385 psi above pump case pressure in neutral at 1500 to 1800 rpm
4. Pressure test point C: 230 to 250 psi above motor case pressure pump. Must be in the stroked position at 1500 to 1800 rpm.

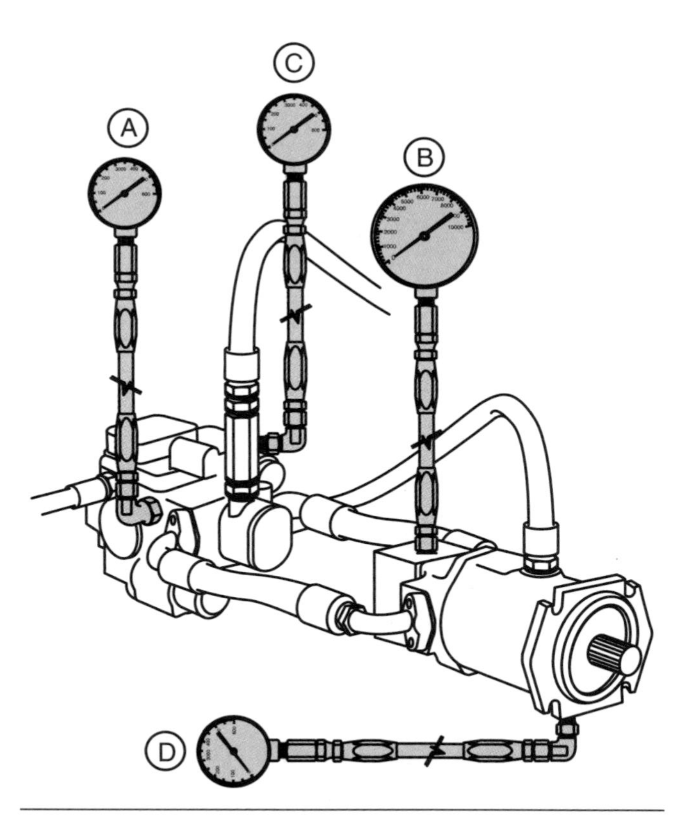

Figure 13-24 Diagram showing typical connection points to test system pressures.

GENERAL TROUBLE-SHOOTING (CLOSED-LOOP) PROCEDURE

The procedures outlined are intended for use as general examples of steps to follow, and should not be used for actual troubleshooting pressures and procedures on any particular hydrostatic drive system. Always refer to the manufacturer's technical documentation for the equipment that you are working on for the proper pressures, procedures and specifications.

Before you look at the troubleshooting charts, you need to familiarize yourself with some basic troubleshooting facts and procedures. Hydrostatic drive systems must maintain defined pressures and flows in various parts of the circuit at different times to operate efficiently. Any changes in these performance values can have an adverse affect on system operation. Certain pressures and flows or restriction in the circuit need to be recorded and monitored. To perform pressure and vacuum reading on a hydrostatic drive circuit, you will need four different gauges with pressure snubbers, as shown in **Figure 13-25.**

To properly identify the correct pressure levels in a hydrostatic closed-loop circuit, you must be able to correctly measure and interpret the values. Be sure that the reading taken is accurate; the pump must be running at maximum rpm. We will discuss the four basic pressures and restrictions that all hydrostatic system operation depends on.

Starting with the charge pump, we may need to measure and record the inlet vacuum by using a vacuum gauge that records values in inches of mercury (Hg). The vacuum gauge used should have a range of 0 to 30 inches of mercury (Hg). Typical vacuum readings measured on the inlet side of the charge pump should be approximately 10 inches of mercury (Hg) at normal operating temperatures. If it is not possible to bring the equipment up to operating temperature, performing this test with cold oil will result in a reading that is slightly higher than the one at operating temperature.

The charge pressure in a typical hydrostatic drive circuit needs to be measured with a quality gauge that has a range from 0 to 400 psi. Typical values for this test are given in at two different operating conditions. To explain the relationship of pressure readings, we will use an example showing the values for a Sundstrand closed-loop hydrostatic drive circuit. The minimum charge pressure is given as a minimum value that should not be less than a given pressure above the case pressure at any time. A reading of 170 psi charge pressure with a case pressure of 40 psi is within the acceptable range of 130 psi above case pressure. The second value we need to examine is the charge pressure when the motor output shaft is turning. This value on this system should be 160 psi above the motor case pressure. Another reading we will check is the charge pressure when the system is in neutral. The minimum reading here should be 190 psi above pump case pressure. The next reading we will check is the system pressure, sometimes called the high pressure. This is measured with a high-pressure gauge that has a range from 0 to 10,000 psi. The gauge is installed in the manifold on the motor housing. This reading should never exceed the maximum value listed in the manufacturer's specifications under any normal operating conditions and temperatures.

The last pressure value is the case pressure. This pressure is measured with a gauge that has a range from 0 to 400 psi, and the reading should not exceed 40 psi under normal operating conditions. The pressure values used in the preceding discussion were

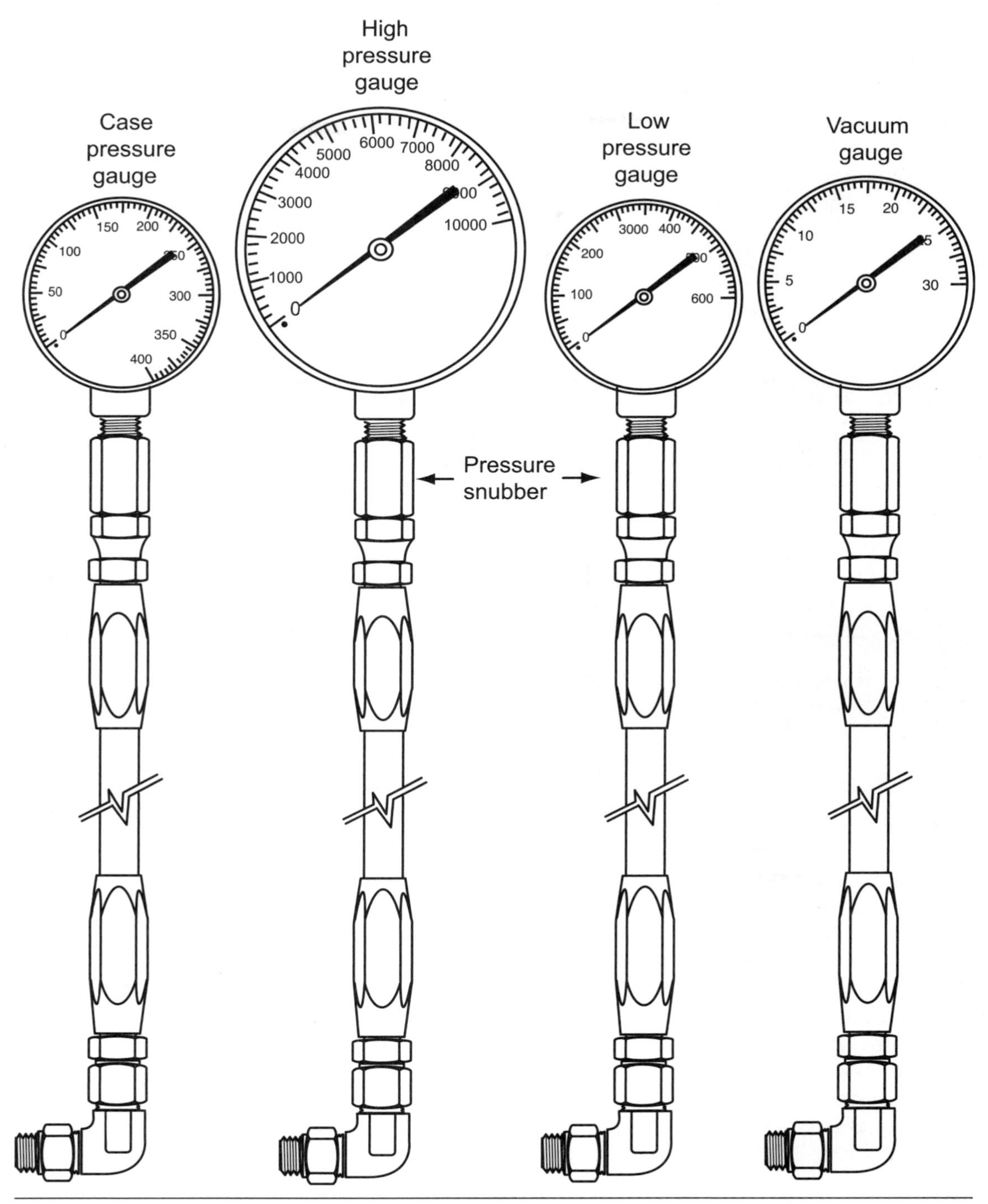

Figure 13-25 Gauges required to test a hydrostatic drive system.

used to show the relationship of each reading to each other reading and may vary somewhat with other manufacturer's systems. Regardless of the values, the four measured values will be identified by the same names on most closed-loop hydrostatic drive circuits, and the relationship of each to each other will be compared in any troubleshooting process. As in all troubleshooting, the process must start at the beginning, in this case, the hydraulic tank.

Testing Charge Pressure

The charge pressure in a hydrostatic drive system is responsible for many jobs within the drive system.

It has to replenish oil that is lost through the case drain, it supplies oil for controlling the servos, and it provides enough oil flow for cooling. It must also maintain a specific pressure within the low pressure side of the hydrostatic system. Testing the charge pressure is a common practice in diagnosing hydrostatic drive system problems. **Photo Sequence 4** outlines a general procedure used in testing a charge pressure. It is by no means intended to replace specific manufacturer's procedures or guidelines. Note that equipment using hydrostatic drives is infinitely variable and that this procedure may not be applicable to all applications.

PHOTO SEQUENCE 4 Testing Charge Pressure in a Hydrostatic Drive

P4-1 This procedure is implemented when the machine has a low performance complaint. The example we are using is on a Caterpillar 248B Skid Steer Loader

This procedure outlines the order in which to perform a charge pressure test on the hydrostatic drive system for this machine. Before any work is started on a machine, proper safety procedure and lock-out systems must be implemented. Before any testing, the machine must be at operating temperature for accurate readings, and oil levels must be at their proper levels.

P4-2 First, raise the boom high enough to properly place the safety arm to lock the boom in the raised position. Remove the holding pin and slowly lower the safety arm into place. **Note:** Do not let the safety arm fall into place; it may damage the cylinder rod.

P4-3 Because the hydrostatic drive system is located below the operator's cab, it must be raised to access the pumps and hydraulic components. To raise the cab, first remove the cab bolts located at the front corners of the cab.

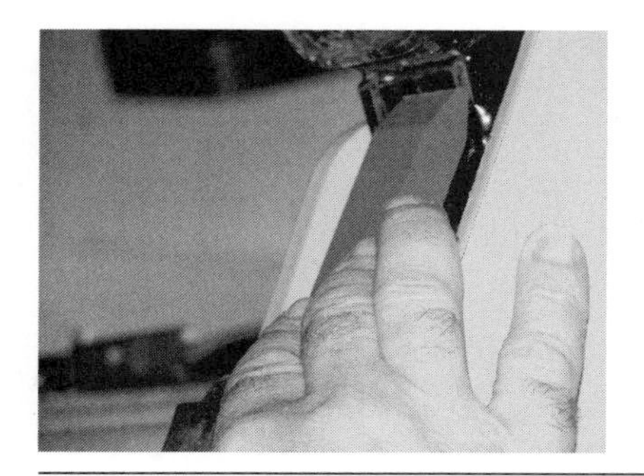

P4-4 and P4-5 This machine is equipped with gas-filled cylinders and will raise with very little effort. With machines not equipped with gas-filled cylinders, an overhead crane is necessary to raise the cab. Once the cab is in the raised position, make sure the safety latch is in place to prevent any sudden drop of the cab.

P4-6 Using the OEM service manual, find the specification for the specific charge pressure setting. (450 ± 30 psi/3100 ± 200 kPa @ low idle) (480 ± 30 psi/3300 ± 200 kPa @ high idle) Locate the pressure tap port for testing the charge pressure.

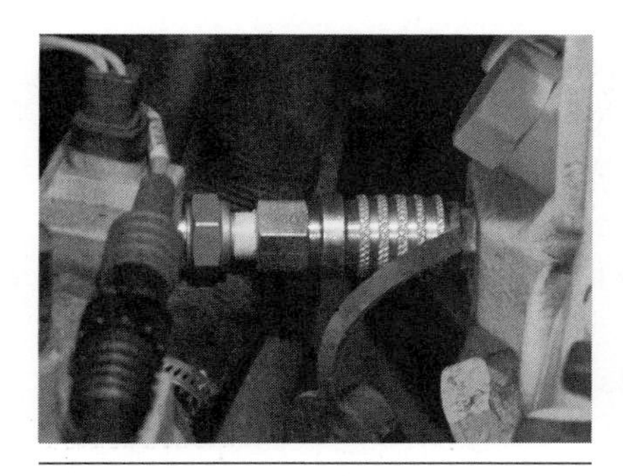

P4-7 Insert a proper pressure gauge or transducer that will fit the range of the charge pressure. (0 to 1000 psi). Failure to do so will damage the gauges.

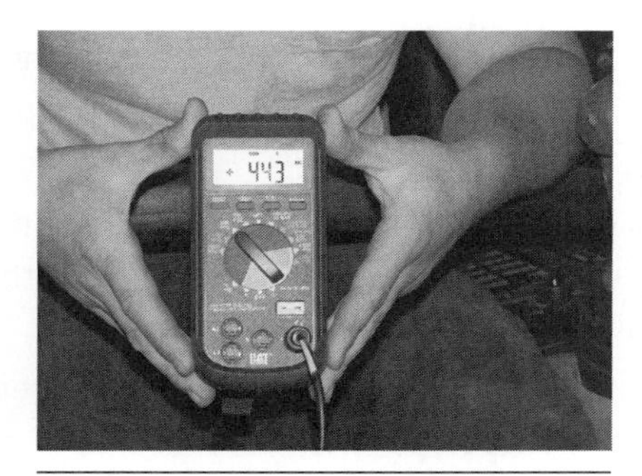

P4-8 Lower the cab and place the pressure gauge in a position that can be read but cannot be damaged from the running engine. Start the engine and take a reading at both low and high idle. Compare your readings with the specification in the service manual. Repeat the test to confirm your readings.

HYDROSTATIC DRIVE (CLOSED-LOOP) TROUBLESHOOTING PROCEDURE

Hydrostatic Drive System Will Not Operate in Either Direction

Note: These troubleshooting guidelines are intended for use as a learning guide and are not intended to be a substitute for the manufacturer's specific troubleshooting guide for a particular application. Always consult the manufacturer's technical documentation for each equipment-specific trouble-shooting guide before attempting to diagnose a system fault.

Cause	Symptoms	Remedy
Hydraulic fluid level low Faulty control linkage Drive coupling disconnected	Below normal charge pressure, or fluctuating charge pressure or both	Find and repair leaks Refill hydraulic tank Check all the linkage for binding. If required, adjust the linkage to pump control arm. Do not move the control arm to accommodate the linkage length Inspect the coupling that drives the pump and check the motor drive coupling for slippage
Inlet line from the charge pump or the filter is plugged or has collapsed	Below normal charge pressure Vacuum test shows high restriction at the charge pump	Clean or replace the filter or replace the inlet line
Charge pump relief valve faulty or stuck open	Below normal charge pressure If the pressure is low when the pump is in neutral, charge pump is faulty If the pressure is low when the system is stroked, the problem is in the manifold charge relief valve	Replace the charge valve assembly
Charge pump drive is severed	No charge pressure in neutral or when stroked Low charge pressure Charge pressure fluctuates rapidly when maximum system pressure is reached Charge pressure drops to zero or near zero Maximum system pressure is below normal high-pressure relief setting	The charge pump should be replaced Check for excessive internal leakage at the pump or motor to determine which one is defective
Pump or motor has internal wear	Hydraulic tank and filter contaminated with brass Pump or motor noisy in operation	Remove the charge relief valve spring in the motor manifold and install a solid pin to replace the spring. This blocks the relief valve in the closed position

(continued)

Cause	Symptoms	Remedy
		Remove the line from the motor case drain connection at the pump case port and install a threaded plug to seal the hole Place a flow meter with the capacity to handle charge pump flow. Attach to the motor case drain line and route the output to the tank Run the equipment at full rpm in neutral and engage the pump control to achieve the highest system pressure you can Observe the flow in gpm coming out of the motor case drain If the flow coming out of the motor is less than 50% of the charge pump flow, the motor has to be replaced Since the pump and motor operate using the same hydraulic oil, the pump should be replaced also If the motor is not damaged, the pump should till be replaced
Control valve linkage is disconnected internally	Charge pressure in neutral will be normal When the control handle is moved, the pump does not stroke	Control linkage at the directional control arm must be disconnected. Move the control arm
Stuck or faulty check valve in the charge circuit	Charge pressure is low or erratic	Check for air in the system Check for damaged charge pressure control valve
Control orifice plugged at the control housing to pump mount	Charge pressure will be normal in neutral, but the pump will not stroke	Remove blockage from the control orifice behind the control housing

Hydrostatic Drive System Operates in One Direction Only

Note: These troubleshooting guidelines are intended for use as a learning guide and are not intended to be a substitute for the manufacturer's specific troubleshooting guide for a particular application. Always consult the manufacturer's technical documentation for each equipment-specific troubleshooting guide before attempting to diagnose a system fault.

Cause	Symptoms	Remedy
Control linkage is faulty	Loss of drive in one direction	Control linkage is out of adjustment or binding
High pressure relief valve sticking or stuck open	System pressure is lower than normal in one direction only	Repair or replace the damaged relief valve
System has one check valve that is faulty	System pressure is low or zero in one direction only The charge pressure may be higher than specification	Repair or replace the 2 check valves located in the pump under the charge pump Check for damaged check valve seats
Faulty directional control valve. Stuck spool or sticking in one direction	Pump will not center itself to neutral when linkage is disconnected	Control valve should be replaced
Motor manifold shuttle valve spool is jammed	Lower than normal system pressure in one direction only	Motor manifold assembly must be replaced

Hydrostatic Drive System Does Not Return to Neutral

Note: These troubleshooting guidelines are intended for use as a learning guide and are not intended to be a substitute for the manufacturer's specific troubleshooting guide for a particular application. Always consult the manufacturer's technical documentation for each equipment-specific troubleshooting guide before attempting to diagnose a system fault.

Cause	Symptoms	Remedy
Control linkage sticky or faulty	System does not return to neutral	Check neutral operation with control linkage disconnected. If it appears normal, the control linkage may be binding Servo cylinders may need adjustment Refer to manufacturer's technical documentation for adjustment instruction Some pumps may need to be factory calibrated
Control valve out of adjustment		The displacement control valve should be replaced or readjusted

System Operating Temperature Runs above 180°F

Note: These troubleshooting guidelines are intended for use as a learning guide and are not intended to be a substitute for the manufacturer's specific troubleshooting guide for a particular application. Always consult the manufacturer's technical documentation for each equipment-specific trouble-shooting guide before attempting to diagnose a system fault.

Cause	Symptoms	Remedy
Low fluid level in the hydraulic system	Low charge pressure reading	Refill hydraulic system with the correct grade of oil
Hydraulic oil cooler plugged		Blow out or power wash cooler air passages
Hydraulic oil by passing the cooler	Check temperature drop of the oil across the hydraulic oil cooler	Repair or replace cooler bypass valve
Plugged filter or suction line	Charge pump vacuum above normal limits system exhibits low charge pressure	Replace hydraulic filter or charge pump inlet line
System exhibits excessive internal leakage	System pressure will be lower than normal in one or both direction	Verify the operation of the high pressure relief valve and replace defective valve
	Charge pressure may be erratic or drop to near zero when the maximum system pressure is reached	Pump or motor may both need replacement

Hydraulic System Noisy in Operation

Note: These troubleshooting guidelines are intended for use as a learning guide and are not intended to be a substitute for the manufacturer's specific troubleshooting guide for a particular application. Always consult the manufacturer's technical documentation for each equipment-specific troubleshooting guide before attempting to diagnose a system fault.

Cause	Symptoms	Remedy
Air in the system	Hydraulic oil in the tank is foaming	Fill hydraulic tank to normal level
Air entering the suction side of the charge pump circuit	Charge pressure is low or erratic	Check for leaks on the suction side of the pump, filter, and hose connections

Sluggish Acceleration and Deceleration

Note: These trouble-shooting guidelines are intended for use as a learning guide and are not intended to be a substitute for the manufacturer's specific trouble shooting guide for a particular application. Always consult the manufacturer's technical documentation for each equipment-specific troubleshooting guide before attempting to diagnose a system fault.

Cause	Symptoms	Remedy
Air entering the system	Poor equipment operating performance	Check the orifice behind the control valve to pump housing for blockage

(continued)

Cause	Symptoms	Remedy
Control orifice is blocked with foreign material		Remove the charge pump and remove plug from the charge pump gauge port Blow air through it to dislodge any foreign material
Internal wear or damage to pump and motor		Remove the charge relief valve spring in the motor manifold and install a solid pin to replace the spring This blocks the relief valve in the closed position Remove the line from the motor case drain connection at the pump case port and install a threaded plug to seal the hole Place a flow meter with the capacity to handle charge pump flow to the motor case drain line and route the output to the tank Run the equipment at full rpm in neutral and engage the pump control to achieve the highest system pressure possible Observe the flow in flow gpm coming out of the motor case drain If the flow coming out of the motor is less than 50% of the charge pump flow, the motor has to be replaced Since the pump and motor operate using the same hydraulic oil the pump should be replaced If the motor shows no sign of damage, then only the pump should be replaced

HYDRAULIC RETARDER SYSTEMS

Large capacity heavy-duty equipment operating on long, steep grades such as pits and quarries often can reach speeds as high as 50 mph. Equipment manufactures design brake systems with average operating conditions in mind. Often these large haulers can benefit from the use of an additional auxiliary brake system (hydraulic retarder), which some equipment manufacturers include on their equipment. As an example, a haulage truck operating in an open pit or mining environment may be required to operate on grades as steep as 20% this will quickly overheat a conventional brake system and lead to premature brake lining wear. The design intent of a hydraulic retarder in this application is to keep the equipment speed in check when operating on steep grades by using the conventional brakes only to bring the vehicle to a full stop. As the name suggests, a retarder uses a fluid to create resistance in an enclosure that consists of two stationary parts forming a housing with a rotating member in the middle (a rotor). The hydraulic retarder can best be defined as a device that can convert the kinetic energy of a moving vehicle into heat energy, as shown in **Figure 13-26.** The resulting heat energy is then sent to an auxiliary cooler, which dissipates it to the atmosphere. Selective use of a hydraulic retarder by an operator can maximize the intervals between service brake overhauls, which in turn will reduce the operating downtime and operating costs of the equipment.

Although there are many different designs and manufacturers of hydraulic retarders, we will focus on two of the most common types used on heavy equipment. There are some key differences in retarder design and mounting location. The integral retarder design typically has two distinctly different design configurations. The retarder may be located either at the front of the transmission between the torque converter housing (input retarder), as shown in **Figure 13-27,** or mounted to the rear (output retarder) of the transmission, as shown in **Figure 13-28.**

To explain the advantage of using this type of auxiliary braking system on heavy equipment, we will use as an example a hypothetical 80,000 pound haulage truck (approximately 36,290 kilograms) operating in an open pit environment. When the truck descends an 8% grade at approximately 30 mph, it could require approximately 500 hp of continuous braking to prevent a speed increase while descending the grade. In our example, this haulage truck's brakes are designed to absorb approximately 250 hp of kinetic energy. If we take into account the truck's rolling resistance and parasitic losses of the engine and power train, (estimated at approximately 100 hp) the truck's braking capacity could fall short by up to 150 hp on this type of grade. If the equipment were required to make an emergency stop, the heat energy dissipation would be a lot higher, and the brake system would most likely not stand up to the job. Although this example is a hypothetical one, you can see the problems that can arise.

In these types of applications, the use of a hydraulic retarder reduces the operating braking temperatures and increases the service life of the brake system. To maintain high efficiency in hydraulic retarder design, the clearance between the housings and rotor must be minimal. Hydraulic retarder efficiency can run from 80% to as high as 115% equalling the efficiency of most diesel engine retarders. Oil temperature and pressure gauges are mounted in the

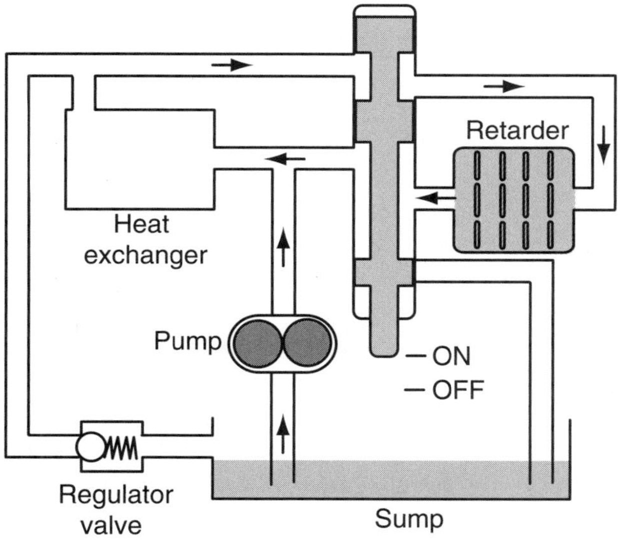

Figure 13-26 Typical hydraulic retarder layout.

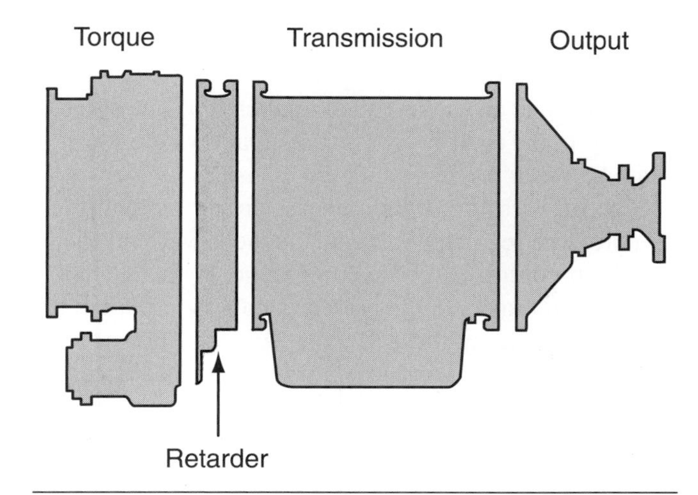

Figure 13-27 Front mounted hydraulic retarder.

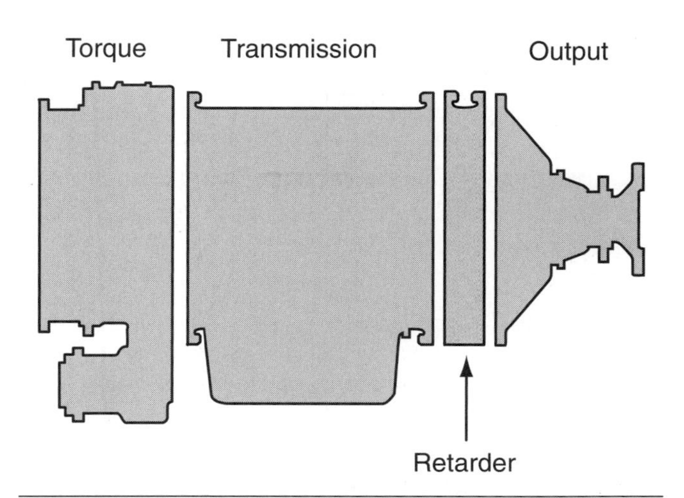

Figure 13-28 Rear mounted hydraulic retarder.

operator's compartment to allow instant monitoring of these two important conditions. We will focus on two common systems in use in these off-road applications: the Allison and the Caterpillar designs. Some equipment manufacturers integrate the controls of the retarder with the powershift transmission controls while others are independently mounted. Some manufacturers use an independent pump and others use the existing transmission pump to provide fluid for retarder operation.

Input Retarders

The front mounted hydraulic retarders or input retarders, as shown in Figure 13-27, are mounted to the front of a transmission between the torque converter and the transmission case. The housings are typically made from cast iron or aluminium alloys and have integrally cast vanes formed on the inside surface. These vanes are sometimes referred to as pockets. The design of the rotor vanes, with respect to width, length, and depth depends on the intended application. The rotor is fastened to the **turbine shaft,** which allows the rotor to rotate between the front and rear housings.

The oil pressure in the retarder circuit is controlled by a pressure regulator valve, while the flow of oil to the retarder is controlled by a control valve. With the engine operating at high idle, the pressure in the retarder circuit depends on the application and can vary from 40 to 120 psi. When the operator activates this circuit, the oil flow enters the retarder near the center of the rotor. The rotor pumps pressurized oil to the pockets in the two stationary members (stators). This causes a high resistance in the retarder, which forces the rotor to slow down, absorb kinetic energy, and turn it into heat energy. Since the rotor is connected to the transmission, the retarding action is transmitted to the drive wheels. To remove the excess heat that is produced when retarder operation is required, engine speed is reduced to an idle to allow the cooling system to absorb the heat from the retarder circuit. The amount of retarder braking will depend on the flow control, which is determined by the operator's use of either a hand or foot control.

Output Retarders

Rear mounted or output retarders, as shown in Figure 13-28, are mounted to the rear of a transmission between the main gear box and the output shaft. The housings are typically made from cast iron or aluminium alloys and have integrally cast vanes (pockets) formed on the inside surface. The rotor is splined to the output shaft, which allows the rotor to turn with the output shaft at the same speed. Operation of the rear-mount retarder is similar in principle to the front-mount design. One additional design feature has been added to this design: a multi-plate **friction clutch.** Although the retarder operates similar to the front-mounted design, it has the ability to quickly apply braking effect through the initial application of the friction brake and while the oil charges the retarder. The main advantage to this design is that the forces generated by the retarder's action do not get transmitted through the transmission components because the retarder is located behind the transmission. Therefore, only the output shaft receives the braking force. Because the rear-mounted retarder turns at output shaft speeds, the braking effect can be enhanced due to varying speeds that can be achieved by selecting different gear ranges.

ONLINE TASKS

1. Use a suitable Internet search engine to research the operation of an off-road hydrostatic drive system. Identify a piece of equipment in your particular location, paying close attention to the hydrostatic drive circuit operation, and identify whether it is an open-loop or closed-loop drive circuit. Outline the reasoning behind the manufacturer's choice of drive circuit design for this particular application.
2. Check out this url: *http://www.deere.com/en_US/golfturf/media/powerpoint/customersupport/2653a_hydraulic_system.ppt* for more information on hydro static drives.

SHOP TASKS

1. Select a piece of equipment in your shop that has a hydrostatic drive system and identify the type of components that make up the drive circuit. Identify whether it is a closed-loop or open-loop design, and outline how oil is directed to this circuit along with the charge pressures used.
2. Select a piece of mobile equipment in your shop. Identify the type and model of the equipment and list all of the major components and the corresponding part numbers of a typical hydrostatic drive circuit. Identify and use the correct parts book for the particular piece of equipment being worked on.

Summary

- The hydrostatic drive system is a fluid drive system that uses hydraulic fluid under pressure to convert diesel engine power to fluid power and then to convert it back to mechanical power at the wheels or tracks.
- The basic differences between a hydrostatic drive system and a hydrodynamic drive system are the velocity of the oil flow and the pressures that are used.
- The integral system uses a pump and motor mounted as a single assembly on the back of a diesel engine.
- The hydrostatic drive system is capable of providing a dynamic braking effect as a means of slowing and stopping the vehicle or even the ability to coast when required.
- Hydraulic fluid will not compress but it will take the shape of any hydraulic system component it happens to flow through.
- The available torque on startup in a typical hydrostatic drive system can be as high as 90%.
- Internal leakage in the pump and motor circuit is required for lubrication and cooling.
- The charge pumps supplies the oil flow for the main pump, as well as oil flow for the lubrication and cooling of a hydrostatic drive system.
- The main pump selected for use in a hydrostatic drive circuit may be a positive displacement pump design with a variable or a fixed displacement.
- The use of a heat exchanger is a requirement if the operating temperature of the hydraulic oil will exceed 180 degrees at the motor case drain.
- The reservoir will be sized to a minimum of half of the charge pump volume in gallons per minute (gpm).
- The displacement control valve controls the swashplate angle in the pump.
- In a closed-loop system, the oil flows from the pump to the motor and is returned to the inlet of the pump. In an open-loop system, the hydraulic oil flows from the pump to the motor and back to the hydraulic tank.
- The term stepless is often used to describe the operation of the hydrostatic drive system.
- Hydrostatic drive circuits cannot tolerate any dirt getting in to the system; the dirt circulates within the components and quickly causes excessive wear in the pump and motor circuit.
- An open-loop hydrostatic drive circuit is a hydraulic drive that may be used in heavy equipment where two or more different drive functions or loads are necessary.
- The selection of hydraulic lines should be in keeping with equipment manufacturer's recommendations for length, diameter, and pressure capabilities.
- Before you perform any maintenance to hydrostatic drive equipment, take note of any potential hazards that may expose you to the possibility of injury or equipment damage.
- The components in the hydrostatic drive system can be damaged quickly from too much heat created by excessive internal leakage.
- The natural tendency of any hydraulic oil that is exposed to excessive oil temperatures is to lose its viscosity.
- The maintenance intervals that are developed by equipment owners are often a result of operating experiences and may differ from manufacturer's recommendations.
- Whenever work is performed on equipment, use approved wheel chocks on both sides of the drive wheels to prevent accidental movement of the equipment.
- The hydraulic retarder can best be described as a device that converts the kinetic energy of a moving vehicle into heat energy.
- The retarder may be located either at the front of the transmission between the torque converter housing (input retarder), or at the rear mounted to the rear of the transmission (output retarder).
- When a haulage truck descends an 8% grade at 30 mph, it could require approximately 500 hp of continuous braking to prevent a speed increase while descending the grade.
- To maintain high efficiency in hydraulic retarder design, the clearance between the housings and rotor has to be minimal. Hydraulic retarder efficiency can run from 80% to as high as 115% equalling the efficiency of most diesel engine retarders.
- The rotor pumps pressurized oil to the pockets in the two stationary members (stators).
- To remove the excess heat that is produced when retarder operation is required, the engine speed is

reduced to an idle allowing the cooling system to absorb the heat from the retarder circuit.

- On steep grades, the use of a hydraulic retarder reduces the operating braking temperatures and increases the service life of the brake system.
- The main advantage of the rear-mounted retarder is that the forces generated by the retarder's action do not get transmitted to the transmission components because the retarder is located behind the transmission and only the output shaft receives the braking force.
- A rear-mounted retarder turns at transmission output shaft speeds; the braking effect can be enhanced due to the speeds that can be achieved by selecting different gear ranges.

Review Questions

1. Which of the following statements best describes what takes place in a hydrostatic drive system?
   A. high flow velocity with low pressure
   B. low flow velocity with high pressure
   C. high flow velocity with low vortex flow
   D. low flow velocity with low rotary flow
2. What happens to the pressure in the drive system on a loader equipped with a hydrostatic drive when the load on the motor increases?
   A. Drive pressure in the system increases.
   B. Drive pressure in the system decreases.
   C. Drive pressure in the system stays the same.
   D. Charge pressure in the system increases.
3. What is the maximum potential torque available at startup on a hydrostatic drive loader?
   A. approximately 100%
   B. approximately 90%
   C. approximately 70%
   D. approximately 50%
4. What takes place in a variable-displacement pump when the swashplate angle is increased?
   A. More oil is displaced per stroke.
   B. Less oil is displaced per stroke.
   C. Charge pressure increases.
   D. Motor speed decreases.
5. What determines the rate and direction of flow in a hydrostatic drive system?
   A. piston diameter
   B. swashplate angle
   C. pump speed
   D. rotor speed
6. Which component supplies makeup oil for the hydrostatic system?
   A. crossover relief valve
   B. swashplate angle.
   C. low-pressure relief valve
   D. charge pump
7. Which component supplies oil to the main pump on a typical hydrostatic drive system?
   A. charge pump
   B. swashplate
   C. low-pressure relief valve
   D. main relief valve
8. Which component directs the low pressure side of the drive loop to the low-pressure charge relief valve?
   A. shuttle valve
   B. swashplate
   C. low-pressure relief valve
   D. charge pump

9. Which of the following systems is capable of offering independent track control and infinite variable speed control?

   A. hydrostatic drive
   B. hydro-dynamic drive
   C. belt drive
   D. chain drive

10. Which component is always used in a closed-loop hydrostatic drive circuit?

    A. charge pump
    B. variable-displacement pump
    C. fixed-displacement pump
    D. fixed-displacement motor

11. Which of the following controls the direction and the amount the swash plate moves?

    A. servo pistons
    B. charge pump
    C. low-pressure relief valve
    D. shuttle valve

12. What function do the main relief valves serve in a hydrostatic drive circuit?

    A. protect against excessive pressure spikes
    B. protect the motor from over speed
    C. protect the charge pump from excessive pressure
    D. maintain maximum torque output at high speeds

13. When a hydrostatic drive system is in neutral, what regulates system pressure?

    A. charge relief valve
    B. shuttle valve
    C. low-pressure relief valve
    D. charge pump

14. What is the main design feature difference between a closed-loop system and an open-loop system?

    A. The return oil from the motor goes directly back to the pump on the close-loop system.
    B. The return oil from the pump goes back to the motor on the closed-loop system.
    C. The return oil from the pump goes back to the tank on the closed-loop system.
    D. The return oil from the pump goes back to the charge pump on the closed-loop system.

15. What takes place in a variable displacement pump when the pistons and barrel rotate with the valve plate at an angle?

    A. The pump is in neutral.
    B. The charge pressure will decrease.
    C. Pistons will stroke.
    D. The motor will stop turning.

16. What takes place when the swashplate in the motor is moved to the maximum angle for maximum torque?

    A. high speed operation
    B. neutral operation
    C. low speed operation
    D. reverse operation

17. Which component determines the rate and direction of flow inside a variable-displacement pump?

    A. swashplate
    B. pistons
    C. control valve
    D. charge pump

18. Which component controls the servo piston movement in a variable-displacement pump?

    A. displacement control valve
    B. charge pump
    C. retraction plate
    D. shuttle valve

19. What takes place when the swashplate in the motor is moved to the minimum angle for minimum torque?
    A. high speed operation
    B. neutral operation
    C. low speed operation
    D. reverse operation
20. What takes place in a hydrostatic drive system when the load on the motor decreases?
    A. The pressure in the system increases.
    B. The pressure in the system decreases.
    C. The pressure in the system remains constant.
    D. The speed of the system decreases.
21. Which type of hydraulic retarder is located at the front of the powershift transmission between the converter and the main gear box?
    A. the input retarder
    B. the friction clutch retarder
    C. the output retarder
    D. the range retarder
22. Which one of the following components is only found in the rear-mounted retarder?
    A. the rotor
    B. the stator
    C. the friction clutch
    D. the pressure regulating valve
23. What determines the degree of braking in a hydraulic retarder?
    A. the oil flow and pressure
    B. the stator speed
    C. the rotor speed
    D. the speed of the turbine
24. Where does the oil enter the hydraulic retarder during activation?
    A. at the outer edge of the stator housing
    B. near the center of the rotor
    C. on both sides of the stator
    D. it is always charged with oil
25. To what component is the rotor mounted in a rear-mounted hydraulic retarder?
    A. turbine shaft
    B. output shaft
    C. stator
    D. countershaft

Driveline Systems

Learning Objectives

After reading this chapter, you should be able to

- Describe the fundamentals of drivelines.
- Describe the components and construction features of drivelines.
- Describe the principles of operation of drivelines.
- Describe the recommended maintenance procedures for drivelines.
- Describe the diagnostic procedures to troubleshoot drivelines.

Key Terms

brinelling
broken-back driveshaft
compound driveline angle
cross
delta-wing
driveshaft
end yoke
galling
high-block
inclinometer
low-block
needle bearings
non-parallel driveshaft
parallel driveshaft
phasing
run-out
slip assembly
slip joint
slip yoke
trunnion
universal joint
weld yoke
Welsh plug
wing-type U-joints
working angle

INTRODUCTION

Driveline technology has evolved considerably in the last 100 years. Even though the concept of the modern **universal joint** was first discovered a few centuries ago, the application of these principles was not used to advantage until early in the twentieth century when a fellow by the name of Clarence Spicer pioneered the development of the modern driveshaft U-joint. Anyone working in the heavy equipment trade these days has come across the name "Spicer," made famous by their pioneering development and manufacturing of the modern driveline and universal joints. The modern driveline and U-joint arrangement shown in **Figure 14-1** has evolved to what could be considered an engineering marvel.

Heavy equipment found at construction sites, open pit mines and quarries these days is capable of hauling hundreds of tons at a time. The power developed by the diesel engine must pass through a series of **driveshafts** and powertrain components on its way to the drive wheels. The torque generated by the use of large gear

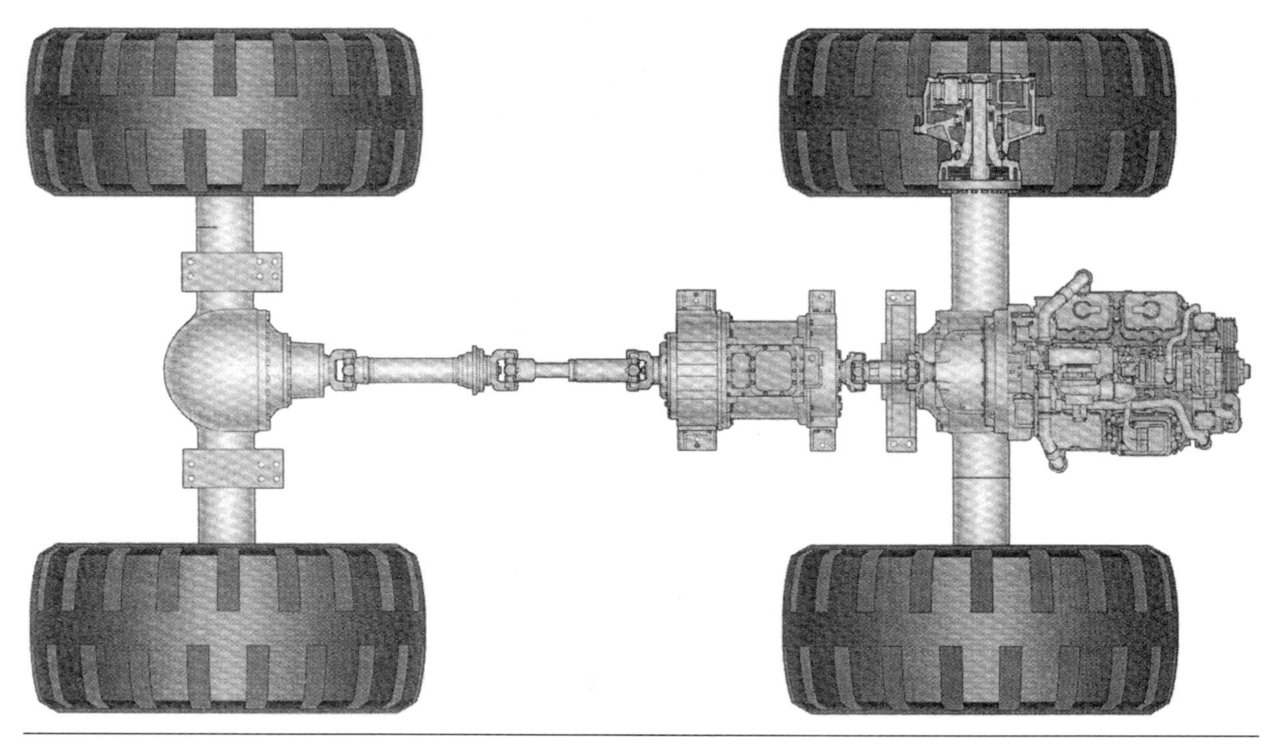

Figure 14-1 Modern driveline and U-joint arrangement. (*Courtesy of Caterpillar*)

reductions found in heavy equipment powertrain components (often as high as 100,000 lb-ft) is generated at the wheel ends. This high torque must be handled by series of drivelines on its way to the final drives. It is not uncommon to find an overall gear reduction of 100 to 1 from the engine to the drive wheels on heavy equipment.

In this example, we will look at the gear reduction of a typical loader used in a quarry operation. Let's assume the engine is capable of producing 1200 lb-ft of torque at the flywheel, and the typical torque converter used in this application has a two to one internal gear reduction between the turbine shaft and the output shaft; this would send 2400 lb-ft of torque to the transmission. A typical four-speed heavy-duty powershift transmission will often use a first gear with a six to one reduction. If you do the math, you will see that the overall torque produced at the output shaft of the transmission is on the order of 14,400 lb-ft. This torque now flows through a differential where it is increased by a four to one reduction to 57,600 lb-ft. The last reduction is a final reduction at the wheels through a four to one planetary gear set, producing an output torque of 230,400 lb-ft. This example shows the torque potential that can be transmitted through the drivelines between powertrain components on heavy equipment.

The transmission of power through a drivetrain as shown in **Figure 14-2** is not necessarily in a straight line; equipment design often necessitates component placement that requires drivelines to transmit power at odd angles. Modern suspension systems pose another problem; final drives are often mounted on flexible suspension systems that cause the drive axles to move when the equipment is driven over uneven terrain. The use of a **slip joint** allows the driveshaft length to shorten and lengthen as the drive axle moves over a roadway. A final hurdle was overcome by the addition of universal joints on both ends; this allowed the driveshaft to adapt to speed changes, produced by driveline angle changes between the powertrain components while it is in operation.

FUNDAMENTALS

Driveline designs can be divided into two categories: parallel driveshaft installations and **broken-back driveshaft** installations **(non-parallel).** Either of these designs may be used on a typical piece of heavy equipment. In a loader that had the torque converter mounted to the back of the engine and the transmission separately mounted rigidly to the frame of the loader, it is common to find a rigidly mounted drive shaft con-

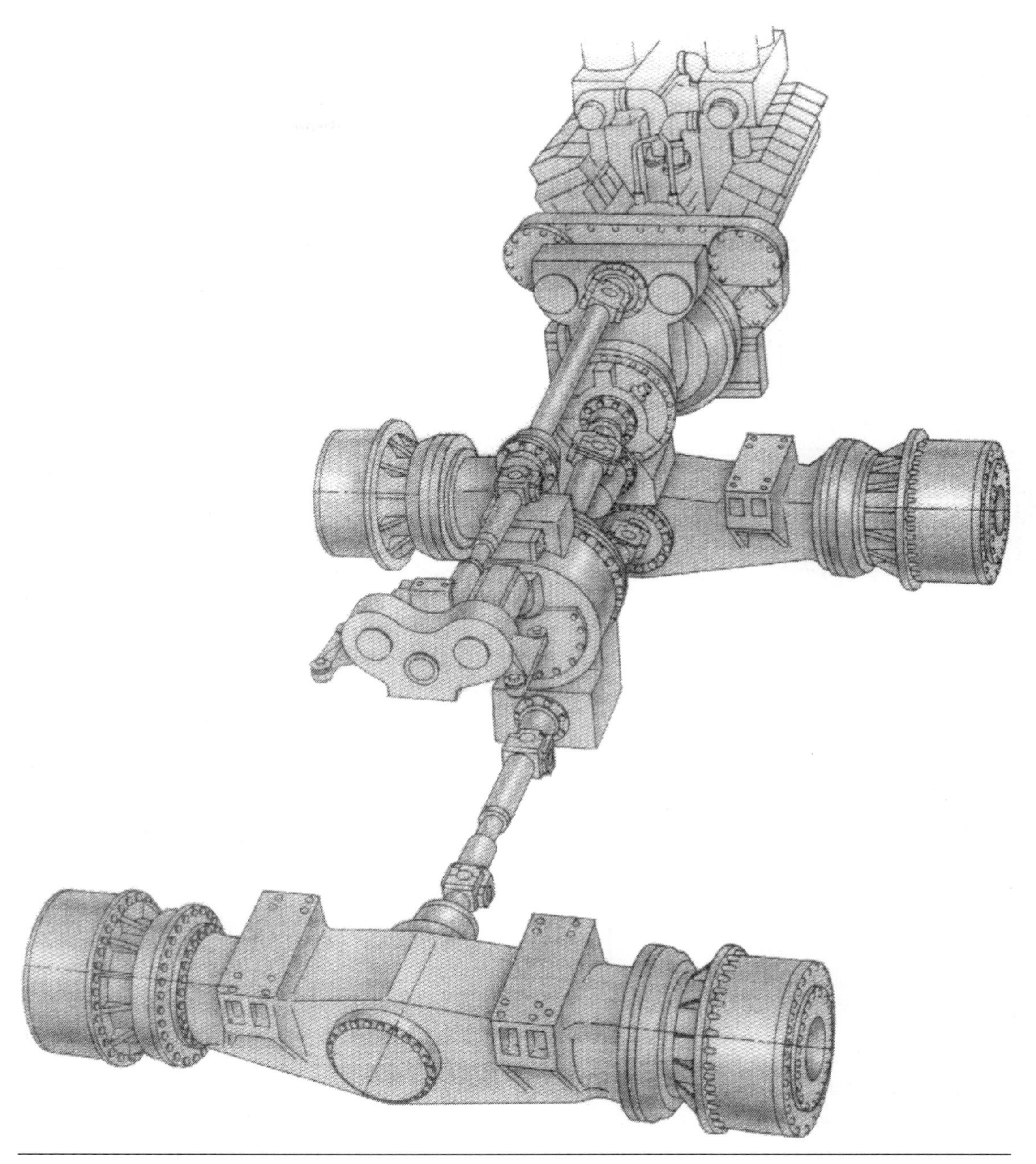

Figure 14-2 Typical heavy equipment driveline arrangement. (*Courtesy of Caterpillar*)

necting the two components with a universal joint at each end. This design is suitable for use where both input and output sources are rigidly mounted.

In an application where one of the components *is* able to move, this design would not be suitable and a slip joint would be required to allow for movement. An example of this is a typical transmission-to-drive axle connection with a suspension mounted drive axle. In **Figure 14-3,** we see a typical off-road application of an axle that is mounted rigidly to the frame and does not have any significant movement other than frame and component flexing. The rigid design is commonly found on heavy equipment and generally has a long trouble-free service life. Not only it is reliable and low cost, but it also is easily maintained since there are no suspension alignment concerns to worry about as the equipment wears.

In applications where the equipment has a live suspension, such as a large haulage truck, the use of a parallel driveshaft arrangement is common. As the

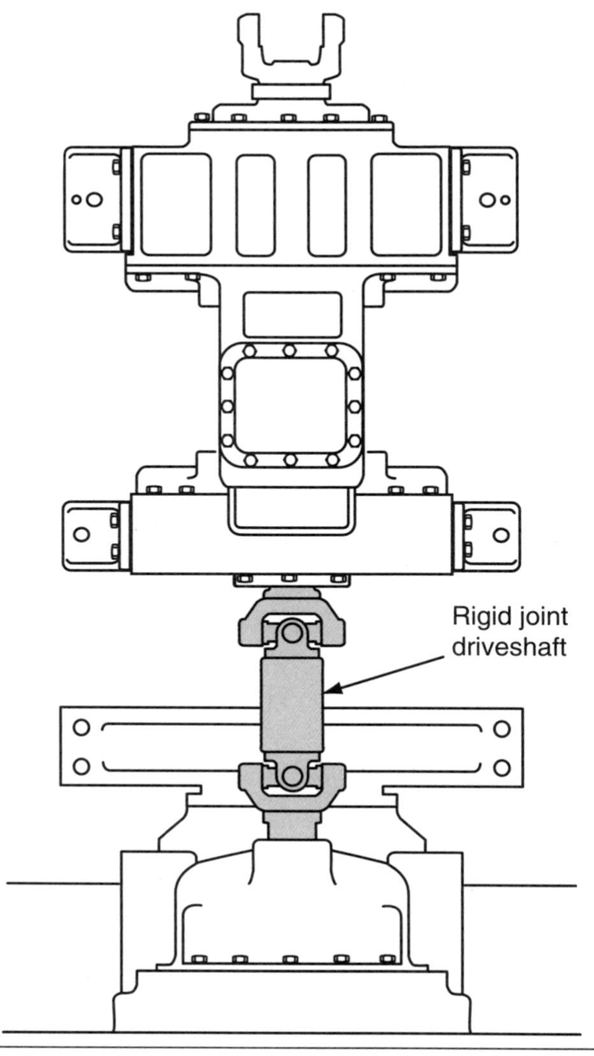

Figure 14-3 Typical rigid joint driveshaft installation found on heavy equipment.

haulage truck moves over uneven terrain the driveshaft is able to change length by the use of a splined slip joint. A parallel design is better suited to this application because the operating angle tends to remain constant as the axles move through their intended design travel. It is important to note that the yokes must remain parallel to each other while in operation, and the **working angles** must remain equal to minimize vibration and wear of the U-joints.

In certain applications, it is better to use a non-parallel design installation of a driveshaft. The yoke angles are not parallel to each other in this installation, but the driveline working angles must remain equal if a vibration-free operation is to be maintained. Due to the design restrictions on some equipment, it becomes necessary not only to have a vertical plane offset, but also to have a horizontal plane offset, as shown in **Figure 14-4.** Due to the inherently slow-speed operation of most heavy equipment, working angles on driveshafts are often much higher that on high-speed vehicles. Vibration does not usually become a problem on heavy equipment; the equipment often operates on rough terrain and the operator would probably not notice vibration until it was extreme.

It is important to note that the life of the driveshaft U-joints decreases as the working angles increase. Driveline speed plays an important role in determining the maximum allowable working angles that can be used. The use of large gear ratios on heavy equipment does not allow the vehicle too much opportunity for high road speeds. Typical road speeds range from between 2 to 25 kph (1 to 15 mph). A universal joint has a maximum speed at which it can operate, this speed is always proportional to the working angle of the installation. Heavy equipment generally does not operate at speeds above 50 kph (30 mph), which limits the maximum speed at which the drivelines turn. This allows equipment manufacturers to use driveline angles of 8 to 12 degrees on large, slow-moving equipment. Typical driveline speeds on final drives can be up to 100 to 800 rpm. **Figure 14-5**

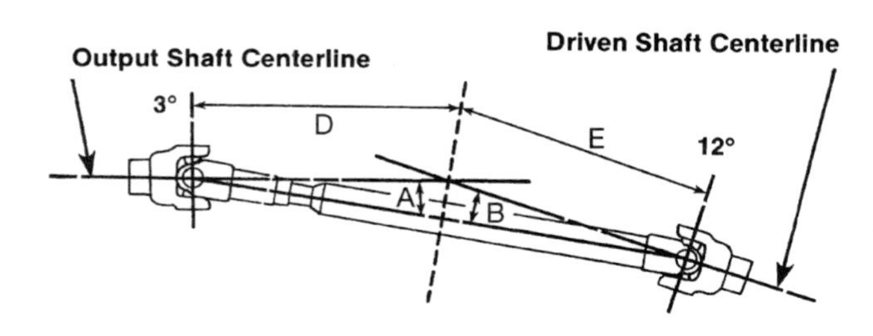

Figure 14-4 Typical non-parallel driveline design installation. Note the unequal yoke angles between the output shaft yoke and the driven shaft yoke. (*Courtesy of International Truck and Engine Corporation*)

Drive Shaft RPM	Maximum Working Angle
4000	4 degrees of 15′
3000	5 degrees of 50′
2000	8 degrees of 40′
1500	11 degrees of 30′
1000	12 degrees of 30′

Figure 14-5 Typical driveline working angle and speed relationship found on heavy equipment.

shows a general example of speed versus the maximum working angles that are often used.

Universal joints connect the drivelines to the powertrain components and carry the full torque that the equipment has the capability of generating, while at the same time allowing the driveshaft to operate at different angles. Because U-joints transmit the full torque that a machine is capable of producing, the **cross** or spider, as some manufacturers call it, is constructed of a high quality forged casting. The four ends, or trunnions, where the **needle bearings** are located have a case-hardened surface that is ground to a fine finish.

Interconnected internal grease passages, as shown in **Figure 14-6,** are drilled in the trunnions to allow the needle bearing to be greased. Although a few manufacturers are starting to use a sealed U-joint that requires no greasing during its service life, the majority of universal joint manufacturers still provide a means of greasing the **trunnion** bearings. Grease fittings are located close to the center of the cross, allowing for periodic lubrication of the U-joint bearings.

Although there are many different cap designs available for heavy-duty universal joints, we will discuss the three most common cap designs. The **low-block** design is a standard type of cap and is suitable for use on smaller equipment that does not regularly achieve high torque levels in service. A good example of this application is a typical 2- to 4-yard loader. The **delta-wing** design is generally used on 4- to 8-yard loaders between the torque converter and the transmission where torque values are minimal. The **high-block** cap (high wing), as shown in **Figure 14-7,** is often found on large equipment between the transmission and the final drive. It is used for heavy-duty application. Its large block design is capable of carrying high torque values that are generated by this type of equipment. A good example of this use is on an 8- to 12-yard loader used in quarries and open pit mines.

As discussed earlier, we know that driveshaft speeds are not very high on a typical piece of heavy equipment; this does not mean that we will ignore working angle geometry. Due to the high torque loads on U-joints used on heavy equipment, we need to ensure that working angles are held to within 1 degree of each other to maintain equal U-joint speed at each end of the driveshaft. This will prevent uneven speed fluctuations, as shown in **Figure 14-8** which can lead to premature U-joint needle bearing failure.

Heavy equipment designs often use some form of driveshaft support when drive shafts pass through equipment bulkheads as shown in **Figure 14-9.** The distance between the engine, transmission, and drive axle on heavy equipment can be substantial, and may require the use of driveshaft supports. The use of a

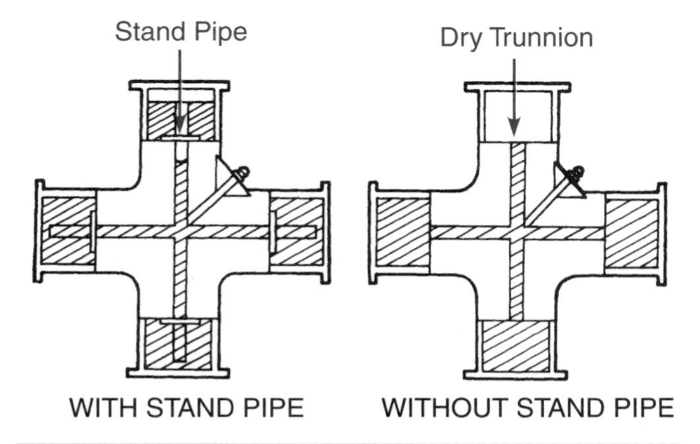

Figure 14-6 Cross-drilled trunnion used to distribute grease to the bearing end. The one on the left uses a check valve (standpipe), which prevents reverse flow of lubricating grease from escaping the bearing end. (*Courtesy of Chicago Rawhide*)

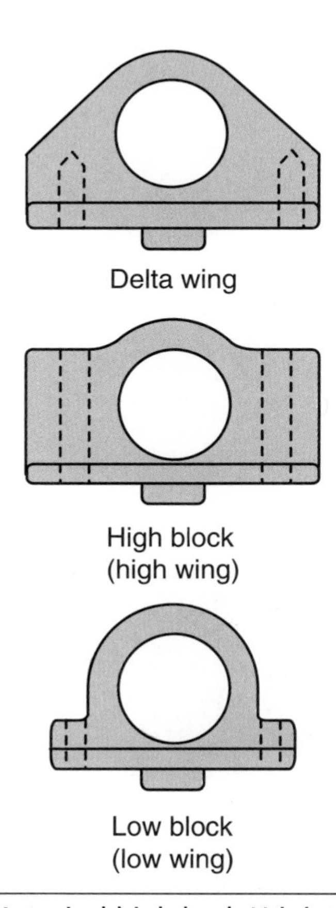

Figure 14-7 A typical high back U-joint design.

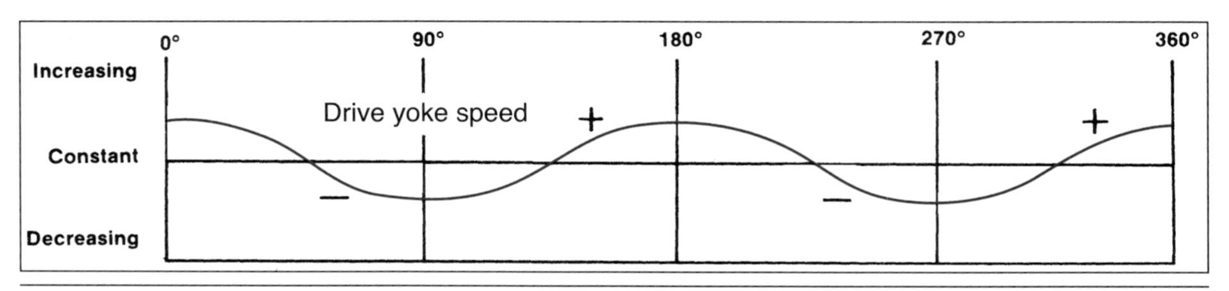

Figure 14-8 Speed fluctuation of a typical driveshaft during one revolution. (*Courtesy of Dana Corporation*)

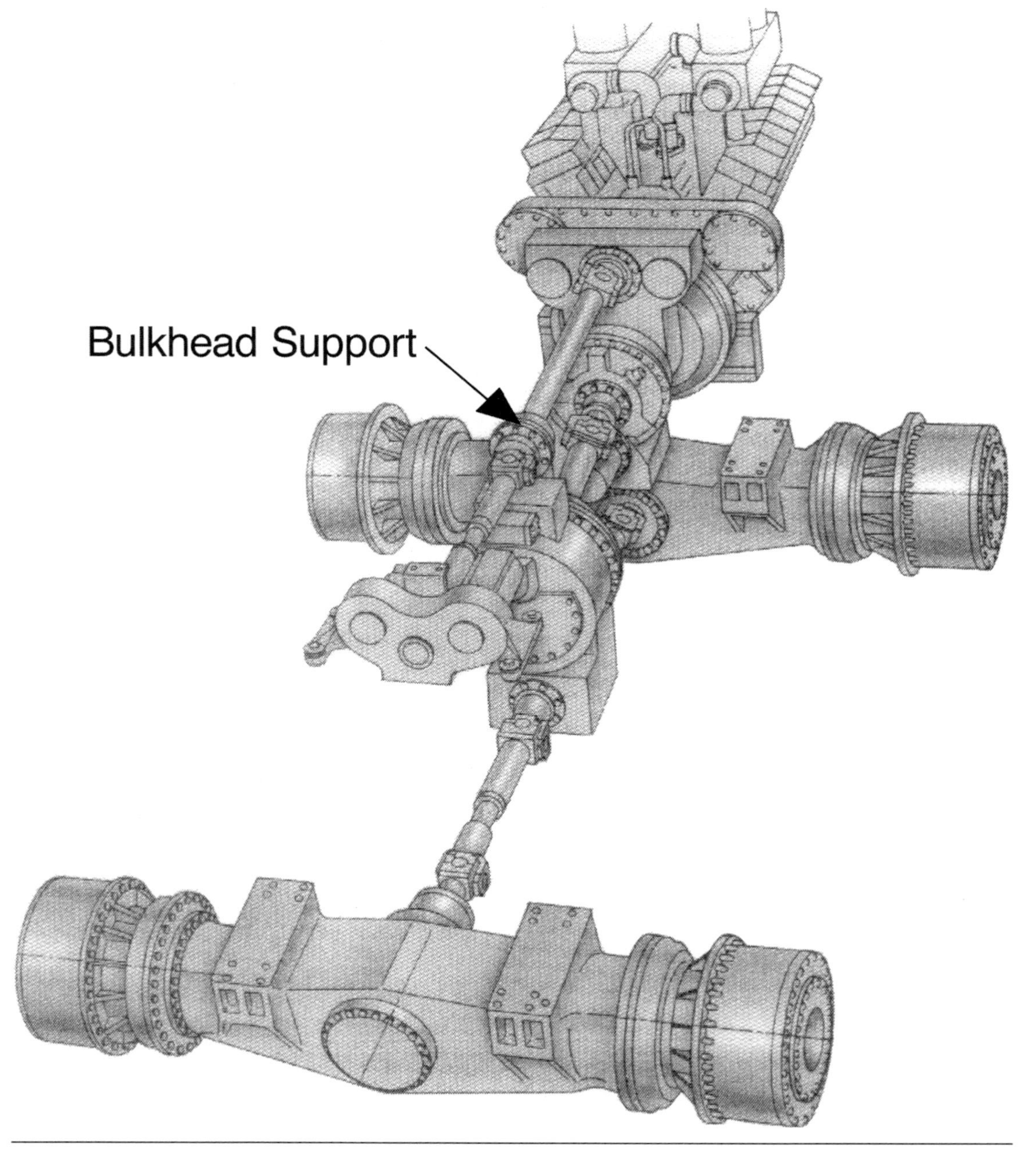

Figure 14-9 Typical driveline bulkhead support-bearing assembly found on heavy equipment. (*Courtesy of Caterpillar*)

bulkhead support bearing (a pillow block bearing) supports the driveshaft and decreases its natural tendency to flex while transferring torque. Flange-type bearings are generally used in this type of installation and are generally flange mounted to the frame. Usually, a lube fitting is provided for lubrication, but on some of the larger units where accessibility to the bearing is limited due to the equipment design, a remote greasing point is provided. On smaller equipment, this support bearing may be sealed, thus making lubrication unnecessary.

CONSTRUCTION FEATURES

Figure 14-10 shows an exploded view of a typical haulage truck drivetrain that consists of three separate driveline assemblies. The coupling shaft assembly connects the transmission to the standard slip assembly through a center bearing to the front axle. The short coupling **slip assembly** connects the two axles. We will look at each of these assemblies and learn how each contributes to the transfer of torque to the drive wheels. As we learned earlier in this chapter, the torque carried through the drivelines on off-road equipment is significantly higher in most cases than what you would normally have with on-road equipment. Because of this off-road drive lines are generally designed to handle the expected slow speeds and high torque found on off-road equipment.

Tubing

The most common tubing used in the construction of heavy-duty drivelines is cold-rolled electric-welded; this is generally considered standard factory build on most off-road equipment. The tubing is manufactured by rolling the steel into a tube shape and welding it together. The fabrication process does not always produce a perfectly true running shaft, but it is generally suitable for low-speed applications like those found between the transmission and the drive axles on most equipment. The wall thickness and driveline diameter determines the torque-carrying capabilities of the driveshaft. A higher quality of construction used to fabricate driveshafts takes the conventional rolled-tube, electric-welded shaft one step farther. The tubing is drawn over a mandrel to ensure that it is straight and runs true. The drawn-over mandrel process produces a much straighter tube and also adds to the strength of the tubes by stress relieving the shaft. The third type of tubing used for driveshaft construction is known as seamless tubing, which is

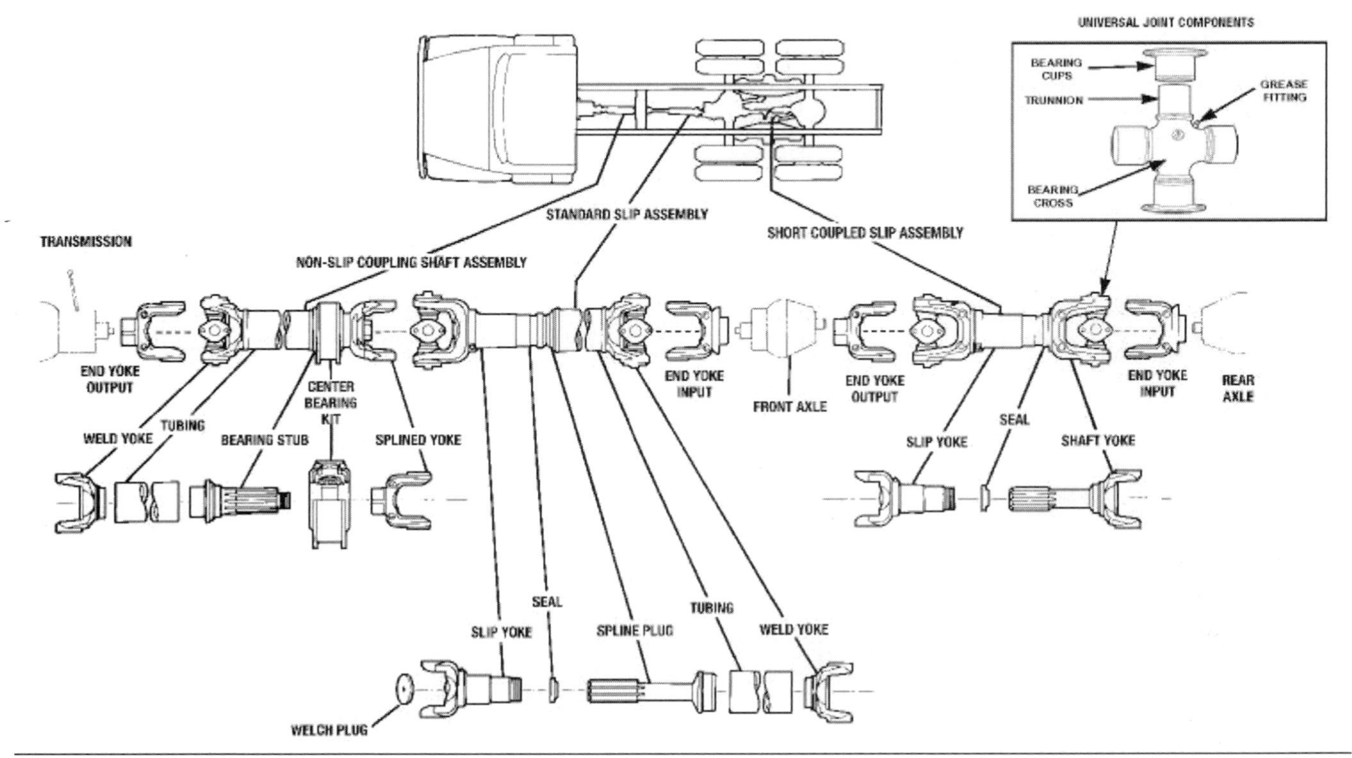

Figure 14-10 Typical exploded view of typical haulage truck driveline configuration. (*Courtesy of Arvin Meritor*)

heavy-wall tubing and is generally used in heavy-duty applications such as large haulage trucks found in open pit mining and quarry operations. Because of the thickness of the tubing, it is suited for slow-speed applications. Factors that must be considered when selecting tubing for a driveshaft are speed, length, operating angle, and torque-carrying capabilities.

Driveshafts

The driveshafts are balanced by the manufacturer but given the slow speeds at which drivelines rotate at on off-road equipment, vibration is seldom a concern. When driveshaft replacement is required, be sure to observe the equipment manufacturer's recommendations and use only the recommended shaft for the application. The use of a driveshaft that is a bigger or smaller in diameter than the one being replaced may lead to a shorter than normal life expectancy of the driveline and U-joints. Oversizing is not necessarily better; a larger rotating mass may lead to overloading of component bearings. Forged yokes are welded to each end of the driveshaft to accept the U-joints, which are required to compensate for the offset that exist between the engine, transmission, and drive axles. The yokes at each end of the driveshaft are connected through the U-joint to a yoke on the output or input shaft of the transmission and drive axles. In most cases, the driveline connection will be a two-piece connection between the components to allow for differences in component mounting variations on equipment and to compensate for driveshaft length changes due to suspension movement during operation.

Slip Joints

The splines on the slip joints are made from high quality hardened steel; the male and female end have matching splines to provide a tight-fitting connection between the two ends of the driveshaft, as shown in **Figure 14-11.** The machined slip joints are welded to the ends of each shaft. The male end of the slip joint has external splines, which are an exact match to the internal splines found on the female end of the drive shaft. When connected together, the splines of each fit snugly yet allow the overall length of the driveshaft to shorten or lengthen during operation. The sliding fit must be held to tight tolerances because the full torque developed must be transmitted through this connection. In some cases, some manufacturers coat the splines with an anti-friction coating, such as nylon or high phosphate grease. When mounted together, the splines are sealed with a threaded dust cap that holds a seal to keep dirt out of the splines on the female end and a blind plug (Welsh plug) or cap at the other end. These require periodic lubrication with high quality grease through a grease fitting located on the female end of the shaft.

Universal joints

Universal joints (U-joints) are necessary on each end of a driveshaft to connect powertrain components together that are not mounted on the same plane. U-joints compensate for speed changes that a driveline experiences when torque is transferred through the powertrain components. **Figure 14-12** shows the parts breakdown of a typical universal joint. The bearing cross, or spider, is constructed from high quality forged steel and has a hardened, machined section on the end of each of the four tips, called trunnions.

These trunnions have bearing caps with internal needle bearings mounted on the ends that allow the caps to rotate. A rubber seal, as shown in **Figure 14-13** is located at the end of each cap to keep dirt and water out of the needle bearing during operation. Needle

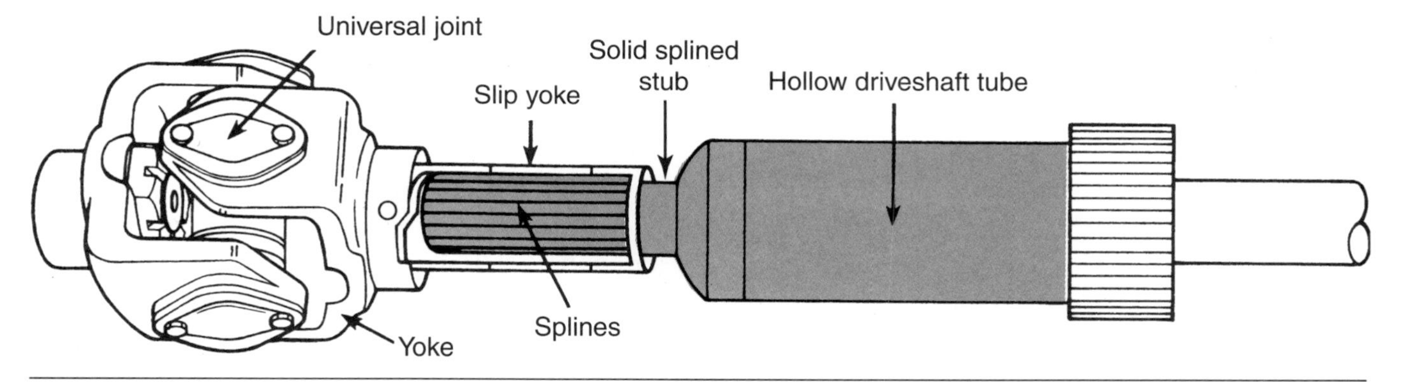

Figure 14-11 Slip joint connection on a typical two-piece driveshaft. (*Courtesy of Chicago Rawhide*)

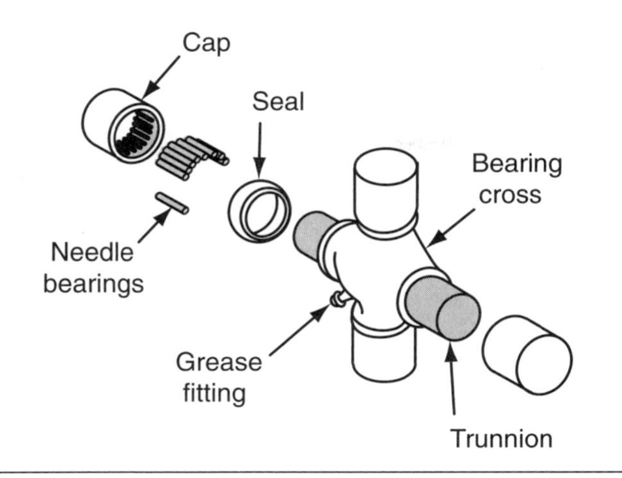

Figure 14-12 Typical universal joint construction showing the major parts.

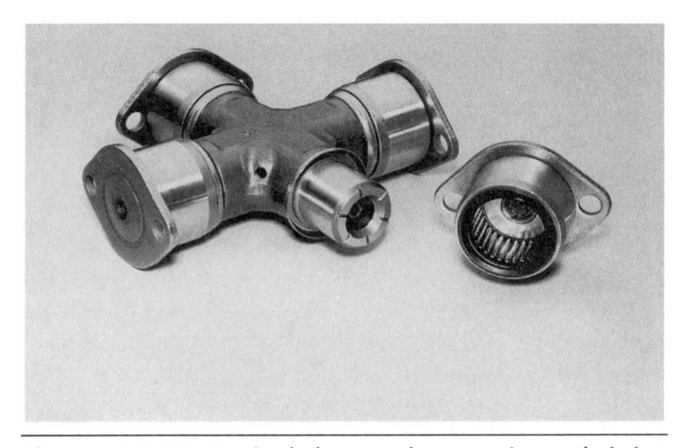

Figure 14-13 Typical heavy-duty universal joint showing one bearing cap removed. (*Courtesy of Chicago Rawhide*)

bearings are ideally suited for use in this application since they require very little radial space and are capable of carrying high radial loads thus giving them a long service life if properly maintained. The bearing caps are mounted to corresponding forged yokes at each end of the driveshaft. Although they come prelubricated, needle bearings must be lubed with high quality grease after initial installation and again at regular service intervals.

A grease fitting is mounted to the bearing cross that has internally drilled passages to allow the grease to reach each needle bearing without removing the caps, as shown in **Figure 14-14.** Some manufacturers locate a standpipe (check valve) at each end of the bearing cross. This check valve prevents the grease from running out of the trunnion when the vehicle is parked or the trunnion is stopped in the vertical position. Without

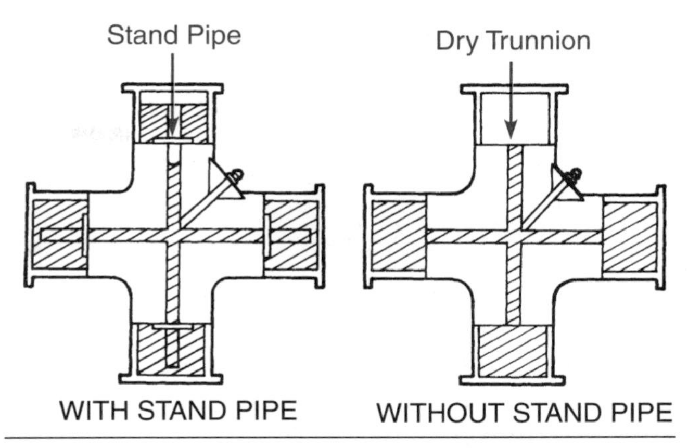

Figure 14-14 A stand-pipe (check valve) which prevents the hot grease from flowing down away from the vertical trunnion. (*Courtesy of Chicago Rawhide*)

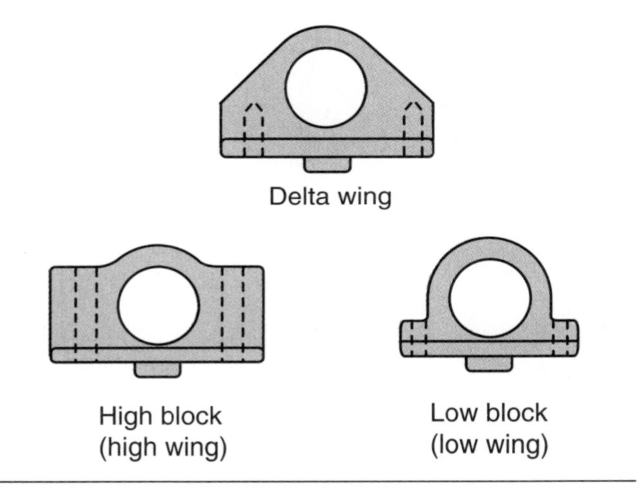

Figure 14-15 Three bearing cap designs used on heavy-duty U-joints.

this check valve action, the bearing would quickly become damaged from lack of lubrication.

Wing Design. U-joints come in many different design configurations. **Figure 14-15** shows four styles that are common on some of the smaller off-road equipment, such as small two-yard front end loaders. The bearing caps come in different design configurations, which depend on the service application of the U-joint. Generally, heavy equipment U-joints come in what is commonly called the wing configuration. Although these are not the only type of bearing caps used on off-road equipment, they are by far the most common. The wing configuration is available in three different styles. **Figure 14-16** shows the heavy-duty high-block style

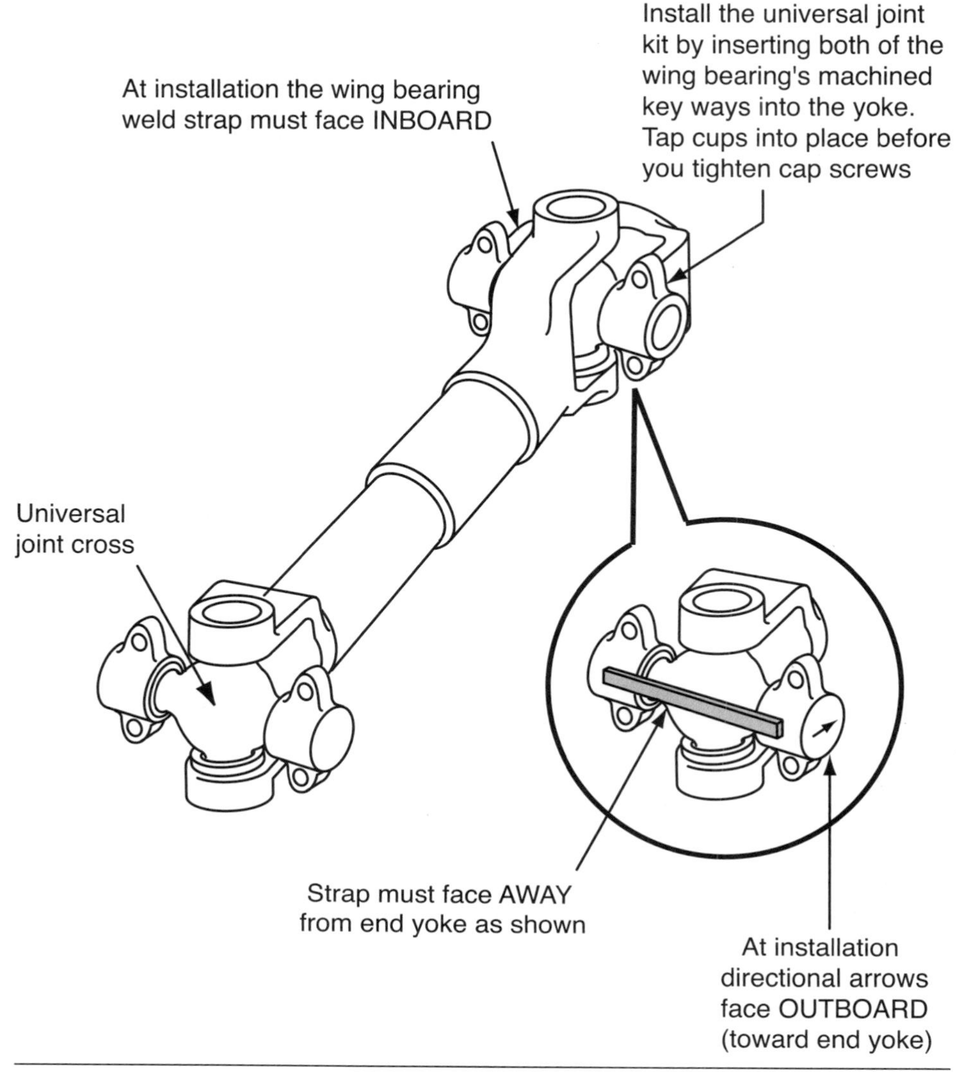

Figure 14-16 Heavy-duty wing-style joint with machined keyways in the yoke.

with a keyway used in severe-duty applications. The low-block and delta-wing bearing caps are found on drivelines that transfer power from the engine to the transmission, where torque is the lowest. Generally, high-block bearing caps are used exclusively on driveshafts between the transmission and the drive axles or off-road equipment.

All wing-style heavy-duty bearing caps have a keyway that fits into a corresponding keyway in the yoke, as shown in Figure 14-16, allowing for an extremely high-torque carrying connection. The wing joints come with special high grade cap screws and should be torqued to the manufacturer's specifications. Some manufacturers recommend that Loc Tite® be applied to the bolt threads before they are torqued to specification. Always follow the manufacturer's recommended installation procedures when installing U-joints on a driveshaft.

OPERATING PRINCIPLES

Off-road equipment has two driveshaft configurations that are used in the majority of equipment. The parallel type is the more popular of the two, and it is used in equipment. When the components are mounted solidly to the frame of the equipment, it is easy to maintain parallel angles of the flanges and yokes. The working angles should be equal because the angles are directly dependent on the speed of the driveshaft.

If the correct working angles for a given speed cannot be met with the component installation parallel,

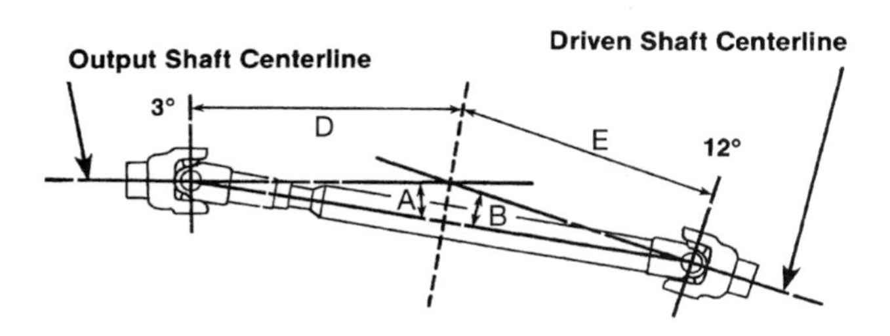

Figure 14-17 Non-parallel driveline installation shows the transmission angle is 3 degrees off vertical and the driven shaft is 12 degrees off vertical. (*Courtesy of International Truck and Engine Corporation*)

each component can be shimmed to achieve the working angle that is required. For example if the transmission center line is offset 3 degrees down from horizontal, and the drive axle is offset 3 degrees up from the horizontal, shimming can be used to reduce the working angles, if necessary.

The non-parallel design is used when the zero vertical angles of the component installation are not able to be met. **Figure 14-17** shows that the transmission is offset 3 degrees from the vertical, and the driven shaft is 12 degrees off vertical. Angles A and B need to be equal to avoid speed differences in the each of the U-joints on the same shaft. If the working angles are held to within 1 degree of each other the driveshaft will slow down at the front U-joint and speed up at the rear U-joint by an equal amount, cancelling out any speed fluctuations. Universal joint life is affected by the angles and speeds at which they operate, and exceeding the working angle and recommended speed of the driveshaft will almost certainly lead to problems. To achieve maximum U-joint life, the working angles must be held to a maximum working angle difference of no more than 1 degree. Off-road equipment operates at slow speeds and working angles are often higher than those found on high-speed on-road equipment. Placement of powertrain components in equipment often requires the use of higher working angles. Fortunately, we have speed (or the lack there of) working in our favor.

Compound Driveline Installations

Off-road equipment component placement can result in drivelines having compound driveline angles. **Figure 14-18** shows a typical driveline installation with the driveline installed with a vertical plane offset. This is by far the most common type of installation and results in a long, trouble-free driveshaft U-joint life.

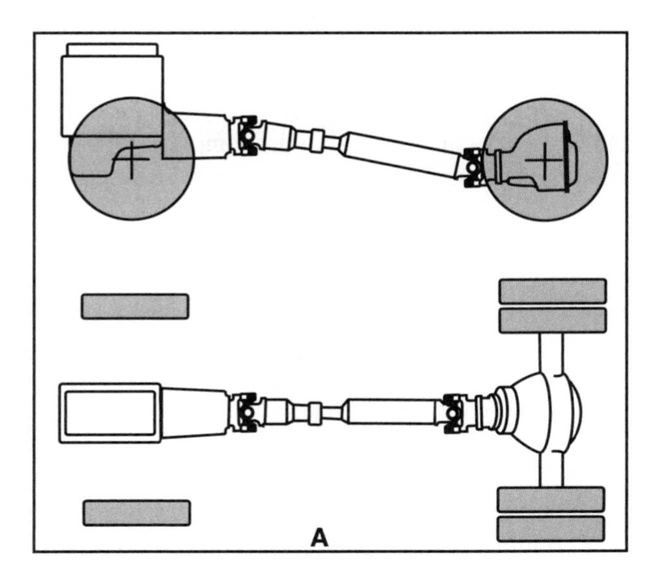

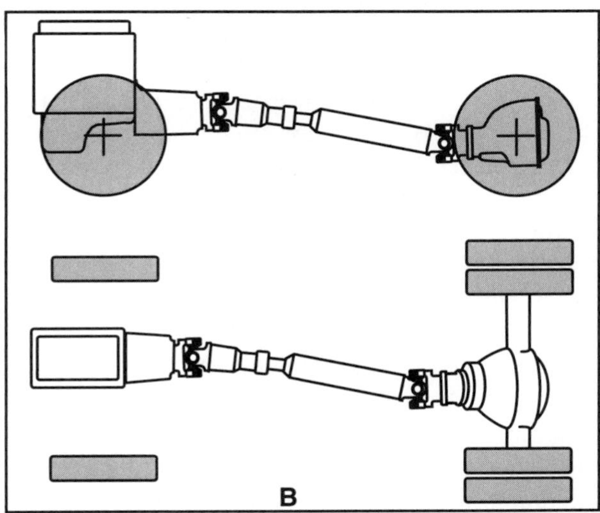

Figure 14-18 Single plane driveline angle installation.

Often, manufacturers will offset the installation in the horizontal plane but maintain a 0-degree vertical plane.

On some equipment it is not possible to install the components so that either a vertical or horizontal offset is achieved. Design limits on some equipment require that the components is installed in such a way that a compound driveline installation is unavoidable.

Compound driveline installations cannot be engineered out of the equipment design, and compromises will always exist. Fortunately, speeds on equipment are low; this keeps the negative effects of operating at high working angles and offsets to a minimum. Vibration on off-road equipment may not be noticeable to the operator, but the effect of vibration on the needle bearings used in the U-joints will take its toll on U-joint life. All U-joints must operate at an angle of at least a ½ degree.

This sets up a speed change that causes the shaft to slow down and speed up during each revolution.

These changes take place at each end of the driveshaft. Speed fluctuations in the driveline cause torsional vibrations to be set up in the drive shaft. These harmful vibrations can have a negative effect on the needle bearing in the U-joints as well as cause problems in the transmission itself and with the drive axles. If the working angles are set up correctly within 1 degree of each other, these speed changes cancel out each other, minimizing this effect.

Driveline Speed

If it were possible to watch a drive shaft U-joint move through an angle from an end view, we could see that the U-joint at the driven end moves through an ellipse. The U-joint moves through each quadrant in a fixed amount of time, in other words the driveshaft increases and decreases in speed are predictable for each revolution. When we examine a conventional two U-joint driveshaft, we can see that if the second U-joint had an equal working angle it would decelerate at the same time that the first U-joint was accelerating and at about the same rate. The acceleration and deceleration rate is directly proportional to the working angle of the U-joints. Since the working angles can change because of suspension movement, the velocity of the driveshaft is in constant change. In order for the driveshaft to cancel out these changes in speed while it transmits torque, the U-joints must be installed in phase with each other. In operation, while the driveshaft turns at the same rpm, the shaft will increase and decrease its speed once each revolution. If we install the driveshaft correctly in phase, the opposing forces generated by this phenomenon will cancel each other. If the driveshaft was installed one tooth out, resulting in an out-of-phase condition, the resulting forces would be similar to a person snapping a long rope; the whipping action would produce a snapping action at the opposite end of the rope. If this can be reproduced at both ends simultaneously, the forces will cancel each other out. This is how driveline **phasing** works; the forces generated at each end of the driveshaft cancel each other out if the driveshaft is correctly assembled.

Phasing

When removing a two-piece driveshaft, look for the phasing marks, as shown in **Figure 14-19.** If none are present, scribe a mark on the two connecting halves

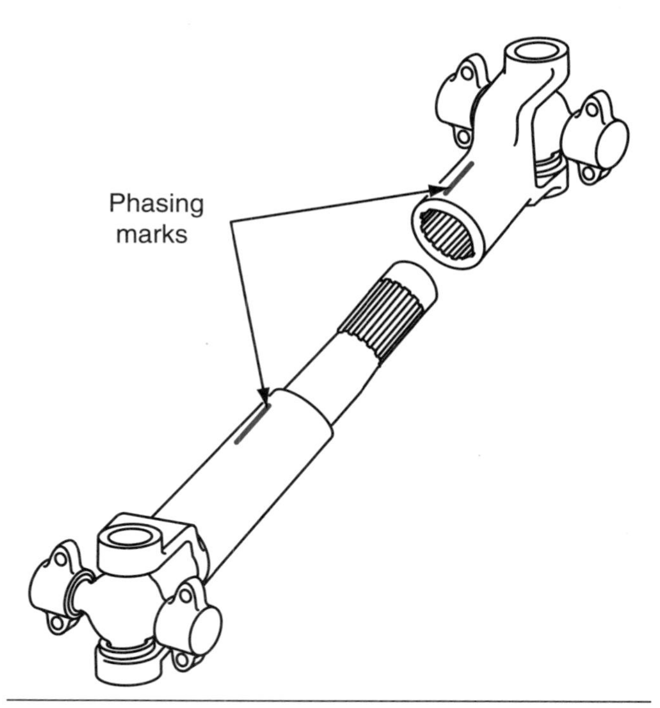

Figure 14-19 Where to place the phasing marks on a driveline.

before removing the driveshaft. This will ensure that the driveshaft is installed correctly when installation occurs. One other factor needs to be considered when determining driveshaft installation parameters: the length and diameter of the driveshaft must be considered (as well as the rpm) when determining installation parameters. Most manufactures restrict driveshaft length to a maximum of 70 inches. The greater the weight and length of the shaft, the greater the radial forces generated. As the speed of the driveshaft goes up, balance becomes more important. The overall length of the driveshaft is measured across the centerline of the U-joints. If proper maintenance has been adhered to, U-joint operating life, as rated by most manufactures, is calculated for a continuous operating load at 3000 rpm for 5000 hours with a 3-degree working angle. If you were to double the working angle, the life expectancy would be reduced by 50%. If you reduce the load by 50% you double the life. Because the drivelines on off-road equipment run at various loads and speeds, it becomes impossible to predict the life of a U-joint.

MAINTENANCE PROCEDURES

This section contains maintenance procedures for drivelines in general. Some procedures require the use of special tools to service drivelines safely. Failure to use the special tools and follow the correct procedures as outlined by the tool manufacturer may result in personal injury or equipment damage. The procedures outlined here are intended for use as general examples of the steps to follow and should not be used for actual maintenance procedures on any particular drive system. Always refer to the manufacturer's technical documentation for the equipment that you are working on for proper procedures and specifications. Before you start any work on equipment, whether it is servicing or repair, you need to become familiar with a few general safety precautions. In many cases, when equipment needs to be serviced or repaired, it is not always necessary to remove the parts. Access to the area being serviced should be high-pressure washed before work begins and after it is completed; this minimizes the chance of contamination and prevent potential slipping conditions as well as minimizing fire hazards from oil on or near hot surfaces.

- To perform certain types of repairs or service work, it may be necessary to have the vehicle raised off the ground on approved safety stands to perform tests and repairs. Always be aware of possible unexpected equipment movement that may pose a safety hazard. To determine whether these procedures must be followed, always consult the manufacturer's service manual for the type of equipment and repairs that you are going to perform.
- When cleaning areas on equipment that may pose a fire hazard, use *only* cleaning agents that are not flammable.
- Use approved wheel chocks on both sides of the drive wheels to prevent accidental movement of the equipment. After performing some types of repairs or service work, starting the equipment for the first time may cause an unexpected loss of drive, which can cause a loss of dynamic braking and lead to an unexpected freewheeling condition on some types of equipment. Always consult the manufacturer's technical documentation for the proper procedures.
- When starting up the equipment for the first time after a repair or service procedure to a driveline, extreme caution must be used when dealing with rotating shafts. Keep your hands clear of possible rotating shafts on initial start-up.
- Always wear approved eye protection and gloves when you perform equipment maintenance or service.

Driveline Inspection

Driveshaft inspection needs to be performed on a regular basis because driveline failures on equipment often lead to a loss of drive, exposing the operator to potential injury and/or damaging to the equipment. When equipment comes in for service or repair you can limit driveline performance issues by performing a few inspection procedures each time the driveshafts are lubed. Driveline vibrations and U-joint failures are often caused by loose **end yokes,** slip spline play, bent driveshaft tubing, and excessive radial play. Also check the drivelines visually for twisting, dents, missing balance weights, and build up of dirt on the shaft leading to balance problems.

Regular inspection of the driveshaft during routine lube intervals will pay off in fewer unexpected breakdowns. The most common cause of U-joint and slip joint problems can be traced to lack of proper lubrication. The importance of proper lubrication must be emphasized, as many service personnel do not follow the recommended procedures. Enough grease must be injected to ensure that the old grease from the U-joint or splines is flushed out because it may contain grit that acts as an abrasive when mixed with the grease.

Tech Tip: Before you attempt to grease driveshaft fittings, clean the nipple carefully. This prevents any dirt from being forced into the bearing during re-lubing.

Lubrication

The importance of using the correct type of lubrication on drivelines cannot be overemphasized on off-road equipment. Lithium soap-based extreme pressure (EP) grease that meets the National Lubricating Grease Institutes (NLGI) classification grade 1 or 2 specifications should be selected. Typically, grade 3 and 4 greases are not recommended for use in cold weather. If the bearings are sealed they are pre-lubed with synthetic grease and do not require any additional greasing. When replacing a U-joint, note that the grease in the U-joint is only there to protect it during shipping and storage. Lubing the U-joint after installation is a service requirement and should never be put off. Lubrication schedules can vary greatly depending on operating conditions. Off-road equipment generally requires that the severe-duty cycle recommendations on lube charts be followed. Off-road classifications are generally considered to apply to equipment that operates on unpaved roadways for 10% or more of the time.

U-Joint and Slip Joint Lubrication

Proper lubrication of the driveshaft U-joints and slip joint is extremely important to the life expectancy of the driveshaft (See Table 14-1). When applying grease to the U-joint it is always a good idea to flush the old grease from the bearing caps. Pump grease through the fitting slowly until the new grease flows from each of the four caps. This ensures that any dirt that has gotten into the caps has been flushed out and replaced with fresh grease. If grease does not flow from all four bearing caps, try moving the driveshaft U-joint from side to side while applying grease.

If this does not resolve the problem, you may have to remove the U-joint to check the condition of the bearing cap and needle bearings. Under no circumstances should this condition be ignored; this will always lead to a driveshaft failure. The price of a U-joint is a small price to pay compared to the cost of a driveshaft. If driveshaft separation occurs while the equipment is moving, extensive damage may occur to not only the driveshaft, but also to any surrounding components, such as transmissions and drive axles. Whenever U-joints are lubed, it makes sense to lube the slip joint also. The same technique used on the U-joints applies to the slip joint.

Flush fresh grease through the grease fitting until the grease comes out the relief hole in the end cap. Block the relief hole and continue to flush the grease through the slip joint until the grease appears at the **slip yoke** seal area. It may become necessary to remove the end cap to perform a complete flushing of the slip joint.

Using the appropriate grease for the application is important. For example, in extremely cold weather the use of a cold-weather-rated grease will ensure that the grease does not freeze in the splines and cause unnecessary wear or damage. Pillow block bearings or hanger bearings usually come sealed. If no grease fitting is present, refer to the manufacturer's recommendations for proper service procedure. If it becomes necessary to replace the hanger bearing, be sure to note the shim pack thickness used under the hanger bearing or pillow block. The shim pack ensures the correct location of the bearing assembly to maintain correct driveshaft angles.

Table 14-1: LUBRICATION SCHEDULE

Type of Service	Miles	Time
City driving	5,000 to 8,000	3 months
On highway	10,000 to 15,000	1 month
Extended haul	50,000	3 months
Severe usage off road	2,000 to 3,000	1 month

U-Joint Replacement

Replacement of universal joints is considered a routine procedure in a service shop and does not require any special tools other than a U-joint puller and press, which should be part of a service shop's tool inventory. In some cases, replacement of the U-joint may require you to support the driveshaft to prevent it from dropping unexpectedly. Before performing this work refer to the manufacturer's service procedure in the technical documentation to ensure that the work is performed correctly.

Tech Tip: To prevent the possibility of a eye injury, always wear safety glasses when performing any work on the equipment. When it becomes necessary to use a hammer to dislodge a bearing cap or U-joint from a yoke, use a soft metal hammer, like copper or brass, to prevent damage to the hardened parts. Serious personal injury or damage can occur if safety procedures are not followed correctly. The equipment should always be blocked with the approved wheel chocks to prevent accidental movement when a driveline is disconnected.

U-Joint Installation

Hardened hammers have a place in a technician's tool box, but they should not be used to remove a U-joint from a yoke. Damage will often result from the use of this tool. If the use of a hammer is required, use a brass or copper hammer. This will prevent any chance of damaging the forward yoke or hardened bearing cups. When a U-joint replacement is performed, it is important to realize that serious damage can occur to the driveshaft assembly if excessive force or blows are used. The price of a U-joint is insignificant compared to the price of a driveshaft. If driveshaft separation is required to service a U-joint, mark the slip splines and tube shaft with a scribe to ensure correct alignment when assembling the two halves together; this will maintain correct phasing. When replacing U-joints with the driveshaft mounted in a vice, be sure not to exert excessive clamping force on the drive tube; this will distort or bend the tube, causing premature driveline or U-joint failure.

There are a number of U-joint mounting designs that are used on equipment, as shown in **Figure 14-20;** only the most common types used will be discussed in detail in this section.

The wing-type U-joint, as shown in **Figure 14-21,** is the most common design used on off-road driveshafts; and the installation procedures will be discussed here in depth. Proper blocking of the equipment is essential to performing the work safely. If you intend to remove the driveshaft, it is necessary to gather the correct blocking or slings to support the shaft once it is disconnected from the yoke. Support the driveshaft with the appropriate blocking or sling, loosen the four cap screws and remove them from the yoke end of the driveline. While supporting the driveline yoke, pry the U-joint from the component end yoke. If a hammer blow is required to loosen the driveline from the end yoke, use only a brass, copper, or soft blow hammer. The use of heat to loosen a U-joint cap from the yoke is discouraged; it will ruin the U-joint. If a slide hammer is required to remove the U-joint from the driveshaft, bear in mind that damage to the driveshaft will occur if excessive force is used. Although it is not always possible to dislodge U-joints from driveshafts without the use of some force, the use of large hammers is discouraged; damage to the driveshaft occurs most often with this method. Wing-style U-joints are generally permanently assembled with steel straps welded by the manufacturer during assembly. Unless you are discarding the U-joint, do not cut these straps; they protect the U-joint from misalignment when fitting to the yoke. Permalube U-joints have no grease fittings and do not require greasing. If a grease fitting is present, then the U-joint must be greased with the appropriate grease following the lube procedure outlined by the manufacturer.

Before installing the new U-joint measure the distance between the yoke ears with a yoke gauge tool to ensure that the correct U-joint is being installed. See **Figure 14-22.** This fit is critical to proper operation and life of the U-joint assembly. To install the U-joint back onto the driveshaft, tap the bearing caps lightly with a soft blow hammer until the keyway is bottomed out in the yoke. Install new cap screws and hand tighten them to the yoke. Do not apply lubrication or anti-seize compound to the yoke saddle of the

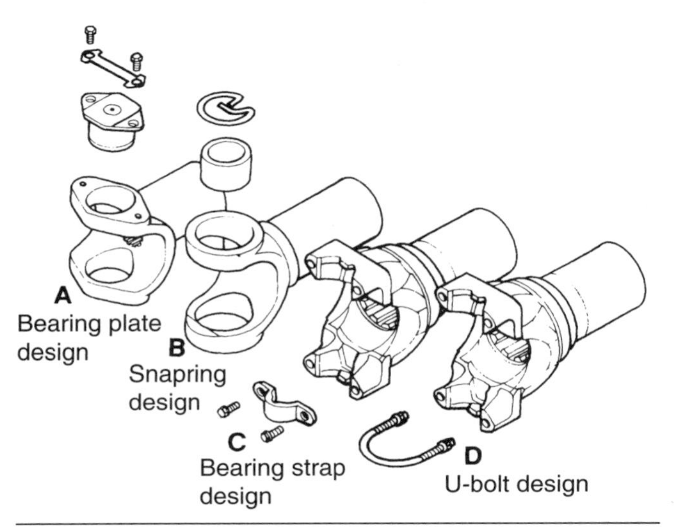

Figure 14-20 Four types of mounting joints used on heavy equipment. (*Courtesy of Spicer Universal Joint Division/Dana Corporation*)

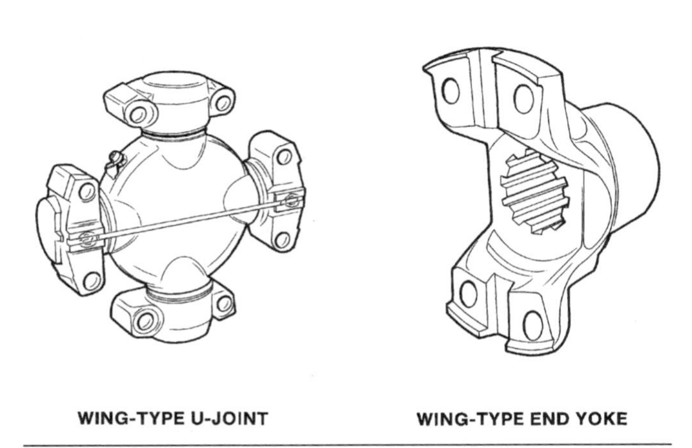

Figure 14-21 Typical wing-type U-joint used on off-road equipment.

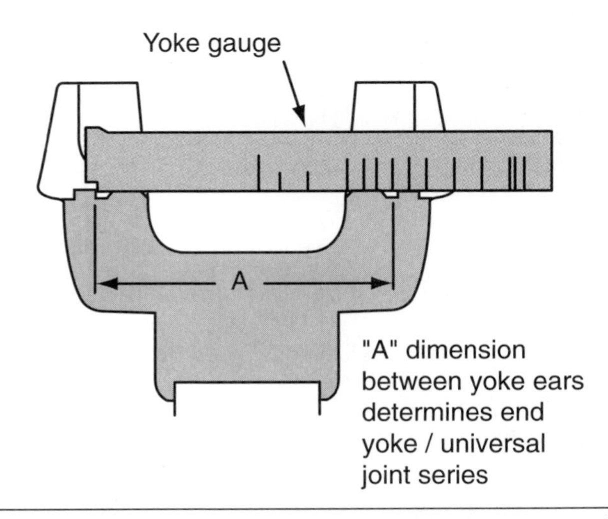

Figure 14-22 Measuring the distance between the yoke ears to determine correct U-joint dimension.

connection. Always use the new bolts that are supplied with the U-joint.

Note: Some manufacturers apply a lock patch to the bolt threads, making it necessary to wrench the bolts in. Do not remove this coating from the threads. Use a torque wrench to alternately tighten the bolts to the proper specifications.

The torque chart shown in **Table 14-2** is provided as an example: refer to the U-joint manufacturer's specifications for the proper bolt torque. After installation, be sure to lubricate the U-joints until the grease flows from all four trunnion bearings. If the grease will not flow from all four trunnions, shake the assembly up and down and side to side while trying to apply grease. If this does not work, loosen the bearing cap capscrews and apply grease until all four trunnion bearings purge grease. Re-torque the capscrews to the correct torque specifications to complete the installation.

Bearing Plate U-Joint Replacement

As with all U-joint removal procedures follow the safety precautions listed below. To prevent any possibility of eye injury, wear safety glasses when performing this type of work. If it becomes necessary to use a hammer to dislodge a bearing cap, or a U-joint from a yoke, use a soft blow hammer. A soft metal hammer, like copper or brass, may be used to prevent damage to the hardened parts. Serious personal injury or equipment damage can occur if safety procedures are not followed correctly. The equipment should always be blocked with the approved wheel chocks to prevent accidental movement when a driveline is disconnected.

- Expose the bolt heads by bending back the lock tabs before removing the bolts that hold the plate to the cap.
- Install a suitable U-joint two-jaw puller, as shown in **Figure 14-23** to remove the cap plates.
- Push the yoke to one side to facilitate the removal of the bearing cap from the yoke bore.
- Slide the yoke to the opposite end and remove the opposite bearing cap from the yoke bore.
- Remove the U-joint from the yoke and repeat the last two steps to completely remove the U-joint.
- Carefully install the new U-joint cross into the yoke with the bearing caps removed.
- Install the bearing caps carefully through the yoke onto the cross joint trunnions, being careful not to dislodge any needle bearings in the process.
- If necessary, use a soft blow hammer to lightly seat the bearing caps in place before hand tightening.
- Use a suitable torque wrench to tighten the bolts using an alternating pattern; then repeat the previous three steps to install the U-joint to the **weld yoke.**
- Always use the new bolts that are supplied with the U-joint. Some manufacturers apply a lock patch to the bolt threads, making it necessary to wrench the bolts in. Do not remove this coating from the threads.

Table 14-2: TORQUE CHART

U-Joint	Name	Description	Torque
Wing style	Capscrew	$1/2 - 20 \times 2 1/2$	115–135 ft-lbs 155–183 nm
	Capscrew	$1/2 - 20 \times 1 1/2$	115–135 ft-lbs 155–183 nm
		3/8-24	40–50 ft-lbs 54–4 nm
		7/16-20	63–83 ft-lbs 85–112 nm
	Jam nut	1/2-H.D.	63–83 lb ft 85–112 nm

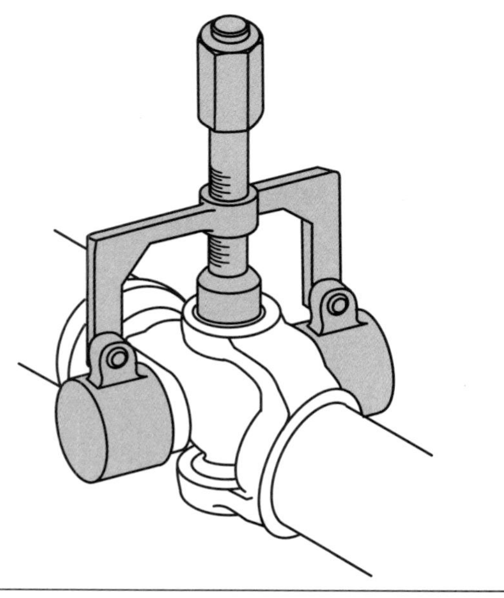

Figure 14-23 Universal joint puller used to remove plate-type U-joint cap plates.

- Refer to the U-joint manufacturer's specifications for the proper bolt torque.
- After installation, be sure to lubricate the U-joints until the grease flows from all four trunnion bearings. If the grease will not flow from all four trunnions, shake the assembly up and down and side to side while trying to apply grease.
- If the above procedure does not work, loosen the bearing cap capscrews and apply grease until all four trunnion bearings purge grease. Re-torque the capscrews to the correct torque specifications to complete the installation.

Snap Ring Design U-Joint Replacement

As outlined in the previous procedure, a technician should follow the safety precautions listed below. To prevent any possibility of eye injury, always wear safety glasses when performing this type of work. If it becomes necessary to use a hammer to dislodge a bearing cap, or a U-joint from a yoke, the use of a soft blow hammer or a copper or brass hammer should always be used to prevent damaging the bearing caps. Serious personal injury or equipment damage can occur if safety procedures are not observed. Equipment should always be blocked with the approved wheel chocks to prevent unexpected movement when a driveline is disconnected.

- Remove the snap rings from the bearing caps on all four U-joints. It may be necessary to lightly tap on each of the bearing caps to relieve the tension on the snap rings.
- Use the correct removal tool for round bushing caps; using a different cap installation tool will damage the bearing caps. The use of the following three approved removal and installation procedures is highly recommended to prevent damage to the cups.

Removing Bearing Caps with a Bridge and Bearing-Cup Receiver Tool

- Place the U-joint into the bearing-cup bushing receiver being careful to align the bearing cups correctly in the receiver.
- Use the press to force the bottom cup out of the yoke, as shown in **Figure 14-24.**
- Turn the shaft end for end and repeat the last procedure until the cup is free of the yoke.
- The U-joint is now free to be removed from the yoke.

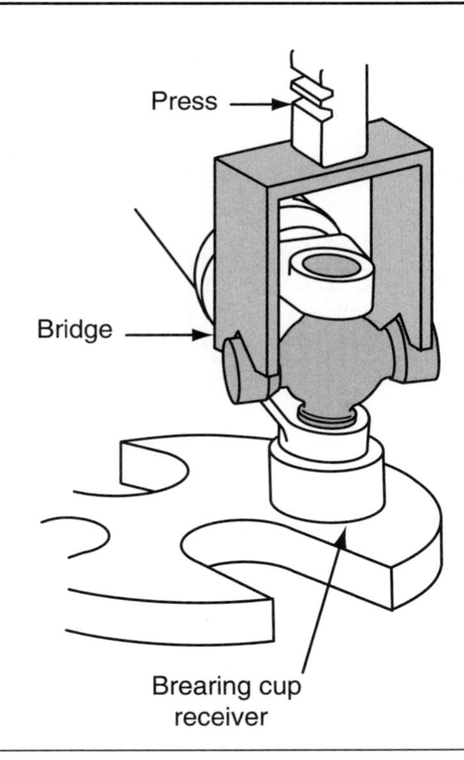

Figure 14-24 Removing bearing caps using a press, bridge, and bearing cup receiver.

Removing a U-joint with a Universal Joint Press

- Carefully place the universal joint press on to the U-joint, as shown in **Figure 14-25.**
- With an appropriate wrench, turn the screw clockwise until the bearing cap becomes free, then turn counterclockwise until the bearing cap can be removed.
- Turn the U-joint end for end and repeat the procedure to remove the other two bearing caps and remove the U-joint from the yoke.
- Using a wrench, turn the adjustment screw clockwise until the bushing loosens, then turn it counterclockwise until the bushing can be removed.
- Turn the U-joint end for end and repeat the procedure for the other two bushings and remove the U-joint from the yoke.

Installation of Replacement Universal Joints

- Carefully remove two bearing cups and thread the joint thought the yoke.
- Replace one bearing cup on the trunnion and position the cross to accept the second bearing cup from the outside of the yoke.

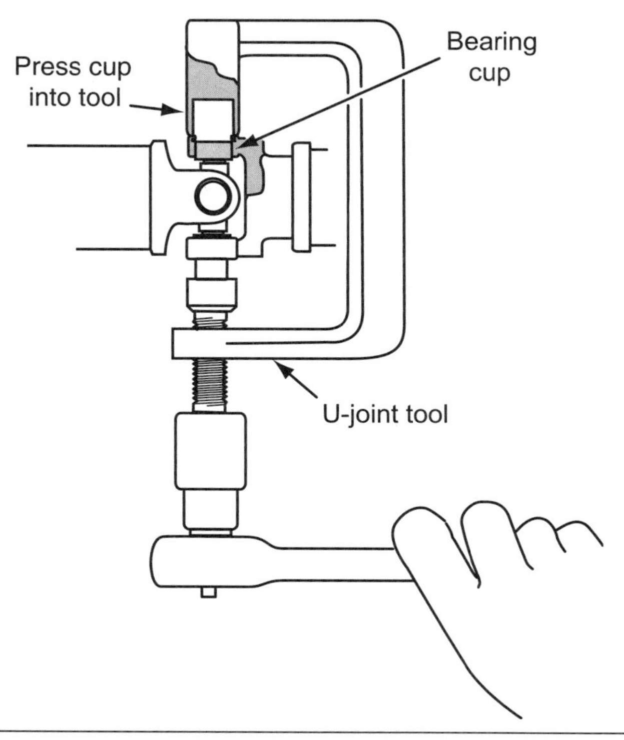

Figure 14-25 Removing bearing caps using a universal joint press

- At this point, you can use either the U-joint press or the yoke bearing-cup installation tool to press the first bushing into the yoke just past the snap ring groove. This will allow you to install the first snap ring in position.
- After the second cup is pushed into the appropriate location, install the second snap ring. Be sure the snap ring is fully seated into position.
- Use a soft blow hammer to strike the yoke ears to ensure the bearing cups are centered and the universal joint is not binding.
- Hand tighten the cap screws on the bearing plate or strap design and torque up to manufacturer's specifications.
- Although the U-joint is factory pre-lubed for shipping purposes, this is inadequate for actual operating conditions. You must be sure to lube it until grease appears at all four bearing caps.

Driveshaft Installation

Once the U-joints have been installed, the driveline should be assembled correctly before installation on the equipment. Examine the driveshaft tubing for any damage that may have occurred while it was disassembled for servicing. Look for things like missing weights, evidence of twisting, or dirt build up, which may cause operating problems. The following is a general list of potential problems to look for when installing a driveline on equipment:

- Dents or irregularities on the tube surface of the driveline need to be examined and a determination on its continued use should be made.
- The splines should not bind when moved in and out in the driveline's normal travel range.
- Verify that the slip joint seal is in good condition and will not become a source of contamination to the slip splines. (If in doubt, replace the seal. It's a small price to pay compared to the price of a driveline.)
- The U-joint crosses must not be binding or have tight spots in their normal travel range.
- Check the yoke flanges carefully for any signs of damage, such as burrs, which will interfere with a proper fit. Often parts are painted to for aesthetic reasons and to prevent rusting. The paint should be removed from the mating surfaces of the yoke; it prevents proper seating of the parts during assembly.
- If there is evidence of driveline twisting on the tube, alignment of the yokes needs to be verified.

Driveline Balance

The problem of driveline vibration is not as apparent on off-road equipment as it is on high-speed road equipment, but it does take its toll on the driveline components. Operators are not likely to complain about vibration on off-road equipment. Therefore, it is important for a technician to spot any problems during servicing that would lead to early driveline component failure. Although the speeds of the drivelines are relatively low due to low road speeds, imbalance can still cause problems for components such as U-joint trunnion bearings, driveshaft hanger bearings, and bulkhead bearings. A badly unbalanced driveline can cause bending movements to occur, which may lead to early driveline failure. If imbalance is suspected on a particular driveline, it may become necessary to have the driveshaft dynamically balanced at a shop that specializes in this type of procedure. Vibration on off-road equipment with an active suspension makes it next to impossible to make balance corrections to a driveline while on the equipment. If vibration damage shows up on off-road drivelines, the working angles should be verified on affected drivelines.

Measuring and Recording Driveline Angles

The following procedure may be used for off-road equipment. The procedures outlined here are intended for use as examples and should not be used as actual procedures on any particular equipment. Always refer to the manufacturer's technical documentation for the equipment that you are working on for proper procedures and specifications. Before you start any work on equipment, whether it is servicing or repair, you need to familiarize yourself with a few general safety precautions. The area being serviced should be high-pressure washed before work begins to minimize the chance of inaccurate measurements. Be sure that an approved set of wheel chocks are in place before you begin measuring.

Tools Required

- A good quality electronic **inclinometer** or spirit-level protractor is needed to measure drive-line angles.
- A tape measure is needed to measure ride height or component height.
- A photocopy of a driveline worksheet is very helpful.

Procedure

- If the equipment has pressurized tires, be sure to inflate them to the recommended values before beginning this procedure. The equipment should be checked on level ground front to rear as well as from side to side.
- It is necessary to place the output yokes on driveshafts that are to be checked in the vertical position.
- Use a magnetic-base protractor or an inclinometer to measure the driveline angles. The use of a high quality electronic inclinometer is recommended for this type of work. Accuracy is very important.
- Before you measure any angles determine the components angles. For example look at the components and determine whether the front of the component is higher than the rear, or the rear is higher than the front.
- Determine the driveline phasing by checking the examples in **Figure 14-26** and record on a sheet of paper.
- The first measurement taken and recorded on a sheet of paper should be the transmission output yoke angle, as shown in **Figure 14-27.** If the yoke is inaccessible, use a flat section of the transmission case, such as a countershaft cover. Record the reading on a sheet of paper.
- Measure the driveline angle by placing the inclinometer on a flat section of the driveline tubing to measure the first two driveshaft readings, as shown in **Figure 14-28.** Be sure that your inclinometer is not resting on any dirt or on a weld joint. Record the data on a sheet of paper.

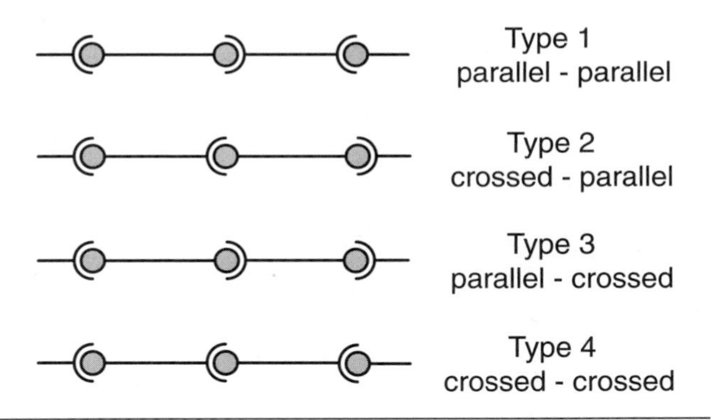

Figure 14-26 Example of a driveline phasing configuration.

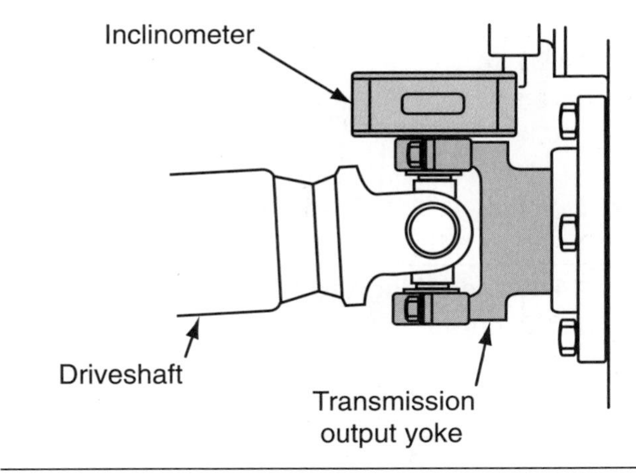

Figure 14-27 Measuring the transmission output yoke angle.

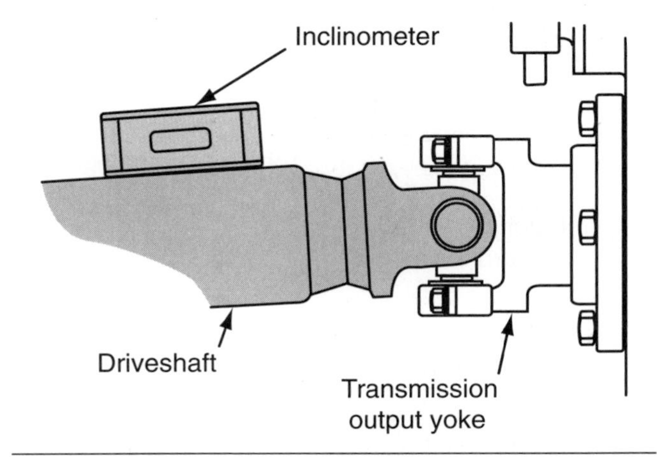

Figure 14-28 Measuring the driveline angle with an inclinometer.

- Place the inclinometer on flat portion of the drive axle or on the output yoke if possible (this will depend on the type of U-joint being used), and record the angle on a sheet of paper.
- Place the inclinometer on the interaxle driveline, as shown in **Figure 14-29,** and record the angle on a sheet of paper. If there is no room to place the inclinometer on the driveline tube, then place it vertically and subtract 90 degrees from your reading.
- Place the inclinometer on the input yoke of the rear axle. (This will depend on the type of U-joint being used.) *Note:* a spacer may be required, or you may have to place the inclinometer on a flat portion of the axle tube to obtain a reading.

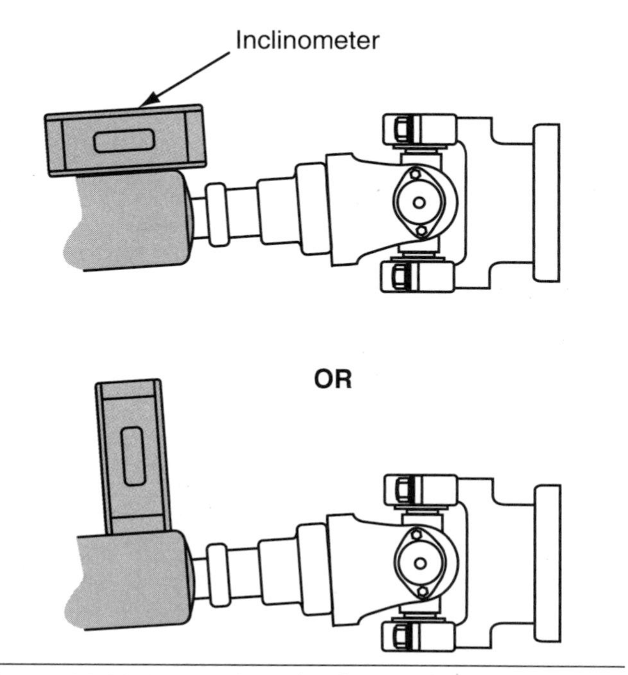

Figure 14-29 Measuring the forward rear drive axle angle on the output yoke.

Making Corrections to the U-Joint Operating Angles

The methods that are required to make changes to the driveline working angles on off-road equipment will depend totally on the type of powertrain used. If the equipment has an active suspension system such as Air Ride®, a slightly different procedure must be followed. Before you proceed to measure the working angles on equipment, check the manufacturer's recommendations for performing this type of driveline evaluation. **Figure 14-30** shows the location of angles that need to be measured.

Equipment that has leaf-spring suspension uses axle shims to change the working angle of the components. These are generally installed on the axle saddle between the leaf springs and the axle. This adjustment tilts the axle to change the working angles on the driveline.

On some off-road equipment, the components are solidly mounted to the frame; in this case the component is shimmed between the mounting points to change the working angles. Equipment that has components that

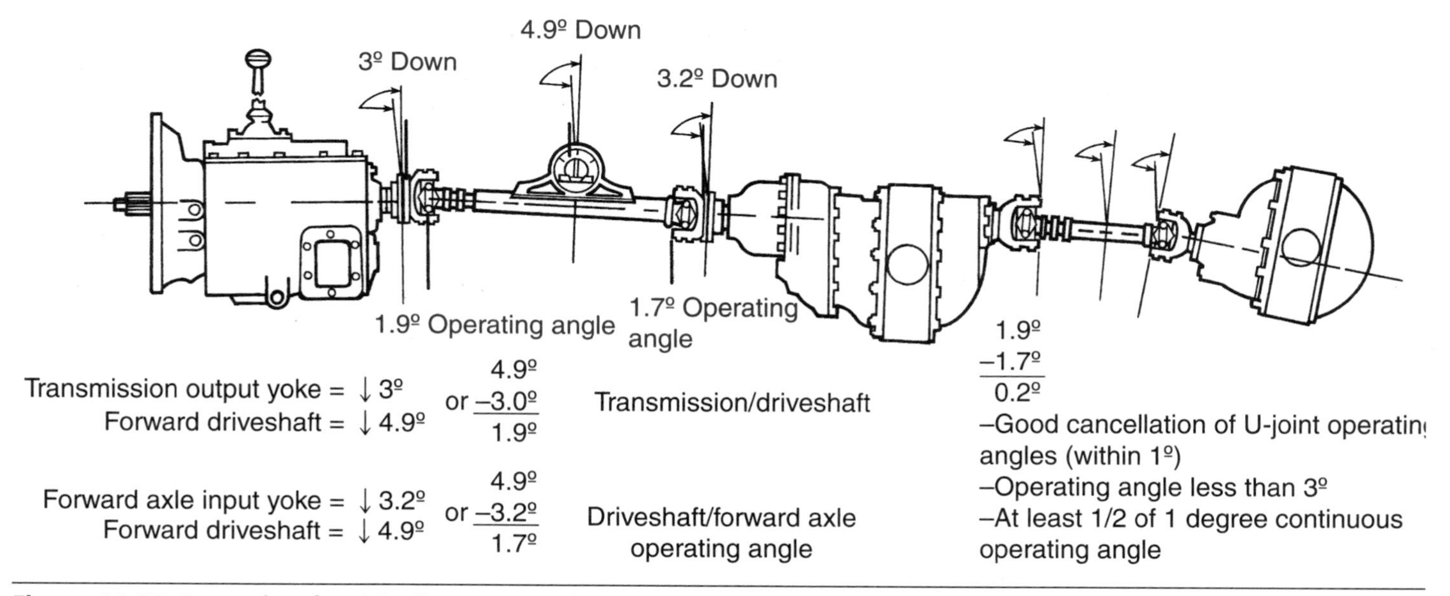

Figure 14-30 Example of a driveline system showing completed measuring procedure along with the calculations necessary to evaluate the working angles. (*Courtesy of Dana Corporation*)

are solidly mounted to the frame, such as front end loaders, generally has very few problems with changes in working angles. The same can't be said for equipment that uses some form of live suspension.

The U-joint operating angles can be changed by things such as worn bushings in the spring hangers or incorrect air bag height dimensions. Changes to the chassis from overloading or wear and tear on the engine and transmission mounts can account for undesirable working angle changes.

Driveline run-out has a negative effect on vibration and driveline component wear, as shown in **Figure 14-31.** Checking the driveline run-out is a relatively simple procedure and can be done in the field with only a few simple tools. The one stipulation that must be adhered to is that the driveline must be clean in the area where the run-out readings are to be performed.

If run-out is detected while measuring driveline uniformity, a maximum tolerance of 0.010 is generally allowable, but as always, refer to the manufacturer's technical documentation for specific applications.

When trying to pinpoint a run-out problem on a driveline, one factor that could play a major role in maintaining ideal working angle is yoke run-out. A dial indicator is mounted to the transmission housing close to the yoke positioned on a machined surface of the yoke near the yoke shoulder, as shown in **Figure 14-32.** Different designs of U-joints will necessitate a creative approach to obtaining this measurement. Take a reading on one end of the yoke, then rotate the yoke 180 degrees and take a second reading. A good rule is not to exceed .005″ run-out. Always check the manufacturer's technical documentation for specifications.

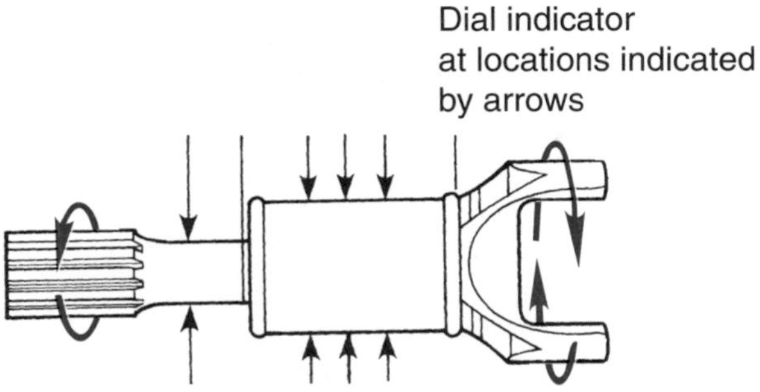

Dial indicator
at locations indicated
by arrows

Figure 14-31 Arrows indicate the locations to perform measurements required to access the driveline run-out condition of the driveline. (*Courtesy of Dana Corporation*)

Vertical Alignment Check of a Yoke

Checking the vertical alignment is performed by using a protractor held against the machined face of the yoke, as shown in **Figure 14-33.** Ensure that the equipment is on level ground and that the component and its yoke are properly torqued. A loose mount will give a false reading, leading to a bad diagnosis. To perform this check, the driveshaft must be disconnected. Access to the yoke face is required, as well as the ability to rotate the yoke to a vertical position to obtain the reading. A protractor with a magnetic base and attached to the yoke should give a reading of 90 degrees minus the component offset angle. The reading should be within 1/2 degree of angle of the component to which the yoke is attached. Should the reading be higher than specified, the yoke must be replaced.

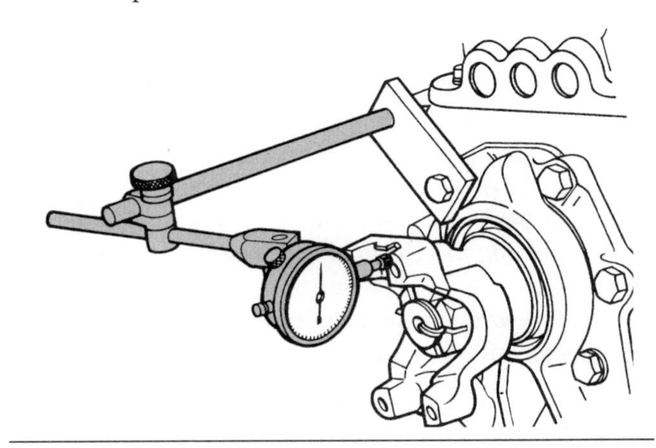

Figure 14-32 When performing these run-out checks, be sure that the pointer is on a machined surface. (*Courtesy of Dana Corporation*)

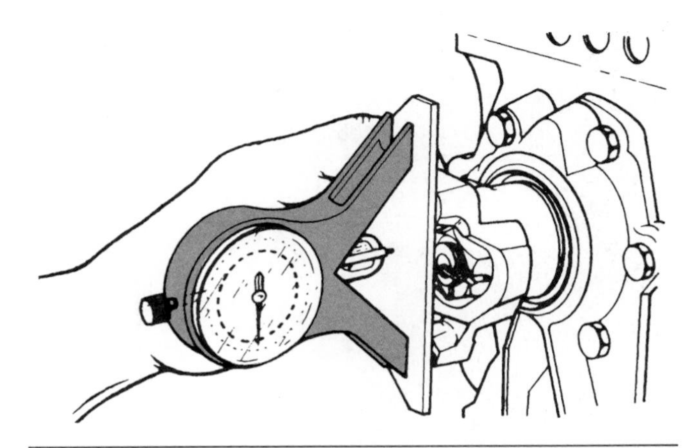

Figure 14-33 Magnetic-based protractor used to check vertical alignment of the yoke. (*Courtesy of International Truck and Engine Corporation*)

FAILURE ANALYSIS

Although a large number of driveline failures can be identified and explained just by performing a detailed visual inspection of the failed parts, certain types of driveline failures require the use of specialized equipment and the knowledge to determine the cause. Often the failures can be attributed to poor equipment operation, which is often the result of operators not being aware of the damage that can occur from certain operating practices. The second issue we need to be aware of is the use of driveline components that do not meet the application requirements for a specific use. Driveline components are manufactured to meet a large variety of applications; the need to be selected on this basis. A third issue to be discussed is the use of poor maintenance practices, which have sometimes become accepted standards in some shops.

We will examine the most common types of failures and attempt to explain what occurred to cause the failure. Driveline failures occur in many different forms, but in the majority of cases, U-joints are the most commons causes of driveline failures. Whenever driveline service is performed, a technician should spend some time inspecting the U-joints and yokes for any signs of problems.

The following list is representative of the conditions a technician should look for when examining a U-joint for re-use:

- **Galling.** This occurs when friction between two surfaces causes metal to be displaced.
- **Spalling.** Fatigue causes flakes of metal to break off, as shown in **Figure 14-34.** This condition is associated with splines as well as trunnion races.
- **Brinelling.** Surface fatigue causes grooves to wear into a surface, as shown in **Figure 14-35.** It is often associated with improper U-joint installation. Watch for false brinelling, which looks like a polished surface rather than actual grooves worn into a surface.
- **Pitting.** This is caused by corrosion that forms little pits in the metal leading to surface wear and parts failure.
- **Cracks.** These are an indication of metal fatigue, which stars out as tiny stress cracks. Eventually this condition will weaken the metal until it finally breaks, as shown in **Figure 14-36.**

Figure 14-35 Example of brinelling on the surface of a U-joint trunnion bearing surface. (*Courtesy of Arvin Meritor*)

Figure 14-34 Example of spalling on a U-joint trunnion bearing surface. (*Courtesy of Arvin Meritor*)

Figure 14-36 Hairline cracks in metal finally lead to complete fracture of the U-joint cross. (*Courtesy of Arvin Meritor*)

Tables 14-3 through **14-9** show various conditions, their possible cause, and suggested corrective actions used to address driveline failures.

Table 14-3: PREMATURE WEAR PROBLEMS		
Condition	**Cause**	**Corrective Action**
Premature U-joint wear	Cross hole misaligned on end yoke	Check the alignment of the cross holes on the end yoke with an alignment bar Replace if misaligned
	Excessive angularity	Measure U-joint operating angles Reduce the angles if excessive
	Lubrication frequency Inadequate or incorrect lube for the application	Check lubrication and application requirement to Ensure correct lube is being used
	Damaged or worn seals	Replace U-joint
Repeated U-joint wear	Excessive heavy operating loads	Use a higher capacity U-joint to carry heavy load
	Operating at excessive angles and speeds	Install a high capacity heavy-duty driveline and U-joint
	Worn or damaged seals	Replace universal joints
	Inadequate lubrication intervals or incorrect grade of lubricant	Lubricate according to manufacturer's recommendations
End galling of trunnion and bearing assembly	Driveline operating at excessive angle	Measure driveline angle and reduce if they are excessive
	Torque load to high for driveline U-joint size	Install a higher capacity driveline U-joint
	Lubrication intervals are inadequate or wrong grade of lubricant being used	Lubricate according to manufacturer's recommendations
Needle rollers brinelled into bearing cup and cross	Extremely heavy running loads	Install a higher capacity driveline U-joint
	Insufficient operating angles	Increase the operating angles to at least 2 degrees
	Normal appearing bearing wear	If brinelling occurs in a small area, it is not necessary to replace the U-joint
	Lubrication intervals are inadequate or wrong grade of lubricant being used	Lubricate according to manufacturer's recommendations
Spalling of needle bearing into trunnion surfaces	Lubrication intervals are inadequate or wrong grade of lubricant being used	Lubricate according to manufacturer's recommendations

(continued)

Table 14-3: (CONTINUED)

Condition	Cause	Corrective Action
	Contamination from dirt	Check U-joints for wear replace if necessary If brinelling occurs in a small area, it is not necessary to replace the U-joint
	Normal bearing wear	Worn components should be replaced
Bearing or cross assembly broken	Excessive torque loads for the size of driveline and U-joints	Replace with a higher capacity driveline U-joint

Table 14-4: YOKE FRACTURE PROBLEMS

Condition	Cause	Corrective Action
Yoke broken or cracked	Yoke lug interference at full jounce or rebound	Use high angle yokes. Check the design application
	Excessive torque loads for the size of driveline and U-joints	Replace with a higher capacity driveline U-joint
	Fatigue from bending caused by high secondary couple loads	Reduce the universal joint running angles

Table 14-5: UNIVERSAL JOINT CENTER PROBLEMS

Condition	Cause	Corrective Action
Trunnion or U-joint cross fracture	Excessive high loading	Overloading or abuse of vehicle
Fractured bushing	Excessive loading of U-joint	Check for excessive driveline torque in low gear. Use high capacity U-joints
	Excessive operating angles	Check for high working angles reduce if necessary
	Worn or damaged parts	Replace parts if worn excessively

Table 14-6: WING-STYLE U-JOINT PROBLEMS

Condition	Cause	Corrective Action
Loose mounting bolts	Dirt or paint on the mounting pads	Check to see if the mounting pads are fretted or loose drive tang
Broken mounting bolts	Overtorqued or undertorqued mounting bolts Fretting on the mounting pads or around bolt holes with fretting on the drive tang	Yoke mounting pads must be free of dirt and bushing must be fully seated before torque is applied

(*continued*)

Table 14-6: (CONTINUED)

Condition	Cause	Corrective Action
	Fretting on the mounting pad or bolt hole the bolt	Loose bolt
	Excessive driveline angularity	Check for excessive driveline operating angle and reduce the angle if necessary

Table 14-7: ROUND BUSHINGS PROBLEMS

Condition	Cause	Corrective Action
Difficult to remove or replace bushings	Yoke bushing hole distorted Corrosion caused by fretting with rust build up	When removing the U-joints be careful not to distort the cup holes when using a hammer on the center cross
New center cross will not install in the yoke	Yoke ears are distorted from abuse and are binding the parts	Yoke must be replaced

Table 14-8: SLIP YOKE SPLINE WEAR PROBLEMS

Condition	Cause	Corrective Action
Seized splines	Not using correct lubrication on splines	Lubricate according to manufacturer's recommendations
	Worn or damaged splines	Splines need to be replaced
	Contamination	Lubricate the splines according to manufacturer's recommendations
Galling	Splines are worn or damaged	Mating splines need to be replaced
	Contamination	Lubricate the splines according to manufacturer's recommendations Check the condition of the spline seal
Outside diameter is worn excessively at the extremes	Wrong lubrication	Should use a larger diameter tube
	Extremely loose outside diameter fit	Spline components need to be replaced
Spline shafts or tube is broken in torsion	Tube size inadequate	Replace with a larger capacity driveline and U-joints

Table 14-9: SHAFT AND TUBE PROBLEMS

Condition	Cause	Corrective Action
Support bearing shaft wear	Driveline is too long for the operating speeds	Replace with a two-piece driveline with a shaft support bearing

(*continued*)

Table 14-9: (CONTINUED)

Condition	Cause	Corrective Action
	Incorrect lube of bearings	Center bearing should be replaced
Support shaft rubber insulator wear	Bending fatigue due to excessive coupling loads	Reduce the U-joint operating angles
	Too heavy a torque load for the U-joint and driveline size	Replace with higher capacity driveline and U-joint
	Support shaft bearing misaligned	Mounting bracket needs to be realigned to frame cross member
Tube circle weld is fractured	Balance weight located in the apex of the weld yoke	Tube needs to be replaced and rebalanced
	Balance weight is to close to the circle weld	Tube needs to be replaced and rebalanced
	Circle weld is defective	Tube needs to be replaced and rebalanced
Shaft is broken from bending	Driveline is to long for the operating speeds	Replace with a two-piece driveline with a shaft support bearing
	Bending fatigue due to excessive coupling loads	Reduce the U-joint operating angle

ONLINE TASKS

1. Use a suitable Internet search engine to research the configuration of an off-road drive-shaft system. Identify a piece of equipment in your particular location, paying close attention to the driveline arrangement, and identify whether it is a parallel or a non-parallel design. Outline the reasoning behind the manufacturer's choice of driveline design for that particular application.
2. Check out this url: http://training.arvinmeritor.com/onlinetraining.asp for more information on driveline configuration and design.

SHOP TASKS

1. Select a piece of equipment in your shop that has a driveline assembly and identify the type of components that make up the driveline arrangement. Identify whether or not the driveline has excessive run-out, and what type of U-joint is used on this arrangement.
2. Select a piece of mobile equipment in your shop that uses a driveline, identify the type and model of the equipment, and list all the major components and the corresponding part numbers of the driveline. Identify and use the correct parts book for this task.

Summary

- A list of the components that make up a typical driveline assembly on off-road equipment is as follows:
  - Yokes
  - U-joints
  - Slip spline
  - Slip assembly
  - Hanger bearings
  - Propeller shafts
- The transmission of power through a drivetrain is not always in a straight line. Equipment design often necessitates component placement that requires drivelines to transmit power at odd angles.

- Driveline designs can be split into two categories: parallel driveshaft installations and broken-back driveshaft (or non-parallel) installation.
- The non-parallel design is used when the zero vertical angles of the component installation cannot be met.
- Off-road equipment component placement can result in drivelines having compound driveline angles.
- Due to the slow speeds at which most heavy equipment operates, working angles on driveshafts are often much higher than on high-speed vehicles.
- Driveline speed plays an important role in determining the maximum number of allowable working angles that will be used. A universal joint has a maximum operating speed which is proportional to the working angle of the installation.
- Because of the torque loads generated on heavy equipment, ensure that working angles are within 1 degree of each other to maintain equal U-joint speed at each end of the driveshaft. This prevents uneven speed fluctuations, which can lead to premature U-joint needle bearing failure.
- The most common tubing used in the construction of heavy-duty drivelines is cold-rolled, electric-welded tubing; this is generally considered to be standard factory build on most off-road equipment.
- In actual operation with the drive yoke rotating at a constant speed, driveline speed will increase and decrease once each revolution.
- Yokes are welded to each end of the driveshaft to accept the U-joints, which are required to compensate for the offset that exist between the engine, transmission, and drive axles.
- The splines on the slip joints are made from high-quality hardened steel. The male and female ends have matching splines to provide a tight fitting connection between the two ends of the driveshaft. The machined slip joints are welded to the ends of each shaft.
- Verify that the slip joint seal is in good condition and will not become a source of contamination to the slip splines. (If in doubt, replace the seals; it is a small price to pay compared to the price of a driveline.)
- If run-out is detected while measuring driveline uniformity, a maximum tolerance of 0.010 is generally allowable. Always refer to the manufacturer's technical documentation for specific applications.
- A driveshaft should not exceed 70 inches in overall length. To prevent whipping at high speed, a hanger bearing may be used.
- The universal joint trunnions have bearing caps with internal needle bearings mounted on the ends that allow the caps to rotate. A seal is located at the end of each cap, which keeps dirt and water out of the needle bearing during operation.
- A grease fitting is mounted to the bearing cross that has internally drilled passages, which allows the grease to reach each needle bearing without removing the caps. Some manufacturer's locate a standpipe (check valve) at each end of the bearing cross.
- The wing-type U-joint is the most common design used on off-road driveshafts. All wing-style, heavy-duty bearing caps have a keyway that fits into a corresponding keyway in the yoke, making for an extremely high torque-carrying connection. The wing joints come with special high grade cap screws and should be torqued to manufacturer's specifications.
- Vibration on off-road equipment may not be noticeable to the operator, but the effect of vibration on the needle bearings used in the U-joints will take its toll on U-joint life. If you were to double the working angle the life expectancy would be reduced by 50%. If you reduce the load by 50%, you double the life of the U-joint.
- Although a U-joint is factory pre-lubed for shipping purposes, this grease is inadequate for actual operating conditions. When applying grease to the U-joint it is always a good idea to flush the old grease from the bearing caps. Pump grease through the fitting slowly until new grease flows from each of the four caps.
- When possible, use a U-joint puller to separate a U-joint from a yoke assembly. When it becomes necessary to use a hammer to dislodge a bearing cap or U-joint from a yoke, use a soft metal hammer (copper of brass) to prevent damage to the hardened parts.
- Before installing the new U-joint measure the distance between the yoke ears with a yoke gauge tool to ensure that the correct U-joint is being installed. This fit is critical to proper operation and life of the U-joint assembly. Do not apply lubrication or anti-seize compound to the yoke saddle of the connection. Always use the new bolts that are supplied with the U-joint.
- Operators are not likely to complain about vibration on off-road equipment. Therefore, it is

important for a technician to spot any problems during servicing that might lead to early driveline component failure.

- Use a magnetic-base protractor or an inclinometer to measure the driveline angles. The use of a high-quality electronic inclinometer is recommended for this type of work. Accuracy is important.
- When trying to pinpoint a run-out problem on a driveline, one factor that could play a role in maintaining ideal working angle is yoke run-out.

Review Questions

1. Why must a driveshaft maintain flexibility while continuously transferring torque between powertrain components?
   A. to allow the driveline to lengthen during cornering
   B. to allow for continuous length changes in reaction to road changes
   C. to allow for contraction during load changes
   D. to allow for speed fluctuations
2. Which of the following consists of a hardened, splined stub, welded to the end of a driveshaft tube?
   A. universal joint
   B. slip yoke
   C. hanger bearing
   D. cardan joint
3. Which component contains a forged journal cross and trunnions?
   A. slip yoke
   B. center support
   C. universal joint
   D. hanger bearing
4. Which statement describes the purpose of a slip yoke on a driveshaft?
   A. slips to dampen clutch engagement and road shock
   B. allows the driveshaft to change length as differentials respond to the road profile
   C. permits the transmission slide back for service
   D. all of the above
5. Which of the following factors would apply to determining driveline diameter and tube wall thickness?
   A. operating torque load
   B. equipment weight
   C. equipment application
   D. all of the above
6. Which of the flowing design features prevent the reverse flow of hot grease in a universal joint?
   A. zerk fittings
   B. trunnions
   C. bearing cap seal
   D. standpipe
7. What term is commonly used to identify the universal joint cross in technical documentation?
   A. spider
   B. trunnion
   C. yoke
   D. spline
8. Which of the angles must be equal on a one-piece parallel joint driveshaft installation with two U-joints?
   A. All working angles on all drive shafts must be the equal.
   B. Angles of both mating yokes must be equal.
   C. Angle of the transmission output shaft and the axle pinion shaft must be equal.
   D. Working angles of its two U-joints must be equal.

9. Which of the following statements indicates a correct installation specification for a set of U-joints on a driveshaft?
   - A. maximum operating angle of 10 degrees
   - B. minimum operating angle of 0 degree
   - C. operating angles each U-joint should be within 3 degree
   - D. operating angles each U-joint should be within 1 degree
10. Which of the following operating applications would be correct when determining more frequent driveline lubrication intervals?
    - A. off-road operation
    - B. pick-up and delivery
    - C. linehaul trucking
    - D. short haul trucking
11. Which one of the failures listed below would likely occur from an incorrectly installed U-joint?
    - A. damage from galling
    - B. damage from spalling
    - C. damage from pitting
    - D. damage from brinelling
12. Which of the statements below would not be considered a cause of driveline vibration?
    - A. driveshaft out-of-balance
    - B. equal operating angles
    - C. improper phasing
    - D. unequal U-joint working angles
13. Which of the statements below would be considered the correct procedure when lubricating U-joints?
    - A. grease until you see grease come out of a trunnion seal
    - B. pump grease until it exits from all four trunnion seals
    - C. apply only two shots of grease per trunnion
    - D. hand pack each bearing cap when first installing U-joint
14. If a driveshaft is out of phase, what must have taken place?
    - A. U-joint has been assembled incorrectly
    - B. transmission yoke has been damaged
    - C. differential carrier yoke has been damaged
    - D. driveshaft has been separated at the slip splines
15. Which of the tools below would be the best method of removing a universal joint from a driveline yoke?
    - A. two-jaw U-joint puller
    - B. 12-pound sledge hammer
    - C. 25-pound slide hammer on yoke
    - D. hydraulic jack on yoke hub
16. When inspecting a U-joint failure, you see metal flakes separating from the bearing surfaces on the cross. Which cause listed below do you suspect is responsible?
    - A. brinelling
    - B. spalling
    - C. galling
    - D. false brinelling

CHAPTER

15 Heavy-Duty Drive Axles

Learning Objectives

After reading this chapter, you should be able to

- Describe the types of heavy-duty drive axles used on off-road equipment.
- Identify the construction features of heavy-duty drive axle assemblies.
- Describe the operation of drive axles used on off-road equipment.
- Identify the different differential designs used on heavy-duty drive axles.
- Identify the components of a heavy-duty off-road differential.
- Describe the operation of a heavy-duty off-road differential.
- Identify the maintenance and adjustment procedures on heavy-duty drive axles.
- Describe the common failures that occur on heavy-duty drive axles and analyze the causes.

Key Terms

amboid gear
banjo housing
bevel gear
carrier assembly
crown gear
differential
differential carrier
differential lock
double reduction axle
full-floating axle
hypoid gear set
interaxle differential
limited slip differential
live axle
no spin differential
pinion gear
planetary gear set
semi-floating axle
single reduction axle
spiral bevel gear set
tandem drive

INTRODUCTION

Heavy-duty drive axles are considered the final phase in a powertrain and are used to deliver torque to the drive wheels. Drive axles can be categorized into three distinct designs. The first design is capable of propelling the vehicle as well as carrying the weight of the load. The second is designed to carry the load and steer the vehicle. The third type is designed to carry and the support the load; this type is found on the rear section of an articulating haulage truck. Drive axles are usually mounted directly to the equipment frame and must be capable of supporting the full weight of the equipment, as well as the torque and shock loads that are generated during operation. Off-road equipment axles fall into two categories, the most common being the axle assembly that only transmits torque to the wheels and has no capability of providing steering to the equipment. The second category provides steering capability.

Articulating steering is used on equipment with an axle that has no steering capability. Articulating wheel loaders have a **differential** in each of the two axle assemblies, while larger articulating haulage trucks can have two rear axles, as shown in **Figure 15-1** with a single axle on the front section. All drive axles (sometimes called **live axles**), whether steering or straight drive, are required to change the direction of the flow of power and reduce the speed while increasing the torque to the drive wheels. The **differential carrier** is required to accommodate the change in wheel speeds when cornering, allowing the outer wheel to travel farther than the inner wheel when a vehicle corners. Axles extend out from the differential carrier to the wheels that drive the wheel hubs. On large equipment with outboard planetary drives, the axles do not carry any weight and are called **full-floating axles.** Axles that are used only to steer the equipment or carry the weight of the load are known as dead axles. Most configurations of off-road equipment use all-wheel drive due to the terrain they operate on; they rarely use a dead axle.

The second category of drive axle assembly used on off-road equipment has steering capabilities and is used on equipment where an articulating frame is not required. Because of the high torque requirement and slow travel speeds, axle assemblies used on off-road equipment come with a conventional differential **carrier assembly** and often use a second speed reduction, through the use of a planetary drive, that can be located inboard or outboard on the axle assembly. On large axle assemblies, the differential carrier is considered the primary reduction with the axles driving a secondary reduction using a planetary drive. Lubrication to both the primary and secondary drives is usually the same type of oil, but the drives often have their own oil level checks.

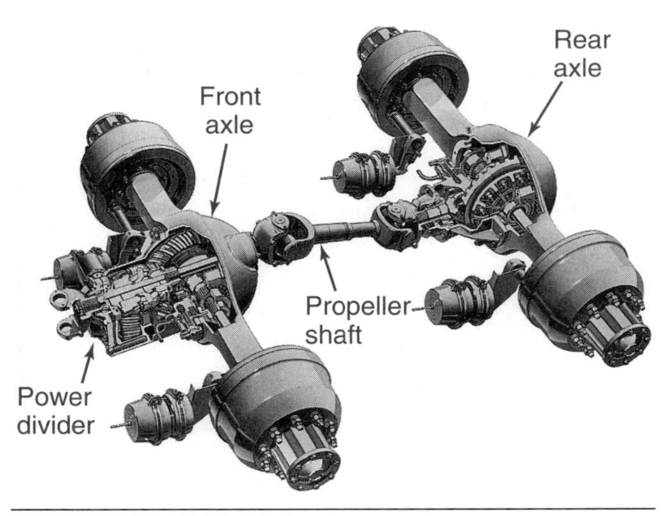

Figure 15-1 Typical tandem drive used on large articulating haulage trucks. (*Courtesy of Arvin Meritor*)

FUNDAMENTALS

Although the drive axle's primary role is to carry the vehicle load, it must also be able to provide a means of propelling the equipment. The role of the differential is crucial to the operation of the drive axle. It must split the torque coming from the transmission and send each wheel the appropriate amount of power to propel the equipment and at the same time allow the wheels to individually rotate at a different speeds when the equipment needs to turn. Each drive axle, or half shafts as they are sometimes called, sends power independently to the wheels. The power flows from the driveshaft to the pinion shaft to the ring gear (or **crown gear**), which changes the flow of power 90 degrees and splits it, sending equal amounts of power to each drive wheel. The design of the crown and **pinion gear** set allows the flow of power to be redirected at a 90° angle. The term *differential* is derived from the term *to differentiate*—to show a difference. The actual speed needs of each wheel can be split by the differential without interrupting the power sent to the wheels.

Tech Tip: One of the problems we are faced with when working with different manufacturer's documentation is the terminology that is used when addressing different parts of a drive axle assembly. For example, some manufacturers call the ring gear a crown gear or a drive axle a half shaft. The same thing occurs with the term driveshaft. Some manufactures use this term to identify the shaft that transmits power from the differential to the wheel ends. Others refer to the driveshaft as the shaft that transmits power to the differential carrier from the transmission. Although we will try to use the same terms for each consistently, you should try to get used to this lack of standardization as it is not likely to change in the near future. Most of us in the trade will work with many different OEM equipment manuals, and we will get used to each manufacturer's technical terms.

CONSTRUCTION FEATURES

The carrier assembly consists of seven major subcomponents as shown in **Figure 15-2:**

- Ring gear
- Input shaft
- Pinion gear
- Differential spider
- Differential side gear
- Differential pinion gear
- Drive axles.

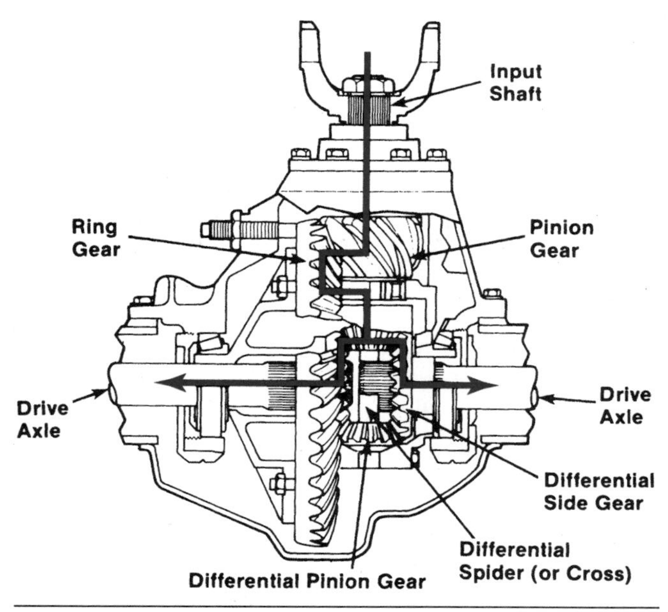

Figure 15-2 Typical differential identifying the main components and showing the flow of power to the drive axles. (*Courtesy of Roadranger Marketing. One great drivetrain from two great companies—Eaton and Dana Corporation*)

To gain an understanding of how the subcomponents work together, we will first need to examine the components that make up the carrier assembly. Figure 15-2 shows the input yoke mounted to the splined pinion shaft and secured in place with a large nut that must be torqued to specification.

Note: The torque on this nut is critical to the pinion shafts tapered roller bearing pre-load, and must be performed properly. The flow of power is transmitted to the crown gear at right angles from the pinion gear, which is in constant mesh with it.

The crown gear is bolted to the differential case halves, as shown in **Figure 15-3.** The breakdown of the parts in the differential assembly is also shown in the exploded view. The four legs of the spider are bolted into corresponding holes in the differential case halves. Four pinion gears are mounted to the spider legs, which are in mesh with the teeth of the two side gears. The differential case is held in place with a tapered roller bearings on each end that are adjusted with a pre-load when

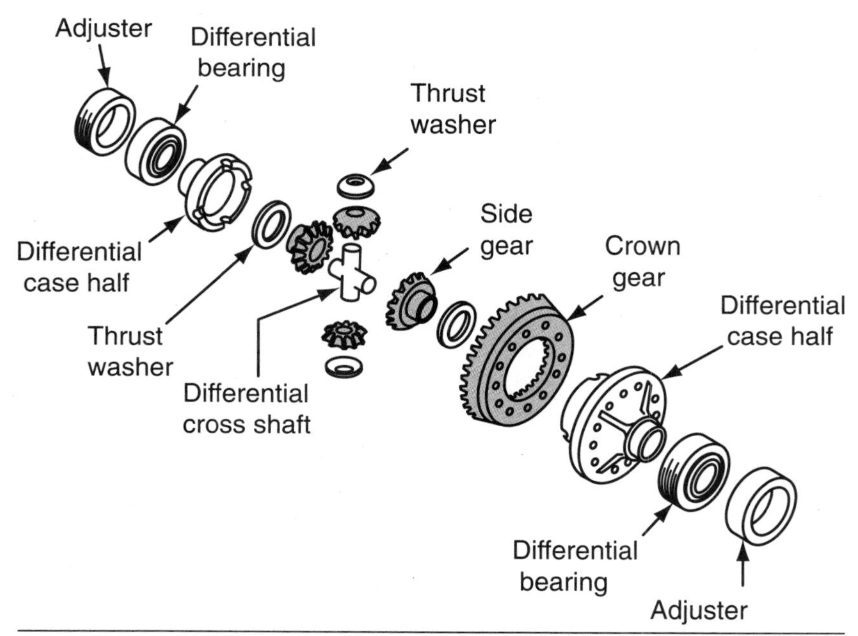

Figure 15-3 Typical breakdown of components in a differential carrier.

the running pattern is set up. The flow of power is sent to the carrier assembly from the crown gear where the flow of power is divided and sent to the axles. The side gears have internal splines which mesh with the internal splines of the axles. The carrier assembly is flange mounted to the **banjo housing.** The axles are inserted into the ends of the banjo housing and mesh internally with the side gears.

OPERATING PRINCIPLES

In the past, solid axles were used extensively and worked well when the equipment moved in a straight line with little or no turning. Tire wear became a problem if the equipment had to be turned frequently. To help prevent this excessive tire scuffing, a method was devised which permitted the wheels to travel at different speeds when cornering and still allow torque to be transferred to the wheels during cornering.

To get a clear understanding of how this differential action works, we will examine what takes place during operation. When the equipment is driven forward in a straight line, torque is transmitted through a driveshaft to the pinion shaft of the differential carrier. The torque is then sent through the ring gear, which changes the direction of power flow by 90 degrees, allowing the axles to be driven at right angles to the input power.

When torque rotates the crown gear, which is bolted to the differential case, it causes the spider to rotate. Differential pinions are mounted to the each one of the four spider shafts. The pinions, which are in mesh with the side gears, rotate thereby allowing the axles to transmit torque to the drive wheels. The differential pinion gears are at right angles to the side gears; when the differential gears rotate with the spider gears, the side gears move along with them. Notice that the side gears are free to rotate, as they are not splined to the differential housing. They are free to rotate independent of the differential housing. If the axles are both going the same speed, then the differential pinion gears will not rotate on their axes, as shown in **Figure 15-4.** The torque applied to the side gears will be equal as long as the equipment is moving in a straight line. There will be no motion inside the differential housing; the differential gears, ring gear, and drive axles all rotate at the same speed.

Differential Operation

If a U-turn is performed with a loader, you would see the differential action that takes place inside a conventional differential. If the inside wheel slows down 10% the outside wheel will speed up 10%.

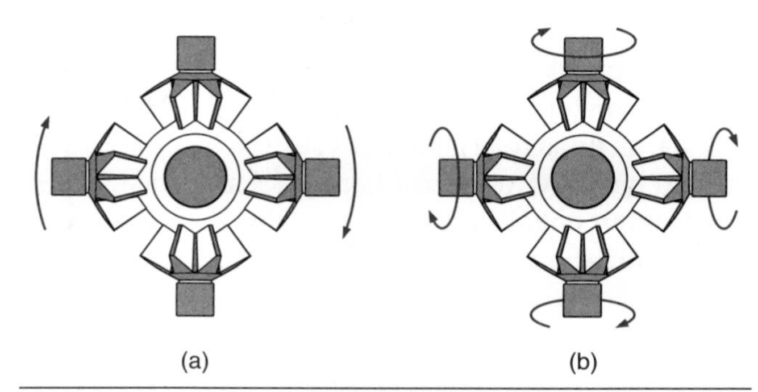

Figure 15-4 (A) Differential action in straight line motion. (B) Differential action during cornering.

As an example, a loader with a 10-foot width makes a 180-degree turn. The inside wheel is traveling on a 10-foot (3.05 m) radius and the outside wheel is traveling on a 20-foot (6.1 m) radius. If you perform the calculations, you will see that the inside wheel actually travels a distance of 31.5 feet (9.6 m), while the outside wheel travels over twice that distance 63 feet (19.2 m). This shows why a differential is necessary on off-road equipment.

Without differential action, the tires would take a real beating—not to mention the strain on the drive axle components. The inside wheel has more resistance when making the turn; this causes unequal torque to be applied to the side gears in the differential. This change in resistance forces one axle to slow down thus forcing the differential pinions to walk around the side gear. This movement in turn, causes the opposite side gear to speed up an equal amount.

There is a downside to this design. If one wheel loses traction, the differential will operate the same as if in a turn. The same amount of torque is sent to both wheels, but it can only be equal to the amount that is required to turn the wheel with the least amount of resistance.

Locking Differential Operation

Both the solid axle and conventional differential have design limitations that do not work well individually on equipment. Therefore, a design had to be developed that combined the benefits of a solid axle with the turning capabilities of a conventional differential. The first locking differentials were designed to be locked and unlocked by the operator of the equipment. The operator would lock the differential whenever he operated in a straight line or when wheel slip occurred, and then unlock it when cornering.

The operator could choose to engage the locking differential anytime road conditions such as mud, snow, ice, or other slippery road conditions presented themselves. The lockout could be engaged mechanically by using air or hydraulic pressure. Depending on the equipment make, when the **differential lock** was activated, a solenoid would send air or hydraulic oil to the shift mechanism, which would act on the shift fork to engage the splines on the differential case. This locked the differential case, gearing, and axle shafts together to provide equal torque to both wheels.

Figure 15-5 shows single reduction differential carrier with lockout. The carrier consists of the following parts:

- Actuator assembly
- Shift fork
- Sensor switch
- Outer splines—axle shaft to shift collar
- Shift collar and differential case splines
- Axle shaft
- Mechanical interface between fork and detent
- Shift collar in the locked position.

Figure 15-5 also shows the differential action with the lock disengaged (A) and engaged (B). When in the engaged position, a sensor actuates a dash light indicating that the differential is operating in the locked position. When the operator disengages the differential lock, the solenoid exhausts the pressure in the shift mechanism, allowing the return spring to move the shift fork and shift collar and disengaging the lock mechanism. The differential can now operate in its normal mode.

This type of locking differential design is often used on motor graders; this traction differential uses a clutch between the left side gear and the differential housing. Depending on the manufacturer of the differential, either oil or air pressure can be used to activate the clutches.

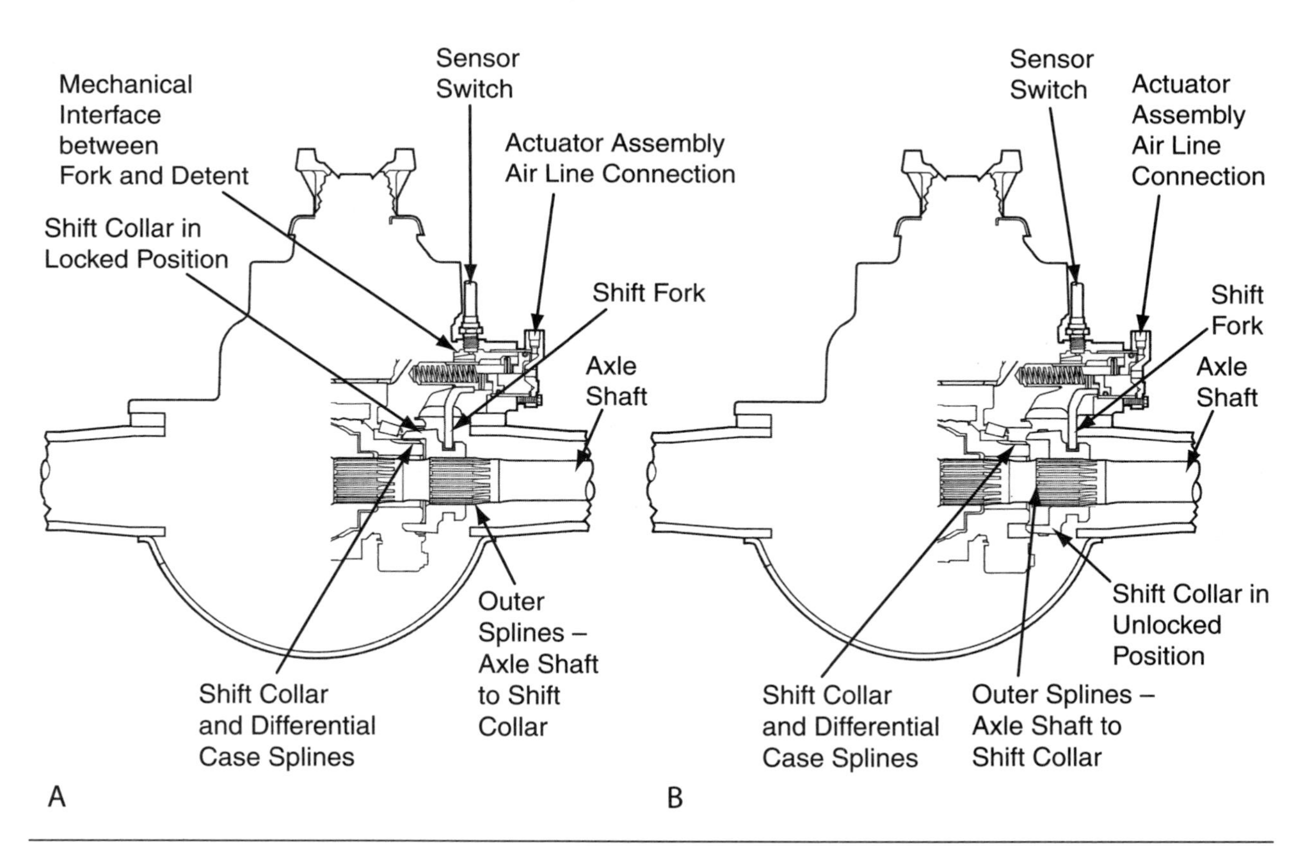

Figure 15-5 A parts breakdown of a singled reduction differential carrier showing (A) lock in the disengaged position, and (B) lock in the engaged position. (*Courtesy of Roadranger Marketing. One great drivetrain from two great companies—Eaton and Dana Corporations*)

Figure 15-6 shows how a typical traction control works. When engaged this design locks the left side gear to the differential case with the use of a clutch pack that in this case, consists of 21 discs. The 10 friction discs are splined to the driver, while the 11 steel discs have tangs on the outside and engage the differential case while spring pressure keeps the discs clamped together. The shift fork is controlled by an actuator that when engaged, forces the internal splines and the splines on the end of the side gear to lock. This action locks the side gear to the differential housing and causes both axles to turn at the same speed.

With this design, the wheels can slip based on a predetermined load, such as when the equipment must negotiate a turn. This slipping is based on the clutch disc's coefficient of friction. The increased torque on the outside wheel forces the plates in the clutch pack to slip and allows the outside side gear to increase its speed, forcing the differential to operate in normal mode. This allows the outside axle to turn faster than the inner axle. This design is available with either an operator-controlled shift controller or with a permanent traction control, which uses a driver that is engaged with the side gear at all times. The latter design can only operate in the conventional differential mode when the clutch discs slip; it will always be in the lockup mode.

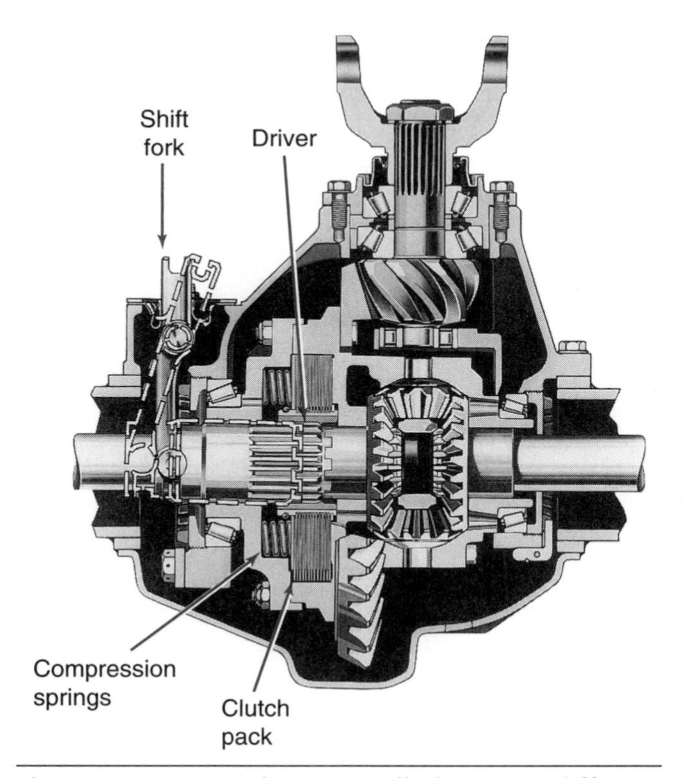

Figure 15-6 Heavy-duty controlled traction differential. (*Courtesy of Roadranger Marketing. One great drivetrain from two great companies—Eaton and Dana Corporations*)

Hydraulic Lock No Spin Operation

Another variation of this design uses hydraulic pressure to activate the clutch discs. The operator can lock out the differential whenever the eqipment is operating in a straight line. When engaged the oil flows into a cavity behind a piston which forces the clutch discs to lock the left side gear to the differential case. This causes the side gears and the spider to rotate together as a unit at the same speed. The differential pinions will lock the other side gear, forcing both axles to rotate at the same speed. This operating mode forces the differential to operate as if it were a solid axle. To release the lockup oil is exhausted from the piston cavity by the operator. The differential can now operate in its normal mode, allowing the side gears to rotate independently of each other as when cornering.

Mechanical Jaw Clutch No Spin Operation

This simple mechanical differential lock is widely used on smaller equipment such as backhoes. A jaw clutch is used to lock the differential, providing equal torque to both wheels. This type of lock is usually engaged with a foot control: the operator steps on a pedal, operated through a mechanical linkage, and constant torque is applied to a lock lever. The lever forces a fork to engage a coupling on the side gear into an adapter on the differential housing. This action pushes the two halves of the jaw clutch together, locking the side gear to the differential housing and providing equal torque distribution to each wheel. Once engagement has taken place, the lock pedal can be released. While the wheels are locked together, the side force generated by torque transfer keeps the jaw clutch engaged. When the torque becomes equal at each wheel, the reduced side force will automatically disengage the jaw clutch. This design is simple and provides long trouble-free life on smaller equipment.

Limited Slip Differential

This design of differential lock, as shown in **Figure 15-7** is commonly used on smaller wheel loaders and backhoe loaders. It is very compact and lends itself to installation in tight spaces such as those found on small equipment. This design takes advantage of gear tooth separating forces to apply the clutches. The **limited slip differential** has been designed to provide equal torque application to both wheels during normal operation. Figure 15-7 shows the two separate clutches, one on each side of the differential housing, connecting the side gear to the differential case. A small amount of side clearance is left in each clutch during assembly to ensure that each clutch engages correctly during operation. Under normal operating conditions with good traction at both wheels, equal torque will be distributed to each wheel during straight line operation. In a conventional differential, if one wheel loses traction and speeds up, the other wheel will slow down in proportion to the speed increase on the free wheel. This affects forward motion with speed and torque losses. If a limited slip differential is used in this same situation, the clutches make it more difficult for this to occur because as more input torque is applied to the differential, greater internal friction is developed in the side clutch. The locking effect takes place due to high internal friction generated by the increase in torque being applied. If a speed difference at the axles occurs, such as during cornering, the separating forces that occur in the differential during this process compress the clutch packs and allow the torque in the faster turning wheel to be redirected to the wheel that has greater traction. During turning, the wheels generate enough torque to overcome the clutch packs, allowing the wheels to change speed just as in a conventional differential.

Unlike a conventional differential spider, the spider used in the limited slip differential is a three-piece design. This design is used on small loaders and backhoes. The longer shaft holds two pinion gears, while each of the short shafts hold one pinion gear each. The ends of each shaft fit a corresponding bore in the connector, as shown in **Figure 15-8.** Dowels are used to fasten the shafts to the rotating housing. Internal slack

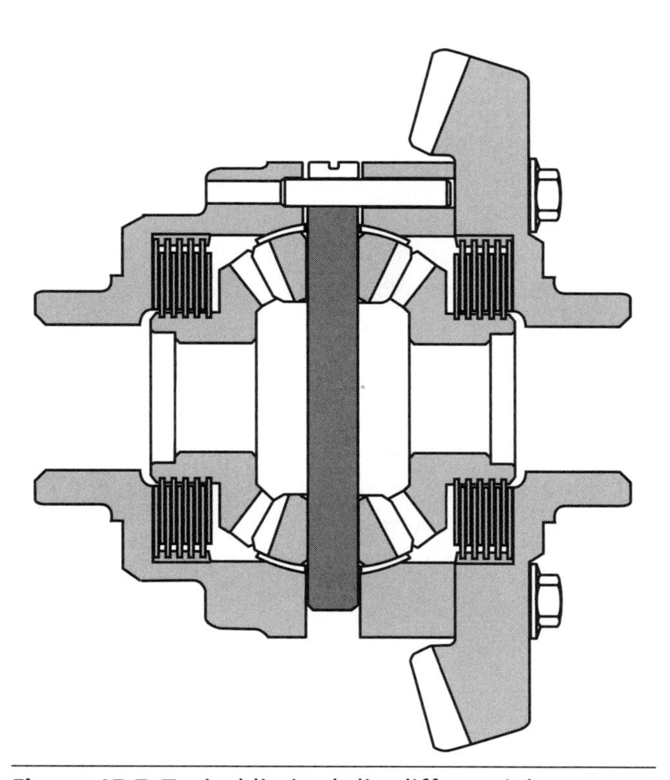

Figure 15-7 Typical limited slip differential.

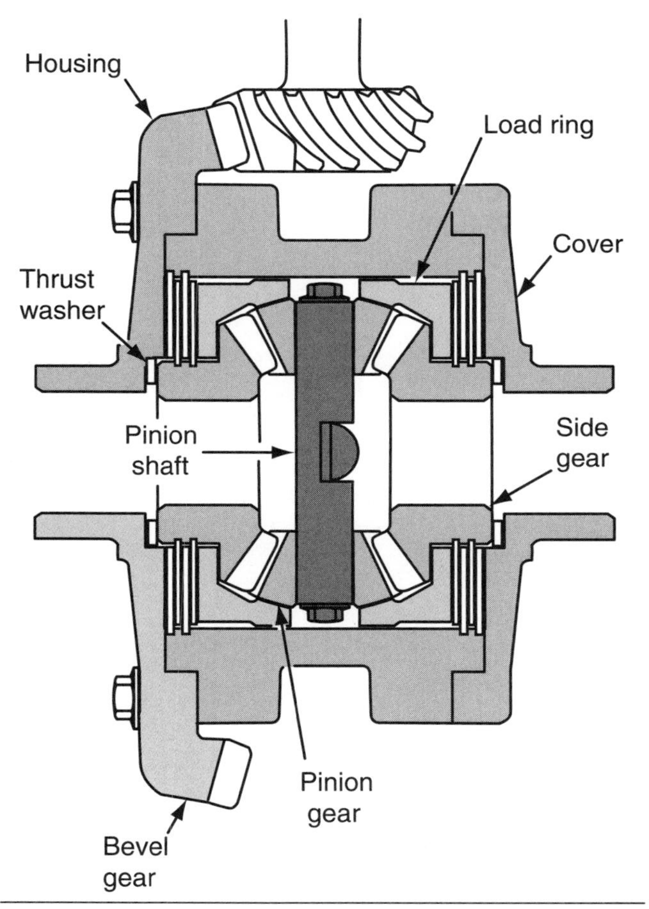

Figure 15-8 Limited slip differential that uses a load ring between the connector housing the differential housing.

is present in the arrangement, which allows the gear tooth separating forces to be transferred to the clutch discs. A side gear sits on top of each connector and meshes with the differential pinion gears. The clutch packs sit against the side gear contact face. The discs are splined to the side gears, while the plates have tangs that connect them to the housing.

Whenever a speed difference is present in the differential, gear tooth separating forces act on the side gears by axially squeezing the clutch packs together. The amount of torque is totally dependent on the input torque applied to the differential. When a clutch pack is engaged on the faster wheel end, the total input torque will increase, allowing the torque to be split when the clutches are engaged. The clearance in the clutch packs is adjusted by the use of shims when they are assembled.

On larger equipment a slightly different design is often used. Load rings are used to provide additional separating forces when a speed difference is present between the wheel ends. This design uses a two-piece spider that is notched in the middle to allow the two shafts to intersect with each other. Both ends of the spider shafts are squared off and mate with wedges in two pressure rings inside the differential housing.

This design uses a pressure ring that holds the side gear in place. The pressure ring has lugs on the outside diameter that fit into corresponding slots in the differential housing. The design is such that the pressure ring is allowed to move axially in the differential housing whenever the clutch packs are engaged. The spider shafts are driven by the pressure rings and have a natural tendency to ride up in the wedge during operation.

During axle speed differences, the separating forces of the gear teeth will force the pressure rings to rotate the spider shafts. The square end of the spider shafts in the wedge-shaped cutout will try to resist the rotational forces that are present, increasing the axial forces on the pressure rings. This force will engage the clutch packs, resulting in torque transfer. Note that the spiders are not connected to the differential housing directly; they are held in place by the pressure ring.

Figure 15-9 shows the discs and plates being installed. The plates have protruding notches on the outside diameter that match notches in the rotating housing, and the discs have internal teeth that mesh with the side gears. Axial forces that are generated by unequal axle speed force the clutch packs to engage. When fully assembled, the side gears sit on both sides of the spider shafts and pinion gears. They will mesh together at all times. The notches in the pressure rings hold the spider shafts in place. The clutch plates sit against the pressure rings, rotating with the differential housing, while the discs rotate with the side gear.

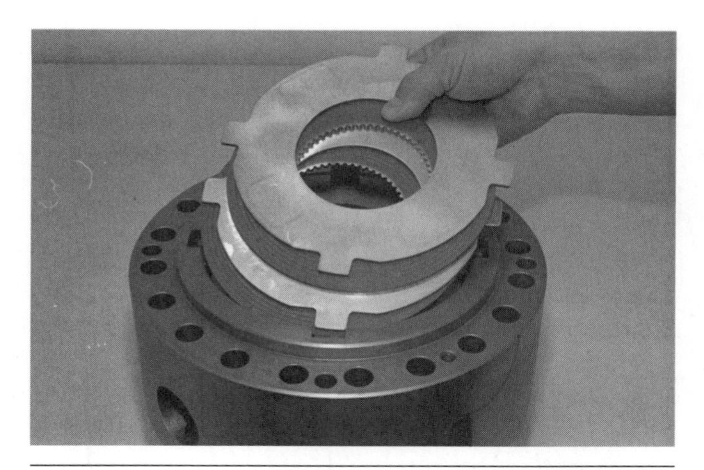

Figure 15-9 Note that the steel plates are held on the outside by the rotating housing and the discs are held by internal teeth to the side gears. (*Courtesy of Caterpillar*)

In **Figure 15-10,** the equipment is operating in a straight line with equal traction on each wheel. The torque is divided equally between the two wheels. The differential pinion gears rotate as a complete unit and do not rotate on their own axis. If wheel slip occurs, separation forces will engage the clutch packs, forcing the faster wheel to transfer torque to the slower wheel. The braking torque will be proportional to the input torque.

No Spin Differential Operation

The **no spin differential** uses jaw clutches that are splined to the side gears. The jaw clutches stay engaged with the side gears as long as the equipment is operated in a straight line. When the equipment turns, the outside wheel will speed up disengaging the outside jaw clutch, allowing the outside wheel to freewheel at a faster speed than the inside wheel. All the torque is sent to the inside wheel for traction. During straight line operation, the no spin splits the power equally between the two wheels and operates much like a solid axle.

Figure 15-11 shows how the spider fits into the jaw clutch, which also has internal teeth that are splined to the side gear. The axles are splined internally and mesh with the side gear. During cornering, the jaw clutches move axially, disengaging from the teeth on the spider. While this takes place, all the drive torque goes to the inner wheel to propel the equipment. The springs are

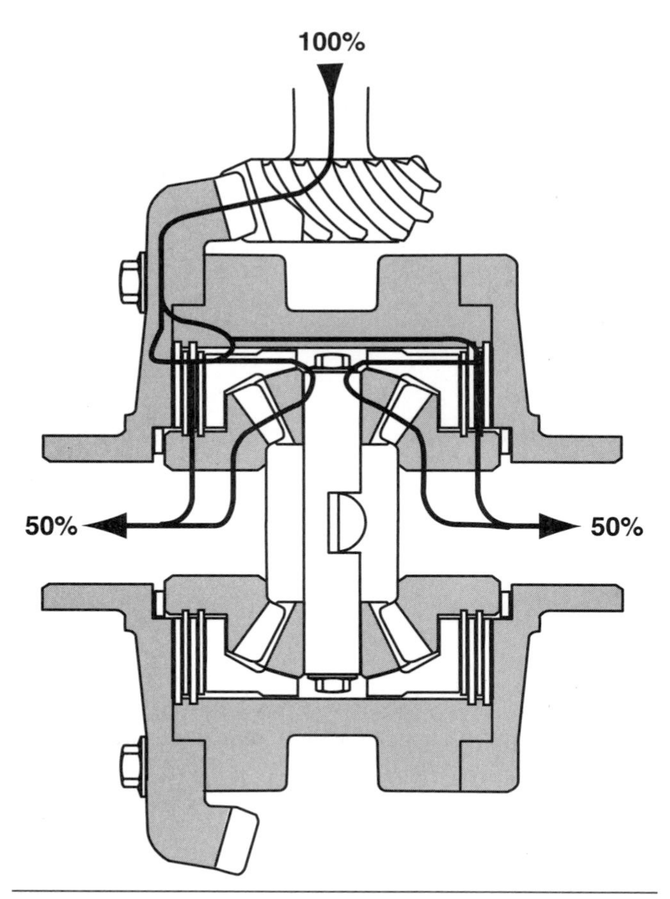

Figure 15-10 Limited slip differential operating with equal axle speed. Torque is divided equally between both wheels.

Figure 15-11 Jaw clutch and side gear. (*Courtesy of Caterpillar*)

used to keep the spider and clutch engaged during straight line operation.

Although the name spider is used to describe this subcomponent, it does not look like a traditional spider in a conventional differential. This unit still uses four cylindrical projections spaced at right angles to each other that fit corresponding spaces in the differential housing, which forces the spider assembly to rotate with the differential housing. There is a center cam ring with square teeth that mesh with the jaw clutch, which is held in place by a snap ring in the center of the spider assembly. One tooth on the spider is longer than the rest and is called the spider key. The spider key mates with a corresponding notch in the cam ring. When the equipment changes direction, the notch feels the force from the spider key. When the equipment is cornering the jaw clutch disengages the faster turning outside wheel, forcing the holdout ring to keep the jaw clutch disengaged until the outside wheel speed becomes synchronized with the spider. When this occurs, the jaw clutch is allowed to engage once again with the spider teeth.

When the jaw clutch and spider are engaged, the holdout ring is driven by the center cam. The holdout ring sits in a groove between the two layers of teeth on the jaw clutch. The only connection between the jaw clutch and the holdout ring is friction. Friction forces the holdout ring out when the jaw clutch disengages, forcing the teeth of the holdout ring to disengage from the center cam. Friction forces the holdout ring around until it contacts the spider key. The spider key will hold the holdout ring so that the teeth of the ring stand on the teeth of the center cam and turn with the spider. The top of the holdout ring will slip into the groove of the jaw clutch. When the jaw clutch is disengaged and turning counterclockwise, the end of the holdout ring contacts the spider key and keeps the holdout ring in the raised position. A slight clockwise movement will allow the teeth of the holdout ring to mesh with the center cam, and to be driven by the center cam.

Figure 15-12 shows the spider with the center cam ring in place and each of the jaw clutches on either side with the holdout rings in place. When the equipment is moving in a straight line with equal traction on both wheels, the outer teeth of the spider mesh with the outer teeth of the jaw clutches. The center cam teeth mesh with the holdout ring as well as with the inner teeth on the jaw clutches. The teeth of the spider drive the jaw clutch when the outer teeth are in mesh, sending equal torque to both wheels. When the equipment corners, one of the jaw clutches will begin to rotate faster due to the larger turning radius of the outer wheel. This forces the jaw clutch teeth to move up in the slot, initiating clutch disengagement.

Whenever the outer teeth of the jaw clutch move up in the slot, the inner teeth will disengage from the cen-

Figure 15-12 Spider/cam ring assembly. Note the holdout rings on the jaw clutches. (*Courtesy of Caterpillar*)

Figure 15-13 Left jaw clutch in the disengaged position. (*Courtesy of Caterpillar*)

ter cam. The faster speed of the jaw clutch forces the inner teeth of the jaw clutch to ride up the teeth of the center cam. When this happens the holdout ring comes out of its groove on the center cam, and the jaw clutch moves back on the splines of the side gear. As long as the jaw clutch is trying to rotate faster that the spider, the two components will remain disengaged. When the wheel speed becomes synchronized, the ground resistance exerts a negative force on the free wheel, causing the jaw clutch to slow down. Friction between the jaw clutch and holdout ring moves the holdout ring back into mesh with the center cam teeth. When this occurs, the spring forces the jaw clutch back into engagement with the spider.

A snap ring holds the cam ring in place and allows the cam ring to turn on the snap ring until it contacts the spider key and notch. The notch is required because the equipment can be driven in both directions. When the spider is rotated in the direction of the arrow, shown in **Figure 15-13** the spider's outer teeth push against the left edge of the bottom jaw clutch slot, driving the jaw clutch in the same direction as the spider. The teeth of the center cam and inner jaw clutch teeth are in mesh and force the jaw clutch to move to the end of the slot. Although the center cam can move, a snap ring holds it in place in the spider. The notch in the cam ring limits the amount of movement when it contacts the spider key. In Figure 15-13, we can see that the spider key is in contact with the right side of the cam ring. The shape of the inner teeth force disengagement to take place if the jaw clutch is forced to rotate at a higher speed that the spider.

Off-road equipment manufacturers often use a no spin or limited slip differential to replace a standard differential. Equipment manufacturers do not recommend using two no spin differentials in both front and rear axles assemblies. In good road conditions, the limited slip differential works well, but on slippery road conditions the maneuverability of the equipment is compromised because the clutch packs do not slip easily. This limits differential action, which increases the turning radius of the equipment. This design also tends to increase operating costs because of drag associated with these types of differentials.

A no spin differential provides better traction than a limited slip differential, and has an increased turning radius because the inside wheel does not slow down as in a standard differential. Only the outside wheel speeds up in a no spin differential. If two no spins were used on a loader, the turning radius would be even worse than with one. Although there are drawbacks to no spin differentials, if good traction is a must, then they are required in some applications. If the use of a no spin or limited slip differential is a requirement, then tire size must be matched on the same axle. Never operate a no spin or limited slip differential with two different tire diameters on the same axle for any length of time; damage will result from the constant clutch action. One tire will turn faster if diameter difference is excessive, forcing jaw clutch disengagement or clutch slippage as is the case with the limited slip differential.

Planetary Differentials

This design is often referred to as an interaxle and is commonly found on articulating, off-road equipment where you often need to separate the drive shafts on a tractor from the trailer. The use of standard differentials between two axle shafts is necessary to allow for speed differences in wheels due to turning or uneven terrain.

Planetary differentials operate much like standard differentials. This design is used to minimize driveline stress as the driveshafts are allowed to turn at different speeds. Like the standard differential, the planetary differential can vary the torque split between each wheel end. In an articulating truck, equal torque can be sent to each axle or be split between the two axles, depending on design. If an articulating haulage truck has three axles, the torque can be split by the planetary differential so that 40% of the torque goes to the tractor and 60% goes to the trailer. Each axle of the trailer will receive 30% of the total torque. Planetary differentials can be locked to provide maximum torque when road conditions are poor.

The planetary differential input torque is provided by a gear in the transmission transfer drive that meshes with the carrier of the planetary differential. The two sets of planetary gear shafts are held by the carrier and drive the planet gears. There is a clutch in the carrier that can be used to lock the planetary differential, if required. When operated in a straight line, the planet gears do not rotate on their own axes; they rotate as a unit, transferring all the torque to the wheels, as shown in **Figure 15-14.**

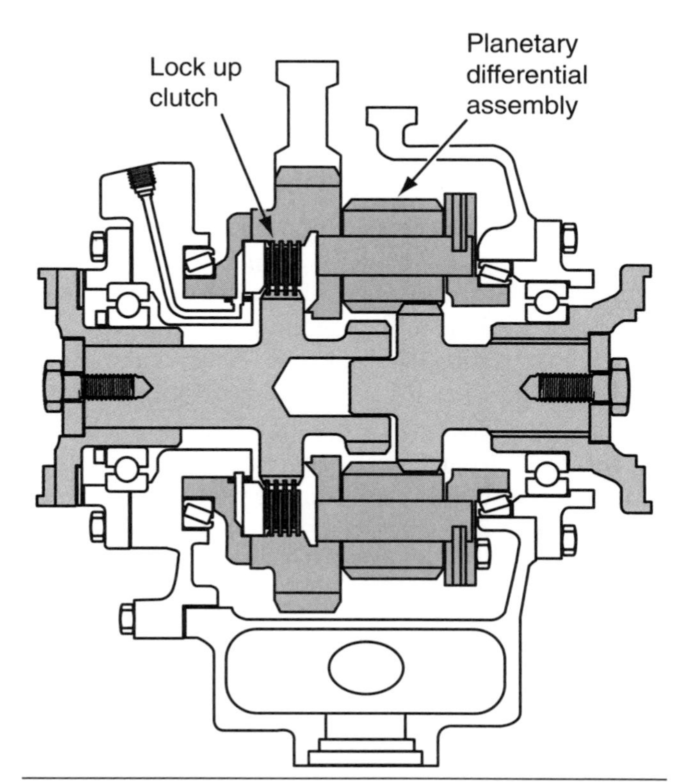

Figure 15-14 Cutaway view shows the lockup clutch location in relation to the planetary differential assembly.

If a change in resistance is felt at one of the wheels, the planet gears on the axle that is turning slower begin to rotate on their own axes. This causes the teeth of the planet gears which are meshed with the other set, to speed up the other axle, allowing the torque to be split between the two axles proportionally. If the differential is placed in the locked position, a clutch pack locks the carrier to one axle shaft, which provides equal torque to each drive wheel.

DIFFERENTIAL GEAR DESIGNS

A more thorough explanation of gear types can be found in Chapter 9, Introduction to Powertrains. Many different designs of gears are used in off-road differentials; we will outline the most popular types used in this chapter. The following is a list of the most common types:

- Spiral bevel gears
- Spur bevel gears
- Hypoid gears
- Amboid gears

The crown and pinion gear sets on off-road equipment often use spiral bevel, hypoid, or **amboid gear** designs. The manufacturer's equipment design and component layout and application will determine the type of gears used. **Bevel gears** are used extensively in the differential where they transfer power at right angles to the drive axles. The teeth can be either straight cut or spiral, depending on the application. Remember, the longer the teeth contact area, the greater the load carrying capacity will be; this rule applies to any gear design.

Spiral bevel gears are set up so that the pinion sits on the centerline axis of the ring gear. This gear design provides good, continuous tooth contact with one tooth in contact at all times during operation. It also generates less noise than the straight cut spiral design. Straight spur gears are used in standard differential gearing found in standard differentials.

Hypoid gears look similar to spiral bevel gears, but they are easily identified by the offset that is present between the pinion gear and ring gear. **Figure 15-15** shows that the pinion gear sits below the centerline of the ring gear, which requires a longer tooth design. This is an advantage in the contact area because there are two teeth in contact in this design compared to one-tooth contact in the **spiral bevel gear sets.** Hypoid gears are considered to be a stronger gear set. Given an equal diameter size of ring gear, the **hypoid gear set**

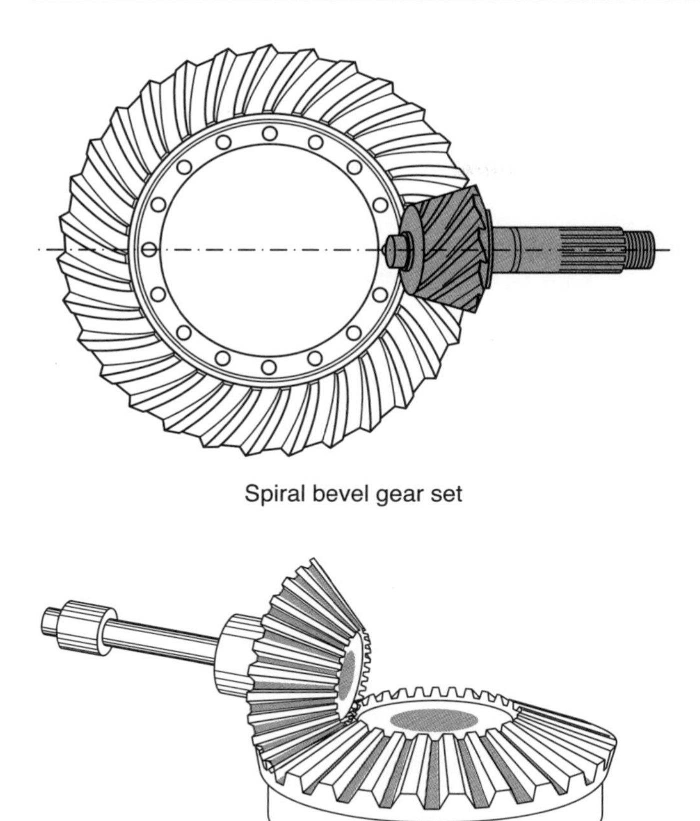

Figure 15-15 Examples of straight cut and spiral bevel gear sets. (*Courtesy of Arvin Meritor*)

will use a bigger pinion due to its greater torque carrying capacity. Hypoid gear sets drive forward on the convex side of the ring gear tooth, which provides greater strength and quieter operation when compared to spiral bevel gears.

Because of the operating characteristics of a hypoid gear set, the use of EP-rated lubrication is a must since the gears slide across each other wiping away any lubrication that is present between the contact teeth. Most equipment manufacturers recommend the use of synthetics with this design because of the superior shear strength of synthetic lubricants.

Amboid gears are similar in design to hypoid gears; the difference is the location of the pinion in relation to the ring gear. Figure 15-15 shows the location of the pinion, which is located above the centerline of the ring gear. This requires that the driveshaft be mounted higher than on the hypoid design, and the drive side of the gear is on the concave side of the ring gear tooth. This arrangement is commonly used on the rear axle of a **tandem drive** installation.

POWER DIVIDERS

Many manufacturers use a tandem drive arrangement on larger off-highway haulage trucks. This allows the equipment to haul heavier loads when compared to single axle designs. This design requires the use of extra gearing to split the drive torque between the two axles. The additional gearing must also allow the axle to operate at different speeds if required.

A power divider performs two different functions. It must be able to divide the driving torque between two axle assemblies, and it must be able to provide differential action between the two axles when necessary.

An interaxle is located in the forward rear axle and uses most of the same components that are found in a conventional differential. **Figure 15-16** shows a cross section of the differential. Note the location of the **interaxle differential.** It is located between the input shaft and the output shaft. Helical gears are often used to send torque to the front differential pinion. A lockout is used on the forward drive axle to lock the interaxle differential thereby providing maximum traction when road conditions are poor. The lockout mechanism is engaged by the operator, who implements a sliding clutch that lock the helical and differential side gears together to give a positive drive to both axles. The lockout should only be applied during slow speed operation and never engaged during wheel slip. This feature should only be used when road conditions are poor; continuous use of this feature will lead to excessive tire wear.

Operation

The flow of power comes from the transmission and enters the input shaft of the forward axle assembly, where it flows to the forward and rear axles. The interaxle differential, which is mounted on the input shaft of the forward differential, operates the same as a conventional differential providing differential action between the forward and rear axle assembly. A conventional differential is located in each of the two axle assemblies, which provide differential action in each of the drive axles on both axle assemblies. **Figure 15-17** shows the flow of power that is sent through the input shaft that is splined to the interaxle differential spider. From here the power is sent through the interaxle differential to the forward and rear axles. The power flows into the

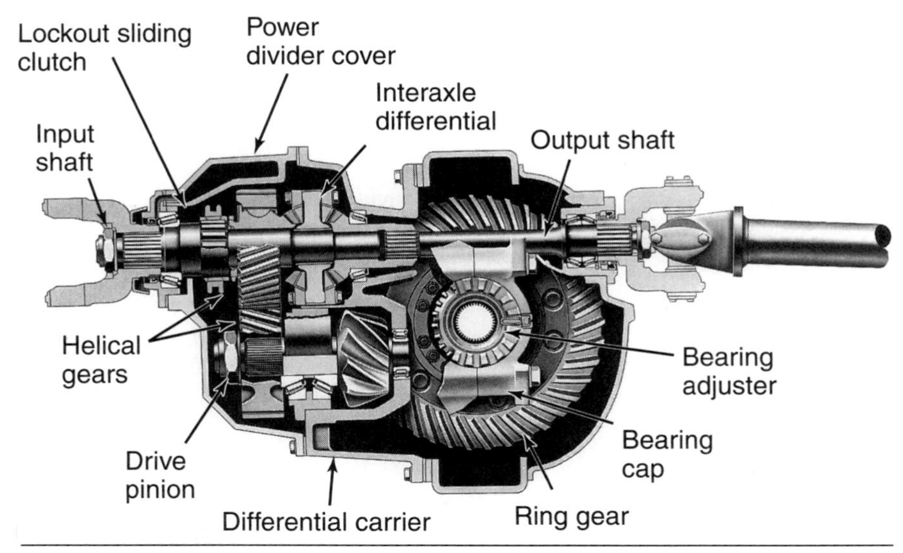

Figure 15-16 Sectional view of an interaxle differential showing how the torque is split between the forward and rear differential. (*Courtesy of Dana Corporation*)

forward axle and is transferred from the spider through the floating helical side gear on the input shaft to the drive pinion and on through the ring gear to the forward rear axle.

The rear axle has the power coming in through the spider to the output shaft side gear, and moving on to the propeller shaft and to the drive pinion and ring gear of the rear axle when the unit is operating in the normal mode (unlocked). If the interaxle is not in operation, the helical side gear and interaxle differential are turning together as a unit. During this phase of operation torque is transferred to the forward and rear differentials equally. The forward axle pinion is driven from the input shaft by the floating helical side gear. The rear axle is driven from the output shaft side gear, through the output shaft side gear, to the output shaft, and to the propeller shaft.

When lockout operation is required, the operator engages the lockout mechanism, which allows a sliding clutch to lock the differential side gear to the helical gear. This locks the interaxle to produce a positive drive to each axle. Whenever there is a wheel speed difference between the forward axle and rear axle, the interaxle differential splits the torque and speed between the two axle assemblies. The same thing can occur whenever there is a speed difference between wheels on the same axle assembly. Each differential is capable of splitting the torque and speed between each wheel on each axle assembly.

The lockout mechanism can be air or hydraulically actuated, depending on the application and manufacturer. The mechanism consists of a shift fork and pushrod assembly and a sliding lockout clutch. The actuating force extends the pushrod and shift fork mechanism, which forces the clutch against a helical side gear. The clutch is splined to the input shaft, which is splined to the differential. When shifted, the clutch locks the helical side gear to the differential to stop any differential action from occurring. If the operator locks the interaxle differential, power is divided equally between the two axle assemblies.

LUBRICATION

A large percentage of interaxles are lubricated by the splash method: oil is distributed by the gears turning and splashing the oil internally. Directional baffles are used to direct oil to critical areas of the differential assembly. Some manufacturers provide a lubrication pump that ensures positive oil distribution to all areas of the interaxle differential. Getting oil to the interaxle differential is essential, and using the splash method does not ensure that oil gets into this area; it is a positive method of oil circulation.

External gear pumps are often chosen for this purpose. **Figure 15-18** illustrates a cutaway of a typical drive axle carrier showing the mounting location of

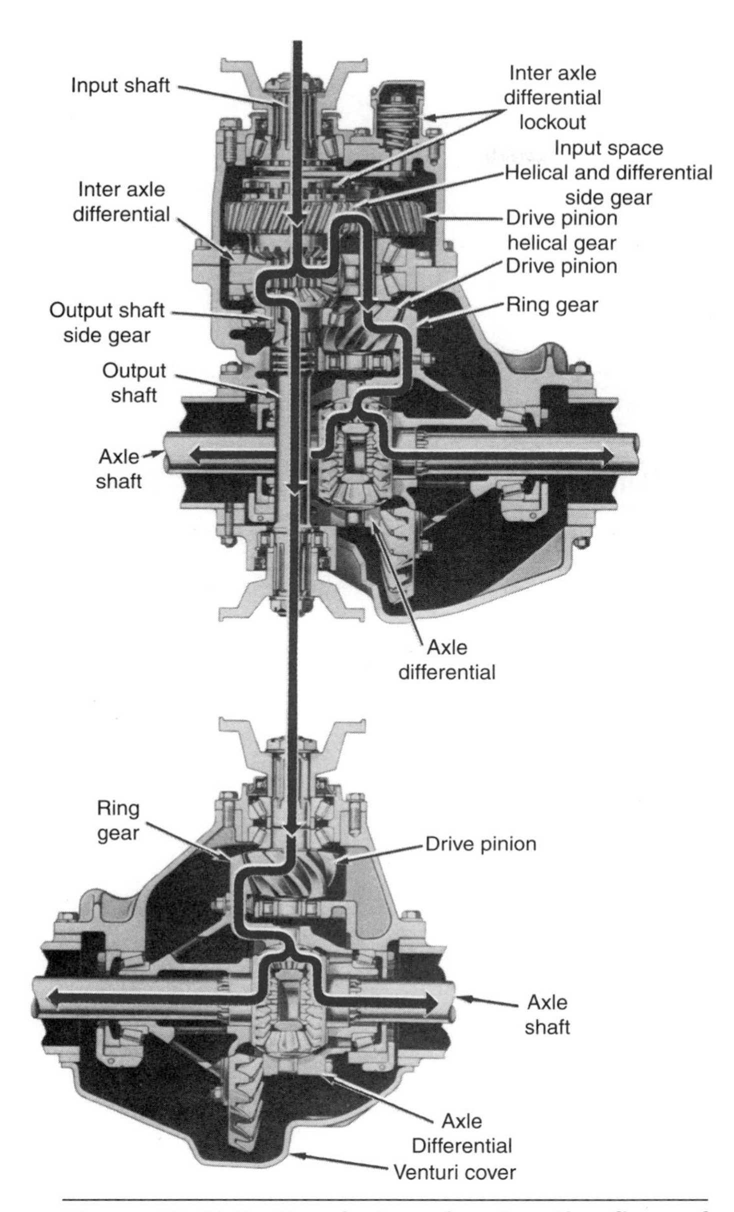

Figure 15-17 Sectional view showing the flow of power through the forward and rear differential. (*Courtesy Eaton Corporation Axle Division*)

the lube pump. The oil pump picks up oil in the sump and sends it through a filter. From there it is pumped through an external oil line in some cases, or it can be routed internally to the top of the carrier assembly where it is distributed to spray nozzles that spray oil on critical areas such as the differential gears and the front pinion bearing assembly. Additional lubrication is provided by rotation of the ring gear, which splashes oil around the inside to lubricate the drive gears and all internal bearings. If the vehicle is equipped with a positive lubrication system, oil filter replacement is

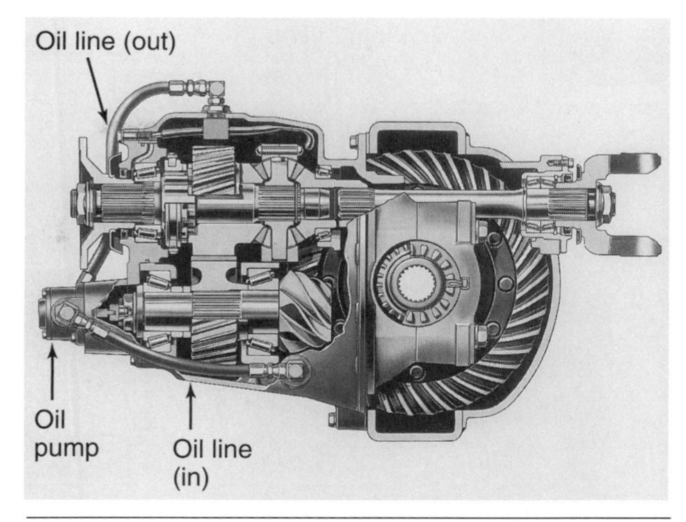

Figure 15-18 Cutaway of a power divider. Note the location of the lube pump. (*Courtesy of Arvin Meritor*)

required on a regular basis. Always refer to the manufacturer's documentation to determine the service interval between oil filter replacements.

DOUBLE REDUCTION AXLES

Off-road equipment is often required to haul large, heavy loads. An additional gear reduction may be required to get the job done. A **double reduction axle** uses two gear sets to provide additional torque, as shown in **Figure 15-19.** The design uses a conventional pinion ring gear to get the first gear reduction. The second gear reduction is provided by a helical spur gear and pinion set. The pinion and ring gear drive work the same as a conventional differential; however, the differential case is not bolted to the ring gear, instead it is mounted to the spur gear. The spur pinion is keyed to a shaft that holds the ring gear to provide drive. The spur pinion and spur gear are in constant mesh. The torque flows into the input shaft and goes through the ring gear to provide the first reduction.

Power now flows to the spur pinion, which is driven directly by the shaft that the ring gear is mounted to. The second reduction is achieved through the spur pinion driving the spur gear mounted to the differential carrier, and power flows through a conventional differential to the drive axles.

Some manufacturers use a **planetary gear set** in place of helical spur gears to achieve double reduction. Input torque is sent to the drive pinion and flows

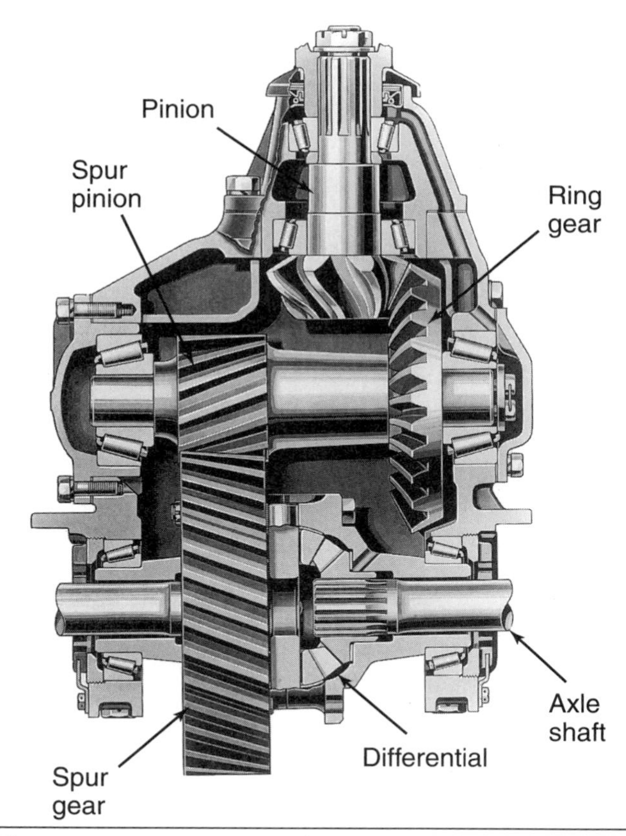

Figure 15-19 Cutaway of a double reduction drive axle assembly. (*Courtesy of Dana Corporation*)

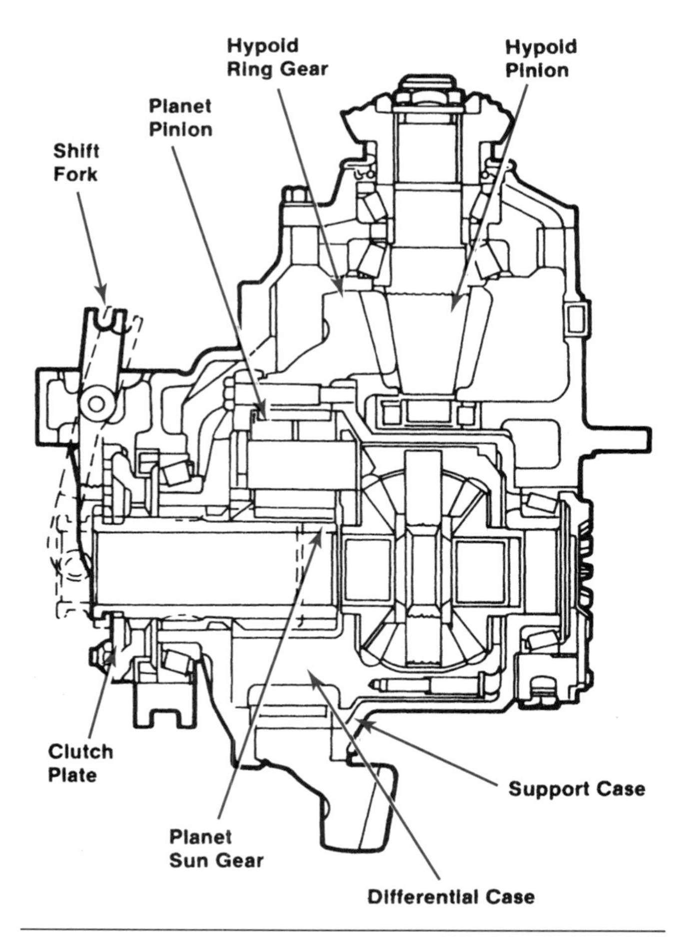

Figure 15-20 A two-speed planetary drive axle. Note the shifting mechanism used to engage the second speed reduction through the internal planetary drive gears. (*Courtesy of Arvin Meritor*)

through the ring gear for the first reduction. The ring gear has internal teeth that hold four planetary idler pinions. These pinions revolve around a stationary sun gear to provide the second gear reduction. Torque is transferred through the planetary gears to the differential and out to the drive axles. The complete flow of power will flow through drive pinion, ring gear, pinion idlers, and out through the differential to the axle shafts.

TWO-SPEED DRIVE AXLES

Two-speed axle assemblies are designed to provide the advantages of two axles in one unit, as shown in **Figure 15-20.** This design is capable of operating as a **single reduction axle,** and can be shifted to a low-speed operating mode by the use of an external shift mechanism, which combines the use of planetary gears to provide a second reduction. Although it is similar in construction to the double reduction axle, this axle can be operated in two different gear ratios independent of each other, while the double reduction can only operate through both gear reductions at the same time. This makes this type of drive axle extremely flexible since the unit is capable of operating in high-speed single reduction, and can be shifted into a low-speed reduction when required. These can be used on both axles of a tandem drive. When operating in high range, the flow of power only goes through the ring gears and pinions on both axles. When operated in low range, the flow of power goes through the planetary gears to produce the second reduction.

Equipment with two-speed axles can take advantage of the extra speed range and use it to get more effective gearing. An example of this is the use of low range in each of the two axle ratios: First gear in low range, then first gear again in high range, and then continue up the gear ranges using two stages for each range.

Tech Tip: If disassembly becomes necessary when servicing equipment, always refer to the manufacturer's service manual to identify which axle design you are dealing with. For example, look at two different axle designs manufactured by the same company. Although they are both classified the same, they have significant differences that must be noted or mistakes can be made when servicing. There are two different types of differential carrier designs used by a manufacturer: a ribbed carrier and a conical carrier assembly. Lubrication for each is quite different and should be noted as such. With the conical differential carrier, oil is picked up by the gear support case and is sent to the pickup trough. The oil is channeled to the carrier passage to the right differential bearing, then passes through the differential and planetary units to the left-hand differential bearing, and then flows back to the reservoir. With the ribbed differential carrier, the oil is picked up by an oil collector drum and sent to the oil distributor; from there it is sent in two different directions. Half flows to the drive pinion bearings, while the other half flows to the right-hand differential bearing, then through the differential to the planetary gears and on to the left-hand differential bearing before returning back to the case.

DRIVE AXLE CONFIGURATIONS

Off-road equipment drive axles, in most cases, come with an extra planetary drive reduction that will generally not be found on highway vehicles. Planetary drives are compact in design and can be made as small as necessary to fit into almost any space. They are inherently capable of carrying large amounts of torque. The design of this drive allows the torque to be split evenly throughout all the gears. The planetary reduction drive may be located either inboard or outboard on the drive axle. Inboard planetary drives sit right next to the differential and receive torque from a short axle shaft that serves as a sun gear or input for the planetary drive. Outboard planetary drives are located in the wheel end of the axle assembly and are driven by a much longer axle from the differential, which sends torque to the input or sun gear of the drive. Each design has its advantages and disadvantages.

The inboard planetary drive has room at the wheel ends for a much larger range of wheel widths. The disadvantage is that the axle needs to be built stronger because it carries the full weight of the load on the wheel ends and absorbs end thrust. The axle has two tapered roller bearings at the wheel end to support the load and is driven by the planet carrier for a final torque increase.

The outboard planetary drive, on the other hand, has the planetary reduction in the wheel ends, which gives the planetary drive its own rigid housing. The axle is driven by the differential sending torque through the axle to the other end, which acts as the sun gear or input on the planetary drive. An advantage of this design is that the axles do not support the weight of the equipment and can be easily removed if the equipment needed to be towed.

Planetary Drive Operation

Planetary drives receive torque from the differential through an axle shaft, which serves as a sun gear (input gear). This shaft has splines at both ends; one end fits in the differential, the other end (sun gear) sends torque to the planetary drive gears. The sun gear meshes with the three planet gears, which are mounted together in a planet carrier. The ring gear is held or locked in place; the planet carrier acts as the output, resulting in a gear reduction. Torque from the sun gear forces the planet pinions to rotate around the inside of the ring gear, which is held stationary. The planet pinion carrier rotates in the same direction as the sun gear, sending engine torque to the wheels at a slower speed. This increases torque.

Regardless of the design, both types of planetary drives operate the same: the sun gear is the input, the ring gear is held, and the planet carrier becomes the output. The one main difference between the two designs is that the axle in the outboard design is a full-floating axle and does not carry any weight, while the inboard design axle carries the full weight of the load through large tapered roller bearings mounted at the wheel end of the axle housing.

Full-Floating Axle

A full-floating axles is used on large, slow moving off-road equipment with a drive wheel on each end of the axle housing. **Figure 15-21** shows two large tapered roller bearings that are mounted to the wheel hub and carry the full weight of the equipment and load. On

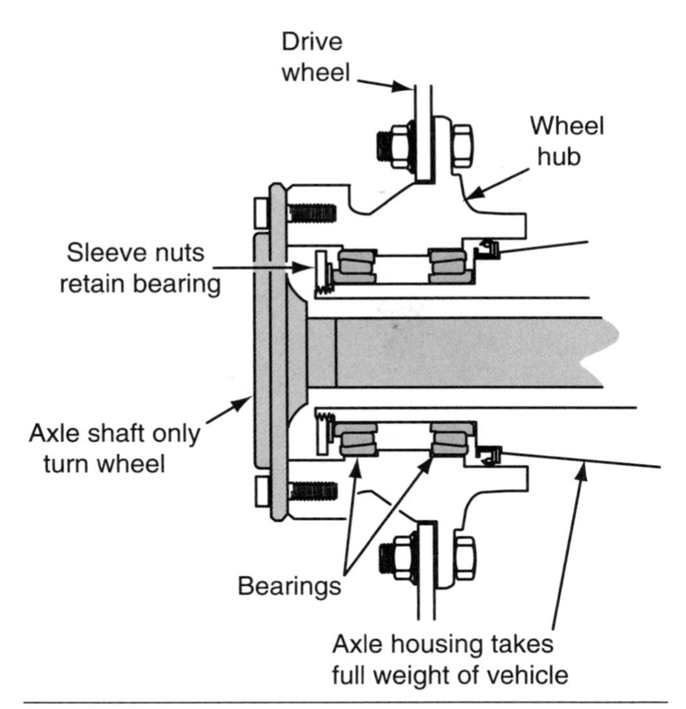

Figure 15-21 Typical heavy-duty full-floating drive axle.

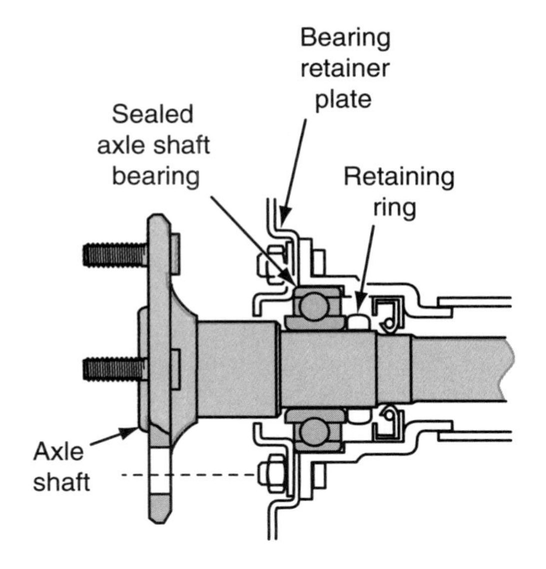

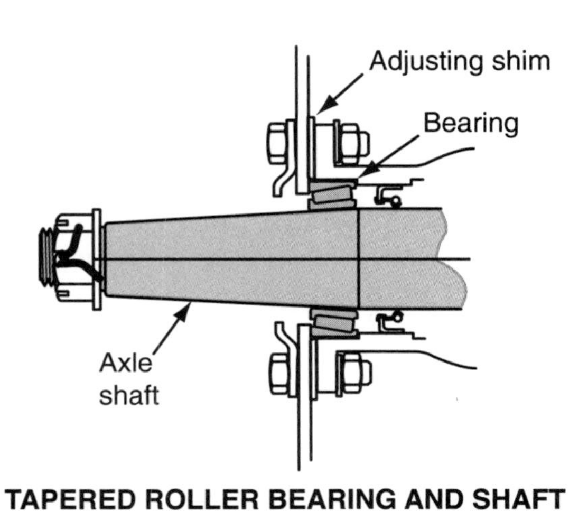

Figure 15-22 Typical heavy-duty semi-floating drive axle.

larger equipment, the bearings are mounted on a spindle that bolts to the axle housing. The axle itself only transmits the torque of the engine through it and does not carry any weight; hence, the term *floating* is used to describe its function. On smaller equipment, the axle is connected to the drive wheels through a bolted flange that can be removed to gain access to the axle without removing the wheels.

On larger equipment with outboard planetary drives, the axle is floating and held in place with a thrust plate that can be removed to gain access to the planetary drive gears for servicing. The axles are easily removed if and when the equipment needs to be towed in a non-run situation. Caution should be used when removing the axles to allow the wheels to free-wheel. The equipment should be secured, either with wheel chocks or a tow hook to another piece of towing equipment to prevent accidental movement.

Semi-Floating Axle

The term *semi-floating* is derived from the way the axle is mounted in the assembly, as shown in **Figure 15-22.** The axle at the differential is splined to the side gear and floats at this end. At the wheel end of the **semi-floating axle,** the shaft carries one or two bearings, depending on the axle load carrying capacity. At the wheel end of the axle, the load is carried by the axle, which also transmits the engine torque. This design forces the axle to handle all of the torque and strain caused by uneven road conditions. This design comes with two types of bearing configurations: tapered roller bearings or heavy-duty ball bearings. If tapered roller bearings are used, a means of adjustment is provided for preloading the bearings, either with shims or a threaded nut which is torqued to a manufacturer's recommended torque specification. One of the main disadvantages of this design is that, if the axle were to break, the wheel could fall off. If that happened to the full-floating design, the result would be only a loss of drive to the wheel end. These are often found on light duty applications were weight and load is often less than 5,000 lbs per axle.

Drive/Steer Axles

Drive/steer axles are used on off-road equipment that does not use an articulating steering arrangement. Some manufacturers use this type of drive/steer configuration because of limitations in size, weight, length, and width of equipment design. The drive axle section of this drive/steer axle is similar to a conventional drive axle. The difference is in the wheel end design.

Kingpins are used to mount the wheel end to the axle housing, allowing the wheel end to turn from side to side. The steering knuckle that supports the kingpin at the top and bottom has bushings that have recessed grooves to retain grease. Each steering knuckle is connected to the other by way of a tie-rod that is threaded at each end to allow wheel adjustment. The ends of the cross tubes have a left-hand thread at one end and a right-hand thread at the other end, which provides a means to adjust the toe in or out during alignments. Engine power is sent through the differential and flows through the Cardin axle shaft universal joint to the outer axle shaft to drive the wheels. The outer ends of the shaft are splined to a drive flange, which is bolted to the wheel hub. Depending on the manufacturer a single- or double-cardan joint allows the equipment to be transfer torque and steer at the same time. If the wheel end is equipped with a planetary drive reduction the outer ends of the axle shafts are splined to a sun gear that transfers torque to a pinion carrier, resulting in the pinion gears walking around the stationary ring gear and providing forward reduction, as shown in **Figure 15-23.**

DRIVE AXLE LUBRICATION

Technicians who perform service and repairs on off-road equipment must understand the importance of following the manufacturer's maintenance procedures. A good number of axle-related failures are caused by using a lubricant that is not ideal for the application. This results in a shorter component life than could otherwise be achieved. Many failures could be prevented by following sound maintenance practices concerning lubrication selection and use. Axle failures can and have produced many undesirable consequences in the past. When lubrication levels are not properly maintained in axles, the life of the bearings and gears will be adversely affected, shortening their life or often leading to catastrophic failures. Regardless of how well the equipment is designed and operated, if it is not properly maintained, premature axle component wear will occur and will most definitely lead to early failure. Although some failures are caused by improper installation of components, technicians need to make themselves aware of correct methods and ensure that the proper replacement parts and tools are used to assemble the components.

Lubrication-related damage may take place even if the correct lubricant is used. Often the intervals between servicing are inadequate for the application, resulting in component damage. Lubricants that are used in axle assemblies have three general functions. They

- Provide an adequate lubrication film quickly and reduce friction between sliding and rolling surfaces.
- Remove excess heat from high friction areas and maintain correct operating temperatures.
- Remove dirt and wear particles from the bearings and other high friction areas.

If and when lubrication damage occurs to axle components, there are generally three basic problem areas that are responsible:

- Contaminated lubrication
- Low lubrication levels
- Incorrect lubricant for the application or depleted additives.

Lubrication practices discussed in this chapter are general examples of what should be done and when it should be performed. In actual practice, maintenance should be performed based on the OEM manufacturer's recommended servicing literature. Many manufacturers allow the use of synthetic lubricants in axles; this extends service intervals significantly. The initial cost of synthetic lubricants is much higher upfront, but in the long run can be significantly less costly. Drain intervals can often be increased by two to three times when compared to conventional crude-based lubricants. Synthetic lubricants have many qualities that conventional oils do not have, such as a much higher boiling point when compared to conventional oil (600°F versus 350°F). Another important factor to consider when deciding on the use of synthetic oil is the amount of cold weather operation to which your equipment will be exposed. Synthetic oil has much better fluidity than conventional oil in extremely cold weather; this could mean the difference between having lubrication instantly or operating for a few minutes before the oil can flow properly.

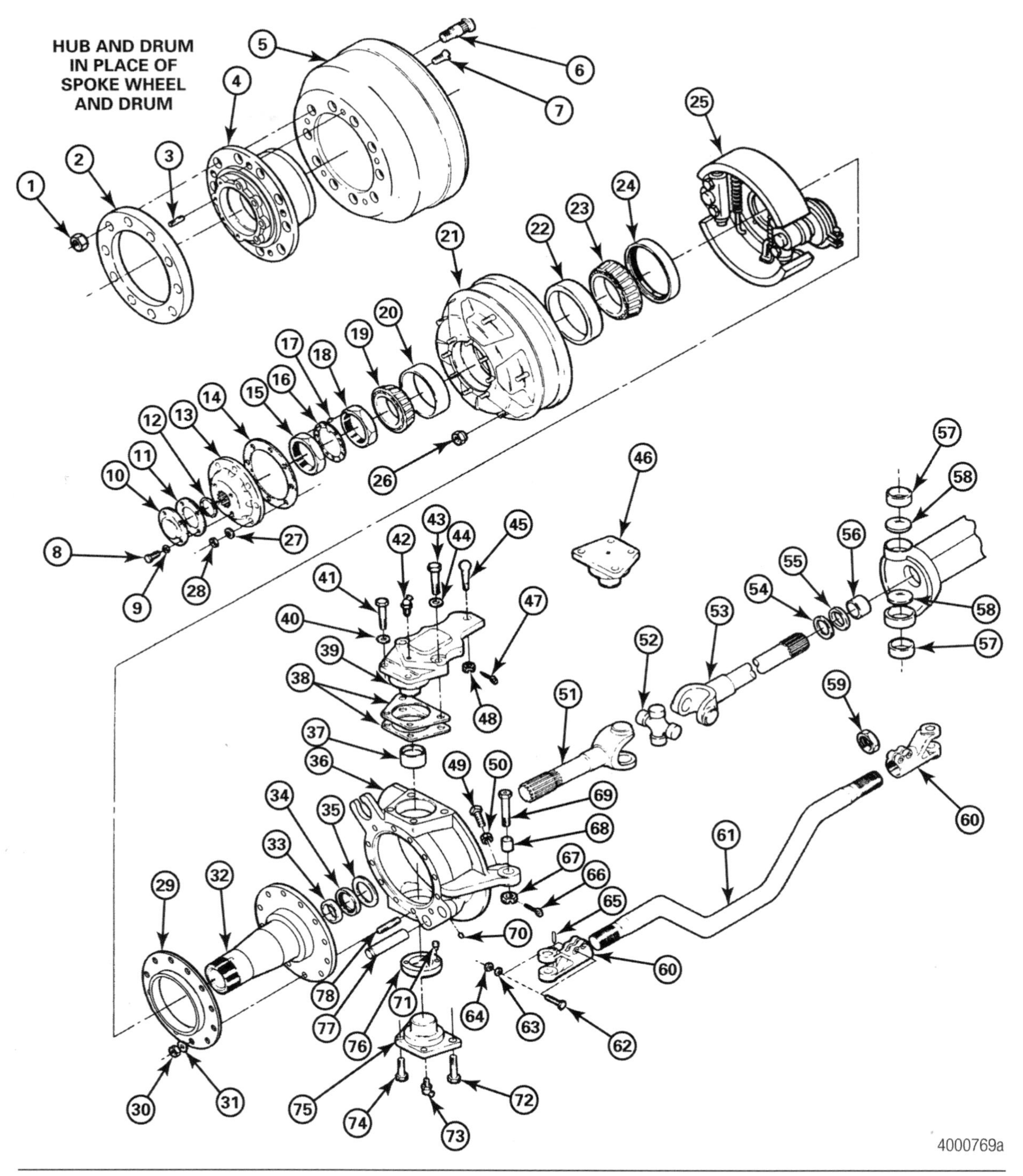

Figure 15-23 Exploded view of a typical heavy-duty drive/steer axle wheel end. (*Courtesy of Arvin Meritor*)

Using the Proper Lubricant

Many a differential has been ruined by a phenomenon called "channeling," which occurs in cold weather: the thickened oil is parted by the rotating ring gear and is too thick to flow back. This results in inadequate lubrication until the oil heats up. Often, by then the gears and bearings have sustained damage from lack of lubrication. A good solution to this type of problem is to use a high quality synthetic lube—one that meets API-GL-5 specifications. Use of synthetic lube also provides the benefit of extended drain intervals; which reduces the cost of labor and ensures better protection. All lubricants used in axle assemblies must meet the American Petroleum Institute (API) and the Society of Automotive Engineers (SAE) Gear Lubrication (GL) standards. Today the use of GL-1, GL-2, GL-3, GL-4, and GL-6 is no longer approved for modern axle assemblies by many manufacturers and should be discontinued. The recommended lubricant to use in drive axles today is API-GL-5, which is rated as an extreme pressure lubricant and is required for use in axle assemblies that use hypoid gears. Using the correct viscosity will be operating-temperature dependent, and should be adhered to. **Table 15-1** shows the proper grade of lubricant for a given operating temperature.

Most OEMs approve the use of synthetic lubricants as long as they meet API-GL-5 classification standards. Due to the wide variety of operating extremes that off-road equipment is subjected to, it is difficult to determine ideal drain intervals for a lot of equipment. Off-road equipment service intervals are usually not measured in miles but are measured in hours of operation.

Contaminated Lubrication

Some equipment shops have a policy of mixing different brands of lubricants; this is not recommended. Not all lubricants are compatible with each other. Every effort should be made to prevent the mixing of different lubricants in axle assemblies. Some additives used by certain manufacturers are not compatible with other manufacturer's additives. Never mix different brands of gear lube. Some synthetic oils, when mixed with petroleum-based oil, will thicken. This can lead to foaming, which can result in early component failure. Water, dirt, and wear particles can cause extensive damage in a short period of time. Water can enter the axle assembly through a faulty shaft seal or through the carrier to a banjo housing joint. Wear particles are generally a result of normal wear and can be minimized by the use of magnetic drain plugs. The magnets collect metal particles that settle to the bottom of the axle housing during shutdown periods. Some manufacturers use an oil pump in the drive axle to distribute oil and a filter in the circuit to filters out dirt particles.

Table 15-1: PROPER GRADE OF LUBRICANT

Ambient Temperature Range	Correct Grade
−40°F to −15°F	75W
−15°F to 100°F	80W–90
−15°F and above	80W–140
10°F and above	85W–140

Results of Lubrication Failures

Whenever a lubricant does not perform up to the equipment's application standards, the life of the internal components will be shortened. Following is a list of causes and effects that may occur as a result of lubrication failures:

- Lubrication that is contaminated with water can cause scoring, etching, or pitting on the running surfaces of the gears and bearings.
- Underfilling the axle assembly may lead to metal-to-metal contact, which can lead to high friction that can cause overheating and results in oil breakdown. This will usually lead to severe scoring and galling that can actually melt metal surfaces. The internal parts will often be blackened from the heat.
- The use of an improper lubricant, which includes the practice of mixing different oils, may lead to surface etching or lubricant breakdown, leading to metal-to-metal contact.
- Using a viscosity that is not suitable for the application temperatures, may result in a lubrication film breakdown, leading to severe scoring and galling. The internal gears will often be blackened from the heat.

Tech Tip: When filling the axle housing and planetary wheel ends, allow enough time for the oil to flow through the components and find its natural level. When the filling is completed, allow a few minutes and then recheck the level again. The lubricant should be warm when drained; this ensures that contaminants are suspended in the oil and get flushed out.

Low Lubricant Levels

Lubrication levels on off-road equipment are generally checked every 250 hours of operation. Oil levels in drive axles must be checked properly; the oil level must be even with the bottom of the oil filler plug hole. A good rule is to replace the oil every 2500 hours of operation or after one year whichever comes first.

Remove the plug in the axle banjo housing; the oil should be even with the bottom of the oil level plug hole, as shown in **Figure 15-24.** Note that if the pinion

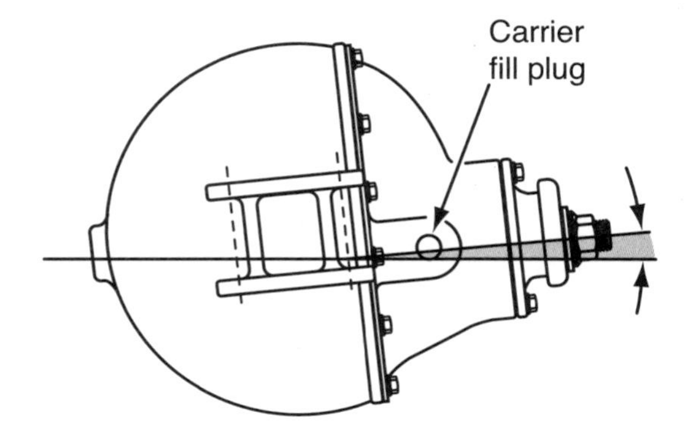

Pinion angle less than 7°
fill to carrier fill plug hole

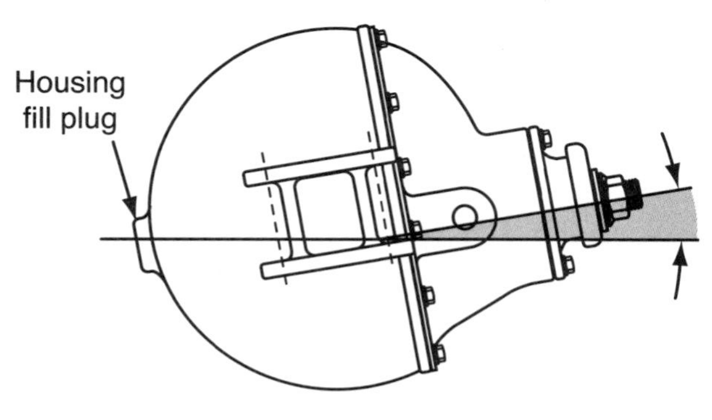

Pinion angle more than 7°
fill to housing fill plug hole

Figure 15-24 Checking the oil level in an axle housing.

angle is more than 7 degrees, a different oil level is used. Refer to Figure 15-24 for details. To properly check the oil levels in a differential drive axle and the planetary wheel ends, the axle first should be run and then allowed to stand for approximately 5 minutes on level ground before checking the oil level. This allows time for the oil to drain back down to the sump. Some axles have seals in the axles separating the oil in the wheel ends from the planetary drive. In this case, the oil level is checked independently in each wheel end and the differential banjo housing. The wheel end seal prevents the oil in the wheel ends from entering the differential banjo housing. When checking the oil level in a top mount or inverted pinion drive axle, a slightly different procedure must be followed. The oil hole on these units is located in the axle banjo housing. Note that some axles have another smaller hole located just below the oil level hole; this hole is for a temperature sensor and should not be used to fill or check the oil level.

Checking the Planetary Wheel Ends

To check the oil level of the planetary wheel ends, the wheel hub oil level plug must be in the down position. Note that you may need to move the vehicle to line up the oil level plug in the correct location. Remove the plug and check the oil level; if it is low remove the fill plug and add lube oil as required. New planetary wheel ends should be drained of initial fill after 75 to 100 hours of operation; this prevents any break-in material from causing excessive wear. Oil should be drained when warm to ensure that any wear particles will drain with the oil and will not have time to settle to the bottom. Some axles do not have an oil check plug at the wheel ends; in this case, the wheel ends of the axle must be raised one at a time to allow oil to flow to the opposite wheel end. After completing this procedure, the oil level in the banjo housing may have to be topped up.

Lubrication Change Intervals

Lubricants used in axle assemblies are necessary to provide lubrication to the internal components and, equally important, they distribute additives throughout the axle and wash away small wear particles that would otherwise cause accelerated wear to the gears and bearings. The oil also carries away excess heat from high friction areas. All these conditions, along with constant exposure to heat, lower the performance properties of the oil. Oil analysis should be used to determine optimum oil change intervals. Guide lines that are given in the manufacturer's documentation are based on average operating conditions. The average temperature of lubricants in axles is between 160°F and 220°F (71°C to 104°C) in summer months. Operating equipment with temperatures above 200°F (104°C) causes the oil to oxidize at a higher rate, depleting the additives that give the oil its increased load-carrying capacity. This reason alone makes more frequent oil change intervals desirable.

DRIVE AXLE INSPECTION

Extensive dirt buildup on the outside of the axle housing can contribute to high lubricating oil temperatures and should be cleaned off regularly. Power washing the axle assembly when dirt is noticeable makes good sense. If the axle is to be repaired, it is mandatory to power wash the area that will be dismantled for servicing. During the inspection procedure, certain off-road equipment axles may require a technician to pressurize the axle center housing and wheel ends to check for leaks before disassembly. An axle housing that has seals that isolate the wheel ends (some wheel ends with internal wet disk brakes) needs to be air-pressure-tested at each wheel end separately in addition to testing the banjo housing. This air-pressure testing procedure requires the use of a 30 psi (206 kPa) air pressure gauge, an air shutoff valve, and an air regulator. Regulated air pressure (12 psi or 82 kPa) is applied to the planetary wheel end fill hole. The pressure should be maintained for 15 seconds. If the pressure does not hold, the wheel hub oil seal is leaking and must be replaced. Pressure check the banjo housing the same way but use the air vent hole in the banjo housing to perform this test. If 12 psi (82 kPa) pressure does not hold, find the source of the leak and make the necessary repairs.

AXLE REPAIR PROCEDURES

The procedures outlined below are intended to be used as a guide and are not intended to replace the manufacturer's technical documentation or repair manuals. Extreme care should be exercised when working on axle assemblies because the subassemblies are, in some cases, very heavy and may require the use of a lifting device. Before beginning work on an axle assembly, the technician should read over the manufacturer's recommended repair procedures carefully to avoid any possibility of damaging the subcomponents through improper disassembly and assembly and to prevent any possibility of injury. Before beginning any repairs, ensure that the vehicle is properly blocked up and the tires and wheels have been removed and stored in a safe location to facilitate the safe removal of the subassemblies. Drain all lubricants from the subassembly that is to be removed.

Planetary Wheel End Removal

1. Once the oil has been drained, match mark the planet cover, spider, and hub, as shown in **Figure 15-25.** Remove the bolts and planet cover.
2. Separate and remove the planetary spider assembly from the wheel hub. The use of puller screws may be necessary to complete this task. The adhesive that seals the spider to the hub may make it necessary to use a pry bar to loosen the spider assembly.
3. Depending on the model of the axle assembly, removing a set screw (if present) may be required before removing the planet pins.
4. In some cases, a press may be required to remove the pins, as shown in **Figure 15-26;** in other cases, they may be drifted out with a block of hardwood or brass drift. *Note*: Do not strike the planet pins with a hammer; they are hardened steel and may splinter and cause personal injury.

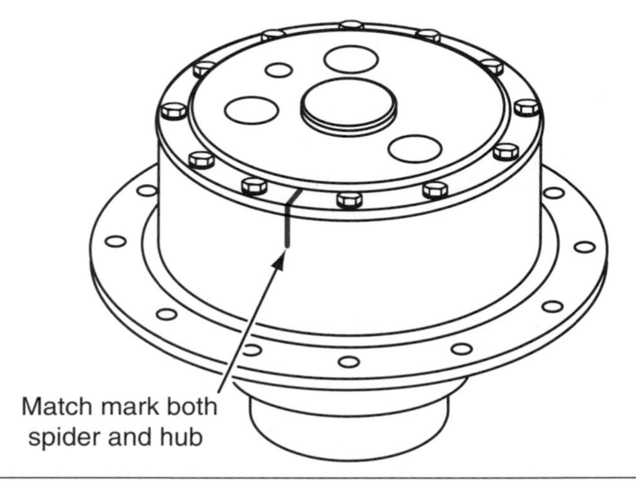

Figure 15-25 Match mark the spider, hub, and planet cover before removing.

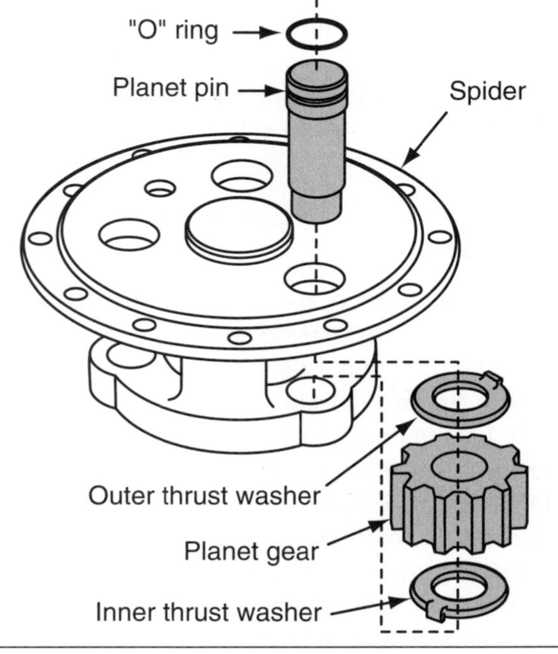

Figure 15-26 Breakdown of the planetary spider.

5. Remove the planet pins, thrust washers, and needle bearing if present.
6. Remove the ring axle shaft and sun gear.
7. Remove the wheel bearing adjusting nut lock. *Note*: Depending on the model, some units may use a double nut with lock and may be locked to the hub in a different manner.
8. Before removing the wheel bearing adjusting nut, use a sling to support the wheel hub, and then remove the adjusting nut.
9. Remove the ring gear hub from the wheel hub. Be careful not to pinch your fingers in the process.
10. On some models, the ring gear to ring gear hub may be bolted together with cap screws and lock plates on the back of the assembly.
11. Remove the outer wheel bearing from the hub.
12. Remove the inner seal and inner bearing from the hub.

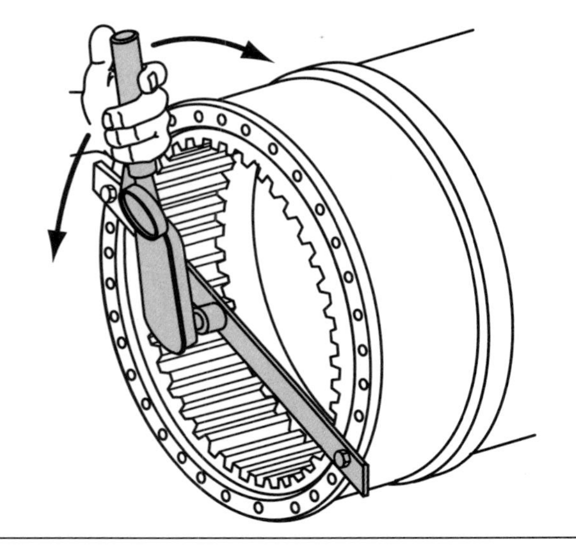

Figure 15-27 Rolling resistance of the wheel end measurement.

Planetary Wheel Ends Installation

1. Install new bearing cups with a suitable driver if the inner wheel bearings are to be replaced.
2. Install a new oil seal in the rear of the hub and lubricate the lip.
3. Depending on the axle model, a suitable lifting device may be required to reinstall the hub assembly on the axle. Note that damage to the inner lip seal may occur if alignment is not maintained during the assembly process.
4. Reinstall the ring gear assembly into the hub and install the adjustment nut to prevent the hub from slipping off the spindle.

Wheel Bearing Adjustment Procedures

Due to the large number of adjustment procedures that are used on different types of wheel ends, it is important to understand that the following adjustment procedures are examples of typical adjustment procedures and should not be used to actually adjust wheel bearing pre-load. Always refer to the manufacturers recommended procedures for a particular type of pre-load wheel end adjustment. **Figure 15-27** show how to check bearing the rolling resistance of the wheel end assembly.

Two basic bearing adjustment procedures that are used by most manufactures will be outlined in this section. The nut and lock plate design and the retainer plate and shim pack design. Before adjustment procedures can be made on any wheel bearing assembly, the tapered bearing and cups must be seated properly in the hub assembly. To bottom out the bearing cups in the hub assembly, the adjusting nut must be torqued up to approximately 500 lb ft (677 Nm) repeatedly until the torque does not advance the adjusting nut any further.

Nut and Lock Plate Adjustment The nut and lock plate design has many variations. Always refer to the manufacturer's installation procedure for the particular type that is being worked on.

1. Loosen the nut a quarter turn until a slight bearing end play is felt and the hub rotates freely.
2. Measure the no-load rolling resistance of the wheel hub assembly. Refer to the **Table 15-2** for the correct values. Now refer to Figure 15-27. *Note*: The rolling resistance is the force being shown on the dial torque wrench while the wheel is turning and is not to be confused with the start-up torque of the wheel, which is always higher.
3. Torque the adjusting nut to the recommended value above the rolling resistance value. Refer to Table 15-2 for the correct value.
4. Once the correct pre-load is reached, apply NeverSeize™ or a similar product to the threads before installing the jam nut against the adjusting nut lock. Be sure that the socket does not contact the outer tangs of the lock when torque is applied to the jam nut. This could shear the inner

Table 15-2: TORQUE CHART

Condition	Rolling Torque Range	Jam Nut Torque
New Tapered Roller Bearing Adjustment	7 to 12 lb-ft (10–16 Nm) torque greater than no-load rolling torque reading	500 lb-ft (677 Nm)
Adjusting Used Bearings	3 to 5 lb-ft (4–6 Nm) torque greater than no-load rolling torque reading	500 lb-ft (677 Nm)

Table 15-3: RECOMMENDED TORQUE VALUE

Cap Screw Size	Recommended Torque Value
1/2–13 Grade 8	90–100 lb-ft (123–135 Nm)
3/4–10 Grade 8	300–330 lb-ft (407–447 Nm)
1 1/4–10 Grade 8	1475–1625 lb-ft (1999.9–2711.8 Nm)
1 1/4–7 Grade 9	1850–2000 lb-ft (2508.4–2711.8 Nm)

lock tang, and torque the jam nut to 500 lb-ft (677 Nm).

5. Recheck the rolling torque value; be sure that it is not above the recommended torque value shown in Table 15-2.
6. If the torque value has not been achieved, repeat the procedure until satisfactory values have been reached.
7. After the correct rolling resistance values have been achieved, bend the lock tangs over the jam nut to prevent loosening.

Plate and Shim Pack Adjustment. The following steps are to be used as examples only. They should not be used as actual procedures. Always refer to the manufacturer's documentation for the recommended adjustment procedures for a particular type of wheel end bearing.

1. Before beginning, measure the retaining plate thickness and record the value.
2. Install the retainer plate without the shims.
3. Tighten the cap screws to the recommended torque values as outlined in **Table 15-3.**
4. Strike the retainer with a soft blow hammer to ensure bearings are seated properly, then loosen the cap screws two turns and rotate the wheel end four or five times.

Table 15-4: EXAMPLE OF CALCULATIONS TO DETERMINE SHIM PACK THICKNESS

Plate Thickness	.994" (25.25 mm)
Depth measured at 3 places, 120 degrees apart. Add together and divide by 3 to get an average reading.	1.127" (28.62 mm) 1.125" (28.57 mm) 1.126" (28.60 mm) 3.378"/3 = 1.126" (28.60 mm)
Subtract Plate Thickness	1.126" (28.60 mm) .994" (25.25 mm) .132" (3.35 mm)
Add .005" (0.2 mm)	.005" (0.12 mm) .137" (3.47 mm)
Select Shim Pack Using Micrometer	**Total Shim Required** .137" (3.47 mm)

5. Measure the rolling resistance using the correct dial indicator torque wrench and adapter.
6. Use a depth micrometer to measure the distance from the face of the retainer plate to the end of the spindle.
7. Subtract the retaining plate thickness (step 1) from the average reading in step 7 and add .005" to the shim pack thickness to get the correct amount of shim.
8. Measure the shim pack thickness and record the value used for the first attempt.
9. Install the shim pack under the retainer plate and torque up the retainer bolts to recommended value in Table 15-3, following the manufacturer's recommended procedures. See **Table 15-4** for an example calculation.
10. Before rechecking the rolling resistance, roll the wheel end over four or five times.

11. The rolling resistance with a new bearing should be 30 to 50 lb-ft (41–67 Nm) above the no-load rolling resistance.
12. Find the rolling resistance for the used bearings (a used bearing is one that has been in use for 300 hours) resistance.
13. If the rolling resistance is high, add a .001" shim and recheck. If the reading is low, remove a .001" shim and recheck.

Differential Removal

Before attempting to lift the equipment, be sure that wheel chocks are placed on the opposite set of wheels to prevent any accidental movement. The following steps to remove a differential from a drive axle assembly are to be used as an example of how this procedure is carried out. Any attempt to remove a differential should be preceded by reading the manufacturers recommended procedures for each specific axle.

1. Use a jack that is rated for the load you are lifting to raise the equipment, and block it with approved stands or blocking.
2. Drain the lubricant from the axle assembly differential housing and disconnect the driveline universal joint from the carrier.
3. Drain the wheel ends and remove the planetary covers and axle shafts from each end.
4. Before removing the differential mounting bolts, place the differential jack under the differential assembly for support.
5. Remove all the cap screws except the two top ones, and take up the weight with the differential jack.
6. Back off the last two fasteners to the end without actually removing them completely. Loosen the differential carrier with a pry bar.
7. Carefully remove the two fasteners and gently slide the differential jack out with the differential.
8. Remove the differential housing gasket, if used, and discard it.
9. After the differential is overhauled or replaced, reinstall the differential following the above procedure in reverse.

Differential Overhaul

The differential carrier should be secured to an overhaul stand before attempting to disassemble it. Inspect the crown and pinion for physical damage before attempting to remove it. If there are broken teeth or excessive wear, it may not be necessary to measure the backlash before removing. If the gear set is determined to be usable, then the backlash must be measured and recorded. The crown and pinion must be reinstalled with the same backlash to maintain the correct tooth contact pattern. If the gear set is damaged and must be replaced, this last step is not required.

1. If the differential carrier has a thrust screw adjustment it must be backed off before disassembly, as shown in **Figure 15-28.**
2. Rotate the differential carrier so that the crown gear is at the top to facilitate easy removal.
3. Match mark each bearing cap to its corresponding carrier leg as shown in **Figure 15-29.** Use a

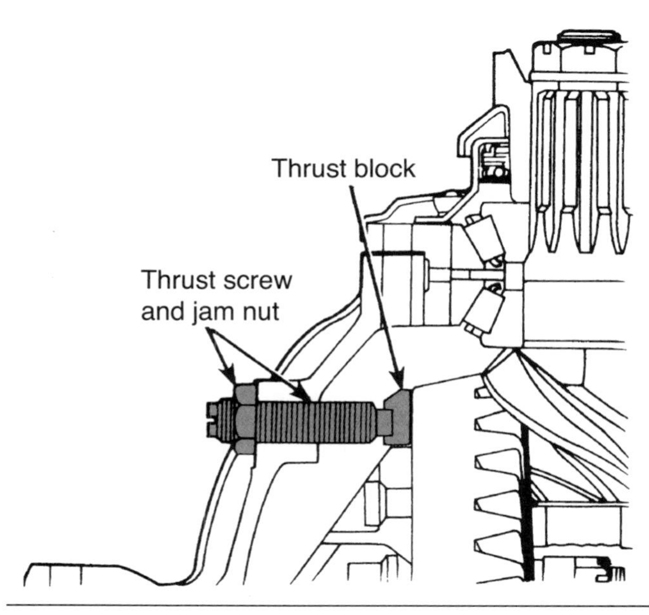

Figure 15-28 Location of the thrust screw assembly. (*Courtesy of Arvin Meritor*)

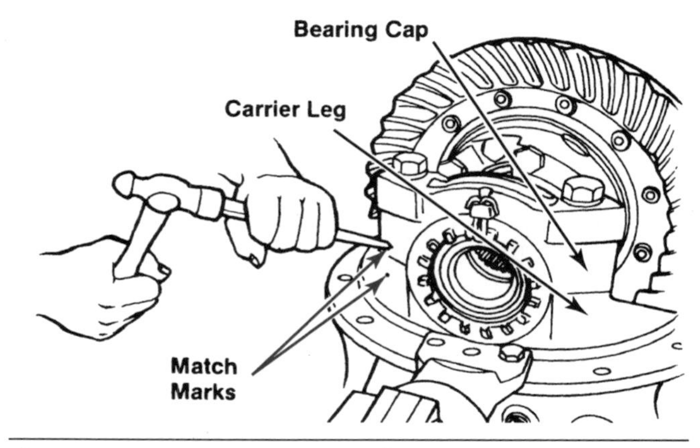

Figure 15-29 How to correctly match mark the caps before removal. (*Courtesy of Arvin Meritor*)

center punch; paint marks will wash off when the parts are cleaned.

4. Remove the cotter pins or lock plates that hold the side bearing adjustment rings in place.
5. Loosen and remove the two adjusting rings from the carrier. If the gear set is to be reused, keep the bearing caps and adjusters together for reuse in their original location.
6. Once the bearing caps are removed, lift the differential crown and pinion out of the housing. (Don't forget to remove the thrust block at this time if used.)

Differential/Crown Gear Disassembly.

1. If match marks are not present on the carrier case, use a center punch to identify correct orientation. The carrier halves must be installed in the same location when reassembled.
2. Separate the case halves and remove the spider, four pinion gears, two side gears, and the thrust washers, as shown in **Figure 15-30.**
3. If you need to replace the differential side bearings a puller is required to remove them without damaging the housing.
4. If the crown gear needs to be replaced, unbolt it from the differential casing. In some cases, it may need to be pressed out of the flange case.

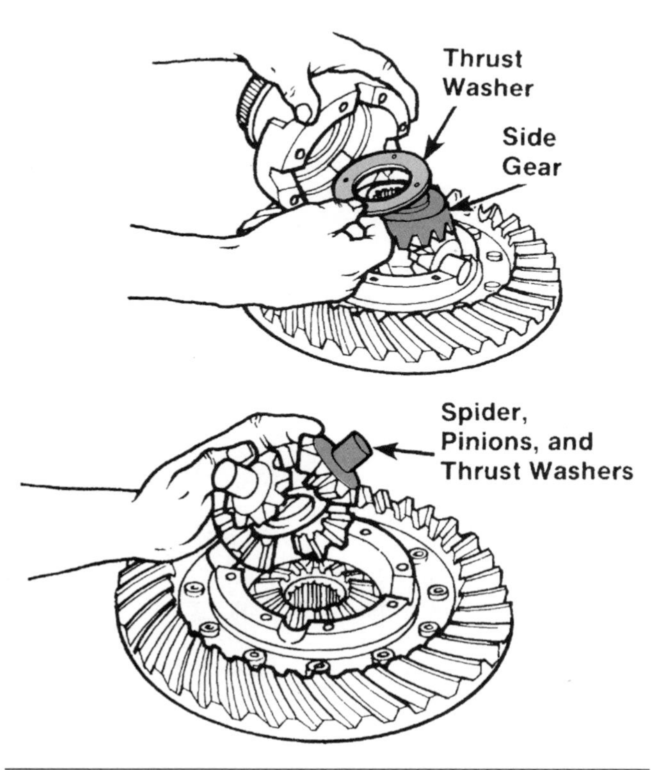

Figure 15-30 Removal of the spider and pinion gears. (*Courtesy of Arvin Meritor*)

Removing the Pinion Drive. The pinion drive on heavy-duty differentials is mounted to a pinion carrier, which houses two tapered roller bearings and the drive pinion and yoke.

1. First remove the pinion nut with an impact. Sometimes a pinion nut will have a cotter pin locking it to the shaft; remove it first.
2. In some cases, such as on some larger differentials, the pinion nut is torqued as high as 1000 lb-ft and must first be loosened with a large $^{3}/_{4}$-inch drive power bar.
3. Refer to **Figure 15-31** to view the procedure for this type of pinion nut removal.
4. Once the pinion yoke is removed, remove the cover and seal assembly along with the gasket from the bearing cage.
5. If the seal needs to be replaced, it must be installed with a press or sleeve driver or seal damage will occur.
6. Carefully remove the drive pinion assembly from the housing; do not damage or loosen the shims, as shown in **Figure 15-32.** Measure the shim pack and record the measurement. It will be needed to calculate the pinion depth when reassembling.

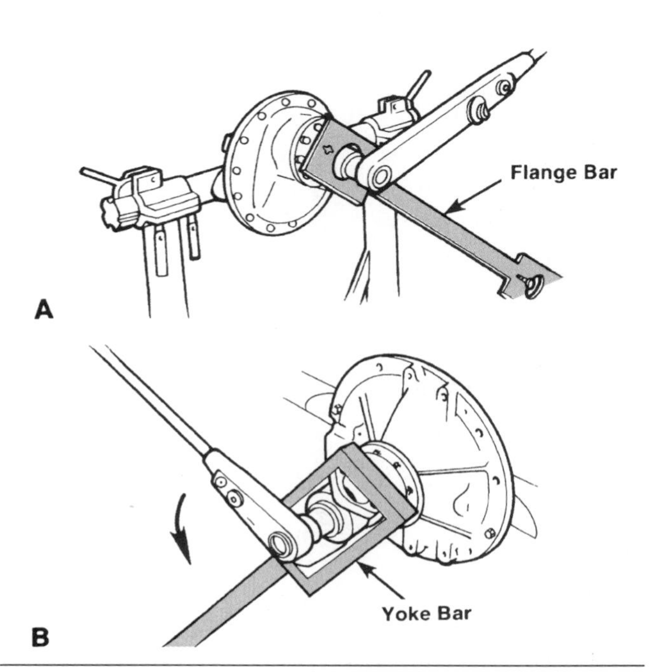

Figure 15-31 Removing the input (A) flange or (B) yoke. (*Courtesy of Arvin Meritor*)

Drive Pinion Bearing Cage Disassembly. On large differentials a press will usually be required to remove the pinion from the bearing carrier, as shown in **Figure 15-33.** As with any type of disassembly, the use of a specific repair manual for the type of differential that is being disassembled should be reviewed before undertaking this type of work.

1. Use a suitable press to remove the pinion from the bearing carrier, as shown in **Figure 15-33.** Be careful not to drop the pinion to the ground once it has cleared the bearing; this could cause damage to the pinion gear faces.

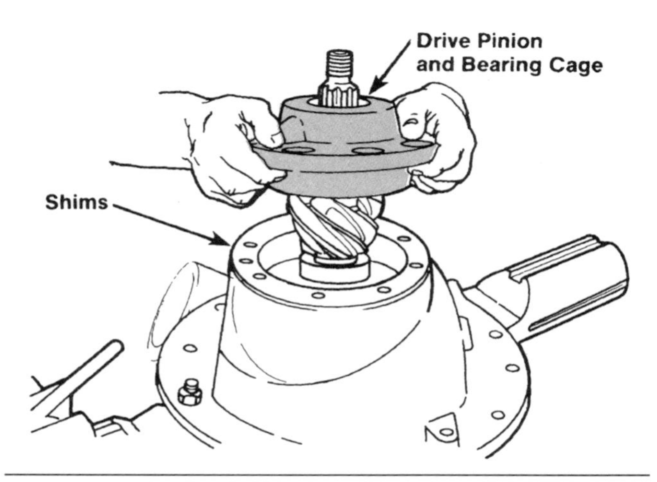

Figure 15-32 Removing the drive pinion and bearing cage. (*Courtesy of Arvin Meritor*)

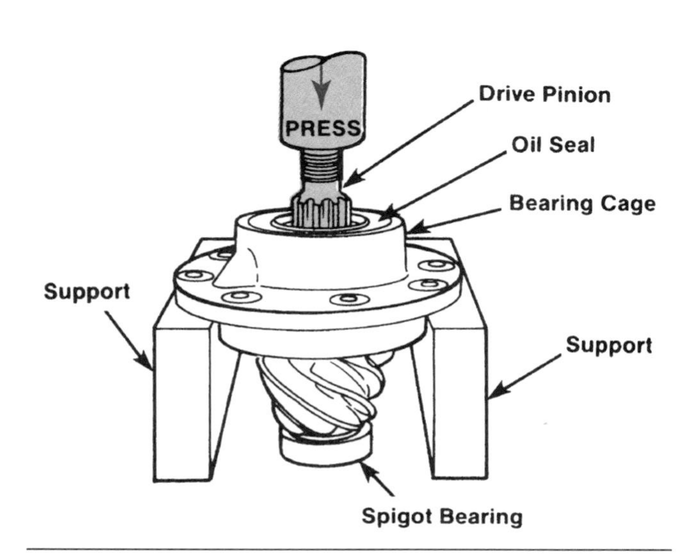

Figure 15-33 Pressing the drive pinion from the bearing cage. (*Courtesy of Arvin Meritor*)

2. If you are replacing the pinion bearings, a press will need to be used to remove the inner bearing from the pinion shaft, as shown in **Figure 15-34.**
3. If the spigot bearing is to be replaced, the snap ring must first be removed and a suitable bearing puller must be used to remove the bearing.

Differential Carrier Assembly

The steps outlined below are basically the reverse steps that are required to disassemble a differential carrier. This procedure is intended to be an example showing the general steps required to assemble a differential. The procedures that are used may vary extensively from differential to differential and should only be attempted after you have consulted the manufacturer's repair manual for a specific product.

1. If the pinion bearings have been replaced, the new bearing cups must be pressed into place in the bearing cage on a press. Installing the cups with a metal hammer is not recommended; the cups are made from hardened, high quality steel and may splinter from impact blows.
2. Press the inner bearing cone onto the drive pinion using a press. Be sure to apply pressure only on the inner race of the cone, not on the roller cage, or damage to the bearing will occur.

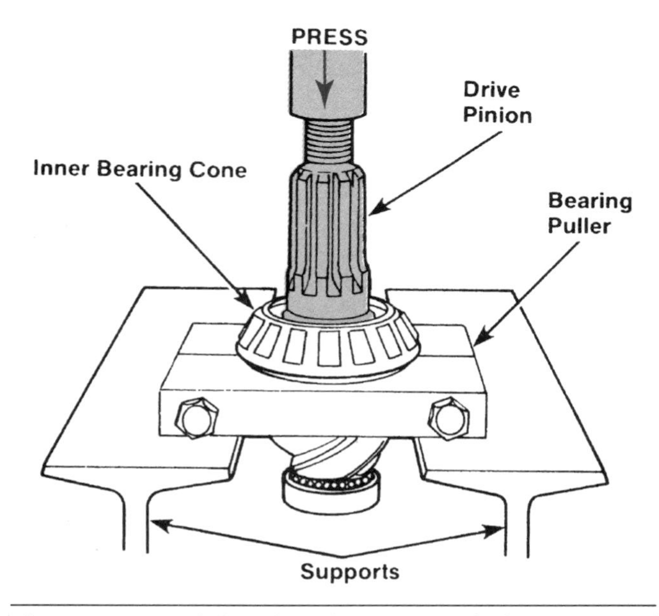

Figure 15-34 Pinion bearing removal. (*Courtesy of Arvin Meritor*)

3. Use a press to install the spigot bearing to the end of the drive pinion and install the snap ring in place, if applicable. Be sure that the bearing is installed correctly. Some bearings have a radius on one shoulder of the inner race. This radius faces toward the drive pinion teeth. The other side may have no radius and will prevent the bearing from seating properly.
4. Lubricate all the bearings and cups with axle lube before installing the drive pinion in the cage.
5. Install the bearing spacers and shims onto the pinion and press the outer bearing into place using a press. Refer to **Figure 15-35** for correct orientation.
6. Set the correct pre-load to the drive pinion assembly. To adjust the bearing pre-load, please refer to the section under testing and adjustment later in this chapter.
7. To adjust the shim pack thickness between the pinion cage and the differential housing, refer to the section under testing and adjusting outlined later in this chapter.

Installing the Drive Pinion Assembly. After completing the pre-load adjustments to the drive pinion assembly, install the assembly into the differential housing as indicated below:

1. If a new crown and pinion is to be installed, follow the calculation procedure outlined in testing and adjusting later in this chapter to determine the shim pack thickness. If the same crown and pinion is reinstalled, then install the shim pack that was removed during disassembly.
2. Be sure to line up the oil holes in the shim pack with the oil holes in the housing using the guide studs as shown in **Figure 15-36.**
3. To install the drive pinion assembly in the housing, it may become necessary to use a soft blow hammer to force the parts together. Install the seal cover and gasket to the assembly. Refer to **Figure 15-37** for the correct orientation.
4. Install and torque all the cap screws to the correct torque values recommended by the manufacturer's specifications.
5. Install the input yoke and torque up to the manufacturer's recommendations; use the yoke or flange bar during the torque procedure.

Installing the Crown Gear. The steps outlined here are shown as examples only. Crown gears, or ring gears as they are sometimes called, may have different installation procedures that are dependent on brand name. Always refer to the manufacturer's technical documentation for the actual procedures and specifications to perform this type of work.

1. To facilitate ease of installation the ring gear should be heated, using water to a temperature of 160 to 180°F for a few minutes. Never expose the gear to any type of flame, such as a torch, to heat the gear; damage to the hardened surface could result. Pressing a cold ring gear into place can result in damage to the case halves from the interference fit. The displaced metal could be

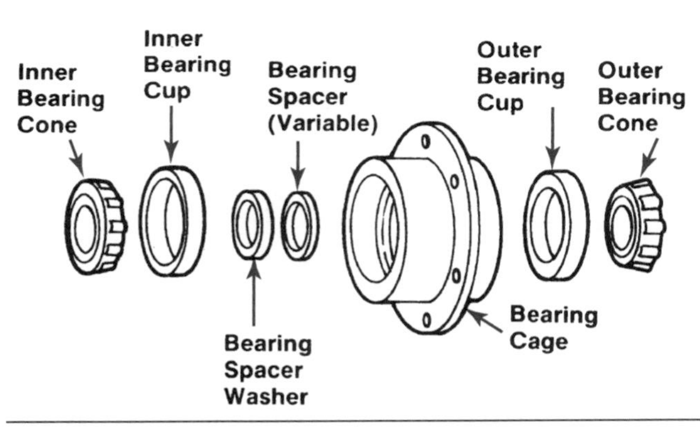

Figure 15-35 Correct orientation of parts in a pinion cage assembly. (*Courtesy of Roadranger Marketing. One great drive train from two great companies—Eaton and Dana Corporations*)

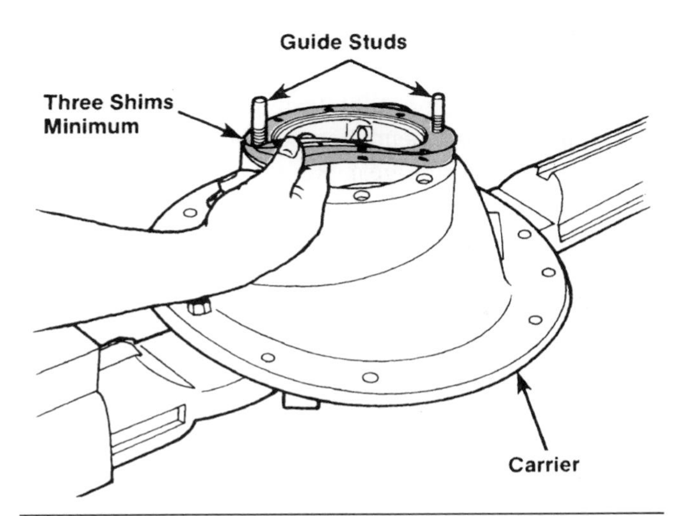

Figure 15-36 Installation of the shim pack before the drive pinion assembly is installed. (*Courtesy of Arvin Meritor*)

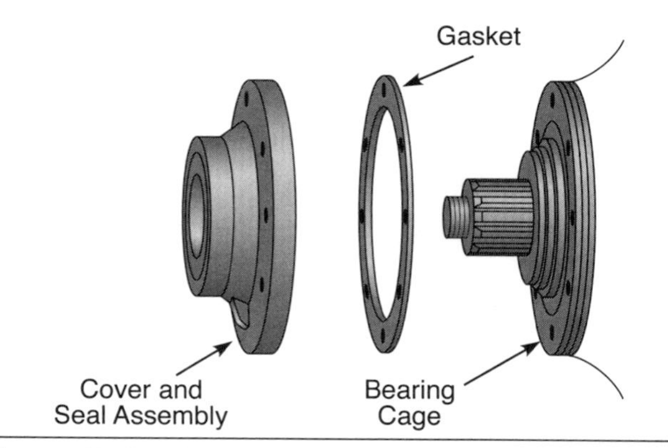

Figure 15-37 Correct orientation of the seal cover and gasket.

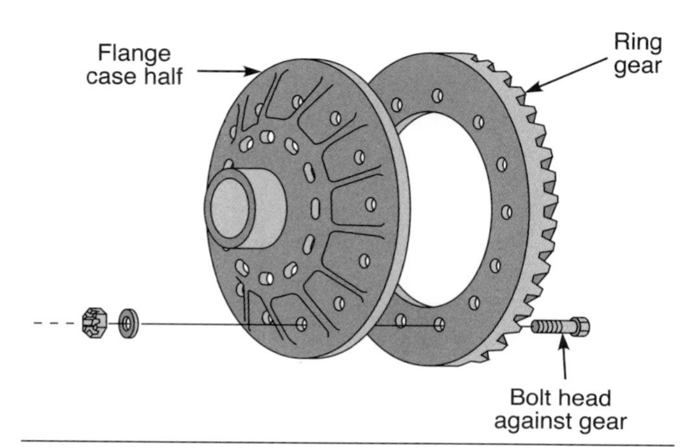

Figure 15-38 Correct orientation of the crown gear.

trapped between the parts and cause excessive ring gear runout.

2. Install and torque the fasteners to the manufacturer's recommendation so that the bolt head is against the ring gear. Refer to **Figure 15-38** for the correct orientation.
3. If new side bearings are to be used, install the bearings to both sides of the case halves by using a press, as shown in **Figure 15-39.**
4. Lubricate both sides of the case halves along with all the internal parts before assembly.
5. Place the case half with the crown gear facing up on a suitable bench to begin assembly.
6. Install one thrust washer and one side gear into the case, as shown in **Figure 15-40.**
7. Install the spider, differential pinions, thrust washers, and side gear. Refer to **Figure 15-41** for correct orientation.

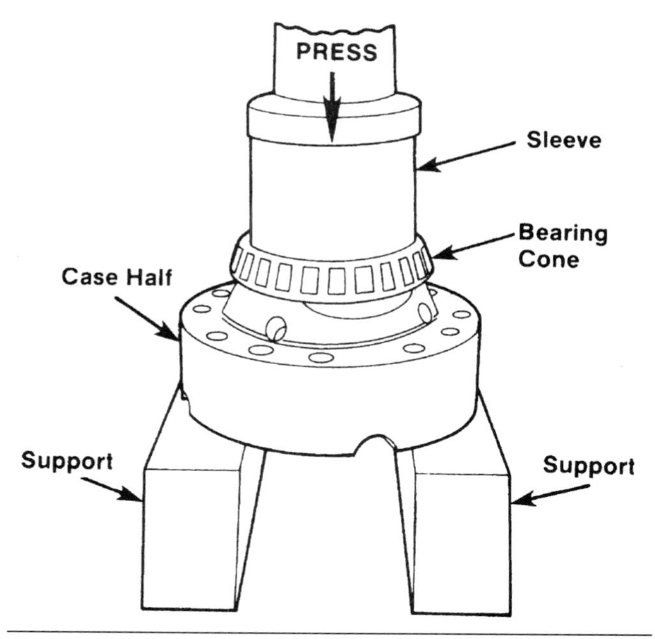

Figure 15-39 Bearing cone installation in case half. (*Courtesy of Arvin Meritor*)

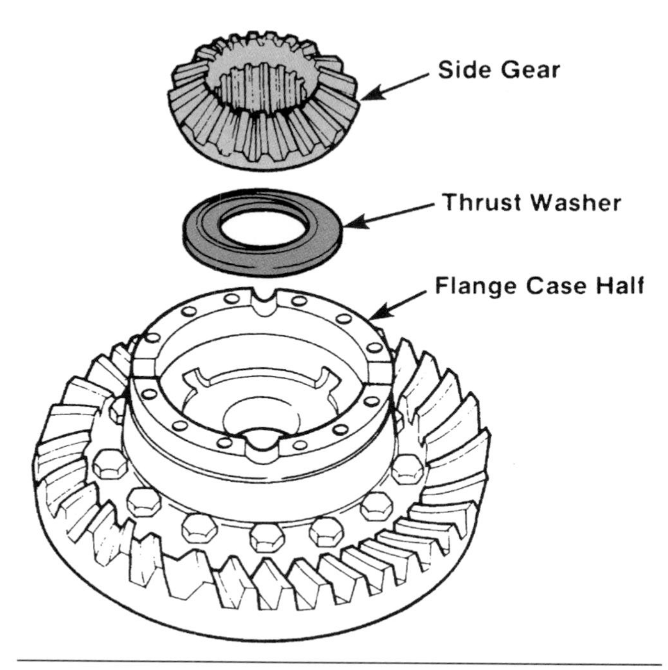

Figure 15-40 Correct orientation of the thrust washer and side gear. Lubricate all internal parts with axle lube before installing into place. (*Courtesy of Arvin Meritor*)

8. Install the other case half into position, being careful of alignment. Tighten the fasteners in a cross pattern to the manufacturer's torque specifications.

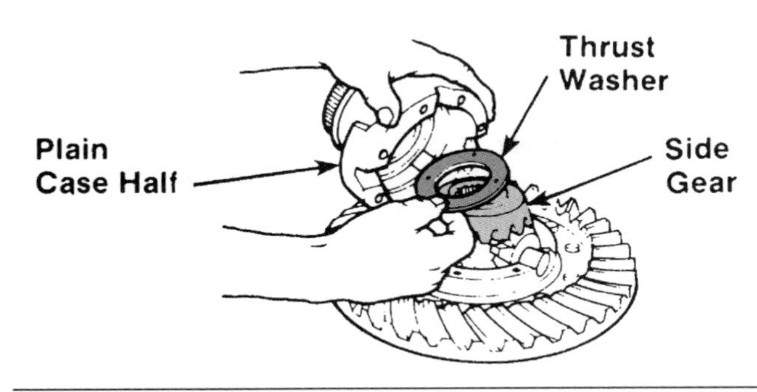

Figure 15-41 Correct orientation of the spider, differential pinions, thrust washer, and side gear. (*Courtesy of Arvin Meritor*)

Check to see that the differential spiders rotate freely when assembled.

Crown Gear Assembly Installation. The following steps are general examples of installation procedures. The actual procedures for this type of installation require that you follow the manufacturer's technical documentation on installing the crown gear assembly into the differential housing.

1. Be sure that the cups and bearing bores of the differential carrier are clean and dry. Apply lube to the inside surface of the cups, and be sure to lube the roller bearings in the process.
2. Install the bearing cups onto the side bearing before installing the crown assembly into the housing. Lower the assembly carefully into the housing, being careful that the cups stay in place on the side bearings.
3. Install the bearing cups following the orientation marks that were made during disassembly, and torque up to specifications.
4. To perform all the necessary adjustments in the right sequence, refer to the testing and adjusting section later in this chapter.
5. Perform these steps in the following: Adjust the pre-load on the differential bearing; Check the runout of the ring gear; Adjust the backlash of the ring gear; Check and adjust the tooth contact pattern; Adjust the thrust screw.

Installing the Differential Carrier into the Axle Housing

After completing the differential carrier assembly, it is now ready to install into the axle housing. The steps outlined here are only examples. You should always consult the manufacturer's technical documentation before attempting to install the carrier into the axle housing; not all installations are exactly the same. The installation procedure is the reverse of the disassembly procedure and is outlined below:

1. Be sure to thoroughly clean all mating surfaces with clean rags and solvent before attempting to assemble the components.
2. Inspect the mating surfaces for burrs or any other abnormality that may interfere with the fitment. Use a file to dress down any burrs that may be present.
3. Install a thread locking compound, such as Loktite® to the studs and torque up to specification.
4. In some cases, the manufacturer does not require a gasket to be installed between the differential carrier and the mounting flange of the banjo housing. Apply a small bead of silicone to the banjo housing surface to form a seal.
5. Install the differential carrier carefully, making sure the mating surfaces are in contact all the way around. Use only one fastener at each corner (at this time) to ensure that the assembly is drawn in evenly.
6. Inspect the mating area carefully to ensure a good fit before installing the rest of the fasteners. Torque all of the fasteners up to the manufacturer's specification.
7. Reinstall the axle, sun gear, and planetary carrier into each wheel end. *Note:* Clean the mating surfaces of the wheel hub and planetary carrier and apply a bead of silicone to the wheel hub contact area before assembling.
8. Be sure to reinstall the planetary carrier to the correct orientation, using the alignment marks that were made during disassembly. Clean the mating surfaces of the wheel hub and planetary carrier and apply a bead of silicone to the wheel hub contact area before assembling.

Axle Assembly Adjustments and Checks

Due to the large number of differential designs, this section will cover the adjustments made to a conventional single reduction differential assembly. The adjustment procedures that are shown in this section are to be used as examples; any adjustments a technician would make on an actual repair will require the use of the manufacturer's shop manual for the differential that is being overhauled.

Pinion Bearing Pre-load Adjustment. Newer differentials carriers use a press-fit outer bearing on the drive pinion. Although not as common, some of the older units use a slip-fit outer bearing that slips over the drive pinion. This section outlines both adjustment procedures.

Press-Fit Method of Adjustment

1. Assemble all the parts that are required to perform this procedure. Refer to **Figure 15-42** to view the correct order of assembly. Be sure to center the spacers between the bearing cones.
2. If a new gear set or new bearings are used, you must choose a nominal spacer based on the manufacturer's specifications. If the original parts are used, use the original spacers that were removed when it was disassembled.
3. Place the cage assembly in a suitable press and apply and hold a load onto the bearings. As pressure is applied, rotate the cage back and forth; this ensures seating the bearings.
4. Hold pressure on the bearing assembly. Use a string that is wound around the bearing cage and attached to a spring scale to measure pre-load on the bearings.
5. The scale should be read while the cage is actually turning; this will avoid getting the start-up torque reading, which would be higher.
6. When using this method of pre-load calculation, the radius (half the diameter of the cage) must be used to perform these calculations.
7. Depending on the specification, it may become necessary to adjust the pinion pre-load, in which case you either add or remove shims or spacers to achieve the desired pre-load.
8. When using this method it is important to select a spacer that is .001" larger than specified to compensate for bearing expansion when it is pressed onto the pinion shaft. This method will result in pre-load that will be correct *most* of the time.

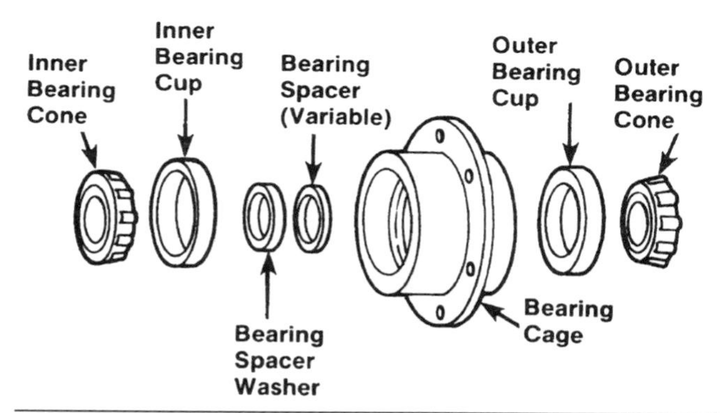

Figure 15-42 Correct orientation of parts in a pinion cage assembly before adjusting the pre-load. (*Courtesy of Roadranger Marketing. One great drive train from two great companies—Eaton and Dana Corporations*)

The following is an example of formulas used to calculate pre-load in Imperial Measurement:

Pull (lb) × radius (inches) = pre-load (in.-lb)
To convert to Newton meters use:
Pre-load (in.-lb) × 0.113 (conversion value) = pre-load (N-m)

The following is an example of formulas used to calculate pre-load in Metric Measurement:

Pull (kg) × radius (cm) = pre-load (kg-cm)
To convert to Newton meters use:
Pre-load (kg-cm) × 0.098 (conversion value) = pre-load (N-m)

In this example we will use actual values and use the formulas to calculate pre-load on a drive pinion. The pull on the scale is 10 pounds and the pinion cage diameter is 10 inches.

The following is an example in Imperial Measurement:

10 lb × 5 in. = 50 in.-lb (pre-load)
Convert to Newton Meters use:
50 in.-lb × 0.113 = 5.65 Nm

The following is an example in Metric Measurement:

4.6 kg × 2.7 cm = 58.42 kg-cm (pre-load)
To convert to Newton meters:
58.42 kg-cm × 0.098 = 5.73 N-m (pre-load)

Yoke method of Pre-load Adjustment. This method uses the pinion bearing assembly, fully assembled. The prior adjustment procedure did not use the drive pinion in the equation.

- Completely assemble the drive pinion cage including the input flange, as shown in **Figure 15-43,** by using a press.
- Assemble the yoke, flange washer, and nut and torque to specification. Note the position of

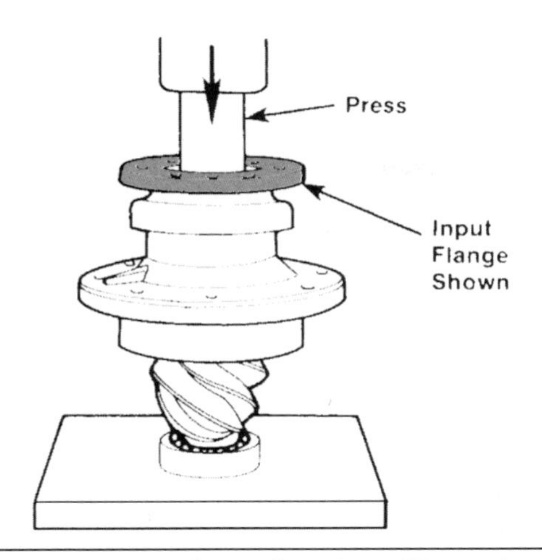

Figure 15-43 Procedure for installing the input flange with a press. (*Courtesy of Arvin Meritor*)

the yoke on the splines of the drive pinion; it must rest against the outer bearing before it is torqued.

- There are two methods that can be used to perform this. If you are working on a small differential assembly (12-inch ring gear or smaller) and have a large vise available, you can place the lip of the pinion carrier in a vise equipped with soft jaws to hold it from turning while you torque up the pinion nut to specification. If you are working on a larger differential assembly, you will temporarily install it in the banjo housing and secure it with two hand-tightened bolts to hold it in place during the torquing procedure. Do not install the shims at this time since it is only a temporary installation.
- Lock the yoke with a flange bar and torque up the pinion nut to specification.
- Place a dial indicator torque wrench with the appropriate socket on the pinion nut and rotate it slowly while checking the reading on the dial.
- If adjustment of the pre-load is necessary to change the spacer thickness, the unit must be disassembled and the entire procedure must be repeated.

Checking/Adjusting Pinion Cage Shim Pack

If the crown and pinion is being reused, start with the shim pack that was removed when the unit was disassembled. If a new gear set is used, then use this procedure to determine the shim pack thickness.

1. Measure the old shim pack with a micrometer and record that number along with the number on the new drive pinion. Refer to **Figure 15-44.**
2. If the number on the old pinion was a plus number, subtract it from the shim pack thickness. If the number was a minus, add it to the shim pack thickness.
3. Check the number on the new pinion and record it for reference later.
4. If the new pinion number is a plus, add the number to the shim pack thickness that was calculated previously. If the number on the new pinion is a minus, subtract it from the previously calculated shim pack.
5. The answer to the previous two steps now yields the new shim pack thickness to be installed with the new gear set.

See **Table 15-5** and **15-6** for examples of the calculations.

Differential Bearing Pre-Load Adjustment

Although there is more than one method for making this adjustment, only the most common one will be shown in this example. As with all adjustments, this example is to be used only as a general guide; always

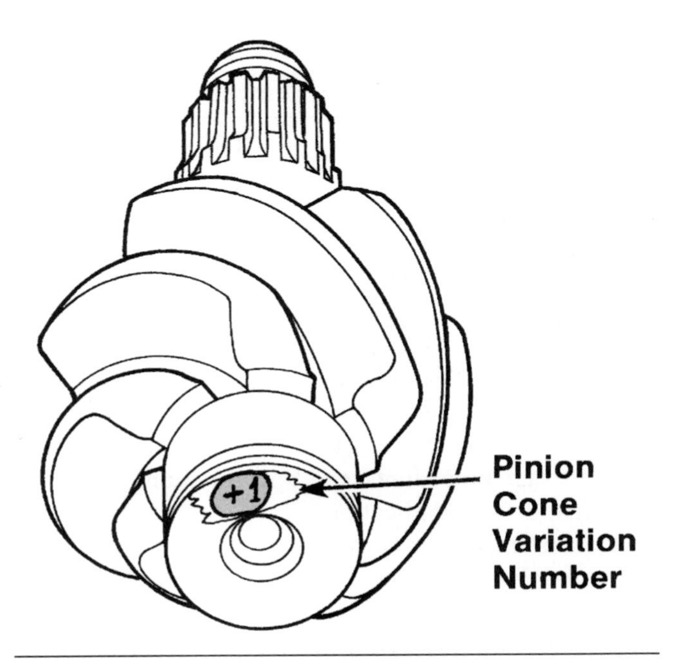

Figure 15-44 Location of the pinion cone variation number. (*Courtesy of Arvin Meritor*)

Table 15-5: CALCULATIONS OF SHIM PACK THICKNESS

	Inches	mm
Old Shim Pack Thickness	.030	.76
Old PC Number PC 2	−.002	−05
Standard Shim Pack Thickness	.028	.71
New PC Number PC + 5	+.005	+.13
New Shim Pack Thickness	.033	.84

	Inches	mm
Old Shim Pack Thickness	.030	.76
Old PC Number PC − 2	+.002	+.05
Standard Shim Pack Thickness	.032	.81
New PC Number PC + 5	+.005	+.13
New shim pack thickness	.037	.94

Table 15-6: CALCULATIONS OF SHIM PACK THICKNESS

	Inches	mm
Old Shim Pack Thickness	.030	.76
Old PC Number PC + 2	−002	−05
Standard Shim Pack Thickness	.028	.71
New PC Number PC − 5	−005	−.13
New Shim Pack Thickness	.023	.58

	Inches	mm
Old Shim Pack Thickness	.030	.76
Old PC Number PC − 2	+.002	+.05
Standard Shim Pack Thickness	.032	.81
New PC Number PC + 5	−005	−.13
New Shim Pack Thickness	.027	.68

consult the manufacturer's technical documentation to get the exact procedures and specifications for each individual design application.

1. It is important to start with a small amount of endplay in the carrier assembly. Place a dial indicator onto a flat surface, such as the mounting flange of the carrier, and place the plunger of the dial indicator so that it sits 90 degrees to the back of the ring gear, as shown in **Figure 15-45.**
2. To achieve a small amount of endplay to begin this procedure, loosen the adjusting ring opposite the dial indicator until a small amount of endplay is registered on the gauge. Use an appropriate T-bar wrench for this adjustment.
3. To adjust the ring gear to the left or right, use a pry bar, as shown in **Figure 15-46** as you read the dial indicator. Use the differential case as a pinch point when performing this adjustment. Tighten the same adjusting ring until no endplay shows on the dial indicator, moving the ring gear to the right or left as required to achieve no end play. Keep repeating until zero endplay is reached.

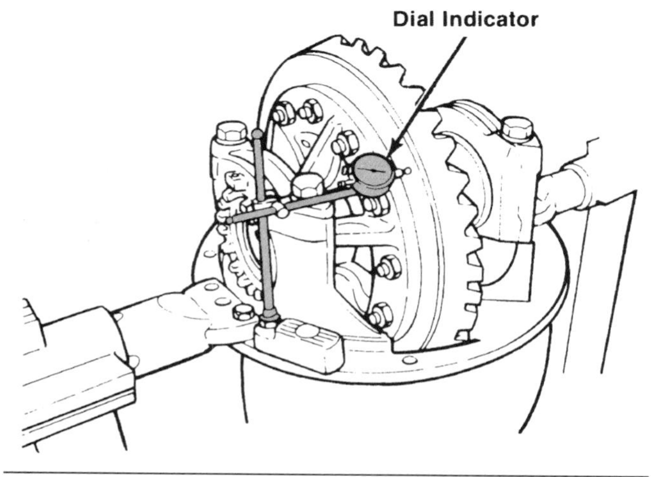

Figure 15-45 Location of the dial indicator used to measure carrier end play. (*Courtesy of Arvin Meritor*)

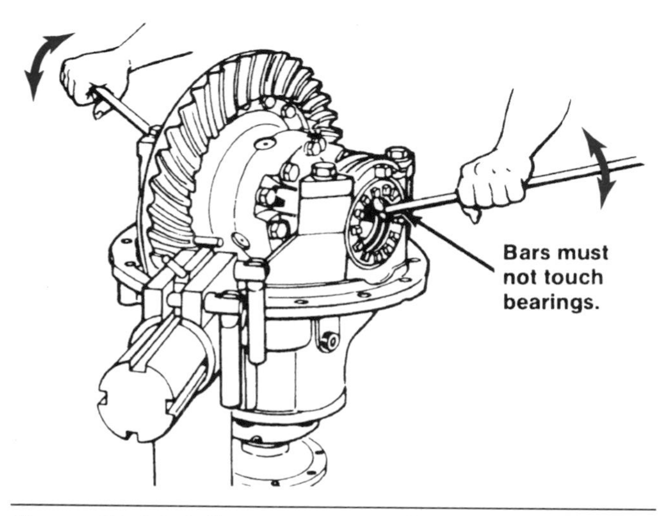

Figure 15-46 Adjustment procedure used to achieve pre-load to the differential assembly bearings. (*Courtesy of Arvin Meritor*)

4. To complete this adjustment procedure, tighten each adjusting ring one notch farther in from zero-lash to pre-load the bearings.

Ring Gear Runout Check

To check the runout on the ring gear assembly, the following steps must be followed carefully or an inaccurate reading will result. Follow the steps below in sequence to measure the ring gear runout.

1. Place a magnetic dial indicator on the flange mount of the differential housing and position the pointer 90 degrees to the back face of the ring gear with a slight pre-load on the pointer.
2. Adjust the dial indicator to zero and slowly rotate the ring gear. Record the total runout during one complete revolution of the ring gear. A nominal value will typically be around .008". But as always, consult the manufacturer's technical documentation for the specific differential that is being adjusted.
3. If excessive runout is recorded, then the differential must be disassembled and checked carefully for any burrs that might have been missed during assembly. The root cause of excessive runout must be corrected as this will lead to a short crown and pinion life.

Ring Gear Backlash Adjustment. If the old ring gear set is being reused, adjust the backlash to the values that were measured before disassembly. If a new ring gear set is used, adjust the backlash according to the manufacturer's specifications. *Note:* Backlash readings should be taken in at least three different places on the ring gear to get an average reading.

1. Place a magnetic dial indicator on the flange mount of the differential housing and position the pointer 90 degrees to the face of a ring gear tooth with at least one revolution of pre-load on the pointer and set it to zero, as shown in **Figure 15-47.**
2. Move the ring gear back and forth slightly and read the dial indicator movement; repeat this in three places spread out evenly around the ring gear. If the reading is not within specifications, adjust the backlash as outlined in following step.
3. Depending on which way the backlash must go, move the adjusting ring one turn at a time by loosening one bearing and tightening the opposite one. This will maintain bearing pre-load. Backlash will increase when the ring gear is moved away from the pinion and decrease when moved toward the pinion. Refer to **Figure 15-48.**

Ring Gear Tooth Contact Adjustment. As with any adjustments on a differential, this one is critical to the life of the gear set and, if not done correctly, can lead to a catastrophic failure. The tooth contact pattern consists of a contact patch that runs lengthwise

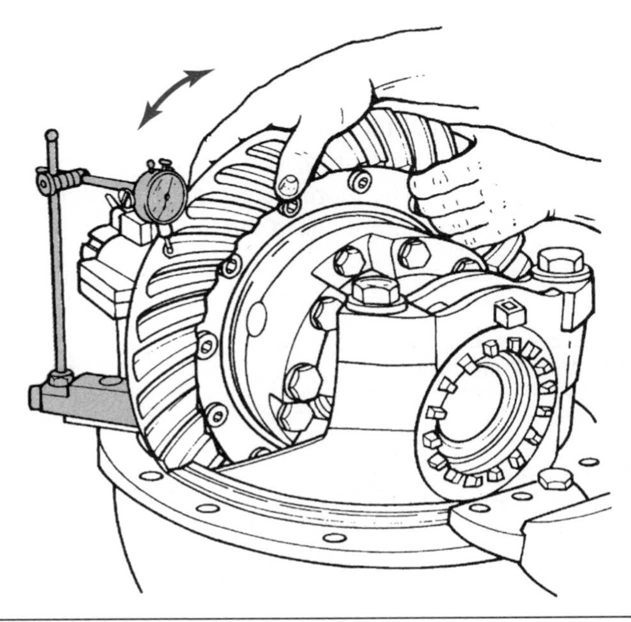

Figure 15-47 How to correctly measure backlash of the ring gear. *Note:* Check in at least 3 places to get an average reading. (*Courtesy of Arvin Meritor*)

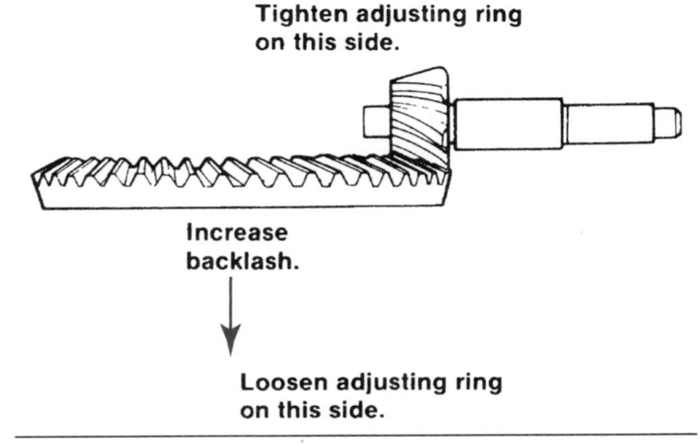

Figure 15-48 What happens when the ring gear is moved away from the pinion. *Note:* Backlash increases. (*Courtesy of Arvin Meritor*)

along the tooth of the ring gear as well as the profile of the tooth.

1. Paint at least 8 to 12 teeth on the ring gear before rolling the gear teeth together to check the pattern. A good pattern, as shown in **Figure 15-49** will have the contact mark centered on the ring gear teeth lengthwise. This test is in an unloaded condition and does not represent what takes place under load. The contact pattern in this case will be about one-half to two-thirds of the tooth.
2. When checking the pattern on a used gear set, the contact will not usually display square as is the case with a new gear set. The contact pattern will have a pocket at the toe of the gear that forms a tail at the root of the tooth. This pattern would look better if a load could be placed on the gear set. The more wear there is on the gear set, the more pronounced the tail will be.

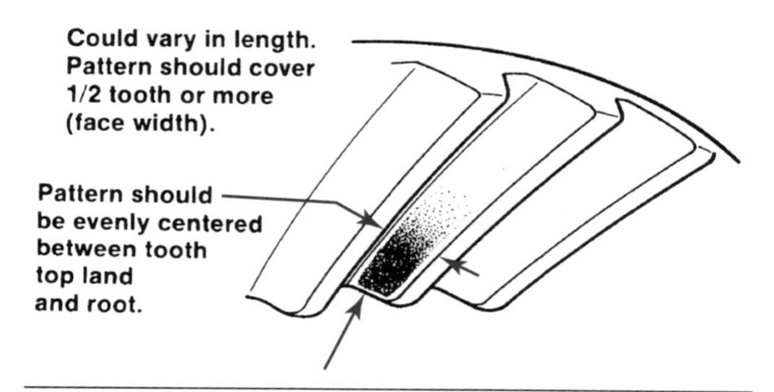

Figure 15-49 Contact pattern on a new gear set. (*Courtesy of Roadranger Marketing. One great drive train from two great companies—Eaton and Dana Corporations*)

3. Before you disassemble the gear set, it is a good idea to perform a tooth contact check and make a sketch of the pattern for future reference. A good pattern will not extend to the toe, and it will be centered evenly along the tooth face between the top land and the root. At no time should the pattern run off the tooth. If required, adjust the contact pattern by moving the ring gear and pinion gear either away from each other or toward each other. Pinion position in relation to ring gear position is determined by the thickness of the shim pack between the pinion cage and the differential housing. This adjustment controls the tooth depth.
4. These adjustments are interrelated because one will affect the other; they must be considered together even though they are two distinct adjustments.

   Whenever these adjustments must be made adjust the pinion first, then adjust the backlash. Continue this process until a satisfactory pattern is achieved. Here is where experience comes in: An experienced technician will set a good pattern in a few minutes, whereas an inexperienced technician will often take quite awhile before figuring it out. **Figure 15-50** shows an example of a contact pattern that needs to be corrected.

Thrust Screw Adjustment

If the differential has a thrust screw, it must be adjusted properly. Follow the steps listed below to complete this procedure correctly. Before beginning this adjustment, the carrier must be rotated in the stand until the ring gear is facing down toward the floor.

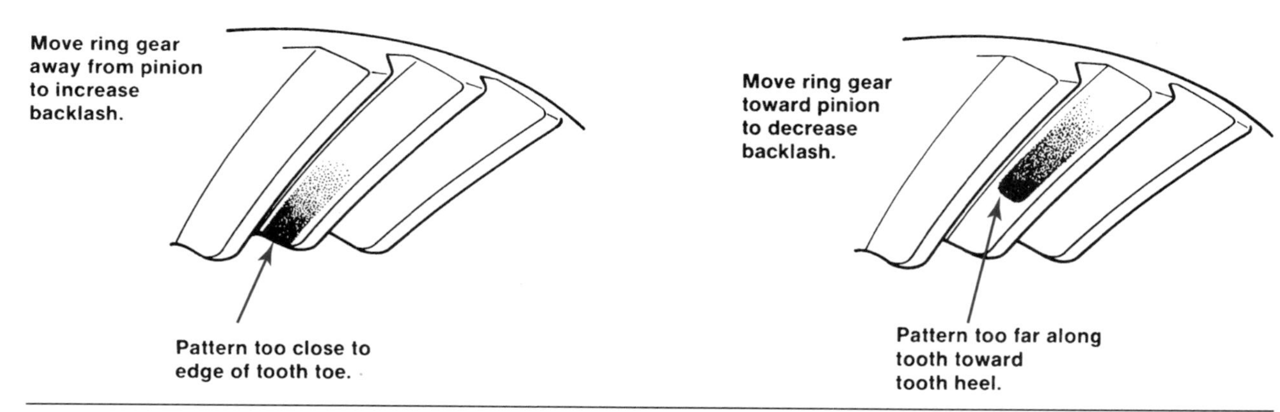

Figure 15-50 Contact pattern too far toward the heel and too close to the toe. (*Courtesy of Roadranger Marketing. One great drive train from two great companies—Eaton and Dana Corporations*)

1. Place the thrust block on the back side of the ring gear and rotate the ring gear until the thrust block and the thrust screw hole are lined up. Install the thrust screw until it contacts the thrust block, and then back out approximately one-half turn.
2. Install the lock nut onto the thrust screw and secure it, making sure that the screw does not turn during this procedure. Torque the jam nut to specifications.

FAILURE ANALYSIS

Drive axles are subject to failure from a wide variety of conditions; some are caused by inappropriate equipment operation and others are caused by poor maintenance practice. A technician's ability to assess the condition and come up with reasonable conclusions rests on his or her ability to understand the limitations of drive axle components and their lubricants. Component damage can lead to expensive repairs and downtime. If a technician replaces the failed components without identifying and correcting the root cause, more problems associated with the failure will occur. It is a considerable challenge for the technician to achieve maximum component service life.

The equipment operators have an important role to play in this effort, and must operate the equipment in a manner that will not stress the axle's components to the point of failure. When a failure occurs on a drive axle, two questions must be answered. What occurred? And how can another failure of this type be prevented from occurring again? There are recognized limitations to failure analysis in the field and, in some cases, a component may need to be analyzed in a lab. A qualified technician can identify most failures that occur and recommend procedures that will prevent the same failure from occurring again. This section will look at some common failures and show what to look for to successfully identify corrective measures.

Failure Types

Axle assemblies fail due to one or more of the following conditions:

- Normal wear
- Improper lubrication
- Fatigue
- Shock loading
- Improper operation

Normal Wear. Drive axles are designed to operate for a reasonable amount of time in the field without any more care than regular maintenance. All components will eventually fail, but there are preventative tasks that can be performed by the operator and service technician that will minimize early failures. During the break-in process, a small amount of wear takes place. In some cases, this wear has a positive effect on component life. Gears that mesh together benefit from this type of wear. Components often come from the manufacturer with a defect in the surface finish. These conditions can be recognized by a qualified technician as normal wear, thus eliminating unnecessary component replacement. Some parts should always be replaced when the axle components are overhauled.

Improper Lubrication. Using the incorrect type of lubricant can have a negative effect on the axle component life. Contamination from things like moisture or dirt can cause scoring and pitting of mating surfaces. Dirt in the lubrication acts like sandpaper, wearing out the mating components and shorting their potential life. **Figure 15-51** shows the damage that can occur when the additive package becomes depleted. The following list addresses more conditions that are likely to cause damage:

Figure 15-51 How depleted additive damage causes ridging on the gear teeth. (*Courtesy of Arvin Meritor*)

- Overheating damage results from high oil operating temperatures, which can be caused by several operating conditions. Operating the equipment with low oil in the axle assemblies will lead to higher operating temperatures.
- Overfilling the axle assembly with oil may also lead to higher than normal operating temperatures.
- Excessive increase in engine power or torque rating and restricted air flow over oil coolers used on some equipment will lead to high oil temperatures.
- Using the wrong grade or viscosity of lubrication may lead to overheating of internal components.
- Depleted additives in the lubricant can cause a wear pattern that is referred to as crow's-feet. These look like scoring lines or ridges on the gear teeth.
- Mixing any lubricants together will most certainly lead to premature failure.

Fatigue (Torsional Bending). This condition is a result of torsional loading and bending of a shaft or component that leads to progressive destruction of the component. The contoured lines often seen on the face of a broken component represent typical characteristics that are called steps. These occur before total breakdown of the part. Refer to **Figure 15-52** for an example of this condition. This condition can be usually attributed to overloading of the axle assembly.

Repeated overloading of the axle shafts can lead to a star-shaped torsional fracture in the splined area of an axle shaft. Refer to **Figure 15-53** to see how this symmetrical star-shaped break looks like.

Shock Loading. Shock loading happens to components that are stressed beyond the strength of the material, which sometimes causes instantaneous breakage, as shown in **Figure 15-54.** This occurrence may not always happen immediately; it sometimes results in a crack, which eventually leads to breakage. Shock impacts can subject axle shafts to high torsional or twisting forces that can lead to torsional shear. When initial shock impact is not great enough to shear the tooth or shaft, a crack will develop and, in a short time, will lead to failure. One of the common causes associated with this type of failure is an operator who changes direction of a loader without bringing the equipment to a complete stop. Losing traction on one side of an axle drive, and then suddenly regaining traction, will often lead to this type of failure. In most cases, this type of failure does not produce immediate catastrophic failures, the failure can take place days or weeks later.

Improper Operation. This operating condition can have a devastating effect on many components in a drive axle assembly. A common contributor to this

Figure 15-52 How torsional fatigue can look after a failure. (*Courtesy of Arvin Meritor*)

Figure 15-53 A star-shaped torsional fatigue fracture. (*Courtesy of Arvin Meritor*)

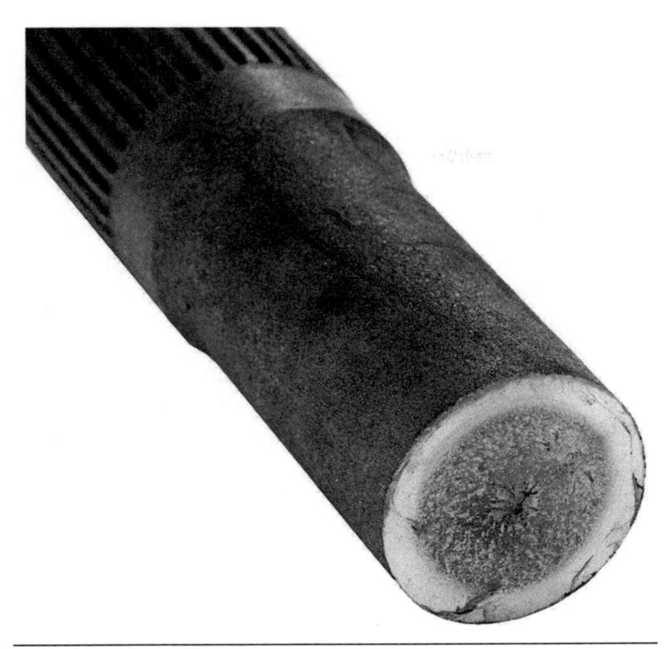

Figure 15-54 How a high impact shock load can lead to a clean break of an axle shaft. (*Courtesy of Arvin Meritor*)

Figure 15-55 Damage on a differential spider. (*Courtesy of Arvin Meritor*)

problem is overloading, which can lead to axle housing damage. Operating equipment that exceeds the Gross Axle Weight Rating (GAWR) can often cause extensive component damage. Identifying these causes should initiate action to prevent further occurrences by showing the operators the results of this type of abuse.

Spinout is another common condition that occurs frequently on equipment such a haulage trucks, as shown in **Figure 15-55.** On a single rear axle haulage truck differential spinout occurs when the equipment loses traction on one side of the drive, which causes one wheel to spin excessively. In the case of a tandem drive, the spinout occurs in the interaxle differential; one axle spins while the other remains stopped. This condition may leads to differential spider and pinion damage or breakage.

ONLINE TASKS

1. Use a suitable Internet search engine to research the configuration of an off-road drive/steer axle system. Identify a piece of equipment in your particular location paying close attention to the drive/steer axle arrangement, and identify whether or not it uses an additional planetary gear reduction in the wheel end. Outline the reasoning behind the manufacturer's choice of this particular design for that particular application.
2. Check out this url: *http://training.arvinmeritor.com/onlinetraining.asp* for more information on drive/steer axle configuration and design.

SHOP TASKS

1. Select a piece of equipment in your shop that has a drive/steer axle assembly and identify the type of components that make up the wheel end arrangement. Identify whether or not the drive axle has excessive wear in the wheel ends and what type of U-joint is used on this arrangement.
2. Select a piece of off-road equipment in your shop that uses a drive/steer axle and identify the type and model of the equipment, listing all the major components and the corresponding part numbers of the drive/steer wheel end assembly. Identify and use the correct parts book for this task.

Summary

- The drive axle's primary role is to carry the vehicle load. It must also be able to provide a means of propelling the equipment.
- Drive axles can be categorized into three distinct designs. The first is capable of propelling the vehicle as well as carrying the weight of the load. The second is designed to carry the load and steer the vehicle. The third is designed to carry the load; this type is found on the rear section of an articulating haulage truck.
- Torque is sent through the ring gear, which changes the direction of power flow 90 degrees, allowing the axles to be driven at a right angle to the input power.
- During a turn, the inside wheel has more resistance when making the turn; this causes unequal torque to be applied to the side gears in the differential. This change in resistance forces one axle to slow down, thus forcing the differential pinions to walk around the side gear.
- A jaw clutch is used to lock the differential, providing equal torque to both wheels. This type of lock is usually engaged with a foot control; the operator steps on a pedal that works through a mechanical linkage, resulting in a constant torque being applied to a lock lever.
- The limited slip differential has been designed to provide equal torque application to both wheels during normal operation. The locking effect takes place due to high internal friction generated by the increase in torque being applied.
- The no spin differential uses jaw clutches that are splined to the side gears. The jaw clutches stay engaged with the side gears as long as the equipment is operated in a straight line. When the equipment turns, the outside wheel speeds up, disengaging the outside jaw clutch and allowing the outside wheel to freewheel at a faster speed than the inside wheel.
- The planetary differential input torque is provided by a gear in the transmission transfer drive that meshes with the carrier of the planetary differential. The two sets of planetary gear shafts are held by the carrier and drive the planet gears.
- A large percentage of interaxles are lubricated by the splash method; oil is distributed by the gears turning and slashing the oil internally. Directional baffles are used to direct oil to critical areas of the differential assembly.
- A power divider's job can be categorized into two different parts. It must be able to divide the driving torque between two axle assemblies, and it must be able to provide differential action between the two axles when required.
- Equipment with two-speed axles can take advantage of the extra speed range and use it to get more effective gearing. An example of this is the use of low range in each of the two axle ratios: First gear in low range, then first gear again in high range, and continue up through the gear ranges.
- Planetary axle drives receive torque from the differential through an axle shaft, which serves as a sun gear (input gear). This shaft has splines at both ends: One to fit in the differential, the other end (sun gear) to send torque to the planetary drive gears. The sun gear meshes with the three planet gears, which are mounted together in a planet carrier. The ring gear is held or locked in place while the planet carrier provides the output, resulting in a gear reduction.
- Full-floating axles are used on large, slow moving off-road equipment with a drive wheel on each end of the axle housing. Two large, tapered roller bearings are mounted to the wheel hub and carry the full weight of the equipment and load.
- Semi-floating axles at the differential are splined to the side gear and float at this end. At the wheel end of the axle, the shaft carries one or two bearings, depending on the axle-load carrying capacity, and carries the full weight of the load at this end.
- When lubrication levels are not properly maintained in axles, the life of the bearings and gears will be adversely affected, shorting their life or leading to catastrophic failures.
- Many a differential has been destroyed by a phenomenon called "channelling." In sub-zero temperatures the cold thickened oil is parted by the rotating ring gear and is too thick to flow back, resulting in inadequate lubrication until the oil heats up.
- To check the oil level of the planetary wheel ends, the wheel hub oil level plug must be in the down

position. You may need to move the vehicle to line up the oil level plug in the correct location.

- If the gear set in a differential is determined to be reusable, then the backlash must be measured and recorded. The crown and pinion must be reinstalled with the same backlash to maintain the correct tooth contact pattern. If the gear set is damaged, it must be replaced.
- If the pinion bearings have been replaced, the new bearing cups must to be pressed into place in the bearing cage on a press. Installing the cups with a metal hammer is not recommended; the cups are made from hardened, high quality steel and may splinter from impact blows.
- If the crown and pinion is to being reused, start with the shim pack that was removed when the unit was disassembled.
- If excessive ring gear runout is recorded, then the differential must be disassembled and checked carefully for any burrs that might have been missed during assembly. The root cause of excessive runout must be corrected since this will lead to a short crown and pinion life.
- Depending on whether the backlash must be increased or decreased, move the adjusting ring one turn at a time by loosening one bearing and tightening the opposite to maintain bearing preload. Backlash will increase when the ring gear is moved away from the pinion and decrease when moved toward the pinion.
- When checking the pattern on a used ring gear and pinion set, the contact will not usually display square as on a new gear set. The contact pattern will have a pocket at the toe of the gear that forms a tail at the root of the tooth. This pattern would look better if a load could be placed on the gear set.

Review Questions

1. What is the final drive section of an articulating haulage truck drive axle usually known as?

   A. rear transaxle
   B. dead axle
   C. differential axle
   D. articulating axle

2. On a single reduction differential what is the drive axle *ring gear* bolted or riveted to?

   A. axle shafts
   B. bearing cage
   C. differential case flange
   D. pinion gear

3. Which of the following describes what takes place in a differential during cornering?

   A. outside wheel rotates faster than the inside wheel
   B. outside wheel rotates slower than the inside wheel
   C. inside wheel is accelerated to the speed of the outside wheel
   D. outside wheel is braked to the speed of the inside wheel

4. Which one of the following statements accurately describes a drive axle configuration that uses two gear sets to provide additional gear reduction and has greater torque at the drive wheels?

   A. single reduction axle
   B. power splitter axle
   C. tandem drive axles
   D. double reduction axle

5. Which one of the following statements is considered accurate?

   A. all two-speed axles use double reduction
   B. double reduction axles are always two-speed axles
   C. all tandem drive axles are double reduction
   D. single reduction axles are seldom used in haulage trucks

6. Which of the following statements best describes an *amboid* gear arrangement?
   A. axis of the drive pinion is at the centerline of the ring gear
   B. axis of the drive pinion is below the centerline of the ring gear
   C. axis of the drive pinion does not identify the amboid gear design
   D. axis of the drive pinion is above the centerline of the ring gear
7. Which of the following components must rotate at the same speed when the differential lock on a single reduction axle is engaged?
   A. differential case and both axles
   B. both axles and the drive pinion
   C. drive pinion and crown gear
   D. drive pinion and both axles
8. Which of the following axle hub arrangements carries the weight of the load on the wheel end of the axle?
   A. semi-floating
   B. full-floating
   C. tandem axles
   D. trailing axles
9. Identify which of the following axle types would use a *kingpin*?
   A. tandem drive axles
   B. steering axles
   C. trailer lift axles
   D. articulating axles
10. What components change the direction of powerflow in a drive axle carrier?
    A. differential gearing
    B. crown and pinion
    C. planetary reduction gearing
    D. axle shafts and side gears
11. Which of the following describes another term commonly used to identify a differential carrier housing?
    A. dead axle
    B. pusher axle
    C. banjo housing
    D. trailing housing
12. Where does the gear reduction actually take place on a single reduction differential carrier?
    A. differential gearing
    B. planetary carrier
    C. propeller shaft and pinion
    D. pinion and crown gearset
13. Which of the following identifies the gear design when a spiral cut pinion and ring gears mesh so that the pinion intersects the ring gear just below its center axis?
    A. amboid
    B. hypoid
    C. spiral bevel
    D. spur gear
14. What do the differential pinions rotate on in a interaxle differential?
    A. spider
    B. ring gear
    C. pinion gear
    D. propeller shaft
15. What connects the steering knuckles on a drive/steer axle?
    A. steering differential arm
    B. pitman arm
    C. tie rod
    D. drag link

CHAPTER 16 Final Drives

Learning Objectives

After reading this chapter, you will be able to:

- Describe the types of final drive systems used on off-road equipment.
- Identify the purpose of final drives used on off-road equipment.
- Identify the final drive system designs used on off-road equipment.
- Identify the construction features of final drive systems.
- Describe the operation of final drive systems on off-road equipment.
- Identify the maintenance and adjustment procedures on final drive systems.
- Describe the common failures that occur on final drive systems and analyze the causes.

Key Terms

backlash
beach marks
bull-type final drive
chain drives
double reduction drive
elongation
final drive
herringbone gears
mechanical seals
pin lock
pinion drive
pinion shaft
planet gears
planetary drive
pre-load
ring gear
roller link
side bar
single reduction drive
spiral fracture
sprocket shaft gear
spur gears
sun gear
viscosity

INTRODUCTION

The term *final drive* as the name suggests, is the final stage in a series of reductions that starts at the engine flywheel, flowing through the torque converter and transmission and ending at the **final drive** (wheels or tracks). **Figure 16-1** shows an example of a typical final drive assembly used on a motor grader. The term can actually be applied to four types of drive systems: **chain drives, pinion drives, planetary drives,** and straight axle drives. Straight axle drives, and planetary drives on wheel ends are covered in detail in another section of this book. Reference to them will be made when discussing final drive designs since some manufacturers include the use of straight axles and planetary drives in their designs.

Final drives accept power from the transmission, reducing its speed and increasing the torque through the

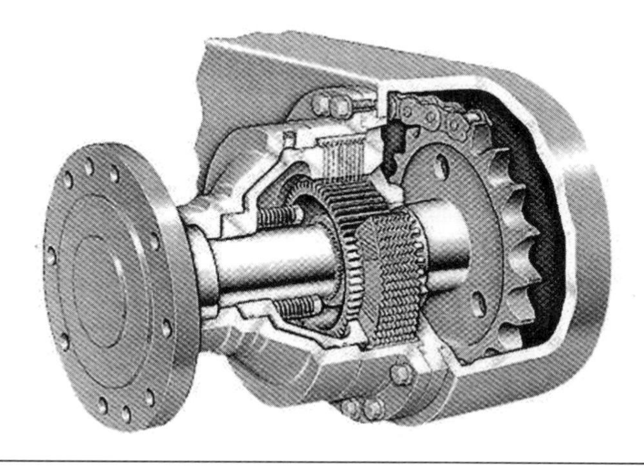

Figure 16-1 Final drive on a Cat 12H motor grader. (*Courtesy of Caterpillar*)

use of various gear and chain drive arrangements. Although the main purpose of a final drive is to increase torque and decrease speed, some manufacturers take advantage of this arrangement to drop the flow of power lower to the ground as is the case of a final drive on a bulldozer or grader. This concept has a number of advantages such as reducing the drive speed and stresses while at the same time increasing the available torque at the drive wheels to move large loads.

Transmissions could be designed to accommodate the speed reductions and torque increases that are required on off-road equipment; but that would require the use of heavier components, such as shafts and gearing, making the transmissions much bulkier and heavier than would otherwise be necessary. By using a final drive to increase torque, the other components used in the powertrain can be much lighter because they do not have to carry huge torque loads. Final drives are designed to support the weight of the machine and have to absorb the torque and shock loads at the wheels or tracks. **Figure 16-2** shows a typical example of a pinion drive used in many final drive applications. In the case of a bulldozer the final drive assembly includes steering clutches that are required to vary final drive speed on each track for steering control.

FUNDAMENTALS

Final drive designs used on off-road equipment can be classified into two basic configurations; single reduction and double reduction. The method used to get these reductions is not limited to just power transfer through gears, although this is the most popular. Chain drives are used extensively on off-road equipment and are quite suitable for this application. Because speeds are quite low, this type of drive is a suitable alternative.

In many cases, straight cut gears are used in final drives to transmit power. A few exceptions are the use of spiral bevel or straight bevel gears, worm gears, or **herringbone gears.** The purpose of a final drive is to change the direction of powerflow and to provide a means of either lowering or raising the output to match equipment design limitations. **Figure 16-3** shows an example of a pinion drive that drops the output lower to the ground to improve ground clearance. Speed reductions through

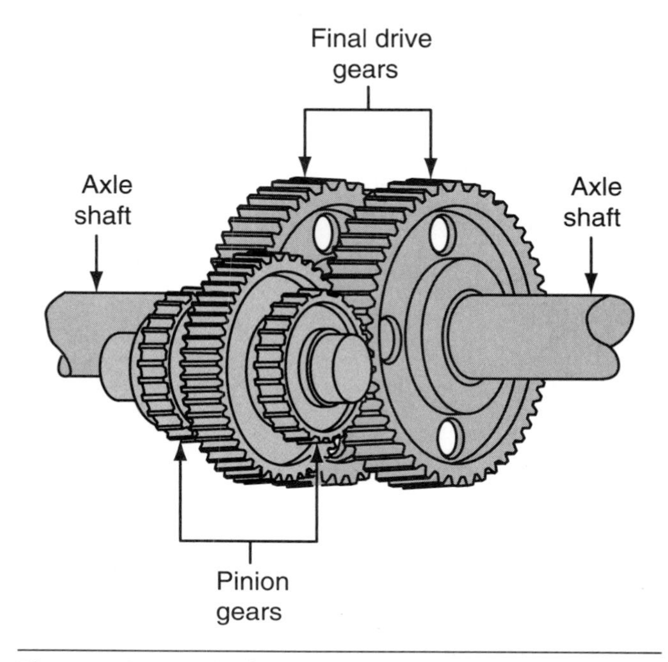

Figure 16-2 Typical pinion drive used in a differential case drive.

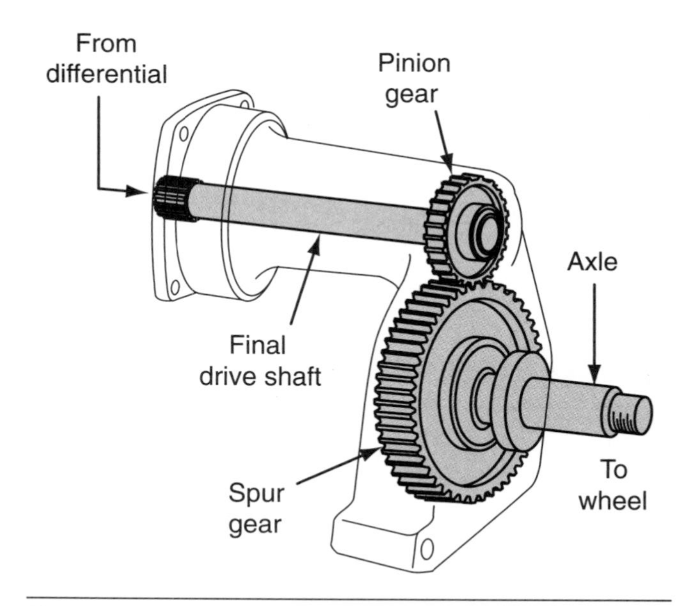

Figure 16-3 Typical pinion drive used in a final drive of a bulldozer.

final drives can vary from quite low, 3 to 1, up to as high as 12 to 1. Most final drives use various gear designs to transmit torque, but they are not limited to this method.

Often manufacturers use chain drives to change speed and torque on equipment, as is the case on many road graders. Chain drives provide an economical means of transmitting torque at low speeds over long distances; e.g., the tandem drives found in road graders.

Another popular method of transmitting torque in a final drive is through the use of worm gears. Worm gear drives allow for a large speed reduction in conjunction with a very large increase in torque to take place in a relatively small area compared to gear trains, which take up more space to get the same results.

CONSTRUCTION FEATURES

Pinion Drives

There are two basic designs of pinion drives used on off-road equipment; the first has the drive assembly located in the differential housing, (Figure 16-2) while the second type has the final reduction incorporated into the outer end of the final drive (**Figure 16-4**).

The pinion drive is possibly the most common of all the final drives used on off-road equipment. Most manufacturers utilize this concept on the majority of small track equipment, such as bulldozers. This drive utilizes a single gear reduction and is often referred to in the trade as a **bull-type final drive.** The **spur gear** and pinion drive design is commonly found on smaller equipment; it has the advantage of containing all the gears in the differential transmission case, making it more compact and cost-effective to service and maintain because one lubrication system is necessary. This design is quite popular on small tractors as the power is transmitted in a straight line to the wheel ends, minimizing cost by requiring fewer parts. The axle shafts generally have two tapered roller bearings that are mounted to the shaft to carry the weight of the machine and absorb end thrust; they are connected to the final drive gears by splines on the ends of the axle shafts. The bearings are generally located at each end of the axle shaft facing each other. These bearings are usually capable of varying the pre-load, either by shimming or by use of an adjusting nut with a locking mechanism to maintain the proper pre-load setting.

Figure 16-4 Typical pinion drive used in a final drive of a bulldozer. Note the use of the straight spur gear design in this application. (*Courtesy of Caterpillar*)

Bull-Type Final Drive

This drive design is more popular on track equipment where ground clearance may be a concern. Locating the pinion drive at the ends of the axle assembly allows the gears to be separate in their own housing, necessitating their own reservoir. If a failure were to occur, the damage would be contained in the outer housing and not take out the rest of the gear train, as in an integral design. This design has the advantage of dropping the powerflow closer to the ground, and is often referred as drop-axle housing. The weight of the equipment is supported by the entire final drive assembly and can supply an increase in torque through the use of only one gear set—a small pinion driving a large bull gear. **Figure 16-5** shows an example of a **single reduction drive** commonly used on bulldozers. On smaller dozers, the housings that contain this gear set are usually made from a one-piece casting and are bolted to the steering clutch housing. The pinion gear is splined to the **pinion shaft** and held in place with roller bearings; the larger final drive gear is usually splined to the track drive sprocket, which is supported by large tapered roller bearings. Lubrication for this design is achieved by using the time-proven splash method; the housing acts as a sump for the lubricating oil and is splashed around by the rotating final drive gear. Newer equipment uses **mechanical seals** to contain the oil in and keep dirt out. This type of seal is better at keeping the dirt out and has been proven to outlast conventional lip-type seals in this application.

Double Reduction Final Drives

On larger track equipment **double reduction drives,** as shown in **Figure 16-6,** are commonly used in the final drive in an elevated position. This concept is utilized in order to reduce the shock loads and implement loads that

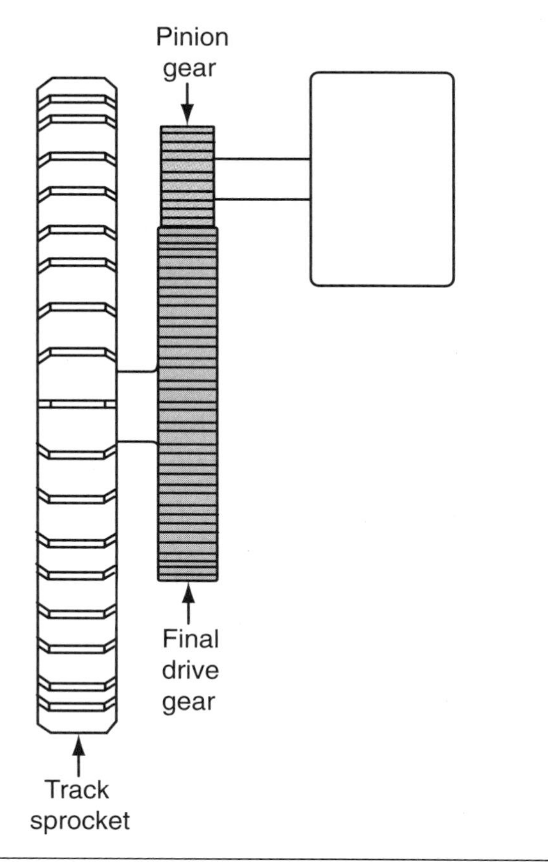

Figure 16-5 Single reduction final drive on a dozer.

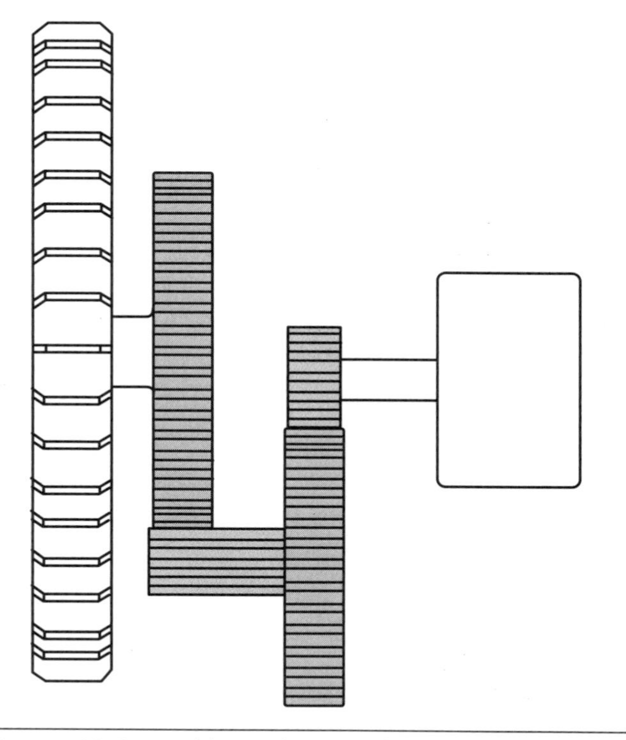

Figure 16-6 Double reduction final drive used on large dozers.

Figure 16-7 Position of planetary final drive used in an elevated design. (*Courtesy of Caterpillar*)

cause gear and bearing misalignment. It also prevents the final drive from coming into contact with work surfaces or any other abrasive material that could accelerate wear.

A double reduction drive has two separate gear reductions in line with each other. They are generally located in a separate housing at the track end of the drive. **Figure 16-7** shows a typical example of a double reduction final drive used on dozers. The housings are often made in two parts with the inner one being part of the steering housing assembly. The outer housing is bolted to the steering housing, which encloses the final drive components.

The bearings used to support the drive pinion are tapered roller bearings capable of supporting the high loads that are generated at this end of the drive. As with most tapered roller bearings, shims are usually used to adjust the bearing pre-load. Most manufacturers also incorporate the brakes and steering clutches into this section of the drive.

The intermediate pinion and gear set are splined together on the same shaft and are usually supported by a large tapered roller bearing at each end (refer to **Figure 16-8**) with the pre-load adjustment set by shimming the bearings. The sprocket shaft gear is splined to the sprocket shaft, which also holds the drive sprocket. The sprocket is mounted to the sprocket hub, which is held in place with either a large nut or a series of bolts. In most cases, newer double reduction drives are pressure lubricated; some of the older units still use the splash method of lubrication.

Planetary Final Drives

Planetary final drive designs also come in single reduction and double reduction configurations and function much the same way as previously noted. The

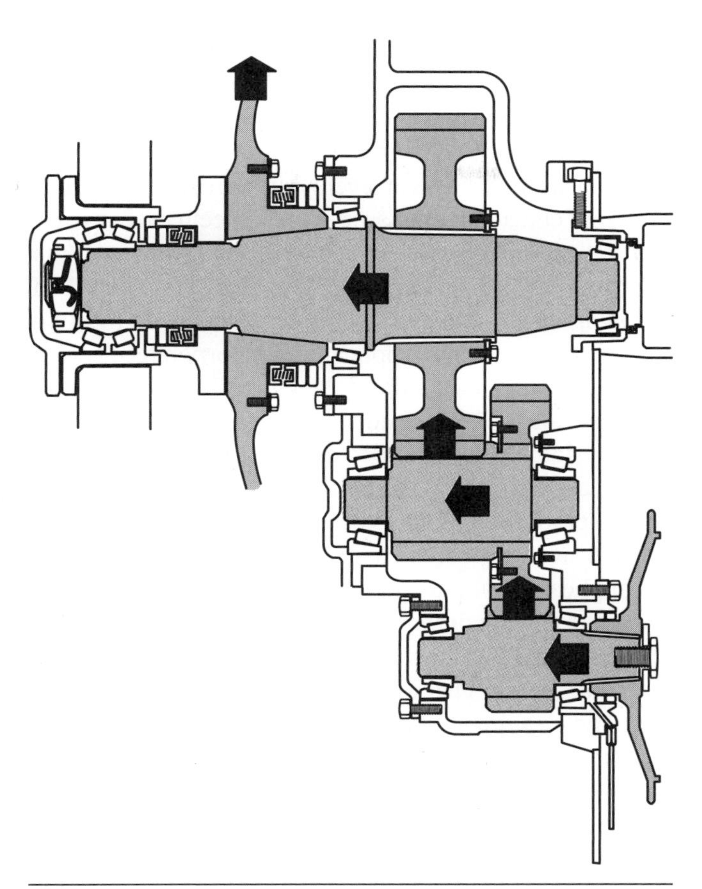

Figure 16-8 Flow of power through a typical final drive on large dozers.

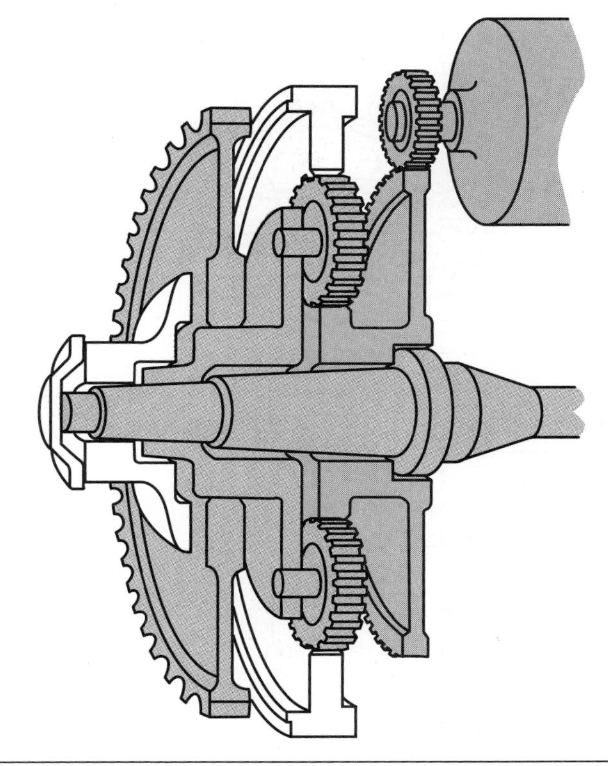

Figure 16-9 Cutaway of a flow of power through a typical planetary final drive.

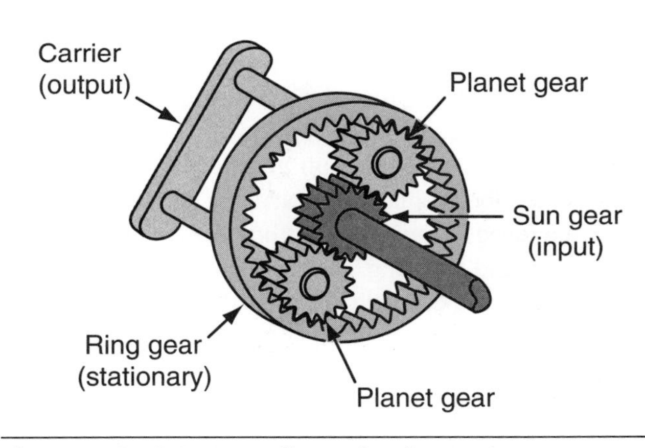

Figure 16-10 Power flows in from the sun gear, transferring torque to the planets which drive the track sprocket.

main difference is the way torque is transmitted through them. **Figure 16-9** shows an example of a typical planetary final drive used on some types of dozers.

These types of final drives are the ones most commonly used on modern equipment for many good reasons, the most important of which is size. Planetary drives are inherently small, especially when compared to conventional pinion drives. This reduces the overall size of the final drive considerably. Not only is the drive more compact, but it can also transmit more torque than a conventional pinion drive of similar size because it is designed to use more than one gear at a time to transfer torque. This design provides the ability to equalize torque and minimize the loads on support bearings and shafts, which translates into longer bearing and component life.

Figure 16-10 shows simple planetary gears; the torque flows from the axle to the **sun gear,** which transfers the torque to the **planet gears** that are mounted together on a carrier. Depending on the size of the final drive, there can be up to four planets mounted together on a carrier revolving around the inside of a stationary **ring gear.** The ring gear is either bolted in or held on splines to prevent it from rotating. This design follows one of four planetary gear laws that states: Whenever the planet carrier is the output and the ring gear is held, a forward reduction drive results. *Note*: The more planets there are, the stronger the gear set.

There are variations of double reduction planetary drives produced by manufacturers to suit many applications. Some manufacturers use a combination of a pinion drive with a planetary drive at the drive sprocket end to drive the track. **Figure 16-11** shows one variation of this

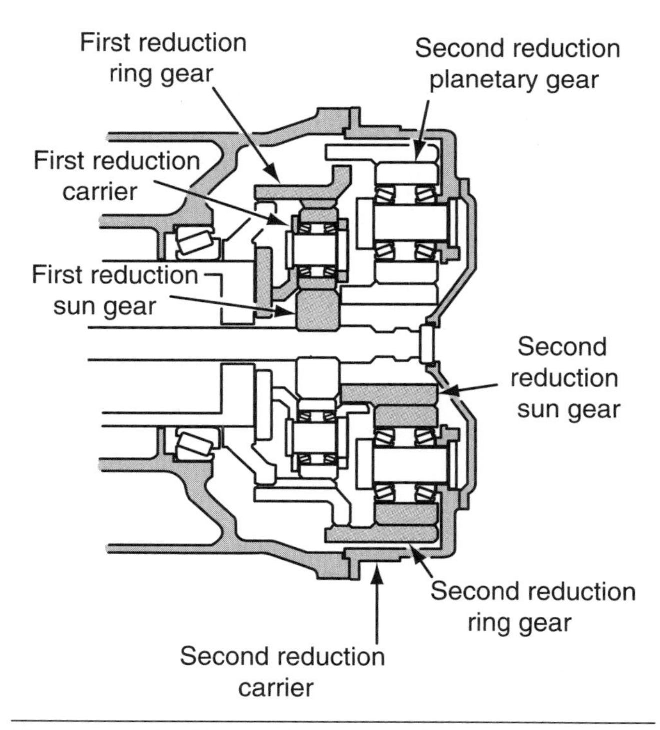

Figure 16-11 Flow of power through a double reduction planetary final drive.

popular drive. Two sets of planetary gears are used in this application with the inboard sun gear receiving torque from the axle. This forces the inboard carrier which is splined to the outboard sun gear, to walk around the ring gear, causing the outboard sun gear to turn at the same speed as the inboard planetary carrier. The second reduction is achieved through the outboard planetary carrier walking around the ring gear, which transmits torque to the track. An incredible amount of torque increase is produced from a light, compact drive. Tapered roller bearings are used to support the weight of the assembly as well as the load.

CHAIN DRIVES

Chain drives are used in many heavy equipment applications, such as road graders, where torque is transferred from the final drive sprocket to the wheel sprockets. Using a chain drive instead of gears for this purpose is an inexpensive and efficient way to transmit torque over a specific distance. Chain drives are capable of transmitting torque for long periods of time without much maintenance so long as dirt is kept out of the equation. The roller chain links mesh with the teeth on the sprockets to provide a positive drive between the front and rear wheels of the tandem drive, the same as would be provided by a gear drive but with no chance of slippage. **Figure 16-12** shows a typical final drive arrangement used on a grader.

To distribute the wear evenly and prevent the **roller link** from contacting the same sprocket teeth each revolution, roller chains have an even number of links and the sprockets they run on have an odd number of teeth. When operating under load, the slack side of the chain must be on the bottom. Some applications incorporate an idler sprocket on the slack side of the chain to take up excessive chain slack, with the added disadvantage of having to make periodic adjustments to the chain tension. A typical grader tandem drive has a drive sprocket that drives two wheel sprockets, one at either end of the tandem drive, with a heavy-duty roller chain. To get long chain life with this application, alignment is critical and must be checked whenever any repair work is carried out.

Figure 16-13 shows torque coming from the axle shaft into the sun gear of the planetary final drive reduction. The two drive sprockets are mounted to the planet carrier and transmit power to the two drive chains. A driven sprocket at each end of the tandem housing is mounted to a wheel spindle. Power is transmitted by the two chains to each of the wheels on either side of the grader's final drive. The wheel spindle housing contains the components that make up the brake system. Lubricating oil in the tandem housing is also used as a means of cooling the brakes. The pivot housing bolted to the tandem housing allows the tandem drive to oscillate. This connection allows the wheels to follow

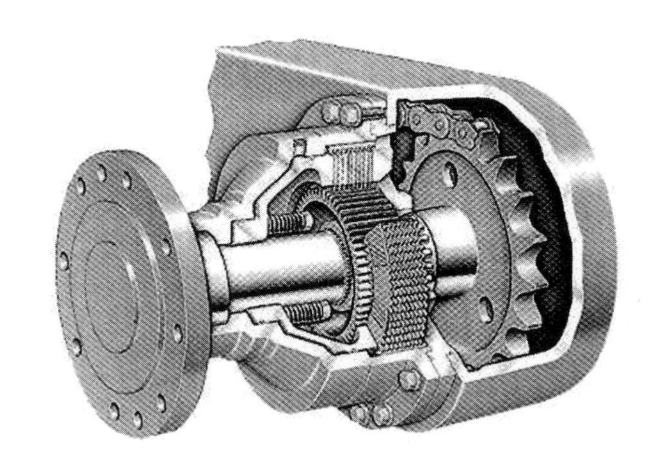

Figure 16-12 Flow of power through a grader tandem final drive. (*Courtesy of Caterpillar*)

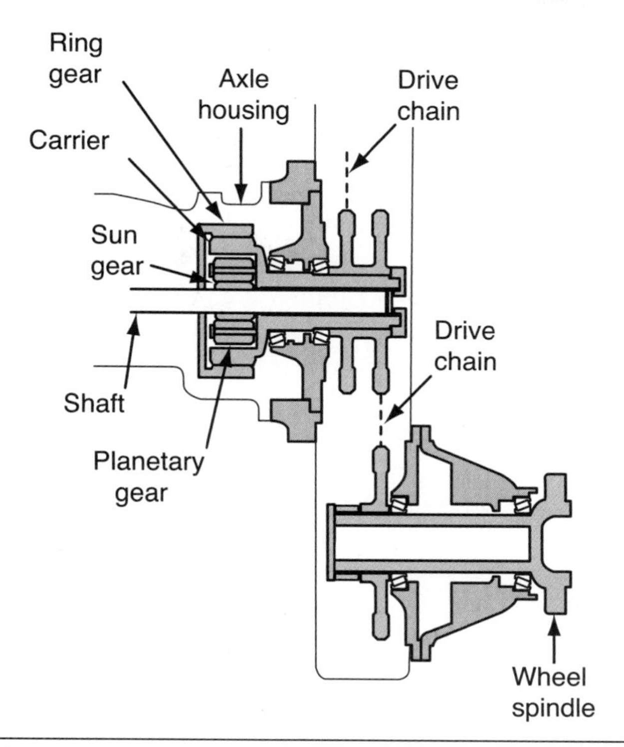

Figure 16-13 Flow of power through a tandem motor grader chain drive.

the road contour, providing traction to all four wheels continuously.

Roller chains are made up of a series of pin links and roller links connected to each other with **side bars;** rollers and bushings are fitted to the pin links and are also held together with side bars. Roller chains come preassembled from the manufacturer with provision for a master link to connect them together. **Figure 16-14** shows the parts that make up typical roller chain construction. (Each pin link interconnects to a roller link, with the pin links consisting of two pins with two side bars, one for each side. The roller links consist of two side bars, two bushings, and two rollers.) These parts are interconnected to form a roller chain. A master link is used to connect them together; the only difference between a pin link and a master link is that the latter can be easily removed to facilitate chain replacement.

Chain Adjustment

Chain tension can vary greatly on final drives, due to equipment application. Always consult the manufacturer's technical documentation for each application to determine the amount of tension or slack that is required. Some manufacturers use an idler sprocket to regulate chain tension; others use a guide shoe. A good rule to use when measuring chain tension is to have about one quarter inch of slack for every foot between shaft centers when the chain tension is removed from the drive side of the chain. The key to preventing chain and sprocket wear is to ensure that sprocket alignment is not compromised and that lubricating oil is removed and filtered or replaced as recommended by the equipment manufacturer. Chain lubrication is generally provided by the drive chains themselves, which run through the oil in the housing and distribute oil along the drive chain's path.

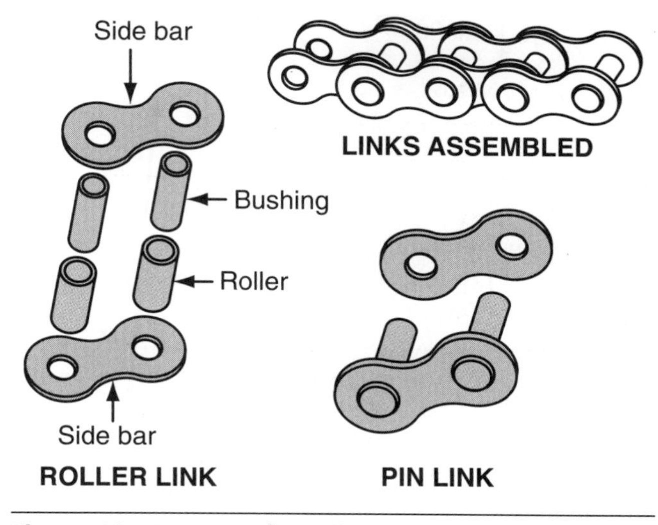

Figure 16-14 Parts of a roller chain.

Chain Wear

To make a determination on drive chain wear, the chain should be checked when mounted in the respective housing. The drive is designed to tolerate normal wear that takes place over a period of time; up to a maximum of 3% elongation is considered normal. There are exceptions to this rule. Always consult the manufacturer's technical documentation before determining the allowable elongation specifications for the equipment that you are checking. Chain replacement is required when the chain reaches elongation specifications. Extensive damage to the tandem drive may result if the chain breaks during operation.

Some manufacturers recommend checking chain length under tension as an indicator of drive chain wear. Follow the manufacturer's recommended chain removal procedure to remove the drive chains. Once the chain is removed, clean it using the manufacturer's approved cleaning procedure, and inspect the chain for cracks, broken rollers and pins, and deformation. If any of these conditions are found the chain must be replaced. After determining the specifications for drive chain

Table 16-1: CHAIN TENSION MEASUREMENTS

Model	Chain Pitch	No. of Chain Links	New Measurement	Maximum Wear	Chain Tension
120H	44.5 mm	14	622.3 mm	641.1 mm	173 kg
135H	1.75 in.		24.50 in.	25.24 in.	381 lbs
12H,140H,	50.8 mm	12	609.6 mm	628.7 mm	277 kg
143H,160H,163H	2.00 in.		24.00 in.	24.72 in.	610 lbs
14H	57.2 mm	11	628.7 mm	647.7 mm	286 kg
	2.25 in.		24.75 in.	25.50 in.	630 lbs
16H	63.5 mm	10	635 mm	654.1 mm	354 kg
	2.50 in.		25.00 in.	25.75 in.	780 lbs
24H*	76.2 mm	10	762 mm	773.4 mm	782 kg
	3.00 in.		30 in.	30.45 in.	1724 lbs

*Manufacturer recommends 1.5% maximum chain elongation. (*Courtesy of Caterpillar.*)

length are higher than allowed, you need to determine whether the sprockets are worn.

CAUTION *Always replace both chains in tandem housings at the same time. Splicing in a new section is not recommended; this will cause improper engagement with the sprockets, leading to sprocket damage.*

To determine chain elongation dimensions, hang the chain from a hoist and attach a scale that is capable of measuring the chain tension that is shown in **Table 16-1.** Attach the free end of the scale to a heavy object, such as the grader. Refer to Table 16-1 for the correct tension to apply to the scale. To achieve accurate dimensions the chain must be placed under tension to displace the oil that may be trapped between the rollers and pins of the chain. Once the correct tension is applied, measure the distance across the number of chain links recommended in Table 16-1 to determine elongation limits. If the chain passes this test for chain wear, it may be reused. Table 16-1 shows an example of measurements for Caterpillar grader final drives recommended chain tension and measurement specifications. Always refer to the manufacturer's specification for the equipment that is being worked on.

In some cases, chain and sprocket wear cannot easily be determined visually without some disassembly of the tandem final drive. Consult the manufacturer's technical documentation to determine the course of action that needs to be followed. Sometimes it is possible to check for chain wear by looking at the chain wrapped around the largest sprocket in the drive that is accessible. The chain links should not ride up the tooth, as shown in **Figure 16-15,** but should sit at or near the bottom of the sprocket. If the sprocket teeth are worn, the chain should be replaced.

If chain elongation is not a problem, it may be possible to reuse the chain if it passes the visual inspection when wrapped around a sprocket. Removing a link may allow you to get additional life out of the chain drive. In some cases, the sprockets can be reversed on some equipment, extending drive life. New links should not be used on old chains; the new links will have a shorter pitch. This will cause shock loading of the chain during operation and should be avoided at all costs. In most cases, it is advisable to change the sprockets when installing a new chain. Any sprocket tooth wear, which may look good to the eye, will shorten the life of the new chain considerably. A determination must be made based on the manufacturer's recommended procedure for each specific application.

To evaluate chain wear and operating conditions, the following guide in **Table 16-2** can be used as an example of common problems and their likely causes.

MAINTENANCE AND LUBRICATION

Modern final drive assemblies are capable of operating for extended periods of time with very little mainte-

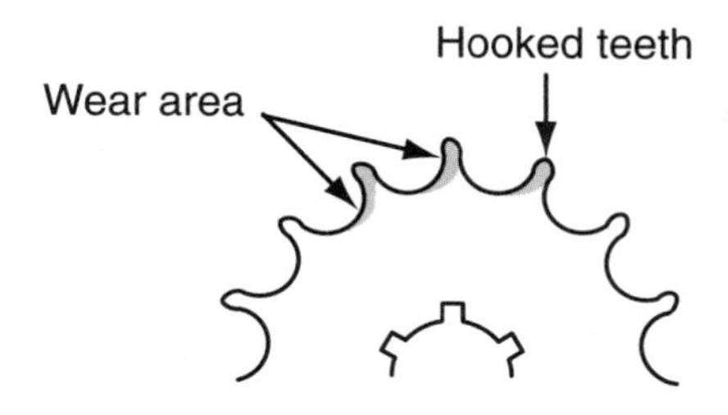

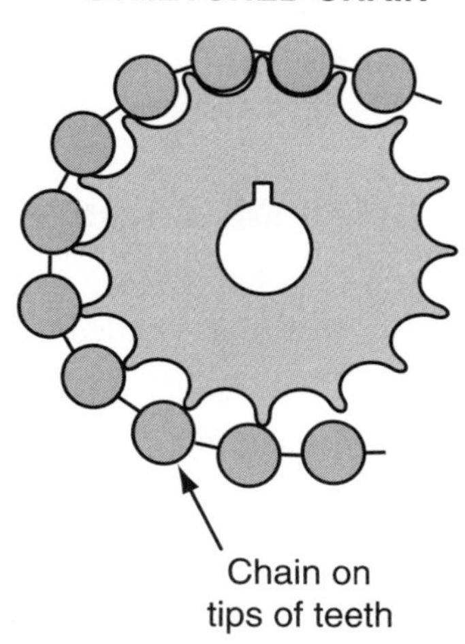

Figure 16-15 Chain link wear and tooth wear.

nance service. Technicians are often required to make a determination as to how much work can be done economically in the field on final drives and where that work will take place. A thorough knowledge of operating principles related to final drives is required to be able to make that assessment. A technician must be able to determine whether to rebuild a final drive assembly or replace it. The importance of maintaining and servicing the final drive systems cannot be understated, as the cost of repairing and replacing the components is time-consuming and expensive. Equipment technicians require a good understanding of modern lubrication requirements for final drive systems since a large percentage of problems stem from improper servicing procedures.

By their very nature, final drives support the weight of the equipment as well as transmitting torque to the tracks or wheels. Oil leaks from final drives invariably lead to low oil levels, which in turn lead to inadequate lubrication conditions and ultimately to early failure. Since the life of a final drive is dependent on using the correct lubricant to aid in reducing friction between moving parts, there must be a means of removing excess heat

Table 16-2: DETERMINING CHAIN WEAR AND OPERATING CONDITIONS

Problem	Likely Cause
Excessive wear on the side bars and link plates	Misalignment
Pins, bushings, or rollers broken	Sudden shock loads Insufficient lubrication Mismatched chain and sprocket Too high an chain operating speed
Chain whip	Incorrect slack adjustment Seized chain joints Chain wear uneven
Chain climbing the sprockets	Worn out chain Excessive chain slack Sprockets worn out Mismatched chain and sprocket
Broken sprocket teeth	Excessive shock loading Chain climbing the sprockets
Chain drive lubrication overheats	Too high operating speed Insufficient lubrication Oil level too high in housing
Chain sticks to sprockets	Mismatched chain and sprocket Incorrect lubricant Excessive chain slack

as well as suspending dirt and wear particles in the lubricant between drain intervals. Effective lubrication is totally dependent on using the correct lubricant and changing it at the appropriate intervals. The recommendations discussed in this chapter are general and provide a guideline only. Always refer to the manufacturer's technical documentation for specific procedures and information as to the correct oil for the equipment you are servicing.

The practice of mixing different brands of oil is discouraged; it often results with compatibility issues. As an example, some oil manufacturer's advertise that their synthetic oil products are fully compatible with any type of gear lube. Experts in the equipment service field caution that mixing lubricants may cause accelerated breakdown of the additives used in the oil, which may lead to increased wear. In some cases **viscosity** changes and oil foaming occur from mixing synthetics and petroleum-based lubricants. Most manufacturers today approve the use of synthetic oils in final drives, as they have a list of impressive qualities that are not found in conventional lubricants. Synthetic lubricants are much more effective in cold weather operations and are also superior in high temperature applications. The ability of the lubricant to cling to the components is much better with synthetics than to conventional oils. This is often referred to as throw-off resistance in manufacturer's lubrication literature. Lube change intervals may sometimes be extended by using synthetic oil, but always refer to the manufacturer's recommendations when determining oil change intervals.

Maintaining correct oil levels in a final drive is critical. Oil foaming or operating with too low an oil level can be prevented. Refer to the manufacturer's service literature for the proper procedures and techniques used to determine the correct oil level.

Some manufacturers use magnetic type drain plugs to capture metal wear particles. Excessive metal particles found on drain plugs can be an indication that things are about to go wrong in the final drive. These symptoms should be investigated before a costly breakdown occurs. Following simple manufacturer's recommendations, such as draining the lubricant when it is warm, will eliminate most particles and dirt because they will be suspended in the oil. Letting the oil cool causes particles to settle to the bottom of the housing; draining cool oil results in only partial particle removal.

Always check lubrication requirements before servicing equipment; conventional gear oils may not be compatible with friction materials that are often used in brakes or steering clutches, which are located internally in many final drives. Some gear oils are not compatible with the seal materials often found in final drives and may cause brake performance problems. Not all multigrade oils will meet the requirements of all manufacturers, as some multi grade oils use viscosity index improvers that use high molecular weight polymers, which can lead to poor viscosity performance. Final drive oil should be able to handle high temperature and load conditions and be able to develop a sufficient coefficient of friction to meet the demands of brakes and steering clutches used in final drives. Choosing the correct oil for this application is critical to brake and steering clutch performance. The lubricant should have good rust and corrosion protection as well as a low rate of oxidation. It must be highly resistant to foaming and protect against copper corrosion.

Tandem Drive Breather Cleaning/Replacement

Although the tandem drives are sealed from the environment, they must have a means of equalizing the pressure that is built up internally during operation. A breather is located on each drive (**Figure 16-16**) to prevent pressure buildup during operation. Periodic cleaning or replacement of this breather is required and should not be neglected; seal damage will result from excessive pressure buildup in the drive if it becomes plugged.

Tandem Drive Oil Change

Before performing any service work to the tandem drive, make sure that it is cleaned thoroughly around the fill plug and drain plug.

1. Place a suitable container under the drain plug and remove the drain plug and oil check plug to drain the oil. Refer to **Figure 16-17** for the location of plugs.
2. It may be necessary to flush the drive with diesel fuel to get rid of slug buildup. To inspect the inside of the drive housing, remove the inspection cover on the top of the drive.
3. Clean and reinstall the drain plug when the oil drain is completed.
4. Refill the tandem drive with an approved lubricant until oil flows from the check plug hole in the outer cover, as shown in **Figure 16-18.**

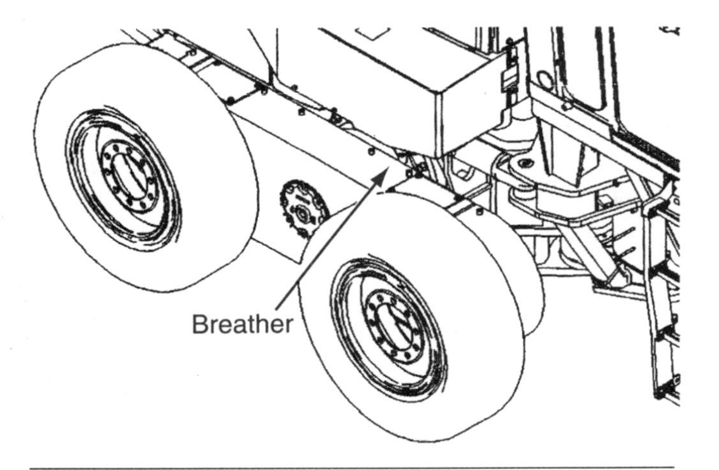

Figure 16-16 Location of the breather. (*Courtesy of Caterpillar*)

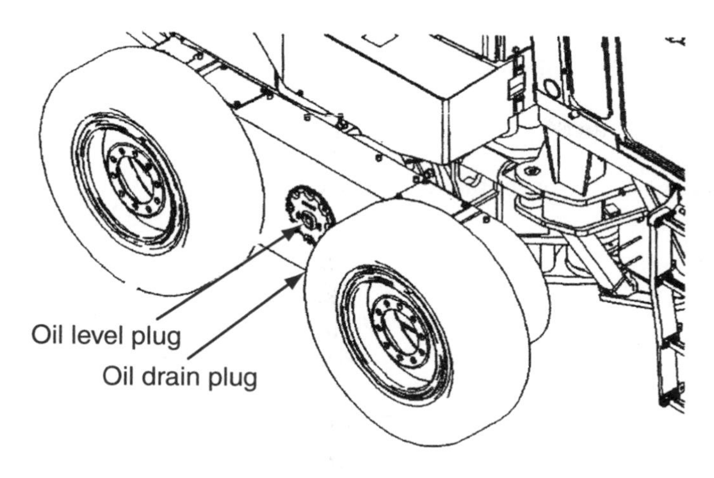

Figure 16-17 Location of oil level and drain plug. (*Courtesy of Caterpillar*)

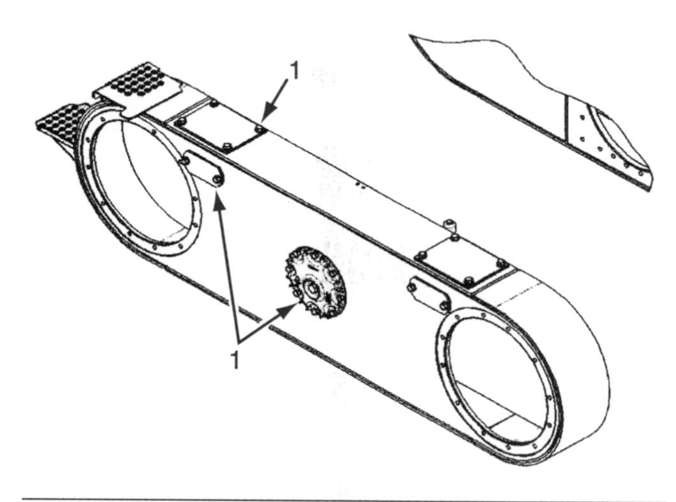

Figure 16-18 Location of the oil level check plug. (*Courtesy of Caterpillar*)

5. Install the inspection cover, operate the equipment, and check the drive housing for any leaks.
6. Recheck the oil level refill until it is even with the bottom of the oil check hole, as shown in Figure 16-18.

FINAL DRIVE BEARING ADJUSTMENT

Whenever an adjustment procedure of this nature must be performed, the equipment must be properly blocked up to ensure that the work can proceed safely. Always refer to the manufacturer's recommended blocking procedure. The following general steps provide a guide to the actual steps performed before checking and adjusting of the final drive can take place. Always refer to the manufacturer's procedures for more specific instructions on your off-road equipment. Be sure to dispose of old final drive oil following locally accepted regulations and procedures. Never reuse the old fluid; dirt may have been accidentally introduced into the oil. Rapid wear would result if the old oil were reused.

1. Be sure to park the grader on a flat, level surface if possible, and make sure that the transmission control lever is in the "park brake on" position.
2. Place the wheel lean pin into the front axle. This will prevent the front of the machine from unexpectedly shifting sideways.
3. If the grader has an articulating frame, make sure to install the articulating lock pin. This will prevent the machine from unexpectedly moving sideways due to the articulating frame.
4. Place wheel chocks at the front and rear tires.
5. Be sure to choose a jack that is rated for lifting the intended weight.
6. Use only approved stands that are rated to support the intended load.
7. Before attempting to raise the back of the grader (tandem drive end), repeat steps 3 through 7.
8. Drain down the oil level in the tandem housing below the bottom of the outer cover.
9. Remove the outer cover and store the cover and bolts in a secure location.
10. Remove the lock that holds the adjusting nut in place and store in a secure location.
11. Refer to the manufacturer's technical documentation to select the correct spanner wrench for the final drive you are attempting to adjust. Refer to **Table 16-3** for the example of a tool selection chart.
12. Using the appropriate torque wrench (**Table 16-4**) turn the tires while adjusting the torque to the

Table 16-3: TOOL SELECTION CHART

Part Number	Description	Quantity
6V-4074	Wrench for 120H	1
	Wrench for 135H	1
9U-5015	Torque Wrench 3/4 in. drive	1

Table 16-4: EXAMPLE OF A BEARING PRE-LOAD FOR FINAL DRIVE CHART

Final Drive Group	Part No for Nut	Adjustment Torque	Final Torque
194-6773 194-6774	2G-6396	170 Nm (125 lb-ft)	102 Nm (75 lb-ft)
8W-4368 202-7593	8D-3540	204 Nm (150 lb-ft)	136 Nm (100 lb-ft)

appropriate torque values also shown in Table 16-4. If the final drive has been disassembled, strike the hub with a large brass punch and hammer to ensure correct seating of the sprockets.

13. Recheck the nut torque *again* and repeat the last two steps, if required, until the torque remains constant.
14. Back off the adjusting nut one lock position.
15. Once again, retighten the adjusting nut to the final torque as specified in Table 16-4.

Note: It is important to differentiate between the rolling torque specifications in the disassembly and the assembly instructions; they are not the same.

1. Once torqued to the final specifications, check to see that the lock will fit against the adjusting nut, then install and torque into place. If it does not fit properly, tighten further until it fits properly. *Note:* Do not back off the adjusting nut to install the lock nut; the resulting torque will be incorrect.
2. Reinstall the outer cover to the tandem housing and refill with new lubricating oil to the correct level.

Differential and Bevel Gear Check

During some point in a technician's career, he or she will be called upon to make a determination as to the condition of a final drive differential without removing it for inspection. The following is a general procedure that will provide a guide to determining excessive **backlash** in the differential bevel gear in a grader final drive. This example is based on a typical procedure used by Caterpillar. Before proceeding with this test, be sure to follow the manufacturer's suggested blocking procedure outlined in the technical documentation for each particular piece of equipment; different types of equipment may have different techniques for to securing the equipment safely. Once the equipment is properly blocked, the general steps outlined below will lead to determining backlash in the final drive.

1. Remove the driveshaft and store in a secure place. Be careful not to drop any of the U-joint bearing caps.
2. Use the appropriate 3/4-drive power bar and socket to remove the yoke from the differential housing.
3. Attach the specialized tooling to the pinion shaft as shown in **Figure 16-19.** Refer to **Table 16-5** for the tool number.
4. Securely attach the magnetic base of the dial indicator to the carrier housing.
5. Perform three evenly spaced readings approximately 120 degrees apart. Refer to **Figure 16-20.** *Note:* Each different model of grader has a slightly different procedure that must be followed to obtain accurate readings. For some models use position "A." If the reading at "2A" is used, you have to divide the results by two. To take readings on some models, position "B" must be used. This example illustrates the importance of getting and using the correct procedures to perform this task. If you were to use any method and specifications

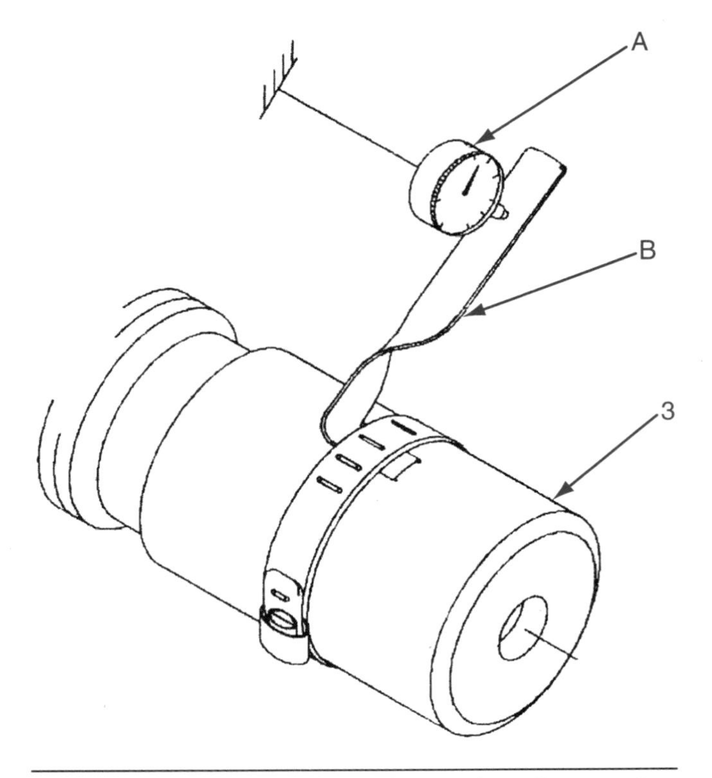

Figure 16-19 Correct setup of special tooling used to measure backlash in a final drive differential assembly. (*Courtesy of Caterpillar*)

Table 16-5: EXAMPLE OF REQUIRED TOOLING

Reference Number	Part Number	Description	Quantity
A	8T-5096	Dial Indicator Gauge Assembly	1
B	127-8072	Backlash Gauge	1

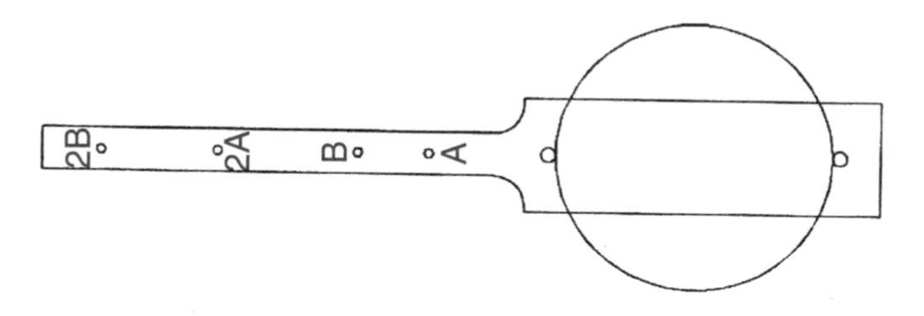

Figure 16-20 Example of special tooling used to measure backlash on the drive pinion of a final drive differential. (*Courtesy of Caterpillar*)

other than those shown in this example for this type of equipment, incorrect values would be the obtained.

6. If readings are to be taken at position "2B" you must divide the readings by two to calculate the correct value.
7. In this example refer to **Table 16-6** for the correct specifications for each application. If the backlash readings are not within parameters, further disassembly will be required to determine the course of action.

Differential Lock Check On a Caterpillar 120H Road Grader

The differential used in this example, shown in **Figure 16-21,** has a lock that is used with a conventional differential to force both tandem drives to be locked together to prevent differential action from occurring during low traction conditions. The differential locking mechanism is activated from the dash by a solenoid-activated hydraulic valve that sends pressure oil from the transmission to the differential lock valve located on the differential carrier. This system uses a series of clutch discs that, when activated, lock the left side gear to the bevel gear, preventing the pinions from turning on the spider shaft, and causing the spider assembly to rotate as a unit. This action forces the two output axles to turn at the same speed, which provides equal power to both tandem drives.

Table 16-6: EXAMPLE OF SPECIFICATIONS USED FOR ADJUSTMENT

Model	Assembled Backlash
12H	0.025 mm + 0.10 – 0.07 mm (0.010 + 0.004 – 0.003 in.)
120H	0.025 mm + 0.10 – 0.07 mm (0.010 + 0.004 – 0.003 in.)
135H	0.025 mm + 0.10 – 0.07 mm (0.010 + 0.004 – 0.003 in.)
140H	0.025 mm + 0.10 – 0.07 mm (0.010 + 0.004 – 0.003 in.)
143H	0.025 mm + 0.10 – 0.07 mm (0.010 + 0.004 – 0.003 in.)
160H	0.025 mm + 0.10 – 0.07 mm (0.010 + 0.004 – 0.003 in.)
163H	0.025 mm + 0.10 – 0.07 mm (0.010 + 0.004 – 0.003 in.)

FAILURE ANALYSIS

Final drives are subject to more extreme torque loading and shock abuse potential than any other component in the powertrain system in off-road equipment. When a final drive failure occurs, it often results in expensive parts replacement and long periods of downtime. Failure analysis can be a complex procedure and may require the use of sophisticated lab equipment to make an accurate determination of the root cause. Fortunately, most failures can be accurately assessed in the field. Determining the root cause of a failure is the primary objective of this section of the book. We will examine the necessary skills that are required to determine the cause of the failure. Often, you can determine the root cause of the failure by careful visual inspection of the failed component.

Why and what led up to the failure may be difficult to determine. A common error in failure analysis may occur when a technician mistakenly assumes that the first damaged component found is responsible for breakdown. In many cases, the damaged component is a result, rather than a cause, of the failure. A proven method of determining the root cause of a failure is to use an investigative approach. The steps listed below are a good starting point to use in developing good diagnostic and analytical skills for failure analysis.

1. Record all the information collected during the assessment stage.
2. Talk to the operator who was using the equipment at the time of the failure.

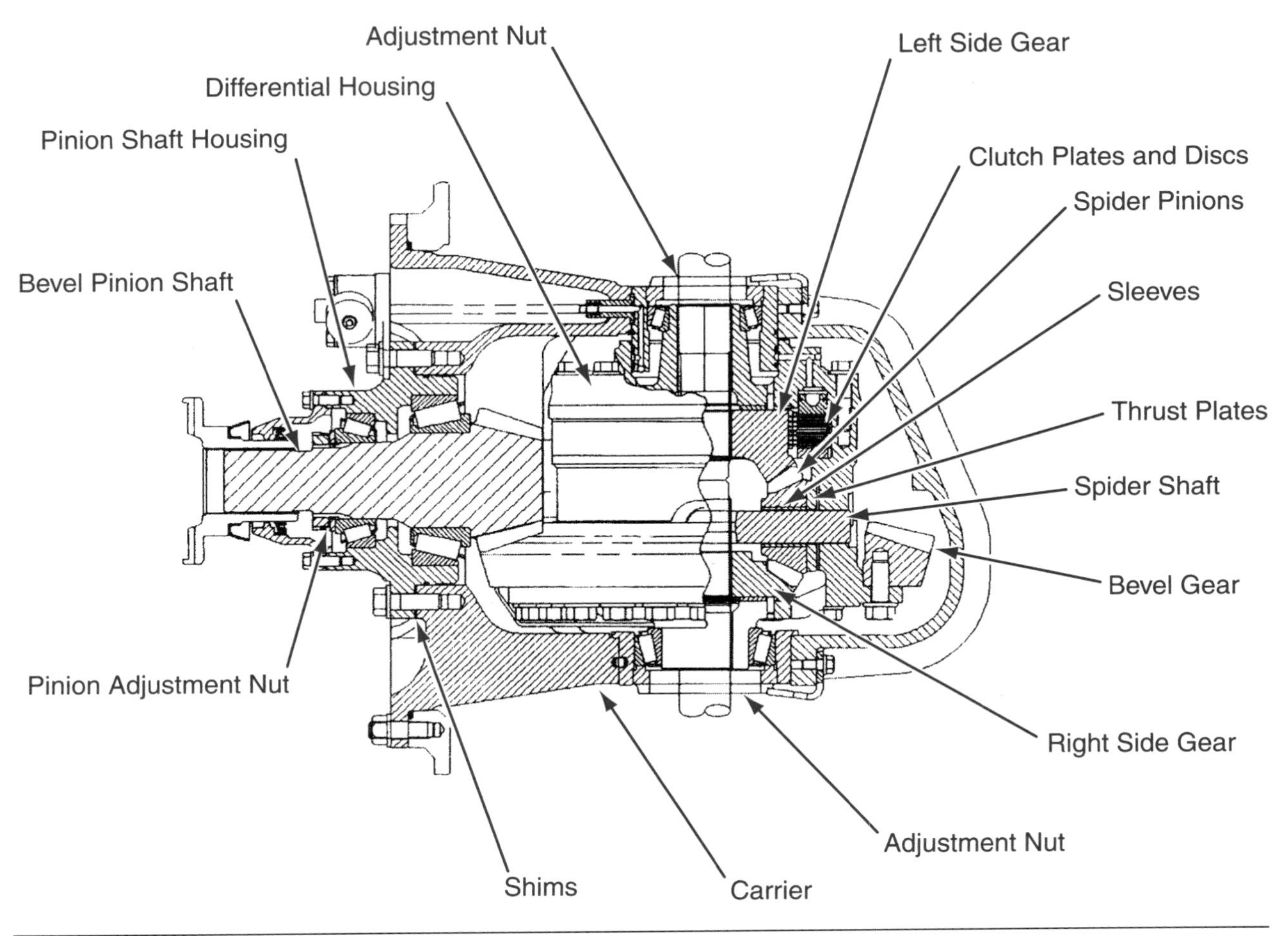

Figure 16-21 Cutaway view of the differential assembly. (*Courtesy of Caterpillar*)

3. Assess whether the failure was instantaneous or occurred gradually.
4. If there were multiple failures, determine which occurred first and what led up to the failure. For example, did a lubricating pump fail first, which lead to a major component failure, or was it the other way around?
5. Was the equipment operating above its designed torque capabilities?
6. Have any repairs been made or parts replaced recently on the equipment in question: If parts have been repaired or adjusted, it would be beneficial to talk to the technician who performed the work.
7. Determine how the equipment was brought to the repair facility. Was it towed or loaded on a trailer and trucked in?
8. Be sure to establish the accurate number equipment hours or length of service, since the last maintenance, and what the equipment was used for.
9. Review the equipments service records to determine whether the equipment received proper service, including the use of recommended lubricants.
10. Review the equipment's overall condition. Look for things like oil leaks, operating abuse, and tire or track damage. Is the equipments exterior condition clean or is it caked with dirt?
11. Be sure not to disturb the failed components: they should be left in their original positions. Do not clean the parts before inspection. If they are left outside, make sure they are not exposed to the environment; cover them up with a tarp or other suitable cover.
12. Begin by inspecting the obvious damage first, keeping in mind that there may be more that one damaged component related to this failure.
13. If possible try to establish the failure type. Was it lubrication-related failure or was it caused by a shock load? Was it an operating-related failure, such as a failure caused by spinout (which is

common in final drive failures)? Try to establish whether equipment abuse led up to the failure.

14. In the case of final drives, if it is possible check and record backlash, end play, and runout. It may even be possible to check tooth contact.

Keep in mind that this information is intended to be used as a guide and not to be absolute. There are quite a few types of failures that can occur; some of them are related to the design, and others to the operational application of the equipment. Final drive components usually fail as a result of the following:

- Normal wear
- Lubrication failures
- Shock loading
- Fatigue failure

Normal Wear

Although final drive components are designed and built to last for the life of the equipment, they eventually wear out during normal use. Initially, the components will have a slightly higher level of wear due to "break-in." When new components mesh for the first time, a certain amount of wear takes place; however, this will benefit the overall life of the equipment. Gears meshing with each other improves their contact patterns during the break-in period. It is important that a technician can recognize what normal wear looks like, and does not get it confused with other types of abnormal wear. Some manufacturers recommend that the lubricant be replaced after an initial break-in period, as this will remove the wear particles from the initial break-in period. Be sure to refer to the manufacturers recommended maintenance procedures to determine the course of action for the equipment in question. Progressive companies use a reputable oil analysis program to establish benchmarks for determining acceptable wear in final drive systems.

Lubrication Failures

Some of the most difficult diagnoses to make are pinpointing the root causes of lubrication failures. Often premature wear occurs from using the wrong type of lubricant or not having enough lubrication. These can be confused or mistaken for premature wear caused by equipment abuse or from operating the equipment for the wrong application. These deficiencies can lead to accelerated bearing and gear wear in final drives. The following list is representative of problems that are associated with lubrication failures:

- Inadequate lubrication is a common problem on off-road equipment; oil leaks from gaskets and seals in final drives may lead to low oil levels and should always be repaired as soon as possible. When lubrication changes are performed on equipment, it is important to be sure that the correct level of oil is present in the final drive. Many final drives rely on the splash method to distribute the lubricant around the inside of the housing. If the oil level was too low, an inadequate amount of oil is circulated, leading to possible metal-to-metal contact that causes overheating and leads to oil breakdown. **Figure 16-22** shows burned lubricant on a differential case. Eventually, the mating components and bearings will become scored, leading to component failure. When this takes place, the internal parts become black and discoloured and the oil generally thickens and smells burnt.
- Using a lubricant with a viscosity that is not suited for the operating temperature range to which the equipment is subjected, will result in shorter component life. Refer to the manufacturer's recommended viscosity rating to determine the correct oil viscosity for the operating environment. Using oil that is not designed for cold weather operation may result in insufficient lubrication due to the thickening of the lubricant and leads to metal-to-metal contact. Mixing different types of oils also has the potential for undesirable viscosity changes. When the correct oil viscosity is

Figure 16-22 Burned lubricant on a differential case. (*Courtesy of Arvin Meritor*)

chosen, oil gets circulated immediately to the bearings and gears, lessening the effect of a dry start up. **Figure 16-23** shows damage caused by using non-EP-rated lubricant.

- Lubrication contamination is probably the most common problem that occurs in heavy equipment. Dirt, water, and even break-in debris can lead to component failure. This often starts as scoring or etching and progresses to pitting of the mating surfaces. Anytime you have dirt or any other abrasive material in the oil, it acts as an abrasive and accelerates the wear of mating parts tenfold. Moisture in the sump, in combination with conventional oil in a final drive, will lead to acid formation caused by the water and small amounts of sulphur found in conventional oil. This leads to corrosion and acid etching of mating surfaces, which eventually leads to component failure.

Shock Loading

Shock loading can often be traced to improper equipment operation. A sudden load applied to a final drive component instantly or continuously over a longer period of time results in cracks or immediate breakage, depending on the severity and frequency of the shock load. When shock loading becomes excessive, gear teeth and shafts are overloaded past the point that the material can handle, resulting in cracking or immediate breakage, as shown in **Figure 16-24.** These types of impacts subject components to large amounts of twisting forces that can eventually lead to fractures. Often shock loading indicators will show up as small curved rings, called **beach marks,** on the surface of the component. If this shock loading continues, it will lead to progressively larger beach marks and eventually result in complete breakage. Shock loading can be distinguished by a rough crystalline surface texture that is evident on the broken parts.

Fatigue Failure

Another common condition that can be mistaken for shock loading (**Figure 16-25**) is fatigue failure, as shown in **Figure 16-26.** This condition occurs when the component in question is subjected to torsional or bending forces often caused by overloading of final drives. There are several types of fatigue that will be discussed briefly in this section. *Surface fatigue* can be defined as a condition that occurs between the surfaces of mating parts that are subjected to high torque loads. It starts with pieces of metal flaking off and progressively leads to pitting and finally to complete breakage. *Bending fatigue*, if severe enough, will also cause mating gears to distort the running patterns, which can move the contact pattern to the ends of teeth and lead to concentrated loading. This type of fatigue will eventually lead to shaft

Figure 16-23 Effects of using oil that is not EP-rated. Extensive wear has taken place on the teeth leaving sharp edges. (*Courtesy of Arvin Meritor*)

1 ROUGH CRYSTALLINE AREA
2 SMEARED AREA

Figure 16-24 Broken teeth that resulted from shock loading. (*Courtesy of Arvin Meritor*)

Figure 16-25 Pinion gear set from a final drive damaged by fatigue due to overloading. (*Courtesy of Arvin Meritor*)

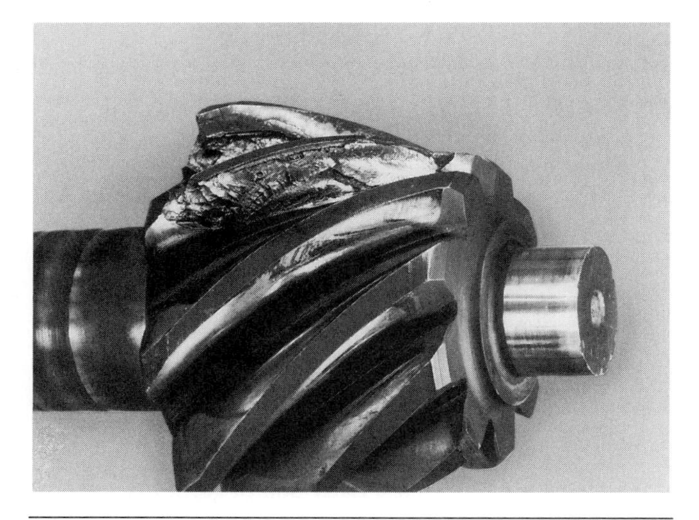

Figure 16-26 Results of overload fatigue that has lead to cracks and spalling. (*Courtesy of Roadranger Marketing. One great drive train from two great companies—Eaton and Dana Corporation*)

breakage as shown in **Figure 16-27.** Bending fatigue will generally cause a spiral-type fracture to occur to the shaft. *Overloading* of axle shafts is a common condition that occurs in final drives. A crack that develops from extremely high loads on axles may progress to the center core of the shaft, eventually breaking it.

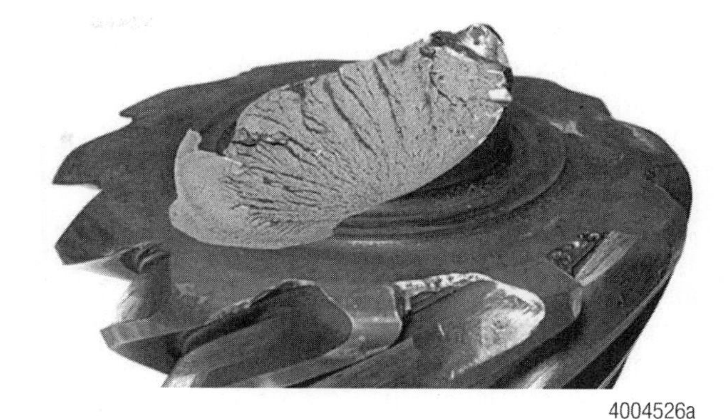

Figure 16-27 Pinion gear broken by shock loading. Note the crystalline appearance and the angle of the break. (*Courtesy of Arvin Meritor*)

Surface pitting is another condition that is caused by overloading on final drive bearings and gears. Surface pitting starts as spalling on the metal surface of mating parts, such a tooth surfaces. The teeth of gears are case hardened to a predetermined depth by the manufacturer. The underlining metal is not as hard as the surface of the teeth; this is necessary to allow the parts to absorb some shock loads. Continually applying heavy loads to the gear teeth surfaces causes the hardened surfaces on the teeth to break down, leading to pitting. The mating surfaces of the teeth flake off to expose a layer of softer metal, which leads to rapid metal breakdown and results in pitting.

ONLINE TASKS

1. Use a suitable Internet search engine to research the configuration of an off-road chain drive system. Identify a piece of equipment in your particular location, paying close attention to the chain drive arrangement, and identify whether or not it uses an additional planetary gear reduction. Outline the reasoning behind the manufacturer's choice of this particular design for that application.
2. Check out this url: *http://science.howstuffworks.com/skid-steer2.htm* for more information on chain drive configuration and design.

SHOP TASKS

1. Select a piece of equipment in your shop that has a chain drive axle assembly and identify the components that make up the chain drive arrangement.

Identify whether the chain has excessive wear, and what type of chain drive link design is used on this arrangement.

2. Select a piece of off-road equipment in your shop that uses a chain drive axle and identify the type and model of the equipment, listing all the major components and the corresponding part numbers of the chain drive assembly. Identify and use the correct parts book for this task.

Summary

- Final drives are the last stage of torque increase and speed reduction before the power is transmitted to the ground through the tires or tracks.
- Final drives accept power from the transmission, reducing its speed and increasing the torque through the use of various gear and chain drive arrangements.
- By using a final drive to increase torque the other components used in the power train can be made much lighter because they will not have to carry huge torque loads.
- Final drives are designed to support the weight of the equipment and absorb the torque and shock loads at the wheels or tracks.
- Final drive designs used on off-road equipment can be classified into two basic configurations: single reduction and double reduction.
- Chain drives are used extensively on off-road equipment and are quite suitable in this application. Speeds are quite low, making chain drives a realistic alternative.
- Worm gear drives allow for a large speed reduction and a very large increase in torque to take place in a relatively small area compared to gear trains, which take up more space to get the same results.
- There are two basic designs of pinion drives used on off-road equipment; the first has the drive assembly located in the differential housing, and the second has the final reduction incorporated into the outer end of the final drive.
- Locating the pinion drive at the ends of the axle assembly allows the gears to be separate in their own housing, necessitating their own reservoir. If a failure were to occur, the damage would be contained in the outer housing.
- Lubrication is achieved in this design by using the time-proven splash method. The housing acts as a sump for the lubricating oil, which is splashed around by the rotating final drive gears.
- On larger track equipment, double reduction drives are commonly used in the final drive in the elevated position.
- The bearings used to support the drive pinion are tapered roller bearings capable of supporting the high loads that are generated at this end of the drive. As with most tapered roller bearings, shims are used to adjust the bearing pre-load.
- Planetary final drive designs come in single reduction and double reduction configurations and basically work much the same way. The main difference is the way in which torque is transmitted through them.
- Some manufacturers use a combination of a pinion drive with a planetary drive at the drive sprocket to drive the track or chain.
- Chain drives are used in many heavy equipment applications, such as road graders, where torque is transferred from the final drive sprocket to the wheel sprockets.
- Chain drives are capable of transmitting torque for long periods of time without much maintenance so long as dirt is kept out of the equation.
- The roller chain links mesh with the teeth on the sprockets to provide a positive drive between the front and rear wheels of the motor grader tandem drive.
- To distribute the wear evenly and prevent the roller links from contacting the same sprocket teeth all the time, roller chains have an even number of links and the sprockets they run on have an odd number of teeth.
- On grader final drives, the two drive sprockets are often mounted to the planet carrier and transmit power to two drive chains (one at each end of the tandem housing). The wheel spindle housing contains the components that make up the brake system. Some manufacturers use an idler sprocket to regulate chain tension; others use a guide shoe.
- A means of allowing the tandem drive to oscillate is provided by the pivot housing that connects to the

tandem housing with bolts. This connection allows the wheels to follow the road contour providing traction to all four wheels continuously.

- A chain drive is designed to tolerate some normal wear over a period of time. Three percent elongation of the chain is considered normal. Most manufacturers recommend checking chain length under tension as an indicator of drive chain wear.
- New chain links should not be used on old chains because the new link will have a shorter pitch, which will cause shock loading of the chain during operation and should be avoided at all costs. In most cases, it is advisable to change the sprockets when installing a new chain. Any sprocket tooth wear, which may look good visually, will shorten the life of the new chain considerably.
- Do not use conventional gear oils. They may not be compatible with friction materials that are used in brakes or steering clutches that are located internally in most final drives.
- Although tandem drives are sealed from the environment, they must have a means of equalizing the pressure that is built up internally during operation. A breather is located on each drive to prevent pressure buildup during operation.
- It may be necessary to flush the tandem drive with diesel fuel to get rid of slug buildup. When a final drive failure occurs, it often results in expensive parts replacement and long periods of downtime.
- When new internal components mesh for the first time, a certain amount of wear takes place, which benefits the overall life of the component. Gears meshing with each other will improve their contact patterns during the break-in period, leading to better load distribution.
- When it comes to assessing overhaul and service intervals, progressive companies use a reputable oil analysis program to establish benchmarks to determining acceptable wear levels in final drive systems.
- Inadequate lubrication is a common problem on off-roads equipment. Oil leaks from gaskets and seals on final drives lead to low oil levels and should always be repaired.
- Shock loading can often be traced to improper equipment operation where a sudden load is applied to a final drive component instantly or continuously over a longer period of time, resulting in cracks or immediate breakage depending on the severity and frequency of the shock load. Shock loading indicators will show up as small curved rings on the surface of the component. These rings are called beach marks.
- Another common condition that can be mistaken for shock loading is fatigue failure. This condition occurs when the components in question are subjected to torsional or bending forces caused by overloading of final drives. *Fatigue* is a condition that occurs between the surfaces of mating parts subjected to high torque loads. It starts with pieces off metal flaking of and progressively leads to pitting and finally complete breakage.
- Lubrication contamination is probably the most common problem that occurs in heavy equipment. Dirt, water, and even break-in debris can lead to component failure if lubrication oil is not changed according to manufacturer's recommendations.
- Bending fatigue, if severe enough, will cause mating gears to distort the running patterns, which transfers the contact pattern to the ends of teeth leading to concentrated loading.

Review Questions

1. On a single reduction final drive on a bulldozer where does the final gear reduction actually take place?
   A. differential gearing
   B. drop axle
   C. propeller shaft
   D. pinion drive
2. Which of the following indicates that the lube level in dozer final drive housing is acceptable?
   A. oil level is exactly even with the bottom of the filler hole
   B. lube level pours out the filler hole when the plug is removed
   C. lube level is one inch below the filler hole
   D. lube level is $1/2$ inch below the filler hole

3. When is the best time to drain the final drive oil during a regularly scheduled lube change?

   A. when it has cooled to ambient temperature
   B. at operating temperature
   C. after cooling for one hour
   D. any time if it's hot out

4. Which of the following types of final drive failures is likely produce a relatively smooth, flat fracture pattern?

   A. shock load
   B. fatigue
   C. lubrication failure
   D. constant torque overload

5. Which of the following are indications of fatigue-type failures in final drive gear components?

   A. pitted and spalled tooth surfaces
   B. cone-shaped fracture
   C. flat fracture pattern
   D. Brinelling

6. Which of the following types of final drive failures can result in a component becoming blackened and discolored?

   A. fatigue loading
   B. lubrication
   C. shock load
   D. torsional loading

7. In a final drive on a dozer if a sun gear in the planetary final drive is driven and the ring gear is stationary which of the following statements would be correct?

   A. The planet pinions will walk around the ring gear.
   B. All answers are correct.
   C. The speed of the planet pinion carrier will be reduced.
   D. The planet pinion carrier will rotate the same direction as the sun gear.

8. Which of the following statements describes the term backlash?

   A. the space between the bearings on the pinion
   B. the clearance between the crown teeth and the pinion teeth
   C. the pre-load on the carrier bearings
   D. the pre-load on the backlash bearing carrier

9. Which type of chain is used in motor grader final drives?

   A. plain
   B. roller
   C. gear
   D. silent

10. Which of the following is done to determine grader drive chain elongation dimensions?

    A. lay the chain flat on a bench measure its length
    B. hang the chain from a hoist, attach a scale, place the chain under tension
    C. hang the chain from a hoist, attach, and measure its overall length
    D. wrap the chain around a sprocket to determine overall wear

11. When replacing a grader drive chain, which of the following is an acceptable practice?

    A. replace only those roller that are worn
    B. replace only stiff rollers
    C. never add new roller links to a used chain
    D. replace the complete chain

12. When checking the backlash on a grader final drive differential, how many readings are required?

    A. 2
    B. 3
    C. 1
    D. 4

Steering Systems

Learning Objectives

After reading this chapter, you should be able to

- Describe the fundamentals of steering systems for off-road equipment.
- Describe the steering components used on off-road steering systems.
- Describe the terms *toe, caster, camber,* and *axle inclination.*
- Describe the fundamentals of front-wheel steer systems.
- Describe the principles of operation of front-wheel steer systems.
- Describe the fundamentals of all-wheel steer systems.
- Describe the principles of operation of all-wheel steer systems.
- Describe the fundamentals of differential steer systems.
- Describe the principles of operation of all-differential steer systems.
- Describe the fundamentals of clutch steer/brake systems on track equipment.
- Describe the principles of operation of clutch steer/brake systems on track equipment.
- Describe the fundamentals of articulating steer systems.
- Describe the principles of operation of articulating steer systems.

Key Terms

articulating steering
camber
caster
coordinated steer
crab steer
differential steer
drive steer
kingpin
pitman arm
steering clutch
tie-rod assembly
toe-in
toe-out

INTRODUCTION

Off-road steering systems enable equipment operators to convert steering wheel or joystick movement, through a series of electro-hydraulic or gear systems, into precise angular control of equipment with very little physical effort. Without some form of assistance, steering heavy equipment would be next to impossible; the effort to articulate or control wheel angles would be extremely arduous. At one time heavy equipment mechanical linkage, along with worm gears, was used extensively to steer equipment. On modern equipment, such as the dozer shown in **Figure 17-1,** the use of electric solenoids and hydraulics make the job of steering effortless compared to how difficult it was in the past.

Steering systems on off-road equipment come in a variety of design configurations, on-road equipment generally involves only the front wheels of the vehicle. This chapter covers the most common forms of steering systems found on off-road equipment, including conventional front drive axle steering systems, **articulating steering** systems, differential steering systems, and **steering clutch**/brake systems commonly found on crawler tractors. In most cases off-road equipment rarely uses axles that are just for steering. This chapter focuses on common types of steering found on heavy equipment; conventional on-road steer axles are covered in detail in *Heavy Duty Truck Systems*, 4th edition by Sean Bennett. No single steering system will work well on all of the different types of equipment used off road, however, all have heavy duty steer systems are generally operated with some form of hydraulic assist.

Although front-wheel **drive steer** is common, off-road equipment such as backhoes often use the front wheels to steer only, as shown in **Figure 17-2,** while the rear wheels propel the equipment. Front-wheel steer used on this type of equipment incorporate the same basic components as are used by on-road equipment. Equipment designed for some applications requires the use of all-wheel drive; in which case the front axle is equipped with a differential as well as steering capability. Other types of equipment, such as front end loaders and off-road haulage trucks, steer by controlling the angular relationship of a two-piece frame. The two sections are connected together by a center hinge pin with hydraulic cylinders attached to both the front and rear sections of the equipment that controls articulation.

Track-based equipment takes a different approach to steering. Manufacturers use a steering clutch/brake combination on each track. These combinations can operate independently of each other to provide directional control. Steering is accomplished by stopping or

Figure 17-1 Crawler tractor showing the steering clutch and brake assembly. (*Courtesy of Caterpillar*)

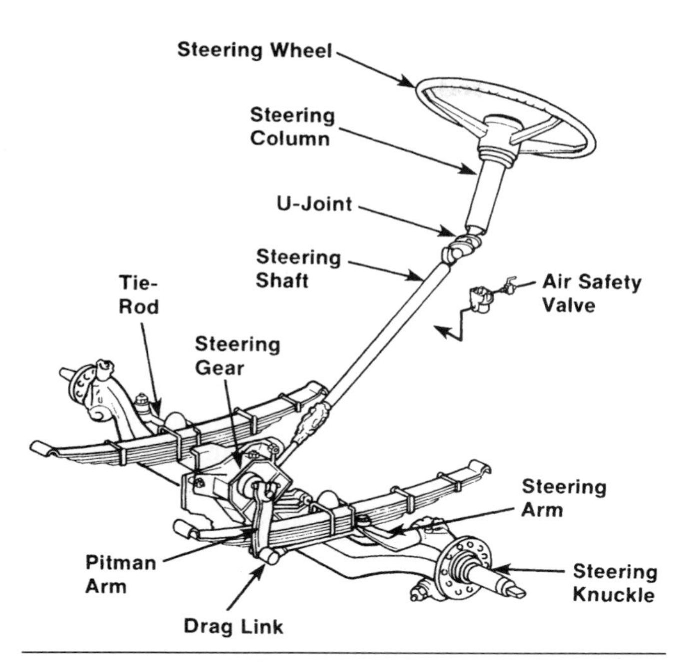

Figure 17-2 Typical front steer axle assembly. (*Courtesy of Air-O-Matic Power Steering. Div. of Sycon Corp.*)

slowing one track with a brake while maintaining forward or reverse motion. Differential steering is used on equipment that requires more precise control of the drive tracks. This system uses a series of planetary gears to provide torque input to the drive tracks, and a third input from a steering motor that can increase the speed of each track. When the speed of one track is increased, the other track slows down proportionally. The term **differential steer** comes from the action that takes place when the equipment is steered; it uses the same principle of operation as a conventional differential—where one side speeds up and the other side slows down an equal amount.

Crab steer is found on equipment such as large forklifts that require precise manuverability in tight corners such as in lumberyards or steel plants. The operator uses this option when it becomes necessary to shift the load left or right and still remain at right angles to the drop-off location. All four wheels turn in the same direction at the same time, allowing the equipment to move to the left or right while maintaining load alignment. Normal operation of the forklift is performed with the front wheels doing the steering, but in cases where side shifting with forward motion is required the operator can switch to crab steering by simply selecting it with a lever in the cab.

Coordinated steering is often used on equipment such as heavy-duty forklifts or specialized purpose-built lifters, such as the Cat TH83 Telehandler. Crab and coordinated steer are often found on purpose-built material handlers such as the Cat TH63 and TH83 as well as conventional front wheel steering. Having the ability to steer in a number of different ways makes this type of equipment well-suited for precision material handling situations like lumberyards and construction sites.

FUNDAMENTALS

Steering off-road equipment is accomplished in a number of different ways; all of them are dependent on input from an operator who controls the steering process with a steering wheel or mechanism. Steering can be done with a pilot-operated steering control system that uses a joystick with an electro-hydraulic control subsystem to actuate steer cylinders. Safe and predictable steering response is dependent on a number of factors that will be explained in detail later in this chapter. Each system will be dealt with separately since each system is uniquely different from the others in construction and operation. Steering on each system is somewhat different from the others, but is dependent on the same basic principles.

The track or wheel geometry and the frame alignment contribute to safe operation and to long component life. A heavy equipment technician must be familiar with the basic principles of operation on the various systems found on off-road equipment, and should rely exclusively on the manufacturer's technical documentation for procedures and specifications. In all cases, the equipment and tools used to perform this type of work must be checked at regular intervals to ensure accuracy. Specialized tools that are not properly maintained and tested for accuracy will result in poor work quality, which may compromise safe equipment operation. When it comes to steering, every effort must be made to ensure that the work performed is accurately done; any unexpected loss of steering during the operation of the equipment may result in a serious injury or equipment damage. In the case of conventional front-wheel steering assemblies, alignment of the steering components will ensure maximum component life. Alignment also has a significant effect on handling and tire life. In pilot-operated hydraulic steering, whether clutch-based or differential steer, a sound knowledge of the basic operation of each system is essential.

FRONT-WHEEL STEER SYSTEM

Off-road steering systems that use front-wheel steer generally use a steering wheel or a "stick" steering control valve and steering linkage. Although mechanical steering was used on older equipment, we will deal with newer systems that use hydraulics to assist the operator in steering control. Whether the equipment comes with a steering wheel or stick steering, some form of hydraulic assist is used on off-road equipment.

Steering Control Arm

The steering lever is attached to the steering knuckle on the driver's side of the equipment at one end and connects to the drag link, which is used to turn the front wheels. The terminology that is used to identify this lever by different manufacturers varies somewhat. Terms such as *steering lever*, *control arm*, and *steering arm* are used to identify this lever. As with all steering components the steering arm is made from high quality drop-forged tempered steel, and should never be heated to straighten out a bend; this could draw out the temper and weaken the arm.

Steering Knuckle

The steering knuckle is the movable portion of the front axle that pivots on a knuckle pin or **kingpin** which provides the steering action as shown in **Figure 17-3.** Kingpins can be either straight or tapered. Tapered pins are sealed and do not generally require lubrication; straight kingpins have grease fittings fitted to the knuckle caps at each end. Refer to the manufacturer's service literature for the correct type of grease and service intervals when maintenance is to be performed on the equipment.

New equipment often comes with sealed steering knuckles and does not require any additional lubrication during its service life. Most steering system knuckles require periodic lubrication and will have grease fittings on the knuckles. As with any part of the steering system, take the weight off the wheel ends when lubricating; this aids in the distribution of grease through the joint. The wheel spindle is mounted to the steering knuckle, which holds the wheel assembly and brake spider. The steering control arm is connected to the top of the left steering knuckle. The Ackerman arm, or cross-steering lever, is connected to the bottom of the steering knuckles of both wheel ends through the tie-rod ends of the cross tube, as shown in **Figure 17-4.**

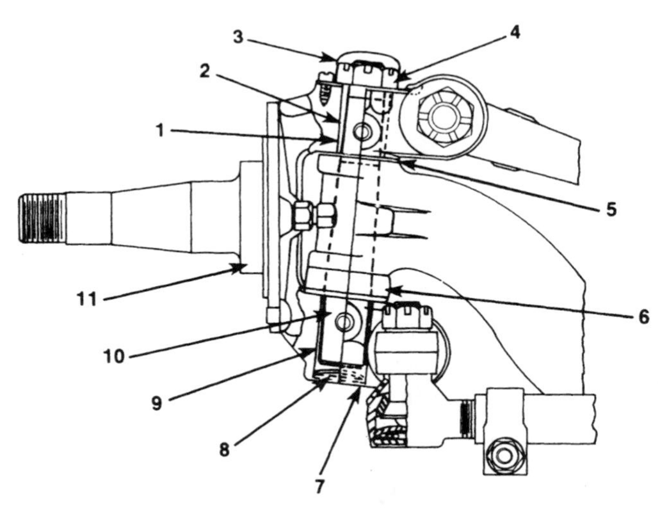

1 Knuckle Upper Bushing
2 Knuckle Pin Sleeve
3 Upper Dust Cap
4 Knuckle Pin Nut
5 Shims
6 Thrust Bearing
7 Expansion Plug Lock Ring
8 Expansion Plug
9 Knuckle Lower Bushing
10 Tapered Knuckle Pin
11 Knuckle/Spindle

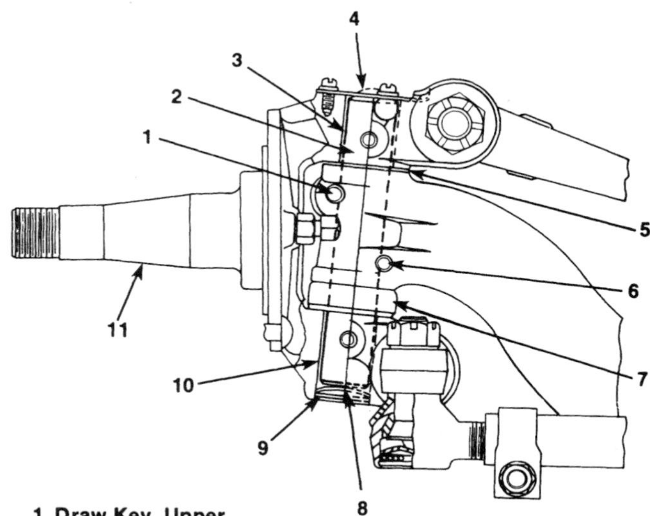

1 Draw Key, Upper
2 Knuckle Pin
3 Knuckle Bushing, Upper
4 Kingpin Cap
5 Shims
6 Draw Key, Lower
7 Thrust Bearing
8 Expansion Plug
9 Expansion Plug Lock Ring
10 Knuckle Bushing Lower
11 Knuckle/Spindle

Figure 17-3 Two styles of kingpins that are used on front-wheel steer equipment. (*Courtesy of Arvin Meritor*)

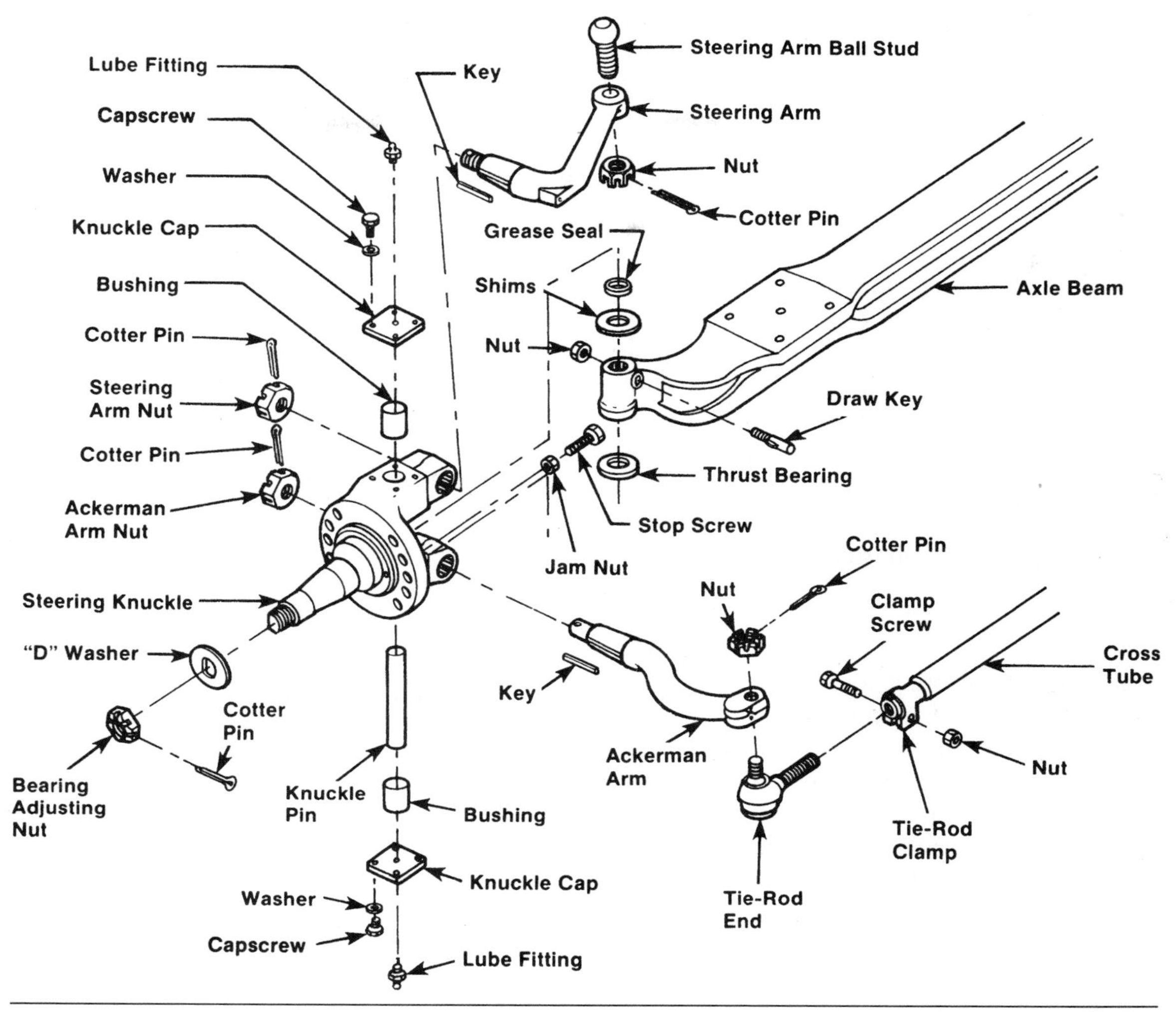

Figure 17-4 Components of a typical front-wheel steer assembly. (*Courtesy of Roadranger Marketing. One great drive train from two great companies—Eaton and Dana Corporations*)

Akerman Arm

The Akerman arm connects the right- and left-hand steering knuckles to the cross tube through tie-rod ball studs. One end of the arm is bolted through the steering knuckle, while the other end is bolted to the tie-rod and held with a castellated nut and a cotter pin to lock it in place. As with any steering system component, never attempt to straighten the Akerman arm if it is bent. The Akerman arm is made from drop-forged steel and should not be heated or bent back into shape. If it is damaged, it must be replaced.

Cross Steering Tube Assembly

The cross tube assembly holds the tie rods at each end. The tie-rods are threaded into the cross tube and connected to an Akerman arm on each steering knuckle. One end is left-hand thread and the other is right-hand thread. The threaded tie-rod end provides a means of adjusting the **toe-in** and **toe-out** settings.

Tie-Rod Assembly

Tie-rods are required to connect the steering knuckles and steering linkage together; this allows the two wheel ends to work in unison to provide steering control to the equipment. The tie-rods are ball sockets that allow the steering linkage to move freely in all directions, preventing binding. Each tie-rod is threaded at each end, which provides a means of connecting them to the cross tube. The cross tube has matching threads at each end to accept the tie-rods; a clamp is used to fasten them in place once proper alignment is achieved. The adjustment defines the toe setting because each tie-rod has opposite threads that are used to lengthen or shorten the cross tube length to achieve optimum toe setting. Once the toe setting is adjusted, tie-rod clamps are used to secure the tie-rod in the correct position.

Drag Link

The drag link is the mechanical connection between the **pitman arm** and the steering lever. Depending on the manufacturer, the drag link can be of one-piece or two-piece construction and is made from drop-forged steel for strength and durability. The connection between the drag link, the pitman arm, and the steering arm is made through ball joints at each end, which absorbs most of the axle motion from the steering gear.

Pitman Arm

The pitman arm or steering gear lever is used on steering systems that use conventional steering gear boxes. The pitman arm changes the rotary motion of the steering gear to linear motion. The length of the pitman arm will affect the steering response of the system; the longer the arm, the quicker the steering response will be. Some manufacturers scribe a timing mark on the arm, which must be lined up with a corresponding mark on the steering gear sector shaft to center the steering properly.

Steering Gear

Although mechanical steering gear systems are a rare find on equipment nowadays, they will be explained briefly. The focus will be primarily on steering systems that use hydraulic pressure from the equipment hydraulic system to steer the equipment through an elaborate steering control valve that is still operated by a conventional steering wheel, such as found on backhoes.

There are two general categories of mechanical steering boxes, the recirculating ball-type and the worm sector shaft-type as shown in **Figures 17-5** and **17-6.** Both systems multiply steering input torque from the operator and change the direction of power from rotary to linear motion. The worm and sector shaft steering uses a worm gear that rotates with the steering column. The sector shaft is located at 90 degrees to the worm shaft and converts the rotary motion of the worm gear into linear movement of the pitman arm.

The recirculating ball-type steering gear uses a worm gear with a ball nut that has internal grooves that match those threads of the worm gear. The ball nut has exterior teeth that mesh with the sector gear teeth. The balls used in the grooves of the ball nut allow the mechanism to move freely without much friction. The ball nut moves up and down on the worm gear, causing the sector shaft to rotate to provide steering control through the pitman arm, which is connected to the steering arm and steering knuckle.

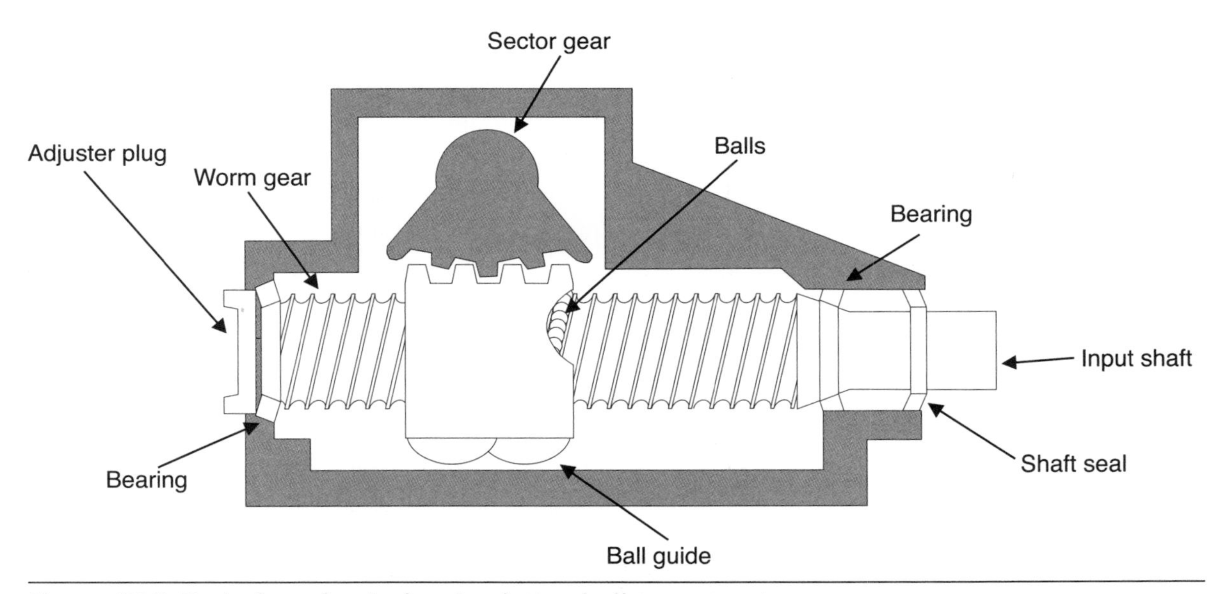

Figure 17-5 Typical mechanical recirculating ball-type steering gear.

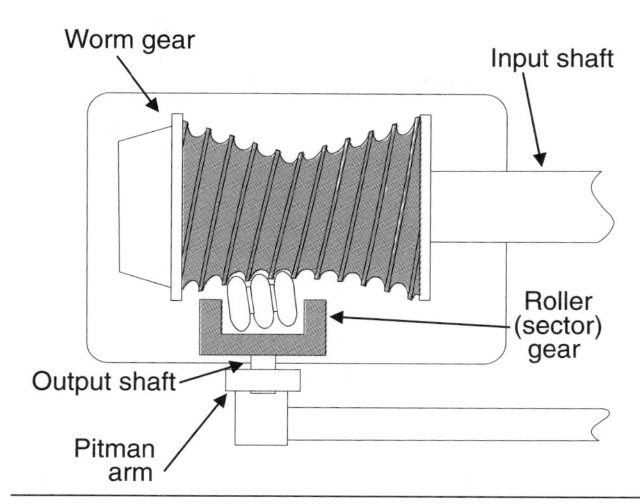

Figure 17-6 Typical mechanical worm sector shaft steering system.

Hydraulic Steering Valves

Although not all manufacturers use the same type of steering valve as that shown in **Figure 17-7,** they all use a similar principle for steering the front wheels of a backhoe. One type of John Deere's systems will be explained in detail to show how steering control is achieved with this design. The steering valve utilizes a rotary spool and sleeve assembly that is housed in a valve body, which has a gerotor mounted at the bottom of the valve body. When the steering valve is stationary, the steering valve reverts to the neutral position. The oil flow through the valve is blocked by the spool and sleeve, which is lined up in such a way that oil flow is stopped. When the operator turns the wheel to the left, the spool is turned in such a way that the oil is allowed to flow in the valve.

The flow of oil goes through the sleeve into spool passages and in the housing to one side of the gerotor, forcing the gerotor gear to rotate. This forces the oil out of the gerotor into outlet passages in the housing and through the sleeve and spool; it then flows out the outlet port marked "L" for left. From there the oil is sent to the rod end of the steering cylinder. The oil from the cylinder is forced back into the steering valve and out to the tank. When the rotation of the steering wheel is stopped, the gerotor gear continues to turn the valve sleeve until the oil flow is blocked, returning the valve to the neutral position. When the operator turns the wheel to the right, the same movement takes place in the valve in the opposite direction. Oil is sent to different passages and exits out the port marked "R" to the steering cylinder.

In the event of engine power or pressure loss, the steering can still be regulated manually. If no oil pressure is available, a cross pin comes in contact with the spool to provide a mechanical means of rotating the gerotor gear. This provides oil flow to the steering cylinder for emergency steering.

The hydraulic system provides oil to the steering system through a pressure control valve that acts as a priority valve. The steering system and backhoe circuit get priority; no oil will be allowed to flow to the loader circuit until the steering and backhoe circuit have 1300 psi of pressure.

STEERING GEOMETRY TERMINOLOGY

Off-road equipment with front-wheel steer utilizes the same concepts and terms that are used to describe conventional on road steering systems. Several common steering system components are discussed for clarification along with descriptions of alignment terms and troubleshooting techniques. Off-road equipment is not generally faced with the same type of steering system problems that are associated with on-road equipment, so alignment procedures may vary and are totally dependent on the road conditions and speed that the equipment will operate on.

Camber

Camber is the amount the front wheels tilt outward or inward at the top. The wheels are said to have positive camber if they are inclined outward at the top; and if the wheels are inclined inward at the top, they have negative or reverse camber. Camber setting can vary considerably on off-road equipment and is determined by the terrain and speed at which the equipment is operating. Camber adjustments are based on the amount of kingpin inclination that is used on the equipment. Incorrect camber can have a negative effect on tire life. Excessive positive camber causes tires to wear more on the outside of the tire, and negative camber will cause the tires to wear more on the inside. Unequal camber angles (from side to side) can have a negative impact on steering, causing the equipment to lead to the left or right. Camber corrections should never be made by bending the axles; this practice can lead to axle weakening and breakage. Camber angle (**Figure 17-8**) is dependent on the manufacturer's axle design and does

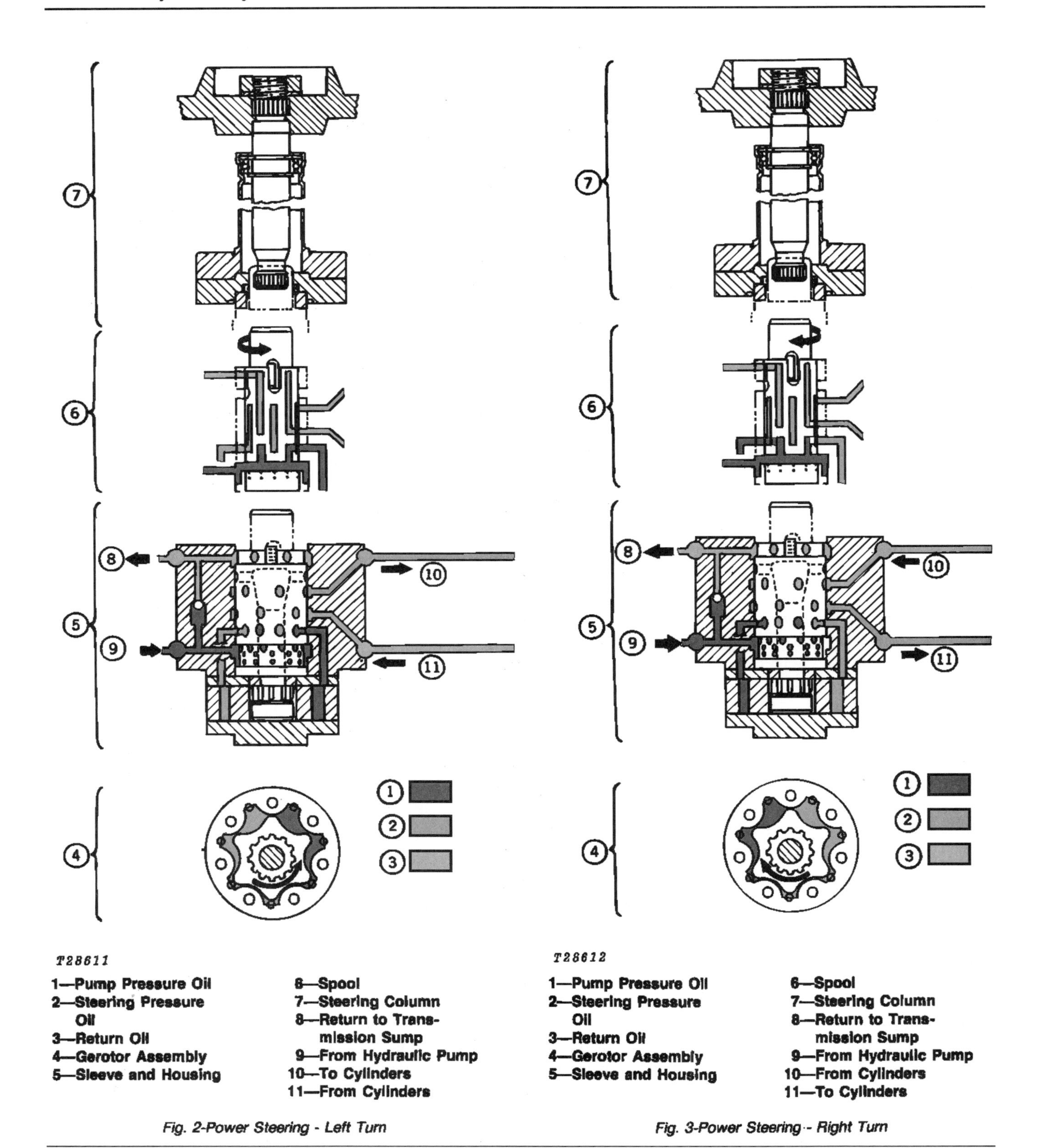

Figure 17-7 Typical hydraulic steering circuit on a John Deere 310 backhoe. (*Courtesy of John Deere and Company*)

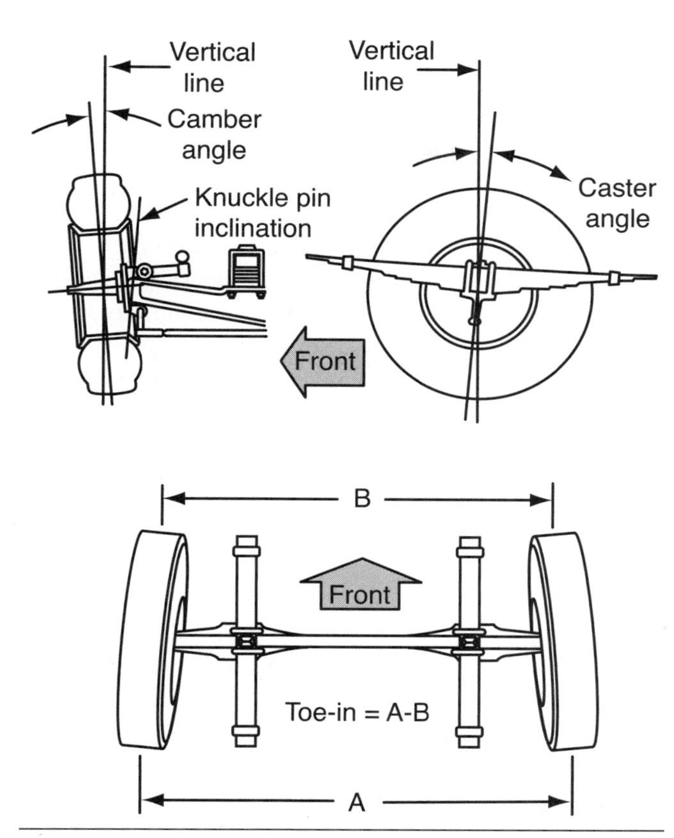

Figure 17-8 Adjustment for front steer off-road equipment.

not normally become a problem unless the axles have been bent in service.

Caster

Caster can be defined as the rearward or forward tilt of the knuckle pin (kingpin), either toward the front or rear of the equipment. It is expressed in degrees and as positive, negative, or zero caster. Heavy equipment steer axles are often set at a positive caster, which generally produces improved tracking and tends to make the steering axle return to a straight steer position more easily after a turn is made. Positive caster (Figure 17-8) is the rearward tilt of the kingpin as viewed from the front of the equipment.

Caster setting has little or no effect on the tire life of off-road equipment. Adjustment is made by placing shims between the front springs and spring seats. Always use the correct width of shim; the width should be equal to the spring width.

Kingpin Inclination

Kingpin inclination is defined as the number of degrees the knuckle pins (kingpins) are inclined inward at the top when viewed from the front of the equipment. Kingpin and caster angles (Figure 17-8) are set up on equipment to produce an optimum footprint contact with the road surface. These two adjustments can also affect the directional stability of the equipment. Adjustment should not have to be altered after the initial setup unless the front axle has been bent, in which case the bent parts should be replaced. No attempt to straighten out bent steering system components should be made under any circumstances.

Toe-In/Toe-Out

Toe-in is the relationship of distance between the front of the tires (when measured between the center lines) to the same measurement at the rear of the tires (Figure 17-8). This difference in distance is required to compensate for the natural tendency of the front wheels to toe-out during normal operation due to the effect of load and speed. The whole concept of toe-in adjustment is based on the premise that ideal toe-in should be zero when the equipment is moving in a straight line during normal operating conditions.

Tech Tip: Whenever a technician checks, tests, or performs any work on a steering system, the following precautions must be observed to ensure safe and correct equipment operation.

- Under no circumstance should a technician attempt to straighten out a steering axle component; always replace the bent component. On front-wheel steer equipment, never attempt to correct camber by bending steering components.
- Under no circumstance should welding be performed on steering components. Welding may change the structural integrity of the metal and weaken the axle housing or components.
- Never use heat to remove a steering component unless you are discarding it. Bent steering components should be replaced and no attempt to straighten them should be made.

DRIVE STEER AXLE

Off-road equipment must be able to travel on roadways that are less than ideal, necessitating the use of all-wheel drive in many instances. Front-drive steer axles must be able not only to support the weight of the equipment, but also to provide drive capability to the front wheels and house the front brakes. The spindle is mounted on the knuckle, which is mounted to the axle housing through the knuckle caps. The knuckle caps are able to swivel on bushings mounted in bores on the axle housing. Steering control is accomplished through the upper knuckle cap, which is connected to the steering arm. Both ends of the wheels are connected together by a cross tube, which has ball ends (tie-rods) on each end to provide a swivel connection between them.

The swivel joints (tie-rod ends) are protected from dirt and road debris with boots or dust seals made from rubber or synthetic materials that require periodic greasing and adjustments. There are numerous configurations for steering linkage for off-road equipment drive steer axles and all are dependent on equipment application.

The linkage on most off-road equipment may be arranged differently from the one that is used as an example in this chapter, but all have the same general parts that are usually located behind the front-drive steer axle. The cross tube and tie-rod ends connect the two wheel ends together and provides a means of maintaining alignment between the wheel ends. The steering arm, which is shown mounted to the wheel end, provides the connection to the powered steering system that is based on equipment application. The tie-rod ends do wear out, and the clearance in the ball and socket assembly will eventually become excessive, leading to poor steering control and excessive tire wear in some cases. In many cases, the operator may not be able to detect wear while operating the equipment.

STEERING SYSTEM MAINTENANCE

It is the technician's roll to inspect the steering system every time the equipment comes in for service. When the equipment does comes in for service the tie-rod ends should be inspected for wear or possible damage from road debris. **Table 17-1** shows examples of the various categories of operation for off-road equipment and should not be used as a service guideline. Always refer to the equipment manufacturer's service recommendations to determine actual intervals for servicing equipment. Table 17-1 shows examples of tie-rod service intervals.

Checking Drive Steer Axle Linkage Condition

Tech Tip: Before attempting to lift and block equipment, make sure it is parked on a level surface. Set the park brake and place wheel chocks in place to prevent unwanted equipment movement. Support the vehicle with approved safety stands. Whenever the steering system is to be checked for wear, the equipment must be raised and blocked in such a way that the steering components are unloaded. Never perform any type of work on equipment that is supported by a jack; always use approved stands to support the equipment.

Performing inspections of the steer axle components is part of a technician's role in servicing equipment.

Table 17-1: TIE-ROD SERVICE INTERVALS

Designation	Application	Inspection Intervals	Lube Intervals	Operating Conditions
Light service on-road	Linehaul only	Each oil change	100,000 miles (161,000 km)	More than 50,000 miles per year, 95% on-road
Medium duty	Heavy haul	20,000 miles (32,200 km)	40,000 miles (64,400 km)	Less than 50,000 miles a year
Harsh duty	Logging, construction	10,000 miles (16,100 km)	10,000 miles (16,100 km)	Low mileage operation Mostly off-road operation
Severe duty	Mining, landfill operations	5,000 miles (100 hours)	5,000 miles (100 hours)	Heavy duty service
Very Severe duty	Mining, logging, construction	48 hours	48 hours	Severe duty 80–100% off-road

Checks should include wear of linkage components, as well as checking clearances between the steering knuckles and axle centers, as shown in **Figure 17-9.** Inspection techniques used to determine the wear in steering knuckles requires that the equipment be jacked and blocked up properly. Try shaking the wheels vertically to see if there is excessive clearance between the knuckle pins and bushings or bearings. Inspect the front axle for any sign of bending or twisting, as well as cracks that may have developed. Refer to the manufacturer's technical documentation for clearance specification for knuckle joints. Check the condition of the front axle fasteners for any signs of movement, which could indicate a loose or broken axel mounting bolts. The wheels should turn freely from stop to stop. Be sure that the steering axle stops are adjusted correctly. Refer to the equipment shop manual for the correct procedures and specifications.

Do not grease the **tie-rod assembly** before the inspection is performed; the grease may take up the wear that is present in the joint. The wheels must be in the straight ahead position. If possible, turn the wheels through a full left to right turn and then back to center; this ensure that the linkage is properly unloaded.

1. Carefully check the tie-rod boots for any damage or cracks. Be sure to inspect the boot seals for breaks. If damage is discovered, the complete tie-rod must be replaced.
2. Be sure to check that the cotter pin is properly installed through the tie-rod end. If the cotter pin is missing, retighten the tie-rod end nut to the correct torque, following the manufacturer's torque charts. Install a new pin. Never back off the nut to install the cotter pin.
3. Check the cross tube carefully for any sign of impact damage. Look for cracks. If in doubt, replace the cross tube with an original equipment tube of the same size and diameter. Never try to straighten the tube after it has been bent. Never use heat on the tube; it may affect the strength of the metal. If the tie-rod clamps are damaged, replace them.
4. Check for excessive play in the tie-rod joint by grasping the cross tube with both hands and pulling and pushing on the cross tube.
5. Check the tie-rod thread depth in the cross tube slot. The threads must be in past the depth of

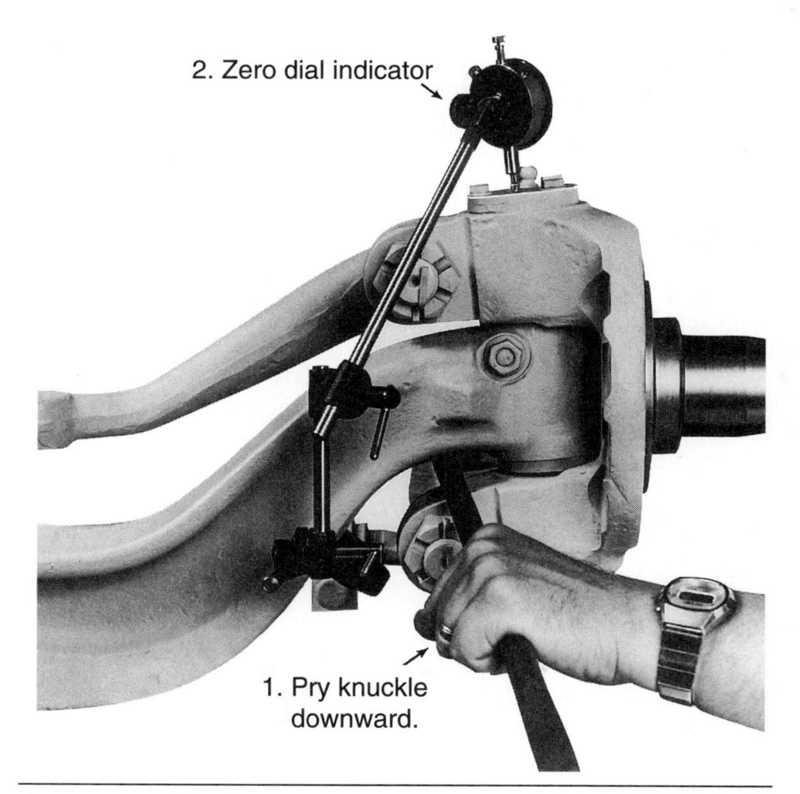

Figure 17-9 Checking steering knuckle for excessive vertical clearance. (*Courtesy of Roadranger Marketing. One great drive train from two great companies—Eaton and Dana Corporations*)

the cross tube slot. Make corrections if the depth is not correct. Note that if this adjustment is made, it is necessary to recheck the toe-in and toe-out setting. Refer to specific manufacturer's recommended procedure to perform this adjustment.

CAUTION *Before proceeding with any type of service or repair work to equipment that has a steer axle assembly, make sure that the equipment park brake has been applied and the rear wheels are blocked with wheel chocks. Use an appropriate jack to lift the equipment so that safety stands can be placed in an appropriate location to support the equipment with the wheels raised off the ground. Under no circumstances should anyone attempt to perform work on a piece of equipment that is supported with a hydraulic jack.*

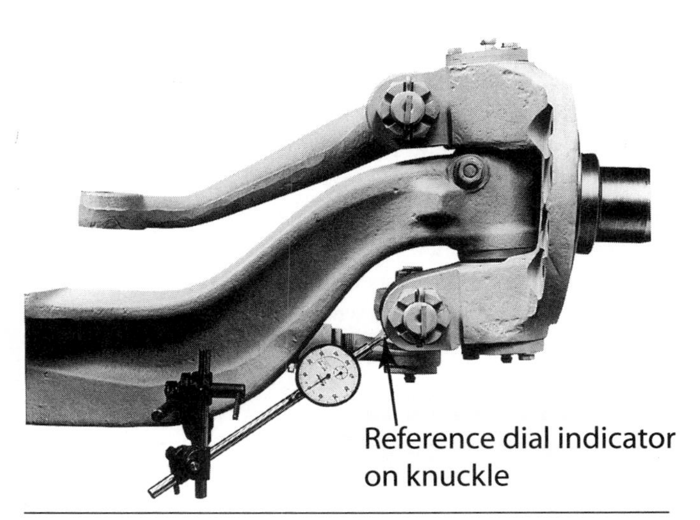

Figure 17-10 Checking lower kingpin bushing clearance using a dial indicator. (*Courtesy of Roadranger Marketing. One great drive train from two great companies—Eaton and Dana Corporations*)

Steering Knuckle Inspection

Steering knuckles can be a high maintenance item. The life of this connection depends on proper service and adjustment procedures. Steering knuckles should be lubricated according to the manufacturer's recommended time intervals, which can vary widely. The lubrication intervals and the type of grease used depend on the service application of the equipment. Steering knuckle wear accounts for a major portion of steering problems found on equipment with this type of steering system. **Figure 17-10** shows the procedure used to check lower kingpin bushing wear on the knuckle. The following procedure can be used as a guide to determine the amount of wear in the steering knuckle joints. Always refer to the manufacturer's recommended inspection procedures and specifications.

Steering Knuckle Clearance Checks. Because of the extreme road conditions that off-road equipment must endure, adequate amounts of lubrication are extremely important to prevent wear in the steering components that rotate or turn frequently such as steering knuckles.

1. With the dial indicator base mounted securely on the axle, as shown in Figure 17-9, set the dial indicator vertically on the end cap with the correct amount of pre-load.
2. Using a suitable pry bar, force the assembly downward and adjust the dial to zero.
3. Use the pry bar to force the assembly up and note the dial indicator reading or lower the front axle to check for excessive clearance. Refer to the manufacturer's recommended clearance in the shop manual to determine clearance limits.
4. To determine kingpin bushing clearances, position the dial indicator on the upper part of the knuckle as shown in Figure 17-9.
5. Vigorously push and pull on the wheel assembly and observe the dial indicator reading.
6. Again, refer to the manufacturer's recommended clearance in the shop manual to determine clearance limits.
7. Check the lower kingpin bushing by following the same procedure used to check the top kingpin bushing clearance.

Lubrication. Steering linkage joints often have grease fittings, although newer equipment will often have sealed joints that are factory pre-lubed and do not require grease during their life span. When performing maintenance work on equipment you may find a grease fitting that refuses to take grease; always replace the fitting immediately. Refer to the manufacturer's recommended lubrication chart to determine the greasing intervals on equipment. Lubrication intervals can vary widely and are dependent on the environment in which the equipment is operating.

DIFFERENTIAL STEER SYSTEMS

Differential steer systems are commonly used on a wide variety of equipment. Although there are differences in designs from manufacturer to manufacturer, the basic operating principles are very similar. The following example used to explain the construction and operating principles is a Caterpillar design, as shown in **Figure 17-11.**

Precise steering control, while maintaining equal power distribution to the drive wheels or tracks is often a requirement on off-road equipment. Differential steer provides the ability to deliver equal amounts of power to both axles while the equipment is driven in a straight line. When the operator need to correct steering direction, a steering motor drives one set of planetary gears that speeds up one axle and slows down the other by an equal amount, hence the term *differential steer*. The system operates on the same principle as a conventional differential, except that it uses planetary gears to achieve the differential action. Steering control is achieved without any change in power input speed, as shown in **Figure 17-12.**

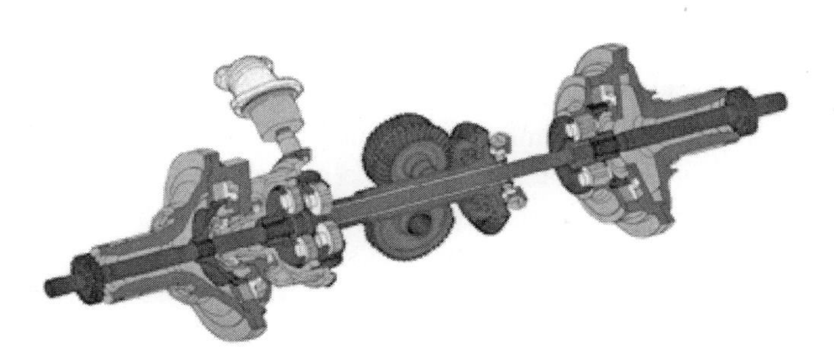

Figure 17-11 Cat typical differential steer assembly. *(Courtesy of Caterpillar)*

Fundamentals

The Caterpillar planetary differential steer system makes use of a drive planetary that sends torque to a planetary on each end of the assembly. The left planetary set is called the steering planetary; the set on the right is called the equalizing planetary. Power to the assembly can come from either the transmission or from a steering motor, which can drive the steering planetary. When the steering controls are activated by the operator, the hydraulic steering motor receives power from the equipment's closed-loop hydraulic system.

In Figure 17-12 we see that all three sun gears connected together by a common center shaft, which

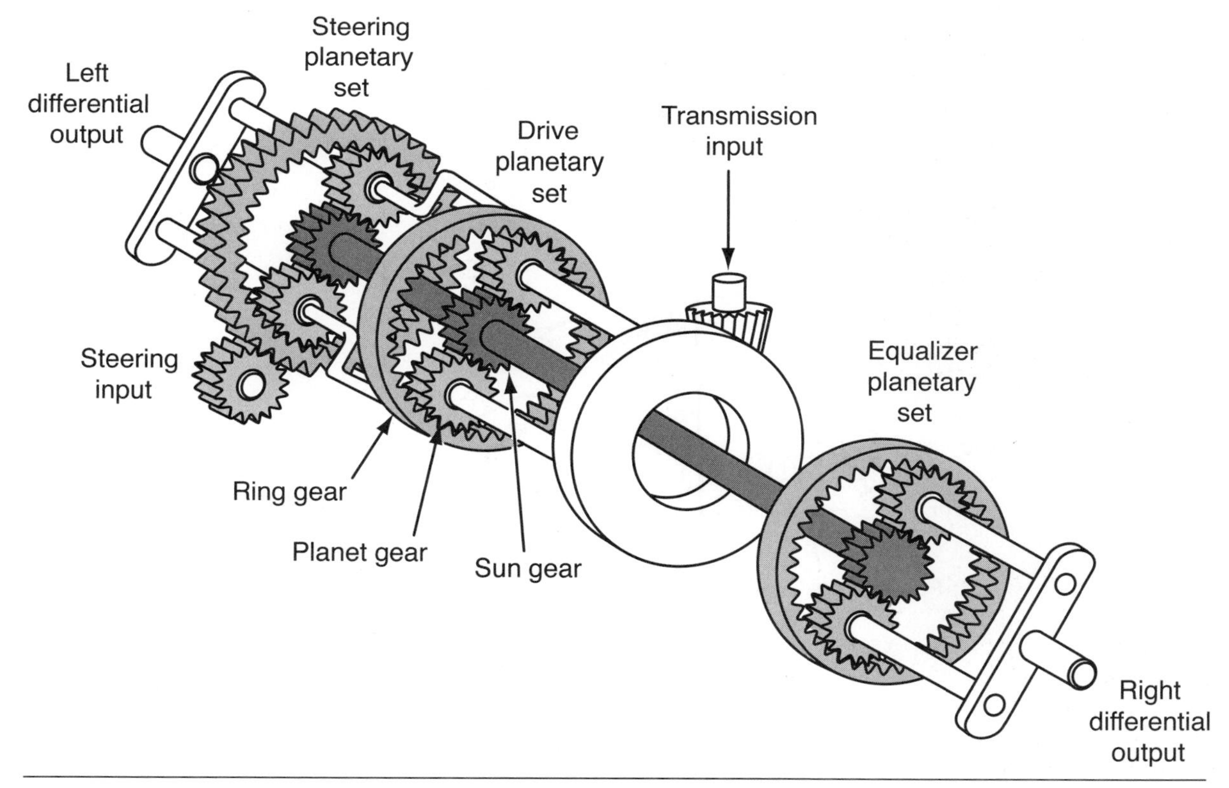

Figure 17-12 Cat differential steer assembly showing the components that make up this drive design.

allows them to turn at the same speed. The equalizing planetary ring gear is bolted to the right brake housing and never rotates in this configuration. The carrier of the steering planetary drives the left axle shaft, while the drive for the right axle comes from the carrier in the equalizing planetary.

With the equipment operating in the straight forward direction, as shown in **Figure 17-13,** the flow of power comes into the steering assembly through a pinion bevel gear set. The planetary drive carrier is connected directly to the bevel gear. Powerflow is split between the drive planetary ring gear and the sun gear. The drive ring gear reduces the speed and increases the torque that is transmitted to the left outer axle. Power flows to the right side of the drive through a drive planetary sun gear/center shaft to the equalizing sun gear, which then drives a set of planets, resulting in a speed reduction and torque increase to the right side drive of the equipment. Equipment engineers design the gear-tooth count to produce equal speeds and torque at both left and right sides of the drive when the equipment is operated in a straight forward direction. While operating in this mode with no other steering inputs, the steering planetary gear set ring gear does not rotate.

In order for the equipment to make a left turn, the operator needs to speed up one track and slow down the other by an equal amount. This results in a track speed differential between the left and right side. This form of speed control from side to side will turn the equipment with extreme precision. Steering motor output is controlled by steering controls, which determine the direction and speed of the steer motor. During a left turn, the steering motor rotates the steering planetary ring gear in the opposite direction of the planetary carrier, as shown in **Figure 17-14.** This operational change causes the ring gear to slow down the speed of the left side drive; at the same time it speeds up the planets that drive the right side sun gear, which drives the equalizing planets on the right side faster thereby producing a left turn. The majority of power to propel the equipment comes from the transmission during a turn. Any input from the steer motor results in a change in speed and direction of the steering planetary ring gear and causes speed differences between tracks.

When the operator needs to turn right, he or she activates the steering controls that cause the steering motor to turn the steering planetary ring gear in the same direction as the carrier. The ring gear input speeds

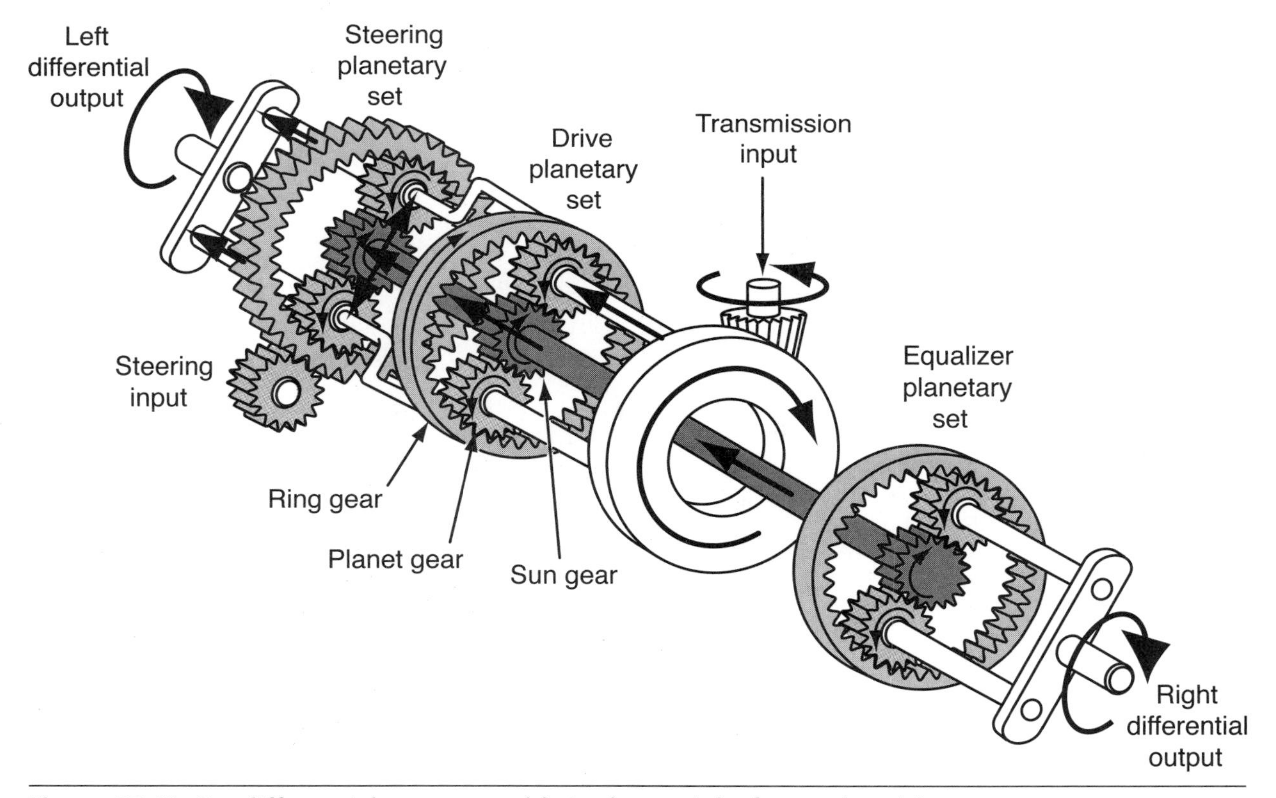

Figure 17-13 Cat differential steer assembly in the straight forward position.

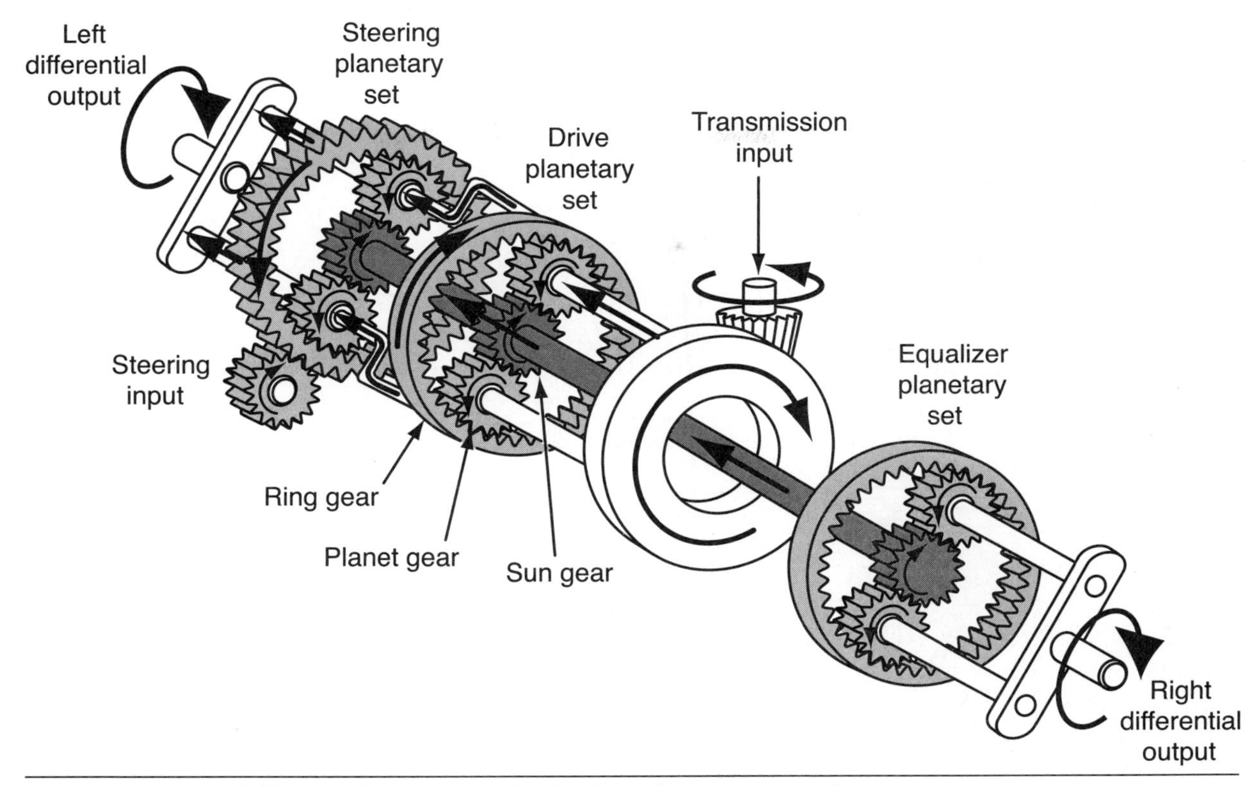

Figure 17-14 Cat differential steer assembly in the left turn mode of operation.

up the left carrier, causing the left axle to turn faster. Because the ring gear and carrier turn together, the planets are forced to slow down. Since the planets now drive the sun gear slower, the right axle slows down and forces the equipment to make a right turn.

This steering maneuver allows the operator to control the counter-rotation procedure with extreme precision. Although this function is not generally used in the actual working operation of the equipment, it does allow the operator to have increased maneuverability in areas were space is at a premium. In order to perform this maneuver the equipment must be placed in neutral because this procedure is controlled entirely by the steering motor input. Ground conditions must be similar under both tracks for this to work properly. The steering motor turns the steering planetary ring gear in the opposite direction, forcing the steering planets to turn in the same direction as the ring gear. Power flows to the sun gear causing it to rotate in the opposite direction of the steering planetary ring gear, which in turn drives the right side track in the opposite direction.

Operation

To fully understand the operation of this steering system, you must understand how the three planetary gear sets interconnect with each other, as shown in **Figure 17-15,** and how direction and speed are manipulated with this design. The main flow of power comes from the transmission, which drives the transmission pinion. Power flows through the crown gear and drive planets, which are connected directly to the crown gear. A hydraulic steering motor, which drives the steering planetary ring gear, is driven by a variable-displacement pump. The drive oil is part of a closed-loop hydraulic system used to power the circuit. This circuit requires a counterbalance valve to provide make-up oil, regulate the pressure, and protect the circuit from excessive pressure spikes and overspeed.

When the operator activates the steering control lever, the variable-displacement pump swashplate angle is altered, forcing the drive pump to rotate the drive planetary ring gear. To achieve straight forward drive, the oil in the steering circuit is blocked from flowing in either direction, which effectively locks the motor in place and prevents the drive planetary ring gear from rotating. The power flows from the transmission to the pinion drive. The planetary drive carrier is connected directly to the bevel gear. Powerflow is split between the drive planetary ring gear and the sun gear. The drive ring gear reduces the speed and increases the torque, which is transmitted to the left outer axle. At the same time

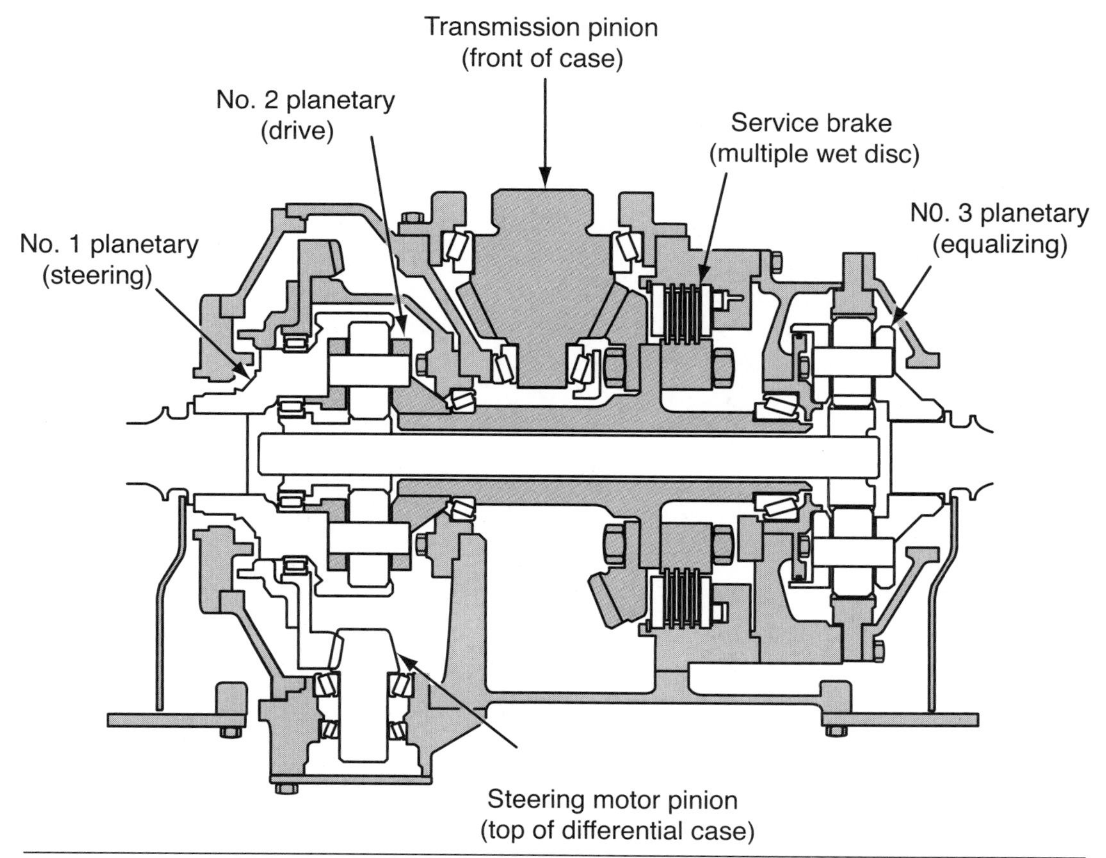

Figure 17-15 Cat differential steer assembly showing the parts that make up this drive system.

power flows to the right side drive through the drive planetary sun gear/center shaft to the equalizing sun gear (which in turn drives a set of planets to produce speed reduction and torque increase equal to the left side drive).

When the operator moves the steering controller to initiate a right turn (refer to **Figure 17-16**) the counterbalance valve is forced into the right steer position. This allows the flow of oil to enter a passage on the right side of counterbalance valve through a metering orifice and to fill the right chamber. Once this chamber is filled, the oil flow forces the check valve off its seat thus allowing oil to flow around the right crossover relief to the steering motor and causing it to rotate. This action forces the oil in the return side to return to the left side of the counterbalance valve, where it must overcome a predetermined spring tension before it forces the stem to shift to the left. This shift opens a cross-drilled port, and allows oil flow to return through the steering control valve. In order to protect the circuit from excessively high pressure, the crossover relief valves can sense excessive pressure on either side of the steering circuit. During normal high-pressure operation, oil is present in the return side crossover relief valve chamber; if excessive pressure develops during normal operation, the left crossover relief valve will opens to allow the excess oil to flow to the return side.

To initiate a left turn (refer to Figure 17-16), the counterbalance valve is forced into the left steer position. This allows the flow of oil to enter a passage on the left side of counterbalance valve through a metering orifice and to fill the left chamber of the counterbalance valve. After this chamber is filled, the oil flow forces a check valve off its seat thus allowing oil to flow around the left crossover relief to the steering motor, which forces it to rotate. This forces the oil in the return side to return to the right side of the counterbalance valve, where it must overcome a predetermined spring tension forcing the stem to shift to the right. The shift exposes a cross-drilled port and allows oil flow to return through the steering control valve. In order to protect the circuit from excessively high pressure, the

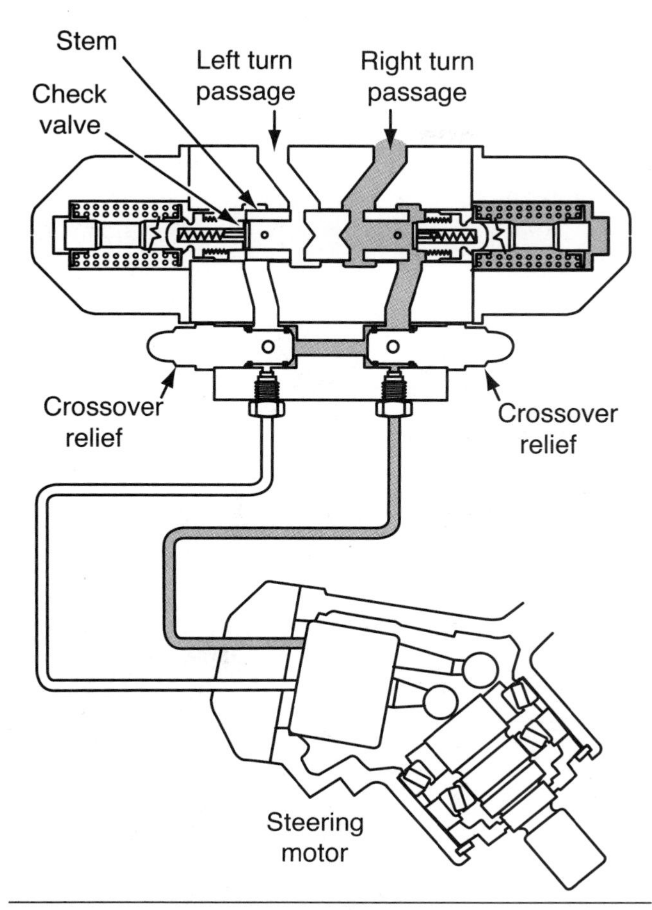

Figure 17-16 Cat counterbalance valve used in the differential steer circuit showing oil flow during a right turn.

crossover relief valves can sense excessive pressure on either side of the steering circuit. During normal high-pressure operation, oil is present in the left side crossover relief valve chamber; if excessive pressure develops during normal operation, the right crossover relief valve opens to allow the excess oil to flow to the return side.

Overspeed can occur on equipment during operation on steep slopes. Steering control can be lost if overspeed occurs during a turn on a steep slope. The counterbalance valve has been designed to prevent this condition from occurring. Operating pressure will decrease rapidly in an overspeed condition. The pressure in the right side of the steer motor drops. This causes the pressure to drop in the right side of the counterbalance valve. When pressure drops below a preset amount, the stem moves to the right to stop the flow of oil to the return side. When this occurs high backpressure builds in the steer motor, which limits overspeed. To protect the circuit from excessive backpressure, the left crossover relief valve forces any excess return oil to flow back to the supply side to prevent any chance of cavitation. There may be some operating conditions that will result in extreme overspeed; in these cases, the steering control valve has a make-up valve that allows make-up oil to enter the supply side of the circuit to maintain a positive oil supply for continued steering control.

STEERING CLUTCH AND BRAKE STEER SYSTEMS

Manufacturers of track-type equipment use independent control mechanisms to steer and brake each track separately, as shown in **Figure 17-17.** There are many design variations for these systems. This section covers the basic operating principles of this design and uses the Caterpillar system as an example.

A brake and a steer clutch are mounted together on each side of the final drive assembly. The steering brake and clutch assemblies are designed as modular units and are bolted together. Depending on the manufacturer and model, the equipment can have a brake assembly design using either band or multi-clutch design. The operating principle is the same in both cases; the braking action is just accomplished by using different hardware designs. In operation, the steering clutch disengages gradually to slow down the track on one side.

When the equipment is required to make a hard turn, the steering clutch is disengaged and the brake is applied, forcing the track to stop turning or slow down depending on the amount of brake application. Power flows in through the pinion and bevel gear set to drive a coupling that is connected to the yoke assembly. The clutch and brake can be controlled by either a pedal or lever. On some track-type equipment, two steering pedals or two steering levers may be used to control the spring-applied hydraulically-released clutches. The clutch outer drum also incorporates the brake drum. To execute a slight turn, the operator partially disengages the clutch on one track to slow it down; this causes the equipment to turn slightly because of the unequal track speeds.

The pedal steer clutch control system (**Figure 17-18**) uses the top valve to control the left side of the equipment and is shown in the hold position. Pressure flows into both control valves; no pressure is directed to the clutch from the left control valve in this position. The oil is diverted to the right side control valve, where it flows past the spool to the right side steer clutch, disengaging it. The right control valve spool is depressed, allowing oil to flow to the steer clutch. The design of this system

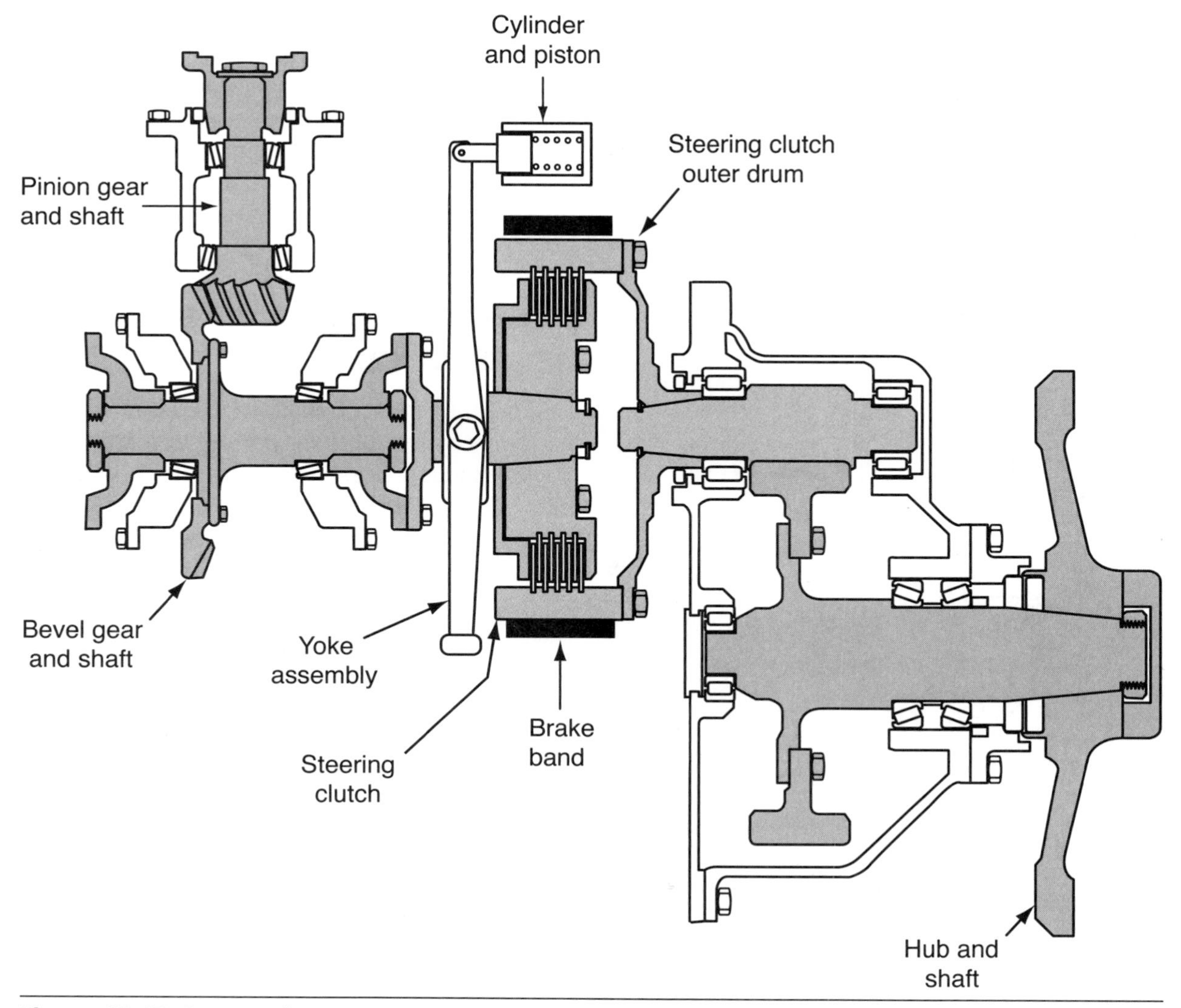

Figure 17-17 Cat crawler tractor steering clutch and brake assembly.

allows only one clutch to be disengaged at one time. The flow of oil to the left steer control valve is sent to the right side control valve where it flows through the right control valve spool to the steering clutch. These control valves are designed so that each also functions as a pressure reducing valve. This allows the operator to get feedback from the pedals; when the pressure increases, the combination of the spring pressure and oil pressure tries to move the spool to the right against an inner spring.

The steering clutch control valve operation has no crossover supply passages in this design; this prevent the operator from disengaging both clutches at the same time. The right side lever is activated, sending oil from the pump to the right steer clutch and disengaging it. Clutch operation is proportional: the more the pedal is depressed, the more the clutch slips. The left side steering control valve blocks oil from flowing to the left steer clutch, maintaining drive speed.

If the equipment uses a steer clutch/brake band combination, when the clutch pedal is depressed, oil is sent to a steering clutch cylinder. This forces the yoke and pressure plate assembly to release some of the clutch spring pressure and allows the clutch discs to slip slightly. If full pressure is allowed into the steering clutch cylinder, the clutch disengages the track from the powerflow completely. When the operator releases the clutch completely, a mechanical brake band is applied to the outer drum to stop or slow down the track.

On newer equipment many manufacturers use a multi-disc brake, shown in **Figure 17-19.** This replaces the drum-style brake used on older equipment. Modern equipment design has done away with levers, in most

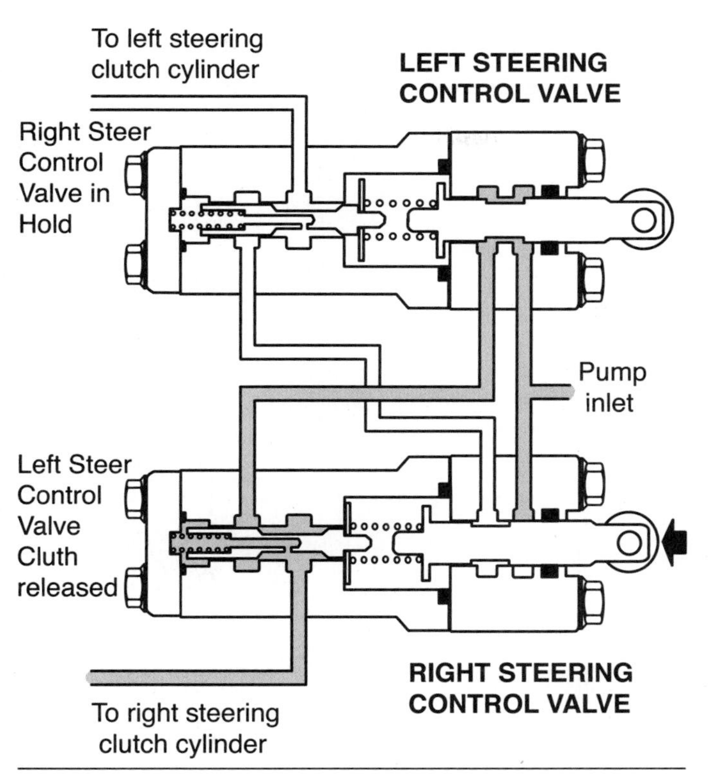

Figure 17-18 Cat crawler tractor clutch pedal steer control valves.

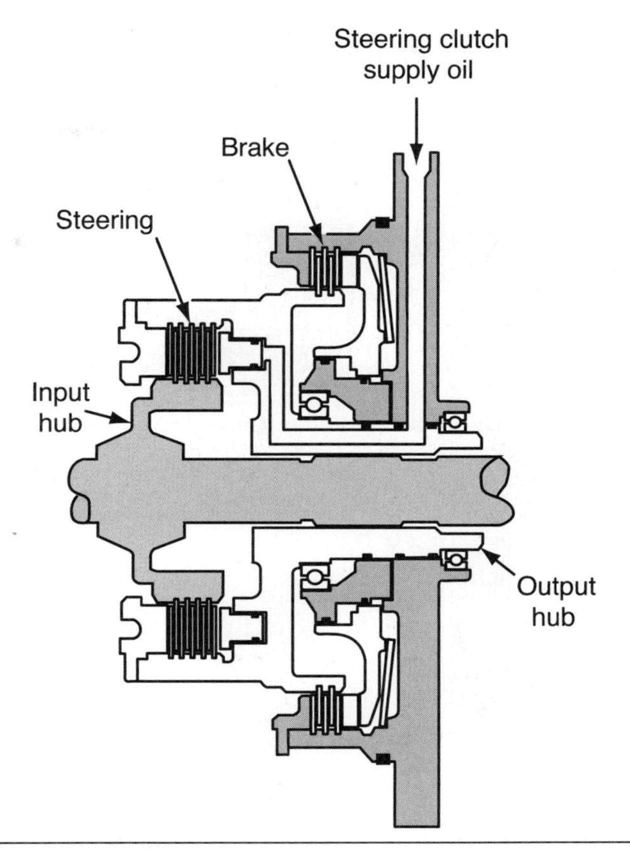

Figure 17-19 Cat crawler tractor multi-disc brake assembly.

cases, and replaced them with fingertip steering controls that use electric/hydraulics to control the flow of oil to the steer clutch brake assembly. The assembly is lubricated and oil cooled with oil that is circulated through an oil cooler with a pump. The steering clutches are applied by hydraulic pressure, and the brake is applied by spring pressure and released with hydraulic pressure. When the operator activates the controls for a hard turn, the steering clutch is fully disengaged and the brake is applied.

The systems are designed to apply the brakes automatically when hydraulic pressure is lost. Power flows from the transmission through a bevel gear to the inner axle shaft to the input hub on each side. The input hub and output hub are connected through the steering clutches. The brake housing is attached to the output hub through the brake clutch and does not rotate. A slight activation of the steering clutch causes the clutches to slip, resulting in a change in track speed. Whenever the brake is applied it forces the output hub to stop rotating it is connected mechanically to the stationary brake housing when the brakes are applied. When the service brakes are activated by the foot pedal, the oil flow to the brakes is gradually reduced and will vary with the amount of pedal travel the operator applies. When the steering control is pulled past detent, brake pressure decreases by 285 psi immediately; this action allows the brakes to be gradually applied from this point on. Full brake engagement occurs at maximum brake pedal travel, at which point the pressure is reduced to approximately 28 psi. This allows Bellville springs to apply the brakes.

Steering/Brake Hydraulic Circuit Operation

To ensure safe operation of the equipment under all possible operating conditions, the steering and brake control circuit gets priority over the equipment hydraulic circuit. The circuit incorporates a priority control valve to ensure that the steering and brake circuit gets oil first, as shown in **Figure 17-20.** If hydraulic pressure is lost for any reason, the steering clutches will be disengaged and the brakes will come on because they are spring-applied. There are four pressure-reducing valves located

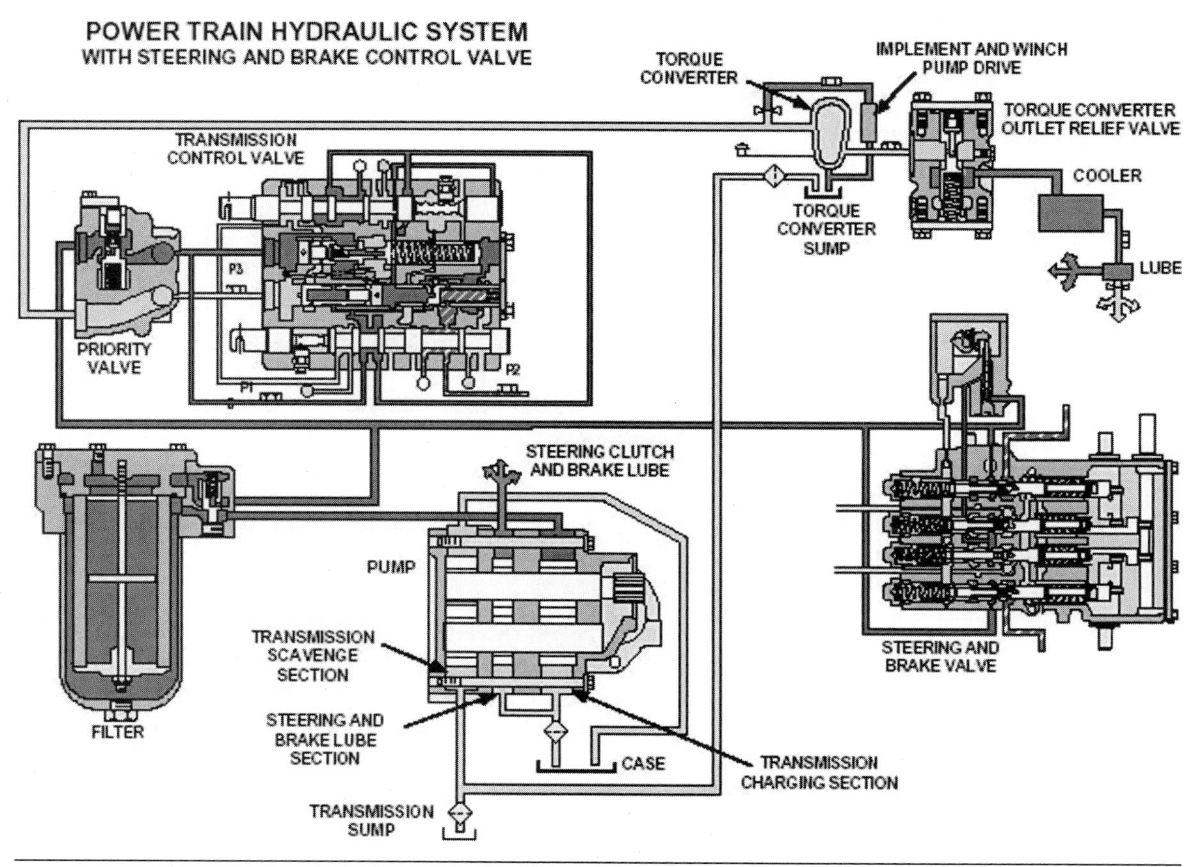

Figure 17-20 Cat crawler tractor powertrain hydraulic system. (*Courtesy of Caterpillar*)

in the steering brake valve control. Since the steer circuit is hydraulically-applied and the brake circuit is spring-applied the oil requirements are quite different for each circuit. The steering brake control valves have four separate circuits: a separate circuit for each steering clutch and brake, and a parking brake circuit for each brake. Whenever the operator applies the park brake, the oil flow to the brakes is cut off. If the operator were to perform a hard turn, the oil to the appropriate steering and brake circuit would be drained. A small correction to the steering, such as when making a large radius turn, would result in a gradual draining of a left or right steering circuit.

Steering and Brake Valve Operation

Figure 17-21 shows straight line operation flow through the steer brake valve. To achieve straight line operation, the brake/steering levers and brake pedals do not require any input from the operator. Oil flows in from the priority flow control valve and enters into the cavities for the steering clutch reducing valves. From there the oil flows through a check valve to the cavities around the brake reducing valves. Whenever the steering clutches or brakes are activated, oil will flow to the clutch cavities. During this process the oil also flows from the outlet cavity and opens a ball check valve and fills a reaction chamber. After maximum clutch pressure is reached, a spool in the reaction chamber moves to the left against spring pressure into the metered position. This restricts the flow of oil to the clutch and is responsible for controlling the clutch pressure. When the clutch pressure in the reaction cavity drops, a spring shifts the spool to the right; during this phase more oil can flow into the clutch until maximum system pressure is achieved. The operation of the brake circuit is quite different from the clutch, as the brakes require oil pressure to release them.

When the operator applies the brakes, cam lobes force the plungers located in the brake spool cavities to compress the springs against the brake spools. Once plunger-spring force is sufficient to overcome the opposing forces, the spools will move to block a port that supplies pump oil through the park brake check valve from flowing to the brakes. At this point the oil from the spring-applied brakes will drain, automatically applying the brakes. When the operator applies the park brake, a

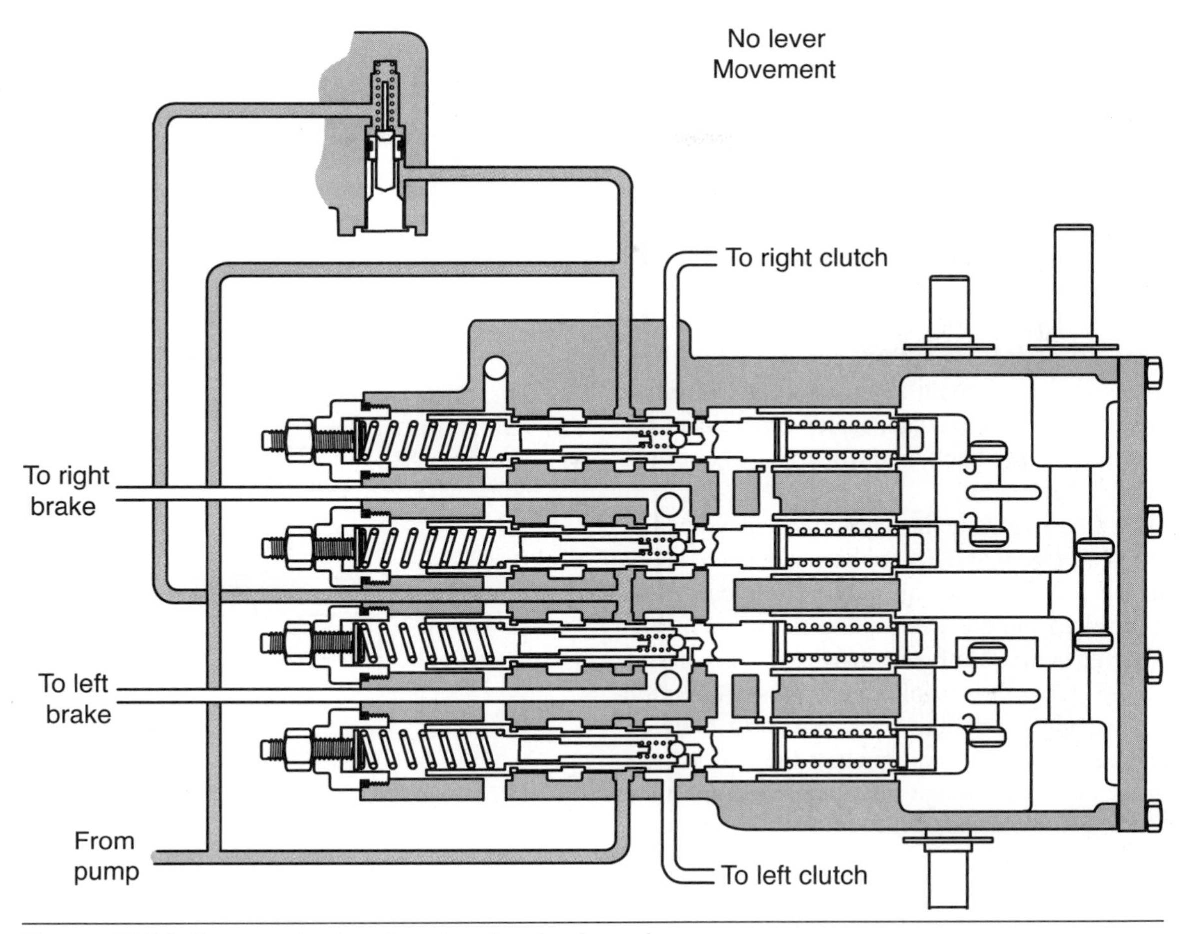

Figure 17-21 Cat crawler tractor steering brake valve.

poppet valve blocks the flow of oil by closing a check valve and preventing oil from flowing to the brake spools. This causes the oil to drain from the brake spool cavity, applying the spring-applied brakes.

This valve design prevents the brakes from applying before the clutch is fully disengaged. When the operator activates the steering control, a lever forces the steering shaft to rotate, causing a cam lobe to shift the steering clutch spool. This action blocks the flow of oil to the clutch and at the same time allows the oil from the clutch to drain. If the operator were to continue to move the lever past its halfway point, a cam lobe on the shaft would contact the brake plunger and start to apply the brakes. In this application, no brake application can occur until the steering clutch is in the fully released position.

ALL-WHEEL STEER

Off-road equipment steering applications must often be more versatile than what can be achieved with front-wheel steer only. This section deals with a steering concept that is common to many manufacturers of specialized material handling equipment. The construction industry benefits from equipment with various steering modes. We will cover the basic concepts of all-wheel steer by using examples that are based on material handlers found in the construction industry. Three steering modes are used in this design: two-wheel steer, circle steer, and crab steer. The front and rear axles have similar designs, both use conventional steering components such as tie-rods and steering cylinders. The tie-rods are connected to each wheel through a steering case on each side; when the operator turns the steering wheel hydraulic oil causes the steering cylinder to move the tie-rods, which changes the steering direction of the wheels. The steering cases are mounted on pivot pins on the axle housing, enabling them to turn from side to side. The basic components that make up this steering system include a variable-displacement steering pump, a selector valve, front and rear steer cylinders, a steering metering pump, and a load-sensing valve bank.

Fundamentals

In normal operating situations, two-wheel steer allows the operator to use the equipment much like any conventional material handling machinery. This mode should be used for most operating conditions, and it should always be used whenever the unit is operated on a roadway. Only the front wheels will steer in this mode. The rear wheels must be in the straight ahead position for safe operation in this mode. When switching between all-wheel steer and conventional front wheel steer some equipment manufacturers install a self-aligning feature that makes it easier and quicker for an operator to switch back and forth between two-wheel and all-wheel steer.

In some work environments, equipment may need to be operated in an area that does not have sufficient space to perform the work that is required. In these cases, the operator can select a circle steer option to enable him to turn in a shorter radius. Both sets of wheels steer in this mode, with the rear wheels turning the opposite direction to the front wheels. Switching back and forth between any of the steering modes requires that the rear wheels first be aligned. On equipment that does not come with the self-aligning feature, the operator must set it up manually before a switch can be made. Equipment that comes with the self-aligning feature uses a mode selector switch along with a master steering selector switch to select the operating mode. These features will be explained in more detail later in this section.

Crab steer gives the equipment the ability to move sideways while moving also forward or backward like a crab, hence the name *crab steer*. The rear wheels are steered in the same direction as the front wheels causing the equipment to move either left or right while advancing. The maneuver is extremely beneficial when working the tight quarters.

STEERING OPERATION

The hydraulic pump that is used to provide oil flow in this circuit is a variable-displacement axial piston pump. A load-sensing system in the valve bank supplies oil to the metering pump, which supplies oil to steer the equipment. The rear steer axle is controlled by a three-position solenoid-operated steer mode selector valve, as shown in **Figure 17-22.** On models that are equipped with the self-aligning rear steer, additional hardware is required to self-align the rear axle.

A proximity switch is located on the right side of the rear axle, with a corresponding target attached to the right side of the steering yoke. When switching to conventional front-wheel steer, a position sensor is used to indicate straight alignment, which allows locking the rear wheels in the required position. This sensor is de-energized when the operator selects crab or circle steer. When operating in front-wheel steer mode, oil is blocked from circulating to the rear wheels by the master steering selector switch, which de-energizes the mode selector valve. Springs center the spool to block the flow of oil to the cylinder locking it in place. The variable-displacement axial piston pump incorporates a pressure and flow compensator valve to meet the demands of the circuit. This valve controls the swashplate angle to regulate the pressure that acts on the actuation piston.

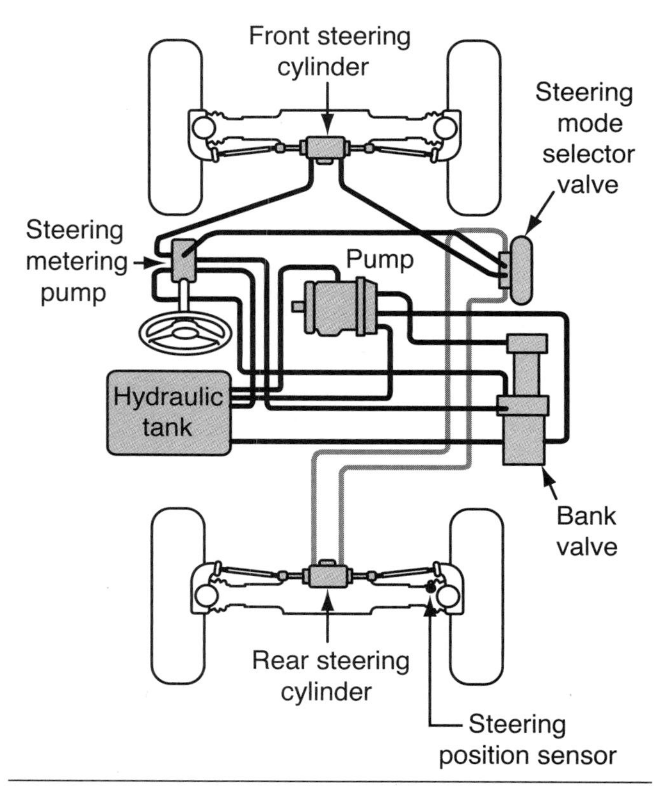

Figure 17-22 Cat telehandler steering system component layout.

As the actuation pressure decreases, the bias spring increases the stroke, sending more oil to the steer circuit. Signal pressure from an implement and supply pressure determine how much oil is sent to the actuator piston in the pump. In this circuit, the pressure difference between the supply pressure and the signal pressure is referred to as the margin pressure. The spring acting on the flow compensator spool determines the actual margin pressure. Supply pressure forces the flow compensator spool to push against a spring and signal pressure on the other side of the spool.

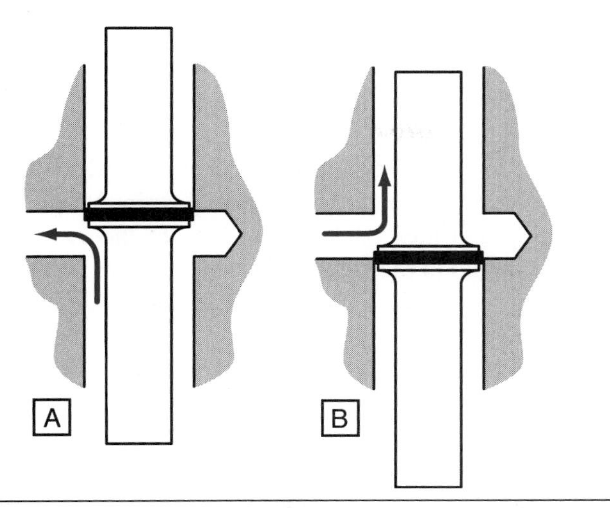

Figure 17-23 Flow compensator spool in the two modes of operation.

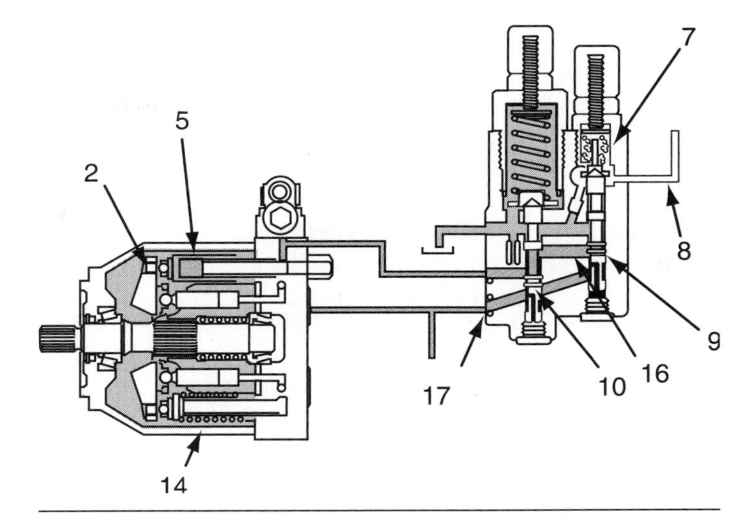

Figure 17-24 Steering circuit in the standby mode.

Position "A" in **Figure 17-23** shows oil flow when the pump pressure is greater than the combination of spring and signal pressure. In this position the oil flow decreases the stroke in the pump, lowering the pressure in the circuit and forcing the flow compensator spool into position "B", which sends oil flow from the actuator piston back to drain. An orifice in the signal passage regulates the speed of the actuator piston by constantly bleeding oil back to drain. The pressure compensator spool and spring will limit maximum system pressure when necessary, for example, if the signal relief valve becomes jammed or was improperly adjusted.

During periods of idle with no steering input, the system goes into standby mode and the only oil flow would be to compensate for internal leakage, as shown in **Figure 17-24.** No-load sensing pressure is present at the spring side of the flow compensator spool (pressure de-strokes the pump, and only a minimal amount of oil is necessary due to internal circuit leakage) and an adequate amount of supply pressure is maintained.

Pump Up-Stroking/De-Stroking

When the operator activates a circuit, the pump immediately gets stroked; signal pressure (along with spring pressure from the flow compensator valve) becomes higher than the pump standby pressure, blocking the flow of oil to the actuator piston and at the same time allowing the oil in the actuator circuit to return to tank through the constant bleed orifice. The bias spring forces the pump to increase its stroke and send more oil to the circuit, increasing the pressure until the flow compensator spool moves into the metering position. An increase in pump stroke can be caused by a number of conditions; a change in engine speed, steering requirements, and implement activation. Although the signal pressure does not change, a drop in supply pressure can also cause the pressure to the actuator piston to drop, causing a stroke increase in the pump.

When oil flow requirements drop, signal pressure, along with spring pressure in the flow compensator spool circuit above the spool, drops to below the pump pressure. This causes the flow compensator spool to move up, forcing the oil pressure to increase to the actuator piston and causing a decrease in stroke length. This action decreases the pump output as well as system pressure. Once the flow requirements are met, the flow compensator spool moves back into the metering position. If no further input from the operator is detected, the system reverts to low pressure standby mode.

A number of activities can activate the de-stroke mode of operation; they include an increase in engine speed, decreased steering demand, or implement activity. Note that the signal pressure does not necessarily have to decrease to de-stroke the pump; an increase in supply pressure will force the flow compensator spool to shift up, causing a pressure increase to the actuator piston and overcoming bias spring pressure, which would de-stroke the pump.

Certain operating conditions often result in what a high-pressure stall, for example, when a cylinder reaches the end of its stroke. To protect the circuit, the signal relief opens to limit maximum signal pressure. Supply pressure at the pump inlet forces the pressure and flow compensator spools to shift up to the metered position to maintain the flow rate. When the circuit goes into stall, the pump stroke increases the flow. The pressure compensator spool acts as a backup to the signal relief valve, limiting maximum system pressure.

Steering Metering Pump Operation

The steering pump in this circuit controls the direction and speed of the turn with two separate sections within the pump; the control section and the metering section, as shown in **Figure 17-25.** The direction control is achieved with a closed center rotary control valve, and the metering is controlled by a gerotor pump. An increase in steering wheel input increases the flow of oil to the steering cylinders. When steering input stops, centering springs position the spool in such away that it blocks the flow of oil to the steering cylinders. Turning the steering wheel to the right forces the spool to turn, overcoming centering spring tension; travel is limited by a drive pin coming in contact with a slot in the spool. This action allows oil to flow from the control section to the metering section through a check valve in the inlet port of the metering section.

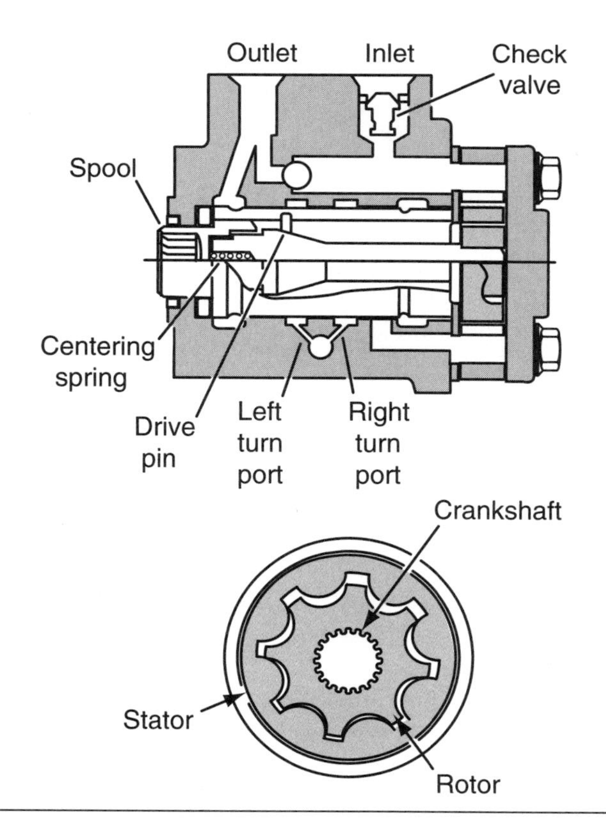

Figure 17-25 Steering metering pump cross section.

Continued turning of the wheel causes the drive pin to turn a driveshaft, forcing the rotor to turn inside the stator. This action causes the oil flow to be metered back to the control section of the pump and out the right turn port to the corresponding steering cylinder. Return oil flows back through the left turn port into the metering pump and then back to the tank. In the event of a pressure spike in the steering circuit, a check valve protects the system from overpressure. If the equipment were to lose power or incur a pump failure, the steering metering pump can still be manually operated. Two check valves in the steering valve allow the equipment to be steered manually. A make-up valve opens between the metering pump and the steering cylinder and a check valve (Figure 17-25) prevents the oil from entering the steering pump.

Two-Wheel Steer Operation

When the operator executes a right turn in two-wheel steer mode, oil flows from the metering pump to the mode selector valve. From there it flows to the right side of the steering cylinder located on the front axle. The pressurized oil enters the right side of the steer cylinder and forces the tie-rods to move to the left, initiating a right turn. The oil from the return side of the steer cylinder is displaced back to the tank. In the event that the operator bottoms out the steering on either side, pressure will rise until the signal relief valve, located in the end of the valve bank, opens to prevent any buildup of excess pressure in the circuit and returning excess oil back to the tank. To perform a left turn, the oil flow is reversed. In the event of an unexpected loss of power while driving, the circuit has provisions for emergency steering capability. The operator can still operate the steering wheel because the metering section of the metering pump will act as a hydraulic pump and provide enough flow to steer the equipment until it comes to a complete stop.

Crab Steer Operation

When the operator wants to steer the equipment slightly to the left while still moving ahead, he or she has to select the appropriate position on the master selector switch to energize a solenoid on the mode selector valve. This action offsets the selector valve spool to the crab steer position. The operator can now rotate the steering wheel counterclockwise. Oil will flow to the metering pump and into the left side of the front steer cylinder, and oil from the right side of the front steer cylinder will flow to the mode selector valve and on to the right side of the rear steering cylinder. Excess oil from the left side of the rear steer cylinder flows back to the tank through the mode selector valve and metering pump. This forces the two axles to steer in the same direction (left) at the same time.

When the operator wants to steer the equipment slightly to the right in tight quarters while still moving ahead, he or she will select the appropriate position on the master selector switch to energize a solenoid on the mode selector valve. This action offsets the selector valve spool to the crab steer position. Turning the steering wheel to the right in the crab steer mode sends pressurized oil through the mode selector valve and on to the left side of the steer cylinder at the rear of the equipment. This forces the oil from the right side of the rear steering cylinder to flow out through the mode selector valve and into the right side of the front steering cylinder. This action forces the front and rear wheels to steer equally to the right at the same time.

Coordinated (Circle) Steer Operation

When the equipment is required to initiate a tight right turn, the appropriate position on the master selector switch is selected to energize a solenoid on the mode selector valve. This places the selector valve spool in the circle steer position. Turning the wheel clockwise sends pressurized oil through the mode selector valve to the right side of the rear steering cylinder. Oil from the left side of the rear steering wheel is forced to flow out through the mode selector valve to the right side of the front steering cylinder. At the same time, oil from the left side of the front steering cylinder is directed back to the tank through the metering pump. This action causes the front wheels of the equipment to steer to the right, which forces the rear cylinder rod to the left and causes the wheels to turn left. To perform a left turn while operating in the circle steer mode, the operator turns the steering wheel to the left, forcing oil to flow from the metering pump into the left side of the front steering cylinder. Oil from the opposite end of the front steering cylinder flows through the mode selector valve to the left side of the rear steering cylinder and forces the rear wheels to steer to the right.

ARTICULATING STEERING

Articulating steering systems, as shown in **Figure 17-26** are the most common form of steering found on off-road equipment. Construction, mining, and forestry industry equipment utilizes this form of steering in many of their off-road applications. Slow speeds and the terrain that the equipment operates on make this system ideal. Since no axle wheel steer system is necessary, the system design is simple and often has very few complex components. A change in steering direction is made by pivoting the unit at the swivel joint that connects the two halves together. Large, specialized bearings are used to absorb the weight and load at this joint. The steering system uses two hydraulic cylinders that are connected to both halves of the equipment, as shown in figure 17-26. When the operator activates the steering to change the direction of travel, either with a steering wheel or joystick, hydraulic oil is sent to one end of one steering cylinder and the opposite end of the other steering cylinder, forcing the equipment to articulate at the joint. Depending on the size and type of equipment, off-road vehicles can come equipped with one or two axles on the load bearing end.

Figure 17-26 Typical articulating steering joint (Cat 535B skidder). (*Courtesy of Caterpillar*)

Fundamentals

Although not all hydraulic steering systems use the same types of components, they do share common components including a hydraulic pump(s), steering cylinder(s), and a steering control valve. Our example uses the steering system from a CAT 992 loader, which is typical of an articulating steering system used on off-road equipment. A tandem pump, or in some cases two separate pumps, is used to provide oil flow for this steering circuit design. This steering system design consists of two individual circuits: the pilot circuit and the steering circuit. The system is protected from overpressure by a relief valve. The steering control valve directs oil to the steering cylinders when the operator turns the steering wheel or joystick. Oil flows to one side of one steer cylinder and to the opposite side of the other steering cylinder.

The steering control valve controls the oil flow to the steering cylinders and allows the oil to flow to the steering system filter and oil cooler when no steering input is present. The oil cooler is protected from excess pressure by a cooler flow-control valve, which limits the pressure by sending excess oil back to the tank. When the operator initiates a turn, oil flows from the steering control valve to the steering cylinders, causing the equipment to turn. During steering wheel or joystick input, oil is allowed to flow to the steering cylinders; however, whenever steering input is stopped, oil flow stops.

The pilot circuit in this example consists of a hand metering unit (HMU) and two neutralizer valves. As long as the equipment is running, oil flow is always available at the HMU. When the steering system is activated, the pilot system is responsible for movement of the spool in the steering control valve. The HMU is made up of two sections: a metering section and a directional control section. The metering section is actually a small gear pump that supplies a metered oil flow to the steering control valve. The more steering input there is, the quicker the oil flows to the steering control valve, which causes the valve spool to move farther and allows more oil flow from the steering pump to the steering cylinders to initiate faster steering response. A neutralizer valve is used in the pilot circuit to stop the oil flow when the equipment reaches maximum steer position in either direction. In the event of an accidental engine power loss, the HMU will not provide emergency steering capabilities. As long as the pilot circuit is able to send oil flow to the steering valve to control the valve spool position, the equipment will have steering capability.

The smaller of the two pumps used in this circuit is also responsible for charging the accumulator through a charge valve located in the circuit. Once the accumulator is charged, the excess oil is sent to the steering control valve where it combines with the oil flow of the main pump. The return oil from the steering circuit flows through a return filter and cooler before returning to the tank. A bypass valve, located at the filter, allows the oil to bypass the cooler and go directly to the tank in the event of a high restriction at the cooler. The HMU control section operates as a closed-center valve, which stops the oil flow through the valve when there is no steering input.

Steering input forces the internal gear to turn inside the outer gear, sending a controlled flow of pilot oil through the valve body to the neutralizer valve. As long as there is steering input, the oil flow will continue to be metered. When steering input stops, springs will force the sleeve back to the neutral position, closing the passage between the metering section and the control section of the HMU. In **Figure 17-27,** the flow of oil is blocked by the valve spool keeping the valve in the neutral position. A bleed slot in the spool allows a small amount of oil to flow to the return. This facilitates a faster response time from the spool, allowing pump pressure to unload more quickly when it returns to neutral position. In the event of an overpressure condition in the neutral position, the pilot valve will open, limiting maximum pressure to the relief valve setting. In order for the spool to return to the neutral position when steering input is ended, the oil that acts on the end of the spool must be forced to flow to the opposite end of the spool through the use of a spring on one end of the spool.

Operation

When steering input is initiated to make a right turn, oil flows from the pilot circuit through the neutralizer valve to the right side (spring side) of the valve spool in the steering control valve. A pressure drop occurs across the metering orifices, forcing the spool to the left. The amount of spool movement is dependent on the metered oil flow from the HMU. Oil can now flow from the pump to the head end of the left steering cylinder and the rod end of the right steering cylinder to initiate a right turn. The quickness of the turn will depend on the continuous input from the HMU.

In the event of an overpressure condition, the pilot valve circuit will bleed off any excess pressure above the relief valve setting to protect the system from overpressure. Oil will force the relief valve off its seat, allowing oil to flow through an internal metered passage and lowering the pressure in the spring chamber. The pressure on the inlet side of the flow control valve moves it to the right so that oil can be dumped through holes in the flow control valve.

To prevent the equipment from bottoming out on the steering stops, a neutralizer valve is used in the pilot circuit to stop the oil flow to the steering valve, as shown in **Figure 17-28.** When the equipment is at the end of a

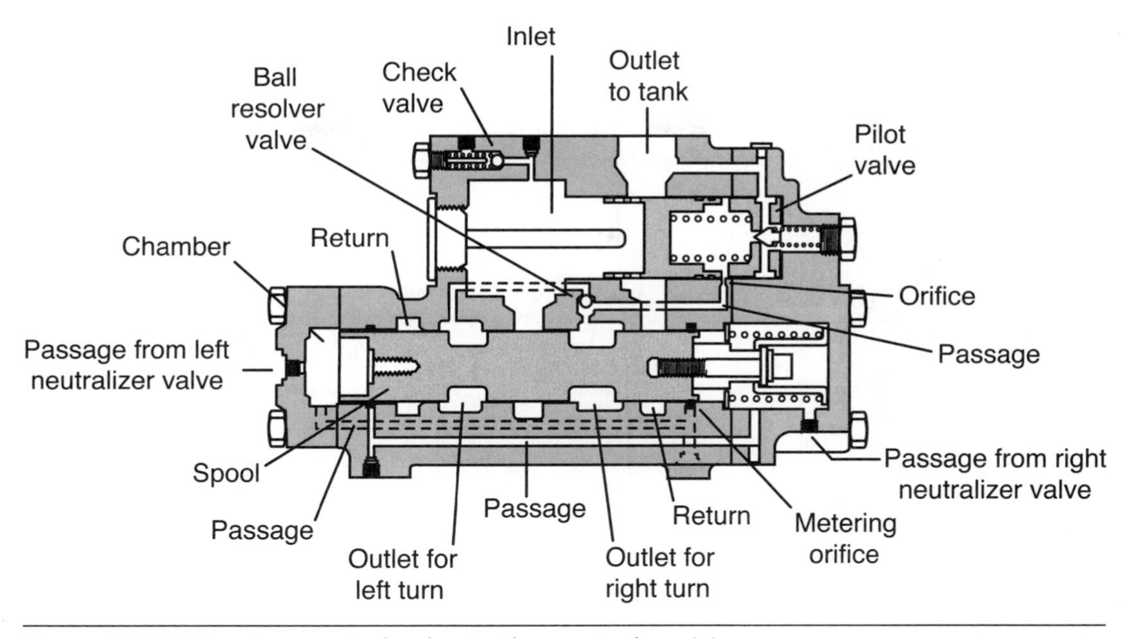

Figure 17-27 Steering control valve in the neutral position.

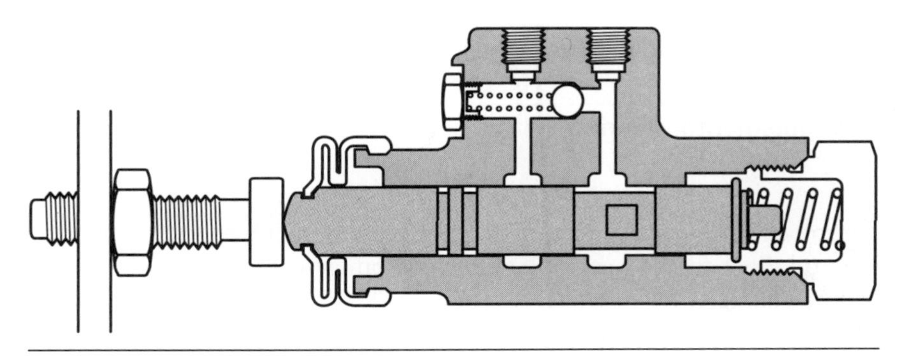

Figure 17-28 Neutralizer valve in the closed position.

right turn, a striker located on the frame comes in contact with the stem of the neutralizer valve and shifts it to the right to allow pilot oil to flow from the inlet to the outlet. This stops the flow of oil to the steering valve spool and returns it to the neutral position.

In order for the equipment to steer to the left, the oil must flow from the opposite end of the steering valve spool through the right neutralizer valve and through a check ball into the steering valve spool passage to initiate a left turn. The minute the striker moves back into the normal operating position, the check ball passage will close, and normal operation of the steering circuit is resumed. This adjustment is set up by the manufacturer, but may need to be readjusted whenever a neutralizer valve or striker needs to be replaced.

In many operating conditions, the use of a supplemental steering system is a necessity for obvious safety reasons, as shown in **Figure 17-29.** The HMU is not capable of providing emergency steering if engine power is lost. The supplemental steering system has a few additional components that are necessary to provide emergency steering capabilities. Addition of a supplemental steering pump, a diverter valve, and a check valve is required.

The steering pump must be mounted to the wheel side of the equipment so that it will turn when the equipment is in motion. This supplementary pump is generally driven by the wheel end driven gears in the transmission. As long as the engine is operating, the main steering pump will supply oil through the check

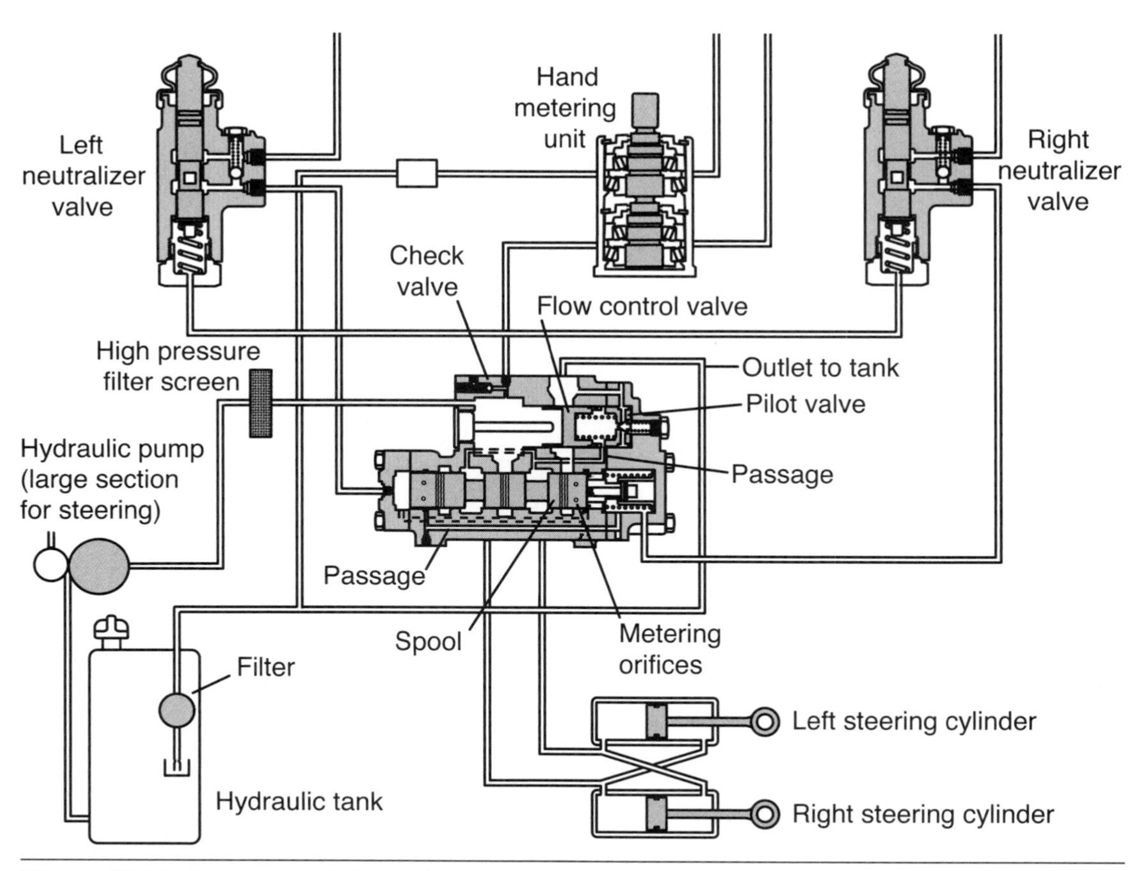

Figure 17-29 Supplemental steering system components.

valve to the steering control valve. The small section of the steering pump sends oil flow to the accumulator charge valve first. After the charge valve circuit is satisfied, the oil is sent through the diverter valve to the steering control valve, combining with the primary oil flow feeding the steering valve. When the equipment is in motion and the engine rpm is above a predetermined level, usually around 1300 rpm, the diverter valve sends the oil flow from the supplemental pump back to the tank.

If the rpm drops below a predetermined speed, the oil flow from the supplemental steering pump and the smaller primary steering pump are joined in the diverter valve and sent to the steering control valve. At this point, the oil flow combines with the primary pump output. As soon as the rpm goes over a predetermined amount the diverter valve will send the supplemental oil pump output back to the tank.

Under normal operating conditions the oil flow for the steering circuit comes from the two sections of the primary steering pump. If a pump failure occurs or engine power is lost, the diverter valve will send the supplemental oil flow back into the steering control valve circuit to provide emergency steering until the equipment comes to a stop.

The diverter valve operation above a predefined rpm (usually around 1300 rpm) is simply to send the oil flow from the supplemental steering pump back to the tank. In **Figure 17-30,** the diverter spool is shifted to the right and oil flows into the diverter spool cavity at the top, around the spool, and back to the tank. During this operating condition, only the oil flow from the small section of the primary pump goes to the steering control valve to combine with the oil flow from the larger primary pump. When the rpm of the equipment drops below a predefined level (usually around 1300 rpm) the diverter spool moves to the left, stopping the flow of oil back to tank. Pressure in the diverter valve passage increases until the check valve in the inlet cavity of the inlet passage opens, allowing oil from the supplemental pump to combine with the oil from the small section of the primary pump. This oil now flows to the steering control valve where it combines with the flow of the larger section of the primary pump.

As long as the equipment is in motion the supplemental pump will supply oil flow to the steering control

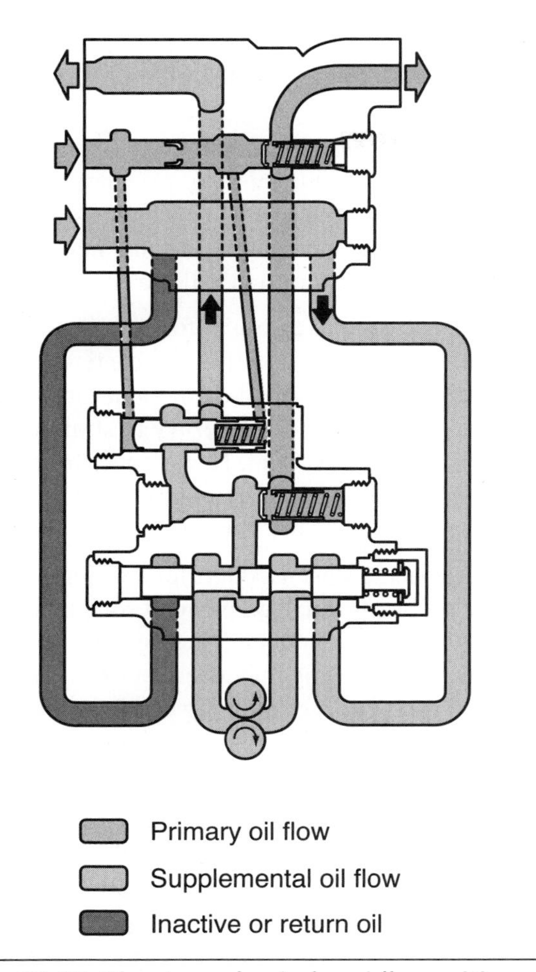

Figure 17-30 Diverter valve in low idle position.

valve. In the event of reverse movement, the reversing spool in the bottom of the diverter valve (Figure 17-30) shifts to the left to allow oil to flow from the supplemental pump, past a check valve, and into the diverter spool cavity.

If the equipment is moving forward, the oil is sent to the left of the reversing spool, shifting it to the right. The oil flow through the diverter valve passage increases until the check valve in the inlet cavity of the inlet passage opens, allowing oil from the supplemental pump to combine with the oil from the small section of the primary pump. This oil now flows to the steering control valve where it joins with the flow of the larger section of the primary pump.

ONLINE TASKS

1. Use a suitable Internet search engine to research the configuration of a typical off-road steering system. Identify a piece of equipment in your particular location, paying close attention to the steering system design and arrangement, and identify whether it has its own separate hydraulic system or uses an existing dump and hoist circuit to supply oil for the steering. Outline the reasoning behind the manufacturer's choice of this design for a particular application.
2. Check out this url: *http://training.arvinmeritor.com/onlinetraining.asp* for more information on steering system configuration and design.

SHOP TASKS

1. Select a piece of equipment in your shop that has a steer axle assembly and identify the type of components that make up the steer axle wheel end assembly. Identify whether or not the kingpins have excessive wear and what type of kingpin bushing design is used on this arrangement.
2. Select a piece of off-road equipment in your shop that uses a conventional steer axle and identify the type and model of the equipment, listing all the major components and the corresponding part numbers of the steer axle assembly. Identify and use the correct parts book for this task.

Summary

- Off-road steering systems enable equipment operators to convert steering wheel or joystick movement, either through a series of electro-hydraulic or gear systems, into precise angular control of equipment with very little physical effort from the equipment operator.
- Use of electric solenoids and hydraulics makes the operator's job of steering effortless when compared to how it was done in the past.
- Equipment designed for off-road applications requires the use of all-wheel drive. In these cases the front axle is equipped with differential as well as steering capabilities.
- Many types of off-road equipment, such as front end loaders and off road haulage trucks, steer by controlling the angular relationship of a two-piece frame that utilizes solid axles with no steering

capabilities mounted on each of the frame components.

- All steering components are generally made from high quality, drop-forged tempered steel and should never be heated to straighten a bend. This could draw out the temper and weaken the arm.
- Tie-rods are ball sockets that allow the steering linkage to move freely in all directions, preventing binding. They connect the steering knuckles and steering linkage together, allowing the two wheel ends to work in unison to provide steering control on the equipment.
- The drag link is a mechanical connection between the pitman arm and the steering lever; the drag link can be either a one- or two-piece design and is generally made from drop-forged steel for strength and durability.
- On front-wheel steer systems, the hydraulic system provides oil to the steering system through a pressure control valve that acts as a priority valve. The steering system gets priority since no oil will be allowed to flow to the loader circuit until the steering system gets a predefined pressure.
- Camber is expressed as the amount the front wheels tilt outward or inward at the top. The wheels are said to have positive camber if the wheels are inclined outward at the top; if the wheels are inclined inward at the top, they have negative or reverse camber.
- Caster is defined as the rearward or forward tilt of the knuckle pin (kingpin) either toward the front or rear of the equipment. It is expressed in degrees and is positive, negative, or zero caster.
- Kingpin inclination can be defined as the number of degrees the knuckle pins (kingpins) are inclined inward at the top when viewed from the front of the equipment.
- Front drive steer axles must be able to support the weight of the equipment and must also provide drive capability to the front wheels as well as house the front brakes.
- Before attempting to lift and block equipment, make sure it is parked on a level surface, set the park brake, and place approved wheel chocks in place to prevent unwanted equipment movement. Follow manufacturer's recommended procedures and only use approved safety stands to support the equipment.
- Do not grease the tie-rod assemblies before the inspection is performed. The grease may take up the wear that is present in the joint.
- Carefully check the tie-rod boots for any damage or cracks. Inspect the tie-rod thread depth in the cross tube slot to be sure its adequate. Be sure that the cotter pin is properly installed through the tie-rod end. If the cotter pin is missing, retighten the tie-rod end nut to the correct torque, following manufacturer's torque charts. Install a new pin. Never back off the nut to install the cotter pin.
- Check the cross tube carefully for any sign of impact damage. Look for cracks. If in doubt, replace the cross tube with an original equipment tube of the same size and diameter. Never try to straighten the tube after it has been bent. Never apply heat to the tube, it may affect the strength of the metal. If the tie-rod clamps are damaged, replace them.
- Steering knuckle wear accounts for a major portion of steering problems found on off-road equipment with this type of steering system. Steering knuckles should be lubricated following manufacturer's recommended time intervals. When performing service work on equipment, you may find a grease fitting refusing to take grease. Replace the fitting immediately.
- Differential steering systems provides the ability to deliver equal amounts of power to both axles while the equipment is driven in a straight line. When the operator needs to correct steering direction, a steering motor drives one set of planetary gears, which speeds up one axle and slows down the other by an equal amount, hence the term *differential steer*.
- The planetary differential steer system contains a drive planetary, which sends torque to a planetary gear set on each end of the assembly. The one on the left is referred to as the steering planetary; the one on the right is called the equalizing planetary. Power to the assembly can come from either the transmission or from a steering motor, which drives the steering planetary ring gear.
- The planetary differential steer system uses a hydraulic steering motor to drive the steering planetary ring gear with oil flow coming from a variable-displacement pump. The drive oil is part of a closed-loop hydraulic system. Steering motor output is controlled by steering controls, which are used to control the direction and speed of the steer motor. This circuit requires a counterbalance valve to provide make-up oil, regulate the pressure, and protect the circuit from excessive pressure spikes and overspeed.
- Steering clutch and brake systems on track-type equipment use independent control mechanisms to steer and brake each track separately. A brake and a

steer clutch are mounted together on each side of the final drive assembly. Depending on the manufacturer and model of the equipment, it can have a brake assembly design using either band or multi-clutch design.

- When the track equipment makes a hard turn, the steering clutch is disengaged and the brake is applied, forcing the track to stop turning or slow down depending on the amount of brake force application. If full pressure is allowed into the steering clutch, the clutch will disengage the track from the powerflow completely.
- Newer equipment has largely done away with mechanical control levers and has replaced them with fingertip steering controls that use electric/hydraulics to control the flow of oil to the steer clutch brake assembly. The assembly is lubricated and oil cooled with oil that is circulated through an oil cooler with a pump.
- To ensure safe operation of the equipment under all possible operating conditions, the steering and brake control circuit gets priority over the equipment hydraulic circuit. The circuit incorporates a priority control valve to ensure that the steering and brake circuit gets oil first. If hydraulic pressure is lost for any reason, the steering clutches will be disengaged and the brakes will come on because they are spring-applied.
- The operation of the brake circuit is quite different from the clutch; the brakes are spring-applied and require oil pressure to release them.
- Equipment that uses all-wheel steer generally incorporates three different steering modes; two-wheel steer, circle steer, and crab steer modes. The front and rear axles have similar designs; both use conventional steering components such as tie-rods and steering cylinders.
- The basic components that make up an all-wheel steering system include a variable-displacement steering pump, a selector valve, front and rear steer cylinders, a steering metering pump, and a load-sensing valve bank.
- The rear steer axle on all-wheel steer systems is controlled by a three-position solenoid-operated steer mode selector valve. On models that are equipped with the self-aligning rear steer, additional hardware is required to self-align the rear axle. A proximity switch is located on the right side of the rear axle with a corresponding sensor target, which is attached to the right side steering yoke.
- The variable-displacement pump used on all-wheel steer systems incorporates a pressure and flow compensator valve to supply oil flow for the circuit. This valve controls the swashplate angle by regulating the pressure that acts on the actuation piston.
- Front-wheel steer is used for most operating conditions on specialized equipment such as a telehandler. When driven on roadways, only the front wheels will steer in this mode; the rear wheels must be in the straight ahead position.
- All-wheel steer equipment has the ability to use the circle steer option, which enables the equipment to turn in a shorter radius. Both sets of wheels steer in this mode; the rear wheels steer in the opposite direction to the front wheels. Switching back and forth between any of the steering modes requires that the rear wheels first be aligned.
- All-wheel steer equipment will go into standby mode during periods of idle with no steering input, and only minimal oil flow will be necessary to compensate for internal leakage. No-load sensing pressure is present at the spring side of the flow compensator spool. Pressure de-strokes the pump, and only a minimal amount of oil is necessary, due to internal circuit leakage, to maintain an adequate amount of supply pressure.
- On all-wheel steer equipment when oil flow requirements drop, signal pressure (along with spring pressure in the flow compensator spool circuit above the spool) drops to below the pump pressure. This causes the flow compensator spool to move up, forcing oil pressure to increase to the actuator piston and causing a decrease in pump stroke length.
- Certain operating conditions on all-wheel steer equipment will often result in a high pressure during stall conditions, for example, when a cylinder reaches the end of its stroke. To protect the circuit, the signal relief will open, limiting maximum signal pressure. Supply pressure at the pump inlet forces the pressure and flow compensator spools to shift up to the metered position to maintain the flow rate. When the circuit goes into stall, the pump stroke increases the flow. Note that the pressure compensator spool acts as a backup to the signal relief valve by limiting maximum system pressure.
- The steering pump in an all-wheel steer circuit controls the direction and speed of the turn with two separate sections within the pump; the control section and the metering section. Direction control

is achieved with a closed-center rotary control valve, while the metering is controlled by a gerotor pump.

- If a material handler loses power or incurs a pump failure, the steering metering pump can still be manually operated. Two check valves in the steering valve allow the equipment to be steered manually. A make-up valve opens between the metering pump and the steering cylinder. A check valve is used to prevent the oil from entering the steering pump.
- In the crab steer mode, if the steering wheel is turned counterclockwise, oil will flow to the metering pump and into the left side of the front steer cylinder, while oil from the right side of the front steer cylinder flows to the mode selector valve and on to the right side of the rear steering cylinder. Excess oil from left side of the rear steer cylinder flows back to the tank through the mode selector valve and metering pump. This forces the two axles to steer in the same direction (left) at the same time.
- In the coordinated (circle) steer mode, turning the wheel clockwise sends pressurized oil through the mode selector valve to the right side of the rear steering cylinder. Oil from the left side of the rear steering wheel is forced to flow out through the mode selector valve to the right side of the front steering cylinder. At the same time, oil from the left side of the front steering cylinder is directed back to the tank through the metering pump. This causes the front wheels to steer to the right, which forces the rear cylinder rod to the left and causes the rear wheels to turn left.
- Articulating steering changes the steering direction by pivoting the unit at the swivel joint that connects the two halves of the equipment together. Large specialized bearings are used to absorb the weight and load at this joint. The steering system uses two hydraulic cylinders to steer the equipment.
- The articulating steering system contains two individual control circuits: the pilot circuit and the steering circuit. The system is protected from overpressure by a relief valve with a steering control valve directing oil to the steering cylinders when the operator turns the steering wheel or joystick. Oil flows to one side of one steering cylinder and to the opposite side of the other steering cylinder to steer the equipment.
- The pilot circuit consists of a hand metering unit (HMU) and two neutralizer valves. As long as the equipment is running, oil flow is always available at the HMU. As long as the pilot circuit is able to supply oil flow to the steering valve to control the valve spool position, the equipment will have steering capability.
- The return oil from the steering circuit flows through a return filter and cooler before returning to the tank. A bypass valve located at the filter causes the oil to bypass the cooler and go directly to tank in the event of a restriction at the cooler.
- The HMU control circuit operates as a closed-center valve, which stops the oil flow through the valve when there is no steering input. Steering input forces the internal gear to turn inside the outer gear, sending a controlled flow of pilot oil through the valve body to the neutralizer valve. As long as there is steering input, the oil flow will continue to be metered.
- In the event of an overpressure condition in the neutral position, the pilot valve will open to limit pressure to the relief valve setting. In order for the spool to return to the neutral position when steering input is ended, the oil that acts on the end of the spool must be forced to flow to the opposite end of the spool by a spring on one end of the spool.
- When steering input is activated to make a right turn, oil flows from the pilot circuit through the neutralizer valve to the right side (spring side) of the valve spool in the steering control valve. A pressure drop occurs across the metering orifices, forcing the spool to the left. The amount of spool movement is dependent on the metered oil flow HMU.
- In the event of an overpressure condition, the pilot valve circuit will bleed off any excess pressure above the relief valve setting to protect the system from overpressure. Oil will force the relief valve off its seat, allowing oil to flow through an internal metered passage and lowering the pressure in the spring chamber.
- To prevent the equipment from bottoming out on the steering stops, a neutralizer valve is used in the pilot circuit to stop the oil flow to the steering valve when the equipment is at the end of a turn. A striker located on the frame comes in contact with the stem of the neutralizer valve, shifting it to the right and allowing pilot oil to flow from the inlet to the outlet.
- In many operating environments, the use of a supplemental steering system is a necessity for

safety reasons. The HMU is not capable of providing emergency steering if engine power is lost.

- The supplemental steering pump must be mounted to the wheel side of the equipment so that it turns whenever the equipment is in motion. This pump is generally driven by the wheel end driven gears in the transmission.
- Diverter valve operation above a predefined rpm (usually around 1300 rpm) sends the oil flow from the supplemental steering pump back to the tank. The diverter spool is shifted to the right and oil flows into the diverter spool cavity at the top, around the spool, and back to the tank.
- When the rpm of the equipment drops below a predefined level, the diverter spool moves to the left, stopping the supplemental flow of oil back to tank. Pressure in the diverter valve passage increases until the check valve in the inlet cavity of the inlet passage opens, allowing oil from the supplemental pump to combine with the oil from the small section of the primary pump.

Review Questions

1. On a two-wheel steer backhoe, what device does a steering knuckle pivot on?
   A. pitman arm
   B. Ackerman arm
   C. kingpin
   D. steering control arm
2. On front-wheel steer equipment, which component is responsible for controlling steering toe-out on turns?
   A. pitman arm
   B. drag link
   C. Ackerman arm
   D. steering control arm
3. If you increase the length of a tie-rod on a front-wheel steer backhoe steering system, which of the following will happen?
   A. toe-out increases
   B. toe-in increases
   C. toe-in decreases
   D. toe remains neutral
4. When the front tires of a backhoe lean outward slightly at the top, which of the following is correct?
   A. positive camber
   B. positive caster
   C. negative camber
   D. negative caster
5. On a backhoe, how is steering knuckle vertical clearance adjusted?
   A. adding or subtracting shims
   B. using the correct tapered wedge shims
   C. selecting the correct steering knuckle
   D. adjusting a stop bolt and jam nut
6. When making a tie-rod end adjustment, which of the following statements should be observed when tightening the fastening clamps?
   A. ensure that the tie-rod end threads extend into the tie-rod beyondthe split slot
   B. ensure the equipment weight is off the wheels
   C. ensure that there is a no space between the clamp and the end e of the crosstub
   D. straighten the cross tube before adjusting the tie-rods with a torch
7. On equipment that uses differential steering, which component is driven by the transmission input shaft?
   A. sun gear
   B. ring gear/planetary carrier
   C. steering motor input gear
   D. equalizing planetary gear set

8. When a right turn is preformed on equipment that uses differential steering, which of the following components initiates the turning process?

   A. hydraulic steering motor
   B. equalizing planetary gear set
   C. steering planetary gear set
   D. sun gear

9. On equipment with differential steer, which of the following components is driven by the hydraulic steering motor?

   A. transmission pinion gear
   B. equalizing planetary gear set
   C. steering planetary ring gear
   D. sun gear

10. How many sun gears are used on equipment with differential steering?

    A. 1
    B. 2
    C. 3
    D. 4

11. Which gear is always held and never rotates in the differential steer system?

    A. drive planetary ring gear
    B. equalizing planetary ring gear
    C. steering planetary ring gear
    D. right planetary drive gear

12. When moving in a straight forward direction, which gear receives equal power with the drive planetary ring gear?

    A. right differential output gear
    B. equalizing planetary ring gear
    C. steering planetary ring gear
    D. sun gear

13. On a track loader, during counterrotation, which component turns the planetary ring gear in the opposite direction?

    A. hydraulic steering motor
    B. equalizing planetary ring gear
    C. steering planetary ring gear
    D. sun gear

14. On a differential steer loader, which component receives oil from the counterbalance valve after a right turn is initiated?

    A. steer motor
    B. left brake
    C. right clutch
    D. right brake

15. How many steering and brake clutches are used on a track loader with a clutch steer/brake system?

    A. 1 of each on each track
    B. 2 of each on each track
    C. 3 of each on each track
    D. 4 of each on each track

16. When a track loader with a steer clutch/brake system makes a hard right turn, which component is disengaged first?

    A. left steering clutch
    B. right steering clutch
    C. left brake
    D. right brake

17. On a track loader with a steer clutch/brake system, what occurs when the operator depresses a clutch pedal?

    A. Oil is sent to the brake piston.
    B. Oil is sent to the steering clutch control valve.
    C. Oil is sent to the brake control valve.
    D. Oil is sent to the steering clutch cylinder.

18. On track loaders that utilize clutches for the steering and braking of the tracks, how many pressure reducing valves are used?

    A. 1 C. 3

    B. 2 D. 4

19. On a track loader with a steer clutch/brake system that is traveling in a straight line, which component receives oil flows from the priority flow control valve?

    A. steering clutch pressure reducing valves C. brake control valve

    B. steering clutch control valve D. steering clutch cylinder

20. When the park brake is applied on a track loader with a steer clutch/brake system, which valve blocks the flow of oil to the brake spools?

    A. pressure reducing valves C. poppet valve

    B. steering clutch control valve D. steering clutch cylinder

21. On equipment with all-wheel steer, which component is responsible for self-aligning the rear axle?

    A. position sensor C. equalizing sensor

    B. pressure sensor D. pressure reducing sensor

22. Which type of hydraulic pump is used for the steering circuit on an all-wheel steer system?

    A. Variable-displacement axial piston pump C. fixed displacement axial piston pump

    B. gear pump D. fixed displacement vane pump

23. What occurs in the all-wheel steering circuit when no-load sensing pressure is present at the spring side of the flow compensator spool?

    A. crab steer mode C. coordinated steer mode

    B. standby mode D. two-wheel steer

24. On equipment with all-wheel steer, under what operating condition will the pressure compensating spool act as a relief valve?

    A. High-pressure stall condition C. high speed mode

    B. standby mode D. low speed mode

25. On equipment with all-wheel steer operating in crab steer mode and steering to the left, which component receives oil flows from the metering pump?

    A. right side front steer cylinder C. right side rear steer cylinder

    B. left side front steer cylinder D. left side front steer cylinder

26. On equipment with all-wheel steer operating in coordinated steer mode and steering to the right, which component receives oil flows from the mode selector valve?

    A. right side front steer cylinder C. right side rear steer cylinder

    B. left side front steer cylinder D. left side front steer cylinder

27. When a Cat 992B loader with a hydraulic steering system performs a right turn, where does the oil flow after it leaves the neutralizer valve?

    A. steering control valve C. right side rear steer cylinder

    B. left side front steer cylinder D. left side front steer cylinder

28. When a Cat 992B loader reaches its safe left or right steering limits, what occurs?

A. Neutralizer valve stops oil flow to the steering cylinders.

B. Neutralizer valve stops oil flow to the steering valve.

C. Pilot valve will relieve excess pressure.

D. Flow control valve will block flow to the steering cylinders.

29. On a Cat 992B loader that comes equipped with supplemental steering circuit, what drives the supplemental steering pump?

A. engine

B. transmission output gear

C. transmission input gear

D. torque converter

30. On a 992B Cat loader, what happens to the oil from the diverter valve at high engine rpm?

A. flows to the steering cylinders valve

B. flows to the steering control valve

C. flows to the neutralizer valve

D. returns to the tank

CHAPTER 18 Track and Undercarriage Systems

Learning Objectives

After reading this chapter, you should be able to

- Describe the fundamentals of track and undercarriage systems.
- Describe the components that make up track and undercarriage systems.
- Describe the principle of operation of track and undercarriage systems.
- Describe the maintenance and repair procedures associated with track and undercarriage systems.
- Describe the troubleshooting procedures for track and undercarriage systems.

Key Terms

Bellville washer
broached
carrier roller
dry joint
elevated sprocket
equalizer bar
extra long (XL) (front) track system
extra long (XR) (rear) track system
galled
grouser
hydraulic track adjuster
I.D.
idler
link pitch
lip seal
low ground pressure (LGP) system
master link
master pin
oscillating undercarriage
pitted
sealed tracks
spalling
sprocket
track guard
track link
track pin
track pitch
track roller
track shoes
wear step

INTRODUCTION

Off-road track equipment operates at slow speeds on terrain that is not generally considered a roadway when compared to on-road applications. Off-road applications, such as heavy equipment operating in a quarry, mine, or forestry application, uses specialized tracks and undercarriages to meet the extreme duty operating cycles associated with this type of environment. Other off-road equipment, such as that which operates in the forestry industry, requires a different specialized track and undercarriage unique to that environment. Because of the numbers of applications required for off-road equipment tracks, only the most popular types of track systems and undercarriages will be covered in this chapter. Always refer to the manufacturer's technical documentation for information on servicing and repairing tracks and undercarriages.

Tracks and undercarriages on off-road equipment are extremely heavy and require the use of specialized equipment to handle them safely. Tracks account for a high percentage of maintenance and repair costs when compared to the rest of the equipment. Because the tracks are in constant contact with the roadway, they must be built to absorb shock loads and designed to operate submerged in muddy or loose soil conditions. As a result of the extreme operating conditions to which the tracks are exposed, wear is inevitable. The technician's job is to evaluate the wear and perform the required maintenance on a regular basis, making the necessary adjustments when required to minimize wear. Good operating and maintenance techniques can have a positive effect on the life of the track.

FUNDAMENTALS

Track and undercarriage systems can be categorized into four distinct designs with a number of variations in each category, as shown in **Figure 18-1.** These are the standard track system, **low ground pressure (LGP) system, extra long (XL) (front) track system**, and **extra long (XR) (rear) track system.** The standard track system is designed to perform well on firm ground and can be equipped with different shoe widths. The LGP system is designed with a wider track, which increases ground contact for better flotation and stability on lose surfaces. Shoe width on the LGP track is generally a lot wider than the standard track. The XL track system has an extra long

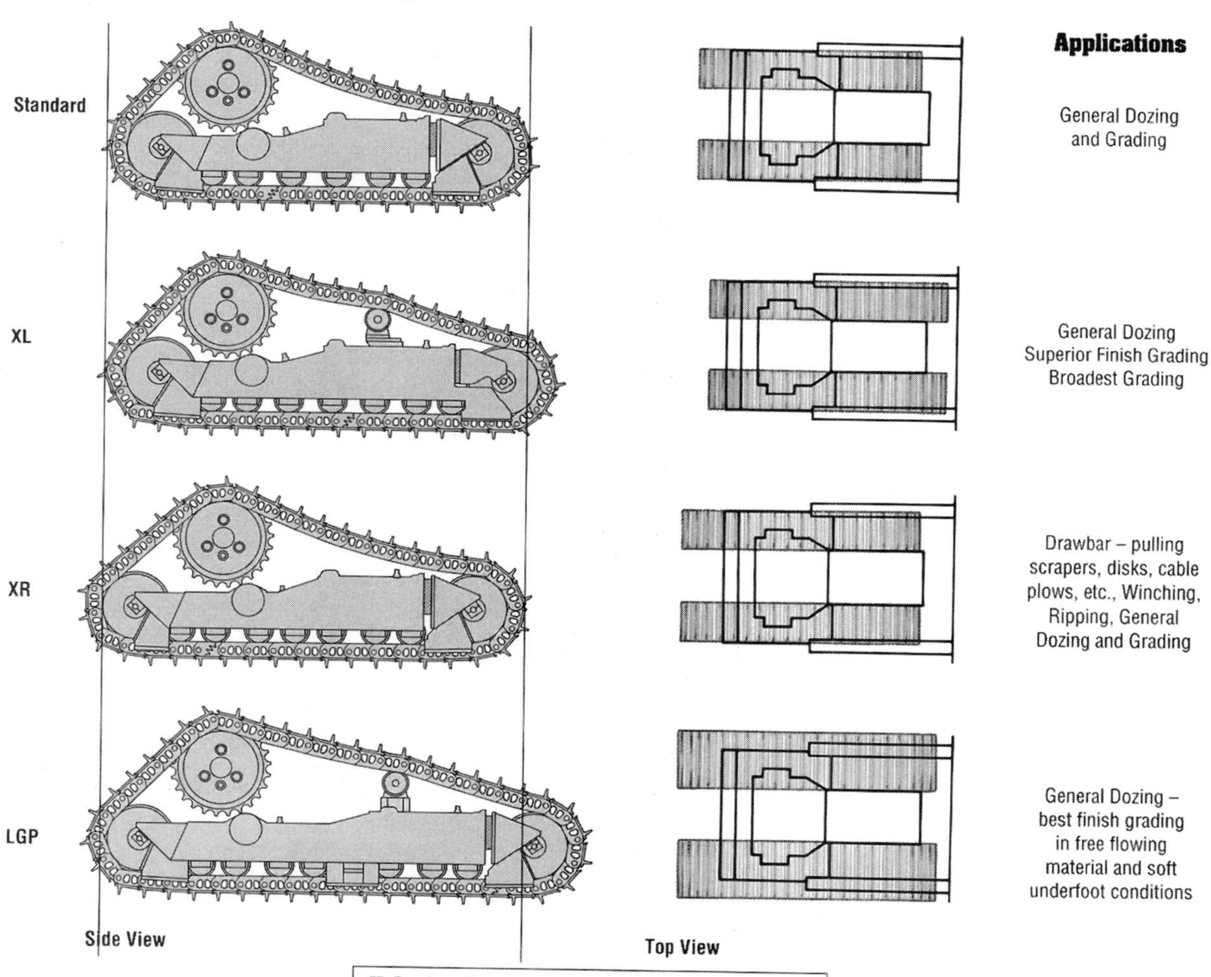

Figure 18-1 Typical undercarriage and track configurations. (*Courtesy of Caterpillar*)

front track design, which provides better balance with extremely good flotation characteristics. It is better suited for finish grading on free flowing material. The XR track system has an extra long rear section, which is more suitable for pulling scrapers, as well as for winching or ripping operations. In the agricultural applications, a belt track is often used to reduce the zone of compaction. This design increases ground contact significantly over conventional LGP track systems.

To gain an understanding of the construction of a track, look at it as a large endless chain with shoes that are held in place with bolts. The chain section of the track is made up of **track links,** pins, and bushings. A considerable number of these interconnect with each other to form the track. The complete track and undercarriage system is made up of more that just the chain section. There are actually a number of components that form the complete track system. The drive **sprocket**, front **idler**, **track rollers**, tension mechanism, roller guards, and the frame are all required to make up a track and undercarriage system.

The final drive on a crawler tractor drives a complete track and undercarriage system on each side of the equipment. Each link of the track is held together with a press-fit pin and bushing to keep each link section in proper alignment with each other link section during operation. This combination of parts is called a link assembly. Pressed together, each section acts like a hinge, which allows the chain flexibility when rotating on the undercarriage. The track links also provide a means for attaching the **track shoes** that are bolted to the links. The parts of each link assembly are induction hardened to provide good wear characteristics. Pins and bushings are machined to provide a smooth bearing between them, which also increases durability. A **master link** or **master pin** completes the track assembly.

The master link, as shown in **Figure 18-2,** is slightly different from the rest of the links. The bushing used in the master link assembly is slightly shorter than the rest of the bushings, and the master pin, as shown in **Figure 18-3,** has a smaller diameter to facilitate easy installation and removal. Some manufacturer's master pins can be identified by a mark or drill hole on the end of the outside face of the pin; other manufacturers use a two-piece master link that bolts together. The master link uses the same pins and bushings as the rest of the track, making splitting the track easier because no special tools are required to disassemble the track. Before attempting any maintenance or repairs to a track and undercarriage system always refer to the manufacturer's technical documentation or service manual.

Track systems also come in what is commonly called a sealed system, where the pin and bushing are lubricated on assembly and are sealed from dirt entering between them. This design eliminates internal wear and is quite popular.

Track shoes are bolted to the link assembly to provide a means of supporting the equipment and better traction and flotation during operation. They are generally constructed from a hardened metal plate and come in many different configurations. Refer to **Table 18-1** for more design features. The operating ground conditions will determine the type of shoes to be used. Operation on highly abrasive soil demands a high quality hardened

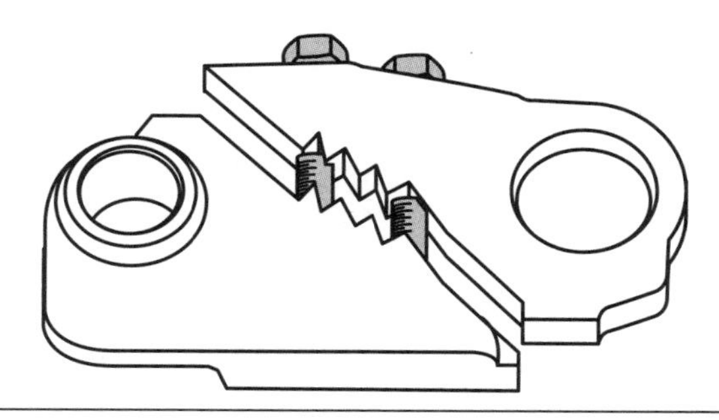

Figure 18-2 Split master link.

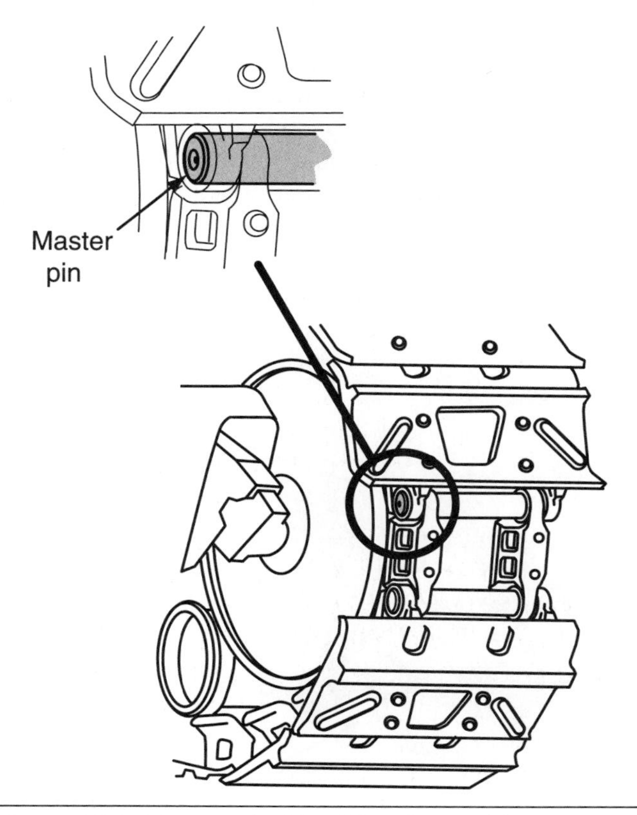

Figure 18-3 Master pin identification mark.

Table 18-1: TRACK SHOE DESIGN FEATURES	
No. 1	The grouser is used to penetrate the ground for traction.
No. 2	The plate surface provides a means for flotation. The wider the plate the more flotation capabilities are available.
No. 3	The design of the leading edge is curved downward. This provides the trailing edge, which is curved upward, interference-free operation. The curve also adds strength to the shoe.
No. 4	These circular cuts are called "link reliefs." They ensure that the leading edge of the shoe will not interfere with the links when bending occurs around the sprocket or idlers during operation.

surface that will resist wear. The number of ridges (called **grousers**) on the track shoe designates the terrain the shoe will operate on, as shown in **Figure 18-4.** The hardness of the surface material on the grousers determines whether they are standard or extreme-service track shoes.

In some applications, the use of rubber shoes is necessary so as not to damage the operating surface. As an example, consider a small excavator that must work from a sidewalk, which should not be damaged, in order to excavate the surrounding area. The width of the track shoes also plays an important role when selecting the appropriate shoe. Working on snow, as in the case of ski hill groomers, requires an extremely wide shoe to maintain maximum flotation. Shoe width should provide the appropriate flotation for the terrain. See **Figure 18-5** and **Table 18-2.**

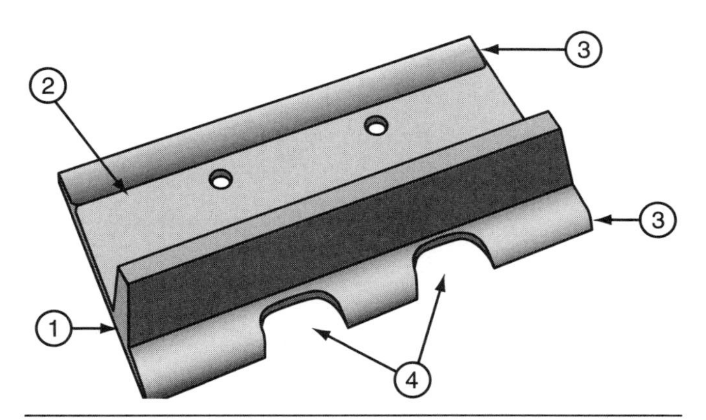

Figure 18-4 Typical track shoe. (*Courtesy of Caterpillar*)

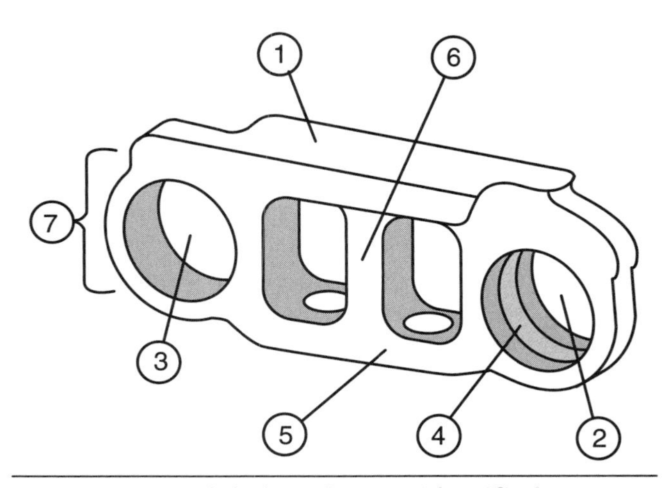

Figure 18-5 Track link with parts identified.

Table 18-2: TRACK LINK PARTS	
No. 1	The track roller runs on this section of the track link.
No. 2	Pin bore.
No. 3	Bushing bore.
No. 4	Counterbore required to hold the sealing rings on sealed or lubricated tracks.
No. 5	The shoe strap has bolt holes to accept mounting bolts for track shoes.
No. 6	Struts are used to help support the link rail.
No. 7	The bushing strap is located at the end of the track link behind the bushing bore.

TRACK AND UNDERCARRIAGE COMPONENTS

Track Links, Pins, and Bushings

Each track section is made up of two track links, a hardened pin, and a bushing. These individual track sections are interconnected to the link assembly. The two track links used in each section has provisions for attaching a track shoe and also provides a rail for the track rollers to maintain accurate track alignment. **Sealed tracks** have a solid pin. Sealed and lubricated tracks have a hollow pin, which provides a path for lubricating the pin and bushing of the next track section. When installing a center-drilled pin, the cross drill hole must be installed toward the rail of the link, which keeps the pin in compression to resist the possibility of crushing. The pins and bushings are press-fit into the links. *Note*: Be sure to follow the correct orientation; there are both right-hand and left-hand links. Clearance between the pins and bushings is minimal, providing only enough clearance for lubrication. In the case of a sealed track, the seal fits over the pin against the track link before installing another link.

Drive Sprockets

The sprocket, as shown in **Figure 18-6,** is not mounted to the track frame but is attached to the final drive. Its job is to transfer drive torque to the track. The teeth on the outside of the sprocket act like gear teeth. They engage the track links and propel the equipment on the continuous track loop. Older equipment often has a one-piece sprocket assembly, the use of individual bolts on segments did not show up until quite recently. On older models, the old sprocket had to be cut off and a new one welded in its place. Many equipment owners replaced the welded version with a weld on adapter ring; a new sprocket could then be bolted on as necessary, saving a considerable amount of time. Operating on different types of terrain requires that the segments be quickly interchangeable.

On modern equipment, changing sprocket segments is quick and easy; position the track so that the segment that requires changing is on the inside (away from the track links), unbolt the old segment, and bolt in and torque the new one. Rotate the track a few feet and repeat the procedure until all the segments are replaced.

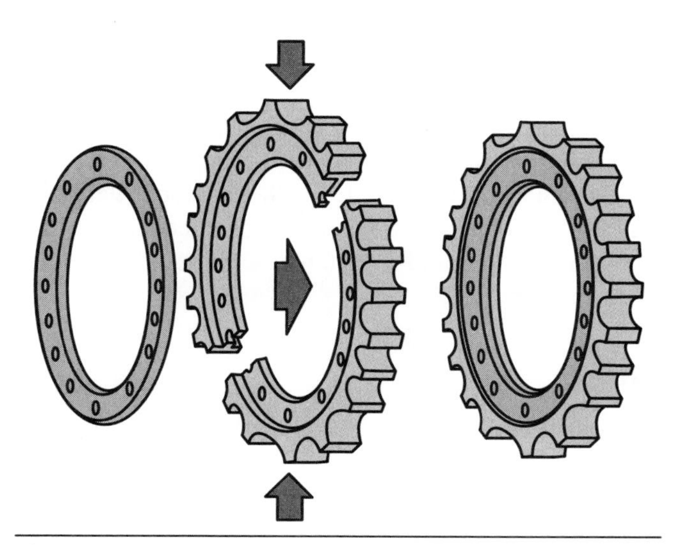

Figure 18-6 Components that make up the drive sprocket.

Tech Tip: When equipment must operate in mud or snow, a special type of segment with a slotted tooth design to prevent foreign material from building up and over-tightening the track is available. To maintain long track life, track tension must be maintained properly during operation.

By design, the sprocket has an odd number of teeth to provide a change in tooth-to-sprocket contact every revolution of the sprocket. This is commonly called "hunting tooth." This design allows for even distribution of wear among the sprocket teeth.

Track Rollers and Carrier Rollers

Modern track design utilizes two different types of track rollers to maintain track alignment. The bottom rollers are called track rollers; they support the weight of the equipment and ensure that the weight of the equipment is distributed evenly over the bottom of the track. The track rollers are spaced closely together, and there are generally a large number of them mounted to the bottom side of the track frame. The track rollers maintain track alignment as well as distribute the weight of the equipment evenly over the length of the track.

There are two types of track roller designs: single flange and double flange. Single flange rollers are used closest to the sprockets. Double flange rollers maximize track stability and alignment. The surface of the track rollers is hardened to the same Rockwell scale as the track links. Lubrication and cooling are provided by oil sealed in the housing. Because of the twin flanges, double flange track rollers are used to maximize track alignment.

There are several different design configurations of track rollers. Always refer to the specific shop manual of the equipment you are going to work on for detailed instructions. **Figure 18-7** shows the parts breakdown of a typical Cat track roller. Refer to the **Table 18-3** for a detailed description of each part.

Carrier rollers are located above the frame rail. Their purpose is to support the weight of the top section of track as it rolls between the idler and the sprocket. These rollers generally have a single flange that aids in controlling track sag and whipping during operation. A secondary function of the carrier rollers is to maintain

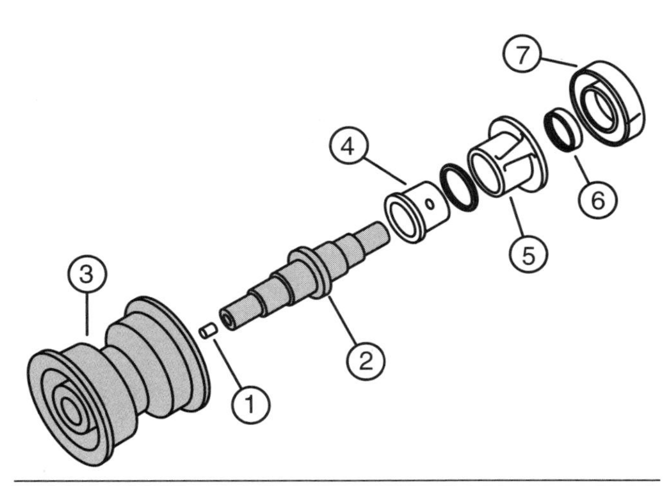

Figure 18-7 Exploded view of the components used in track rollers.

Table 18-3: TRACK ROLLER PARTS	
No. 1	A plug is used in the end of the support shaft to seal in the lubricating oil.
No. 2	The support shaft is used to support the roller shell, which can rotate freely on the shaft.
No. 3	The shell has a hardened outer surface that rolls on the track links.
No. 4	A bronze bushing is used between the shaft and shell to provide an excellent wear surface.
No. 5	This bushing is made from cast and is used to support the bearing to the end collar.
No. 6	Rigid seals are used to keep the dirt out and lubrication in.
No. 7	The end collar holds the roller seals in place and has provision for mounting the rollers to the track frame.

alignment on the upper portion of the track between the idler and sprocket. These rollers also have a hardened surface that is equal in hardness to the track link surfaces; this can have a slightly negative effect on the life of the track links. Some equipment manufacturers' cantilever mount the carrier rollers to minimize material buildup.

Figure 18-8 shows an exploded view of a typical carrier roller. **Table 18-4** identifies the parts.

The retaining ring on a carrier roller fastens the end collar to the roller shell and is held in position on the

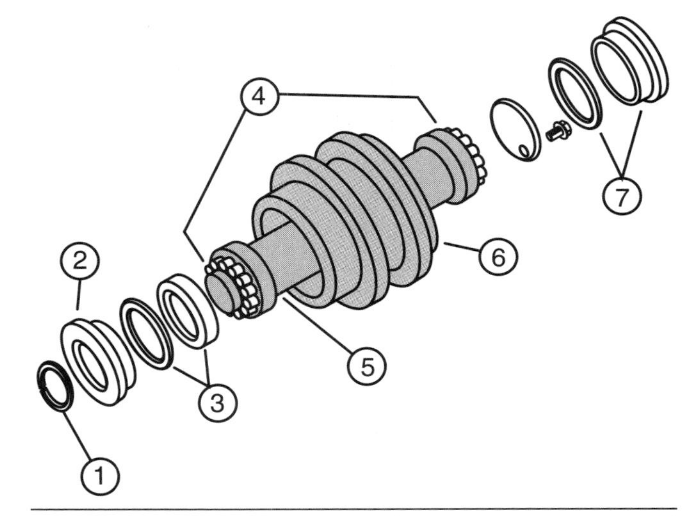

Figure 18-8 Exploded view of the components used in Cat carrier roller.

Table 18-4: PARTS OF A CARRIER ROLLER	
No. 1	Retaining ring clips
No. 2	End collar
No. 3	Seals
No. 4	Two roller bearings and cups
No. 5	Support shaft
No. 6	Carrier shell
No. 7	Plug and O-ring

support shaft by a groove. The bearings are held in place in the carrier shell by the end collar and seals, which also prevent oil from leaking out of and dirt getting into the bearings. The two roller bearing assemblies allow the carrier roller to roll freely on the shaft. The shaft has mounting points to secure the carrier roller to the track frame. A plug and O-ring seal the lubrication opening in the shaft. The retainer plate, which mounts to the shaft, holds the tapered roller bearings inside the carrier housing. A roller cover provides a seal on one end and is fastened with bolts to the end of the carrier shell.

Idlers

Depending on the track design, the equipment could have one or two idlers per track section. If the equipment has an elevated sprocket system there will be two idlers: a front idler and a back idler. Equipment that has a conventional oval track requires only one idler. The purpose of an idler(s) is to help guide the track through the track rollers and to help support part the weight of the equipment. **Figure 18-9** shows an example of a typical idler assembly. Besides providing alignment, the idlers job is to maintain the correct tension and slack on the track. The tension on the track is adjusted by moving the idler back and forth on the track frame. This is accomplished with the use of hydraulic pressure to compress a high-tension spring mechanism to tighten the track.

Equipment with oval tracks can have a two-position idler, with one position that is used when drawbar work is required. This position minimizes the amount of track that is in contact with the ground, resulting in less track wear. When equipment must operate with heavy implements attached to the front, the use of the lower mount for the idler places more track on the roadway making the equipment more stable but causing more track wear. Too much track tension is a contributing factor to accelerated wear because of increased loading on the track components.

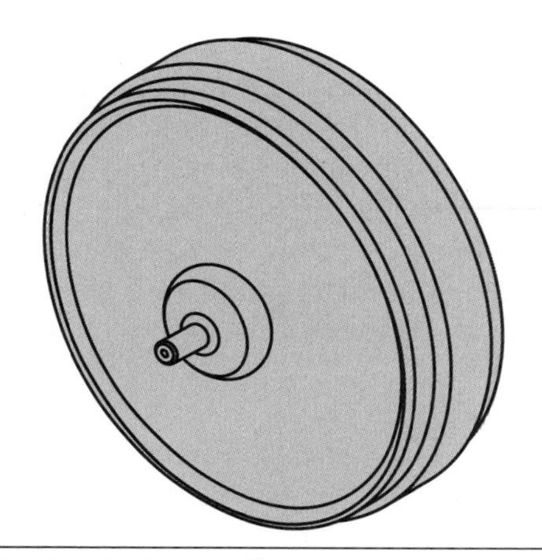

Figure 18-9 Typical Cat idler assembly.

Figure 18-10 shows a breakdown of a typical idler assembly, **Table 18-5** identifies the components that make up the assembly.

The bearing used between the idler shaft and shell in Figure 18-10 is a plain bimetallic bearing with excellent load and wear characteristics. A retainer ring holds the end collar to the idler shell. The end collar holds the seals and the other parts inside the assembly. The outer circumference of the idler has a raised center section on which the track rides. The surface hardness of the idler, as with all track components, is the same as the track links.

The tensioning mechanism connects to the idler to maintain the correct track tension. In older equipment, the tensioning mechanism used a recoil spring with a mechanically adjusted rod to adjust track tension. More recently manufactured equipment uses hydraulic pressure to balance spring tension.

Seals

Tracks operate in extremely abrasive conditions and would quickly wear out the pins and bushings if they were not protected with seals. Seals keep lubrication in and dirt out, providing a much longer service life. Not

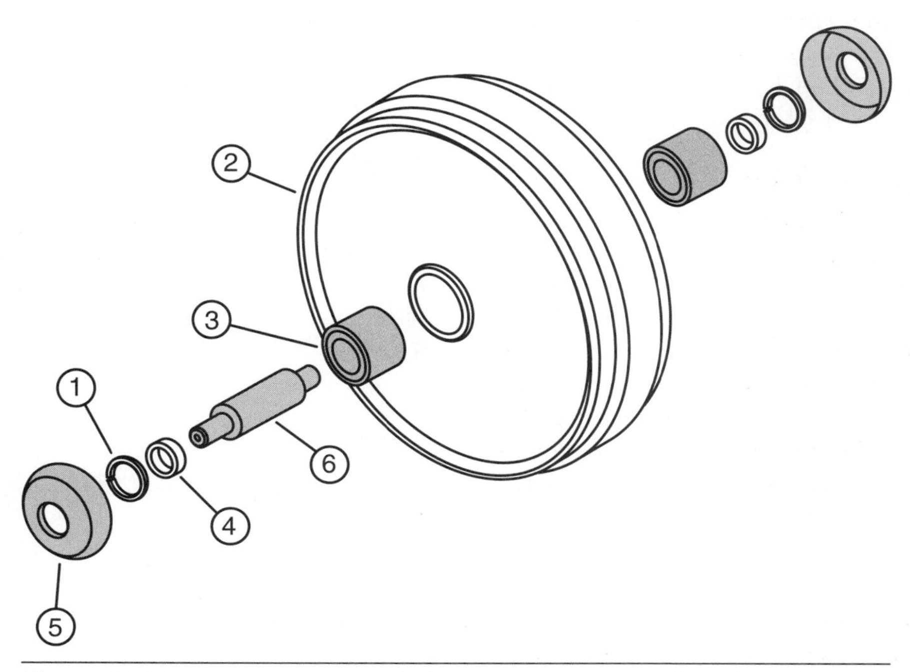

Figure 18-10 Expanded view of the components used in the idler.

Table 18-5: COMPONENTS OF AN IDLER ASSEMBLY	
No. 1	Retaining ring
No. 2	Idler shell
No. 3	Idler shaft
No. 4	Seals
No. 5	End collar
No. 6	Idler bearing

only do the seals keep dirt out, but they also carry a certain amount of side load, preventing unnecessary wear of the track links counterbore. Conventional track seals use **Bellville washers** that are capable of maintaining constant sealing pressure as they wear, which helps keep dirt out.

Some manufacturers use rigid seals which consist of two individual parts like those used on lubricated tracks. These are made up of a **lip seal** and stiff wear-resistant material on the inside called the "can." The seal shape integrity is maintained by the extremely durable urethane material from which it is constructed. The seal is held against the bushing by the load ring. These seals are capable of withstanding operating temperatures as high as 170°F (70°C). The seal is protected from thrust loads by a steel thrust ring, which limits seal compression.

The cost of this type of seal system is considerably more that the conventional Bellville washer design, but the result is a much longer pin and bushing life. Over the long haul, this type of seal system yields lower operating costs. Wear is generally not an issue with this type of seal system, and operation is quieter due to the lubricated pins and bushings. Frictional losses are reduced with a lubricated track than with a conventional track, resulting in lower operating fuel costs.

Track Guards

To keep the tracks from getting packed with dirt, roller guards are often used to keep rocks and foreign debris from getting between the track rollers and track links. The **track guard** also helps guide the track. Good practice requires that the area around the track guard be cleaned regularly, especially after operating in muddy conditions. Operating with packed debris in the track mechanism results in a too tight track condition, which greatly accelerates wear. Track tension should be checked periodically when operating in mud, sand, or snow.

The frame is the backbone of the track assembly; it holds the front idler in place and the track rollers and guards are bolted to the underside of the frame. The recoil assembly and the track carrier roller(s) are all mounted to the frame. Generally the frame causes very few problems when compared to the rest of the track components.

Split Master Links/Master Pin and Bushing

The split master link design is used by many manufacturers because it requires no specialized equipment to remove and install. Half of the link is installed on one end of a link assembly and the other half to the opposite end. The links are serrated and bolted together when the track is assembled. A split master link has about the same life span as the rest of the links in that track.

Another popular type of track connection method requires the use of a master pin and link. The master pin is longer than a normal **track pin** and has a machined step on one end. The master bushing is a little shorter than the rest of the bushings and does not fit all the way into the link bores. When track separation is required, the master pin must be removed first, then the master bushing is forced between the master pin track sections.

Equalizer Bars

The track frames on each side of the equipment are connected together through an **equalizer bar** near the front of the machine, as shown in **Figure 18-11,** and are mounted to the sprocket that is part of the final drive at the rear of the machine. All the weight of the equipment is carried by the track rollers and frame. With this type of mounting, each side can move

Figure 18-11 Equalizer bar in the undercarriage. (*Courtesy of Caterpillar*)

independently on uneven terrain to maintain contact with the ground. Diagonal braces are used to maintain track alignment when on uneven terrain. One end of each brace is welded to the track frame while the other end is mounted to the sprocket shaft and swivels on a bearing. These bearings must be lubricated at regular service intervals.

Often older track loading equipment does not have a pivoting equalizer bar, therefore, independent track movement is not possible. Older equipment often has a rigid frame and is not able to operate on uneven terrain effectively. **Figure 18-12** shows the undercarriage of a typical track loader with a rigid frame. This type of equipment is best suited to operation on flat, hard surfaces such as those found in quarries and pits.

Some manufacturers have added an **oscillating undercarriage**, as shown in **Figure 18-13,** as an alternative to independent track movement. This design has the added benefit of being able to pivot from front to rear, which can increase stability when the equipment is operating on uneven ground. The track frames have an equalizer bar mounted between to allow each track frame to pivot independently on a pivot shaft. With this design, an idler swing link is used to permit the idler to move horizontally to absorb any shock loads it may encounter during operation. This ensures that the correct amount of track tension is always maintained. Pins are used to prevent the equalizer bar from having excessive oscillating travel. Rubber pads are used to dampen vibration between the equalizer bar and the main frame.

Figure 18-12 Track loader with a rigid track system. (*Courtesy of Caterpillar*)

Figure 18-13 Track loader with an oscillating undercarriage. (*Courtesy of Caterpillar*)

Elevated Sprocket

Improvement in track equipment design has led to the inception of the elevated track design, which offers a few major advantages over a conventional track sprocket arrangement. The major benefit of this design is that it isolates the final drive from excessive shock loads. The position of the final drive also creates a better balance arrangement that also provides better traction, as shown in **Figure 18-14.** This design requires the use of two idlers and a separate roller frame for the front and rear.

There is always a down side to any design, and this one is no exception. Compared to a conventional undercarriage design, which has one travel loaded flex point, the **elevated sprocket** undercarriage has three flex points where loading takes place during operation. Therefore, in the long run, the conventional undercarriage with only one load point during forward travel will likely last longer.

The two rear track frames are connected together with a pivot shaft; this allows the shock loads to be absorbed by the main frame. The front roller frames fit

Figure 18-14 Elevated sprocket track dozer. (*Courtesy of Caterpillar*)

inside the rear roller frames and are connected to each other with an equalizer bar that is pinned in the center with limited movement from side to side. A recoil spring provides track tension. Some manufacturers build tractors with a suspended undercarriage, which can improve ground contact as the track can oscillate in the roller frames providing better ground contact at the bottom of the track. This type of specialized equipment comes with longer, wider tracks that increase stability.

Specialized equipment, such as pipe layers, use an elevated sprocket track but have a unique setup in that the roller frame does not oscillate. This unique design allows the shock loads to be absorbed by the undercarriage. The roller frames are connected to the main frame two-point mounting system for good stability required in pipe laying applications. A pivot shaft is still used to connect the right and left rear roller frames, allowing ground shock to be transmitted directly to the main frame. The equalizer bar is replaced with a hard bar, which connects the two front roller frames together. The front roller frames slide in and out the same way they do on a conventional elevated sprocket track, and track tension is maintained with a recoil spring. A counterweight is mounted on the right side of the equipment to offset the boom bolted on the left side of the equipment.

Track Shoes

Track shoes are used to provide a gripping function on many different surfaces, as shown in **Figure 18-15.** Many manufacturers make a wide variety of track shoes, and they are rated by a vertical designation to indicate the shoe's ability to withstand abrasion and a horizontal designation to indicate impact resistance. Impact resistance is determined by the hardness of the material. A high impact can be defined as hard concrete or rock, with continuous exposure to six-inch bumps. The opposite end of the spectrum is low impact, which can be defined as completely penetrable material such as landscaping soil, with low potential for high bumps. The width of the track shoe can be tailored to suit the application. The shoe's ability to provide flotation, a large contact area, bending resistance, penetration, and self-cleaning are all points that must be considered.

The standard single grouser shoe is probably the most popular shoe in use. It is well-suited for dozer and drawbar applications and provides excellent life when used in landscaping loam, gravel, or clay applications. This shoe is best for operation in low and moderate impact and abrasion conditions. These shoes have good

Figure 18-15 Typical track shoe designs. (*Courtesy of Caterpillar*)

penetration with excellent resistance to bending and wear and are available in a variety of widths. In applications that require working on high impact material, an extreme service grouser is available. This shoe has a much harder surface for high impact conditions. When the links outlast the shoes, you probably need extreme service grousers.

For operating conditions that require constant turns in tight places, such as on excavations, the use of a double grouser shoe will work best. See **Figure 18-16.** The use of two short grousers instead of one tall grouser causes less ground disturbance and makes the shoes extremely rigid, which also makes them highly resistant to bending. This shoe is commonly found on hydraulic excavators where less ground penetration is desirable.

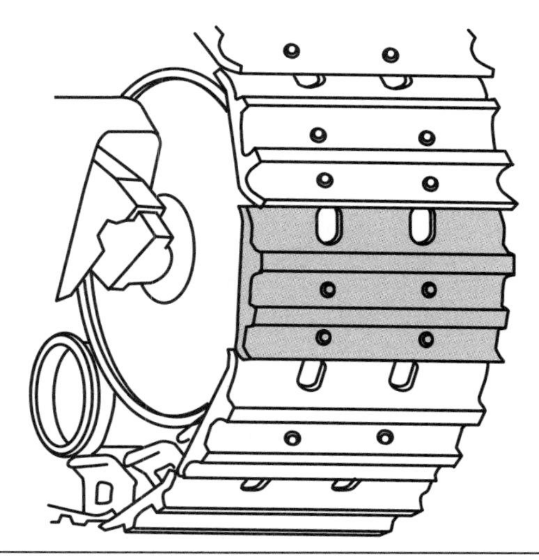

Figure 18-16 Typical double grouser shoe.

When equipment must be operated in loose, muddy ground the use of a self-cleaning, low ground pressure shoe, as shown in **Figure 18-17,** is desirable. These shoes should never be used in high impact or abrasive conditions. As the track rotates around the sprocket or idler, they separate slightly to allow the dirt to drop out. This design has very high flotation capability on loose, muddy ground.

In operating conditions where packing of the track becomes a problem, the use of an open-center grouser, as shown in **Figure 18-18,** is recommended. This design allows the dirt to be pushed out by the sprocket as the track rotates around the sprocket. This lessens the possibility of the track becoming too tight by forcing out dirt that tends to pack around the inside of the track.

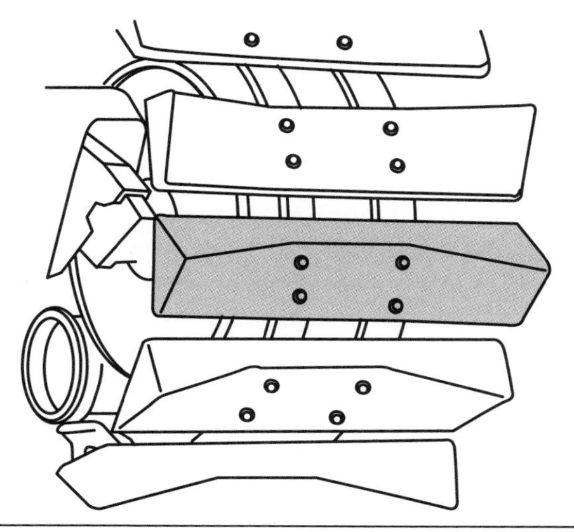

Figure 18-17 Typical self-cleaning low ground pressure shoe.

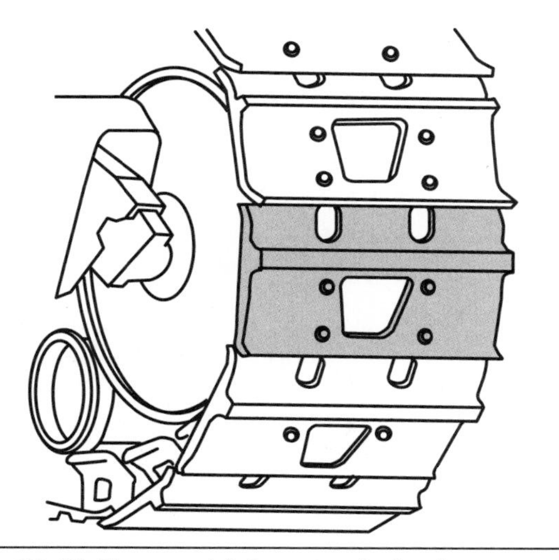

Figure 18-18 Trapezoidal center hole shoe.

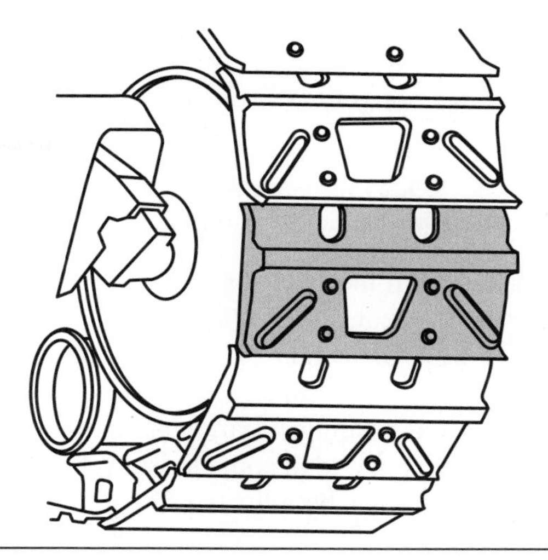

Figure 18-19 Chopper shoe.

Operation in snow also requires the use of an open-center grouser shoe because snow will quickly pack inside the track, causing the track too become to tight. Some manufacturers supply an all-purpose grouser shoe with three small grousers. This type of shoe is very versatile and can be used on hard smooth surfaces as well as in soft muddy conditions.

Some manufacturers make a grouser shoe that has the ability to cut and chop the surface of the ground. This design (shown in **Figure 18-19**) is intended to be used where debris can get lodged in the shoes, such as in demolition applications and at landfill sites. The shoe looks a lot like a single open-center grouser but has additional diagonal side grousers that help chop up the material it is operating on. Other specialty shoes also come in a variety of materials including rubber which is often used in applications on smooth finished surfaces that would otherwise be damaged by conventional steel shoes. The downside to rubber shoes is their short life span when compared to steel shoes.

UNDERCARRIAGE OPERATION AND TERMINOLOGY

Well-designed track components work and wear together and should wear out at about the same time. If components such a links wear out before the shoes, the type of link design used for that particular application should be examined. Track terminology used to describe the operation and wear characteristics of track systems and will be used in this section. Some common terminology is similar between manufacturers. **Link pitch** is a term used to define the distance between the bushing bore center and the pin bore center. This dimension does not generally change during the life of the track.

Track Pitch

Track pitch on the other hand, does change and is an indication of wear. *Track pitch* can be described as the distance from the center of one pin to the center of an adjacent pin in an adjoining track link section. When the track pins and bushings wear, the length of the track is increased, causing it to become loose. When the track tension can no longer be adjusted, the pins and bushings need to be turned or replaced depending on the track design. *Sprocket pitch* can be defined as the distance between the centerline of one bushing sitting in a sprocket tooth to the center of the adjacent bushing sitting two teeth over in the sprocket.

On a new track, the pitch of the track and sprocket are the same dimension. During the life of the track, the track pitch increases due to wear, but the link and sprocket pitch will not change. As the track pitch increases the teeth of the sprocket do not mesh properly with the track, causing excessive wear to take place, as shown in **Figure 18-20.** If the bushings are allowed to wear excessively, all the related track components will wear rapidly and eventually lead to sprocket jumping, causing excessive sprocket wear. When the bushings wear externally, the worn section covers about one-third of the surface of the bushing. For the most part, use of a lubricated and sealed track design eliminates this problem. Once the sprocket has excessive wear, rotating the pins and bushings will cause continued wear at a higher rate than normal: the only solution is to replace the sprocket.

Track wear occurs when the parts that are in contact with each other during heavy loading slide against each other. Wear on the track components is accelerated when you factor in dirt between the components, horsepower, and the weight of the equipment. Operation of the equipment in reverse at a high rate of speed can cause accelerated wear in the pins and bushings. This also applies when the equipment is operated with a tight track due to dirt packing between the sprocket teeth and bushings. External wear on the bushings along with internal wear of both pins and bushings indicates that the pins and bushings need to be turned.

Look for wear lines on the drive side of the sprocket teeth; this indicates the need to rotate the pins and bushings. In order to rotate the pins and bushings successfully, the wear must not extend completely around the pins. On a conventional track with a master pin and bushing, faster wear occurs first at the master pin since the pin is shorter and the bushing does not always extend into the counterbore of the master link. Years of equipment operation have shown that replacing the master pin on a regular basis helps minimize the wear on the rest of the pins and bushings and minimizes sprocket tooth wear. On tracks that use the split master link, this excessive wear is not a problem because the pins are identical throughout the track.

On a new track during forward operation there are two types of loads imposed on the parts; these occur as the two track sections hinge, as shown in **Figure 18-21.** When the bushing enters or leaves the sprocket teeth in forward, the bushing rotates slightly as it leaves the sprocket. The opposite occurs in reverse; when the bushing enters the sprocket it rotates slightly. This movement between the bushing and sprocket is similar to that of gear teeth when they mesh.

When the bushing enters the sprocket in forward, the first bushing to come into contact with the sprocket takes about 85% of the load, as shown in **Figure 18-22.** The rest of the load is absorbed by the next three bush-

Figure 18-20 Sprocket pitch dimension measured center to center between two adjacent bushings.

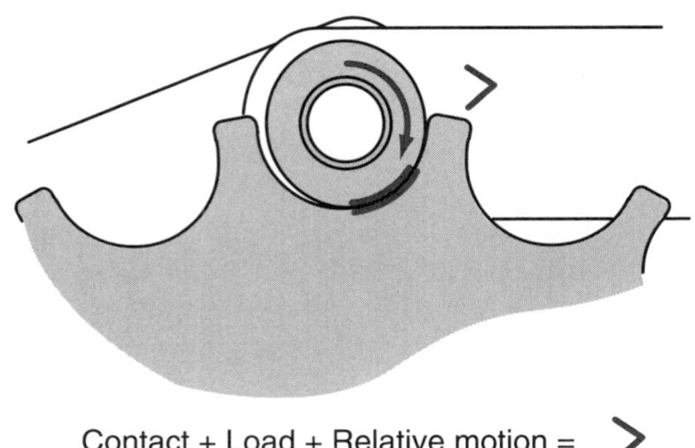

Figure 18-21 How wear occurs in a track.

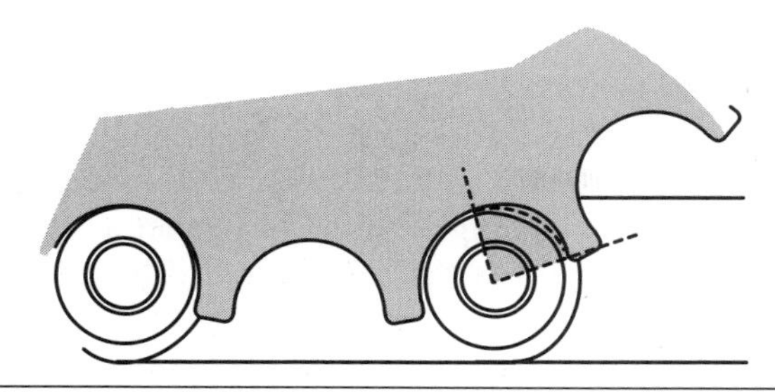

Figure 18-22 How wear occurs in the first track pin and bushing during operation in reverse.

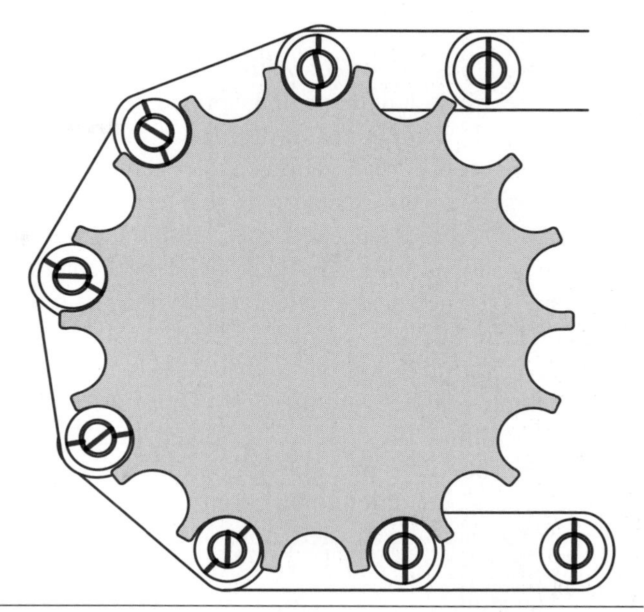

Figure 18-23 Relative motion between the pins, bushings, and sprocket as it rotates

ings. When there is a mismatch between the sprocket and track pitch during operation, the bushing will slide and turn as it seats in the sprocket, causing wear on the forward drive side. An internally lubed track has an advantage over a conventional track; it prevents wear from taking place in this situation. In reverse the first bushing comes into contact with the sprocket teeth about 30 to 60 degrees from the center of the forward drive side. This forces the bushing to rotate through the tooth root area and over to the reverse drive side of the tooth. The bushing wears externally and causes excessive wear to the sprocket tooth.

Some manufacturers incorporate a design that allows the bushings to rotate; this reduces the wear significantly. To show how the relative motion of the sprocket, pin, and bushing occurs, refer to **Figure 18-23.** The lines show how much movement occurs as the sprocket engages the pins and bushings during operation. When the equipment moves in the forward direction, the pin enters the sprocket first; the first bushing is then picked up by the forward drive side of the sprocket loading it with approximately 80% of the load. Because the pin and bushing are lubricated, no negligible wear takes place.

In reverse, if the track is in good condition and has been properly adjusted, the bushing end of the track enters the sprocket first. The forward drive side of the track picks up the first bushing and rotates it as it moves to the reverse drive side of the sprocket; this can cause more wear than travel in the forward direction. If the track has been improperly adjusted and is too tight, the load will be increased significantly. During forward operation, with good track tension and no significant packing occurring, no load is present at the top of the sprocket. With a tight track, loading will occur, causing excessive wear. In reverse, there is normally some wear while operating with a properly adjusted track. If the track tension is tighter than recommended, the wear factor increases significantly.

Unfortunately, some degree of packing occurs on track equipment during operation on certain types of terrain. Excessive packing can increase the track tension significantly and causes excessive wear. Track adjustment should always be carried out after operating in an area to establish a normal amount of packing in the tracks. When packing gets between the track and sprocket, the bushings sit on top of the packing and cause the tension on the track to increase significantly. This increases the sprocket pitch and results in a mismatch between the track and sprocket pitch. The contact between the track and sprocket occurs in the wrong area.

In the forward direction the bushing is picked up by the reverse side and slides over to the forward side. This prevents the sprocket from driving the bushing until it moves to the forward drive side. This sliding action dramatically increases the wear on the sprocket and external sides of the bushings. When operating in reverse with a packed track, the sprocket engages the bushing closer to the reverse drive side of the sprocket, causing excessive wear from the sliding action.

Inevitably, track wear will occur over time, causing the track pitch to increase significantly. In forward, the bushing will contact the drive side of the tooth higher up on the tip, initiating a sliding action between the two parts and causing excess wear. In reverse, the bushing contacts the sprocket tooth higher on the forward drive side. As the sprocket rotates, the bushings rotate through the bottom of the sprocket tooth and over to the reverse drive side of the tooth, again increasing the wear significantly.

When the track pitch is increased due to wear, packing can have a positive effect on the overall track pitch. Packing the track lengthens the sprocket pitch, and when the track wear increases the track pitch and is equal to the sprocket pitch, good contact is maintained. The down side to this scenario is that abrasive wear occurs rapidly from the packed material in the track. To minimize wear, packing the track must be kept to a minimum; too much packing tightens the track so much that significant bushing wear will occur.

When operating in conditions that are less than ideal, track packing can be minimized by choosing a shoe design that allows material to be forced out through holes in the shoe. It is probably a good idea not to use roller guards in conditions that promote packing. The undercarriage must be cleaned more often when operating in conditions that cause excessive track packing to occur. Adjusting the track correctly for the operating conditions is beneficial to track life; a loose track *will* cause excessive wear to roller flanges and sprocket teeth. If the track tension is left loose, the track roller frame can experience significant damage. The importance of selecting the correct track shoe cannot be over emphasized because the wrong shoe can promote excessive wear. For instance, wider shoes are excellent in muddy conditions because they provide a greater degree of flotation. But keep in mind that selecting a shoe wider than is necessary is not desirable. A wider shoe is more susceptible to bending and breakage, and using a wide shoe in a high impact operating condition will lead to excessive track wear and undercarriage damage.

Roller Wear

The rollers on track equipment do not all wear at the same rate or equally; they should be rotated, if possible, to extend the life of the rollers. Rollers in certain positions will wear more on one flange than another. The flanges on the lead rollers of the track must be kept in relatively good condition as they guide the track onto the sprocket. The track rollers have a surface hardness that is about the same as the track links, and generally wear at about the same rate. Unless broken from impact damage, the rollers are replaced at about the same time as the track links, pins, and bushings. Carrier rollers can last significantly longer than the track rollers because they only support the weight of the track and are used to guide the track. Track rollers help support the weight of the equipment.

A typical track system can have a number of roller arrangements depending on the track type. For example a XL track can have seven track rollers and generally arranged so that the front roller (located at the front of the dozer) is a single flange roller, the second roller is a double flange roller, the third roller is a single flange roller, and so on, ending with the last roller as a single flange roller. The number of track rollers depends on the track design and can vary from manufacturer to manufacturer. The tightness of a track can affect the life of the carrier rollers and has a negative effect on idler life.

MAINTENANCE AND ADJUSTMENTS

It may be impossible to stop track wear altogether on a track equipped machine, but wear can be slowed to a great extent by good maintenance practice and operation. This section helps you better understand the wear and adjustments that take place on a track and undercarriage system. Always follow the manufacturer's repair manuals to carry out maintenance and repair procedures on track systems as they may vary from the example that is presented here. Because of varying operating environments, no one maintenance practice will work for all applications. It is imperative that you consult the manufacturer's guide for equipment service recommendations when considering the maintenance and adjustments to be made to the track system. Operating track equipped machines in conditions that are abrasive leads to a short life of components that rub together in an abrasive environment. With older track systems the pins and bushings are a constant source of maintenance, but in modern sealed and lubricated systems, pin and bushing wear has been virtually eliminated. Track systems and undercarriages wear whenever they are operated; the severity of the wear depends on the environment. Understanding what causes tracks to wear allows a technician to maintain and adjust the track system to optimum specifications.

One of the most important considerations is the selection of the shoe. Never use a shoe that is wider than necessary; added width will cause accelerated wear and lead to structural problems on the track and undercarriage. Track shoes that are wider than necessary will affect the life of track links, rollers, and the idler flange. Even the weight of the wider shoes will cause an increase in wear and that wear will increase significantly on rough terrain.

It cannot be overemphasized how much correct track adjustment affects track life. Not only is track tension easily corrected, it is also easily adjusted on modern track systems. Track tension adjustment must be performed on the terrain that the equipment will work on. Certain conditions, like mud, will pack the track,

Figure 18-24 Track adjusted too tight. *(Courtesy of Caterpillar)*

more quickly than other conditions, such as crushed rock. In some instances, a certain amount of track packing is necessary before a track adjustment is performed. Always refer to the manufacturer's recommended procedures for track adjustment. Operation with too tight a track, as shown in **Figure 18-24,** leads to excessive bushing and sprocket wear, which is caused by increased loads on components. Increased loads from a tight track increase wear on the idlers and carrier rollers because of the excessive tension placed on them. High-speed operation should be avoided as much as possible; the increase in track revolutions accelerates the wear on track components tremendously. If the equipment is to be operated in conditions that will lead to excessive track packing, the roller guards should be removed to aid in expelling the material, which will help minimize packing.

Good Maintenance Practice

The following list is a quick reference for inspection and maintenance that reduces damage from wear.

- Pick the correct shoe for the application; use the narrowest shoe you can, keeping in mind the importance of maintaining flotation.
- For track adjustment, be sure to perform the adjustments in the conditions under which the equipment will operate. Always follow the manufacturer's recommended track adjustment procedures.
- Get in the habit of inspecting the equipment for loose bolts and components. Check for leaks and excessive wear.
- Whenever track components need to be re-torqued, follow the manufacturer's recommended torque procedures.
- The use of roller guards should be avoided in conditions such as mud or snow which can lead to excessive track packing.
- When inspecting and servicing equipment, be sure to inquire about the operating environment. The type of work that the equipment is to perform dictates the wear severity and location. For example, when the equipment is performing drawbar work or ripping, the weight is shifted to the back of the equipment, increasing the wear on the rear rollers and sprockets.

Generally track wear can be measured by using either of the procedures shown in **Figures 18-25** and **18-26.**

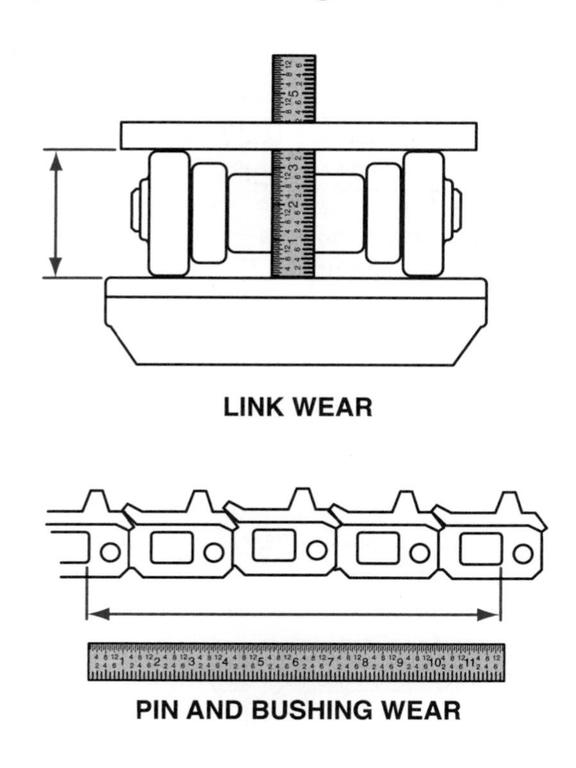

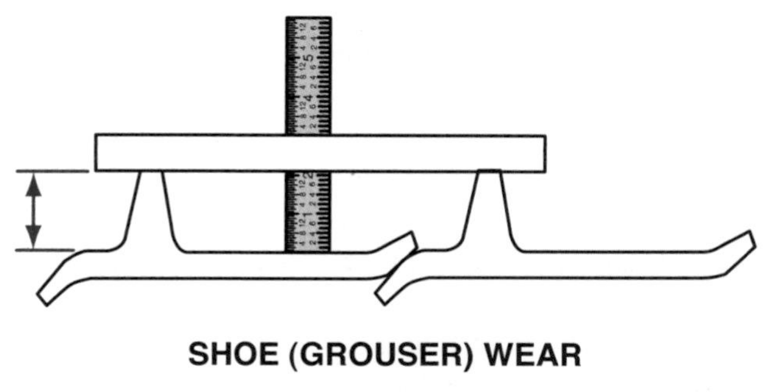

Figure 18-25 Track measuring technique using measuring tools.

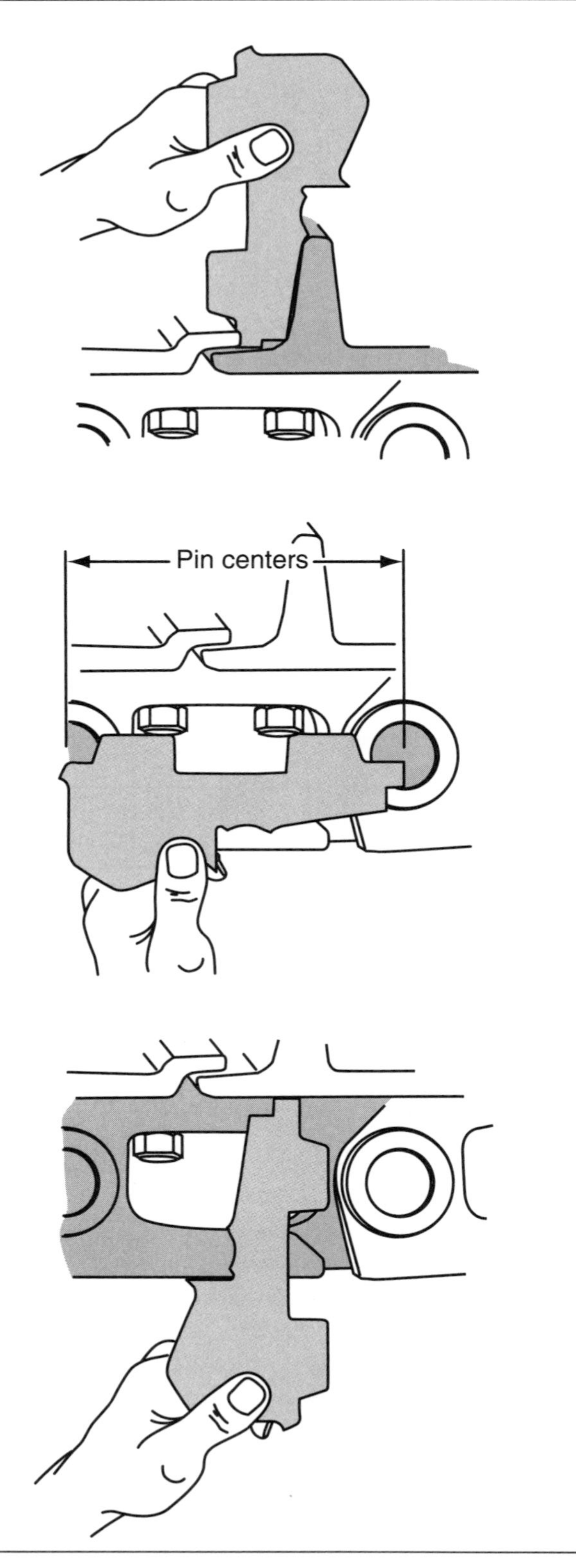

Figure 18-26 How gauges are used to access track wear.

The use of specialized measuring tools is one method used by some manufacturers. The other equally popular method is to use specialized wear gauges. Measuring gauges include tape measures, depth gauges, and vernier callipers. Readings can be compared to the manufacturer's wear limit recommendations to determine whether or not replacement is necessary. Some manufacturers supply special wear gauges to measure track component wear. The problem with some types of tools is that they do not give you an actual measurement; rather, they give a visual indication that is subjective.

Tech Tip: Before performing any type of wear evaluation or taking measurements of the tracks, make sure the equipment is properly blocked and tagged with the engine off. Never attempt to perform any measuring of track components while the equipment is in operation. Some measurements require that the track be tightened before performing any measurements.

Before attempting to make any determination as to the amount of pin and bushing wear on a track, the track must be tightened by placing an old track pin, as shown in **Figure 18-27,** between the sprocket teeth and the closest link. Back up the equipment until the slack is taken up. Note that the pin should be at approximately the 12 o'clock position. Figure 18-25 shows how to measure across several links to determine wear. When using this measurement technique, do not include the master pin in any of the measurements. If you have measured across four links, then you must divide the value by four and compare it to the manufacturer's specification to determine actual wear.

The use of a wear gauge is another common method used by some manufacturers to determine the pin and bushing wear. Note that the track must first be cleaned before any wear measurements can be accurately made. Figure 18-26 shows how to use a wear gauge to determine pin and bushing wear.

Shoe wear is determined by measuring of the height of the grouser (Figure 18-25). The shoe bolts should be checked for tightness when checking for wear. The most common cause of shoe looseness is that they were not torqued correctly in the first place. Always refer to the manufacturer's recommended torque procedure for the equipment in question.

When installing new shoes on a track, it is good practise to re-check the bolt torque after approximately 50 to 100 hours of operation. Initial torque specifications are generally given in lb-ft or Nm, and require that an additional torque turn be applied to the bolts to ensure the bolts maximum clamping force is achieved.

Figure 18-27 How to tension the track to check pin and bushing wear. (*Courtesy of Caterpillar*)

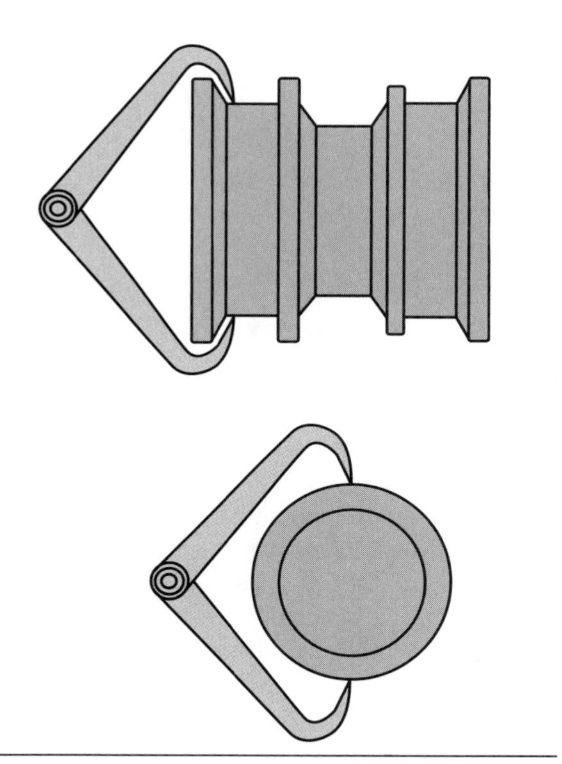

Figure 18-28 Measurement procedure for roller and bushing wear

Roller and bushing wear can be checked by using several different methods. The procedure used in **Figure 18-28** is for clarification purposes; always refer to the manufacturer's recommended procedure for this check. To check the rollers for wear by using the gauge method (**Figure 18-29**), be sure to use the correct gauge for the equipment in question. Use of an incorrect gauge may lead to expensive repair work that may not be necessary.

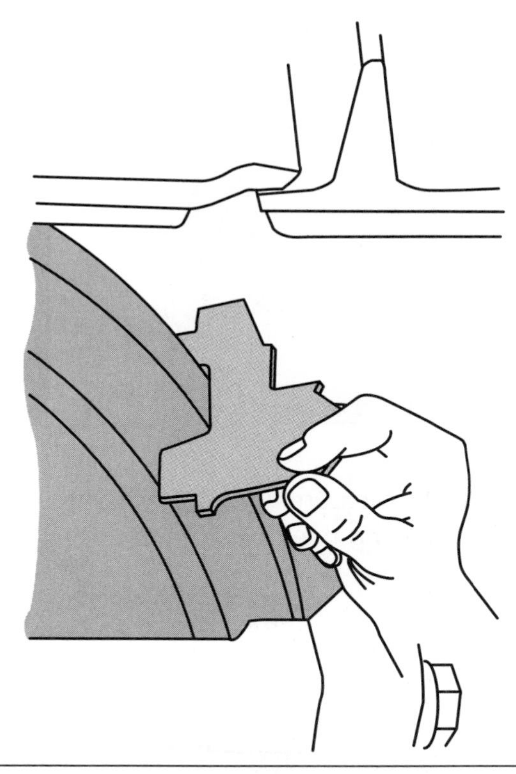

Figure 18-29 Measurement procedure to determine front idler flange wear using a wear gauge.

Track rollers will wear on the outer portion of the flanges and need to be checked accurately by using an accepted method to determine maximum wear. The most common method of checking roller wear requires two measurements to be taken to determine wear. **Figure 18-30** shows how to perform this procedure accurately. After taking the readings, subtract measurement C from measurement B to determine the roller radius. If the equipment has roller guards in place, the radius of the rollers must be measured with a ruler.

Front idlers tend to wear on the side and outer edges of the flanges, so proper idler wear measurements are needed to determine the wear on the flanges. This measurement can be made in one of two ways: the use of a specialized gauge, as shown in Figure 18-29, or with a straight edge and ruler, as shown in **Figure 18-31.** Always refer to the manufacturer's specifications to determine the allowable limits.

Determining sprocket wear with a gauge is recommended by many manufacturers. The round section of the manufacturer-supplied gauge goes into the root of the sprocket, as shown in **Figure 18-32.** Use the manufacturer's specifications for the specific sprocket to determine the wear.

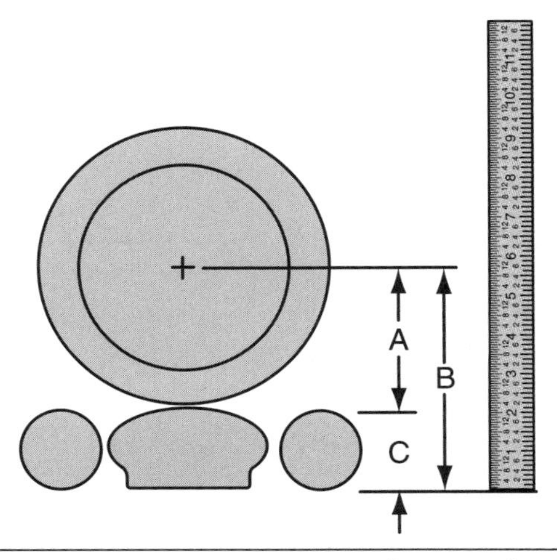

Figure 18-30 Measurement procedure to determine roller wear using the indirect methods with a ruler.

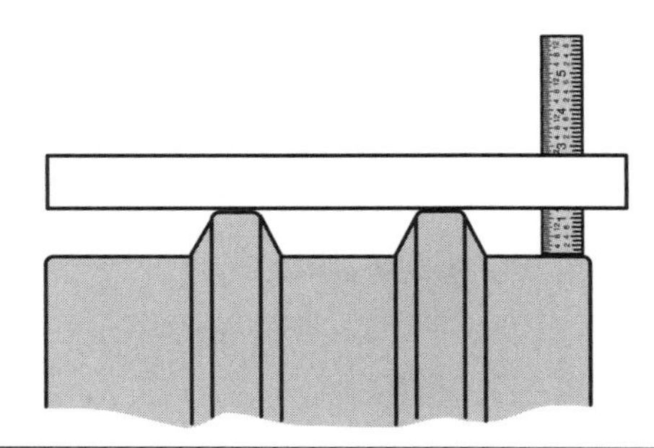

Figure 18-31 Measurement procedure to determine front idler flange wear using a straight edge and ruler.

CAUTION *Under no circumstance should anyone attempt to repair a track without having been given the proper instructions and special training that is required to perform this type of work safely. The track and its components are often very heavy and bulky to handle by hand. The only way to avoid potential accidents when working on track equipment is to familiarize yourself with the manufacturer's recommended procedures and to follow basic safety rules and precautions. A technician's first responsibility is to ensure that the work can be performed safely and that no one is exposed to any unnecessary hazards. No manufacturer can anticipate all potential hazards and repair procedures. Therefore, it becomes the responsibility of the technician to make sure that the procedures and techniques used to perform any repairs or maintenance to the equipment that are not covered in the technical documentation will not cause personal injury or damage to the equipment.*

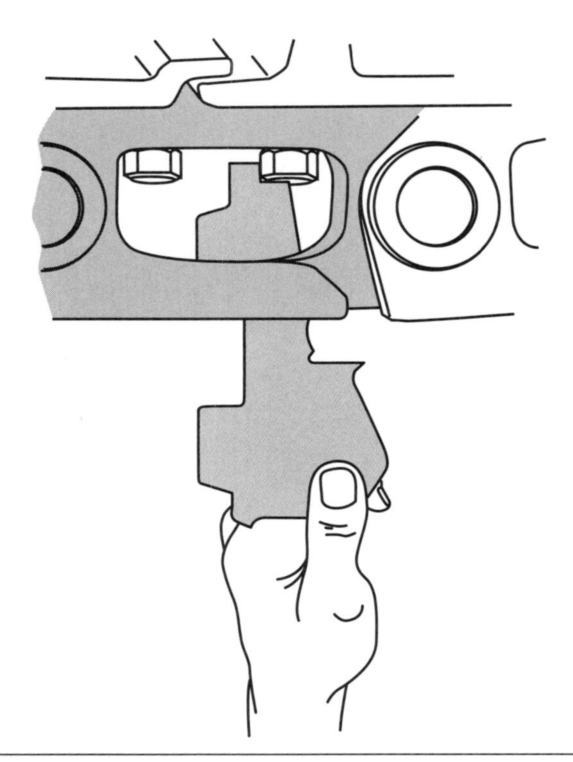

Figure 18-32 Measurement procedure to determine sprocket wear using a manufacturer-supplied wear gauge.

Track Adjustment

Correct track adjustment is necessary to maintain long track life with low operating costs. Nothing shortens the life of a track faster than incorrect adjustments. If the track is adjusted tight, excessive loads are placed on the mating components, accelerating track component wear. If the track is too loose, it will cause track whip at higher speeds, leading to excessive wear of undercarriage components. To keep these problems to a minimum, a technician must perform these adjustments according to the manufacturer's recommendations. It must be stressed that this explanation of track adjustment is shown as an example and may vary based on the age and make of the equipment. This procedure is for Caterpillar D3 to D9 track-type tractors and loaders. Not all equipment uses the same techniques; some excavators and loaders use different measurements and, in some cases, different procedures. Refer to the manufacturer's recommended procedures for the equipment in question.

1. To ensure that the adjustments have been performed correctly, the equipment must be located in the actual working environment. Do not clean the track to make the adjustments; in some cases the track will pack slightly and any

attempt to clean it may lead to an adjustment that will become too tight when the equipment is back in operation. **Figure 18-33** indicates that A is the track adjustment mechanism location and B is the location of the front idler bearing assembly.

2. Before beginning any adjustment procedure, make sure you are wearing approved safety glasses and gloves, if required. Clean any debris from the top of the adjustment mechanism plate before opening. After wiping the fitting, use a grease gun to adjust the track until the track idler is at its farthest forward position. Ensure that the relief valve remains closed.

CAUTION *At any point in the adjustment procedure, never attempt to remove any packing material between any of the track components. Once this step is completed the track between the idler and the carrier roller will be tight and look almost straight.*

1. Clean off the track frame and place a mark behind the rear edge of the idler bearing. The distance will depend on whether the track has one carrier roller per side or more than one. Tracks with one carrier roller per side will use 3/8" (10 mm) as the correct dimension for this step; tracks with two or more carrier rollers will use 1/2 (13 mm).
2. Before performing this step, make sure you are wearing safety glasses. Once step 3 is completed on both tracks, open the relief valve slowly.

Figure 18-33 Cat dozer in the correct adjustment condition. Track tension should be adjusted in the work environment. (*Courtesy of Caterpillar*)

3. Locate a drawbar or track pin and insert it in the sprocket tooth near a link on both tracks, as shown in **Figure 18-34.**
4. Ensure that the relief valve is in the open position on each track; start the equipment, and move the equipment in the reverse direction until the front idler backs up one-half inch or more. This should place the pin between the track link and sprocket in the 12 o'clock position, as shown in **Figure 18-35.** At this point, move the machine forward slightly and remove the track pin from the sprocket.
5. Ensure that the hydraulic relief valve is closed on each track adjustment mechanism. Adjust the **hydraulic track adjuster** until the mark on each roller frame is visible. The track will sag between the idler and the carrier roller. This sag

Figure 18-34 Technician inserting a track pin in a sprocket tooth. (*Courtesy of Caterpillar*)

Figure 18-35 Pin in the proper location (12 o'clock). (*Courtesy of Caterpillar*)

will provide the correct amount of track tension for the application in which the equipment is operating. Secure the track adjustment protection plates and return the equipment to service.

Track Removal

This general overview of the track installation procedure is to be used as an example and is not intended to replace the manufacturer's shop manual and supplemental bulletins. In many cases, it may be necessary to remove the track from the equipment to perform major repair work to the track. There are two methods that are used to connect tracks together: the master link and the master pin. A general summary of the track removal process on a master link track will be given so that the procedures involved in this type of work are well understood. Never attempt to perform any track repairs to equipment unless you have been properly trained and have the necessary equipment and manufacturer's recommended procedures on hand to perform the work safely.

The split master link uses a procedure that is unique to that design; the technician must follow the manufacturer's recommended procedures.

Tech Tip: High pressure may be present in the adjusting cylinder in the form of high grease pressure. Never expose yourself to the possibility of injury by visually inspecting to see if the grease is released. Visually inspect the rod to see if it has moved into the front support of the track roller frame. When removing the track from the equipment, do not stand anywhere near the front of the equipment; this will minimize any chance of injury if something went wrong. The track is very heavy and can weigh upward of 1500 pounds.

1. The equipment must be placed in a suitable position so that when the track is removed it is on flat ground. *Note*: The track must be loosen by opening the relief valve located under a protection cover on the roller frame. If the track will not slacken on its own, place a piece of round steel bar stock on the inside of the links between the bushing and the sprocket and drive the equipment in reverse slowly until the track loosens.
2. Place the track in position so that the master link is on the centerline of the rear idler, as shown if **Figure 18-36.** Place suitable blocking under the grouser bar, as shown just under the master link, and remove the four bolts (6) that hold the master shoe in place. Separate the track by removing the bolts that hold the master link together.
3. This step requires that one end of a chain (7) be fastened to the end of the top section of track, as shown in **Figure 18-37,** and the other end be fastened to a wheel loader. Start the unit and rotate the final drive in the forward direction. Allow the track to move forward so that it is pulling the wheel loader in a forward direction. Be sure that there is enough room in the front of the equipment to completely remove the track.
4. In the event that it may become necessary to raise the equipment to remove the track, use the tools listed in **Table 18-6.** Use tool A (hydraulic jack) from the table to raise the side of the equipment. Place tool B (stand) into position under the equipment. Due to the track weight, the use of the appropriate lifting devices is necessary. Use extreme caution when removing the track.

Figure 18-36 Correct location for the track blocking. (*Courtesy of Caterpillar*)

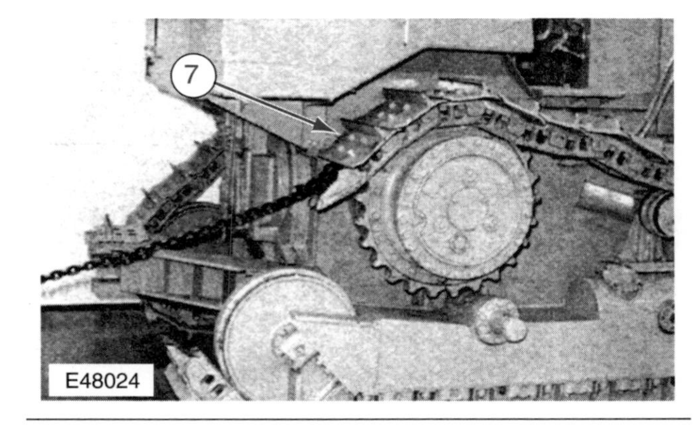

Figure 18-37 Correct procedure to remove the track. (*Courtesy of Caterpillar*)

Table 18-6: TOOLING FOR TRACK SEPARATION

Examples of Required Tooling for Track Separation	A	B
8S-7610 Base	2	
8S-7650 Cylinder	2	
9U-6600 Hydraulic jack	2	
8S-7640 Stand		2
8S-7611 Tube		2
8S-7615 Pin		2

Track Installation

1. This general overview of the track installation procedure is for an example purposes and is not intended to replace the manufacturer's shop manual and supplemental bulletins. The manufacturer's procedures go into greater detail and should be used when performing this type of repair work. Never attempt to perform any track repairs to equipment unless you have been properly trained and have the necessary equipment and manufacturer's recommended procedures on hand to perform the work safely. The same specialty tools that are necessary to separate the track are required to reconnect the track.
2. The track must be positioned and aligned under the track roller frame in such a way that the links will engage the track sprocket and roller flanges correctly. Once the track is in position, carefully lower the unit onto the track. Visually inspect that the track is properly aligned with the track roller flanges.
3. Connect the same chain that was used to remove the track to the end of the split track on one end and connect the other end of the chain to a loader or other suitable equipment that is capable of pulling the chain over the front idler. Have the loader pull the chain over the front idler and carrier rollers until it engages the sprocket teeth.
4. Caution should be exercised whenever the equipment is operated to ensure that all personnel working in and around the equipment are not in danger of being caught up in any of the moving parts. At this point, the equipment must be started and the final drive must be rotated in reverse to pull the track together so that the master link can be reconnected.
5. Remove the pull chain and carefully align the two halves of the master link (labeled 5 and 6) so that the bolts can be installed and torqued correctly, as shown in **Figure 18-38.** Follow the manufacturer's recommended installation procedure to install and torque the master link bolts correctly. Reinstall the shoe on the master link and tighten the bolts following the manufacturer's recommended procedure.
6. Close the relief and fill valves, as shown in **Figure 18-39** on the track adjuster located on the track frame and perform an initial track adjustment. Follow the track adjustment procedure as outlined in the manufacturer's technical documentation. *Note*: If the track assembly has been cleaned due to the repair work performed during removal, the track adjustment procedure must

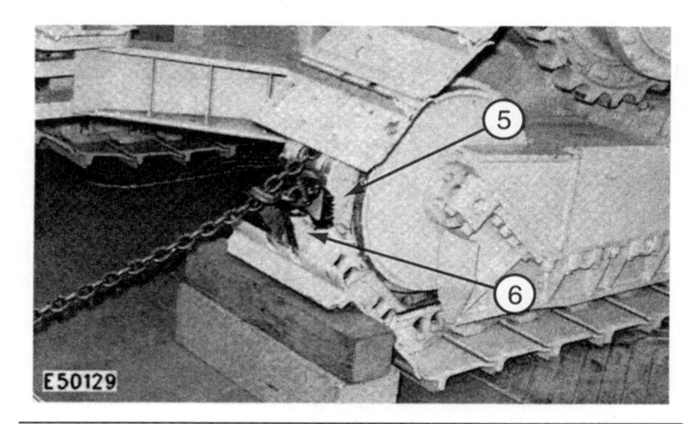

Figure 18-38 Correct procedure used to align the master link. (*Courtesy of Caterpillar*)

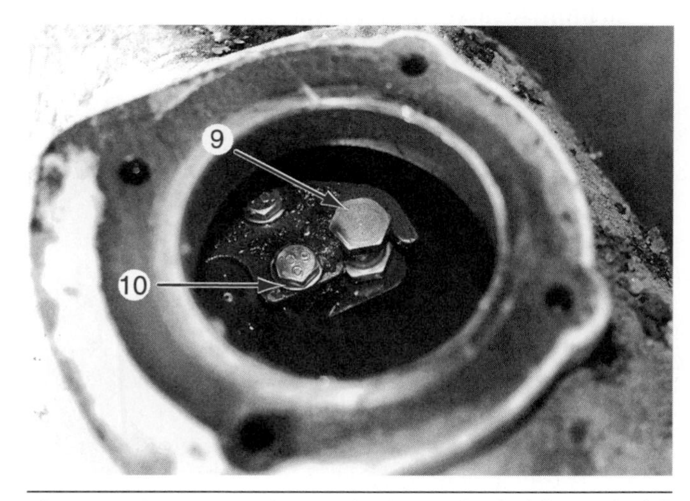

Figure 18-39 Location of the relief (9) and fill valves (10). (*Courtesy of Caterpillar*)

be repeated once the equipment operates at a worksite long enough to be sure that the track has a minimal amount of packing. This ensures that the track is properly adjusted. Replace the protective cover and O-ring on the track frame and torque to specifications.

Track Disassembly

In some cases, it may become necessary to remove and replace a track link component without removing the track from the equipment. The following procedure outlines these techniques in general and is only used to clarify the process. When performing this type of repair, always consult the manufacturer's shop manual for the particular type of equipment that is being repaired. To perform this work, it is necessary to work with hydraulic pressures that can cause personal injury if proper procedures are not followed. Never inspect hydraulic lines for leaks with your hands; a small pin hole could easily cause serious injury.

1. To replace a track component, position the blocking (labeled as 1) between the front idler and the carrier roller, as shown in **Figure 18-40.** Completely loosen the track by relieving all hydraulic pressure at the track adjustment relief valve located in the track frame. High pressure may be present in the adjusting cylinder in the form of high grease pressure. Never expose yourself to the possibility of injury by visually inspecting to see if the grease is released.
2. Use a suitable hoist to pull the weight of the track section to be worked on, as shown in **Figure 18-40,** and use the appropriate blocking (**Figure 18-41**) to support the weight of the track. Under no circumstance should anyone work under a suspended load; the track is heavy and, if it were to fall, it could cause severe injury or even the loss of a limb. Remove the carrier roller fastening bolts (3) from the carrier track roller frame (2), and then remove the bolts and bracket (4) that secure it to the track frame.
3. Place the appropriate blocks to support the track when it is disassembled. Place two additional blocks under the section of the track that is to be worked on. Remove the three track shoes that are located over the pins (4 and 5) that are to be removed. Because of the weight of the special hydraulic tool required to remove the pins, a hoist attached to the Tooling (A) is necessary to hold it in the correct alignment, as shown in **Figure 18-42.**

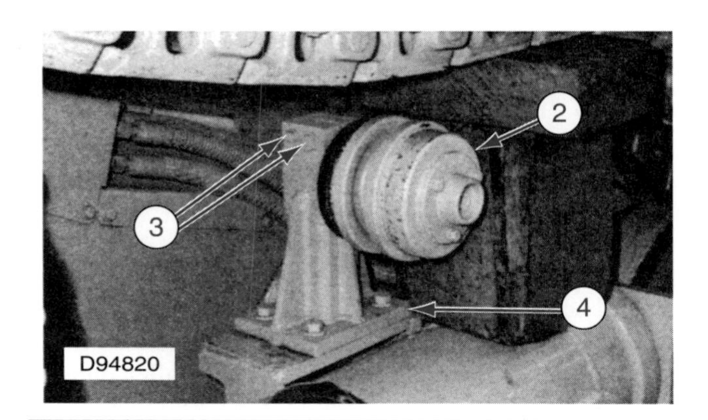

Figure 18-41 Procedure for removing the carrier roller. (*Courtesy of Caterpillar*)

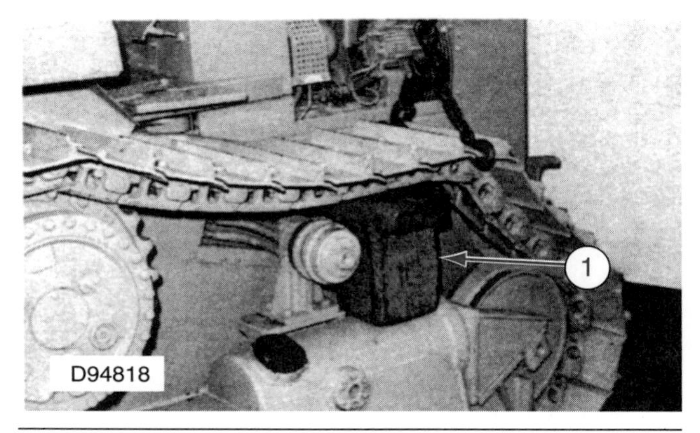

Figure 18-40 Procedure for blocking the track to remove the carrier roller. (*Courtesy of Caterpillar*)

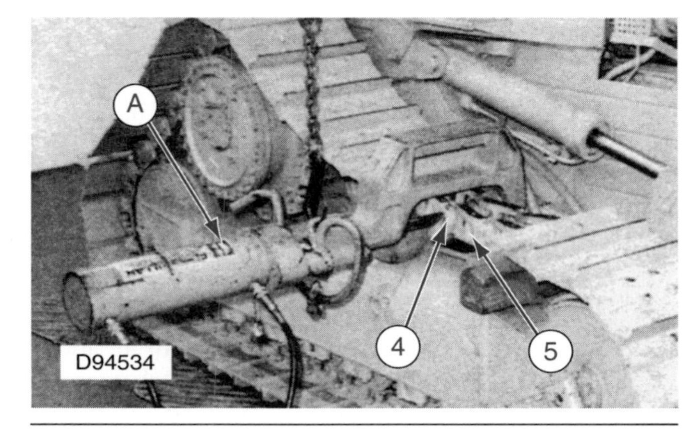

Figure 18-42 Procedure used to support the track before disassembly. (*Courtesy of Caterpillar*)

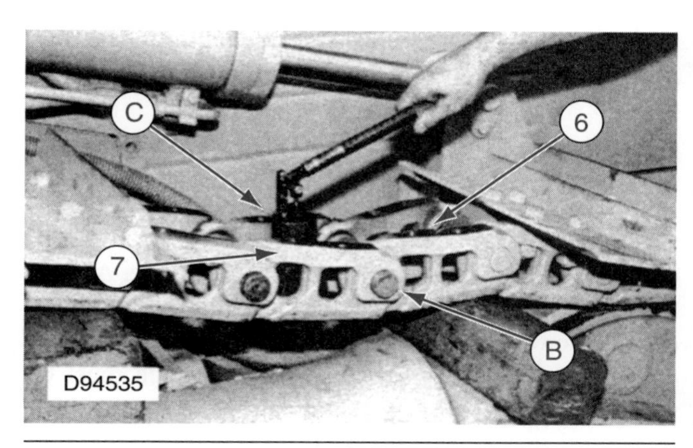

Figure 18-43 Procedure used to remove a link with a special hydraulic jack. (*Courtesy of Caterpillar*)

Table 18-7: TOOLING FOR TRACK REPLACEMENT

Disassembly Tools	A	B	C	D
9U-6600 Hand hydraulic pump	1			1
5P-2380 Track tool	1			1
8S-9978 Ram adapter	1			
5P- 2188 Adapter	1			
5F-9447 Pilot pin		2		
Felco Hydraulic jack (20-ton)			1	
8S-9975 Ram adapter				1
7S-8469 Bushing adapter				1
2S-8235 Bushing pusher				1

4. Use the specialized hydraulic jack, as shown in **Figure 18-43,** to spread the link assembly so it can be removed from the track. Be careful not to damage the seal faces with the jack when forcing them apart. Use the tool to push the pin from the track and install tool B pilot pin (refer to **Table 18-7**) to keep the track together until the second pin is removed. Place the hydraulic removal tool against the second pin and remove it; install the second pilot pin. Remove the seals and rings and clean up the link assembly. Use the bushing adapter with the ram adapter to remove the bushings.

Drive Sprocket Removal and Installation

The following is an outline of the procedure used to remove and replace a drive sprocket. It is only to be used as an example. Always refer to the specific manufacturer's technical documentation when replacing the segments on track equipment. Manufacturers of modern track equipment generally use one of two different sprockets designs, the older being a one-piece unit that is bolted on to the final drive and the newer being a segment-type or multi-piece design.

The newer (segment-type) is the easiest to replace since all that is required is to place the segment that needs to be replaced so that it does not contact any of the track links and unbolt it from the final drive. Before installing a new sprocket segment, be sure to wash and remove any paint or debris on the parts that may interfere with good contact between the mating parts. Locate the segment into position on the final drive, as shown in **Figure 18-44,** and install the lubricated bolts so that the head of each bolt is against the final drive. Torque the nuts as outlined in the manufacturer's technical documentation.

Figure 18-44 Segment-type sprocket in the correct position for removal and installation. (*Courtesy of Caterpillar*)

On equipment that uses a one-piece segment, the track must be disassembled before the sprocket can be replaced. Follow the track disassembly procedures outlined in the manufacturer's technical documentation to split the track at the sprocket before attempting to replace the sprocket.

Track Roller Removal and Installation

The following is an outline of the procedure used to remove and replace a track roller. It is only intended as an example. Always refer to the specific manufacturer's technical documentation when replacing the track rollers on track equipment. Before any track roller can be replaced the track must be loosened by following the manufacturer's recommended procedure as outlined in the section on track disassembly. **Table 18-8** provides an example of tools that are required in this example and may differ from the tools actually required for a specific type of track system. Use the electric hydraulic jack to lift one side of the equipment off the ground and block by using the approved stands. Place the lifting fork under the track roller to be removed and secure with a hoist, as shown in **Figure 18-45.**

Next, remove the bolts (3) that secure the caps (1) on the roller to the track frame. Remove the track roller (2) from the track assembly. *Note*: The weight of the roller may vary somewhat (50 to 120 lbs), but they are not light and will require care when removing them.

Reinstalling the track roller requires that you follow the procedure outlined in reverse order. Always follow the manufacturer's technical documentation when performing this type of repair procedure on equipment.

Table 18-8: TOOLS FOR TRACK ROLLER REMOVAL

Required Tools for Track Roller Removal	A	B	C
8S-7610 Base	2		
8S-7650 Cylinder	2		
3S-6224 Electric hydraulic pump or jack	2		
8S-7640 Stand		2	
8S-7611 Tube		2	
8S-7615 Pin		2	
FT-1309 Lifting fork			1

Idler Removal and Installation

The following procedure is a summary of the steps used to remove and replace an idler assembly. This procedure is only intended to be used an example. Always refer to the specific manufacturer's technical documentation when replacing the idler on track equipment.

Before an idler can be replaced, the track must be separated by following the manufacturer's recommended procedure as outlined in the technical documentation on track separation. The idler on dozers is generally too heavy to handle by hand; a suitable hoist should be used to secure the idler, as shown in **Figure 18-46,** to safely hold it in place before attempting to remove the mounting hardware. Once the idler is secure, remove the bolts and caps. Carefully remove the shim pack (4) and identify it for correct replacement. Move the retaining blocks (5) toward the roller frame. Use the hoist to lift the idler from its mounting position.

To perform an overhaul to the idler, refer to the manufacturer's technical documentation for the detailed procedure and the tools required to perform this type of work.

To reinstall the idler, place the blocks in position on the dowels on the front roller frame. Use a hoist to lower the idler into position. Reinstall the correct shim pack between the front roller frame and the blocks, as shown in Figure 18-46. Carefully align the holes (6) in the idler shaft with the pin in the caps. Reinstall the mounting caps and bolts and torque up to specifications. Follow the manufacturer's recommended procedure to reconnect the track and make adjustments according to specifications.

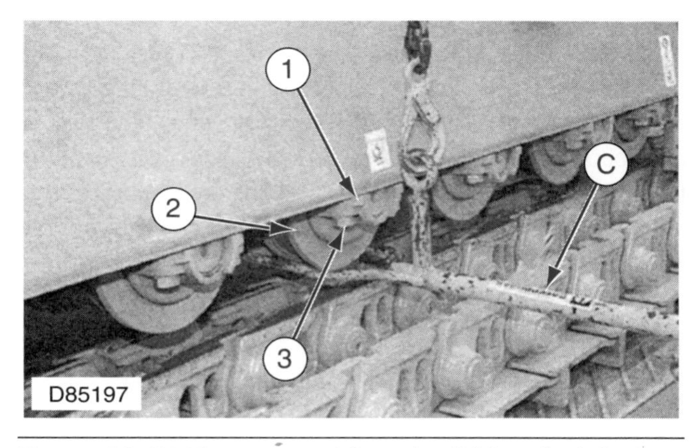

Figure 18-45 How to position the lifting fork before attempting to remove and install a track roller. (*Courtesy of Caterpillar*)

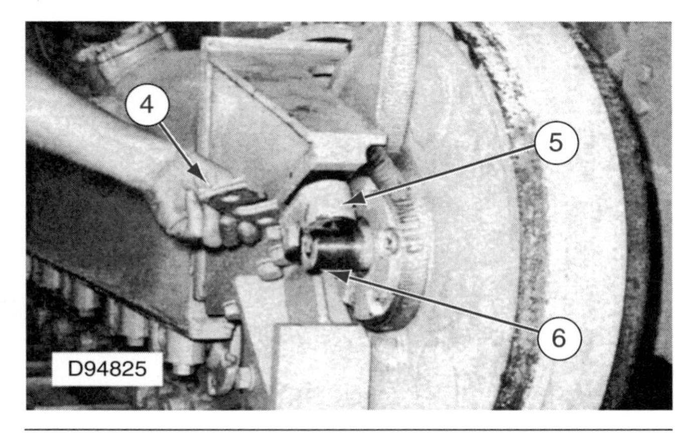

Figure 18-46 Shim removal process. Identify the shim packs for installation. (*Courtesy of Caterpillar*)

Track Component Inspection

Once track components have been removed from the track, it is important to examine the condition of the parts to determine if they are reusable. If the equipment has been operating in an abrasive environment, the links can become worn out, resulting in stretching and cracking. This section outlines the inspection procedures used to determine the usability of the components in question. As always, refer to the manufacturer's technical documentation for the recommended procedures for inspecting the equipment being repaired. It cannot be overstated how important it is to assess the usability of the components since the cost of replacing all the components would lead to excessive costs. Technicians should always adhere to the manufacturer's guidelines when it comes to reusability of parts. Replace only the parts that are worn or damaged beyond use. Before any of the track parts can be properly inspected, they must be cleaned.

When new links or other track components are used for the repair, always be sure to remove the paint around the bores of a link or other components that have mating surfaces, such as track shoes and fasteners. A layer of paint prevents a proper fit and interferes with good shoe retention. An angle grinder with a wire brush or an abrasive wheel will clean up any paint in these areas quickly. Caution should be exercised whenever a high-speed grinder is used. To prevent the possibility of injury, always wear a face shield and be sure no other personnel are in the immediate area when performing this operation.

Figure 18-47 Area on a link that is susceptible to cracking.

Figure 18-48 Crack in the pin boss.

Track Link Assessment

Links can develop cracks over time; these cracks can show up near the bushing strap, in which case the links are not to be reused. There are numerous spray compounds that can be used to detect cracks in the track components. However, if none are available in the field, spray paint can be used to detect any cracks that are not easily visible. One area that needs to be examined carefully is the link strut area. Links that have developed cracks near the strut, as shown in **Figure 18-47,** should not be reused. Failure to detect cracks in this area could lead to a catastrophic track failure during operation, resulting in extensive damage to the track and undercarriage.

The toenail sections, as well as the pin boss area of the link, are other areas that are prone to developing cracks. **Figures 18-48** shows the area in question that needs to be carefully examined for cracks. Over time the bushing bores tend to become oval shaped; this results in excessive pin-to-bushing clearances and affects the sealing capabilities of the fit. When examining the condition of the pin, if you notice that the surface is rusted all the way across the bore, it can safely be assumed that the press fit is no longer present and the link should be discarded. Since the bushing bore does not do any sealing, slight scoring is acceptable as long as a press fit is still attainable.

Often the link rail surfaces (**Figure 18-49**) will have pieces break off the edges; this is known as **spalling** and does not necessarily mean that the link needs to be discarded. If the link has less than one-third of the rail surface broken off, it can be reused safely; if the spalling is significantly deep and located in the bushing area, the link should be discarded.

Figure 18-49 Link with minor rail spalling. This link can be reused.

Figure 18-50 Link with rolled metal up over the pin boss. It may be reused but must be ground off. (*Courtesy of Caterpillar*)

Link counterbore damage usually results from a seal failure and, depending on the extent of damage, the link can be reused. Often a lightly **pitted** or corroded link counterbore can be cleaned up enough to reuse. An exception to this is when the equipment has been allowed to operate with a **dry joint.** The counterbore sidewall quickly becomes damaged from contact with the bushing if allowed to run dry. The track seal load ring is supported by the counterbore sidewall; if excessive wear is present in this area, the load ring will fail in a short time. Often links with minor damage to the counterbore sidewalls can be reused as long as the counterbore height is greater than the load ring height when installed.

A link counterbore with thrust ring wear can be reused as long as the depths of the wear do not exceed a specified amount. Refer to the manufacturer's specifications for this information as it may vary. Often track roller flanges will cause metal on the sides of the links to be rolled over the track pin, as shown in **Figure 18-50.** This must be removed with a grinder before attempting to remove the pins. Occasionally, the roller flanges can cause metal to roll over the surface of the link where the track press tool sits for assembly. This metal must be removed before assembly is attempted.

Figure 18-51 Link bore with major scoring that extends the full width of the bore. It should not be reused. (*Courtesy of Caterpillar*)

Track Pin Assessment

Sometimes during disassembly, small imperfections in the pin ends can cause minor scoring in the link pin bores. As long as the scoring does not extend the full width of the bushing bore, the link can be reused. If it extends the full width of the bore, a path for lubricant to leak out is created and the link should not be reused. In **Figure 18-51** the link bore is scored from one end to the other and would provide a leak path for the lubrication. It should not be reused.

Track pin damage comes in a variety of forms and some knowledge is required to be able to access the usability of the used pins. Pin damage, as shown in **Figure 18-52,** can generally be categorized into three types: spalling, galling, and step wear. Because the pins are heat treated, spalling takes the form of flaking or breaking off of metal in small pieces. This condition can weaken the pin significantly, and most manufacturers recommend that the pins not be reused if they show signs of significant spalling.

Figure 18-52 Track pin with minor galling damage. This pin may be reused.

Figure 18-53 Track pin with a 10-degree chamfer ground on the end. This pin may be reused.

Whenever making determinations on the reusability of track pins or other track components, always refer to the manufacturer's technical documentation for the exact parameters. Parameters differ significantly on some types of equipment; larger equipment often has less tolerance for wear in areas like track pins. Metal transfer on pins is caused by the slight movement of the pin in the bore during operation. If you encounter pins with this type of damage, it is best not to attempt to reuse them since the metal that is fused to the pin will damage the link bore and cause scoring or broaching as the pin is pressed into the link bore.

Pins that develop some form of **wear step** damage generally come from the wet joint design. Minor wear under .005" can be tolerated for reuse in most cases, but checking the specific manufacturers recommendations is always recommended to determine wear tolerances. In the case of pin end damage, it is generally permitted to grind a 10-degree chamfer on the end of the pin, as shown in **Figure 18-53,** to remove any burrs caused by contact with the roller flanges. If this rough edge is not removed properly, most likely it will lead to broaching or scoring to the link bore.

Some equipment manufacturers use positive pins retention (PPR). These pins have a small groove on the end that accepts a retainer ring. With this pin design you cannot chamfer the ends; it would weaken the retention ring groove. These pins can have some damage to the ends, but no more than 25% of the groove should be damaged or broken.

Bushings and Counterbore Assessment

Cracks that develop in the bushing area of the link (**Figure 18-54**) need to be identified. A crack in this area will allow the lubricant to leak out of the joint, resulting in a dry joint before very long. It is important to perform a thorough inspection to ensure that no links with cracks in this area are reused.

On sealed tracks it is important to ensure that the area where the track seal fits in the link counterbore is free from any excessive wear grooves. These could affect the integrity of the seal fit. Bushing groove gauges are available from the manufacturers to measure

Figure 18-54 Track link with a small crack in the bushing area. This link should not be reused.

acceptable wear on the ends. A specialized gauge, as shown in **Figure 18-55,** is used on the ends to check whether or not the seal groove is deep enough to be visible with the gauge in place. If it is, then the bushing should be discarded.

If a pin shows signs of galling, then the bushings that held that pin will most likely have the same problem. A lack of lubrication on mating surfaces causes galling, which generally occurs at the same time on both mating surfaces. On certain types of equipment, slightly **galled** bushings can be reused. As always, refer to the manufacturer's recommended usability procedures to determine if the parts are to be reused. On some equipment if galling does not occur for more than one-half inch laterally or for more than 90 degrees of circumference, it may be reused.

If the bushings develop a slight movement in the link bore, metal transfer will occur. Metal will transfer from the link bushing **I.D.** (inside diameter) because the pin is made from a harder material than the bushing. Pins that develop metal transfer on them should not be reused; the pin-to-pin bore must seal in the oil on wet systems. This does not matter on the bushing-to-bushing bore fit as this connection does not have to seal oil in. Some metal transfer is considered acceptable on bushings for reuse by some manufacturers. Check with the specific manufacturers replacement recommendations for the particular type of equipment being worked on.

Thrust rings are used on sealed track systems to provide the correct amount of seal compression to the track pin assembly; they must not be reused if any cracks or chips are present, as shown in **Figure 18-56.** A slight amount of wear can be tolerated on either side of the thrust rings. When assessing the reusability of the rings, if you can still see the lubrication slots on both sides of the thrust rings, then they are generally reusable. If galling is present on the thrust faces that goes all the way around on either side, the rings should not be reused.

Some manufacturers use a special track system. This system differs from conventional track systems in that the pin assembly comes preassembled. The pin, bushings, and seals are supplied as a complete unit, which is assembled by the manufacturer and cannot be disassembled for inspection. The track is replaced as a complete unit. In most cases, if there is minor broaching on the link bore, it is generally reusable. The track link needs to be inspected by following the procedure used for conventional sealed tracks.

Rotating Pins and Bushing

On certain track systems it may be desirable to rotate the track pins and bushings, as shown in **Figure 18-57,** once a nominal track pitch dimension has been exceeded. Because the track pins and bushings wear in a predictable location, they can be rotated 180 degrees to bring track pitch back to acceptable dimensions. This procedure requires the use of a specialized track press, which can be portable or can require that the track be taken to a shop that has the equipment. Note that this procedure may not always be economical; the cost of

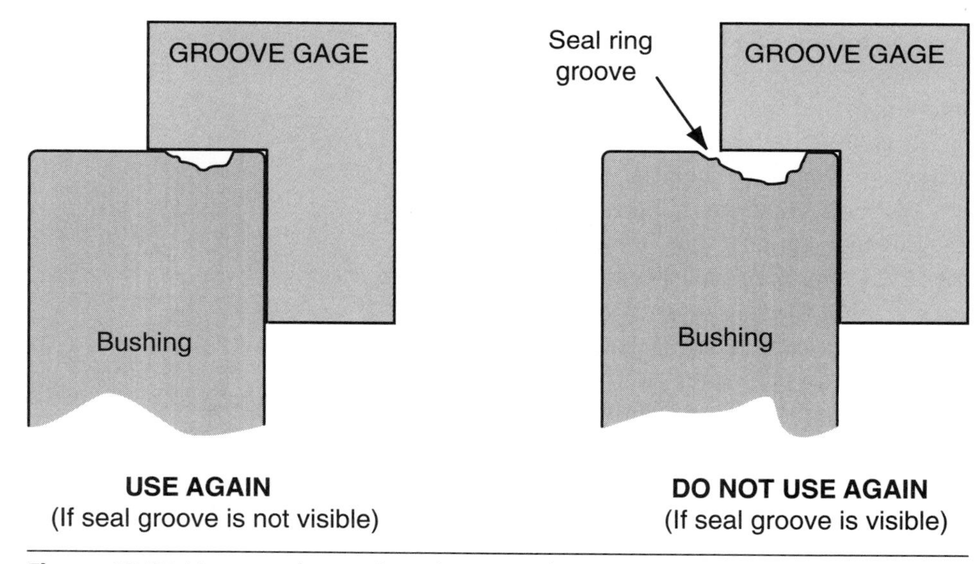

Figure 18-55 How to determine the wear limits on bushing grooves using a manufacturers-supplied groove gauge.

Figure 18-56 Thrust ring with a crack through it. Thrust rings with cracks should never be reused. (*Courtesy of Caterpillar*)

labor to perform this work may exceed the cost of replacing the pins and bushing with new ones.

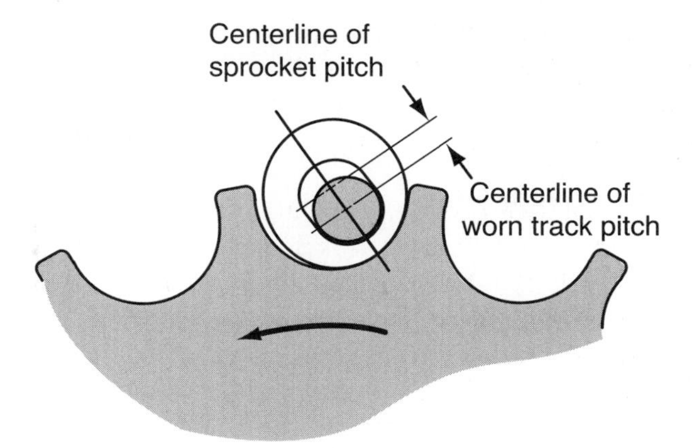

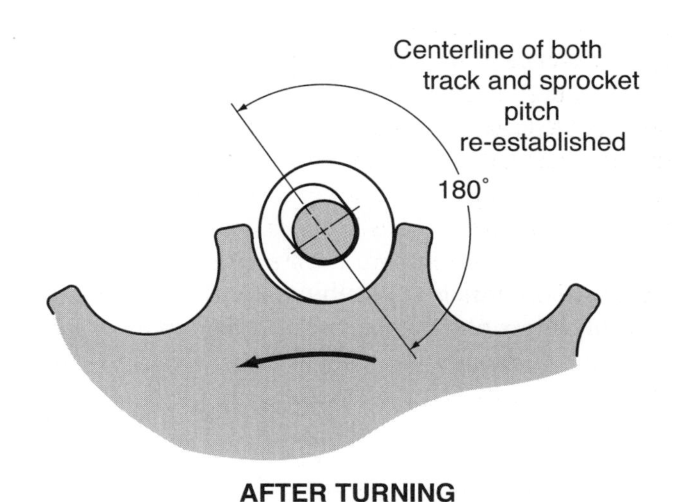

Figure 18-57 How track link pitch can be restored by rotating the pins and bushing 180 degrees.

GENERAL GUIDE TO TROUBLE-SHOOTING TRACK PROBLEMS

Table 18-9 provides a quick reference guide to troubleshooting track problems:

ONLINE TASKS

1. Use a suitable Internet search engine to research the configuration of a typical off-road track loader undercarriage. Select a piece of equipment in your particular location, paying close attention to the track design and arrangement, and identify whether it has an additional gear reduction at the drive sprocket. Outline the reasoning behind the manufacturer's choice of this particular design for a particular application.
2. Check out this url: *http://www.tpub.com/content/engine/14081/css/14081_229.htm* for more information on drive track configuration and design.

SHOP TASKS

1. Select a piece of equipment in your shop that has a drive track assembly and identify the type of components that make up the drive track major component assembly. Identify whether the track pins have excessive wear and what type of track pin bushing design is used on this arrangement.
2. Select a piece of off-road equipment in your shop that uses a drive track and identify the type and model of the equipment. List all of the major components and the corresponding part numbers of the track assembly. Identify and use the correct parts book for this task.

Table 18-9: TROUBLESHOOTING TRACK PROBLEMS

Problem	Possible Cause
Accelerated wear on one side of the final drive sprocket teeth	Track misalignment Loose track Sprocket misaligned due to improper installation
Accelerated wear on one side of the idler(s) or rollers	Track misalignment Idler misaligned due to improper installation Track adjusted improperly or too loose
Accelerated wear to track pins, bushings, and track links	Track adjustment too tight Misaligned track Excessively worn sprocket teeth Excessive high-speed operation
Excessive wear to the final drive bearings	Track adjusted too tight Misaligned track
Operator complains of loss of drawbar power	Improper track adjustment Track adjusted too tight
Operator complains that the dozer drifts either to the right or left while being driven in a straight direction	Improper track adjustment One track is tighter than the other Track misalignment
Frequent packing of the track during operation	Improper track adjustment Track adjusted too loose
The track is noisy during operation	Improper track adjustment Track adjusted too loose
The track whips excessively during operation	Improper track adjustment Track adjusted too loose Idler seized in the retracted position
The track is thrown off during operation	Improper track adjustment Track adjusted too loose

Summary

- Heavy equipment operating in a quarry, mine, or forestry application uses specialized tracks and undercarriages to meet the extreme-duty operating cycles associated with this type of environment.
- Track and undercarriage systems can be categorized into four distinct designs with a number of variations in each category. These designs are the standard track system low ground pressure (LGP) system, extra long (XL) (front) track system, and extra long (XR) (rear) track system.
- The chain section of the track is made up of track links, pins, and bushings. A number of these, interconnecting with each other, form the track.
- There are several components that combine together to form a track system. The drive sprocket, front idler, track rollers, tension mechanism, roller guards, and frame are all required to make up a track and undercarriage system.

- Each individual track section is made up of two track links, a hardened pin, and bushing. These individual track sections are interconnected from link to link. The two track links used in each section have provisions for attaching a track shoe, and also provide a rail that helps the track rollers maintain track alignment during operation.
- Each link of the track is held together with a press-fit pin and bushing, which keep each link section in proper alignment with each other link section during operation. This is called a link assembly. Pressed together, each section acts like a hinge, which allows the chain flexibility when rotating on the undercarriage.
- Sealed tracks have a solid pin. Sealed and lubricated tracks have a hollow pin to provide a path for lubricating the pin and bushing of the next track section.
- Track shoes are bolted to the link assembly to provide a means of supporting the equipment and to provide traction and flotation during operation. They are generally constructed from a hardened metal plate and come in many different configurations.
- Track shoes are used to provide a gripping function on a wide variety of surfaces. Many manufacturers make an assortment of track shoes that are available. They are rated by a vertical designation, which indicates the shoe's ability to withstand abrasion, and a horizontal designation, which indicates impact resistance.
- When dozers must be operated in loose, muddy ground conditions, the use of a self-cleaning, low ground pressure shoe is desirable. These shoes should never be used in high impact or abrasive conditions. As the track rotates around the sprocket or idler, they separate slightly allowing the dirt to drop out. This design has very high flotation capability on loose, muddy ground.
- Tracks utilize two different types of track rollers to maintain track alignment. The bottom rollers support the weight of the equipment and ensure that the weight of the equipment is distributed evenly over the bottom of the track.
- Carrier rollers are located above the frame rail. Their purpose is to support the weight of the top section of track as it rolls between the idler and the sprocket.
- Rollers have a hardened surface that is equal in hardness to the track link surfaces, This can have a slightly negative effect on the life of the track links.
- The purpose of an idler(s) is to help guide the track through the track rollers and to support part the weight of the equipment. Besides providing alignment, the idler's job is to maintain the correct tension and slack on the track.
- Depending on the track design, a dozer could have one or two idlers per track section. If the equipment had an elevated sprocket system, there will be two idlers: a front idler and a back idler. Equipment that has a conventional oval track requires only one idler.
- Tracks operate in extremely abrasive conditions and would quickly wear out any pins and bushings if they were not protected with seals. Not only do the seals keep dirt out but they also carry a certain amount of side load, preventing unnecessary wear of the track link's counterbore. Conventional track seals use Bellville washers that are used because they are capable of maintaining constant sealing pressure to keep dirt out as they wear.
- A tensioning mechanism connects to the idler to maintain the correct track tension. The tensioning mechanism uses either a recoil spring with a mechanically adjusted rod or a hydraulic tensioning cylinder to adjust track tension.
- On some dozers, the track frames on each side of the equipment are connected together through an equalizer bar near the front of the machine and are mounted to the sprocket, which is part of the final drive at the rear of the machine. All the weight of the equipment is carried by the track rollers and frame. The mounting on each side can move independently over uneven terrain to maintain contact with the ground.
- Improvement in track design has led to the inception of the elevated track design which has a few advantages over the conventional track sprocket arrangement. The major benefit of this design is that it isolates the final drive from excessive shock loads. The position of the final drive also creates a better balance arrangement, which provides better traction.
- The elevated track design requires the use of two idlers and a separate roller frame for the front and

rear. There is always a downside to any design. Compared to a conventional undercarriage design, which has one travel loaded flex point, the elevated sprocket undercarriage has three flex points where loading takes place during operation.

- On a track during forward operation, there are two type of loads imposed on the parts; these loads occur as the two track sections hinge. When the bushing enters or leaves the sprocket teeth in forward, the bushing rotates slightly as it leaves the sprocket. The opposite occurs in reverse; when the bushing enters the sprocket, it rotates slightly. This movement between the bushing and sprocket is similar to that of gear teeth when they mesh.
- When the bushing enters the sprocket in forward, the first bushing to come in contact with the sprocket takes about 85% of the load. The rest of the load is absorbed by the next three bushing. When there is a mismatch between the sprocket and track pitch during operation, the bushing will slide and turn as it seats in the sprocket, causing wear on the forward drive side.
- An internally lubed track has an advantage over a conventional track; it prevents wear between the pins and bushings during operation.
- When packing gets between the track and sprocket, the bushings sit on top of the packing, causing the tension on the track to increase significantly. This increases the sprocket pitch and causes a mismatch between the track and sprocket pitch.
- When the track pitch is increased due to wear, packing can have a positive effect on the overall track pitch. Packing the track lengthens the sprocket pitch; and when the track wear increases the track pitch and is equal to the sprocket pitch, good contact is maintained. The downside to this scenario is that abrasive wear occurs rapidly from the packed material in the track.
- When operating a dozer in conditions that are less than ideal, track packing can be minimized by choosing a shoe design that allows material to be forced out through holes in shoe. It is probably a good idea not to use roller guards in conditions that promote packing.
- Adjusting the track correctly for the operating conditions will lengthen the track life because a loose track will cause excessive wear to roller flanges and sprocket teeth. If the track tension is too loose, the track roller frame can experience significant damage.
- The rollers on track equipment do not all wear at the same rate equally and if possible, should be rotated to extend the life of the rollers. Rollers in certain positions will wear more on one flange than on another. The flanges on the lead rollers of the track must be kept in relatively good condition because they guide the track onto the sprocket.
- On older track systems the pins and bushings were a constant source of maintenance; in modern sealed and lubricated systems, pin and bushing wear is all but eliminated. Track systems and undercarriages wear whenever they are operated; the severity of the wear depends on the environment they are working in.
- Never use a shoe that is wider than necessary; the added width causes accelerated wear and will lead to structural problems on the track and undercarriage. Track shoes that are wider than necessary will also affect the life of track links, rollers, and the idler flange. The weight of the wider shoes will cause an increase in wear, which increases significantly on rough terrain.
- Track wear can be measured using two different procedures. The use of specialized measuring tools is one method used by some manufacturers. The other equally popular method is the use of specialized wear gauges. Measuring gauges include tape measures, depth gauges, and vernier callipers.
- Before performing any type of wear evaluation or measurements to the tracks, be sure the equipment is properly blocked and tagged with the engine off. Never attempt to perform any measuring of the track components while the equipment is in operation. Some measurements require that the track be tightened before performing any measurements.
- To ensure that the track adjustments have been performed correctly, the equipment must be located in the actual working environment. Do not clean the track to make the adjustments; in some cases, the track will pack slightly and any attempt to clean it may lead to an adjustment that will become too tight when the equipment is back in operation.

- Correct track adjustment is necessary to maintain long track life with low operating costs. Nothing shortens the life of a track faster than incorrect adjustments. If the track is adjusted too tight, excessive loads are placed on the mating components, which accelerates track component wear. If the track is too loose, it will cause track whip at higher speeds and lead to excessive wear of undercarriage components.
- Never attempt to perform any track repairs to equipment and manufacturer's recommended procedures on hand to perform the work safely.

Review Questions

1. On a track loader, which of the following would likely cause excessive wear on one side of the track idlers and rollers?

   A. track misalignment
   B. track too loose
   C. track too tight
   D. excessive pin and bushing wear

2. On a track loader, which one of the following would likely cause the track to jump the sprocket?

   A. track too tight
   B. track too loose
   C. track misalignment
   D. excessive track link wear

3. On a track loader, which one of the following would likely cause excessive wear to the final drive bearing?

   A. track misalignment
   B. track too loose
   C. track too tight
   D. excessive pin and bushing wear

4. Which of the following defines the term Hunting Tooth design when referring to track equipment?

   A. Every other sprocket tooth engages a different track bushing during operation.
   B. Every sprocket tooth engages the same track bushing during operation.
   C. Every other sprocket tooth engages the same track bushing during operation.
   D. Sprockets have an even number of teeth to match an even number of links.

5. When adjusting track sag, which of the following dimensions is considered normal?

   A. 3.5 inches
   B. 1.5 inches
   C. .25 inches
   D. 4.5 inches

6. When pins and bushings wear on tracks, how many degrees do you rotate the pins and bushings to restore track pitch?

   A. 120 degrees
   B. 90 degrees
   C. 180 degrees
   D. 360 degrees

7. What takes place when the track pins and bushings wear?

   A. Track pitch increases
   B. Track pitch decreases
   C. Track tension increases
   D. Track pitch remains constant

8. Which of the following is an accepted method of measuring pin and bushing wear?

   A. Use a tape measure to measure the distance across several links from one pin edge to the same edge on another pin, then divide that number by the number of links.

   B. Use a tape measure to measure the distance across two links from one pin edge to the same edge on the other pin, then divide that number by the number of links.

   C. Check the wear pattern on the sprocket teeth to determine excessive pin and bushing wear.

   D. Measure the overall track length to determine pin and bushing wear.

9. When pins and bushings wear out on a track, what happens to the wear line on the mating sprocket?

   A. wear line will be higher

   B. wear line will be lower

   C. wear line will be the same

   D. track tension can compensate for pin wear

10. On a track loader, what is the purpose of a split master link?

    A. required to compensate for track wear

    B. a two-piece link used to separate the track

    C. used to identify the location of the master pin

    D. can be adjusted to compensate for track wear

CHAPTER

19 Suspension Systems

Learning Objectives

After reading this chapter, you should be able to

- Describe the fundamentals of off-road suspension systems.
- Identify the components that make up off-road suspension systems.
- Describe the principles of operation for off-road suspension systems.
- Outline the maintenance and repair procedures associated with off-road suspension systems.
- Identify the basic troubleshooting procedures for off-road suspension systems.

Key Terms

cross tube
cylinder pressure
drive bushing
equalizing beam suspension
jounce
leaf spring
leaf-spring system
load cushion
nitrogen-charged accumulators
oil/pneumatic cylinder
oscillating hitch
rebound
saddle assembly
solid rubber suspension
spring hangers
spring packs
suspension cylinder
torque rods
U-bolt
un-sprung weight

INTRODUCTION

The primary purpose of an off-road suspension systems (**Figure 19-1**) is to support the total weight of the equipment. If no suspension system were used, the shock loads encountered during operation would be transmitted directly to the equipments frame. These jolts would soon cause metal fatigue, resulting in cracks in the frame structure. Equally important is the comfort of the operator, who in many cases must remain at the controls for hours on end. Modern suspension systems play a major role in providing stability to the equipment during operation on rough terrain, as well as absorbing shock loads and providing a comfortable ride for the operator. A suspension system that performs well when both loaded and unloaded generally has to compromise some performance at both ends of the spectrum. A well-designed suspension system will perform well with or without a load.

Off-road suspension systems come in a wide variety of designs, some can be as simple as those found on many types of slow moving LHDs (load haul and dump equipment), which consist of drive axles that are mounted directly to the frame with only the tires to absorb ground impact. Conventional **leaf-spring systems** are still quite common on many haulage truck applications in pits and quarries. Complex systems may incorporate nitrogen-filled shocks that are monitored

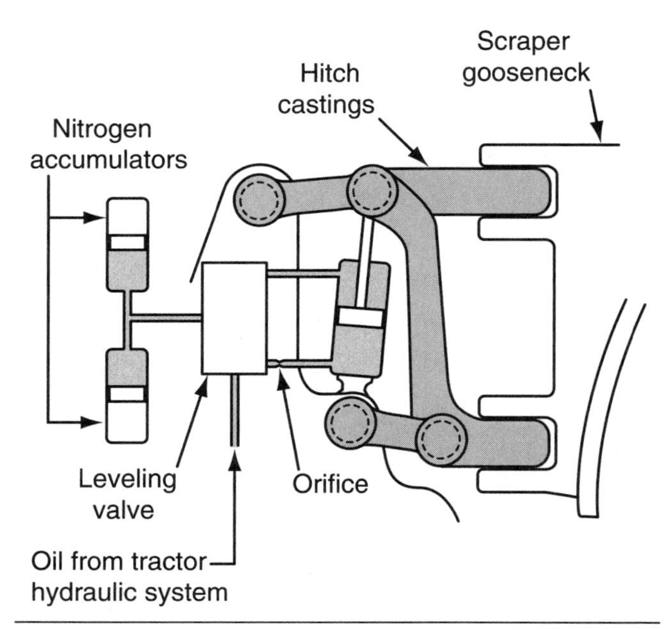

Figure 19-1 Caterpillar cushion hitch suspension system used on a 637E Scraper.

and regulated by the operator, or an onboard ECM (Electronic Control Module) that monitors and controls the operation of the suspension system.

Terminology associated with heavy-duty suspension system design and performance is expressed in commonly used descriptive terms such as, sprung and un-sprung weight, jounce, **rebound** and oscillation. **Un-sprung weight** is described as the total weight of the axle plus all the components that are mounted to it, including all the brake hardware and suspension components. Sprung weight is defined as the load that the suspension must support during normal operation. The term **jounce** is used to identify the spring position in its maximum compressed travel position. Some manufacturers refer to the rubber blocks used to prevent axle to frame contact as "jounce blocks." The term **rebound** describes what happens to the energy that is stored in a spring; it kicks back. On many types of equipment, jounce and rebound are controlled with dampening devices such as shock absorbers and springs.

Suspension work on off-road equipment often involves the use of heavy lifting devices because the suspension system components are usually heavy. When working with **suspension cylinders,** be mindful of the fact that the cylinder may be under pressure and can move suddenly without warning. Never place yourself in a position that could cause you to become pinned between a tire and the frame by the sudden movement of a suspension component.

When working on suspension systems that use gas/suspension cylinders, be sure that all the nitrogen in the suspension cylinder has been released before checking the oil level. Never remove any plugs or valves from a suspension cylinder if the rod is not fully retracted. Make sure all the nitrogen pressure is dissipated. Under no circumstances should you work under a suspended load without using approved blocking and lifting procedures. If a jack or hoist is to be used, be sure that it has the capacity to lift the load. When working around heavy equipment always wear the required personal protective equipment based on job conditions.

Tech Tip: The equipment that you work on will often be large enough that you will be required to climb up to a location where the work will actually be performed. Always use the three-point contact rule when mounting or dismounting the equipment: Maintain contact with two hands and one foot, or two feet and one hand. Never attempt to mount or dismount moving equipment. Do not carry tools with you when mounting or dismounting the equipment. Use a hand line or other safe means to pull the tools and components up to where you are working.

Use only approved lifting devices when lifting components, and always follow approved lifting procedures. Even when using the correct sling or chain to lift, you may still run into a problem if the proper lifting techniques are not followed. **Figure 19-2** shows what happens to the load capacity of a chain or sling when it is used at different angles.

- Figure 19-2A shows the load rating capacity of the chain in a vertical position. Load capacity is 100% of its workload rating.
- Figure 19-2B shows the load rating capacity of the chain at a 60° angle from vertical. Load capacity is reduced to 86% of its workload rating.
- Figure 19-2C shows the load rating capacity of the chain at a 45° angle from vertical. Load capacity is reduced to 70% of its workload rating.
- Figure 19-2D shows the load rating capacity of the chain at a 30° angle from vertical. Load capacity is reduced to 50% of its workload rating.

Always exercise caution when working with hydraulic suspension systems that contain high pressures. A technician is often required to locate hydraulic leaks on these systems. When checking for hydraulic

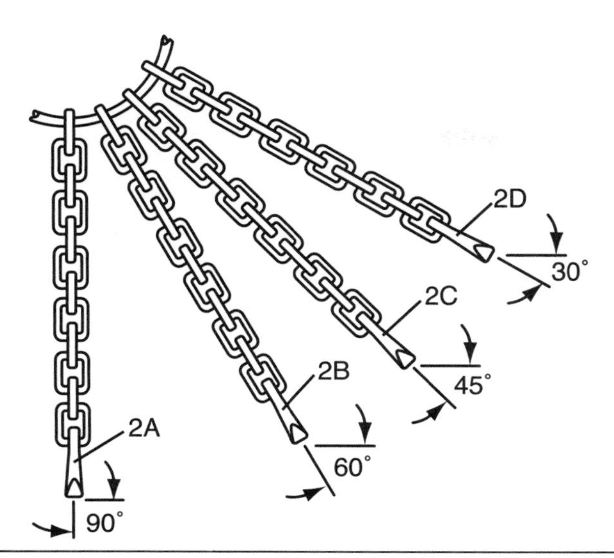

Figure 19-2 Lifting devices used at different angles will affect the load lifting capabilities. Refer to load lifting charts to determine the correct slings and angles.

leaks, never run your hands over the lines and components. Any high-pressure leaks can penetrate the skin leading to serious injury. If oil penetrates the skin, immediate medical attention is mandatory. Operating temperatures on hydraulic suspension systems can often be over 212°F (100°C).

Appropriate protective equipment must be used when handling hot oil or components. Hydraulic tanks are often pressurized to prevent dirt contamination and require special procedures to check the oil level safely. On equipment that uses high-pressure nitrogen in the suspension systems use of a mixture of soap and water sprayed around any potential leaks is the only safe way to locate the problem area. Highly compressed nitrogen has the potential to cause serious injury to exposed skin.

FUNDAMENTALS

Suspension systems on off-road equipment are designed to absorb road and load impacts by preventing them from reaching the frame. Off-road haulage trucks often use independent suspension cylinders that are capable of independent rebound rates, as shown in **Figure 19-3.** This allows the load stresses to be absorbed by the suspension instead of the frame. Off-road equipment operating on rough terrain benefits from independent suspension systems that allow the axles to oscillate from side to side and absorb torsional and bending forces.

Figure 19-3 How the suspension cylinder is incorporated into the front suspension in a typical haulage truck. (*Courtesy of Caterpillar*)

Since distinct categories of suspension systems are in use on off-road equipment: independent oil/pneumatic suspension systems used on haulage equipment, and solid axle suspension systems used on LHD equipment. These systems will be discussed in depth in this chapter. Although designs differ from manufacturer to manufacturer, the basic operating principles apply to each design.

Oil/Pneumatic Suspension Systems

Equipment that is used for haulage operation in pits and quarries often uses independent oil/pneumatic suspension systems. These systems on newer haulage trucks are often complex and require a complete understanding of both hydraulics and electrical/electronics fundamentals to service. These large trucks use a suspension cylinder at each front wheel end (instead of a conventional kingpin) as part of the steering linkage. These **oil/pneumatic cylinders** act as shock absorbers in the suspension. The front suspension cylinder barrels are attached to the frame of the equipment at the top and are fastened to the steering knuckle at the bottom, as shown in **Figure 19-4.** When the equipment is in

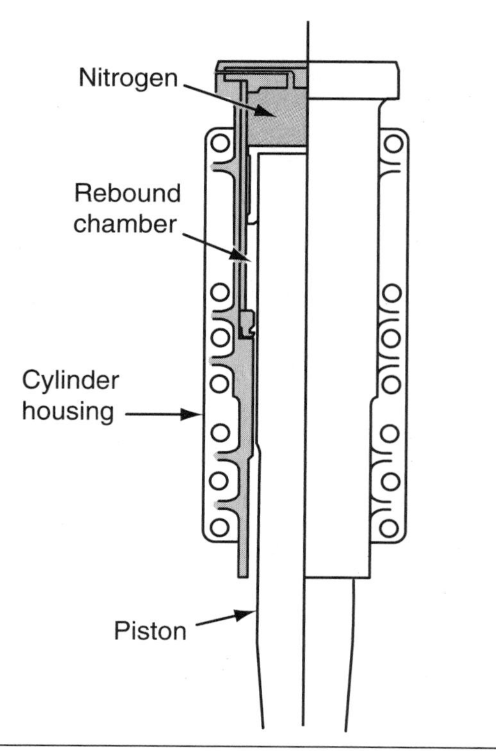

Figure 19-4 Cutaway of a suspension cylinder in a typical haulage truck.

motion, the front wheels can move up and down with any changes in the roadway.

The front suspension cylinders contain nitrogen gas in chamber 2 (**Figure 19-5**), which reacts to the wheel movement during operation. As the wheel moves up, the gas charge in chamber 2 is compressed, forcing oil to flow to chamber 6 through orifice 4 and ball check 5. When the front suspension cylinder moves down during operation (**Figure 19-6**), the compressed nitrogen gas forces the rod to move down. The oil flows from cavity 6 through orifice 4 to chamber 2. At the same time, the increase in pressure forces check ball 5 to close off the port, forcing all the oil to flow through orifice 4 on its way to chamber 2. Orifice 4 is slowly closed off as the rod extends downward to prevent the rod from bottoming out in the cylinder.

The rear suspension is similar to the front in that it also uses oil/pneumatic suspension cylinders. The cylinder barrels are connected to the equipment frame at the top with the stem end of the cylinder mounted to a ball joint at the front of the rear axle housing. This allows the suspension cylinder to pivot on the differential housing during operation. This is completely opposite of the connection on the front of the equipment. The rear suspension cylinders must support the entire weight of the rear of the haulage truck, including the load.

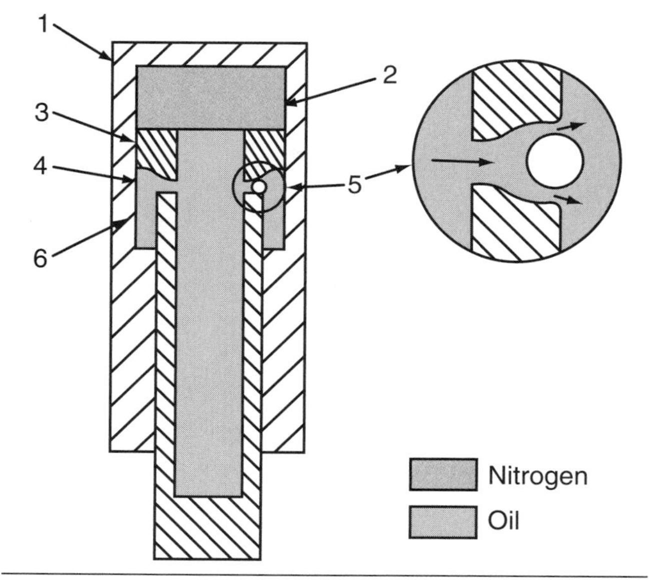

Figure 19-5 Illustration shows a cutaway of a front suspension cylinder with the rod moving up. 1. Cylinder, 2. Nitrogen chamber, 3. Cylinder rod, 4. Orifices (2), 5. Ball check. (*Courtesy of Caterpillar*)

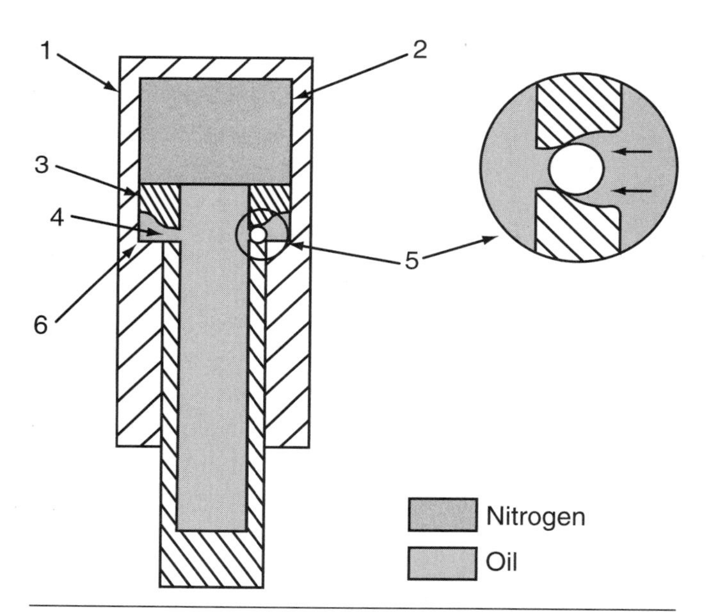

Figure 19-6 Cutaway of a front suspension cylinder with the rod moving down: 1. Cylinder, 2. Nitrogen chamber, 3. Cylinder rod, 4. Orifices (2), 5. Ball check. (*Courtesy of Caterpillar*)

During operation, the rod is able to react to terrain changes and move up and down in the housing. The compressed nitrogen in cavity 2 provides a cushion during operation, while the ability to absorb shock loads is controlled by the orifice 5. Check ball 6 controls the flow rate of the oil between chambers 2 and 4. When one of the wheel ends of the rear axle moves up during operation, the cylinder housing is forced to move up on the rod (**Figure 19-7**). As a result of this motion, the nitrogen pressure in cavity 2 increases rapidly to provide a cushioning action. This increase in nitrogen pressure also forces the oil in cavity 2 to flow to cavity 4 by way of orifice 6 and ball check 5.

When the rear suspension cylinder housing changes direction and moves down, as shown in **Figure 19-8,** the combined weight of the vehicle and load along with the nitrogen in cavity 2 forces the rod to move out. The oil in cavity 4 is forced to flow to chamber 2 through orifice 6 as check valve 5 closes due to reverse oil flow. Because of the restrictive flow of oil from cavity 4 to cavity 2, a shock absorber action is produced in the cylinder. Oil flow from chamber 2 to cavity 4 is much faster during down movement of the rod, and slows down considerably in the opposite direction because ball check 5 closes (producing a much slower flow rate).

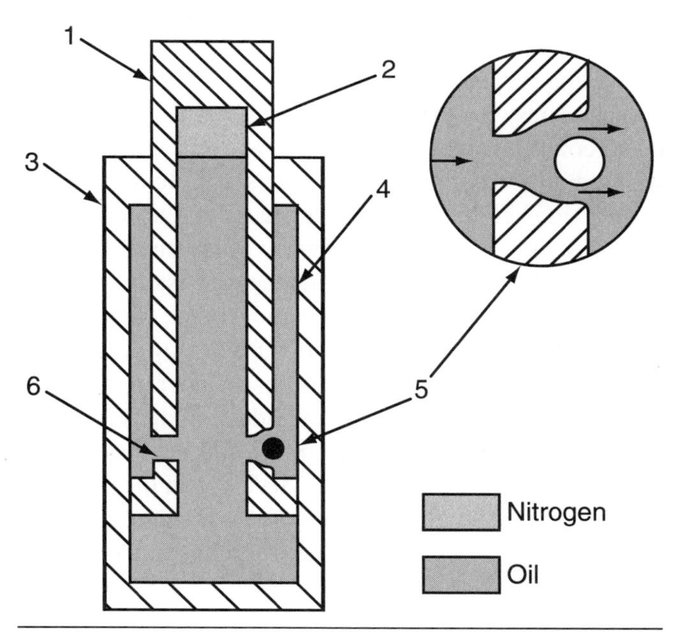

Figure 19-7 Cutaway of a rear suspension cylinder with the cylinder moving up: 1. Rod, 2. Nitrogen chamber, 3. Cylinder, 4. Cavity (2), 5. Ball check, 6. Orifices (2). (*Courtesy of Caterpillar*)

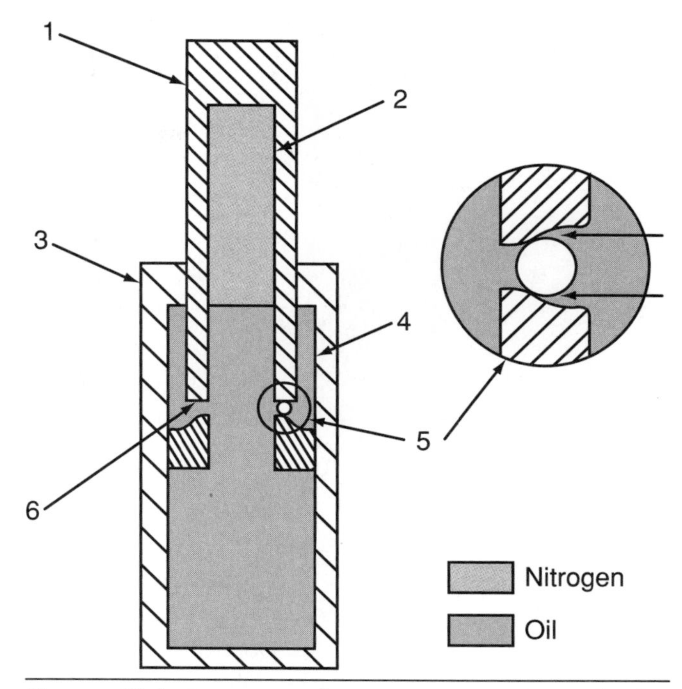

Figure 19-8 Cutaway of a rear suspension cylinder with the cylinder moving down: 1. Rod, 2. Nitrogen chamber, 3. Cylinder, 4. Cavity (2), 5. Ball check, 6. Orifices (2). (*Courtesy of Caterpillar*)

ARTICULATING TRUCK SUSPENSION SYSTEM

Large articulating haulage trucks, as shown in **Figure 19-9** incorporate a suspension system that will often integrate the suspension system circuit into the vehicle's hydraulic system which, in our example, provides oil to the hoist, steering, brake, and fan drive circuits. A common hydraulic tank provides oil for all hydraulic circuits on the equipment. On older models, accumulators were used to supply stored hydraulic pressure for suspension pressure. The system was a closed system and was isolated from the rest of the hydraulic system through the use of control valves. Newer equipment often uses a similar circuit, shown in **Figure 19-10** that is able to utilize excess oil from a brake or hydraulic circuit to control ride height with the use of directional control valves.

Refer to **Figure 19-11** for the following suspension circuit identifiers. The suspension frame is connected to the front axle on both sides by mounting bolts. The suspension frame is mounted to each side of the **oscillating hitch** by a pin on the upper side at both ends of the front axle. At the bottom it is attached to the front frame

Figure 19-9 Terex TA40 articulating off-road haulage truck. (*Courtesy of Terex*)

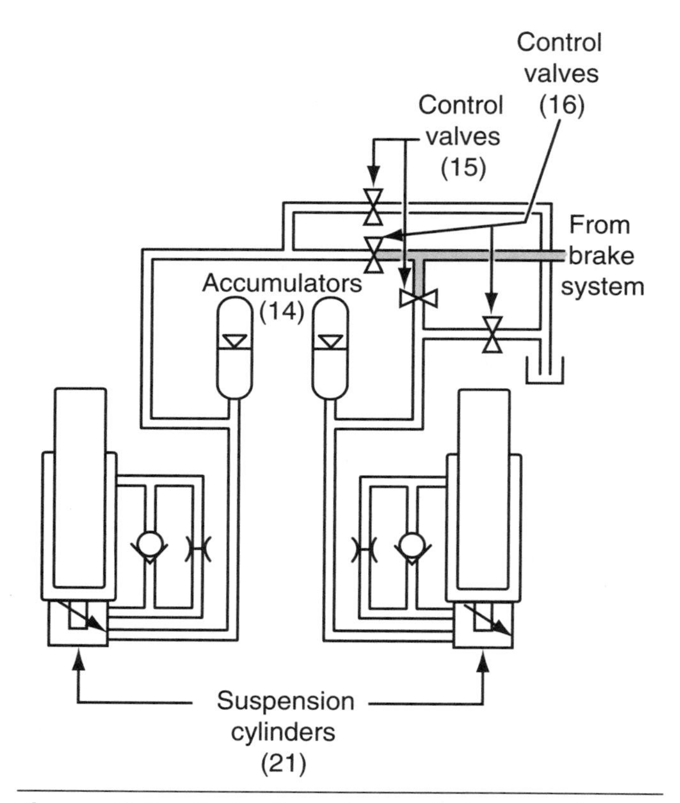

Figure 19-10 Illustration of a typical suspension circuit on an articulating truck.

through the lower suspension cylinders mounting points along with rod. Pins, located at the bottom of the suspension cylinders, connect the cylinders to the suspension frame. Through the use of suspension frame pivots connected to the oscillating hitch, independent vertical motion can be achieved. The dampening action of the suspension cylinders is controlled through the use of **nitrogen-charged accumulators.**

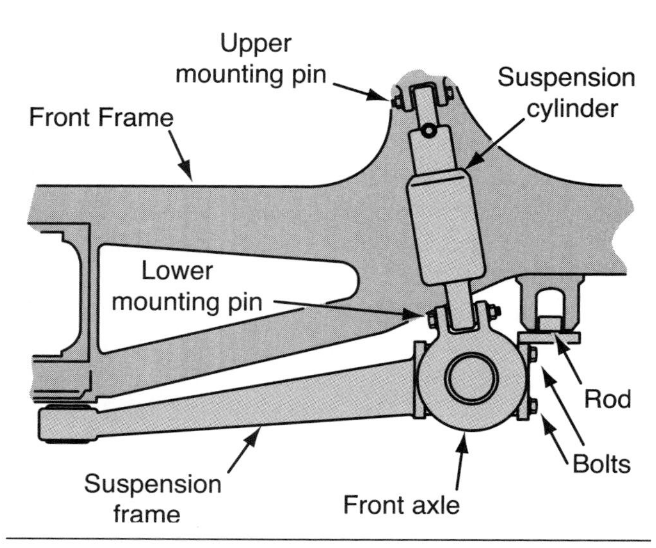

Figure 19-11 Typical suspension setup on an articulating haulage truck.

Suspension System Hydraulic Circuit Operation

In order to raise the suspension system up to the desired operational height, two control valves must be activated, as shown in **Figure 19-12,** with the engine running so that oil from the main hydraulic circuit can flow to the suspension cylinders. Once the correct ride height is reached the operator closes control valves 9, which isolate the suspension cylinders so that they can operate independently. To lower the ride height the operator will open control valves 10, also shown in Figure 19-12, which drains off oil from the suspension cylinders lowering the ride height. The nitrogen charge in the accumulators acts like a shock absorber by cushioning the ride during operation. As the load on the suspension cylinders changes with vertical suspension movement, the pressure in each independent circuit varies up or down, depending on which way the suspension cylinder is moving. The gas in the suspension cylinders increases when compressed during piston-up travel and decreases during piston-down travel. If a leak develops in the nitrogen side of the accumulator, the operator will notice a change in ride height on the side where the nitrogen leak has occurred.

Independent Gas-Charged Suspension Cylinder Maintenance

A large number of haulage trucks come equipped with independent gas suspension systems. Each suspen-

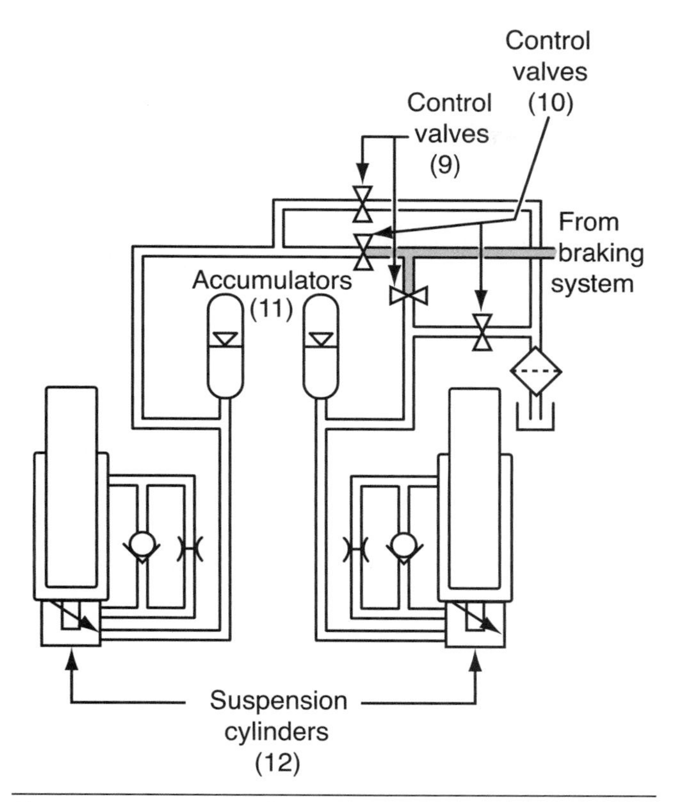

Figure 19-12 Typical suspension hydraulic circuit used on a CAT 725 articulating truck.

sion cylinder works independently from every other cylinder; if one cylinder develops a leak, it would not affect the operation of the other three suspension cylinders. Gas, charged suspension systems minimize road shock transfer to the truck frame. Each wheel is connected to the truck frame by a gas/hydraulic suspension cylinder. The suspension cylinders act as the kingpins for the steering system on the front axle. The suspension cylinders also provide the spring and shock absorber action for the rear suspension system and dampen suspension oscillations.

A key advantage to independent gas/hydraulic suspension systems is their simplicity of design and their ability to absorb changing load and road conditions. The more modern suspension systems have taken the place of conventional mechanical leaf-type suspension systems. Gas/hydraulic suspension systems require very little maintenance or adjustment when compared to the older conventional leaf-spring systems. Properly serviced gas/hydraulic systems often operate for long periods of time between breakdowns. The combination of this suspension system with modern truck seat suspension provides the operator a smooth ride and minimal driver fatigue.

Tech Tip: Be sure to carefully read over the manufacturer's service procedures before attempting to service the ride suspension system. Although nitrogen itself is not flammable, it can cause serious injuries due to the high pressures that are used. This pressure must be released in a safe manner. Be aware of pinch points as changes to ride height will be affected when the nitrogen is released to the atmosphere. Never attempt to check the oil level in a suspension cylinder until all the nitrogen has been bled off. Never remove any valves or plugs from the suspension cylinder unless the rod is in the fully retracted position.

Before making any adjustment to a suspension system, the technician must consider that the pressures in each suspension cylinder can be affected by changes in ride height due to leaks at opposite ends of the equipment. For example, when the left rear suspension cylinder ride height is low, it is likely that the right front suspension cylinder is riding high because of the shift in load weight. A detailed visual inspection of each cylinder is necessary to determine the source of the problem. Visually inspect for oil leaks at all connections on the suspension cylinders, valves, and all connecting hoses and pipes. If the vehicle is equipped with an oil distribution manifold and corresponding pressure sensors, check for leaks around them. Common sources of leaks are the O-rings used on the distribution manifold. Nitrogen may have leaked out of the suspension cylinders, which will affect ride height. Finding these leaks is often very difficult and requires the use of a soap and water solution to detect an external leak around the charge valve. Be sure to check for a loose valve body; this is often responsible for the nitrogen leak. If a loose valve body is discovered, be sure to use a torque wrench to tighten it back to the manufacturer's recommended torque value.

When checking suspension cylinder operation, look for excessive cylinder stem travel. If the chromed area of the stem is clean, this is a good indication of excessive travel. Discuss the ride quality with the operator to see if there has been any noticeable change. With the equipment empty, check the strut height to see if the struts are fully extended; this is usually an indication of a suspension problem. To correctly diagnose a suspension system problem it is imperative that the technician remember that suspension cylinders operate independently of each other. Note that suspension cylinders are to be serviced and charged in pairs and cannot be serviced individually.

Conditions for servicing the suspension cylinders may vary somewhat from machine to machine, but generally will follow these guidelines. Service is required when

- A suspension cylinder shows signs of external leakage.
- The weight of the truck changes substantially (more than 5000 lb).
- The struts are fully extended when the truck is empty.
- The operator complains of a rougher than normal ride.
- The suspension **cylinder pressure** varies more than 50 psi (345 kPa) from cylinder to cylinder.
- The ride height varies more than .500" from the estimated ride height of the equipment.

Note that these dimensions are temperature sensitive so the air temperature difference between a shop or building where the adjustments are performed must be no greater than 20°F (11°C). **Table 19-1** lists examples of dimensions that would typically be used on haulage trucks. These dimensions are equipment specific and a shop manual for the particular type of equipment that is to be adjusted must be consulted.

In some cases, a manufacturer will recommend that a gauge block be fabricated to check the measurements during the height adjustment procedures. An example of a typical drawing showing the fabrication dimensions is shown in **Figure 19-13.** Measurements depending on the manufacturing brand of the components. Always refer to the equipment manufacturer's technical documentation for specific instructions on the tools required to perform this adjustment procedure.

Table 19-2 is an example of a manufacturer's list of tools required to charge suspension system components on a typical haulage truck. These are specific to the example discussed in this chapter.

Tools and equipment may vary due to the type of equipment being serviced. Always refer to the specific manufacturer's shop manual for the exact tools and procedures required for this type of work.

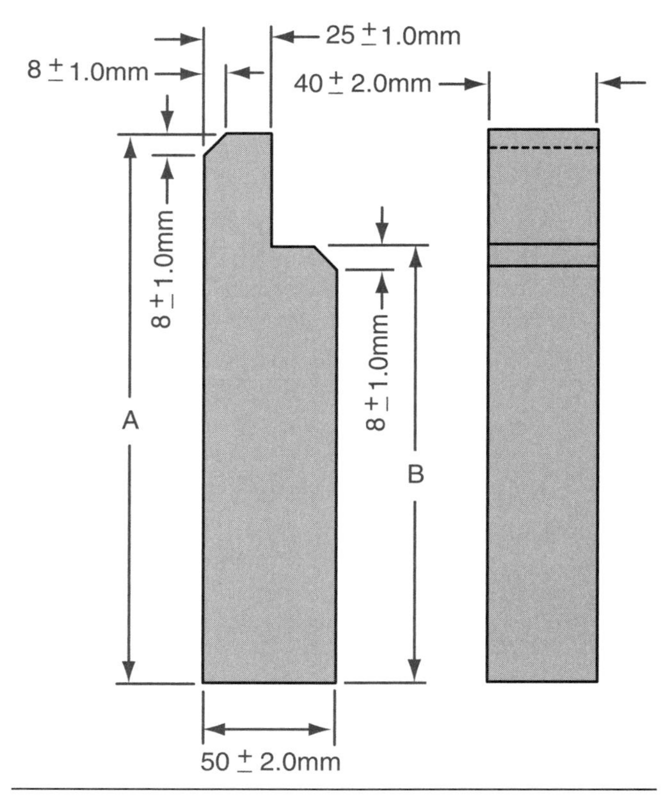

Figure 19-13 CAT suspension cylinder height adjustment tool dimensions.

Table 19-1: TYPICAL DIMENSIONS FOR HAULAGE TRUCKS			
Gauge Block Dimensions (Front Suspension Cylinders)			
FT Gauge Block	**Tractor or truck**	**Gauge Block Length A**	**Gauge Block Length B**
1680	771C, 775B	8.50 in. (216.0 mm)	6.69 in. (170.0 mm)
Rear Suspension Cylinder Nitrogen Charge Dimensions (Distance Below the Index Line)			
Quarry Truck	**First Charge Line**	**Second Charge Line**	
771C	5.0 in. (128 mm)	5.5 in. (140 mm)	
775B	4.9 in. (124 mm)	5.4 in. (137 mm)	

Table 19-2: EXAMPLE OF TOOLS REQUIRED FOR NITROGEN CHARGING EQUIPMENT

Part No.	Consists of	Item Description	Number Needed
FT1680		Gauge	2
7S5437		Nitrogen charge Group	1
	1S8941	Hose assembly	1
	7S5106	Chuck	1
	7S8713	Gauge	1
	8S4600	Nipple	2
	8S1506	Coupling assembly	2
	1J3914	Hose assembly	1
	7S8712	Gauge	2
	7S8714	Gauge	1
	2D7325	Tee	1
	1S8937	Valve	1
	2S5244	Nipple	1
	8S4599	Coupling assembly	1
	8S1505	Regulator assembly	1
	5P3041	Nipple assembly	1
5P8610		Auxiliary fitting group	
	1S8941	Hose assembly	1
	7S5106	Chuck	1
	8S7169	Coupling	2
	2J9803	Hose assembly	2
	1S8937	Valve	2
	5P8998	Pipe nipple	2
	2D7325	Tee	1
	3D8884	90° elbow	1

Note: Auxiliary Fitting Group 5P8610 is used with 7S5437 when charging two cylinders at the same time. 7S8712 gauge is used to check the cylinder pressure before charging.

Adjusting Rear Gas-Charged Suspension Cylinders

Tech Tip: When releasing the pressure in the rear suspension cylinders, the ride height will decrease quickly. The chance that you may be pinned between the tire and a fender or frame component is possible. Make sure that you stand in a safe location before releasing any pressure from the suspension cylinders.

Because both rear cylinders will be charged together, remove both charge valve caps. Install a special chuck (7S5106) listed in Table 19-2 on each of the charge valve as shown in **Figure 19-14.** Release all the nitrogen and oil from the suspension cylinder by turning the check valve clockwise. Once the cylinders have bottomed out, allow pressure in the suspension cylinders to equalize by leaving the check valves open for approximately five minutes. Once the pressure has been equalized close the check valves by turning counterclockwise. Before attaching the oil fill unit to begin the fill cycle, make sure the oil lines are full of oil by cycling the oil refill unit with the ends of the lines submerged in oil.

Before the charging procedure can begin, a rule must be placed on the cylinder, as shown in **Figure 19-15,** so that the cylinder extension can be measured accurately. A suitable magnet, such as a base for a dial indicator, should be used to hold the rule in place. The rule should be placed in a position that is parallel to the centerline of the suspension cylinder, as shown in Figure 19-15. A mark must be placed on the rule (position 3) that corresponds to the top edge of the suspension cylinder head. Once this is done mark a

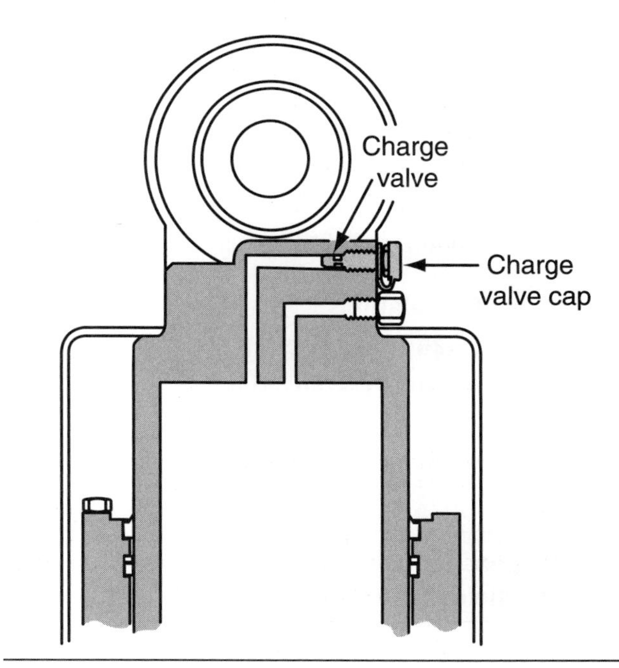

Figure 19-14 CAT rear suspension cylinder showing the location of the charge valve: 1. Charge valve cap, 2. Charge valve.

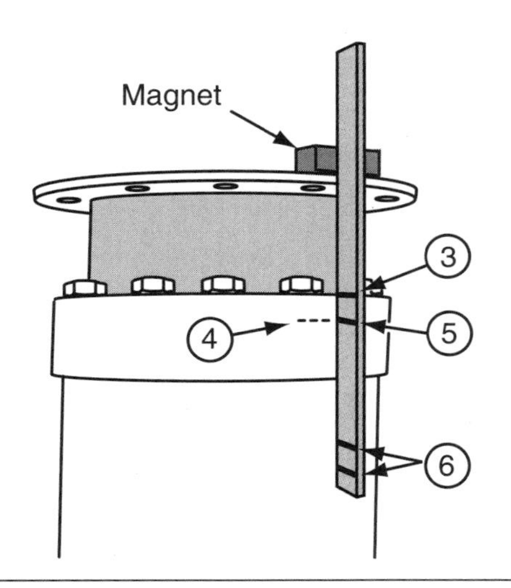

Figure 19-15 CAT rear suspension cylinder showing the location of the rule and marks.

corresponding line on the cylinder (line 4) that matches line 3. Another line (5) must be placed one inch below line 3 on the rule. Two more lines must be placed below line 5. The dimensions are shown on the Nitrogen Charge Chart, **Table 19-3.** Line 6 is shown in Figure 19-15. Remember, if the temperature difference is greater than 20°F (11°C), then a correction must be made. For every 10°F (5.5°C) .100" (2.54 mm) must be added to the dimension from the example shown in Table 19-3.

When performing these calculations, temperature differences must be averaged out to the nearest 10°F (5.5°C).

- Shop temperature averaged out to 10°F (5.5°C)
- Outside average temperature is −40°F (−40°C)

Following is an example of how to perform the calculations used to compensate for temperature difference.

Difference Calculations

10 °C − (−40 °C) = 50 °C or
[50 °F − (−40 °F) = 90 °F]

Table 19-3: EXAMPLE OF NITROGEN CHARGE DIMENSIONS

Rear Suspension Cylinders (Distance Below the Index Line)

Quarry Truck	First Charge Line	Second Charge Line
771C	5.02 in. (127.5 mm)	5.52 in. (140.2 mm)
775B	4.87 in. (123.7 mm)	5.37 in. (136.4 mm)

Extra length required for temperature difference, if required

Metric 50°C × (2.54 mm/5.5°C) = 23.1 mm
Imperial [90°F × (0.10 in./10°F) = 0.90 in.]

Total distance for the first nitrogen charge line dimension

Metric 127.5 mm + 23.1 mm = 163.3 mm
Imperial 5.02 in. + 0.90 in. = 5.92 in.

Total distance for the second nitrogen charge line dimension

Metric 140.2 mm + 23.1 mm = 163.3 mm
Imperial 5.52 in. + 0.90 in. = 6.42 in.

Be sure to mark the results from these calculations on the rule located on the suspension cylinder before beginning the charging process.

Rear Suspension Cylinder Oil Charge Procedure

Once the calculations have been made and the data transferred to the rule on the suspension cylinders, the oil refill unit must be connected to the suspension cylinders as shown in **Figure 19-16.** Before starting the charging procedure make sure the gate valves are open on the lines, and then slowly open the check valves by turning them counterclockwise. Adjust the sir pressure connected to the oil refill unit to 125 psi (860 kPa). Observe the movement of the cylinders; if one cylinder starts to travel at a faster rate than the other, close the gate valve to the faster cylinder so that it does

Figure 19-16 CAT oil refill unit connected to the rear suspension cylinders. (*Courtesy of Caterpillar*)

not go past line 5 on the rule. Continue filling the other cylinder until it reaches line 5 in Figure 19-15. When both cylinders (line 5) are even with the index (line 4) the oil level is correct. Close the check valves on the cylinders and close the air and gate valves before disconnecting the oil lines.

Rear Suspension Cylinder Nitrogen Charge Procedure

> *Tech Tip:* Nitrogen gas is the only gas that should be used to charge suspension cylinders. Other gases such as oxygen are highly explosive and must never be used to charge suspension cylinders. Be sure that the gas cylinder to be used is indeed nitrogen and not oxygen; the cylinders can look very similar. The fitting used to connect the nitrogen bottles to the suspension cylinders should never be used on other types of compressed gas bottles such as oxygen.

The following procedures are intended for explanation purposes only. Always refer to the specific technical documentation for the equipment that you will be servicing.

Install the quick disconnect couplings used to charge the suspension cylinders with oil to the nitrogen charge lines of the nitrogen charge kit. Recheck the line 4-to-line 5 alignments on each cylinder to make sure it is still lined up correctly before beginning the nitrogen charge procedure. This next step will require that the rear suspension cylinders be charged with nitrogen until line 4 is lined up with line 6 (**Figure 19-17**). Before connecting the charge lines to the cylinders, make sure that the charge valves are fully closed; this will prevent the possibility of oil from the suspension cylinders, which is under pressure, getting into the nitrogen charge lines.

Once the previous step has been completed, connect the charge lines to the rear suspension cylinders, as shown in **Figure 19-18.** Adjust the nitrogen regulator located on the charge bottle to 350 psi (2400 kPa). Slowly open the gate valves, and turn the chucks clockwise on both suspension cylinders to begin the charging process. Continue to charge the suspension cylinders until line 4 aligns with line 6, as shown in Figure 19-17.

If the cylinders do not reach the correct height at the same time, close the gate valve on the suspension cylinder that has reached the correct height dimensions first and allow the other cylinder to charge until the correct height is reached. Once charging dimensions have been reached, close both gate valves before closing the check valves at the suspension cylinders. Remove the nitrogen charge hoses from the suspension cylinders and install the protective caps. Torque the caps to the recommended torque values outlined in the technical documentation and remove the steel rules and magnets.

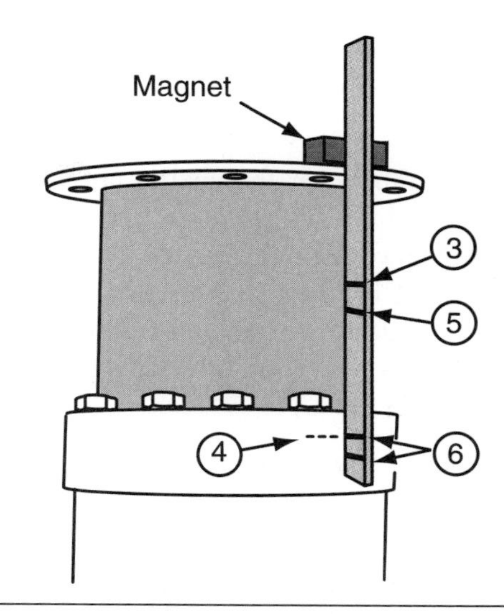

Figure 19-17 Height dimension measuring procedure used to charge rear suspension cylinders on a CAT quarry truck.

Figure 19-18 CAT nitrogen charge unit connected to the rear suspension cylinders. (*Courtesy of Caterpillar*)

Front Suspension Cylinder Charge Procedures

Tech Tip: Whenever the charge pressure in a suspension cylinder is released, the ride height of the vehicle will be reduced quickly. Be sure to stand in a safe location so that the rapid change in ride height does not trap you between a fender or frame component and the tire.

Although the charge procedures are somewhat similar on the front and rear suspension cylinders, there are some differences that need to be identified. The location of the charge ports and the measuring techniques used on the front suspension cylinder is not the same as those used to adjust the height of the rear cylinders. To bleed of the oil in the front suspension cylinders, remove the charge valve caps, as shown in **Figure 19-19,** and install the chucks from the oil charge unit to the charge valves of both suspension cylinders.

Install a suitable drain line to each connection so that the oil is drained into an appropriate container. While standing in a safe location, slowly open the check valves on the charge valves by turning in a clockwise direction, allowing the oil and nitrogen to bleed off until the cylinder reaches the bottom of its travel. Note that pressurized oil and escaping nitrogen gas can pose a safety hazard to anyone who comes in contact with either one. Give the pressure in the cylinders a chance to equalize by allowing the valves to remain open for approximately five minutes before closing them. Once this step is complete, place the charge line ends in a suitable container and cycle the oil refill pump until the lines are full of charge oil. Reattach the lines to the charge valves on the front suspension cylinders.

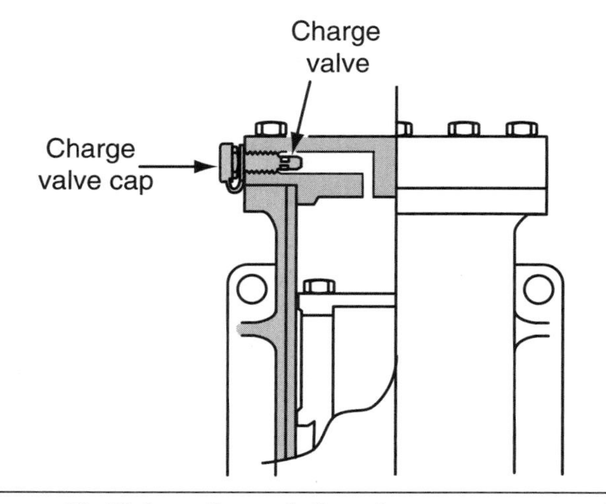

Figure 19-19 Location of the charge ports on the front suspension cylinders on a CAT quarry truck.

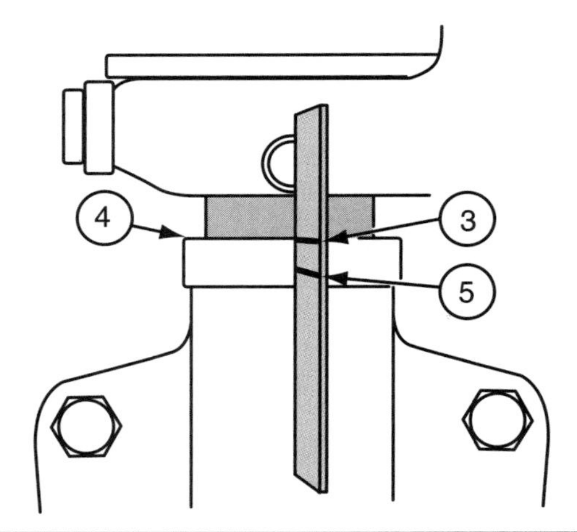

Figure 19-20 Height dimension measuring procedure used to charge the front suspension cylinders.

The rule and its corresponding height adjustment lines are not the same for the front suspension cylinders and rear suspension cylinders. **Figure 19-20** shows the location of the rule with the appropriate scribed lines in place. However, the magnet used for the rear suspension cylinder measurements can be reused for the front cylinders. Mark up line 3 on the rule to correspond with the top edge of the suspension cylinder's head. Place a second mark (line 4) on the cylinder head edge. A third line (5) must be placed 1 in. (25.4 mm) in this case below line 3 on the rule. Once these steps are complete, you are ready to begin the oil charge procedure steps.

Oil Charge Procedure for the Front Suspension Cylinders

The following procedures are intended to be used for demonstration purposes only. Always refer to the specific technical documentation for the equipment that you will be servicing.

The charge procedure for the front cylinders is similar to the procedure used to charge the rear suspension cylinders. The same specialized tools are required to perform this next step. Attach the check valves and lines from the oil charge unit to the front suspension cylinder charge valves. Adjust the air pressure to 125 psi (860 kPa) and make sure the gate valves are in the open position.

Open the check valves slowly by turning in a clockwise direction. Continue injecting oil until line 5 reaches line 4. If one cylinder reaches the line first, close the gate valve to that cylinder and allow the remaining cylinder to fill until it reaches the correct height. Close the gate valves and regulated air pressure before bleeding the air pressure to zero. If required, reuse of the quick disconnects is allowed to charge the suspension cylinders with nitrogen if they are not included with the nitrogen charge kit.

Nitrogen Charge Procedure for the Front Suspension Cylinders

The use of gauge blocks to set the correct ride height on the front suspension cylinders in this example is shown to clarify the steps required to perform this procedure. **Figure 19-21** shows an example of gauge blocks that are used to set the ride height. This requirement may vary depending on the manufacturer and model of equipment. Suspension dimensions will be affected by temperature differences between the adjustment location and the operating location. For example, if the temperature difference between the shop where the adjustment is performed and the outside temperature where the equipment is operating is greater than 20°F (11°C), a different calculated dimension is required for extra shim thickness and must be calculated based on the temperature difference. An example of this calculation is provided for a typical haulage truck. The procedures and dimensions for different manufacturer's equipment or even different models by the same manufacturer can vary substantially. Always consult the manufacturer's specific technical documentation for the equipment in question.

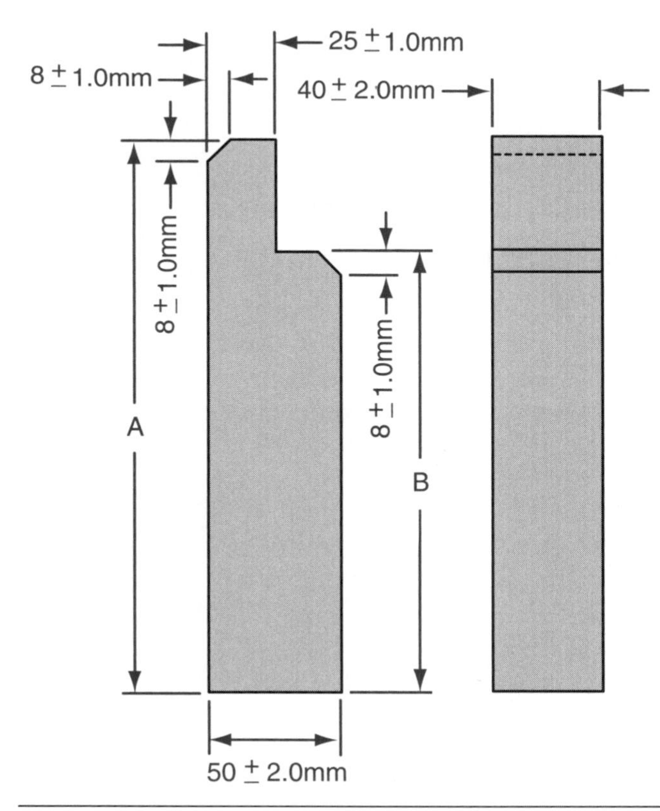

Figure 19-21 Dimensions to fabricate the gauge used to measure the front suspension cylinder height on a CAT quarry truck.

Calculations. The following is an example of calculations used to compensate for temperature differences greater than 20°F (11°C). Before beginning the calculations, temperature differences must be rounded off to the nearest 10°F (5.5°C).

Workshop Temperature 10°F (5.5°C)
Outside Air Temperature −40°F (−40°C)

Calculated temperature difference

Metric 10°C − (−40°C) = 50°C
Imperial −50°F − (−40°F) = 90°F

Shim Thickness

Metric 50°C × (5.1 mm/5.5°C) = 46.4 mm
Imperial 90°F × (.20 in./10°F) = 1.80 in.

Gauge plus Shim Thickness

Metric 170.0 mm + 46.4 mm = 216.4 mm
Imperial 6.69 in. + 1.80 in. = 8.49 in.

Procedures. The following procedures are intended to be used for example purposes only. Refer to the specific equipment technical documentation for the actual service procedures.

Be sure that the charge valves are fully closed before connecting the nitrogen charge lines to the suspension cylinders to prevent oil from entering the nitrogen charge lines. The first step is to adjust the nitrogen regulated pressure to 600 psi (4150 kPa). Next, open the gate valves on the high-pressure lines, allowing the nitrogen to flow to the charge valves. Slowly open the charge valve check valves by turning clockwise on both suspension cylinders to allow nitrogen to enter the cylinders. If one of the cylinders extends faster that the other and has reached slightly higher that the correct height, close the gate valve to that cylinder and continue to allow the other cylinder to reach the same height before closing the remaining gate valve. In this example, continue to inject nitrogen into the suspension cylinders until they have reached the same height (6.69 in.).

Close off both gate valves first, and then close the charge valves on the suspension cylinders. Remove the lines at the quick disconnects. With the gauge blocks and shims in place, slowly open the charge valves to release some of the nitrogen until the cylinder comes to rest on the gauge block. At this point, close the charge valve.

Reconnect the nitrogen charge lines to the suspension cylinders and adjust the regulated nitrogen pressure to 350 psi (2400 kPa). Once the previous steps have been completed, open the gate valves and charge valves and allow the nitrogen to flow into the suspension cylinders for approximately five minutes to equalize the pressure. After the pressure has been equalized, close the charge valves and gate valves and close the nitrogen supply at the tank.

Disconnect the nitrogen charge kit from the suspension cylinders, reinstall the protective valve caps, and torque to specifications. If the gauge blocks are not easy to remove it may be necessary to start the vehicle, raise the truck body, and steer the wheels from side to side to facilitate removal.

REUSE OF GAS/HYDRAULIC SUSPENSION SYSTEM COMPONENTS

The suitability to reuse components in a suspension system can be determined by following the manufacturer's technical documentation. Welding and re-machining of the seal lands is often done to the head of the rear suspension cylinder housing. The rod assemblies are often sent out for re-chroming or polishing to restore the finish. The cylinders themselves can often be honed out or re-tubed. Guidelines that provide the dimensional specifications can be found in the technical documentation of the shop manual for any particular piece of equipment. Always follow the guidelines provided by the manufacturer to determine the reusability of suspension components.

When determining the reuse of suspension cylinders, you must take into consideration the overall length of the cylinder, the bore dimensions, the bore finish, and the internal diameter of the connecting eyes. The cylinder rods must meet specific tolerances as well. The overall length of the cylinder rod, as well as the outside diameter of the rod must be within specifications. The rod eye dimensions must be within specifications; often the eye will have a replaceable bushing.

LOAD, HAUL, AND DUMP RIDE CONTROL SUSPENSION SYSTEMS

Off-road load, haul, and dump equipment, as shown in **Figure 19-22**, does not utilize any suspension system other than the tires to absorb road shock. Because of the slow operating speeds of this type of equipment and the specific work that this equipment performs, conventional suspension systems are not necessary. The front axles (bucket end) on this type of equipment are bolted solidly to the frame with the only shock absorbing capability coming from the tires themselves.

The rear axles are bolted to an oscillating axle or rear axle trunnion bearing support, as shown in **Figure 19-23,** which pivots on large bushings running lengthwise along the length of the equipment. This

Figure 19-22 Typical load, haul, and dump vehicle. (*Courtesy of Caterpillar*)

Figure 19-23 Typical loader with rear oscillating axle during operation on uneven ground.

allows the equipment to maintain road contact during operation on uneven ground.

Some manufacturers offer an option for a load ride control. This allows the loader bucket to ride on a column of oil in the hoist cylinders that are connected hydraulically to an accumulator through a control valve, allowing road shock to be absorbed by the accumulator. A control switch located on the dump hoist control lever is used to activate the ride control circuit.

Figure 19-24 shows the ride control circuit in the inactive position, solenoid 7, which is a normally closed valve blocking the flow of oil from the pilot circuit to the diverter valve 8. When the lift control valve solenoid is energized, and the lift control valve in the hold position, pilot oil flows to the diverter valve, forcing the spool inside to shift to the left. With the diverter valve in this position and the control valve is in the hold position, the oil from the head end and

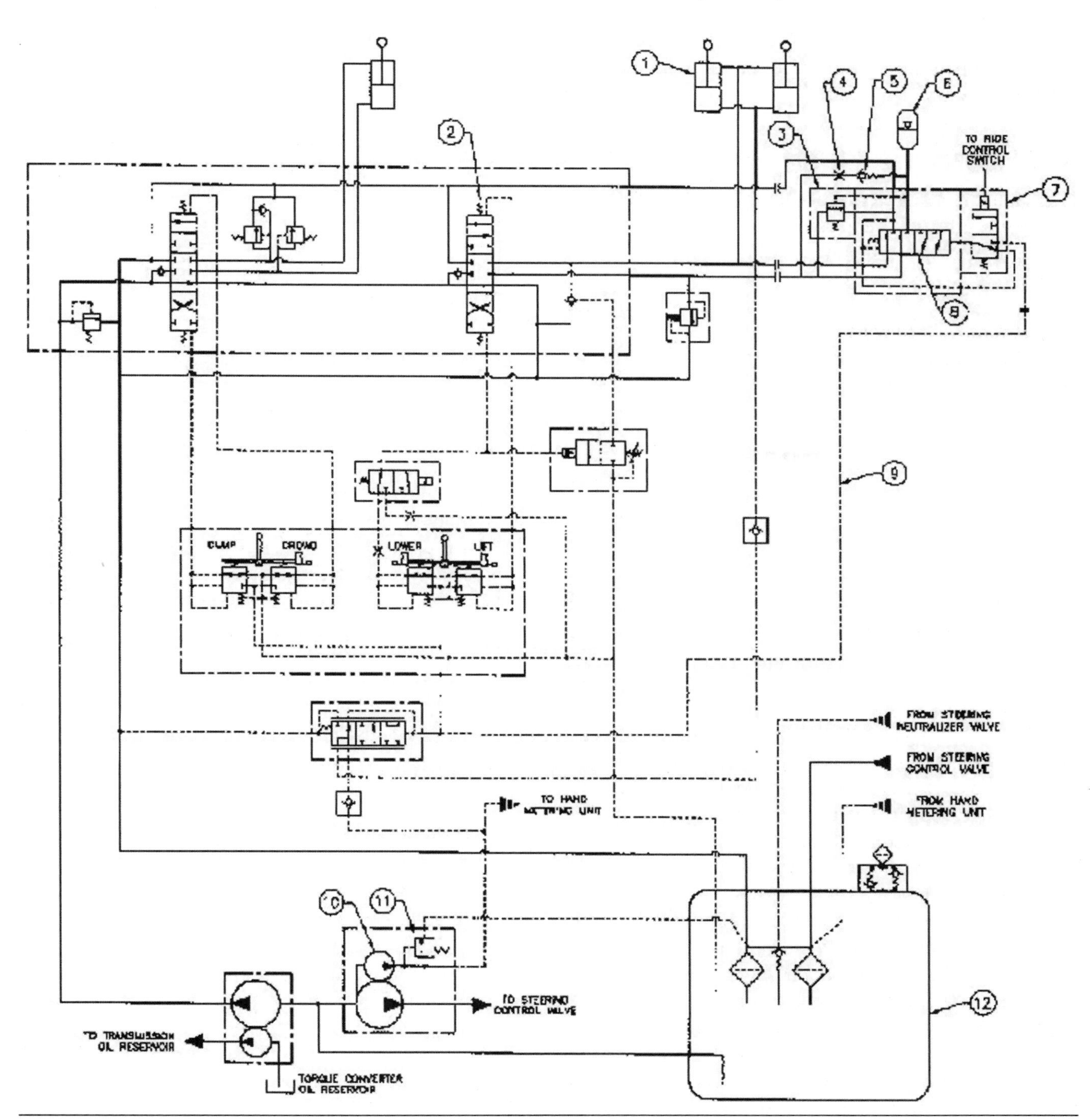

Figure 19-24 Ride control circuit used on a load, haul, and dump loader.

the rod end of the lift cylinder flows through the right side of the diverter valve. The oil from the head end of the lift cylinder flows directly into accumulator 6. The rod end oil flows back to the tank through the diverter valve.

When the equipment is operated on rough ground with a full bucket, the combined weight of the bucket and load act on the oil in the lift cylinder's head end, which is open to the accumulator. The nitrogen pre-charged accumulator absorbs the shock loads, acting like a shock absorber. The rod end receives make up oil since it is open through the diverter valve into the lift cylinders.

RIDE CONTROL SUSPENSION SYSTEM TESTING AND ADJUSTING

Tech Tip: When working on any high-pressure hydraulic system, always use a suitable device, such as a piece of cardboard, to check for hydraulic leaks. A high-pressure leak can penetrate the skin easily, causing serious injury. If hydraulic fluid is injected into the skin, immediate medical attention is required. Sudden movement of the equipment with components or valves disconnected can cause a high-pressure discharge to occur. Never place yourself in a position that will cause you to be in a direct line of fire of a potential hydraulic oil discharge. Be prepared to collect any hydraulic fluid spill that may escape during the testing and adjustment procedures in a suitable container. Disposal of hydraulic fluids must be performed in accordance to local laws and policies.

The ride control circuit can be serviced by using the following procedure. Start the equipment and turn on the ride control option. With the joystick control lever, lower the bucket to the ground and place the lever in the float position. This step relieves all the hydraulic pressure in the ride control accumulator. Once this step is completed the equipment can be shut down and the hydraulic lines to the accumulator can be safely removed.

If you determine that the accumulator needs to be replaced, you must open the gas discharge valve on the accumulator one turn to release all the compressed nitrogen gas in the accumulator gas chamber.

Tech Tip: Compressed nitrogen is the only gas that should be used to charge the ride control accumulator. Other compressed gases such as oxygen are highly explosive and must never be used to charge accumulators. Be sure that the gas cylinder to be used is indeed nitrogen and not oxygen; the cylinders can look very similar in some cases. The fitting used to connect the nitrogen bottles to the accumulator should never be used on other types of compressed gas bottles such as oxygen.

Charging the Accumulator

The following procedures are used for example purposes only. Refer to the specific equipment technical documentation for the actual service procedures. Be sure that the charge valve is fully closed before connecting the nitrogen charge lines to the accumulator to prevent oil from entering the nitrogen charge lines.

The equipment must be started, warmed up, and left running to perform this procedure. To charge the accumulator with dry nitrogen the following steps must be performed in sequence:

1. Locate the ride control accumulator and remove the guard that protects the gas charge valve and remove the cap.
2. Before you can attach the nitrogen charge adapter to the valve, the T handle must be turned all the way out (counterclockwise) before connecting the chuck to the accumulator.
3. See **Figure 19-25.** Connect the chuck (5) to the accumulator charge valve.
4. This step requires that the gas in the accumulator to be discharged to the atmosphere. Connect the quick disconnects (3) and (13) together and place the open end of the hose in such a way that the discharge will not harm anyone.
5. Open the T handle (4) and valve (1) and raise the lift arms of the equipment. This forces the gas in the accumulator to be expelled to the atmosphere. Be sure that the open end of the hose will not discharge toward anyone.
6. Place the needle valve (1) in the open position. Turn the T handle (4) on the gas chuck (5) clockwise all the way in.
7. Ensure that no one is positioned near the bucket end of the equipment and that adequate headroom is available to raise the bucket to its maximum

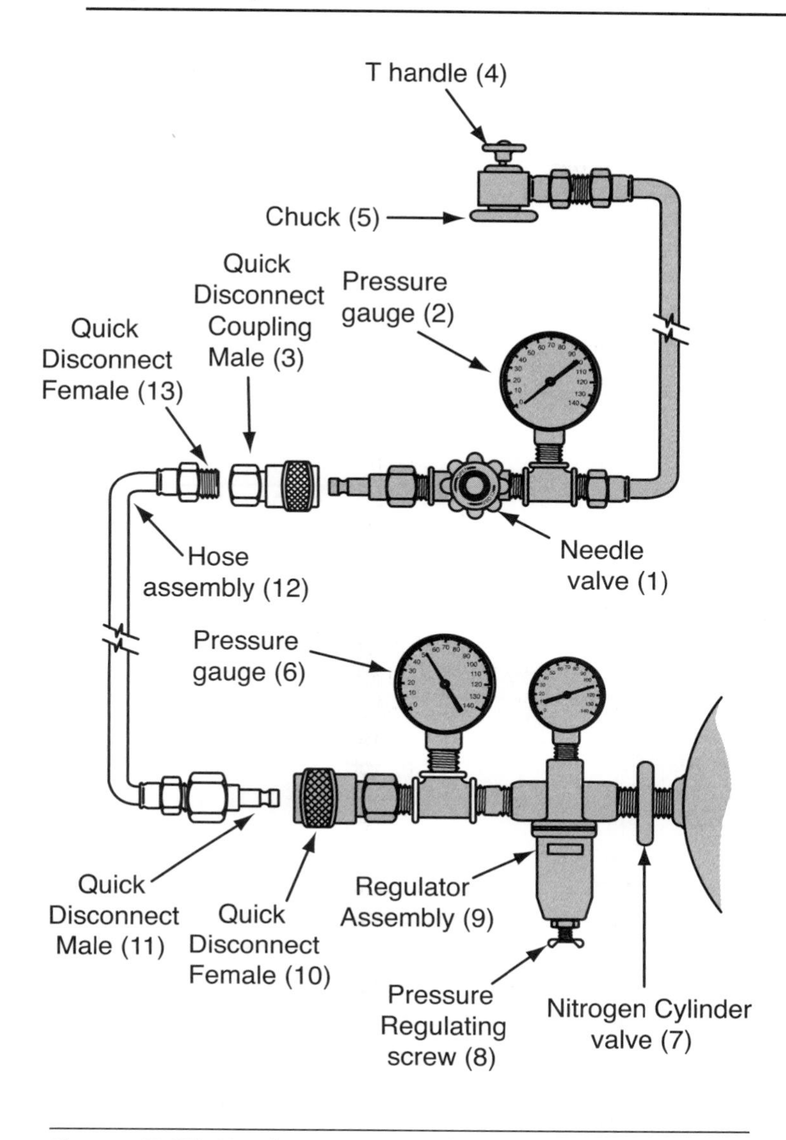

Figure 19-25 Hardware required to charge the ride control accumulator.

height. Lift arms must be raised to their maximum height and held for five seconds. This step ensures that all the gas has been vented from the accumulator and the accumulator piston has moved all the way up.

8. Close the needle valve (1) after all the gas pressure in the accumulator has been released to the atmosphere. Place the joystick in the hold position. Lower the bucket to the ground, place the joystick in the float position, and move the ride control switch to the on position.
9. Connect the quick disconnect coupling (10) to the quick disconnect coupling (11) and attach the hose to the nitrogen cylinder.
10. Adjust the nitrogen pressure (9) to 00 psi + −5 psi (2070 + −35kPa). This pressure is based on an ambient temperature of 70°F (21°C). To compensate for temperature differences consult the technical documentation for the equipment being serviced. *Note*: If the equipment has been in operation, the temperature of the accumulator may be higher than the ambient temperature. Use the accumulator temperature if it is higher than the ambient temperature.
11. Open the needle valve (1) and observe the time it takes the gauge to reach charge pressure. If the gauge reads the charge pressure instantly the gas charge was not relieved correctly and all of the previous steps must be repeated. If the gauge takes 10 to 20 seconds to register the charge pressure, then the gas has been successfully removed from the accumulator.
12. Do not remove the line from the accumulator before turning off the nitrogen gas supply. This will ensure that the accumulator piston is in the bottomed out position, and the accumulator is completely filled with nitrogen.
13. Place the ride control switch in the off position raise the lift arms, and hold the control lever in the raise position for about five seconds. This ensures that the pressure reaches and maintains the system relief valve setting, which should register on the gas pressure gauge. If a lower pressure is observed on the gauge then the accumulator has not been properly charged and the charge procedure must be redone.
14. If the hydraulic system relief pressure was reached, lower the bucket to the ground and place the ride control switch in the on position. Observe that the nitrogen gas pressure drops instantly to charge pressure. If the nitrogen gas pressure has not dropped to the charge pressure, repeat the charge procedure until you get these results.
15. Remove the charge hardware and install the cap and protective cover back onto the accumulator.

Installing a New Accumulator

When installing a new accumulator on the ride control system, certain steps must be observed so that all the air is removed from the nitrogen end of the accumulator. Pour approximately two quarts (1.9 litres) of SAE hydraulic oil into the nitrogen end of the accumulator. The oil is necessary to ensure that all the air from the nitrogen end of the accumulator is removed and to provide lubrication to the upper seal of the accumulator piston.

1. Connect the nitrogen gas chuck to the accumulator charge valve and connect the male and female quick disconnects together (3 and 13), as shown in Figure 19-25. Place the open end of the hose in a suitable container to collect the oil.
2. Open the needle valve (1) and screw the T handle on the gas chuck in a clockwise direction until it is fully open. Make sure that the accumulator is in the upright position and move the piston to the nitrogen end of the accumulator.
3. The air and oil will be vented out of the accumulator as the piston moves up. When the oil flow stops, close the needle valve completely and allow any remaining oil to drain from the hose.
4. Reconnect the quick disconnect couplings (10 and 11) together. Connect the nitrogen charge hardware to the nitrogen bottle and adjust the pressure to 300 psi (2070 kPa). Be sure to check the ambient temperature chart and compensate if required.
5. Charge the accumulator by opening the needle valve (1) with 300 psi (2070 kPa). Close the gas chuck (5) and remove the nitrogen charge hardware from the accumulator. The accumulator is now ready to install.

Checking Charge Pressure

The charge pressure in the ride control accumulator may need to be checked periodically or whenever a problem in the performance of the system is suspected. It is not necessary to follow all the steps outlined in the charge procedure just to check the pressure in the system. The following steps are outlined so that the charge pressure can be checked quickly. Start the equipment up and place the ride control switch on the joystick in the on position. Lower the bucket to the ground and place the joystick in the float position. This will ensure that all the hydraulic pressure is relieved in the ride control accumulator.

1. Shut down the equipment and remove the guard and protective cap from the accumulator.
2. Close the needle valve (1) on the nitrogen charge hardware and connect the chuck to the accumulator charge valve. Be sure to turn the T handle completely counterclockwise before connecting the chuck to the accumulator charge valve.

CHARGING PRESSURE AND TEMPERATURE RELATIONSHIP FOR THE ACCUMULATOR	
Temperature °C (°F)	Pressure kPa (psi)*
–7 (20)	1790 (260)
–1 (30)	1835 (266)
4 (40)	1875 (272)
10 (50)	1915 (278)
16 (60)	1955 (284)
21 (70)	2070 (300)
27 (80)	2110 (306)
32 (90)	2150 (312)
38 (100)	2200 (318)
43 (110)	2250 (324)
49 (120)	2300 (330)

*Nominal allowable pressure tolerance equals ± 35 kPa (± 5 psi)

Figure 19-26 Temperature compensating chart used to charge an accumulator. (*Courtesy of Caterpillar*)

3. Open the T handle by turning clockwise and read the gauge (2) to see if the accumulator pressure is within specifications. If the equipment has been in operation then the temperature charts, shown in **Figure 19-26,** must be consulted to determine the correct pressure reading.

EQUALIZING BEAM SUSPENSION SYSTEMS

Tandem drive applications that are used on articulating haulage trucks often use **equalizing beam suspension** systems, as shown in **Figure 19-27;** these systems have proven to be up to the task of operating in rough service applications. To maintain good ground contact, the equalizing beams are allowed to flex on pivots. This provides a means of balancing the loads imposed on the drive axles while still maintaining good ground contact.

There are two distinct types of equalizing beam suspension designs: the leaf-spring equalizing beam, as shown in **Figure 19-28,** and the solid rubber equalizing beam, as shown in **Figure 19-29.** The latter beam is the most common on off-road equipment. This design is ideally suited to off-road equipment as the layout actually lowers the center of gravity by placing the beam below the centerline of the axle. Torque rods are provid-

Figure 19-27 Articulating haulage truck.

ed to absorb additional torque generated by the axle and road shock.

Leaf-Spring Suspension

The less popular of the two suspension designs used on off-road equipment utilizes **leaf springs** on each equalizer beam. The springs are mounted to saddle mounts at the top of each equalizer beam and are connected to the frame by **spring hangers.** The spring assemblies pivot on spring pins attached to the hanger brackets. The ends of the springs are not rigidly attached to the spring hangers, therefore providing a means of moving back and forth during spring flexing. The design of the equalizing beam suspension allows the load to be equally distributed between the axles while compensating for road irregularities. The starting and stopping forces, known as windup torque, are absorbed by the torque rods. A **cross tube** is used to connect the equalizing beams together; this reduces potentially damaging load transfers that occur during operation.

SOLID RUBBER SUSPENSION SYSTEMS

This design seems to be the most popular type of equalizing beam suspension on off-road haulage trucks. A **solid rubber suspension** incorporates solid rubber **load cushions** instead of leaf springs to absorb road shock. A **saddle assembly** on each side of the suspension

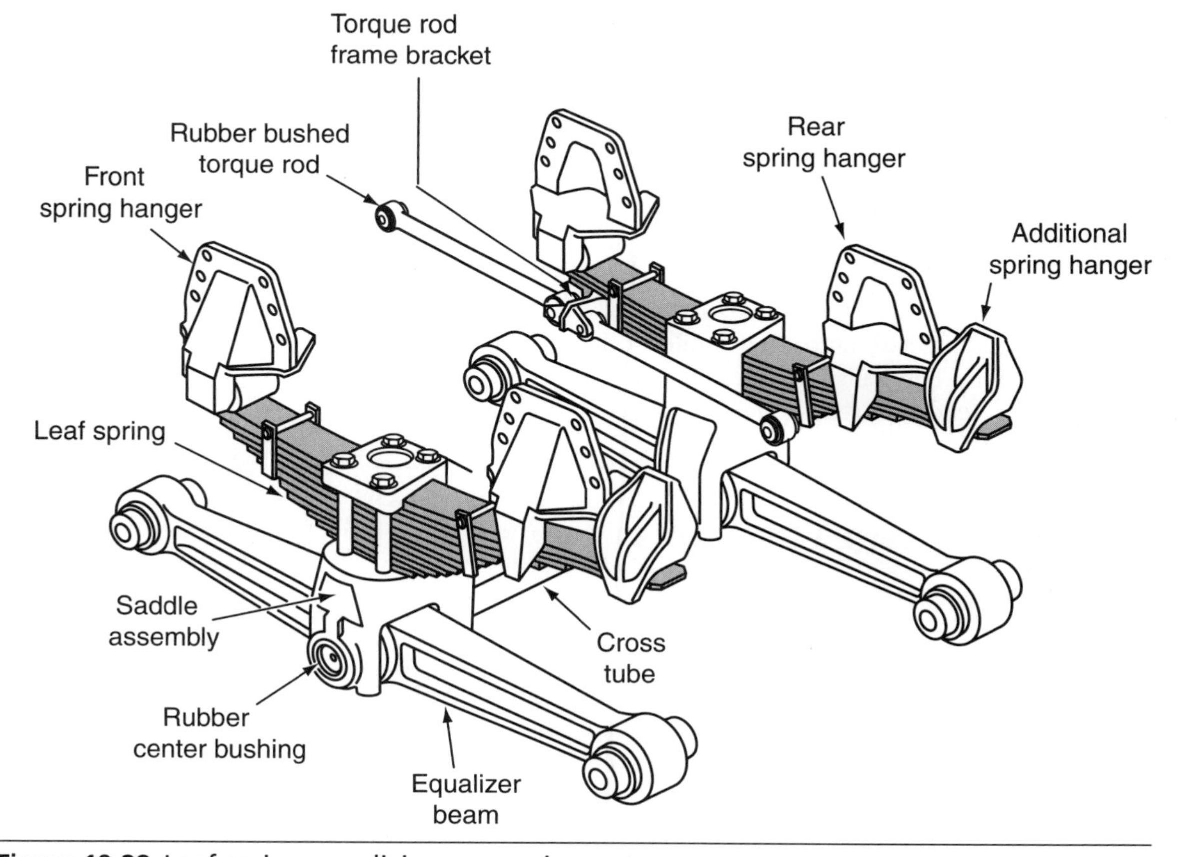

Figure 19-28 Leaf-spring equalizing suspension system.

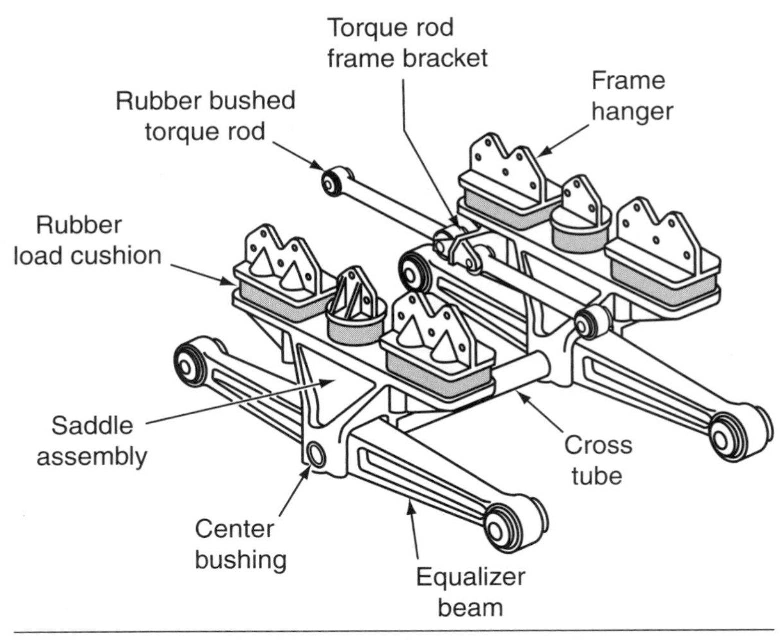

Figure 19-29 Solid rubber equalizing suspension system.

system holds the rubber load cushions. **Figure 19-30** shows a typical rubber cushion equalizer beam suspension system. Each of the four rubber blocks is connected to the axle assembly by drive pins. The operational loads generated by cornering and braking are all absorbed by the suspension through these pins. When the suspension is operating with no load, the outer edges of the spring load cushions are in contact with the saddle. As the load increases on the vehicle, the load cushion is compressed, further absorbing the additional loads. The alignment is maintained by the drive pins, which are encased in rubber bushing. The vertical **drive bushings** are used on each drive pin to allow vertical movement during loading and unloading while still maintaining suspension alignment. This design works well when loaded, but tends to give a rough ride when operating without a load.

SERVICING SUSPENSION SYSTEMS

Off-road equipment operates in severe-duty service, making inspection of the suspension systems on a regular basis essential. The equipment should be powerwashed before regular inspection and servicing. Powerwashing the suspension system allows detailed inspection of the suspension components, which includes looking for cracks or physical damage. Rubber bushings should be inspected for deterioration and cracking. If the rubber components show any signs of deterioration or cracking, they should be replaced as

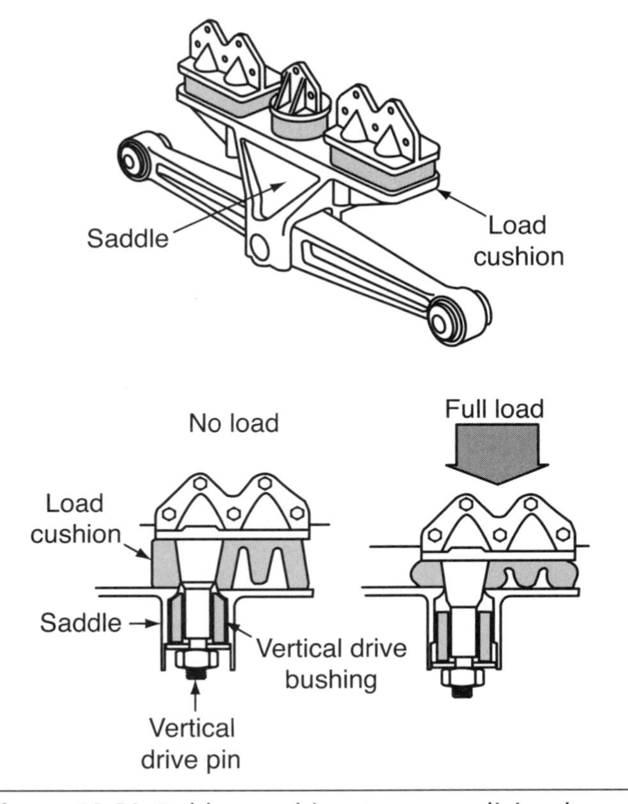

Figure 19-30 Rubber cushion-type equalizing beam suspension system.

soon as possible. Suspension system bushings are not easily removed; often specialized tools are required to remove them properly. The use of a torch is not recommended because the fumes produced by the burning rubber are toxic. A specialized hydraulic press is often required to properly remove the bushing and sleeve assemblies. In some cases it is easier to replace the complete equalizer arm than to attempt to replace the bushing without specialized equipment. Pressures used to replace these components often reach 50 tons.

The load cushions on equalizer beam suspension systems are easily assessed because they are visible for a thorough inspection. The compounds that the blocks are made from may often be damaged by cleaning solvents that are used to clean the area before repairs are performed on the suspension system. Be sure to use only manufacturer-approved cleaning products on these suspension components. The blocks are usually made from synthetic materials that may be susceptible to damage from petroleum products, such as grease or oil. Visual inspection is all that is necessary to inspect these components.

Check to see if the un-sprung height has changed significantly. Continued operation of equipment with equalizer beam suspensions that have worn-out load cushions should not be allowed; damage to the suspension system components, including the frame itself, can result and lead to premature failure and expensive repair bills.

U-bolts that secure the **spring packs** must be checked for tightness on a regular basis. If the bolts are corroded, cleaning them and freeing them up before checking the torque is mandatory, as the corrosion will affect the re-torque values. Whenever new fasteners are installed on suspension systems, it is highly recommended that the fasteners be re-torqued after an initial run-in period.

If a grease fitting on a spring pin refuses to take grease during a scheduled maintenance service, try taking the weight off the spring before changing the grease fitting, or try using a hand grease gun since it is capable of higher operating pressures than air-powered systems. Lubricating the suspension pins is critical to increasing the life span and controlling the operating costs of equipment and should always be a priority during servicing.

Figure 19-31 shows how wear to the equalizer bracket can occur when the spring ends shift. Look for loose U-bolts or problems with the U-bolt seating whenever these conditions are encountered. Wear can be caused by improperly mounted springs on the axle housing, indicating a need for a suspension alignment check.

Whenever a broken or cracked spring is found in a spring pack, most manufacturers recommend that the complete spring pack be replaced. Never apply grease or paint to the springs; this will affect the springs' dampening ability. The same thing applies if the springs are allowed to rust severely; anything that affects the dampening characteristics of the springs should not be tolerated. The springs should be disassembled and cleaned or replaced if cleaning is not possible.

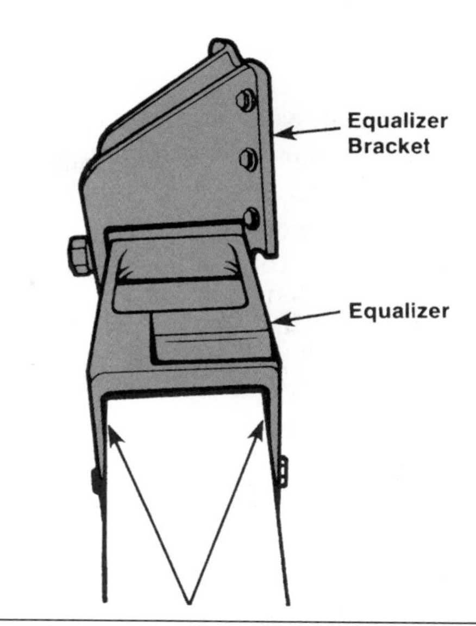

Figure 19-31 Misaligned spring rubbing against an equalizer bracket. (*Courtesy of International Truck and Engine Corporation*)

The spring and bracket bushings are generally considered a high maintenance item and should be inspected for cracks and gouging on a regular basis. If the bushings are free from damage, then they must be checked to see if they meet dimension specifications. Always refer to the manufacturer's recommended dimensions and clearances to determine the usability of the parts. This general rule can be used if specifications are not readily available: A maximum of .020" clearance between the bushing and pin can be tolerated; if this limit is exceeded, then the bushings or pins or both may need to be replaced.

Following manufacturer's recommended service procedures will result in lower repair costs along with less downtime.

TROUBLESHOOTING PROCEDURES

Tech Tip: Before any maintenance or repair work begins on suspension systems, the technician must read the service manual and become familiar with the procedures and tools required to perform the work in question. Suspension system components and their related parts are often heavy and, in some cases, are under tension due to the weight of the equipment. Failure to follow the manufacturer's recommended disassembly procedures may result in personal injury or equipment damage. Proper training is essential when it comes to this type of work. The equipment must be properly blocked and chocked before any work begins.

The general guidelines provided in this chapter are examples and are not intended be used as a complete troubleshooting guide. It is always advisable to follow the specific troubleshooting procedures that are outlined in the manufacturer's technical documentation; however, a general guideline helps to resolving general complaints that are similar on a wide variety of equipment suspension systems. Be careful when trying to access the nature of the symptoms; similar but different complaints can often be attributed to the same symptoms.

The following troubleshooting chart (**Table 19-4**) can be used to assess the most common types of complaints that may develop on leaf-spring suspension systems.

ASSEMBLY, DISASSEMBLY, AND REPLACEMENT PROCEDURES

General Guidelines for Leaf-Spring Replacement

CAUTION *Caution should be exercised whenever replacing springs. The leafs are under a great deal of tension and could expand unexpectedly if the center bolt were to break during re-torquing, if the U-bolts were not in place. If a spring pin is to be removed be sure that the equipment is blocked properly so that it does not drop unexpectedly and cause personal injury.*

These guidelines are shown for example purposes and should never be used to replace the manufacturer's procedures. The specific replacement procedures will differ from equipment to equipment but, in general, will follow the same basic guidelines as outlined below:

1. Place the appropriate chocks under the wheels to prevent the equipment from moving unexpectedly.
2. Use an approved lifting device to raise the equipment to relive the weight from the springs, and place safety stands under the frame.
3. If shocks are used, disconnect and remove them.
4. Remove the U-bolts and retainers along with the spring bumpers.
5. If the pins are lubricated, remove the grease fittings and lubricators.
6. Remove the nuts that secure the spring shackle pins.
7. Remove the bracket pin and shackle pin by sliding the spring off the bracket pin.
8. To install the new spring, fasten the pivot end of the spring to the frame bracket first.
9. Fasten the shackle or slipper end by aligning it with the other frame bracket.
10. Install the U-bolts to secure the springs to the axle, but do not apply the final torque values until the load is on the springs.
11. Remove the frame supports to place the springs under tension, and torque the U-bolts and other hardware to the manufacturer's recommended values.
12. Whenever replacing spring components, make sure to re-check the torque values after a predefined operating time or distance.

Guidelines for Equalizer Replacement

Tech Tip: Before any suspension repair work can begin, the technician must read over the procedures in the appropriate service manual and become familiar with both the procedures and tools required to perform the work in question. Suspension system components and their related parts are often heavy, and in some cases, are under tension due to the weight of the equipment. Failure to follow the manufacturers recommended disassembly procedures may result in personal injury or equipment damage. Proper training is essential. The use of a torch to burn out the bushings is not recommended on suspension systems, because the fumes given off by the burning bushings can be noxious and may lead to the potential to start a fire.

These guidelines are shown for example purposes and should not be used to replace the manufacturer's technical documentation. The specific replacement procedures will differ from equipment to equipment but, in general, will follow the same basic guidelines outlined below.

Suspension equalizer replacement requires that the front wheels of the equipment be properly chocked to

Table 19-4: LEAF-SPRING SUSPENSION SYSTEM TROUBLESHOOTING GUIDE		
Symptoms	**Cause**	**Corrective Action**
Equipment leans to one side	Broken spring leaves	Replace the spring assembly
	Spring assembly weak	Replace the spring assembly
	Bent or twisted from rail	Correct the frame rail bend or twist
	Spring load capacities are not the same on both sides of the equipment	Replace the spring assemblies with the recommended assemblies
Equipment wanders	Broken spring leaves	Replace the spring assembly
	Wheels not properly aligned	Redo the wheel alignment according to manufacturer's recommendations
	Incorrect caster settings	Adjust caster to manufacturers recommended settings
	Steering gear not centered	Center the steering
	Drive axle alignment not properly set	Align the drive axles according to manufacturer's recommendations
Equipment suspension bottoms out during operation	Equipment overloaded	Follow equipment recommended load capacity rating
	Broken spring leaves	Spring assembly should be replaced
	Weak Spring assemblies	Spring assembly should be replaced
Spring breakage occurs regularly	Equipment overloaded or extremely rough service duty	Equipment load should be reduced to match operating conditions
	Insufficient center bolt torque	Re-torque center bolt to manufacturers recommended values
	Improperly torqued U-bolt nuts	Re-torque the U-bolt nuts to manufacturer's recommended values
	Worn-out or damaged spring pin bushings allowing spring endplay	Replace the spring pins and bushings
Excessive spring noise	Improperly torqued spring center bolt or U-bolts allows excessive spring movement	Inspect components for damage and replace any damaged parts before re-torquing to manufacturer's recommended values
	Loose, bent, or broken spring shackle	Replace the damaged components as required and torque the fasteners to the manufacturer's recommended values
	Spring pin bushing worn out or damaged causing excessive spring endplay	Replace the worn or damaged spring pins

prevent any unexpected movement of the equipment during the replacement procedure. Use an appropriate floor jack to raise the rear of the equipment so that the axles can be supported on safety stands in such a way that the weight is removed from the leaf springs. Once the equipment is properly blocked, remove the wheels from the axles and store in a safe location.

1. If the equipment being worked on has a tandem axle, the cotter pin from the outboard end of each spring retainer should be removed. Now, the retainer pins should be removed.
2. Loosen and remove the nut from the equalizer cap and tube assembly. Remove the inboard bearing washer, bolt, and outboard bearing washer as shown in **Figure 19-32.**
3. Place an appropriate bar between the tube assembly and equalizer cap and pinch the equalizer cap to remove the weight of the equalizer. Position a short piece of pipe through the inboard equalizer cap and tube assembly bolt hole. Gently tap the cap and tube assembly out of the equalizer.

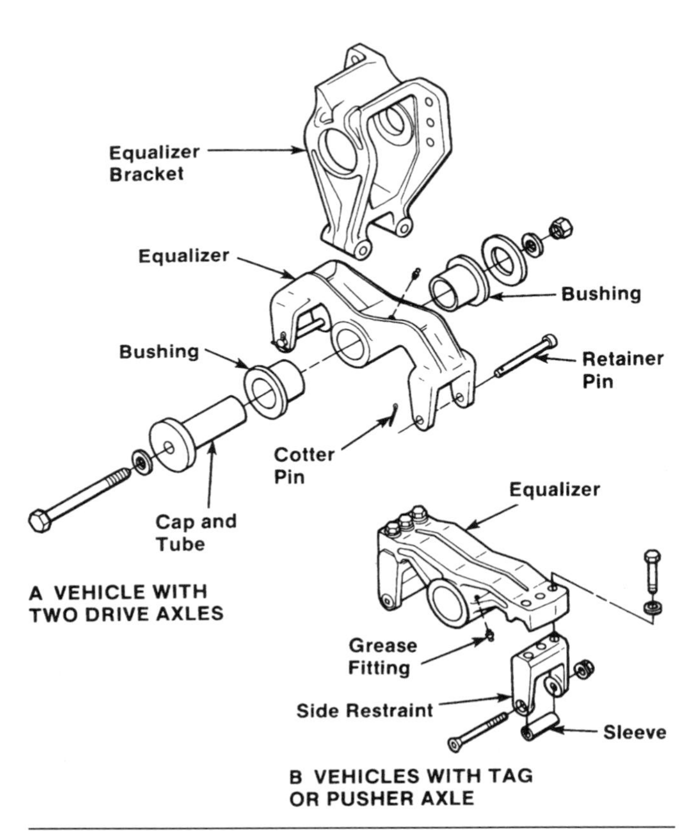

Figure 19-32 Exploded view of a typical equalizer assembly. *(Courtesy of Freightliner, LLC)*

4. At this point, the equalizer bushings should be cleaned so that they can be inspected for damage and wear. If the bushings show evidence of damage or the dimensions are excessive, they should be replaced. Refer to the technical documentation for the equipment in question for the correct specifications.
5. Before installing the bushing back into the equalizer, lubricate the inside of the bushing with the appropriate chassis grease and then install the equalizer back into the equalizer bracket.
6. Before installing the equalizer cap and tube assembly, apply grease to the appropriate area and insert the cap and tube assembly into the equalizer through the bracket. Force the equalizer cap and tube assembly partway into the equalizer and install the inboard wear washers between the inboard equalizer bushing and the bracket. Complete the installation by pushing the cap and tube assembly the rest of the way into the equalizer bracket.
7. Install the equalizer outer bearing washer on to the equalizer cap and tube assembly bolt *before* installing it into the cap and tube assembly.
8. Install the inboard bearing washer, locknut, and tube assembly bolt, and torque the nut to the manufacturer's recommended specifications.
9. Reinstall the wheels that were removed and raise the equipment up to remove the safety stands from under the frame and axles. Lower the equipment to the floor.
10. Before the equipment can be returned to service, the rear axle alignment must be checked and adjusted, following manufacturer's recommended procedures.

Equalizer Beam Bushing Service

CAUTION *Always exercise caution when working with portable hydraulic systems that use high pressures. If a hydraulic leak is suspected, never run your hands over the lines to locate the leak. A high-pressure leak can penetrate the skin, causing serious injury. If any oil penetrates the skin, immediate medical attention is necessary. Operating temperatures on hydraulic suspension systems can often be over 212°F (100°C). The appropriate protective equipment must be used when handling portable hydraulic equipment.*

When any type of replacement work on suspension systems is considered, the area must be thoroughly powerwashed and cleaned so that the components can be inspected for physical damage and fatigue cracks. If the rubber bushings show any signs of cracking or deterioration, they should be replaced. In most cases, specialized service tools are required to replace the equalizer beam bushings. If however, the specialized tools are not readily available, then a portable hydraulic press and the appropriate steel piping can be used to replace the equalizer beam bushings.

The load cushions or jounce blocks should be inspected for signs of distortion or damage. These blocks act as shock load insulators and are important to the operation of the suspension system. Before performing cleaning and assembly operations in this area be sure not to expose the load cushions to any type of chemicals or petroleum products; they are often made from synthetic material which can be easily damaged by these products. When visually inspecting the load cushions, check that the blocks are still bonded to the metal insert and are not about to separate. If the unloaded height of the load cushion is shorter than $\frac{1}{4}$ inch below that of a new one, then it must be replaced.

Reconditioning Equalizing Beam Suspension Systems

As for all major repair work, the equipment must be thoroughly cleaned before any work can begin. Once the equipment is clean, a complete visual inspection must be made to determine the extent of the repair work and the parts that may be required to complete the work. In some cases it is important to be sure that the replacement parts are available before the work actually gets underway. No one can afford to have their equipment torn apart and waiting for replacement parts to arrive. Planning ahead is in everyone's best interest.

Tech Tip: Before any major suspension repair work can begin, the technician should carefully read over the procedures in the service manual and become familiar with the procedures and tools required to perform the work in question. Failure to follow the manufacturer's recommended disassembly procedures may result in personal injury or equipment damage. Proper training is essential. The use of a torch to burn out the bushings is not recommended on suspension systems, because the fumes given off by the burning bushing material can be noxious and can lead to an equipment fire. Because a hammer and various punches are used in the removal and installation procedure, safety glasses and a face shield should be used.

Equalizer beams are manufactured from different materials. This can pose a problem if the installation procedure is too aggressive and uses too much force. Aluminium and cast steel equalizer beams can be easily damaged during the installation. Care must be exercised during installation so as not to overstress the metal. Nodular iron equalizer beams must be installed in a specific way so that the markings indicating "up" are all facing in the correct direction. An arrow or the word UP is cast into the side of the equalizer beam. Reinforced gate pads are built into each end and the middle section of the beams.

Once a thorough inspection has been completed and the parts required to complete the repairs are in place, it is time to prepare the equipment for repair work. Locate the equipment on a clean hard surface that is reasonably flat. Place wheel chocks on both axles of the equipment. Remove all driveshafts from the rear axles and mark their location. It is always a good to mark and reinstall the parts in the same location as they were removed from.

Next, remove the four saddle caps that are used to lock the center bushing of the equalizer beams to the saddle assemblies (**Figure 19-33**). The **torque rods** must be disconnected from the axle housing by loosening the lock nuts and driving the shaft from the bushing and axle housing brackets. A suitable hoist is required to lift the truck frame enough so that there is room for the lower part of the saddle to clear the top of the axle housing. Visually ensure that the axles are still blocked properly so that they do not pivot on the wheels. Use safety stands or approved blocking to support the equipment frame.

Leaf-Spring Equalizer Beam Disassembly

As with all disassembly procedures, always read over the manufacturer's technical documentation before attempting this repair procedure. Although there are several design variations of equalizer beam mounting configurations, only one of the three most common types, the bolt-type beam end mounting, will be used as an example. This mounting is shown in

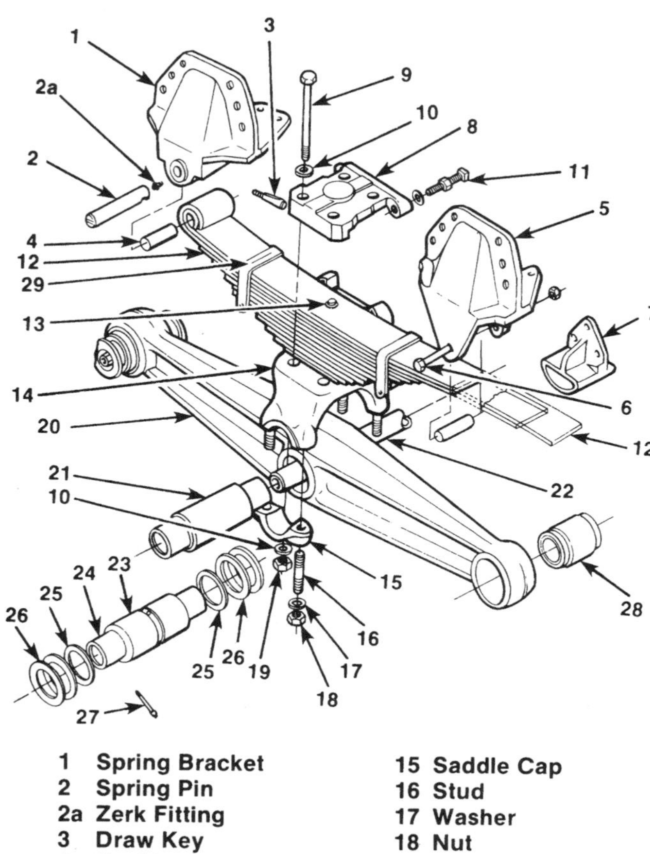

1 Spring Bracket
2 Spring Pin
2a Zerk Fitting
3 Draw Key
4 Bushing
5 Spring Bracket
6 Rebound Bolt with Spacer
7 Spring Bracket (extended leaf only)
8 Top Pad
9 Top Pad Bolt
10 Washer
11 Setscrew
12 Leaf Spring Assembly
13 Center Bolt
14 Saddle Assembly
15 Saddle Cap
16 Stud
17 Washer
18 Nut
19 Nut
20 Equalizer Beam
21 Center Bushing
22 Cross Tube
23 Center Bushing Bronze
24 Sleeve
25 Seals
26 Washer
27 Zerk Fitting
28 End Bushing
29 Part Number

Figure 19-33 Exploded view of an equalizing beam suspension assembly. (*Courtesy of International Truck and Engine Corporation*)

Figure 19-34. This section explains the procedure used to disassemble the leaf-spring equalizer beam suspension.

1. The equalizer beam must be disconnected from the axle housing before the equalizer beam end bolt can be removed.
2. Drive the bushing adapters and sleeve out of the bushing and axle housing brackets. The notches in the adapters will help locate and keep the chisel in place during this operation. To wedge out the adapter, drive the chisel into one side first then into the other.

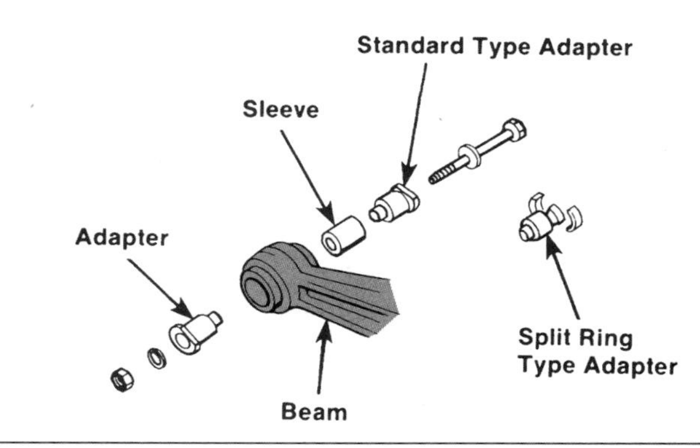

Figure 19-34 Exploded view of a three-piece adapter bolt-type beam end mounting. (*Courtesy of International Truck and Engine Corporation*)

3. Remove the first adapter and proceed to remove the opposite adapter with a hammer and drift.
4. Now, the equalizer beams can usually be separated from the cross tube. The cross tube pivots on the inner sleeve of the center bushing and slides back and forth approximately three inches. This results in the cross tube having a polished appearance near the center bushing, which is considered normal.
5. Once the cross tubes are separated, remove the saddle assemblies, springs, or cushions.
6. The spring and saddle assembly must be supported on a floor jack and safety stand before the spring-aligned set screws can be loosened. Next, remove the spring top saddle bolts and nuts. The top pad can now be removed safely. Lower the floor jack to remove the saddle.
7. To remove the spring assembly, place the floor jack under the spring and loosen the locknut on the spring pin draw key, as shown in Figure 19-33.
8. Caution must be exercised in this step. The threads on the draw key can be easily damaged. Back off the nut just enough to protect the draw key threads before tapping the nut with a soft blow hammer to loosen the draw key.
9. After the draw key has been removed, drive the spring pin out of the spring bracket and lower the spring assembly from the frame using the jack.

Leaf-Spring Equalizer Beam Assembly

1. Position the spring in the spring saddle in such a way that the bolt head of the spring center bolt is positioned in the hole in the saddle.

2. Place the spring top pad over the cup on the main spring leaf, and install the bolts and nuts that secure the saddle to the top pad. Seat the nuts on the bolts, but do not torque them yet.
3. To align the spring in the saddle, tighten the spring aligning set screws to the recommended torque, then tighten the aligning screw locknuts in place before torquing the saddle to top pad bolt nuts to the manufacturer's specifications.
4. Position a jack on the spring and saddle so that the assembly sits in the front and rear spring mounting brackets, and align the spring eye with the spring pin bore in the front bracket.
5. Install the spring pin while aligning the draw key slot in the pin with the draw key bore in the bracket, install the draw key, lock washer, and nut, and torque up to manufacturer's recommended values.
6. Replace the zerk grease fitting in the spring pin and grease with the recommended lubricant.

Solid Rubber Equalizer Beam Disassembly

- Begin by removing the four drive pin bushing retainer caps (Figure 19-34) before removing the drive pin nuts.
- Next, remove the load cushions and saddle from the vertical drive pins. This step may require that the drive pin nuts be cut to facilitate removal. Depending on the difficulty of the removal, it may be necessary to replace the drive pin bushing.
- Remove the torque rods from the cross member frame bracket and install the new spring, cushion, and saddle assemblies.

Solid Rubber Equalizer Beam Assembly

1. Assemble the drive pin bushings to the saddle and install the bearing caps in place *before* installing the rubber load cushions on the saddle.
2. Lubricate the drive pins with multipurpose grease on the frame brackets, before assembling the frame brackets to the saddle. Install the drive pin nuts and washers with the flanges facing down and tighten them snugly. If the frame hangers have been removed, the nuts must be torqued to the manufacturer's specifications after they have been assembled to the chassis frame.
3. Install the preassembled saddles, load cushions, and frame hangers to the chassis frame, and install the mounting bolts, lock washers, and nuts in place. Torque the mounting bolts to the manufacturer's recommended torque values *before* torquing the drive pin nuts.

Cross Shaft Installation

Note: This procedure applies to both the leaf-spring and rubber spring equalizer beam suspensions. The final torque values on the mounting nuts should not be applied until the wheels have been installed and lowered to the ground.

1. Before inserting the bushing adapters, align the beam end bushing with the hanger brackets. Lubricate the adapters with multipurpose grease, and place the flat surface of each adapter in the vertical position.
2. With a hammer, lightly tap the adapters into the bushing until the adapter flanges seat against the outer face of the hanger brackets. Install the bolt and locknut in place. *Do not* torque up to specifications yet. When installing rubber bushed equalizing beams to the axles and saddles, do not tighten the nuts until the torque rods are installed and the equalizer beams are level with the frame. This reduces the chance of creating torsional windup in the suspension.

Reinstalling Equalizing Beam Suspension Assemblies

1. One end of the cross tube must be positioned into one end of the center sleeve of the equalizer beam. Ensure that the tube seats correctly into the sleeve. Now, install the other equalizer beam into the cross tube and locate the beam in the spring saddle.
2. At this point, the axle assemblies are ready to be positioned under the center of the saddle; make sure that the outer bushings are positioned in such a way that they line up with the center of the saddle legs. The frame can now be lowered carefully. Be sure that the saddles center on the beam end bushings. Install the saddle caps and nuts. *Do not* apply the final torque until the torque rods are installed and the equalizer beams are level with the frame.

3. Make sure that the cross shaft is free to float in the bushing assemblies on each side; this ensures correct alignment of the left and right equalizer beams.
4. Torque the saddle caps to ensure that the sleeve is secure to the bushing. This tightening procedure has no effect on the pivot of the cross shaft.
5. Connect the torque rods to the axle brackets and the frame brackets. During the torque sequence, tap the bracket with a hammer to seat the tapered torque rod stud into the bracket. If the torque rod ends have straddle mounts with two holes, as shown in **Figure 19-35,** a spacer is required between the bracket and cross member to get the correct axle adjustment. Once these steps are complete, torque the shaft nuts and equalizer beam end fittings to the manufacturer's recommended torque values.
6. Reinstall the axle shafts in the correct locations as per markings during disassembly. Reconnect all the brake lines, drive shafts, and hardware that were removed during the disassembly procedure.

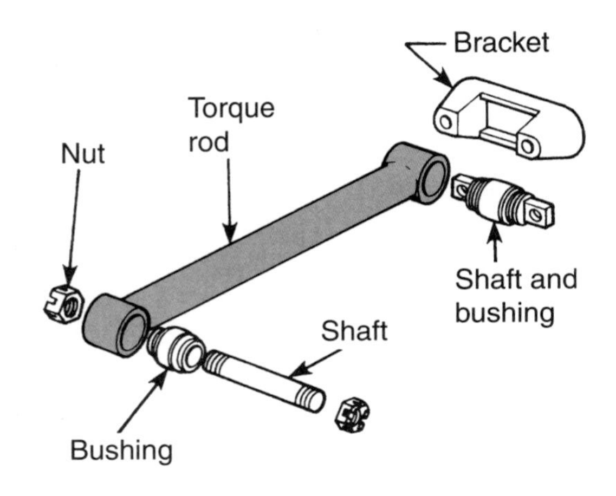

Figure 19-35 Exploded view of a torque rod assembly. (*Courtesy of International Truck and Engine Corporation*)

ONLINE TASKS

1. Use a suitable Internet search engine to research the configuration of a typical off-road suspension system. Identify a piece of equipment in your particular location, paying close attention to the suspension system design and arrangement, and identify whether it has automated ride height compensation ability. Outline the reasoning behind the manufacturer's choice of design for this particular application.
2. Check out this url: *http://www.meritorhvs.com/Product_CVS.aspx?product_id=47&top_nav_str=hvs* for more information on suspension system configuration and design.

SHOP TASKS

1. Select a piece of equipment in your shop that has a chain assembly and identify the type of components that make up the suspension system assembly. Identify whether or not the suspension components have excessive wear and what type of design is used on this arrangement.
2. Select a piece of off-road equipment in your shop that uses a conventional spring suspension system and identify the type and model of the equipment. List all the major components and the corresponding part numbers of the major suspension system components. Identify and use the correct parts book for this task.

Summary

- An off-road suspension system's primary purpose is to support the total weight of the equipment. If no suspension system were used, the shock loads encountered during operation would be transmitted directly to the equipment frame, which would quickly lead to metal fatigue and cracks in the frame structure.
- Off-road suspension systems come in a wide variety of designs, some can be as simple as those found on many types of slow moving LHDs (load haul dump equipment). An LHD consists of a drive axle mounted directly to the frame with only the tires to absorb ground impact. Conventional leaf-spring systems are still used on many haulage truck

applications in pits and quarries. Some complex equipment systems use nitrogen-filled shocks that are monitored and regulated by the operator, or an onboard ECM that monitors and controls the operation of the suspension system.

- Un-sprung weight is the total weight of the axle plus all the components that are mounted to it, including all the brake hardware and suspension components. Sprung weight is defined as the load that the suspension must support during normal operation.
- When working on suspension systems that use gas/suspension cylinders, be sure that all the nitrogen in the suspension cylinder has been released before checking the oil level. Never remove any plugs or valves from a suspension cylinder if the rod is not fully retracted and all the nitrogen pressure is dissipated.
- Large haulage trucks use a gas/suspension cylinder at each front wheel end instead of a conventional kingpin as part of the steering linkage. These oil/pneumatic cylinders act as shock absorbers in the suspension.
- Large articulating haulage trucks incorporate a suspension system that is often integrated into the hydraulic system circuit to provide ride height adjustment.
- A key advantage of independent gas/hydraulic suspension systems is their simplicity of design and ability to absorb changing load and road conditions. The gas/hydraulic suspension systems take the place of conventional mechanical leaf-type suspension systems on off-road haulage trucks.
- To correctly diagnose a suspension system problem it is imperative that the technician take into consideration that suspension cylinders operate independently of each other.
- Before making any adjustment to a suspension system, the technician must consider that the pressures in each suspension cylinder can be affected by changes in ride height due to leaks at opposite ends of the equipment. For example, when the left rear suspension cylinder ride height is low, it is likely that the right front suspension cylinder is riding high, due to a shift in load weight.
- Nitrogen gas is the only gas that should be used to charge suspension cylinders. Other gases, such as oxygen, are highly explosive and must never be used to charge suspension cylinders. Be sure that the gas cylinder to be used is indeed nitrogen and not oxygen; the cylinders can look very similar.
- Pressurized oil and escaping nitrogen gas can pose a safety hazard to anyone who comes in contact with either one. Be sure to allow the pressure in the suspension cylinders to equalize by allowing the valves to remain open for approximately five minutes before closing them.
- When charging, if one cylinder reaches the height line first, close the gate valve to that cylinder and allow the remaining cylinder to fill until it reaches the correct height.
- Be sure that the charge valves are fully closed before connecting the nitrogen charge lines to the suspension cylinders to prevent oil from entering the nitrogen charge lines.
- Suspension dimensions can be affected by temperature differences between the adjustment location and the operating location. For example if the temperature difference between the shop where the adjustment is performed and the outside temperature where the equipment is operating is greater that 20°F (11°C), a different calculated dimension is required for extra shim thickness. It must be calculated based on the temperature difference.
- Tandem drive applications that are used on articulating haulage trucks often use equalizing beam suspension systems; these systems have been proven to be up to the task of operating in rough service applications.
- The design of the equalizing beam suspension allows the load to be equally distributed between the axles while compensating for road irregularities. The starting and stopping forces, known as windup torque, are absorbed by the torque rods. A cross tube is used to connect the equalizing beams together; this reduces potentially damaging load transfers that occur during operation.
- A specialized hydraulic press is often required to properly remove and install the bushings and sleeve assemblies in most equalizer beam suspension systems.
- Caution should be exercised whenever replacing suspension springs. The leafs are under a great deal of tension and could expand unexpectedly if the center bolt were to break during re-torquing, if the U-bolts were not in place.

- If a grease fitting on a spring pin refuses to take grease during a scheduled maintenance service, try taking the weight off the spring before changing the grease fitting, or try using a hand grease gun, which is capable of higher operating pressures than air-powered systems.
- Exercise caution when working with portable hydraulic press systems that use high pressures. If a hydraulic leak is suspected, never run your hands over the lines to locate the leak. A high-pressure leak can penetrate the skin, causing serious injury. If any oil penetrates the skin, immediate medical attention is necessary. Operating temperatures on hydraulic suspension systems can often be over 212°F (100°C).

Review Questions

1. Which of the following best describes the term un-sprung weight?
   A. the total weight of the axle including all components that are mounted to it
   B. the total weight of the each end of an axle
   C. the total weight of the equipment
   D. the rebound rate of the equipment
2. Which of the following symptom would be the result of a broken spring pack center bolt?
   A. erratic steering
   B. U-bolt failure
   C. bouncing front end
   D. broken leaf spring
3. If you discovered that a spring pack had a single cracked leaf, which of the following should be done?
   A. The complete spring pack should be replaced.
   B. Dismantle the spring pack and replace the broken leaf.
   C. If the crack is small, weld it.
   D. Do nothing, it is only a crack.
4. While lubricating a spring pack assembly you find that the bracket spring pin will not take grease. Which of the following should be done first?
   A. change the grease fitting on the bracket spring pin
   B. replace the bracket spring pin
   C. remove the weight from the spring and try to grease again
   D. heat it up with a torch to unplug the grease passage
5. How are the springs loaded in a spring pack under tension?
   A. U-Bolts
   B. centerbolt
   C. spring pins
   D. torque rod
6. What term is used to describe the greatest compressed condition of a suspension spring?
   A. jounce
   B. rebound
   C. deflection rate
   D. un-sprung load
7. What purpose does the torque arm serve on a suspension system?
   A. to control axle torque
   B. to control rebound rates
   C. to control jounce
   D. to control the suspension height

8. Which of the following terms describe an *equalizing beam* suspension?

   A. torsion bar
   B. walking beam
   C. air spring
   D. multi-leaf variable rate

9. Which of the following functions can be performed by torque rods?

   A. dampen axle windup caused by driveline torque
   B. align axles
   C. dampen axle twist under braking
   D. all of the above

10. What do large quarry trucks often use in the front suspension to replace the kingpin?

    A. suspension cylinder
    B. torque rods
    C. coil spring
    D. leaf spring

11. How are the rear cylinder barrels connected to the equipment frame on a large quarry truck?

    A. torque rod
    B. kingpin
    C. cylinder stem
    D. spring pin

12. Where does the adjustable ride control on a haulage truck get its oil supply?

    A. from the brake or hydraulic circuit
    B. from an independent hydraulic system
    C. does not need an external supply source
    D. none of the above

13. What would happen if one cylinder were to develop a gas leak in a suspension cylinder on a haulage truck with an independent gas charged system?

    A. would not affect the other three suspension cylinders
    B. all suspension cylinders would experience a pressure drop
    C. pressure would rise in the remaining three cylinders
    D. ride height would be affected

14. What procedure must be observed when charging a suspension cylinder with nitrogen?

    A. compensate for temperature if required
    B. drain the oil from the cylinder first
    C. support the equipment to keep it level
    D. take the weight of the suspension cylinder

15. When installing a new accumulator on the ride control system, what is the first step to be taken?

    A. fill the accumulator with oil first
    B. purge the accumulator of all air
    C. check the pressure in the accumulator
    D. confirm that hydraulic pressure is present

Glossary

Absolute pressure a pressure scale with its zero point at absolute vacuum. Expressed as pounds per square inch absolute (psia).

Accumulator (hydraulics) a device used to store potential energy. May refer to a pressure storage device in a hydraulic circuit or electrical energy storage devices.

Active suspension a hydraulic suspension system used on off-road equipment and large haulage trucks. The system has the capability of making changes to the ride height.

Actuator device that produces mechanical motion in response to an electrical, air, or hydraulic signal.

Adaptive suspension a suspension designed to adjust to load and road conditions; this usually means an air suspension system. Adaptive suspensions can either be non-electronic or electronically controlled. The term *active suspension* is also used.

Aeration the blending of a gas into a liquid such as the mixing of air into oil by the churning action of rotating gears.

Aftercooler describes a variety of different heat exchangers used on engines, transmissions, and air brake systems. An aftercooler on an air brake system is used to condense water from moisture-saturated air leaving the compressor by cooling it.

Air dryer a device, usually located downstream from the air compressor, designed to remove moisture from the air system. Air dryers use two principles to eliminate air from an enclosed brake circuit: cooling/condensation and dessicant media.

Air/hydraulic master cylinder an air-actuated hydraulic master cylinder that enables a pneumatic brake control circuit to actuate a hydraulic brake circuit.

Air-over-hydraulic brakes brake system that uses an air control circuit to actuate a hydraulic slave circuit.

Air-over-hydraulic intensifier converts control circuit air pressure (from the brake treadle valve) into hydraulic slave circuit pressure to actuate the foundation brakes.

Air spring suspension any one of a number of different types of pneumatic suspension.

Allen wrench hex-shaped wrench designed to fit into recessed hex Allen screws.

Alloy mixture of metallic elements to create a specific metal. Steel is an alloy of iron and carbon.

Amboid gear set a gear arrangement in which the axis of the drive pinion gear is set above the radial center line of the crown gear.

American National Standards Institute (ANSI) administrator and coordinator of U.S. private sector, voluntary standardization. ANSI standards are developed by consensus.

Analog use of physical variables to represent numbers such as voltage or length.

Annulus the largest gear member of a simple planetary (epicyclical) gear set. Located outside of the planetary gears.

Anti-corrosion agent a solution or additive that inhibits chemical corrosion.

Application pressure the hydraulic pressure that is used to apply the service or park brake on a hydraulic brake system.

Articulating steering a steering design that uses hydraulic cylinders connected to the front and rear sections of a loader frame. The cylinders are connected together through a hinge pin, which articulates the front and rear sections of equipment to achieve steering control. The axles used on this equipment are non-steer axles.

Axial load a pure tension or compression load acting along the long axis of a straight structural member or shaft.

Axial runout the amount of wobble on a shaft or rotor when viewed from the front or back.

Axle shaft a shaft that is splined to the differential at one end and transmits torque to the wheel ends.

Back pressure the potential pressure that is created on the return side of a hydraulic circuit. This added resistance to flow increase the pressure required to move the load.

Backlash the lost motion or looseness (play) between the faces of meshing gears or threads.

Baffle a separating plate used in a hydraulic tank to separate the inlet side from the outlet side.

Band type clutch a device used to connect or disconnect (engage or disengage) two rotatable parts so that they revolve, as one unit or separately, by means of a friction band wrapped around a rotating flywheel.

Banjo housing the housing of an axle assembly that contains the differential, brake, and drive components.

Bar metric unit of pressure measurement roughly equivalent to one unit of atmospheric pressure.

Beach marks the term used to describe shock loading on gears which shows up as small curved rings on the surface of the component.

Belleville springs see *diaphragm/bevel spring.*

Belleville washer a type of cone-shaped washer that provides a pre-load on the part to which it is. Used on older crawler tracks to provide a seal to keep dirt from getting between the pins and bushing.

Bernoulli's Principle states that the sum of the kinetic and potential energy in a moving, incompressible fluid will remain the same: so, in a fluid moving in a steady state in through a circuit, if the potential energy decreases, the kinetic energy must increase so that the sum of the two remains the same.

Bevel gear a gear design used in differential spider gears in a conventional differential to split the flow of power proportionally between wheel ends.

Boom circuit part of an excavator implement hydraulic circuit. The boom arm connects to and pivots on the upper cab structure. The boom circuit is actuated by double-acting hydraulic cylinders.

Boyle's Law states that for a given, confined quantity of gas, pressure is inversely related to the volume so that as one value goes up the other has to go down.

Brake application pressure the hydraulic pressure that is used to apply the service brakes. This pressure is modulated by the foot brake valve and can vary from zero to the maximum allowable brake application pressure limit.

Brake chambers foundation brake actuators that convert pneumatic pressure into linear force to actuate slack adjusters.

Brake pads the friction faces and plate assembly that is used in disc brakes calipers to stop a vehicle.

Brake pressure regulated by either a charge valve or a pressure regulating valve, which is often part of the foot brake valve assembly.

Brake pump a hydraulic pump that is used to charge the brake circuit with hydraulic oil. Used to apply the service brake or release the spring-applied park brake.

Brake shoes the friction faces and lever assembly that is forced against the drum in a brake system to slow down or stop a vehicle.

Brake valve a hydraulic valve that is foot operated. Modulates pressure applied to the wheel ends of a vehicle to slow down or stop it quickly or smoothly.

Brinelling grooves worn into a cross by the needle rollers and caused by a lack of lubrication. Can also be caused by excessive loading.

Broken-back driveshaft a driveshaft installation where the U-joint angles are equal, but the yokes are not parallel.

Bucket circuit part of an excavator implement hydraulic circuit. The bucket connects to and pivots on the stick; the bucket circuit is actuated by a double-acting hydraulic cylinder.

Bull-type final drive a term used to describe a drive that utilizes a single gear reduction. Often found in a final drive assembly of a bull dozer.

Burst pressure the pressure at which a fluid power device is rated to fail or rupture when subjected to overpressure. Usually $2\frac{1}{2}$ to 4 times specified working pressure.

Bypass an optional or secondary route for fluid flow in a circuit.

Calipers a disc brake component that is used in conjunction with brake pads to apply force on a rotor to slow down or stop a vehicle.

Cam an eccentric or oval shaped component.

Cam ring an eccentric liner, usually stationary, within which mobile pump members rotate.

Camber the tilt of a wheel assembly when viewed from the front of the vehicle from true vertical. A wheel that is leaning inward at the top from true vertical is said to have negative camber. A wheel that is leaning outward at the top from true vertical is said to have positive camber.

Cardan universal joint a device in which a cross connects the yokes of a driveshaft.

Carrier assembly the drive axle assembly the holds the gearing that changes the flow of power at right angles to propel the wheels.

Carrier rollers located above the frame rails. Their purpose is to support the weight of the top section of track as it rolls between the idler and the sprocket. These rollers generally have a single flange, which aids in controlling track sag and whipping during operation.

Case drain (hydrostatic drive) a term used to identify the return oil flowing back to the hydraulic tank from the case or housing of a piston pump or motor. The oil flow is a result of internal leakage built into the piston pump required for internal lubrication.

Caster the tilt of the kingpin in either the forward or backward direction when viewed from the side. Positive caster is when the kingpin leans to the rear of the centerline. Negative caster is when the kingpin leans forward of the centerline.

Cavitation bubble collapse erosion in hydraulic or engine coolant circuits.

Centerbolt a bolt that is used to hold a multi-leaf spring pack together.

Centrifugal clutch a clutch that utilizes a centrifugal force to apply pressure against a friction disc in proportion to the rotational speed of the unit.

Ceramic facing a clutch disk facing made from a mixture of ceramics and copper or iron.

Chain drive a term used to describe the final drive design used on a grader tandem drive.

Chain hoist a ratcheting block and chain lift device. Either manually or electrically actuated.

Charge check valve a check valve that is internally located in the charge valve. Prevents the flow of oil in the charge valve from returning back to the tank.

Charge pressure the pressure in an accumulator used in a hydraulic system that has fluid under pressure stored in it.

Charge pump a positive displacement pump that is used to supply oil flow to a piston pump in a hydrostatic drive circuit.

Charge relief valve a relief valve that is located internally in a brake charge valve. Used to ensure that maximum brake pressure is never exceeded.

Charge valve an automatic regulating valve that is capable of regulating pressure in a brake circuit between a high and a low setting.

Check valve a valve in a fluid circuit used to permit flow in one direction only.

Cherry picker a portable shop hoist device consisting of a hydraulically-raised lift arm or boom.

Chock the blocking of a wheel on equipment before performing any work. A triangular metal object that looks like a wedge is used to chock the equipment from moving.

Closed-center a hydraulic circuit in which the pump is not unloaded to the reservoir when in the center or neutral position.

Coefficient of friction the coefficient of friction can be *static* or *kinetic*. The coefficient of *static* friction is defined as the ratio of the maximum static friction force between the surfaces in contact to the normal force. The coefficient of *kinetic* friction is defined as the ratio of the kinetic friction force between the surfaces in contact to the normal force. Both static and kinetic coefficients of friction depend on the pair of surfaces in contact; their values are usually determined experimentally. For a given pair of surfaces, the coefficient of static friction is larger than that of kinetic friction.

Coil spring a coil of wire that contributes to the motive force of a spring. In extension and torsion springs, all the coils are active coils. In compression springs, only the coils that show daylight between them are active coils. A spiral-shaped spring that can be compressed or extended without permanent deformation.

Collar shift has parallel shafts fitted with gears in constant mesh. If gear speeds are different grinding will occur. Used for low gear on some older model transmissions.

Come-along a cable or chain hook, hand-actuated, linear ratcheting device used for lifting or applying separation pressure.

Compensator control a component that is used on a variable-displacement pump or motor to control the displacement as required to compensate for pressure changes.

Compound angle a term used to describe the angular relationship of a driveline on both horizontal and vertical planes.

Compressibility the measure of the volume change in a fluid when subjected to pressure.

Compressor any of a number of mechanically driven devices used to compress a gas. In air brake systems, compressors create the potential energy required to actuate foundation brakes.

Conductor in electricity, any material that permits electrical current to flow; any element with less than four electrons in its valence shell. In hydraulics, enclosed circuits or pipes that route hydraulic media through the circuit.

Cone-type clutch a cone clutch serves the same purpose as a disk or plate clutch. However, instead of mating two spinning disks, the cone clutch uses two conical surfaces to transmit friction and torque.

Constant-mesh a transmission that contains helical gears. In this type of transmission, certain countershaft are constantly in mesh with the main shaft gears. The main shaft meshing gears are arranged so that they cannot move endwise. They are supported by roller bearings that allow them to rotate independently of the main shaft.

Cooler any of a number of different types of heat exchanger.

Cooling circuit a specialized oil circuit often used in a hydraulic brake system to directs a metered amount of oil through a cooler to remove excess heat.

Coordinated steer both front and rear steer axles turn in or out in opposite directions, allowing the equipment to turn in a tighter circle.

Counterbalance valve a valve that creates resistance to flow in one direction while permitting unrestricted flow in the opposite direction.

Coupler device that links two components or circuits such as the fittings used in hydraulic circuits. In most cases, couplers have both a male and a female component.

Coupling any one of a number of different types of mechanical connection. Used to transfer fluid, linear, or rotary force.

Crab steer used on equipment that requires precise steering control in confined spaces. All four wheels turn in the same direction at the same time, allowing the equipment to move to the left or right while maintaining load alignment.

Crack pressure the pressure at which a pressure-actuated valve or device opens to pass fluid.

Cracking pressure a condition that occurs in a hydraulic system when a pressure reaches a predefined level allowing excess flow to bypass a circuit.

Cross the "body" of a universal joint.

Cross tube used to connect the equalizing beams together. Reduces potentially damaging load transfers that occur during operation on haulage truck with equalizer bean suspensions.

Crown gear (differential) the larger gear in a differential that is driven at right angles by a pinion gear. Some manufacturers refer to it as a ring gear.

Cylinder (hydraulic) a component that is capable of converting fluid energy into mechanical motion.

Cylinder pressure describes the existing nitrogen charge pressure found in suspension cylinders on off-road equipment.

Dampen to buffer or insulate.

Dampened discs clutch friction discs with built-in coaxial springs used to dampen driveline oscillation and shock loads. Many of today's clutches use dampened discs due to the run slow, gear fast, drivelines used with high torque engines.

Delta wing a particular design of bearing cap commonly found on heavy-duty driveshaft U-joints.

Diaphragm/bevel spring like coil spring equipped clutches, diaphragm springs operate under indirect pressure using a retainer and lever arrangement to exert pressure on the pressure plate. Also called a Belleville-type spring.

Differential drive axle gearing used to transmit torque at right angles while at the same time capable of varying speed to each wheel during cornering.

Differential carrier the component that contains the differential gearing in an axle assembly.

Differential hydraulic cylinder a hydraulic cylinder that has opposed piston areas which are not of equal size.

Differential lock a term used to describe what takes place when the both axles on a drive axle are locked together mechanically.

Differential side gear the side gears mesh with the pinions and have internal splines that connect to the axles.

Differential spider the cross that is used in a conventional differential to hold the four pinion bevel gears.

Differential steer this design uses a series of planetary gears to provide torque input to the drive tracks, with a third input from a steering motor that can increase the speed of each track independently. When the speed of one track is increased, the other track slows down proportionally.

Direct drive gearing of a transmission so that one revolution of the engine produces one revolution of the transmission's output shaft. The final drive ration of a direct drive transmission is 1:1.

Directional control valve a valve that functions to direct or block flow through selected channels or circuits.

Directional valve a hydraulic valve that can control the direction of oil flow in a circuit or stop it completely.

Disc brake vehicle brake system in which squeeze pressure is applied to a rotor by a caliper. Both pneumatic and hydraulic disc brakes are used on off-highway vehicles.

Disk-and-plate-type clutch a device used to break torque flow between an engine and transmission and to apply/release application force to other units. Utilizes a friction disc sandwiched between a flywheel and pressure plate.

Displacement the amount of fluid that passes through a pump or motor during one revolution or stroke.

Displacement control valve a valve that can vary the displacement of oil flowing in a hydrostatic drive circuit.

Double-acting cylinder a hydraulic cylinder that is capable of converting hydraulic energy into mechanical motion in either direction.

Double check valve used when a component or circuit must receive or be controlled by the higher of two possible sources of fluid pressure. The valve is usually designed to bias the higher source pressure.

Double reduction axle a differential that uses two different gear sets to achieve two speeds at the axles.

Double reduction drive the term used to describe a final drive that has two separate gear reductions in line to each other. Generally located in a separate housing at the track end of the drive of a bull dozer.

Drift movement or wander from a preset position or value. Can be applied to hydraulic implement position or engine rpm.

Drive bushings vertical bushings on each drive pin that allow vertical movement during loading and unloading while still maintaining suspension alignment.

Drive gear a gear that drives another gear or causes another gear to turn.

Driveline the power transmitting components that connect the transmission output yoke to the drive axle.

Driven gear a gear that is being driven or caused to turn by another gear that is meshed with it.

Driven member a gear that is driven by another gear.

Driveshaft the shaft that connects the powertrain components together to transmit torque to the wheels.

Drive steer type of drive axle commonly found on heavy equipment. Provides the capability to steer and propel the equipment at the same time.

Drivetrain the group of components that generate power and deliver it to the road surface, water, or air. This includes the engine, transmission, driveshafts, differentials, and the final drive.

Drum the rotating component of a drum brake system that is fixed to the wheel assembly, which rotates around stationary set of brake shoes. Brake action is achieved when the shoes are forced out against the inside of the drums by mechanical force that is controlled through air or hydraulic pressure.

Drum brakes a brake system used on equipment that uses a rotating drum attached to a wheel assembly that rotates around stationary friction shoes.

Dryer reservoir module (DRM) an integrated reservoir, dryer, governor, and pressure protection pack assembly used in some vehicle air supply circuits.

Dry joint a condition associated with track systems where by the counterbore sidewall becomes damaged quickly from contact with the bushing when allowed to run dry.

Dryseal coupler threaded fitting in which the roots and crests of the opposing threads contact before the flanks contact and a seal is created by crushing the thread crests. Dryseal couplers are usually intended for one-off usage and usually do not require thread seal compound.

Dry-type clutch separates the engine from the transmission. There is a dry connection between the engine and transmission. It is not bathed in fluid that robs it of some energy. Since the surfaces of a wet clutch can be slippery, stacking multiple clutch disks can compensate for slippage.

Dynamic friction kinetic (or dynamic) friction occurs when two objects are moving relative to each other and rub together. The coefficient of kinetic friction is typically denoted as μ_k, and is usually less than the coefficient of static friction.

Dual circuit application valve the *foot valve* used with FMVSS 121 compliant service brakes and sometimes with off-highway equipment. Masters service applications from primary and secondary circuits. Also known as *treadle valve*.

Electro-hydraulic valve a hydraulic valve, controlled by a electrical signal, which controls the flow.

Electro-kinetic friction applies to the forces restricting the movement of an object that is already in motion on a relatively smooth hard surface.

Electro-magnetic clutch an electrically-actuated clutch used for on-off cycles. Usually spring-loaded (mechanically) to engage and electrically energized to release, but the opposite can also be true.

Elevated sprocket this sprocket is not mounted to the track frame but is attached to the final drive. The elevated position keeps the final drive out of the dirt. If the equipment has an elevated sprocket system, there are two idlers, a front idler and a back idler. Equipment with a conventional oval track requires only one idler.

Elongation the wear that takes place on crawler tracks. As track components wear, the track experiences elongation.

End yoke the name of the output yoke on a powertrain component that connects to the driveline yoke.

Endplay axial play in a shaft wheel or rotor. A dial indicator is used to measure the endplay.

Equalizer bar or beam a term used to describe a beam that is used in a tandem equalizer beam suspension system.

Equalizing beam suspension a suspension system used on off-road equipment that lowers the center of gravity of the axle load.

Expanding shoe-type clutch a form of clutching device that utilizes a set of brake shoe-type friction material within a drum used to apply and release engine torque to a specific drive-transmission.

Extra long (XL) (Front) track system: the XL track system has an extra long front track design that provides better balance with extremely good flotation characteristics, which is better suited for finish grading on free flowing material.

Extra long (XR) (Rear) track system: the XR track system has an extra long rear section that is more suitable for pulling scrapers, as well as for winching or ripping operations. In the agricultural applications, a belt track is often used to reduce the zone of compaction. This design increases ground contact significantly over conventional LGP track systems.

False brinelling a condition that is a result of manufacturing. It appears as a polished surface usually associated with bearing surfaces. Often mistaken for a defect.

Fatigue wear damage that occurs to metal and components after many hours of operation under heavy loads.

Fatigue failures catastrophic failures that occur due to old age or excessive wear.

Ferrule a compression ring used to create a seal between the line and coupling nut. Ferrules used in pneumatic and hydraulic circuits can be made out of malleable plastic, brass, copper, or steels. Many ferrules deform to conform with the fitting when torqued and should not be reused.

Filter a component that captures contaminants from hydraulic fluid during operation.

Final drive see *input shaft* and *output shaft.*

Fitting a mechanical coupling device used to attach or connect lines and hoses; usually consists of both a male and a female member.

Fixed-displacement (motor) a hydraulic drive motor that accept an equal amount of oil per revolution from a hydraulic pump regardless of speed.

Fixed-displacement (pump) displaces an equal amount of oil each stroke or revolution, regardless of speed.

Flaring capable of being spread outward. A *flaring* tool is used to expand tubing. Also describes the aerodynamic panels used on some vehicles.

Flow continuous movement of a *fluid* within or outside of a circuit. The term can also be used to describe the movement of electrons in an electrical circuit.

Flow control valve used in hydraulic circuits to regulate the speed of actuators. Function by defining a flow area. Rated by operating pressure and capacity. This valve can be used in meter-in or meter-out applications; two general types are used: bleed-off and fixed orifice.

Flow rate the volume of hydraulic oil that passes through a component during a specific time.

Fluid coupling a hydrodynamic device used to transmit rotating mechanical power from one source to another.

Flywheel runout a deviation of specified rotational travel between the engine flywheel and Bell housing. Runout is measured with a dial indicator.

Foot valve a service brake application valve used to apply the brakes and located in the operator's compartment.

Force push or pull effort measured in units of weight. Force represents the ability to do work but this ability does not necessarily result in any work done: for instance, you can push all you like on a bulldozer but it is unlikely to move the vehicle. The formula for force (F) is calculated by multiplying pressure (P) by the area (A) it acts on. $F = P \times A$.

Friction the resistance to relative motion between two surfaces in contact with each other.

Friction clutch the driven disc used in a hydraulic clutch pack of a powershift transmission.

Friction disc a type of disc used in a wet disc brake system that has a friction material bonded to both surfaces, which is forced to rub against steel discs in the wheel end of the wet disc brake system by either hydraulic pressure or spring pressure.

Full-floating axle an axle that does not carry any weight but transmits torque to the drive wheels.

Full flow filter a filter which is capable of accepting the full flow of a hydraulic circuit through it during circuit operation.

Galling transfer or displacement of metal. Usually occurs between two metal parts that come in contact with each other. Can be caused by a lack of lubrication, or overloading.

Gear pitch the number of teeth per given unit of pitch diameter in a gear.

Gear ratio the ratio of an input shaft speed to the speed of the output shaft.

Gears any of a number of different types of toothed or spiraled shafts designed to either impart or receive torque from another.

Gear set two or more gears that intermesh.

Grouser the ridges on a track shoe that designate the terrain the shoe will operate on. The hardness of the surface material on the grousers determines whether they are standard or extreme service track shoes.

Head the vertical height of a column of liquid above a given point; expressed in linear units (feet). Head can be used to express gauge pressure. Head pressure is equal to the height of a liquid in a column multiplied by the density of the liquid.

Head pressure equal to the height of a liquid in a column multiplied by the density of the liquid. See *head*.

Heat exchanger a component that transfers heat from one medium (i.e., fluid to fluid) to another.

Helical cut gear a gear with teeth cut at some angle other than at a right angle across the face of the gear, thus permitting more than one tooth to be engaged at all times and also providing a smoother and quieter operation than the spur gear.

Herringbone gears a gear design used on heavy-duty final drive applications. A gear arrangement in which the drive gears are shaped like chevrons. They provide high torque capability with no axial loading.

High block a heavy-duty design of bearing cap used on off-road equipment U-joints.

High brake pressure any brake pressure above a predefined value in a brake system that is necessary to stop the equipment safely.

High pressure relief valve a regulating valve used in a hydraulic circuit to protect the system from over pressure.

Hoist any of a number of different types of lifting devices including chain falls, cherry pickers, scissor jacks, and cranes.

Horizontal parallel or level with the horizon.

Hydraulic analyzer a hydraulic circuit diagnostic instrument consisting of a pressure gauge(s), flow meter, gate (choke) valve, and thermometer. May also be equipped with additional instruments.

Hydraulic pump a component in a hydraulic system that converts mechanical motion into fluid power.

Hydraulically-released clutch a hydraulic release mechanism comprised of a master cylinder and slave cylinder used to engage and disengage a clutch through a medium of hydraulic fluid.

Hydraulic track adjuster a hydraulic actuator that is used to adjust the track tension on a crawler track by increasing or decreasing the hydraulic pressure in the actuator. The actuator adjusts the position of the idler to change track tension.

Hydraulics the science of using fluid power circuits that use liquids, usually specially formulated oils, in confined circuits to transmit power or multiply human or machine effort.

Hydrodynamic drive in a hydrodynamic drive system, the hydraulic oil flows at high speed and relatively low pressure to provide power to the drive wheels.

Hydrodynamics the science of forces acting on or exerted by liquids; usually, but not always, in confined circuits.

Hydrokinetics the study of fluid behavior under flow.

Hydrometer an instrument used to measure the specific gravity of a liquid, usually coolant or battery electrolyte.

Hydropneumatics term used to describe a compound pneumatic and hydraulic circuit: an example is air-over-hydraulic brakes.

Hydroscopic a term used to describe a fluid's ability to absorb moisture. A feature of a good brake fluid is its ability to absorb moisture to prevent corrosion.

Hydrostatic drive uses a positive displacement hydraulic pump and motor to transfer power through rotary motion to the drive wheels. In a hydrostatic drive system, the oil is subjected to high pressures at a relatively slow speed.

Hydrostatics the study of fluid behavior and pressure exerted by liquids at rest.

Hygroscopic capable of absorbing moisture. A characteristic of some brake fluids. This is an advantage for hydraulic brake fluid because it inhibits water droplets that can cause corrosion, freeze, or boil.

Hygroscopic breather hydraulic reservoir breather capable of eliminating moisture and airborne particulate through the vent.

Hypoid gear set a gear set used in a differential that has the pinion mounted below the centerline of the crown gear.

ID acronym that refers to the inside diameter of circular opening in a component.

Idler the front idler on a crawler track is used to align, support and guide the track as it rotates. It also can be used to provide a means of tensioning the track.

Impeller/pump an impeller is a rotating component of a pump, usually made of aluminum, which transfers energy from the engine fluid being pumped by forcing the fluid outward from the center of rotation.

Inclinometer a device used to measure the angles of drivelines and flanges in a powertrain.

Inflammable capable of being *inflamed* or burned.

Input retarder a type of brake mechanism used on a driveline between a main transmission housing and a torque converter. Absorbs torque produced by the engine and converters it to heat energy.

Input shaft (final drive) a shaft that transmits torque from a power source (transmission) to a final drive.

Interaxle differential a gear design used to transmit power from the front axle to the rear axle on a tandem drive axle arrangement.

International Standards Organization (ISO) an international standards governing body that describes itself as "non-governmental," but many ISO standards become legally enforced.

Jack a hydraulic or air-over-hydraulic vertical lifting device.

Jaw clutch a simple mechanical lock used on drive axles to lock the differential, providing positive drive to both wheels.

Joint Industry Committee (JIC) organization that sets compatibility standards for hydraulic circuits.

Jounce (travel) the greatest compressed position of a suspension system.

Kilopascal (kPa) metric unit of pressure used mostly in engineering in North America. One hundred kPa is equivalent to 14.56 psi.

Kinetic energy the energy of motion.

Kingpin a linkage pin that allows the steering knuckle to pivot from side to side. Usually made from high quality steel.

kPa kilo-Pascal. Metric unit of pressure measurement. 101.3 kPa = 14.7psi = 1atm

Laminar flow streamlined flow through a hydraulic circuit. A hydraulic circuit should be designed to minimize friction caused by sharp angles and internal obstructions. Also, a desirable flow condition that can occur in a hydraulic system that minimizes friction.

Land raised surface beside a groove such as is found on a spool valve.

Laws of leverage the mechanical advantage obtained by the use of a lever or combination of levers. The laws of leverage can be derived using Newton's law of motion.

Leaf-spring system a series of springs used in equipment suspension systems to provide shock absorbing capabilities. A single spring is called a leaf spring.

Limited slip differential used in a differential to provide equal torque to both drive wheels during normal operation.

Linear actuator device that converts hydraulic energy into linear motion such as COE cab lift cylinders or dump box rams. Also, a hydraulic cylinder which converts hydraulic energy into linear motion.

Liner the stationary external pump member used in gear and vane pumps; also known as a *cam ring*.

Link pitch the distance between the bushing bore center and the pin bore center. This dimension does not generally change during the life of the track.

Lip seal a type of dynamic seal that is generally made from synthetic material. It uses a spring on the inside of the lip to form a tight seal on a shaft.

Live axle a drive axle that propels the equipment. Often used in conjunction with a dead axle in a tandem axle configuration.

Load cushion when a suspension is operating with no load, the outer edges of the spring load cushions are in contact with the saddle. As the load increases on the vehicle, the load cushion is compressed, further absorbing the additional loads.

Load proportioning valve (LPV) a valve used in equipment with hydraulic brakes to sense load at the rear axle and adjust the brake pressure accordingly.

Low brake pressure any brake pressure below a predefined value in a brake system that is necessary to stop the equipment safely.

Low block a light-duty bearing cap design found on light-duty off-road equipment driveline U-joints.

Low ground pressure system (LGP) track equipment that uses the low ground pressure system uses a wider track which increases ground contact for better flotation and stability on lose surfaces. Shoe width on the LGP track is generally a lot wider than on the standard track.

Machinist's rule a steel ruler used to take shop linear measurements and may act as a straight edge.

Magnetic-type clutch the torque is produced by the magnetic field across the air gap between the two sections of the clutch causing them to rotate as one unit.

Main pressure the system pressure standard used as system pressure to manage the hydraulic circuits in an automatic transmission.

Manifold a fluid conductor that allows for multiple connections.

Manometer a single tube bent into a U-shape and mounted to a linear measuring scale. It may be filled to the zero on the calibration scale with either water (H_2O) or mercury (Hg), depending on the pressures range to be measured.

Manual control valve a hydraulic control device that is actuated by the operator of the equipment such as a foot valve or lever which controls directional control.

Master cylinder a hydraulic brake circuit that develops brake pressure to apply the brakes at the wheel end of the vehicle. In a hydraulic brake system, the dual circuit application control valve.

Master link in order to assemble or disassemble some crawler tracks, a master link is required to complete the crawler track assembly.

Master pin in order to assemble or disassemble some crawler tracks, a master pin is required to complete the crawler track assembly.

Mechanical linkage a series of rigid links connected with joints to form a closed chain, or a series of closed chains. Each link has two or more joints, and the joints have various degrees of freedom to allow motion between the links.

Mechanical seal a seal that uses two highly machined metal faces that rotate on each other to provide a seal between two rotating parts. Usually found at the wheel end of a drive axle to seal the oil in the axle housing and rotating wheel end.

Meter to regulate or control the amount of fluid in a circuit.

Metering the precise measuring of fluid flow.

Metering valve a valve used on a vehicle to regulate the brake pressure to the front and rear brakes. Improves brake balance.

Micrometer a precise measuring instrument typically consisting of a thimble, barrel, and anvil. Used to make ID, OD, and thickness measurements.

Modulate to regulate.

Modulated brake pressure varies by the amount of pressure applied to the foot pedal. The foot brake valve varies (modulates) the pressure to the service brake.

Modulated foot valve a foot-operated brake valve that is capable of either raising or lowering the brake pressure by increasing the travel of the actuator.

Modulating valve a valve that regulates or proportions. In air brake systems, it may have *inversion valve* features.

Modulation a lowering or raising condition that occurs in a brake system wheel end pressure that is controlled by the brake valve.

Motor a rotary device that can change hydraulic energy into mechanical energy.

Multiple countershaft any transmission that utilizes more than one countershaft to supply the gear ratios necessary for the transmission output shaft.

Multiple countershaft transmission a housing containing at least two countershafts having parallel axes of rotation and being fixed against all movement except rotative with respect to the housing.

National Institute for Automotive Service Excellence (ASE) body that defines certification standards and manages certification testing for the motive power trades including truck, auto, schoolbus, automachinists, etc.

Needle bearing the term used to describe the type of bearing used in universal joint cups that rotate against the universal joint trunnion during driveshaft rotation.

Needle cup the part of a cross which fits over the trunnion and holds the needle rollers.

Needle rollers (bearing) cylindrical bearings positioned inside the bore of the needle cup.

Nitrogen-charged accumulator a dampening device used on suspension systems on off-road haulage trucks. Nitrogen-charged accumulators provide a dampening action on the suspension cylinders.

Non-constant-mesh entails sliding the teeth of the gear pairs together, and hoping they engage at some point.

Non-parallel driveshaft the driveline arrangement where the companion flanges of a transmission and drive axle are not parallel to each other, but the working angles of the U-joints are equal.

Non-slip shaft assembly a driveline that is a fixed length. Consists of a weld yoke, tubing, center bearing, and an end yoke.

No-spin differential used on off-road equipment to provide a positive drive to both wheels during straight line operation. When cornering, the outside wheel is disengaged by a jaw clutch allowing the outside wheel to speed up while the inside wheel supplies the drive.

OD acronym for the outside diameter of a circular object such as a bushing.

Occupational Safety and Health Administration (OSHA) federal body that sets standards and monitors employee health and safety in the workplace.

Oil/pneumatic cylinders act as a shock absorbers in the suspension of heavy equipment. The front suspension cylinder barrels are attached to the frame of the equipment at the top and are fastened to the steering knuckle at the bottom.

Open center a hydraulic circuit in which pump delivery fluid is allowed to recirculate freely to the reservoir in the 'center' (neutral) position.

Organic facing clutch a clutch disc make from organic materials such as glass, mineral wool, and carbon.

Orifice a restriction in a circuit which limits flow.

Original equipment manufacturer (OEM) used to describe a chassis manufacturer or a manufacturer of major chassis subcomponents such as engines, transmissions or brakes.

Oscillate movement or cycles with rhythmic regularity. See *oscillation*. Can refer to back and forth movement.

Oscillating hitch used on off-road equipment to provides independent vertical motion of suspension frame pivots.

Oscillating undercarriage this design has the added benefit of being able to pivot from front to rear, which can increase stability when the equipment is operating on uneven ground. The track frames have an equalizer bar mounted in between them to allow each track frame to pivot independently on a pivot shaft.

Oscillation the act of oscillating. The natural dynamics of frame rails, the vibrations of drivelines, wheel dynamics, and electrical wave pulses.

Output retarder a hydrodynamic device located on the output shaft of a transmission that converts power to heat energy. Hydraulic driveline brakes mounted on the rear of a transmission that are designed to apply retarding force directly to the driveline.

Output shaft (final drive) a shaft in a transmission that directs torque to a final drive input shaft.

Over-center clutch used where the internal linkage is necessarily different to provide an "over-center" action to lock the clutch plate with the flywheel.

Overdrive gearing of a transmission so that one revolution of the engine produces more than one revolution of the transmission's output shaft.

Oxidation the result of oxygen reacting with a substance such as a metal or oil. Rust is the result of oxidizing iron.

Packing general term used to describe sealing media used in movable hydraulic components.

Parallel drivehaft a driveline arrangement that has all the yokes and companion flanges parallel to each other. The working angles of the U-joints must also be equal.

Park brake valve an electric hydraulic valve that is used in a brake circuit to activate the park brakes.

Pascal's Law states that pressure applied to a confined liquid is transmitted undiminished in all directions and acts with equal force on all equal areas and at right angles to those areas.

Phasing the relationship of the end yokes on each end of a two-piece driveline to each other.

Pilot any of a number of devices that receive signals and then effect an action. Can also be known as a relay.

Pilot line a control or signal line to an actuator. In hydraulic systems, a pilot line often uses a lower pressure value.

Pilot pressure a hydraulic valve used to provide lower pressure in a circuit that is used to control another circuit, such as the dump/hoist circuit on a loader.

Pilot valve a secondary valve that is used to control the operation of another valve.

Pilot-operated relief valve consists of a main relief valve through which the main pressure line passes, and a smaller relief valve that acts as a trigger to control the main relief valve. Used to handle large volumes of fluid with little pressure differential.

Pin link pins that join the individual links together on a roller chain.

Pinion drive a term used to describe the type of drive used on some types of heavy equipment. Uses two bevel spur gears, a pinion gear, and a ring gear to transmit the flow of power at right angles through a final drive in a powertrain.

Pinion gear (differential) a short stub shaft with gear teeth machined on one end that mates to a crown gear in a differential. The crown and pinion gears often come in a set.

Pinion shaft a term used to describe a shaft that is used in the final drive of a bull dozer. The pinion shaft is held in place with roller bearings, while the larger final drive gear is usually splined to the track drive sprocket, which is also supported by large tapered roller bearings.

Pitch diameter the diameter of the pitch circle. Operating pitch diameter is the pitch diameter at which the gears operate.

Pitman arm the connection used between the steering gear and the drag link in a conventional steering system.

Pitting surface irregularities on metal that are caused by corrosion or from abrasive wear.

Plain bevel gears of conical form designed to operate on intersecting axes. Most commonly the shafts will intersect at a 90° angle. Straight bevel gears (as opposed to spiral bevel gears) have teeth that are straight, radially, from the gear's centerpoint.

Planetary drive a term used to describe a type of drive that consists of at least one ring gear, a sun gear, and two or more planet gears. Used to transmit torque to wheel end or final drive.

Planet gear the name given to the gears (usually 3) that rotate around the inside of a ring gear on the final drive wheel end of a heavy-duty drive axle.

Planetary carrier typically, the planet gears are mounted on a movable arm or carrier, which itself may rotate relative to the sun gear. Holds one or more peripheral planet gears, of the same size, meshed with the sun gear and ring gear.

Planetary gear set a gear set consisting of three elements: a sun gear, a planetary carrier (containing 3 or more planetary gears), and a ring gear (annulus). The three separate elements intermesh, and the hold/release status of each determines the output ratios. Also known as an *epicyclic gear set.*

Planetary pinion gears small gears fitted into a framework called the planetary carrier.

Planetary shift transmission any transmission configuration that utilizes planetary gearing.

Pneumatic brakes air brakes.

Pneumatically-controlled clutch a pneumatic release mechanism comprised of a treadle valve and slave cylinder used to engage and disengage a clutch through a medium of compressed air.

Pneumatic suspension an air suspension.

Pneumatics a branch of fluid power physics dealing with pressure and gas dynamics.

Pop-off valve a term used to describe the safety pressure relief valve used on air brake systems. Most pop-off valves are set to trip at 150 psi and are located on the supply (wet) tank in the system. See *safety pressure relief valve.*

Positive displacement a characteristic of a pump or motor that has its inlet positively sealed from its outlet so that no fluid can recirculate within the component: another way of saying this is: what goes into the pump per cycle is what comes out.

Positive displacement pump a fluid pump where the inlet and outlet are not connected hydraulically.

Pounds per square inch (psi) standard unit of pressure used in North America: the force acting over a sectional square inch. One psi is equivalent to 6.891 kPa.

Power the rate of performing work.

Powerflow the mechanical flow path that conducts motive power through a system or component. For instance, the powerflow through a transmission in a specific gear ratio would be the sequential routing of the torque flow path through those gears that are actually engaged.

Powershift transmissions in which the shifting or selecting of a gear ratio does not require breaking torque from the engine or the synchronization of gears.

Powertrain the components that make up the drive system on equipment. It includes the engine, torque converter, transmission, and final drive.

Pre-charge pressure the pressure of nitrogen gas used in an accumulator before hydraulic pressure is added.

Pre-load a load placed on a tapered roller bearing to maintain good tooth contact between gear teeth in a powertrain. Also, a manufacturer's suggested predetermined force that should be applied to a bearing assembly to maintain a specified shaft or gear alignment.

Pressure the force applied over unit area. Usually expressed in pounds per square inch (psi) or kilopascals (kPa).

Pressure differential the difference in pressure values between two points of a circuit or component.

Pressure differential valve a warning light used in a brake circuit to alert the operator that there is a difference in brake pressure in from one side to the other in a dual circuit brake system.

Pressure drop a change in pressure between two points in a hydraulic line caused by the energy lost to maintain flow in the circuit.

Pressure plate applies pressure to the clutch disc for the transfer of torque to the transmission. The pressure plate, when coupled with the clutch disc and flywheel, makes and breaks the flow of power from the engine to the transmission.

Pressure protection valve a normally closed, pressure sensitive control valve. Some are externally adjustable. Can be used to delay the charging of a circuit or reservoir, or to isolate a circuit or reservoir in the event of a pressure drop-off.

Pressure reducing valve a valve used in a circuit that lowers the pressure to a value that is necessary to operate an additional circuit at reduced pressure.

Pressure relief valve a pressure limiting device that trips at a predetermined pressure to define circuit pressure or act as a safety device. A *safety pop-off valve* is one type of pressure relief valve.

Pressure switch an electric hydraulic valve used in a hydraulic brake circuit to monitor the brake pressure. When the pressure becomes too low, it turns on a warning light in the operator's compartment.

Preventative maintenance (PM) scheduled maintenance downtime designed to eliminate unscheduled equipment breakdowns.

Preventative maintenance inspection (PMI) the inspection that accompanies a PM schedule.

Prime mover the initial mechanical source of motive power. In the context of a hydraulic circuit, the means used to drive the hydraulic pump would be properly the prime mover, although the term is sometimes used to describe the pump itself.

Priority valve a protective device on hydraulic (automatic transmissions) and pneumatic circuits. Ensures that the control system upstream from a valve or subcircuit has sufficient pressure to function.

Propeller shaft driveshaft.

Proportioning valve a hydraulic brake valve used on vehicles equipped with front disc and rear drum brakes. Its function is to reduce application pressure to the rear drum brakes to prevent premature lock-up under severe braking.

psi pounds per square inch. Standard measure of pressure values used in North America. One psi is equivalent to 6.9 kPa (metric measure of pressure).

psia pounds per square inch absolute. Measure of pressure values used in North America that factors in atmospheric pressure of 14.7 psi.

Pull-type clutch type of clutch that pulls the release bearing toward the transmission for release.

Pump any of a number of different devices used to move or displace fluids.

Push-type clutch type of clutch that is released by a pedal push. The release bearing is pushed toward the flywheel for release.

Quick-connect coupler see *quick-release coupler*.

Quick-release coupler any of a number of different types of hydraulic or pneumatic connecters with spring-loaded seal couplings designed for fast connect/disconnect.

Quick-release valve an air circuit valve acts like a Tee fitting when air is applied to it from source, but exhausts air at the valve when source air is removed. The objective is to speed release times. Used on various air brake and other vehicle pneumatic circuits.

Rack and pinion gear a straight length of square or rectangular steel bar with teeth on one side, across which a pinion is driven. Technically a gear rack is a spur gear with an infinite pitch diameter.

Radial extending from the center of an axis or center of a circle. An example is the spokes on a wheel.

Radial load a load acting perpendicular or 90 degrees to a shaft.

Radial play eccentric movement or runout in a rotating component. See *radius*.

Radial runout the wobble of a turning wheel as viewed from the side of the vehicle. Used to describe one type of runout on rotors, shafts, and wheels.

Radius a straight line extending from the center of a circle to its circumference. The radius of a circle is half its diameter.

Ram a hydraulic (or pneumatic) linear actuator. Examples are a dump box and COE cab lift rams.

Ratio valve an air brake valve used on some pre-ABS front axle brakes to proportion, on a graduated scale, application pressures to the front brakes.

Rebound occurs when the energy that is stored in a spring kicks back. On many types of equipment, rebound is controlled with dampening devices such as shock absorbers and springs.

Reciprocating motion the back and forth motion of pistons in a positive-displacement hydraulic piston pump.

Regulator a device that limits system or subsystem pressure in a fluid power circuit.

Relay valve in air brake systems, a valve that receives a control signal, then uses a local supply of air to send out to actuator devices. Most relay valves replicate the signal pressure valve to the out-circuit, but some are designed to amplify or modulate the signal.

Release bearing unit within the clutch used to disengage the clutch; also known as throw-out bearing.

Relief valve a hydraulic valve that sends excess oil back to the hydraulic tank.

Reservoir any of a number of different types of storage chamber, both pressurized and non-pressurized, used to store fluids such as hydraulic oil or the compressed air in the air system. A tank or container that is used to store hydraulic oil.

Residual pressure the potential energy contained in a hydraulic circuit when it is inactive. Safety precautions should be taken when working around hydraulic equipment.

Reverse modulation a condition that occurs in a spring-applied brake circuit that modulates the brake pressure in what would normally be the opposite of what takes place in conventional hydraulic brakes. Example: when the brake pedal is depressed in a spring-applied brake, the pressure is lowered in proportion to the travel of the brake pedal.

Rigid anything that cannot flex or bend.

Rigid disc steel clutch plates to which friction linings, or facings, are bonded or riveted.

Ring gear spur gear such as that used on a flywheel for engine cranking, or the outer member of a planetary gear set.

Roller link the link of a chain drive that holds the roller in place. Each link interconnects to another to form a chain.

Root mean square (rms) term used to mean averaging when performing machining operations and also used to describe AC voltage measurement by giving it an equivalent DC value.

Rotary actuator a device that converts fluid flow into incremental rotary motion as opposed to continuous cycle rotary motion.

Rotary flow fluid movement, such as that produced by a paddle wheel in water or a propeller in air, that results in centrifugal force. See *rotary oil flow*.

Rotary oil flow caused by the centrifugal force applied to the fluid as the converter rotates. Occurs as turbine speed increases within the torque converter.

Rotate to turn or spin.

Rotating clutch pack a multi-disk-and-plate-type clutch, when engaged, will rotate to provide an input within a transmission.

Rotation the motion of a gear, shaft, or wheel.

Rotor any of a number of different components that rotate. In air brakes, the rotor turns with the wheel and has braking force applied to it. In an alternator, the rotor provides the magnetic fields required to create current flow.

Runout any kind of wobble in a rotating component. Runout may be axial or radial. Using the example of a wheel on a vehicle, axial runout is wobble as viewed from the front or rear of the vehicle, and radial runout is wobble viewed from the side of the vehicle.

Saddle assembly an assembly on each side of the suspension system that holds rubber load cushions. Each of the four rubber blocks is connected to the axle assembly by drive pins.

Safety pressure relief valve the pressure relief valve used on air brake systems, more commonly called a *pop-off valve*. Usually set to trip at 150 psi and located on the supply (wet) tank in the system.

Scissor jack an air-over-hydraulic portable floor hoist that lifts equipment using a scissor action.

Sealed tracks a pin design used on crawler tracks that uses a solid pin between the links. Sealed and lubricated tracks have a hollow pin, which provides a path for lubricating the pin and bushing of the next track section.

Semi-floating axle the term semi-floating is derived from the way the axle is mounted in the assembly. The ale at the differential is splined to the side gear, and floats at this end. At the wheel end of the axle, the shaft carries one or two bearings depending on the axle load carrying capacity.

Sensor energy conversion device used to signal a condition or state. Sensors are used to signal speed, pressure, temperature, etc.

Sequence valve a hydraulic valve that is capable of reducing pressure in sequence to two different circuits.

Service bulletin see *technical service bulletin* (TSB).

Service literature most kinds of electronically accessed service instructions and the data bases that retain them. Replaces hard copy service manuals and TSBs. Most OEMs no longer make hard (paper) copy versions of service literature available.

Service manual a hard (paper) copy book that outlines service repair and troubleshooting procedures. Today most service manuals are electronic due to lower cost and ease of updating. Consequently, the term *service literature* is more often used.

Servo a component actuated by another. One type of servo is a hydraulic slave piston that is actuated by a control valve.

Servo brake a braking action that occurs due to the action of one shoe on the other in a drum brake circuit. Example: the primary shoe of a drum brake transmits a force to the secondary shoe and forces it to wedge into the drum providing braking action.

Servo control a hydraulic valve that can control the flow of oil in proportion to the input signal.

Shift bar housing transmission subassembly that houses shift rails, shift yolks, detent balls and springers, interlock balls, and pin and neutral shaft.

Shift fork or yoke Y-shaped components located between gears on the main shaft that, when actuated, cause the gears to engage or disengage via the sliding clutches. Shift forks are located between low and reverse, first and second, and third and fourth gears in a standard main section.

Shift valves hydraulic valves that direct hydraulic pressure to the clutch packs to engage a specific gear ratio. The shift valve determines when to shift from one gear to the next.

Shock absorbers hydraulic devices used to dampen vehicle spring oscillations.

Shuttle valve a prioritizing valve with two inlet ports and a single outlet port. The ball or spool shuttles to direct the higher of the two inlet pressure values to the outlet. Also, a valve located in a circuit that allows flow to a component from two different sources.

Side bar a connector. Roller chains are made up a series of pin links and roller links connected to each other with side bars.

Sight glass allows a technician to see the flow of a liquid in a line (such as fuel or refrigerant) for diagnostic purposes.

Single-acting cylinder a hydraulic cylinder designed to produce thrust in one direction only: returned by gravity or spring force.

Single-countershaft transmission any transmission that utilizes a single countershaft to supply the gear ratio's necessary for the transmission output shaft.

Single-phase main the standard 110-120 V-AC household electrical supply.

Single reduction axle a drive axle that utilizes a single gear reduction that is achieved through the use of a crown and pinion gear set.

Single reduction drive a term used to describe a final drive that reduces the gear ration through the drive only once. An example is a pinion drive used on small bull dozer final drives.

Slave cylinder a device in the hydraulic clutch system that is activated by hydraulic force, disengaging the clutch. See wheel cylinders.

Slave valve any air or hydraulic actuator. A spool-type hydraulic valve that uses pressure from a pilot valve to move the position of its spool.

Sliding gear gears that may slide along the shaft to which they are attached. This motion is controlled by the shift forks. They are splined to the shaft so they always rotate at shaft speed. A sliding gear has protrusions (dogs) on its sides which are designed to engage slots in the freewheeling gears and cause it to transmit power. Only one sliding gear can be engaged to a freewheeling gear at any particular time; this is what determines which gear the transmission is currently in.

Slip assembly the male/female splined section of a driveshaft that fits together and can lengthen or shorten during operation.

Slip joint the splined part of a driveshaft that is needed to accommodate the changes in length experienced during operation.

Slip yoke a part of a driveline assembly that provides a means for the length of a driveshaft to change during operation.

Slop widely used slang term for endplay: see *endplay*.

Slug see *slug volume*.

Slug volume the fluid volume of a pump chamber or engine cylinder swept by the plunger/piston in one cycle.

Snap ring a retaining split ring with radial spring pressure. Snap rings are usually fitted to grooves and are used to retain, shims, springs, bearings, and other components.

Socket a female cavity to which a pin or shaft is inserted. Electrical connectors are equipped with sockets to which terminal pins are inserted to make a connection.

Solid rubber suspension a solid rubber cushion-type equalizing beam suspension system used on off-road haulage truck.

Solvent substance that dissolve other substances.

Spalling a condition that occurs on metal parts when chips or flakes of metal separate from fatigue as opposed to wear.

Specification an assembly, engineering or technical detail or value.

Spike a sudden surge in hydraulic pressure or electrical voltage. Also, a slang term meaning the tractor steering column-mounted trailer service brake valve.

Spiral bevel gear set a helical gear set that includes a crown and pinion that mesh at right angles with the pinion gear mounted on the centerline of the crown gear.

Spline radial lands that either protrude from, or are milled into, a shaft to permit it engaging with female splines for the purpose of transmitting torque.

Spool valve commonly used hydraulic valve in which a sliding spool within a valve body is used to open and close hydraulic subcircuits. A cylindrical part that moves from side to side in a hydraulic valve to control oil flow in a circuit.

Spring hangers specialized connectors. Springs are mounted to the top of each equalizer beam onto saddle mounts and are connected to the frame by spring hangers. The spring assemblies pivot on spring pins attached to the hanger brackets.

Spring packs a series of springs bound together with a center bolt in an elliptical shape. They provide a dampening action to the equipment during operation.

Sprocket a wheel with toothed cogs designed to engage with a chain or mesh with a similar shaped device. The sprocket is not mounted to the track frame but is attached to the final drive. Its job is to transfer drive torque to the track. The teeth on the outside of the sprocket act like gear teeth on a gear.

Sprocket shaft gear the sprocket shaft gear is splined to the sprocket shaft, which also holds the drive sprocket. The sprocket is mounted to sprocket hub of a final drive on a bull dozer.

Spur bevel gear set a straight cut gear set that includes a crown and pinion that mesh at right angles with the pinion gear mounted on the centerline of the crown gear.

Spur gears cylindrical gears with teeth that are straight and parallel, and operate on parallel axes.

Static charge accumulated electrical charge potential not flowing in circuit.

Static discharge arc that occurs when static electrical charge finds a flow path.

Static friction occurs when the two objects are not moving relative to each other. The coefficient of static friction is typically denoted as μ_r. The initial force to get an object moving is often dominated by static friction. S friction is, in most cases, higher than the kinetic friction.

Static head measured pressure in a circuit in which there is no fluid flow; normally expressed in feet of water.

Stationary clutch pack a multi-disk-and-plate-type clutch. When engaged it will hold stationary one of the three planetary gear sets (ring gear, sun gear, planetary gears).

Stator located between the torque converter pump/impeller and turbine. It redirects the oil flow from the turbine back into the impeller in the direction of impeller rotation with minimal loss of speed or force.

Steel disc a type of disc used in a wet disc brake system that is made from steel material which is forced to rub against friction discs in the wheel end of the wet disc brake system by either hydraulic pressure or spring pressure.

Steering clutch a clutch design found on track equipment. Provides the capability to slow down each track independently to achieve direction control.

Stick circuit part of an excavator implement hydraulic circuit. The stick arm connects to and pivots on the boom and has the bucket connected to it. The stick circuit is actuated by double-acting hydraulic cylinders.

Straight cut spur the leading edges of the teeth are aligned parallel to the axis of rotation. These gears can only mesh correctly if they are fitted to parallel axles.

Sump a reservoir.

Sun gear the central gear of a planetary gear set, usually located on axle of a wheel end planetary drive.

Supply pressure pressure that is common to a circuit such as the pressure that comes from a charge valve to charge the accumulators of a brake circuit.

Supply tank in a vehicle air brake circuit, the supply tank is the first reservoir in the circuit to receive air from the air compressor. The supply tank feeds the primary, secondary, and any auxiliary air tanks in the circuit.

Surge a temporary rise in pressure or fluid level in a circuit.

Suspension cylinder off-road haulage truck use suspension cylinders to help absorb road shock during operation. Some suspension systems contain high hydraulic and nitrogen pressures.

Swashplate a stationary canted (angled) plate used in an axial piston pump on which pistons in a rotating cylinder ride: as the pump cylinder rotates, the pistons reciprocate in the barrel bores. The swash angle can be varied to provide variable displacement.

Swing circuit the means of hydraulically rotating the upper structure on a typical excavator circuit usually through 360 degrees.

Synchronizer performs two functions. It brings components that are rotating at different speeds to one synchronized speed to ensure that the main shaft and the main shaft gear to be locked to it are rotating at the same speed, and it actually locks these components together.

System pressure a common pressure in a fluid circuit. An example is the main hydraulic dump and hoist pressure found in a hydraulic circuit. Usually refers to the range of pneumatic circuit pressures between cut-out and cut-in pressure values in an FMVSS 121 compliant air brake system. Can refer to other fluid power circuit pressures such as those in transmission hydraulics and engine lube circuits.

Tandem drive a two-axle drive used on off-road haulage trucks to provide four-wheel drive at the rear of the equipment.

Technical service bulletin (TSB) service literature either on-line or hard copy that updates or modifies a repair procedure.

Tee fitting a fluid circuit valve usually set up with a single source but distributes to two branches of the circuit; often T-shaped.

Temper the heat treating process used to harden (or in some cases soften) materials. Term is usually applied to heat treatments of middle and high alloy steels.

Tensile strength the highest stress a material can withstand without separating, expressed in psi. In steels, tensile strength typically exceeds *yield strength* by about 10%.

Three-phase main high potential electrical supply used for shop machinery available in 220 V-AC and 450-600 V-AC depending on region.

Tie rod a rod or crosstube that connects both wheels indirectly to a steering box, which provides steering input to the wheels.

Toe-in a condition that has the front of a pair of tires on the same axle point inward as viewed from the front of the equipment.

Toe-out a condition that has the front of a pair of tires on the same axle point outward as viewed from the front of the equipment.

Torque the measure of force applied to a lever arm. Normally expressed in lb.-ft. (pound-feet) or lb.-in. (pound-inch).

Torque converter lockup clutch improves cruising power transmission efficiency. The application of the clutch locks the turbine to the pump, causing all power transmission to be mechanical, thus eliminating losses associated with fluid drive.

Torque multiplication condition when the input torque from an engine is less than the output torque produced through a torque converter. Accomplished through the use of a stator.

Torque rods used on suspension systems to absorb additional torque generated by the axle and road shock.

Torricelli's tube Torricelli discovered the concept of atmospheric pressure by inverting a tube filled with mercury into a bowl of the liquid, then observing that the column of mercury in the tube fell until atmospheric pressure acting on the surface balanced against the vacuum created in the tube. His experiment proved that vacuum in the column would support 29.92 inches of mercury at sea level.

Torsion twisting, especially when one end of a beam or body is held.

Torsional failures usually a total failure of a shaft or part in a powertrain, resulting in loss of drive.

Toxic poisonous.

Toxicity having poisonous characteristics.

Track circuit the drive and steer hydraulic circuit used in bulldozers, excavators, and other track-driven equipment.

Track guards used to keep rocks and foreign debris from getting in between the track rollers and track links.

Track link one element of the chain section of the track. Ttrack links, pins, and bushings interconnect with each other to form the track.

Track pin connects the links and bushings together on a track to form a chain.

Track pitch the distance from the center of one pin to the center of an adjacent pin in an adjoining track link section. When the track pins and bushings wear, the length of the track increases and the track becomes loose.

Track roller the bottom rollers on a crawler track are called track rollers. They support the weight of the equipment and ensure that the weight of the equipment is distributed evenly over the bottom of the track.

Track shoes bolted to the link assembly to provide a means of supporting the equipment and to provide traction and flotation during operation of track systems. They are generally constructed from a hardened metal plate and come in many different configurations.

Transfer case an additional gearbox located between the main transmission and the rear axle. Its function is to transfer torque from a machine's transmission to the front and rear driving axles.

Treadle valve a dual circuit service brake actuation valve that meters air from the primary and secondary service reservoirs to the service lines and brake chambers. Also known as a *foot valve.*

Trip in technology, this term is used to mean to open. Examples: a safety valve is *tripped* when its specified pressure is achieved, or a circuit breaker is *tripped* when its specified current is exceeded.

Trunnion the machined surfaces of the universal joint cross. The bearing cups fit into this cross.

Tube hydraulic line or passage.

Tube-type clutch a clutch fitted with a tube whose inflation causes the clutch to engage, and deflation, to disengage.

Turbine the output (driven) member of a torque converter splined to the input shaft of the transmission.

Two-way valve a direction control valve used in a hydraulic system to divide the oil flow into two flow paths.

Two-way check valve a shuttle valve that is supplied with air from two sources and has a single outlet. The valve is designed to shuttle and bias air flow from the higher of the two sources to the outlet.

U-bolt a U-shaped steel bolt with threads at both end that is used to fasten a multi-leaf spring pack to the axles of equipment.

Underdrive any drivetrain arrangement in which an output speed ratio is less than the input speed.

Underwriter's Laboratory (UL) organization that ensures product compliance to standards of safety.

Universal joint (U-joint) a joint that provides a flexible coupling that allows power to be transmitted from one shaft to another.

Unsprung weight the weight of the equipment that is not supported by the suspension system.

Vacuum the absence of matter and therefore the absence of pressure. The term is often used to refer to a partial vacuum, which means any pressure condition below atmospheric.

Valve a mechanical device used to control flow in a fluid power circuit.

Vapor a gas. One of the three states of matter, the other two being solid and liquid.

Variable-displacement (motor) a hydraulic motor that is only capable of accepting an equal volume of oil per revolution or stroke from a hydraulic pump.

Variable-displacement (pump) a hydraulic pump that is capable of varying the displacement per stroke or revolution regardless of the speed.

Variable-pressure reducing valve a pressure reducing valve that can be adjusted to specific pressures required in hydraulic brake systems.

Vent a breather device.

Vertical exactly upright or plumb.

Vibration rhythmic or arrhythmic movement or oscillation. In vehicle technology, usually refers to an undesirable condition. Example: *driveline vibration.*

Viscosity the shear resistance of a liquid. Often used to mean oil thickness or resistance to flow. Viscosity ratings are assigned by SAE (Society of Automotive Engineers)

Viscosity index (VI) a measure of the viscosity-to-temperature performance of a fluid.

Volume geometric space within a chamber. The term is often applied to mean pump output, in which case it is represented in gpm.

Vortex eddy force caused by spiraling fluid. A moving vehicle creates a *vortex* behind it as it cuts through the air. Term is used to describe both gas and liquid dynamics.

Vortex oil flow circular oil flow in a torque converter that occurs as oil is forced from the impeller to the turbine, then back to the impeller during operation.

Wear step track pins develop a specific form of damage called wear step. It is generally associated with the wet joint design. Minor wear under .005″ can be tolerated for reuse in most cases.

Weld yoke a term used to describe a heavy-duty yoke design used on a typical heavy equipment driveshaft application.

Welsh plug a plate or cup used to seal the hole in the throat of a slip yoke.

Wet tank commonly used slang term for supply tank. See *supply tank.*

Wet-type clutch a wet clutch contains many plates that are in the oil spray of the transmission and oil gets between the plates. A wet-type friction clutch plate keeps lubricant on a frictional slide face, stabilizes frictional torque transmission, and suppresses shuddering noise. A wet clutch is immersed in a cooling lubricating fluid, which also keeps the surfaces clean and gives smoother performance and longer life.

Wheel cylinder a hydraulic piston-type valve that is used to exert force in a drum brake system to force shoes out against the drums of a vehicle to slow down or stop it.

Wheel end pressure the hydraulic pressure found at the wheel end in a fluid brake circuit. This pressure is generally modulated by the brake valve.

Wing a bearing cap design. For example, a delta wing is a common bearing cap design.

Wing-type U-joint a U-joint cup design that is held in place by bolts that secure it to a yoke on a driveshaft or powertrain component.

Work when effort or force produces an observable result (movement), work results. In a hydraulic circuit, this means moving a load. To produce work in a hydraulic circuit, there must be flow. Work is measured in units of force multiplied by distance, for example, pound-feet.

Working angle the angle defined by the intersection of the centerlines of two driveshafts connected by a universal joint.

Worm gear worm gear sets run on non-intersecting, perpendicular shaft axes and provide high ratios of reduction. The worm is a threaded cylindrical shaft that drives the worm gear (also referred to as the worm wheel). Worm gears look somewhat similar to helical gears except that they have a curved throat recessed in their face to allow the worm access to the flanks of the gear teeth.

Yield strength the highest stress a material can withstand without permanent deformation, expressed in psi. In steels, yield strength is typically about 10% less than *tensile strength.*

Yoke the parts that are interconnected by a universal on a driveshaft.

Zerk fitting a type of grease nipple that uses a spring-loaded ball to seal an orifice. It accepts grease from a grease gun but does not allow dirt or other foreign material to enter the component.

Index

C

N

O

P

Q

R

T